Transistor Substitution Handbook

by

The Howard W. Sams
Engineering Staff

Howard W. Sams & Co., Inc.
4300 WEST 62ND ST. INDIANAPOLIS, INDIANA 46268 USA

SEVENTEENTH EDITION
FIRST PRINTING—1978

International Standard Book Number: 0-672-21515-2
Library of Congress Catalog Card Number: 78-54456

Printed in the United States of America.

SECTION 1

Making Transistor Substitutions

Both American and foreign transistors are included in this handbook. Section 1 lists these transistors in numerical and alphabetical order, placing the registered 2N numbers near the beginning of the list.

Substitutes for these transistors come under two headings: computer-selected substitutes and general-purpose replacements. The computer-selected substitutes were obtained by giving the computer the physical dimensions, seven of the most critical electrical parameters (polarity; material; maximum power-, voltage-, and current-handling capabilities; typical gain; and frequency response) of over 19,000 bipolar transistors—including rf power types. Those that matched within certain tolerances were listed in this book.

General-purpose replacement transistors follow the computer-selected substitutes. These substitutes are those recommended by the manufacturer or supplier of the general-purpose replacement transistor and are marked with an asterisk (*). Further information on these types can be found in Section 2.

In critical applications, it is advisable to check the specifications of the substitute against those of the transistor being replaced. These specifications can be found in the Howard W. Sams *Transistor Specifications Manual.*

Detailed descriptions of several lines taken from this handbook are shown for reference on the page preceding the first page of substitutions.

Callout	To Replace	Substitute This Type	Callout
	2A	*PTC109,*GE-51,*TR-17,*HEPG0005,*SK3006,*HF35,*ECG126,*WEP635,*276-2005	
1	2B	SEE 2N52	
2	*2C*	*PTC109,*TR-05,*HEPG0001,*SK3005,*ECG100,*WEP254,*RT-118	
	2D	*PTC109,*TR-05,*HEPG0001,*SK3005,*ECG100,*WEP254,*RT-118	
	2E	*PTC109,*GE-51,*TR-17,*HEPG0001,*SK3006,*HF35,*ECG126,*WEP635	
	2F	*PTC109,*GE-51,*TR-17,*HEPG0002,*HF35,*ECG160,*WEP637,*276-2005	
	2G	*GE-51,*TR-17,*HEPG0003,*SK3006,*HF35,*ECG126,*WEP635	
	2N27	*PTC102,*GE-2,*TR-05,*SK3005,*ECG100,*276-2004	
	2N28	*TR-08,*HEPG0011,*SK3011,*ECG101,*WEP641,*RT-119	
	2N29	*TR-09,*HEPG0011,*SK3011,*ECG101,*WEP641,*RT-119	
	2N34	2N59A,2N59B,2N59C,2N60A,2N60B,2N60C,2N381,2N518,2N526,2N526A,2N597, 2N651,2N651A,2N652A,2N1008B,2N1057,2N1057+, 2N1187,2N1192,2N1348, 2N1349,2N1350,2N1351,2N1373,2N1375,2N1447,2N1448,2N1451,2N1452,2N1925, 2N1954,2N1956,2N1957, 2N2956,MA100,MA882, MM404A,SF.T243, *PTC109, *GE-80,*TR-85,*HEPG0005,*SK3004,*AT30M,*ECG102A,*WEP250, *276-2005, *RT-121	5
3	*2N34+*	2N59A,2N59B,2N59C,2N382,2N383,2N461,2N527,2N527A,2N597,2N651,2N651A, NKT227,*GE-80,*HEPG0005,*SK3004,*ECG102A	
	2N34A	2N59A,2N59B,2N59C,2N60A,2N60B,2N60C,2N382,2N461,2N526,2N526A,2N527A, 2N597,2N651,2N651A,2N652A,2N1008A,2N1008B, 2N1125,2N1187,2N1192,2N1348, 2N1349,2N1350,2N1351,2N1371,2N1375,2N1377,2N1448,2N1452,2N1925,2N1926, 2N1954,2N1956, 2N1957,2N2956,2N2957,MA882,MA887,MM404A,*PTC109,*GE-80, *TR-85, *HEPG0005,*ECG102A,*WEP250,*276-2005, *RT-121	6
	2N35	2N377A,2N1000,2N1012,2N1299,2N1473,*PTC108,*GE-8,*TR-08,*HEPG0011,*SK3010, *NR5,*ECG103A,*WEP274,*276-2001, *RT-122	
	2N35+	2N214,*GE-8,*HEPG0011,*SK3010,*ECG103A	
	2N227	M.P.2N226	7
4	GE-1	REFER TO SECTION 2	
	GE-2	REFER TO SECTION 2	
	GE-3	REFER TO SECTION 2	
	2N5916†	2N5917†,2N6202,*PTC144,*HEPS3005,*276-2020	
	2N5917†	2N5916†,2N6202,*PTC144,*HEPS3005,*276-2020	8
	2N5918†	*276-2020	
	2N5919	2N5919A†,2N6204,*276-2020	
	2N5919A†	2N5919,2N6204,*276-2020	

1. Cross-reference to a type that has superseded the type being replaced.

2. Italic type indicates this type is no longer manufactured.

3. Two slightly different transistors have been manufactured using the same type number, so a plus sign (+) has been added to distinguish between them. Refer to Howard W. Sams *Transistor Specifications Manual* for details.

4. General-purpose replacement transistors. Section 2 contains further information on these types.

5. Computer-selected substitutes.

6. General-purpose replacement transistors—marked with an asterisk (*)—as recommended by the manufacturer of the replacement.

7. A matched pair of 2N226's.

8. Where † sign follows type number, it signifies a transistor whose ceramic body contains beryllium oxide. **Warning:** *Do Not* crush, grind, or abrade these portions, since the dust resulting from such action may be hazardous if inhaled. Disposal should be by burial.

To Replace	Substitute This Type
2A	*PTC109,*GE-51,*TR-17,*HEPG0005,*SK3006,*HF35,*ECG126,*WEP635,*276-2005
2B	SEE 2N52
2C	*PTC109,*TR-05,*HEPG0001,*SK3005,*ECG100,*WEP254,*RT-118
2D	*PTC109,*TR-05,*HEPG0001,*SK3005,*ECG100,*WEP254,*RT-118
2E	*PTC109,*GE-51,*TR-17,*HEPG0001,*SK3006,*HF35,*ECG126,*WEP635
2F	*PTC109,*GE-51,*TR-17,*HEPG0002,*HF35,*ECG160,*WEP637,*276-2005
2G	*GE-51,*TR-17,*HEPG0003,*SK3006,*HF35,*ECG126,*WEP635
2N27	*PTC102,*GE-2,*TR-05,*SK3005,*ECG100,*WEP254,*276-2004,*RT-118
2N28	*TR-08,*HEPG0011,*SK3011,*ECG101,*WEP641,*RT-119
2N29	*TR-09,*HEPG0011,*SK3011,*ECG101,*WEP641,*RT-119
2N34	2N59A,2N59B,2N59C,2N60A,2N60B,2N60C,2N381,2N518,2N526,2N526A,2N597, 2N651,2N651A,2N652A,2N1008B,2N1057,2N1057+,2N1187,2N1192,2N1348, 2N1349,2N1350,2N1351,2N1373,2N1375,2N1447,2N1448,2N1451,2N1452,2N1925, 2N1954,2N1956,2N1957,2N2956,2SB57A,2SB226,2SB227,2SB304A,2SB427, 2SB428,MA100,MA882,MA887,MM404A,SF.T243,*PTC109,*GE-80,*TR-85, *HEPG0005,*SK3004,*AT30M,*ECG102A,*WEP250,*276-2005,*RT-121
2N34+	2N59A,2N59B,2N59C,2N382,2N383,2N461,2N527,2N527A,2N597,2N651,2N651A, 2N652,2N652A,2N1008A,2N1008B,2N1124,2N1125,2N1187,2N1188,2N1189,2N1349, 2N1371,2N1375,2N1377,2N1448,2N1449,2N1452,2N1926,2N1954,2N1955,2N1956, 2N1957,2N1997,2N2956,2N2957,2SB77A,2SB136A,2SB227,2SB428,MA882,MA887, MM404A,NKT217,NKT227,*GE-80,*HEPG0005,*SK3004,*ECG102A,*WEP250
2N34A	2N59A,2N59B,2N59C,2N60A,2N60B,2N60C,2N382,2N461,2N526,2N526A,2N527A, 2N597,2N651,2N651A,2N652A,2N1008A,2N1008B,2N1125,2N1187,2N1192,2N1348, 2N1349,2N1350,2N1351,2N1371,2N1375,2N1377,2N1448,2N1452,2N1925,2N1926, 2N1954,2N1956,2N1957,2N2956,2N2957,2SB57A,2SB226,2SB227,2SB304A, 2SB428,MA882,MA887,MM404A,*PTC109,*GE-80,*TR-85,*HEPG0005,*ECG102A, *WEP250,*276-2005,*RT-121
2N35	2N377A,2N1000,2N1012,2N1299,2N1473,*PTC108,*GE-8,*TR-08,*HEPG0011, *SK3010,*NR5,*ECG103A,*WEP641B,*276-2001,*RT-122
2N35+	2N214,*PTC108,*GE-8,*HEPG0011,*SK3010,*ECG103A,*WEP641B
2N36	2N60,2N61,2N64,2N109,2N131,2N131A,2N133,2N133A,2N188,2N188A,2N188A+, 2N196,2N206,2N217,2N241,2N241A,2N241A+,2N270,2N320,2N321,2N363,2N407, 2N408,2N414,2N414A,2N414B,2N414C,2N415,2N415A,2N416,2N422,2N422A, 2N450,2N466,2N502,2N502A,2N502B,2N567,2N568,2N569,2N570,2N572,2N582, 2N584,2N610,2N654,2N838,2N980,2N1115,2N1171,2N1175,2N1265/5,2N1303, 2N1305,2N1309A,2N1313,2N1372,2N1383,2N1414,2N1415,2N1681,2N1726, 2N1727,2N1728,2N1746,2N1747,2N1748,2N1788,2N1790,2N1865,2N1867,2N2048, 2N2168,2N2587,2N2717,2N2795,2N3323,2N3324,2N3325,2N3412,2N3883,2N6365, 2N6365A,2SA169,2SA171,2SA173,2SA204,2SA206,2SA207,2SA209,2SA212, 2SA406,2SA407,2SB103,2SB114,2SB220,2SB221,2SB262,2SB299,2SB328, ASY26,ASZ21,MM380,NKT224,SF.T222,SF.T228,SYL792,*GE-51,*TR-85, *HEPG0005,*SK3003,*AT20M,*ECG102A,*WEP250,*AF125,*276-2004,*RT-121
2N36+	2N60,2N61,2N64,2N131,2N131A,2N133,2N133A,2N188,2N188A,2N188A+,2N196, 2N206,2N215,2N241A+,2N269,2N270,2N270+,2N283,2N320,2N361,2N363,2N408, 2N408+,2N414,2N414A,2N414B,2N414C,2N422,2N450,2N466,2N502,2N502A, 2N502B,2N567,2N568,2N580,2N580+,2N582,2N584,2N610,2N611,2N615,2N838, 2N980,2N1115,2N1171,2N1175,2N1303,2N1305,2N1313,2N1372,2N1383,2N1414, 2N1415,2N1524,2N1728,2N1746,2N1747,2N1748,2N1790,2N1865,2N1867,2N2048, 2N2168,2N2587,2N2717,2N2795,2N3883,2N6365,2N6365A,2SA79,2SA138,2SA139, 2SA169,2SA171,2SA173,2SA204,2SA206,2SA207,2SA209,2SA212,2SA325, 2SA326,2SA351,2SA352,2SA353,2SA354,2SA406,2SA407,2SB33,2SB34,2SB37, 2SB38,2SB59,2SB60A,2SB66,2SB75,2SB79,2SB89,2SB95,2SB103,2SB114,2SB156A, 2SB171,2SB185,2SB220,2SB221,2SB262,2SB263,2SB288,2SB299,2SB328,2SB365, 2SB475,ASY24B,ASY26,ASZ21,MM380,NKT214,NKT224,SF.T222,SF.T227,SF.T228, SF.T319,SF.T322,SF.T352,*GE-51,*TR-85,*HEPG0005,*SK3003,*ECG102A, *WEP250
2N37	2N61,2N64,2N104,2N111A,2N112A,2N123A,2N130A,2N131,2N131A,2N133,2N133A, 2N185,2N187,2N187A,2N188,2N188A,2N196,2N197,2N198,2N206,2N215,2N269, 2N283,2N316,2N317,2N319,2N320,2N361,2N363,2N396A,2N403,2N404,2N405, 2N406,2N413,2N428,2N502B,2N519A,2N565,2N566,2N579,2N579+,2N580, 2N580+,2N611,2N613,2N614,2N615,2N979,2N980,2N1017,2N1018,2N1303,2N1313,

To Replace	Substitute This Type
(2N112)	2N615,2N705,2N710,2N828,2N828A,2N964,2N968,2N1017,2N1018,2N1280,2N1303, 2N1313,2N1347,2N2048,2N2401,2N2717,2N2860,2N6365A,2SA56,2SA205,2SA208, 2SA405,2SA406,2SA411,ASZ21,MM380,SF.T227,*PTC109,*GE-52,*TR-17, *HEPG0009,*SK3005,*HF12M,*ECG126,*WEP635,*AC188/01,*276-2005
2N112+	2N315A,2N316A,2N396A,2N404A,2N428,2N1017,2N1018,2N1313,2SA205,2SA406, *PTC109,*GE-52,*HEPG0009,*SK3005,*ECG126,*WEP635
2N112A	2N111A,2N315A,2N316A,2N396A,2N404A,2N413,2N428,2N524A,2N525A,2N1017, 2N1018,2N1303,2N1313,2SA205,2SA406,SF.T226,*PTC109,*TR-05,*HEPG0005, *SK3005,*HF12M,*ECG100,*WEP254,*AC188/01,*276-2004,*RT-118
2N113	2N303,2N559,2N582,2N584,2N711A,2N711B,2N829,2N838,2N972,2N1171,2N1282, 2N2048,2N2401,2N2402,2N2587,2N2717,2N3883,2N6365,2N6365A,2SA169,2SA207, 2SA406,2SA407,ASZ21,MM380,*GE-1,*TR-05,*HEPG0009,*SK3004,*HF12H, *276-2005,*RT-118
2N113+	2N1171,2SA207,2SA406,2SA407,*GE-1,*TR-05,*HEPG0009,*SK3004
2N114	2N711B,2N829,2N838,2N964A,2N972,2N985,2N2022,2N2273,2N2928,2N3371,*GE-1, *TR-05,*HEPG0002,*SK3005,*HF20H,*ECG100,*WEP254,*276-2005,*RT-118
2N114+	*GE-1,*HEPG0002,*SK3005,*ECG100,*WEP254,*RT-118
2N115	*GE-3,*TR-01
2N117	2N160,2N160A,2N332,2N471,2N472,2N472A,2N474A,2N475A,2N1149,2N1276, 2SC387A,2SC478,2SC945,*PTC126,*GE-60,*TR-86,*HEPS0014,*SK3124,*SN80, *ECG123,*WEP53,*BF167,*276-2011,*RT-100
2N117+	2N472,2N472A,2N475A,2N1149,2N2521,2SC478,2SC945,TIS126,*GE-60,*HEPS0014, *ECG123,*WEP53
2N118	2N118A,2N161,2N161A,2N333,2N334,2N474,2N475,2N479A,2N480A,2N542A,2N783, 2N834A,2N842,2N1150,2N1277,2N1387,2N2368,2N3663,2N3825,2N3843,2N3843A, 2N3983,2N3984,2N3985,2N4292,2N4293,2SC120,2SC121,2SC124,2SC321, 2SC322,2SC382,2SC455,2SC717,BF224J,BF225J,BSX25,D16G6,EN918,GET706, KT218,MPS834,MPS918,MPS6540,MPS6542,MPS6543,MPS6545,MPS6546, MPS6547,MPS6548,MPSH19,MPSH20,MPSH37,MT1038,MT1038A,MT1039, MT1060,PET3001,SE1010,SE5025,TIS47,TIS125,ZT22,ZT42,*PTC126,*GE-60,*TR-86, *HEPS0014,*SK3124,*SN80,*ECG123,*WEP53,*BF167,*276-2011,*RT-100
2N118+	2N334,2N475,2N480A,2N717,2N842,2N844,2N1150,2N1151,2N2432A,2N2522,2S102, 2SC302,BF224J,BF225J,MPS6544,MPS6545,ZT22,ZT42,*GE-60,*HEPS0014, *ECG123,*WEP53,*RT-100
2N118A	2N162,2N162A,2N479,2N480,2N543A,2N708A,2N784,2N784A,2N840,2N1151,2N1152, 2N1278,2N2318,2N2319,2N2387,2N2432,2N2432A,2N2729,2N3340,2N3563,2N3564, 2N3605A,2N3606A,2N3688,2N3689,2N3690,2N3844,2N3844A,2N3854A,2N3983, 2N4137,2N4294,2N4295,2N5029,2N5030,2N5130,2SC54,2SC204,2SC302,2SC356, 2SC380-R,2SC380A-R,2SC387AG,2SC394-O,2SC400-R,2SC455,2SC468,2SC595, 2SC601,2SC689,2SC724,2SC752G-R,2SC963,2SC1123,2SC1126,2SC1128, 2SC1129,2SC1240,2SC1274,BF224J,BF225J,BF255,BFY39-1,EN914,EN2369A, EN3009,EN3011,EN3013,EN3014,GET708,GET914,GET2369,GET3013,GET3014, GET3646,KT218,MM1941,MPS6507,MPS6511,MPSA10-RED,MPSA20-RED,MPSH19, MT1060A,PET3001,PT2760,SE3001,SE3002,SE5001,SE5002,SE5003,SE5006,TIS22, TIS45,TIS46,TIS48,TIS49,TIS51,TIS52,TIS55,TIS129,ZT23,ZT43,*PTC126,*GE-60, *TR-86,*HEPS0014,*SK3124,*ECG123,*WEP53,*BF167,*276-2011,*RT-100
2N118A+	2N480,2N543A,2N708A,2N718,2N929,2N1152,2N1674,2N2387,2N2845,2N2847, 2N3301,2N3693,2N3826,2N3858A,2N3862,2N3946,2N3973,2N3975,2N4140,2N4966, 2N4994,2N5027,2N5824,2S103,2SC283,2SC318,2SC335,2SC336,2SC620,2SC725, 2SC815,2SC838,2SC839,2SC896,2SC938,BFY39-1,BSY93,EN697,GET2221, MM3903,PET1001,SE1001,TIS22,TIS87,ZT23,ZT43,*GE-60,*HEPS0014,*ECG123, *WEP53
2N119	2N336,2N479,2N480,2N543A,2N708,2N708A,2N784A,2N843,2N929,2N957,2N1152, 2N1278,2N1674,2N2242,2N2318,2N2319,2N2369,2N2369A,2N2387,2N2501,2N2885, 2N3009,2N3011,2N3013,2N3014,2N3340,2N3563,2N3564,2N3605A,2N3606A, 2N3688,2N3689,2N3690,2N3691,2N3693,2N3845,2N3845A,2N3855A,2N3858, 2N3862,2N4123,2N4137,2N4138,2N4294,2N4295,2N4418,2N4420,2N4421,2N4422, 2N4966,2N4968,2N4996,2N4997,2N5029,2N5030,2N5136,2N5137,2N5824,2N6566, 2SC67,2SC122,2SC169,2SC204,2SC318,2SC335,2SC356,2SC380-R,2SC380A-R, 2SC394-O,2SC400-O,2SC460,2SC461,2SC468,2SC595,2SC601,2SC601N,2SC620, 2SC634,2SC639,2SC710,2SC712,2SC713,2SC724,2SC752G-R,2SC764,2SC838,

To Replace	Substitute This Type
(2N119)	2SC839,2SC924,2SC963,2SC964,2SC1123,2SC1128,2SC1129,2SC1187,2SC1188, 2SC1189,2SC1215,2SC1240,2SC1274,2SC1319,2SC1320,2SC1372,2SC1686, 2SC1687,2SC1688,A321,BF121,BF123,BF125,BF127,BF198,BF199,BF241,BF255, BF311,BF377,BF378,BFY39-1,EN708,EN914,EN2369A,EN3009,EN3011,EN3013, EN3014,GET708,GET914,GET2369,GET3013,GET3014,GET3646,MPS3646, MPS3693,MPS6512,MPS6574-YEL,MPS6576-YEL,MPSA10-RED,MPSA20-RED, MPSH10,MPSH11,MT1060A,PET1001,PET2001,PT2760,SE1001,SE2001,SE3001, SE3005,SE5001,SE5002,SE5003,SE5006,TIS22,TIS45,TIS46,TIS48,TIS49,TIS51, TIS52,TIS55,TIS87,TIS129,TP3705,ZT23,ZT43,ZT708,*PTC126,*GE-60,*TR-86, *HEPS0014,*SK3124,*SN80,*ECG123,*WEP53,*BF167,*276-2011,*RT-100
2N119+	2N480,2N543A,2N708A,2N718,2N843,2N929,2N1152,2N1674,2N2387,2N2845, 2N2847,2N3301,2N3693,2N3826,2N3858A,2N3862,2N3903,2N3946,2N3973,2N3974, 2N3975,2N3976,2N4140,2N4966,2N4994,2N5027,2N5824,2S103,2SC283,2SC318, 2SC335,2SC336,2SC620,2SC725,2SC815,2SC838,2SC839,2SC896,2SC938, 2SC1189,2SC1320,2SC1688,BFY39-1,BSY93,EN697,GET2221,MM3903,MPS3693, MPS6565,MPS6576-YEL,PET1001,SE1001,TIS22,TIS87,TP3705,ZT23,ZT43,*GE-60, *HEPS0014,*ECG123,*WEP53
2N120	2N841,2N930,2N1153,2N2349,2N2388,2N3692,2N3694,2N3860,2N4124,2N4419, 2N4967,2N5209,2N5826,2SC367G-Y,2SC372-Y,2SC372G-Y,2SC380-Y,2SC380A-Y, 2SC394-GR,2SC400,2SC400-GR,2SC538A,2SC644,2SC711,2SC711A,2SC733-Y, 2SC735-Y,2SC752G-Y,2SC847,2SC848,2SC900,2SC912M,2SC1085,2SC1175, 2SC1285,2SC1364,2SC1632,2SC1634,2SC1684,2SC1727,2SC1849,2SC1853, 2SC1854,2SC1992,BC107,BC107A,BC108,BC108A,BC123,BC167A,BC171A, BC237A,BC238A,BCY58A,BCY58B,BCY59A,BCY59B,BSW88,BSW89,BSX38,BSX79, EN930,MPS3694,MPS6514,MPS6573,MPS6574-SIL,MPS6575,MPS6576-SIL, MPS-A10,MPS-A20,MPS-K20,MPS-K21,MPS-K22,MPSA10-BLU,MPSA10-YEL, MPSA20-BLU,MPSA20-YEL,NPSA20,PBC107A,PET1002,PET2002,SE1002,SE2002, SE4001,TIS23,TZ81,*PTC123,*GE-39,*TR-86,*HEPS0014,*SK3124,*SN80, *ECG123,*WEP53,*BC107B,*276-2011,*RT-100
2N120+	2N117+,2N332,2N472,2N472A,2N475A,2N1149,2N2520,2N2610,*GE-39,*HEPS0014, *ECG123,*WEP53
2N122	*GE-32,*TR-23
2N123	2N417,2N598,2N1284,2N1307,2N1309,2N1354,2N1355,2N1356,2N1357,2N1892, 2N1998,2SA210,ASY27,*PTC109,*GE-1,*TR-05,*HEPG0002,*SK3005,*HF6M, *ECG100,*WEP254,*AC188/01,*276-2005,*RT-118
2N123/5	2N417,2N598,2N1284,2N1307,2N1309,2N1354,2N1355,2N1356,2N1357,2N1892, 2N1998,ASY27,*PTC109,*GE-51,*HEPG0002,*AC188/01,*276-2005
2N123+	2N417,2N598,2N1284,2N1307,2N1309,2N1354,2N1355,2N1356,2N1357,2N1892, 2N1998,ASY27,*GE-1,*HEPG0002,*SK3005,*ECG100,*WEP254
2N123A	2N315A,2N316A,2N317A,2N396A,2N404,2N428,2N1017,2N1018,2N1313,2N1347, 2N1942,2SA205,2SA406,SF.T227,*PTC109,*GE-51,*TR-17,*HEPG0002,*SK3006, *ECG126,*WEP635,*AC188/01,*276-2005
2N124	2N212,2N529/N,2N530/N,2N531/N,2N585,2N634,2N1891,*PTC108,*GE-5,*TR-08, *HEPG0011,*SK3011,*NR5,*ECG101,*WEP641,*AC187/01,*RT-119
2N124+	2N124,2N182,2N212,2N529/N,2N530/N,2N531/N,2N585,2N634,2N1891,*GE-5, *TR-08,*HEPG0011,*SK3011,*ECG101,*WEP641
2N125	2N94A,2N312,2N377,2N446,2N533/N,2N557,2N576,2N634,2N635,2N1090,2N1995, 2N1996,2SC129,*PTC108,*GE-5,*TR-08,*HEPG0011,*SK3011,*NR5,*ECG101, *WEP641,*AC187/01,*RT-119
2N125+	2N94A,2N125,2N169,2N182,2N183,2N312,2N377,2N446,2N533/N,2N557,2N576, 2N634,2N635,2N1090,2N1995,2N1996,2SC11,2SC129,2SC180,2SC181,*GE-5, *TR-08,*HEPG0011,*SK3011,*ECG101,*WEP641
2N126	2N388,2N446,2N558,2N634A,2N636,2N1091,2N1114,2N1304,2N1306,2N1624,2SC90, 2SC91,2SC129,ASY29,*PTC108,*GE-5,*TR-08,*HEPG0011,*SK3011,*NR5, *ECG103A,*WEP641B,*AC187/01,*RT-122
2N126+	2N78,2N78A,2N126,2N169A,2N184,2N446,2N558,2N636,2N1091,2N1114,2N1304, 2N1306,2SC90,2SC91,2SC129,2SC181,ASY29,*GE-5,*TR-08,*HEPG0011,*SK3011, *ECG103A,*WEP641B
2N128	2SA251,*GE-51,*TR-17,*HEPG0003,*SK3008,*JR30X,*ECG126,*WEP635,*AF125
2N129	*GE-51,*TR-17,*HEPG0003,*SK3008,*JR30X,*ECG126,*WEP635,*AF125
2N130	2N104,2N111A,2N112A,2N130A,2N396A,2N404,2N404A,2N413,2N413A,2N426,

To Replace	Substitute This Type
(2N155)	*WEP232,*276-2006,*RT-127
2N156	TI-156,*DS503,*GE-3,*TR-01,*HEPG6003,*SK3009,*PT25,*WEP230,*276-2006, *RT-124
2N157	*PTC105A,*GE-25,*TR-01,*HEPG6003,*SK3009,*PT40,*ECG121,*WEP232,*OC28, *276-2006,*RT-127
2N157A	*PTC105A,*GE-25,*TR-01,*HEPG6005,*SK3009,*PT40,*ECG121,*WEP232,*OC28, *276-2006,*RT-127
2N158	2N158A,*DS503,*GE-3,*TR-01,*HEPG6005,*SK3009,*PT40,*WEP232,*RT-127
2N158A	*DS503,*GE-3,*TR-01,*HEPG6012,*SK3009,*PT40,*RT-127
2N160	2N160A,2N472,2N472A,2N475A,2N1149,2N1276,2SC478,2SC945,*PTC123,*GE-60, *TR-86,*HEPS0014,*SK3124,*SN80,*ECG123,*WEP53,*BF167,*276-2013
2N160A	2SC945,*PTC123,*GE-60,*TR-86,*HEPS0014,*SK3124,*ECG123,*WEP53,*BF167, *276-2013,*RT-100
2N161	2N161A,2N334,2N475,2N480A,2N717,2N783,2N834A,2N842,2N844,2N1150,2N1277, 2N2368,2S102,2SC120,2SC121,2SC124,2SC321,2SC322,2SC382,BF224J,BF225J, BSX25,GET706,KT218,MPS834,MPS6544,MPS6545,MPSH20,MPSH37,TIS47, TIS125,ZT22,ZT42,*PTC123,*GE-60,*TR-86,*HEPS0014,*SK3124,*SN80,*ECG123, *WEP53,*BF167,*276-2013,*RT-100
2N161A	2N717,2N783,2N834A,2S102,2SC321,2SC322,BSX25,GET706,MPS834,MPSH37, ZT22,ZT42,*PTC123,*GE-60,*TR-86,*HEPS0014,*SK3124,*ECG123,*WEP53, *BF167,*276-2013,*RT-100
2N162	2N162A,2N480,2N543A,2N708,2N708A,2N718,2N784A,2N840,2N929,2N957,2N1151, 2N1152,2N1278,2N1674,2N2242,2N2369,2N2369A,2N2387,2N2432A,2N2522, 2N2845,2N2885,2N3009,2N3013,2N3014,2N3301,2N3605A,2N3606A,2N3688, 2N3689,2N3690,2N3693,2N3826,2N3946,2N3973,2N3975,2N4137,2N4140,2N4295, 2N4420,2N4421,2N4422,2N4994,2N5029,2S103,2SC54,2SC67,2SC283,2SC302, 2SC335,2SC336,2SC468,2SC601,2SC601N,2SC689,2SC725,2SC752G-R,2SC815, 2SC838,2SC839,2SC896,2SC938,2SC1240,2SC1274,2SC1686,BF198,BF241, BFY39-1,BSY93,EN697,EN708,EN914,EN2369A,EN3009,EN3013,EN3014,GET708, GET914,GET2221,GET2369,GET3013,GET3014,GET3646,MPS3646,MPS6512, MPSA10-RED,MPSA20-RED,PET1001,SE1001,SE5001,SE5002,SE5003,SE5006, TIS22,TIS45,TIS46,TIS48,TIS49,TIS52,TIS55,TIS129,ZT23,ZT43,ZT708,*PTC123, *GE-60,*TR-70,*HEPS0014,*SK3039,*SN80,*ECG107,*WEP56,*BF167,*276-2013, *RT-108
2N162A	2N708,2N708A,2N718,2N784A,2N929,2N957,2N2242,2N2387,2N2432A,2N2522, 2N2845,2N2885,2N3013,2N3014,2N3301,2N3605A,2N3606A,2N3946,2N3973, 2N3975,2N4140,2N4295,2N4420,2N4421,2N4422,2S103,2SC54,2SC67,2SC283, 2SC302,2SC335,2SC336,2SC468,2SC601,2SC601N,2SC689,2SC725,2SC752G-R, 2SC815,2SC838,2SC839,2SC938,2SC1274,BFY39-1,BSY93,EN697,EN708, EN914,EN3013,EN3014,GET708,GET914,GET2221,GET2369,GET3013,GET3014, GET3646,MPS3646,TIS22,TIS45,TIS46,TIS52,TIS55,ZT23,ZT43,ZT708,*PTC123, *GE-60,*TR-70,*HEPS0014,*SK3039,*ECG107,*WEP56,*BF167,*276-2013,*RT-108
2N163	2N162,2N162A,2N480,2N543A,2N708,2N708A,2N718,2N784A,2N840,2N929,2N957, 2N1151,2N1152,2N1278,2N1674,2N2242,2N2369,2N2369A,2N2387,2N2432A, 2N2522,2N2845,2N2885,2N3009,2N3013,2N3014,2N3301,2N3605A,2N3606A, 2N3688,2N3689,2N3690,2N3693,2N3826,2N3946,2N3973,2N3975,2N4137,2N4140, 2N4295,2N4420,2N4421,2N4422,2N4994,2N5029,2S103,2SC54,2SC67,2SC283, 2SC302,2SC335,2SC336,2SC468,2SC601,2SC601N,2SC689,2SC725,2SC752G-R, 2SC815,2SC838,2SC839,2SC896,2SC938,2SC1240,2SC1274,2SC1686,BF198, BF241,BFY39-1,BSY93,EN697,EN708,EN914,EN2369A,EN3009,EN3013,EN3014, GET708,GET914,GET2221,GET2369,GET3013,GET3014,GET3646,MPS3646, MPS6512,MPSA10-RED,MPSA20-RED,PET1001,SE1001,SE5001,SE5002,SE5003, SE5006,TIS22,TIS45,TIS46,TIS48,TIS49,TIS52,TIS55,TIS129,ZT23,ZT43,ZT708, *PTC123,*GE-60,*TR-86,*HEPS0014,*SK3124,*SN80,*ECG123,*WEP53,*BF167, *276-2013,*RT-100
2N163A	2N162A,2N708,2N708A,2N718,2N784A,2N929,2N957,2N2242,2N2387,2N2432A, 2N2522,2N2845,2N2885,2N3013,2N3014,2N3301,2N3605A,2N3606A,2N3946, 2N3973,2N3975,2N4140,2N4295,2N4420,2N4421,2N4422,2S103,2SC54,2SC67, 2SC283,2SC302,2SC335,2SC336,2SC468,2SC601,2SC601N,2SC689,2SC725, 2SC752G-R,2SC815,2SC838,2SC839,2SC938,2SC1274,BFY39-1,BSY93,EN697, EN708,EN914,EN3013,EN3014,GET708,GET914,GET2221,GET2369,GET3013,

To Replace	Substitute This Type
(2N163A)	GET3014,GET3646,MPS3646,TIS22,TIS45,TIS46,TIS52,TIS55,ZT23,ZT43,ZT708, *PTC123,*GE-60,*TR-86,*HEPS0014,*SK3124,*ECG123,*WEP53,*276-2013, *RT-100
2N164	2N94A,2N164A,2N312,2N377,2N385,2N439,2N439A,2N440,2N440A,2N446,2N533/N, 2N557,2N576,2N635,2N636,2N1090,2N1091,2N1304,2N1605,2N1995,2N1996, 2SC90,2SC128,2SC129,ASY28,*PTC108,*GE-5,*TR-08,*HEPG0011,*SK3011, *NR10,*ECG101,*WEP641,*AC187/01,*RT-119
2N164A	2N94A,2N312,2N377,2N385,2N439,2N439A,2N440,2N440A,2N446,2N533/N,2N557, 2N576,2N635,2N636,2N1090,2N1091,2N1304,2N1605,2N1995,2N1996,2SC90, 2SC128,2SC129,ASY28,*PTC108,*GE-5,*TR-08,*HEPG0011,*SK3011,*NR10, *ECG101,*WEP641,*AC187/01,*RT-119
2N165	2N78,2N78A,2N169A,2N440,2N440A,2N446,2N558,2N636,2N797,2N1091,2N1114, 2N1304,2N1306,2SC90,2SC91,2SC129,ASY28,ASY29,*PTC108,*GE-7,*TR-08, *HEPG0011,*SK3011,*NR10,*ECG101,*WEP641,*AC187/01,*RT-119
2N166	2N94A,2N164,2N164A,2N216,2N312,2N377,2N445,2N531/N,2N532/N,2N533/N, 2N557,2N576,2N585,2N585+,2N634,2N635,2N679,2N1090,2N1302,2N1605,2N1891, 2N1995,2N1996,*PTC108,*GE-5,*TR-08,*HEPG0011,*SK3011,*NR10,*ECG101, *WEP641,*AC127,*276-2001,*RT-119
2N167	2N357A,2N358A,2N377A,2N439,2N439A,2N440,2N440A,2N1000,2SC128,ASY28, SF.T298,*PTC108,*GE-6,*TR-08,*HEPG0011,*SK3011,*NR10,*ECG101,*WEP641, *RT-119
2N167A	2N440,2N440A,SF.T298,*GE-6,*TR-08,*HEPG0011,*SK3011,*ECG101,*WEP641, *RT-119
2N168	2N78,2N78A,2N146,2N165,2N169A,2N388,2N440,2N440A,2N446A,2N558,2N634A, 2N636,2N797,2N1091,2N1114,2N1304,2N1306,2N1308,2N1624,2SC90,2SC91, ASY29,*GE-5,*TR-08,*HEPG0011,*SK3011,*NR5,*ECG101,*WEP641,*AC187/01, *RT-119
2N168+	2N292,2N292A,2N634,2N1366,2N1891,*GE-5,*HEPG0011,*SK3011,*WEP641
2N168A	2N78,2N78A,2N146,2N165,2N169A,2N388,2N440,2N440A,2N446A,2N558,2N634A, 2N636,2N797,2N1091,2N1114,2N1304,2N1306,2N1308,2N1624,2SC90,2SC91, ASY29,*GE-5,*TR-08,*HEPG0011,*SK3011,*NR5,*ECG101,*WEP641,*AC187/01, *RT-119
2N168A+	2N94A,2N145,2N167,2N169,2N169A+,2N312,2N377,2N439,2N439A,2N446,2N533/N, 2N557,2N576,2N635,2N1090,2N1198,2N1217,2N1995,2N1996,2SC128,2SC129, 2SC180,2SC181,ASY28,OC139,*GE-5,*HEPG0011,*SK3011,*ECG101,*WEP641
2N169	2N312,2N440,2N440A,2N557,2N635,2N636,2N1091,2N1996,2SC128,2SC129,ASY28, *PTC108,*GE-7,*TR-08,*HEPG0011,*SK3011,*NR10,*ECG101,*WEP641,*RT-119
2N169+	2N94A,2N145,2N164,2N164A,2N167,2N169,2N169A+,2N312,2N377,2N439,2N439A, 2N446,2N532/N,2N533/N,2N557,2N576,2N635,2N1090,2N1198,2N1217,2N1605, 2N1995,2N1996,2SC11,2SC128,2SC129,2SC180,2SC181,2SD75,2SD77,ASY28, OC139,*PTC108,*GE-7,*TR-08,*HEPG0011,*SK3011,*ECG101,*WEP641
2N169A	2N440,2N440A,2N1091,2N1306,2SC91,ASY29,*PTC108,*GE-5,*TR-08,*HEPG0011, *SK3011,*NR10,*ECG101,*WEP641,*RT-119
2N169A+	2N440,2N440A,2N1091,2SC129,2SC181,ASY28,*GE-5,*TR-08,*HEPG0011,*SK3011, *ECG101,*WEP641
2N170	2N94A,2N125,2N164,2N164A,2N169,2N312,2N377,2N531/N,2N532/N,2N533/N, 2N557,2N576,2N634,2N635,2N1090,2N1605,2N1891,2N1995,2N1996,*GE-7,*TR-08, *HEPG0011,*SK3011,*NR5,*ECG101,*WEP641,*AC187/01,*RT-119
2N172	2N94A,2N164A,2N312,2N377,2N439,2N439A,2N446,2N532/N,2N533/N,2N557, 2N576,2N585+,2N635,2N1090,2N1302,2N1605,2N1995,2N1996,2SC89,2SC128, 2SC129,ASY28,*PTC108,*GE-5,*TR-08,*HEPG0011,*SK3011,*NR5,*ECG101, *WEP641,*AC187/01,*RT-119
2N172+	*TR-08,*HEPG0011,*SK3011,*ECG101,*WEP641
2N173	2N1099,*GE-4,*TR-03,*HEPG6006,*SK3012,*PT501,*ECG105,*WEP233,*276-2006
2N174	2N1100,*GE-4,*TR-03,*HEPG6006,*SK3012,*PT501,*ECG105,*WEP233,*276-2006
2N174A	*GE-4,*TR-03,*HEPG6006,*SK3012,*PT515,*ECG105,*WEP233,*276-2006
2N175	2N132,2N132A,2N220,2N415,2N415A,2N416,2N422A,2N466,2N654,2N654+,2N1125, 2N1192,2N1305,2N1307,2N1309A,2N1348,2N1349,2N1350,2N1351,2N1352,2N1356, 2N1374,2N1376,2N1381,2N1382,2N1383,2N1681,2SA207,2SB187,2SB188,2SB299, 2SB303,40269,ASY27,SF.T223,*PTC109,*GE-51,*TR-85,*HEPG0005,*SK3003, *AT10H,*ECG102A,*WEP250,*AF125,*276-2007,*RT-121

To Replace	Substitute This Type
2N175+	2N711B,2N796,2N829,2N964A,2N972,2N973,2N974,2N982,2N983,2N2022,2N2273, 2N2928,2N3320,2N3371,2SA52,2SA53,2SA79,2SA138,2SA139,2SA350,2SA351, 2SA352,2SA353,2SA354,2SA355,2SA408,2SA409,2SA417,2SA451,2SB33,2SB34, 2SB60A,2SB75,2SB77,2SB91,2SB114,2SB115,2SB154,2SB184,2SB187,2SB188, 2SB199,2SB263,2SB303,2SB328,2SB329,2SB364,40359,40488,AFY15,MM404,OC44, OC45,OC47,SF.T307,SF.T308,SF.T317,SF.T320,SF.T323,SF.T353,*GE-51,*TR-85, *HEPG0005,*SK3003,*ECG102A,*WEP250
2N176	2N242,2N380,2N1168,2N1359,2N1360,*PTC105A,*GE-3,*TR-01,*HEPG6013, *SK3009,*PT25,*ECG104,*WEP230,*AD149,*276-2004,*RT-124
2N178	2N663,*PTC105A,*GE-3,*TR-01,*HEPG6003,*SK3009,*PT25,*ECG104,*WEP230, *OC28,*276-2006,*RT-124
2N179	*PTC105A,*GE-16,*TR-01,*HEPG6003,*SK3009,*ECG104,*WEP230,*276-2006, *RT-124
2N180	2N59A,2N59B,2N60A,2N60B,2N360,2N381,2N414,2N414A,2N414B,2N414C,2N415, 2N415A,2N416,2N422A,2N461,2N466,2N518,2N567,2N568,2N569,2N571,2N654, 2N654+,2N838,2N1008A,2N1125,2N1130,2N1171,2N1175,2N1175A,2N1192, 2N1305,2N1307,2N1309A,2N1348,2N1349,2N1350,2N1351,2N1352,2N1356,2N1371, 2N1373,2N1375,2N1377,2N1415,2N1447,2N1448,2N1451,2N1452,2N1681, 2N1707,2N2956,2N2957,2N3323,2N3324,2N3325,2N6365,2SA204,2SA207,2SA407, 2SB221,2SB222,2SB299,MM404A,NKT224,NKT226,SF.T223,*GE-53,*TR-85, *HEPG0005,*SK3004,*AT30H,*ECG102A,*WEP250,*276-2005,*RT-121
2N180+	*GE-53,*HEPG0005,*SK3004,*ECG102A,*WEP250
2N181	2N1008A,2N1125,2N1373,2N1375,2N1377,AFY11,*PTC135,*GE-53,*TR-85, *HEPG0005,*SK3004,*AT30H,*ECG102A,*WEP250,*276-2005,*RT-121
2N181+	*GE-53,*HEPG0005,*SK3004,*ECG102A,*WEP250
2N182	2N183,2N576A,2N587,*PTC108,*GE-5,*TR-08,*HEPG0011,*SK3011,*NR5,*ECG101, *WEP641,*276-2001,*RT-119
2N182+	2SC180,*PTC108,*GE-5,*TR-08,*SK3011,*ECG101,*WEP641,*AC127
2N183	2N440,2N440A,2N576A,2N1091,2SC129,ASY28,*PTC108,*GE-5,*TR-08,*HEPG0011, *SK3011,*NR5,*ECG101,*WEP641,*276-2001,*RT-119
2N183+	2SC181,*PTC108,*GE-5,*TR-08,*HEPG0011,*SK3011,*ECG101,*WEP641, *AC187/01
2N184	2N1306,2N1308,2SC91,ASY29,*PTC108,*GE-6,*TR-08,*HEPG0011,*SK3011,*NR10, *ECG101,*WEP641,*RT-119
2N184+	*PTC108,*GE-6,*TR-08,*HEPG0011,*SK3011,*ECG101,*WEP641
2N185	2N61,2N317A,2N403,2N404,2N413,2N565,2N611,2N653,2N1017,2N1303,2N1313, 2N1413,2N1414,2N1942,2SA205,2SA206,2SA406,2SB185,2SB220,2SB220A,ASY26, SF.T222,SF.T227,*PTC135,*GE-52,*TR-85,*HEPG0005,*SK3004,*AT30M, *ECG102A,*WEP250,*AC188/01,*RT-121
2N185+	*GE-52,*TR-85,*HEPG0005,*SK3004,*ECG102A,*WEP250
2N186	2N111A,2N112A,2N186A,2N186A+,2N187,2N187A,2N315A,2N316A,2N317A,2N319, 2N396A,2N404,2N404A,2N413,2N413A,2N426,2N427,2N428,2N573,2N1017,2N1018, 2N1191,2N1413,2SA205,2SB219,2SB219A,SF.T221,SF.T226,*PTC102,*GE-53, *TR-85,*HEPG0005,*SK3004,*AT30M,*ECG102A,*WEP250,*276-2007,*RT-121
2N186+	2N111A,2N112A,2N186,2N186A,2N186A+,2N187,2N187A,2N315A,2N316A,2N317A, 2N319,2N396A,2N404,2N404A,2N413,2N413A,2N426,2N427,2N428,2N573,2N1017, 2N1018,2N1191,2N1413,2SA205,2SB219,2SB219A,SF.T221,SF.T226,*PTC102, *GE-53,*TR-85,*HEPG0005,*SK3004,*ECG102A,*WEP250
2N186A	2N187A,2N319,2N396A,2N573,2N1018,2N1191,2N1413,2SA205,2SB219,2SB219A, SF.T221,*PTC102,*GE-53,*TR-85,*HEPG0005,*SK3004,*AT30M,*ECG102A, *WEP250,*276-2004,*RT-121
2N186A+	2N186A,2N187A,2N319,2N396A,2N573,2N1018,2N1191,2N1413,2SA205,2SB219, 2SB219A,SF.T221,*GE-53,*TR-85,*HEPG0005,*SK3004,*ECG102A,*WEP250
2N187	2N61,2N61A,2N187A,2N188,2N188A,2N188A+,2N320,2N361,2N363,2N404,2N413, 2N414,2N414A,2N414B,2N414C,2N505,2N653,2N1017,2N1171,2N1191,2N1303, 2N1313,2N1414,2N1478,2SA204,2SA205,2SA206,2SA406,2SA407,2SB220, 2SB220A,ASY26,SF.T222,SF.T227,*PTC102,*GE-53,*TR-85,*HEPG0005,*SK3004, *AT30M,*ECG102A,*WEP250,*276-2007,*RT-121
2N187+	2N61,2N61A,2N187,2N187A,2N188,2N188A,2N188A+,2N320,2N361,2N363,2N404, 2N413,2N414,2N414A,2N414B,2N414C,2N505,2N653,2N1017,2N1171,2N1191, 2N1303,2N1313,2N1414,2N1478,2SA204,2SA205,2SA206,2SA406,2SA407,2SB220,

To Replace	Substitute This Type
(2N187A)	2SB220A,ASY26,SF.T222,SF.T227,*PTC102,*GE-53,*TR-85,*HEPG0005,*SK3004,*ECG102A,*WEP250
2N187A	2N188A,2N320,2N414B,2N414C,2N653,2N1191,2N1414,2N1478,2SA205,2SA206,2SA406,2SA407,2SB220,2SB220A,SF.T222,*PTC102,*GE-53,*TR-05,*HEPG0005,*SK3004,*AT30M,*ECG102,*WEP631,*276-2004,*RT-120
2N187A+	2N61,2N61A,2N187A,2N188A,2N188A+,2N320,2N361,2N363,2N414B,2N414C,2N653,2N1191,2N1313,2N1414,2N1478,2SA205,2SA206,2SA406,2SA407,2SB220,2SB220A,SF.T222,*GE-53,*TR-05,*HEPG0005,*SK3004,*ECG102,*WEP631
2N188	2N60,2N60A,2N61,2N61A,2N188A,2N188A+,2N241,2N241A,2N241A+,2N320,2N321,2N363,2N414,2N414A,2N414B,2N414C,2N415,2N415A,2N416,2N422A,2N505,2N569,2N570,2N572,2N654,2N1171,2N1175,2N1192,2N1303,2N1305,2N1309A,2N1313,2N1348,2N1349,2N1350,2N1351,2N1383,2N1414,2N1415,2N1478,2N1681,2SA204,2SA206,2SA207,2SA406,2SA407,2SB220,2SB220A,2SB221,2SB221A,2SB326,2SB328,ASY26,SF.T222,SF.T228,*PTC102,*GE-53,*TR-05,*HEPG0005,*SK3004,*AT30H,*ECG102,*WEP631,*RT-120
2N188+	2N59A,2N60,2N60A,2N188A+,2N241,2N241A,2N241A+,2N321,2N360,2N414,2N414A,2N414B,2N414C,2N415,2N415A,2N416,2N422A,2N505,2N520A,2N569,2N570,2N572,2N654,2N654+,2N1128,2N1171,2N1175,2N1175A,2N1192,2N1305,2N1309A,2N1348,2N1349,2N1350,2N1351,2N1374,2N1382,2N1383,2N1415,2N1478,2N1681,2N1706,2N1707,2SA204,2SA206,2SA207,2SA407,2SB221,2SB221A,2SB326,2SB328,ACY23,ACY32,ASY26,SF.T228,*PTC102,*GE-53,*HEPG0005,*SK3004,*ECG102,*WEP631
2N188A	2N241A,2N241A+,2N320,2N321,2N414B,2N414C,2N422A,2N654,2N1175,2N1192,2N1348,2N1349,2N1350,2N1351,2N1383,2N1414,2N1415,2N1478,2SA206,2SA207,2SA406,2SA407,2SB220,2SB220A,2SB221,2SB221A,2SB326,SF.T222,*PTC102,*GE-53,*TR-05,*HEPG0005,*SK3004,*AT30H,*ECG102,*WEP631,*276-2005,*RT-120
2N188A+	2N59A,2N60,2N60A,2N241A,2N241A+,2N321,2N360,2N414B,2N414C,2N422A,2N654,2N654+,2N1175,2N1175A,2N1192,2N1348,2N1349,2N1350,2N1351,2N1374,2N1382,2N1383,2N1415,2N1478,2N1681,2N1706,2N1707,2SA206,2SA207,2SA407,2SB221,2SB221A,2SB326,*PTC102,*GE-53,*TR-05,*HEPG0005,*SK3004,*ECG102,*WEP631
2N189	2N61,2N61A,2N131A,2N133A,2N187,2N187A,2N188,2N188A,2N188A+,2N206,2N215,2N269,2N320,2N361,2N363,2N404,2N413,2N414,2N414A,2N414B,2N414C,2N502B,2N505,2N567,2N568,2N582,2N584,2N653,2N1017,2N1171,2N1191,2N1303,2N1313,2N1372,2N1413,2N1414,2N1450,2N1478,2N1495,2N1499B,2N2587,2N2795,2N2955,2N3883,2N6365,2N6365A,2SA204,2SA205,2SA206,2SA212,2SA406,2SA407,2SB185,2SB220,2SB220A,ASY26,MM380,NKT221,NKT224,NKT228,SF.T222,SF.T227,*GE-53,*TR-05,*HEPG0005,*SK3004,*AT30N,*ECG102,*WEP631,*276-2005,*RT-120
2N189+	2N186A,2N186A+,2N187A,2N319,2N396A,2N573,2N1018,2N1191,2N1413,2N2381,2SA205,2SB219,2SB219A,NKT225,SF.T221,*GE-53,*HEPG0005,*SK3004,*ECG102,*WEP631
2N190	2N60,2N60A,2N61,2N61A,2N188,2N188A,2N188A+,2N241,2N241A,2N241A+,2N320,2N321,2N363,2N414,2N414A,2N414B,2N414C,2N415,2N415A,2N416,2N422A,2N466,2N502A,2N502B,2N505,2N569,2N570,2N572,2N582,2N584,2N654,2N838,2N1171,2N1175,2N1192,2N1303,2N1305,2N1309A,2N1313,2N1348,2N1349,2N1350,2N1351,2N1383,2N1414,2N1415,2N1478,2N1495,2N1681,2N2587,2N2795,2N2956,2N3323,2N3324,2N3325,2N3883,2N6365,2N6365A,2SA204,2SA206,2SA207,2SA212,2SA406,2SA407,2SB103,2SB114,2SB220,2SB220A,2SB221,2SB221A,2SB326,2SB328,ASY26,MM380,SF.T222,SF.T228,*GE-53,*TR-05,*HEPG0005,*SK3004,*AT30M,*ECG102,*WEP631,*276-2004,*RT-120
2N190+	2N61,2N61A,2N187A,2N188A,2N188A+,2N320,2N361,2N363,2N414B,2N414C,2N653,2N1191,2N1313,2N1414,2N1478,2N1495,2N3883,2SA205,2SA206,2SA406,2SA407,2SB185,2SB220,2SB220A,MM380,SF.T222,*GE-53,*HEPG0005,*SK3004,*ECG102,*WEP631
2N191	2N59A,2N132A,2N241,2N241A,2N321,2N360,2N362,2N415,2N415A,2N416,2N467,2N520A,2N569,2N570,2N571,2N572,2N654,2N654+,2N1128,2N1175A,2N1274,2N1307,2N1309A,2N1352,2N1356,2N1370,2N1374,2N1376,2N1381,2N1382,2N1706,2N1707,2N1808,2N1892,2N2273,2N2956,2N2957,2N3323,2N3324,2N3325,2N3371,2SB115,2SB116,2SB186,2SB187,2SB188,2SB222,2SB222A,2SB327,

To Replace	Substitute This Type
(2N191)	2SB329,40269,ASY27,SF.T223,*GE-53,*TR-05,*HEPG0005,*SK3004,*AT30H, *ECG102,*WEP631,*276-2002,*RT-120
2N191+	2N59A,2N60,2N60A,2N188A+,2N241A,2N241A+,2N321,2N360,2N414B,2N414C, 2N422A,2N654,2N654+,2N1175,2N1175A,2N1192,2N1348,2N1349,2N1350,2N1351, 2N1374,2N1382,2N1383,2N1415,2N1478,2N1681,2N1706,2N1707,2N3883,2SA206, 2SA207,2SA407,2SB221,2SB221A,2SB326,*GE-53,*HEPG0005,*SK3004,*ECG102, *WEP631
2N192	2N359,2N417,2N508A,2N521A,2N598,2N599,2N655,2N655+,2N1193,2N1309, 2N1354,2N1355,2N1357,2N1808,2N1892,2N1998,2N2635,2SB117,2SB223,2SB223A, MA1703,*GE-53,*TR-05,*HEPG0005,*SK3004,*AT30H,*ECG102,*WEP631,*RT-120
2N192+	2N59A,2N241A,2N321,2N360,2N362,2N422A,2N654,2N654+,2N1008A,2N1175A, 2N1192,2N1349,2N1356,2N1381,2N1382,2N1681,2N1706,2N1707,2SB222,2SB222A, 2SB327,SF.T223,*GE-53,*HEPG0005,*SK3004,*ECG102,*WEP631
2N193	2N94,2N194A,2N233A,*PTC108,*GE-5,*TR-08,*HEPG0011,*SK3011,*NR5, *ECG103A,*WEP641B,*AC187/01,*RT-122
2N193+	2N94A,2N357A,2N358A,2N377,2N385,2N439,2N439A,2N440,2N440A,2N576,2N635, 2N636,2N679,2N1302,2N1304,2N1605,2N1995,2N1996,2SC128,2SC129,2SD11, 2SD38,2SD75,2SD77,AC127,ASY28,SF.T298,*PTC108,*GE-5,*HEPG0011,*SK3011, *ECG103A,*WEP641B
2N194	2N94,2N193,2N194A,2N212,2N233A,*PTC108,*GE-5,*TR-08,*HEPG0011,*SK3011, *NR10,*ECG103A,*WEP641B,*AC187/01,*RT-122
2N194+	*GE-5,*TR-08,*HEPG0011,*SK3011,*ECG103A,*WEP641B
2N194A	2N94,2N193,2N212,2N233A,*PTC108,*GE-5,*TR-08,*HEPG0011,*SK3011,*NR10, *ECG103A,*WEP641B,*AC187/01,*RT-122
2N194A+	2N94,*PTC108,*GE-5,*TR-08,*HEPG0011,*SK3011,*ECG103A,*WEP641B
2N195	2N522A,2N523A,2N1316,2N1317,2N1379,2N2613,2SB400,*GE-52,*TR-05, *HEPG0005,*SK3004,*AT20M,*ECG102,*WEP631,*AC126,*276-2005,*RT-120
2N196	2N60A,2N60B,2N61A,2N61B,2N363,2N381,2N414,2N414A,2N414B,2N414C,2N415, 2N415A,2N416,2N422A,2N466,2N505,2N518,2N568,2N570,2N572,2N654,2N1171, 2N1175,2N1192,2N1303,2N1305,2N1309A,2N1313,2N1348,2N1349,2N1350, 2N1351,2N1373,2N1414,2N1415,2N1447,2N1451,2N1681,2SA204,2SA206,2SA207, 2SA406,2SA407,2SB103,2SB220,2SB221,2SB299,ASY26,NKT224,SF.T222, *GE-53,*TR-05,*HEPG0005,*SK3004,*AT20M,*ECG102,*WEP631,*276-2004, *RT-120
2N197	2N61A,2N61B,2N196,2N215,2N361,2N363,2N381,2N413,2N414,2N414A,2N414B, 2N414C,2N466,2N505,2N518,2N568,2N1017,2N1171,2N1191,2N1303,2N1313, 2N1348,2N1350,2N1351,2N1373,2N1414,2N1446,2N1447,2N1451,2SA204,2SA205, 2SA206,2SA207,2SA406,2SA407,2SB220,2SB299,2SB397,ASY26,NKT224,SF.T222, SF.T227,*GE-53,*TR-05,*HEPG0005,*SK3004,*AT20M,*ECG102,*WEP631, *276-2004,*RT-120
2N198	2N61A,2N61B,2N104,2N111A,2N112A,2N196,2N197,2N215,2N283,2N361,2N363, 2N396A,2N404A,2N413,2N428,2N465,2N566,2N573,2N1017,2N1018,2N1056, 2N1191,2N1303,2N1313,2N1413,2N1446,2SA205,2SA406,2SB219,2SB220,2SB397, NKT225,SF.T221,SF.T222,SF.T226,SF.T227,*GE-53,*TR-05,*HEPG0005,*SK3004, *AT20M,*ECG102,*WEP631,*276-2004,*RT-120
2N199	2N413A,2N425,2N426,2N427,2N460,2N464,2N1408,*GE-53,*TR-05,*HEPG0005, *SK3004,*AT20M,*ECG102,*WEP631,*276-2004,*RT-120
2N200	2N404A,2N465,2N524,2N524A,2N525,2N525A,2N573,2N650,2N650A,2N1056,2N1191, 2N1446,2SB224,MA886,SF.T226,*GE-2,*TR-05,*HEPG0005,*AT20M,*ECG102, *WEP631,*276-2004,*RT-120
2N204	2N205,2N464,MA885,*GE-2,*TR-05,*HEPG0011,*SK3004,*AT20M,*ECG102, *WEP631,*276-2001,*RT-120
2N205	2N204,2N464,MA885,*GE-2,*TR-05,*HEPG0011,*SK3004,*AT20M,*ECG102, *WEP631,*276-2001,*RT-120
2N206	2N381,2N414,2N414A,2N414B,2N414C,2N415,2N415A,2N416,2N422A,2N466,2N505, 2N518,2N525,2N525A,2N526,2N526A,2N597,2N650,2N650A,2N651,2N651A,2N652A, 2N654,2N1171,2N1192,2N1303,2N1305,2N1309A,2N1313,2N1348,2N1349,2N1350, 2N1351,2N1373,2N1447,2N1478,2N1681,2SA204,2SA206,2SA207,2SA406,2SA407, 2SB57A,2SB225,2SB226,2SB299,2SB304A,2SB326,2SB427,ASY26,MA887,SF.T222, *PTC135,*GE-53,*TR-05,*HEPG0005,*SK3004,*AT30M,*ECG102,*WEP631, *276-2005,*RT-120

To Replace	Substitute This Type
2N207	2N207A,2N207B,2N582,2N584,2SA209,2SA212,2SA251,SF.T228,*PTC109,*GE-52, *TR-05,*HEPG0005,*SK3003,*JR30,*ECG102,*WEP631,*AC126,*RT-120
2N207A	2N207,2N207B,2N582,2N584,2SA209,2SA212,2SA251,SF.T228,*PTC109,*GE-52, *TR-05,*HEPG0005,*SK3003,*JR30,*ECG102,*WEP631,*AC126,*276-2004,*RT-120
2N207B	2N207,2N207A,2N582,2N584,2SA209,2SA212,2SA251,SF.T228,*PTC109,*GE-52, *TR-05,*HEPG0005,*SK3003,*ECG102,*WEP631,*AC126,*276-2004,*RT-120
2N207B+	2N705,2N710,2N779,2N779A,2N828,2N828A,2N934,2N964,2N965,2N966,2N967, 2N968,2N969,2N970,2N980,2N2401,2N2487,2N2488,2N2489,2N2717,2N2795, 2N2860,2N3322,2SA56,2SA405,2SA410,2SA411,2SA450,ASZ21,MM380,*GE-52, *TR-05,*HEPG0005,*SK3003,*ECG102,*WEP631
2N211	2N94,2N193,2N194,2N194A,2N212,2N233,2N233A,2N529/N,*GE-5,*TR-08, *HEPG0011,*SK3011,*NR5,*ECG103A,*WEP641B,*AC127,*RT-122
2N211+	2N94,2N193,2N194,2N194A,2N212,2N233,2N233A,2N529/N,*GE-5,*HEPG0011, *SK3011,*ECG103A,*WEP641B
2N212	2N634,2N1891,*PTC108,*GE-5,*TR-08,*HEPG0011,*SK3011,*NR10,*ECG103A, *WEP641B,*AC187/01,*RT-122
2N212+	2N94A,2N357A,2N358A,2N377,2N385,2N439,2N439A,2N440,2N440A,2N576,2N635, 2N636,2N1304,2N1605,2N1995,2N1996,2SC128,2SC129,2SD75,2SD77,ASY28, SF.T298,*PTC108,*GE-5,*TR-08,*HEPG0011,*SK3011,*ECG103A,*WEP641B
2N213	2N213A,*PTC134,*GE-8,*TR-08,*HEPG0011,*SK3010,*NA20,*ECG103A,*WEP641B, *276-2001,*RT-129
2N213+	2N213,2N213A,2N388A,*PTC134,*GE-8,*HEPG0011,*SK3010,*ECG103A,*WEP641B
2N213A	*PTC134,*GE-8,*TR-08,*HEPG0011,*SK3010,*NA20,*ECG103A,*WEP641B, *276-2001,*RT-122
2N213A+	2N213,2N213A,2N388A,*PTC108,*GE-8,*HEPG0011,*SK3010,*ECG103A,*WEP641B
2N214	*PTC108,*GE-8,*TR-08,*HEPG0011,*SK3010,*NA20,*ECG103A,*WEP641B, *276-2001,*RT-122
2N214+	2N213,2N213A,*PTC134,*GE-8,*HEPG0011,*SK3010,*ECG103A,*WEP641B
2N215	2N381,2N414,2N414A,2N414B,2N414C,2N466,2N518,2N525,2N525A,2N526,2N526A, 2N650,2N650A,2N653,2N1171,2N1192,2N1303,2N1305,2N1313,2N1348,2N1350, 2N1351,2N1373,2N1446,2N1447,2N1478,2SA204,2SA206,2SA207,2SA406,2SA407, 2SB57A,2SB225,2SB226,2SB299,2SB304A,2SB326,2SB397,2SB427,ASY26,MA886, SF.T222,SF.T227,*PTC109,*GE-52,*TR-85,*HEPG0005,*SK3004,*AT30H,*ECG102A, *WEP250,*AC126,*276-2005,*RT-121
2N215+	2SB37,2SB75A,*PTC109,*GE-52,*TR-85,*HEPG0005,*SK3004,*ECG102A,*WEP250
2N216	2N94A,2N356,2N356A,2N357,2N357A,2N358,2N358A,2N377,2N439,2N439A,2N557, 2N576,2N585+,2N635,2N679,2N1090,2N1302,2N1605,2N1995,2N1996,2SC89, 2SC128,2SC129,ASY28,*PTC108,*GE-5,*TR-08,*HEPG0011,*SK3011,*NR5, *ECG103A,*WEP641B,*AC187/01,*RT-122
2N216+	2N94A,2N145,2N164,2N164A,2N169,2N172,2N183,2N216,2N312,2N377,2N439, 2N439A,2N445,2N446,2N532/N,2N533/N,2N557,2N576,2N585+,2N635,2N679, 2N1090,2N1198,2N1217,2N1302,2N1605,2N1995,2N1996,2SC11,2SC89,2SC128, 2SC129,2SC179,2SC180,2SC181,2SD75,2SD77,ASY28,*GE-5,*HEPG0011, *SK3011,*ECG103A,*WEP641B
2N217	2N109,2N415,2N415A,2N416,2N422A,2N467,2N654,2N654+,2N1008A,2N1125, 2N1307,2N1309A,2N1349,2N1352,2N1356,2N1374,2N1376,2N1381,2N1382,2N1681, 2N1808,2N1892,2SB186,2SB187,2SB188,2SB327,40269,ASY27,SF.T223,*PTC135, *GE-53,*TR-85,*HEPG0008,*SK3004,*AT30H,*ECG102A,*WEP250,*276-2005, *RT-121
2N217+	2SB370A,2SB431,*PTC135,*GE-53,*TR-85,*HEPG0008,*SK3004,*ECG102A, *WEP250
2N218	2N139,2N414,2N414B,2N414C,2N415,2N415A,2N416,2N582,2N584,2N1171,2N1305, 2N1313,2N1681,2SA204,2SA206,2SA207,2SA209,2SA212,2SA251,2SA406,2SA407, ASY26,SF.T228,*PTC135,*GE-50,*TR-17,*HEPG0003,*SK3005,*HF12H,*ECG160, *WEP637,*AC188/01,*276-2007
2N218+	2N140+,2SA251,SF.T307,*PTC109,*GE-50,*HEPG0003,*SK3005,*ECG160,*WEP637
2N219	2N140,2N303,2N829,2N964A,2N972,2N985,2N1307,2N2022,2N2273,2N2699,2N3320, 2SA210,40359,SF.T229,*PTC107,*GE-50,*TR-17,*HEPG0005,*SK3005,*HF12H, *ECG160,*WEP637,*276-2005
2N219+	2N140+,2SA251,SF.T307,*PTC109,*GE-50,*TR-17,*HEPG0005,*SK3005,*ECG160, *WEP637

To Replace	Substitute This Type
2N220	2N132,2N132A,2N175,2N415,2N415A,2N416,2N422A,2N466,2N654,2N654+,2N1125, 2N1192,2N1305,2N1307,2N1309A,2N1348,2N1349,2N1350,2N1351,2N1352,2N1356, 2N1374,2N1376,2N1381,2N1382,2N1383,2N1681,2SA207,2SB187,2SB188,2SB299, 2SB303,40269,ASY27,SF.T223,*PTC109,*GE-51,*TR-85,*HEPG0005,*SK3004, *AT10H,*ECG102A,*WEP250,*AF125,*276-2007,*RT-121
2N220+	2N711B,2N796,2N829,2N964A,2N972,2N973,2N974,2N982,2N983,2N2022,2N2273, 2N2928,2N3320,2N3371,2SA52,2SA53,2SA79,2SA138,2SA139,2SA350,2SA351, 2SA352,2SA353,2SA354,2SA355,2SA408,2SA409,2SA417,2SA451,2SB33,2SB60A, 2SB75,2SB77,2SB91,2SB114,2SB115,2SB154,2SB184,2SB187,2SB188,2SB263, 2SB303,2SB328,2SB329,2SB364,40359,40488,AFY15,MM404,SF.T307,SF.T308, SF.T317,SF.T320,SF.T323,SF.T353,*GE-51,*TR-85,*HEPG0005,*SK3004, *ECG102A,*WEP250
2N223	2N359,2N417,2N467,2N508,2N508A,2N521A,2N598,2N599,2N631,2N655,2N655+, 2N827,2N1129,2N1284,2N1309,2N1317,2N1354,2N1355,2N1357,2N1705,2N1808, 2N1892,2N1998,2N2635,2SB186,2SB223,2SB223A,2SB327,NKT222,NKT223, *GE-53,*TR-85,*HEPG0005,*SK3003,*AT30H,*ECG102A,*WEP250,*AC188/01, *276-2005,*RT-121
2N223+	2N241A,2N321,2N467,2N654,2N654+,2N1008,2N1175A,2N1356,2N1374,2N1376, 2N1381,2N1382,2N1705,2N1706,2N1707,2N2706,2SB22,2SB186,2SB187,2SB188, 2SB222,2SB222A,2SB327,2SB431,2SB496,NKT211,NKT213,NKT216,SF.T223, SF.T323,SF.T353,*GE-53,*HEPG0005,*SK3003,*ECG102A,*WEP250
2N224	2N59,2N59A,2N362,2N369,2N417,2N508A,2N520A,2N571,2N598,2N609,2N631, 2N632,2N654+,2N655,2N1125,2N1128,2N1129,2N1130,2N1175A,2N1193,2N1274, 2N1307,2N1309,2N1352,2N1354,2N1355,2N1356,2N1357,2N1370,2N1374,2N1376, 2N1381,2N1382,2N1706,2N1707,2N1808,2N1892,2N1998,2SB186,2SB187,2SB188, 2SB222,2SB222A,2SB223,2SB223A,2SB327,2SB329,ASY27,MM404,MM404A, SF.T223,*PTC135,*GE-53,*TR-85,*HEPG0005,*SK3004,*AT30H,*ECG102A, *WEP250,*AC188/01,*276-2005,*RT-121
2N224+	2N59,2N59A,2N60,2N60A,2N217+,2N241A,2N270,2N321,2N360,2N362,2N369, 2N415,2N415A,2N416,2N422A,2N520A,2N570,2N571,2N572,2N609,2N632,2N654, 2N654+,2N1008A,2N1125,2N1128,2N1130,2N1175,2N1175A,2N1192,2N1274, 2N1305,2N1307,2N1309A,2N1349,2N1352,2N1356,2N1370,2N1374,2N1376,2N1381, 2N1382,2N1383,2N1415,2N1681,2N1706,2N1707,2N2953,2SB22,2SB38,2SB56, 2SB89,2SB187,2SB188,2SB221,2SB221A,2SB222,2SB222A,2SB304,2SB327, 2SB328,2SB329,2SB496,ACY23,ACY32,ASY27,MM404,MM404A,NKT211,SF.T223, SF.T323,SF.T353,*GE-53,*HEPG0005,*SK3004,*ECG102A,*WEP250
2N225	M.P.2N224
2N226	2N59,2N59A,2N60,2N60A,2N241,2N241A,2N241A+,2N270,2N321,2N360,2N362, 2N369,2N414,2N414A,2N414B,2N414C,2N415,2N415A,2N416,2N422A,2N505, 2N520A,2N567,2N568,2N569,2N570,2N571,2N572,2N609,2N610,2N632,2N654, 2N654+,2N1124,2N1125,2N1128,2N1130,2N1171,2N1175,2N1175A,2N1192, 2N1274,2N1305,2N1307,2N1309A,2N1348,2N1349,2N1350,2N1351,2N1352, 2N1356,2N1370,2N1372,2N1374,2N1376,2N1381,2N1382,2N1383,2N1415,2N1478, 2N1681,2N1706,2N1707,2N3883,2SA204,2SA207,2SA407,2SB187,2SB188,2SB221, 2SB221A,2SB222,2SB222A,2SB299,2SB304,2SB326,2SB328,2SB329,ASY27, MM404,MM404A,NKT221,NKT228,SF.T223,SF.T228,*PTC135,*GE-53,*TR-85, *HEPG0005,*SK3004,*AT30H,*ECG102A,*WEP250,*AC188/01,*276-2005, *RT-121
2N226+	2N59,2N59A,2N60,2N60A,2N109+,2N217+,2N241A,2N270,2N321,2N360,2N362, 2N369,2N414,2N414A,2N414B,2N414C,2N415,2N415A,2N416,2N422A,2N505, 2N520A,2N568,2N570,2N571,2N572,2N609,2N610,2N632,2N654,2N654+, 2N1008A,2N1124,2N1125,2N1128,2N1130,2N1171,2N1175,2N1175A,2N1192, 2N1274,2N1305,2N1307,2N1309A,2N1348,2N1349,2N1350,2N1351,2N1352, 2N1356,2N1370,2N1372,2N1374,2N1376,2N1381,2N1382,2N1383,2N1415,2N1478, 2N1681,2N1706,2N1707,2N2953,2N3883,2SA204,2SA207,2SA407,2SB22,2SB38, 2SB56,2SB89,2SB187,2SB188,2SB221,2SB221A,2SB222,2SB222A,2SB299, 2SB304,2SB326,2SB328,2SB329,2SB475,2SB496,ACY23,ACY32,ASY27,MM404, MM404A,NKT211,NKT221,NKT228,SF.T223,SF.T228,SF.T323,SF.T353,*GE-53, *HEPG0005,*SK3004,*ECG102A,*WEP250
2N227	M.P.2N226
2N228	2N214,2N385A,2N1012,2N1299,*PTC108,*GE-8,*TR-08,*HEPG0011,*SK3010,*NA30,

To Replace	Substitute This Type
(2N228)	*ECG103A,*WEP641B,*AC127,*276-2001,*RT-122
2N228+	2N214,*PTC108,*GE-8,*TR-08,*HEPG0011,*SK3010,*ECG103A,*WEP641B
2N229	2N306,2N445,2N446,*PTC108,*GE-59,*TR-08,*HEPG0011,*SK3010,*NA30, *ECG103A,*WEP641B,*AC127,*276-2001,*RT-122
2N229+	ASY29,*PTC108,*GE-59,*HEPG0011,*SK3010,*ECG103A,*WEP641B
2N231	2N128,2N345,2N501A,2N559,2N588A,2N711,2N711A,2N768,2N769,2N779,2N779A, 2N794,2N795,2N828,2N828A,2N934,2N964,2N965,2N966,2N967,2N968,2N969, 2N970,2N971,2N984,2N1427,2N2259,2N2401,2N2402,2N2451,2N2487,2N2488, 2N2860,2N3322,2SA15,2SA16,2SA56,2SA202,2SA251,2SA405,2SA410,2SA411, 2SA412,2SA450,*GE-51,*TR-17,*HEPG0008,*SK3008,*AF125,*276-2005
2N232	*GE-51,*TR-17,*HEPG0003,*SK3008,*ECG160,*WEP637,*276-2002
2N233	2N94,2N193,2N194A,2N212,2N233A,*PTC108,*GE-5,*TR-08,*HEPG0011,*SK3011, *NR5,*ECG103A,*WEP641B,*AC127,*RT-122
2N233+	2N212,2N356,2N357,2N357+,2N358,2N529/N,2N530/N,2N531/N,2N532/N,2N585, 2N634,2N1891,*PTC108,*GE-5,*TR-08,*HEPG0011,*SK3011,*ECG103A,*WEP641B
2N233A	2N94,2N193,2N194A,2N212,*PTC108,*GE-5,*TR-08,*HEPG0011,*SK3011,*NR5, *ECG103A,*WEP641B,*AC187/01,*RT-122
2N233A+	2N94A,2N356,2N357,2N357+,2N358,2N377,2N445,2N530/N,2N531/N,2N532/N, 2N533/N,2N557,2N576,2N585,2N585+,2N634,2N679,2N1605,2N1891,2N1995, 2SC180,2SD75,*PTC108,*GE-5,*TR-08,*HEPG0011,*SK3011,*ECG103A, *WEP641B
2N234	*PTC105A,*GE-16,*TR-01,*HEPG6003,*SK3009,*ECG121,*WEP232,*276-2006, *RT-127
2N234A	*PTC138,*GE-3,*TR-01,*HEPG6003,*SK3009,*PT25,*ECG121,*WEP232,*OC28, *276-2006,*RT-127
2N235	2N235A,2N236A,2N257-GRN,2N399,2N2836,AD149,CDT1310,*GE-16,*TR-01, *HEPG6003,*SK3009,*ECG121,*WEP232,*276-2006,*RT-127
2N235A	2N236A,2N2836,AD131,AD149,AUY19,CDT1311,*PTC138,*GE-3,*TR-01,*HEPG6003, *SK3009,*PT40,*ECG121,*WEP232,*OC28,*276-2006,*RT-127
2N235B	2N420,2N457A,2SB138,ASZ16,ASZ17,CDT1311,*PTC138,*GE-3,*TR-01,*HEPG6003, *SK3009,*ECG121,*WEP232,*OC28,*276-2006,*RT-127
2N236	*PTC105A,*GE-3,*TR-01,*HEPG6005,*SK3009,*ECG121,*WEP232,*276-2006, *RT-127
2N236A	2N235A,2N420,2SB138,AD149,CDT1311,*PTC138,*GE-3,*TR-01,*HEPG6003, *SK3009,*PT40,*ECG121,*WEP232,*OC28,*276-2006,*RT-127
2N236B	2N235A,2N235B,2N236A,2N420,2N2836,2SB127A,2SB138,AD131,AD149,AUY19, CDT1311,*PTC138,*GE-3,*TR-01,*HEPG6003,*SK3009,*PT40,*ECG121,*WEP232, *OC28,*276-2006,*RT-127
2N237	2N60B,2N60C,2N61B,2N61C,2N381,2N518,2N1057,2N1057+,2N1373,2N1447, 2N1451,2N1954,2N1956,2N1957,*PTC109,*GE-2,*TR-05,*HEPG0005,*SK3004, *AT20H,*ECG102,*WEP631,*276-2005,*RT-120
2N238	2N60,2N61,2N64,2N109,2N131,2N131A,2N133,2N133A,2N175,2N188,2N188A, 2N188A+,2N196,2N206,2N217,2N220,2N241,2N241A,2N241A+,2N270,2N270+, 2N320,2N321,2N363,2N407,2N408,2N408+,2N414,2N414A,2N414B,2N414C, 2N415,2N415A,2N416,2N422,2N422A,2N450,2N466,2N502,2N502A,2N502B,2N567, 2N568,2N569,2N570,2N572,2N582,2N584,2N610,2N654,2N838,2N980,2N1115, 2N1171,2N1175,2N1265/5,2N1303,2N1305,2N1309A,2N1313,2N1372,2N1383, 2N1414,2N1415,2N1524,2N1681,2N1726,2N1727,2N1728,2N1746,2N1747,2N1748, 2N1788,2N1790,2N1865,2N1867,2N2048,2N2168,2N2587,2N2717,2N2795,2N3323, 2N3324,2N3325,2N3412,2N3883,2N6365,2N6365A,2SA79,2SA138,2SA139,2SA169, 2SA171,2SA173,2SA204,2SA206,2SA207,2SA209,2SA212,2SA325,2SA326,2SA351, 2SA352,2SA353,2SA354,2SA406,2SA407,2SB33,2SB34,2SB37,2SB38,2SB56, 2SB59,2SB60A,2SB66,2SB75,2SB89,2SB95,2SB103,2SB114,2SB156A,2SB171, 2SB220,2SB221,2SB262,2SB263,2SB288,2SB299,2SB328,2SB365,2SB475,AFY15, ASY24B,ASY26,ASZ21,MM380,NKT213,NKT214,NKT216,NKT224,SF.T222,SF.T228, SF.T319,SF.T320,SF.T322,SF.T352,SYL792,*GE-51,*TR-85,*HEPG0005,*SK3004, *AT20H,*ECG102A,*WEP250,*AF125,*276-2002,*RT-121
2N240	2SA251,*GE-50,*TR-17,*HEPG0003,*SK3008,*JR30,*ECG160,*WEP637
2N241	2N59A,2N241A,2N321,2N360,2N362,2N415,2N415A,2N416,2N520A,2N571,2N654, 2N654+,2N1175A,2N1307,2N1309A,2N1352,2N1356,2N1381,2N1382,2N1706, 2N1707,2N1808,2N1892,2SB222,2SB222A,2SB327,2SB329,ASY27,SF.T223,

To Replace	Substitute This Type
(2N241)	*PTC102,*GE-53,*TR-05,*HEPG0005,*SK3004,*AT30H,*ECG102,*WEP631,*RT-120
2N241+	2N59A,2N60,2N60A,2N241,2N241A,2N241A+,2N321,2N360,2N362,2N415,2N415A, 2N416,2N422A,2N520A,2N571,2N654,2N654+,2N1175,2N1175A,2N1192,2N1305, 2N1307,2N1309A,2N1349,2N1352,2N1356,2N1381,2N1382,2N1383,2N1415, 2N1681,2N1706,2N1707,2SB221,2SB221A,2SB222,2SB222A,2SB327,2SB328, 2SB329,ASY27,SF.T223,*PTC102,*GE-53,*HEPG0005,*SK3004,*ECG102, *WEP631
2N241A	2N321,2N654,2N654+,2N1008A,2N1175A,2N1356,2N1381,2N1382,2N1706,2N1707, 2SB222,2SB222A,2SB327,SF.T223,*PTC102,*GE-53,*TR-05,*HEPG0005,*SK3004, *AT30H,*ECG102,*WEP631,*276-2005,*RT-120
2N241A+	2N241A,2N321,2N422A,2N654,2N654+,2N1008A,2N1175,2N1175A,2N1192,2N1349, 2N1356,2N1381,2N1382,2N1383,2N1415,2N1706,2N1707,2SB221,2SB221A,2SB222, 2SB222A,2SB327,SF.T223,*PTC102,*GE-53,*TR-05,*HEPG0005,*SK3004,*ECG102, *WEP631
2N242	2N380,2N4243,2N4246,*PTC901,*GE-76,*TR-01,*HEPG6003,*SK3009,*PT40, *ECG104,*WEP230,*276-2006,*RT-124
2N243	2N342,2N342A,2N342B,2N497,2N497A,2N1943,2N3512,*PTC125,*GE-47,*TR-63, *HEPS5026,*BF338,*276-2012
2N244	2N339,2N340,2N343,2N343A,2N343B,2N656,2N656A,2N1338,2N1613,2N1613A, 2N1837A,2N2106,2N2107,2N2224,2N2846,2N3015,2N3110,2N3299,2N3326,2N3554, 2SC32,2SC152,2SC594-R,2SC1218,2SC1385,BC341-6,BSY51,BSY92,PT1558, SE8002,*PTC125,*GE-47,*TR-63,*HEPS5026,*BF338,*276-2012
2N245	2N342,2N342A,2N342B,2N497A,2N707A,*PTC125,*GE-63,*TR-63,*BF338,*276-2008
2N246	2N339,2N340,2N343,2N343A,2N343B,2N656A,2N1613A,2N2106,2N2107,*PTC125, *GE-63,*TR-63,*BF338,*276-2008
2N247	2N2588,*PTC107,*GE-1,*TR-17,*HEPG0009,*SK3007,*JR30X,*ECG126,*WEP635, *276-2005
2N247/33	2N1395,2N1396,2N1397,2N2588,*PTC145,*TR-17,*HEPG0003,*SK3006,*ECG126, *WEP635
2N248	2N1195,2N1789,*GE-51,*TR-17,*HEPG0003,*SK3007,*JR100,*ECG126,*WEP635, *AF125
2N249	2N1495,2N2173,MM380,*GE-53,*TR-05,*HEPG0005,*SK3005,*AT30M,*ECG100, *WEP254,*276-2004,*RT-118
2N249+	*GE-53,*HEPG0005,*ECG100
2N250	*PTC138,*GE-49,*TR-01,*HEPG6003,*SK3009,*PT25,*ECG104,*WEP230,*276-2006, *RT-124
2N250A	2N251A,2N456B,2N457B,2N3611,2N3612,2N3613,2N3614,2N4243,2N4244,2N4246, 2N4247,MP2060,MP2061,*PTC105A,*GE-76,*TR-01,*HEPG6003,*SK3009,*ECG104, *WEP230,*276-2006,*RT-124
2N251	2N1293,2N1295,*GE-25,*TR-01,*HEPG6005,*SK3009,*PT40,*ECG121,*WEP232, *OC28,*276-2006,*RT-127
2N251A	2N457B,2N458B,2N3612,2N3614,2N3617,2N4242,2N4243,2N4245,2N4246,MP2061, MP2062,*PTC105A,*GE-76,*TR-01,*HEPG6005,*SK3009,*ECG121,*WEP232, *276-2006,*RT-127
2N252	2N123A,2N139,2N218,2N269,2N302,2N316,2N317,2N396A,2N404,2N428,2N500, 2N502B,2N532/P,2N533/P,2N579,2N579+,2N580,2N580+,2N581,2N583,2N779, 2N779A,2N828,2N828A,2N964,2N968,2N979,2N980,2N1017,2N1018,2N1280, 2N1313,2N1347,2N1499A,2N1499B,2N1742,2N1743,2N1744,2N1745,2N1787, 2N1868,2N2048,2N2169,2N2170,2N2401,2N2487,2N2488,2N2489,2N2672,2N2717, 2N2795,2N2796,2N2860,2N6365A,2SA56,2SA205,2SA251,2SA405,2SA406,2SA411, ASZ21,MM380,SF.T227,*GE-51,*TR-17,*HEPG0008,*SK3005,*HF6M,*ECG126, *WEP635,*AF125,*276-2014
2N252+	*GE-51,*TR-17,*HEPG0008,*SK3005,*ECG126,*WEP635
2N253	2N94A,2N312,2N356,2N357,2N358,2N377,2N446,2N532/N,2N533/N,2N557,2N576, 2N585+,2N635,2N1090,2N1302,2N1605,2N1995,2N1996,2SC89,2SC129,*GE-5, *TR-08,*HEPG0011,*SK3011,*NR5,*ECG101,*WEP641,*AC127,*276-2001,*RT-119
2N253+	*GE-5,*TR-08,*HEPG0011,*SK3011,*ECG101,*WEP641
2N254	2N182,2N438,2N438A,2N585,2N634,2N1891,2N1994,*GE-5,*TR-08,*HEPG0011, *SK3011,*NR5,*ECG101,*WEP641,*AC187/01,*276-2001,*RT-119
2N254+	*GE-5,*TR-08,*HEPG0011,*SK3011,*ECG101,*WEP641
2N255	2N155,2SB137,*PTC138,*GE-3,*TR-01,*HEPG6003,*SK3009,*PT12,*ECG104,

To Replace	Substitute This Type
(2N255)	*WEP230,*OC28,*276-2006,*RT-124
2N255A	2N234A,2N256A,*PTC138,*GE-25,*TR-01,*HEPG6003,*SK3009,*PT25,*ECG104, *WEP230,*OC28,*276-2006,*RT-124
2N256	CDT1310,CDT1319,*PTC105A,*GE-3,*TR-01,*HEPG6003,*SK3009,*PT25,*ECG104, *WEP230,*OC28,*276-2006,*RT-124
2N256A	2SB149-N,*PTC138,*GE-25,*TR-01,*HEPG6003,*SK3009,*PT25,*ECG104,*WEP230, *OC28,*276-2006,*RT-124
2N257	2N456,2N457,2N638,*PTC138,*GE-3,*TR-01,*HEPG6003,*SK3009,*PT40,*ECG104, *WEP230,*OC28,*276-2006,*RT-124
2N257-BLK	2N297A,2N392,2N1168,2N1534,2N1539,2N1540,2N1544,2N1544A,2N1545,2N1545A, 2N2064A,2N2143,2N2143A,2N2144,2N2144A,CDT1310,CDT1311,CDT1319, *PTC105A,*GE-25,*HEPG6003,*WEP230,*276-2006
2N257-GRN	2N257-BLK,2N297A,2N1168,2N1534,2N1535,2N1539,2N1540,2N2064A,2N2143, 2N2143A,2N2144,2N2144A,2SB246,2SB247,2SB248,2SB248A,2SB251A,CDT1310, CDT1311,CDT1319,*PTC114,*GE-25,*TR-16,*HEPG6003,*WEP230,*276-2006
2N257-WHT	2N669,*PTC901,*GE-76,*HEPG6003,*WEP230,*276-2006
2N258	
2N259	
2N263	2N916,2N916A,2N3859A,2N3903,2N3947,2N3974,2N3976,2N4227,2N5825,2S104, 2SC366G,2SC366G-O,2SC899,2SC907A,2SC933,2SC943,2SC944,2SC965, 2SC1007,2SC1189,2SC1320,2SC1453,2SC1688,BFY39-2,EN916,MPS3693, MPS6565,MPS6576-BLUE,MPS6576-GREEN,MPS6576-YEL,TP3705,ZT24,ZT44, *PTC123,*GE-39,*TR-95,*HEPS0014,*SK3039,*ECG108,*WEP56,*276-2013, *RT-113
2N264	2N337,2N475,2N717,2N840,2N842,2N844,2N1151,2N2432A,2N2522,2SC302,BF224J, BF225J,*PTC123,*GE-39,*TR-95,*HEPS0014,*SK3039,*ECG108,*WEP56,*BF167, *276-2013,*RT-113
2N265	2N359,2N508A,2N521A,2N522A,2N599,2N655+,2N1193,2N1309,2N1316,2N1354, 2N1355,2N1357,2N1379,2N1998,2N1999,2N2613,2N2635,2SB223,2SB223A,MA1703, *GE-53,*TR-05,*HEPG0005,*SK3003,*AT30H,*RT-120
2N265+	2N359,2N417,2N508A,2N521A,2N522A,2N598,2N599,2N655,2N655+,2N1193, 2N1309,2N1316,2N1354,2N1355,2N1357,2N1379,2N1998,2N1999,2N2613,2N2635, 2SB117,2SB223,2SB223A,MA1703,*GE-53,*HEPG0005,*SK3003
2N267	*PTC109,*GE-52,*TR-17,*HEPG0002,*SK3005,*ECG160,*WEP637,*276-2005
2N268	2N1159,2N1536,2N1537,2N2140,2N2140A,2N2141,2N2141A,2N2145,2N2145A, 2N2146,2N2146A,2SB252,2SB253,CDT1312,CDT1313,*PTC105A,*GE-25,*TR-01, *HEPG6005,*SK3009,*PT40,*ECG121,*WEP232,*276-2006,*RT-127
2N268A	2N268,2N561,2N665,2N1536,2N1537,2N2140,2N2140A,2N2141,2N2141A,2SB249, 2SB252,2SB253,CDT1321,CDT1322,SF.T240,*PTC105A,*GE-25,*TR-01,*HEPG6005, *SK3009,*PT40,*ECG121,*WEP232,*276-2006,*RT-127
2N269	2N414,2N582,2N584,2N1171,2N1349,2N1350,2N1351,2SA206,2SA207,2SA406, 2SA407,*GE-1,*TR-17,*HEPG0003,*SK3005,*HF6M,*ECG160,*WEP637
2N269+	*GE-1,*HEPG0003,*SK3005,*ECG160,*WEP637
2N270	2N1008A,2N1125,2N1374,2N1376,2N1381,2N1382,*PTC135,*GE-53,*TR-05, *HEPG0005,*SK3004,*AT30H,*ECG102,*WEP631,*AC188/01,*RT-120
2N270+	2N109,2N217,2N217+,2N270,2N415,2N415A,2N416,2N422A,2N467,2N654,2N654+, 2N1008A,2N1125,2N1192,2N1305,2N1307,2N1309A,2N1349,2N1352,2N1356, 2N1374,2N1376,2N1381,2N1382,2N1383,2N1681,2SB22,2SB38,2SB56,2SB77, 2SB79,2SB89,2SB187,2SB188,2SB327,40269,ACY23,ACY32,ASY27,SF.T223, SF.T323,*PTC109,*HEPG0005,*SK3004,*WEP631
2N271	2N302,2N303,2N829,2N972,2N973,2N974,2N1204A,2N1281,2N1282,2N1344,2N3883, MM380,SF.T228,SYL792,*GE-1,*TR-05,*HEPG0005,*SK3005,*HF12H,*ECG100, *WEP254,*AC188/01,*276-2004,*RT-118
2N271+	2N381,2N414,2N414A,2N414B,2N414C,2N525A,2N526A,2N650,2N653,2N1171, 2N1192,2N1303,2N1305,2N1313,2N1348,2N1350,2N1351,2N1478,2SA204,2SA206, 2SA207,2SA406,2SA407,ASY26,*PTC109,*GE-1,*HEPG0005,*SK3005,*ECG100, *WEP254
2N271A	2N302,2N303,2N829,2N972,2N973,2N974,2N1204A,2N1281,2N1282,2N1344,2N3883, MM380,SF.T228,SYL792,*GE-1,*TR-05,*HEPG0005,*SK3005,*HF12H,*ECG100, *WEP254,*AC188/01,*276-2005,*RT-118
2N272	2N359,2N417,2N508A,2N521A,2N522A,2N598,2N599,2N631,2N655,2N655+,2N1129,

To Replace	Substitute This Type
(2N272)	2N1193,2N1309,2N1316,2N1354,2N1355,2N1357,2N1379,2N1808,2N1892,2N1998, 2N2635,2SB186,2SB223,2SB223A,MA1703,NKT222,NKT223,*GE-53,*TR-05, *HEPG0005,*SK3003,*AT30H,*ECG102,*WEP631,*276-2005,*RT-120
2N272+	2N59,2N241A,2N321,2N360,2N362,2N415,2N415A,2N416,2N467,2N520A,2N569, 2N571,2N654,2N654+,2N1008,2N1128,2N1175A,2N1274,2N1307,2N1309A,2N1352, 2N1356,2N1370,2N1374,2N1376,2N1381,2N1382,2N1706,2N1707,2N1808,2N1892, 2N3323,2N3324,2N3325,2N3371,2SB22,2SB56,2SB77,2SB186,2SB187,2SB188, 2SB222,2SB222A,2SB327,2SB329,2SB364,2SB431,2SB496,40269,ACY23,ACY32, ASY27,NKT211,NKT213,NKT216,SF.T125,SF.T125P,SF.T223,SF.T323,SF.T353, *GE-53,*HEPG0005,*SK3003,*ECG102,*WEP631
2N273	2N44,2N44+,2N44A,2N111A,2N112A,2N315A,2N316A,2N396A,2N404A,2N413A, 2N425,2N426,2N427,2N428,2N460,2N464,2N465,2N524,2N524A,2N573,2N1018, 2N1056,2N1408,2N2100,2N2381,2N2382,2SB219,2SB219A,MA885,NKT225, SF.T221,SF.T226,*GE-53,*TR-05,*HEPG0005,*SK3004,*AT30H,*ECG100,*WEP254, *276-2004,*RT-118
2N273+	2N43,2N43+,2N43A,2N43A+,2N381,2N525,2N525A,2N586,2N1057,2N1057+, 2N1373,2N1495,2N1924,2SB225,2SB427,SF.T144,SF.T243,*GE-53,*HEPG0005, *SK3004,*ECG100,*WEP254
2N274	2N384,2N2588,*PTC107,*GE-1,*TR-17,*HEPG0003,*SK3007,*JR30X,*ECG160, *WEP637
2N277	2N278,*GE-4,*TR-03,*HEPG6004,*SK3012,*PT501,*ECG105,*WEP231,*276-2006
2N278	*PTC106,*GE-4,*TR-03,*HEPG6004,*SK3012,*PT501,*ECG105,*WEP231,*276-2006
2N279	2N61,2N111A,2N112A,2N123A,2N185,2N187A,2N188A,2N269,2N280,2N284,2N315A, 2N316A,2N317A,2N319,2N320,2N361,2N363,2N368,2N396A,2N402,2N403,2N404, 2N413,2N428,2N519A,2N565,2N566,2N579,2N579+,2N580,2N580+,2N611,2N613, 2N614,2N615,2N653,2N1017,2N1018,2N1204,2N1204A,2N1303,2N1313,2N1347, 2N1413,2N1450,2N1942,2N2048,2N6365A,2SA205,2SA208,2SA406,2SB185,2SB219, 2SB219A,2SB220,2SB220A,NKT225,SF.T221,SF.T222,SF.T227,*PTC107,*GE-52, *TR-85,*HEPG0005,*SK3004,*AT30H,*ECG102A,*WEP250,*AC188/01,*276-2004, *RT-121
2N280	2N60,2N61,2N109,2N188A,2N188A+,2N217,2N241A,2N241A+,2N270,2N320,2N321, 2N363,2N407,2N408,2N414,2N414A,2N414B,2N414C,2N415,2N415A,2N416,2N422, 2N422A,2N450,2N466,2N567,2N568,2N569,2N570,2N572,2N582,2N584,2N610, 2N654,2N838,2N1115,2N1171,2N1175,2N1204A,2N1303,2N1305,2N1309A,2N1313, 2N1372,2N1383,2N1414,2N1415,2N1478,2N1681,2N2048,2N2271,2N2587,2N3323, 2N3324,2N3325,2N3883,2N6365,2N6365A,2SA204,2SA206,2SA207,2SA209, 2SA212,2SA406,2SA407,2SB103,2SB220,2SB220A,2SB221,2SB221A,2SB299, 2SB304,2SB326,2SB328,ASY26,NKT221,NKT224,NKT228,SF.T222,SF.T228, SYL792,*PTC107,*GE-52,*TR-85,*HEPG0005,*SK3004,*AT30H,*ECG102A, *WEP250,*AC188/01,*276-2005,*RT-121
2N281	2N60,2N109,2N217,2N270,2N414,2N414A,2N414B,2N414C,2N415,2N415A,2N416, 2N450,2N567,2N568,2N569,2N570,2N572,2N582,2N584,2N610,2N654,2N1115, 2N1171,2N1281,2N1282,2N1305,2N1344,2N1372,2N1383,2N1478,2N1681,2N2271, 2SA204,2SA206,2SA207,2SA209,2SA212,2SA407,2SB103,2SB221,2SB221A, 2SB299,2SB304,2SB326,2SB328,ASY26,MA113,MA115,NKT221,NKT224,NKT228, SF.T228,SYL792,*PTC135,*GE-52,*TR-85,*HEPG0005,*SK3003,*AT30H, *ECG102A,*WEP250,*AC126,*276-2004,*RT-121
2N282	M.P.2N281
2N283	2N60A,2N60B,2N61A,2N61B,2N215,2N381,2N414,2N414A,2N414B,2N414C,2N422, 2N466,2N505,2N518,2N525,2N525A,2N526,2N526A,2N567,2N568,2N650,2N650A, 2N653,2N1171,2N1175,2N1192,2N1303,2N1305,2N1313,2N1348,2N1350,2N1351, 2N1373,2N1414,2N1415,2N1446,2N1447,2N1451,2N1478,2SA204,2SA206,2SA207, 2SA406,2SA407,2SB57A,2SB103,2SB220,2SB220A,2SB221,2SB221A,2SB225, 2SB226,2SB299,2SB304,2SB304A,2SB326,2SB397,2SB427,ASY26,MA886,NKT221, NKT224,NKT228,SF.T222,SF.T227,*PTC135,*GE-52,*TR-85,*HEPG0005,*SK3003, *AT20M,*ECG102A,*WEP250,*AC188/01,*276-2005,*RT-121
2N284	2N60A,2N60B,2N61A,2N61B,2N381,2N414,2N414A,2N414B,2N414C,2N422A,2N505, 2N518,2N525,2N525A,2N526,2N526A,2N567,2N568,2N650,2N650A,2N651A,2N1171, 2N1175,2N1192,2N1303,2N1305,2N1313,2N1348,2N1349,2N1350,2N1351,2N1373, 2N1414,2N1415,2N1446,2N1447,2N1451,2N1478,2N1681,2SA204,2SA206,2SA207, 2SA406,2SA407,2SB57A,2SB220,2SB220A,2SB221,2SB221A,2SB225,2SB226,

To Replace	Substitute This Type
(2N284)	2SB299,2SB304,2SB304A,2SB326,2SB427,ASY26,MA886,NKT221,NKT224,NKT228, SF.T222,SF.T227,*PTC135,*GE-53,*TR-85,*HEPG0005,*SK3004,*AT20M, *ECG102A,*WEP250,*276-2004,*RT-121
2N284A	2N60C,2N61C,2N1187,2N1924,2N1925,2N1954,2N1956,2N1957,ASY23,MA100, MA881,SF.T243,*DS-26,*GE-2,*TR-85,*HEPG0005,*SK3004,*AT30M,*ECG102A, *WEP250,*276-2004,*RT-121
2N285	2N257-WHT,2N392,2N1137,2N1138,2SB318,*PTC901,*GE-16,*TR-01,*HEPG6005, *SK3009,*ECG104,*WEP230,*276-2006,*RT-124
2N285A	2N235A,2N236A,2N399,AD149,*PTC138,*GE-3,*TR-01,*HEPG6003,*SK3009,*PT40, *ECG104,*WEP230,*OC28,*276-2006,*RT-124
2N285B	2N257-BLK,2N4241,CDT1310,CDT1319,*PTC138,*GE-3,*TR-01,*HEPG6003,*SK3009, *ECG104,*WEP230,*AD149,*276-2006,*RT-124
2N290	2SB354,*GE-4,*TR-01,*HEPG6006,*SK3012,*ECG104,*WEP230,*276-2006,*RT-124
2N291	2N59,2N59A,2N60,2N60A,2N241A,2N241A+,2N321,2N360,2N362,2N414B,2N414C, 2N422A,2N609,2N610,2N654,2N654+,2N1008A,2N1175,2N1175A,2N1192,2N1348, 2N1349,2N1350,2N1351,2N1356,2N1372,2N1374,2N1376,2N1381,2N1382,2N1383, 2N1415,2N1478,2N1681,2N1706,2N1707,2N3883,2SA207,2SA407,2SB221, 2SB221A,2SB222,2SB222A,2SB304,2SB326,NKT221,NKT228,SF.T223,*GE-53, *TR-05,*HEPG0005,*SK3004,*AT30H,*ECG102,*WEP631,*AC188/01,*276-2007, *RT-120
2N291+	2N60,2N60A,2N61,2N61A,2N188A,2N188A+,2N241A+,2N320,2N361,2N363,2N414B, 2N414C,2N610,2N611,2N653,2N1175,2N1192,2N1313,2N1348,2N1350,2N1351, 2N1372,2N1383,2N1414,2N1415,2N1478,2N1495,2N2173,2N3883,2SA206,2SA207, 2SA406,2SA407,2SB220,2SB220A,2SB221,2SB221A,2SB304,2SB326,MM380, NKT221,NKT228,SF.T222,SF.T322,*GE-53,*HEPG0005,*SK3004,*ECG102, *WEP631
2N292	*GE-7,*TR-08,*HEPG0011,*SK3011,*NR5,*ECG101,*WEP641,*AC187/01,*RT-119
2N292+	2N94A,2N292,2N292A,2N377,2N533/N,2N557,2N576,2N634,2N1366,2N1891,2N1995, 2SC128,2SC180,*GE-7,*TR-08,*HEPG0011,*SK3011,*ECG101,*WEP641
2N292A	2N292,2N576,2N634,2N1366,2N1891,2N1995,*GE-5,*TR-08,*HEPG0011,*ECG101, *WEP641,*AC187/01,*RT-119
2N293	*GE-7,*TR-08,*HEPG0011,*SK3011,*NR10,*ECG101,*WEP641,*AC187/01, *276-2001,*RT-119
2N293+	2N557,2N576,2N1366,2SC128,*GE-7,*TR-08,*HEPG0011,*SK3011,*ECG101, *WEP641
2N296	2N1293,2N1295,2N2836,AD131,AD132,AUY19,AUY20,*GE-25,*TR-01,*HEPG6003, *SK3020,*PT50,*ECG104,*WEP230,*BDY20,*276-2006,*RT-124
2N297	2N457,2N458,2N561,2N1160,2N1530,2N1531,2SB86,2SB87,ASZ15,CDT1320, CDT1321,*PTC105A,*GE-25,*TR-01,*HEPG6005,*SK3009,*ECG121,*WEP232, *276-2006,*RT-127
2N297A	2N375,2N618,2N1546,2N1546A,*PTC105A,*GE-25,*TR-01,*HEPG6005,*SK3009, *PT50,*ECG121,*WEP232,*276-2006,*RT-127
2N299	2N2258,2N2999,*GE-51,*TR-17,*HEPG0003,*SK3006,*JR30X,*ECG160,*WEP637
2N300	2N501,2N741,2N846A,2N960,2N961,2N962,2N963,2N2400,2SA449,*GE-51,*TR-17, *HEPG0003,*SK3006,*JR30X,*ECG160,*WEP637
2N301	2N257-BLK,2N257-GRN,2N301A,2N2143,2N2143A,2N2144,2N2144A,*PTC105A, *GE-25,*TR-01,*HEPG6003,*SK3009,*PT40,*ECG121,*WEP232,*276-2006,*RT-127
2N301A	2N1159,2N2144,2N2144A,2N2145,2N2145A,*PTC105A,*GE-25,*TR-01,*HEPG6005, *SK3009,*PT40,*ECG121,*WEP232,*276-2006,*RT-127A
2N302	2N303,2N414,2N829,2N972,2N1171,2N1204A,2N1281,2N1282,2N1313,2N1344, 2N3883,2SA204,2SA206,2SA207,2SA406,2SA407,*PTC109,*GE-1,*TR-05, *HEPG0005,*SK3003,*HF12H,*ECG100,*WEP254,*276-2005,*RT-118
2N303	2N972,2N985,*PTC102,*GE-2,*TR-05,*HEPG0005,*SK3003,*HF20H,*ECG102, *WEP631,*276-2005,*RT-120
2N306	*PTC108,*GE-59,*TR-08,*HEPG0011,*SK3010,*NA30,*ECG103A,*WEP641B,*AC127, *276-2001,*RT-129
2N306+	2N184,2N385,2N388,2N440,2N440A,2N445A,2N446A,2N634A,2N636,2N647,2N1091, 2N1114,2N1304,2N1306,2N1624,2N2354,2SC86,2SC89,2SC90,2SC91,2SC129, 2SC179,2SC181,2SD11,2SD21,2SD22,2SD25,2SD30,2SD38,2SD186,2SD187, 2SD195,ASY29,NKT713,*PTC108,*GE-59,*HEPG0011,*SK3010,*ECG103A, *WEP641B

To Replace	Substitute This Type
2N307	2N242,2N456B,2N4244,*PTC901,*GE-76,*TR-01,*HEPG6003,*SK3009,*PT25,*276-2006,*RT-127
2N307A	*PTC901,*GE-76,*TR-01,*HEPG6003,*SK3009,*PT25,*ECG121,*WEP232,*276-2006,*RT-127
2N307B	2N235A,2N236A,2N285B,2N399,2N401,2N1291,2SB426A,AD130,AD149,AD150,ADY27,*PTC105A,*GE-3,*TR-01,*HEPG6003,*SK3009,*ECG121,*WEP232,*OC28,*276-2006,*RT-127
2N308	2N61,2N111A,2N112A,2N123A,2N187,2N187A,2N188,2N188A,2N269,2N316,2N317,2N319,2N320,2N361,2N363,2N396A,2N404,2N413,2N428,2N502B,2N579,2N579+,2N580,2N580+,2N614,2N615,2N979,2N980,2N1017,2N1018,2N1303,2N1313,2N1347,2N1499A,2N1499B,2N1742,2N1743,2N1744,2N1745,2N1789,2N1866,2N1867,2N1868,2N2048,2N2489,2N2672,2N2717,2N2795,2N2796,2N6365A,2SA205,2SA208,2SA406,2SB185,2SB219,2SB220,ASZ21,MM380,SF.T221,SF.T222,SF.T227,*GE-51,*TR-05,*HEPG0009,*SK3007,*AT20M,*ECG100,*WEP254,*AF125,*276-2005,*RT-118
2N308+	*GE-51,*HEPG0009,*SK3007,*ECG100,*WEP254
2N309	2N61,2N111A,2N112A,2N123A,2N187,2N187A,2N188,2N188A,2N269,2N316,2N317,2N319,2N320,2N361,2N363,2N396A,2N404,2N413,2N428,2N502B,2N579,2N579+,2N580,2N580+,2N614,2N615,2N979,2N980,2N1017,2N1018,2N1303,2N1313,2N1347,2N1499A,2N1499B,2N1742,2N1743,2N1744,2N1745,2N1789,2N1866,2N1867,2N1868,2N2048,2N2489,2N2672,2N2717,2N2795,2N2796,2N6365A,2SA205,2SA208,2SA406,2SB185,2SB219,2SB220,ASZ21,MM380,SF.T221,SF.T222,SF.T227,*GE-51,*TR-05,*HEPG0009,*SK3007,*AT20M,*ECG100,*WEP254,*AF125,*276-2005,*RT-118
2N309+	*GE-1,*HEPG0009,*SK3007,*ECG100,*WEP254
2N310	2N60,2N61,2N188,2N188A,2N188A+,2N241,2N241A,2N241A+,2N320,2N321,2N363,2N414,2N414A,2N414B,2N414C,2N415,2N415A,2N416,2N422A,2N450,2N466,2N502,2N502A,2N502B,2N569,2N570,2N572,2N582,2N584,2N654,2N838,2N980,2N1115,2N1171,2N1175,2N1265/5,2N1303,2N1305,2N1309A,2N1313,2N1383,2N1414,2N1415,2N1681,2N1726,2N1727,2N1728,2N1746,2N1747,2N1748,2N1788,2N1790,2N1865,2N1867,2N2048,2N2168,2N2587,2N2717,2N2795,2N3323,2N3324,2N3325,2N3412,2N6365,2N6365A,2SA169,2SA171,2SA173,2SA204,2SA206,2SA207,2SA209,2SA212,2SA406,2SA407,2SB103,2SB114,2SB220,2SB221,2SB262,2SB328,ASY26,ASZ21,MM380,SF.T222,SF.T228,SYL792,*GE-51,*TR-05,*HEPG0009,*SK3007,*AT20M,*ECG102,*WEP631,*276-2005,*RT-120
2N310+	*GE-51,*HEPG0009,*SK3007,*ECG102,*WEP631
2N311	2N60,2N61,2N109,2N217,2N270,2N363,2N414,2N414A,2N414B,2N414C,2N415,2N415A,2N416,2N450,2N567,2N568,2N569,2N610,2N654,2N1115,2N1171,2N1281,2N1282,2N1303,2N1305,2N1313,2N1344,2N1372,2N1383,2N1681,2SA204,2SA206,2SA207,2SA406,2SA407,2SB220,2SB221,2SB299,2SB328,ASY26,MA113,MA115,NKT224,SF.T222,SF.T228,SYL792,*PTC109,*GE-50,*TR-05,*HEPG0002,*SK3005,*HF6M,*ECG100,*WEP254,*AC126,*RT-118
2N312	2N183,2N440,2N440A,2N557,2N635,2N636,2N1091,2SC129,ASY28,*PTC108,*GE-6,*TR-08,*HEPG0011,*SK3011,*NR10,*ECG101,*WEP641,*276-2001,*RT-119
2N312+	2N358,2N358A,2N635,2SC129,ASY28,*PTC108,*GE-6,*HEPG0011,*SK3011,*ECG101,*WEP641
2N315	2N315A,2N316,2N316A,2N317,2N317A,2N396A,2N426,2N427,2N428,2N1018,*PTC109,*GE-1,*TR-05,*HEPG0005,*SK3005,*HF12M,*ECG100,*WEP254,*AC188/01,*276-2004,*RT-118
2N315A	2N316A,*GE-2,*TR-05,*HEPG0005,*SK3004,*ECG100,*WEP254,*AC188/01,*276-2004,*RT-118
2N315B	*PTC109,*GE-53,*TR-05,*HEPG0005,*SK3005,*ECG100,*WEP254,*276-2002,*RT-118
2N316	2N316A,2N317,2N317A,2N1017,*GE-1,*TR-17,*HEPG0002,*SK3005,*HF12M,*ECG160,*WEP637
2N316A	*GE-2,*TR-05,*HEPG0005,*SK3004,*ECG102,*WEP631,*276-2004,*RT-120
2N317	2N317A,2N1017,*GE-1,*TR-05,*HEPG0002,*SK3005,*HF20M,*ECG100,*WEP254,*RT-118
2N317A	*GE-51,*TR-05,*HEPG0002,*SK3005,*ECG100,*WEP254,*RT-118
2N319	2N320,2SB220,2SB220A,SF.T222,*PTC135,*GE-2,*TR-05,*HEPG0005,*SK3003,*AT20M,*ECG102,*WEP631,*AC188/01,*276-2007,*RT-120

To Replace	Substitute This Type
2N320	2N321,2N1478,2SB221A,2SB326,*PTC135,*GE-2,*TR-05,*HEPG0005,*SK3003, *AT20H,*ECG102A,*WEP631,*AC188/01,*276-2004,*RT-120
2N321	2SB222,2SB222A,2SB327,SF.T223,*PTC135,*GE-2,*TR-05,*HEPG0005,*SK3003, *AT20H,*ECG102,*WEP631,*AC188/01,*276-2005,*RT-120
2N322	2N60,2N61,2N188A,2N188A+,2N241A,2N241A+,2N320,2N321,2N323,2N363,2N414, 2N414A,2N414B,2N414C,2N415,2N415A,2N416,2N422A,2N450,2N466,2N569,2N654, 2N1115,2N1171,2N1175,2N1303,2N1305,2N1309A,2N1313,2N1383,2N1414, 2N1415,2N1478,2N1681,2SA204,2SA206,2SA207,2SA406,2SA407,2SB220, 2SB220A,2SB221,2SB221A,2SB326,2SB328,ASY26,SF.T222,SF.T228,SYL792, *PTC102,*GE-2,*TR-05,*HEPG0005,*SK3003,*AT20H,*ECG102,*WEP631, *AC188/01,*276-2005,*RT-120
2N323	2N241A,2N321,2N324,2N360,2N362,2N415,2N415A,2N416,2N508,2N520A,2N571, 2N654,2N654+,2N1008,2N1175A,2N1307,2N1309A,2N1352,2N1356,2N1381,2N1382, 2N1705,2N1706,2N1707,2N1808,2N1892,2SB222,2SB222A,2SB327,2SB329,40269, ASY27,SF.T223,SF.T229,*PTC102,*GE-2,*TR-05,*HEPG0005,*SK3004,*AT20H, *ECG102,*WEP631,*AC188/01,*276-2005,*RT-120
2N324	2N417,2N508,2N598,2N1307,2N1356,2N1705,2N1808,2N1892,40269,ASY27,SF.T223, SF.T229,*PTC102,*GE-53,*TR-05,*SK3003,*AT20H,*ECG102,*WEP631,*AC188/01, *276-2005,*RT-120
2N325	2N1291,AD148,NKT452,*PTC105A,*GE-49,*TR-01,*SK3009,*PT40,*ECG104, *WEP230,*OC28,*276-2006,*RT-124
2N326	*TR-91
2N327	2N327A,2N327B,2N1232,2N1439,2N1440,2N2601,HA7534,SHA7534,*PTC127,*GE-65, *TR-05,*SK3005,*SP70,*ECG100,*WEP254,*BC327,*276-2022,*RT-118
2N327A	*GE-65,*TR-05,*HEPS0032,*SK3114,*ECG100,*WEP254,*BC327,*276-2022,*RT-118
2N327B	2N1232,*GE-82,*TR-88,*HEPS0032,*SK3025,*ECG129,*WEP242,*BC327,*276-2023, *RT-115
2N328	2N328A,2N328B,2N330A,2N721,2N721A,2N1230,2N1441,2N2393,2N3219,2N3979, 2N4008,2S322,FT1746,HA7533,SHA7533,TIS138,*PTC127,*GE-65,*TR-88,*SK3114, *SP70,*ECG129,*WEP242,*BC327,*276-2022,*RT-115
2N328A	*GE-65,*TR-88,*HEPS0032,*SK3114,*ECG129,*WEP242,*BC327,*276-2022,*RT-115
2N328B	2N1233,*GE-82,*TR-88,*HEPS0032,*SK3025,*ECG129,*WEP242,*BC327,*276-2023, *RT-115
2N329	2N329A,2N329B,2N722,2N722A,2N1231,2N1442,2N1443,2N1469A,2N2394,2N2425, 2N2946A,2N3039,2N3120,2N3121,2N3219,2N3829,2N3840,2N3979,2N5041,2N5042, 2N5230,2N5231,2N5365,2SA523,2SA558,2SA559A,2SA732,A5T4248,BC360-6, BSW72,EN1132,FT1746,FT5041,GI-3703,HA7537,MPS404A,MPS6516,MPSA70-RED, NPS6516,SHA7537,TIS38,TIS137,TW135,ZT183,ZT283,*PTC127,*GE-65,*TR-88, *HEPS0032,*SK3025,*SP70,*ECG129,*WEP242,*BC327,*276-2022,*RT-115
2N329A	*GE-65,*TR-88,*HEPS0032,*SK3025,*ECG129,*WEP242,*BC327,*276-2022,*RT-115
2N329B	2N2425,2N3840,2N3910,2N3913,2N4981,2N4982,2N5231,*GE-82,*TR-88,*HEPS0032, *SK3025,*ECG129,*WEP242,*BC327,*276-2023,*RT-115
2N330	2N328A,2N328B,2N330A,2N721,2N721A,2N722A,2N1233,2N1441,2N1442,2N1443, 2N2393,2N2602,2N2603,2N3039,HA7538,SHA7538,*PTC127,*GE-65,*TR-88, *SK3025,*SP70,*ECG129,*WEP242,*BC327,*276-2022,*RT-115
2N330A	2N328A,2N328B,2N1232,2N1233,2N1441,HA7534,HA7538,SHA7534,SHA7538, *GE-65,*TR-88,*HEPS0032,*SK3025,*ECG129,*WEP242,*BC327,*276-2022, *RT-115
2N331	2N381,2N414B,2N414C,2N422A,2N526,2N526A,2N527A,2N597,2N651,2N651A, 2N652A,2N654,2N654+,2N1008A,2N1124,2N1125,2N1192,2N1348,2N1349,2N1350, 2N1351,2N1356,2N1373,2N1375,2N1377,2N1447,2N1448,2N1478,2SA207,2SA407, 2SB57A,2SB226,2SB227,2SB304A,2SB326,2SB427,2SB428,MA887,SF.T223, *PTC109,*GE-53,*TR-17,*HEPG0005,*SK3008,*AT20H,*ECG160,*WEP637, *AC188/01,*276-2005
2N332	2N472,2N472A,2N475A,2N1149,2SC478,2SC945,*PTC123,*GE-39,*TR-86, *HEPS0014,*SK3124,*SN80,*ECG123,*WEP53,*BF167,*276-2013,*RT-100
2N332A	2N497,2N742,2N756,2N756A,2N757,2N757A,2N759B,2N3512,*PTC123,*GE-81, *TR-86,*HEPS0014,*SK3124,*ECG123,*WEP53,*BC337,*276-2012,*RT-100
2N333	2N334,2N475,2N717,2N842,2N844,2N1151,2N2432A,2N2522,2S102,2SC302,BF224J, BF225J,*PTC123,*GE-39,*TR-86,*HEPS0014,*SK3124,*SN80,*ECG123,*WEP53, *BF167,*276-2013,*RT-100

To Replace	Substitute This Type
2N333A	2N334A,2N335A,2N551,2N758,2N758A,2N759,2N759A,2N760B,2N761,2N1987, 2N3122,2N3295,2N5188,2SC30,2SC31,2SC46,2SC116,2SC152,PT2540,SE8001, *PTC123,*GE-81,*TR-86,*HEPS0014,*SK3124,*ECG123,*WEP53,*BC337, *276-2012,*RT-100
2N334	2N480,2N543A,2N717,2N840,2N842,2N844,2N1151,2N1152,2N2387,2N2432A, 2N2522,2N3973,2N3975,2SC302,2SC725,BF224J,BF225J,BFY39-1,BSY93,ZT23, ZT43,*PTC123,*GE-39,*TR-86,*HEPS0014,*SK3124,*SN80,*ECG123,*WEP53, *BF167,*276-2013,*RT-100
2N334A	2N656,2N758,2N758A,2N759,2N759A,2N760B,2N1410,2N1410A,2N1987,2N3122, 2N3295,2N5188,2SC30,2SC32,2SC46,2SC116,2SC152,2SC594-R,2SC1166-R, 2SC1360,BSY92,*PTC123,*GE-81,*TR-86,*HEPS0014,*SK3124,*ECG123,*WEP53, *BC337,*276-2012,*RT-100
2N334B	2N497,2N742,2N756A,2N759B,2N3512,A777,*PTC123
2N335	2N480,2N543A,2N708A,2N718,2N929,2N1152,2N1674,2N2387,2N2845,2N2847, 2N3301,2N3693,2N3826,2N3858A,2N3862,2N3946,2N3973,2N3975,2N4140,2N4966, 2N4994,2N5027,2N5824,2S103,2SC318,2SC335,2SC336,2SC620,2SC725,2SC815, 2SC838,2SC839,2SC896,2SC938,BFY39-1,BSY93,EN697,GET2221,MM3903, PET1001,SE1001,TIS22,TIS87,ZT23,ZT43,*PTC123,*GE-39,*TR-86,*HEPS0014, *SK3124,*SN80,*ECG123,*WEP53,*BF167,*276-2013,*RT-100
2N335A	2N335B,2N336A,2N551,2N656,2N759,2N759A,2N760B,2N1410,2N1410A,2N1987, 2N2846,2N3015,2N3122,2N3299,2N3326,2N3554,2N3641,2N3642,2SC30,2SC32, 2SC46,2SC152,2SC594-R,2SC1166-R,2SC1360,BSY92,*PTC123,*GE-81,*TR-86, *HEPS0014,*SK3124,*ECG123,*WEP53,*BC337,*276-2012,*RT-100
2N335B	2N656,2N718A,2N735,2N735A,2N1565,2N2846,2N2848,2N3015,2N3109,2N3299, 2N3326,2N3554,2N3641,2N3642,2N4960,2N4961,2N4962,2N4963,2N5380,2N6538, 2SC32,2SC594-R,2SC1166-R,A5T3903,BSY92,FT3567,FT3568,FT3722,SE8002, TN-3903,*PTC123,*GE-81,*TR-86,*HEPS0014,*SK3124,*ECG123,*WEP53,*BF338, *276-2012,*RT-100
2N336	2N543,2N841,2N916,2N916A,2N2161,2N3859A,2N3903,2N3947,2N3974,2N3976, 2N4227,2N5825,2S104,2SC366G,2SC366G-O,2SC899,2SC933,2SC943,2SC944, 2SC965,2SC1189,2SC1320,2SC1453,2SC1688,BFY39-2,EN916,MPS3693, MPS6565,MPS6576-BLUE,MPS6576-GREEN,MPS6576-YEL,TP3705,TZ82,ZT24, ZT44,*PTC123,*GE-39,*TR-86,*HEPS0014,*SK3124,*SN80,*ECG123,*WEP53, *BF167,*276-2013,*RT-100
2N336A	2N760,2N1704,2N2846,2N2848,2N3015,2N3299,2N4960,2N4962,2N5380,2N6538, 2SC352,2SC594-O,2SC651,2SC1166-O,A5T3903,TIS105,TN-3903,*PTC123,*GE-81, *TR-86,*HEPS0014,*SK3124,*ECG123,*WEP53,*BC337,*276-2012,*RT-100
2N337	2N480,2N543A,2N717,2N840,2N842,2N844,2N1151,2N1152,2N2387,2N2432A, 2N2522,2N3973,2N3975,2SC302,2SC725,BF224J,BF225J,BFY39-1,BSY93,ZT23, ZT43,*PTC123,*GE-39,*TR-70,*HEPS0014,*SK3039,*SN80,*ECG107,*WEP56, *BF167,*276-2013,*RT-108
2N337A	2N656,2N2846,2N2848,2N3015,2N3299,2N3326,2N3554,2N3641,2N3642,2SC32, 2SC352,2SC594-R,2SC651,2SC1164-O,2SC1166-R,2SC1199,2SC1360,2SC1412, BSY92,TIS105,*PTC123,*GE-20,*TR-95,*HEPS0014,*ECG108,*WEP56,*BC337, *276-2009,*RT-113
2N338	2N543,2N841,2N916,2N916A,2N2161,2N3859A,2N3903,2N3947,2N3974,2N3976, 2N4227,2N5825,2S104,2SC366G,2SC366G-O,2SC899,2SC933,2SC943,2SC944, 2SC965,2SC1007,2SC1189,2SC1320,2SC1453,2SC1688,BFY39-2,EN916,MPS3693, MPS6565,MPS6576-BLUE,MPS6576-GREEN,MPS6576-YEL,TP3705,TZ82,ZT24, ZT44,*PTC123,*GE-39,*TR-21,*HEPS0014,*SK3124,*SN80,*BF167,*276-2013, *RT-100
2N338A	2N760,2N760A,2N1986,2N3643,2N4960,2N4962,2N5380,2N6538,2SC594,2SC594-O, 2SC852,2SC1166-O,2SC1999,A5T3903,BC141,TN-3903,*PTC123,*GE-20,*TR-86, *HEPS0014,*SK3124,*ECG123,*WEP53,*BC337,*276-2009,*RT-100
2N339	2N343,2N343A,2N343B,2N656A,2N1613A,2N2106,2N2107,7A31,*PTC125,*GE-63, *TR-25,*HEPS5014,*BF338,*276-2008
2N339A	2N340A,2N717,2N734,2N734A,2N758A,2N759A,2N760B,2N844,2N2395,2N2514, 2N2522,2N3078,2S102,2SC30,BF237,MPS6544,*PTC123,*GE-63,*TR-25, *HEPS5014,*BF338,*276-2008
2N340	2N657A,2N3498,*PTC125,*GE-27,*TR-78,*HEPS5026,*BF338,*276-2008
2N340A	2N734,2N734A,2N845,2N912,2N2460,2N2464,2N2514,2N3078,BFY80,*PTC125,

To Replace	Substitute This Type
(2N340A)	*GE-27,*TR-78,*HEPS3010,*BF338,*276-2008
2N341	2N1613B,2N3500,MM3008,*PTC117,*GE-27,*TR-78,*HEPS5026,*BF338,*276-2008
2N341A	2N739A,2N2518,2N4068,2N5184,2SC273,2SC856,2SC1596,40354,*PTC117,*GE-27, *TR-78,*HEPS3021,*BF338,*276-2008
2N342	2N342A,2N342B,2N497A,2N707A,*PTC125,*GE-63,*TR-25,*HEPS0005,*BF338, *276-2008
2N342A	2N342B,2N498A,*PTC125,*GE-27,*TR-78,*HEPS5026,*BF338,*276-2008
2N342B	2N498A,MM3000,*PTC125,*GE-18,*BF338,*276-2012
2N343	2N343A,2N343B,2N656A,2N1613A,2N2106,2N2107,*PTC125,*GE-63,*TR-25, *HEPS0005,*BF338,*276-2008
2N343A	2N343,2N343B,2N656A,2N1613A,2N2106,2N2107,*PTC125,*GE-63,*TR-87, *HEPS0005,*SK3024,*ECG128,*WEP243,*BF338,*276-2008,*RT-114
2N343B	2N656A,2N1613A,2N2106,2N2107,*PTC125,*GE-63,*HEPS0005,*BF338,*276-2008
2N344	2N240,*GE-51,*TR-17,*HEPG0003,*SK3006,*JR100,*ECG126,*WEP635,*AF125
2N345	2SA408,2SA409,*GE-51,*TR-17,*HEPG0003,*SK3006,*JR100,*ECG126,*WEP635, *AF125
2N346	*GE-51,*TR-17,*HEPG0003,*SK3006,*JR100,*ECG126,*WEP635,*AF125
2N350	2N351,*PTC138,*GE-3,*TR-01,*HEPG6003,*SK3009,*PT40,*ECG121,*WEP232, *OC28,*276-2006,*RT-127
2N350A	2N351A,2N457B,2N1535A,2N4243,*PTC105A,*GE-25,*TR-01,*HEPG6003,*SK3009, *PT40,*ECG121,*WEP232,*276-2006,*RT-127
2N351	NKT402,NKT404,*PTC138,*GE-3,*TR-01,*HEPG6003,*SK3009,*PT40,*ECG121, *WEP232,*OC28,*276-2006,*RT-127
2N351A	2N376A,2N457B,2N1535A,2N4243,2N4246,*PTC105A,*GE-25,*TR-01,*HEPG6003, *SK3009,*PT40,*ECG121,*WEP232,*276-2006,*RT-127
2N352	*GE-16,*TR-01,*HEPG6005,*SK3009,*ECG121,*WEP232,*276-2006,*RT-127
2N353	*GE-16,*TR-01,*HEPG6005,*SK3009,*ECG121,*WEP232,*276-2006,*RT-127
2N354	*HEPG0001
2N355	*PTC127
2N356	2N356A,2N357,2N357+,2N357A,2N358,2N358A,*DS-72,*GE-7,*TR-08,*HEPG0011, *SK3011,*NR5,*ECG101,*WEP641,*AC187/01,*276-2001,*RT-119
2N356A	2N357A,2N358A,*DS-72,*GE-5,*TR-08,*HEPG0011,*SK3011,*ECG101,*WEP641, *AC187/01,*276-2001,*RT-119
2N357	2N357+,2N357A,2N358,2N358A,*DS-72,*GE-6,*TR-08,*HEPG0011,*SK3011,*NR10, *ECG101,*WEP641,*276-2001,*RT-119
2N357+	2N357,2N357A,2N358,2N358A,*GE-6,*HEPG0011,*SK3011,*ECG101,*WEP641
2N357A	2N358A,SF.T298,*DS-72,*GE-5,*TR-08,*HEPG0011,*SK3011,*ECG101,*WEP641, *276-2001,*RT-119
2N358	2N358A,*DS-72,*GE-6,*TR-08,*HEPG0011,*SK3011,*NR10,*ECG101,*WEP641, *276-2001,*RT-119
2N358+	2N358,2N358A,*GE-6,*HEPG0011,*SK3011,*ECG101,*WEP641
2N358A	SF.T298,*DS-72,*GE-5,*TR-08,*HEPG0011,*SK3011,*ECG101,*WEP641,*276-2001, *RT-119
2N359	2N1194,2N1316,2N1999,MA1703,MA1704,*PTC135,*GE-2,*TR-05,*HEPG0005, *SK3004,*AT30H,*ECG102,*WEP631,*AC188/01,*276-2005,*RT-120
2N360	2N382,2N383,2N527,2N527A,2N598,2N652,2N654+,2N655,2N1008A,2N1193, 2N1356,2N1448,2N1449,2N1452,2N1707,2N1997,2SB222,2SB222A,2SB227,2SB327, SF.T223,*PTC102,*GE-2,*TR-05,*HEPG0005,*SK3004,*AT30H,*ECG102,*WEP631, *276-2005,*RT-120
2N361	2N381,2N414B,2N414C,2N525,2N525A,2N526,2N526A,2N1313,2N1348,2N1350, 2N1351,2N1446,2N1447,2N1478,2SA206,2SA207,2SA406,2SA407,2SB221A,2SB225, 2SB226,2SB326,*PTC102,*GE-2,*TR-05,*HEPG0005,*SK3004,*AT30H,*ECG102, *WEP631,*276-2005,*RT-120
2N362	2N508A,2N598,2N599,2N655,2N655+,2N1008A,2N1193,2N1354,2N1355,2N1356, 2N1357,2N1998,2SB223,2SB223A,2SB327,SF.T223,*PTC102,*GE-2,*TR-05, *HEPG0005,*SK3004,*AT20M,*ECG102,*WEP631,*AC188/01,*276-2004,*RT-120
2N363	2N60A,2N60B,2N381,2N414B,2N414C,2N422A,2N525,2N525A,2N526,2N526A,2N597, 2N650,2N651,2N652A,2N654,2N1175,2N1192,2N1313,2N1348,2N1349,2N1350, 2N1351,2N1415,2N1447,2N1451,2N1478,2N1681,2SA206,2SA207,2SA406,2SA407, 2SB57A,2SB220,2SB220A,2SB221,2SB221A,2SB225,2SB226,2SB326,SF.T222, *PTC102,*GE-2,*TR-05,*HEPG0005,*SK3004,*AT20M,*ECG102,*WEP631,

To Replace	Substitute This Type
(2N363)	*276-2005,*RT-120
2N364	*PTC108,*GE-59,*TR-08,*HEPG0011,*SK3010,*NA30,*ECG103,*WEP641A,*RT-122
2N364+	2N364,*GE-59,*HEPG0011,*SK3010,*ECG103,*WEP641A
2N365	2N356A,2N357A,2N358A,2N377A,2N439,2N439A,2N576A,2N587,2N1000,2N1473, 2N1605A,2SC128,ASY28,*PTC108,*GE-59,*TR-08,*HEPG0011,*SK3010,*NA30, *ECG103,*WEP641A,*RT-122
2N365+	2N356A,2N365,2N438,2N438A,2N439,2N439A,2N576A,2N587,2N1605A,2N1994, 2SC128,2SD75A,*GE-59,*TR-08,*HEPG0011,*SK3010,*ECG103,*WEP641A
2N366	2N213,2N213A,2N385A,2N445A,2N446A,2N447A,2N1012,2N1993,*PTC108,*GE-59, *TR-08,*HEPG0011,*SK3010,*NA30,*ECG103,*WEP641A,*RT-122
2N366+	2N366,2N385A,2N440,2N440A,2N445A,2N446A,2N1012,2N1299,2SD38,SF.T298, *GE-59,*HEPG0011,*SK3010,*ECG103,*WEP641A
2N367	2N413A,2N425,2N563,2N1408,*PTC135,*GE-53,*TR-05,*HEPG0005,*SK3004, *AT30N,*ECG102,*WEP631,*276-2004,*RT-120
2N368	2N61A,2N61B,2N381,2N413,2N414,2N414A,2N414B,2N414C,2N422,2N518,2N525, 2N525A,2N567,2N568,2N650,2N650A,2N653,2N1017,2N1171,2N1191,2N1303, 2N1313,2N1373,2N1414,2N1446,2N1478,2SA204,2SA205,2SA206,2SA406,2SA407, 2SB220,2SB220A,2SB225,2SB397,2SB427,ASY26,MA886,NKT221,NKT224,NKT228, SF.T222,SF.T227,*PTC135,*GE-53,*TR-05,*HEPG0005,*SK3004,*AT30M,*ECG102, *WEP631,*276-2004,*RT-120
2N368+	2N61A,2N61B,2N215,2N283,2N284,2N315A,2N316A,2N368,2N413,2N524,2N524A, 2N525,2N525A,2N565,2N566,2N650,2N650A,2N653,2N1017,2N1191,2N1303, 2N1313,2N1413,2N1414,2N1446,2SA205,2SA206,2SA406,2SB172,2SB220, 2SB220A,2SB224,2SB225,2SB288,2SB397,ASY26,MA886,NKT214,SF.T222, SF.T227,*GE-53,*HEPG0005,*SK3004,*ECG102,*WEP631
2N369	2N59A,2N59B,2N382,2N383,2N417,2N461,2N467,2N508A,2N571,2N654+,2N655, 2N1008A,2N1125,2N1175A,2N1193,2N1307,2N1309,2N1352,2N1354,2N1355, 2N1356,2N1357,2N1371,2N1375,2N1377,2N1448,2N1449,2N1452,2N1707,2N1892, 2SB222,2SB223,NKT222,NKT223,NKT226,SF.T223,*PTC109,*GE-52,*TR-05, *HEPG0005,*SK3004,*AT30H,*ECG102,*WEP631,*AC126,*276-2007,*RT-120
2N369+	2N59A,2N59B,2N60A,2N60B,2N369,2N382,2N415,2N415A,2N416,2N422A,2N461, 2N467,2N526,2N526A,2N527A,2N569,2N570,2N571,2N572,2N597,2N651,2N651A, 2N652A,2N654,2N654+,2N1125,2N1175,2N1175A,2N1192,2N1305,2N1307, 2N1309A,2N1349,2N1352,2N1356,2N1371,2N1375,2N1377,2N1415,2N1448,2N1452, 2N1681,2N1707,2N2614,2N2953,2SB37,2SB38,2SB56,2SB57A,2SB59,2SB66, 2SB77A,2SB89A,2SB103,2SB173,2SB175,2SB176,2SB221,2SB221A,2SB222, 2SB222A,2SB226,2SB227,2SB304,2SB304A,2SB327,2SB428,ACY23,ACY32,MA887, NKT211,NKT213,NKT216,NKT226,SF.T125P,SF.T223,*GE-52,*HEPG0005,*SK3004, *ECG102,*WEP631
2N370	*PTC107,*GE-50,*TR-17,*HEPG0003,*SK3007,*JR100,*ECG126,*WEP635
2N371	*PTC107,*GE-50,*TR-17,*HEPG0003,*SK3007,*JR100,*ECG126,*WEP635
2N372	*PTC107,*GE-50,*TR-17,*HEPG0003,*SK3007,*JR100,*ECG126,*WEP635
2N373	2N1023,2N1066,2N1397,2N2496,2N2588,2N3783,2N3784,2SA246,*PTC107,*GE-50, *TR-17,*HEPG0003,*SK3007,*JR100,*ECG126,*WEP635
2N374	2N1023,2N1066,2N1397,2N2496,2N2588,2N3783,2N3784,2SA246,*PTC107,*GE-50, *TR-17,*HEPG0003,*SK3007,*JR100,*ECG126,*WEP635
2N375	2N618,2N1362,2N1363,*PTC105A,*GE-76,*TR-01,*HEPG6005,*SK3009,*PT50, *ECG104,*WEP230,*276-2006,*RT-124
2N376	NKT402,NKT404,NKT405,*PTC138,*GE-76,*TR-01,*HEPG6003,*SK3009,*PT40, *ECG104,*WEP230,*276-2006,*RT-124
2N376A	2N1545,2N4246,*PTC105A,*GE-25,*TR-01,*HEPG6003,*SK3009,*ECG104,*WEP230, *276-2006,*RT-124
2N377	2N357A,2N358A,2N377A,2N439,2N439A,2N576A,2N1000,2N1995,2SC128,2SC129, ASY28,*PTC108,*GE-8,*TR-08,*HEPG0011,*SK3011,*NR10,*ECG101,*WEP641, *276-2001,*RT-119
2N377A	2N1000,*PTC134,*GE-6,*TR-08,*HEPG0011,*SK3011,*NR10,*ECG101,*WEP641, *276-2001,*RT-119
2N378	2N242,2N380,*PTC901,*GE-76,*TR-01,*HEPG6003,*SK3009,*PT150,*ECG104, *WEP230,*276-2006,*RT-124
2N379	2N458B,2N1021,2N1021A,2N4242,2N4245,2N5156,*PTC105A,*GE-76,*TR-01, *HEPG6003,*SK3009,*PT150,*ECG121,*WEP232,*276-2006,*RT-127

To Replace	Substitute This Type
2N380	*PTC105A,*GE-76,*TR-01,*HEPG6003,*SK3009,*PT150,*ECG104,*WEP230, *276-2006,*RT-124
2N381	*GE-2,*TR-05,*HEPG0005,*SK3004,*AT30H,*ECG102,*WEP631,*276-2005,*RT-120
2N382	2N383,*GE-2,*TR-05,*HEPG0005,*SK3004,*AT30H,*ECG102,*WEP631,*276-2005, *RT-120
2N383	*GE-2,*TR-05,*HEPG0005,*SK3004,*AT30H,*ECG102,*WEP631,*276-2005,*RT-120
2N384	2N2588,*PTC107,*GE-51,*TR-17,*HEPG0008,*SK3007,*JR100,*ECG160,*WEP637
2N385	2N377A,2N385A,2N440,2N440A,2N1114,2N1299,2N1304,2N1306,2SC129,ASY28, ASY29,SF.T298,*PTC108,*GE-8,*TR-08,*HEPG0011,*SK3011,*NR10,*ECG101, *WEP641,*276-2001,*RT-119
2N385A	2N388A,*PTC108,*TR-08,*HEPG0011,*SK3011,*ECG101,*WEP641,*276-2001, *RT-119
2N388	2N388A,2N634A,2N1306,2N1308,2N1624,ASY29,*PTC108,*GE-8,*TR-08,*HEPG0011, *SK3011,*NR5,*ECG101,*WEP641,*AC187/01,*276-2001,*RT-119
2N388A	*DS-72,*GE-7,*TR-08,*HEPG0011,*SK3011,*NR5,*ECG101,*WEP641,*276-2001, *RT-119
2N389	*HEPS5004
2N389A	2N389,2N424,2N424A,*HEPS5004
2N392	2N1546,2N1546A,SF.T250,*PTC105A,*GE-16,*TR-01,*HEPG6005,*SK3009,*PT150, *ECG104,*WEP230,*276-2006,*RT-124
2N393	*GE-51,*TR-17,*HEPG0003,*SK3008,*JR30X,*ECG126,*WEP635
2N394	2N1281,2N1282,2N1344,SF.T228,*PTC109,*GE-1,*TR-05,*HEPG0005,*SK3005, *HF6M,*ECG100,*WEP254,*AC188/01,*276-2005,*RT-118
2N394A	*PTC109,*TR-05,*HEPG0005,*SK3005,*ECG100,*WEP254,*276-2005,*RT-118
2N395	2N396,2N397,*PTC135,*GE-1,*TR-05,*HEPG0005,*SK3005,*HF6M,*ECG100, *WEP254,*AC188/01,*276-2005,*RT-118
2N396	*PTC135,*GE-1,*TR-05,*HEPG0005,*SK3005,*HF6M,*ECG100,*WEP254,*AC188/01, *276-2005,*RT-118
2N396A	2N1018,2SA406,*GE-1,*TR-05,*HEPG0005,*SK3005,*HF6M,*ECG100,*WEP254, *276-2005,*RT-118
2N397	*GE-2,*TR-05,*HEPG0002,*SK3005,*HF12H,*ECG100,*WEP254,*276-2007,*RT-117
2N398	2N398A,2N2042,2N2042A
2N398A	2N2042,2N2042A
2N398B	*HEPG6012
2N399	2N235A,2N236A,2N257-GRN,2N285A,AD149,*PTC138,*GE-3,*TR-01,*HEPG6003, *SK3009,*PT40,*ECG104,*WEP230,*AD149,*276-2006,*RT-124
2N400	2N257-WHT,2N419,2N4241,*PTC114,*GE-25,*TR-01,*HEPG6003,*SK3009,*PT40, *ECG104,*WEP230,*AD149,*276-2006,*RT-124
2N401	2N235A,2N236A,2N257-GRN,2N399,2N2836,2SB138,AD149,CDT1310,CDT1311, *PTC138,*GE-3,*TR-01,*HEPG6003,*SK3009,*PT40,*ECG104,*WEP230,*OC28, *276-2006,*RT-124
2N402	2N61,2N61A,2N396A,2N403,2N573,2N611,2N613,2N653,2N1018,2N1191,2N1313, 2N1413,2SA205,2SA406,2SB185,2SB219,2SB219A,2SB220,2SB220A,SF.T221, SF.T222,*PTC135,*GE-53,*TR-05,*HEPG0005,*SK3003,*AT20M,*ECG102, *WEP631,*AC188/01,*276-2004,*RT-120
2N403	2N61,2N61A,2N414B,2N414C,2N610,2N611,2N653,2N1191,2N1313,2N1348,2N1350, 2N1351,2N1372,2N1414,2N1478,2SA205,2SA206,2SA207,2SA406,2SA407,2SB220, 2SB220A,2SB326,NKT221,NKT228,SF.T222,*PTC135,*GE-53,*TR-05,*HEPG0005, *SK3003,*AT20M,*ECG102,*WEP631,*AC188/01,*276-2004,*RT-120
2N404	2N1017,2N1171,2SA207,2SA407,*GE-53,*TR-05,*HEPS0032,*SK3004,*HF6M, *ECG102,*WEP631,*276-2006,*RT-120
2N404A	*GE-2,*TR-05,*HEPS0032,*SK3004,*ECG102,*WEP631,*276-2021,*RT-120
2N405	2N61,2N187A,2N188A,2N188A+,2N215,2N320,2N361,2N363,2N404,2N406,2N413, 2N1017,2N1303,2N1313,2N1413,2N1414,2SA205,2SA206,2SA406,2SB185,2SB220, ASY26,SF.T222,SF.T227,*PTC107,*GE-52,*TR-05,*HEPG0005,*SK3003,*AT20M, *ECG102,*WEP631,*AC188/01,*276-2004,*RT-120
2N405+	2N215+,2N406+,2SB185,NKT214,SF.T322,*GE-52,*HEPG0005,*SK3003,*ECG102, *WEP631
2N406	2N61,2N187A,2N188A,2N188A+,2N215,2N320,2N361,2N363,2N404,2N405,2N413, 2N1017,2N1303,2N1313,2N1413,2N1414,2SA205,2SA206,2SA406,2SB185,2SB220, ASY26,SF.T222,SF.T227,*PTC107,*GE-52,*TR-85,*HEPG0005,*SK3004,*AT20M,

To Replace	Substitute This Type
(2N406)	*ECG102A,*WEP250,*AC188/01,*276-2004,*RT-123
2N406+	2N215+,2N405+,2SB185,NKT214,SF.T322,*GE-52,*TR-85,*HEPG0005,*ECG102A, *WEP250
2N407	2N59,2N109,2N217,2N241A,2N321,2N360,2N362,2N369,2N415,2N415A,2N416, 2N422A,2N467,2N520A,2N569,2N571,2N609,2N632,2N654,2N654+,2N1008,2N1009, 2N1128,2N1130,2N1175A,2N1274,2N1307,2N1309A,2N1352,2N1356,2N1370, 2N1374,2N1376,2N1381,2N1382,2N1681,2N1706,2N1707,2N1808,2N1892,2N2271, 2N3323,2N3324,2N3325,2SB186,2SB187,2SB188,2SB222,2SB222A,2SB327, 2SB329,40269,ASY27,NKT222,NKT223,NKT226,SF.T223,SYL792,*PTC107,*GE-52, *TR-85,*HEPG0005,*SK3003,*AT20M,*ECG102A,*WEP250,*AC188/01,*276-2006, *RT-121
2N407+	2N217+,2N270+,2N408+,2N654+,2N838,2N3323,2N3324,2N3325,2SB34,2SB38, 2SB56,2SB59,2SB75,2SB77,2SB89,2SB156A,2SB187,2SB188,2SB263,2SB328, 2SB329,2SB364,2SB496,ACY23,ACY32,NKT211,NKT213,NKT216,NKT226,SF.T323, *GE-52,*TR-85,*HEPG0005,*SK3003,*ECG102A,*WEP250
2N408	2N59,2N60,2N109,2N217,2N241A,2N241A+,2N270,2N321,2N360,2N362,2N369, 2N407,2N415,2N415A,2N416,2N422A,2N467,2N520A,2N569,2N571,2N609,2N632, 2N654,2N654+,2N838,2N1008,2N1115,2N1128,2N1130,2N1175,2N1175A,2N1274, 2N1305,2N1307,2N1309A,2N1352,2N1356,2N1370,2N1374,2N1376,2N1381,2N1382, 2N1383,2N1415,2N1681,2N1706,2N1707,2N2271,2N3323,2N3324,2N3325,2SB187, 2SB188,2SB221,2SB221A,2SB222,2SB222A,2SB304,2SB327,2SB328,2SB329, 40269,ASY27,NKT226,SF.T223,SYL792,*PTC107,*GE-52,*TR-85,*HEPG0005, *SK3003,*AT20M,*ECG102A,*WEP250,*AC188/01,*276-2004,*RT-123
2N408+	2N217+,2N407+,2N838,2N3323,2N3324,2N3325,2SB56,2SB59,2SB75,2SB77, 2SB156A,2SB187,2SB188,2SB263,2SB328,2SB329,2SB364,2SB496,ACY23,ACY32, NKT211,NKT213,NKT216,NKT226,SF.T323,*GE-52,*TR-85,*HEPG0005,*SK3003, *ECG102A,*WEP250
2N409	2N140,2N219,2N302,2N303,2N410,2N411,2N412,2N521,2N559,2N582,2N584,2N711, 2N711A,2N711B,2N794,2N795,2N829,2N972,2N973,2N974,2N1281,2N1282,2N1300, 2N1301,2N1344,2N1683,2N2048,2N2401,2N2402,2N2717,2N2795,2N3320,2N3322, 2N3883,2SA169,2SA171,2SA451,ASZ21,*PTC107,*GE-50,*TR-17,*HEPG0003, *SK3005,*HF12H,*ECG126,*WEP635
2N409+	2N559,2N711,2N711A,2N711B,2N794,2N795,2N829,2N972,2N973,2N974,2N2401, 2N2402,2N2717,2N2795,2N3320,2N3322,2SA12,2SA13,2SA15,2SA16,2SA138, 2SA139,2SA201,2SA202,2SA412,2SA451,AFY15,ASZ21,SF.T307,*GE-50, *HEPG0003,*SK3005,*ECG126,*WEP635
2N410	2N140,2N219,2N302,2N303,2N409,2N411,2N412,2N521,2N559,2N582,2N584,2N711, 2N711A,2N711B,2N794,2N795,2N829,2N972,2N973,2N974,2N1281,2N1282,2N1300, 2N1301,2N1344,2N1683,2N2048,2N2401,2N2402,2N2717,2N2795,2N3320,2N3322, 2N3883,2SA169,2SA171,2SA451,ASZ21,*PTC107,*GE-50,*TR-17,*HEPG0003, *SK3005,*HF12H,*ECG126,*WEP635,*276-2007
2N410+	2N559,2N711,2N711A,2N711B,2N794,2N795,2N829,2N972,2N973,2N974,2N2401, 2N2402,2N2717,2N2795,2N3320,2N3322,2SA12,2SA13,2SA15,2SA16,2SA138, 2SA139,2SA201,2SA202,2SA412,2SA451,AFY15,ASZ21,SF.T307,*PTC107,*GE-50, *TR-17,*HEPG0003,*SK3005,*ECG126,*WEP635
2N411	2N303,2N412,2N796,2N829,2N964A,2N972,2N973,2N974,2N985,2N1683,2N2022, 2N2273,2N2699,2N3320,2SA210,40359,SF.T229,*PTC107,*GE-50,*TR-17, *HEPG0003,*SK3005,*HF12H,*ECG126,*WEP635
2N411+	2N796,2N829,2N964A,2N972,2N973,2N974,2N985,2N2022,2N2273,2N2699,2N3320, 40359,AFY15,SF.T308,*PTC107,*GE-50,*TR-17,*HEPG0003,*SK3005,*ECG126, *WEP635
2N412	2N303,2N411,2N796,2N829,2N964A,2N972,2N973,2N974,2N985,2N1683,2N2022, 2N2273,2N2699,2N3320,2SA210,40359,SF.T229,*PTC107,*GE-50,*TR-17, *HEPG0003,*SK3005,*HF12H,*ECG126,*WEP635
2N412+	2N796,2N829,2N964A,2N972,2N973,2N974,2N985,2N2022,2N2273,2N2699,2N3320, 40359,AFY15,SF.T308,*PTC107,*GE-50,*TR-17,*HEPG0003,*SK3005,*ECG126
2N413	2N381,2N414,2N414B,2N414C,2N525A,2N1017,2N1171,2N1303,2N1313,2N1348, 2N1350,2N1351,2N1478,2SA204,2SA205,2SA206,2SA207,2SA406,2SA407,ASY26, *GE-1,*TR-05,*HEPG0005,*SK3005,*HF6M,*ECG100,*WEP254,*AC188/01, *276-2004,*RT-118
2N413A	2N111A,2N112A,2N315A,2N316A,2N396A,2N404A,2N413,2N426,2N427,2N428,

To Replace	Substitute This Type
(2N413A)	2N524A,2N1017,2N1018,2N1191,2SA205,SF.T226,*GE-53,*TR-05,*HEPG0005, *SK3005,*HF6M,*ECG100,*WEP254,*AC188/01,*276-2004,*RT-118
2N414	2N1171,2N1307,2N1309A,2N1349,2N1350,2N1351,2SA204,2SA207,2SA407,*PTC109, *GE-1,*TR-05,*HEPG0001,*SK3005,*HF12M,*ECG100,*WEP254,*RT-118
2N414A	2N381,2N414,2N414B,2N414C,2N415,2N415A,2N416,2N526A,2N527A,2N597,2N651, 2N654,2N1171,2N1192,2N1305,2N1307,2N1309A,2N1348,2N1349,2N1350,2N1351, 2N1356,2N1478,2N1681,2SA204,2SA207,2SA407,*PTC109,*GE-1,*TR-05, *HEPG0005,*SK3005,*HF12H,*ECG100,*WEP254,*AC188/01,*276-2005,*RT-118
2N414B	2N414C,2N526A,2N527A,2N1348,2N1349,2N1350,2N1351,2N1356,2SA207,2SA407, *PTC109,*GE-1,*TR-05,*HEPG0002,*SK3005,*ECG100,*WEP254,*AC188/01, *RT-118
2N414C	2N414B,2N526A,2N527A,2N1348,2N1349,2N1350,2N1351,2N1356,2SA207,2SA407, *PTC109,*GE-1,*TR-05,*HEPG0002,*SK3005,*ECG100,*WEP254,*AC188/01
2N415	2N383,2N415A,2N416,2N1307,2N1309A,2N1356,2N1892,*PTC109,*GE-1,*TR-05, *HEPG0002,*SK3005,*HF12H,*ECG100,*WEP254,*AC188/01,*RT-118
2N415A	2N383,2N415,2N416,2N1307,2N1309A,2N1356,2N1892,*PTC109,*GE-1,*TR-05, *HEPG0002,*SK3005,*HF12H,*ECG100,*WEP254,*AC188/01,*RT-118
2N416	2N383,2N415,2N415A,2N1307,2N1309A,2N1356,2N1892,*PTC109,*GE-1,*TR-05, *HEPG0002,*SK3005,*HF12H,*ECG100,*WEP254,*AC188/01,*RT 118
2N417	2N599,2N1309,2N1316,2N1357,*GE-1,*TR-05,*HEPG0002,*SK3005,*HF20H, *ECG100,*WEP254,*RT-118
2N418	CDT1313,KR6503,*PTC105A,*GE-25,*TR-01,*HEPG6005,*SK3009,*PT50,*ECG121, *WEP232,*276-2006,*RT-127
2N419	2N400,*PTC114,*GE-3,*TR-01,*HEPG6003,*SK3009,*PT50,*ECG104,*WEP230, *OC28,*276-2006,*RT-124
2N420	2N420A,2SB138,CDT1311,CDT1312,KR6501,*PTC138,*GE-25,*TR-01,*HEPG6005, *SK3009,*PT50,*ECG121,*WEP232,*276-2006,*RT-127
2N420A	2N418,CDT1313,KR6503,*PTC105A,*GE-25,*TR-01,*HEPG6005,*SK3009,*PT50, *ECG121,*WEP232,*276-2006,*RT-127
2N422	2N381,2N518,2N526,2N526A,2N527A,2N597,2N651,2N651A,2N652A,2N1008A, 2N1125,2N1192,2N1309A,2N1348,2N1349,2N1350,2N1351,2N1373,2N1375,2N1377, 2N1447,2N1448,2SB57A,2SB226,2SB227,2SB304A,2SB427,2SB428,MA887, *PTC135,*GE-53,*TR-05,*HEPG0005,*SK3003,*AT20H,*ECG102,*WEP631,*AC128, *276-2005,*RT-120
2N422A	2N382,2N383,2N527,2N527A,2N597,2N651,2N652,2N652A,2N1008A,2N1189,2N1349, 2N1448,2N1449,2N1997,2SB227,*PTC135,*GE-53,*TR-05,*HEPG0005,*ECG102, *WEP631,*276-2004,*RT-120
2N424	*HEPS5004
2N424A	2N424,*HEPS5004
2N425	2N315A,2N316A,2N426,2N427,2N428,2N524A,2N1018,*GE-1,*TR-05,*HEPG0005, *SK3005,*HF3M,*ECG100,*WEP254,*AC188/01,*276-2004,*RT-118
2N426	2N315A,2N316A,2N427,2N428,2N1017,2N1018,*GE-1,*TR-05,*HEPG0005,*SK3005, *HF6M,*ECG100,*WEP254,*276-2004,*RT-118
2N427	2N316A,2N428,2N1017,2N1018,*GE-1,*TR-05,*HEPG0002,*SK3005,*HF6H,*ECG100, *WEP254,*276-2004,*RT-118
2N428	2N1017,2N1018,*GE-1,*TR-05,*HEPG0002,*SK3005,*HF12H,*ECG100,*WEP254, *276-2004,*RT-118
2N438	2N438A,*PTC134,*GE-7,*TR-08,*HEPG0011,*SK3011,*NR5,*ECG101,*WEP641, *276-2001,*RT-119
2N438A	2N438,*GE-7,*TR-05,*HEPG0011,*SK3011,*NR5,*ECG100,*WEP254,*276-2001, *RT-118
2N439	2N439A,2N440,2N440A,*DS-72,*GE-7,*TR-08,*HEPG0011,*SK3011,*NR5,*ECG101, *WEP641,*276-2001,*RT-119
2N439A	2N439,2N440,2N440A,*GE-7,*TR-05,*HEPG0011,*SK3011,*NR5,*ECG100,*WEP254, *276-2001,*RT-118
2N440	2N440A,*DS-72,*GE-6,*TR-08,*HEPG0011,*SK3011,*NR10,*ECG101,*WEP641, *276-2001,*RT-119
2N440A	2N440,*GE-6,*TR-05,*HEPG0011,*SK3011,*NR10,*ECG100,*WEP254,*RT-118
2N441	2N442,2N443,*GE-4,*TR-03,*HEPG6004,*SK3012,*PT501,*ECG105,*WEP233, *276-2006
2N442	2N443,*GE-4,*TR-03,*HEPG6006,*SK3012,*PT501,*ECG105,*WEP233,*276-2006

To Replace	Substitute This Type
2N443	2N174,2N1099,*GE-4,*TR-03,*HEPG6006,*SK3012,*PT501,*ECG105,*WEP233, *276-2006
2N444	*PTC108,*GE-59,*TR-08,*HEPG0011,*SK3011,*NR5,*ECG101,*WEP641,*AC127, *276-2001,*RT-119
2N444+	2N356,2N356A,2N357,2N357+,2N358,2N438,2N438A,2N444,2N585,2N634,2N1891, 2N1994,2SD19,*PTC108,*GE-59,*SK3011,*ECG101,*WEP641
2N444A	2N576A,2N587,2N1000,2N1473,2N1605A,*PTC108,*GE-59,*TR-08,*HEPG0011, *SK3010,*ECG103,*WEP641A,*276-2001,*RT-122
2N445	2N446,2N576,*PTC108,*GE-7,*TR-08,*HEPG0011,*SK3010,*NR5,*ECG101, *WEP641,*AC127,*276-2001,*RT-119
2N445+	2N356,2N356A,2N357,2N357A,2N358,2N358A,2N377,2N439,2N439A,2N445,2N446, 2N557,2N576,2N585+,2N635,2N679,2N1090,2N1302,2N1605,2N1995,2N1996, 2SC89,2SC128,2SC129,ASY28,*PTC108,*GE-7,*HEPG0011,*SK3010,*ECG101, *WEP641
2N445A	2N385A,2N440,2N440A,2N446A,2N1012,2N1299,*PTC108,*GE-5,*TR-08,*HEPG0011, *SK3010,*ECG101,*WEP641,*AC187/01,*276-2001,*RT-119
2N446	ASY29,*PTC108,*GE-54,*TR-08,*HEPG0011,*SK3011,*NR5,*ECG101,*WEP641, *AC187/01,*276-2001,*RT-119
2N446+	2N357A,2N358A,2N440,2N440A,2N446,2N635,2N636,2N1090,2N1091,2N1114, 2N1304,2N1306,2N1996,2SC90,2SC129,ASY28,ASY29,SF.T298,*PTC108, *HEPG0011,*SK3011,*ECG101,*WEP641
2N446A	2N388A,2N447A,*PTC108,*GE-5,*TR-08,*HEPG0011,*SK3010,*ECG103,*WEP641A, *AC187/01,*276-2001,*RT-122
2N447	2N357A,2N358A,2N377,2N385,2N439,2N439A,2N440,2N440A,2N585+,2N635,2N636, 2N679,2N1090,2N1091,2N1302,2N1304,2N1605,2N1996,2SC89,2SC90,2SC128, 2SC129,2SD11,ASY28,SF.T298,*PTC108,*GE-5,*TR-08,*HEPG0011,*SK3011, *ECG101,*WEP641,*276-2001,*RT-119
2N447A	2N388A,*GE-8,*TR-08,*HEPG0011,*SK3010,*ECG103,*WEP641A,*276-2001,*RT-122
2N447B	2N214,2N228,2N366,2N385,2N385A,2N388,2N440,2N440A,2N445A,2N446A,2N634A, 2N1012,2N1114,2N1299,2N1304,2N1306,2N1624,2N2426,2SD11,2SD21,2SD22, 2SD186,2SD187,ASY29,*TR-08,*HEPG0011,*SK3010,*ECG103,*WEP641A, *276-2001,*RT-122
2N448	2N94A,2N164,2N164A,2N169,2N172,2N312,2N377,2N439,2N439A,2N531/N, 2N532/N,2N533/N,2N557,2N576,2N585,2N585+,2N634,2N635,2N1090,2N1302, 2N1605,2N1995,2N1996,2SC128,ASY28,*GE-7,*TR-08,*HEPG0011,*SK3011,*NR5, *ECG101,*WEP641,*AC127,*276-2001,*RT-119
2N448+	2N94A,2N167,2N169,2N169A+,2N312,2N377,2N439,2N439A,2N533/N,2N557,2N576, 2N634,2N635,2N1090,2N1891,2N1995,2N1996,2SC128,2SC180,OC139,*GE-7, *TR-08,*HEPG0011,*SK3011,*ECG101,*WEP641
2N449	2N145,2N169A,2N440,2N440A,2N558,2N636,2N797,2N1091,2N1114,2N1303,2SC129, ASY28,ASY29,*GE-7,*TR-08,*HEPG0011,*SK3011,*NR10,*ECG101,*WEP641, *276-2001,*RT-119
2N449+	2N145,2N167A,2N169A,2N440,2N440A,2N558,2N636,2N797,2N1091,2N1114,2N1306, 2SC91,2SC129,2SC181,ASY28,ASY29,OC140,*GE-7,*TR-08,*HEPG0011,*SK3011, *ECG101,*WEP641
2N450	2N414,2N415,2N415A,2N416,2N1115,2N1171,2N1305,2N1307,2N1309A,2N1356, 2N1681,2SA204,2SA207,2SA407,ASY27,SF.T228,SYL792,*PTC109,*GE-1,*TR-05, *HEPG0002,*SK3005,*HF6M,*ECG100,*WEP254,*AC188/01,*276-2007,*RT-117
2N450+	2N417,2N521A,2N522A,2N598,2N599,2N1284,2N1309,2N1316,2N1317,2N1354, 2N1355,2N1357,2N1998,*GE-1,*HEPG0002,*SK3005,*ECG100,*WEP254
2N456	2N457,2SB85,2SB86,*PTC105A,*GE-25,*TR-01,*HEPG6003,*SK3009,*PT40, *ECG104,*WEP230,*276-2006,*RT-124
2N456A	2N457A,2N2143,2N2143A,2N2144,2N2144A,2N3611,2N3612,2N3613,2N3614,2N5890, 2N5891,2N5894,2N5895,MP2060,MP2061,*PTC105A,*GE-16,*TR-01,*HEPG6003, *SK3009,*ECG104,*WEP230,*276-2006,*RT-124
2N456B	2N457B,*PTC901,*GE-3,*TR-01,*HEPG6003,*SK3009,*ECG104,*WEP230,*276-2006, *RT-124
2N457	2N458,2SB86,2SB87,*PTC105A,*GE-25,*TR-01,*HEPG6005,*SK3009,*PT50, *ECG104,*WEP230,*276-2006,*RT-124
2N457A	2N458A,2N2144,2N2144A,2N2145,2N2145A,2N3612,2N3614,2N3617,2N5891,2N5892, 2N5895,2N5896,2SB339,MP2061,MP2062,*PTC105A,*GE-3,*TR-01,*HEPG6005,

To Replace	Substitute This Type
(2N457A)	*SK3009,*ECG104,*WEP230,*276-2006,*RT-124
2N457B	2N458B,*PTC901,*TR-01,*HEPG6005,*SK3009,*ECG104,*WEP230,*276-2006, *RT-124
2N458	2SB87,2SB88,*PTC105A,*GE-25,*TR-01,*HEPG6005,*SK3009,*PT50,*ECG104, *WEP230,*276-2006,*RT-124
2N458A	2N2145,2N2145A,2N2146,2N2146A,2N3617,2N3618,2N5892,2N5896,2SB339, MP2062,MP2063,*PTC105A,*GE-3,*TR-01,*HEPG6005,*SK3009,*ECG104,*WEP230, *276-2006,*RT-124
2N458B	2N1021A,*PTC901,*GE-3,*TR-01,*HEPG6005,*SK3009,*ECG121,*WEP232, *276-2006,*RT-127
2N459	*PTC105A,*GE-76,*TR-01,*HEPG6005,*SK3012,*PT50,*ECG121,*WEP232,*276-2006, *RT-127
2N459A	2N1021A,2N1022A,*TR-01,*HEPG6018,*SK3009,*ECG121,*WEP232,*RT-127
2N460	2N524,2N524A,2SB224,*GE-2,*TR-05,*HEPG0005,*SK3004,*AT30M,*ECG102, *WEP631,*276-2004,*RT-120
2N461	2N382,2N383,2N527,2N652,2N1189,2N1449,2N1997,2N2000,*GE-1,*TR-05, *HEPG0005,*SK3004,*AT30M,*ECG102,*276-2004,*RT-120
2N462	*GE-80,*TR-05,*HEPG0005,*SK3004,*AT30M,*ECG102,*WEP631,*AC128,*276-2005, *RT-120
2N463	
2N464	2N465,*GE-53,*TR-05,*HEPG0005,*SK3004,*AT30M,*ECG102,*WEP631,*276-2004, *RT-120
2N465	*GE-2,*TR-05,*HEPG0005,*SK3004,*AT30H,*ECG102,*WEP631,*276-2004,*RT-120
2N466	2N382,2N422A,2N526,2N526A,2N527A,2N597,2N651,2N651A,2N652A,2N1008A, 2N1192,2N1309A,2N1348,2N1349,2N1350,2N1351,2N1375,2N1377,2N1447,2N1448, 2SB57A,2SB226,2SB227,2SB428,MA887,*PTC135,*GE-53,*TR-05,*HEPG0005, *SK3003,*AT30H,*ECG102,*WEP631,*AC128,*276-2005,*RT-120
2N467	2N383,2N527,2N598,2N652,2N1185,2N1189,2N1193,2N1449,2N1997,2N1998, 2N2171,MA888,*PTC135,*GE-53,*TR-05,*HEPG0005,*SK3003,*AT20H,*ECG102, *WEP631,*AC126,*276-2005,*RT-120
2N470	2N471,2N479A,2N835,2N958,2N1082,2N1387,2N1682,2N2214,2N2719,2N3825, 2N3843,2N3843A,2N3984,2N3985,2N4292,2N4293,2S131,2SC171,2SC172,2SC385A, 2SC387A,D16G6,MPS918,MPSH07,MPSH08,ZT20,ZT40,*PTC126,*GE-86,*TR-86, *HEPS0014,*SK3124,*SN80,*ECG123,*WEP53,*BF365,*276-2011,*RT-100
2N471	2N472,2N479A,2N480A,2N757,2N834,2N839,2N1140,2N1387,2N3825,2N3843, 2N3843A,2N3984,2N3985,2N4292,2N4293,2SC382,2SC387A,2SC945,BSX25,D16G6, GET706,MPS918,MPS6545,MPS6546,MPS6547,MPSH07,MPSH08,TIS125,TIS126, ZT22,ZT42,*PTC126,*GE-60,*TR-24,*HEPS0014,*SN80,*BF173,*276-2013
2N471A	2N471,2N472,2N472A,2N474A,2N475A,2N756,2N4072,MPSH34,*PTC126,*GE-60, *TR-86,*HEPS0014,*SK3124,*ECG123,*WEP53,*BF173,*276-2013,*RT-100
2N472	2N742,2N757,2N757A,2N759B,2N839,2N2310,2N2521,2SC945,MPS6544,MPS6545, TIS126,ZT22,ZT42,*PTC123,*GE-212,*TR-21,*HEPS0014,*SK3122,*SN80, *ECG123A,*WEP736,*BC107B,*276-2013,*RT-102
2N472A	2N472,2N742,2N756,2N756A,2N757,2N757A,2N759B,2N839,2N2521,2SC945,TIS126, *PTC123,*GE-212,*TR-21,*HEPS0014,*SK3124,*ECG123A,*WEP736,*BC107B, *276-2013,*RT-102
2N473	2N474,2N478,2N479,2N706,2N706A,2N706B,2N709A,2N743,2N784,2N947,2N959, 2N988,2N989,2N1139,2N1386,2N2205,2N2318,2N2319,2N2432,2N2571,2N2572, 2N2711,2N2713,2N2729,2N2921,2N2926-BRN,2N3010,2N3544,2N3844,2N3844A, 2N3854,2N3854A,2N4294,2N5030,2N5187,2N5399,2SC55,2SC63,2SC68,2SC98, 2SC204,2SC356,2SC387AG,2SC395A-R,2SC400-R,2SC455,2SC595,2SC717, 2SC718,2SC719,2SC724,2SC1126,2SC1231,40519,EN706,EN918,EN3011,GI-2711, GI-2713,GI-3605,GI-3606,GI-3607,MM1748,MM1941,MPS2713,MPS2926-BRN, MPS6507,MPS6511,MPS6540,MPS6542,MPS6548,MPSH19,MT1038,MT1038A, MT1039,MT1060,PET3001,PET8101,SE3001,SE3002,TIS44,TIS51,ZT21,ZT41, ZT203P,ZT403P,*PTC126,*GE-86,*TR-86,*HEPS0014,*SK3124,*ECG123,*WEP53, *BC107B,*276-2011,*RT-100
2N474	2N475,2N479,2N480,2N758,2N759,2N783,2N784,2N834A,2N840,2N842,2N2318, 2N2319,2N2368,2N2432,2N2432A,2N2710,2N2729,2N3261,2N3510,2N3605A, 2N3606A,2N3648,2N3793,2N3844,2N3844A,2N3854A,2N4294,2N5030,2SC52, 2SC54,2SC204,2SC302,2SC321,2SC322,2SC356,2SC387AG,2SC394-O,2SC400-R,

To Replace	Substitute This Type
(2N474)	2SC455,2SC468,2SC576,2SC595,2SC601,2SC689,2SC717,2SC724,2SC752G-R, 2SC1126,40405,BF224J,BF225J,EN914,EN918,EN2369A,EN3009,EN3011,EN3013, EN3014,GET708,GET914,GET3013,GET3014,GET3646,GI-3793,KT218,MM1941, MPS834,MPS6507,MPS6511,MPS6540,MPS6542,MPS6543,MPS6548,MPSA10-RED, MPSA20-RED,MPSH19,MPSH20,MPSH37,MT1038,MT1038A,MT1039,MT1060, PET3001,SE3001,SE3002,TIS45,TIS46,TIS47,TIS49,TIS51,TIS52,TIS55,TIS84,TIS108, ZT23,ZT43,ZT203P,ZT403P,ZT2368,*PTC126,*GE-61,*TR-21,*HEPS0014,*SK3122, *SN80,*ECG123A,*WEP736,*BC107B,*276-2013,*RT-102
2N474A	2N471,2N472,2N472A,2N475A,2N756,2N834,2SC387A,2SC945,MPSH07,MPSH08, MPSH34,TIS126,*PTC126,*GE-60,*TR-21,*HEPS0014,*SK3124,*ECG123A, *WEP736,*BF173,*276-2013,*RT-102
2N475	2N480,2N717,2N758,2N758A,2N759,2N759A,2N760B,2N840,2N842,2N844,2N2312, 2N2314,2N2395,2N2432A,2N2522,2N3973,2N3975,2SC30,2SC302,2SC725,BF224J, BF225J,BSY93,MPSA05,ZT23,ZT43,*PTC123,*GE-212,*TR-21,*HEPS0014, *SK3122,*SN80,*ECG123A,*WEP736,*BC107B,*276-2013,*RT-102
2N475A	2N472,2N472A,2N742,2N756,2N756A,2N759B,2N2521,2SC945,MPSH34,TIS126, *PTC123,*GE-212,*TR-21,*HEPS0014,*SK3124,*ECG123A,*WEP736,*BC107B, *276-2013,*RT-102
2N476	*PTC126,*GE-86,*TR-86,*HEPS0014,*SK3124,*ECG123,*WEP53,*BF365,*276-2011, *RT-100
2N477	*PTC126,*GE-60,*TR-86,*HEPS0014,*SK3124,*ECG123,*WEP53,*BF173,*276-2013, *RT-100
2N478	2N479,2N709,2N709A,2N744,2N753,2N1386,2N1992,2N2206,2N2318,2N2319, 2N2331,2N2475,2N2569,2N2570,2N2656,2N2711,2N2713,2N2784,2N2922, 2N2926-RED,2N3011,2N3298,2N3340,2N3394,2N3563,2N3564,2N3605,2N3606, 2N3607,2N3633,2N3845,2N3845A,2N3855,2N3855A,2N3858,2N3959,2N3960, 2N4294,2N4418,2N4968,2N4996,2N4997,2N5030,2N5126,2N5127,2N5128, 2N5129,2N5136,2N5137,2N5200,2N5201,2N5399,2S095A,2S501,2S512,2SC99, 2SC168,2SC204,2SC300,2SC301,2SC337,2SC356,2SC400-O,2SC460,2SC461, 2SC528,2SC595,2SC710,2SC712,2SC713,2SC718,2SC719,2SC724,2SC836, 2SC922,2SC924,2SC1215,2SC1231,2SC1293,2SC1372,2SC1393,2SC1394, 2SC1395,2SC1396,2SC1789,2SC1988,40231,40234,A321,BC170A,BF121,BF123, BF125,BF127,BF255,BF377,BF378,EN706,EN744,EN3011,MPS2713,MPS2926-RED, MPS6568,MPSH02,MPSH10,MPSH11,MPSH30,MPSH31,MT1060A,PT720,SE3001, SE3005,TIS51,TIS64A,ZT21,ZT41,ZT2205,*PTC126,*GE-86,*TR-86,*HEPS0014, *SK3124,*SN80,*ECG123,*WEP53,*BF365,*276-2011,*RT-100
2N479	2N338A,2N480,2N708,2N708A,2N760,2N784A,2N843,2N929,2N957,2N2242,2N2318, 2N2319,2N2331,2N2369,2N2369A,2N2387,2N2501,2N3009,2N3011,2N3013,2N3014, 2N3340,2N3563,2N3564,2N3605A,2N3606A,2N3646,2N3688,2N3689,2N3690, 2N3691,2N3693,2N3845,2N3845A,2N3855A,2N3858,2N3862,2N4123,2N4137, 2N4294,2N4295,2N4418,2N4420,2N4421,2N4422,2N4966,2N4968,2N4996,2N4997, 2N5029,2N5030,2N5136,2N5137,2N5824,2SC67,2SC169,2SC204,2SC318,2SC335, 2SC356,2SC380-R,2SC380A-R,2SC394-O,2SC400-O,2SC460,2SC461,2SC468, 2SC595,2SC601,2SC601N,2SC620,2SC639,2SC710,2SC712,2SC713,2SC724, 2SC752G-R,2SC764,2SC836,2SC838,2SC839,2SC852,2SC922,2SC924,2SC941-R, 2SC963,2SC964,2SC1123,2SC1128,2SC1129,2SC1187,2SC1188,2SC1189, 2SC1215,2SC1240,2SC1274,2SC1319,2SC1320,2SC1372,2SC1393,2SC1394, 2SC1395,2SC1396,2SC1686,2SC1687,2SC1688,A321,BF121,BF123,BF125,BF127, BF198,BF199,BF241,BF255,BF311,BF377,BF378,BFY39-1,EN708,EN914,EN2369A, EN3009,EN3011,EN3013,EN3014,GET708,GET914,GET2369,GET3013,GET3014, GET3646,MPS3646,MPS3693,MPS6512,MPS6574-YEL,MPS6576-YEL,MPSA10-RED, MPSA20-RED,MPSH10,MPSH11,MT1060A,PET1001,PET2001,PT2760,SE1001, SE2001,SE3001,SE3005,SE5001,SE5002,SE5003,SE5006,TIS45,TIS46,TIS48,TIS49, TIS51,TIS52,TIS55,TIS64A,TIS87,TIS129,TP3705,ZT23,ZT43,ZT708,*PTC126,*GE-60, *TR-86,*HEPS0014,*SK3124,*SN80,*ECG123,*WEP53,*BF173,*276-2013,*RT-100
2N479A	2N474,2N475,2N480A,2N758,2N759,2N783,2N784,2N834A,2N842,2N1140,2N1387, 2N2368,2N2427,2N2432,2N2432A,2N2729,2N3663,2N3825,2N3843,2N3843A, 2N3854A,2N3983,2N3984,2N3985,2N4292,2N4293,2SC52,2SC54,2SC302,2SC321, 2SC322,2SC382,2SC387AG,2SC400-R,2SC455,2SC689,2SC717,2SC1126,BF224J, BF225J,BSX25,D16G6,EN918,GET706,KT218,MM1941,MPS834,MPS918,MPS6507, MPS6511,MPS6540,MPS6542,MPS6543,MPS6545,MPS6546,MPS6547,MPS6548,

To Replace	Substitute This Type
(2N479A)	MPSH19,MPSH20,MPSH32,MPSH37,MT1038,MT1038A,MT1039,MT1060,PET3001, SE1010,SE3002,SE5025,TIS47,TIS84,TIS108,TIS125,ZT22,ZT42,*PTC126,*GE-60, *TR-86,*HEPS0014,*SK3124,*ECG123,*WEP53,*BF173,*276-2013,*RT-100
2N480	2N338A,2N708A,2N718,2N760,2N843,2N929,2N2387,2N2396,2N2845,2N2847, 2N3301,2N3641,2N3642,2N3693,2N3826,2N3858A,2N3862,2N3903,2N3946,2N3973, 2N3974,2N3975,2N3976,2N4140,2N4962,2N4966,2N4994,2N5027,2N5380,2N5824, 2S103,2SC318,2SC335,2SC336,2SC620,2SC725,2SC815,2SC838,2SC839,2SC852, 2SC896,2SC938,2SC1189,2SC1320,2SC1688,BFY39-1,BSY93,EN697,FT3641, FT3642,GET2221,MM3903,MPS3693,MPS6565,MPS6576-YEL,PET1001,SE1001, TIS87,TN-3903,TP3705,ZT23,ZT43,*PTC123,*GE-212,*TR-86,*HEPS0014,*SK3124, *SN80,*ECG123,*WEP53,*BC107B,*276-2013,*RT-100
2N480A	2N475,2N717,2N758,2N758A,2N759,2N759A,2N760B,2N842,2N844,2N2395,2N2432A, 2N2522,2S102,2SC30,2SC302,BF224J,BF225J,MPS6544,MPS6545,ZT22,ZT42, *PTC123,*GE-212,*TR-21,*HEPS0014,*SK3122,*ECG123A,*WEP736,*BC107B, *276-2013,*RT-102
2N481	2N60,2N241A,2N241A+,2N320,2N321,2N323,2N450,2N482,2N483,2N1115,2N1281, 2N1282,2N1344,2N1383,2SB328,SF.T228,SYL792,*PTC109,*GE-52,*TR-05, *HEPG0005,*SK3005,*HF12H,*ECG100,*WEP254,*AC126,*276-2004,*RT-118
2N482	2N60,2N241A,2N241A+,2N321,2N323,2N324,2N362,2N450,2N483,2N484,2N520A, 2N1008,2N1115,2N1281,2N1282,2N1344,2N1381,2N1382,2N1383,2N1706,2SB328, 2SB329,40269,ASY27,SF.T228,SF.T229,SF.T237,SYL792,*PTC109,*GE-52,*TR-05, *HEPG0005,*SK3005,*HF6H,*ECG100,*WEP254,*AC126,*276-2004,*RT-118
2N483	2N321,2N324,2N484,2N508,2N520A,2N1705,2N1706,2N1808,40269,ASY27,SF.T229, SF.T237,*PTC109,*GE-1,*TR-05,*HEPG0005,*SK3005,*HF6H,*ECG100,*WEP254, *AC188/01,*276-2004,*RT-118
2N484	2N508,2N1284,ASY27,SF.T229,*PTC109,*GE-1,*TR-05,*HEPG0002,*SK3005, *HF12M,*ECG100,*WEP254,*AC188/01,*276-2005,*RT-118
2N485	2N450,2N486,2N1115,2N1281,2N1282,2N1344,SF.T228,SYL792,*PTC109,*GE-1, *TR-05,*HEPG0005,*SK3005,*HF12H,*ECG100,*WEP254,*AC188/01,*276-2004, *RT-118
2N486	SF.T229,*PTC109,*GE-2,*TR-05,*HEPG0001,*SK3005,*HF20H,*ECG100,*WEP254, *RT-118
2N487	2N2630,2N2718,*PTC107,*GE-1,*TR-05,*HEPG0001,*SK3004,*ECG102,*WEP631, *RT-120
2N495	2N726,2N858,2N860,2N861,2N923,2N939,2N978,2N2391,2N2969,2N2971,2N4008, 2N6567,2S322,2S322A,2S3210,2S3220,A5T3638,MM4052,MPSH85,TIS138,TP3638, *PTC127,*GE-65,*TR-20,*HEPS0013,*SK3114,*EP25,*ECG106,*WEP52,*BC177, *276-2022,*RT-126
2N496	2N726,2N860,2N861,2N862,2N2391,A5T3638,TP3638,TP5142,*PTC127,*GE-65, *TR-20,*HEPS0013,*SK3114,*EP25,*ECG106,*WEP52,*BC177,*276-2021,*RT-126
2N497	*PTC125,*GE-18,*TR-87,*HEPS5014,*ECG128,*WEP243,*276-2008,*RT-114
2N497A	*PTC125,*GE-63,*TR-87,*HEPS5014,*ECG128,*WEP243,*276-2008,*RT-114
2N498	*PTC144,*GE-27,*TR-87,*HEPS0014,*ECG128,*WEP243,*276-2008,*RT-114
2N498A	*PTC144,*GE-32,*TR-21,*HEPS3019,*276-2008
2N499	2N499A,2N2587,2N2955,2N6365,2N6365A,*PTC107,*GE-51,*TR-17,*HEPG0008, *SK3006,*JR200,*ECG126,*WEP635,*276-2009
2N499A	2N2587,2N2955,2N6365,2N6365A,*PTC107,*TR-17,*HEPG0009,*SK3006,*ECG126, *WEP635,*276-2005
2N500	2N705,2N710,2N779,2N779A,2N828,2N828A,2N964,2N968,2N1499B,2N2048,2N2169, 2N2170,2N2487,2N2488,2N2489,2N2795,2N2796,2N2860,2SA56,2SA405,2SA411, *PTC107,*GE-51,*TR-17,*HEPG0008,*SK3008,*JR200,*ECG126,*WEP635
2N501	2N501A,2N705,2N710,2N779,2N779A,2N828,2N828A,2N964,2N968,2N979,2N980, 2N2487,2N2488,2N2489,2N2795,2N2796,2N2860,2SA56,2SA405,2SA411,*PTC107, *GE-51,*TR-17,*HEPG0003,*SK3006,*JR200,*ECG126,*WEP635
2N501/18	2N828,2N968,*PTC145,*TR-17,*HEPG0003,*SK3006,*ECG126,*WEP635
2N501A	2N559,2N711B,2N779,2N779A,2N829,2N838,2N984,2N2402,2N2795,*PTC107, *GE-51,*TR-17,*HEPG0003,*SK3006,*JR200,*ECG126,*WEP635
2N502	2N502A,2N982,*PTC145,*GE-51,*TR-17,*HEPG0009,*SK3006,*JR200,*ECG126, *WEP635,*276-2005
2N502A	*PTC145,*GE-51,*TR-17,*HEPG0009,*SK3006,*JR200,*ECG126,*WEP635,*276-2005
2N502B	2N502A,2N2587,*PTC145,*TR-17,*HEPG0009,*SK3006,*ECG126,*WEP635

To Replace	Substitute This Type
2N503	*PTC107,*GE-51,*TR-05,*HEPG0002,*SK3005,*JR200,*ECG100,*WEP254,*RT-118
2N504	*PTC107,*GE-1,*TR-17,*HEPG0003,*SK3008,*JR100,*ECG126,*WEP635
2N505	*GE-2,*TR-05,*HEPG0002,*SK3005,*HF12M,*ECG100,*WEP254,*276-2005,*RT-118
2N506	2N60A,2N60B,2N60C,2N61A,2N61B,2N61C,2N381,2N505,2N518,2N525,2N525A, 2N526,2N526A,2N597,2N650,2N650A,2N651,2N651A,2N652A,2N1057,2N1057+, 2N1187,2N1192,2N1348,2N1349,2N1350,2N1351,2N1373,2N1447,2N1451,2N1495, 2N1924,2N1925,2N1954,2N1956,2N1957,2N2956,2SB57A,2SB225,2SB226, 2SB304A,2SB427,MA100,MA882,MA887,SF.T243,*GE-80,*TR-85,*HEPG0005, *SK3004,*AT30M,*ECG102A,*WEP250,*AC126,*276-2004,*RT-121
2N506+	2N43,2N43A,2N61A,2N61B,2N61C,2N381,2N505,2N518,2N525,2N525A,2N586, 2N650,2N650A,2N1057,2N1186,2N1191,2N1373,2N1446,2N1495,2N1924,2N2955, 2SB225,2SB289,2SB427,MA100,MA206,MA881,MA886,SF.T243,*GE-80,*TR-85, *HEPG0005,*SK3004,*ECG102A,*WEP250
2N507	2N377A,2N444A,2N576A,2N587,2N1000,2N1312,2N1473,2N1605A,2N1672,2N1672A, 2SD75A,2SD77A,*GE-8,*TR-08,*HEPG0011,*SK3010,*NA30,*ECG103A,*WEP641B, *AC127,*276-2001,*RT-122
2N508	2N417,2N521A,2N522A,2N598,2N599,2N1284,2N1309,2N1316,2N1317,2N1354, 2N1355,2N1357,2N1808,2N1892,2N1998,*PTC135,*GE-53,*TR-05,*HEPG0005, *SK3003,*AT30H,*ECG102,*WEP631,*AC188/01,*276-2005,*RT-120
2N508A	2N599,2N655+,2N1190,2N1316,2N1354,2N1355,2N1357,2N1998,2N1999,2N2171, 2N3427,2SB223,2SB223A,*PTC135,*GE-2,*TR-05,*HEPG0005,*SK3004,*ECG102, *WEP631,*276-2005,*RT-120
2N509	*PTC107,*GE-51,*TR-17,*ECG160,*WEP637
2N511	2N511A,2N512,2N512A,2N513,2N513A,*PTC901,*GE-3,*TR-01,*SK3009,*PT250, *ECG104,*WEP230,*RT-124
2N511A	2N511B,2N512A,2N512B,2N513A,2N513B,*PTC901,*GE-3,*TR-01,*SK3009,*PT250, *ECG104,*WEP230,*RT-124
2N511B	2N512B,2N513B,*PTC901,*GE-3,*TR-01,*SK3009,*ECG104,*WEP230,*RT-124
2N512	2N511,2N511A,2N512A,2N513,2N513A,*PTC901,*GE-3,*TR-01,*SK3009,*PT250, *ECG104,*WEP230,*RT-124
2N512A	2N511A,2N511B,2N512B,2N513A,2N513B,*PTC901,*GE-3,*TR-01,*SK3009,*PT250, *ECG104,*WEP230,*RT-124
2N512B	2N511B,2N513B,*PTC901,*GE-3,*TR-01,*SK3009,*ECG104,*WEP230,*RT-124
2N513	2N511,2N511A,2N512,2N512A,2N513A,*PTC901,*GE-3,*TR-01,*SK3009,*PT250, *ECG104,*WEP230,*RT-124
2N513A	2N511A,2N511B,2N512A,2N512B,2N513B,*PTC901,*GE-3,*TR-01,*SK3009,*PT250, *ECG104,*WEP230,*RT-124
2N513B	2N511B,2N512B,*PTC901,*GE-3,*TR-01,*SK3009,*PT250,*ECG104,*WEP230, *RT-124
2N514	2N514A,*PTC901,*GE-3,*TR-01,*SK3009,*PT250,*ECG121,*WEP232,*RT-127
2N514A	2N514B,*PTC901,*GE-3,*TR-01,*SK3009,*PT250,*ECG121,*WEP232,*RT-127
2N514B	*PTC901,*GE-3,*TR-01,*SK3009,*PT250,*ECG121,*WEP232,*RT-127
2N515	2N94,2N193,2N194,2N194A,2N212,2N233A,*GE-5,*TR-08,*HEPG0011,*SK3011, *NR5,*ECG103A,*WEP641B,*AC187/01,*276-2001,*RT-122
2N515+	2N94,2N193,2N194,2N194A,2N212,2N233A,2N1367,*GE-5,*TR-08,*HEPG0011, *SK3011,*WEP641B
2N516	2N94,2N193,2N194,2N194A,2N212,2N233A,*GE-5,*TR-08,*HEPG0011,*SK3010, *NR5,*ECG103A,*WEP641B,*AC187/01,*276-2001,*RT-122
2N516+	2N212,2N1366,*GE-5,*TR-08,*HEPG0011,*SK3010,*WEP641B
2N517	2N94,2N193,2N194,2N194A,2N212,2N233A,*GE-5,*TR-08,*HEPG0011,*SK3011, *NR5,*ECG103A,*WEP641B,*AC187/01,*276-2001,*RT-122
2N517+	2N182,2N212,2N438,2N438A,2N585,2N634,2N1366,2N1891,2N1994,*GE-5,*TR-08, *HEPG0011,*SK3011,*WEP641B
2N518	*GE-51,*TR-05,*HEPG0002,*SK3005,*HF12H,*ECG100,*WEP254,*276-2005,*RT-118
2N519	2N61,2N111A,2N112A,2N315A,2N316,2N316A,2N317,2N317A,2N396A,2N403,2N404, 2N413,2N428,2N519A,2N520,2N565,2N566,2N579,2N579+,2N580,2N580+,2N611, 2N613,2N653,2N1017,2N1018,2N1280,2N1303,2N1313,2N1347,2N1942,2SA205, 2SA208,2SA406,2SB219,2SB219A,2SB220,2SB220A,MA112,SF.T221,SF.T222, SF.T227,*PTC109,*GE-2,*TR-05,*HEPG0005,*SK3005,*HF6M,*ECG100,*WEP254, *AC188/01,*276-2004,*RT-118
2N519A	2N61,2N61A,2N104,2N111A,2N112A,2N215,2N396A,2N403,2N404,2N404A,2N413,

To Replace	Substitute This Type
(2N519A)	2N428,2N565,2N573,2N611,2N613,2N1017,2N1018,2N1191,2N1303,2N1313,2N1413, 2SA205,2SA406,2SB185,2SB219,2SB220,NKT225,SF.T221,SF.T222,SF.T226, SF.T227,*PTC135,*GE-50,*TR-05,*HEPG0005,*SK3005,*ECG100,*WEP254, *AC188/01,*276-2004,*RT-118
2N520	2N404,2N413,2N414,2N414B,2N414C,2N580,2N580+,2N1017,2N1171,2N1280, 2N1281,2N1303,2N1313,2N1344,2N1478,2N1942,2SA204,2SA205,2SA206,2SA207, 2SA209,2SA406,2SA407,2SB326,ASY26,SF.T227,SF.T228,*PTC109,*GE-2,*TR-05, *HEPG0005,*SK3005,*HF6M,*ECG100,*WEP254,*AC188/01,*276-2005,*RT-118
2N520A	2N417,2N508A,2N571,2N654+,2N655,2N1193,2N1307,2N1309,2N1354,2N1355, 2N1356,2N1357,2N1707,2N1808,2N1892,2SB222,2SB223,40269,ASY27,SF.T223, *PTC109,*GE-50,*TR-05,*HEPG0005,*SK3005,*ECG100,*WEP254,*AC188/01, *276-2005,*RT-118
2N521	2N1282,2N1307,2SA210,SF.T229,*PTC109,*GE-1,*TR-05,*HEPG0005,*SK3005, *HF12M,*ECG100,*WEP254,*276-2005,*RT-118
2N521A	2N522A,2N1316,*GE-50,*TR-05,*HEPG0002,*SK3005,*ECG100,*WEP254,*276-2005, *RT-118
2N522	2N522A,2N523,2N1309,*GE-2,*TR-05,*HEPG0002,*SK3003,*HF20H,*ECG100, *WEP254,*276-2005,*RT-118
2N522A	*GE-50,*TR-05,*HEPG0002,*SK3005,*ECG100,*WEP254,*276-2005,*RT-118
2N523	*GE-2,*TR-05,*HEPG0002,*SK3003,*HF20H,*ECG100,*WEP254,*276-2005,*RT-118
2N523A	*GE-50,*TR-05,*HEPG0002,*SK3005,*ECG100,*WEP254,*276-2005,*RT-118
2N524	2N524A,2N525,2N525A,2SB225,*GE-2,*TR-05,*HEPG0005,*SK3004,*AT30M, *ECG102,*WEP631,*276-2004,*RT-120
2N524A	2N525A,*GE-2,*TR-05,*HEPG0005,*SK3004,*AT30M,*ECG102,*WEP631,*276-2004, *RT-120
2N525	2N525A,2N526,2N526A,2N597,2SB225,2SB226,*GE-2,*TR-05,*HEPG0005,*SK3004, *AT30M,*ECG102,*WEP631,*276-2004,*RT-120
2N525A	2N526A,2N597,*GE-2,*TR-05,*HEPG0005,*SK3004,*AT30M,*ECG102,*WEP631, *276-2004,*RT-120
2N526	2N526A,2N527A,2N597,2N1997,2SB226,2SB227,*GE-2,*TR-05,*HEPG0005,*SK3004, *AT30H,*ECG102,*WEP631,*276-2005,*RT-120
2N526A	2N527A,*GE-2,*TR-05,*HEPG0005,*SK3004,*AT30H,*ECG102,*WEP631,*276-2004, *RT-120
2N527	*GE-2,*TR-05,*HEPG0005,*SK3004,*AT30H,*ECG102,*WEP631,*276-2005,*RT-120
2N527A	*TR-05,*HEPG0005,*SK3004,*ECG102,*WEP631,*276-2005,*RT-120
2N528	
2N529	M.P. NPN † PNP
2N529/N	2N182,2N212,2N438,2N438A,2N530/N,2N531/N,2N585,2N634,2N1891,2N1994, *PTC108,*GE-5,*TR-08,*NA20,*AC127
2N529/P	2N111A,2N112A,2N123A,2N315,2N396A,2N425,2N426,2N427,2N428,2N530/P, 2N531/P,2N579+,2N1018,2N1347,2SA208,*PTC109,*GE-50,*TR-05,*AC126
2N530	M.P. NPN † PNP
2N530/N	2N94A,2N182,2N212,2N438,2N438A,2N529/N,2N531/N,2N532/N,2N533/N,2N576, 2N585,2N634,2N1891,2N1994,2N1995,*PTC108,*GE-5,*NR10,*AC187/01
2N530/P	2N111A,2N112A,2N123A,2N315,2N316,2N317,2N396A,2N426,2N427,2N428, 2N531/P,2N532/P,2N533/P,2N579,2N579+,2N1018,2N1347,2SA208,*PTC109, *GE-50,*TR-17,*AC188/01
2N531	M.P. NPN † PNP
2N531/N	2N94A,2N182,2N183,2N377,2N439,2N439A,2N530/N,2N532/N,2N533/N,2N557, 2N576,2N585,2N634,2N1605,2N1891,2N1994,2N1995,2SC128,*PTC108,*GE-5, *TR-08,*NR10,*AC187/01
2N531/P	2N111A,2N112A,2N123A,2N269,2N316,2N317,2N396A,2N404,2N413,2N428,2N520, 2N532/P,2N533/P,2N579,2N579+,2N580,2N580+,2N583,2N615,2N1017,2N1018, 2N1280,2N1347,2SA205,2SA208,SF.T227,*PTC109,*GE-50,*TR-82,*SK3123, *ECG176,*AC188/01
2N532	M.P. NPN † PNP
2N532/N	2N94A,2N182,2N183,2N312,2N377,2N439,2N439A,2N531/N,2N533/N,2N557,2N576, 2N634,2N635,2N1090,2N1605,2N1891,2N1995,2N1996,2SC128,*PTC108,*GE-5, *NR10,*AC187/01
2N532/P	2N269,2N316,2N317,2N404,2N533/P,2N579,2N580,2N580+,2N583,2N1017,2N1280, 2N1313,2SA205,2SA406,SF.T227,*PTC109,*GE-50,*TR-17,*AC188/01

To Replace	Substitute This Type
2N533	M.P. NPN † PNP
2N533/N	2N94A,2N182,2N183,2N312,2N377,2N439,2N439A,2N446,2N532/N,2N557,2N576, 2N634,2N635,2N1090,2N1605,2N1995,2N1996,2SC128,2SC129,ASY28,*PTC108, *GE-5,*TR-08,*NR10,*AC187/01
2N533/P	2N269,2N404,2N414,2N414B,2N414C,2N450,2N580,2N580+,2N582,2N583,2N584, 2N1017,2N1171,2N1280,2N1281,2N1313,2N1344,2SA171,2SA173,2SA204,2SA205, 2SA206,2SA209,2SA212,2SA406,2SA407,ASY26,SF.T227,SF.T228,*PTC109, *GE-50,*AC188/01
2N534	*GE-51,*TR-85,*HEPG0003,*SK3123,*AT30M,*ECG102A,*WEP250,*RT-121
2N535	2N417,2N535A,2N535B,2N536,2N655,2N1309,2N1316,2N1354,2N1355,2N1357, 2N2613,*GE-52,*TR-85,*HEPG0005,*SK3003,*AT20M,*ECG102A,*WEP250, *AC188/01,*276-2005,*RT-121
2N535A	2N417,2N535,2N535B,2N536,2N655,2N1309,2N1316,2N1354,2N1355,2N1357, 2N2613,*GE-52,*TR-85,*HEPG0005,*SK3003,*AT20M,*ECG102A,*WEP250, *AC188/01,*276-2005,*RT-121
2N535B	2N417,2N535,2N535A,2N536,2N655,2N1309,2N1316,2N1354,2N1355,2N1357, 2N2613,*GE-52,*TR-85,*HEPG0005,*SK3003,*AT20M,*ECG102A,*WEP250, *AC188/01,*276-2005,*RT-121
2N536	2N1316,2N2613,*DS26,*GE-52,*TR-85,*HEPG0005,*SK3003,*HF3H,*ECG102A, *WEP250,*AC126,*276-2002,*RT-121
2N536+	2SB400,*GE-52,*HEPG0005,*SK3003,*ECG102A,*WEP250
2N537	*GE-2,*TR-17,*HEPG0003,*ECG160,*WEP637
2N538	2N539,*DS-503,*GE-3,*TR-01,*SK3009,*PT40,*ECG121,*WEP232,*RT-127
2N538A	2N539A,2N1261,2N1262,*DS-503,*GE-3,*TR-01,*SK3009,*PT40,*ECG121,*WEP232, *RT-127
2N539	2N540,*DS-503,*GE-3,*TR-01,*HEPG6012,*SK3009,*PT40,*ECG104,*WEP230, *RT-124
2N539A	2N540A,2N1202,2N1262,2N1263,*DS-503,*GE-3,*TR-01,*HEPG6006,*SK3009,*PT40, *ECG104,*WEP230,*276-2006,*RT-124
2N540	*DS-503,*GE-3,*TR-01,*SK3009,*PT40,*ECG104,*WEP230,*RT-124
2N540A	2N1202,2N1263,*DS-503,*GE-3,*TR-01,*SK3009,*PT40,*ECG104,*WEP230,*RT-124
2N541	2N542,2N2712,2N2714,2N2924,2N2926-YEL,2N3241A,2N3392,2N3856,2N3856A, 2N3859,2N3860,2N4251,2N5131,2N5133,2N5224,2S502,2SC400,2SC400-Y,2SC454, 2SC458,2SC847,2SC848,2SC934,2SC966,2SC1085,2SC1281,2SC1363,2SC1631, 2SC1633,2SC1854,2SD392,40232,40233,40398,40400,A5T3572,BC108A,BC168A, BC170B,BC172A,BC238A,BC548A,BCY58A,MPS2714,MPS2926-GRN,MPS2926-YEL, PBC108A,SE4001,TN80,*PTC126,*GE-60,*TR-21,*HEPS0014,*SK3122,*SN80, *ECG123A,*WEP736,*BF365,*276-2011,*RT-102
2N541A	*PTC126
2N542	2N543,2N841,2N930,2N2388,2N3241A,2N3242A,2N3692,2N3694,2N3856A,2N3859, 2N3860,2N4074,2N5209,2N5825,2N5826,2SC367G,2SC367G-Y,2SC372-Y, 2SC372G-Y,2SC380-Y,2SC380A-Y,2SC394-GR,2SC400,2SC400-Y,2SC454,2SC458, 2SC733-Y,2SC735-Y,2SC752G-Y,2SC847,2SC848,2SC876,2SC933,2SC941-Y, 2SC966,2SC1085,2SC1175,2SC1216,2SC1285,2SC1364,2SC1632,2SC1634, 2SC1727,2SC1854,2SC1999,BC107A,BC108A,BC123,BC167A,BC171A,BC237A, BC238A,BC547A,BC548A,BCY58A,BCY59A,BF240,BFY39-2,BSW88,BSW89,BSX38, BSX79,EN930,MPS6514,MPS6573,MPS6574-GREEN,MPS6574-SIL,MPS6575, MPS6576-GREEN,MPS6576-SIL,MPS-A10,MPS-A20,MPS-K20,MPS-K21,MPS-K22, MPSA10-BLU,MPSA10-GRN,MPSA10-YEL,MPSA20-BLU,MPSA20-GRN,MPSA20-YE L,NPSA20,PBC107A,PET1002,PET2002,SE1002,SE2002,SE4001,TN80,TZ81, *PTC126,*GE-60,*TR-21,*HEPS0014,*SK3122,*SN80,*ECG123A,*WEP736, *BC107B,*276-2013,*RT-102
2N542A	2N479,2N480,2N543A,2N758,2N759,2N783,2N784,2N840,2N2318,2N2319,2N2368, 2N2387,2N2432,2N2432A,2N2729,2N3340,2N3605A,2N3606A,2N3663,2N3844, 2N3844A,2N3854A,2N3983,2N4137,2N4294,2N5030,2SC52,2SC54,2SC204, 2SC302,2SC321,2SC322,2SC356,2SC380-R,2SC380A-R,2SC387AG,2SC394-O, 2SC400-R,2SC455,2SC468,2SC595,2SC601,2SC689,2SC717,2SC724,2SC752G-R, 2SC1126,BF224J,BF225J,BF255,BFY39-1,EN914,EN918,EN2369A,EN3009,EN3011, EN3013,EN3014,GET708,GET914,GET3013,GET3014,GET3646,KT218,MM1941, MPS6507,MPS6511,MPSA10-RED,MPSA20-RED,MPSH19,MT1038,MT1038A, MT1039,MT1060,PET3001,PT2760,SE3001,SE3002,TIS45,TIS46,TIS49,TIS51,TIS52,

To Replace	Substitute This Type
(2N542A)	TIS55,TIS84,TIS129,ZT23,ZT43,*PTC126,*GE-60,*TR-86,*HEPS0014,*SK3124, *ECG123,*WEP53,*BF173,*276-2013,*RT-100
2N543	2N760A,2N909,2N930B,2N956,2N2483,2N2484A,2N2645,2N3302,2N3643,2N3827, 2N3859A,2N3904,2N3947,2N4141,2N4995,2N5028,2N5209,2N5381,2N5825,2N5826, 2SC366G,2SC366G-Y,2SC734-Y,2SC875,2SC876,2SC933,2SC943,2SC979-Y, 2SC980G-Y,2SC1007,2SC1175,2SC1364,2SC1632,2SC1634,2SC1999,BC107A, BC174A,BC182A,BC182L,BC190A,BC237A,BC547A,BSX79,EN956,EN1711, EN2484,FT3643,GET930,GET2222,GET2222A,MM3904,TN-3904,TZ82,*PTC123, *GE-212,*TR-21,*HEPS0014,*SK3122,*SN80,*ECG123A,*WEP736,*BC107B, *276-2013,*RT-100
2N543A	2N708A,2N718,2N718A,2N2396,2N2845,2N2847,2N3301,2N3641,2N3642,2N3826, 2N3858A,2N3862,2N3903,2N3946,2N3973,2N3974,2N3975,2N3976,2N4140,2N4962, 2N4966,2N4994,2N5027,2N5380,2N5824,2S103,2SC318,2SC335,2SC336,2SC383, 2SC473,2SC620,2SC714,2SC725,2SC815,2SC838,2SC839,2SC896,2SC938, 2SC979-R,2SC980,2SC980G-R,2SC1189,2SC1320,2SC1688,BSY93,EN697, EN718A,EN915,EN1613,FT3641,FT3642,GET929,GET2221,GET2221A,MM3903, MPS6565,TN-3903,TP3705,*PTC123,*GE-212,*TR-21,*HEPS0014,*SK3122, *ECG123A,*WEP736,*BC107B,*276-2013,*RT-100
2N544	2N3783,2N3784,A1383,*PTC107,*GE-50,*TR-17,*HEPG0003,*SK3008,*JR30X, *ECG126,*WEP635,*AF125
2N544/33	2N2588,2N3783,2N3784,A1383,*PTC107,*GE-50,*TR-17,*HEPG0003,*SK3008, *ECG126,*WEP635,*AF125
2N545	2N1613A,2N2106,2N2107,2N2193,2N2193A,2N2193B,2N2218A,2N2868,2N3110, 2N3678,2N3722,2N6375,2N6376,A5T2193,BC140-6,BSY53,ZT1613,*PTC144,*GE-63, *276-2012
2N546	2N548,2N550,MM3724,MM3725,SE8041,SE8042,*PTC144,*GE-63,*TR-86, *HEPS0014,*SK3124,*ECG123,*WEP53,*276-2012,*RT-100
2N547	2N545,2N549,2N1613A,2N2106,2N2107,2N2193,2N2193A,2N2193B,2N2218A, 2N2868,2N3110,2N3678,2N3722,2N6375,2N6376,A5T2193,BC140-6,BSY53,ZT1613, *PTC144,*GE-63,*TR-86,*HEPS0014,*SK3124,*ECG123,*WEP53,*276-2012, *RT-100
2N548	2N546,2N550,MM3724,MM3725,SE8041,SE8042,*PTC136,*GE-47,*TR-86, *HEPS0014,*SK3124,*ECG123,*WEP53,*BC337,*276-2012,*RT-100
2N549	2N545,2N547,2N1613A,2N2106,2N2107,2N2193,2N2193A,2N2193B,2N2218A, 2N2868,2N3110,2N3678,2N3722,2N6375,2N6376,A5T2193,BC140-6,BSY53,ZT1613, *PTC144,*GE-63,*TR-86,*HEPS0014,*SK3124,*ECG123,*WEP53,*276-2012, *RT-100
2N550	2N546,2N548,MM3724,MM3725,SE8041,SE8042,*PTC136,*GE-47,*TR-86, *HEPS0014,*SK3124,*ECG123,*WEP53,*BC337,*276-2012,*RT-100
2N551	2N545,2N547,2N549,2N656,2N1613,2N1613A,2N1837,2N1837A,2N2106,2N2107, 2N2218A,2N3110,2N3678,BSY53,*PTC123,*GE-81,*TR-86,*HEPS0014,*SK3124, *ECG123,*WEP53,*276-2012,*RT-100
2N552	2N546,2N548,2N550,2N2237,2N4264,SE8041,SE8042,*PTC123,*GE-81,*TR-86, *HEPS0014,*SK3124,*ECG123,*WEP53,*BC337,*276-2012,*RT-100
2N553	2N1536,2N1536A,2N1537,2N1537A,2N1541,2N1542,2N2145,2N2145A,2N2146, 2N2146A,2N3617,2N3618,2SB152A,2SB249,2SB253,*PTC105A,*GE-3,*TR-01, *HEPG6005,*SK3009,*PT50,*ECG121,*WEP232,*276-2006,*RT-127
2N554	2N555,*PTC105A,*GE-25,*TR-01,*HEPG6003,*SK3009,*PT12,*ECG121,*WEP232, *276-2006,*RT-127
2N555	2N663,2N1291,*PTC138,*GE-76,*TR-01,*HEPG6003,*SK3009,*PT25,*ECG121, *WEP232,*276-2006,*RT-127
2N556	2N357A,2N358A,2N377,2N377A,2N385,2N439,2N439A,2N440,2N440A,2N445A, 2N585+,2N679,2N1000,2N1012,2N1090,2N1091,2N1299,2N1302,2N1304,2N1473, 2N1605A,2N2426,2SC89,2SC90,2SC128,2SC129,2SD11,2SD20,2SD21,ASY28, SF.T298,*PTC108,*GE-8,*TR-08,*HEPG0011,*SK3011,*NR10,*ECG101,*WEP641, *AC187/01,*276-2001,*RT-119
2N557	*PTC108,*GE-6,*TR-08,*HEPG0011,*SK3011,*NR10,*ECG101,*WEP641,*276-2001, *RT-119
2N558	*PTC108,*GE-6,*TR-08,*HEPG0011,*SK3011,*NR10,*ECG101,*WEP641,*276-2001, *RT-119
2N559	2N2022,*GE-51,*TR-17,*HEPG0003,*ECG160,*WEP637

To Replace	Substitute This Type
2N560	*PTC125,*GE-81,*TR-86,*HEPS0014,*ECG123,*WEP53,*BF338,*276-2012,*RT-100
2N561	2SB87,2SB88,*GE-25,*TR-01,*HEPG6005,*SK3009,*PT50,*ECG121,*WEP232, *276-2006,*RT-127
2N563	2N315A,2N316A,2N425,2N426,2N427,2N428,2N460,2N524,2N524A,2N1018, 2SB219A,MA885,SF.T143,*GE-53,*TR-05,*HEPG0005,*SK3004,*AT30N,*ECG102, *WEP631,*276-2004,*RT-120
2N564	2N315A,2N316A,2N425,2N426,2N427,2N428,2N460,2N524,2N524A,2N1018, 2SB219A,MA885,*PTC135,*GE-53,*TR-05,*HEPG0005,*SK3004,*AT30N,*ECG102, *WEP631,*276-2004,*RT-120
2N565	2N381,2N525,2N525A,2N567,2N568,2N650,2N650A,2N653,2N1017,2N1171,2N1303, 2N1313,2N1446,2N1478,2SB220A,2SB225,2SB427,MA886,NKT221,NKT228, SF.T144,*PTC135,*GE-53,*TR-05,*HEPG0005,*SK3004,*AT30M,*ECG102, *WEP631,*276-2005,*RT-120
2N566	2N381,2N525,2N525A,2N568,2N650,2N650A,2N653,2N1017,2N1171,2N1303,2N1313, 2N1446,2N1478,2SB220A,2SB225,2SB427,MA886,NKT221,NKT228,*PTC135, *GE-53,*TR-05,*HEPG0005,*SK3004,*AT30M,*ECG102,*WEP631,*276-2007, *RT-120
2N567	2N381,2N461,2N526,2N526A,2N527A,2N568,2N569,2N571,2N597,2N651,2N651A, 2N652A,2N1008A,2N1171,2N1305,2N1307,2N1309A,2N1348,2N1349,2N1350, 2N1351,2N1447,2N1448,2N1451,2N1452,2N1478,2N1707,2SB57A,2SB221A, 2SB222A,2SB226,2SB227,2SB304,2SB304A,2SB326,2SB427,2SB428,MA887, NKT211,NKT221,NKT228,SF.T125P,SF.T144,*PTC135,*GE-53,*TR-08,*HEPG0011, *SK3004,*AT30H,*ECG103A,*WEP641B,*276-2002,*RT-122
2N568	2N381,2N461,2N526,2N526A,2N527A,2N571,2N597,2N651,2N651A,2N652A, 2N1008A,2N1171,2N1305,2N1307,2N1309A,2N1348,2N1349,2N1350,2N1351, 2N1447,2N1448,2N1451,2N1452,2N1478,2N1707,2SB57A,2SB221A,2SB222A, 2SB226,2SB227,2SB304,2SB304A,2SB326,2SB427,2SB428,MA887,NKT221, NKT228,*PTC135,*GE-53,*TR-05,*HEPG0005,*SK3004,*AT30H,*ECG102, *WEP631,*276-2005,*RT-120
2N569	2N382,2N383,2N461,2N527,2N527A,2N571,2N597,2N651,2N652,2N652A,2N1008A, 2N1189,2N1307,2N1309A,2N1448,2N1449,2N1452,2N1707,2N1892,2N1997, 2SB222A,2SB227,2SB327,2SB428,2SB431,2SB461,MA887,NKT211,SF.T125P, *PTC135,*GE-53,*TR-05,*HEPG0005,*SK3003,*AT20H,*ECG102,*WEP631, *AC188/01,*276-2005,*RT-120
2N570	2N382,2N383,2N461,2N527,2N527A,2N571,2N572,2N597,2N651,2N652,2N652A, 2N1008A,2N1189,2N1307,2N1309A,2N1448,2N1449,2N1452,2N1707,2N1892, 2N1997,2SB222A,2SB227,2SB327,2SB428,2SB461,MA887,*PTC135,*GE-53, *TR-05,*HEPG0005,*SK3003,*AT20H,*ECG102,*WEP631,*AC188/01,*276-2005, *RT-120
2N571	2N382,2N383,2N527,2N598,2N599,2N652,2N1307,2N1309,2N1449,2N1892,2N1997, 2N1998,2N2541,2SB223A,2SB327,*PTC135,*GE-52,*TR-05,*HEPG0005,*SK3003, *ECG100,*WEP254,*276-2007,*RT-118
2N572	2N382,2N383,2N461,2N527,2N527A,2N570,2N571,2N597,2N651,2N652,2N652A, 2N1008A,2N1189,2N1307,2N1309A,2N1448,2N1449,2N1452,2N1707,2N1892, 2N1997,2SB222A,2SB227,2SB327,2SB428,2SB461,MA887,*PTC135,*GE-53, *TR-05,*HEPG0005,*SK3003,*AT10H,*ECG100,*WEP254,*AC188/01,*276-2005, *RT-118
2N573	2N524A,2N525A,2N650,2N650A,2N1186,2N1924,*GE-2,*TR-05,*HEPG0007,*SK3004, *ECG102,*WEP631,*RT-120
2N574	2N574A,*GE-4,*TR-03,*HEPG6006,*SK3004,*PT515,*WEP233,*276-2006
2N574A	*GE-4,*TR-03,*HEPG6010,*SK3004,*PT515
2N575	2N575A,*GE-4,*TR-03,*SK3012,*PT515,*ECG105,*WEP233
2N575A	*GE-4,*TR-03,*SK3012,*PT515,*ECG105,*WEP233
2N576	*DS-72,*GE-5,*TR-08,*HEPG0011,*SK3011,*NR5,*ECG101,*WEP641,*276-2001, *RT-119
2N576A	*DS-72,*GE-5,*TR-08,*HEPG0011,*SK3011,*NR5,*ECG101,*WEP641,*276-2001, *RT-119
2N578	2N425,2N426,2N427,2N578+,*PTC135,*GE-1,*TR-05,*HEPG0005,*SK3005,*HF6M, *ECG100,*WEP254,*AC188/01,*276-2004,*RT-118
2N578+	2N426,2N427,*PTC135,*GE-1,*HEPG0005,*SK3005,*ECG100,*WEP254
2N579	2N580,2N580+,2N1017,2N1171,2N1313,2N1942,2SA209,*PTC135,*GE-1,*TR-05,

To Replace	Substitute This Type
(2N579)	*HEPG0005,*SK3005,*HF6M,*ECG100,*WEP254,*AC188/01,*276-2004,*RT-118
2N579+	2N316A,2N317A,2N428,2N580,2N580+,2N1017,2N1018,*GE-1,*HEPG0005,*SK3005, *ECG100,*WEP254
2N580	2N580+,2N1171,*GE-1,*TR-05,*HEPG0005,*SK3005,*HF12H,*ECG100,*WEP254, *276-2004,*RT-118
2N580+	*GE-1,*HEPG0005,*SK3005,*ECG100,*WEP254
2N581	2N123A,2N269,2N315A,2N316,2N316A,2N317,2N317A,2N396A,2N404,2N428,2N579, 2N579+,2N580,2N580+,2N583,2N1017,2N1018,2N1313,2N1347,2N1942,2SA205, 2SA406,SF.T227,*PTC109,*GE-1,*TR-05,*HEPG0005,*SK3005,*HF6M,*ECG100, *WEP254,*AC188/01,*276-2004,*RT-118
2N582	2N584,2N1309A,2SA207,2SA407,*GE-2,*TR-05,*HEPG0002,*SK3003,*HF20H, *ECG100,*WEP254,*276-2005,*RT-118
2N583	2N269,2N404,2N414,2N414B,2N414C,2N450,2N580,2N580+,2N582,2N584,2N1017, 2N1171,2N1313,2N1942,2SA204,2SA205,2SA206,2SA207,2SA209,2SA212,2SA406, 2SA407,ASY26,SF.T227,SF.T228,*PTC109,*GE-1,*TR-17,*HEPG0005,*SK3005, *HF6M,*ECG126,*WEP635,*AC188/01,*276-2004
2N583+	*PTC109,*GE-1,*HEPG0005,*SK3005,*ECG126,*WEP635
2N584	2N582,2N1309A,2SA207,2SA407,*DS-26,*GE-2,*TR-17,*HEPG0002,*SK3003, *HF20H,*ECG126,*WEP635,*276-2005
2N584+	*GE-2,*HEPG0002,*SK3003,*ECG126,*WEP635
2N585	2N356A,2N357A,2N358A,2N438,2N438A,2N439,2N439A,2N585+,2N587,2N1000, 2N1090,2N1302,2N1891,2N1994,2N1995,2SC128,*PTC108,*GE-8,*TR-08, *HEPG0011,*SK3011,*NR5,*ECG101,*WEP641,*276-2001,*RT-119
2N585+	2N356A,2N357A,2N358A,2N439,2N439A,2N1000,2N1090,2N1302,2N1995,2SC89, 2SC128,2SC129,ASY28,*PTC108,*GE-5,*TR-08,*HEPG0011,*SK3011,*ECG101, *WEP641
2N586	*GE-1,*TR-05,*HEPG0005,*SK3005,*HF12M,*ECG100,*WEP254,*276-2005,*RT-118
2N587	*DS-72,*GE-5,*TR-08,*HEPG0011,*SK3011,*NR5,*ECG101,*WEP641,*276-2001, *RT-119
2N588	*PTC107,*GE-51,*TR-17,*HEPG0008,*SK3006,*JR200,*ECG126,*WEP635
2N588A	2N501A,2N559,2N711A,2N711B,2N779,2N779A,2N829,2N838,2N980,2N984,2N2401, 2N2402,2N2587,2N2795,2N6365,2N6365A,*PTC107,*TR-17,*HEPG0003,*SK3006, *ECG126,*WEP635
2N591	2N109+,2N217+,2N838,2N2188,2N2614,2N2953,2N2956,2N2957,2N3323,2N3324, 2N3325,2SB37,2SB56,2SB59,2SB66,2SB77A,2SB173,2SB175,2SB176,40261, ACY23,ACY32,MM404A,NKT213,NKT216,NKT226,*GE-53,*TR-85,*HEPG0009, *SK3004,*AT30H,*ECG102A,*WEP250,*276-2005,*RT-123
2N591/5	2N59A,2N59B,2N60A,2N60B,2N360,2N382,2N422A,2N461,2N467,2N526,2N526A, 2N527A,2N597,2N651,2N651A,2N652A,2N1125,2N1175,2N1175A,2N1192,2N1309A, 2N1349,2N1371,2N1375,2N1377,2N1415,2N1448,2N1452,2N2956,2N2957,2N3323, 2N3324,2N3325,2SB57A,2SB226,2SB227,2SB304A,2SB428,MA887,MM404A, *PTC109,*GE-53,*TR-85,*ECG102A,*WEP250,*276-2005,*RT-121
2N592	2N111A,2N112A,2N396A,2N413,2N428,2N1017,2N1018,2SA205,*GE-50,*TR-05, *SK3005,*HF3M,*ECG100,*WEP254,*AC188/01,*RT-118
2N593	*GE-53,*TR-05,*SK3005,*HF3M,*ECG100,*WEP254,*RT-118
2N594	2N439,2N439A,2N585,2N585+,2N1090,2N1302,2N1995,2N1996,2SC89,2SC128, 2SC129,ASY28,*PTC108,*GE-5,*TR-08,*SK3011,*NR5,*ECG101,*WEP641, *AC187/01,*RT-119
2N595	2N439,2N439A,2N440,2N440A,2N585+,2N1090,2N1091,2N1302,2N1304,2N1996, 2SC89,2SC90,2SC128,2SC129,ASY28,*PTC108,*GE-5,*TR-08,*SK3011,*NR5, *ECG101,*WEP641,*AC127,*RT-119
2N596	2N1091,2N1304,2N1306,2SC89,2SC90,2SC91,2SC129,ASY29,*PTC108,*GE-5, *TR-08,*SK3011,*NR5,*ECG101,*WEP641,*AC127,*RT-119
2N597	2N1997,*GE-1,*TR-05,*HEPG0005,*SK3005,*HF12M,*ECG100,*WEP254,*276-2005, *RT-118
2N597+	2N597,*GE-1,*HEPG0005,*SK3005,*ECG100,*WEP254
2N598	2N1998,*GE-1,*TR-05,*HEPG0005,*SK3005,*HF12M,*ECG100,*WEP254,*276-2005, *RT-118
2N599	*GE-1,*TR-05,*HEPG0005,*SK3005,*HF12M,*ECG100,*WEP254,*276-2005,*RT-118
2N600	*GE-51,*TR-05,*SK3005,*ECG100,*WEP254,*RT-118
2N601	*GE-51

To Replace	Substitute This Type
2N602	2N404,2N580,2N580+,2N582,2N584,2N1017,2N1171,2N2048,2N3883,2N6365, 2N6365A,2SA207,2SA406,2SA407,*PTC107,*GE-50,*TR-17,*HEPG0002,*SK3008, *JR100,*ECG126,*WEP635,*276-2004
2N603	2N838,2N2956,2N3323,2N3324,2N3325,2N6365,2N6365A,*PTC107,*GE-51,*TR-17, *HEPG0002,*SK3008,*JR100,*ECG126,*WEP635,*276-2005
2N604	2N838,2N2188,2N2956,2N2957,2N3323,2N3324,2N3325,*PTC107,*GE-51,*TR-17, *HEPG0002,*SK3008,*JR100,*ECG126,*WEP635,*276-2005
2N605	2N123,2N303,2N324,2N415,2N415A,2N416,2N617,2N711B,2N829,2N838,2N964A, 2N972,2N985,2N1115,2N1282,2N1305,2N1307,2N1356,2N1681,2N1705,2N2022, 2N2928,2N3371,2SA169,2SA210,40269,40359,ASY27,SF.T223,SF.T229,SYL792, *GE-50,*TR-17,*HEPG0002,*SK3008,*JR100,*ECG126,*WEP635,*AC188/01, *276-2005
2N606	2N417,2N508,2N521A,2N523,2N617,2N827,2N1284,2N1307,2N1309,2N1353,2N1354, 2N1355,2N1356,2N1357,2N1705,2N1808,2N1892,2N2699,2N2928,2N3371,40269, 40359,ASY27,SF.T223,*GE-50,*TR-17,*HEPG0002,*SK3008,*JR100,*ECG126, *WEP635,*AC188/01,*276-2005
2N607	2N359,2N417,2N508,2N521A,2N522A,2N523,2N827,2N1284,2N1309,2N1316,2N1317, 2N1353,2N1354,2N1355,2N1357,2N1808,2N1892,2N2613,2N2635,2N2699,*GE-50, *TR-17,*HEPG0002,*SK3008,*JR100,*ECG126,*WEP635,*AC188/01,*276-2005
2N608	2N359,2N521A,2N522A,2N523,2N523A,2N827,2N1284,2N1309,2N1316,2N1317, 2N1353,2N1354,2N1355,2N1357,2N2613,2N2635,2N3995,*GE-50,*TR-17, *HEPG0002,*SK3008,*JR100,*ECG126,*WEP635,*AC188/01,*276-2005
2N609	2N59,2N59A,2N508A,2N598,2N654+,2N655,2N1008A,2N1175A,2N1193,2N1354, 2N1355,2N1356,2N1357,2N1374,2N1376,2N1381,2N1382,2N1707,2N1998,2SB222, 2SB222A,2SB223,2SB223A,2SB327,SF.T223,*PTC109,*GE-53,*TR-05,*HEPG0005, *SK3004,*AT20H,*ECG102,*WEP631,*AC188/01,*276-2005,*RT-120
2N610	2N59,2N59A,2N60,2N60A,2N422A,2N609,2N654,2N654+,2N1008A,2N1175,2N1175A, 2N1192,2N1348,2N1349,2N1350,2N1351,2N1356,2N1374,2N1376,2N1381,2N1382, 2N1383,2N1415,2N1681,2N1707,2SA207,2SB221,2SB221A,2SB222,2SB222A, 2SB304,2SB326,SF.T223,*PTC109,*GE-53,*TR-05,*HEPG0005,*SK3004,*AT20H, *ECG102,*WEP631,*AC188/01,*276-2005,*RT-120
2N611	2N60,2N60A,2N61,2N61A,2N414B,2N414C,2N422A,2N610,2N1175,2N1192,2N1313, 2N1348,2N1349,2N1350,2N1351,2N1372,2N1383,2N1414,2N1415,2N1478,2N1681, 2SA206,2SA207,2SA406,2SA407,2SB220,2SB220A,2SB221,2SB221A,2SB304, 2SB326,NKT221,NKT228,SF.T222,*PTC109,*GE-53,*TR-05,*HEPG0005,*SK3004, *AT20H,*ECG102,*WEP631,*AC188/01,*276-2004,*RT-120
2N612	2N396A,2N402,2N403,2N573,2N613,2N1018,2N1191,2N1413,2SA205,2SB219, 2SB219A,SF.T221,*PTC109,*GE-53,*TR-05,*HEPG0005,*SK3004,*AT20M,*ECG102, *WEP631,*AC188/01,*276-2004,*RT-120
2N613	2N61,2N61A,2N403,2N611,2N653,2N1191,2N1313,2N1413,2N1414,2SA205,2SA406, 2SB220,2SB220A,SF.T222,*PTC109,*GE-53,*TR-05,*HEPG0005,*SK3004,*AT20M, *ECG102,*WEP631,*AC188/01,*276-2004,*RT-120
2N614	2N315A,2N316A,2N317A,2N404,2N413,2N579,2N580,2N580+,2N615,2N653,2N1017, 2N1303,2N1313,2N1942,2SA205,2SA406,2SB220,2SB220A,SF.T222,SF.T227, *PTC109,*GE-52,*TR-05,*HEPG0005,*SK3005,*HF6M,*ECG100,*WEP254, *AC188/01,*276-2004,*RT-118
2N615	2N414,2N414B,2N414C,2N1171,2N1303,2N1305,2N1313,2N1478,2N1681,2SA204, 2SA206,2SA207,2SA209,2SA406,2SA407,2SB221A,2SB326,ASY26,SF.T227, SF.T228,SYL792,*PTC109,*GE-2,*TR-05,*HEPG0005,*SK3005,*HF6M,*ECG100, *WEP254,*AC188/01,*276-2004,*RT-118
2N616	*GE-1,*TR-05,*HEPG0005,*SK3005,*HF12H,*276-2004,*RT-118
2N617	2N417,2N521A,2N523,2N599,2N1284,2N1309,2N1354,2N1355,2N1357,2N1892, *PTC135,*GE-1,*TR-05,*HEPG0005,*SK3005,*ECG100,*WEP254,*AC188/01, *276-2004,*RT-118
2N618	2N1363,*PTC105A,*GE-16,*TR-01,*HEPG6005,*SK3009,*PT50,*276-2006,*RT-127
2N619	*GE-61,*TR-95,*SK3039,*ECG108,*WEP56,*276-2013,*RT-113
2N620	*GE-61,*TR-95,*SK3039,*ECG108,*WEP56,*276-2013,*RT-113
2N621	*GE-61,*TR-95,*SK3039,*ECG108,*WEP56,*276-2013,*RT-113
2N622	*GE-61,*TR-86,*HEPS5025,*ECG123,*WEP53,*276-2013,*RT-100
2N623	2N1499B,2N2955,2N6365,2N6365A,*PTC107,*GE-20,*TR-17,*HEPG0009,*ECG126, *WEP635

To Replace	Substitute This Type
2N624	2N1066,2N1224,2N1225,2N2496,2N3281,2N3282,AFY37,AFY42,MM5000,MM5001, MM5002,*GE-1,*TR-17,*HEPG0003,*SK3008,*HF20M,*ECG126,*WEP635
2N625	*GE-20,*TR-08,*NR10,*ECG103,*WEP641A,*RT-122
2N626	*GE-21
2N627	2N628,MP1549,MP1549A,MP1550,MP1550A,*PTC105A,*GE-76,*TR-01,*HEPG6013, *SK3009,*PT150,*ECG121,*WEP232,*276-2006,*RT-127
2N628	2N629,MP1550,MP1550A,MP1551,MP1551A,*PTC105A,*GE-76,*TR-01,*HEPG6013, *SK3009,*PT150,*ECG121,*WEP232,*276-2006,*RT-127
2N629	2N630,MP1551,MP1551A,MP1552,MP1552A,*PTC105A,*GE-76,*TR-01,*HEPG6013, *SK3009,*PT150,*ECG121,*WEP232,*276-2006,*RT-127
2N630	MP1552,MP1552A,*PTC105A,*GE-3,*TR-35,*HEPG6018,*SK3009,*PT150,*ECG179, *WEPG6001,*276-2006,*RT-147
2N631	2N359,2N508A,2N599,2N655+,2N1316,2N1354,2N1355,2N1357,2N1379,2N1998, 2N1999,2SB223,2SB223A,MA1703,MA1704,*PTC135,*GE-53,*TR-05,*HEPG0005, *SK3003,*AT20H,*ECG102,*WEP631,*AC188/01,*276-2005,*RT-120
2N632	2N382,2N383,2N417,2N461,2N508A,2N527,2N571,2N598,2N599,2N652,2N655, 2N655+,2N1008A,2N1189,2N1193,2N1307,2N1309,2N1352,2N1354,2N1355,2N1356, 2N1357,2N1371,2N1377,2N1449,2N1892,2N1997,2N1998,2SB223,2SB223A, 2SB327,MA888,SF.T223,*PTC102,*GE-53,*TR-05,*HEPG0005,*SK3004,*AT30H, *ECG102,*WEP631,*AC188/01,*276-2005,*RT-120
2N633	2N59A,2N59B,2N60A,2N60B,2N360,2N381,2N422,2N422A,2N461,2N466,2N518, 2N526,2N526A,2N527A,2N597,2N651,2N651A,2N652A,2N1008A,2N1124,2N1125, 2N1175,2N1175A,2N1192,2N1309A,2N1348,2N1349,2N1350,2N1351,2N1371, 2N1373,2N1375,2N1377,2N1415,2N1447,2N1448,2N1451,2N1452,2SB57A,2SB226, 2SB227,2SB304A,2SB427,2SB428,MA887,*PTC102,*GE-53,*TR-05,*HEPG0005, *SK3004,*AT30H,*ECG102,*WEP631,*AC128,*RT-120
2N634	2N357A,2N358A,2N439,2N439A,2N576,2N635,2N1891,2N1995,2N1996,*PTC108, *GE-8,*TR-08,*HEPG0011,*SK3011,*NR10,*ECG101,*WEP641,*AC187/01, *276-2001,*RT-119
2N634A	2N1306,2N1308,*GE-8,*TR-08,*HEPG0011,*SK3011,*NR10,*ECG101,*WEP641, *AC187/01,*276-2001,*RT-119
2N635	2N358A,2N440,2N440A,2N636,*PTC108,*GE-6,*TR-08,*HEPG0011,*SK3011,*NR10, *ECG101,*WEP641,*276-2001,*RT-119
2N635A	2N636A,*GE-6,*TR-08,*HEPG0011,*SK3011,*NR10,*ECG101,*WEP641,*276-2001, *RT-119
2N636	*PTC108,*GE-5,*TR-08,*HEPG0011,*SK3011,*NR10,*ECG101,*WEP641,*276-2001, *RT-119
2N636A	*GE-6,*TR-08,*HEPG0011,*SK3011,*NR10,*ECG101,*WEP641,*276-2001,*RT-119
2N637	2N457A,2N458A,ASZ15,ASZ17,KR6500,*PTC138,*GE-76,*TR-01,*HEPG6005, *SK3009,*PT40,*ECG104,*WEP230,*276-2006,*RT-124
2N637A	2N637B,2N1021,2N1021A,2N1022A,2N3616,2N3618,2N5156,*PTC105A,*GE-76, *TR-01,*HEPG6005,*SK3009,*PT50,*ECG104,*WEP230,*276-2006,*RT-124
2N637B	2N1021,2N1021A,2N1022A,2N3616,2N3618,*PTC105A,*GE-3,*TR-01,*HEPG6005, *SK3009,*PT50,*ECG104,*WEP230,*276-2006,*RT-124
2N638	2N637,2SB86,2SB87,ASZ15,ASZ17,*GE-76,*TR-01,*HEPG6005,*SK3009,*PT40, *ECG104,*WEP230,*276-2006,*RT-124
2N638A	2N637A,2N637B,2N638B,2N3616,2N5156,*PTC105A,*GE-76,*TR-01,*HEPG6005, *SK3009,*PT50,*ECG104,*WEP230,*276-2006,*RT-124
2N638B	2N637B,2N3616,2N5156,*PTC105A,*GE-3,*TR-01,*SK3009,*PT50,*ECG104, *WEP230,*RT-124
2N639	2N456,2SB85,*PTC138,*GE-25,*TR-01,*HEPG6005,*SK3009,*PT40,*ECG104, *WEP230,*OC28,*276-2006,*RT-124
2N639A	2N297,2N457,2N458,2N561,2N638,2N639B,2SB86,2SB87,2SB358,ASZ15,CDT1320, CDT1321,*PTC138,*GE-25,*TR-01,*HEPG6005,*SK3009,*PT50,*ECG104,*WEP230, *OC28,*276-2006,*RT-124
2N639B	2N297,2N457,2N458,2N561,2N638,2N639A,2SB86,2SB87,2SB358,ASZ15,CDT1320, CDT1321,*PTC122,*GE-25,*TR-01,*HEPG6005,*SK3009,*ECG104,*WEP230, *276-2006,*RT-124
2N640	2N2588,A1383,*PTC107,*GE-51,*TR-17,*HEPG0009,*SK3008,*JR100,*ECG126, *WEP635,*AF125,*276-2005
2N641	2N2588,A1383,*PTC107,*GE-1,*TR-17,*HEPG0009,*SK3008,*JR100,*ECG126,

To Replace	Substitute This Type
(2N641)	*WEP635,*AF125,*276-2005
2N642	2N2588,A1383,*PTC107,*GE-51,*TR-17,*HEPG0009,*SK3007,*JR100,*ECG126, *WEP635,*AF125,*276-2005
2N643	*PTC145,*GE-53,*TR-17,*HEPG0002,*SK3007,*JR100,*ECG126,*WEP635,*276-2004
2N644	*PTC145,*GE-51,*TR-17,*HEPG0002,*SK3007,*JR100,*ECG126,*WEP635,*276-2004
2N645	*PTC145,*GE-51,*TR-17,*HEPG0002,*SK3007,*JR100,*ECG126,*WEP635,*276-2004
2N646	*PTC108,*GE-8,*TR-08,*HEPG0011,*SK3010,*ECG103,*WEP641A,*RT-122
2N647	2SC86,2SC179,2SC181,2SD77,2SD186,2SD187,NKT713,*PTC108,*GE-5,*TR-08, *HEPG0011,*SK3010,*NA30,*ECG103A,*WEP641B,*AC127,*276-2001,*RT-122
2N649	2N647,2N797,2N2354,2SC86,2SC179,2SC181,2SD25,2SD27,2SD33,2SD61,2SD62, 2SD77,2SD96,2SD186,2SD187,2SD195,AC127,NKT713,*PTC108,*GE-5,*TR-08, *HEPG0011,*SK3010,*NA30,*ECG103A,*WEP641B,*AC187/01,*RT-122
2N650	2N525A,2N526A,2N597,2N651,2N652A,*GE-2,*TR-05,*HEPG0005,*SK3004,*AT30H, *ECG102,*WEP631,*276-2007,*RT-120
2N650A	2N525A,2N526A,2N597,2N650,2N651,2N651A,2N652A,2N1187,*GE-2,*TR-05, *HEPG0005,*SK3004,*ECG102,*WEP631,*276-2005,*RT-120
2N651	2N527A,2N597,2N652,2N1997,*GE-2,*TR-05,*HEPG0005,*SK3004,*AT30H,*ECG102, *WEP631,*276-2005,*RT-120
2N651A	2N527A,2N597,2N651,2N652,2N652A,2N1188,2N1997,*GE-2,*TR-05,*HEPG0005, *SK3004,*ECG102,*WEP631,*276-2005,*RT-120
2N652	2N1997,*GE-2,*TR-05,*HEPG0007,*SK3004,*AT30H,*ECG102,*WEP631,*276-2005, *RT-120
2N652A	2N527A,2N597,2N651,2N652,2N1188,2N1997,*GE-2,*TR-05,*HEPG0005,*SK3004, *ECG102,*WEP631,*276-2005,*RT-120
2N653	2N525A,2N526A,2N650,*GE-53,*TR-05,*HEPG0005,*SK3004,*AT30H,*276-2007, *RT-120
2N654	2N527A,2N597,2N651,2N652,2N1997,*GE-2,*TR-05,*HEPG0005,*SK3004,*AT30H, *ECG102,*WEP631,*RT-120
2N654+	2N382,2N383,2N527,2N527A,2N598,2N652,2N655,2N1008A,2N1448,2N1449,2N1997, 2N1998,2SB227,2SB327,SF.T223,*PTC135,*HEPG0005,*SK3004,*ECG102, *WEP631
2N655	2N598,2N1998,2N3427,*GE-2,*TR-05,*HEPG0007,*SK3004,*AT30H,*ECG102, *WEP631,*RT-120
2N655+	2N599,2N1316,2N1999,2N2171,2N3427,2N3428,*PTC135,*GE-2,*HEPG0007, *SK3004,*ECG102,*WEP631
2N656	2N1837A,*PTC125,*GE-18,*TR-87,*HEPS5014,*SK3024,*ECG128,*WEP243, *276-2008,*RT-114
2N656A	2N2106,2N2107,*PTC125,*GE-18,*TR-25,*HEPS5014,*SK3024,*276-2012,*RT-114
2N657	*PTC144,*GE-32,*TR-05,*HEPS5026,*SK3024,*ECG128,*WEP254,*276-2012, *RT-118
2N657A	2N4000,2N4001,*PTC144,*GE-32,*HEPS3019,*276-2012
2N658	2N659,*GE-48,*TR-82,*HEPG0005,*HF6M,*ECG176,*WEP238,*BC327,*276-2022, *RT-127
2N659	*GE-48,*TR-82,*HEPG0005,*SK3005,*HF12H,*ECG176,*WEP238,*BC327,*276-2022, *RT-127
2N660	2N661,*GE-1,*TR-82,*HEPG0005,*SK3005,*HF20H,*ECG176,*WEP238,*276-2005, *RT-127
2N661	*DS-25,*GE-1,*TR-82,*HEPG0005,*SK3005,*HF20H,*ECG176,*WEP238,*276-2005, *RT-127
2N662	2N660,*DS-25,*GE-1,*TR-82,*HEPG0005,*SK3005,*HF12H,*ECG176,*WEP238, *276-2004,*RT-127
2N663	*PTC105A,*GE-25,*TR-01,*HEPG6003,*SK3009,*PT40,*ECG121,*WEP232,*276-2006, *RT-127
2N665	*PTC105A,*GE-25,*TR-01,*HEPG6005,*SK3009,*PT50,*ECG121,*WEP232,*276-2006, *RT-127
2N669	*DS503,*GE-25,*TR-01,*HEPG6003,*SK3009,*PT40,*ECG121,*WEP232,*276-2006, *RT-127
2N670	*GE-53,*HEPG6011,*AT100H,*276-2006
2N671	*GE-53,*AT100H
2N672	*GE-53,*TR-82,*HEPG6011,*AT100H,*ECG176,*WEP238,*276-2006,*RT-127
2N673	*GE-53,*AT100H

To Replace	Substitute This Type
2N674	*PTC102,*GE-53,*TR-05,*HEPG6010,*SK3005,*ECG100,*WEP254,*276-2002, *RT-100
2N675	*GE-53,*276-2006
2N677	2N677A,2N1162,2N1162A,2N1554,2N1554A,CTP1504,*PTC901,*GE-76,*TR-35, *HEPG6005,*SK3009,*PT150,*ECG179,*WEPG6001,*276-2006,*RT-147
2N677A	2N1164,2N1164A,2N1554,2N1554A,2N1555,2N1555A,CQT940B,CQT940BA,CTP1503, CTP1504,*PTC901,*GE-76,*TR-35,*HEPG6005,*SK3009,*ECG179,*WEPG6001, *276-2006,*RT-147
2N677B	2N677C,2N1166,2N1166A,2N1556,2N1556A,CTP1500,*DS503,*GE-76,*TR-05, *HEPG6005,*SK3009,*ECG179,*WEPG6001,*276-2006,*RT-147
2N677C	2N1166,2N1166A,2N1556,2N1556A,CTP1500,*DS503,*GE-16,*TR-01,*HEPG6005, *SK3009,*ECG121,*WEP232,*276-2006,*RT-127
2N678	2N678A,2N1558,2N1558A,*PTC901,*GE-76,*TR-35,*HEPG6003,*SK3009,*PT150, *ECG179,*WEPG6001,*276-2006,*RT-147
2N678A	2N1558,2N1558A,2N1559,2N1559A,*PTC901,*GE-76,*TR-35,*HEPG6005,*SK3009, *ECG179,*WEPG6001,*276-2006,*RT-147
2N678B	2N678C,2N1560,2N1560A,*DS-503,*GE-76,*TR-01,*HEPG6005,*SK3009,*ECG121, *WEP232,*276-2006,*RT-127
2N678C	2N1560,2N1560A,*DS-503,*GE-16,*TR-01,*HEPG6005,*SK3009,*ECG121,*WEP232, *276-2006,*RT-127
2N679	2N356A,2N357A,2N358A,2N377,2N377A,2N439,2N439A,2N576A,2N1000,2N1302, 2N1473,2N1995,2SC128,2SC129,ASY28,*PTC108,*GE-8,*TR-08,*HEPG0011, *SK3011,*NR5,*ECG101,*WEP641,*AC187/01,*276-2001,*RT-119
2N680	*PTC109,*GE-52,*TR-17,*HEPG0003,*SK3004,*AT20M,*ECG126,*WEP635,*AC126, *276-2007
2N694	*PTC107,*GE-52,*TR-17,*HEPG0003,*SK3008,*ECG126,*WEP635
2N695	*PTC107,*GE-51,*TR-17,*HEPG0003,*SK3006,*HF35,*ECG126,*WEP635
2N696	BSY34,*PTC144,*GE-81,*TR-86,*HEPS0014,*SK3124,*SN80,*ECG123,*WEP53, *BC337,*276-2012,*RT-100
2N696A	*PTC144,*GE-47,*HEPS3011,*276-2012
2N697	*PTC144,*GE-81,*TR-86,*HEPS0014,*SK3024,*SN80,*ECG128,*WEP53,*BC337, *276-2012,*RT-100
2N697A	*PTC144,*GE-47,*TR-86,*HEPS3011,*SK3024,*ECG123,*WEP53,*276-2008,*RT-100
2N698	*PTC125,*GE-18,*HEPS3019,*SN80,*BF338,*276-2012
2N699	2N699A,2N699B,2N2102,2N2102A,2N2243,2N2243A,2N2405,2N2443,2N3020, 2N3036,2SC510-O,A5T2243,BSY85,ZT91,ZT92,ZT93,ZT2102,*PTC144,*GE-18, *TR-78,*HEPS5026,*SK3024,*SN80,*WEP712,*BF338,*276-2012,*RT-110
2N699A	2N699B,2N1342,2N2443,2N4925,2SC686,*PTC125,*GE-18,*TR-78,*HEPS3019, *SK3024,*SN80,*WEP712,*BF338,*276-2008,*RT-110
2N699B	2N2443,*PTC125,*GE-18,*TR-78,*HEPS3019,*SK3024,*WEP712,*BF338,*276-2008
2N700	*GE-51,*TR-17,*HEPG0003,*JR200,*ECG160,*WEP637
2N700/18	2N700,*GE-51,*TR-17,*HEPG0003,*ECG160,*WEP637
2N700A	*DS-56,*GE-51,*TR-17,*HEPG0003,*JR200,*ECG160,*WEP637
2N702	2N703,2N1708A,2N4264,2N5772,A5T4123,BSY62,BSY63,*PTC123,*GE-61,*TR-95, *HEPS0011,*SK3039,*EN10,*ECG108,*WEP56,*BC107B,*276-2016,*RT-113
2N703	2N6426,2N6427,2SC1788,2SC1851,A5T4123,*PTC123,*GE-61,*TR-95,*HEPS0011, *SK3039,*EN10,*ECG108,*WEP56,*BC107B,*276-2016,*RT-113
2N705	*PTC145,*GE-51,*TR-17,*HEPG0003,*JR200,*ECG160,*WEP637
2N705A	2N705,2N710,2N711A,2N828,2N828A,2N964,2N968,2N2401,2N2402,2N2587,2N2860, 2N6365,2N6365A,2SA56,2SA405,2SA411,2SB185,MM380,NKT224,*GE-51,*TR-17, *HEPG0003,*JR200,*ECG160,*WEP637
2N706	2N706A,2N706B,2N783,2N784,2N2205,2N2318,2N2319,2N2368,2N2656,2N2710, 2N3261,2N3510,2N3511,2N3605A,2N3606A,2N3648,2N4137,2N5030,2N5187,2SC68, 2SC204,2SC321,2SC322,2SC356,2SC576,2SC595,40405,GET708,GET914, GET3013,GET3014,GET3646,KT218,MM1941,MPSH19,MT1038,MT1038A,MT1039, PT2760,ZT2368,*PTC136,*GE-210,*TR-95,*HEPS0011,*SK3039,*SN60,*ECG108, *WEP56,*BC107B,*276-2016,*RT-113
2N706/46	2N959,*PTC136,*GE-210,*TR-53,*HEPS0011,*BC107B,*276-2009
2N706/51	2N959,*GE-210,*TR-53,*BC107B,*276-2015
2N706A	2N706,2N706B,2N783,2N784,2N2205,2N2318,2N2319,2N2368,2N2656,2N2710, 2N3261,2N3510,2N3511,2N3605A,2N3606A,2N3648,2N4137,2N5030,2N5187,2SC68,

To Replace	Substitute This Type
(2N706A)	2SC204,2SC321,2SC322,2SC356,2SC576,2SC595,40405,GET708,GET914,GET3013, GET3014,GET3646,KT218,MM1941,MPSH19,MT1038,MT1038A,MT1039,PT2760, ZT2368,*PTC136,*GE-210,*TR-95,*HEPS0011,*SK3039,*SN60,*ECG108,*WEP56, *BC107B,*276-2016,*RT-113
2N706A/46	2N959,*PTC136,*GE-210,*TR-53,*HEPS0011,*BC107B,*276-2011
2N706A/51	2N959,*PTC126,*GE-210,*TR-53,*HEPS0011,*BC107B,*276-2011
2N706B	2N706,2N706A,2N783,2N784,2N2205,2N2318,2N2319,2N2368,2N2656,2N2710, 2N3261,2N3510,2N3511,2N3605A,2N3606A,2N3648,2N4137,2N5030,2N5187,2SC68, 2SC204,2SC321,2SC322,2SC356,2SC576,2SC595,40405,GET708,GET914, GET3013,GET3014,GET3646,KT218,MM1941,MPSH19,MT1038,MT1038A,MT1039, PT2760,ZT2368,*PTC136,*GE-210,*TR-95,*HEPS0011,*SK3039,*ECG108,*WEP56, *BC107B,*276-2016,*RT-113
2N706B/46	2N959,*PTC136,*GE-210,*TR-53,*HEPS0011,*BC107B,*276-2009
2N706B/51	2N959,*PTC123,*GE-210,*TR-53,*BC107B,*276-2011
2N706C	2N743A,2N744A,*GE-210,*TR-95,*HEPS0011,*SK3039,*ECG108,*WEP56,*BC337, *276-2009,*RT-113
2N707	*GE-212,*TR-95,*HEPS0011,*SK3039,*ECG108,*WEP56,*BC107B,*276-2016, *RT-113
2N707A	*PTC117,*GE-20,*TR-86,*HEPS0011,*SK3124,*ECG123,*WEP53,*BF338,*276-2009, *RT-100
2N708	2N914,2N916,2N916A,2N2221,2N2242,2N2501,2N2539,2N2845,2N2847,2N3013, 2N3014,2N3115,2N3210,2N3211,2N3301,2N3646,2N3946,2N4013,2N4962,2N5144, 2N5380,2N5772,2SC67,2SC577,2SC639,2SC764,A5T3903,A5T4123,BSW82, FT3641,FT3642,MM3903,TIS133,TIS134,TN-3903,ZT708,*PTC136,*GE-17,*TR-21, *HEPS0011,*SN60,*ECG123A,*WEP736,*276-2011,*RT-102
2N708/46	*PTC136,*GE-17,*HEPS0011,*276-2009
2N708/51	MM3903,*GE-61,*TR-51,*BF173,*276-2016
2N708A	2N2845,2N2847,2N3301,2N3946,2N4962,2N5380,A5T3903,FT3641,FT3642,GET2221, GET2221A,MM3903,TN-3903,*PTC136,*GE-17,*TR-21,*HEPS0011,*SK3122, *ECG123A,*WEP736,*276-2009,*RT-102
2N709	2N2475,2N2784,2N3959,2N3960,2N5200,2N5201,*PTC133,*GE-17,*TR-95, *HEPS0011,*EN10,*ECG108,*WEP56,*276-2038,*RT-113
2N709A	2N2784,2N3633,2N3959,2N3960,2N5200,2N5201,2SC718,2SC1231,MT1060A, *PTC126,*GE-17,*TR-95,*HEPS0011,*SK3039,*ECG108,*WEP56,*276-2009, *RT-113
2N710	2N705,*PTC145,*GE-51,*TR-17,*HEPG0003,*JR200,*ECG160,*WEP637
2N710A	2N726,2N860,2N861,2N862,2N978,2N2274,2N2276,2N2373,2N2391,2N2411,2N2969, 2N2971,2N6567,GET3638,MPSH85,TP3638,TP5142,*PTC127,*GE-51,*TR-17, *HEPG0003,*ECG160,*WEP637
2N711	2N711A,2N711B,2N829,2N964A,2N972,2N973,2N974,2N985,2N2402,2SA451, *PTC145,*GE-51,*TR-17,*HEPG0003,*JR200,*ECG160,*WEP637
2N711A	2N711B,2N829,2N838,2N964A,2N972,2N985,2N2402,*PTC145,*GE-51,*TR-17, *HEPG0003,*JR200,*ECG160,*WEP637
2N711B	2N838,2N3323,2N3324,2N3325,*PTC145,*GE-51,*TR-17,*HEPG0003,*JR200, *ECG160,*WEP637
2N715	2N716,*PTC117,*GE-20,*TR-86,*HEPS0011,*SK3124,*ECG123,*WEP53,*276-2009, *RT-100
2N716	*PTC117,*GE-20,*TR-25,*HEPS0025,*276-2009
2N717	2N735,2N735A,2N758A,2N759A,2N760B,2N2395,2N2515,2N2522,2S103,2SC1166-R, *PTC123,*GE-210,*TR-95,*HEPS0011,*SK3039,*EN10,*ECG108,*WEP56,*BC337, *276-2009,*RT-113
2N718	2N718A,2N736,2N736A,2N736B,2N2396,2N2516,2N4962,2N4963,2N5380,2N6538, 2S104,2SC1166-O,A5T3903,FT3567,FT3568,TIS96,TIS99,TN-3903,*PTC123,*GE-210, *TR-95,*HEPS0015,*SK3122,*EN10,*ECG108,*WEP56,*BC337,*276-2009,*RT-113
2N718A	2N736A,2N736B,2N870,2N910,2N2465,*PTC123,*GE-20,*TR-95,*HEPS3020, *SK3122,*EN10,*ECG108,*WEP56,*BF338,*276-2009,*RT-113
2N719	2N738,2N739A,2N2518,2N4390,2SC1596,40354,TIS101,*PTC125,*GE-18,*HEPS0005, *ECG194,*BF338,*276-2012
2N719A	*PTC144,*GE-18,*HEPS3019,*ECG194,*BF338,*276-2012
2N720	2N720A,2N739,2N740,2N740A,2N2519,SE7016,*PTC125,*GE-18,*HEPS0005, *ECG128,*WEP243,*BF338,*276-2008

To Replace	Substitute This Type
2N720A	2N740A,*PTC125,*GE-18,*HEPS3019,*ECG128,*WEP243,*BF338,*276-2008
2N721	2N721A,2N722A,2N2393,2N2603,*PTC127,*GE-82,*TR-20,*HEPS0013,*SK3114,*SP70,*ECG159,*WEP242,*BC327,*276-2021,*RT-115
2N721A	*PTC127,*GE-82,*TR-20,*HEPS0013,*SK3114,*SP70,*ECG159,*WEP242,*BC327,*276-2021,*RT-115
2N722	2N722A,2N2394,2N2603,2N3581,2N6067,2SA661-R,A5T5448,A8T3703,*PTC127,*GE-82,*TR-20,*HEPS0013,*SK3114,*SP70,*ECG159,*WEP242,*BC327,*276-2023,*RT-115
2N722A	2SA661-R,*PTC127,*GE-82,*TR-20,*HEPS0013,*SK3114,*SP70,*ECG159,*WEP242,*BC327,*276-2021,*RT-115
2N725	*PTC145,*GE-51,*TR-17,*HEPG0003,*SK3008,*HF12M,*ECG160,*WEP637
2N726	2N2391,2N2411,2SA558,TIS137,*PTC131,*GE-22,*TR-20,*HEPS0013,*SP70,*ECG106,*WEP52,*276-2034,*RT-126
2N727	2N869,2N869A,2N2392,2N2412,2N3829,2N3905,2N5382,2SA467,2SA509-O,2SA509G-O,A5T3905,BSW72,MM3905,TIS37,TN-3905,*PTC131,*GE-65,*TR-20,*HEPS0013,*SP70,*ECG106,*WEP52,*BC177,*276-2034,*RT-126
2N728	2N729,2N2615,2N2616,2N2712,2N2714,2N2923,2N2926-ORG,2N3241A,2N3393,2N3856,2N3856A,2N3859,2SC388A,2SC934,2SD392,A5T3571,A5T3572,BC108A,BC168A,BC170B,BC172A,BC238A,BCY58A,MPS2714,MPS2926-ORG,MPS2926-YEL,MPS6568A,MPS6569,MPS6570,TIS62A,TIS63A,TIS86,*PTC126,*GE-81,*TR-63,*HEPS0011,*SK3124,*SN80,*ECG123,*WEP53,*BC337,*276-2030,*RT-100
2N729	2N916,2N916A,2N2615,2N2616,2N3241A,2N3856A,2N3859,2N4873,2N5825,2SC367G,2SC367G-Y,2SC735-Y,2SC933,2SC1175,2SC1216,2SC1999,BC107A,BC108A,BC167A,BC171A,BC237A,BC238A,BCY58A,BCY59A,BF240,BFY39-2,MPS6513,MPS6574-BLUE,MPS6574-GREEN,MPS6576-BLUE,MPS6576-GREEN,MPSA20-GRN,MPSA20-WHT,TIS62A,TIS63A,TIS86,ZT44,*PTC126,*GE-81,*TR-86,*HEPS0011,*SK3124,*SN80,*ECG123,*WEP53,*BC337,*276-2030,*RT-100
2N730	2N758A,2N3737,2N6375,*PTC144,*GE-20,*TR-25,*HEPS3020,*BC337,*276-2009
2N731	2N718A,2N2897,2N2900,2N4962,2N4963,A5T2193,FT3567,FT3568,TN-3903,*PTC144,*GE-20,*HEPS3020,*BC337,*276-2009
2N734	2N2464,*PTC123,*GE-81,*TR-25,*HEPS5026,*276-2008
2N734A	2N2464,*PTC123,*GE-81,*TR-25,*BF338,*276-2012
2N735	2N718A,2N735A,2N870,2N911,2N2465,2N4963,BSW84,FT3567,FT3568,FT3722,SE8012,TIS135,TIS136,*PTC123,*GE-81,*HEPS5026,*276-2008
2N735A	2N718A,2N870,2N911,2N2465,2N4963,TIS135,TIS136,*PTC123,*GE-81,*TR-25,*HEPS0005,*BF338,*276-2008
2N736	2N736B,2N871,2N910,2N956,2N2466,2N2645,2N6540,2N6541,2SC1998,BC546A,BSW85,FT3569,TIS95,TIS98,*PTC123,*GE-81,*HEPS0005,*276-2008
2N736A	2N736B,*PTC125,*GE-81,*TR-78,*HEPS5026,*BF338
2N736B	*PTC125,*GE-81,*HEPS5026,*BF338
2N738	2N739A,2N4390,40354,TIS101,*PTC125,*GE-18,*TR-78,*HEPS5025,*SK3045,*ECG154,*WEP712,*BF338,*276-2008,*RT-110
2N738A	*PTC125,*GE-18,*BF338,*276-2012
2N739	2N720A,2N740,2N740A,*PTC125,*GE-18,*TR-78,*HEPS5025,*SK3045,*ECG154,*WEP712,*BF338,*276-2008,*RT-110
2N739A	*PTC125,*GE-27,*TR-78,*HEPS5025,*SK3045,*ECG154,*WEP712,*BF338,*RT-110
2N740	2N740A,2SC1438,2SC1439,A5T4410,*PTC125,*GE-18,*TR-78,*HEPS0005,*SK3045,*ECG154,*WEP712,*BF338,*276-2008,*RT-110
2N740A	*PTC125,*GE-27,*TR-78,*HEPS5025,*SK3045,*ECG154,*WEP712,*BF338,*RT-110
2N741	2N741A,2N828,2N828A,2N964,2N968,2N2630,2N2860,2SA411,*PTC145,*GE-51,*TR-17,*HEPG0003,*SK3006,*JR200,*ECG160,*WEP637
2N741A	*PTC145,*GE-51,*TR-17,*HEPG0003,*JR200,*ECG160,*WEP637
2N742	*PTC123,*GE-17,*TR-86,*HEPS0011,*SK3124,*ECG123,*WEP53,*BC337,*276-2009,*RT-100
2N742A	2N742,2N757A,2N759B,*PTC123,*GE-81,*TR-86,*HEPS0011,*SK3124,*ECG123,*WEP53,*276-2012,*RT-100
2N743	2N2318,2N2319,2N2656,2N5187,2SC68,2SC204,2SC395A-R,2SC595,*PTC136,*GE-20,*TR-95,*HEPS0011,*SK3039,*EN10,*ECG108,*WEP56,*BC107B,*276-2016,*RT-113
2N743A	2N744A,*GE-17,*TR-95,*HEPS0025,*SK3039,*EN10,*ECG108,*276-2038,*RT-113

To Replace	Substitute This Type
2N744	2N4418,2S095A,2S512,2SC395A,2SC395A-O,2SC735-O,PT720,ZT2205,ZT2938, *PTC136,*GE-20,*TR-95,*HEPS0011,*SK3039,*EN10,*ECG108,*WEP56,*BC107B, *276-2016,*RT-113
2N744A	*HEPS0025,*276-2038
2N745	BF224J,BF225J,*PTC123,*GE-20,*TR-21,*HEPS0011,*SK3122,*ECG123A,*WEP736, *276-2009,*RT-102
2N746	2N3903,2SC899,2SC1688,MPS3693,MPS6565,MPS6576-BLUE,MPS6576-GREEN, MPS6576-YEL,*PTC123,*GE-20,*TR-21,*HEPS0011,*SK3122,*ECG123A,*WEP736, *276-2009,*RT-102
2N747	2N4123,2N4294,2N4295,2N4418,2N4420,2N4421,2N4422,2SC570,2SC624,2SC1123, 2SC1128,2SC1215,2SC1293,2SC1789,MPS2369,MPS3394,MPS3646,MPS6512, MPS6574-YEL,MPSA10-RED,MPSA20-RED,MPSH10,MPSH11,NPS6512,TIS45, TIS46,TIS48,TIS49,TIS51,TIS52,TIS55,*PTC126,*GE-20,*TR-21,*HEPS0011, *SK3122,*ECG123A,*WEP736,*276-2009,*RT-102
2N748	2N3793,2N3825,2N4292,2N4293,2SC1126,BF224J,BF225J,MPS834,MPS918, MPS6507,MPS6511,MPS6540,MPS6542,MPS6543,MPS6545,MPS6546,MPS6547, MPS6548,MPS6567,MPSH19,MPSH20,MPSH32,MPSH37,TIS47,*PTC126,*GE-20, *TR-21,*HEPS0011,*SK3122,*ECG123A,*WEP736,*276-2009,*RT-102
2N749	2N751,2N1388,2N1390,2SC302-M,2SC307-M,2SC590N,*GE-18,*TR-21,*HEPS0011, *SK3122,*ECG123A,*WEP736,*276-2009,*RT-102
2N751	2N749,2N1388,2N1390,2SC302-M,2SC307-M,2SC590N,*GE-18,*TR-21,*HEPS0011, *SK3122,*ECG123A,*WEP736,*276-2009,*RT-102
2N753	2N2242,2N2501,2N3013,2N3014,2N3211,2N4123,2N4418,2S512,2SC67,2SC577, 2SC601N,2SC639,2SC735-O,2SC764,A5T4123,BSW82,PT720,ZT2205,ZT2938, *PTC126,*GE-61,*TR-21,*HEPS0011,*SK3122,*SN60,*ECG123A,*WEP736, *BC107B,*276-2016,*RT-102
2N754	2N717,2N735,2N735A,2N758A,2N759A,2N760B,2N844,2N2395,2N2515,2N2522, 2N3973,2N3975,2S103,2SC383,2SC979-R,2SC1166-R,BSY93,*PTC125,*GE-81, *TR-21,*HEPS0011,*SK3122,*ECG123A,*WEP736,*BF338,*276-2012,*RT-102
2N755	2N719,2N738,2N739A,2N845,2N2460,2N2464,2N2518,2N4390,2N5184,2SC979A-R, BSY79,*PTC125,*GE-18,*HEPS5026,*BF338,*276-2012
2N756	2N742,2N756A,2N757,2N757A,2N759B,*PTC123,*GE-210,*TR-95,*HEPS0011, *SK3039,*EN10,*ECG108,*WEP56,*BC337,*276-2008,*RT-113
2N756A	2N742,2N757A,2N759B,*PTC123,*GE-81,*TR-86,*HEPS0011,*SK3124,*ECG123, *WEP53,*BF338,*276-2008,*RT-100
2N757	2N757A,2N758,2N758A,*PTC123,*GE-210,*TR-95,*HEPS0011,*SK3039,*EN10, *ECG108,*WEP56,*BC337,*276-2008,*RT-113
2N757A	2N758A,*PTC123,*GE-81,*TR-86,*HEPS0011,*SK3124,*ECG123,*WEP53,*BF338, *276-2008,*RT-100
2N758	2N758A,2N759,2N759A,2N760B,*PTC125,*GE-210,*TR-95,*HEPS0011,*SK3039, *EN10,*ECG108,*WEP56,*BC337,*276-2008,*RT-113
2N758A	2N759A,2N760B,*PTC125,*GE-81,*TR-86,*HEPS0011,*SK3124,*ECG123,*WEP53, *BF338,*276-2008,*RT-100
2N758B	*PTC125,*GE-81,*TR-86,*HEPS0011,*SK3124,*ECG123,*WEP53,*BF338,*276-2008, *RT-100
2N759	2N759A,2N760B,*PTC125,*GE-81,*TR-95,*HEPS0011,*SK3039,*SN60,*ECG108, *WEP56,*BC337,*276-2008,*RT-113
2N759A	2N760B,*PTC125,*GE-81,*TR-86,*HEPS0011,*SK3124,*ECG123,*WEP53,*BF338, *276-2008,*RT-100
2N759B	2N758A,*PTC125,*GE-81,*TR-86,*HEPS0011,*SK3124,*ECG123,*WEP53,*BF338, *276-2008,*RT-100
2N760	2N760A,*PTC125,*GE-81,*TR-95,*HEPS0011,*SK3039,*SN60,*ECG108,*WEP56, *BC337,*276-2008,*RT-113
2N760A	*PTC125,*GE-81,*TR-86,*HEPS0011,*SK3124,*ECG123,*WEP53,*276-2008,*RT-100
2N760B	2N759A,*PTC125,*GE-81,*TR-86,*HEPS0011,*SK3124,*ECG123,*WEP53,*276-2008, *RT-100
2N761	2N656,2N758A,2N759A,2N760B,*PTC123,*GE-17,*TR-95,*HEPS0011,*SK3039, *EN10,*ECG108,*WEP56,*BC337,*276-2009,*RT-113
2N762	2N6428,2N6428A,2N6539,*GE-17,*TR-95,*HEPS0011,*SK3039,*EN10,*ECG108, *WEP56,*BC337,*276-2009,*RT-113
2N768	2N711A,2N769,2N779,2N779A,2N828,2N964,2N965,2N968,2N969,2N984,2N2401,

To Replace	Substitute This Type
(2N768)	2N2402,2N2487,2N2488,2N2489,2N2795,2N2860,2N3322,2SA411,2SA450,*PTC145, *GE-51,*TR-17,*HEPG0001,*JR200,*ECG160,*WEP637
2N769	2N983,2N3320,2N3322,*PTC145,*GE-51,*TR-17,*HEPG0001,*JR200,*ECG160, *WEP637
2N770	2N706,2N706A,2N706B,2N743,2N959,2N2205,2N2214,2N2719,2N3825,2SC68, MPS706,MPS706A,MPS918,MPS6540,MPS6542,MPS6543,MPS6548,MT1038, MT1038A,MT1039,SE1010,SE5025,*PTC132,*GE-20,*TR-86,*HEPS0011,*SK3124, *ECG123,*WEP53,*276-2009,*RT-100
2N771	2N744,2N784,2N988,2N989,2N2206,2N2318,2N2319,2N2656,2N2921,2N2922, 2N3011,2N3544,2N3564,2N3691,2N4294,2N5030,2N5126,2N5187,2N5418,2SC55, 2SC99,2SC204,2SC356,2SC394-O,2SC395A-R,2SC400-R,2SC595,2SC718,2SC719, 2SC724,2SC1123,2SC1128,2SC1231,EN706,EN744,EN3011,MM1941,MPS6507, MPS6511,MPS6568,MPSH30,MPSH31,NT3000,PET2001,PT720,PT2760,TIS51, ZT2205,*PTC132,*GE-20,*TR-86,*HEPS0011,*SK3124,*ECG123,*WEP53, *276-2009,*RT-100
2N772	2N706,2N706A,2N706B,2N783,2N784,2N834A,2N959,2N2205,2N2368,2N2921, 2N3261,2N3544,2N3793,2N3854A,2N5187,2SC54,2SC55,2SC68,2SC321,2SC322, 2SC400-R,2SC455,2SC576,2SC689,40405,GI-3793,KT218,MM1941,MPS706, MPS706A,MPS834,MPS6507,MPS6511,MPS6540,MPS6542,MPS6543,MPS6548, MPSH19,MPSH20,MPSH37,MT1038,MT1038A,MT1039,PET3001,SE1010,SE5025, TIS47,ZT403P,ZT2368,*PTC132,*GE-20,*TR-86,*HEPS0011,*SK3124,*ECG123, *WEP53,*276-2009,*RT-100
2N773	*PTC123,*GE-20,*TR-86,*HEPS0011,*SK3124,*ECG123,*WEP53,*276-2009,*RT-100
2N774	*PTC123,*GE-20,*TR-86,*HEPS0011,*SK3124,*ECG123,*WEP53,*276-2009,*RT-100
2N775	*PTC123,*GE-20,*TR-86,*HEPS0011,*SK3124,*ECG123,*WEP53,*276-2009,*RT-100
2N776	*PTC123,*GE-20,*TR-86,*HEPS0011,*SK3124,*ECG123,*WEP53,*276-2009,*RT-100
2N777	*PTC123,*GE-20,*TR-86,*HEPS0011,*SK3124,*ECG123,*WEP53,*276-2009,*RT-100
2N778	*PTC123,*GE-20,*TR-86,*HEPS0011,*SK3124,*ECG123,*WEP736,*276-2009, *RT-100
2N779	2N779A,2N829,2N838,2N972,2N983,2N984,2N2795,2N3320,*PTC145,*GE-51,*TR-17, *HEPG0003,*ECG160,*WEP637
2N779A	2N779,2N829,2N838,2N972,2N983,2N984,2N2795,2N3320,*PTC145,*GE-51,*TR-17, *HEPG0003,*JR200,*ECG160,*WEP637
2N779B	2N829,2N838,2N972,*PTC145,*GE-51,*TR-17,*HEPG0003,*ECG160,*WEP637
2N780	*PTC125,*GE-212,*TR-70,*HEPS0011,*SK3039,*EN10,*ECG107,*WEP56,*BC107B, *276-2008,*RT-108
2N781	2N829,2N972,2SA417,NKT221,NKT228,*PTC109,*GE-1,*TR-17,*HEPG0003,*SK3004, *JR200,*ECG160,*WEP637,*AC188/01,*276-2005
2N782	2N828,2N968,2N969,2N970,*PTC109,*GE-2,*TR-17,*HEPG0003,*SK3123,*JR200, *ECG160,*WEP637
2N783	2N2220,2N2710,2N3261,2N3510,2N3511,2N3605A,2N3606A,2N3648,2N3973,2N3975, 2SC321,2SC322,2SC576,40405,GET708,GET914,GET3013,GET3014,GET3646, MPS6532,ZT83,ZT113,*PTC136,*GE-210,*TR-21,*HEPS0011,*SK3122,*EN10, *ECG123A,*WEP735,*BC107B,*276-2016,*RT-102
2N784	2N708,2N708A,2N784A,2N914,2N2242,2N2318,2N2319,2N2481,2N2710,2N3011, 2N3013,2N3014,2N3210,2N3261,2N3510,2N3511,2N3605A,2N3606A,2N3646, 2N3647,2N3648,2N5772,2SC67,2SC204,2SC356,2SC577,2SC595,2SC1274, BSW82,GET708,GET914,GET2369,GET3013,GET3014,GET3646,ZT708,*PTC136, *GE-210,*TR-21,*HEPS0011,*SK3122,*EN10,*ECG123A,*WEP735,*276-2016, *RT-102
2N784A	2N708,2N708A,2N914,2N2221,2N2242,2N2481,2N2501,2N2539,2N2845,2N2847, 2N3013,2N3014,2N3115,2N3210,2N3211,2N3301,2N3646,2N3647,2N3946,2N4013, 2N4962,2N5027,2N5144,2N5380,2N5772,2SC67,2SC577,2SC639,2SC764,A5T3903, A5T4123,BSW82,FT3641,FT3642,GET2221,GET2369,GI-3641,MM3903,TN-3903, ZT708,*PTC136,*GE-20,*TR-21,*HEPS0011,*SK3122,*EN10,*ECG123A,*WEP735, *276-2009,*RT-102
2N784A/46	2N3013,2N3014,2N3647,MM3903,*PTC123,*GE-210,*TR-64,*HEPS0011,*BC337, *276-2016
2N784A/51	2N3013,2N3014,2N3647,MM3903,*PTC123,*GE-210,*TR-64,*BC107B,*276-2016
2N789	2N2521,2SC478,2SC945,TIS126,*PTC123,*GE-39,*TR-86,*HEPS0011,*SK3124, *ECG123,*WEP53,*276-2013,*RT-100

To Replace	Substitute This Type
2N790	2N717,2N840,2N844,2N2387,2N2432A,2N2522,2N3973,2N3975,2S103,2SC302, 2SC725,BF224J,BF225J,BFY39-1,BSY93,ZT43,*PTC123,*GE-20,*TR-86,*HEPS0011, *SK3124,*ECG123,*WEP53,*276-2009,*RT-100
2N791	2N718,2N843,2N916,2N916A,2N929,2N2847,2N3693,2N3826,2N3858A,2N3862, 2N3903,2N3946,2N3974,2N3976,2N4227,2N4966,2N4994,2N5027,2N5824,2S104, 2SC318,2SC366G-O,2SC620,2SC815,2SC899,2SC938,2SC944,2SC1189,2SC1320, 2SC1453,2SC1688,EN916,MM3903,MPS3693,MPS6565,MPS6576-BLUE, MPS6576-YEL,PET1001,SE1001,TIS87,TP3705,ZT44,*PTC123,*GE-20,*TR-86, *HEPS0011,*SK3124,*ECG123,*WEP53,*276-2009,*RT-100
2N792	2N708A,2N718,2N843,2N929,2N2845,2N2847,2N3301,2N3693,2N3826,2N3858A, 2N3862,2N3903,2N3946,2N3974,2N3976,2N4140,2N4966,2N4994,2N5027,2N5824, 2S103,2SC318,2SC335,2SC336,2SC366G-O,2SC620,2SC815,2SC838,2SC839, 2SC896,2SC938,2SC1189,2SC1320,2SC1688,GET2221,MM3903,MPS3693, MPS6565,MPS6576-YEL,PET1001,SE1001,TIS22,TIS87,TP3705,*PTC123,*GE-20, *TR-86,*HEPS0011,*SK3124,*ECG123,*WEP53,*276-2009,*RT-100
2N793	2N2523,2N3117,2N5210,2N5827,2N6112,2SC828A,2SC1222,2SC1313,2SC1345, 2SC1416,2SC1416A,BC183L,BC184L,BC413C,BC414C,BCY59C,BCY59D,FT107C, GET2484,MPS-A09,MPSA18,SE4020,*GE-20,*TR-86,*HEPS0011,*SK3124, *ECG123,*WEP53,*276-2009,*RT-100
2N794	2N711,2N711A,2N711B,2N795,2N796,2N829,2N964A,2N972,2N973,2N974,2N985, 2N2402,2SA451,*PTC145,*GE-51,*TR-17,*HEPG0003,*JR100,*ECG160,*WEP637
2N795	2N796,*GE-51,*TR-17,*HEPG0003,*JR100,*ECG160,*WEP637
2N796	2N827,2N2699,*GE-51,*TR-17,*HEPG0003,*JR100,*ECG160,*WEP637
2N797	*GE-7,*TR-08,*HEPG0011,*SK3010,*NR5,*ECG103,*WEP641A,*276-2001,*RT-122
2N799	*PTC109,*GE-1,*TR-05,*HEPG0005,*SK3004,*HF12H,*ECG102,*WEP631, *AC188/01,*276-2007,*RT-120
2N800	*GE-1,*TR-85,*HEPG0005,*SK3123,*HF12H,*ECG102A,*WEP250,*AC188/01, *276-2007,*RT-121
2N801	*GE-1,*TR-05,*HEPG0005,*SK3005,*HF6H,*ECG100,*WEP254,*AC188/01, *276-2005,*RT-118
2N802	*GE-1,*TR-05,*HEPG0005,*SK3005,*HF6H,*ECG100,*WEP254,*AC188/01,*RT-118
2N803	*DS-25,*GE-1,*TR-84,*HEPG0005,*SK3005,*HF12H,*ECG158,*WEP238,*AC188/01, *276-2005
2N804	*DS-25,*GE-1,*TR-84,*HEPG0005,*SK3005,*HF12H,*ECG158,*WEP238,*AC188/01
2N805	*DS-25,*GE-1,*TR-84,*HEPG0005,*SK3005,*HF20H,*ECG158,*WEP238,*276-2005
2N806	*DS-25,*GE-1,*TR-84,*HEPG0006,*SK3005,*HF20H,*ECG158,*WEP238
2N807	*DS-25,*GE-1,*TR-85,*HEPG0005,*SK3005,*HF20H,*ECG102A,*WEP250,*276-2005, *RT-121
2N808	*DS-25,*GE-1,*TR-85,*HEPG0005,*SK3005,*HF20H,*ECG102A,*WEP250,*RT-121
2N809	*PTC109,*GE-1,*TR-05,*HEPG0005,*SK3005,*HF12H,*ECG100,*WEP254,*276-2005, *RT-118
2N810	*GE-1,*TR-05,*HEPG0005,*SK3005,*HF12H,*ECG100,*WEP254,*RT-118
2N811	*GE-1,*TR-05,*HEPG0005,*SK3005,*HF12H,*ECG100,*WEP254,*276-2005,*RT-118
2N812	*GE-1,*TR-05,*HEPG0005,*SK3005,*HF12H,*ECG100,*WEP254,*RT-118
2N813	*DS-25,*GE-1,*TR-05,*HEPG0005,*SK3005,*HF20H,*ECG102A,*WEP250,*276-2005, *RT-121
2N814	*DS-25,*GE-1,*TR-05,*HEPG0007,*SK3005,*HF20H,*ECG102A,*WEP250,*RT-121
2N815	*PTC108,*GE-8,*TR-05,*HEPG0005,*SK3011,*NR5,*ECG102A,*WEP250,*AC187/01, *276-2005,*RT-121
2N816	*GE-8,*TR-85,*HEPG0005,*SK3011,*NR5,*ECG102A,*WEP250,*AC187/01,*RT-121
2N817	*GE-6,*TR-84,*HEPG0005,*SK3011,*NR5,*ECG158,*WEP238,*276-2004
2N818	*GE-6,*TR-84,*HEPG0005,*SK3011,*NR5,*ECG158,*WEP238
2N819	*DS-72,*GE-6,*TR-85,*HEPG0005,*SK3011,*NR5,*ECG102A,*WEP250,*AC187/01, *276-2004,*RT-121
2N820	*DS-72,*GE-6,*TR-85,*HEPG0005,*SK3011,*NR5,*ECG102A,*WEP250,*AC187/01, *RT-121
2N821	*DS-72,*GE-6,*TR-08,*HEPG0011,*SK3011,*NR10,*ECG101,*WEP641,*RT-119
2N822	*DS-72,*GE-6,*TR-08,*HEPG0011,*SK3011,*NR10,*ECG101,*WEP641,*RT-119
2N823	*PTC108,*GE-5,*TR-08,*HEPG0011,*SK3011,*NR10,*ECG101,*WEP641,*RT-119
2N824	*PTC108,*GE-5,*TR-08,*HEPG0011,*SK3011,*NR10,*ECG101,*WEP641,*RT-119
2N825	*GE-1,*TR-05,*HEPG0005,*SK3011,*HF12H,*ECG102,*WEP631,*AC188/01,

To Replace	Substitute This Type
(2N825)	*276-2005,*RT-120
2N826	*GE-1,*TR-05,*HEPG0005,*SK3004,*HF12H,*ECG102,*WEP631,*AC188/01,*RT-120
2N827	*GE-53,*TR-17,*HEPG0003,*ECG160,*WEP637
2N828	2N968,*PTC145,*GE-51,*TR-17,*HEPG0003,*JR200,*ECG160,*WEP637
2N828A	2N828,2N968,*PTC145,*TR-17,*HEPG0003,*ECG160,*WEP637
2N829	2N972,2N985,*PTC145,*GE-51,*TR-17,*HEPG0003,*ECG160,*WEP637
2N834	2N834A,2SC321,2SC322,2SC576,40405,MPS834,MPSH37,*PTC136,*GE-20,*TR-95, *HEPS0011,*SN80,*ECG108,*WEP56,*276-2016,*RT-113
2N834A	2N2710,2N3648,2SC321,2SC322,2SC576,*PTC133,*TR-21,*HEPS0011,*SK3122, *ECG123A,*WEP736,*276-2038,*RT-102
2N835	2N834,MPSH07,MPSH08,*PTC136,*GE-20,*TR-21,*HEPS0011,*SK3122,*EN10, *ECG123A,*WEP736,*BC107B,*276-2016,*RT-102
2N837	2N711,2N711A,2N711B,2N794,2N795,2N829,2N2401,2N2402,2SA417,2SA451, 2SB185,*PTC109,*GE-52,*TR-17,*HEPG0003,*AT20M,*ECG160,*WEP637,*AC126, *276-2006
2N838	2N2956,2N2957,*PTC145,*GE-53,*TR-17,*HEPG0003,*ECG160,*WEP637
2N839	2N717,2N757,2N757A,2N758,2N758A,2N842,2N844,2N2395,2S102,BF224J,BF225J, MPS6544,MPS6545,ZT42,*PTC123,*GE-212,*TR-21,*HEPS0011,*SK3122,*EN10, *ECG123A,*WEP736,*BC107B,*276-2013,*RT-102
2N840	2N708A,2N718,2N929,2N2387,2N2396,2N2845,2N2847,2N3301,2N3641,2N3642, 2N3858A,2N3862,2N3946,2N3973,2N3975,2N4140,2N4994,2N5027,2N5824,2S103, 2SC318,2SC896,2SC1166-R,BFY39-1,BSY93,EN697,FT3641,FT3642,GET2221, MM3903,PET1001,TIS87,TIS105,ZT43,*PTC123,*GE-212,*TR-21,*HEPS0011, *SK3122,*EN10,*ECG123A,*WEP736,*BC107B,*276-2013,*RT-102
2N841	2N909,2N929A,2N930B,2N2483,2N2484,2N2484A,2N3302,2N3904,2N3947,2N4141, 2N4995,2N5028,2N5209,2N5381,2N5826,2SC366G-Y,2SC538A,2SC1166-Y,2SC1175, 2SC1364,2SC1850,2SC1991,2SC1992,A5T3904,BC107,BC107A,BC167A,BC171A, BC174A,BC182A,BC182L,BC190A,BC237A,BC547A,BCY59A,BCY59B,BSX79, FT3643,GET2222,MM3904,MPS3694,MPS6566,MPS6575,MPS6576-SIL,PET1002, TN-3904,*PTC123,*GE-212,*TR-95,*HEPS0011,*SK3039,*EN10,*ECG108,*WEP56, *BC107B,*276-2013,*RT-113
2N842	2N717,2N758,2N758A,2N759,2N759A,2N760B,2N840,2N844,2N2312,2N2314,2N2395, 2N2522,2N3973,2N3975,2SC302,2SC1166-R,BF224J,BF225J,BFY39-1,BSY93, MPSA05,ZT43,ZT83,ZT113,*PTC123,*GE-212,*TR-21,*HEPS0011,*SK3122,*EN10, *ECG123A,*WEP736,*BC107B,*276-2013,*RT-102
2N843	2N760,2N760A,2N841,2N916,2N916A,2N3416,2N3643,2N3859A,2N3903,2N3947, 2N3974,2N3976,2N4013,2N4227,2N4951,2N4962,2N5368,2N5380,2N5825,2N6538, 2S104,2SC366G,2SC366G-O,2SC850,2SC933,2SC943,2SC968,2SC970, 2SC1166-O,2SC1210,2SC1211,2SC1247,2SC1318,2SC1852,2SC1999,A5T3903, BFY39-2,MPS3693,MPS6565,MPS6576-BLUE,MPS6576-GREEN,MPS6576-YEL, MPS-A05,TN-3903,TP3705,ZT44,ZT81,ZT82,ZT84,ZT111,ZT112,ZT114,*PTC123, *GE-212,*TR-95,*HEPS0011,*SK3039,*EN10,*ECG108,*WEP56,*BC107B,*276-2013, *RT-113
2N844	2N735,2N735A,2N758A,2N759A,2N760B,2N2515,2N2522,2N3973,2N3975,2S103, 2SC383,2SC979-R,2SC1166-R,BSY93,MPSA05,MPSA06,ZT83,ZT113,*PTC123, *GE-81,*TR-21,*HEPS0011,*SK3122,*ECG123A,*WEP736,*276-2008,*RT-102
2N845	2N738,2N739A,2N2460,2N2464,2N2518,2N4390,2SC979A-R,ZT86,ZT116,*PTC125, *GE-18,*HEPS5026,*276-2008
2N846	*PTC107,*GE-52,*TR-17,*HEPG0003,*ECG160,*WEP637
2N846A	2N828,2N828A,2N964,2N968,2N2487,2N2488,2N2489,2N2630,2N2796,2SA411, *PTC145,*GE-51,*TR-17,*HEPG0003,*JR200,*ECG160,*WEP637
2N846B	2N828,2N828A,2N964,2N968,2N2630,2SA411,*PTC145,*GE-2,*TR-17,*HEPG0003, *SK3123,*ECG160,*WEP637
2N849	*GE-61,*TR-95,*HEPS0016,*SK3039,*EN10,*ECG108,*WEP56,*BC107B,*276-2016, *RT-113
2N850	*GE-61,*TR-95,*HEPS0020,*SK3039,*EN10,*ECG108,*WEP56,*BC107B,*276-2016, *RT-113
2N851	*GE-20,*TR-95,*HEPS0016,*SK3039,*ECG108,*WEP56,*BC107B,*276-2016,*RT-113
2N852	*GE-20,*TR-95,*HEPS0016,*SK3039,*EN10,*ECG108,*WEP56,*BC107B,*276-2016, *RT-113
2N858	*GE-65,*TR-20,*HEPS0013,*SK3114,*SP70,*ECG159,*WEP242,*BC177,*276-2022,

To Replace	Substitute This Type
(2N858)	*RT-115
2N859	2N2946A,2N3840,2N4981,2N5231,*GE-65,*TR-20,*HEPS0013,*SK3114,*SP70, *ECG159,*WEP242,*BC177,*276-2022,*RT-115
2N860	2N2969,2N2971,2N4008,MM4052,*GE-65,*TR-20,*HEPS0013,*SK3114,*SP70, *ECG159,*WEP242,*BC177,*276-2022,*RT-115
2N861	2N2969,2N4008,*GE-65,*TR-20,*HEPS0013,*SK3114,*SP70,*ECG159,*WEP242, *BC177,*276-2022,*RT-115
2N862	2N861,2N2969,2N2971,MM4052,*GE-65,*TR-20,*HEPS0013,*SK3114,*SP70, *ECG159,*WEP242,*BC177,*276-2022,*RT-115
2N863	2N2944A,2N2945A,TW135,ZT152,*GE-65,*TR-20,*HEPS0013,*SK3114,*SP70, *ECG159,*WEP242,*BC177,*276-2022,*RT-115
2N864	2N865,2N2280,2N4006,2N4007,2N5141,ZT152,*PTC127,*GE-65,*TR-20,*HEPS0013, *SK3114,*SP70,*ECG159,*WEP242,*BC177,*276-2021,*RT-115
2N865	ZT152,*GE-65,*TR-20,*HEPS0013,*SK3118,*SP70,*ECG106,*WEP52,*BC177, *276-2021,*RT-126
2N866	2N758,*PTC123,*GE-20,*TR-86,*HEPS0011,*SK3124,*ECG123,*WEP53,*276-2009, *RT-100
2N867	2N760,2N2331,2N3646,2N3736,A5T5450,A8T3705,D33D24,SE8040,TN62,*PTC123, *GE-20,*TR-86,*HEPS0011,*SK3124,*ECG123,*WEP53,*276-2009,*RT-100
2N869	2N4034,2N4035,2N4060,2N5382,2N5383,A5T3905,A5T3906,BC558A,BC559A, BCY78A,MM3906,TN-3905,TN-3906,*PTC127,*GE-65,*TR-20,*HEPS0013,*SP70, *ECG106,*WEP52,*276-2024,*RT-126
2N869A	2N3829,*TR-20,*HEPS0013,*ECG106,*WEP52,*276-2023,*RT-126
2N870	2N720A,2N740A,2N910,2N2465,*PTC123,*GE-18,*HEPS5026,*ECG128,*WEP243, *BF338,*276-2008
2N871	*PTC123,*GE-18,*HEPS5026,*ECG128,*WEP243,*276-2008
2N902	2N2521,2SC478,2SC945,TIS126,*HEPS0016
2N903	2N717,2N840,2N844,2N2387,2N2432A,2N2522,2N3973,2N3975,2S103,2SC302, 2SC725,BF224J,BF225J,BFY39-1,BSY93,ZT43,*HEPS0016
2N904	2N718,2N843,2N916,2N916A,2N929,2N2847,2N3693,2N3826,2N3858A,2N3862, 2N3903,2N3946,2N3974,2N3976,2N4227,2N4966,2N4994,2N5027,2N5824,2S104, 2SC318,2SC366G-O,2SC620,2SC815,2SC899,2SC938,2SC944,2SC965,2SC1189, 2SC1320,2SC1453,2SC1688,EN916,MM3903,MPS3693,MPS6565,MPS6576-BLUE, MPS6576-YEL,PET1001,SE1001,TIS87,TP3705,ZT44,*HEPS0016
2N905	2N708A,2N718,2N843,2N929,2N2845,2N2847,2N3301,2N3693,2N3826,2N3858A, 2N3862,2N3903,2N3946,2N3974,2N3976,2N4140,2N4966,2N4994,2N5027,2N5824, 2S103,2SC318,2SC335,2SC336,2SC366G-O,2SC620,2SC815,2SC838,2SC839, 2SC896,2SC938,2SC1189,2SC1320,2SC1688,GET2221,MM3903,MPS3693, MPS6565,MPS6576-YEL,PET1001,SE1001,TIS22,TIS87,TP3705,*HEPS0016
2N906	2N2523,2N3117,2N5210,2N5827,2N6112,2SC828A,2SC1222,2SC1313,2SC1345, 2SC1416,2SC1416A,BC183L,BC184L,BC413C,BC414C,BCY59C,BCY59D,FT107C, GET2484,MPS-A09,MPSA18,SE4020,*HEPS0030
2N909	2N929A,2N930A,2N930B,2N956,2N2645,2N4409,2N5381,2N6540,2SC1166-GR, A5T3904,A5T4409,FT3643,TIS95,TIS98,TN-3904,*PTC123,*GE-210,*TR-86, *HEPS0011,*ECG123,*WEP56,*BC337,*276-2009,*RT-100
2N910	2N740A,2N871,2N2466,*PTC123,*GE-18,*HEPS0005,*ECG128,*WEP243,*BF338, *276-2008
2N911	2N720A,2N870,2N2465,*PTC123,*GE-18,*HEPS5026,*ECG128,*WEP243,*BF338, *276-2008
2N912	2N2464,*PTC123,*GE-18,*HEPS5026,*ECG128,*WEP243,*BF338,*276-2012
2N914	2N916,2N916A,2N2221,2N2501,2N2539,2N2845,2N2847,2N3115,2N3210,2N3211, 2N3301,2N3646,2N4013,2N4962,2N5144,2SC577,BSW82,FT3641,FT3642,TIS133, TIS134,TN-3903,ZT708,*PTC136,*GE-17,*TR-95,*HEPS0011,*SN60,*ECG108, *WEP56,*276-2009,*RT-113
2N914/46	*PTC136,*GE-17,*TR-64,*HEPS0011,*276-2009
2N914/51	2N3013,2N3014,MM3903,*GE-61,*TR-95,*HEPS0016,*SK3039,*ECG108,*WEP56, *BF173,*276-2016,*RT-113
2N915	2SC1998,BC546A,GET2222A,*GE-20,*HEPS0025,*ECG123A,*WEP736
2N916	2N916A,2N3643,2N3947,2N5381,2SC1991,2SC1999,A5T3904,BC547A,BCY59A, GET2222,MM3904,TN-3904,*PTC136,*GE-17,*TR-95,*HEPS0011,*SK3039,*EN10, *ECG108,*WEP56,*276-2009,*RT-113

To Replace	Substitute This Type
2N916A	2N916,2N3643,2N3947,2N5381,2SC1991,2SC1999,A5T3904,BC547A,BCY59A, GET2222,MM3904,TN-3904,*PTC136,*GE-17,*TR-95,*HEPS0011,*SK3039,*ECG108, *WEP56,*276-2009,*RT-113
2N917	*PTC121,*GE-60,*TR-83,*HEPS0016,*SK3019,*EN10,*ECG161,*WEP709,*BF173, *276-2011,*RT-113
2N917A	*GE-86,*TR-83,*HEPS0016,*SK3117,*ECG161,*WEP709,*BF173,*276-2011,*RT-113
2N918	2N5851,2N5852,MT1061,*PTC132,*GE-86,*TR-95,*HEPS0017,*SK3018,*EN10, *ECG108,*WEP56,*276-2011,*RT-113
2N918/51	2N2729,MT1038,MT1038A,MT1039,MT1060,*GE-86,*276-2011
2N919	2N2318,2N2656,2N2710,2N3510,2N3511,2N3605A,2N3606A,2N3648,2SC204, 2SC321,2SC322,2SC576,GET708,GET914,GET3013,GET3014,GET3646,*PTC136, *GE-47,*TR-21,*HEPS0011,*SK3122,*EN10,*ECG123A,*WEP736,*276-2038, *RT-102
2N920	2N2242,2N2501,2N3013,2N3014,2N3211,2SC67,2SC577,2SC639,2SC764,A5T4123, BSW82,PT720,*PTC136,*GE-47,*TR-21,*HEPS0011,*SK3122,*EN10,*ECG123A, *WEP736,*276-2038,*RT-102
2N921	BSX48,BSX49,*GE-47,*TR-21,*HEPS0011,*SK3122,*EN10,*ECG123A,*WEP736, *276-2038,*RT-102
2N922	2N6538,A5T3903,TIS110,TIS133,TIS134,*GE-17,*TR-21,*HEPS0011,*SK3122,*EN10, *ECG123A,*WEP736,*276-2009,*RT-102
2N923	2N925,2N938,2N939,*GE-65,*TR-20,*HEPS0013,*SK3114,*SP70,*ECG159,*WEP242, *BC177,*276-2022,*RT-115
2N924	2N940,2N2946A,2N3219,2N3840,2N3979,*GE-65,*TR-20,*HEPS0013,*SK3114,*SP70, *ECG159,*WEP242,*BC177,*276-2022,*RT-115
2N925	2N927,*GE-65,*TR-20,*HEPS0013,*SK3114,*SP70,*ECG159,*WEP242,*BC177, *276-2022,*RT-115
2N926	2N928,*GE-65,*TR-20,*HEPS0013,*SK3114,*SP70,*ECG159,*WEP242,*BC177, *276-2022,*RT-115
2N927	*GE-82,*TR-20,*HEPS0013,*SK3114,*SP70,*ECG159,*WEP242,*276-2021,*RT-115
2N928	2N3841,*GE-82,*TR-20,*HEPS0013,*SK3114,*SP70,*ECG159,*WEP242,*276-2021, *RT-115
2N929	2N718,2N760,2N916,2N916A,2N2396,2N2847,2N3858A,2N3903,2N3946,2N3974, 2N3976,2N4227,2N4962,2N5027,2N5380,2N5824,2N6538,2S104,2SC318, 2SC366G-O,2SC1166-O,A5T3903,EN697,MM3903,TN-3903,TP3705,ZT44,*PTC123, *GE-212,*TR-21,*HEPS0011,*SK3122,*SN80,*ECG123A,*WEP736,*BC107B, *276-2013,*RT-102
2N929A	2N930A,BC546B,TIS94,TIS97,*PTC123,*GE-81,*TR-21,*HEPS0011,*SK3122,*SN80, *ECG123A,*WEP736,*BC337,*276-2012,*RT-102
2N930	2N909,2N929A,2N930A,2N930B,2N2388,2N2484,2N2484A,2N2586,2N3302,2N3904, 2N5381,2N5826,2N6112,2SC538A,2SC587,2SC587A,2SC1166-GR,2SC1364, 2SC1850,2SC1992,A5T3904,A157B,BC107,BC107B,BC167B,BC171B,BC174B, BC182B,BC182L,BC183L,BC190B,BC237B,BC413B,BC414B,BCY59B,BFY39-3, BSX79,FT3643,GI-3566,TN-3904,TP3704,*PTC123,*GE-212,*TR-21,*HEPS0015, *SK3122,*SN80,*ECG123A,*WEP736,*BC107B,*276-2013,*RT-102
2N930A	2N6428,2N6428A,2N6539,BC546B,TIS94,TIS97,*GE-81,*TR-21,*HEPS0011,*SK3039, *EN10,*ECG123A,*WEP736,*BC337,*276-2012,*RT-113
2N930B	2N929A,2N930A,2N956,2N2645,2N5381,2N6540,A5T3904,TIS95,TIS98,TN-3904, *PTC123,*GE-210,*BC337,*276-2012
2N934	2N828,2N968,2N969,2N970,*PTC145,*GE-2,*TR-17,*HEPG0003,*ECG160,*WEP637
2N935	2N925,2N927,2N936,2N945,*GE-65,*TR-20,*HEPS0013,*SK3114,*SP70,*ECG159, *WEP242,*BC327,*276-2023,*RT-115
2N936	2N925,2N927,*GE-65,*TR-20,*HEPS0013,*SK3114,*SP70,*ECG159,*WEP242, *BC327,*276-2023,*RT-115
2N937	2N926,2N928,2N3060,*GE-65,*TR-20,*HEPS0013,*SK3114,*SP70,*ECG159, *WEP242,*BC327,*276-2023,*RT-115
2N938	*GE-65,*TR-20,*HEPS0013,*SK3114,*ECG159,*WEP242,*BC177,*276-2022,*RT-115
2N939	2N2946A,*GE-21,*TR-20,*HEPS0013,*SK3114,*SP70,*ECG159,*WEP242,*BC177, *276-2023,*RT-115
2N940	2N2946,2N2946A,2N3840,2N3910,2N3913,2N4981,*GE-65,*TR-20,*HEPS0013, *SK3114,*SP70,*ECG159,*WEP242,*BC177,*276-2022,*RT-115
2N941	*GE-65,*TR-20,*HEPS0013,*SK3114,*SP70,*ECG159,*WEP242,*BC177,*276-2022,

To Replace	Substitute This Type
(2N941)	*RT-115
2N942	2N941,2N3677,*GE-65,*TR-20,*HEPS0013,*SK3114,*SP70,*ECG159,*WEP242, *BC177,*276-2022,*RT-115
2N943	2N938,2N944,2N945,*GE-65,*TR-20,*HEPS0013,*SK3114,*SP70,*ECG159,*WEP242, *BC177,*276-2022,*RT-115
2N944	2N938,2N943,2N945,*GE-65,*TR-20,*HEPS0013,*SK3114,*SP70,*ECG159,*WEP242, *BC177,*276-2022,*RT-115
2N945	*GE-65,*TR-20,*HEPS0013,*SK3114,*SP70,*ECG159,*WEP242,*BC177,*276-2022, *RT-115
2N946	*GE-82,*TR-20,*HEPS0013,*SK3114,*SP70,*ECG159,*WEP242,*276-2023,*RT-115
2N947	2N2318,2N2656,2N3011,2SC55,2SC204,MPSH02,MPSH19,PT2760,*PTC136,*GE-10, *TR-21,*HEPS0011,*SK3122,*SP70,*ECG123A,*WEP736,*BC337,*276-2009, *RT-102
2N955	*DS81,*GE-11,*TR-08,*HEPG0011,*SK3011,*NR700,*ECG101,*WEP641,*RT-119
2N955A	*DS81,*GE-11,*TR-08,*HEPG0011,*SK3011,*NR700,*ECG101,*WEP641,*RT-119
2N956	2N871,*PTC123,*GE-81,*HEPS0015,*276-2008
2N957	2N916,2N916A,2N2242,2N2501,2N2847,2N3013,2N3014,2N3903,2N3946,2N3974, 2N3976,2N4123,2N4227,2N4962,2N5027,2N5380,2SC67,2SC169,2SC601N,2SC620, 2SC639,2SC764,2SC815,2SC899,2SC944,MM3903,TN-3903,*PTC132,*GE-61, *TR-21,*HEPS0015,*SK3122,*ECG123A,*WEP736,*BF365,*276-2016,*RT-102
2N958	2N2214,*HEPS0025
2N959	2N2319,2N3261,2N3510,2N3511,2N3648,2N5187,2SC356,40405,*HEPS0020
2N960	2N2630,*PTC145,*GE-51,*TR-17,*HEPG0003,*JR200,*ECG160,*WEP637
2N961	2N960,2N2630,*PTC145,*GE-51,*TR-17,*HEPG0003,*JR200,*ECG160,*WEP637
2N962	2N960,2N961,2N963,2N2630,*PTC145,*GE-51,*TR-17,*HEPG0003,*JR200,*ECG160, *WEP637
2N963	2N960,2N961,2N962,2N2630,*PTC145,*GE-51,*TR-17,*HEPG0003,*JR200,*ECG160, *WEP637
2N964	2N828,2N968,2SA411,*PTC145,*GE-51,*TR-17,*HEPG0003,*JR200,*ECG160, *WEP637
2N964A	2N827,2N2699,*PTC145,*GE-2,*TR-17,*HEPG0003,*ECG160
2N965	2N828,2N964,2N968,2N969,2N2402,2N2860,2SA411,*PTC145,*GE-51,*TR-17, *HEPG0003,*JR200,*ECG160,*WEP637
2N966	2N828,2N964,2N965,2N967,2N968,2N969,2N970,2N2402,2N2860,2SA411,2SA450, *PTC145,*GE-51,*TR-17,*HEPG0003,*JR200,*ECG160,*WEP637
2N967	2N828,2N964,2N965,2N966,2N968,2N969,2N970,2N2402,2N2860,2SA411,2SA450, *PTC145,*GE-51,*TR-17,*HEPG0003,*JR200,*ECG160,*WEP637
2N968	2N828,*PTC145,*GE-51,*TR-17,*HEPG0003,*JR200,*ECG160,*WEP637
2N969	2N828,2N968,*PTC145,*GE-51,*TR-17,*HEPG0003,*JR200,*ECG160,*WEP637
2N970	2N828,2N968,2N969,*PTC145,*GE-51,*TR-17,*HEPG0003,*JR200,*ECG160, *WEP637
2N971	2N828,2N968,2N969,2N970,*PTC145,*GE-51,*TR-17,*HEPG0003,*JR200,*ECG160, *WEP637
2N972	2N985,*PTC145,*GE-51,*TR-17,*HEPG0003,*JR200,*ECG160,*WEP637
2N973	2N972,2N985,*PTC145,*GE-51,*TR-17,*HEPG0003,*JR200,*ECG160,*WEP637
2N974	2N972,2N973,2N985,*PTC145,*GE-51,*TR-17,*HEPG0003,*JR200,*ECG160, *WEP637
2N975	2N972,2N973,2N974,2N985,*PTC145,*GE-51,*TR-17,*HEPG0003,*JR200,*ECG160, *WEP637
2N976	2N5332,MPS6535,*PTC131,*GE-51,*TR-17,*HEPG0003,*JR200,*ECG160,*WEP637
2N977	*PTC145,*GE-2,*TR-17,*HEPG0003,*SK3123,*ECG160,*WEP637
2N978	2N721,2N722,2N722A,2N2393,2N2394,2N3341,2SA558,A5T4248,EN1132,TIS137, *PTC127,*GE-82,*TR-20,*HEPS0013,*SK3006,*JR200,*ECG106,*WEP52,*BC327, *276-2021
2N979	2N980,2N2489,2N2795,2N2796,*PTC131,*GE-1,*TR-17,*HEPG0003,*JR200, *ECG160,*WEP637
2N980	2N838,2N2795,2N3323,2N3324,2N3325,*PTC131,*GE-51,*TR-17,*HEPG0003, *ECG160,*WEP637
2N982	*PTC127,*GE-51,*TR-17,*HEPG0003,*SK3006,*JR200,*ECG160,*WEP637
2N983	2N982,2N3320,*PTC127,*GE-51,*TR-17,*HEPG0003,*SK3006,*JR200,*ECG160, *WEP637

To Replace	Substitute This Type
2N984	2N829,2N838,2N964A,2N972,2N982,2N983,2N985,2N3320,*PTC131,*GE-51,*TR-17, *HEPG0003,*SK3006,*JR200,*ECG160,*WEP637
2N985	*PTC145,*GE-2,*TR-17,*HEPG0003,*JR200,*ECG160,*WEP637
2N987	*PTC145,*GE-51,*TR-17,*HEPG0003,*SK3006,*JR200,*ECG160,*WEP637
2N988	2N744,2N989,2N2318,2N2319,2N2656,2N3011,2N5030,2N5200,2N5201,2N5418, 2SC204,2SC595,2SC718,2SC1231,MPS6568,MPSH02,MPSH30,MPSH31,PT720, PT2760,ZT2205,*PTC136,*GE-20,*TR-95,*HEPS0011,*SK3039,*EN10,*ECG108, *WEP56,*BC107B,*276-2016,*RT-113
2N989	2N744,2N988,2N2318,2N2319,2N2656,2N3011,2N5030,2N5200,2N5201,2N5418, 2SC204,2SC595,2SC718,2SC1231,MPS6568,MPSH02,MPSH30,MPSH31,PT720, PT2760,ZT2205,*PTC136,*GE-20,*TR-95,*HEPS0011,*SK3039,*EN10,*ECG108, *WEP56,*BC107B,*276-2016,*RT-113
2N990	2N991,2N993,*PTC107,*GE-51,*TR-17,*HEPG0008,*SK3007,*JR200,*ECG160, *WEP637,*276-2005
2N991	2N990,2N993,*PTC107,*GE-50,*TR-17,*HEPG0003,*SK3006,*ECG160,*WEP637
2N992	2N990,2N991,2N993,*PTC107,*GE-50,*TR-17,*HEPG0001,*SK3006,*ECG160, *WEP637,*276-2006,*RT-183
2N993	2N990,2N991,*PTC107,*GE-51,*TR-17,*HEPG0008,*SK3005,*JR200,*ECG160, *WEP637,*276-2007
2N995	2N869A,2N995A,2N3209,2N3545,2N3576,2N3829,2N4060,2N5354,2SA741,TIS37, *PTC127,*GE-65,*TR-20,*HEPS0013,*SK3118,*SP70,*ECG106,*WEP52,*BC327, *276-2024,*RT-126
2N995A	2N869,2N869A,2N995,2N3209,2N3545,2N3576,2N3829,2N4060,2N5354,2SA741, A5T4125,TIS37,*PTC127,*GE-65,*TR-20,*HEPS0013,*SK3118,*ECG106,*WEP52, *276-2024,*RT-126
2N996	2N869A,2N995,2N995A,2N3209,2N3248,2N3545,2N3546,2N3576,2N4060,2N5056, 2N5057,2N5354,2N5771,2SA741,A5T4260,A5T4261,MM2894,*PTC127,*GE-65, *TR-20,*HEPS0013,*ECG106,*WEP52,*BC327,*276-2024
2N997	*276-2009
2N998	
2N999	
2N1000	*GE-6,*TR-08,*HEPG0011,*SK3011,*NR10,*ECG101,*WEP641,*276-2001,*RT-119
2N1003	*PTC145,*GE-51,*TR-17,*HEPG0009,*SK3006,*ECG126,*WEP635
2N1004	2N383,2N1449,*PTC145,*GE-51,*TR-17,*HEPG0009,*ECG126,*WEP635
2N1005	2N835,2N958,2N1472,2N2214,2N2719,2N3825,2N3843,2N3843A,2N3984,2N3985, 2N4292,2N4293,2S131,2SC171,2SC172,2SC387A,D16G6,MPS918,ZT20,ZT40, *PTC126,*GE-214,*TR-95,*HEPS0016,*SK3039,*EN10,*ECG108,*WEP635,*BF183, *276-2011,*RT-113
2N1006	2N728,2N729,2N2475,2N2569,2N2570,2N2615,2N2616,2N2923,2N2926-ORG,2N3298, 2N3393,2N3633,2N3845,2N3845A,2N3856,2N3856A,2N3858,2N3959,2N3960, 2N4418,2N4996,2N4997,2N5132,2N5200,2N5201,2S095A,2S512,2SC388A, 2SC400-O,2SC458,2SC459,2SC460,2SC528,2SC619,2SC633,2SC645,2SC710, 2SC712,2SC713,2SC924,2SC925,2SC1215,2SC1359,2SC1372,2SC1789,40398, 40400,BF123,BF254,BF377,BF378,BSY61,GI-2712,GI-2714,GI-2716,MPS2926-ORG, MPS3563,MPS6568A,MPS6569,MPS6570,MPSH10,MPSH11,PET3002,PET4001, SE3005,TIS86,*PTC126,*GE-214,*TR-24,*HEPS0014,*SK3122,*ECG123A, *WEP736,*BF183,*276-2011,*RT-102
2N1007	2N257-WHT,*GE-3,*TR-01,*HEPG6003,*SK3009,*PT25,*ECG104,*WEP230,*OC28, *276-2006,*RT-124
2N1008	*PTC135,*GE-2,*TR-05,*HEPG0005,*SK3004,*AT20H,*ECG102,*WEP631, *AC188/01,*276-2005,*RT-120
2N1008A	*PTC135,*GE-2,*TR-05,*HEPG0005,*SK3004,*AT20H,*ECG102,*WEP631,*276-2005, *RT-120
2N1008B	*GE-2,*TR-05,*HEPG0005,*SK3004,*AT30H,*ECG102,*WEP631,*276-2005,*RT-120
2N1009	2N1008A,*PTC135,*GE-52,*TR-05,*HEPG0005,*SK3003,*ECG102,*WEP631, *276-2005,*RT-120
2N1010	2N377,2N445,2N557,2N576,2N585,2N585+,2N634,2N635,2N679,2N1090,2N1302, 2N1605,2N1995,2N1996,*PTC108,*GE-7,*TR-08,*HEPG0011,*SK3011,*NA30, *ECG103A,*WEP641B,*AC127,*RT-122
2N1010+	2SC11,2SC179,2SC180,2SC181,2SD75,2SD77,*PTC108,*GE-7,*TR-08,*SK3011, *ECG103A,*WEP641B

To Replace	Substitute This Type
2N1011	2N665,2N1536,2N1537,2N1541,2N1542,2N1971,2N2140,2N2140A,2N2141,2N2141A, 2N2145,2N2145A,2N2146,2N2146A,2SB152A,2SB249,2SB253,2SB339,ASZ18, *PTC105A,*GE-25,*TR-01,*HEPG6005,*SK3009,*PT150,*ECG121,*WEP232, *276-2006,*RT-127
2N1012	*GE-6,*TR-08,*HEPG0011,*SK3011,*NA30,*ECG101,*WEP641,*RT-119
2N1014	*GE-25,*TR-01,*HEPG6005,*SK3009,*PT40,*ECG121,*WEP232,*276-2006,*RT-127
2N1015	
2N1015A	
2N1015B	2N1016B
2N1015C	2N1015D,2N1016C,2N1016D
2N1015D	2N1015E,2N1016D
2N1015E	
2N1015F	
2N1016	
2N1016A	
2N1016B	
2N1016C	2N1016D
2N1016D	
2N1016E	
2N1016F	
2N1017	*GE-51,*TR-17,*HEPG0002,*SK3011,*HF20H,*ECG160,*WEP232,*276-2004
2N1018	*GE-51,*TR-17,*HEPG0002,*SK3011,*JR100,*ECG160,*WEP637,*276-2004
2N1021	2N1021A,2N1022A,*DS503,*GE-3,*TR-01,*HEPG6005,*SK3009,*PT50,*ECG121, *WEP232,*276-2006,*RT-127
2N1021A	2N1022A,*DS503,*GE-3,*TR-01,*HEPG6005,*SK3009,*ECG121,*WEP232,*276-2006, *RT-127
2N1022	*DS503,*GE-3,*TR-01,*HEPG6005,*SK3009,*PT50,*ECG121,*WEP232,*276-2006, *RT-127
2N1022A	*DS503,*GE-3,*TR-01,*HEPG6005,*SK3009,*ECG121,*WEP232,*276-2006,*RT-127
2N1023	2N2588,*PTC107,*GE-51,*TR-17,*HEPG0008,*SK3006,*JR100,*ECG160,*WEP637
2N1024	2N1219,2N1230,2N4008,2N6567,MM4052,*GE-65,*TR-20,*HEPS0012,*SK3114, *SP70,*ECG106,*WEP52,*BC177,*276-2022,*RT-126
2N1025	2N938,2N1232,2N1440,2N1441,2N1474A,*GE-65,*TR-20,*HEPS0012,*SK3114,*SP70, *ECG159,*WEP242,*BC177,*276-2022,*RT-115
2N1026	2N939,2N1233,2N1474A,2N2946A,*GE-65,*TR-20,*HEPS0012,*SK3114,*SP70, *ECG159,*WEP242,*BC177,*276-2022,*RT-115
2N1026A	SEE 2N1469
2N1027	2N1231,2N4007,2N4008,*GE-65,*TR-20,*HEPS0012,*SK3114,*SP70,*ECG159, *WEP242,*BC177,*276-2022,*RT-115
2N1028	*GE-65,*TR-20,*HEPS0012,*SK3114,*SP70,*ECG159,*WEP242,*BC177,*276-2021, *RT-115
2N1029	2N1550,2N1550A,*PTC901,*GE-16,*TR-01,*HEPG6003,*SK3009,*ECG121,*WEP232, *276-2006,*RT-127
2N1029A	2N1550,2N1550A,2N1551,2N1551A,*GE-3,*TR-01,*HEPG6005,*SK3009,*ECG121, *WEP232,*276-2006,*RT-127
2N1029B	2N1552,2N1552A,CQT940A,CQT1076,DTG-1200,*GE-3,*TR-01,*HEPG6005,*SK3009, *ECG121,*WEP232,*276-2006,*RT-127
2N1029C	2N1552,2N1552A,CQT940A,CQT1076,DTG-1200,*GE-3,*TR-01,*HEPG6005,*SK3009, *ECG121,*WEP232,*276-2006,*RT-127
2N1030	2N1146A,2N1558A,*PTC901,*GE-16,*TR-01,*HEPG6003,*SK3009,*ECG121, *WEP232,*276-2006,*RT-127
2N1030A	2N1146A,2N1146B,2N1558A,2N1559,2N1559A,*PTC901,*GE-16,*TR-01,*HEPG6005, *SK3009,*ECG121,*WEP232,*276-2006,*RT-127
2N1030B	2N1146C,2N1560,2N1560A,*GE-16,*TR-01,*HEPG6005,*SK3009,*ECG121,*WEP232, *276-2006,*RT-127
2N1030C	*GE-16,*TR-01,*HEPG6005,*SK3009,*ECG121,*WEP232,*276-2006,*RT-127
2N1031	2N511A,2N512A,2N513A,2N1031A,2N1163,2N1163A,CRT1544,MP1554,MP1554A, *PTC901,*GE-76,*TR-01,*HEPG6003,*SK3009,*PT150,*ECG121,*WEP232, *276-2006,*RT-127
2N1031A	2N511A,2N511B,2N512A,2N512B,2N513A,2N513B,2N1165,2N1165A,CRT1544, CRT1545,CTP3503,MP1554,MP1554A,MP1555,MP1555A,*PTC901,*GE-76,*TR-01,

To Replace	Substitute This Type
(2N1031A)	*HEPG6005,*SK3009,*PT150,*ECG121,*WEP232,*276-2006,*RT-127
2N1031B	2N1031C,2N1167,2N1167A,2N2445,CRT1553,CTP3500,MP1556,MP1556A,*DS503, *GE-76,*TR-01,*HEPG6005,*SK3009,*PT250,*ECG121,*WEP232,*276-2006, *RT-127
2N1031C	2N1167,2N1167A,2N2445,CRT1553,CTP3500,MP1556,MP1556A,*DS503,*GE-3, *TR-01,*HEPG6005,*SK3009,*PT250,*ECG121,*WEP232,*276-2006,*RT-127
2N1032	2N1032A,2N1147A,MP1558,MP1558A,*PTC901,*GE-76,*TR-01,*HEPG6003,*SK3009, *PT150,*ECG121,*WEP232,*276-2006,*RT-127
2N1032A	2N1147A,2N1147B,MP1558,MP1558A,MP1559,MP1559A,*PTC901,*GE-76,*TR-01, *HEPG6005,*SK3009,*PT150,*ECG121,*WEP232,*276-2006,*RT-127
2N1032B	2N1032C,2N1147C,MP1560,MP1560A,*DS503,*GE-76,*TR-01,*HEPG6005,*SK3009, *PT250,*ECG121,*WEP232,*276-2006,*RT-127
2N1032C	2N1147C,MP1560,MP1560A,*DS503,*TR-01,*HEPG6005,*SK3009,*PT250,*ECG121, *WEP232,*276-2006,*RT-127
2N1034	2N1036,2N3061,2N3911,2N3912,2N3914,2N3915,*GE-65,*TR-88,*HEPS0012, *SK3114,*SP70,*ECG129,*WEP242,*BC177,*276-2022,*RT-115
2N1035	2N329A,2N329B,2N926,2N928,2N937,2N1233,2N1442,2N1443,2N1475,2N2425, 2N3060,2N3346,2N4982,BCY31,HA7538,SHA7538,*GE-65,*TR-88,*HEPS0012, *SK3114,*SP70,*ECG129,*WEP242,*BC177,*276-2022,*RT-115
2N1036	2N3061,2N3911,2N3912,2N3914,2N3915,*GE-65,*TR-88,*HEPS0012,*SK3114,*SP70, *ECG129,*WEP242,*BC177,*276-2022,*RT-115
2N1037	2N329A,2N329B,2N926,2N928,2N937,2N1035,2N1233,2N1442,2N1443,2N1475, 2N2425,2N3060,2N3346,2N4982,BCY31,HA7538,SHA7538,*GE-65,*TR-88, *HEPS0012,*SK3114,*SP70,*ECG129,*WEP242,*BC177,*276-2022,*RT-115
2N1038	2N1039,2N2564,2N2565,*DS503,*GE-2,*TR-82,*HEPG6011,*SK3123,*AT100M, *ECG176,*WEP238,*276-2006,*RT-136
2N1038-1	2N1039-1,2N1042-1,2N1043-1,2N2560,2N2561
2N1038-2	2N1039-2,2N1042,2N1043
2N1039	2N1040,2N2565,2N2566,*DS503,*GE-3,*HEPG6011,*AT100M,*276-2006,*RT-136
2N1039-1	2N1040-1,2N1043-1,2N1044-1,2N2561,2N2562
2N1039-2	2N1040-2,2N1043,2N1044
2N1040	2N1041,2N2566,2N2567,*DS503,*GE-3,*TR-01,*HEPG6012,*SK3009,*AT100H, *ECG104,*WEP230,*RT-124
2N1040-1	2N1041-1,2N1044-1,2N1045-1,2N2562,2N2563
2N1040-2	2N1041-2,2N1044,2N1045
2N1041	2N2567,*DS503,*GE-3,*TR-01,*HEPG6012,*SK3009,*AT100H
2N1041-1	2N1045-1,2N2563
2N1041-2	2N1045
2N1042	2N1038-2,2N1039-2,2N1043,2N2556,2N2557,*DS26,*GE-2,*TR-17,*AT100H,*ECG126, *WEP635
2N1042-1	2N1038-1,2N1039-1,2N1043-1,2N2560,2N2561
2N1042-2	2N1043-2,2N2564,2N2565
2N1043	2N1039-2,2N1040-2,2N1044,2N2557,2N2558,*DS26,*GE-2,*TR-17,*AT100H,*ECG126, *WEP635
2N1043-1	2N1039-1,2N1040-1,2N1044-1,2N2561,2N2562
2N1043-2	2N1044-2,2N2565,2N2566
2N1044	2N1040-2,2N1041-2,2N1045,2N2558,2N2559,*GE-2,*TR-05,*SK3004,*AT100H, *ECG102,*WEP631,*RT-120
2N1044-1	2N1040-1,2N1041-1,2N1045-1,2N2562,2N2563
2N1044-2	2N1045-2,2N2566,2N2567
2N1045	2N1041-2,2N2559,*GE-2,*TR-05,*SK3004,*AT100H,*ECG102,*WEP631,*RT-120
2N1045-1	2N1041-1,2N2563
2N1045-2	2N2567
2N1046	2N1046A,2N1046B,MP1612,*DS503,*GE-25,*TR-01,*HEPG6005,*SK3009,*HO300, *ECG121,*WEP232,*276-2006,*RT-127
2N1046A	2N1046B,MP1612A,MP1612B,*DS503,*GE-25,*TR-01,*HEPG6005,*SK3009,*PT40, *276-2006,*RT-127
2N1046B	2N1046A,MP1612A,MP1612B,*DS503,*GE-25,*TR-01,*HEPG6005,*SK3009,*PT40, *ECG121,*WEP232,*276-2006,*RT-127
2N1047	2N1047C,*GE-66
2N1047A	2N1047,2N1047B,2N1047C,2N1690

To Replace	Substitute This Type
2N1047B	2N1047,2N1047C,2N1690
2N1047C	
2N1048	2N1048C
2N1048A	2N1048,2N1048B,2N1048C,2N1691
2N1048B	2N1048,2N1048C,2N1691
2N1048C	
2N1049	2N1049C,*GE-66
2N1049A	2N1049,2N1049B,2N1049C
2N1049B	2N1049,2N1049C
2N1049C	
2N1050	2N1050C,*GE-32
2N1050A	2N1050,2N1050B,2N1050C
2N1050B	2N1050,2N1050C
2N1050C	
2N1052	*GE-32,*TR-78,*HEPS5025,*SK3045,*ECG154,*WEP712,*276-2012,*RT-110
2N1053	MM3009,*GE-32,*TR-78,*HEPS5025,*SK3045,*ECG154,*WEP712,*276-2012,*RT-110
2N1054	*GE-32,*TR-78,*HEPS0014,*SK3045,*SN80,*ECG154,*WEP712,*276-2012,*RT-110
2N1055	2N657,2N699B,2N1613B,2N1889,2N1893,2N1974,2N3108,2N3498,2SC49,2SC69, BSY55,BSY87,MM3008,SE8010,*PTC144,*GE-32,*HEPS5026,*276-2012
2N1056	2N1924,*GE-2,*TR-05,*HEPG0005,*SK3004,*AT20M,*ECG102,*WEP631,*276-2004, *RT-120
2N1057	2N381,2N526,2N526A,2N527A,2N597,2N1008B,2N1057+,2N1925,2N1926,2SB57A, 2SB226,2SB227,2SB304A,2SB427,2SB428,SF.T243,*GE-2,*TR-05,*HEPG0005, *SK3004,*AT20M,*ECG102,*WEP631,*276-2005,*RT-120
2N1057+	2N381,2N526,2N526A,2N527A,2N597,2N1008B,2N1057,2N1925,2N1926,2SB57A, 2SB226,2SB227,2SB304A,2SB427,2SB428,SF.T243,*HEPG0005,*SK3004,*ECG102, *WEP631
2N1058	2N1891,*GE-5,*TR-08,*HEPG0011,*SK3011,*NR10,*ECG103A,*WEP641B, *AC187/01,*276-2001,*RT-122
2N1058+	2N94A,2N357,2N357A,2N358,2N358A,2N377,2N439,2N439A,2N557,2N576,2N635, 2N1090,2N1198,2N1605,2N1995,2N1996,2SC128,2SC129,2SC180,2SC181,2SD75, 2SD77,ASY28,*TR-08,*HEPG0011,*SK3011,*ECG103A,*WEP641B
2N1059	*PTC134,*GE-59,*TR-08,*HEPG0011,*SK3010,*NA30,*ECG103A,*WEP641B,*AC127, *276-2001,*RT-122
2N1059+	2N214,*PTC108,*HEPG0011,*SK3010,*ECG103A,*WEP641B
2N1060	2N742,2N742A,2N757,2N757A,2N759B,2N2310,2N2521,ZT42,*GE-63,*TR-95, *SK3039,*ECG108,*WEP56,*RT-113
2N1065	2N2956,*PTC107,*GE-51,*TR-17,*HEPG0002,*SK3008,*HF20H,*ECG160,*WEP637, *276-2005
2N1066	2N1225,2N1396,2N1397,2N2588,*PTC107,*GE-51,*TR-17,*HEPG0003,*SK3006, *JR100,*ECG160,*WEP637
2N1067	*GE-28,*TR-76,*HEPS5014,*276-2020
2N1068	*GE-28,*TR-76,*HEPS5014,*276-2020
2N1069	2N1070,2N1487,2N2305,2SD124A,BDY23A,HST9801,HST9802,ZT1487,*PTC173, *GE-77,*TR-59,*HEPS7002,*SK3027,*HN100,*ECG130,*WEP247,*BDY20, *276-2020,*RT-131
2N1070	2N1069,2N1487,2N2305,2SD124A,BDY23A,HST9801,HST9802,ZT1487,*PTC173, *GE-77,*TR-59,*HEPS7002,*SK3027,*HN100,*ECG130,*WEP247,*BDY20, *276-2020,*RT-131
2N1072	*GE-14,*HEPS3010
2N1073	*PTC105A,*GE-76,*TR-01,*HEPG6003,*SK3034,*PT150,*ECG121,*WEP232, *276-2006,*RT-127
2N1073A	2N2691,CTP3500,CTP3503,MP2526,*PTC105A,*GE-76,*TR-01,*HEPG6005,*SK3034, *PT250,*ECG121,*WEP232,*276-2006,*RT-127A
2N1073B	MP2527,MP2528,*DS-503,*GE-25,*TR-27,*HEPG6008,*SK3034,*ECG127,*WEP235
2N1074	*HEPS0025
2N1075	*HEPS0025
2N1076	*HEPS0025
2N1077	*HEPS0025
2N1078	2N1328,*GE-54,*TR-01,*SK3009,*ECG121,*WEP232,*RT-127
2N1079	2N1080

To Replace	Substitute This Type
2N1080	2N1079
2N1081	*GE-63,*TR-21,*HEPS0014,*SK3122,*ECG123A,*WEP736,*276-2009,*RT-102
2N1082	2N474,2N706,2N706A,2N706B,2N783,2N784,2N834A,2N959,2N1140,2N1387,2N2205, 2N2368,2N2427,2N2432,2N2719,2N2729,2N3261,2N3544,2N3793,2N3825,2N3843, 2N3843A,2N3854A,2N4292,2N4293,2N5187,2SC52,2SC54,2SC55,2SC63,2SC68, 2SC171,2SC172,2SC321,2SC322,2SC382,2SC387AG,2SC400-R,2SC455,2SC576, 2SC689,2SC717,2SC1126,40405,BSX25,EN918,GET706,GI-3793,KT218,MM1941, MPS706,MPS706A,MPS834,MPS918,MPS6507,MPS6511,MPS6540,MPS6542, MPS6543,MPS6546,MPS6547,MPS6548,MPSH19,MPSH20,MPSH32,MPSH37, MT1038,MT1038A,MT1039,MT1060,PET3001,SE1010,SE3002,SE5025,TIS44,TIS47, TIS84,TIS108,TIS125,ZT203P,ZT403P,ZT2368,*PTC126,*GE-20,*TR-21,*HEPS0014, *SK3122,*ECG123A,*WEP736,*276-2009,*RT-102
2N1086	2N146,2N147,2N558,*GE-6,*TR-08,*HEPG0011,*SK3011,*NR10,*ECG101,*WEP641, *AC187/01,*RT-119
2N1086+	2N145,2N169,2N312,2N557,2N576,2N635,2N1996,*GE-6,*HEPG0011,*SK3011, *ECG101,*WEP641
2N1086A	2N146,2N147,2N558,*GE-6,*TR-08,*HEPG0011,*SK3011,*NR10,*ECG101,*WEP641, *AC187/01,*RT-119
2N1086A+	2N145,2N169,2N312,2N557,2N576,2N635,2N1996,*GE-6,*HEPG0011,*SK3011, *ECG101,*WEP641
2N1087	2N146,2N147,2N558,*GE-6,*TR-08,*HEPG0011,*SK3011,*NR10,*ECG101,*WEP641, *AC187/01,*RT-119
2N1087+	2N145,2N169,2N312,2N557,2N576,2N635,2N1996,*GE-6,*HEPG0011,*SK3011, *ECG101,*WEP641
2N1090	2N357A,2N358A,2N439,2N439A,2N440,2N440A,2N1091,2SC90,SF.T298,*GE-6, *TR-08,*HEPG0011,*SK3011,*NR10,*ECG101,*WEP641,*AC187/01,*276-2001, *RT-119
2N1091	2N440,2N440A,2SC91,*GE-6,*TR-08,*HEPG0011,*SK3011,*NR10,*ECG101, *WEP641,*276-2001,*RT-119
2N1092	*GE-47,*TR-87,*HEPS5014,*SK3024,*SN80,*ECG128,*WEP243,*276-2012,*RT-114
2N1093	2N383,2N598,2N599,2N1307,2N1309,2N1449,2N1892,2N1998,*PTC135,*GE-53, *TR-17,*HEPG0002,*HF12H,*ECG160,*WEP637,*AC188/01,*276-2007
2N1094	*GE-52,*TR-17,*HEPG0002,*ECG160,*WEP637,*276-2004
2N1097	2N59,2N60,2N241A,2N241A+,2N321,2N360,2N362,2N414B,2N414C,2N654,2N654+, 2N1008,2N1281,2N1282,2N1356,2N1374,2N1376,2N1381,2N1382,2N1383,2N1478, 2N1681,2N1706,2N1707,2SA207,2SA407,2SB221,2SB221A,2SB222,2SB222A, 2SB326,SF.T223,*PTC102,*GE-2,*TR-05,*HEPG0005,*SK3003,*AT30H,*ECG102, *WEP631,*AC188/01,*276-2005,*RT-120
2N1098	2N60,2N61,2N188A,2N188A+,2N241A,2N241A+,2N320,2N321,2N363,2N414B, 2N414C,2N654,2N1097,2N1281,2N1282,2N1313,2N1383,2N1478,2N1681,2SA206, 2SA207,2SA406,2SA407,2SB220,2SB220A,2SB221,2SB221A,2SB326,SF.T222, *PTC102,*GE-2,*TR-05,*HEPG0005,*SK3003,*AT30H,*ECG102,*WEP631, *AC188/01,*276-2005,*RT-120
2N1099	*GE-4,*TR-03,*HEPG6006,*SK3012,*PT515,*ECG105,*WEP233,*276-2006
2N1100	*PTC106,*GE-4,*TR-08,*HEPG6006,*SK3012,*PT515,*ECG105,*WEP233,*276-2006
2N1101	2N306,2N576,*PTC108,*GE-59,*HEPG0011,*SK3010,*NA30,*ECG103A,*WEP641B, *AC127,*276-2001,*RT-122
2N1101+	2N306,2N576,2N1101,AC127,*PTC108,*GE-59,*HEPG0011,*SK3010,*ECG103A, *WEP641B
2N1102	2N576A,2N1473,2N1605A,*PTC108,*GE-59,*TR-08,*HEPG0011,*SK3010,*NA30, *ECG103A,*WEP641B,*276-2001,*RT-122
2N1102+	2N576A,2N1102,2N1473,2N1605A,*PTC108,*GE-59,*HEPG0011,*SK3010,*ECG103A, *WEP641B
2N1103	2N480,2N543A,2N708A,2N718,2N840,2N929,2N1151,2N1152,2N1674,2N2387, 2N2432A,2N2522,2N2845,2N3301,2N3693,2N3826,2N3946,2N3973,2N3975,2N4140, 2N4994,2S103,2SC302,2SC335,2SC336,2SC725,2SC815,2SC838,2SC839,2SC896, 2SC938,BFY39-1,BSY93,EN697,GET2221,PET1001,SE1001,TIS22,ZT23,ZT43, *PTC123,*GE-11,*TR-21,*HEPS0014,*SK3124,*ECG123A,*WEP736,*276-2009, *RT-102
2N1104	2N263,2N338,2N543,2N841,2N843,2N916,2N916A,2N2161,2N3859A,2N3903,2N3947, 2N3974,2N3976,2N4227,2N5825,2S104,2SC366G,2SC366G-O,2SC899,2SC933,

To Replace	Substitute This Type
(2N1104)	2SC943,2SC944,2SC965,2SC1007,2SC1189,2SC1320,2SC1453,2SC1688,BFY39-2, EN916,MPS3693,MPS6565,MPS6576-BLUE,MPS6576-GREEN,MPS6576-YEL, TP3705,ZT24,ZT44,*PTC123,*GE-39,*TR-86,*HEPS0014,*SK3124,*ECG123, *WEP53,*276-2013,*RT-100
2N1105	*PTC125,*HEPS3020
2N1106	*PTC144,*HEPS5026
2N1107	*GE-51,*TR-17,*HEPG0008,*SK3005,*JR30X,*ECG126,*WEP635,*AF125,*276-2002
2N1108	*GE-51,*TR-17,*HEPG0008,*SK3008,*JR30X,*ECG126,*WEP635,*AF125,*276-2002
2N1108+	2N499,2N499A,2N501A,2N502B,2N559,2N588A,2N779,2N779A,2N828,2N828A, 2N964,2N968,2N980,2N1280,2N1499B,2N1742,2N1743,2N1744,2N1745,2N1787, 2N1868,2N2048,2N2169,2N2401,2N2487,2N2488,2N2489,2N2672,2N2717,2N2795, 2N2860,2N6365A,2SA56,2SA251,2SA405,2SA411,ASZ21,MM380,*GE-1, *HEPG0008,*SK3008,*ECG126,*WEP635
2N1109	*GE-51,*TR-17,*HEPG0008,*SK3007,*JR30X,*ECG126,*WEP635,*AF125,*276-2002
2N1109+	2N500,2N501,2N741,2N846A,2N960,2N979,2N1195,2N1499A,2N1500,2N1742, 2N1743,2N1744,2N1745,2N1864,2N1868,2N2170,2N2718,2N2796,*HEPG0008, *SK3007,*ECG126,*WEP635
2N1110	*GE-51,*TR-17,*HEPG0008,*SK3008,*JR30X,*ECG126,*WEP635,*AF125,*276-2002
2N1110+	2N499,2N499A,2N500,2N501,2N501A,2N502B,2N588A,2N779,2N779A,2N828, 2N828A,2N964,2N968,2N979,2N980,2N1280,2N1499A,2N1499B,2N1742,2N1743, 2N1744,2N1745,2N1787,2N1868,2N2048,2N2169,2N2170,2N2401,2N2487,2N2488, 2N2489,2N2672,2N2717,2N2795,2N2796,2N2860,2N6365A,2SA56,2SA251,2SA405, 2SA411,ASZ21,MM380,*GE-51,*HEPG0008,*SK3008,*ECG126,*WEP635
2N1111	*GE-51,*TR-17,*HEPG0008,*SK3007,*JR30X,*ECG126,*WEP635,*AF125,*276-2002
2N1114	2N440,2N440A,2N1306,2N1308,ASY29,SF.T298,*PTC108,*GE-8,*TR-08,*HEPG0011, *SK3011,*NR10,*ECG101,*WEP641,*RT-119
2N1115	2N415,2N415A,2N416,2N1305,2N1307,2N1309A,2N1356,2N1681,ASY27,SYL792, *PTC109,*GE-1,*TR-17,*HEPG0002,*SK3005,*HF6M,*ECG160,*WEP637, *AC188/01,*276-2005
2N1115A	*GE-2,*TR-17,*HEPG0005,*SK3005,*ECG160,*WEP637,*276-2005
2N1116	2N2594,*PTC144,*GE-63,*TR-21,*HEPS0014,*SK3122,*ECG123A,*WEP736, *276-2012,*RT-102
2N1117	2N1116,2N2594,*PTC144,*GE-63,*TR-21,*HEPS0014,*SK3122,*ECG123A,*WEP736, *276-2012,*RT-102
2N1118	2N861,2N1118A,2N2162,2N2968,2N2969,2N2970,2N2971,2N4008,MM4052,*GE-65, *TR-20,*HEPS0012,*SK3114,*SP70,*ECG159,*WEP242,*BC177,*276-2022,*RT-115
2N1118A	2N861,2N1118,2N2162,2N2968,2N2969,2N2970,2N2971,2N4008,MM4052,*PTC131, *GE-65,*TR-20,*HEPS0012,*SK3114,*SP70,*ECG159,*WEP242,*BC177,*276-2022, *RT-115
2N1119	2N941,2N942,2N2378,*GE-65,*TR-20,*HEPS0012,*SK3114,*SP70,*ECG159, *WEP242,*BC177,*276-2021,*RT-115
2N1120	2N1165,2N1165A,2N1167,2N1167A,2N2445,MP1555,MP1555A,MP1556,MP1556A, MP2145A,MP2146A,*DS503,*GE-76,*TR-01,*HEPG6005,*SK3009,*PT250,*ECG121, *WEP232,*276-2006,*RT-127
2N1121	2N78,2N78A,2N145,2N165,2N169A,2N440,2N440A,2N446,2N558,2N636,2N797, 2N1091,2N1114,2N1198,2N1217,2N1304,2N1306,2SC90,2SC91,2SC129,ASY28, ASY29,*PTC108,*GE-7,*TR-08,*HEPG0011,*SK3011,*NR10,*ECG101,*WEP641, *AC187/01,*276-2001,*RT-119
2N1121+	2N145,2N167A,2N169,2N169A,2N169A+,2N312,2N440,2N440A,2N558,2N635,2N636, 2N797,2N1091,2N1114,2N1306,2N1996,2SC129,2SC181,ASY28,ASY29,OC140, *GE-7,*TR-08,*HEPG0011,*SK3011,*ECG101,*WEP641
2N1122	2N501,2N741,2N846A,2N960,2N961,2N962,2N963,2N979,2N2400,2N2796,*GE-51, *TR-17,*HEPG0003,*SK3005,*JR30,*ECG160,*WEP637
2N1122A	2N501,2N741,2N846A,2N960,2N979,2N2796,*GE-51,*TR-17,*HEPG0003,*SK3005, *JR30,*ECG160,*WEP637
2N1123	*GE-53,*AT100H
2N1124	*GE-80,*TR-05,*HEPG0005,*SK3004,*AT100H,*ECG102,*WEP631,*276-2005, *RT-120
2N1124+	2N1008A,2N1008B,2N1124,*GE-80,*HEPG0005,*ECG102,*WEP631
2N1125	*GE-80,*TR-05,*HEPG0005,*SK3004,*AT100H,*ECG102,*WEP631,*276-2005, *RT-120

To Replace	Substitute This Type
2N1125+	2N1008A,*GE-80,*HEPG0005,*ECG102,*WEP631
2N1126	*GE-53,*TR-05,*HEPG6011,*SK3004,*ECG102,*WEP631,*AC128,*RT-120
2N1127	*GE-53,*HEPG6011,*SK3004,*AC128,*RT-120
2N1128	2N520A,2N571,2N598,2N654+,2N655,2N1008A,2N1307,2N1309,2N1706,2N1707, 2N1808,2N1892,2N1998,2SB222A,2SB223A,2SB327,SF.T223,*PTC102,*GE-53, *TR-05,*HEPG0005,*SK3004,*AT30H,*ECG102,*WEP631,*AC188/01,*276-2005, *RT-120
2N1128+	2N521A,2N571,2N598,2N599,2N655,2N655+,2N1008A,2N1307,2N1309,2N1808, 2N1892,2N1998,2SB223A,2SB327,2SB431,NKT211,NKT218,SF.T223,SF.T323, *GE-53,*HEPG0005,*SK3004,*ECG102,*WEP631
2N1129	2N521A,2N522A,2N599,2N655+,2N1309,2N1316,2N1998,2N1999,2SB223A,MA1703, MA1704,*PTC135,*GE-53,*TR-05,*HEPG0005,*SK3004,*AT30H,*276-2005,*RT-120
2N1129+	2N521A,2N522A,2N599,2N655+,2N1316,2N1999,MA1703,MA1704,NKT218,*GE-53, *HEPG0005,*SK3004
2N1130	2N382,2N383,2N461,2N527,2N571,2N598,2N599,2N652,2N655,2N655+,2N1008A, 2N1189,2N1307,2N1309,2N1449,2N1892,2N1997,2N1998,2SB223A,2SB327,MA888, SF.T223,*PTC135,*GE-80,*TR-85,*HEPG0005,*SK3004,*AT30H,*ECG102A, *WEP250,*276-2005,*RT-121
2N1130+	2N383,2N527,2N598,2N599,2N652,2N655,2N655+,2N1185,2N1189,2N1309,2N1449, 2N1892,2N1997,2N1998,2N2171,2SB223A,2SB327,2SB370A,2SB431,MA888, NKT218,*GE-80,*HEPG0005,*SK3004,*ECG102A,*WEP250
2N1131	2N1131A,2N3245,2N3468,2N3762A,2N3763,2N5022,40538,BC143,*PTC141,*GE-21, *TR-20,*HEPS0012,*SK3114,*SP70,*ECG159,*WEP242,*BC327,*276-2021,*RT-115
2N1131/51	*GE-21,*BC327
2N1131A	2N3763,92PE77B,92PE77C,40538,*PTC127,*GE-21,*TR-20,*HEPS0012,*SK3114, *SP70,*ECG159,*WEP242,*BC327,*276-2021,*RT-115
2N1131A/51	*GE-21,*BC327
2N1132	2N1132A,2N1132B,2N2800,2N2904,2N2904A,2N3081,2N3245,2N3468,2N3762A, 2N3763,2N4030,2N5865,2SA498-R,2SA503-O,2SA503-Y,2SA546,2SA571,2SA717, BC161-10,BC161-6,MM3726,*PTC141,*GE-21,*TR-20,*HEPS0012,*SK3114,*SP70, *ECG159,*WEP242,*BC327,*276-2021,*RT-115
2N1132/51	*GE-21,*BC327
2N1132A	2N2904,2N2904A,2N3763,2N4030,2N4031,2N4404,2N5865,2SA503-O,2SA503-Y, 2SA512-O,2SA512-R,2SA516,2SA546,2SA560,2SA571,2SA717,BC161-10,BC161-6, *PTC127,*GE-21,*TR-20,*HEPS0012,*SK3114,*SP70,*ECG159,*WEP242,*BC327, *276-2021,*RT-115
2N1132A/51	*GE-21,*BC327
2N1132B	2N3081,2N4036,*PTC127,*GE-67,*TR-20,*HEPS0012,*SK3114,*ECG159,*WEP242, *276-2021,*RT-115
2N1132B/51	*GE-67
2N1135	*TR-20,*HEPS0012,*SK3114,*ECG159,*WEP242,*276-2021,*RT-115
2N1135A	*TR-20,*HEPS0012,*SK3114,*ECG159,*WEP242,*276-2021,*RT-115
2N1136	2N457A,2N458A,2N1137,2SB339,ASZ16,ASZ18,KR6501,*PTC105A,*GE-76,*TR-01, *HEPG6005,*SK3009,*PT50,*ECG121,*WEP232,*276-2006,*RT-127
2N1136A	2N1022,2N1136B,2N1137A,2N1137B,KR6503,*PTC105A,*GE-76,*TR-01,*HEPG6005, *SK3009,*PT50,*ECG121,*WEP232,*276-2006,*RT-127
2N1136B	2N1022,2N1137B,KR6503,*PTC105A,*GE-16,*TR-01,*HEPG6018,*SK3009,*PT50, *ECG127,*WEP235,*276-2006
2N1137	2N1138,ASZ16,KR6501,*PTC105A,*GE-16,*TR-01,*HEPG6005,*SK3009,*PT50, *ECG121,*WEP232,*276-2006,*RT-127
2N1137A	2N1137B,2N1138A,2N1138B,KR6503,*PTC105A,*GE-16,*TR-01,*HEPG6005,*SK3009, *PT50,*ECG121,*WEP232,*276-2006,*RT-127
2N1137B	2N1138B,KR6503,*PTC105A,*GE-16,*TR-01,*HEPG6018,*SK3009,*PT50,*ECG127, *WEP235,*276-2006
2N1138	*PTC138,*GE-16,*TR-01,*HEPG6005,*SK3009,*ECG121,*WEP232,*276-2006, *RT-127
2N1138A	2N1138B,*DS503,*GE-16,*TR-01,*HEPG6005,*SK3009,*PT50,*ECG121,*WEP232, *276-2006,*RT-127
2N1138B	*DS503,*GE-16,*TR-92,*HEPG6005,*SK3009,*PT50,*276-2006
2N1139	2N702,2N2320,2SC53,2SC87,TN81,*PTC123,*GE-17,*TR-21,*HEPS0014,*SK3122, *ECG123A,*WEP736,*BC337,*276-2009,*RT-102

To Replace	Substitute This Type
2N1140	2N758,2N758A,2N759,2N759A,2N760B,2N1987,2N3122,2N3295,2N5188,2SC30, 2SC31,2SC46,2SC47,2SC52,2SC116,BSY91,SE8001,*PTC123,*GE-20,*TR-25, *HEPS0014,*BC337,*276-2009
2N1141	*TR-17,*HEPG0002,*ECG160,*WEP637,*276-2005
2N1141A	*TR-17,*HEPG0002,*ECG160,*WEP637,*276-2005
2N1142	2N1141,*TR-17,*HEPG0009,*ECG160,*WEP637
2N1142A	*TR-17,*HEPG0002,*ECG160,*WEP637,*276-2005
2N1143	2N1141,2N1142,2N2929,*TR-17,*HEPG0002,*ECG160,*WEP637,*276-2005
2N1143A	*TR-17,*HEPG0002,*ECG160,*WEP637,*276-2005
2N1144	2N59,2N60,2N241A,2N241A+,2N303,2N321,2N323,2N324,2N360,2N362,2N414, 2N414A,2N414B,2N414C,2N415,2N415A,2N416,2N450,2N520A,2N569,2N571,2N654, 2N654+,2N711A,2N711B,2N829,2N838,2N964A,2N972,2N985,2N1008,2N1097, 2N1115,2N1128,2N1144+,2N1171,2N1273,2N1274,2N1281,2N1282,2N1305,2N1307, 2N1344,2N1352,2N1356,2N1370,2N1374,2N1376,2N1381,2N1382,2N1383, 2N1478,2N1681,2N1706,2N1707,2N2402,2N3371,2N3883,2N6365,2SA204,2SA207, 2SA407,2SB187,2SB188,2SB221,2SB221A,2SB222,2SB222A,2SB326,2SB328, 2SB329,40269,ASY27,SF.T223,SF.T228,SF.T229,SF.T237,SYL792,*GE-52,*TR-17, *HEPG0002,*SK3003,*AT20H,*ECG160,*WEP637,*AC126,*276-2005
2N1144+	2N59,2N60,2N241A,2N241A+,2N321,2N360,2N362,2N654,2N654+,2N1008,2N1282, 2N1356,2N1374,2N1376,2N1381,2N1382,2N1383,2N1681,2N1705,2N1706,2N1707, 2SB187,2SB188,2SB221,2SB221A,2SB222,2SB222A,2SB327,SF.T223,*GE-52, *TR-17,*HEPG0002,*SK3003,*ECG160,*WEP637
2N1145	2N59,2N60,2N241A,2N241A+,2N303,2N321,2N323,2N324,2N360,2N362,2N414, 2N414A,2N414B,2N414C,2N415,2N415A,2N416,2N450,2N520A,2N569,2N571,2N654, 2N654+,2N711A,2N711B,2N829,2N838,2N964A,2N972,2N985,2N1008,2N1097, 2N1115,2N1128,2N1144+,2N1171,2N1273,2N1274,2N1281,2N1282,2N1305,2N1307, 2N1344,2N1352,2N1356,2N1370,2N1374,2N1376,2N1381,2N1382,2N1383, 2N1478,2N1681,2N1706,2N1707,2N2402,2N3371,2N3883,2N6365,2SA204,2SA207, 2SA407,2SB187,2SB188,2SB221,2SB221A,2SB222,2SB222A,2SB326,2SB328, 2SB329,40269,ASY27,SF.T223,SF.T228,SF.T229,SF.T237,SYL792,*GE-52,*TR-17, *HEPG0002,*SK3003,*AT20H,*ECG160,*WEP637,*AC126,*276-2005
2N1145+	2N59,2N60,2N241A,2N241A+,2N321,2N360,2N414B,2N414C,2N654,2N654+, 2N1008,2N1097,2N1144+,2N1281,2N1282,2N1374,2N1382,2N1383,2N1478,2N1681, 2N1706,2N1707,2SA207,2SA407,2SB221,2SB221A,2SB326,*PTC102,*GE-52, *TR-17,*HEPG0002,*SK3003,*ECG160,*WEP637
2N1146	2N1146A,*PTC901,*GE-76,*TR-01,*HEPG6003,*SK3009,*PT150,*ECG104,*WEP230, *276-2006,*RT-124
2N1146A	2N1146B,*PTC901,*GE-76,*TR-01,*HEPG6005,*SK3009,*ECG104,*276-2006,*RT-124
2N1146B	2N1146C,*PTC901,*GE-76,*TR-01,*HEPG6005,*SK3009,*ECG104,*WEP230, *276-2006,*RT-124
2N1146C	*DS503,*GE-3,*TR-01,*HEPG6005,*SK3009,*ECG104,*WEP230,*276-2006,*RT-124
2N1147	2N1147A,*PTC901,*GE-76,*TR-01,*HEPG6003,*SK3009,*PT150,*ECG104,*WEP230, *276-2006,*RT-124
2N1147A	2N1147B,*PTC901,*GE-76,*TR-01,*HEPG6005,*SK3009,*PT150,*ECG104,*WEP230, *276-2006,*RT-124
2N1147B	2N1147C,*PTC901,*GE-76,*TR-01,*HEPG6005,*SK3009,*ECG104,*WEP230, *276-2006,*RT-124
2N1147C	*DS503,*GE-3,*TR-01,*HEPG6005,*SK3009,*ECG104,*WEP230,*276-2006,*RT-124
2N1149	2N472,2N839,2N2521,2SC478,2SC945,TIS126,*PTC123,*GE-39,*TR-86,*HEPS0014, *SK3124,*SN80,*ECG123,*WEP53,*BF167,*276-2013,*RT-100
2N1150	2N475,2N480A,2N717,2N842,2N844,2N1151,2N2432A,2N2522,2S102,2SC302, BF224J,BF225J,MPS6544,MPS6545,ZT22,ZT42,*PTC123,*GE-39,*TR-86, *HEPS0014,*SK3124,*SN80,*ECG123,*WEP53,*BF167,*276-2013,*RT-100
2N1151	2N480,2N708A,2N718,2N840,2N929,2N1152,2N1674,2N2387,2N2432A,2N2522, 2N2845,2N3301,2N3693,2N3826,2N3946,2N3973,2N3975,2N4140,2N4994,2S103, 2SC283,2SC302,2SC335,2SC336,2SC725,2SC815,2SC838,2SC839,2SC896, 2SC938,BFY39-1,BSY93,EN697,GET2221,PET1001,SE1001,TIS22,ZT23,ZT43, *PTC123,*GE-39,*TR-86,*HEPS0014,*SK3124,*SN80,*ECG123,*WEP53,*BF167, *276-2013,*RT-100
2N1152	2N480,2N708A,2N718,2N843,2N929,2N1674,2N2387,2N2845,2N2847,2N3301, 2N3693,2N3826,2N3858A,2N3862,2N3903,2N3946,2N3973,2N3974,2N3975,2N3976,

To Replace	Substitute This Type
(2N1152)	2N4140,2N4966,2N4994,2N5027,2N5824,2S103,2SC283,2SC318,2SC335,2SC336, 2SC620,2SC725,2SC815,2SC838,2SC839,2SC896,2SC938,2SC1189,2SC1320, 2SC1688,BFY39-1,BSY93,EN697,GET2221,MM3903,MPS3693,MPS6565,MPS6576-YEL,PET1001,SE1001,TIS22,TIS87,TP3705,ZT23,ZT43,*PTC123,*GE-39,*TR-86, *HEPS0014,*SK3124,*SN80,*ECG123,*WEP53,*BF167,*276-2013,*RT-100
2N1153	2N909,2N930,2N2388,2N2484,2N2484A,2N2586,2N3302,2N3694,2N3827,2N3904, 2N4287,2N4967,2N4995,2N5826,2SC366G-Y,2SC538A,2SC587,2SC587A,2SC711A, 2SC1000-GR,2SC1000G-GR,2SC1175,2SC1364,2SC1380-GR,2SC1380A-GR, 2SC1632,2SC1634,2SC1681-GR,2SC1685,2SC1850,2SC1992,A157B,BC107, BC107B,BC123,BC167B,BC171B,BC174B,BC182B,BC182L,BC190B,BC237B, BC413B,BC414B,BCY59B,BFY39-3,BSX79,EN2484,GI-3566,MPS3694,MPS6566, MPS6575,MPS6576-SIL,PBC107A,PBC107B,PET1002,SE1002,TIS23,TIS24,TP3704, *PTC123,*GE-39,*TR-86,*HEPS0014,*SK3124,*SN80,*ECG123,*WEP53,*BC107B, *276-2013,*RT-100
2N1154	2N497,2N3512,7A30,PT2540,SE8001,*PTC125,*GE-47,*TR-63,*HEPS0005,*BF338, *276-2012
2N1155	2N498,MM3000,*PTC125,*GE-18,*TR-25,*HEPS0005,*BF338,*276-2012
2N1156	MM3001,*PTC125,*GE-18,*HEPS5025,*BF338,*276-2012
2N1157	2N1157A,*GE-4,*TR-03,*SK3012,*PT501,*ECG105,*WEP233
2N1157A	*GE-4,*TR-03,*SK3012,*PT501,*ECG105,*WEP233
2N1158	2N1158A,2N2099,*PTC145,*TR-17,*HEPG0002,*ECG160,*WEP637
2N1158A	2N741A,2N2099,*PTC145,*TR-17,*HEPG0002,*ECG160
2N1159	*PTC105A,*GE-16,*TR-01,*HEPG6005,*SK3009,*PT50,*ECG121,*WEP232,*276-2006, *RT-127
2N1160	2N3615,2N3616,*PTC105A,*GE-76,*TR-01,*HEPG6005,*SK3009,*PT50,*ECG121, *WEP232,*276-2006,*RT-127
2N1162	2N1162A,*PTC901,*GE-76,*TR-01,*HEPG6001,*SK3009,*PT250,*ECG104,*WEP230, *RT-124
2N1162A	*PTC901,*GE-76,*TR-01,*HEPG6001,*SK3009,*ECG104,*WEP230,*RT-124
2N1163	2N511A,2N512A,2N513A,2N1163A,CRT1544,*PTC901,*GE-76,*TR-01,*HEPG6014, *SK3009,*PT250,*ECG121,*WEP232,*RT-127
2N1163A	2N511A,2N512A,2N513A,CRT1544,*PTC901,*GE-76,*TR-01,*HEPG6014,*SK3009, *ECG121,*WEP232,*RT-127
2N1164	2N1164A,2N1166,2N1166A,*PTC901,*GE-76,*TR-01,*HEPG6001,*SK3009,*PT250, *ECG104,*WEP230,*RT-124
2N1164A	2N1166A,*PTC901,*GE-76,*TR-01,*HEPG6001,*SK3009,*ECG104,*WEP230,*RT-124
2N1165	2N1165A,2N1167,2N1167A,*PTC901,*GE-76,*TR-01,*HEPG6009,*SK3009,*PT250, *ECG104,*WEP230,*RT-124
2N1165A	2N1167,2N1167A,*PTC901,*GE-76,*TR-01,*HEPG6009,*SK3009,*ECG121,*WEP232, *RT-127
2N1166	2N1166A,*DS503,*GE-3,*TR-01,*HEPG6001,*SK3009,*PT250,*ECG104,*WEP230, *RT-124
2N1166A	*DS503,*GE-16,*TR-01,*HEPG6001,*SK3009,*ECG121,*WEP232,*RT-127
2N1167	*DS503,*GE-25,*TR-01,*HEPG6009,*SK3009,*PT250
2N1167A	2N1167,*DS503,*GE-25,*TR-01,*HEPG6009,*SK3009
2N1168	*PTC105A,*GE-16,*TR-01,*HEPG6005,*SK3009,*PT50,*ECG121,*WEP232,*276-2006, *RT-127
2N1169	*GE-5,*TR-08,*NR10,*ECG103,*WEP641A,*AC187/01,*RT-122
2N1170	2N587,*GE-5,*TR-08,*ECG103,*WEP641A,*RT-122
2N1171	2N1349,*PTC102,*GE-1,*TR-05,*HEPG0005,*SK3005,*HF12M,*ECG100,*WEP254, *276-2004,*RT-118
2N1172	*DS503,*GE-3,*TR-01,*SK3009,*PT50,*ECG104,*WEP230,*RT-124
2N1173	*GE-8,*TR-08,*HEPG0011,*SK3010,*ECG103A,*WEP641B,*276-2001,*RT-122
2N1174	2N2374,*GE-2,*TR-05,*HEPG0005,*SK3010,*ECG102,*WEP631,*276-2005,*RT-120
2N1175	2N382,2N422A,2N461,2N526,2N526A,2N527A,2N597,2N651,2N652A,2N1008A, 2N1175A,2N1192,2N1349,2N1415,2N1448,2N1452,2SB57A,2SB226,2SB227, *PTC135,*GE-53,*TR-05,*HEPG0005,*SK3004,*HF6H,*ECG102,*WEP631, *276-2005,*RT-120
2N1175A	2N382,2N383,2N461,2N527,2N527A,2N598,2N652,2N1008A,2N1189,2N1193,2N1448, 2N1449,2N1452,2N1997,2N1998,2SB227,*PTC135,*GE-53,*TR-05,*HEPG0005, *SK3004,*HF6H,*ECG102,*WEP631,*276-2005,*RT-120

TRANSISTOR SUBSTITUTES

To Replace	Substitute This Type
2N1176	*PTC135,*GE-53,*TR-05,*HEPG0005,*SK3005,*AT100H,*ECG100,*WEP254, *AC188/01,*276-2004,*RT-118
2N1176A	2N524,2N524A,2N525,2N525A,2N650,2N650A,2N1176B,2N1186,2N1446,2N1924, 2SB224,MA881,MA886,*PTC135,*GE-80,*TR-05,*HEPG0005,*SK3005,*AT100H, *ECG100,*WEP254,*276-2004,*RT-118
2N1176B	2N1186,2N1614,2N1924,MA881,*TR-05,*HEPG0005,*SK3005,*AT100H,*ECG100, *WEP254,*276-2004,*RT-118
2N1177	*PTC107,*GE-51,*TR-17,*HEPG0009,*SK3007,*JR100,*ECG160,*WEP637,*276-2005
2N1178	*PTC107,*GE-51,*TR-17,*HEPG0009,*SK3007,*JR100,*ECG160,*WEP637,*276-2005
2N1179	2N3783,2N3784,*PTC107,*GE-51,*TR-17,*HEPG0009,*SK3007,*JR100,*ECG160, *WEP637,*276-2005
2N1180	2N1396,2N1397,2N2588,2N3783,2N3784,2SA246,AF202S,AFY39,*DS-56,*GE-51, *TR-17,*HEPG0009,*SK3006,*JR100,*ECG126,*WEP635,*276-2005
2N1182	2N457B,*PTC901,*GE-76,*TR-01,*HEPG6005,*SK3014,*ECG121,*WEP232, *276-2006,*RT-127
2N1183	2N1183A,*DS503,*TR-82,*HEPG6011,*SK3123,*AT100H,*ECG176,*WEP238, *276-2006,*RT-127
2N1183A	2N1183B,*DS503,*TR-82,*HEPG6011,*SK3123,*AT100H,*ECG176,*WEP238, *276-2006,*RT-127
2N1183B	*DS503,*TR-82,*HEPG6012,*AT100H,*ECG176,*WEP238,*RT-127
2N1184	2N1184A,*DS503,*TR-82,*HEPG6011,*SK3123,*AT100H,*ECG176,*WEP238, *276-2006,*RT-127
2N1184A	2N1184B,*DS503,*TR-82,*HEPG6011,*AT100H,*ECG176,*WEP238,*276-2006, *RT-127
2N1184B	*DS503,*TR-82,*HEPG6012,*AT100H,*ECG176,*WEP238,*RT-127
2N1185	2N3427,2N3428,*GE-1,*TR-05,*HEPG0005,*SK3005,*HF3H,*ECG100,*WEP254, *276-2005,*RT-118
2N1186	*GE-1,*TR-05,*HEPG0005,*SK3005,*HF3H,*ECG100,*WEP254,*276-2004,*RT-118
2N1187	2N1188,*GE-1,*TR-05,*HEPG0005,*SK3005,*HF3H,*ECG100,*WEP254,*276-2005, *RT-118
2N1188	*GE-1,*TR-05,*HEPG0005,*SK3005,*HF3H,*ECG100,*WEP254,*276-2005,*RT-118
2N1189	2N1185,2N2000,*GE-53,*TR-05,*HEPG0005,*SK3004,*ECG102,*WEP631,*276-2005, *RT-120
2N1190	2N3428,*GE-53,*TR-05,*HEPG0005,*SK3004,*ECG102,*WEP631,*276-2005,*RT-120
2N1191	2N525A,2N650,2N1348,2N1350,2N1351,SF.T243,*GE-2,*TR-05,*HEPG0005,*SK3004, *AT30H,*ECG102,*WEP631,*276-2007,*RT-120
2N1192	2N527A,2N597,2N651,2N652,2N1349,2N1997,*GE-2,*TR-05,*HEPG0005,*SK3004, *AT20M,*ECG102,*WEP631,*RT-120
2N1193	2N3427,*GE-53,*TR-05,*HEPG0007,*SK3004,*AT30H,*ECG102,*WEP631,*RT-120
2N1194	MA1702,*GE-2,*TR-05,*HEPG0007,*SK3004,*AT30H,*ECG102,*WEP631,*RT-120
2N1195	*PTC145,*GE-51,*TR-17,*HEPG0002,*JR100,*ECG160,*WEP637,*276-2005
2N1196	2N1197,*PTC127,*GE-82,*TR-20,*HEPS5023,*ECG159,*WEP242,*276-2021,*RT-115
2N1197	2N1196,*PTC127,*GE-82,*TR-88,*HEPS5023,*SK3025,*ECG129,*WEP242,*276-2021, *RT-115
2N1198	2N357A,2N358A,2N377A,2N440,2N440A,2N1000,2N1090,2N1091,2N1114,2N1304, 2N1306,2SC90,2SC129,ASY28,ASY29,SF.T298,*GE-5,*TR-08,*HEPG0011,*SK3011, *NR10,*ECG103,*WEP641A,*RT-122
2N1199	2N835,2N958,2N1472,ZT202P,ZT402P,*PTC123,*GE-212,*TR-21,*HEPS0025, *SK3122,*NR10,*ECG123A,*WEP736,*BC107B,*276-2016,*RT-102
2N1199A	*PTC132,*TR-21,*HEPS0011,*SK3122,*ECG123A,*WEP736,*276-2009,*RT-102
2N1200	2N1199,*PTC123,*GE-11,*TR-21,*HEPS0014,*SK3124,*ECG123A,*WEP736, *276-2009,*RT-102
2N1201	2N1199,*PTC123,*GE-11,*TR-86,*HEPS0014,*SK3124,*ECG123,*WEP53,*276-2009, *RT-100
2N1202	2N1263,*GE-16,*TR-03,*SK3012,*PT40,*ECG105,*WEP233
2N1203	*GE-4,*TR-03,*SK3012,*ECG105,*WEP233
2N1204	2N1204A,*GE-51,*TR-17,*HEPG0005,*SK3006,*ECG126,*WEP635,*276-2004
2N1204A	*TR-17,*HEPG0005,*SK3006,*ECG126,*WEP635,*276-2004
2N1206	BFX55,HST4453,HST9005,*PTC144,*GE-63,*TR-25,*HEPS5014,*BF338,*276-2008
2N1207	2N2435,*GE-27,*TR-78,*HEPS5026,*SK3045,*ECG154,*WEP712,*BF338,*276-2008, *RT-110

To Replace	Substitute This Type
2N1208	2N1212,2N3487,ST400,*HEPS5004
2N1209	ST400,ST401,*HEPS5004
2N1210	2N1211,*HEPS5004
2N1211	*HEPS5004
2N1212	2N1208,2N3487,ST400,*HEPS5004
2N1213	*PTC109,*GE-2,*TR-17,*HEPG0002,*SK3006,*ECG126,*WEP635,*276-2005
2N1214	*PTC109,*GE-2,*TR-17,*HEPG0002,*SK3006,*ECG126,*WEP635,*276-2005
2N1215	*PTC109,*GE-2,*TR-17,*HEPG0002,*SK3006,*ECG126,*WEP635,*276-2005
2N1216	*PTC109,*GE-2,*TR-17,*HEPG0002,*SK3006,*ECG126,*WEP635,*276-2005
2N1217	2N440,2N440A,2N636,2N797,2N1091,2N1114,2N1198,2N1304,2N1306,2SC90, 2SC129,ASY28,ASY29,*GE-5,*TR-08,*HEPG0011,*SK3010,*NR10,*ECG103, *WEP641A,*276-2001,*RT-122
2N1217+	2N167,2N169A+,2N440,2N440A,2N635,2N636,2N1090,2N1091,2N1114,2N1304, 2N1306,2N1996,2SC90,2SC129,2SC181,ASY28,ASY29,*PTC108,*GE-5,*TR-08, *ECG103,*WEP641A
2N1218	2N1294,*TR-01,*HEPS5014,*ECG121,*WEP232,*276-2018,*RT-127
2N1219	2N4008,MM4052,*GE-65,*TR-20,*HEPS0012,*SK3114,*SP70,*ECG159,*WEP242, *BC177,*276-2022,*RT-115
2N1220	2N327B,2N1230,MM4052,*GE-65,*TR-20,*HEPS0012,*SK3114,*SP70,*ECG159, *WEP242,*BC177,*276-2022,*RT-115
2N1221	2N1219,2N4008,MM4052,*GE-65,*TR-20,*HEPS0012,*SK3114,*SP70,*ECG159, *WEP242,*BC177,*276-2022,*RT-115
2N1222	2N327B,2N1220,2N1230,MM4052,*GE-65,*TR-20,*HEPS0012,*SK3114,*SP70, *ECG159,*WEP242,*BC177,*276-2022,*RT-115
2N1223	2N327B,*GE-65,*TR-20,*HEPS0012,*SK3114,*SP70,*ECG159,*WEP242,*BC177, *276-2022,*RT-115
2N1224	2N1066,2N1225,2N1226,2N1395,2N1396,2N1397,2N2588,*PTC107,*GE-51,*TR-17, *HEPG0009,*SK3007,*JR100,*ECG126,*WEP635,*276-2005
2N1225	2N1066,2N1396,2N1397,2N2588,*PTC107,*GE-51,*TR-17,*HEPG0008,*SK3006, *JR100,*ECG126,*WEP635
2N1226	*DS25,*GE-51,*TR-17,*HEPG0008,*SK3008,*JR100,*ECG126
2N1227	2N669,*PTC114,*GE-25,*TR-01,*HEPG6003,*SK3009,*PT40,*ECG104, /EP230, *276-2006,*RT-124
2N1228	2N6567,MM4052,*GE-22,*TR-88,*HEPS0012,*SK3025,*SP70,*ECG129,*WEP242, *BC327,*276-2023,*RT-115
2N1229	2N2945A,2N4007,TW135,*GE-22,*TR-88,*HEPS0012,*SK3025,*SP70,*ECG129, *WEP242,*BC327,*276-2023,*RT-115
2N1230	2N4008,*GE-82,*TR-88,*HEPS0012,*SK3025,*SP70,*ECG129,*WEP242,*BC327, *276-2023,*RT-115
2N1231	2N2946A,*GE-82,*TR-88,*HEPS0012,*SK3025,*SP70,*ECG129,*WEP242,*BC327, *276-2023,*RT-115
2N1232	*PTC103,*GE-82,*TR-88,*HEPS0012,*SK3025,*SP70,*ECG129,*WEP242,*276-2021, *RT-115
2N1233	2N4982,*GE-82,*TR-88,*HEPS0012,*SK3025,*SP70,*ECG129,*WEP242,*276-2021, *RT-115
2N1234	*TR-88,*SK3025,*ECG129,*WEP242,*RT-115
2N1235	2N1260
2N1238	HA7522,SHA7522,*GE-67,*TR-20,*HEPS0032,*SP70,*ECG106,*WEP52,*276-2025, *RT-126
2N1239	HA7526,SHA7526,*GE-67,*TR-20,*HEPS0032,*SP70,*ECG106,*WEP52,*276-2025, *RT-126
2N1240	HA7523,SHA7523,*GE-67,*TR-20,*HEPS0032,*SP70,*ECG106,*WEP52,*276-2025, *RT-126
2N1241	HA7527,SHA7527,*GE-67,*TR-20,*HEPS0032,*SP70,*ECG106,*WEP52,*276-2025, *RT-126
2N1242	HA7524,SHA7524,*GE-67,*TR-28,*HEPS5022
2N1243	HA7528,SHA7528,*GE-67,*TR-28,*HEPS5022
2N1244	HA7525,SHA7525
2N1245	2SB137,2SB149-N,*PTC138,*GE-16,*TR-01,*HEPG6003,*SK3009,*ECG104,*WEP230, *276-2006,*RT-124
2N1246	2N4241,*PTC138,*GE-16,*TR-01,*HEPG6003,*SK3009,*ECG104,*WEP230,*276-2006,

To Replace	Substitute This Type
(2N1246)	*RT-124
2N1247	2N473,2N3662,GI-3605,GI-3606,GI-3607,PET8101,*PTC126,*GE-214,*TR-86, *HEPS0014,*SK3124,*SN80,*ECG123,*WEP53,*BF183,*276-2011,*RT-100
2N1248	2N470,2N1586,2SC385A,*PTC126,*GE-214,*TR-86,*HEPS0014,*SK3124,*NR10, *ECG123,*WEP53,*BF183,*276-2011,*RT-100
2N1249	2N473,2N1247,2N1417,2N1589,2N3662,2SC386A,2SC812,GI-3605,GI-3606,GI-3607, PET8101,*PTC126,*GE-214,*TR-86,*HEPS0014,*SK3124,*SN80,*ECG123,*WEP53, *BF183,*276-2011,*RT-100
2N1250	AMF112,AMF113,*HEPS5004
2N1251	2N635A,2N636A,*DS66,*GE-5,*TR-08,*HEPG0011,*SK3010,*NA20,*ECG103A, *WEP641B,*AC127,*RT-122
2N1251+	2N388,2N446A,2N447A,2N634A,2N635A,2N1251,2N1308,2N1624,2N1993,2SD23, 2SD167,2SD170,2SD170A,*HEPG0011,*ECG103A,*WEP641B
2N1252	*PTC144,*GE-81,*TR-87,*HEPS0014,*SK3024,*ECG128,*WEP243,*BC337,*276-2012, *RT-114
2N1252A	2N707A,2N3512,92PE37B,92PE37C,BSX59,BSX61,*PTC144,*HEPS3011
2N1253	MM1803,*PTC144,*GE-81,*TR-87,*HEPS0014,*SK3024,*ECG128,*WEP243,*BC337, *276-2008,*RT-114
2N1253A	2N2194,2N2194A,2N2194B,2N3110,2N3253,2N3444,2N3735,2N3830,2N3831,2N5188, 2N5189,BC142,ZT94,*PTC144,*GE-63,*TR-25,*HEPS3011
2N1254	2N721,2N722,2N722A,2N978,2N1255,2N1256,2N1257,2N1259,2N1991,2N2393, 2N2394,2N3341,2SA558,A5T4248,EN1132,TIS137,*PTC127,*GE-65,*TR-88, *HEPS0012,*SK3025,*SP70,*ECG129,*WEP242,*BC177,*276-2022,*RT-115
2N1255	2N722,2N722A,2N1257,2N2394,2N3250,2N3341,2N3581,2N3703,2N3829,2N3905, 2N5041,2N5042,2N5382,2N5448,2N6067,2SA467,2SA499-O,2SA500-O,2SA522A, 2SA523,2SA532,2SA558,2SA838,A5T3905,A5T5448,A8T3703,EN1132,FT5041, GI-3703,MM3905,MPS3703,TIS37,TN-3905,TP3703,*PTC127,*GE-65,*TR-88, *HEPS0012,*SK3025,*SP70,*ECG129,*WEP242,*BC177,*276-2022,*RT-115
2N1256	2N721,2N722,2N722A,2N1257,2N1259,2N2393,2N2394,2N2603,2SA558,2SA594-R, 2SA661-R,A5T4248,EN1132,*PTC127,*GE-65,*TR-88,*HEPS0012,*SK3025,*SP70, *ECG129,*WEP242,*BC177,*276-2022,*RT-115
2N1257	2N722,2N722A,2N2394,2N2603,2N3250,2N3250A,2N3581,2N3703,2N3905,2N4142, 2N5041,2N5042,2N5382,2N5448,2N6067,2SA467,2SA499-O,2SA522A,2SA523, 2SA523A,2SA532,2SA539,2SA544,2SA552,2SA558,2SA594-R,2SA594N,2SA661-R, A5T3905,A5T5448,A8T3703,BC325,EN1132,FT5041,GET2904,GET2905,GI-3703, MM3905,MPS3703,TN-3905,TP3703,TZ551,*PTC127,*GE-65,*TR-88,*HEPS0012, *SK3025,*SP70,*ECG129,*WEP242,*BC177,*276-2022,*RT-115
2N1258	2N3250,2N3581,2N3702,2N3703,2N3905,2N4034,2N4060,2N5382,2N5447,2N5448, 2N6067,2SA467,2SA467G-O,2SA467G-Y,2SA499-O,2SA499-Y,2SA500-O,2SA500-Y, 2SA532,2SA561-O,2SA561-Y,2SA562-O,2SA562-Y,2SA642,2SA659,2SA710, 2SA711,2SA838,A5T3905,A5T5447,A5T5448,A8T3702,A8T3703,A177,BC177, BC177A,BC202,BC203,BC213L,BC251A,BC257A,BC261A,BC307A,BC308A, BC415-6,BC415A,BC416-6,BC416A,BC557A,BC558A,BC559A,BC560A,BCY78A, BCY79A,FT3645,GI-3702,MM3905,MM3906,MPS3702,MPS3703,TIS37,TN-3905, TP3644,TP3702,TP3703,*PTC127,*GE-65,*TR-88,*HEPS0012,*SK3025,*SP70, *ECG129,*WEP242,*BC177,*276-2022,*RT-115
2N1259	2N722,2N722A,2N2394,2N2603,2N3250,2N3250A,2N3581,2N3703,2N4142,2N5448, 2N6067,2SA499-O,2SA522A,2SA523A,2SA532,2SA539,2SA544,2SA552,2SA561-O, 2SA594-O,2SA594-R,2SA594N,2SA661-O,2SA661-R,A5T4249,A5T5448,A8T3703, BC177,BC325,EN1132,GET2904,GET2905,GI-3703,MPS3703,TP3703,TZ551, *PTC127,*GE-65,*TR-88,*HEPS0012,*SK3025,*SP70,*ECG129,*WEP242,*BC177, *276-2022,*RT-115
2N1260	2N1235
2N1261	2N1262,*GE-4,*TR-03,*SK3012,*PT40,*ECG105,*WEP233
2N1262	2N1202,2N1263,*GE-4,*TR-03,*SK3012,*PT40,*ECG105,*WEP233
2N1263	2N1202,*GE-4,*TR-03,*SK3012,*PT40,*ECG105,*WEP233
2N1264	2N700A,2N990,2N991,2N993,2N2654,2N2671,2N3127,2N3279,2N3280,2N3281, 2N3282,2N3283,2N3284,2SA341,2SA342,AF239,AF240,AFY12,AFY37,AFY42, GM290A,GM378A,MM1139,*GE-51,*TR-05,*HEPG0005,*SK3006,*HF3M,*ECG100, *WEP254,*AF125,*RT-118
2N1265	2N59,2N60,2N123,2N132A,2N241,2N241A,2N270,2N303,2N321,2N323,2N324,2N362,

To Replace	Substitute This Type
(2N1265)	2N450,2N502,2N520A,2N521,2N582,2N584,2N609,2N610,2N711,2N711A,2N711B, 2N794,2N795,2N796,2N829,2N964A,2N972,2N973,2N974,2N982,2N983,2N984, 2N985,2N1097,2N1115,2N1128,2N1265/5,2N1273,2N1274,2N1281,2N1282,2N1344, 2N1345,2N1370,2N1372,2N1374,2N1376,2N1381,2N1382,2N1383,2N1683,2N1706, 2N2168,2N2271,2N2273,2N2402,2N3320,2N3371,2N3412,2N3883,2SA79,2SA209, 2SA210,2SA212,2SA417,2SA451,2SB22,2SB34,2SB75,2SB77,2SB89,2SB156, 2SB156A,2SB187,2SB188,2SB199,2SB263,2SB328,2SB329,2SB365,2SB475, 2SB496,40269,ASY27,MA113,MA115,MA117,MA287,MM404,NKT121,NKT122, NKT123,NKT127,NKT128,NKT129,SF.T228,SF.T229,SF.T237,SF.T307,SF.T308, SF.T323,SF.T353,SYL792,*GE-52,*TR-05,*HEPG0005,*SK3003,*AT10M,*ECG102, *WEP631,*AC126,*276-2005,*RT-120
2N1265/5	2N59,2N123,2N132A,2N415,2N415A,2N416,2N467,2N520A,2N569,2N570,2N571, 2N572,2N654,2N654+,2N1175A,2N1274,2N1307,2N1309A,2N1352,2N1356,2N1370, 2N1374,2N1376,2N1381,2N1382,2N1707,2N1808,2N1892,2SA210,2SB186,2SB187, 2SB188,2SB222,2SB222A,2SB327,2SB329,40269,ASY27,SF.T223,*PTC109,*GE-52, *TR-85,*AC126,*276-2005
2N1266	2N315,2N529/P,2N530/P,2N578,2N578+,2N741,2N741A,2N960,2N961,2N962, 2N963,2N1343,2N2630,2N2718,*GE-50,*TR-85,*HEPG0005,*SK3005,*AT10M, *ECG102A,*WEP250,*AF125,*276-2005,*RT-121
2N1266+	2N140,2N219,2N302,2N303,2N409,2N410,2N411,2N412,2N450,2N485,2N486,2N521, 2N559,2N582,2N584,2N711,2N711A,2N711B,2N794,2N795,2N829,2N972,2N973, 2N974,2N1115,2N1281,2N1282,2N1300,2N1301,2N1344,2N1524,2N1683,2N2048, 2N2401,2N2402,2N2717,2N2795,2N3320,2N3322,2N3883,2SA12,2SA13,2SA15, 2SA16,2SA138,2SA139,2SA169,2SA171,2SA201,2SA202,2SA209,2SA351,2SA352, 2SA353,2SA354,2SA412,2SA451,40488,AFY15,ASZ21,MM380,SF.T228,SF.T307, SF.T319,SF.T320,SYL792,*GE-50,*TR-85,*HEPG0005,*SK3005,*ECG102A, *WEP250
2N1267	ZT202P,ZT402P,*PTC123,*GE-212,*TR-86,*HEPS0011,*SK3124,*ECG123,*WEP53, *BC107B,*276-2013,*RT-100
2N1268	2N835,2N1409,2N1682,2N2719,MPS918,MPSH07,MPSH08,*PTC123,*GE-212,*TR-24, *HEPS0011,*BC107B,*276-2013
2N1269	2N744,2N784,2N988,2N989,2N1840,2N2206,2N2309,2N2318,2N2319,2N2320, 2N2571,2N2572,2N2656,2N3011,2N3394,2N3544,2N3854A,2N3855A,2N4274, 2N5030,2N5187,2N5399,2N5418,2SC53,2SC55,2SC87,2SC99,2SC204,2SC343, 2SC356,2SC395A-R,2SC595,2SC718,2SC1123,2SC1128,2SC1231,BC170A, MM1941,MPS6507,MPS6511,MPS6568,MPSH02,MPSH30,MPSH31,PET2001, PT720,PT2760,ZT204P,ZT404P,ZT2205,*PTC123,*GE-212,*TR-24,*HEPS0011, *EN10,*BC107B,*276-2013
2N1270	*PTC123,*GE-212,*TR-24,*HEPS0011,*EN10,*BC107B,*276-2016
2N1271	2N835,2N1682,2N2719,MPS918,MPSH07,MPSH08,*PTC123,*GE-212,*TR-24, *HEPS0011,*BC107B,*276-2016
2N1272	2N744,2N784,2N988,2N989,2N2206,2N2318,2N2319,2N2320,2N2656,2N3011, 2N3394,2N3544,2N3855A,2N4274,2N5030,2N5187,2N5399,2N5418,2SC53,2SC55, 2SC87,2SC99,2SC204,2SC343,2SC356,2SC395A-R,2SC595,2SC718,2SC1123, 2SC1128,2SC1231,MM1941,MPS6507,MPS6511,MPS6568,MPSH02,MPSH30, MPSH31,PET2001,PT720,PT2760,ZT2205,*PTC123,*GE-212,*TR-24,*HEPS0011, *EN10,*BC107B,*276-2016
2N1273	2N417,2N508A,2N521A,2N523,2N571,2N599,2N655,2N655+,2N1274,2N1284, 2N1307,2N1309,2N1352,2N1353,2N1354,2N1355,2N1356,2N1357,2N1370,2N1376, 2N1381,2N1808,2N1892,2SB186,2SB187,2SB188,2SB223,2SB223A,2SB327, 2SB329,ASY27,SF.T223,*PTC109,*GE-52,*TR-05,*HEPG0005,*SK3004,*AT10M, *ECG102,*WEP631,*AC188/01,*276-2004,*RT-120
2N1274	2N417,2N508A,2N521A,2N571,2N598,2N599,2N655,2N655+,2N1008A,2N1193, 2N1307,2N1309,2N1352,2N1354,2N1355,2N1356,2N1357,2N1370,2N1376,2N1381, 2N1808,2N1892,2N1998,2SB186,2SB187,2SB188,2SB223,2SB223A,2SB327, 2SB329,ASY27,SF.T223,*PTC135,*GE-53,*TR-05,*HEPG0005,*SK3004,*AT20M, *ECG102,*WEP631,*276-2007,*RT-120
2N1275	2N1234,2N1476,2N3842,HA7535,SHA7535,*GE-21,*TR-88,*SK3025,*ECG129, *WEP242,*RT-115
2N1276	2N472,2SC478,2SC945,*PTC123,*GE-60,*TR-86,*HEPS0014,*SK3124,*SN80, *ECG123,*WEP53,*BF167,*276-2013,*RT-100

To Replace	Substitute This Type
2N1277	2N475,2N480A,2N717,2N783,2N834A,2N842,2N844,2N1150,2N2368,2S102,2SC120, 2SC121,2SC124,2SC321,2SC322,2SC382,BF224J,BF225J,BSX25,GET706,KT218, MPS834,MPS6544,MPS6545,MPSH20,MPSH37,TIS47,TIS125,ZT22,ZT42,*PTC123, *GE-60,*TR-86,*HEPS0014,*SK3124,*SN80,*ECG123,*WEP53,*BF167,*276-2013, *RT-100
2N1278	2N480,2N708,2N708A,2N718,2N784A,2N843,2N929,2N957,2N1152,2N1674,2N2242, 2N2369,2N2369A,2N2387,2N2501,2N2845,2N2847,2N2885,2N3009,2N3013,2N3014, 2N3301,2N3605A,2N3606A,2N3688,2N3689,2N3690,2N3693,2N3826,2N3858A, 2N3862,2N3903,2N3946,2N3973,2N3974,2N3975,2N3976,2N4123,2N4137,2N4140, 2N4295,2N4420,2N4421,2N4422,2N4966,2N4994,2N5027,2N5029,2N5824,2S103, 2SC67,2SC122,2SC169,2SC318,2SC335,2SC336,2SC468,2SC601,2SC601N, 2SC620,2SC634,2SC639,2SC725,2SC752G-R,2SC764,2SC815,2SC838,2SC839, 2SC896,2SC938,2SC1187,2SC1188,2SC1189,2SC1240,2SC1274,2SC1319, 2SC1320,2SC1686,2SC1687,2SC1688,BF198,BF199,BF241,BFY39-1,BSY93,EN697, EN708,EN914,EN2369A,EN3009,EN3013,EN3014,GET708,GET914,GET2221, GET2369,GET3013,GET3014,GET3646,MM3903,MPS3646,MPS3693,MPS6512, MPS6565,MPS6576-YEL,MPSA10-RED,MPSA20-RED,PET1001,SE1001,SE5001, SE5002,SE5003,SE5006,TIS22,TIS45,TIS46,TIS48,TIS49,TIS52,TIS55,TIS87,TIS129, TP3705,ZT23,ZT43,ZT708,*PTC123,*GE-60,*TR-86,*HEPS0014,*SK3124,*SN80, *ECG123,*WEP53,*BF167,*276-2013,*RT-100
2N1279	2N841,2N909,2N930,2N1153,2N2388,2N2483,2N2484,2N2484A,2N3302,2N3694, 2N3827,2N3859A,2N3904,2N3947,2N4074,2N4141,2N4967,2N4995,2N5028,2N5209, 2N5825,2N5826,2SC366G-Y,2SC367G-Y,2SC538A,2SC711A,2SC752G-Y,2SC933, 2SC943,2SC1175,2SC1285,2SC1364,2SC1632,2SC1634,2SC1685,2SC1850, 2SC1992,BC107,BC107A,BC123,BC167A,BC171A,BC182A,BC182L,BC237A, BCY59A,BCY59B,BFY39-2,BSX79,EN930,EN2484,GET2222,MM3904,MPS3694, MPS6514,MPS6566,MPS6575,MPS6576-GREEN,MPS6576-SIL,MPS-A10,MPS-A20, MPS-K20,MPS-K21,MPS-K22,MPSA10-BLU,MPSA10-GRN,MPSA10-YEL, MPSA20-BLU,MPSA20-GRN,MPSA20-YEL,NPSA20,PBC107A,PET1002,SE1002, TIS23,TIS24,TZ81,TZ82,*PTC123,*GE-39,*TR-86,*HEPS0014,*SK3124,*SN80, *ECG123,*WEP53,*BC107B,*276-2013,*RT-100
2N1280	*GE-1,*TR-05,*HEPG0005,*SK3005,*HF6H,*ECG100,*WEP254,*276-2007,*RT-118
2N1281	2N660,2N662,2N1282,*GE-1,*TR-05,*HEPG0005,*SK3005,*HF12H,*ECG100, *WEP254,*276-2007,*RT-118
2N1282	2N660,*GE-1,*TR-17,*HEPG0002,*SK3005,*HF12H,*ECG160,*WEP637,*276-2005, *RT-118
2N1284	2N521A,2N522A,2N523A,2N599,2N1316,2N1317,2N1998,2N1999,2N2001,2N2541, MA1704,*PTC135,*GE-1,*TR-05,*HEPG0005,*SK3005,*HF6H,*ECG100,*WEP53, *AC188/01,*276-2005,*RT-118
2N1285	*PTC145,*GE-51,*TR-17,*HEPG0003,*SK3006,*ECG160,*WEP637
2N1287	*GE-52,*TR-05,*HEPG0005,*SK3004,*ECG102,*WEP631,*276-2004,*RT-120
2N1287A	*GE-52,*TR-05,*HEPG0005,*SK3004,*ECG102,*WEP631,*276-2005,*RT-120
2N1288	*GE-8,*TR-08,*HEPG0011,*SK3011,*ECG101,*WEP641,*RT-119
2N1289	*DS-72,*GE-5,*TR-08,*HEPG0011,*SK3011,*NR10,*ECG101,*WEP641,*RT-119
2N1291	2N235A,2N236A,2N285A,2N399,AD149,*PTC138,*GE-3,*TR-01,*HEPG6003,*SK3009, *PT50,*ECG104,*WEP230,*OC28,*276-2006,*RT-124
2N1292	*GE-3,*TR-01,*SK3009,*ECG121,*WEP232,*276-2006,*RT-127
2N1293	2N420,2N665,2N1011,2N1136,2N1295,2N1971,2N2836,2SB127A,2SB128,2SB138, AD131,AD132,AUY19,AUY20,*PTC138,*GE-25,*TR-01,*HEPG6005,*SK3009,*PT50, *ECG121,*WEP232,*OC28,*276-2006,*RT-127
2N1294	2N1296,*GE-16,*TR-01,*SK3014,*ECG121,*WEP232,*276-2006,*RT-127
2N1295	2N418,2N420A,2N665,2N1011,2N1136A,2N1136B,2N1297,2SB215,2SB295, *PTC105A,*GE-25,*TR-01,*HEPG6005,*SK3009,*PT50,*ECG121,*WEP232, *276-2006,*RT-127
2N1296	2N1298,*DS-503,*GE-3,*TR-01,*SK3014,*PT40,*ECG121,*WEP232,*276-2006, *RT-127
2N1297	2N418,2N1136B,2SB128A,2SB295,AD163,*GE-25,*TR-01,*HEPG6005,*SK3009, *PT50,*ECG121,*WEP232,*276-2006,*RT-127
2N1298	*GE-54
2N1299	2N377A,2N385A,*PTC108,*GE-6,*TR-08,*HEPG0011,*SK3011,*NR10,*ECG101, *WEP641,*276-2001,*RT-119

To Replace	Substitute This Type
2N1300	2N711,2N711A,2N711B,2N794,2N795,2N829,2N972,2N973,2N974,2N1204A,2N1301, 2N1683,2N2048,2N2401,2N2402,2N3883,2SA451,*PTC145,*GE-1,*TR-17, *HEPG0002,*HF12M,*ECG160,*WEP637,*276-2005
2N1301	2N795,2N1204A,2N1683,*PTC145,*GE-1,*TR-17,*HEPG0002,*HF12M,*ECG160, *WEP637,*276-2005
2N1302	2N439,2N439A,2N440,2N440A,2N1012,2N1304,*GE-8,*TR-08,*HEPG0011,*SK3011, *NA20,*WEP641,*276-2001,*RT-119
2N1303	2N525A,2N526A,2N597,2N1305,2N1309A,2N1348,2N1349,2N1350,2N1351,*GE-1, *TR-84,*HEPG0005,*SK3004,*AT30M,*ECG158,*WEP238,*276-2007
2N1304	2N440,2N440A,2N1306,*GE-8,*TR-08,*HEPG0011,*SK3011,*NA20,*ECG101, *WEP641,*276-2001,*RT-119
2N1305	2N1307,2N1309A,2N1349,*GE-1,*TR-84,*HEPG0005,*AT30H,*ECG158,*WEP238, *276-2007
2N1306	2N1308,*GE-5,*TR-08,*HEPG0011,*SK3011,*NA30,*ECG101,*WEP641,*276-2001, *RT-119
2N1307	2N1309,*DS-25,*GE-1,*TR-84,*HEPG0005,*AT30H,*ECG158,*WEP238,*276-2007
2N1308	*DS-72,*GE-5,*TR-08,*HEPG0011,*SK3011,*NA30,*ECG101,*WEP641,*RT-119
2N1309	*GE-1,*TR-05,*HEPG0005,*AT30H,*ECG100,*WEP254,*276-2002,*RT-118
2N1309A	*TR-17,*HEPG0002,*ECG160,*276-2005
2N1310	*DS-72,*GE-5,*TR-08,*SK3011,*NR5,*ECG101,*WEP641,*RT-119
2N1311	2N1310,*DS-72,*GE-5,*TR-08,*SK3011,*NR5,*ECG101,*WEP641,*RT-119
2N1312	2N1311,*PTC108,*GE-7,*TR-08,*SK3010,*NR5,*ECG103,*WEP641A,*RT-122
2N1313	2N1348,2N1349,2N1350,2N1351,*GE-1,*TR-17,*HEPG0005,*SK3008,*HF6M, *ECG126,*WEP635,*276-2005
2N1314	2N242,2N307,2N456A,2N1534,2N1539,2N2137,2N2137A,2N2138,2N2138A,2N2142, 2N2142A,2N2143,2N2143A,2N3611,2N5889,2N5890,2SB246,2SB248,2SB250, 2SB251,MP2060,OC27,*PTC114,*GE-25,*TR-01,*HEPG6003,*SK3009,*PT40, *ECG104,*WEP230,*276-2006,*RT-124
2N1316	*GE-1,*TR-05,*HEPG0002,*SK3005,*HF12H,*ECG100,*WEP254,*276-2005,*RT-118
2N1317	2N1316,*GE-1,*TR-05,*HEPG0002,*SK3005,*HF12M,*ECG100,*WEP254,*276-2005, *RT-118
2N1318	2N1317,*GE-1,*TR-05,*HEPG0002,*SK3005,*HF12H,*ECG100,*WEP254,*276-2005, *RT-118
2N1319	2N425,2N426,2N427,*GE-1,*TR-05,*HEPG0005,*SK3005,*HF6M,*ECG100,*WEP254, *AC188/01,*276-2004,*RT-118
2N1320	
2N1321	
2N1322	
2N1323	2N1325
2N1324	
2N1325	2N1327
2N1326	
2N1327	
2N1328	
2N1329	
2N1330	2N1332,*SK3124
2N1331	2N1333
2N1332	2N1334
2N1333	
2N1334	
2N1335	2N699B,2N1336,2N1337,2N1342,2N1893A,2N2102,2N2102A,2N2405,2N3020,BSY55, BSY85,MM2258,ZT91,ZT92,ZT93,ZT2102,*PTC125,*GE-18,*276-2008
2N1336	2N699B,2N1335,2N1337,2N1342,2N1893A,2N2102,2N2102A,2N2405,2N3020,BSY55, BSY85,MM2258,ZT91,ZT92,ZT93,ZT2102,*PTC125,*GE-18,*276-2008
2N1337	2N699B,2N1335,2N1336,2N1342,2N1893A,2N2102,2N2102A,2N2405,2N3020,BSY55, BSY85,MM2258,ZT91,ZT92,ZT93,ZT2102,*PTC125,*GE-18,*276-2008
2N1338	2N1613,2N1837A,2N2218A,2N3107,2N3109,2N3119,2N3725,2N4047,2N4961, 2SC97A,2SC108A-O,2SC307,2SC1218,40408,40635,BSY53,BSY83,BSY87,TN53, ZT1613,*PTC125,*GE-18,*TR-86,*SN80,*ECG123,*WEP53,*276-2008,*RT-100
2N1339	2N699B,2N1335,2N1336,2N1337,2N1340,2N1341,2N1342,2N1893A,2N2102,2N2102A, 2N2405,2N3020,BSY55,BSY85,MM2258,ZT91,ZT92,ZT93,ZT2102,*PTC125,*GE-18,

To Replace	Substitute This Type
(2N1339)	*276-2008
2N1340	2N699B,2N1335,2N1336,2N1337,2N1339,2N1341,2N1342,2N1893A,2N2102,2N2102A, 2N2405,2N3020,BSY55,BSY85,MM2258,ZT91,ZT92,ZT93,ZT2102,*PTC125,*GE-18, *276-2008
2N1341	2N699B,2N1335,2N1336,2N1337,2N1339,2N1340,2N1342,2N1893A,2N2102,2N2102A, 2N2405,2N3020,BSY55,BSY85,MM2258,ZT91,ZT92,ZT93,ZT2102,*PTC125,*GE-18, *276-2008
2N1342	2N1893A,2N3020,MM2260,*PTC125,*276-2008
2N1343	2N426,2N427,*PTC135,*GE-1,*TR-05,*HEPG0005,*SK3005,*HF6M,*ECG100, *WEP254,*AC188/01,*276-2004,*RT-118
2N1344	2N660,2N662,2N1171,2N1281,2N1282,*GE-1,*TR-05,*HEPG0005,*SK3005,*HF12H, *ECG100,*WEP254,*276-2005,*RT-118
2N1345	2N660,2N661,2N1318,*GE-1,*TR-05,*HEPG0002,*SK3005,*HF12M,*ECG100, *WEP254,*276-2005,*RT-118
2N1346	2N523A,*GE-1,*TR-05,*HEPG0002,*SK3005,*HF12H,*ECG100,*WEP254,*276-2005, *RT-118
2N1347	2N315A,2N316A,2N317A,2N396A,2N404,2N428,2N1017,2N1018,2N1313,2N1942, 2SA205,2SA406,SF.T227,*PTC109,*GE-1,*TR-05,*HEPG0005,*SK3005,*HF6H, *ECG100,*WEP254,*AC188/01,*276-2005,*RT-118
2N1348	2N1349,2N1350,2N1351,*GE-2,*TR-05,*HEPG0005,*SK3004,*HF6H,*ECG102, *WEP631,*276-2005,*RT-120
2N1349	*GE-1,*TR-05,*HEPG0005,*SK3005,*HF12H,*ECG100,*WEP254,*276-2005,*RT-118
2N1350	2N1349,2N1351,*GE-1,*TR-05,*HEPG0005,*SK3005,*HF12H,*ECG100,*WEP254, *276-2005,*RT-118
2N1351	2N1349,2N1350,*GE-1,*TR-05,*HEPG0005,*SK3005,*HF12H,*ECG100,*WEP254, *276-2005,*RT-118
2N1352	2N382,2N383,2N417,2N527,2N598,2N599,2N652,2N655,2N655+,2N1008A,2N1189, 2N1193,2N1307,2N1309,2N1354,2N1355,2N1356,2N1357,2N1449,2N1892,2N1997, 2N1998,2SB327,MA888,SF.T223,*PTC135,*GE-53,*TR-05,*HEPG0005,*SK3003, *HF3H,*ECG102,*WEP631,*AC188/01,*276-2005,*RT-120
2N1353	2N523,2N599,2N655+,2N1316,2N1317,2N1354,2N1355,2N1357,2N1999,2SB223, 2SB223A,MA1703,MA1704,*PTC135,*GE-2,*TR-05,*HEPG0005,*SK3003,*HF6M, *ECG102,*WEP631,*AC188/01,*276-2005,*RT-120
2N1354	2N599,2N1316,2N1355,2N1357,2N1998,2N1999,2N2171,2N3427,*GE-1,*TR-05, *HEPG0005,*SK3005,*HF6M,*ECG100,*WEP254,*AC188/01,*RT-118
2N1355	2N599,2N1316,2N1357,2N1999,2N2171,*GE-1,*TR-05,*HEPG0005,*SK3005,*HF6M, *ECG100,*WEP254,*AC188/01,*RT-118
2N1356	2N383,2N598,2N599,2N1354,2N1355,2N1357,2N1998,*GE-1,*TR-05,*HEPG0005, *SK3005,*HF12H,*ECG100,*WEP254,*AC188/01,*276-2005,*RT-118
2N1357	2N599,2N1316,*GE-1,*TR-05,*HEPG0005,*SK3005,*HF12H,*ECG100,*WEP254, *276-2005,*RT-118
2N1358	2N174A,2N1100,*GE-4,*TR-03,*HEPG6006,*SK3012,*PT515,*ECG105,*WEP233, *276-2006
2N1358A	2N1412,2SB334,*GE-4,*TR-03,*SK3012,*ECG105,*WEP233,*276-2006
2N1359	*GE-76,*TR-01,*HEPG6005,*SK3009,*PT40,*ECG121,*WEP232,*276-2006,*RT-127
2N1360	*PTC105A,*GE-16,*TR-01,*HEPG6005,*SK3009,*PT40,*ECG121,*WEP232,*276-2006, *RT-127
2N1362	2N1364,*PTC105A,*GE-16,*TR-01,*HEPG6005,*SK3009,*PT40,*ECG121,*WEP232, *276-2006,*RT-127
2N1363	2N1365,*PTC105A,*GE-4,*TR-01,*HEPG6005,*SK3009,*PT40,*ECG121,*WEP232, *276-2006,*RT-127
2N1364	*DS-520,*GE-16,*TR-01,*HEPG6005,*SK3009,*ECG121,*WEP232,*276-2006,*RT-127
2N1365	*DS-520,*GE-16,*TR-01,*HEPG6017,*SK3009,*ECG121,*WEP232,*276-2006,*RT-127
2N1366	*PTC108,*GE-5,*TR-08,*HEPG0011,*SK3011,*ECG101,*WEP641,*276-2001,*RT-119
2N1367	*PTC108,*GE-5,*TR-08,*HEPG0011,*SK3011,*ECG101,*WEP641,*276-2001,*RT-119
2N1370	2N417,2N508A,2N521A,2N571,2N598,2N599,2N655,2N655+,2N1008A,2N1193, 2N1307,2N1309,2N1352,2N1354,2N1355,2N1356,2N1357,2N1376,2N1381,2N1808, 2N1892,2N1998,2SB223,2SB223A,2SB327,2SB329,ASY27,SF.T223,*PTC135, *GE-53,*TR-05,*HEPG0005,*SK3004,*AT30H,*ECG102,*WEP631,*276-2004, *RT-120
2N1371	2N382,2N383,2N461,2N527,2N652,2N1188,2N1189,2N1377,2N1449,2N1997,MA883,

To Replace	Substitute This Type
(2N1371)	MA888,*GE-2,*TR-05,*HEPG0005,*SK3004,*AT30H,*ECG102,*WEP631,*276-2004, *RT-120
2N1372	2N1008A,2N1374,2N1376,2N1381,2N1382,2N1478,*PTC135,*GE-53,*TR-05, *HEPG0005,*SK3004,*AT30M,*ECG102,*WEP631,*276-2004,*RT-120
2N1373	2N597,2N1375,2N1377,*GE-2,*TR-05,*HEPG0005,*SK3004,*AT30M,*ECG102, *WEP631,*276-2004,*RT-120
2N1374	2N598,2N1008A,2N1376,2N1381,2N1382,2N1998,*PTC135,*GE-53,*TR-05, *HEPG0007,*SK3004,*AT30H,*ECG102,*WEP631,*RT-120
2N1375	2N1377,2N1997,*GE-2,*TR-05,*HEPG0005,*SK3004,*AT30H,*ECG102,*WEP631, *276-2004,*RT-120
2N1376	2N598,2N599,2N1008A,2N1381,2N1998,*PTC135,*GE-53,*TR-05,*HEPG0005, *SK3004,*AT30H,*ECG102,*WEP631,*276-2005,*RT-120
2N1377	2N1997,*GE-2,*TR-05,*HEPG0005,*SK3004,*AT30H,*ECG102,*WEP631,*276-2005, *RT-120
2N1378	2N1379,2N1380,*PTC135,*GE-53,*TR-05,*HEPG0005,*SK3004,*AT10H,*ECG102, *WEP631,*AC188/01,*276-2005,*RT-120
2N1379	2N1999,*PTC135,*GE-53,*TR-05,*HEPG0005,*SK3004,*AT30H,*ECG102,*WEP631, *276-2005,*RT-120
2N1380	2N1378,2N1379,*PTC135,*GE-53,*TR-05,*HEPG0005,*SK3004,*AT10H,*ECG102, *WEP631,*AC188/01,*276-2004,*RT-120
2N1381	2N598,2N599,2N1008A,2N1998,*PTC135,*GE-53,*TR-05,*HEPG0005,*SK3004, *AT30H,*ECG102,*WEP631,*276-2004,*RT-120
2N1382	2N598,2N1008A,2N1998,*PTC135,*GE-2,*TR-05,*HEPG0005,*SK3004,*AT30H, *ECG102,*WEP631,*RT-120
2N1383	2N654,2N654+,2N1008A,2N1192,2N1349,2N1356,2N1381,2N1382,2SB327,SF.T223, *PTC135,*GE-53,*TR-05,*HEPG0005,*SK3004,*AT30H,*ECG102,*WEP631, *276-2004,*RT-120
2N1384	2N1495,*DS56,*GE-51,*TR-17,*HEPG0002,*SK3008,*JR100,*ECG126,*WEP635
2N1385	*GE-51,*TR-17,*HEPG0002,*ECG160,*WEP637
2N1386	2N702,2N703,2N708,2N753,2N784A,2N914,2N2206,2N2237,2N2242,2N2309,2N2318, 2N2319,2N2320,2N2331,2N2369,2N2369A,2N2481,2N2501,2N2656,2N2710,2N3009, 2N3011,2N3013,2N3014,2N3137,2N3210,2N3211,2N3298,2N3394,2N3508,2N3510, 2N3511,2N3605A,2N3606A,2N3646,2N3647,2N3648,2N3855A,2N3858,2N4123, 2N4137,2N4274,2N4275,2N4418,2N5029,2N5030,2N5136,2N5418,2N5769,2N5772, 2S512,2SC67,2SC204,2SC356,2SC577,2SC595,2SC601,2SC601N,2SC639, 2SC764,2SC1068,2SC1123,2SC1128,2SC1240,2SC1274,2SC1293,2SC1387, 2SC1988,A5T4123,BSW82,GET708,GET914,GET2369,GET3013,GET3014,GET3646, MPS2369,MPS6512,MPS6574-YEL,MPSA10-RED,MPSA20-RED,MPSH10,MPSH11, MT1060A,PET2001,PT720,PT2760,TI-484,ZT80,ZT110,ZT204P,ZT404P,ZT708, ZT2205,ZT2369,*PTC126,*GE-61,*TR-86,*HEPS0014,*SK3124,*ECG123,*WEP53, *BC107B,*276-2013,*RT-100
2N1387	2N758,2N759,2N783,2N784,2N834A,2N1140,2N1839,2N1987,2N2236,2N2368, 2N2427,2N2729,2N3261,2N3854A,2N5236,2SC47,2SC52,2SC54,2SC87,2SC188, 2SC302,2SC321,2SC322,2SC576,2SC689,2SC1164-R,40405,BF224J,BF225J, BSX25,GET706,KT218,MM1941,MPS834,MPS918,MPS6507,MPS6511,MPS6540, MPS6542,MPS6543,MPS6545,MPS6546,MPS6547,MPS6548,MPSH19,MPSH20, MPSH32,MPSH37,MT1038,MT1038A,MT1039,MT1060,TI-483,TIS84,TIS108,ZT22, ZT42,ZT203P,ZT403P,ZT2368,*PTC126,*GE-61,*TR-86,*HEPS0014,*SK3124, *ECG123,*WEP53,*BC107B,*276-2013,*RT-100
2N1388	2N749,2N751,2N1390,2SC302-M,2SC307-M,2SC590N,*PTC133,*GE-212,*TR-86, *HEPS0014,*SK3124,*ECG123,*WEP53,*BC107B,*276-2008,*RT-100
2N1389	*PTC123,*GE-212,*TR-86,*HEPS0014,*SK3124,*ECG123,*WEP53,*276-2013, *RT-100
2N1390	2N749,2N751,2N1388,2SC302-M,2SC307-M,2SC590N,*PTC139,*GE-61,*HEPS0011, *276-2013
2N1391	2N4138,2N6566,*GE-214,*TR-08,*HEPG0011,*ECG103,*WEP641A,*BF183,*276-2031, *RT-122
2N1395	2N1396,2N1397,*PTC145,*GE-51,*TR-05,*HEPG0003,*SK3006,*JR100,*ECG100, *WEP254,*RT-118
2N1396	2N1397,*PTC145,*GE-51,*TR-17,*HEPG0003,*SK3006,*JR100,*ECG126,*WEP635
2N1397	2N1396,*PTC145,*GE-51,*TR-17,*HEPG0003,*SK3006,*JR100,*ECG126,*WEP635

TRANSISTOR SUBSTITUTES

To Replace	Substitute This Type
2N1398	*GE-51,*TR-17,*SK3006,*ECG126,*WEP635
2N1399	*GE-51,*TR-17,*SK3006,*ECG126,*WEP635
2N1400	*GE-51,*TR-17,*SK3006,*ECG126,*WEP635
2N1401	*GE-51,*TR-17,*SK3006,*ECG126,*WEP635
2N1402	*GE-51,*TR-17,*HEPG0009,*SK3006,*ECG126,*WEP635
2N1403	2N2381,2N2630,*PTC145,*GE-2,*TR-17,*HEPG0002,*SK3123,*ECG160,*WEP637, *276-2005
2N1404	*GE-2,*TR-05,*HEPG0005,*HF6H,*ECG102,*WEP631,*276-2005,*RT-120
2N1404A	2N111A,2N112A,2N396A,2N404A,2N413,2N426,2N427,2N428,2N1017,2N1018, 2SA205,SF.T226,*PTC109,*TR-17,*HEPG0002,*ECG160,*WEP637,*276-2005
2N1405	*PTC107,*GE-50,*TR-17,*HEPG0003,*JR200,*ECG160,*WEP637,*276-2005
2N1406	*PTC107,*GE-50,*TR-17,*HEPG0003,*JR200,*ECG160,*WEP637
2N1407	*PTC107,*GE-50,*TR-17,*HEPG0003,*JR200,*ECG160,*WEP637
2N1408	2N1614,MA885,*GE-2,*TR-17,*HEPG0002,*AT30N,*ECG160,*WEP637,*276-2005
2N1409	2N1409A,2N2883,2N2884,2SC116,BFR36,BSY91,*PTC123,*GE-81,*TR-17, *HEPG0002,*ECG160,*WEP637,*BC337,*276-2008
2N1409A	2N706C,2N743A,2N2883,2N2884,2SC116,BFR36,*PTC125,*GE-47,*TR-63,*BF338, *276-2008
2N1410	2N656,2N780,2N1410A,2N2846,2N3015,2N3299,2N3326,2N3554,2SC32,2SC594-R, 2SC651,2SC1166-R,2SC1360,BSY92,*PTC123,*GE-81,*TR-17,*HEPG0002, *ECG160,*WEP637,*BC337,*276-2008
2N1410A	2N656,2N780,2N2846,2N3015,2N3299,2N3326,2N3554,2SC32,2SC594-R,2SC651, BSY92,*PTC125,*GE-47,*TR-63,*BF338,*276-2008
2N1411	2N501A,2N559,2N588A,2N768,2N769,2N779,2N779A,2N828,2N828A,2N964,2N965, 2N966,2N967,2N968,2N969,2N970,2N971,2N1427,2N2259,2N2401,2N2451,2N2487, 2N2488,2N2860,2N3322,2SA56,2SA251,2SA405,2SA410,2SA411,2SA450,*GE-51, *TR-17,*HEPG0003,*JR100,*ECG160,*WEP637
2N1412	2N1100,2SB334,*GE-4,*TR-03,*HEPG6006,*PT515,*ECG105,*WEP233,*276-2006
2N1413	2N381,2N525,2N525A,2N650,2N650A,2N1191,2N1373,2N1414,2N1446,2SB225, 2SB427,MA886,*PTC135,*GE-53,*TR-05,*HEPG0005,*SK3004,*AT30M,*ECG102, *WEP631,*276-2004,*RT-120
2N1414	2N381,2N422A,2N526,2N526A,2N597,2N651,2N651A,2N652A,2N1175,2N1192, 2N1348,2N1349,2N1350,2N1351,2N1415,2N1447,2N1451,2SB57A,2SB225,2SB226, 2SB427,MA887,*PTC135,*GE-53,*TR-05,*HEPG0005,*SK3004,*AT30M,*ECG102, *WEP631,*276-2004,*RT-120
2N1415	2N382,2N422A,2N461,2N526,2N526A,2N527A,2N597,2N651,2N652,2N652A, 2N1008A,2N1175A,2N1192,2N1349,2N1448,2N1452,2N1997,2SB226,2SB227, *PTC135,*GE-53,*TR-05,*HEPG0005,*SK3004,*AT30H,*ECG102,*WEP631, *276-2005,*RT-120
2N1416	M.P.2N223
2N1417	2N473,2N474,2N706,2N706A,2N706B,2N743,2N748,2N784,2N947,2N959,2N1082, 2N1387,2N1418,2N2205,2N2432,2N2571,2N2572,2N2719,2N2729,2N2926-BRN, 2N3544,2N3825,2N3843,2N3843A,2N3854,2N3854A,2N4292,2N4293,2N5187, 2S131,2SC55,2SC63,2SC68,2SC98,2SC137,2SC171,2SC172,2SC387AG, 2SC395A-R,2SC400-R,2SC455,2SC467,2SC717,2SC812,2SC1126,40519,EN918, GI-3605,GI-3606,GI-3607,MM1748,MM1941,MPS706,MPS706A,MPS918, MPS2926-BRN,MPS6507,MPS6511,MPS6539,MPS6540,MPS6542,MPS6548, MPSH19,MT1038,MT1038A,MT1039,MT1060,PET3001,PET8101,SE1010,SE3002, SE5025,TIS44,ZT20,ZT40,ZT203P,ZT403P,*PTC126,*GE-39,*TR-21,*HEPS0014, *SK3124,*SN80,*ECG123A,*WEP736,*BC107B,*276-2011,*RT-102
2N1418	2N474,2N475,2N748,2N783,2N784,2N834A,2N842,2N1387,2N2368,2N2432,2N2432A, 2N2729,2N3261,2N3793,2N3825,2N3843,2N3843A,2N3854A,2N4292,2N4293,2SC54, 2SC302,2SC321,2SC322,2SC382,2SC387AG,2SC400-R,2SC455,2SC565,2SC576, 2SC689,2SC717,2SC1126,40405,BF224J,BF225J,BSX25,EN918,GET706,GI-3793, KT218,MM1941,MPS834,MPS918,MPS6507,MPS6511,MPS6540,MPS6542,MPS6543, MPS6545,MPS6546,MPS6547,MPS6548,MPS6567,MPSH19,MPSH20,MPSH37, MT1038,MT1038A,MT1039,MT1060,PET3001,SE1010,SE3002,SE5025,TIS47,TIS125, ZT22,ZT42,ZT203P,ZT403P,ZT2368,*PTC126,*GE-39,*TR-21,*HEPS0014,*SK3124, *SN80,*ECG123A,*WEP736,*BC107B,*276-2013,*RT-102
2N1419	*PTC105A,*GE-16,*TR-01,*HEPG6005,*SK3009,*ECG104,*WEP230,*276-2006, *RT-124

To Replace	Substitute This Type
2N1420	2N1507,2N1972,2N2192,2N2192A,2N2192B,2N3300,A5T2192,ZT1711,*PTC123, *GE-81,*TR-21,*HEPS0014,*ECG123A,*WEP736,*BC337,*276-2008,*RT-102
2N1420A	2N1711A,BSY84,ZT1711,*GE-47,*HEPS3011,*ECG123A,*WEP736,*276-2008
2N1421	*HEPS5000
2N1422	*PTC906,*TR-59,*HEPS5000,*ECG130,*WEP247,*RT-131
2N1423	180T2,BDY23B,*PTC173,*TR-59,*HEPS5000,*ECG130,*WEP247,*RT-131
2N1424	ST400,*HEPS5000
2N1425	2N1023,2N1066,2N1224,2N1225,2N2092,2N2496,2N2671,2N3127,2N3281,2N3282, 2N3783,2N3784,2SA246,*PTC107,*GE-50,*TR-17,*HEPG0003,*SK3008,*JR30, *ECG126,*WEP635,*AF125
2N1426	2N384,2N987,2N2084,2N2588,A1383,SF.T357P,*PTC145,*GE-50,*TR-17,*HEPG0003, *SK3008,*JR30X,*ECG126,*WEP635,*AF125
2N1427	2N2451,2SA251,*GE-2,*TR-17,*HEPG0003,*SK3003,*JR100,*ECG126,*WEP635
2N1428	2N976,2N5142,2N5143,2N5332,2SA559,MPSL07,MPSL08,*PTC127,*TR-20, *HEPS0012,*SK3114,*ECG159,*WEP242,*276-2021,*RT-115
2N1429	2N863,2N864,2N2163,2N2164,2N2280,2N4006,2N4007,2N5141,*PTC127,*GE-210, *TR-88,*HEPS0012,*SK3025,*HF35,*ECG129,*WEP242,*BC337,*276-2009,*RT-115
2N1430	MP2833,MP2834,*GE-25,*TR-01,*HEPG6005,*SK3014,*PT40,*ECG121,*WEP232, *276-2006,*RT-127
2N1431	2N1059,2N2354,2SD72,2SD186,2SD187,ASY29,*PTC108,*GE-59,*TR-08,*HEPG0011, *SK3010,*NA30,*ECG103A,*WEP641B,*AC127,*276-2001,*RT-122
2N1431+	2N2354,2SD27,2SD30,2SD170A,2SD186,2SD187,ASY29,*PTC108,*GE-59,*SK3010, *ECG103A,*WEP641B
2N1432	2N1226,*GE-51,*TR-05,*HEPG0003,*SK3004,*AT30M,*ECG102,*WEP631,*RT-120
2N1433	2N539A,2N540A,2N1202,2N1262,2N1263,*GE-4,*TR-03,*SK3012,*PT50,*ECG105, *WEP233
2N1434	*GE-4,*TR-03,*SK3012,*PT50,*ECG105,*WEP233
2N1435	*GE-4,*TR-03,*SK3012,*PT50,*ECG105,*WEP233
2N1436	2N104,2N111A,2N112A,2N123A,2N130A,2N186,2N186A,2N186A+,2N279,2N315, 2N315A,2N316,2N316A,2N319,2N396A,2N402,2N413A,2N425,2N426,2N427,2N428, 2N500,2N519,2N519A,2N530/P,2N531/P,2N532/P,2N563,2N564,2N579+,2N581, 2N612,2N613,2N614,2N741,2N741A,2N846A,2N960,2N979,2N1018,2N1347, 2N1499A,2N1500,2N1742,2N1743,2N1744,2N1745,2N1864,2N1868,2N2170, 2N2381,2N2630,2N2718,2N2796,2SA208,2SB219,2SB219A,NKT225,SF.T221, *GE-50,*TR-17,*HEPG0002,*ECG160,*WEP637,*276-2005
2N1437	*GE-3
2N1438	*GE-3
2N1439	2N1232,2N1440,2N1441,HA7534,SHA7534,*GE-82,*TR-20,*HEPS0012,*SK3114, *SP70,*ECG159,*WEP242,*BC327,*276-2021,*RT-115
2N1440	2N1232,*GE-82,*TR-20,*HEPS0012,*SK3114,*SP70,*ECG159,*WEP242,*BC327, *276-2021,*RT-115
2N1441	2N1232,2N1233,*GE-82,*TR-20,*HEPS0012,*SK3114,*SP70,*ECG159,*WEP242, *BC327,*276-2023,*RT-115
2N1442	2N1233,2N1443,2N3840,2N3910,2N3913,2N4982,*GE-82,*TR-20,*HEPS0012, *SK3114,*SP70,*ECG159,*WEP242,*BC327,*276-2023,*RT-115
2N1443	2N1233,2N1442,2N3840,2N3910,2N3913,2N4982,*GE-82,*TR-20,*HEPS0012, *SK3114,*SP70,*ECG159,*WEP242,*BC327,*276-2023,*RT-115
2N1444	2N757A,2N2618,*PTC123,*GE-63,*TR-25,*HEPS3011
2N1445	2N4001,*GE-32,*HEPS3019,*276-2012
2N1446	2N1447,*GE-2,*TR-05,*HEPG0005,*SK3004,*AT30M,*ECG102,*WEP631,*276-2004, *RT-120
2N1447	2N1448,*GE-2,*TR-05,*HEPG0005,*SK3004,*AT30M,*ECG102,*WEP631,*276-2005, *RT-120
2N1448	2N1449,*GE-2,*TR-05,*HEPG0005,*SK3004,*AT30H,*ECG102,*WEP631,*276-2005, *RT-120
2N1449	2N2171,*GE-2,*TR-05,*HEPG0005,*SK3004,*AT30H,*ECG102,*WEP631,*276-2005, *RT-120
2N1450	2N43,2N43A,2N43A+,2N61A,2N61B,2N284,2N361,2N363,2N381,2N413,2N414, 2N414A,2N414B,2N414C,2N422,2N466,2N505,2N518,2N525,2N525A,2N567,2N568, 2N650,2N650A,2N653,2N1017,2N1057,2N1057+,2N1171,2N1191,2N1303,2N1313, 2N1348,2N1350,2N1351,2N1373,2N1414,2N1446,2N1447,2N1451,2N1478,2N1495,

To Replace	Substitute This Type
(2N1450)	2N2955,2N6365,2N6365A,2SA204,2SA205,2SA206,2SA207,2SA406,2SA407,2SB220, 2SB220A,2SB225,2SB299,2SB326,2SB427,ASY26,MA886,NKT221,NKT224, NKT228,SF.T222,SF.T227,*PTC102,*GE-52,*TR-05,*HEPG0005,*SK3004,*AT30M, *ECG102,*WEP631,*AC188/01,*276-2006,*RT-120
2N1450+	2N1495,2N2955,2N6365,2N6365A,*PTC102,*GE-52,*SK3004,*WEP631
2N1451	2N382,2N461,2N526,2N526A,2N527A,2N597,2N651,2N652A,2N1447,2N1448,2N1452, 2N1925,2N1926,2SB57A,2SB226,2SB227,*GE-2,*TR-05,*HEPG0005,*SK3004, *AT30M,*ECG102,*WEP631,*276-2004,*RT-120
2N1452	2N382,2N383,2N461,2N527,2N527A,2N652,2N1448,2N1449,2N1997,2N2000,2SB227, *GE-2,*TR-05,*HEPG0005,*SK3004,*AT30H,*ECG102,*WEP631,*276-2005,*RT-120
2N1453	
2N1454	
2N1455	
2N1456	
2N1457	
2N1458	
2N1461	
2N1462	
2N1463	
2N1464	
2N1465	
2N1466	
2N1469	2N940,2N1469A,2N2946,2N2946A,2N3840,2N3910,2N3913,2N4981,*GE-65,*TR-05, *HEPG0005,*SK3005,*SP70,*ECG100,*WEP254,*BC177,*276-2022,*RT-118
2N1469A	
2N1470	*PTC130,*GE-72,*TR-05,*HEPS5000,*SK3005,*ECG100,*WEP254,*276-2020, *RT-118
2N1471	2N1317,MA1703,MA1704,*PTC135,*GE-1,*TR-05,*HEPG0005,*SK3005,*HF6H, *ECG100,*WEP254,*AC188/01,*276-2005,*RT-118
2N1472	2N834,2N835,2N958,2N2214,2N2719,2N3825,GET706,MPS918,*PTC123,*GE-212, *TR-87,*HEPS0011,*SK3024,*EN10,*ECG128,*WEP243,*BC107B,*276-2016, *RT-114
2N1473	*PTC134,*GE-7,*TR-08,*HEPG0011,*SK3011,*NR5,*ECG101,*WEP641,*RT-119
2N1474	2N1233,2N1474A,*GE-82,*TR-20,*HEPS0012,*SK3114,*SP70,*ECG159,*WEP242, *276-2021,*RT-115
2N1474A	2N1232,2N1233,*GE-82,*TR-20,*HEPS0012,*SK3114,*SP70,*ECG159,*WEP242, *276-2021,*RT-115
2N1475	2N4982,*GE-82,*TR-20,*HEPS0012,*SK3114,*SP70,*ECG159,*WEP242,*276-2021, *RT-115
2N1476	2N3842
2N1477	
2N1478	2N597,*GE-2,*TR-05,*HEPG0005,*HF12H,*ECG102,*WEP631,*AC188/01,*276-2005, *RT-120
2N1479	ZT1479,*GE-46,*TR-87,*HEPS5014,*SK3024,*ECG128,*WEP243,*276-2020,*RT-114
2N1480	ZT1480,*GE-28,*TR-87,*HEPS5014,*SK3024,*ECG128,*WEP243,*276-2020,*RT-114
2N1481	*GE-46,*TR-87,*HEPS5014,*SK3024,*ECG128,*WEP243,*276-2020,*RT-114
2N1482	*TR-87,*HEPS5014,*SK3024,*ECG128,*WEP243,*276-2020,*RT-114
2N1483	2N1485,ZT1483,ZT1485,*GE-66,*HEPS5014,*SK3024,*276-2020
2N1484	2N1486,S2N1486,ZT1484,ZT1486,*GE-66,*HEPS5000,*SK3530,*276-2020
2N1485	ZT1485,*GE-66,*HEPS5014,*SK3024,*276-2020
2N1486	ZT1486,*GE-66,*HEPS5000,*SK3530,*276-2020
2N1487	2N1489,2SD15,2SD163,BDY23A,HST9201,HST9205,HST9206,HST9801,HST9802, ZT1487,ZT1489,*PTC173,*GE-77,*TR-59,*HEPS7002,*SK3510,*ECG280,*WEP247, *BDY20,*RT-131
2N1488	2N1490,2N3446,2SD16,2SD118-R,2SD119-R,2SD161,2SD164,2SD177,BDY24A, HST9202,HST9203,HST9207,HST9208,HST9803,HST9804,ZT1488,ZT1490,*PTC173, *GE-75,*TR-59,*HEPS7002,*SK3510,*ECG280,*WEP247,*BDY20,*RT-131
2N1489	2SD15,180T2,BDY23B,ZT1489,*PTC173,*GE-77,*TR-59,*HEPS7002,*SK3510, *ECG280,*WEP247,*BDY20,*RT-131
2N1490	2N3446,2N3448,2SD16,2SD110-R,2SD111-R,2SD118-R,2SD119-R,181T2,BDY24B, ZT1490,*PTC129A,*GE-75,*TR-59,*HEPS7002,*SK3510,*ECG130,*WEP247,

To Replace	Substitute This Type
(2N1490)	*BDY20,*RT-131
2N1491	2N749,2N751,2N1388,2N1390,2N1492,2SC302-M,2SC307-M,2SC590N,*PTC136, *GE-17,*TR-65,*HEPS0014,*SK3048,*EN10,*ECG195A,*276-2009,*RT-102
2N1492	2N749,2N751,2N1388,2N1390,2N1491,2SC302-M,2SC307-M,2SC590N,*PTC136, *GE-17,*TR-51,*HEPS0014,*SK3048,*ECG195A,*276-2009,*RT-102
2N1493	2N749,2N751,2N1388,2N1390,2SC302-M,2SC307-M,2SC590N,*TR-78,*HEPS5026, *ECG154,*WEP712,*276-2012,*RT-110
2N1494	*DS56,*GE-51,*JR100
2N1494A	
2N1495	*GE-51,*TR-17,*HEPG0002,*JR100,*ECG160,*WEP637,*276-2005
2N1495A	2N2382
2N1496	*JR100
2N1499	2N61,2N104,2N111A,2N112A,2N123A,2N130A,2N131A,2N133A,2N185,2N187, 2N187A,2N188,2N188A,2N200,2N206,2N215,2N269,2N283,2N284,2N315A,2N316, 2N316A,2N317,2N317A,2N319,2N320,2N361,2N363,2N368,2N396A,2N402,2N403, 2N404,2N413,2N428,2N519A,2N565,2N566,2N579,2N579+,2N580,2N580+,2N611, 2N613,2N614,2N615,2N653,2N979,2N980,2N1017,2N1018,2N1204,2N1204A,2N1303, 2N1313,2N1347,2N1413,2N1499A,2N1499B,2N1942,2N2048,2N2489,2N2795, 2N2796,2SA205,2SA208,2SA406,2SB185,2SB219,2SB219A,2SB220,2SB220A, NKT225,SF.T221,SF.T222,SF.T227,*PTC107,*GE-51,*TR-17,*HEPG0001,*SK3006, *JR100,*ECG126,*WEP635
2N1499A	2N979,2N980,2N1204,2N1204A,2N1499B,2N2048,2N2489,2N2795,2N2796,*PTC145, *GE-51,*TR-17,*HEPG0001,*SK3006,*ECG126,*WEP635
2N1499B	2N1495,2N2955,*PTC145,*GE-51,*TR-17,*HEPG0001,*SK3006,*ECG126,*WEP635
2N1500	2N500,2N846A,2N960,2N2170,2N2381,2N2630,2N2718,2N2796,*PTC107,*GE-51, *TR-17,*HEPG0002,*SK3006,*JR100,*ECG126,*WEP635,*276-2005
2N1501	2N539,2N539A,2N540,2N540A,2N1202,2N1262,2N1263,2N1502,*GE-4,*TR-01, *SK3012,*PT40,*ECG104,*WEP230,*RT-124
2N1502	2N539,2N539A,2N540,2N540A,2N1202,2N1262,2N1263,2N1501,*GE-4,*TR-01, *SK3012,*PT40,*ECG104,*WEP230,*RT-124
2N1504	
2N1505	2N1506,2N1710,2N4350,2SC302-M,2SC307-M,2SC590N,2SC845,SRF1002,XB401, *PTC144,*GE-47,*TR-95,*HEPS3001,*SK3039,*EN10,*ECG108,*WEP56,*276-2008, *RT-113
2N1506	2N1506A,2N1709,2N1710,2N2874,2N3118,2N3866,2SC302-M,2SC307-M,2SC547, 2SC555,40305,SRF11101,*PTC144,*GE-47,*TR-95,*HEPS3001,*SK3039,*EN10, *ECG108,*WEP56,*276-2030,*RT-113
2N1506A	2N1709,2N2874,2SC302-M,2SC307-M,2SC547,40305,SRF11101,*PTC144,*GE-17, *TR-95,*HEPS3011,*SK3039,*EN10,*ECG108,*WEP56,*RT-113
2N1507	2N1420,2N1972,2N2192,2N2192A,2N2192B,2N3300,A5T2192,ZT1711,*GE-81,*TR-95, *HEPS3020,*SK3039,*EN10,*ECG108,*WEP56,*BC337,*276-2008,*RT-113
2N1508	2N5148,MM1893,*PTC144,*HEPS3019
2N1509	2N5413,2N5414,MM2270,MM3053,*PTC144,*HEPS3020
2N1510	2N1310,*DS-75,*GE-6,*TR-08,*SK3011,*NR10,*ECG101,*WEP641,*RT-119
2N1511	2N1513,*HEPS5004
2N1512	2N1514,2SD54,*HEPS5004
2N1513	*HEPS5004
2N1514	*HEPS5004
2N1516	2N2495,2N2654,2N3279,2N3280,2N3283,2N3284,2SA327,*PTC107,*GE-2,*TR-17, *HEPG0003,*SK3006,*ECG160,*WEP637
2N1518	2N1520,2N1522,2SB478,*GE-4,*TR-03,*HEPG6004,*SK3012,*PT515,*ECG213, *WEP233,*276-2006
2N1519	2N1521,2N1523,2SB479,2SB480,ADY26,*GE-4,*TR-03,*HEPG6006,*SK3012,*PT515, *ECG213,*WEP233,*276-2006
2N1520	2N1522,*GE-4,*TR-03,*HEPG6004,*SK3012,*PT515,*ECG213,*WEP233,*276-2006
2N1521	2N1523,*GE-4,*TR-03,*HEPG6006,*SK3012,*PN350,*ECG213,*WEP233,*276-2006
2N1522	*TR-03,*HEPG6002,*SK3012,*PT515,*ECG213,*WEP233,*276-2006
2N1523	*GE-4,*TR-03,*HEPG6002,*SK3012,*PT515,*ECG105,*WEP233,*276-2006
2N1524	2N838,2N2188,2N2273,2N2956,2N2957,2N3323,2N3324,2N3325,2N6365,2SA353, 2SA354,2SA355,*PTC107,*GE-50,*TR-17,*HEPG0003,*SK3008,*JR100,*ECG160, *WEP233,*AF125,*RT-183

To Replace	Substitute This Type
2N1525	2N838,2N1524,2N2188,2N2273,2N2956,2N2957,2N3323,2N3324,2N3325,2N6365, 2SA353,2SA354,2SA355,*PTC107,*GE-50,*TR-17,*HEPG0003,*SK3008,*JR100, *ECG160,*WEP637,*AF125
2N1526	2N2635,*PTC145,*GE-50,*TR-17,*HEPG0009,*SK3007,*JR100,*ECG126,*WEP635, *AF125,*276-2005,*RT-188
2N1527	2N1526,2N2635,*PTC145,*GE-50,*TR-17,*HEPG0003,*SK3008,*JR100,*ECG160, *WEP637,*AF125
2N1528	*PTC126,*GE-63,*TR-21,*HEPS0014,*SK3124,*ECG123A,*WEP736,*276-2009, *RT-102
2N1529	2N1530,2N2138,2N2138A,2N2139,2N2139A,2N5890,2N5891,*PTC105A,*GE-76, *TR-01,*HEPG6003,*SK3009,*PT50,*ECG121,*WEP232,*276-2006,*RT-127
2N1529A	2N350A,2N1530A,2N2063A,*PTC105A,*GE-76,*TR-01,*HEPG6003,*SK3009, *ECG121,*WEP232,*276-2006,*RT-127
2N1530	2N1531,2N2139,2N2139A,2N2140,2N2140A,2N3615,*PTC105A,*GE-76,*TR-01, *HEPG6005,*SK3009,*PT50,*ECG121,*WEP232,*276-2006,*RT-127
2N1530A	2N1531A,2N2065A,*PTC105A,*GE-76,*TR-01,*HEPG6005,*SK3009,*ECG121, *WEP232,*276-2006,*RT-127
2N1531	2N1532,2N2140,2N2140A,2N2141,2N2141A,2N3615,2N3616,*PTC105A,*GE-76, *TR-01,*HEPG6005,*SK3009,*PT50,*ECG121,*WEP232,*276-2006,*RT-127
2N1531A	2N1532A,2N5156,*PTC105A,*GE-76,*TR-01,*HEPG6005,*SK3009,*ECG121, *WEP232,*276-2006,*RT-127
2N1532	2N3616,*PTC105A,*GE-3,*TR-01,*HEPG6005,*SK3009,*PT150,*ECG121,*WEP232, *276-2006,*RT-127
2N1532A	2N5156,*PTC105A,*GE-3,*TR-01,*HEPG6005,*SK3009,*ECG121,*WEP232, *276-2006,*RT-127
2N1533	*DS503,*GE-3,*TR-01,*HEPG6018,*SK3009,*PT150,*ECG121,*WEP232,*RT-127
2N1534	2N380,2N1168,2N2143,2N2143A,2N2144,2N2144A,*PTC105A,*GE-76,*TR-01, *HEPG6003,*SK3009,*PT150,*ECG121,*WEP232,*276-2006,*RT-127
2N1534A	2N250A,2N251A,2N350A,2N351A,2N376A,2N380,2N456B,2N457B,2N1535A, 2N2063A,2N4243,2N4244,*PTC105A,*GE-76,*TR-01,*HEPG6003,*SK3009,*ECG121, *WEP232,*276-2006,*RT-127
2N1535	2N1536,2N2139,2N2139A,2N2140,2N2140A,2N2144,2N2144A,2N2145,2N2145A, *PTC105A,*GE-76,*TR-01,*HEPG6005,*SK3009,*PT50,*ECG121,*WEP232, *276-2006,*RT-127
2N1535A	2N457B,2N458B,2N1536A,2N2065A,2N4242,2N4243,*PTC105A,*GE-76,*TR-01, *HEPG6005,*SK3009,*ECG121,*WEP232,*276-2006,*RT-127
2N1536	2N1537,2N2140,2N2140A,2N2141,2N2141A,2N2145,2N2145A,2N2146,2N2146A, *PTC105A,*GE-76,*TR-01,*HEPG6005,*SK3009,*PT50,*ECG121,*WEP232, *276-2006,*RT-127A
2N1536A	2N1537A,2N4242,2N5156,*PTC105A,*GE-76,*TR-01,*HEPG6005,*SK3009,*ECG121, *WEP232,*276-2006,*RT-127
2N1537	2N1538,*PTC105A,*GE-3,*TR-01,*HEPG6005,*SK3009,*PT150,*ECG121,*WEP232, *276-2006,*RT-127
2N1537A	2N5156,*PTC105A,*GE-3,*TR-01,*HEPG6005,*SK3009,*ECG121,*WEP232, *276-2006,*RT-127
2N1538	*DS503,*GE-3,*TR-01,*HEPG6018,*SK3009,*PT150,*ECG121,*WEP232,*276-2006, *RT-127
2N1539	2N1168,2N1534,2N1540,2N1544,2N1545,2N2143,2N2143A,2N2144,2N2144A,2N3613, 2N3614,2N4243,2N4244,2N4246,2N4247,2N5894,2N5895,MP2060,MP2061, *PTC105A,*GE-76,*TR-01,*HEPG6003,*SK3009,*PT50,*ECG121,*WEP232, *276-2006,*RT-127
2N1539A	2N250A,2N251A,2N351A,2N376A,2N456B,2N457B,2N1168,2N1540A,2N1544, 2N1544A,2N1545,2N1545A,2N4243,2N4244,2N4246,2N4247,TI-3027,TI-3028, *PTC105A,*GE-76,*TR-01,*HEPG6003,*SK3009,*ECG121,*WEP232,*276-2006, *RT-127
2N1540	2N1541,2N1545,2N1546,2N1546A,2N2144,2N2144A,2N2145,2N2145A,2N3614, 2N3617,2N4242,2N4243,2N4245,2N4246,*PTC105A,*GE-76,*TR-01,*HEPG6005, *SK3009,*PT50,*ECG121,*WEP232,*RT-127
2N1540A	2N457B,2N458B,2N1541A,2N1545,2N1545A,2N1546,2N1546A,2N4242,2N4243, 2N4245,2N4246,*PTC105A,*GE-76,*TR-01,*HEPG6005,*SK3009,*ECG121, *WEP232,*276-2006,*RT-127

To Replace	Substitute This Type
2N1541	2N1542,2N1546,2N1546A,2N1547,2N1547A,2N2145,2N2145A,2N2146,2N2146A, 2N3617,2N3618,2N4242,2N4245,*PTC105A,*GE-76,*TR-01,*HEPG6005,*SK3009, *PT50,*ECG121,*WEP232,*276-2006,*RT-127
2N1541A	2N1542A,2N1546,2N1546A,2N1547,2N1547A,2N4242,2N4245,*PTC105A,*GE-76, *TR-01,*HEPG6005,*SK3009,*ECG121,*WEP232,*276-2006,*RT-127
2N1542	2N1543,2N1547,2N1547A,2N1548,2N3618,*PTC105A,*GE-3,*TR-01,*HEPG6005, *SK3009,*PT150,*ECG121,*WEP232,*276-2006,*RT-127
2N1542A	2N1543A,2N1547,2N1547A,2N1548,*PTC105A,*GE-3,*TR-01,*HEPG6005,*SK3009, *ECG121,*WEP232,*276-2006,*RT-127
2N1543	2N1548,*DS503,*GE-3,*TR-01,*HEPG6018,*SK3009,*PT150,*ECG121,*WEP232, *RT-127
2N1543A	*GE-16,*TR-01,*SK3009,*ECG121,*WEP232,*RT-127
2N1544	2N669,2N1168,2N1545,2N3613,2N3614,2N4246,2N4247,TI-3027,TI-3028,*PTC105A, *GE-76,*TR-01,*HEPG6003,*SK3009,*PT50,*ECG121,*WEP232,*276-2006,*RT-127
2N1544A	2N669,2N1168,2N1544,2N1545,2N1545A,2N3613,2N3614,2N4246,2N4247,TI-3027, TI-3028,*PTC105A,*GE-76,*TR-01,*HEPG6003,*SK3009,*ECG121,*WEP232, *276-2006,*RT-127
2N1545	2N1546,2N1546A,2N3614,2N4245,2N4246,*PTC105A,*GE-76,*TR-01,*HEPG6013, *SK3009,*PT50,*ECG121,*WEP232,*276-2006,*RT-127
2N1545A	2N1545,2N1546,2N1546A,2N3614,2N4245,2N4246,*PTC105A,*GE-76,*TR-01, *HEPG6013,*SK3009,*ECG121,*WEP232,*276-2006,*RT-127
2N1546	2N1546A,2N1547,2N1547A,2N4245,*PTC105A,*GE-76,*TR-01,*HEPG6005,*SK3009, *PT50,*ECG121,*WEP232,*276-2006,*RT-127
2N1546A	2N1547A,*PTC105A,*GE-76,*TR-01,*HEPG6005,*SK3009,*ECG121,*WEP232, *276-2006,*RT-127
2N1547	2N1547A,2N1548,*PTC105A,*GE-3,*TR-01,*HEPG6005,*SK3009,*PT150,*ECG121, *WEP232,*276-2006,*RT-127
2N1547A	*PTC105A,*GE-3,*TR-01,*HEPG6005,*SK3009,*ECG121,*WEP232,*276-2006, *RT-127
2N1548	*DS503,*GE-3,*TR-01,*HEPG6018,*SK3009,*PT150,*ECG121,*WEP232,*RT-127
2N1549	2N1549A,2N1550,2N1550A,*PTC901,*GE-76,*TR-01,*HEPG6003,*SK3009,*PT50, *ECG121,*WEP232,*276-2006,*RT-127
2N1549A	2N1549,2N1550,2N1550A,*PTC901,*GE-76,*TR-01,*HEPG6003,*SK3009,*ECG121, *WEP232,*276-2006,*RT-127
2N1550	2N1550A,2N1551,2N1551A,*PTC901,*GE-76,*TR-01,*HEPG6005,*SK3009,*PT50, *ECG121,*WEP232,*276-2006,*RT-127
2N1550A	2N1550,2N1551,2N1551A,*PTC901,*GE-76,*TR-01,*HEPG6005,*SK3009,*ECG121, *WEP232,*276-2006,*RT-127
2N1551	2N1551A,2N1552,2N1552A,*PTC901,*GE-76,*TR-01,*HEPG6005,*SK3009,*PT50, *ECG121,*WEP232,*276-2006,*RT-127
2N1551A	2N1551,2N1552,2N1552A,*PTC901,*GE-76,*TR-01,*HEPG6005,*SK3009,*ECG121, *WEP232,*276-2006,*RT-127
2N1552	2N1552A,CQT1076,*DS503,*GE-3,*TR-01,*HEPG6005,*SK3009,*PT150,*ECG121, *WEP232,*276-2006,*RT-127
2N1552A	2N1552,CQT1076,*DS503,*GE-3,*TR-01,*HEPG6005,*SK3009,*ECG121,*WEP232, *276-2006,*RT-127
2N1553	2N1553A,2N1554,2N1554A,2N1557,2N1557A,2N1558,2N1558A,CTP1504,CTP1508, *PTC901,*GE-76,*TR-01,*HEPG6003,*SK3009,*PT150,*ECG121,*WEP232, *276-2006
2N1553A	2N1553,2N1554,2N1554A,2N1557,2N1557A,2N1558,2N1558A,CTP1504,CTP1508, *PTC901,*GE-76,*TR-01,*HEPG6003,*SK3009,*ECG121,*WEP232,*276-2006, *RT-127
2N1554	2N1558A,2N1559A,CQT940B,CQT940BA,CTP1503,CTP1504,*PTC901,*GE-16,*TR-01, *HEPG6005,*SK3009,*PT150,*ECG121,*WEP232,*276-2006,*RT-127
2N1554A	2N1554,2N1555,2N1555A,2N1558A,2N1559,2N1559A,CQT940B,CQT940BA,CTP1503, CTP1504,*PTC901,*GE-76,*TR-01,*HEPG6005,*SK3009,*ECG121,*WEP232, *276-2006,*RT-127
2N1555	2N1555A,2N1556A,2N1559,2N1559A,2N1560,2N1560A,CQT940B,CQT940BA, *PTC901,*GE-76,*TR-01,*HEPG6005,*SK3009,*PT150,*ECG121,*WEP232, *276-2006,*RT-127
2N1555A	2N1555,2N1556A,2N1559,2N1559A,2N1560,2N1560A,CQT940B,CQT940BA,*PTC901,

To Replace	Substitute This Type
(2N1555A)	*GE-76,*TR-01,*HEPG6005,*SK3009,*ECG121,*WEP232,*276-2006,*RT-127
2N1556	2N1166A,2N1556A,2N1560,2N1560A,*DS503,*GE-3,*TR-01,*HEPG6005,*SK3009, *PT40,*ECG121,*WEP232,*276-2006,*RT-127
2N1556A	2N1560,2N1560A,*DS503,*GE-3,*TR-01,*HEPG6005,*SK3009,*PT40,*ECG121, *WEP232,*276-2006,*RT-127
2N1557	2N1557A,2N1558,2N1558A,*PTC901,*GE-76,*TR-01,*HEPG6003,*SK3009,*PT150, *ECG121,*WEP232,*276-2006,*RT-127
2N1557A	2N1557,2N1558,2N1558A,*PTC901,*GE-76,*TR-01,*HEPG6003,*SK3009,*ECG121, *WEP232,*276-2006,*RT-127
2N1558	2N1558A,2N1559,2N1559A,*PTC901,*GE-16,*TR-01,*HEPG6005,*SK3009,*PT150, *ECG121,*WEP232,*276-2006,*RT-127
2N1558A	2N1559,2N1559A,*PTC901,*GE-16,*TR-01,*HEPG6005,*SK3009,*ECG121,*WEP232, *276-2006
2N1559	2N1559A,2N1560,2N1560A,*PTC901,*GE-76,*TR-01,*HEPG6005,*SK3009,*PT150, *ECG121,*WEP232,*276-2006,*RT-127
2N1559A	2N1559,2N1560,2N1560A,*PTC901,*GE-76,*TR-01,*HEPG6005,*SK3009,*ECG121, *WEP232,*276-2006,*RT-127
2N1560	2N1560A,*DS503,*GE-3,*TR-01,*HEPG6005,*SK3009,*PT150,*ECG121,*WEP232, *276-2006,*RT-127
2N1560A	2N1560,*DS503,*GE-3,*TR-01,*HEPG6005,*SK3009,*ECG121,*WEP232,*276-2006, *RT-127
2N1561	*GE-51,*TR-17,*HEPG0003,*JR200,*ECG160,*WEP637
2N1562	2N1561,*GE-51,*TR-17,*HEPG0003,*JR200,*ECG160,*WEP637
2N1564	2N1975,2N3108,2N3110,2SC345,PT1558,*PTC123,*GE-81,*TR-25,*HEPS5026, *276-2012
2N1565	2N1613,2N1613A,2N1837,2N1837A,2N1889,2N1974,2N2380,2N2380A,2N2479, 2N3107,2N3109,2N3498,2N3725,2N4047,2N4924,2N4961,2SC307,2SC308,2SC353, 2SC1218,BSY87,SE8002,SE8010,TIS135,TIS136,*PTC123,*GE-81,*TR-25, *HEPS5026,*276-2012
2N1566	2N1711,2N1711A,2N1890,2N1973,2N2049,2N3499,2N6540,2N6541,2SC1509,BSY88, TIS95,TIS98,ZT1711,*PTC123,*GE-81,*HEPS5026,*PT150,*276-2008
2N1566A	*PTC125,*GE-81,*TR-78,*HEPS5026,*BF338
2N1572	2N698,2SC59,2SC149,2SC589,BF178,MM3001,SF.T186,*PTC125,*GE-18,*TR-78, *HEPS5025,*SK3045,*NA30,*ECG154,*WEP712,*BF338,*276-2012,*RT-110
2N1573	2N699A,2N699B,2N1342,2N1613B,2N3500,2N4925,2SC49,2SC470-4,2SC470-5, 2SC590,2SC686,2SC1217,MM3008,TIS101,*PTC125,*GE-18,*TR-78,*HEPS5025, *SK3045,*HF6H,*ECG154,*WEP712,*BF338,*276-2008,*RT-110
2N1574	2N1711B,2N3501,2N4410,A5T4410,*PTC125,*GE-18,*TR-78,*HEPS5025,*SK3045, *ECG154,*WEP712,*276-2008,*RT-110
2N1585	*GE-8,*TR-08,*HEPG0011,*SK3011,*ECG101,*WEP641,*RT-119
2N1586	2N470,2N471,2N835,2N958,2N1472,2N1587,2N2214,2SC387A,*PTC126,*GE-60, *TR-70,*HEPS0014,*SK3124,*SN80,*ECG107,*WEP56,*BF167,*276-2011,*RT-108
2N1587	2N161,2N161A,2N471,2N472,2N472A,2N834,2N839,2N1149,2SC387A,2SC478, 2SC945,TIS126,*PTC126,*GE-60,*TR-70,*HEPS0014,*SK3124,*SN80,*ECG107, *WEP56,*BF167,*276-2011,*RT-108
2N1588	2N2521,*PTC123,*GE-212,*TR-21,*HEPS0005,*SK3122,*ECG123A,*WEP736, *BC107B,*276-2016,*RT-102
2N1589	2N118A,2N478,2N479,2N542A,2N706,2N706A,2N706B,2N709A,2N743,2N784,2N947, 2N959,2N988,2N989,2N1386,2N1590,2N2205,2N2318,2N2319,2N2432,2N2656, 2N2711,2N2713,2N2729,2N2921,2N2926-BRN,2N3010,2N3340,2N3544,2N3605, 2N3606,2N3607,2N3662,2N3663,2N3844,2N3844A,2N3854,2N3854A,2N3983, 2N4294,2N5030,2N5130,2N5399,2SC55,2SC63,2SC68,2SC98,2SC204,2SC356, 2SC387AG,2SC400-R,2SC455,2SC595,2SC668,2SC674,2SC717,2SC718,2SC719, 2SC724,2SC1023,2SC1126,2SC1231,EN706,EN918,EN3011,GI-2711,GI-2713, GI-2715,GI-3605,GI-3606,GI-3607,MM1748,MM1941,MPS2713,MPS2926-RED, MPS6507,MPS6511,MPSH19,MT1038,MT1038A,MT1039,MT1060,PET3001, PET8101,SE3001,SE3002,TIS18,TIS44,TIS51,ZT21,ZT41,*PTC126,*GE-60,*TR-70, *HEPS0014,*SK3124,*SN80,*ECG107,*WEP56,*BF167,*276-2011,*RT-108
2N1590	2N118A,2N118A+,2N119+,2N162,2N162A,2N163,2N479,2N480,2N542A,2N543A, 2N783,2N784,2N840,2N1151,2N1152,2N1278,2N2318,2N2319,2N2368,2N2387, 2N2432,2N2432A,2N2729,2N3340,2N3605A,2N3606A,2N3663,2N3844,2N3844A,

To Replace	Substitute This Type
(2N1590)	2N3854A,2N3983,2N4137,2N4294,2N5030,2N5130,2SC54,2SC120,2SC121,2SC124, 2SC204,2SC283,2SC302,2SC321,2SC322,2SC356,2SC380-R,2SC380A-R, 2SC387AG,2SC394-O,2SC400-R,2SC455,2SC468,2SC595,2SC601,2SC689, 2SC717,2SC724,2SC752G-R,2SC1126,BF224J,BF225J,BFY39-1,EN914,EN918, EN2369A,EN3009,EN3011,EN3013,EN3014,GET708,GET914,GET3013,GET3014, GET3646,KT218,MM1941,MPS6507,MPS6511,MPSA10-RED,MPSA20-RED,MPSH19, MT1038,MT1038A,MT1039,MT1060,PET3001,PT2760,SE3001,SE3002,TIS45, TIS46,TIS49,TIS51,TIS52,TIS55,TIS129,ZT23,ZT43,*PTC126,*GE-60,*TR-70, *HEPS0014,*SK3124,*SN80,*ECG107,*WEP56,*BF167,*276-2011,*RT-108
2N1591	2N717,2N844,2N2515,2N2522,2N3973,2N3975,2S103,2SC284,2SC383,2SC714, 2SC725,2SC979-R,2SC980G-R,BSY93,*PTC123,*GE-212,*TR-21,*HEPS0014, *SK3020,*SN80,*ECG123A,*WEP736,*BC107B,*276-2016,*RT-102
2N1592	2N541,2N542,2N1593,2N2712,2N2714,2N2923,2N2926-ORG,2N3393,2N3856, 2N3856A,2N3859,2N4251,2N5131,2N5132,2S502,2SC281,2SC400,2SC400-Y, 2SC454,2SC458,2SC459,2SC645,2SC847,2SC848,2SC912M,2SC925,2SC930NP, 2SC934,2SC1085,2SC1359,2SC1363,2SC1631,2SC1633,2SC1854,2SD392,93T6, 40398,40400,BC108A,BC130A,BC168A,BC170B,BC172A,BC238A,BCY58A,GI-2714, GI-2716,MPS2714,MPS2926-GRN,MPS2926-YEL,PBC108A,PET4001,SE4001, *PTC126,*GE-60,*TR-70,*HEPS0014,*SK3124,*SN80,*ECG107,*WEP56,*BF365, *276-2011,*RT-108
2N1593	2N542,2N543,2N841,2N930,2N1153,2N1279,2N2349,2N2388,2N3692,2N3694, 2N3856A,2N3859,2N4074,2N5209,2N5825,2SC123,2SC281,2SC367G,2SC367G-Y, 2SC372-Y,2SC372G-Y,2SC380-Y,2SC380A-Y,2SC394-GR,2SC400,2SC400-Y, 2SC454,2SC458,2SC459,2SC645,2SC733-Y,2SC735-Y,2SC752G-Y,2SC847,2SC848, 2SC907,2SC912M,2SC925,2SC930NP,2SC933,2SC965,2SC1085,2SC1175, 2SC1216,2SC1285,2SC1359,2SC1364,2SC1632,2SC1634,2SC1682-BL, 2SC1682-GR,2SC1727,2SC1854,BC107A,BC108A,BC123,BC129A,BC130A,BC167A, BC171A,BC237A,BC238A,BCY58A,BCY59A,BF240,BFY39-2,BSW88,BSW89,BSX38, BSX79,EN930,MPS6513,MPS6573,MPS6574-GREEN,MPS6574-SIL,MPS6575, MPS6576-GREEN,MPS6576-SIL,MPSA10-BLU,MPSA10-GRN,MPSA20-BLU, MPSA20-GRN,PBC107A,PET1002,PET2002,PET4001,SE1002,SE2002,SE4001, TIS23,*PTC126,*GE-60,*TR-70,*HEPS0014,*SK3124,*SN80,*ECG107,*WEP56, *BF365,*276-2011,*RT-108
2N1594	2N909,2N2483,2N2516,2N3827,2N3859A,2N3877,2N3877A,2N3904,2N3947,2N4141, 2N4946,2N4995,2N5028,2S104,2SC366G,2SC366G-Y,2SC734-Y,2SC943,2SC979-Y, 2SC980G-Y,BC174A,BC182A,BC190A,EN956,EN1711,GET930,GET2222,GET2222A, MM3904,TZ82,*PTC123,*GE-212,*TR-21,*HEPS0014,*SK3124,*SN80,*ECG123A, *WEP736,*BC107B,*276-2016,*RT-102
2N1605	2N357A,2N358A,2N377,2N377A,2N439,2N439A,2N576A,2N1000,2N1473,2N1605A, 2N1995,2SC128,2SC129,ASY28,*PTC108,*GE-8,*TR-08,*HEPG0011,*SK3011, *NR15,*ECG101,*WEP241,*276-2001,*RT-119
2N1605A	2N576A,2N1473,*PTC108,*GE-5,*TR-08,*HEPG0011,*SK3011,*ECG101,*WEP241, *276-2001,*RT-119
2N1606	2N941,2N942,2N1119,*TR-88,*HEPS0012,*SK3114,*ECG129,*WEP242,*276-2021, *RT-115
2N1607	2N941,2N942,*TR-20,*HEPS0012,*SK3114,*ECG159,*WEP242,*276-2021,*RT-115
2N1608	*TR-20,*HEPS0012,*SK3114,*ECG159,*WEP242,*276-2021,*RT-115
2N1609	*GE-16,*AT100H
2N1610	*GE-16,*AT100H
2N1611	*GE-3,*AT100H
2N1612	*GE-3,*AT100H
2N1613	2N1837A,2N1889,2N2193,2N2193A,2N2193B,2N3107,2N3109,40635,BC140-10, BC141-10,BSY53,BSY83,BSY87,ZT1613,*PTC125,*GE-216,*TR-87,*HEPS5026, *SK3024,*ECG128,*WEP243,*276-2008
2N1613A	*PTC125,*GE-27,*TR-78,*HEPS3011,*BF338,*276-2008
2N1613B	*PTC125,*GE-27,*TR-78,*HEPS3019,*BF338,*276-2008
2N1614	*DS-26,*GE-2,*TR-84,*HEPG0005,*AT30M,*ECG158,*WEP242,*276-2004
2N1614+	2N1924,2N1925,SF.T243,*GE-2,*HEPG0005,*ECG158,*WEP242
2N1615	2N657,*PTC144,*GE-32,*HEPS5026,*276-2012
2N1616	2N1616A,2N1617,2N1617A,2N3487,HST9901,HST9902,ST400,STT2655,*HEPS5004, *276-2020

To Replace	Substitute This Type
2N1616A	2N1617A,2N3487,ST400,*HEPS5004
2N1617	2N1617A,2N1618,2N1618A,2N3487,2N3488,HST9902,HST9903,STT2654,STT2655, TI-1135,TI-1136,TI-1155,*HEPS5004,*276-2020
2N1617A	2N1618A,2N3487,2N3488,*HEPS5004
2N1618	2N1618A,2N3488,2N3492,HST9903,HST9904,ST86020,ST91057,STT2653,STT2654, TI-1135,TI-1136,TI-1155,*HEPS5004
2N1618A	2N3488,2N3492,*HEPS5004
2N1620	2S024,TI-1126,TI-1145,TI-1146,*HEPS5004
2N1622	2N1310,*GE-54,*TR-08,*HEPG0011,*SK3011,*ECG101,*WEP641,*276-2001,*RT-119
2N1623	2N328A,2N328B,2N330A,2N926,2N928,2N937,2N1035,2N1232,2N1441,2N1474, 2N1474A,2N3345,2N3346,BCY30,BCY31,HA7534,SHA7534,*GE-65,*TR-88, *HEPS0012,*SK3114,*SP70,*ECG129,*WEP242,*BC177,*276-2022,*RT-115
2N1624	2N388,2N388A,2N634A,2N1306,2N1308,ASY29,*PTC108,*GE-6,*TR-08,*HEPG0011, *SK3011,*NR10,*ECG101,*WEP641,*276-2001,*RT-119
2N1631	2N1632,2N1633,2N1634,2N1635,2N1636,2N1637,2N1638,2N1639,*PTC107,*GE-51, *TR-17,*HEPG0001,*SK3008,*JR100,*ECG126,*WEP635,*AF125
2N1632	2N1631,2N1633,2N1634,2N1635,2N1636,2N1637,2N1638,2N1639,*PTC107,*GE-51, *TR-17,*HEPG0001,*SK3008,*JR100,*ECG126,*WEP635,*AF125
2N1633	2N1631,2N1632,2N1634,2N1635,2N1636,2N1637,2N1638,2N1639,*PTC107,*GE-51, *TR-17,*HEPG0002,*SK3008,*NR100,*ECG160,*WEP637,*276-2005
2N1634	2N1631,2N1632,2N1633,2N1635,2N1636,2N1637,2N1638,2N1639,*PTC107,*GE-51, *TR-17,*HEPG0002,*SK3008,*JR100,*ECG160,*WEP637,*AF125,*276-2005
2N1635	2N1631,2N1632,2N1633,2N1634,2N1636,2N1637,2N1638,2N1639,*PTC107,*GE-51, *TR-17,*HEPG0001,*SK3008,*JR100,*ECG126,*WEP635,*AF125
2N1636	2N1631,2N1632,2N1633,2N1634,2N1635,2N1637,2N1638,2N1639,*PTC107,*GE-51, *TR-17,*HEPG0001,*SK3008,*JR100,*ECG126,*WEP635,*AF125
2N1637	2N1631,2N1632,2N1633,2N1634,2N1635,2N1636,2N1638,2N1639,*PTC107,*GE-51, *TR-17,*HEPG0001,*SK3008,*JR100,*ECG126,*WEP635,*AF125
2N1638	2N1631,2N1632,2N1633,2N1634,2N1635,2N1636,2N1637,2N1639,*PTC107,*GE-51, *TR-17,*HEPG0008,*SK3007,*JR100,*ECG160,*WEP637,*AF125,*276-2005
2N1639	2N1631,2N1632,2N1633,2N1634,2N1635,2N1636,2N1637,2N1638,*PTC107,*GE-51, *TR-17,*HEPG0008,*SK3008,*JR100,*ECG160,*WEP637,*AF125,*276-2005
2N1640	2N938,2N943,2N944,2N945,2N1025,2N1439,2N1641,2N2474,2N3677,HA7530, SHA7530,*GE-65,*TR-20,*SK3118,*SP70,*ECG106,*WEP52,*BC177,*276-2022, *RT-126
2N1641	2N923,2N925,2N938,2N1025,2N1230,2N1441,2N1642,2N3345,HA7533,MM4052, SHA7533,*GE-65,*TR-20,*SK3118,*SP70,*ECG106,*WEP52,*BC177,*276-2022, *RT-126
2N1642	2N939,2N1026,2N1230,2N1441,2N3345,2N3346,2N4008,2N6567,MM4052,*GE-65, *TR-20,*SK3118,*SP70,*ECG106,*WEP52,*BC177,*276-2022,*RT-126
2N1643	2N923,2N938,2N1025,2N1219,2N1220,2N1230,2N1641,2N1642,HA7533,MM4052, SHA7533,*GE-210,*TR-20,*HEPS0012,*SK3114,*SP70,*ECG159,*WEP242,*BC337, *276-2009,*RT-115
2N1644	2N2909,*PTC144,*GE-27,*TR-21,*HEPS0014,*SK3122,*ECG123A,*WEP736,*BF338, *276-2009,*RT-102
2N1645	
2N1646	2N559,2N705,2N710,2N828,2N964,2N968,2N2048,2N2402,2N2860,2SA56,2SA405, 2SA411,*GE-51,*TR-17,*HEPG0003,*JR100,*ECG160,*WEP637
2N1647	*HEPS5000
2N1648	
2N1649	*HEPS5000
2N1650	
2N1651	*PTC147,*GE-76,*TR-35,*HEPG6005,*SK3014,*PT250,*ECG179,*WEPG6001, *276-2006,*RT-147
2N1652	2N1653,2N2636,2N2637,2N2638,*DS503,*GE-3,*TR-01,*HEPG6005,*SK3014,*PT250, *ECG121,*WEP232,*276-2006,*RT-127
2N1653	*DS503,*GE-3,*TR-01,*HEPG6009,*SK3009,*PT250,*ECG121,*WEP232,*RT-127
2N1654	2N1476,2N1477,2N1656,*BF338
2N1655	*BF338
2N1656	2N2551,*BF338
2N1658	

To Replace	Substitute This Type
2N1658/13	*HEPG6012
2N1659	
2N1659/13	*HEPG6011
2N1660	
2N1661	
2N1662	
2N1663	2N744,2N784,2N988,2N989,2N2206,2N2318,2N2319,2N2656,2N2921,2N2922, 2N3011,2N3544,2N3564,2N3691,2N3855A,2N4294,2N5030,2N5126,2N5127,2N5187, 2N5418,2SC55,2SC99,2SC204,2SC356,2SC394-O,2SC395A-R,2SC400-R,2SC595, 2SC718,2SC719,2SC724,2SC1123,2SC1128,2SC1231,EN706,EN744,EN3011, MM1941,MPS6507,MPS6511,MPS6568,MPSH30,MPSH31,NT3000,PET2001, PT720,PT2760,TIS51,ZT2205,*PTC132,*GE-20,*TR-21,*HEPS0011,*SK3122, *ECG123A,*WEP736,*276-2009,*RT-102
2N1664	*GE-2,*TR-05,*HEPG0006,*SK3005,*ECG100,*WEP254,*RT-118
2N1665	2N2587,2N6365,*GE-52,*TR-17,*HEPG0002,*ECG160,*WEP637,*276-2005
2N1667	2N2144,2N2144A,2N2145,2N2145A,2SB339,2SB538,ASZ16,ASZ18,KR6501, *PTC105A,*GE-25,*TR-01,*HEPG6003,*SK3009,*PT40,*ECG121,*WEP232, *276-2006,*RT-127
2N1668	2N2139,2N2139A,2N2140,2N2140A,2N2144,2N2144A,2N2145,2N2145A,2SB339, 2SB433,ASZ17,ASZ18,KR6500,*PTC105A,*GE-25,*TR-01,*HEPG6003,*SK3009, *PT40,*ECG104,*WEP230,*276-2006,*RT-124
2N1669	2N2145,2N2145A,2N2146,2N2146A,2SB339,2SB484,ASZ18,*PTC105A,*GE-25, *TR-01,*HEPG6003,*SK3009,*PT40,*ECG104,*WEP230,*276-2006,*RT-124
2N1670	*GE-1,*TR-17,*HEPG0002,*SK3005,*ECG160,*WEP637,*276-2005
2N1672	2N576A,2N587,2N1000,2N1312,2N1473,2N1605A,2N1672A,*PTC108,*GE-6,*TR-08, *HEPG0011,*NR5,*ECG103A,*WEP641B,*RT-122
2N1672A	2N576A,2N587,2N1000,2N1312,2N1473,2N1605A,2N1672,*PTC108,*TR-08, *HEPG0011,*ECG103A,*WEP641B,*RT-122
2N1673	2N987,2N1395,2N1396,2N1397,2N2084,2N2588,A1383,*PTC107,*GE-1,*TR-85, *HEPG0005,*SK3005,*ECG102A,*WEP250,*AF125,*276-2005,*RT-121
2N1674	2N718,2N760,2N843,2N916,2N916A,2N929,2N2161,2N2396,2N2845,2N2847,2N3301, 2N3693,2N3826,2N3858A,2N3862,2N3903,2N3946,2N3974,2N3976,2N4227,2N4962, 2N4966,2N4994,2N5027,2N5380,2N5824,2SC318,2SC335,2SC336,2SC366G-O, 2SC620,2SC815,2SC838,2SC839,2SC852,2SC899,2SC938,2SC944,2SC1189, 2SC1320,2SC1453,2SC1688,EN697,EN916,FT3641,FT3642,MM3903,MPS3693, MPS6565,MPS6576-BLUE,MPS6576-YEL,PET1001,SE1001,TIS87,TN-3903,TP3705, ZT24,ZT44,*PTC123,*GE-212,*TR-21,*HEPS0014,*SK3122,*EN10,*ECG123A, *WEP736,*BC107B,*276-2013,*RT-102
2N1675	
2N1676	2N864,2N865,2N1677,2N2164,2N2280,2N2894A,2N3546,2N4006,2N4208,2N4257, 2N4257A,2N4258,2N4258A,2N4313,2N4872,2N5055,2N5056,2N5057,2N5141, 2N5292,EN2894A,MM2894,MPSL07,MPSL08,*PTC127,*GE-65,*TR-20,*HEPS0012, *SK3114,*SP70,*ECG159,*WEP242,*BC177,*276-2021,*RT-115
2N1677	2N864,2N865,2N1676,2N2164,2N2280,2N2894A,2N3546,2N4006,2N4208,2N4257, 2N4257A,2N4258,2N4258A,2N4313,2N4872,2N5055,2N5056,2N5057,2N5141, 2N5292,EN2894A,MM2894,MPSL07,MPSL08,*PTC127,*GE-65,*TR-20,*HEPS0012, *SK3118,*SP70,*ECG159,*WEP242,*BC177,*276-2021,*RT-115
2N1678	*GE-51,*TR-17,*HEPG0002,*SK3008,*ECG160,*WEP637,*276-2005
2N1679	2N3665,*PTC144
2N1680	2N2297,2N3945,2SC781,*PTC144
2N1681	2N383,2N1349,2N1356,*PTC109,*GE-1,*TR-85,*HEPG0005,*HF12M,*ECG102A, *WEP250,*AC188/01,*276-2005,*RT-121
2N1682	2SC652,2SC998,MPSH07,MPSH08,*PTC136,*GE-17,*TR-86,*HEPS0014,*SK3124, *ECG123,*WEP254,*276-2009,*RT-100
2N1683	2N796,2N2699,*GE-1,*TR-05,*HEPG0005,*SK3005,*HF12M,*ECG100,*WEP254, *276-2005,*RT-118
2N1684	*PTC109,*GE-1,*TR-05,*HEPG0005,*SK3005,*HF12M,*ECG100,*WEP254, *AC188/01,*276-2007,*RT-118
2N1685	*PTC108,*GE-6,*TR-08,*HEPG0011,*SK3011,*NR5,*ECG101,*WEP641,*276-2001, *RT-119
2N1690	2N1049,2N1049C,*HEPS3020

To Replace	Substitute This Type
2N1691	2N1050,2N1050C,*GE-32,*HEPS3019
2N1692	
2N1693	
2N1694	2N356,2N356A,2N357,2N357+,2N358,2N377,2N438,2N438A,2N557,2N576,2N585, 2N585+,2N634,2N1605,2N1891,2N1994,2N1995,2SC128,2SC180,2SD75,*PTC108, *GE-5,*TR-08,*HEPG0011,*SK3011,*NR5,*ECG101,*WEP641,*AC187/01, *276-2001,*RT-119
2N1699	*DS56,*GE-51,*TR-17,*JR100,*ECG126,*WEP635
2N1700	2N1481,2N2297,2N3945,2N5413,2N5414,BSX46,MM2270,SDT4305,SDT4306, SDT4311,SDT4312,ZT1481,ZT1700,*PTC144,*GE-46,*TR-87,*HEPS5014,*SK3512, *ECG128,*WEP243,*276-2020,*RT-114
2N1701	2N1485,ZT1485,ZT1701,*GE-66,*HEPS5014,*SK3530,*276-2020
2N1702	2N1489,2N3232,2N3445,2N3713,2SD15,2SD211,180T2,BDY23B,HST9201,HST9205, HST9206,HST9801,HST9802,KSP1171,ZT1489,*PTC129A,*GE-77,*TR-59, *HEPS7002,*SK3510,*ECG130,*WEP247,*BDY20,*RT-131
2N1703	2N1513,*GE-19,*TR-59,*HEPS5004,*ECG130,*WEP247,*RT-131
2N1704	2N760,2N760A,2N1986,2N3643,2N4960,2N4962,2N5380,2N6538,2SC594,2SC594-O, 2SC1166-O,2SC1999,A5T3903,BC141,TN-3903,*PTC123,*GE-81,*TR-21, *HEPS0014,*SK3122,*SN80,*ECG123A,*WEP736,*BC337,*276-2012,*RT-102
2N1705	2N598,2N599,2N661,2N1317,2N1998,2N2541,*PTC135,*GE-2,*TR-05,*HEPG0005, *SK3004,*HF6H,*ECG102,*WEP631,*AC188/01,*276-2005,*RT-120
2N1706	2N598,2N660,2N661,2N1707,2N1998,2SB222A,2SB223A,2SB327,*PTC135,*GE-2, *TR-05,*HEPG0005,*SK3004,*HF6H,*ECG102,*WEP631,*AC188/01,*276-2005, *RT-120
2N1707	2N382,2N383,2N527,2N527A,2N598,2N652,2N1448,2N1449,2N1997,2N1998, 2SB222A,2SB223A,2SB227,2SB327,*PTC135,*GE-2,*TR-05,*HEPG0005,*SK3004, *HF6H,*ECG102,*WEP631,*276-2005,*RT-120
2N1708	2N2319,*GE-20,*TR-21,*HEPS0011,*SK3122,*SN80,*ECG123A,*WEP736,*276-2038, *RT-102
2N1708A	2N5845A,*GE-20,*TR-21,*HEPS0011,*SK3122,*ECG123A,*WEP736,*276-2016, *RT-102
2N1709	2SC302-M,2SC307-M,*GE-66,*HEPS3010
2N1710	2N1709,2SC302-M,2SC307-M,2SC590N,*GE-28,*TR-76,*HEPS3010
2N1711	2N1890,BSY54,BSY88,ZT1711,*GE-18,*TR-87,*HEPS0015,*SK3024,*ECG128, *WEP243,*276-2008
2N1711A	*PTC125,*HEPS3020,*ECG128,*WEP243,*276-2008
2N1711B	*PTC125,*HEPS3019,*276-2008
2N1713	2N2083,*HEPG0009,*ECG160,*WEP637
2N1714	2N698,2N3108,2N4000,2SC1069,BC141-6,*PTC144,*GE-63,*HEPS3020,*276-2012
2N1715	*GE-32,*HEPS3019,*276-2012
2N1716	2N699B,2N1889,2N2102,2N2102A,2N2243,2N2243A,2N2405,2N2443,2N3036,2N3107, BC141-10,BSY85,ZT91,ZT92,ZT93,ZT2102,*PTC144,*GE-63,*HEPS3020,*276-2012
2N1717	2N3020,*GE-32,*HEPS3019,*276-2012
2N1718	*GE-66,*HEPS3020
2N1719	*HEPS3019
2N1720	*GE-66,*HEPS3020
2N1721	*HEPS3019
2N1722	
2N1722A	
2N1723	
2N1724	ST86020,ST86021,ST91057,ST91058,STT2650,STT2651,STT2652,STT2653
2N1724A	
2N1725	STT2652
2N1726	2N982,2N2273,2N3323,2N3324,2N3325,2N3412,*PTC107,*GE-51,*TR-17,*HEPG0002, *SK3006,*JR200,*ECG160,*WEP637,*276-2005
2N1727	2N982,2N1726,2N1748A,2N2273,2N3323,2N3324,2N3325,2N3412,*PTC107,*GE-51, *TR-17,*HEPG0002,*SK3006,*JR200,*ECG160,*WEP637,*276-2005
2N1728	2N502,2N502A,2N838,2N982,2N1726,2N1727,2N1748A,2N1788,2N2273,2N3323, 2N3324,2N3325,2N3412,*PTC107,*GE-51,*TR-17,*HEPG0002,*SK3006,*JR200, *ECG160,*WEP637,*276-2005
2N1729	2N315A,2N316A,2N317A,2N428,2N653,2N1017,2N1018,2N1303,2N1313,*GE-53,

To Replace	Substitute This Type
(2N1729)	*TR-05,*HEPG0005,*SK3005,*ECG100,*WEP254,*276-2006,*RT-118
2N1730	2N356A,2N357A,2N358A,2N438,2N438A,2N439,2N439A,2N1302,2N1891,2N1994, 2N1995,*PTC108,*GE-8,*TR-08,*HEPG0011,*SK3011,*ECG101,*WEP641,*RT-119
2N1731	2N1348,2N1350,2N1351,*GE-53,*TR-05,*HEPG0005,*SK3005,*ECG100,*WEP254, *276-2004,*RT-118
2N1732	2N439,2N439A,*GE-54,*TR-08,*HEPG0011,*SK3011,*ECG101,*WEP641,*276-2001, *RT-119
2N1742	*PTC145,*GE-51,*TR-17,*HEPG0003,*JR200,*ECG160,*WEP637
2N1743	2N502B,2N1742,2N1744,2N1745,2N6365A,*PTC107,*GE-51,*TR-05,*HEPG0005, *SK3005,*JR200,*ECG100,*WEP254,*276-2004,*RT-118
2N1744	2N502B,2N1742,2N1743,2N1745,2N6365A,*PTC107,*GE-51,*TR-05,*HEPG0005, *SK3005,*JR200,*ECG100,*WEP254,*276-2004,*RT-118
2N1745	2N502B,2N1742,2N1743,2N1744,2N6365A,*PTC107,*GE-51,*TR-17,*HEPG0003, *SK3006,*JR200,*ECG160,*WEP637
2N1746	2N838,2N982,2N1726,2N1788,2N2168,2N2273,2N3323,2N3324,2N3325,2N3412, 2N3883,2N6365,*PTC107,*GE-51,*TR-17,*HEPG0003,*SK3006,*JR200,*ECG160, *WEP637
2N1747	2N502,2N502A,2N838,2N982,2N1865,2N2168,2N2273,2N3323,2N3324,2N3325, 2N6365,*PTC107,*GE-51,*TR-17,*HEPG0003,*SK3011,*JR200,*ECG160,*WEP637
2N1748	2N838,2N1788,2N2273,2N2956,2N2957,2N3323,2N3324,2N3325,*PTC107,*GE-51, *TR-17,*HEPG0002,*SK3006,*JR200,*ECG160,*WEP637,*276-2005
2N1748A	2N2273,2N2957,*PTC107,*TR-17,*HEPG0002,*SK3006,*ECG160,*WEP637, *276-2005
2N1749	2N2188,2N2190,2N2956,2N2957,*PTC145,*GE-51,*TR-17,*HEPG0002,*SK3006, *JR100,*ECG160,*WEP637,*276-2005
2N1750	*PTC107,*GE-51,*TR-17,*HEPG0003,*SK3007,*ECG160,*WEP637
2N1751	*TR-35,*PT250,*ECG179,*WEPG6001,*RT-147
2N1752	*PTC145,*GE-51,*TR-17,*HEPG0002,*JR100,*ECG160,*WEP637,*276-2005
2N1753	*GE-52,*TR-17,*HEPG0002,*ECG160,*WEP637,*276-2005
2N1754	2N779,2N779A,2N828,2N964,2N965,2N968,2N969,2N984,2N2048,2N2168,2N2169, 2N2402,2N2487,2N2488,2N2489,2N2795,2N2860,2N3322,2SA411,*PTC109,*GE-51, *TR-17,*HEPG0002,*SK3005,*JR100,*ECG160,*WEP637,*276-2005
2N1755	2N1756,*DS503,*GE-3,*TR-01,*SK3009,*PT40,*ECG104,*WEP230,*RT-124
2N1756	2N1757,*DS503,*GE-3,*TR-01,*SK3009,*PT40,*ECG104,*WEP230,*RT-124
2N1757	2N1758,*DS503,*GE-3,*TR-01,*SK3009,*PT50,*ECG104,*WEP230,*RT-124
2N1758	*DS503,*GE-3,*TR-01,*SK3009,*PT50,*ECG104,*WEP230,*RT-124
2N1759	2N1760,2N3154,2N3155,*DS-503,*GE-3,*TR-01,*SK3009,*PT40,*ECG104,*WEP230, *RT-124
2N1760	2N1761,2N3155,2N3156,*DS-503,*GE-3,*TR-01,*SK3009,*PT40,*ECG104,*WEP230, *RT-124
2N1761	2N1762,2N3156,2N3157,*DS-503,*GE-3,*TR-01,*SK3009,*PT50,*ECG104,*WEP230, *RT-124
2N1762	2N3157,*DS-503,*GE-3,*TR-01,*SK3009,*ECG104,*WEP230,*RT-124
2N1768	*GE-66,*HEPS5000,*276-2020
2N1769	*GE-66,*HEPS5000,*276-2020
2N1779	*PTC108,*GE-6,*TR-08,*HEPG0011,*SK3011,*NR5,*ECG101,*WEP641,*RT-119
2N1780	*PTC108,*GE-5,*TR-08,*HEPG0011,*SK3011,*NR5,*ECG101,*WEP641,*AC187/01, *RT-119
2N1781	*PTC108,*GE-5,*TR-08,*HEPG0011,*SK3011,*NR5,*ECG101,*WEP641,*AC187/01, *276-2001,*RT-119
2N1782	*GE-2,*TR-17,*HEPG0003,*SK3005,*HF12M,*ECG160,*WEP637,*AC188/01
2N1783	*PTC109,*GE-1,*TR-08,*HEPG0011,*SK3005,*HF12H,*ECG101,*WEP641, *AC188/01,*276-2002,*RT-119
2N1784	*GE-1,*TR-17,*HEPG0003,*SK3005,*HF12M,*ECG160,*WEP637
2N1785	2N711,2N711A,2N711B,2N796,2N829,2N964A,2N972,2N973,2N974,2N982,2N983, 2N984,2N985,2N1683,2N1726,2N1746,2N1748,2N2022,2N2168,2N2273,2N2402, 2N3320,2N3412,2N3883,2SA408,2SA409,2SA451,*PTC107,*GE-51,*TR-17, *HEPG0002,*SK3006,*JR200,*ECG160,*WEP637,*276-2005
2N1786	2N500,2N741,2N741A,2N828A,2N846A,2N979,2N1204,2N1499A,2N1742,2N1743, 2N1744,2N1745,2N1787,2N1868,2N2170,2N2400,2N2630,2N2718,2N2796,2SA449, *PTC107,*GE-51,*TR-17,*HEPG0002,*SK3006,*JR200,*ECG160,*WEP637,

To Replace	Substitute This Type
(2N1786)	*276-2005
2N1787	2N502B,2N559,2N705,2N710,2N711A,2N779,2N779A,2N828,2N828A,2N964,2N968, 2N980,2N984,2N1204A,2N1280,2N1499B,2N1728,2N1746,2N1747,2N1865,2N2048, 2N2168,2N2169,2N2401,2N2402,2N2487,2N2488,2N2489,2N2587,2N2795,2N2860, 2N3883,2N6365,2N6365A,2SA56,2SA251,2SA405,2SA411,*PTC107,*GE-51,*TR-17, *HEPG0002,*SK3006,*JR200,*ECG160,*WEP637,*276-2005
2N1788	2N2956,2N2957,2N3323,2N3324,2N3325,*PTC145,*GE-51,*TR-17,*HEPG0002, *SK3006,*JR200,*ECG160,*WEP637,*276-2005
2N1789	2N1495,2N1866,2N1867,2N2955,*PTC145,*GE-51,*TR-17,*HEPG0002,*SK3006, *JR200,*ECG160,*WEP637,*276-2005
2N1790	2N1788,2N2956,2N2957,2N3323,2N3324,2N3325,*PTC145,*GE-51,*TR-17, *HEPG0002,*JR200,*ECG160,*WEP637,*276-2005
2N1808	2N598,2N599,2N1309,2N1316,2N1892,2N1998,2N2001,*GE-8,*TR-08,*HEPG0011, *SK3011,*NR5,*ECG101,*WEP641,*RT-119
2N1809	2N1816,2N1823,2N1830
2N1810	2N1817,2N1824,2N1831
2N1811	2N1812,2N1818,2N1819,2N1825,2N1826,2N1832,2N1833
2N1812	2N1813,2N1819,2N1826,2N1833
2N1813	2N1814
2N1814	
2N1816	2N1809,2N1823,2N1830
2N1817	2N1810,2N1824,2N1831
2N1818	2N1811,2N1812,2N1819,2N1825,2N1826,2N1832,2N1833
2N1819	2N1812,2N1813,2N1826,2N1833
2N1820	2N1813,2N1814,2N1821,2N1827,2N1828,2N1834,2N1835
2N1821	2N1814,2N1828,2N1835
2N1823	2N1809,2N1816,2N1830
2N1824	2N1810,2N1817,2N1831
2N1825	2N1811,2N1812,2N1818,2N1819,2N1826,2N1832,2N1833
2N1826	2N1812,2N1813,2N1819,2N1833
2N1827	2N1813,2N1814,2N1828,2N1834,2N1835
2N1828	2N1814,2N1835
2N1830	2N1809,2N1816,2N1823
2N1831	2N1810,2N1817,2N1824
2N1832	2N1811,2N1812,2N1818,2N1819,2N1825,2N1826,2N1833
2N1833	2N1812,2N1813,2N1819,2N1826
2N1834	2N1813,2N1814,2N1835
2N1835	2N1814
2N1837	2N1837A,*PTC144,*GE-81,*HEPS3011
2N1837A	*PTC144,*HEPS3011
2N1837B	2N1837A,*PTC144,*HEPS3011
2N1838	2N2270,2N2537,2N3053,2N3724,2N3724A,2N4960,2SC503-Y,A5T5450,A8T3705, BC341-10,TIS110,TIS133,TIS134,*PTC123,*GE-81,*TR-21,*HEPS0014,*SK3122, *ECG123A,*WEP736,*BC337,*276-2030,*RT-102
2N1839	2N1958,2N1958A,2N2217,2N2477,2N5188,2SC189,2SC254,40539,BSX48,BSX49, ZT696,*PTC123,*GE-27,*TR-21,*HEPS0014,*SK3122,*ECG123A,*WEP736,*BF338, *276-2009,*RT-102
2N1840	2N1708A,2N2237,2N2320,2SC343,2SC504-O,BC340-6,BSY81,SE8041,SE8042, *PTC123,*GE-27,*TR-21,*HEPS0014,*SK3122,*ECG123A,*WEP736,*BF338, *276-2009,*RT-102
2N1853	2N522A,2N523A,2N599,2N1316,2N1317,2N2635,*PTC145,*GE-2,*TR-05,*HEPG0005, *SK3005,*AT20M,*ECG100,*WEP254,*276-2004,*RT-118
2N1854	*PTC145,*GE-2,*TR-05,*HEPG0005,*SK3004,*AT20M,*ECG102,*WEP631,*276-2004, *RT-120
2N1864	2N741A,2N979,2N1499A,2N1742,2N1743,2N1744,2N1745,2N1789,2N1868,2N2381, 2N2718,2N2796,*PTC107,*GE-51,*TR-17,*HEPG0002,*SK3008,*JR200,*ECG160, *WEP637,*276-2005
2N1865	2N502,2N502A,2N838,2N982,2N1747,2N2168,2N2273,2N3323,2N3324,2N3325, 2N6365,*PTC107,*GE-51,*TR-17,*HEPG0003,*SK3008,*JR200,*ECG160,*WEP637
2N1866	2N1495,2N1867,2N2955,*PTC145,*GE-51,*TR-17,*HEPG0002,*SK3008,*JR200, *ECG160,*WEP637,*276-2005

To Replace	Substitute This Type
2N2064	2SB449,NKT405,*PTC105A,*GE-76,*TR-01,*HEPG6003,*SK3009,*PT40,*ECG104, *WEP230,*276-2006,*RT-124
2N2064A	2N376A,2N1168,2N1539A,2N1540A,2N1544,2N1544A,2N1545,2N1545A,2N3613, 2N3614,2N4246,2N4247,MP2060,MP2061,TI-3027,TI-3028,*PTC105A,*GE-76,*TR-01, *HEPG6003,*SK3009,*ECG104,*WEP230,*276-2006,*RT-124
2N2065	*PTC105A,*GE-76,*TR-01,*HEPG6005,*SK3009,*PT50,*ECG104,*WEP230,*276-2006, *RT-124
2N2065A	2N458B,2N1021A,2N1536A,2N1537A,2N3615,2N3616,2N3617,2N3618,2N4242, 2N5156,*PTC105A,*GE-76,*TR-01,*HEPG6005,*SK3009,*ECG104,*WEP230, *276-2006,*RT-124
2N2066	*PTC105A,*GE-76,*TR-01,*HEPG6005,*SK3009,*PT50,*ECG104,*WEP230,*276-2006, *RT-124
2N2066A	2N1541A,2N1542A,2N1546,2N1546A,2N1547,2N1547A,2N4245,*PTC105A,*GE-76, *TR-01,*HEPG6005,*SK3014,*ECG104,*WEP230,*276-2006,*RT-124
2N2067	2N1755,2N1756,2N1759,2N1760,2N3158,2N3159,*DS-503,*GE-3,*TR-01,*SK3009, *PT40,*ECG104,*WEP230,*RT-124
2N2068	*PT50
2N2069	*PTC901,*GE-16,*TR-01,*SK3009,*ECG104,*WEP230,*RT-124
2N2070	*PTC901,*GE-16,*TR-01,*HEPG6018,*SK3009,*ECG104,*WEP230,*RT-124
2N2071	*PTC901,*GE-3,*HEPG6018
2N2072	*PTC901,*GE-3,*HEPG6018
2N2075	2N174A,2N1099,2N2075A,2N2079,2N2079A,2N2154,2N2154A,2N2155,2N2155A, 2N2158,2N2158A,2N2159,2N2159A,*DS501,*GE-4,*TR-03,*HEPG6006,*SK3012, *PT515,*ECG105,*WEP233,*276-2006
2N2075A	2N174A,2N1099,2N2075,2N2079,2N2079A,2N2154,2N2154A,2N2155,2N2155A, 2N2158,2N2158A,2N2159,2N2159A,*GE-4,*TR-03,*HEPG6006,*SK3012,*ECG105, *WEP233,*276-2006
2N2076	2N174A,2N1099,2N2075,2N2075A,2N2076A,2N2079,2N2079A,2N2080,2N2080A, 2N2154,2N2154A,2N2155,2N2155A,2N2158,2N2158A,2N2159,2N2159A,*GE-4, *TR-03,*HEPG6006,*SK3012,*PT515,*ECG105,*WEP233,*276-2006
2N2076A	2N174A,2N1099,2N2075,2N2075A,2N2076,2N2079,2N2079A,2N2080,2N2080A, 2N2154,2N2154A,2N2155,2N2155A,2N2158,2N2158A,2N2159,2N2159A,*GE-4, *TR-03,*HEPG6006,*SK3012,*ECG105,*WEP233,*276-2006
2N2077	2N2076,2N2076A,2N2077A,2N2080,2N2080A,2N2081,2N2081A,2N2153,2N2153A, 2N2154,2N2154A,2N2157,2N2157A,2N2158,2N2158A,2N2491,*GE-4,*TR-03, *HEPG6004,*SK3012,*PT515,*ECG105,*WEP233,*276-2006
2N2077A	2N2076,2N2076A,2N2077,2N2080,2N2080A,2N2081,2N2081A,2N2153,2N2153A, 2N2154,2N2154A,2N2157,2N2157A,2N2158,2N2158A,2N2491,*GE-4,*TR-03, *HEPG6004,*SK3012,*ECG105,*WEP233,*276-2006
2N2078	2N1980,2N2077,2N2077A,2N2078A,2N2081,2N2081A,2N2082,2N2082A,2N2152, 2N2152A,2N2153,2N2153A,2N2156,2N2156A,2N2157,2N2157A,2N2491,*GE-4, *TR-03,*HEPG6004,*SK3012,*PT501,*ECG105,*WEP233,*276-2006
2N2078A	2N1980,2N2077,2N2077A,2N2078,2N2081,2N2081A,2N2082,2N2082A,2N2152, 2N2152A,2N2153,2N2153A,2N2156,2N2156A,2N2157,2N2157A,2N2491,*GE-4, *TR-03,*HEPG6004,*SK3012,*ECG105,*WEP233,*276-2006
2N2079	2N2079A,2N2158,2N2158A,2N2159,2N2159A,*GE-4,*TR-03,*HEPG6006,*SK3012, *PT515,*ECG105,*WEP233,*276-2006
2N2079A	2N2079,2N2158,2N2158A,2N2159,2N2159A,*GE-4,*TR-03,*HEPG6006,*SK3012, *ECG105,*WEP233,*276-2006
2N2080	2N2079,2N2079A,2N2080A,2N2158,2N2158A,2N2159,2N2159A,*GE-4,*TR-03, *HEPG6006,*SK3012,*PT515,*ECG105,*WEP233,*276-2006
2N2080A	2N2079,2N2079A,2N2080,2N2158,2N2158A,2N2159,2N2159A,*GE-4,*TR-03, *HEPG6006,*SK3012,*ECG105,*WEP233,*276-2006
2N2081	2N2080,2N2080A,2N2081A,2N2157,2N2157A,2N2158,2N2158A,*GE-4,*TR-03, *HEPG6004,*SK3012,*PT515,*ECG105,*WEP233,*276-2006
2N2081A	2N2080,2N2080A,2N2081,2N2157,2N2157A,2N2158,2N2158A,*GE-4,*TR-03, *HEPG6004,*SK3012,*ECG105,*WEP233,*276-2006
2N2082	2N2081,2N2081A,2N2082A,2N2156,2N2156A,2N2157,2N2157A,*GE-4,*TR-03, *HEPG6004,*SK3012,*PT501,*ECG105,*WEP233,*276-2006
2N2082A	2N2081,2N2081A,2N2082,2N2156,2N2156A,2N2157,2N2157A,*GE-4,*TR-03, *HEPG6004,*SK3012,*ECG105,*WEP233,*276-2006

To Replace	Substitute This Type
2N2083	*PTC109,*GE-52,*TR-17,*HEPG0009,*ECG160,*WEP637
2N2084	*PTC145,*GE-51,*TR-17,*HEPG0003,*SK3006,*JR100,*ECG160,*WEP637
2N2085	*GE-59,*TR-08,*HEPG0011,*SK3011,*ECG101,*WEP641,*RT-119
2N2086	SF.T186,*PTC144,*GE-18,*HEPS3019
2N2087	*GE-81,*HEPS3019
2N2089	2N2092,2N2671,*PTC107,*GE-50,*TR-17,*HEPG0001,*SK3006,*ECG126,*WEP635
2N2090	2N2089,2N2092,2N2671,*PTC107,*GE-50,*TR-17,*HEPG0001,*SK3006,*JR100, *ECG126,*WEP635,*RT-188
2N2091	2N2089,2N2092,2N2671,*PTC107,*GE-50,*TR-17,*HEPG0001,*SK3006,*JR100, *ECG160,*WEP637,*RT-183
2N2092	2N2671,*PTC107,*GE-50,*TR-17,*HEPG0001,*SK3007,*JR100,*ECG126,*WEP635
2N2093	*PTC107,*GE-50,*TR-17,*HEPG0003,*SK3008,*JR100,*ECG160,*WEP637
2N2095	*GE-51,*TR-21,*SK3122,*ECG123A,*WEP736,*RT-102
2N2096	*GE-20,*TR-21,*SK3122,*ECG123A,*WEP736,*RT-102
2N2097	*DS56,*GE-51,*JR100
2N2098	2N1645,*PTC107,*GE-51,*TR-17,*HEPG0002,*ECG160,*WEP637
2N2099	2N2100,2N2381,*PTC135,*GE-51,*TR-17,*HEPG0002,*ECG160,*WEP637
2N2100	2N2382,*PTC135,*GE-51,*TR-17,*HEPG0002,*ECG160,*WEP637
2N2100A	2N1495,*PTC135,*HEPG0005
2N2101	2N1208,2N1212,ST400
2N2102	2N2102A,2N2405,*PTC144,*GE-66,*TR-87,*HEPS3019,*SK3024,*ECG128,*WEP243, *276-2008,*RT-114
2N2102A	2N2102,2N2405,*PTC144,*GE-27,*TR-78,*HEPS3019,*BF338,*276-2008
2N2104	*PTC127,*TR-20,*HEPS0012,*SK3114,*ECG159,*WEP242,*276-2021,*RT-115
2N2105	*PTC127,*TR-20,*HEPS0012,*SK3114,*ECG159,*WEP242,*276-2021,*RT-115
2N2106	*PTC125,*GE-63,*TR-87,*HEPS5014,*SK3024,*SN80,*ECG128,*WEP243,*BF338, *276-2008,*RT-114
2N2107	2N2106,*PTC117,*GE-63,*TR-87,*HEPS5014,*SK3024,*SN80,*ECG128,*WEP243, *BF338,*276-2008,*RT-114
2N2108	*PTC125,*GE-63,*TR-87,*HEPS5014,*SK3024,*SN80,*ECG128,*WEP243,*BF338, *276-2008,*RT-114
2N2109	2N2116,2N2123,2N2130
2N2110	2N2117,2N2124,2N2131
2N2111	2N2112,2N2118,2N2119,2N2125,2N2126,2N2132,2N2133
2N2112	2N2113,2N2119,2N2126,2N2133
2N2113	2N2114
2N2114	
2N2116	2N2109,2N2123,2N2130
2N2117	2N2110,2N2124,2N2131
2N2118	2N2111,2N2112,2N2119,2N2125,2N2126,2N2132,2N2133
2N2119	2N2112,2N2113,2N2126,2N2133
2N2120	2N2113,2N2114,2N2121,2N2127,2N2128,2N2134,2N2135
2N2121	2N2114,2N2128,2N2135,*SK3025,*ECG159
2N2123	2N2109,2N2116,2N2130
2N2124	
2N2125	2N2126
2N2126	
2N2127	2N2113,2N2114,2N2128,2N2134,2N2135
2N2128	2N2114,2N2135
2N2130	2N2109,2N2116,2N2123
2N2131	2N2110,2N2117,2N2124
2N2132	2N2111,2N2112,2N2118,2N2119,2N2125,2N2126,2N2133
2N2133	2N2112,2N2113,2N2119,2N2126
2N2134	2N2113,2N2114,2N2135
2N2135	2N2114
2N2137	2N2137A,2N2138,2N2138A,2N2142,2N2142A,2N2143,2N2143A,*PTC114,*GE-16, *TR-01,*HEPG6003,*SK3009,*PT40,*ECG104,*WEP230,*276-2006,*RT-124
2N2137A	2N2138A,2N2142A,2N2143A,*PTC114,*GE-16,*TR-01,*HEPG6003,*SK3009,*ECG104, *WEP230,*276-2006,*RT-124
2N2138	2N2138A,2N2139,2N2139A,2N2143,2N2143A,2N2144,2N2144A,*PTC105A,*GE-3, *TR-01,*HEPG6003,*SK3009,*PT40,*ECG104,*WEP230,*276-2006,*RT-124

To Replace	Substitute This Type
2N2138A	2N2139A,2N2143A,2N2144A,*PTC105A,*GE-16,*TR-01,*HEPG6003,*SK3009, *ECG104,*WEP230,*276-2006,*RT-124
2N2139	2N2139A,2N2140,2N2140A,2N2144,2N2144A,2N2145,2N2145A,*PTC105A,*GE-16, *TR-01,*HEPG6005,*SK3009,*PT50,*ECG104,*WEP230,*276-2006,*RT-124
2N2139A	2N2140A,2N2144A,2N2145A,*PTC105A,*GE-16,*TR-01,*HEPG6005,*SK3009, *ECG104,*WEP230,*276-2006,*RT-124
2N2140	2N2140A,2N2141,2N2141A,2N2145,2N2145A,2N2146,2N2146A,*PTC105A,*GE-16, *TR-01,*HEPG6005,*SK3009,*PT50,*ECG104,*WEP230,*276-2006,*RT-124
2N2140A	2N2141A,2N2145A,2N2146A,*PTC105A,*GE-16,*TR-01,*HEPG6005,*SK3009, *ECG104,*WEP230,*276-2006,*RT-124
2N2141	2N2141A,2N2146,2N2146A,*PTC105A,*GE-16,*TR-01,*HEPG6005,*SK3009,*ECG104, *WEP230,*276-2006,*RT-124
2N2141A	2N2146A,*PTC105A,*GE-16,*TR-01,*HEPG6005,*SK3009,*PT50,*ECG104,*WEP230, *276-2006,*RT-124
2N2142	2N2142A,2N2143,2N2143A,*PTC114,*GE-16,*TR-01,*HEPG6003,*SK3009,*PT40, *ECG104,*WEP230,*276-2006,*RT-124
2N2142A	2N2143A,*PTC114,*GE-16,*TR-01,*HEPG6003,*SK3009,*ECG104,*WEP230, *276-2006,*RT-124
2N2143	2N2143A,2N2144,2N2144A,*PTC105A,*GE-16,*TR-01,*HEPG6003,*SK3009,*PT40, *ECG104,*WEP230,*276-2006,*RT-124
2N2143A	2N2144A,*PTC105A,*GE-16,*TR-01,*HEPG6003,*SK3009,*ECG104,*WEP230, *276-2006,*RT-124
2N2144	2N2144A,2N2145,2N2145A,*PTC105A,*GE-16,*TR-01,*HEPG6005,*SK3009,*PT50, *ECG104,*WEP230,*276-2006,*RT-124
2N2144A	2N2145A,*PTC105A,*GE-16,*TR-01,*HEPG6005,*SK3009,*ECG104,*WEP230, *276-2006,*RT-124
2N2145	2N2145A,2N2146,2N2146A,*PTC105A,*GE-3,*TR-01,*HEPG6005,*SK3009,*PT50, *ECG104,*WEP230,*276-2006,*RT-124
2N2145A	2N2146A,*PTC105A,*GE-3,*TR-01,*HEPG6005,*SK3009,*ECG104,*WEP230, *276-2006,*RT-124
2N2146	2N2146A,*PTC105A,*GE-3,*TR-01,*HEPG6005,*SK3009,*PT50,*ECG104,*WEP230, *276-2006,*RT-124
2N2146A	*PTC105A,*GE-3,*TR-01,*HEPG6005,*SK3009,*ECG104,*WEP230,*276-2006, *RT-124
2N2147	*PTC138,*GE-16,*TR-01,*HEPG6005,*SK3014,*PT40,*ECG104,*WEP230,*276-2006, *RT-124
2N2148	40421,*PTC138,*GE-16,*TR-01,*HEPG6005,*SK3014,*PT40,*ECG121,*WEP232, *276-2006,*RT-127
2N2150	2N4115
2N2151	2N4115
2N2152	2N2152A,2N2153,2N2153A,2N2156,2N2156A,2N2157,2N2157A,*GE-4,*TR-03, *HEPG6004,*SK3012,*PT515,*ECG213,*WEP233,*276-2006
2N2152A	2N2152,2N2153,2N2153A,2N2156,2N2156A,2N2157,2N2157A,*GE-4,*TR-03, *HEPG6004,*SK3012,*ECG213,*WEP233,*276-2006
2N2153	2N2153A,2N2154,2N2154A,2N2157,2N2157A,2N2158,2N2158A,*GE-4,*TR-03, *HEPG6010,*SK3012,*PT515,*ECG213,*WEP233
2N2153A	2N2153,2N2154,2N2154A,2N2157,2N2157A,2N2158,2N2158A,*GE-4,*TR-03, *HEPG6010,*SK3012,*ECG213,*WEP233
2N2154	2N2154A,2N2155,2N2155A,2N2158,2N2158A,2N2159,2N2159A,*GE-4,*TR-03, *HEPG6010,*SK3012,*PT515,*ECG213,*WEP233
2N2154A	2N2154,2N2155,2N2155A,2N2158,2N2158A,2N2159,2N2159A,*GE-4,*TR-03, *HEPG6010,*SK3012,*ECG213,*WEP233
2N2155	2N2155A,2N2159,2N2159A,*GE-4,*TR-03,*HEPG6010,*SK3012,*PT515,*ECG105, *WEP233
2N2155A	2N2155,2N2159,2N2159A,*TR-03,*SK3012,*ECG105,*WEP233
2N2156	2N2156A,2N2157,2N2157A,*GE-4,*TR-03,*HEPG6010,*SK3012,*PT501,*ECG213, *WEP233
2N2156A	2N2156,2N2157,2N2157A,*GE-4,*TR-03,*HEPG6010,*SK3012,*ECG213,*WEP233
2N2157	2N2157A,2N2158,2N2158A,*GE-4,*TR-03,*HEPG6010,*SK3012,*PT515,*ECG213, *WEP233
2N2157A	2N2157,2N2158,2N2158A,*GE-4,*TR-03,*HEPG6010,*SK3012,*ECG213,*WEP233

To Replace	Substitute This Type
2N2158	2N2158A,2N2159,2N2159A,*GE-4,*TR-03,*HEPG6010,*SK3012,*PT515,*ECG213, *WEP233
2N2158A	2N2158,2N2159,2N2159A,*GE-4,*TR-03,*HEPG6010,*SK3012,*ECG213,*WEP233
2N2159	2N2159A,*GE-4,*TR-03,*HEPG6006,*SK3012,*PT515,*ECG105,*WEP233,*276-2006
2N2159A	2N2159,*GE-4,*TR-03,*HEPG6006,*SK3012,*ECG105,*WEP233,*276-2006
2N2161	2N736,2N736A,2N736B,2N760A,2N915,2N2483,2N2516,2N3077,2N3416,2N3569, 2N3643,2N3859A,2N3877,2N3947,2N4141,2N4227,2N4437,2N4946,2N4951,2N5028, 2N5107,2N5209,2N5368,2N5825,2N6004,2S104,2SC366G,2SC366G-Y,2SC734-Y, 2SC850,2SC875,2SC876,2SC899,2SC933,2SC943,2SC944,2SC968,2SC970, 2SC979-Y,2SC980G-Y,2SC1175,2SC1213A,2SC1220,2SC1247,2SC1277,2SC1318, 2SC1453,2SC1781,2SC1998,2SC1999,BC107A,BC174A,BC182A,BC190A,BC237A, FT3569,GET930,GET2222,GET2222A,GI-3643,MM3904,MPS-A05,TZ82,ZT84,ZT89, ZT114,ZT119,*PTC123,*TR-21,*HEPS0014,*SK3122,*ECG123A,*WEP736, *276-2009,*RT-102
2N2162	2N4008,*GE-65,*TR-20,*HEPS0012,*SK3114,*SP70,*ECG159,*WEP242,*BC177, *276-2022,*RT-115
2N2163	2N1917,2N2162,2N4007,*GE-65,*TR-20,*HEPS0012,*SK3114,*SP(0,*ECG159, *WEP242,*BC177,*276-2022,*RT-115
2N2164	*GE-65,*TR-20,*HEPS0012,*SK3118,*SP70,*ECG159,*WEP242,*BC177,*276-2021, *RT-115
2N2165	*GE-65,*TR-20,*HEPS0012,*SK3114,*SP70,*ECG159,*WEP242,*BC177,*276-2022, *RT-115
2N2166	*GE-65,*TR-20,*HEPS0012,*SK3114,*SP70,*ECG159,*WEP242,*BC177,*276-2022, *RT-115
2N2167	*GE-65,*TR-20,*HEPS0012,*SK3114,*SP70,*ECG159,*WEP242,*BC177,*276-2021, *RT-115
2N2168	2N982,*PTC145,*GE-51,*TR-17,*HEPG0002,*JR200,*ECG160,*WEP637
2N2169	2N2168,*PTC145,*GE-51,*TR-17,*HEPG0002,*JR200
2N2170	2N779,2N779A,2N828,2N828A,2N964,2N968,2N2169,2N2487,2N2488,2N2489, 2N2795,2N2796,2SA411,*PTC145,*GE-51,*TR-17,*HEPG0002,*JR200,*ECG160, *WEP637
2N2171	*GE-1,*TR-05,*HEPG0005,*SK3005,*HF12H,*ECG100,*WEP254,*276-2005,*RT-118
2N2172	2N1204,2N1204A,2N1280,2N1942,2N2173,MM380,*GE-1,*TR-05,*HEPG0005, *SK3005,*ECG100,*WEP254,*276-2005,*RT-118
2N2173	*PTC156,*GE-2,*TR-85,*HEPG0005,*AT100H,*ECG102A,*WEP250,*276-2004, *RT-121
2N2175	2N865,2N2176,2N2944A,2N3058,2N3977,2N5229,2S304,2S324,ZT152,*PTC127, *GE-65,*TR-20,*HEPG0006,*SK3118,*SP70,*ECG106,*WEP52,*BC177,*276-2021, *RT-126
2N2176	2N865,2N2944A,2N3058,2N3977,2N5229,ZT152,*PTC127,*GE-65,*TR-20,*HEPG0006, *SK3118,*SP70,*ECG106,*WEP52,*BC177,*276-2021,*RT-126
2N2177	2N863,2N865,2N2175,2N2176,2N2178,2N2280,2N2944A,2N3217,2N3342,2N3977, 2N4006,2N5141,2N5229,2S304,2S324,ZT152,*PTC127,*GE-65,*TR-20,*HEPG0006, *SK3118,*SP70,*ECG106,*WEP52,*BC177,*276-2021,*RT-126
2N2178	2N863,2N865,2N2176,2N2280,2N2944A,2N3217,2N3977,2N4006,2N5141,2N5229, ZT152,*PTC127,*GE-65,*TR-20,*HEPS0013,*SK3114,*SP70,*ECG159,*WEP242, *BC177,*276-2021,*RT-115
2N2180	2SA252,2SA408,2SA409,*GE-52,*TR-17,*HEPG0003,*ECG126,*WEP635
2N2181	2N941,2N942,2N3677,*GE-65,*TR-20,*HEPS0012,*SK3114,*ECG159,*WEP242, *276-2022
2N2182	M.P.2N2181
2N2183	2N941,2N942,2N2181,2N3677,*GE-21,*TR-20,*HEPS0013,*SK3114,*ECG159, *WEP242,*276-2023,*RT-115
2N2184	M.P.2N2183
2N2185	*PTC131,*GE-65,*TR-20,*HEPS0013,*SK3114,*SP70,*ECG159,*WEP242,*BC177, *276-2022,*RT-115
2N2186	M.P.2N2185
2N2187	M.P.2N2185
2N2188	2N2189,2N2190,2N2191,2N2957,*PTC107,*GE-51,*TR-17,*HEPG0002,*SK3006, *JR200,*ECG160,*WEP637,*276-2005
2N2189	2N2191,*PTC145,*GE-51,*TR-17,*HEPG0002,*SK3006,*JR200,*ECG160,*WEP637,

To Replace	Substitute This Type
(2N2189)	*276-2005
2N2190	2N2191,*GE-51,*TR-17,*HEPG0002,*SK3006,*JR200,*ECG160,*WEP637,*276-2005
2N2191	*GE-51,*TR-17,*HEPG0002,*SK3006,*JR200,*ECG160,*WEP637,*276-2005
2N2192	2N2192A,2N2192B,2N3300,ZT1711,*GE-47,*TR-87,*HEPS3001,*SK3039,*EN10, *ECG128,*WEP243,*276-2008,*RT-113
2N2192A	2N2192,2N2192B,2N3300,ZT1711,*GE-47,*TR-95,*HEPS3001,*SK3039,*EN10, *ECG128,*WEP56,*276-2008,*RT-113
2N2192B	2N2192,2N2192A,2N3300,ZT1711,*GE-47,*HEPS3001,*ECG128,*276-2008
2N2193	2N2193A,2N2193B,*PTC144,*GE-32,*TR-87,*HEPS3011,*SK3039,*EN10,*ECG128, *WEP243,*276-2008,*RT-113
2N2193A	2N2193,2N2193B,*PTC144,*GE-32,*TR-95,*HEPS3011,*SK3039,*EN10,*ECG108, *WEP56,*276-2008,*RT-113
2N2193B	2N2193,2N2193A,*PTC144,*GE-32,*TR-95,*HEPS3011,*SK3039,*EN10,*ECG108, *WEP56,*276-2008,*RT-113
2N2194	2N2194A,2N2194B,2N3110,2N3252,2N3253,2N3444,2N3735,2N5188,2N5262,2SC97, 2SC580,BC140-6,BSX60,*PTC144,*GE-47,*TR-21,*HEPS0014,*SK3122,*SN80, *ECG123A,*WEP736,*276-2008,*RT-102
2N2194A	2N2194,2N2194B,2N3110,2N3252,2N3253,2N3444,2N3735,2N5188,2N5262,2SC97, 2SC580,BC140-6,BSX60,*PTC144,*GE-47,*TR-21,*HEPS0014,*SK3122,*SN80, *ECG123A,*WEP736,*276-2008,*RT-102
2N2194B	2N2194,2N2194A,2N3110,2N3252,2N3253,2N3444,2N3735,2N5188,2N5262,2SC97, 2SC580,BC140-6,BSX60,*PTC144,*GE-47,*TR-21,*HEPS0014,*SK3122,*SN80, *ECG123A,*WEP736,*276-2008,*RT-102
2N2195	2N1958A,2N2194,2N2194A,2N2194B,2N2195A,2N2195B,2N3122,2N5188,2N5189, ZT94,*PTC144,*GE-47,*TR-95,*HEPS3001,*SK3024,*EN10,*ECG108,*WEP56, *276-2008,*RT-113
2N2195A	2N1958A,2N2194,2N2194A,2N2194B,2N2195,2N2195B,2N3122,2N5188,2N5189,ZT94, *PTC144,*GE-47,*TR-87,*HEPS3001,*SK3024,*EN10,*ECG128,*WEP243, *276-2008,*RT-113
2N2195B	2N1958A,2N2194,2N2194A,2N2194B,2N2195,2N2195A,2N3122,2N5188,2N5189,ZT94, *PTC144,*GE-47,*TR-95,*HEPS3001,*SK3024,*ECG108,*WEP56,*276-2008, *RT-113
2N2196	*GE-17,*TR-95,*HEPS3020,*SK3039,*EN10,*ECG224,*WEP56,*RT-113
2N2197	*GE-17,*TR-95,*HEPS3020,*SK3039,*EN10,*ECG108,*WEP56,*RT-113
2N2198	2N2313,EN718A,EN870,*PTC125,*GE-81,*TR-86,*HEPS5026,*SK3124,*ECG123, *WEP53,*276-2012,*RT-100
2N2199	2N741,2N741A,2N960,2N2099,2N2200,*PTC145,*GE-51,*TR-17,*HEPG0002,*JR100, *ECG160,*WEP637,*276-2005
2N2200	2N741,2N741A,2N960,2N2099,2N2199,*PTC145,*GE-51,*TR-17,*HEPG0002,*JR100, *ECG160,*WEP637,*276-2005
2N2201	*HEPS3019
2N2202	*HEPS3019
2N2203	*HEPS3019
2N2204	*GE-12,*TR-81,*HEPS3019,*SK3021,*PN66,*ECG124,*WEP240,*RT-128
2N2205	2N706,2N706A,2N706B,2N783,2N784,2N2318,2N2319,2N2368,2N2656,2N2710, 2N3261,2N3510,2N3511,2N3605A,2N3606A,2N3648,2N4137,2N5030,2N5187,2SC68, 2SC204,2SC321,2SC322,2SC356,2SC576,2SC595,40405,GET708,GET914, GET3013,GET3014,GET3646,KT218,MM1941,MPSH19,MT1038,MT1038A,MT1039, PT2760,ZT2368,*PTC136,*GE-210,*TR-21,*HEPS0011,*SK3124,*SN80,*ECG123A, *WEP736,*BC107B,*276-2016,*RT-102
2N2206	2N3009,2N3013,2N3014,ZT110,*PTC136,*GE-210,*TR-21,*HEPS0011,*SK3122, *EN10,*ECG123A,*WEP736,*BC107B,*276-2016,*RT-102
2N2207	*DS41,*GE-51,*TR-17,*HEPG0009,*SK3006,*HF75,*ECG126,*WEP635,*276-2005, *RT-188
2N2208	*PTC107,*GE-1,*TR-17,*HEPG0001,*SK3006,*ECG126,*WEP635,*276-2002
2N2209	*PTC109,*GE-2,*TR-05,*HEPG0005,*SK3005,*ECG100,*WEP254,*276-2004,*RT-118
2N2210	2N1412,2SB334,*PTC106,*GE-4,*TR-03,*HEPG6006,*SK3012,*PT515,*ECG105, *WEP233,*276-2006
2N2211	*HEPG6018
2N2212	MP2833,MP2834,*PTC122,*GE-25,*TR-01,*HEPG6005,*SK3009,*PT250,*ECG121, *WEP232,*276-2006,*RT-127

TRANSISTOR SUBSTITUTES

To Replace	Substitute This Type
2N2214	2N959,40405,*HEPS0015
2N2216	*GE-27,*TR-78
2N2217	2N3735,2N5188,2N5262,2SC97,2SC580,BSX60,*GE-47,*HEPS3001,*ECG123A, *WEP736,*276-2038
2N2217/51	2N3737,*GE-20,*HEPS0025,*276-2016
2N2218	2N2218A,2N2537,2N2848,2N3725A,2N4960,2N4961,2SC97A,*GE-47,*TR-86, *HEPS3011,*SK3124,*ECG123,*WEP53,*276-2038,*RT-100
2N2218/51	2N5581,*GE-20,*276-2016
2N2218A	2N3725A,2N4961,*TR-86,*HEPS3011,*SK3124,*ECG123,*WEP53,*RT-100
2N2219	2N2219A,2N2538,2N3300,*GE-47,*TR-21,*HEPS3011,*276-2038
2N2219/51	*GE-20,*HEPS0005,*276-2016
2N2219A	*GE-18,*TR-21,*HEPS3011,*SK3024,*ECG123A,*WEP736,*276-2009,*RT-172
2N2220	2N3737,2N6375,*PTC136,*GE-20,*TR-21,*HEPS0015,*SK3122,*ECG123A,*WEP736, *276-2009,*RT-102
2N2221	2N2221A,2N2539,2N4962,2N4963,TN-3903,*PTC136,*GE-20,*TR-21,*HEPS0015, *SK3122,*ECG123A,*WEP736,*276-2009,*RT-102
2N2221A	2N4963,*GE-20,*ECG123A,*WEP736,*276-2009
2N2222	2N2222A,2N2540,A5T2222,FT3643,TIS109,TIS111,TN-3904,*PTC136,*GE-20,*TR-21, *HEPS0015,*SK3122,*ECG123A,*WEP736,*276-2009,*RT-102
2N2222A	*GE-20,*HEPS3001,*276-2009
2N2223	2N2223A,*GE-18,*276-2008
2N2223A	2N2223,*GE-18,*276-2008
2N2224	2N2218,2N2218A,2N2537,2N2846,2N2848,2N3015,2N3299,2N3725,2N3725A,2N4047, 2N4960,2N4961,2SC97A,2SC307,2SC1218,*GE-47,*HEPS3011
2N2226	
2N2227	
2N2228	2N2229
2N2229	
2N2230	
2N2231	
2N2232	2N2233
2N2233	
2N2234	HST5906,*PTC906,*GE-20,*TR-86,*HEPS3024,*SK3124,*ECG123,*WEP53,*RT-110
2N2235	HST5901,*PTC906,*GE-20,*TR-86,*HEPS3020,*SK3124,*ECG123,*WEP736,*RT-110
2N2236	*PTC123,*GE-81,*TR-21,*HEPS0014,*SK3122,*SN80,*ECG123A,*WEP736,*BC337, *276-2008,*RT-102
2N2237	2N2270,2N3724,2N3724A,2N4046,2N4960,2N5145,2N6427,TIS110,TIS133,TIS134, *PTC123,*GE-81,*TR-21,*HEPS0014,*SK3122,*SN80,*ECG123A,*WEP637,*BC337, *276-2008,*RT-102
2N2238	*PTC145,*GE-52,*TR-17,*HEPG0002,*ECG160,*WEP736,*276-2005
2N2239	*PTC195,*GE-63,*HEPS3020,*276-2008
2N2242	2N916,2N916A,2N2221,2N2501,2N2539,2N2847,2N3115,2N3211,2N4013,2N4951, 2N4962,2N5144,2N5368,2SC577,BSW82,TIS133,TIS134,TN-3903,*PTC136,*GE-210, *TR-21,*HEPS0011,*SK3122,*EN10,*ECG123A,*WEP736,*276-2009,*RT-102
2N2243	2N699B,2N2102,2N2102A,2N2243A,2N2405,2N2443,2N3020,2N3036,BSY85, *PTC144,*GE-32,*HEPS3019,*276-2008
2N2243A	2N699B,2N2102,2N2102A,2N2243,2N2405,2N2443,2N3020,2N3036,BSY85,*PTC144, *GE-32,*HEPS3019,*276-2008
2N2244	2N760,2N6427,TIS133,TIS134,*PTC123,*GE-17,*TR-21,*HEPS0011,*SK3122,*EN10, *ECG123A,*WEP736,*BC337,*276-2009,*RT-102
2N2245	2N3242A,2N6426,2SC970,BC547A,*PTC123,*GE-17,*TR-21,*HEPS0011,*SK3122, *EN10,*ECG123A,*WEP736,*BC337,*276-2009,*RT-102
2N2246	BC547B,BC550B,*PTC123,*GE-47,*TR-21,*HEPS0011,*SK3122,*EN10,*ECG123A, *WEP736,*BC337,*276-2009,*RT-102
2N2247	2N718A,2N736A,2N736B,2N4962,2N4963,2N5380,2N6538,A5T3903,TIS96,TIS99, TIS135,TIS136,TN-3903,*PTC123,*GE-81,*TR-21,*HEPS0011,*SK3122,*EN10, *ECG123A,*WEP736,*BC337,*276-2008,*RT-102
2N2248	2N956,2N2645,2N5381,2N6540,2N6541,A5T3904,BC546A,TIS95,TIS98,TN-3904, *PTC123,*GE-81,*TR-21,*HEPS0011,*SK3122,*EN10,*ECG123A,*WEP736,*BC337, *276-2008,*RT-102
2N2249	*GE-17,*TR-21,*HEPS0011,*SK3122,*EN10,*ECG123A,*WEP736,*276-2009,*RT-102

To Replace	Substitute This Type
2N2250	2N760,*PTC123,*GE-17,*TR-21,*HEPS0011,*SK3122,*EN10,*ECG123A,*WEP736, *BC337,*276-2009,*RT-102
2N2251	2N3242A,BC547A,*PTC123,*GE-17,*TR-21,*HEPS0011,*SK3122,*EN10,*ECG123A, *WEP736,*BC337,*276-2009,*RT-102
2N2252	A5T3565,BC547B,BC550B,*PTC123,*GE-47,*TR-21,*HEPS0011,*SK3122,*EN10, *ECG123A,*WEP736,*BC337,*276-2009,*RT-102
2N2253	2N718A,2N736A,2N736B,2N4962,2N4963,2N5380,2N6538,A5T3903,TIS96,TIS99, TN-3903,*PTC123,*GE-81,*TR-21,*HEPS0011,*SK3122,*EN10,*ECG123A,*WEP736, *BC337,*276-2008,*RT-102
2N2254	2N956,2N2645,2N5381,2N6540,2N6541,A5T3904,BC546A,TIS95,TIS98,TN-3904, *PTC123,*GE-81,*TR-21,*HEPS0011,*SK3122,*EN10,*ECG123A,*WEP736,*BC337, *276-2008,*RT-102
2N2255	2N6428,2N6428A,2N6539,BC546B,TIS94,TIS97,*PTC123,*GE-47,*TR-21,*HEPS0011, *SK3122,*EN10,*ECG123A,*WEP736,*BC337,*276-2008,*RT-102
2N2256	*PTC136,*GE-10,*TR-21,*HEPS0011,*SK3122,*EN10,*ECG123A,*WEP736,*BC107B, *276-2016,*RT-102
2N2257	2N709A,2N743,2N988,2N989,2N5186,2SC98,2SC395A-R,2SC718,2SC1231,40519, MM1748,MPS2713,MPS2926-RED,*PTC136,*GE-10,*TR-21,*HEPS0011,*SK3122, *EN10,*ECG123A,*WEP637,*BC107B,*276-2016,*RT-102
2N2258	2N741,2N741A,2N960,2N961,2N962,2N963,2N2630,*PTC145,*GE-51,*TR-17, *HEPG0003,*JR100,*ECG160,*WEP637
2N2259	2N828,2N964,2N965,2N966,2N967,2N968,2N969,2N970,2N971,2N2402,2N2860, 2SA410,2SA411,2SA450,*PTC145,*GE-51,*TR-17,*HEPG0003,*JR100,*ECG160, *WEP637
2N2266	2N2267,2N2268,2N2269,*GE-4,*TR-03,*SK3012,*PT40,*ECG105,*WEP233
2N2267	2N2269,*GE-4,*TR-03,*SK3012,*PT40,*ECG105,*WEP233
2N2268	2N2266,2N2267,2N2269,*GE-4,*TR-03,*SK3012,*PT40,*ECG105,*WEP233
2N2269	2N2267,*GE-4,*TR-03,*SK3012,*PT40,*ECG105,*WEP233
2N2270	*GE-63,*TR-87,*HEPS3001,*SK3024,*ECG128,*WEP243,*276-2008,*RT-114
2N2271	2SB461,*PTC135,*GE-53,*TR-05,*HEPG0005,*SK3004,*AT20H,*ECG102,*WEP631, *AC188/01,*276-2005,*RT-120
2N2272	2N916,2N916A,2N2222,2N2540,2N3116,2N3227,2N3643,2N4951,2N4952,2N5107, 2N5368,2N5369,2N6010,A5T2222,BSW83,EN2219,GI-3643,TIS111,TN-3904,*PTC136, *GE-18,*TR-21,*HEPS0014,*SK3122,*ECG123A,*WEP736,*276-2009,*RT-102
2N2273	2N2957,*PTC145,*GE-51,*TR-17,*HEPG0003,*JR200,*ECG160,*WEP637
2N2274	2N859,2N2945A,2N2946A,2N4284,2N4285,*GE-65,*TR-20,*HEPS0013,*SK3114, *SP70,*ECG159,*WEP242,*BC177,*276-2022,*RT-115
2N2275	M.P.2N2274
2N2276	2N861,2N2274,2N2278,2N2945A,2N4007,2N4284,TW135,*GE-65,*TR-20,*HEPS0013, *SK3114,*SP70,*ECG159,*WEP242,*BC177,*276-2022,*RT-115
2N2277	M.P.2N2276
2N2278	2N861,2N2945A,2N4007,2N4284,2N5229,2N5230,TW135,*GE-65,*TR-20,*HEPS0013, *SK3114,*SP70,*ECG159,*WEP242,*BC177,*276-2022,*RT-115
2N2279	M.P.2N2278
2N2280	2N865,2N4006,2N4007,ZT152,*GE-65,*TR-20,*HEPS0013,*SK3114,*SP70,*ECG159, *WEP242,*BC177,*276-2021,*RT-115
2N2281	M.P.2N2280
2N2282	*GE-3,*TR-01,*SK3009,*AT100M,*ECG104,*WEP230,*RT-124
2N2283	*RT-124
2N2284	
2N2285	*PTC147,*GE-76,*TR-35,*HEPG6005,*SK3014,*PT150,*ECG179,*WEPG6001, *276-2006,*RT-147
2N2286	2N2287,*TR-35,*HEPG6005,*SK3014,*PT150,*ECG179,*WEPG6001,*276-2006, *RT-147
2N2287	*TR-01,*HEPG6005,*SK3014,*PT150,*ECG121,*WEP232,*276-2006,*RT-127
2N2288	*GE-16,*TR-01,*HEPG6003,*SK3009,*PT150,*ECG121,*WEP232,*276-2006,*RT-127
2N2289	MP1612,*DS503,*GE-3,*TR-01,*HEPG6005,*SK3009,*PT150,*ECG121,*WEP232, *276-2006,*RT-127
2N2290	MP1612A,MP1612B,*DS503,*GE-3,*TR-27,*HEPG6005,*SK3009,*PT150,*ECG127, *WEP232,*276-2006
2N2291	*PTC122,*GE-25,*TR-01,*HEPG6003,*SK3009,*PT150,*ECG121,*WEP232,*276-2006,

To Replace	Substitute This Type
(2N2291)	*RT-127
2N2292	2N2832,MP1612,*PTC122,*GE-25,*TR-01,*HEPG6005,*SK3009,*PT150,*ECG121,*276-2006,*RT-127
2N2293	2N2833,2N2834,MP1612A,MP1612B,*PTC122,*GE-25,*TR-01,*HEPG6005,*SK3009,*PT150,*ECG121,*WEP232,*276-2006,*RT-127
2N2294	*GE-25,*TR-01,*HEPG6003,*SK3009,*PT250,*ECG121,*WEP232,*276-2006,*RT-127
2N2295	2N2691,MP2832,*DS503,*GE-25,*TR-01,*HEPG6005,*SK3009,*PT250,*ECG121,*WEP232,*276-2006,*RT-127
2N2296	2N2212,MP2833,MP2834,*DS503,*GE-25,*TR-01,*HEPG6005,*SK3009,*PT250,*ECG121,*WEP232,*276-2006,*RT-127
2N2297	*PTC144,*GE-27,*TR-78,*HEPS3019,*SK3104,*ECG128,*WEP243,*BF338
2N2297/51	2N2461,2N2516,*GE-213,*TR-51,*HEPS0005,*BF338,*276-2009
2N2303	2N2801,2N2905,2N2905A,2N3503,2N3671,2N4032,2N4037,2N4890,2SA498-O,2SA498-Y,2SA503-GR,2SA503-Y,2SA571,2SA580,2SA684,2SA817,2SA891,40537,40634,A5T2907,A5T3505,A5T3645,BC161,BC161-10,BC161-16,BC361-10,TIS112,*PTC127,*GE-82,*TR-20,*HEPS0012,*SK3114,*SP70,*ECG159,*WEP242,*BC327,*276-2021,*RT-115
2N2304	2N1485,ZT1485,*HEPS5014,*SK3024,*276-2020
2N2305	2N1487,2N1489,2SD15,2SD163,180T2,BDY23A,BDY23B,HST9201,HST9205,HST9206,HST9801,HST9802,SDT7731,SDT7732,ZT1487,ZT1489,*PTC173,*GE-77,*TR-59,*HEPS7002,*SK3027,*HN100,*ECG130,*WEP247,*BDY20,*RT-131
2N2306	*HEPS5014
2N2308	2N1484,2N1486,S2N1486,ZT1484,ZT1486,*HEPS5000
2N2309	2N1708A,2N1838,2N1941,2N2237,2N2330,2N3724,2N3724A,2N4046,2N5145,2N6427,2SC306,2SC482-Y,2SC504-Y,2SC509-O,2SC509G-O,A5T5450,A8T3705,BC340-10,BSY81,TI-484,TIS133,TIS134,TN61,*PTC123,*GE-81,*TR-21,*HEPS0014,*SK3122,*SP70,*ECG123A,*WEP736,*BC337,*276-2008,*RT-102
2N2310	*PTC125,*GE-81,*TR-21,*HEPS0014,*SK3122,*ECG123A,*WEP736,*BF338,*276-2008,*RT-102
2N2311	*PTC144,*GE-27,*TR-78,*HEPS5026,*BF338,*276-2008
2N2312	2N2351,2N2351A,A5T2193,*PTC125,*GE-81,*TR-21,*HEPS0014,*SK3122,*ECG123A,*WEP736,*BF338,*276-2012,*RT-102
2N2313	*PTC144,*GE-27,*TR-78,*HEPS5026,*BF338,*276-2012
2N2314	2N2312,2N2352,2N2352A,*PTC123,*GE-62,*TR-21,*HEPS0014,*SK3122,*ECG123A,*WEP736,*BC337,*276-2016,*RT-102
2N2315	2N2317,2N2351,2N2351A,2N2389,2N2900,A5T2193,TIS110,*PTC123,*GE-62,*TR-21,*HEPS0014,*SK3122,*ECG123A,*WEP736,*BC337,*276-2016,*RT-102
2N2316	2N2364,2N2364A,2N2898,2N2899,2N3037,A5T2243,*PTC125,*GE-18,*HEPS5026,*BF338,*276-2008
2N2317	2N2351,2N2351A,2N2389,A5T2193,*PTC123,*GE-81,*HEPS3020,*BF338,*276-2008
2N2318	2N708,2N708A,2N784A,2N2242,2N2501,2N3011,2N3013,2N3014,2N3605A,2N3606A,2N3646,2N5772,2SC67,2SC204,2SC639,2SC764,A5T4123,GET708,GET914,GET2369,GET3013,GET3014,GET3646,ZT708,*PTC126,*GE-17,*TR-21,*HEPS0011,*SK3122,*SN80,*ECG123A,*WEP736,*276-2009,*RT-102
2N2319	2N3013,2N3014,2N4123,2N4418,2N5772,A5T4123,*PTC126,*GE-61,*TR-21,*HEPS0014,*SK3122,*ECG123A,*WEP736,*BF173,*276-2016,*RT-102
2N2320	2N4264,2N5772,A5T4123,BSY63,*PTC136,*GE-215,*TR-21,*HEPS0014,*SK3122,*ECG123A,*WEP736,*276-2038,*RT-102
2N2330	*PTC144,*GE-47,*TR-86,*SK3124,*SN80,*ECG123,*WEP53,*276-2030,*RT-100
2N2331	2N3241A,2SC1999,A5T4123,*PTC123,*GE-17,*TR-21,*HEPS0011,*SK3122,*SN80,*ECG123A,*WEP736,*BC337,*276-2009,*RT-102
2N2332	*GE-65,*TR-20,*HEPS0013,*SK3114,*SP70,*ECG159,*WEP242,*BC177,*276-2022,*RT-115
2N2333	*GE-65,*TR-20,*HEPS0013,*SK3114,*SP70,*ECG159,*WEP242,*BC177,*276-2022,*RT-115
2N2334	*GE-65,*TR-20,*HEPS0013,*SK3114,*SP70,*ECG159,*WEP242,*BC177,*276-2022,*RT-115
2N2335	*GE-65,*TR-20,*HEPS0013,*SK3114,*SP70,*ECG159,*WEP242,*BC177,*276-2022,*RT-115
2N2336	*GE-65,*TR-20,*HEPS0013,*SK3114,*SP70,*ECG159,*WEP242,*BC177,*276-2022,*RT-115

To Replace	Substitute This Type
2N2337	*GE-65,*TR-20,*HEPS0013,*SK3114,*SP70,*ECG159,*WEP242,*BC177,*276-2022, *RT-115
2N2338	*HEPS5004
2N2339	2N1768,*GE-66,*HEPS5014,*276-2020
2N2340	KP3446,*PTC195,*HEPS3020
2N2341	*PTC123,*HEPS3020
2N2342	*PTC195,*HEPS3019
2N2343	*PTC195,*HEPS3019
2N2349	2N909,2N930,2N2388,2N2484,2N2484A,2N2586,2N3302,2N3566,2N3694,2N3827, 2N3904,2N4287,2N4967,2N4995,2N5826,2SC366G-Y,2SC367G-Y,2SC538A,2SC587, 2SC587A,2SC711A,2SC752G-Y,2SC1000-GR,2SC1000G-GR,2SC1175,2SC1285, 2SC1364,2SC1380-GR,2SC1380A-GR,2SC1632,2SC1634,2SC1681-GR,2SC1685, 2SC1850,2SC1992,A157B,BC107,BC107B,BC123,BC167B,BC171B,BC182B,BC182L, BC237B,BC413B,BC414B,BCY59B,BFY39-3,BSX79,EN2484,GI-3566,MPS3694, MPS6514,MPS6520,MPS6566,MPS6575,MPS6576-SIL,MPS-A10,MPS-A20,MPS-K20, MPS-K21,MPS-K22,MPSA10-BLU,MPSA10-YEL,MPSA17,MPSA20-BLU,MPSA20-YEL, NPSA20,PBC107A,PBC107B,PET1002,SE1002,TIS23,TIS24,TP3704,TP3706,TZ81, *PTC123,*GE-46,*TR-21,*HEPS0014,*SK3122,*SN80,*ECG123A,*WEP736, *276-2009,*RT-102
2N2350	2N2350A,*GE-47,*HEPS3020,*BC337,*276-2009
2N2350A	2N2350,*GE-47,*HEPS3020,*BC337,*276-2009
2N2351	2N2351A,*PTC144,*GE-18,*HEPS3011,*276-2008
2N2351A	2N2351,*PTC144,*HEPS3011,*276-2008
2N2352	2N2352A,2N3737,*PTC144,*GE-47,*HEPS3020,*BC337,*276-2009
2N2352A	2N2352,2N3737,*PTC144,*GE-47,*HEPS3020,*BC337,*276-2009
2N2353	2N2352,2N2352A,2N2353A,*PTC144,*GE-47,*TR-21,*HEPS0014,*SK3122,*ECG123A, *WEP736,*BC337,*276-2016,*RT-102
2N2353A	2N2352,2N2352A,2N2353,*PTC144,*GE-47,*TR-21,*HEPS0014,*SK3122,*ECG123A, *WEP736,*BC337,*276-2016,*RT-102
2N2354	2N2430,2SD27,2SD30,2SD170A,2SD186,2SD187,AC186,ASY29,*PTC108,*GE-59, *TR-08,*HEPG0011,*SK3124,*ECG103,*WEP641A,*276-2001,*RT-122
2N2356A	
2N2357	*PTC901,*GE-3,*TR-01,*SK3009,*PT250,*ECG121,*WEP232,*276-2006,*RT-127
2N2358	2N2359,*DS503,*GE-3,*TR-01,*SK3009,*PT250,*ECG121,*WEP232,*276-2006, *RT-127
2N2359	
2N2360	2N2398,*PTC145,*GE-51,*TR-17,*HEPG0002,*JR100,*ECG160,*WEP637,*276-2005
2N2361	2N2360,2N2362,2N2398,2N2399,*PTC107,*GE-51,*TR-17,*HEPG0002,*JR100, *ECG160,*WEP637,*276-2005
2N2362	2N2360,2N2361,2N2398,2N2399,*PTC107,*GE-51,*TR-17,*HEPG0002,*JR100, *ECG160,*WEP637,*276-2005
2N2363	*PTC145,*GE-51,*TR-17,*HEPG0003,*SK3006,*JR100,*ECG126,*WEP635
2N2364	2N2364A,2N2898,2N2899,*PTC144,*GE-18,*HEPS3019,*BF338,*276-2012
2N2364A	2N2364,2N2898,2N2899,*PTC144,*GE-18,*HEPS3019,*BF338,*276-2012
2N2368	2SC576,ZT2368,*GE-20,*TR-21,*HEPS0011,*SK3122,*ECG123A,*WEP736, *276-2009,*RT-102
2N2369	2N2369A,2N3862,2N4873,2SC577,2SC764,ZT2369,ZT2369A,*TR-21,*HEPS0011, *SK3122,*ECG123A,*WEP736,*276-2038,*RT-102
2N2369A	2N2369,2N3862,2N4873,2SC577,2SC764,ZT2369,ZT2369A,*PTC133,*TR-21, *HEPS0011,*SK3122,*ECG123A,*WEP736,*276-2038,*RT-102
2N2370	2N1024,2N1220,2N1228,2N2371,BCY33,MM4052,SHA7532,*GE-65,*TR-20, *HEPS0012,*SK3114,*SP70,*ECG159,*WEP242,*BC177,*276-2022,*RT-115
2N2371	2N1024,2N1219,2N1228,2N3401,2N6567,BCY33,MM4052,SHA7532,*GE-65,*TR-20, *HEPS0012,*SK3114,*SP70,*ECG159,*WEP242,*BC177,*276-2022,*RT-115
2N2372	2N2373,MM4052,*GE-65,*TR-20,*HEPS0013,*SK3114,*SP70,*ECG159,*WEP242, *BC177,*276-2022,*RT-115
2N2373	2N6567,MM4052,*GE-65,*TR-20,*HEPS0013,*SK3114,*SP70,*ECG159,*WEP242, *BC177,*276-2022,*RT-115
2N2374	*GE-2,*TR-05,*HEPG0005,*SK3004,*ECG102,*WEP631,*276-2005,*RT-120
2N2375	2N597,2N1997,*GE-1,*TR-05,*HEPG0005,*SK3004,*HF12M,*ECG102,*WEP631, *276-2005,*RT-120

To Replace	Substitute This Type
2N2376	M.P.2N2375
2N2377	2N2945A,2N4284,2N4285,2N5230,TW135,*GE-65,*TR-20,*HEPS0013,*SK3114, *SP70,*ECG159,*WEP242,*BC177,*276-2022,*RT-115
2N2378	2N860,2N862,*GE-65,*TR-20,*HEPS0013,*SK3114,*SP70,*ECG159,*WEP242, *BC177,*276-2021,*RT-115
2N2379	2N1100,2N2493,*GE-4,*HEPG6006
2N2380	2N1837,2N1837A,2N2218A,2N2380A,2N2479,2N3498,2N3678,2N3722,2N3723, 2N3725,2N3725A,2N4047,2N4961,2SC97A,2SC108A-O,2SC307,2SC349,2SC560, 2SC1218,40635,BSY53,BSY83,BSY87,SE8010,TIS135,TIS136,TN53,*PTC123, *GE-81,*HEPS0015,*276-2009
2N2380A	2N1837,2N1837A,2N2218A,2N2380,2N2479,2N3498,2N3678,2N3722,2N3723,2N3725, 2N3725A,2N4047,2N4961,2SC97A,2SC108A-O,2SC307,2SC349,2SC560,2SC1218, 40635,BSY53,BSY83,BSY87,SE8010,TIS135,TIS136,TN53,*PTC123,*GE-81,*TR-78, *HEPS3011,*BF338
2N2381	2N2382,*GE-51,*TR-17,*HEPG0002,*ECG160,*WEP637,*276-2005
2N2382	*GE-51,*TR-17,*HEPG0002,*ECG160,*WEP637,*276-2005
2N2387	2N2396,MM3903,*GE-212,*TR-21,*HEPS0025,*SK3122,*SN80,*ECG123A,*WEP736, *BC107B,*276-2013,*RT-102
2N2388	2N6112,*GE-212,*TR-21,*HEPS0015,*SK3122,*SN80,*ECG123A,*WEP736,*BC107B, *276-2013,*RT-102
2N2389	*GE-81,*TR-86,*HEPS3020,*SK3124,*SN80,*ECG123,*WEP53,*276-2008,*RT-100
2N2390	*GE-81,*TR-86,*HEPS3020,*SK3124,*SN80,*ECG123,*WEP53,*276-2008,*RT-100
2N2391	*HEPS0013
2N2392	2N3829,MM3905,*HEPS0013
2N2393	*GE-82,*TR-20,*HEPS0019,*SK3114,*SP70,*ECG159,*WEP242,*BC327,*RT-115
2N2394	*GE-82,*TR-20,*HEPS0019,*SK3114,*SP70,*ECG159,*WEP242,*BC327,*RT-115
2N2395	*GE-20,*TR-20,*HEPS0025,*SN60,*ECG106,*WEP52,*BC337,*276-2009,*RT-126
2N2396	2N2389,2N2900,*GE-20,*TR-86,*HEPS0025,*SK3124,*SN80,*ECG123,*WEP53, *BC337,*276-2009,*RT-100
2N2397	*GE-210,*TR-86,*HEPS0013,*SK3124,*SN80,*ECG123,*WEP53,*BC107B,*276-2016, *RT-100
2N2398	2N2360,*PTC145,*GE-51,*TR-17,*HEPG0002,*JR100,*ECG160,*WEP637,*276-2005
2N2399	2N2360,2N2361,2N2362,2N2398,*PTC107,*GE-51,*TR-17,*HEPG0002,*JR100, *ECG160,*WEP637,*276-2005
2N2400	2N828,2N828A,2N964,2N965,2N966,2N967,2N968,2N969,2N970,2N2401,2N2860, 2SA410,2SA411,2SA450,*PTC145,*GE-1,*TR-17,*HEPG0003,*JR100,*ECG160, *WEP637
2N2401	2N711A,2N711B,2N829,2N838,2N972,2N2402,*PTC145,*GE-1,*TR-17,*HEPG0003, *JR100,*ECG160,*WEP637
2N2402	2N838,*PTC145,*GE-51,*TR-17,*HEPG0003,*JR100,*ECG160,*WEP637
2N2405	*GE-27,*TR-87,*HEPS3019,*SK3024,*ECG282,*WEP243,*BF338,*RT-114
2N2410	2N2218,2N2218A,2N2537,2N2846,2N2848,2N3015,2N3299,2N3725A,2N4960,2N4961, 2SC97A,*GE-47,*TR-65,*HEPS3001,*276-2012
2N2410/51	*GE-47,*TR-65,*276-2038
2N2411	2SA558,*PTC131,*GE-65,*TR-20,*HEPS0013,*SK3114,*SP70,*ECG159,*WEP242, *BC177,*276-2034,*RT-115
2N2412	2N869,2N869A,2N3829,2N3905,2N4402,2N5382,2SA467,2SA509-O,2SA509G-O, A5T3905,A5T4402,BSW72,MM3905,TN-3905,*PTC127,*GE-65,*TR-20,*HEPS0013, *SK3114,*SP70,*ECG159,*WEP242,*BC177,*276-2034,*RT-115
2N2413	2N708,2N914,2N916,2N916A,2N2221,2N2242,2N2501,2N2539,2N2845,2N2847, 2N3013,2N3014,2N3115,2N3210,2N3211,2N3301,2N3646,2N3903,2N3946,2N4013, 2N4123,2N4227,2N4962,2N5027,2N5144,2N5380,2N5772,2SC67,2SC577,2SC639, 2SC764,A5T3903,A5T4123,BSW82,FT3641,FT3642,MM3903,MPS6530,TN-3903, ZT84,ZT708,*PTC136,*GE-20,*TR-21,*HEPS0011,*SK3122,*EN10,*ECG123A, *WEP736,*BC107B,*276-2016,*RT-102
2N2414	
2N2415	2N2998,AFY39,*GE-51,*TR-17,*HEPG0003,*ECG160,*WEP637
2N2416	2N2415,2N2998,AFY39,TI-400,TIXM101,*GE-51,*TR-17,*HEPG0003,*ECG160, *WEP637
2N2423	2N1021A,2N1022A,2N1537A,2N1542A,2N1543A,2N3618,*PTC105A,*GE-16,*TR-01, *HEPG6005,*SK3009,*PT150,*ECG121,*WEP232,*276-2006,*RT-127

To Replace	Substitute This Type
2N2424	*GE-65,*TR-20,*HEPS0012,*SK3114,*SP70,*ECG159,*WEP242,*BC327,*276-2023, *RT-115
2N2425	*GE-65,*TR-20,*HEPS0012,*SK3114,*SP70,*ECG159,*WEP242,*BC327,*276-2023, *RT-115
2N2426	2N1012,*PTC108,*GE-8,*TR-08,*HEPG0011,*SK3011,*ECG101,*WEP641,*RT-119
2N2427	2N758,2N758A,2N759,2N759A,2N760B,MPSA05,MPSH32,TIS84,TIS108,*PTC123, *GE-20,*TR-21,*HEPS0011,*SK3122,*SN60,*ECG123A,*WEP736,*BC337, *276-2009,*RT-102
2N2428	2SB273,*PTC135,*GE-52,*TR-85,*HEPG0006,*SK3004,*AT20M,*ECG102A,*WEP250, *AC126,*276-2007,*RT-121
2N2429	*PTC135,*GE-53,*TR-85,*HEPG0005,*SK3004,*AT20M,*ECG102A,*WEP250, *276-2004,*RT-121
2N2430	*PTC134,*GE-54R,*TR-08,*HEPG0011,*SK3010,*NA20,*ECG103A,*WEP641B, *AC127,*276-2001,*RT-122
2N2431	*PTC156,*GE-53,*TR-84,*HEPG6011,*SK3004,*AT20M,*ECG158,*WEP238, *276-2006
2N2432	2N2432A,*GE-212,*TR-21,*HEPS0011,*SK3122,*ECG123A,*WEP736,*BC107B, *276-2013,*RT-102
2N2432A	*GE-212,*TR-51,*HEPS0025,*BC107B,*276-2013
2N2433	HST5002,HST5003,SDT5002,SDT5003,*PTC144,*GE-27,*TR-78,*HEPS3020,*BF338
2N2434	*HEPS3020
2N2435	HST5004,*PTC144,*GE-27,*TR-78,*HEPS3019,*BF338
2N2436	
2N2437	*PTC144,*GE-27,*TR-78,*HEPS3019,*BF338
2N2438	HST5008,*PTC144,*GE-27,*TR-78,*BF338
2N2439	2N2435,HST5003,*PTC144,*GE-27,*TR-78,*HEPS3019
2N2440	2N2436,*HEPS3019
2N2443	*PTC117,*GE-27,*TR-78,*HEPS5026,*ECG128,*WEP243,*BF338,*276-2012
2N2444	2N1907,*HEPG6014,*PT150
2N2445	2N1167,2N1167A,MP1556,MP1556A,MP1560,MP1560A,*HEPG6009,*PT250
2N2446	2N1160,2N3615,*PTC105A,*GE-16,*TR-01,*HEPG6005,*SK3009,*ECG121,*WEP232, *276-2006,*RT-127
2N2447	*GE-2,*TR-85,*HEPG0005,*SK3004,*ECG102A,*WEP250,*276-2007,*RT-121
2N2448	*GE-2,*TR-85,*HEPG0005,*SK3004,*AT30M,*ECG102A,*WEP250,*276-2007,*RT-121
2N2449	*PTC135,*GE-53,*TR-85,*HEPG0007,*SK3004,*AT30H,*ECG102A,*WEP250,*AC128, *RT-121
2N2450	*GE-53,*TR-85,*HEPG0007,*SK3004,*AT30H,*ECG102A,*WEP250,*AC128,*RT-121
2N2451	*PTC109,*GE-51,*TR-17,*HEPG0003,*SK3006,*HF6M,*ECG126,*WEP635
2N2453	2N2453A,2N2903,2N2903A
2N2453A	
2N2455	*PTC145,*GE-51,*TR-17,*HEPG0002,*JR100,*ECG160,*WEP637,*276-2005
2N2456	*PTC145,*GE-51,*TR-17,*HEPG0003,*JR100,*ECG126,*WEP635
2N2459	*PTC125,*HEPS0005
2N2460	*PTC144,*HEPS0005
2N2461	2N2462,*PTC144,*HEPS0005
2N2462	*HEPS0005
2N2463	*PTC125,*GE-27,*TR-78,*HEPS0005,*BF338
2N2464	*PTC144,*HEPS0005
2N2465	2N2466,*PTC144,*GE-54,*HEPS0005
2N2466	*TR-21,*HEPS0005,*SK3122,*AT100H,*ECG123A,*WEP736,*276-2009,*RT-102
2N2467	*HEPG6011,*276-2006
2N2468	*GE-51,*TR-05,*HEPG6012,*SK3004,*AT100H,*ECG102,*WEP631,*RT-120
2N2469	*GE-2,*TR-05,*SK3004,*AT100H,*ECG102,*WEP631,*RT-120
2N2472	*GE-20,*HEPS3019,*276-2008
2N2473	*GE-20,*HEPS3019
2N2474	2N923,2N925,2N938,2N943,2N944,2N945,2N1025,2N1439,2N1641,HA7530,SHA7530, *GE-65,*TR-30,*BC177,*276-2022
2N2475	2N3959,2N3960,BF123,TIS86,*PTC126,*GE-61,*HEPS0016,*BF173,*276-2038
2N2475/46	2N5200,2N5201,BF123,MT1060A,*PTC126,*GE-61,*HEPS0014,*BF173,*276-2038
2N2475/51	2N5200,2N5201,BF123,MT1060A,*PTC126,*GE-61,*HEPS0011,*BF173,*276-2038
2N2476	2N3512,2N5189,BSX59,BSX61,*PTC136,*GE-47,*TR-21,*HEPS0014,*SK3122,

To Replace	Substitute This Type
(2N2476)	*ECG123A,*WEP736,*276-2038,*RT-102
2N2477	2N2217,2N3735,2N5188,2N5262,2N6375,2N6376,2SC97,2SC255,2SC580,BSX60, *PTC136,*GE-47,*TR-21,*HEPS0014,*SK3122,*ECG123A,*WEP736,*276-2038, *RT-102
2N2478	*GE-18,*HEPS3019
2N2479	2N1837,2N1837A,2N2218A,2N3725,2N3725A,2N4047,2N4961,2SC97A,2SC108A-O, 2SC307,2SC560,2SC1218,SE8010,TIS135,TIS136,*PTC123,*GE-81,*HEPS3011
2N2480	D2T2218A,D2T2219A
2N2480A	2N2223,2N2223A,2N2480,2N2652,2N2652A,2N6502,D2T2218A,D2T2219A
2N2481	2N708A,2N914,2N2221,2N2501,2N2539,2N2845,2N2847,2N3115,2N3210,2N3211, 2N3301,2N3646,2N3647,2N4013,2N4962,2N5144,2SC577,BSW82,FT3641,FT3642, GI-3641,TIS133,TIS134,TN-3903,ZT708,*PTC136,*GE-20,*TR-21,*HEPS0011, *SK3122,*EN10,*ECG123A,*WEP736,*276-2009,*RT-102
2N2482	*GE-54,*TR-08,*HEPG0011,*SK3010,*ECG103,*WEP641A,*RT-122
2N2483	2N956,2N2390,2N2484,2N2484A,2N2645,2N5381,2N6540,2N6541,A5T3904,BC546A, MM3904,TIS95,TIS98,TN-3904,*PTC123,*GE-81,*TR-21,*HEPS0025,*SK3122, *ECG123A,*WEP736,*BF338,*276-2008,*RT-102
2N2484	2N3117,BC546B,TIS94,TIS97,*PTC123,*GE-81,*TR-21,*HEPS0005,*SK3122, *ECG123A,*WEP736,*276-2008,*RT-102
2N2484A	2N2484,BC546B,TIS94,TIS97,*PTC123,*GE-63,*HEPS0005,*276-2008
2N2485	
2N2486	2N2485
2N2487	2N779,2N779A,2N828,2N964,2N968,2N984,2N2488,2N2489,2N2795,2SA411, *PTC145,*GE-51,*TR-17,*HEPG0003,*SK3006,*JR30,*ECG126,*WEP635
2N2488	*DS25,*GE-51,*TR-17,*HEPG0003,*SK3006,*JR30,*ECG126,*WEP635
2N2489	2N2795,*PTC145,*GE-51,*TR-17,*HEPG0003,*SK3006,*JR30,*ECG126,*WEP635
2N2490	2N1099,2N2492,2N2493,*GE-4,*TR-03,*HEPG6006,*SK3012,*PT501,*ECG105, *WEP233,*276-2006
2N2491	2N174A,2N1099,2N2075,2N2075A,2N2076,2N2076A,2N2153,2N2153A,2N2154, 2N2154A,*GE-4,*TR-03,*HEPG6006,*SK3012,*PT501,*ECG105,*WEP233,*276-2006
2N2492	2N174A,2N2493,*GE-4,*TR-03,*HEPG6006,*SK3012,*PT501,*ECG105,*WEP233, *276-2006
2N2493	*GE-4,*TR-03,*HEPG6006,*SK3012,*PT501,*ECG105,*WEP233,*276-2006
2N2495	*PTC145,*GE-51,*TR-17,*HEPG0002,*SK3006,*HF75,*ECG160,*WEP637,*276-2005
2N2496	*PTC107,*GE-51,*TR-17,*HEPG0002,*SK3006,*HF75,*ECG160,*WEP637,*276-2005
2N2501	2N3947,2SC1999,*PTC136,*GE-17,*TR-21,*HEPS0011,*SK3122,*EN10,*ECG123A, *WEP736,*276-2009,*RT-102
2N2509	2N720A,2N3037,*PTC125,*GE-18,*TR-78,*HEPS0005,*SK3045,*ECG154,*WEP712, *BF338,*276-2012,*RT-110
2N2510	*GE-20,*TR-78,*HEPS0005,*SK3045,*ECG154,*WEP712,*276-2012,*RT-110
2N2511	*GE-20,*HEPS0005,*ECG154,*WEP712,*276-2012
2N2512	*TR-17,*ECG160,*WEP637
2N2514	*PTC123,*GE-81,*TR-25,*HEPS3011,*BF338,*276-2012
2N2515	2N2389,*PTC123,*GE-81,*HEPS3011
2N2516	*PTC125,*HEPS3011
2N2517	*PTC125,*GE-18,*HEPS3019,*BF338,*276-2012
2N2518	*PTC125,*HEPS3019
2N2519	*PTC125,*HEPS3019
2N2520	*PTC125,*GE-81,*TR-21,*HEPS0014,*SK3122,*ECG123A,*WEP736,*BF338, *276-2008,*RT-102
2N2521	*PTC125,*GE-81,*TR-21,*HEPS0014,*SK3122,*ECG123A,*WEP736,*276-2009, *RT-102
2N2522	*PTC125,*GE-63,*TR-21,*HEPS0014,*SK3122,*ECG123A,*WEP736,*276-2009, *RT-102
2N2523	BC546B,*GE-210,*TR-21,*HEPS0014,*SK3122,*ECG123A,*WEP736,*BC337, *276-2008,*RT-102
2N2524	*GE-210,*TR-21,*HEPS0014,*SK3122,*ECG123A,*WEP736,*BC337,*276-2008, *RT-102
2N2525	*HEPS3019
2N2526	2N1555,2N1555A,2N1556A,2N4242,CQT940A,CQT940B,CQT940BA,CTP1500, CTP1503,*PTC105A,*GE-76,*TR-27,*HEPG6015,*PT150,*ECG127,*WEP235

To Replace	Substitute This Type
2N2527	2N2528,*GE-25,*TR-27,*HEPG6007,*PT150,*ECG127,*WEP235
2N2528	*GE-25,*TR-27,*HEPG6017,*ECG127,*WEP235,*276-2006
2N2529	2N2521,2SC945,TIS126,*PTC123,*GE-39,*TR-21,*HEPS0011,*SK3124,*SN80, *ECG123A,*WEP736,*BF167,*276-2013,*RT-102
2N2530	2N717,2N839,2N842,2N844,2N2521,2S102,MPS6544,MPS6545,TIS126,ZT42, *PTC123,*GE-39,*TR-21,*HEPS0011,*SK3124,*SN80,*ECG123A,*WEP736,*BF167, *276-2013,*RT-102
2N2531	2N708A,2N718,2N840,2N929,2N2387,2N2522,2N2845,2N3301,2N3693,2N3826, 2N3946,2N3973,2N3975,2N4140,2N4994,2S103,2SC302,2SC335,2SC336,2SC725, 2SC815,2SC838,2SC839,2SC896,2SC938,BFY39-1,BSY93,EN697,GET2221, PET1001,SE1001,TIS22,ZT43,*PTC123,*GE-39,*TR-21,*HEPS0011,*SK3124,*SN80, *ECG123A,*WEP736,*BF167,*276-2013,*RT-102
2N2532	2N841,2N916,2N916A,2N2483,2N3859A,2N3947,2N4141,2N5028,2N5209,2N5825, 2S104,2SC366G,2SC366G-Y,2SC899,2SC933,2SC943,2SC944,2SC965,2SC1175, 2SC1453,BC107A,BC167A,BC171A,BC174A,BC182A,BC190A,BC237A,BCY59A, BFY39-2,EN916,GET2222,MM3904,MPS6576-BLUE,MPS6576-GREEN,PBC107A, ZT44,*PTC123,*GE-39,*TR-21,*HEPS0011,*SK3124,*SN80,*ECG123A,*WEP736, *BF167,*276-2013,*RT-102
2N2533	2N717,2N840,2N842,2N844,2N2387,2N2522,2N3973,2N3975,2SC302,2SC725, BF224J,BF225J,BFY39-1,BSY93,ZT43,*PTC123,*GE-39,*TR-21,*HEPS0011, *SK3124,*SN80,*ECG123A,*WEP736,*BF167,*276-2013,*RT-102
2N2534	2N841,2N843,2N916,2N916A,2N3859A,2N3903,2N3947,2N3974,2N3976,2N4227, 2N5825,2S104,2SC366G,2SC366G-O,2SC899,2SC933,2SC943,2SC944,2SC965, 2SC1189,2SC1320,2SC1453,2SC1688,BFY39-2,EN916,MPS3693,MPS6565, MPS6576-BLUE,MPS6576-GREEN,MPS6576-YEL,TP3705,ZT44,*PTC123,*GE-39, *TR-21,*HEPS0011,*SK3124,*ECG123A,*WEP736,*BF167,*276-2013,*RT-102
2N2535	2N2536
2N2536	
2N2537	2N3725A,2N4960,2N4961,*GE-47,*HEPS3001,*276-2038
2N2538	2N2219,2N2219A,2N3300,*GE-18,*HEPS3001
2N2539	2N3643,2N4962,2N4963,TN-3903,*PTC136,*GE-20,*TR-21,*HEPS0011,*SK3122, *ECG123A,*WEP736,*276-2009,*RT-102
2N2540	A5T2222,TIS109,TIS111,TN-3904,*PTC136,*GE-20,*TR-21,*HEPS0011,*SK3122, *ECG123A,*WEP736,*276-2009,*RT-102
2N2541	*GE-53,*TR-82,*HEPG6011,*ECG176,*WEP238,*RT-127
2N2551	*GE-27,*TR-78,*BF338,*276-2012
2N2552	2N1038-1,2N1039-1,2N1042-1,2N1043-1,2N2553,2N2560,2N2561,*AT100M
2N2553	2N1039-1,2N1040-1,2N1043-1,2N1044-1,2N2554,2N2561,2N2562,*AT100M
2N2554	2N1040-1,2N1041-1,2N1044-1,2N1045-1,2N2555,2N2562,2N2563,*AT100M
2N2555	2N1041-1,2N1045-1,2N2563
2N2556	2N1038-2,2N1039-2,2N1042,2N1043,2N2557,*AT100H
2N2557	2N1039-2,2N1040-2,2N1043,2N1044,2N2558,*AT100H
2N2558	2N1040-2,2N1041-2,2N1044,2N1045,2N2559
2N2559	2N1041-2,2N1045
2N2560	2N1038-1,2N1039-1,2N1042-1,2N1043-1,2N2552,2N2553,2N2561
2N2561	2N1039-1,2N1040-1,2N1043-1,2N1044-1,2N2553,2N2554,2N2562
2N2562	2N1040-1,2N1041-1,2N1044-1,2N1045-1,2N2554,2N2555,2N2563
2N2563	2N1041-1,2N1045-1,2N2555
2N2564	2N1038,2N1039,2N2565,*GE-80,*TR-05,*HEPG6011,*SK3004,*AT100H,*ECG102, *WEP631,*RT-120
2N2565	2N1039,2N1040,2N2566,*GE-3,*TR-05,*HEPG6011,*SK3004,*AT100H,*ECG102, *WEP631,*RT-120
2N2566	2N1040,2N1041,2N2567,*HEPG6012
2N2567	2N1041,*HEPG6012
2N2568	
2N2569	2N2331,2N2570,2N2923,2N3241A,2N3393,2N4418,2SC395A,2SC395A-Y,2SC509-O, 2SC509G-O,2SC735-O,2SC934,2SC1317,2SC1788,2SC1851,2SD392,BC170B, ZT2938,*PTC123,*GE-212,*TR-21,*HEPS0011,*SK3122,*ECG123A,*WEP736, *BC107B,*276-2016,*RT-102
2N2570	2N2331,2N2569,2N2923,2N3241A,2N3393,2N4418,2SC395A,2SC395A-Y,2SC509-O, 2SC509G-O,2SC735-O,2SC934,2SC1317,2SC1788,2SC1851,2SD392,BC170B,

To Replace	Substitute This Type
(2N2570)	ZT2938,*PTC123,*GE-212,*TR-21,*HEPS0011,*SK3122,*ECG123A,*WEP736,*BC107B,*276-2016,*RT-102
2N2571	2N2572,*GE-20,*TR-21,*HEPS0011,*SK3122,*ECG123A,*WEP736,*276-2009
2N2572	2N2571,*GE-20,*TR-21,*HEPS0011,*SK3122,*ECG123A,*WEP736,*276-2009
2N2580	2N2580M,2N2582
2N2580M	
2N2581	2N2583
2N2582	
2N2583	
2N2584	
2N2585	
2N2586	2N930A,2N2511,2N2523,2N3117,2N6428,2N6428A,2N6539,BC546B,TIS94,TIS97,*PTC123,*GE-212,*TR-21,*HEPS0011,*SK3122,*EN10,*ECG123A,*WEP736,*276-2008,*RT-102
2N2587	*PTC145,*GE-51,*TR-17,*HEPG0003,*ECG160,*WEP637
2N2588	*PTC107,*GE-52,*TR-17,*HEPG0009,*ECG126,*WEP635
2N2589	
2N2590	*PTC141,*GE-21,*276-2022
2N2591	*PTC141,*GE-21,*HEPS0005,*276-2021
2N2592	2N2593,*GE-21,*HEPS0005,*276-2022
2N2593	*GE-21,*HEPS0005,*276-2021
2N2594	*PTC144,*GE-32,*TR-87,*HEPS3011,*SK3024,*ECG128,*WEP243,*BF338,*276-2008
2N2595	2N2590,2N2591,*PTC127,*GE-82,*TR-20,*HEPS0013,*SK3114,*ECG159,*WEP242,*276-2021,*RT-115
2N2596	2N2592,*PTC127,*GE-82,*TR-20,*HEPS0013,*SK3114,*ECG159,*WEP242,*276-2021,*RT-115
2N2597	2N2593,*PTC127,*GE-82,*TR-20,*HEPS0013,*SK3114,*ECG159,*WEP242,*276-2023,*RT-115
2N2598	2N2599A,*GE-21,*TR-20,*SK3114,*ECG159,*WEP242,*276-2021,*RT-115
2N2599	2N2599A,2N2600A,*GE-21,*TR-20,*SK3114,*ECG159,*WEP242,*276-2021,*RT-115
2N2599A	2N2599,*SK3114,*276-2021
2N2600	2N2600A,*GE-21,*TR-20,*SK3114,*ECG159,*WEP242,*276-2021,*RT-115
2N2600A	2N2600,*SK3114
2N2601	*PTC127,*GE-82,*TR-20,*HEPS0013,*SK3114,*ECG159,*WEP242,*276-2021,*RT-115
2N2602	*PTC127,*GE-82,*TR-20,*HEPS0013,*SK3114,*ECG159,*WEP242,*276-2021,*RT-115
2N2603	*PTC127,*GE-82,*TR-20,*HEPS0013,*SK3114,*ECG159,*WEP242,*276-2021,*RT-115
2N2604	2N2596,2N3580,A5T2604,*PTC127,*GE-82,*TR-20,*HEPS0013,*SK3114,*ECG159,*WEP242,*BC327,*276-2021,*RT-115
2N2605	2N2605A,A5T2605,*PTC127,*GE-82,*TR-20,*HEPS0013,*SK3114,*ECG159,*WEP242,*BC327,*276-2022,*RT-115
2N2605A	*PTC127,*GE-48,*TR-20,*HEPS0013,*SK3114,*ECG159,*WEP242,*BC327,*276-2022,*RT-115
2N2610	2N117+,2N332,2N472,2N472A,2N475A,2N1149,2SC12,2SC478,2SC945,*PTC123,*GE-39,*TR-21,*HEPS0014,*SK3122,*EN10,*ECG123A,*WEP736,*BF167,*276-2013,*RT-102
2N2611	*GE-27,*HEPS3019
2N2612	*DS-503,*GE-3,*TR-01,*HEPG6018,*SK3009,*PT40,*ECG104,*WEP230,*RT-124
2N2613	*DS-26,*GE-2,*TR-85,*HEPG0005,*SK3004,*AT10H,*ECG102A,*WEP250,*RT-121
2N2613+	*SK3004,*ECG102A
2N2614	*DS-26,*GE-2,*TR-85,*HEPG0005,*SK3004,*AT20H,*ECG102A,*WEP250,*RT-121
2N2615	2N2616,2N4873,2SC1216,BF240,TIS62A,TIS63A,TIS86,*PTC126,*GE-61,*HEPS0020,*SK3018,*BF173,*276-2038
2N2616	2N2615,2N3227,2N4873,2SC1216,TIS86,ZT2369A,*PTC126,*GE-61,*TR-95,*HEPS0016,*SK3019,*EN10,*ECG108,*WEP56,*276-2038,*RT-113
2N2617	2N924,2N940,2N2945A,2N2946A,2N3218,2N3219,2N3978,2N3979,2S323,*GE-212,*TR-21,*HEPS0013,*SK3004,*ECG123A,*WEP736,*RT-102
2N2618	*PTC136,*GE-47,*TR-78,*HEPS5026,*SK3045,*ECG154,*WEP712,*BC337,*RT-110
2N2618/46	*PTC136,*GE-20,*HEPS0011,*BC337,*276-2009

To Replace	Substitute This Type
2N2621	2N741,2N741A,2N960,2N2099,2N2381,2N2630,2N2718,*PTC145,*GE-2,*TR-17,*HEPG0002,*SK3123,*ECG160,*WEP637,*276-2005
2N2622	2N2099,2N2100,2N2381,*PTC145,*GE-2,*TR-17,*HEPG0002,*SK3123,*ECG160,*WEP637,*276-2005
2N2623	2N2100,2N2381,2N2382,*PTC145,*GE-2,*TR-17,*HEPG0002,*SK3123,*ECG160,*WEP637,*276-2005
2N2624	2N741,2N741A,2N960,2N2099,2N2381,2N2630,2N2718,*PTC145,*GE-2,*TR-17,*HEPG0002,*SK3123,*ECG160,*WEP637,*276-2005
2N2625	2N2099,2N2100,2N2381,*PTC145,*GE-2,*TR-17,*HEPG0002,*SK3123,*ECG160,*WEP637,*276-2005
2N2626	2N2100,2N2381,2N2382,*PTC145,*GE-2,*TR-17,*HEPG0002,*SK3123,*ECG160,*WEP637,*276-2005
2N2627	2N741,2N741A,2N960,2N2099,2N2381,2N2630,2N2718,*PTC145,*GE-2,*TR-17,*HEPG0002,*SK3123,*ECG160,*WEP637,*276-2005
2N2628	2N2099,2N2100,2N2381,*PTC145,*GE-2,*TR-17,*HEPG0002,*SK3123,*ECG160,*WEP637,*276-2005
2N2629	*PTC145,*GE-2,*TR-17,*HEPG0002,*SK3123,*ECG160,*WEP637,*276-2005
2N2630	*GE-22,*TR-17,*HEPG0003,*SK3123,*ECG160,*WEP637,*276-2021
2N2631	2N749,2N751,2N1388,2N1390,2SC302-M,2SC307-M,2SC590N,*HEPS3010
2N2632	2N2633,2N2880,2N3998,2N4115,HST6014,HST6313,HST6413,SDT6014,SDT6313,STT6313,STT6413,*GE-66,*HEPS5004
2N2633	2N2634,2N4115,STT2802
2N2634	STT2802
2N2635	*PTC145,*GE-53,*TR-17,*HEPG0003,*JR200,*ECG160,*WEP637
2N2636	2N2637,2N2638,*DS503,*TR-35,*HEPG6009,*SK3009,*PT250,*ECG179,*WEPG6001,*RT-147
2N2637	2N2636,2N2638,*DS503,*TR-35,*HEPG6009,*SK3009,*PT250,*ECG179,*WEPG6001,*RT-147
2N2638	2N2636,2N2637,*TR-35,*HEPG6009,*PT250,*ECG179,*WEPG6001,*RT-147
2N2639	2N2640,2N2641,2N2722,*GE-212,*TR-95,*HEP729(2),*SK3039,*ECG108,*276-2013,*RT-113
2N2640	2N2639,2N2641,2N2722,*GE-212,*TR-95,*HEP729(2),*SK3039,*276-2013,*RT-113
2N2641	2N2639,2N2640,2N2722,*GE-212,*TR-95,*HEP729(2),*SK3039,*276-2013,*RT-113
2N2642	2N2643,2N2644,2N2722,2N2915A,2N2919A,*GE-212,*TR-95,*HEP728(2),*SK3039,*276-2013,*RT-113
2N2643	2N2642,2N2644,2N2722,2N2915A,2N2919A,*GE-212,*TR-95,*HEP728(2),*SK3039,*276-2013,*RT-113
2N2644	2N2642,2N2643,2N2722,2N2915A,2N2919A,*GE-212,*TR-95,*HEP728(2),*SK3039,*276-2013,*RT-113
2N2645	2N871,2N956,*PTC123,*GE-81,*HEPS0005,*ECG123A,*WEP736,*276-2008
2N2648	*DS26,*GE-2,*TR-82,*HEPG0002,*AT100H,*ECG176,*WEP238,*276-2005,*RT-127
2N2649	2N749,2N751,2N1388,2N1390,2N1491,2N1709,2N2781,2N2782,2N2783,2N2874,2SC302-M,2SC307-M,2SC547,2SC590N,40305,SRF11101,*HEPS3001
2N2650	2N749,2N751,2N1388,2N1390,2SC302-M,2SC307-M,2SC590N
2N2651	2N2710,2N3510,2N3648,2SC576,*PTC136,*GE-20,*TR-95,*HEPS0016,*SK3039,*EN10,*ECG108,*WEP56,*276-2009,*RT-113
2N2652	2N2223,2N2223A,2N2652A,*GE-18,*276-2008
2N2652A	2N2223,2N2223A,2N2652,*GE-18,*276-2008
2N2654	2N2495,2N3279,2N3280,2N3281,2N3282,2N3283,2N3284,*PTC145,*GE-51,*TR-17,*HEPG0008,*SK3006,*JR200,*ECG160,*WEP637,*276-2005
2N2655	2N2202,*GE-32,*HEPS3019,*276-2011
2N2656	2N708,2N784A,2N914,2N2242,2N2481,2N2501,2N3011,2N3013,2N3014,2N3210,2N3211,2N3511,2N3646,2N3647,2N5772,2SC67,2SC577,2SC639,2SC764,A5T4123,BSW82,GET2369,PT720,ZT708,*PTC136,*GE-210,*TR-21,*HEPS0011,*SK3122,*EN10,*ECG123A,*WEP736,*276-2009,*RT-102
2N2657	2N2658,*GE-66,*HEPS3002
2N2658	*HEPS3002
2N2659	2N2660,2N2665,2N2666,*HEPG6011,*276-2006
2N2660	2N2661,2N2666,2N2667,*HEPG6012
2N2661	2N2667,*HEPG6012
2N2662	2N2663,2N2668,2N2669,*HEPG6011

To Replace	Substitute This Type
2N2663	2N2664,2N2669,2N2670,*HEPG6012
2N2664	2N2670,*HEPG6012
2N2665	2N2666,*HEPG6011,*276-2006
2N2666	2N2667,*HEPG6012
2N2667	*HEPG6012
2N2668	2N2669
2N2669	2N2670
2N2670	
2N2671	*PTC107,*GE-51,*TR-17,*HEPG0003,*SK3006,*JR200,*ECG160,*WEP637
2N2672	2N2955,2N3883,2N6365,2N6365A,*PTC107,*GE-51,*TR-17,*HEPG0001,*SK3006, *JR200,*ECG160,*WEP637,*276-2005
2N2672A	2N43,2N43A,2N43A+,2N61A,2N61B,2N196,2N197,2N215,2N237,2N283,2N284, 2N331,2N361,2N363,2N381,2N413,2N414,2N414A,2N414B,2N414C,2N422,2N466, 2N505,2N518,2N567,2N568,2N633,2N1017,2N1057,2N1057+,2N1171,2N1191, 2N1303,2N1313,2N1348,2N1350,2N1351,2N1373,2N1414,2N1446,2N1447,2N1451, 2N2955,2SA204,2SA205,2SA206,2SA207,2SA406,2SA407,2SB220,2SB299, 2SB397,ASY26,NKT224,SF.T222,SF.T227,*PTC145,*TR-17,*HEPG0002,*ECG160, *WEP637
2N2673	2N2521,*PTC123,*GE-212,*TR-21,*HEPS0011,*SK3122,*EN10,*ECG123A,*WEP736, *BC107B,*276-2010,*RT-102
2N2674	2N2395,2N2514,*PTC123,*GE-212,*TR-21,*HEPS0013,*SK3122,*EN10,*ECG123A, *WEP736,*BC107B,*276-2010,*RT-102
2N2675	2N2396,2N2515,2N2522,2SC896,*PTC123,*GE-212,*TR-21,*HEPS0011,*SK3122, *EN10,*ECG123A,*WEP736,*BC107B,*276-2010,*RT-102
2N2676	BC182A,BC546A,MM3904,*PTC123,*GE-212,*TR-21,*HEPS0011,*SK3122,*ECG123A, *WEP736,*BC107B,*276-2010,*RT-102
2N2677	2N2387,2N2395,2N2522,*PTC123,*GE-61,*TR-21,*HEPS0011,*SK3122,*ECG123A, *WEP736,*BC107B,*276-2013,*RT-102
2N2678	*PTC123,*GE-61,*TR-21,*HEPS0011,*SK3122,*ECG123A,*WEP736,*BC107B, *276-2013,*RT-102
2N2691	*TR-35,*HEPG6009,*ECG179,*WEPG6001,*RT-147
2N2691A	*TR-35,*HEPG6009,*ECG179,*WEPG6001,*RT-147
2N2692	*GE-212,*TR-21,*HEPS0011,*SK3122,*ECG123A,*WEP736,*BC107B,*276-2013, *RT-102
2N2693	*GE-61,*TR-21,*HEPS0011,*SK3122,*ECG123A,*WEP736,*BC107B,*276-2013, *RT-102
2N2694	*GE-61,*TR-51,*HEPS0025,*BC107B,*276-2013
2N2695	A5T3638A,A5T4402,*PTC127,*GE-22,*TR-20,*HEPS0013,*SK3114,*SP70,*ECG159, *WEP242,*BC327,*276-2024,*RT-115
2N2696	2N869,2N869A,2N2695,2N5811,2SA509-O,2SA509G-O,2SA695,A5T3638A,A5T4402, BC192,BSW72,D29E1,SE8540,TN-3905,TQ62,*PTC127,*GE-22,*TR-20,*HEPS0013, *SK3114,*SP70,*ECG159,*WEP242,*BC327,*276-2024,*RT-115
2N2697	2N2698,HST4553,HST4554,SDT4553,SDT4554,*HEPS5000
2N2698	HST4554,SDT4554,*HEPS5000
2N2699	2N827,*GE-66,*TR-08,*HEPG0011,*SK3011,*ECG101,*WEP641,*RT-119
2N2706	*PTC135,*GE-53,*TR-84,*HEPG0005,*SK3004,*HF3H,*ECG158,*WEP238,*276-2007
2N2707	M.P.2N2706,2N2430
2N2708	TC3114,*DS-74,*GE-17,*TR-95,*HEPS0016,*SK3019,*EN10,*ECG108,*WEP56, *276-2038,*RT-113
2N2709	2N327A,2N327B,2N328A,2N330A,2N925,2N927,2N936,2N1232,2N1439,2N1440, 2N1441,2N1474A,2N1623,2N3345,BCY30,HA7534,SHA7534,*GE-21,*TR-20, *HEPS0012,*SK3114,*ECG159,*WEP238,*276-2021,*RT-115
2N2710	2N3511,2N3648,2SC577,*PTC136,*TR-95,*HEPS0016,*SK3039,*ECG108,*WEP56, *276-2038,*RT-113
2N2711	2N2318,2N2656,2N3011,2SC204,PT720,*PTC132,*GE-10,*TR-95,*HEPS0015, *SK3124,*EN40,*ECG108,*WEP56,*BC107B,*276-2015,*RT-113
2N2712	2N5224,A5T4124,BC548A,BCY58A,BCY58B,*PTC132,*GE-17,*TR-95,*HEPS0015, *SK3124,*EN40,*ECG108,*WEP56,*BC107B,*276-2015,*RT-113
2N2713	2N2318,2N2331,2N2656,2N3011,2SC204,PT720,*GE-210,*TR-21,*HEPS0015, *SK3124,*EN40,*ECG123A,*WEP736,*BC107B,*276-2015,*RT-102
2N2714	2N3241A,2SC1317,2SC1851,A5T4124,BCY58A,BCY58B,*GE-210,*TR-21,*HEPS0015,

To Replace	Substitute This Type
(2N2714)	*SK3124,*EN40,*ECG123A,*WEP736,*BC107B,*276-2015,*RT-102
2N2715	2N744,2N753,2N2318,2N2319,2N2331,2N2569,2N2570,2N2656,2N2711,2N2713, 2N2922,2N2926-RED,2N3011,2N3394,2N3605,2N3606,2N3607,2N4418,2S095A, 2S512,2SC99,2SC168,2SC204,2SC300,2SC301,2SC356,2SC395A-O,2SC400-O, 2SC460,2SC461,2SC528,2SC595,2SC724,2SC924,2SC1372,BC170A,EN744, EN3011,MPS2713,MPS2926-RED,PT720,TIS51,ZT41,ZT2205,*PTC132,*GE-86, *TR-95,*HEPS0015,*SK3124,*EN40,*ECG108,*WEP56,*BF365,*276-2015,*RT-113
2N2716	2N2712,2N2714,2N2924,2N2926-YEL,2N3241A,2N4124,2N4419,2N5224,2N5998, 2N6000,2SC372G-Y,2SC395A-Y,2SC400,2SC400-Y,2SC454,2SC735-Y,2SC934, 2SC1085,2SC1317,2SC1363,2SC1684,2SC1849,2SC1854,2SD392,BC108,BC108A, BC168A,BC172A,BC238A,BC548A,BCY58A,BCY58B,BSW88,BSW89,BSX38, MPS2714,MPS2926-GRN,MPS2926-YEL,PBC108A,*PTC132,*GE-211,*TR-95, *HEPS0016,*SK3124,*EN40,*ECG108,*WEP56,*BF365,*276-2015,*RT-113
2N2717	2N838,ASZ21,*PTC107,*GE-51,*TR-17,*HEPG0003,*ECG160,*WEP637
2N2718	2N2381,*HEPG0005
2N2719	2N706,2N706A,2N706B,2N783,2N784,2N834A,2N2205,2N2368,2N3261,2N5187, 2SC68,2SC321,2SC322,2SC576,40405,GET706,KT218,MM1941,MPS706,MPS706A, MPS834,MPS6540,MPS6543,MPS6548,MPSH19,MPSH32,MPSH37,MT1038, MT1038A,MT1039,ZT2368,*PTC136,*GE-20,*TR-21,*HEPS0011,*SK3122, *ECG123A,*WEP736,*276-2009,*RT-102
2N2720	2N2721,*GE-20
2N2721	2N2720,*GE-20
2N2722	*GE-20,*HEP729(2)
2N2726	*GE-27,*HEPS3021,*BF338,*276-2008
2N2727	*GE-27,*HEPS3021,*BF338,*276-2008
2N2728	*GE-4,*TR-03,*HEPG6002,*SK3012,*PT501,*ECG105,*WEP233
2N2729	MT1060A,*PTC126,*GE-61,*TR-95,*HEPS0016,*SK3018,*EN10,*ECG108,*WEP56, *276-2038,*RT-113
2N2730	*GE-4,*TR-03,*SK3012,*PT501,*ECG105,*WEP233
2N2731	2N2730,*GE-4,*TR-03,*HEPG6002,*SK3012,*PT501,*ECG105,*WEP233
2N2732	2N2731,*GE-4,*TR-03,*HEPG6002,*SK3012,*PT501,*ECG105,*WEP233
2N2733	
2N2734	2N2733
2N2735	2N2734
2N2736	
2N2737	2N2736
2N2738	2N2737
2N2739	
2N2740	
2N2741	2N2742
2N2742	2N2743
2N2743	2N2744
2N2744	
2N2745	2N2739,2N2751
2N2746	2N2740,2N2752
2N2747	2N2741,2N2742,2N2748,2N2753,2N2754
2N2748	2N2742,2N2743,2N2749,2N2754,2N2755
2N2749	2N2743,2N2744,2N2750,2N2755,2N2756
2N2750	2N2744,2N2756
2N2751	2N2739,2N2745
2N2752	2N2740,2N2746
2N2753	2N2741,2N2742,2N2747,2N2748,2N2754
2N2754	2N2742,2N2743,2N2748,2N2749,2N2755
2N2755	2N2743,2N2744,2N2749,2N2750,2N2756
2N2756	2N2744,2N2750
2N2757	
2N2758	
2N2759	2N2760
2N2760	2N2761
2N2761	2N2762
2N2762	

To Replace	Substitute This Type
2N2763	2N2757,2N2769,2N2775
2N2764	2N2758,2N2770,2N2776
2N2765	2N2759,2N2760,2N2766,2N2771,2N2772,2N2777,2N2778
2N2766	2N2760,2N2761,2N2767,2N2772,2N2773,2N2778,2N2779
2N2767	2N2761,2N2762,2N2768,2N2773,2N2774,2N2779,2N2780
2N2768	2N2762,2N2774,2N2780
2N2769	2N2757,2N2763,2N2775
2N2770	2N2758,2N2764,2N2776
2N2771	2N2759,2N2760,2N2765,2N2766,2N2772,2N2777,2N2778
2N2772	2N2760,2N2761,2N2766,2N2767,2N2773,2N2778,2N2779
2N2773	2N2761,2N2762,2N2767,2N2768,2N2774,2N2779,2N2780
2N2774	2N2762,2N2768,2N2780
2N2775	2N2757,2N2763,2N2769
2N2776	2N2758,2N2764,2N2770
2N2777	2N2759,2N2760,2N2765,2N2766,2N2771,2N2772,2N2778
2N2778	2N2760,2N2761,2N2766,2N2767,2N2772,2N2773,2N2779
2N2779	2N2761,2N2762,2N2767,2N2768,2N2773,2N2774,2N2780
2N2780	2N2762,2N2768,2N2774
2N2781	2N749,2N751,2N1388,2N1390,2N1491,2N2782,2N2783,2SC302-M,2SC307-M, 2SC590N,*HEPS3021
2N2782	2N749,2N751,2N1388,2N1390,2N1491,2N2781,2N2783,2SC302-M,2SC307-M, 2SC590N,*HEPS3021
2N2783	2N749,2N751,2N1388,2N1390,2N1491,2N2781,2N2782,2SC302-M,2SC307-M, 2SC590N,*TR-17,*HEPS3021,*ECG126,*WEP635
2N2784	2N3633,2N3959,2N3960,2N5200,2N5201,*PTC120,*GE-17,*TR-95,*HEPG0016, *SK3039,*EN10,*ECG108,*WEP56,*276-2011,*RT-113
2N2784/46	2N5200,2N5201,*PTC126
2N2784/51	2N5200,2N5201,*HEPS0024
2N2785	
2N2786	*TR-17,*HEPG0002,*ECG160,*WEP637,*276-2005
2N2786A	*TR-17,*HEPG0002,*ECG160,*WEP637,*276-2005
2N2787	2N3737,2N6375,2N6376,*GE-20,*HEPS3011,*276-2009
2N2788	2N2218A,2N3725A,2N4961,2SC97A,*GE-18,*HEPS3011
2N2789	2N2219A,*GE-18,*HEPS3011
2N2790	2N3737,2N6375,*GE-20,*HEPS3001,*276-2009
2N2791	2N4963,*GE-20,*HEPS3001,*276-2009
2N2792	*GE-20,*HEPS3001,*276-2009
2N2793	*GE-4,*TR-03,*HEPG6002,*SK3012,*ECG105,*WEP633
2N2795	2N838,2N2956,*PTC145,*GE-51,*TR-17,*HEPG0003,*ECG160,*WEP637
2N2796	2N2795,*PTC145,*GE-51,*TR-17,*HEPG0003,*ECG160,*WEP637
2N2797	2N1495,2N2956,*PTC145,*GE-51,*TR-17,*HEPG0002,*ECG160,*WEP637,*276-2005
2N2798	*PTC107,*GE-51,*TR-17,*HEPG0002,*ECG160,*WEP637,*276-2005
2N2799	2N1495,2N1499B,2N2955,*PTC145,*GE-51,*TR-17,*HEPG0002,*ECG160,*WEP637, *276-2005
2N2800	2N3245,2N3468,2N3762A,2N3763,2N4030,2N5865,2SA571,2SA717,MM3726, *PTC141,*GE-48,*TR-20,*HEPS0012,*SK3114,*ECG159,*WEP242,*276-2021, *RT-115
2N2801	2N4032,2SA684,MPS4355,*GE-48,*TR-20,*HEPS0012,*SK3114,*SP70,*ECG159, *WEP242,*276-2021,*RT-115
2N2802	2N2803,2N2804,2N2805,2N2806,2N2807,*GE-65,*TR-20,*HEP715(2),*SK3118, *ECG106,*WEP242,*276-2022,*RT-126
2N2803	2N2802,2N2804,2N2805,2N2806,2N2807,*GE-65,*TR-20,*HEP715(2),*SK3118, *ECG106,*WEP52,*276-2022,*RT-126
2N2804	2N2802,2N2803,2N2805,2N2806,2N2807,*GE-65,*TR-20,*HEP715(2),*SK3118, *ECG106,*WEP52,*276-2022,*RT-126
2N2805	2N2802,2N2803,2N2804,2N2806,2N2807,*GE-65,*TR-20,*HEP715(2),*SK3118, *ECG106,*WEP52,*276-2022,*RT-126
2N2806	2N2802,2N2803,2N2804,2N2805,2N2807,*GE-65,*TR-20,*HEP715(2),*SK3118, *ECG106,*WEP52,*276-2022,*RT-126
2N2807	2N2802,2N2803,2N2804,2N2805,2N2806,*GE-65,*TR-20,*HEP715(2),*SK3118, *ECG106,*WEP52,*276-2022,*RT-126

To Replace	Substitute This Type
2N2808	2N2808A,2N2809A,2N3570,2N4252,2N4253,2N5053,2N5054,2N6304,2N6305, 2SC1747,MT1061A,*PTC133,*HEPS0020
2N2808A	2N2808,2N2809A,2N3570,2N4252,2N4253,2N5053,2N5054,2N6304,2SC1747, MT1061A,*PTC133,*HEPS0020
2N2809	2N2808,2N2808A,2N2809A,2N3570,2N3600,2N4252,2N4253,2N5053,2N5054,2N6304, 2N6305,2SC1747,MT1061A,*PTC132,*HEPS0020
2N2809A	2N2808,2N2808A,2N3570,2N4252,2N4253,2N5053,2N5054,2N6304,2N6305,2SC1747, MT1061A,*PTC133,*HEPS0020
2N2810	2N2808,2N2808A,2N2809,2N2809A,2N2810A,2N3570,2N3571,2N3600,2N4252, 2N4253,2N5053,2N5054,2N6304,2N6305,2SC1747,MT1061A,*PTC132,*HEPS0020
2N2810A	2N2808,2N2808A,2N2809A,2N3570,2N3571,2N4252,2N4253,2N5053,2N5054,2N6304, 2N6305,2SC1747,MT1061A,*PTC133,*HEPS0020
2N2811	2N4301,HST6011,HST6012,HST9902,HST9903,KSP1151,KSP1152,*HEPS5004
2N2812	HST6013,HST6014,KSP1154,KSP1155,*HEPS5004
2N2813	2N4115,HST7150,HST9904,KSP1153
2N2814	2N4115,2N5329,HST7140,HST7154,KSP1156
2N2815	2N2816,2N2819,2N2820,2N2823,2N2824,163-06,163-08,164-06,164-08,STC1728, STC1733
2N2816	2N2820,2N2824,163-08,163-10,163-12,164-08,164-10,164-12,STC1733
2N2817	2N2818,2N2821,2N2822,2N2825,163-14,163-16,163-18,164-14,164-16,164-18, STC1738,STC3706
2N2818	2N2822,163-18,163-20,163-22,163-24,164-18,164-20,164-22,164-24,STC3706
2N2819	2N2820,2N2823,2N2824,STC1728,STC1733
2N2820	2N2824,STC1733
2N2821	2N2822,2N2825,STC1738,STC3706
2N2822	STC3706
2N2823	2N2824,STC1728,STC1733
2N2824	STC1733
2N2825	STC1738,STC3706
2N2826	*HEPG6011
2N2827	*HEPG6011
2N2828	2N2829,*GE-66,*HEPS5000,*276-2020
2N2829	2N2828,*GE-66,*HEPS5000,*276-2020
2N2831	2N744A,*GE-17,*TR-21,*HEPS0011,*SK3024,*EN10,*ECG123A,*WEP736,*276-2009, *RT-102
2N2832	MP1612,*TR-01,*HEPG6014,*ECG121,*WEP232,*RT-127
2N2833	2N2834,MP1612A,MP1612B,*HEPG6009
2N2834	MP1612A,MP1612B,*HEPG6009
2N2836	2N665,2N1535,2N1536,2N2144,2N2144A,2N2145,2N2145A,2SB251A,2SB253, CDT1311,CDT1312,*PTC105A,*GE-25,*TR-01,*HEPG6005,*SK3009,*PT40, *ECG121,*WEP232,*276-2006,*RT-127
2N2837	2N3764A,2N3765,2N4026,2N5815,2N5821,*PTC141,*GE-48,*TR-20,*HEPS0013, *SK3114,*SP70,*ECG159,*WEP242,*276-2021,*RT-126
2N2838	2N4028,2N5817,2N5819,2N5823,2SA817,D29E10,D29E5,D29E6,*GE-48,*TR-20, *HEPS0013,*SK3114,*SP70,*ECG159,*WEP242,*276-2023,*RT-115
2N2845	2N915,2N2847,2N3301,2N3946,2N4962,2N4963,2N5380,A5T3903,FT3641,FT3642, MM3903,TN-3903,*PTC136,*GE-20,*TR-21,*HEPS0013,*SK3025,*ECG123A, *WEP736,*276-2009,*RT-102
2N2846	2N2848,2N3015,2N3299,2N4960,2N4961,*GE-47,*TR-87,*HEPS3001,*SK3024, *ECG128,*WEP243,*276-2038,*RT-114
2N2847	2N915,2N3643,2N4962,2N4963,2N5380,2SC1998,A5T3903,TN-3903,*PTC121, *GE-210,*TR-21,*HEPS0013,*SK3025,*ECG123A,*WEP736,*276-2009,*RT-102
2N2848	2N4960,2N4961,*PTC136,*GE-47,*TR-87,*HEPS3001,*SK3024,*ECG128,*WEP243, *276-2038,*RT-114
2N2849	2N1973,2N2443,2N3107,SDT4456,TN624,TN624-1,*TR-87,*HEPS5000,*SK3024, *ECG128,*WEP243,*276-2008,*RT-114
2N2849-1	2N1973,2N2443,2N2849,2N3107,SDT4456,TN624,TN624-1,*HEPS5000,*276-2008
2N2849-2	TN624-2,*HEPS5000
2N2849-3	TN624-3,*HEPS5000
2N2850	2N699A,2N699B,2N1613B,2N1889,2N1974,2N2851,2N2890,2N3108,2N3421,2SC292, 2SC293,SDT4454,SE7002,*TR-87,*HEPS5000,*SK3024,*ECG128,*WEP243,

To Replace	Substitute This Type
(2N2850)	*276-2008,*RT-114
2N2850-1	2N699A,2N699B,2N1613B,2N1889,2N1974,2N2850,2N2851,2N2890,2N3108,2N3421, 2SC292,2SC293,SDT4454,SE7002,*HEPS5000,*276-2008
2N2850-2	*HEPS5000
2N2850-3	*HEPS5000
2N2851	2N699A,2N699B,2N1613B,2N1889,2N1974,2N2850,2N2890,2N3108,2N3421,2SC292, 2SC293,SDT4454,SE7002,*TR-87,*HEPS5000,*SK3024,*ECG128,*WEP243, *276-2008,*RT-114
2N2851-1	2N699A,2N699B,2N1613B,2N1889,2N1974,2N2850,2N2851,2N2890,2N3108,2N3421, 2SC292,2SC293,SDT4454,SE7002,*HEPS5000,*276-2008
2N2851-2	*HEPS5000
2N2851-3	*HEPS5000
2N2852	2N698,2N1975,2N3419,SDT4452,*TR-87,*HEPS5000,*SK3024,*ECG128,*WEP243, *276-2008,*RT-114
2N2852-1	2N698,2N1975,2N2852,2N3419,SDT4452,*HEPS5000,*276-2008
2N2852-2	
2N2852-3	*HEPS5000
2N2853	2N2848,2N2854,2N3109,2N3420,2N4960,2N4961,SDT9008,*GE-4,*TR-87,*HEPS5014, *SK3024,*ECG128,*WEP243,*276-2008,*RT-114
2N2853-1	2N2848,2N2853,2N2854,2N3109,2N3420,2N4960,2N4961,SDT9008,*GE-46, *HEPS5014,*ECG210,*WEPS3020,*276-2008
2N2853-2	*HEPS5000
2N2853-3	*HEPS5000
2N2854	2N2848,2N3109,2N4960,2N4961,SDT4455,SDT9011,*TR-87,*HEPS5014,*SK3024, *ECG128,*WEP243,*276-2008,*RT-114
2N2854-1	2N2848,2N2854,2N3109,2N4960,2N4961,SDT4455,SDT9011,*GE-46,*HEPS5014, *ECG210,*WEPS3020,*276-2008
2N2854-2	*HEPS5000
2N2854-3	*HEPS5000
2N2855	2N1613A,2N2846,2N2853,2N3015,2N3110,2N3299,2N3326,2N3420,2SC291,SDT4453, SDT9005,SE8002,*GE-4,*TR-87,*HEPS5014,*SK3024,*ECG128,*WEP243, *276-2008,*RT-114
2N2855-1	2N1613A,2N2846,2N2853,2N2855,2N3015,2N3110,2N3299,2N3326,2N3420,2SC291, SDT4453,SDT9005,SE8002,*GE-46,*HEPS5014,*ECG210,*WEPS3020,*276-2008
2N2855-2	*HEPS5000
2N2855-3	*HEPS5000
2N2856	2N3418,2N5188,SDT4451,SDT4483,SDT9002,SE8001,*GE-4,*TR-87,*HEPS5014, *SK3024,*ECG128,*WEP243,*276-2008,*RT-114
2N2856-1	2N2856,2N3418,2N5188,SDT4451,SDT4483,SDT9002,SE8001,*GE-46,*HEPS5014, *ECG210,*WEPS3020,*276-2008
2N2856-2	*HEPS5000
2N2856-3	*HEPS5000
2N2857	2N3570,2N4252,2N4253,2N6304,2SC1747,MT1061A,*PTC133,*GE-17,*TR-83,*EN10, *ECG316,*WEP56,*276-2011,*RT-113
2N2858	2N2859,*HEPS5000
2N2859	2N2911
2N2860	*PTC145,*GE-2,*TR-17,*HEPG0003,*ECG160,*WEP637
2N2861	2N869,2N2412,2N3040,2N3905,2N4034,2N4402,2N5382,2SA467,2SA467G-O, 2SA509-O,2SA509G-O,2SA562-O,2SA719,2SA890,A5T3905,A5T4402,BC192, BSW72,CS9012,GI-3702,MM3905,MPS6562,TN-3905,TQ62,*PTC127,*GE-65,*TR-20, *HEPS0013,*SK3114,*SP70,*ECG159,*WEP242,*BC177,*276-2022,*RT-115
2N2862	2N869,2N869A,2N2412,2N2861,2N3040,2N3829,2N3905,2N4402,2N5382,2SA467, 2SA467G-O,2SA509-O,2SA509G-O,2SA562-O,A5T3905,A5T4402,BC192,BSW72, CS9012,GI-3702,MM3905,MPS6562,TN-3905,TQ62,*PTC127,*GE-65,*TR-20, *HEPS0013,*SK3114,*ECG159,*WEP242,*BC177,*276-2022,*RT-115
2N2863	2SC1384,*GE-47,*HEPS3011,*SN60,*276-2030
2N2864	2SC1384,*GE-47,*HEPS3011,*276-2030
2N2865	2N2708,2N3570,2N3571,2N3572,2N4252,2N4253,2N6304,2N6305,TC3114,*PTC132, *GE-86,*TR-83,*HEPS0016,*SK3019,*EN10,*ECG161,*WEP719,*276-2011,*RT-113
2N2866	
2N2867	2N4115

To Replace	Substitute This Type
2N2868	2N2193,2N2193A,2N2193B,2N2270,2N3109,BC140-10,BSY83,ZT1613,ZT2270, *PTC144,*GE-47,*HEPS3010,*ECG123A,*WEP736,*276-2008
2N2869	2N2869/2N301,2N2870,2N2870/2N301A,2SB471,2SB472,KR6501,*PTC105A,*GE-25, *TR-01,*HEPG6005,*SK3009,*PT40,*ECG121,*WEP232,*276-2006,*RT-127
2N2869/2N301	2N2869,2N2870,2N2870/2N301A,2SB471,2SB472,KR6501,*PTC105A,*GE-25,*TR-01, *SK3009,*ECG121,*WEP232,*276-2006,*RT-127
2N2870	2N2870/2N301A,2SB472,2SB484,KR6503,*PTC105A,*GE-25,*TR-01,*HEPG6005, *SK3009,*PT40,*ECG121,*WEP232,*276-2006,*RT-127
2N2870/2N301A	2N2870,2SB472,2SB484,KR6503,*PTC105A,*GE-25,*276-2006
2N2871	
2N2872	
2N2873	AFY37,*GE-52,*TR-17,*HEPG0003,*ECG160,*WEP637
2N2874	2N751,2N1709,2SC302-M,2SC307-M,2SC590N,*HEPS3010
2N2875	*GE-69,*HEPS5006
2N2876	*GE-66
2N2877	2N2879,HST6309,HST6310,HST6409,HST6410,SDT6309,SDT6310,STT2804, STT2805,STT6309,STT6310,STT6409,STT6410,*GE-66,*HEPS5004
2N2878	2N2880,HST6312,HST6313,HST6412,HST6413,SDT6312,SDT6313,*GE-66, *HEPS5004
2N2879	2N4115,HST6310,HST6409,HST6410,SDT6310,STT2803,STT2804,STT6310,STT6410, *HEPS5004
2N2880	2N4115,HST6313,HST6413,SDT6313,*HEPS5004
2N2881	2N3203,2N3204,*HEPS5013
2N2882	
2N2883	2N743A,2N744A,2N2884,BSX48,BSX49,*GE-17,*TR-95,*HEPS3013,*SK3039,*EN10, *ECG108,*WEP56,*276-2038,*RT-113
2N2884	2N743A,2N744A,2N2883,BSX48,BSX49,*GE-17,*TR-95,*HEPS3013,*SK3039,*EN10, *ECG108,*WEP56,*276-2038,*RT-113
2N2885	2N3013,2N3014,MM3903
2N2886	2N2194,2N2194A,2N2194B,2N2217,2N3122,2N3253,2N3831,2N5188,2SC1052, 2SC1053,2SC1064,2SC1072,2SC1072A,2SC1150,2SD149,2SD329,40539,BSX48, BSX49,*PTC125,*GE-47,*TR-25,*HEPS0014,*BF338,*276-2012
2N2887	*HEPS3019
2N2890	2N699A,2N699B,2N1613B,2N1889,2N1974,2N2443,2N2891,2N3107,SDT4454, SE7002,*GE-18,*HEPS3019,*BF338,*276-2012
2N2891	2N1973,2N2443,SDT4456,*GE-18,*HEPS3019,*BF338,*276-2012
2N2892	2N2880,2N2893,2N3998,2N4115,2N5002,2N5083,2N5284,2N5328,HST6313,HST6413, SDT6313,STT2803,STT2804,STT6313,STT6413,*HEPS5004
2N2893	2N2880,2N3998,2N3999,2N4116,2N5004,2N5285,2N5328,2N5348,2N5349,2N5479, 2N5480,2N5730,HST6316,HST6416,SDT6316,STT6316,STT6416,*HEPS5004
2N2894	2N869A,2N3012,2N3209,2N3546,2N3576,2N3640,2N4258A,2N4423,MM2894,*TR-20, *HEPS0013,*SK3114,*SP70,*ECG159,*WEP242,*276-2009,*RT-115
2N2894A	2N3546,2N4258A,2N5056,2N5057,2N5292,2N5771,A5T4260,A5T4261,*TR-20, *HEPS0019,*SK3114,*ECG159,*WEP242,*276-2024,*RT-115
2N2895	2N2896,2N2898,2N2899,2N3701,*PTC144,*TR-87,*HEPS0005,*SK3024,*ECG128, *WEP243,*276-2009
2N2896	2N2899,2N3700,*HEPS0005
2N2897	2N2900,40458,*GE-47,*TR-87,*HEPS0025,*SK3024,*ECG128,*WEP243,*BC337
2N2898	2N2899,*PTC144,*HEPS3021
2N2899	*HEPS3021
2N2900	*GE-47,*HEPS3020,*BC337
2N2901	*GE-11,*TR-21
2N2902	2N1050C,*HEPS3019
2N2903	2N2453,2N2453A,2N2903A
2N2903A	2N2453,2N2453A,2N2903
2N2904	2N2904A,2N4030,2N4031,2N4404,2SA571,2SA717,*PTC127,*GE-21,*TR-88, *HEPS5022,*SK3025,*SP70,*ECG129,*WEP242,*276-2021,*RT-115
2N2904A	2N2904,2N4030,2N4031,2N4404,2SA571,2SA717,*PTC127,*GE-21,*TR-88, *HEPS5022,*SK3025,*SP70,*ECG129,*WEP242,*276-2021,*RT-115
2N2905	2N2905A,2N4032,2N4033,2N4405,2SA684,*PTC127,*GE-48,*TR-88,*HEPS5022, *SK3025,*SP70,*ECG129,*WEP242,*276-2021,*RT-115

To Replace	Substitute This Type
2N2905A	2N2905,2N4032,2N4033,2N4405,2SA684,*PTC127,*GE-67,*TR-88,*HEPS5022, *SK3025,*SP70,*ECG129,*WEP242,*276-2021,*RT-115
2N2906	2N2906A,2N3485,2N3485A,2N4026,2N4027,*PTC127,*GE-21,*TR-20,*HEPS0013, *SK3114,*SP70,*ECG159,*WEP243,*276-2023,*RT-115
2N2906A	2N2906,2N3485,2N3485A,2N4026,2N4027,*PTC127,*GE-21,*TR-20,*HEPS0013, *SK3114,*ECG159,*WEP243,*276-2023,*RT-115
2N2907	2N2907A,2N3486,2N3486A,2N3505,2N4028,2N4029,A5T2907,A5T3505,A5T3645, TIS112,*PTC127,*GE-48,*TR-20,*HEPS0013,*SK3114,*SP70,*ECG159,*WEP243, *276-2023,*RT-115
2N2907A	2N2907,2N3486,2N3486A,2N3505,2N4028,2N4029,A5T2907,A5T3505,A5T3645, TIS112,*PTC127,*GE-67,*TR-20,*HEPS0013,*SK3114,*SP70,*ECG159,*WEP243, *276-2023,*RT-115
2N2908	*HEPS5004
2N2909	*PTC144,*GE-27,*TR-78,*HEPS3010,*BF338
2N2910	*GE-20,*HEPS0025
2N2911	
2N2912	
2N2913	2N2915,2N2915A,2N2917,2N2919,2N2919A,*HEPS0025
2N2914	2N2916,2N2916A,2N2918,2N2920,2N2920A,2N3680,*HEPS0005
2N2915	2N2913,2N2915A,2N2917,2N2919,2N2919A,*HEPS0025
2N2915A	2N2919A,*HEPS0005
2N2916	2N2914,2N2916A,2N2918,2N2920,2N2920A,2N3680
2N2916A	2N2920A,2N3680,*HEPS0030
2N2917	2N2913,2N2915,2N2915A,2N2919,2N2919A,*HEPS0005
2N2918	2N2914,2N2916,2N2916A,2N2920,2N2920A,2N3680,*HEPS0005
2N2919	2N2919A,*HEPS0005
2N2919A	*HEPS0005
2N2920	2N2920A,2N3680
2N2920A	2N3680
2N2921	2N708,2N784A,2N914,2N957,2N2242,2N2318,2N2319,2N2481,2N2656,2N2710, 2N2922,2N3011,2N3013,2N3014,2N3210,2N3211,2N3510,2N3511,2N3605A, 2N3606A,2N3646,2N3647,2N3648,2N4295,2N4420,2N4421,2N4422,2SC67,2SC169, 2SC204,2SC300,2SC301,2SC356,2SC461,2SC468,2SC577,2SC595,2SC601, 2SC601N,2SC724,2SC752G-R,2SC764,2SC1274,BSW82,EN708,EN914,EN3011, EN3013,EN3014,GET708,GET914,GET2369,GET3013,GET3014,GET3646,MPS3646, PT720,TIS45,TIS46,TIS51,TIS52,TIS55,ZT708,ZT2205,*PTC132,*GE-10,*TR-95, *HEPS0015,*SK3124,*EP35,*ECG108,*WEP56,*BC107B,*276-2016,*RT-113
2N2922	2N2501,2N3211,2N4123,2N4418,2S512,2SC169,2SC300,2SC301,2SC372G-O, 2SC400-O,2SC459,2SC460,2SC461,2SC536,2SC619,2SC639,2SC735-O, 2SC752G-O,2SC764,2SC1276,2SC1372,BSY61,ZT2938,*PTC132,*GE-10,*TR-95, *HEPS0015,*SK3124,*EP35,*ECG108,*WEP56,*BC107B,*276-2016,*RT-113
2N2923	2N2924,2N3227,2N3241A,2N3242A,2N5224,2N5998,2N6000,2N6426,2SC1317, 2SC1788,2SC1851,BC548A,BCY58A,BSW83,BSX38,*PTC132,*GE-212,*TR-21, *HEPS0015,*SK3124,*EN40,*ECG123A,*WEP736,*BC107B,*276-2016,*RT-102
2N2924	2N2925,2N4256,2N4954,2N5371,2N5998,2N6002,2N6008,2SC1849,A5T4124,BC548B, BC549B,BCY58B,BCY58C,*PTC121,*GE-212,*TR-21,*HEPS0015,*SK3124,*EN40, *ECG123A,*WEP736,*BC107B,*276-2016,*RT-102
2N2925	2N6002,2N6008,BC548B,BC549B,BCY58D,*PTC139,*GE-212,*TR-21,*HEPS0015, *SK3039,*EN40,*ECG123A,*WEP736,*BC107B,*276-2016,*RT-102
2N2926	2N2925,2N2926-GRN,2N3391,2N3391A,2N3395-WHT,2N3396-WHT,2N3397-WHT, 2N3398-WHT,2N3415,2N3565,2N3721,2N3900,2N3900A,2N4256,2N4286,2N4419, 2N5172,2N6002,2N6008,2SC373G,2SC400-GR,2SC732-GR,2SC733-GR,2SC735-GR, 2SC1684,2SC1849,2SC1993,91T6,98T2,A158B,A159B,BC108,BC108B,BC109B, BC168B,BC170C,BC172B,BC173B,BC238B,BC239B,BC338,BC548B,BC549B, BCY58B,BCY58C,GI-2925,GI-3391,GI-3391A,GI-3395,GI-3398,GI-3403,GI-3415, GI-3900,GI-3900A,MPS2925,MPS2926,MPS3391,MPS3391A,MPS3395,MPS3721, MPS5172,PBC108B,PBC109B,PET4002,*PTC139,*GE-212,*TR-21,*HEPS0015, *SK3018,*ECG123A,*WEP736,*BC107B,*276-2031,*RT-102
2N2926-BRN	2N2318,2N2656,2N2711,2N2713,2N2926-RED,2N3011,2SC204,PT720,*PTC132, *GE-212,*TR-51,*BC107B,*276-2015
2N2926-GRN	2N2925,2N3900,2N3900A,2N3901,2N6002,2N6008,BC548B,BC549B,BCY58C,

To Replace	Substitute This Type
(2N2926-GRN)	BCY58D,*PTC139,*GE-212,*TR-51,*HEPS0015,*ECG123A,*WEP736,*BC107B,*276-2015,*RT-102
2N2926-ORG	2N2712,2N2714,2N2923,2N2924,2N2926-YEL,2N3241A,2N5224,2N5998,2N6000,2SC1317,2SC1788,2SC1851,BC548A,BCY58A,BSX38,*PTC132,*GE-212,*TR-51,*HEPS0015,*ECG123A,*WEP736,*BC107B,*276-2015,*RT-102
2N2926-RED	2N2331,2N3011,PT720,*PTC132,*GE-212,*TR-51,*HEPS0015,*BC107B,*276-2015
2N2926-YEL	2N2924,2N2925,2N2926-GRN,2N3900,2N3900A,2N4256,2N5998,2N6002,2N6008,2SC1849,A5T4124,BC548B,BC549B,BCY58B,BCY58C,*PTC121,*GE-212,*TR-51,*HEPS0015,*BC107B,*276-2015
2N2927	2N3244,2N3467,2N5042,2SA504-Y,BC360-10,BC360-6,SE8541,SE8542,*PTC127,*GE-48,*TR-20,*HEPS0012,*SK3114,*SP70,*ECG159,*WEP242,*276-2021,*RT-115
2N2928	*GE-2,*TR-17,*HEPG0003,*SK3123,*ECG160,*WEP637
2N2929	*TR-17,*HEPG0002,*ECG160,*WEP637,*276-2005
2N2931	2N2932,2N2933,*PTC101,*GE-11,*TR-21,*HEPS0015,*SK3124,*ECG123A,*WEP736,*276-2016,*RT-102
2N2932	*PTC101,*GE-11,*TR-21,*HEPS0015,*SK3124,*ECG123A,*WEP736,*276-2016,*RT-102
2N2933	2N2931,2N2932,*PTC101,*GE-11,*TR-21,*HEPS0015,*SK3124,*ECG123A,*WEP736,*276-2016
2N2934	2N2935,*PTC101,*GE-11,*TR-21,*HEPS0015,*SK3124,*ECG123A,*WEP736,*276-2016,*RT-102
2N2935	*PTC101,*GE-11,*TR-21,*HEPS0015,*SK3124,*ECG123A,*WEP736,*276-2016,*RT-102
2N2936	2N2920,2N2920A,2N2937,2N3680,*GE-20,*HEPS0025
2N2937	2N2920,2N2920A,2N2936,2N3680,*GE-20
2N2938	*PTC136,*GE-20,*TR-21,*HEPS0011,*SK3122,*ECG123A,*WEP736,*276-2038,*RT-102
2N2939	*HEPS3011
2N2940	2N4943,BSY86,*HEPS3021
2N2941	*HEPS3019
2N2942	*GE-51,*TR-17,*HEPG0002,*ECG160,*WEP637,*276-2005
2N2943	2N838,2N1495,2N2956,2N3323,2N3324,2N3325,*PTC145,*GE-51,*TR-17,*HEPG0002,*ECG160,*WEP637,*276-2005
2N2944	*GE-82,*TR-20,*HEPS0013,*SK3114,*SP70,*ECG159,*WEP242,*BC327,*276-2023,*RT-115
2N2944A	*GE-82,*TR-20,*SK3114,*ECG159,*WEP242,*BC327,*276-2023,*RT-115
2N2945	2N4980,*GE-22,*TR-20,*HEPS0013,*SK3114,*ECG159,*WEP242,*BC327,*276-2023,*RT-115
2N2945A	*GE-82,*TR-20,*HEPS0013,*SK3114,*ECG159,*WEP242,*BC327,*276-2023,*RT-115
2N2946	2N3840,2N3910,2N3911,2N4981,*GE-82,*TR-20,*HEPS0013,*SK3114,*ECG159,*WEP242,*BC327,*276-2023,*RT-115
2N2946A	2N3840,*GE-82,*TR-20,*HEPS0013,*SK3114,*ECG159,*WEP242,*BC327,*276-2023,*RT-115
2N2947	*PTC906,*GE-66,*HEPS3020,*HN100
2N2948	2N2947,*PTC906,*GE-19,*HEPS3020,*HN100
2N2949	2N1958A,2N2217,2N2477,2N3737,2N5188,2SC255,2SC1721,BC142,PT2540,*PTC123,*GE-63,*HEPS3001,*SK3048,*ECG195A,*276-2018
2N2950	
2N2951	2N2952,2N4875,2N4876,2N5421,2SC302-M,2SC307-M,2SC590N,2SC730,2SC731,2SC844,2SC854,2SC855,2SC872,2SC890,2SC908,2SC991,2SC992,2SC1002,2SC1083,2SC1256,2SC1303,2SC1464,2SC1479,2SC1820,2SC1947,SRF1001,*DS-66,*GE-63,*TR-21,*HEPS0014,*SK3122,*ECG123A,*WEP736,*276-2009,*RT-102
2N2952	2N2951,2N4875,2N4876,2N5421,2SC302-M,2SC307-M,2SC590N,2SC730,2SC731,2SC844,2SC854,2SC855,2SC872,2SC890,2SC908,2SC991,2SC992,2SC1002,2SC1083,2SC1256,2SC1303,2SC1464,2SC1479,2SC1820,2SC1947,SRF1001,*PTC123,*GE-63,*TR-21,*HEPS0011,*SK3122,*ECG123A,*WEP736,*276-2009,*RT-102
2N2953	*GE-51,*TR-05,*HEPG0005,*SK3004,*HF35,*ECG102,*WEP631,*276-2005,*RT-120
2N2954	TC3114,*PTC136,*GE-20,*HEPS0025,*276-2011
2N2955	*PTC145,*GE-51,*TR-17,*HEPG0003,*ECG160,*WEP637

To Replace	Substitute This Type
2N2956	2N2957,*PTC145,*GE-51,*TR-17,*HEPG0003,*ECG160,*WEP637
2N2957	*PTC145,*GE-51,*TR-17,*HEPG0003,*ECG160,*WEP637
2N2958	2N2218,2N2218A,2N2537,2N2848,2N3725A,2N4960,2N4961,2SC97A,*PTC136, *GE-47,*TR-21,*HEPS0014,*SK3122,*ECG123A,*WEP736,*276-2038,*RT-102
2N2959	2N2219,2N2219A,2N2538,2N3300,A5T2222,TIS109,TIS111,*PTC136,*GE-47,*TR-21, *HEPS0014,*SK3122,*ECG123A,*WEP736,*276-2038,*RT-102
2N2960	2N2961,*PTC136,*GE-18,*TR-21,*HEPS0014,*SK3122,*ECG123A,*WEP736, *276-2038,*RT-102
2N2961	2N2960,*PTC136,*GE-18,*TR-21,*HEPS0014,*SK3122,*ECG123A,*WEP736, *276-2038,*RT-102
2N2962	*HEPG0008
2N2963	
2N2964	*HEPG0005
2N2965	*HEPG0005
2N2966	*PTC145,*GE-2,*TR-05,*HEPG0002,*ECG100,*WEP254,*RT-118
2N2967	2N709,2N2475,2N2784,2N3633,2N3959,2N3960,2N5200,2N5201,2SC99,*PTC126, *HEPS0011
2N2968	2N2162,2N2969,2N4008,*GE-65,*TR-20,*HEPS0012,*SK3118,*SP70,*ECG106, *WEP52,*BC177,*276-2022,*RT-126
2N2969	2N4008,*GE-65,*TR-20,*HEPS0013,*SK3118,*SP70,*ECG106,*WEP52,*BC177, *276-2022,*RT-126
2N2970	2N2968,2N2969,2N2971,2N4008,MM4052,*GE-65,*TR-20,*HEPS0012,*SK3118, *SP70,*ECG106,*WEP52,*BC177,*276-2022,*RT-126
2N2971	2N2969,2N4008,MM4052,*GE-65,*TR-20,*HEPS0013,*SK3118,*SP70,*ECG106, *WEP52,*BC177,*276-2022,*RT-126
2N2972	2N2974,2N2976,2N2978,*HEPS0025
2N2973	2N2975,2N2977,2N2979,*HEPS0005
2N2974	2N2972,2N2976,2N2978,*GE-20,*HEPS0025
2N2975	2N2973,2N2977,2N2979,*HEPS0005
2N2976	2N2972,2N2974,2N2978,*HEPS0025
2N2977	2N2973,2N2975,2N2979,*HEPS0005
2N2978	*HEPS0025
2N2979	*HEPS0005
2N2980	2N2981,2N2982,*HEPS0005
2N2981	2N2980,2N2982,*HEPS0005
2N2982	2N2980,2N2981,*HEPS0005
2N2983	2N2984,*HEPS3019,*276-2008
2N2984	*HEPS3021,*276-2008
2N2985	2N2986,*HEPS3019,*276-2008
2N2986	*HEPS3021,*276-2008
2N2987	2N4300,*GE-32,*HEPS3019,*276-2008
2N2988	*GE-32,*HEPS3019,*276-2008
2N2989	*GE-32,*HEPS3019,*276-2008
2N2990	*GE-32,*HEPS3019,*276-2008
2N2991	*HEPS3019
2N2992	*HEPS3019
2N2993	*HEPS3019
2N2994	*HEPS3019
2N2996	2N2997,*GE-51,*TR-17,*HEPG0003,*SK3006,*ECG160,*WEP637
2N2997	*GE-51,*TR-17,*HEPG0003,*ECG160,*WEP637
2N2998	*GE-51,*TR-17,*HEPG0003,*ECG160,*WEP637
2N2999	*TR-17,*ECG160,*WEP637
2N3009	2N916,2N916A,2N2369,2N2369A,2N2501,2N3013,2N3014,2N3211,2N3862,2N4013, 2N4873,2N5144,2N5769,2SC67,2SC577,2SC639,2SC764,ZT2369,ZT2369A, *PTC136,*GE-20,*TR-21,*HEPS0011,*SK3122,*SN80,*ECG123A,*WEP736, *276-2009,*RT-102
2N3010	2N709,2N709A,2N2475,2N2784,2N3633,2N3959,2N3960,2N5200,2N5201,2SC718, 2SC1231,MT1060A,*PTC126,*GE-61,*TR-95,*HEPS0016,*SK3039,*SN80,*ECG108, *WEP56,*276-2038,*RT-113
2N3011	2N2501,2N3013,2N3014,2N3211,2N3646,2N5772,2SC67,2SC577,2SC639,2SC764, *PTC136,*GE-20,*TR-95,*HEPS0016,*SK3039,*SN60,*ECG123A,*WEP56,

To Replace	Substitute This Type
(2N3011)	*276-2009,*RT-113
2N3012	2N3546,2N3640,2N4258A,*GE-21,*TR-20,*HEPS0013,*SK3114,*SP70,*ECG159, *WEP242,*276-2023,*RT-115
2N3013	2N916,2N916A,2N2501,2N3014,2N3211,2N4013,2N5144,2SC67,2SC577,2SC639, 2SC764,*PTC136,*GE-20,*TR-21,*HEPS0016,*SK3122,*SN80,*ECG123A,*WEP736, *276-2009,*RT-102
2N3014	2N916,2N916A,2N2501,2N3013,2N3211,2N4013,2N5144,2SC67,2SC577,2SC639, 2SC764,*PTC136,*GE-20,*TR-21,*HEPS0016,*SK3122,*SN80,*ECG123A,*WEP736, *276-2009,*RT-102
2N3015	2N2848,2N3299,2N4960,2N4961,*GE-47,*TR-86,*HEPS3001,*EN10,*ECG123, *WEP53,*276-2038,*RT-100
2N3016	*HEPS3002
2N3017	*HEPS5004
2N3018	*HEPS5004
2N3019	*GE-18,*HEPS3019,*ECG128,*WEP243
2N3020	*GE-27,*TR-52,*HEPS3019,*ECG128,*WEP243
2N3021	2N3022,*GE-69,*TR-52,*HEPS5006
2N3022	2N3023,KSP2021,*GE-69,*HEPS5006
2N3023	KSP2021,KSP2022,*GE-69,*HEPS5006
2N3024	2N3025,*GE-69,*HEPS7003
2N3025	2N3026,*GE-69,*HEPS7003
2N3026	2N5605,2N5613,*GE-69,*HEPS7003
2N3033	*PTC144,*276-2012
2N3034	*PTC125,*GE-81,*TR-53,*276-2012
2N3035	*PTC123,*GE-210,*TR-95,*SK3039,*EN10,*ECG108,*WEP56,*BC107B,*276-2013, *RT-113
2N3036	2N2443,*GE-18,*HEPS3019,*276-2008
2N3037	2N2364,2N2364A,2N2898,2N2899,*GE-18,*HEPS0005,*276-2008
2N3038	2N3057,*GE-18,*HEPS0005,*276-2008
2N3039	2N3485,2N3485A,*GE-82,*TR-28,*HEPS0019,*BC327,*276-2023
2N3040	2N3485,2N3485A,2N3673,*GE-82,*HEPS0019,*BC327,*276-2023
2N3043	2N3044,2N3045,*GE-17,*TR-24,*HEP728(2)
2N3044	2N3043,2N3045,*GE-17,*TR-24,*HEP728(2)
2N3045	2N3043,2N3044,*GE-17,*TR-24,*HEP728(2)
2N3046	2N3043,2N3044,2N3045,2N3047,2N3048,*GE-17,*TR-24,*HEP729(2)
2N3047	2N3043,2N3044,2N3045,2N3046,2N3048,*GE-17,*TR-24,*HEP729(2)
2N3048	2N3043,2N3044,2N3045,2N3046,2N3047,*GE-17,*TR-24,*HEP729(2)
2N3049	2N3050,2N3051,*GE-21,*HEP716(2)
2N3050	2N3049,2N3051,*GE-21,*HEP716(2)
2N3051	2N3049,2N3050,*GE-21,*HEP716(2)
2N3052	*GE-63,*TR-25,*HEPS0025
2N3053	*GE-63,*TR-87,*HEPS3011,*SK3024,*ECG128,*WEP243,*276-2030,*RT-114
2N3054	2N3878,2SD242,2SD257,KSP1025,*PTC112,*GE-66,*TR-81,*HEPS5019,*SK3026, *PN26,*ECG175,*WEP245,*276-2020,*RT-197
2N3054A	*PTC169
2N3055	2N3236,2N6270,2N6271,2N6272,2N6273,108T2,BUY53A,BUY54A,HST9202,HST9203, *PTC140,*GE-14,*TR-59,*HEPS7004,*SK3027,*PN350,*ECG181,*WEP247,*BDY20, *RT-131
2N3055A	2N3235,BUY51A,BUY52A,HST9201
2N3055C	2N3772C
2N3055SD	2N3055,2N3236,2N5629,2N5630,2N6254,2N6270,2N6271,2N6272,2N6273,2SD341, 2SD492,108T2,BUY53A,BUY54A,HST9202,HST9203,IR3772,SDT9202,SDT9203
2N3056	2N2898,*PTC144,*GE-18,*HEPS3019
2N3056A	*HEPS3019
2N3057	*GE-18,*HEPS3019
2N3057A	*HEPS3019
2N3058	2N2944A,2N3059,*PTC127,*GE-82,*TR-20,*HEPS0013,*SK3114,*SP70,*ECG159, *WEP242,*BC327,*276-2023,*RT-115
2N3059	2N2944,*GE-82,*TR-20,*HEPS0013,*SK3114,*ECG159,*WEP242,*BC327,*276-2023, *RT-115
2N3060	2N3063,2N4982,*GE-82,*TR-20,*HEPS0013,*SK3114,*ECG159,*WEP242,*276-2023,

To Replace	Substitute This Type
(2N3060)	*RT-115
2N3061	2N3063,*GE-82,*HEPS0032
2N3062	2N3063,2N3065,*GE-21,*TR-20,*HEPS0013,*SK3114,*ECG159,*WEP242,*276-2023,*RT-115
2N3063	*GE-21
2N3064	*GE-21
2N3065	*GE-21
2N3072	2N2904,2N2904A,2N4030,2N4031,2N4404,2SA503-Y,2SA537,2SA560,2SA571,2SA717,BC361-10,BC361-6,*PTC127,*GE-67,*TR-20,*HEPS0013,*SK3114,*ECG159,*WEP242,*276-2023,*RT-115
2N3073	2N2906,2N2906A,2N3485,2N3485A,2N4026,2N4027,2N5372,2SA697,BSW74,TZ551,*PTC127,*GE-82,*TR-20,*HEPS0013,*SK3114,*ECG159,*WEP242,*276-2023,*RT-115
2N3074	*GE-51,*TR-17,*HEPG0003,*SK3006,*HF75,*ECG160,*WEP637
2N3075	*GE-51,*TR-05,*HEPG0005,*SK3006,*HF75,*ECG100,*WEP254,*276-2007,*RT-118
2N3076	
2N3077	2N736B,2N871,2N910,2N956,2N2390,2N2462,2N2466,2N2516,2N2645,2N3038,*PTC123,*GE-81,*TR-87,*HEPS5026,*SK3124,*ECG128,*WEP243,*BF338,*276-2008,*RT-114
2N3078	2N2460,2N2464,*PTC123,*GE-81,*TR-87,*HEPS5026,*SK3124,*ECG128,*WEP243,*BF338,*276-2008,*RT-114
2N3079	
2N3080	
2N3081	*PTC127,*GE-67,*TR-20,*HEPS0012,*SK3114,*SP70,*ECG159,*WEP242,*276-2021,*RT-115
2N3107	2N1973,*PTC125,*GE-18,*HEPS3002,*ECG128,*WEP712,*BF338,*276-2008
2N3108	2N699B,2N1613B,2N1889,2N1974,*PTC125,*GE-18,*HEPS3002,*ECG128,*WEP712,*BF338,*276-2008
2N3109	2N1973,2N3107,*PTC125,*GE-18,*HEPS3001,*ECG128,*WEP243,*BF338,*276-2008
2N3110	2N1613A,2N1889,2N1974,2N3108,*PTC125,*GE-18,*TR-25,*HEPS3001,*ECG128,*WEP243,*BF338,*276-2008
2N3114	2N1342,2N1893A,2N3500,2N3712,2N4925,2SC310,2SC1217,MM2259,MM2260,MM3009,SE7001,*GE-32,*TR-78,*HEPS3019,*SK3104,*ECG154,*WEP712,*276-2012,*RT-110
2N3115	2N2221,2N2221A,2N2539,2N4962,2N4963,TN-3903,*PTC136,*GE-20,*TR-21,*HEPS0014,*SK3122,*EN10,*ECG123A,*WEP736,*276-2009,*RT-102
2N3116	2N2222,2N2222A,2N2540,A5T2222,FT3643,TIS109,TIS111,TN-3904,*PTC136,*GE-20,*TR-21,*HEPS0014,*SK3122,*EN10,*ECG123A,*WEP736,*276-2009,*RT-102
2N3117	2N6428,2N6428A,2N6539,*GE-18,*TR-87,*HEPS0015,*SK3124,*ECG128,*WEP243,*276-2008,*RT-114
2N3118	2N1506,2N1506A,2N1709,2N3866,2SC302-M,2SC547,2SC555,40305,SRF11101,*GE-18,*HEPS3019
2N3119	*GE-27,*TR-78,*HEPS3019
2N3120	2SA544,2SA552,2SA594-R,2SA594N,*PTC127,*GE-48,*TR-88,*HEPS0014,*SK3025,*ECG129,*WEP242,*276-2022,*RT-115
2N3121	2N3250,2N3250A,2N6067,GET2904,GET2905,TZ551,*PTC127,*GE-82,*TR-20,*HEPS0012,*SK3114,*ECG159,*WEP242,*276-2021,*RT-115
2N3122	2N656,2N1613A,2N2846,2N3015,2N3299,2N3326,2N3554,2SC32,2SC594-R,BSY92,*PTC125,*GE-47,*TR-21,*HEPS0014,*SK3122,*ECG123A,*WEP736,*BF338,*276-2008,*RT-102
2N3123	*HEPS3011
2N3127	*PTC107,*TR-17,*HEPG0003,*ECG160,*WEP637
2N3133	2N2904,2N2904A,2N4030,2SA537,2SA571,2SA717,*PTC127,*GE-21,*TR-88,*HEPS0012,*SK3025,*SP70,*ECG129,*WEP242,*276-2021,*RT-115
2N3134	2N2905,2N2905A,2N3503,2N4032,2SA684,A5T2907,A5T3505,A5T3645,TIS112,*PTC127,*GE-48,*TR-88,*HEPS0012,*SK3025,*SP70,*ECG129,*WEP242,*276-2021,*RT-115
2N3135	2N2906,2N2906A,2N3485,2N3485A,2N4026,*PTC127,*GE-21,*TR-20,*HEPS0012,*SK3114,*SP70,*ECG159,*WEP242,*276-2023,*RT-115
2N3136	2N2907,2N2907A,2N3486,2N3486A,2N3505,2N4028,A5T2907,A5T3505,A5T3645,TIS112,*PTC127,*GE-48,*TR-20,*HEPS0012,*SK3114,*SP70,*ECG159,*WEP242,

To Replace	Substitute This Type
(2N3136)	*276-2023,*RT-115
2N3137	2N3326,2N5769,2SC651,TIS105,*PTC136,*GE-11,*TR-70,*HEPS0016,*SK3039, *SN60,*ECG107,*WEP56,*276-2038,*RT-108
2N3139	
2N3140	*GE-66,*HEPS5000
2N3141	
2N3142	*GE-66,*HEPS5000
2N3143	
2N3144	*GE-66,*HEPS5000
2N3145	
2N3146	*GE-3
2N3147	*GE-3
2N3149	2N3150
2N3150	
2N3151	
2N3153	*GE-212,*TR-17,*HEPG0003,*ECG160,*WEP637,*BC107B,*276-2013
2N3154	2N3155
2N3155	2N3156
2N3156	2N3157
2N3157	
2N3158	2N3159
2N3159	2N3160
2N3160	2N3161
2N3161	
2N3163	2N3164,2N3175,2N3176,2N3187,2N3188,*HEPS7003
2N3164	2N3165,2N3176,2N3177,2N3188,2N3189,*HEPS5005,*276-2007
2N3165	2N3166,2N3177,2N3178,2N3189,2N3190,*HEPS5005
2N3166	2N3178,2N3190,*HEPS5006
2N3167	2N3168,2N3179,2N3180,2N3191,2N3192,*HEPS5005
2N3168	2N3169,2N3180,2N3181,2N3192,2N3193,*HEPS5005,*276-2007
2N3169	2N3170,2N3181,2N3182,2N3193,2N3194,*HEPS5005
2N3170	2N3182,2N3194,*HEPS5005
2N3171	2N3172,2N3183,2N3184,2N3195,2N3196,*PTC172,*GE-74,*TR-29,*HEPS7003
2N3172	2N3173,2N3184,2N3185,2N3196,2N3197,*PTC172,*GE-74,*TR-29,*HEPS7003
2N3173	2N3174,2N3185,2N3186,2N3197,2N3198,*PTC172,*GE-74,*HEPS5006
2N3174	2N3186,2N3198,*PTC172,*GE-74,*HEPS5005
2N3175	2N3176,2N3187,2N3188,*HEPS5005
2N3176	2N3177,2N3188,2N3189,*HEPS5005,*276-2007
2N3177	2N3178,2N3189,2N3190,*HEPS5005
2N3178	2N3190,*HEPS5005
2N3179	2N3180,2N3191,2N3192,*HEPS5005
2N3180	2N3181,2N3192,2N3193,*HEPS5005,*276-2007
2N3181	2N3182,2N3193,2N3194,*PTC172,*HEPS5005
2N3182	2N3194,*HEPS5005
2N3183	2N3184,2N3195,2N3196,*GE-74,*TR-29,*HEPS7003
2N3184	2N3185,2N3196,2N3197,*PTC172,*GE-74,*TR-29,*HEPS7003
2N3185	2N3186,2N3197,2N3198,*PTC172,*GE-74,*HEPS5005
2N3186	2N3198,*PTC172,*GE-74,*HEPS5005
2N3187	2N3175,2N3176,2N3188,*HEPS5005
2N3188	2N3176,2N3177,2N3189,*HEPS5005,*276-2007
2N3189	2N3177,2N3178,2N3190,*HEPS5005
2N3190	2N3178,*HEPS5005
2N3191	2N3179,2N3180,2N3192,*HEPS5005
2N3192	2N3180,2N3181,2N3193,*HEPS5005,*276-2007
2N3193	2N3181,2N3182,2N3194,*HEPS5005
2N3194	2N3182,*HEPS5005
2N3195	2N3183,2N3184,2N3196,*PTC172,*GE-74,*TR-29,*HEPS7003
2N3196	2N3184,2N3185,2N3197,*PTC172,*GE-74,*TR-29,*HEPS7003
2N3197	2N3185,2N3186,2N3198,*PTC172,*GE-74,*HEPS5005
2N3198	2N3186,*PTC172,*GE-74,*HEPS5005
2N3199	2N3200,2N3205,2N3206,*GE-69,*HEPS5006,*276-2027

To Replace	Substitute This Type
2N3200	2N3201,2N3206,*GE-69,*HEPS5006
2N3201	2N3207,*HEPS5006
2N3202	2N3203,*GE-84,*TR-88,*HEPS5013,*SK3025,*ECG129,*WEP242,*276-2025,*RT-115
2N3203	2N3204,*TR-88,*HEPS5013,*SK3025,*ECG129,*WEP242,*276-2025,*RT-115
2N3204	*HEPS3003
2N3205	2N3199,2N3200,2N3206,*GE-69,*276-2027
2N3206	2N3200,2N3201,*GE-69,*HEPS5006
2N3207	*HEPS5006
2N3208	2N3202,2N3203,*GE-84,*TR-88,*HEPS5013,*SK3025,*ECG129,*WEP242,*276-2025, *RT-115
2N3209	2N869A,2N3576,2N3829,*GE-21,*TR-20,*HEPS0013,*SK3118,*SP70,*ECG106, *WEP52,*276-2021,*RT-126
2N3210	2N914,2N916,2N916A,2N2221,2N2501,2N2539,2N2845,2N2847,2N3115,2N3211, 2N3301,2N3646,2N4013,2N4962,2N5144,2SC577,BSW82,FT3641,FT3642,TIS133, TIS134,TN-3903,ZT708,*PTC136,*GE-20,*TR-21,*HEPS0014,*SK3122,*ECG123A, *WEP736,*276-2009,*RT-102
2N3211	2N2501,2N4013,2N5144,*PTC136,*GE-20,*TR-21,*HEPS0014,*SK3122,*ECG123A, *WEP736,*276-2009,*RT-102
2N3212	*DS-503,*GE-25,*TR-01,*SK3009,*ECG121,*WEP232,*RT-127
2N3213	2N3212,*DS-503,*GE-16,*TR-01,*SK3009,*ECG121,*WEP232,*RT-127
2N3214	2N3213,*DS-503,*GE-16,*TR-01,*SK3009,*ECG121,*WEP232,*RT-127
2N3215	2N3214,*DS-503,*GE-16,*TR-01,*SK3009,*ECG121,*WEP232,*RT-127
2N3217	2N2944A,2N2945A,2N3218,2N3977,2N3978,*GE-22,*TR-20,*HEPS0013,*SK3114, *SP70,*ECG159,*WEP242,*BC327,*276-2023,*RT-115
2N3218	2N2945A,2N2946,2N2946A,2N3978,*GE-22,*TR-20,*HEPS0013,*SK3114,*SP70, *ECG159,*WEP242,*BC327,*276-2023,*RT-115
2N3219	2N2946A,2N3979,*GE-82,*TR-20,*HEPS0013,*SK3114,*SP70,*ECG159,*WEP242, *BC327,*276-2023,*RT-115
2N3224	*PTC141,*GE-21,*HEPS3032
2N3225	*PTC141,*HEPS3032
2N3226	2N5034,2N5067,40514,40542,B170001-BLK,B170001-BRN,B170010,B170019, HST9201,HST9205,M5A,M10A,*PTC129A,*GE-77,*TR-59,*HEPS7002,*HN100, *ECG130,*WEP247,*BDY20,*RT-131
2N3227	*PTC136,*TR-95,*HEPS0016,*SK3039,*ECG108,*WEP56,*276-2038,*RT-113
2N3229	*GE-66
2N3232	2N3233,2N3236,*GE-77,*TR-59,*HEPS7002,*ECG130,*WEP247,*RT-131
2N3233	2N3236,*GE-75,*TR-59,*HEPS5004,*SK3027,*ECG130,*WEP247,*276-2020
2N3234	*TR-61,*HEPS5020,*ECG130,*WEP247
2N3235	*PTC140,*GE-77,*HEPS7004
2N3236	*PTC173,*GE-75,*HEPS7004
2N3237	2N6570,2N6571,2SD55,IR3772,*PTC173,*GE-75,*HEPS7000
2N3238	2N3239,2SD55,*GE-75,*HEPS7004
2N3239	2N3238,2SD55,*PTC173,*GE-75,*HEPS7004
2N3240	*TR-52
2N3241A	2N3242A,*GE-17,*TR-21,*HEPS0014,*SK3124,*SN80,*ECG123A,*WEP736,*BC337, *276-2009,*RT-102
2N3242A	*GE-20,*TR-21,*HEPS0014,*SK3124,*ECG123A,*WEP736,*BC337,*276-2009, *RT-102
2N3244	2N3467,2N3762,*GE-67,*TR-77,*HEPS3028
2N3245	2N3468,2N3762A,2N3763,2N5022,MM3726,*GE-67,*TR-20,*HEPS3032,*SK3118, *SP70,*ECG106,*WEP52,*RT-126
2N3248	*PTC127,*GE-22,*TR-20,*HEPS0013,*SK3114,*ECG159,*WEP242,*276-2023, *RT-115
2N3249	2N4058,NPS404,*PTC127,*GE-21,*TR-20,*HEPS0013,*SK3114,*ECG159,*WEP242, *276-2023,*RT-115
2N3250	2N3250A,*PTC127,*GE-21,*TR-20,*HEPS0013,*SK3114,*SP70,*ECG159,*WEP242, *276-2023,*RT-115
2N3250A	*PTC127,*GE-21,*TR-20,*HEPS0013,*SK3114,*ECG159,*WEP242,*276-2009, *RT-115
2N3251	2N3251A,*PTC127,*GE-21,*TR-20,*HEPS0013,*SK3114,*SP70,*ECG159,*WEP242, *276-2023,*RT-115

To Replace	Substitute This Type
2N3251A	*PTC127,*TR-20,*HEPS0013,*SK3114,*ECG159,*WEP242,*276-2009,*RT-115
2N3252	*GE-63,*TR-65,*HEPS3001,*276-2038
2N3253	2N3735,*GE-216,*TR-76,*HEPS3001,*276-2014
2N3261	2N3510,2N3511,2N3647,2N3648,ZT708,*PTC136,*GE-20,*TR-21,*HEPS0014, *SK3122,*ECG123A,*WEP736,*276-2016,*RT-102
2N3262	*TR-87,*HEPS3019,*SK3024,*ECG128,*WEP243
2N3263	
2N3264	2N3263
2N3265	
2N3266	2N3265
2N3268	*PTC144,*GE-27,*TR-86,*HEPS0014,*SK3124,*ECG123,*WEP53,*BF338,*276-2009, *RT-100
2N3279	2N3280,*PTC107,*GE-51,*TR-17,*HEPG0003,*ECG160,*WEP637
2N3280	2N3279,*PTC107,*GE-51,*TR-17,*HEPG0003,*ECG160,*WEP637
2N3281	2N3282,*PTC107,*GE-51,*TR-17,*HEPG0003,*ECG160,*WEP637
2N3282	2N3281,*GE-51,*TR-17,*HEPG0003,*ECG160,*WEP637
2N3283	2N3127,2N3279,2N3280,2N3284,*PTC107,*GE-50,*TR-17,*HEPG0003,*ECG160, *WEP637
2N3284	2N3127,2N3279,2N3280,2N3283,*PTC107,*GE-50,*TR-17,*HEPG0003,*ECG160, *WEP637
2N3285	2N3286,*PTC107,*GE-50,*TR-17,*HEPG0003,*ECG160,*WEP637
2N3286	2N3285,*PTC107,*GE-50,*TR-17,*HEPG0003,*ECG160,*WEP637
2N3287	2N3288,2N3337,2N3338,2N3339,*PTC132,*GE-61,*TR-83,*HEPS0016,*SK3018, *ECG161,*WEP719,*276-2016,*RT-113
2N3288	2N3287,2N3337,2N3338,2N3339,*PTC132,*GE-61,*TR-83,*HEPS0016,*SK3018, *ECG161,*WEP719,*276-2016,*RT-113
2N3289	2N918,2N3287,2N3288,2N3290,2N3337,2N3338,2N3339,2N5851,2N5852,MT1061, *PTC132,*GE-86,*TR-83,*HEPS0016,*SK3019,*EN10,*ECG161,*WEP719,*BC107B, *276-2011,*RT-113
2N3290	2N918,2N3287,2N3288,2N3289,2N3337,2N3338,2N3339,2N5851,2N5852,MT1061, *PTC132,*GE-86,*TR-83,*HEPS0016,*SK3019,*EN10,*ECG161,*WEP719,*BC107B, *276-2011,*RT-113
2N3291	2N3292,*PTC132,*GE-61,*TR-83,*HEPS0016,*SK3019,*EN10,*ECG161,*WEP719, *BC107B,*276-2016,*RT-113
2N3292	2N3291,*PTC132,*GE-61,*TR-83,*HEPS0016,*SK3019,*EN10,*ECG161,*WEP719, *BC107B,*276-2016,*RT-113
2N3293	2N3291,2N3292,2N3294,2N6599,2N6600,*PTC132,*GE-86,*TR-83,*HEPS0016, *SK3019,*EN10,*ECG161,*WEP719,*BC107B,*276-2015,*RT-113
2N3294	2N3291,2N3292,2N3293,2N6599,2N6600,*PTC132,*GE-86,*TR-83,*HEPS0016, *SK3019,*EN10,*ECG161,*WEP719,*BC107B,*276-2015,*RT-113
2N3295	2N2217,2N5188,BSX49,*GE-63,*HEPS3008
2N3296	2N2950,*GE-28,*TR-76
2N3297	2N2947,*PTC906,*GE-19,*HN100
2N3298	2N2501,2N2615,2N3241A,2N4123,2N4418,2N4873,2S512,2SC639,2SC735-O, 2SC1216,A5T4123,MPS6513,MPS6574-BLUE,MPS6574-GREEN,MPS6574-YEL, MPSH10,MPSH11,ZT80,ZT110,ZT2369A,ZT2938,*PTC136,*GE-10,*TR-76, *HEPS5026,*BC107B,*276-2016
2N3299	2N2848,2N3015,2N4960,2N4961,*GE-47,*TR-64,*HEPS3008,*276-2038
2N3300	BCY65,*GE-47,*HEPS3008,*ECG123A,*WEP736,*276-2038
2N3301	2N915,2N2847,2N3946,2N4962,2N4963,2N5380,A5T3903,MM3903,TN-3903,*PTC136, *GE-210,*TR-21,*HEPS0015,*SK3122,*ECG123A,*WEP736,*276-2009,*RT-102
2N3302	2N4409,A5T4409,FT3643,*PTC136,*GE-210,*TR-21,*HEPS0015,*SK3122,*ECG123A, *WEP736,*276-2009,*RT-102
2N3303	2N3137,2N5065,*HEPS3001,*276-2038
2N3304	2N2894A,2N3012,2N3546,2N3639,2N3640,2N4207,2N4208,2N4209,2N4257A, 2N4258A,2N4872,2N5056,2N5057,2N5292,2N5771,A5T4260,A5T4261,MPS3639, MPS3640,MPSL07,MPSL08,*PTC131,*GE-21,*TR-20,*SP70,*ECG106,*WEP52, *276-2021,*RT-126
2N3307	*PTC127,*GE-21,*TR-20,*HEPS0013,*SK3118,*SP70,*ECG159,*WEP242,*276-2034, *RT-115
2N3308	2N3307,*PTC127,*GE-21,*TR-20,*HEPS0013,*SK3118,*SP70,*ECG159,*WEP242,

To Replace	Substitute This Type
(2N3308)	*276-2034,*RT-115
2N3309	BSX48,BSX49,MPS3725,*GE-215
2N3311	2N3312,2N3314,2N3315,*DS501,*GE-4,*TR-03,*HEPG6004,*SK3012,*ECG105, *WEP233,*276-2006
2N3312	2N3313,2N3315,*DS501,*GE-4,*TR-03,*HEPG6004,*SK3012,*ECG105,*WEP233, *276-2006
2N3313	*DS501,*GE-4,*TR-03,*HEPG6006,*SK3012,*ECG105,*WEP233,*276-2006
2N3314	2N3315,*DS501,*GE-4,*TR-03,*HEPG6004,*SK3012,*ECG105,*WEP233,*276-2006
2N3315	*DS501,*GE-4,*TR-03,*HEPG6004,*SK3012,*ECG105,*WEP233,*276-2006
2N3316	*HEPG6006,*276-2006
2N3317	*GE-65,*TR-20,*HEPS0013,*SK3114,*SP70,*ECG159,*WEP242,*BC177,*276-2022, *RT-115
2N3318	*GE-65,*TR-20,*HEPS0013,*SK3114,*SP70,*ECG159,*WEP242,*BC177,*276-2022, *RT-115
2N3319	*GE-65,*TR-20,*HEPS0013,*SK3114,*SP70,*ECG159,*WEP242,*BC177,*276-2021, *RT-115
2N3320	*PTC145,*GE-51,*TR-17,*HEPG0003,*ECG160,*WEP637
2N3321	*PTC145,*GE-51,*TR-17,*HEPG0003,*ECG160,*WEP637,*276-2021
2N3322	2N3320,*PTC145,*GE-51,*TR-17,*HEPG0003,*ECG160,*WEP637,*276-2021
2N3323	2N2956,2N2957,2N3324,2N3325,*PTC145,*GE-50,*TR-17,*HEPG0003,*ECG160, *WEP637
2N3324	2N2956,2N2957,2N3323,2N3325,*PTC145,*GE-51,*TR-17,*HEPG0008,*ECG160, *WEP637,*276-2005
2N3325	2N2956,2N2957,2N3323,2N3324,*PTC145,*GE-51,*TR-17,*HEPG0003,*ECG160, *WEP637
2N3326	*HEPS3001,*ECG123A,*WEP736,*276-2014
2N3337	2N3338,2N3339,2SC563A,*PTC136,*GE-61,*HEPS0025,*ECG161,*WEP56,*276-2016
2N3338	2N3337,2N3339,2SC563A,*PTC136,*GE-61,*TR-83,*HEPS0025,*SK3117,*ECG161, *WEP709,*276-2016,*RT-113
2N3339	2N3337,2N3338,2SC563A,*PTC136,*GE-61,*TR-83,*HEPS0025,*SK3117,*ECG161, *WEP709,*276-2016,*RT-113
2N3340	2N5772,A5T4123,*PTC123,*GE-210,*TR-21,*HEPS0011,*SK3122,*ECG123A, *WEP736,*BC337,*276-2009,*RT-102
2N3341	2N2394,2N3581,2N6067,A5T3905,A5T5448,TIS37,*PTC127,*GE-82,*TR-20, *HEPS0011,*SK3114,*SP70,*ECG159,*WEP242,*BC327,*276-2023,*RT-115
2N3342	2N1917,2N1918,2N2945A,2N4284,2N4285,2N5230,*GE-65,*TR-20,*HEPS0012,*SP70, *ECG106,*WEP52,*BC177,*276-2022,*RT-126
2N3343	2N939,2N1026,2N1219,2N1231,2N1917,2N1918,2N2946A,2N3344,2N4008,2N6567, *GE-65,*TR-88,*HEPS0012,*SP70,*ECG129,*WEP242,*BC177,*276-2022,*RT-115
2N3344	2N939,2N940,2N1026,2N1231,2N1469,2N1469A,2N2946A,2N3346,2N4008,2N6567, *GE-65,*TR-88,*HEPS0012,*SK3025,*SP70,*ECG129,*WEP242,*BC177,*276-2022, *RT-115
2N3345	2N1232,2N1474A,2N3346,*GE-65,*TR-88,*HEPS0012,*SK3025,*SP70,*ECG129, *WEP242,*BC177,*276-2022,*RT-115
2N3346	2N1233,*GE-65,*TR-20,*HEPS0012,*SK3114,*SP70,*ECG159,*WEP242,*BC177, *276-2022,*RT-115
2N3347	2N3348,2N3349,2N3350,2N3351,2N3352,*HEPS0019,*276-2034
2N3348	2N3347,2N3349,2N3350,2N3351,2N3352,*HEPS0019,*276-2034
2N3349	2N3347,2N3348,2N3350,2N3351,2N3352,*HEPS0019,*276-2034
2N3350	2N3347,2N3348,2N3349,2N3351,2N3352,*HEPS0019,*276-2034
2N3351	2N3347,2N3348,2N3349,2N3350,2N3352,*HEPS0019,*276-2034
2N3352	2N3347,2N3348,2N3349,2N3350,2N3351,*HEPS0019,*276-2034
2N3371	2N2957,*GE-51,*TR-17,*HEPG0003,*ECG160,*WEP637
2N3375	*GE-28,*HEPS3020
2N3388	2N2443,2N3923,2SC686,*PTC117,*GE-27,*TR-78,*SK3045,*ECG154,*WEP712, *BF338,*276-2012,*RT-110
2N3389	2SC154C,2SC1048,A5T5059,*PTC117,*GE-27,*TR-78,*SK3045,*ECG154,*WEP712, *BF338,*276-2012,*RT-110
2N3390	2N3398-BLU,BC548C,BC549C,MPS3390,*PTC139,*GE-212,*TR-21,*HEPS0024, *SK3124,*EN40,*ECG123A,*WEP736,*BC337,*276-2031,*RT-102
2N3391	2N3391A,2N6002,2N6008,BCY58D,*PTC139,*GE-212,*TR-21,*HEPS0015,*SK3124,

To Replace	Substitute This Type
(2N3391)	*EN40,*ECG123A,*WEP736,*BC107B,*276-2016,*RT-102
2N3391A	2N3391,2N6002,2N6008,BCY58D,*PTC139,*GE-212,*TR-21,*HEPS0015,*SK3124, *EN40,*ECG123A,*WEP736,*BC107B,*276-2016,*RT-102
2N3392	2N2924,2N2925,2N3391,2N3391A,2N4256,2N4954,2N5172,2N5371,2N5998,2N6002, 2N6008,2SC1849,A5T4124,BC338,BC548B,BC549B,BCY58B,BCY58C,MPSA16, TP3706,*PTC121,*GE-212,*TR-21,*HEPS0015,*SK3124,*EN40,*ECG123A, *WEP736,*BC107B,*276-2016,*RT-102
2N3393	2N2923,2N2924,2N3227,2N3241A,2N3242A,2N3392,2N5224,2N5998,2N6000,2N6426, 2SC509-Y,2SC509G-Y,2SC1317,2SC1788,2SC1851,BC548A,BCY58A,BSW83, BSX38,*PTC101,*GE-212,*TR-86,*HEPS0015,*SK3122,*EN40,*ECG123,*WEP53, *BC107B,*276-2016,*RT-105
2N3394	2N2331,2N2501,2N2923,2N3211,2N3393,2N6427,2SC509-O,2SC509G-O,2SC639, 2SC764,A5T4123,*PTC101,*GE-212,*TR-21,*HEPS0015,*SK3124,*EN40,*ECG123A, *WEP736,*BC107B,*276-2016,*RT-102
2N3395-WHT	2N3391,2N3391A,2N3396-WHT,2N3397-WHT,2N3398-BLU,2N3398-WHT,2N6002, 2N6008,A5T3391,A5T3391A,A7T3391,A7T3391A,A8T3391,A8T3391A,BCY58D, MPS3391,MPS3391A,PET4003,TP3566,*PTC139,*GE-212,*TR-51,*SK3156, *BC107B,*276-2031
2N3395-YEL	2N2924,2N2925,2N3391,2N3391A,2N3392,2N3395-WHT,2N3396-WHT,2N3396-YEL, 2N3397-WHT,2N3397-YEL,2N3398-WHT,2N3398-YEL,2N3415,2N4256,2N4954, 2N5172,2N5371,2N5998,2N6002,2N6008,2SC1849,40397,A5T3391,A5T3391A, A5T3392,A5T4124,A5T5172,A7T3391,A7T3391A,A7T3392,A7T5172,A8T3391, A8T3391A,A8T3392,A8T5172,BC338,BC548B,BC549B,BCY58B,BCY58C,MPS3391, MPS3391A,MPSA16,MPSA17,PET4002,TIS90-GRY,TIS90M-GRY,TIS92-GRY, TIS92M-GRY,TP3566,TP3706,TZ81,*PTC121,*GE-212,*TR-51,*BC107B,*276-2031
2N3396-ORG	2N2923,2N2924,2N3227,2N3241A,2N3242A,2N3392,2N3393,2N3395-YEL, 2N3396-YEL,2N3397-ORG,2N3397-YEL,2N3398-ORG,2N3398-YEL,2N3414,2N3509, 2N4074,2N5224,2N5998,2N6000,2N6426,2SC509-Y,2SC509G-Y,2SC847,2SC848, 2SC849,2SC966,2SC967,2SC1246,2SC1317,2SC1788,2SC1851,40398,A5T3392, A7T3392,A8T3392,BC548A,BCY58A,BSW83,BSX38,PET4001,TIS90-BLU,TIS90-VIO, TIS90M-BLU,TIS90M-VIO,TIS92-BLU,TIS92-VIO,TIS92M-BLU,TIS92M-VIO,TN80,TZ81, *PTC101,*GE-212,*TR-51,*BC107B,*276-2031
2N3396-WHT	2N3391,2N3391A,2N3395-WHT,2N3397-WHT,2N3398-BLU,2N3398-WHT,2N6002, 2N6008,A5T3391,A5T3391A,A7T3391,A7T3391A,A8T3391,A8T3391A,BCY58D, MPS3391,MPS3391A,PET4003,TP3566,*PTC139,*GE-212,*TR-51,*SK3156, *BC107B,*276-2031
2N3396-YEL	2N2924,2N2925,2N3391,2N3391A,2N3392,2N3395-WHT,2N3395-YEL,2N3396-WHT, 2N3397-WHT,2N3397-YEL,2N3398-WHT,2N3398-YEL,2N3415,2N4256,2N4954, 2N5172,2N5371,2N5998,2N6002,2N6008,2SC1849,40397,A5T3391,A5T3391A, A5T3392,A5T4124,A5T5172,A7T3391,A7T3391A,A7T3392,A7T5172,A8T3391, A8T3391A,A8T3392,A8T5172,BC338,BC548B,BC549B,BCY58B,BCY58C,MPS3391, MPS3391A,MPSA16,MPSA17,PET4002,TIS90-GRY,TIS90M-GRY,TIS92-GRY, TIS92M-GRY,TP3566,TP3706,TZ81,*PTC121,*GE-212,*TR-51,*BC107B,*276-2031
2N3397-ORG	2N2923,2N2924,2N3227,2N3241A,2N3242A,2N3392,2N3393,2N3395-YEL, 2N3396-ORG,2N3396-YEL,2N3397-YEL,2N3398-ORG,2N3398-YEL,2N3414,2N3509, 2N4074,2N5224,2N5998,2N6000,2N6426,2SC509-Y,2SC509G-Y,2SC847,2SC848, 2SC849,2SC966,2SC967,2SC1246,2SC1317,2SC1788,2SC1851,40398,A5T3392, A7T3392,A8T3392,BC548A,BCY58A,BSW83,BSX38,PET4001,TIS90-BLU,TIS90-VIO, TIS90M-BLU,TIS90M-VIO,TIS92-BLU,TIS92-VIO,TIS92M-BLU,TIS92M-VIO,TN80,TZ81, *PTC101,*GE-212,*TR-51,*BC107B,*276-2031
2N3397-RED	2N2331,2N2501,2N2923,2N3211,2N3393,2N3394,2N3396-ORG,2N3397-ORG, 2N3398-ORG,2N3398-RED,2N6427,2N6566,2SC509-O,2SC509G-O,2SC639,2SC764, A5T4123,MPS6560,PET4001,TIS90-GRN,TIS90-YEL,TIS90M-GRN,TIS90M-YEL, TIS92-GRN,TIS92-YEL,TIS92M-GRN,TIS92M-YEL,*PTC101,*GE-212,*TR-51, *BC107B,*276-2031
2N3397-WHT	2N3391,2N3391A,2N3395-WHT,2N3396-WHT,2N3398-BLU,2N3398-WHT,2N6002, 2N6008,A5T3391,A5T3391A,A7T3391,A7T3391A,A8T3391,A8T3391A,BCY58D, MPS3391,MPS3391A,PET4003,TP3566,*PTC139,*GE-212,*TR-51,*SK3156, *BC107B,*276-2031
2N3397-YEL	2N2924,2N2925,2N3391,2N3391A,2N3392,2N3395-WHT,2N3395-YEL,2N3396-WHT, 2N3396-YEL,2N3397-WHT,2N3398-WHT,2N3398-YEL,2N3415,2N4256,2N4954,

To Replace	Substitute This Type
(2N3397-YEL)	2N5172,2N5371,2N5998,2N6002,2N6008,2SC1849,40397,A5T3391,A5T3391A, A5T3392,A5T4124,A5T5172,A7T3391,A7T3391A,A7T3392,A7T5172,A8T3391, A8T3391A,A8T3392,A8T5172,BC338,BC548B,BC549B,BCY58B,BCY58C,MPS3391, MPS3391A,MPSA16,MPSA17,PET4002,TIS90-GRY,TIS90M-GRY,TIS92-GRY, TIS92M-GRY,TP3566,TP3706,TZ81,*PTC121,*GE-212,*TR-51,*BC107B,*276-2031
2N3398-BLU	2N5526,BC548C,BC549C,*PTC139,*GE-212,*SK3156,*BC337,*276-2031
2N3398-ORG	2N2923,2N2924,2N3227,2N3241A,2N3242A,2N3392,2N3393,2N3395-YEL, 2N3396-ORG,2N3396-YEL,2N3397-ORG,2N3397-YEL,2N3398-YEL,2N3414,2N3509, 2N4074,2N5224,2N5998,2N6000,2N6426,2SC509-Y,2SC509G-Y,2SC847,2SC848, 2SC849,2SC966,2SC967,2SC1246,2SC1317,2SC1788,2SC1851,40398,A5T3392, A7T3392,A8T3392,BC548A,BCY58A,BSW83,BSX38,PET4001,TIS90-BLU,TIS90-VIO, TIS90M-BLU,TIS90M-VIO,TIS92-BLU,TIS92-VIO,TIS92M-BLU,TIS92M-VIO,TN80,TZ81, *PTC101,*GE-212,*TR-51,*BC107B,*276-2031
2N3398-RED	2N2331,2N2501,2N2923,2N3211,2N3393,2N3394,2N3396-ORG,2N3397-ORG, 2N3397-RED,2N3398-ORG,2N6427,2N6566,2SC509-O,2SC509G-O,2SC639,2SC764, A5T4123,MPS6560,PET4001,TIS90-GRN,TIS90-YEL,TIS90M-GRN,TIS90M-YEL, TIS92-GRN,TIS92-YEL,TIS92M-GRN,TIS92M-YEL,*PTC101,*GE-212,*TR-51, *BC107B,*276-2031
2N3398-WHT	2N3391,2N3391A,2N3395-WHT,2N3396-WHT,2N3397-WHT,2N3398-BLU,2N6002, 2N6008,A5T3391,A5T3391A,A7T3391,A7T3391A,A8T3391,A8T3391A,BCY58D, MPS3391,MPS3391A,PET4003,TP3566,*PTC139,*GE-212,*TR-51,*SK3156, *BC107B,*276-2031
2N3398-YEL	2N2924,2N2925,2N3391,2N3391A,2N3392,2N3395-WHT,2N3395-YEL,2N3396-WHT, 2N3396-YEL,2N3397-WHT,2N3397-YEL,2N3398-WHT,2N3415,2N4256,2N4954, 2N5172,2N5371,2N5998,2N6002,2N6008,2SC1849,40397,A5T3391,A5T3391A, A5T3392,A5T4124,A5T5172,A7T3391,A7T3391A,A7T3392,A7T5172,A8T3391, A8T3391A,A8T3392,A8T5172,BC338,BC548B,BC549B,BCY58B,BCY58C,MPS3391, MPS3391A,MPSA16,MPSA17,PET4002,TIS90-GRY,TIS90M-GRY,TIS92-GRY, TIS92M-GRY,TP3566,TP3706,TZ81
2N3399	2N3127,2N3281,2N3282,AFY37,AFY42,MM5000,MM5001,MM5002,*PTC107,*GE-50, *TR-17,*HEPG0003,*ECG160,*WEP637
2N3401	2N924,2N939,2N1026,2N1231,2N2946A,2N3219,2N3979,2N4008,2N6567,HA7537, SHA7537,*GE-65,*TR-20,*HEPS0012,*SK3114,*SP70,*ECG159,*WEP242,*BC177, *276-2022,*RT-115
2N3402	*PTC178,*GE-81,*TR-25,*HEPS0015,*SK3124,*SN80,*BC337,*276-2008,*RT-102
2N3403	*GE-47,*TR-21,*HEPS0015,*SK3124,*SN80,*ECG192,*BC337,*276-2008,*RT-102
2N3404	*DS-66,*GE-81,*TR-21,*HEPS0015,*SK3124,*EN30,*ECG192,*WEP736,*BC337, *276-2008,*RT-102
2N3405	*GE-47,*TR-21,*HEPS0015,*SK3124,*EN30,*BC337,*276-2008,*RT-102
2N3412	2N982,2N3323,2N3324,2N3325,*PTC145,*GE-51,*TR-17,*ECG160,*WEP637, *276-2005
2N3414	2N3227,2N3241A,2N3242A,2N3509,2N5998,2N6000,2N6426,2SC509-Y,2SC509G-Y, 2SC967,2SC1246,2SC1317,2SC1788,2SC1851,2SD400,BC338-16,BC338-25,BSW83, GI-3706,PET3706,TN60,TN80,*PTC170,*GE-210,*TR-21,*HEPS0033,*SK3122, *SN80,*ECG123A,*WEP736,*BC337,*276-2014,*RT-102
2N3415	2N6002,2N6008,BC338-40,*GE-47,*TR-21,*HEPS0024,*SK3124,*EN30,*ECG123A, *WEP736,*BC337,*276-2014,*RT-102
2N3416	2N760A,2N909,2N956,2N2222,2N2222A,2N2350,2N2350A,2N2390,2N2484A,2N2540, 2N2645,2N3116,2N3302,2N3704,2N4952,2N5107,2N5369,2N5449,2N5582,2N5818, 2N5822,2N6004,2N6010,2N6014,2SC850,2SC968,2SC970,2SC1247,2SC1318, 2SC1781,2SC1813,2SC1852,40458,A5T2192,A5T2222,A5T5449,A8T3704,BCW34, BSW85,D33D25,D33D26,D33D30,GI-3643,GI-3704,PET3704,TIS109,TIS111,TN-3904, *PTC170,*GE-210,*TR-21,*HEPS0015,*SK3124,*EN10,*ECG123A,*WEP736, *BC337,*276-2008,*RT-102
2N3417	2N4953,2N5370,2N5376,2N6006,2N6012,2N6016,D33D27,*GE-47,*TR-21,*HEPS0015, *SK3124,*EN10,*ECG123A,*WEP736,*BC337,*276-2008,*RT-102
2N3418	*GE-28,*TR-76,*HEPS3002,*276-2012
2N3419	*276-2012
2N3420	*GE-28,*TR-76,*HEPS3002,*276-2012
2N3421	*276-2012
2N3426	2N5065,*GE-47,*TR-65,*HEPS3001,*BC337,*276-2038

To Replace	Substitute This Type
2N3427	2N3428,*GE-53,*TR-05,*HEPG0006,*SK3004,*ECG102,*WEP631,*276-2005,*RT-120
2N3428	MA1702,*GE-53,*TR-05,*HEPG0006,*SK3004,*ECG102,*WEP631,*276-2005,*RT-120
2N3429	
2N3430	
2N3431	2N3432
2N3432	2N3433
2N3433	153-30,154-30
2N3434	153-30,154-30
2N3439	*GE-32,*SK3103,*ECG128,*WEP243,*276-2008
2N3440	*GE-32,*TR-81,*HEPS3021,*SK3044,*ECG124,*WEP240,*RT-128
2N3441	*TR-81,*HEPS5012,*SK3131,*ECG286,*WEP241
2N3442	2N3773,2N6262,43104,IR3773,*PTC118,*HEPS5020,*SK3079,*ECG284
2N3444	2N3253,2N3735,*PTC144,*HEPS3002,*276-2010
2N3445	2N3232,2N3233,2N3236,2N3446,HST9202,HST9206,HST9207,*PTC140,*GE-14, *TR-59,*HEPS7002,*SK3027,*ECG130,*WEP247,*RT-131
2N3446	HST9202,HST9203,HST9207,HST9208,*PTC173,*GE-14,*TR-59,*HEPS5004, *SK3027,*ECG130,*WEP247,*276-2020,*RT-131
2N3447	2N3448,*PTC140,*GE-14,*TR-59,*HEPS7002,*SK3027,*ECG130,*WEP247,*RT-131
2N3448	*PTC173,*GE-14,*TR-59,*HEPS7004,*SK3027,*ECG130,*WEP247,*RT-131
2N3461	*TR-25,*HEPG6011,*276-2006
2N3467	2N3244,2N3762,*GE-67,*TR-77,*HEPS3028,*ECG129,*276-2006
2N3468	2N3245,2N3762A,2N3763,2N5022,MM3726,*PTC144,*GE-67,*HEPS3032,*ECG129, *WEPS3031
2N3469	*GE-28,*TR-87,*HEPS5014,*SK3024,*ECG128,*WEP243,*276-2018,*RT-114
2N3470	
2N3471	
2N3472	2N3473
2N3473	
2N3474	
2N3475	
2N3476	2N3477
2N3477	
2N3478	2SC817,2SC988,2SC988A,2SC988B,2SC1117,2SC1260,2SC1529,40915,*DS-74, *GE-86,*TR-83,*HEPS0016,*EN10,*ECG316,*WEP719,*276-2011,*RT-113
2N3485	2N3485A,*PTC127,*GE-21,*TR-20,*HEPS0012,*SK3114,*SP70,*ECG159,*WEP242, *276-2023,*RT-115
2N3485A	2N3485,*PTC127,*GE-21,*TR-20,*HEPS0012,*SK3114,*ECG159,*WEP242,*276-2021, *RT-115
2N3486	2N3486A,A5T2907,A5T3505,A5T3645,TIS112,*PTC127,*GE-48,*TR-20,*HEPS0012, *SK3114,*SP70,*ECG159,*WEP242,*276-2023,*RT-115
2N3486A	2N3486,A5T2907,A5T3505,A5T3645,TIS112,*PTC127,*GE-67,*TR-20,*HEPS0012, *SK3114,*ECG159,*WEP242,*276-2021,*RT-115
2N3487	2N3488
2N3488	2N3492
2N3489	
2N3490	2N3491
2N3491	
2N3492	
2N3493	S1297,*PTC132,*GE-60,*TR-95,*HEPS0016,*SK3039,*ECG108,*WEP56,*BF183, *276-2011,*RT-113
2N3494	*PTC127,*TR-88,*HEPS5023,*SK3025,*ECG129,*WEP242,*276-2022,*RT-115
2N3495	2SA712,A5T5400,*276-2021
2N3496	*PTC127,*GE-21,*TR-20,*HEPS0013,*SK3114,*ECG159,*WEP242,*276-2023, *RT-115
2N3497	A5T5400,*276-2022
2N3498	*PTC144,*GE-20,*TR-86,*SN80,*ECG123,*WEP53,*276-2009,*RT-100
2N3499	*PTC144,*GE-18,*276-2008
2N3500	*GE-32,*HEPS5026,*SK3104
2N3501	*GE-18,*HEPS5026,*SK3104
2N3502	2N2905,2N2905A,2N3503,2N4032,2SA684,*PTC127,*GE-48,*TR-88,*HEPS0012, *SK3025,*ECG129,*WEP242,*276-2021,*RT-115

To Replace	Substitute This Type
2N3503	2N2905,2N2905A,2N4032,2N4033,2N4405,2SA684,*PTC127,*GE-67,*TR-88, *HEPS0012,*SK3025,*ECG129,*WEP242,*276-2021,*RT-115
2N3504	2N2907,2N2907A,2N3486,2N3486A,2N3505,2N4028,A5T2907,A5T3504,A5T3505, A5T3644,A5T3645,TIS112,*PTC127,*GE-48,*TR-20,*HEPS5022,*SK3114,*ECG159, *WEP242,*276-2021,*RT-115
2N3505	2N2907,2N2907A,2N3486,2N3486A,2N4028,2N4029,A5T2907,A5T3505,A5T3645, TIS112,*PTC127,*GE-67,*TR-20,*HEPS0012,*SK3114,*ECG159,*WEP242, *276-2021,*RT-115
2N3506	*GE-46,*TR-87,*HEPS3002,*SK3024,*ECG128,*WEP243,*276-2008,*RT-114
2N3507	2N5335,*GE-28,*TR-87,*HEPS3002,*SK3024,*ECG128,*WEP243,*276-2008,*RT-114
2N3508	2N3647,2N6427,TIS110,TIS133,TIS134,*PTC123,*GE-210,*TR-95,*HEPS0016, *SK3039,*ECG108,*WEP56,*BC337,*276-2009,*RT-113
2N3509	TIS111,*PTC123,*GE-210,*TR-95,*HEPS0016,*SK3124,*ECG108,*WEP56,*BC337, *276-2009,*RT-113
2N3510	2N2501,2N3211,2N3511,2N3647,2N3648,2N5144,ZT708,*PTC136,*GE-20,*TR-21, *HEPS0011,*SK3122,*ECG123A,*WEP736,*276-2009,*RT-102
2N3511	*PTC136,*GE-20,*TR-95,*HEPS0016,*SK3039,*ECG108,*WEP56,*276-2009,*RT-113
2N3512	2N707A,*GE-47,*TR-64,*276-2014
2N3527	2N4980,*GE-82,*TR-20,*HEPS0013,*SK3019,*EN10,*ECG159,*WEP242,*BC327, *276-2023,*RT-115
2N3543	2SC765
2N3544	2N2369,2N2369A,2N2710,2N4137,2N5029,2N5769,2SC577,2SC601,2SC601N, 2SC689,2SC1123,2SC1128,MM1941,MPS2369,MPS6507,ZT2369,*PTC133,*TR-95, *HEPS0016,*SK3039,*ECG108,*WEP56,*276-2038,*RT-113
2N3545	2N869A,2N3576,2N3829,*PTC127,*GE-22,*TR-20,*HEPS0013,*SK3114,*ECG159, *WEP242,*276-2023,*RT-115
2N3546	2N5056,2N5057,2N5771,A5T4260,A5T4261,*TR-20,*HEPS0019,*SK3122,*ECG159, *WEP242,*276-2023,*RT-115
2N3547	2N3548,*PTC127,*GE-67,*TR-20,*HEPS0013,*SK3114,*ECG159,*WEP242,*276-2023, *RT-115
2N3548	2N3549,2N3550,*PTC127,*GE-48,*TR-20,*HEPS0013,*SK3114,*ECG159,*WEP242, *BC327,*276-2023,*RT-115
2N3549	2N3550,*GE-21,*TR-20,*HEPS0013,*SK3114,*ECG159,*WEP242,*276-2023,*RT-115
2N3550	2N3549,*GE-21,*TR-20,*HEPS0013,*SK3114,*ECG159,*WEP242,*BC327,*276-2023, *RT-115
2N3551	2N3552
2N3552	
2N3553	2SC302-M,2SC307-M,2SC590N,40605,*GE-28,*HEPS3001,*276-2014
2N3554	2N1837A,2N2224,2N2846,2N2848,2N3015,2N3299,2N3326,2N3725,2N4047,2N4960, 2N4961,2SC307,2SC594-O,2SC1218,*GE-215,*TR-65,*HEPS3001,*SK3197, *276-2030
2N3563	2N2369,2N2369A,2N2615,2N2616,2N3862,2N4295,2N4996,2N4997,2N5029,2SC601N, 2SC639,2SC764,2SC836,2SC922,2SC1123,2SC1128,2SC1187,2SC1188,2SC1189, 2SC1215,2SC1240,2SC1319,2SC1320,2SC1393,2SC1394,2SC1395,2SC1396, 2SC1687,2SC1688,BF123,BF199,BF311,BF377,BF378,MPS3563,MPSH10,MPSH11, MT1060A,SE3005,TIS48,TIS64A,TIS87,*PTC132,*GE-86,*TR-70,*HEPS0020, *SK3018,*EN10,*ECG229,*WEP56,*BF173,*276-2011,*RT-108
2N3564	2N2369,2N2369A,2N2501,2N3009,2N3011,2N3013,2N3014,2N3646,2N3688,2N3689, 2N3690,2N3862,2N4295,2N4420,2N4421,2N4422,2N4996,2N4997,2N5029,2SC67, 2SC169,2SC335,2SC380-O,2SC380A-O,2SC601N,2SC639,2SC752G-O,2SC764, 2SC836,2SC922,2SC1187,2SC1188,2SC1189,2SC1274,2SC1319,2SC1320, 2SC1393,2SC1394,2SC1395,2SC1396,2SC1687,2SC1688,BF121,BF123,BF125, BF127,BF198,BF199,BF241,BF311,GET2369,MPS3646,MT1060A,SE3005,SE5001, SE5002,SE5003,SE5006,TIS48,TIS87,*PTC132,*GE-20,*TR-70,*HEPS0020,*SK3018, *EN10,*ECG229,*WEP56,*BF173,*276-2011,*RT-108
2N3565	2N4286,2N4287,BC547B,BC550B,BCY58C,BCY58D,BCY59C,BCY59D,MPSA16, MPSA18,PET4002,*PTC139,*GE-212,*TR-21,*HEPS0015,*SK3122,*SN80, *ECG123A,*WEP736,*BC107B,*276-2031,*RT-102
2N3566	2N930A,2N2523,2N2586,2N3117,2N5827,2N6112,2N6428,2N6428A,2N6539,2SC587, 2SC587A,2SC1166-GR,A157B,BC183L,BC184L,BC413C,BC414C,BC547B,BC550B, BCY59C,BCY59D,GI-3566,MPSA16,MPSA18,TIS94,TIS97,TP3566,TP3704,*PTC121,

To Replace	Substitute This Type
(2N3566)	*GE-64,*TR-21,*HEPS0015,*SK3124,*SN80,*ECG123A,*WEP736,*BC107B, *276-2013,*RT-102
2N3567	2N718A,2N870,2N911,2N1837,2N2221A,2N2317,2N2351,2N2351A,2N2380,2N2380A, 2N2389,2N2479,2N3056,2N3568,2N4014,2N4963,2N5581,2SC349,2SC497-O, 2SC498-O,A5T2193,BSW84,EN1613,FT3567,FT3568,FT3722,SE8012,TN54, *PTC170,*GE-20,*TR-87,*HEPS0015,*SK3024,*SN80,*ECG128,*WEP243, *276-2009,*RT-114
2N3568	2N718A,2N870,2N911,2N1837,2N2221A,2N2317,2N2351,2N2351A,2N2380,2N2380A, 2N2389,2N2479,2N3056,2N3567,2N4014,2N4963,2N5581,2SC349,2SC497-O, 2SC498-O,A5T2193,BSW84,EN1613,FT3567,FT3568,FT3722,SE8012,TN54, *PTC170,*GE-81,*TR-87,*HEPS0015,*SK3024,*SN80,*ECG128,*WEP243, *276-2008,*RT-114
2N3569	2N871,2N956,2N2222A,2N2390,2N2645,2N3057,2N5582,2SC497-Y,2SC498-Y,BSW85, EN1711,SE6021,SE6021A,*PTC170,*GE-20,*TR-87,*HEPS0015,*SK3024,*SN80, *ECG128,*WEP243,*276-2009,*RT-114
2N3570	2SC1747,*PTC133,*TR-95,*SK3018,*ECG108,*WEP56,*RT-113
2N3571	2N3570,2N3572,*PTC133,*GE-86,*TR-95,*HEPS0017,*SK3018,*ECG108,*WEP56, *276-2011,*RT-113
2N3572	*PTC132,*GE-86,*TR-95,*HEPS0017,*SK3018,*ECG108,*WEP56,*276-2037,*RT-113
2N3576	2N869A,2N3829,*GE-20,*TR-20,*HEPS0013,*SN60,*ECG106,*WEP52,*276-2023, *RT-126
2N3579	2N2591,*PTC127,*GE-82,*TR-20,*HEPS0013,*SK3114,*ECG159,*WEP242,*276-2023, *RT-115
2N3580	2N2592,2N2593,*PTC127,*GE-82,*TR-20,*HEPS0013,*SK3114,*ECG159,*WEP242, *276-2023,*RT-115
2N3581	*PTC127,*GE-82,*TR-20,*HEPS0013,*SK3114,*SP70,*ECG159,*WEP242,*BC327, *276-2023,*RT-115
2N3582	*PTC127,*GE-82,*TR-20,*HEPS0013,*SK3114,*SP70,*ECG159,*WEP242,*BC327, *276-2023,*RT-115
2N3583	*DS-41,*GE-12,*TR-81,*HEPS3021,*ECG286,*WEP240,*RT-128
2N3584	*GE-66,*TR-81,*ECG286,*WEP240
2N3585	*GE-66,*TR-81,*ECG124,*WEP240
2N3588	2N3783,2N3784,AF200,AF201,AF202,AF202S,AFY39,MM5000,MM5001,MM5002, *GE-50,*TR-17,*HEPG0003,*SK3006,*HF35,*ECG160,*WEP637
2N3589	*GE-32,*HEPS3021
2N3590	*GE-32,*HEPS3021
2N3591	*GE-32,*HEPS3021,*276-2008
2N3592	*GE-32,*HEPS3021,*276-2008
2N3593	*GE-32,*HEPS3021,*276-2008
2N3594	*GE-32,*HEPS3021,*276-2008
2N3595	*HEPS3021
2N3596	*HEPS3021
2N3597	2N3598,HST8002,HST8015,HST8301,KSP1254,SDT8002,SDT8015,SDT8301
2N3598	2N3599,2N5733,HST8002,HST8003,HST8015,HST8016,HST8301,HST8302,KSP1254, KSP1255,SDT8002,SDT8003,SDT8015,SDT8016,SDT8301,SDT8302
2N3599	2N5733,HST8003,HST8016,HST8302,KSP1255,SDT8003,SDT8016,SDT8302
2N3600	2N2808,2N2808A,2N2809,2N2809A,2N3570,2N4252,2N4253,2N5053,2N5054,2N6304, 2N6305,2SC567,2SC568,2SC1747,MT1061A,*PTC132,*GE-86,*TR-95,*HEPS0016, *EN10,*ECG108,*WEP56,*276-2011,*RT-113
2N3605	2N744,2N2656,2N3011,2N3606,2N3607,2N4418,2S095A,2S512,2SC168,2SC395A-O, 2SC735-O,2SC1372,EN744,PT720,ZT2205,*GE-20,*TR-21,*HEPS0015,*SK3019, *EN10,*ECG123A,*WEP736,*BC107B,*276-2011,*RT-102
2N3605A	2N708,2N708A,2N784A,2N914,2N2221,2N2242,2N2481,2N2501,2N2539,2N2710, 2N2845,2N2847,2N3013,2N3014,2N3115,2N3210,2N3211,2N3301,2N3510,2N3511, 2N3606A,2N3646,2N3647,2N3648,2N3903,2N3946,2N4123,2N4140,2N4962, 2N5027,2N5144,2N5380,2N5772,2SC67,2SC577,2SC639,2SC764,2SC1274, A5T3903,A5T4123,BSW82,GET708,GET914,GET2221,GET2369,GET3013,GET3014, GET3646,GI-3641,GI-3642,MM3903,MPS6530,TN-3903,ZT83,ZT708,*GE-20,*TR-64, *HEPS0025,*276-2016
2N3606	2N744,2N2656,2N3011,2N3605,2N3607,2N4418,2S095A,2S512,2SC168,2SC395A-O, 2SC735-O,2SC1372,EN744,PT720,ZT2205,*GE-20,*TR-21,*HEPS0015,*SK3019,

To Replace	Substitute This Type
(2N3606)	*EN10,*ECG123A,*WEP736,*BC107B,*276-2011,*RT-102
2N3606A	2N708,2N708A,2N784A,2N914,2N2221,2N2242,2N2481,2N2501,2N2539,2N2710, 2N2845,2N2847,2N3013,2N3014,2N3115,2N3210,2N3211,2N3301,2N3510,2N3511, 2N3605A,2N3646,2N3647,2N3648,2N3903,2N3946,2N4123,2N4140,2N4962, 2N5027,2N5144,2N5380,2N5772,2SC67,2SC577,2SC639,2SC764,2SC1274, A5T3903,A5T4123,BSW82,GET708,GET914,GET2221,GET2369,GET3013,GET3014, GET3646,GI-3641,GI-3642,MM3903,MPS6530,TN-3903,ZT83,ZT708,*GE-20,*TR-64, *HEPS0025,*276-2016
2N3607	2N744,2N2656,2N3011,2N3605,2N3606,2N4418,2S095A,2S512,2SC168,2SC395A-O, 2SC735-O,2SC1372,EN744,PT720,ZT2205,*GE-20,*TR-21,*HEPS0011,*SK3019, *EN10,*ECG123A,*WEP736,*BC107B,*276-2011,*RT-102
2N3611	2N3612,2N4243,2N4244,MP2060,MP2061,*PTC105A,*GE-76,*TR-01,*HEPG6003, *SK3009,*PT40,*ECG121,*WEP232,*276-2006,*RT-127
2N3612	2N3615,2N3617,2N4242,2N4243,*PTC105A,*GE-76,*TR-01,*HEPG6005,*SK3009, *ECG121,*WEP232,*276-2006,*RT-127
2N3613	2N3614,2N4246,2N4247,MP2060,MP2061,*PTC105A,*GE-76,*TR-01,*HEPG6003, *SK3009,*ECG121,*WEP232,*276-2006,*RT-127
2N3614	2N4245,2N4246,*PTC105A,*GE-76,*TR-01,*HEPG6005,*SK3009,*ECG121,*WEP232, *276-2006,*RT-127
2N3615	2N3616,2N3617,2N3618,2N4242,*PTC105A,*GE-76,*TR-01,*HEPG6005,*SK3009, *ECG121,*WEP232,*276-2006,*RT-127
2N3616	2N3618,*PTC105A,*GE-16,*TR-01,*HEPG6018,*SK3009,*ECG121,*WEP232, *276-2006,*RT-127
2N3617	2N3618,2N4242,2N4245,*PTC105A,*GE-76,*TR-01,*HEPG6005,*SK3009,*ECG121, *WEP232,*276-2006,*RT-127
2N3618	*PTC105A,*GE-16,*TR-01,*HEPG6018,*SK3009,*ECG121,*WEP232,*276-2006, *RT-127
2N3619	*GE-28,*TR-76,*HEPS3002
2N3620	2N3624,*GE-28,*TR-76
2N3621	2N3622,2N3625,2N3626,2N3629,2N3630,B144003,B144004,B144006,B144007, *GE-66
2N3622	2N3621,2N3625,2N3626,2N3629,2N3630,B144003,B144004,B144006,B144007, *GE-66
2N3623	*GE-28,*TR-76,*HEPS3002
2N3624	2N3620,*GE-28,*TR-76
2N3625	2N3621,2N3622,2N3626,2N3629,2N3630,B144003,B144004,B144006,B144007, *GE-66
2N3626	2N3621,2N3622,2N3625,2N3629,2N3630,B144003,B144004,B144006,B144007, *GE-66
2N3627	*GE-66,*HEPS3002
2N3628	*GE-28,*TR-76
2N3629	2N3630,*GE-66
2N3630	2N3629,*GE-66
2N3632	2N4933,40665,*GE-66,*HEPS3007
2N3633	2N3960,2N5200,2N5201,*GE-66,*TR-95,*HEPS0016,*SK3039,*ECG108,*WEP56, *276-2011,*RT-113
2N3634	2N3636
2N3635	2N3637
2N3636	
2N3637	
2N3638	2N869A,2N2695,2N2696,2N2927,2N3244,2N3467,2N4402,2N5023,2N5041,2N5042, 2N5811,2SA504-O,2SA504-Y,2SA509-O,2SA509G-O,2SA695,A5T3638A,A5T4402, BC360-10,BC360-6,BSW72,MPS3638,MPS6533,SE8540,SE8541,SE8542,TN-3905, TQ61,TQ62,ZT180,ZT280,*PTC103,*GE-22,*TR-20,*HEPS0019,*SK3025,*EP25, *ECG159,*WEP242,*BC327,*276-2034,*RT-115
2N3638A	2N869,2N4402,2N4403,2N6001,2SA504-GR,2SA504-Y,2SA509-O,2SA509-Y, 2SA509G-O,2SA509G-Y,2SA683,2SA695,2SA719,2SA890,2SB544,A5T3638A, A5T4402,A5T4403,BC360-10,BC360-16,BSW73,CS9012,GI-3638A,MPS3638A, MPS6534,TN-3905,TN-3906,ZT180,ZT187,ZT280,ZT287,*PTC103,*GE-82,*TR-20, *HEPS0019,*SK3025,*EP25,*ECG159,*WEP242,*276-2034,*RT-115
2N3639	2N3640,2N5771,*PTC131,*GE-21,*TR-20,*HEPS0019,*SK3118,*EP25,*ECG159,

To Replace	Substitute This Type
(2N3639)	*WEP242,*276-2021,*RT-115
2N3640	2N5771,*PTC131,*GE-21,*TR-20,*HEPS0019,*SK3114,*EP25,*ECG159,*WEP242, *276-2021,*RT-115
2N3641	2N2846,2N2848,2N3015,2N3299,2N3326,2N3642,2N4960,2N4961,2N4962,2N4963, 2N5380,2N6538,2SC594-O,A5T3903,FT3722,TIS96,TIS99,TN-3903,*PTC101, *GE-210,*TR-21,*HEPS0015,*SK3122,*EN10,*ECG123A,*WEP736,*BC337, *276-2016,*RT-102
2N3642	2N2846,2N2848,2N3015,2N3299,2N3326,2N3641,2N4960,2N4961,2N4962,2N4963, 2N5380,2N6538,2SC594-O,A5T3903,FT3722,TIS96,TIS99,TN-3903,*PTC101, *GE-210,*TR-21,*HEPS0015,*SK3024,*EN10,*ECG123A,*WEP736,*BC337, *276-2016,*RT-102
2N3643	2N2960,2N2961,2N3300,2N4409,2N5381,2SC1991,2SC1998,A5T3904,A5T4409, BC546A,TN-3904,*PTC121,*GE-210,*TR-21,*HEPS0015,*SK3122,*ECG123A, *WEP736,*276-2009,*RT-102
2N3644	2N2905,2N2905A,2N2907,2N2907A,2N3486,2N3486A,2N3502,2N3503,2N3504, 2N3505,2N3645,2N3671,2N3672,2N3673,2N4028,2N4032,2N6005,2SA684,2SA720, 2SA891,A5T2907,A5T3504,A5T3505,A5T3644,A5T3645,BC361-10,FT3644,FT3645, GI-3644,TIS112,*PTC103,*GE-48,*TR-20,*HEPS0019,*SK3114,*EP25,*ECG159, *WEP242,*276-2021,*RT-115
2N3645	2N2905,2N2905A,2N2907,2N2907A,2N3486,2N3486A,2N3503,2N3505,2N3671, 2N3672,2N3673,2N4028,2N4029,2N4032,2N4033,2SA684,2SA720,2SA891,A5T2907, A5T3505,A5T3645,BC361-10,FT3644,TIS112,*PTC103,*GE-67,*TR-88,*HEPS5022, *EP25,*ECG129,*WEP242,*276-2021,*RT-115
2N3646	2N5772,*PTC132,*GE-20,*TR-21,*HEPS0011,*SK3019,*EN10,*ECG123A,*WEP736, *BF173,*276-2016,*RT-102
2N3647	*PTC136,*GE-20,*TR-86,*HEPS0015,*SN80,*ECG123,*WEP53,*276-2009,*RT-100
2N3648	*PTC136,*GE-20,*TR-95,*HEPS0016,*SK3039,*SN80,*ECG108,*WEP56,*276-2009, *RT-113
2N3659	HST5055,HST5056,HST5555,HST5556,PT2524,PT2525,*GE-32,*HEPS3021
2N3660	2N3661,*GE-84,*HEPS5013
2N3661	*HEPS5013
2N3662	2N2729,2N3544,2N3663,2SC387AG,2SC718,2SC719,2SC1231,EN918,MM1941, MPS6507,MT1038,MT1038A,MT1039,MT1060,PET3001,PET8101,TIS18,*PTC132, *GE-11,*TR-70,*HEPS0016,*SK3039,*EN10,*ECG107,*WEP56,*276-2011,*RT-108
2N3663	2N2368,2N2729,2SC387AG,2SC689,EN918,MM1941,MPS6507,MT1038,MT1038A, MT1039,MT1060,PET3001,TIS84,TIS129,*PTC132,*GE-11,*TR-70,*HEPS0016, *SK3018,*SN80,*ECG107,*WEP56,*276-2011,*RT-108
2N3664	2N3961,*276-2017
2N3665	*HEPS3019,*ECG128,*WEPS3021
2N3666	*HEPS3019,*ECG128,*WEPS3021
2N3667	2N3055A,2N3235,2N3771,2N5302,2N6371,2SD231,2SD249,1561-0403,1561-0404, 1582-0403,1582-0404,1582-0405,1582-0603,1582-0604,1582-0605,B170002-BLK, B170002-BRN,B170020,HST9201,HST9205,IR3771,MJ2801,MJ3771,SDT9201, SDT9205,*PTC140,*GE-77,*TR-59,*HEPS7004,*SK3036,*HN100,*ECG130, *WEP247,*RT-131
2N3671	2N2905,2N2905A,2N3503,2N4032,2N4033,2N4405,2SA684,A5T2907,A5T3505, A5T3645,TIS112,*PTC127,*GE-21,*TR-88,*HEPS0012,*SK3025,*ECG129,*WEP242, *276-2021,*RT-115
2N3672	2N2907,2N2907A,2N3486,2N3486A,2N3505,2N4028,2N4029,A5T2907,A5T3505, A5T3645,FT3644,TIS112,*PTC127,*GE-21,*TR-20,*HEPS0019,*SK3114,*ECG159, *WEP242,*276-2024,*RT-115
2N3673	2N3486,2N3486A,A5T2907,A5T3505,A5T3645,TIS112,*PTC127,*GE-21,*TR-20, *HEPS0019,*SK3114,*ECG159,*WEP242,*276-2024,*RT-115
2N3675	2N2858,2N2859,2N3676,*GE-66,*HEPS3002,*276-2020
2N3676	2N2858,2N2859,2N3675,*HEPS3002
2N3677	*GE-82,*TR-88,*HEPS0013,*SK3025,*ECG129,*WEP242,*BC327,*276-2023,*RT-115
2N3678	2N2218A,2N3722,2N3723,2N3725A,2N4961,*HEPS3001,*ECG123A,*WEP736, *276-2014
2N3680	
2N3681	2N3571,2N3572,2N5031,2N5032,2N6665,2SC611,2SC653,*PTC132,*GE-11, *HEPS0017,*276-2011

To Replace	Substitute This Type
2N3688	2N2369,2N2369A,2N2501,2N3009,2N3013,2N3014,2N3646,2N3689,2N3690,2N3862, 2N4295,2N4420,2N4421,2N4422,2N5029,2SC67,2SC169,2SC335,2SC336,2SC601N, 2SC639,2SC752G-O,2SC764,2SC1187,2SC1188,2SC1189,2SC1274,2SC1319, 2SC1320,2SC1687,2SC1688,BF198,BF199,BF241,GET2369,MPS3646,SE5001, SE5002,SE5003,SE5006,TIS48,TIS87,*PTC132,*GE-60,*TR-95,*HEPS0015,*SK3039, *EN10,*ECG108,*WEP56,*276-2016,*RT-113
2N3689	2N2369,2N2369A,2N2501,2N3009,2N3013,2N3014,2N3646,2N3688,2N3690,2N3862, 2N4295,2N4420,2N4421,2N4422,2N5029,2SC67,2SC169,2SC335,2SC336,2SC601N, 2SC639,2SC752G-O,2SC764,2SC1187,2SC1188,2SC1189,2SC1274,2SC1319, 2SC1320,2SC1687,2SC1688,BF198,BF199,BF241,GET2369,MPS3646,SE5001, SE5002,SE5003,SE5006,TIS48,TIS87,*PTC132,*GE-60,*TR-95,*HEPS0025,*SK3039, *EN10,*ECG108,*WEP56,*276-2016,*RT-113
2N3690	2N2369,2N2369A,2N2501,2N3009,2N3013,2N3014,2N3646,2N3688,2N3689,2N3862, 2N4295,2N4420,2N4421,2N4422,2N5029,2SC67,2SC169,2SC335,2SC336,2SC601N, 2SC639,2SC752G-O,2SC764,2SC1187,2SC1188,2SC1189,2SC1274,2SC1319, 2SC1320,2SC1687,2SC1688,BF198,BF199,BF241,GET2369,MPS3646,SE5001, SE5002,SE5003,SE5006,TIS48,TIS87,*PTC132,*GE-60,*TR-95,*HEPS0025,*SK3018, *ECG108,*WEP56,*276-2016,*RT-113
2N3691	2N916,2N916A,2N957,2N2242,2N2369,2N2369A,2N2501,2N3009,2N3013,2N3014, 2N3693,2N3862,2N4123,2N4873,2SC67,2SC169,2SC372-O,2SC372G-O,2SC380-O, 2SC380A-O,2SC394-Y,2SC536,2SC601N,2SC620,2SC639,2SC735-O,2SC752G-O, 2SC764,2SC899,2SC964,2SC965,2SC1187,2SC1188,2SC1189,2SC1276,2SC1319, 2SC1320,2SC1686,2SC1687,2SC1688,BF198,BF199,BF241,EN916,MPS3693, MPS6574-BLUE,MPS6574-YEL,MPS6576-BLUE,MPS6576-YEL,PET1001,PET2001, SE1001,SE2001,TIS87,*PTC132,*GE-60,*TR-95,*HEPS0016,*SK3018,*SN80, *ECG108,*WEP56,*BF365,*276-2016,*RT-113
2N3692	2N3242A,2N3694,2SC373,2SC373G,2SC735-GR,2SC1084,2SC1992,A157B,BC107, BC107B,BC123,BC167B,BC171B,BC237B,BC413B,BC414B,BCY58B,BCY59B, BSW88,BSW89,BSX38,BSX79,MPS3694,MPS6514,MPS6520,MPS6573, MPS6574-SIL,MPS6575,MPS6576-SIL,PET1002,PET2002,SE1002,SE2002,*PTC132, *GE-210,*TR-95,*HEPS0025,*SK3124,*EN10,*ECG108,*WEP56,*BC107B, *276-2016,*RT-113
2N3693	2N916,2N916A,2N2847,2N3826,2N3862,2N3903,2N3946,2N3974,2N3976,2N4227, 2N4962,2N4994,2N5027,2N5380,2SC620,2SC815,2SC899,2SC944,2SC965, 2SC1189,2SC1320,2SC1688,EN916,MM3903,MPS3693,MPS6565,MPS6576-BLUE, MPS6576-YEL,PET1001,SE1001,TIS87,TN-3903,*PTC132,*GE-210,*TR-95, *HEPS0025,*SK3122,*SN80,*ECG123A,*WEP56,*BC107B,*276-2011,*RT-113
2N3694	2N3302,2N3827,2N3904,2N4995,2N5381,2SC1992,A157B,BC107,BC107B,BC123, BC167B,BC171B,BC174B,BC190B,BC237B,BC413B,BC414B,BCY59B,BSX79, MPS3694,MPS6566,MPS6575,MPS6576-SIL,PET1002,SE1002,TN-3904,*PTC132, *GE-210,*TR-21,*HEPS0025,*SK3122,*SN80,*ECG123A,*WEP736,*BC107B, *276-2013,*RT-102A
2N3700	*TR-78,*HEPS3019,*SK3045,*ECG154,*WEP712,*RT-110
2N3701	2N2896,2N2899,*HEPS3019,*ECG154,*WEPS3021
2N3702	2N2838,2N2907,2N2907A,2N3251,2N3251A,2N3486,2N3486A,2N3504,2N3505, 2N3672,2N3673,2N3962,2N5373,2N5374,2N5375,2N5379,2N5383,2N5447,2N5817, 2N5819,2N6005,2SA659,2SA661-GR,2SA661-Y,2SA720,2SA891,A5T2907,A5T3504, A5T3505,A5T3644,A5T3645,A5T3906,A5T4403,A5T5447,A8T3702,BC327-16, BC327-25,BCW35,BCY79,BCY79A,BCY79B,BSW73,D29E5,D29E6,GET2906, GET2907,MM3906,TIS112,TN-3906,TQ60,TZ552,TZ553,*PTC103,*GE-82,*TR-20, *HEPS0019,*SK3114,*EP35,*ECG159,*WEP242,*BC177,*276-2034,*RT-303
2N3703	2N2838,2N2906,2N2906A,2N3250,2N3250A,2N3485,2N3485A,2N3672,2N3673, 2N5372,2N5448,2N5815,2N5817,2N5821,2N5823,2SA661-O,A5T5448,A8T3703, BSW74,D29E10,D29E5,GET2904,GET2905,TZ551,*PTC103,*GE-82,*TR-20, *HEPS0019,*SK3114,*EP35,*ECG159,*WEP242,*BC177,*276-2023,*RT-115
2N3704	2N2222,2N2222A,2N2540,2N3302,2N5449,2N5818,2N6112,A5T2222,A5T5449, A8T3704,D33D26,TIS109,TIS111,TN-3904,*GE-20,*TR-21,*HEPS0015,*SK3024, *ECG123A,*WEP736,*BC337,*276-2014,*RT-102
2N3705	2N915,2N2539,2N2897,2N2900,2N4962,2N5450,2N5822,40458,A5T5450,A8T3705, D33D25,D33D30,TIS110,TN54,TN-3903,*GE-20,*TR-21,*HEPS0030,*SK3124,*EN40, *ECG123A,*WEP736,*BC337,*276-2014,*RT-102

To Replace	Substitute This Type
2N3706	2N5451,2N6012,2N6112,A5T5451,A8T3706,BC337-40,D33D27,*DS-66,*GE-47,*TR-21, *HEPS0015,*SK3122,*ECG123A,*WEP736,*BC337,*276-2014,*RT-102
2N3707	2N3708,A5T3565,A5T3707,A5T3708,A8T3707,A8T3708,BC547B,BC550B,BCY58B, BCY58C,BCY59B,BCY59C,MPSA16,MPSA17,PET4002,*PTC101,*GE-212,*TR-21, *HEPS0015,*SK3122,*EN40,*ECG123A,*WEP736,*BC107B,*276-2013,*RT-102
2N3708	2N3711,A5T3565,A5T3708,A5T3711,A8T3708,A8T3711,BC547B,BC550B,BCY58D, BCY59D,PET4003,*TR-21,*HEPS0033,*SK3124,*ECG123A,*WEP736,*276-2010
2N3708-BLU	2N3708,2N3711,A5T3565,A5T3708,A5T3711,A8T3708,A8T3711,BC547B,BC550B, BCY58C,BCY58D,BCY59C,BCY59D,PET4002,PET4003,*PTC139,*GE-212,*TR-51, *BC107B,*276-2013
2N3708-BRN	2N760,2N2501,2N3709,2N6566,A5T3709,A8T3709,ZT708,*PTC101,*GE-61,*TR-51, *BC107B,*276-2013
2N3708-GRN	2N3707,2N3708,A5T3565,A5T3707,A5T3708,A8T3707,A8T3708,BC547B,BC550B, BCY58B,BCY58C,BCY59B,BCY59C,MPSA16,MPSA17,PET4002,*PTC139,*GE-212, *TR-51,*BC107B,*276-2013
2N3708-ORG	2N3241A,2N3242A,2N3710,2N4074,2SC1999,A5T3710,A8T3710,BC547A,BCY58A, BCY59A,PET4001,*PTC101,*GE-61,*TR-51,*BC107B,*276-2013
2N3708-RED	2N760,2N2501,2N3709,2N6566,2SC1999,A5T3709,A8T3709,PET4001,*PTC101, *GE-61,*TR-51,*BC107B,*276-2013
2N3708-VIO	BC550C,PET4003,*PTC139,*GE-212,*BC107B,*276-2013
2N3708-YEL	2N3242A,2N3707,2N3710,A5T3707,A5T3710,A8T3707,A8T3710,BC547A,BCY58B, BCY59B,MPSA17,*PTC101,*GE-61,*TR-51,*BC107B,*276-2013
2N3709	2N3241A,2N4074,2SC1999,A5T3709,A8T3709,BCY58A,BCY59A,PET4001,*PTC101, *GE-61,*TR-21,*HEPS0015,*SK3124,*EN40,*ECG123A,*WEP736,*BC107B, *276-2013,*RT-102
2N3710	2N3707,A5T3707,A5T3710,A8T3707,A8T3710,BCY58B,BCY58C,BCY59B,BCY59C, MPSA16,MPSA17,PET4002,*PTC101,*GE-61,*TR-21,*HEPS0015,*SK3124,*EN40, *ECG123A,*WEP736,*BC107B,*276-2013,*RT-102
2N3711	A5T3711,A8T3711,BC550C,BCY58D,BCY59D,PET4003,*PTC139,*GE-212,*TR-21, *HEPS0005,*SK3124,*EN40,*ECG123A,*WEP736,*BC107B,*276-2013,*RT-102
2N3712	2N1342,2N1893A,2N3500,2N4925,2SC1217,MM2259,MM2260,*GE-32,*TR-78, *HEPS3019,*SK3045,*ECG154,*WEP712,*276-2012,*RT-110
2N3713	2N3714,*PTC173,*GE-14,*TR-59,*HEPS7002,*SK3036,*ECG130,*WEP247,*RT-131
2N3714	*PTC173,*GE-14,*TR-59,*HEPS7004,*SK3036,*ECG130,*WEP247,*RT-131
2N3715	2N3716,*PTC173,*GE-14,*TR-59,*HEPS7002,*SK3036,*ECG130,*WEP247,*RT-131
2N3716	2N5039,*PTC173,*GE-14,*TR-59,*HEPS7004,*SK3036,*ECG284,*WEP247,*RT-131
2N3719	2N3720,*GE-29,*TR-88,*HEPS5013,*SK3025,*ECG129,*WEP242,*276-2025,*RT-115
2N3720	*TR-88,*HEPS5013,*SK3025,*ECG129,*WEP242,*276-2025,*RT-115
2N3721	2N3390,2N3391,2N3391A,2N3395-WHT,2N3396-WHT,2N3397-WHT,2N3398-BLU, 2N3398-WHT,2N3415,2N3900,2N3900A,2N3901,2N6002,2N6008,A5T3391, A5T3391A,A7T3391,A7T3391A,A8T3391,A8T3391A,BC548B,BC549B,BCY58D, MPS3390,MPS3391,MPS3391A,PET4003,*PTC101,*GE-62,*TR-70,*HEPS0015, *SK3124,*EN10,*ECG56,*WEP720,*BC337,*276-2010,*RT-108
2N3722	2N3723,2N3725A,2N4961,*HEPS3019
2N3723	*GE-27,*TR-55,*HEPS3019,*SK3047,*ECG123A
2N3724	2N3724A,*GE-47,*HEPS3019,*276-2038
2N3724A	*GE-215,*HEPS3020,*276-2038
2N3725	2N3725A,*GE-18,*HEPS3019
2N3725A	*HEPS3020
2N3730	*GE-25,*TR-27,*HEPG6007,*SK3034,*ECG127,*WEP235,*276-2006
2N3731	*PTC122,*GE-25,*TR-27,*HEPG6008,*SK3035,*ECG127,*WEP235
2N3732	*GE-25,*TR-27,*HEPS5005,*SK3034,*ECG127,*WEP235,*OC28
2N3733	2N5016,MSA7505,MSA8505
2N3734	*GE-215,*HEPS3020,*276-2038
2N3735	*HEPS3020
2N3736	*GE-20,*HEPS3020,*276-2038
2N3737	*HEPS3020
2N3738	2N3739,*PTC104,*GE-32,*TR-81,*HEPS5011,*SK3021,*ECG124,*WEP240,*RT-128
2N3739	*PTC104,*GE-32,*TR-81,*HEPS5011,*SK3021,*ECG124,*WEP240,*RT-128
2N3740	2N3741,*PTC113,*GE-69,*TR-58,*HEPS5018,*ECG218,*276-2025,*RT-133
2N3740A	2N3741A,KSP2024,KSP2025,*PTC113,*GE-69,*TR-58,*HEPS5018,*ECG218,

To Replace	Substitute This Type
(2N3740A)	*276-2025,*RT-133
2N3741	*PTC113,*TR-58,*HEPS5018,*ECG292,*WEPS5007,*276-2027
2N3741A	KSP2025,*PTC113,*TR-58,*HEPS5018,*ECG218,*WEPS5007,*276-2025
2N3742	*GE-27,*HEPS5024,*ECG154,*WEP712,*276-2008
2N3743	*GE-228,*ECG154,*WEP712
2N3744	2N3745,1718-0602,1718-0605,1718-0802,1718-0805,KSP1031,SDT6408,SDT6409, *GE-66
2N3745	2N3746,1718-0802,1718-0805,1718-1002,1718-1005,KSP1031,KSP1032,SDT6409, SDT6410,*GE-66
2N3746	1718-1002,1718-1005,1718-1202,1718-1205,KSP1032,SDT6410
2N3747	2N3748,KSP1034,SDT6411,SDT6412,*GE-66
2N3748	2N3749,2N3996,KSP1034,KSP1035,SDT6412,SDT6413,*GE-66
2N3749	2N3996,KSP1035,SDT6413
2N3750	2N3751,SDT6414,SDT6415
2N3751	2N3752,SDT6415,SDT6416
2N3752	SDT6416
2N3762	MM3726,*GE-29,*TR-88,*HEPS5013,*SK3025,*ECG129,*WEP242,*276-2025,*RT-115
2N3762A	2N3763,MM3726
2N3763	2N4406,*TR-88,*HEPS5013,*SK3025,*ECG129,*WEP242,*276-2025,*RT-115
2N3764	*TR-88,*HEPS5013,*SK3025,*ECG129,*WEP242,*276-2023,*RT-115
2N3764A	2N3765
2N3765	*TR-88,*HEPS5013,*SK3025,*ECG129,*WEP242,*276-2022,*RT-115
2N3766	2N3767,2N5428,2N5430,2N5606,2N5610,2SC1444,2SC1445,KSP1024,KSP1025, *PTC112,*TR-81,*HEPS5019,*ECG175,*WEP241,*276-2017
2N3767	2N3878,2N5430,2N5610,2SC1445,KSP1025,*PTC148,*TR-81,*HEPS5012,*SK3538, *SK3538,*ECG175,*WEP241,*276-2013
2N3771	2N5302,2N5685,BUY51A,BUY52A,IR3771,MJ3771,*PTC175,*GE-75,*TR-36, *HEPS7000,*SK3036,*ECG181,*WEPS7000,*RT-149
2N3772	2N6270,2N6271,2N6272,2N6273,2N6570,2N6571,108T2,BUY53A,BUY54A,IR3772, MJ3772,RCS258,STC2221,STC2225,*PTC175,*GE-75,*TR-36,*HEPS7004,*SK3036, *ECG181,*WEPS7000,*RT-149
2N3772C	2N3055C,109T2
2N3773	2N6259,109T2,1561-1410,1561-1415,1561-1608,1561-1610,1561-1615,1561-1808, 1561-1810,1561-1815,1561-2008,1561-2010,43104,IR3773,STC2222,STC2223, STC2226,STC2227,*HEPS7000,*SK3535,*ECG284,*WEPS7000
2N3774	2N3202,2N3203,2N3775,STC5610,*GE-84,*TR-88,*HEPS5013,*SK3025,*ECG129, *WEP242,*276-2025,*RT-115
2N3775	2N3203,2N3204,2N3776,STC5610,*TR-88,*HEPS5013,*SK3025,*ECG129,*WEP242, *276-2025,*RT-115
2N3776	2N3204,2N3777,STC5611,*TR-88,*HEPS3032,*SK3025,*ECG129,*WEP242,*RT-115
2N3777	STC5611,STC5612,*HEPS3032
2N3778	2N3774,2N3775,2N3779,*GE-84,*TR-88,*HEPS5013,*SK3025,*ECG129,*WEP242, *276-2025,*RT-115
2N3779	2N3775,2N3776,2N3780,*TR-81,*HEPS5013,*ECG175,*WEP241,*276-2025
2N3780	2N3776,2N3777,2N3781,*TR-88,*HEPS3032,*SK3025,*ECG129,*WEP242,*RT-115
2N3781	2N3777,*TR-88,*HEPS3032,*SK3025,*ECG129,*WEP242,*RT-115
2N3782	*GE-83,*TR-88,*HEPS5013,*SK3025,*ECG129,*WEP242,*276-2017,*RT-115
2N3783	2N3784,*GE-51,*TR-17,*HEPG0003,*ECG160,*WEP637
2N3784	2N3783,*GE-51,*TR-17,*HEPG0003,*ECG160,*WEP637
2N3785	2N3783,2N3784,*GE-51,*TR-17,*HEPG0003,*ECG160,*WEP637
2N3788	2SD393,*PTC129A,*GE-36,*TR-67,*HEPS5020
2N3789	2N3790,*PTC149,*HEPS7003,*ECG219,*WEPS7001
2N3790	*PTC172,*HEPS7003,*ECG219,*WEPS7001
2N3791	2N3792,*PTC149,*HEPS7003,*ECG219,*WEPS7001
2N3792	*PTC172,*HEPS7003,*ECG219,*WEPS7001
2N3793	2N708A,2N914,2N957,2N2221,2N2481,2N2710,2N2845,2N3115,2N3210,2N3261, 2N3301,2N3510,2N3511,2N3646,2N3647,2N3648,2SC577,BSW82,GI-3641,GI-3642, GI-3793,ZT83,ZT113,ZT708,*PTC170,*GE-210,*TR-21,*HEPS0016,*SK3124, *ECG123A,*WEP736,*BC337,*276-2016,*RT-102
2N3794	2N3706,2N4953,2N4954,2N5370,2N5371,2N5376,2N5377,2N5451,2N6006,2N6012, 2N6112,BC337,D33D27,GI-3794,MPS3706,*GE-210,*TR-21,*HEPS0015,*SK3124,

To Replace	Substitute This Type
(2N3794)	*ECG123A,*WEP736,*BC337,*276-2016,*RT-102
2N3798	2N3251A,2N3962,2N3963,2N3965,2N4355,2N5373,2N5374,2SA661-GR,BC556B, BSW75,TZ552,TZ553,*PTC127,*GE-67,*TR-20,*HEPS0019,*SK3114,*ECG159, *WEP242,*276-2021,*RT-115
2N3798A	2SA850,*PTC127,*SK3114
2N3799	*GE-21,*TR-20,*HEPS0019,*SK3114,*ECG159,*WEP242,*276-2021,*RT-115
2N3799A	*SK3114
2N3806	2N3808,2N3810
2N3807	2N3809,2N3811
2N3808	2N3806,2N3810
2N3809	2N3807,2N3811
2N3810	2N3806,2N3808
2N3811	2N3807,2N3809
2N3812	2N3814,2N3816,*HEP715(2),*SK3114
2N3813	2N3815,2N3817,*HEPS0019,*SK3114
2N3814	2N3812,2N3816,*HEP715(2),*SK3114
2N3815	2N3813,2N3817,*SK3114
2N3816	2N3812,2N3814,*HEP715(2),*SK3114
2N3817	2N3813,2N3815
2N3818	*GE-66
2N3825	2N783,2N784,2N834A,2N2368,2N3261,2SC54,2SC302,2SC321,2SC322,2SC400-R, 2SC576,2SC689,40405,GET706,KT218,MPS834,MPS6532,MPS6540,MPSH20, MPSH32,MPSH37,MT1038,MT1038A,MT1039,PET3001,TIS47,ZT2368,*PTC132, *GE-210,*TR-70,*HEPS0011,*SK3039,*ECG107,*WEP56,*BC107B,*276-2015, *RT-108
2N3826	2N915,2N2847,2N3903,2N3946,2N3974,2N3976,2N4227,2N4962,2N4963,2N4994, 2N5027,2N5380,2SC473,2SC815,2SC944,2SC979-O,2SC980G-O,EN915,MM3903, MPS6565,TN-3903,*PTC121,*GE-210,*TR-70,*HEPS0005,*SK3018,*ECG107, *WEP56,*BC107B,*RT-108
2N3827	2N3302,2N3904,2N4995,2N5381,BC174B,BC190B,MPS6566,TN-3904,*PTC121, *GE-210,*TR-70,*HEPS0005,*SK3018,*ECG229,*WEP56,*BC107B,*RT-108
2N3828	2N916,2N916A,2N2501,2N3947,2N4013,2N4873,2SC639,2SC933,2SC1216,2SC1999, ZT81,ZT82,ZT84,ZT2369A,*PTC121,*GE-20,*TR-21,*HEPS0015,*SK3122,*EN10, *ECG123A,*WEP736,*276-2016,*RT-102
2N3829	*GE-21,*TR-20,*HEPS0013,*SK3114,*EN40,*ECG159,*WEP242,*276-2023,*RT-115
2N3830	2N3735,*HEPS3001,*276-2014
2N3831	2N3735,*GE-215,*HEPS3001,*276-2014
2N3832	*PTC126,*GE-86,*TR-95,*HEPS0016,*SK3039,*ECG108,*WEP56,*276-2011,*RT-113
2N3833	2N3834,2N3835
2N3834	2N3833,2N3835
2N3835	2N3833,2N3834
2N3839	2N2857,2N3570,2N3600,2N4252,2N4253,2N6304,2N6305,2SC1747,40294,MT1061A, *PTC132,*GE-86,*ECG316
2N3840	2N4981,*GE-82,*TR-20,*HEPS0013,*SK3114,*ECG159,*WEP242,*BC327,*276-2023, *RT-115
2N3841	
2N3842	
2N3843	2N758,2N759,2N783,2N784,2N834A,2N2368,2N3261,2N3793,2N3825,2N3843A, 2N3854A,2SC54,2SC302,2SC321,2SC322,2SC400-R,2SC455,2SC576,2SC689, 40405,BSX25,GET706,GI-3793,KT218,MPS834,MPS6532,MPS6540,MPS6545, MPSH20,MPSH32,MPSH37,MT1038,MT1038A,MT1039,PET3001,SE1010,TIS47, ZT403P,ZT2368,*PTC101,*GE-212,*TR-21,*HEPS0011,*SK3122,*EN40,*ECG123A, *WEP736,*BC107B,*276-2016,*RT-102
2N3843A	2N758,2N759,2N783,2N784,2N834A,2N2368,2N3261,2N3793,2N3825,2N3843, 2N3854A,2SC54,2SC302,2SC321,2SC322,2SC400-R,2SC455,2SC576,2SC689, 40405,BSX25,GET706,GI-3793,KT218,MPS834,MPS6532,MPS6540,MPS6545, MPSH20,MPSH32,MPSH37,MT1038,MT1038A,MT1039,PET3001,SE1010,TIS47, ZT403P,ZT2368,*PTC101,*GE-212,*TR-21,*HEPS0011,*SK3122,*EN40,*ECG123A, *WEP736,*BC107B,*276-2016,*RT-102
2N3844	2N708,2N708A,2N784A,2N914,2N957,2N2242,2N2318,2N2319,2N2369,2N2369A, 2N2481,2N2710,2N3009,2N3011,2N3013,2N3014,2N3210,2N3211,2N3510,2N3511,

To Replace	Substitute This Type
(2N3844)	2N3564,2N3605A,2N3606A,2N3646,2N3647,2N3648,2N3688,2N3689,2N3690,2N3691, 2N3693,2N3844A,2N3855A,2N3862,2N4137,2N4294,2N4295,2N4420,2N4421, 2N4422,2N4969,2N5029,2N5030,2SC67,2SC169,2SC204,2SC335,2SC356,2SC394-O, 2SC461,2SC468,2SC577,2SC595,2SC601,2SC601N,2SC724,2SC752G-R,2SC764, 2SC1274,BFY39-1,BSW82,EN708,EN914,EN2369A,EN3009,EN3011,EN3013, EN3014,GET708,GET914,GET2369,GET3013,GET3014,GET3646,MPS2369, MPS3646,MPS6512,MPSA20-RED,PET1001,PET2001,PT2760,SE1001,SE5001, SE5002,SE5003,SE5006,TIS45,TIS46,TIS48,TIS49,TIS51,TIS52,TIS55,ZT404P,ZT708, ZT2369,*PTC101,*GE-212,*TR-21,*HEPS0015,*SK3122,*EN40,*ECG123A, *WEP736,*BC107B,*276-2016,*RT-102
2N3844A	2N708,2N708A,2N784A,2N914,2N957,2N2242,2N2318,2N2319,2N2369,2N2369A, 2N2481,2N2710,2N3009,2N3011,2N3013,2N3014,2N3210,2N3211,2N3510,2N3511, 2N3564,2N3605A,2N3606A,2N3646,2N3647,2N3648,2N3688,2N3689,2N3690, 2N3691,2N3693,2N3844,2N3855A,2N3862,2N4137,2N4294,2N4295,2N4420,2N4421, 2N4422,2N4969,2N5029,2N5030,2SC67,2SC169,2SC204,2SC335,2SC356,2SC394-O, 2SC461,2SC468,2SC577,2SC595,2SC601,2SC601N,2SC724,2SC752G-R,2SC764, 2SC1274,BFY39-1,BSW82,EN708,EN914,EN2369A,EN3009,EN3011,EN3013, EN3014,GET708,GET914,GET2369,GET3013,GET3014,GET3646,MPS2369, MPS3646,MPS6512,MPSA20-RED,PET1001,PET2001,PT2760,SE1001,SE5001, SE5002,SE5003,SE5006,TIS45,TIS46,TIS48,TIS49,TIS51,TIS52,TIS55,ZT404P,ZT708, ZT2369,*PTC101,*GE-212,*TR-24,*HEPS0015,*SK3122,*ECG123A,*WEP736, *BC107B,*276-2016,*RT-102
2N3845	2N916,2N916A,2N2331,2N2501,2N3241A,2N3845A,2N3856A,2N4013,2N4123, 2N4418,2N4873,2N5144,2SC318,2SC367G,2SC367G-O,2SC372-O,2SC372G-O, 2SC394-Y,2SC400-O,2SC400-Y,2SC454,2SC458,2SC459,2SC460,2SC536,2SC619, 2SC620,2SC639,2SC712,2SC713,2SC735-O,2SC752G-O,2SC933,2SC965,2SC1210, 2SC1216,2SC1276,2SC1372,2SC1999,BFY39-2,EN916,MPS3693,MPS3705, MPS6513,MPS6574-BLUE,MPS6574-GREEN,MPS6574-YEL,MPS6576-BLUE, MPS6576-GREEN,MPS6576-YEL,MPSA20-GRN,MPSA20-WHT,PET3002,SE2001, SE3005,TP3705,ZT81,ZT111,ZT2369A,*PTC132,*GE-212,*TR-70,*HEPS0015, *SK3039,*EN40,*ECG107,*WEP56,*BC107B,*276-2016,*RT-108
2N3845A	2N916,2N916A,2N2331,2N2501,2N3241A,2N3845,2N3856A,2N4013,2N4123,2N4418, 2N4873,2N5144,2SC318,2SC367G,2SC367G-O,2SC372-O,2SC372G-O,2SC394-Y, 2SC400-O,2SC400-Y,2SC454,2SC458,2SC459,2SC460,2SC536,2SC619,2SC620, 2SC639,2SC712,2SC713,2SC735-O,2SC752G-O,2SC933,2SC965,2SC1210, 2SC1216,2SC1276,2SC1372,2SC1999,BFY39-2,EN916,MPS3693,MPS3705, MPS6513,MPS6574-BLUE,MPS6574-GREEN,MPS6574-YEL,MPS6576-BLUE, MPS6576-GREEN,MPS6576-YEL,MPSA20-GRN,MPSA20-WHT,PET3002,SE2001, SE3005,TP3705,ZT81,ZT111,ZT2369A,*PTC132,*GE-212,*TR-70,*HEPS0015, *SK3039,*EN40,*ECG107,*WEP56,*BC107B,*276-2016,*RT-108
2N3846	*GE-17,*TR-70,*SK3039,*EN40,*ECG107,*WEP56,*RT-108
2N3847	
2N3850	HST6310,HST6409,HST6410,SDT6310
2N3851	2N3850
2N3852	HST6308,HST6309,HST6408,SDT6308,SDT6309,*GE-28
2N3853	2N3852,*GE-28
2N3854	2N2318,2N2656,2N2711,2N2713,2N2926-BRN,2N2926-RED,2N3011,2N3394,2N3854A, 2N3855,2N3855A,2N4274,2N5418,2SC204,PET2001,PT720,PT2760,*PTC132, *GE-212,*TR-95,*HEPS0015,*SK3039,*EN40,*ECG108,*WEP56,*BC107B, *276-2015,*RT-113
2N3854A	2N708,2N708A,2N784A,2N914,2N2242,2N2318,2N2369,2N2369A,2N2481,2N2710, 2N3009,2N3011,2N3013,2N3014,2N3210,2N3510,2N3511,2N3605A,2N3606A, 2N3646,2N3647,2N3648,2N3855A,2N4137,2N4274,2N4275,2N5769,2N5772,2SC67, 2SC204,2SC302,2SC577,BSW82,GET708,GET914,GET2369,GET3013,GET3014, GET3646,PET1001,PET2001,PT2760,ZT708,ZT2369,*PTC132,*GE-212,*TR-95, *HEPS0015,*SK3039,*EN40,*ECG108,*WEP56,*BC107B,*276-2016,*RT-113
2N3855	2N2331,2N3394,2N3855A,2N3959,2N3960,2N5418,PET2001,PT720,*PTC132, *GE-212,*TR-70,*HEPS0015,*SK3018,*EN40,*ECG107,*WEP56,*BC107B,*276-2015, *RT-108
2N3855A	2N916,2N916A,2N2242,2N2331,2N2369,2N2369A,2N2501,2N3009,2N3013,2N3014, 2N3211,2N3862,2N4013,2N4873,2N5144,2N5769,2N6427,2SC67,2SC577,2SC639,

To Replace	Substitute This Type
(2N3855A)	2SC764,2SC1210,A5T4123,BSW82,PET1001,PET2001,TIS133,TIS134,ZT2369, ZT2369A,*PTC132,*GE-212,*TR-70,*HEPS0015,*SK3039,*EN40,*ECG107,*WEP56, *BC107B,*276-2016,*RT-108
2N3856	2N2712,2N2714,2N2924,2N2926-YEL,2N3241A,2N3392,2N3856A,2N5224,2N5419, 2N5998,2N6000,2SC1317,2SC1788,2SC1851,BC548A,BCY58A,BSX38,PET2002, *PTC132,*GE-212,*TR-70,*HEPS0015,*SK3039,*EN40,*ECG107,*WEP56,*BC107B, *276-2015,*RT-108
2N3856A	2N3227,2N3241A,2N3242A,2N5998,2N6000,2N6004,2N6426,2SC1175,2SC1317, 2SC1851,2SC1999,BC547A,BC548A,BCY58A,BCY59A,BSW83,BSX38,BSX79, NPSA20,PET1002,PET2002,*PTC132,*GE-212,*TR-70,*HEPS0015,*SK3039,*EN40, *ECG107,*WEP56,*BC107B,*276-2016,*RT-108
2N3857	*GE-82,*TR-20,*HEPS0012,*SK3114,*EN40,*ECG159,*WEP242,*BC327,*276-2021, *RT-115
2N3858	2N916,2N916A,2N2331,2N2501,2N3241A,2N3856A,2N3859,2N4013,2N4873,2N5144, 2N5824,2N5825,2N6427,2SC509-O,2SC509G-O,2SC639,2SC1210,2SC1999, A5T4123,TIS133,TIS134,TP3705,ZT2369A,*PTC101,*GE-212,*TR-21,*HEPS0015, *SK3122,*EN40,*ECG123A,*WEP736,*BC107B,*276-2016,*RT-102
2N3858A	2N736A,2N736B,2N2516,2N3859A,2N4014,2N4962,2N4963,2N5380,2N6538, 2SC1998,A5T3903,TIS96,TIS99,TIS135,TIS136,TN-3903,*PTC170,*GE-81,*TR-21, *HEPS0015,*SK3122,*EN40,*ECG123A,*WEP736,*BF338,*RT-102
2N3859	2N3227,2N3241A,2N3242A,2N3860,2N5825,2N5826,2N5998,2N6000,2N6004,2N6426, 2SC509-Y,2SC509G-Y,2SC1175,2SC1317,2SC1813,2SC1849,2SC1851,A5T4124, BC547A,BC548A,BCY58A,BCY58B,BCY59A,BCY59B,BSW83,BSX38,BSX79, NPSA20,PET1002,PET2002,*PTC101,*GE-212,*TR-21,*HEPS0015,*SK3124,*EN40, *ECG123A,*WEP736,*BC107B,*276-2016,*RT-102
2N3859A	2N3947,2N5381,2N6540,2N6541,A5T3904,BC546A,BCW34,MM3904,TIS95,TIS98, TN-3904,*PTC170,*GE-81,*TR-21,*HEPS0015,*SK3122,*EN40,*ECG123A,*WEP736, *RT-102
2N3860	2N4256,2N4954,2N5371,2N5826,2N5827,2N5998,2N6002,2N6006,2N6008,2N6112, 2SC1849,A5T4124,BC337,BC547B,BC548B,BC549B,BC550B,BCY58B,BCY58C, BCY59B,BCY59C,MPSA16,MPSA17,TP3704,TP3706,*PTC101,*GE-212,*TR-70, *HEPS0011,*SK3039,*EN40,*ECG107,*WEP56,*BC107B,*276-2016,*RT-108
2N3861	
2N3862	*PTC133,*GE-20,*TR-21,*HEPS0025,*SK3122,*SN80,*ECG123A,*WEP736, *276-2038,*RT-102
2N3863	2N3055,2N3055SD,2N3235,2N3236,2N3446,2N3448,2N3713,2N3714,2N5632,2N6254, 2SD41,2SD341,2SD492,*PTC140,*GE-77,*TR-59,*HEPS7002,*SK3036,*SN80, *ECG130,*WEP247,*RT-131
2N3864	2N3055,2N3055SD,2N3236,2N3446,2N3448,2N3714,2N5632,2N5633,2N6254,2SD341, 2SD492,*PTC118,*GE-75,*TR-59,*HEPS7004,*SK3036,*ECG130,*WEP247,*RT-131
2N3865	*PTC118,*TR-61,*HEPS5020
2N3866	2N3866A,2SC302-M,2SC555,*GE-28,*TR-87,*HEPS3008,*SK3195,*ECG311, *WEP243,*276-2009,*RT-114
2N3866A	2N2485,2N2486,*HEPS3008,*ECG311
2N3867	*GE-29,*TR-88,*HEPS5013,*SK3025,*ECG129,*WEP242,*276-2025,*RT-115
2N3868	2N6303,*TR-88,*HEPS5013,*SK3025,*ECG129,*WEP242,*276-2025,*RT-115
2N3877	2N2462,2N2466,2N3877A,2N6540,2N6541,2SC1781,2SC1998,BC546A,BSW85, GET2222A,TIS95,TIS98,*GE-81,*TR-87,*HEPS0005,*SK3024,*EN40,*ECG128, *WEP243,*276-2010,*RT-114
2N3877A	2N2462,2N2466,2N6540,2N6541,2SC1998,BC546A,TIS95,TIS98,*GE-81,*HEPS5025
2N3878	*GE-66,*HEPS5012,*SK3562,*ECG175,*276-2017
2N3879	KSP1163,KSP1166,*HEPS5012,*SK3562,*276-2019
2N3883	*PTC145,*TR-17,*HEPG0005,*ECG160,*WEP637,*276-2004
2N3900	2N3900A,2N3901,2N6002,2N6008,BCY58D,*PTC139,*GE-212,*TR-21,*HEPS0015, *SK3124,*ECG123,*BC107B,*276-2015,*RT-102
2N3900A	2N3900,2N3901,2N6002,2N6008,BCY58D,*PTC139,*GE-212,*TR-21,*HEPS0015, *SK3124,*ECG123A,*WEP736,*BC107B,*276-2015,*RT-102
2N3901	BC548C,BC549C,*PTC139,*GE-10,*TR-21,*HEPS0015,*SK3124,*ECG123A, *WEP736,*BC107B,*276-2015,*RT-102
2N3902	2N5805,DTS-423,IR413,*GE-73,*TR-67,*HEPS5021,*ECG163,*WEP740
2N3903	2N3947,2N4014,2N4962,2N4963,2N5380,2SC943,A5T3903,TN-3903,*GE-210,*TR-53,

To Replace	Substitute This Type
(2N3903)	*HEPS0015,*ECG123A,*WEP736,*276-2016
2N3904	2N2222A,2N4401,2N5381,A5T3904,TIS111,TN-3904,*GE-20,*TR-87,*HEPS0015, *SK3122,*EN10,*ECG123A,*WEP245,*276-2016,*RT-102A
2N3905	2N2906,2N2906A,2N3250,2N3250A,2N3485,2N3485A,2N3672,2N3673,2N4228, 2N5382,2SA467,2SA603,2SA718,2SA720,2SA891,A5T3905,BC212L,MM3905, TN-3905,*PTC103,*GE-21,*TR-20,*HEPS0019,*SK3114,*EP25,*ECG159,*WEP242, *276-2034,*RT-115
2N3906	2N3251,2N3251A,2N5383,2N6005,A5T3906,MM3906,TN-3906,*PTC103,*GE-48, *TR-20,*HEPS0019,*SK3025,*EP25,*ECG159,*276-2034,*RT-115
2N3910	*GE-82,*BC327
2N3911	*GE-82,*TR-20,*HEPS0032,*SK3114,*SP70,*BC327,*276-2021,*RT-115
2N3912	*GE-82,*TR-20,*HEPS0032,*SK3114,*SP70,*BC327,*276-2021,*RT-115
2N3913	2N3910,*GE-82,*BC327
2N3914	2N3911,*GE-82,*TR-20,*HEPS0032,*SK3114,*SP70,*ECG159,*WEP242,*BC327, *276-2023,*RT-115
2N3915	2N3912,*GE-82,*TR-20,*HEPS0032,*SP70,*ECG106,*WEP52,*BC327,*276-2023, *RT-126
2N3916	*GE-32,*HEPS3019
2N3917	*PTC906,*GE-19,*HEPS5000,*HN100
2N3918	*PTC906,*GE-19,*HEPS5000,*HN100
2N3919	HST5904,HST6905,*PTC906,*GE-28,*HEPS7004
2N3920	HST5914,*HEPS7004
2N3923	*PTC117,*GE-27,*TR-78,*HEPS5024,*SK3045,*ECG154,*WEP712,*BF338,*276-2012, *RT-110
2N3924	2SC302-M,2SC307-M,2SC590N,*GE-28,*TR-86,*HEPS3013,*EN40,*ECG123,*WEP53, *276-2009,*RT-100
2N3925	*GE-28
2N3926	2N3927,2N4932,2N5424,2N5424A,2SC831,40282,SRF12213,SRF32214,SRF52214, *PTC128,*GE-28,*TR-66
2N3927	2N4932,2N5102,2N5424,2N5424A,2SC831,40282,SRF52214,*PTC128,*GE-66,*TR-66
2N3928	*HEPS3002
2N3929	
2N3930	2N3931,*ECG159,*WEP242
2N3931	*ECG159,*WEP242
2N3932	2N2808,2N2808A,2N2809,2N2809A,2N2857,2N3570,2N3600,2N3839,2N3933,2N4252, 2N4253,2N5053,2N5054,2N5852,2N6304,2N6305,2SC567,2SC568,2SC583,2SC602, 2SC1747,40294,40295,40413,A485,A490,BFX89,BFY90,MT1061A,ZT2857,*PTC132, *GE-86,*TR-95,*HEPS0016,*ECG108,*WEP56,*276-2015,*RT-113
2N3933	TC3114,*PTC132,*TR-95,*HEPS0016,*SK3039,*ECG108,*WEP56,*276-2038,*RT-113
2N3945	*PTC144,*GE-215,*HEPS3010
2N3946	2N2221A,2N4014,2N4962,2N4963,2N5380,A5T3903,MM3903,TIS135,TIS136,TN-3903, *PTC136,*GE-210,*HEPS5026,*276-2009
2N3947	2N2222A,2N5381,2N5582,A5T3904,MM3904,TIS111,TN-3904,*PTC136,*GE-20, *HEPS5026,*276-2009
2N3948	2SC302-M,2SC731,2SC872,2SC890,2SC908,2SC1001,2SC1002,2SC1165,BLX89, *GE-28,*TR-87,*HEPS3013,*SK3024,*ECG128,*WEP243,*276-2038
2N3950	
2N3959	2N3960,2N5770,*HEPS0024,*276-2013
2N3960	2N3959,2N5770
2N3961	*GE-28,*HEPS3001
2N3962	2N3963,2N3965,*PTC127,*GE-67,*TR-20,*HEPS5022,*SK3114,*ECG159,*WEP717, *276-2021,*RT-115
2N3963	*PTC127,*GE-21,*TR-20,*SK3114,*ECG159,*WEP717,*276-2021,*RT-115
2N3964	2N3965,*GE-48,*TR-20,*HEPS0019,*SK3114,*ECG159,*WEP717,*BC327,*276-2024, *RT-115
2N3965	*GE-21,*TR-20,*HEPS0019,*SK3114,*ECG159,*WEP717,*RT-115
2N3973	2N2221,2N2221A,2N2539,2N2845,2N2847,2N3115,2N3301,2N3974,2N3975,2N3976, 2N4962,2N4963,2N5581,BSW84,GET2221,GET2221A,GI-3641,TIS110,TIS136, TN-3903,*GE-210,*TR-21,*HEPS0025,*SK3122,*EN40,*WEP735,*BC337,*276-2009, *RT-102
2N3974	2N915,2N2539,2N3976,2N4014,2N4951,2N4962,2N4963,2N5368,2SC1318,2SC1852,

To Replace	Substitute This Type
(2N3974)	TIS110,TIS135,TIS136,TN-3903,*GE-210,*TR-21,*HEPS0025,*SK3122,*EN40, *WEP735,*BC337,*276-2009,*RT-102
2N3975	2N2221,2N2221A,2N2539,2N2845,2N2847,2N3115,2N3301,2N3973,2N3974,2N3976, 2N4962,2N4963,2N5581,BSW84,GET2221,GET2221A,GI-3641,TIS110,TIS136, TN-3903,*GE-210,*TR-21,*HEPS0025,*SK3122,*EN40,*ECG123A,*WEP735, *BC337,*276-2009,*RT-102
2N3976	2N915,2N2539,2N3974,2N4014,2N4951,2N4962,2N4963,2N5368,2SC1318,2SC1852, TIS110,TIS135,TIS136,TN-3903,*GE-210,*TR-21,*HEPS0015,*SK3024,*EN40, *ECG123A,*WEP735,*BC337,*276-2009,*RT-102
2N3977	2N2944A,2N2945A,*GE-22,*TR-20,*HEPS0032,*SK3114,*SP70,*ECG159,*WEP717, *BC327,*276-2023,*RT-115
2N3978	2N2945A,2N2946,2N2946A,2N3218,*GE-22,*TR-20,*HEPS0032,*SK3114,*SP70, *ECG159,*WEP717,*BC327,*276-2023,*RT-115
2N3979	2N2946A,2N3219,*GE-82,*TR-20,*HEPS0032,*SK3114,*SP70,*ECG159,*WEP717, *BC327,*276-2023,*RT-115
2N3983	2N708,2N708A,2N784A,2N2318,2N2319,2N2729,2N3011,2N3564,2N3605A,2N3606A, 2N3646,2N3688,2N3689,2N3690,2N4137,2N4294,2N4295,2N4420,2N4421,2N4422, 2N5029,2N5030,2SC54,2SC204,2SC302,2SC335,2SC380-R,2SC380A-R,2SC387AG, 2SC400-R,2SC468,2SC595,2SC601,2SC689,2SC724,2SC752G-R,2SC838,2SC839, 2SC1123,2SC1126,2SC1128,2SC1129,2SC1240,2SC1274,BF224J,BF225J,EN708, EN914,EN2369A,EN3009,EN3011,EN3013,EN3014,GET708,GET914,GET2369, GET3013,GET3014,GET3646,KT218,MM1941,MPS3646,MPS6507,MPS6512, MPSH19,MT1060A,PET3001,PT2760,SE5001,SE5002,SE5003,SE5006,TIS45,TIS46, TIS48,TIS49,TIS51,TIS52,TIS55,TIS84,TIS129,ZT708,*PTC132,*GE-86,*TR-95, *HEPS0016,*SK3039,*ECG108,*WEP56,*BC107B,*276-2011,*RT-113
2N3984	2N834A,2N2368,2N2729,2N3983,2N3985,2N4292,2N4293,2SC54,2SC302,2SC321, 2SC322,2SC387AG,2SC400-R,2SC689,2SC1126,BF224J,BF225J,EN918,KT218, MM1941,MPS834,MPS918,MPS6507,MPS6540,MPS6542,MPS6543,MPS6546, MPS6547,MPS6548,MPSH19,MPSH20,MPSH32,MPSH37,MT1038,MT1038A, MT1039,MT1060,PET3001,TIS47,TIS84,TIS108,TIS125,*PTC132,*GE-86,*TR-95, *HEPS0016,*SK3018,*ECG108,*WEP56,*BC107B,*276-2011,*RT-113
2N3985	2N834A,2N2368,2N2729,2N3983,2N3984,2N4292,2N4293,2SC54,2SC302,2SC321, 2SC322,2SC387AG,2SC400-R,2SC689,2SC1126,BF224J,BF225J,EN918,KT218, MM1941,MPS834,MPS918,MPS6507,MPS6540,MPS6542,MPS6543,MPS6546, MPS6547,MPS6548,MPSH19,MPSH20,MPSH32,MPSH37,MT1038,MT1038A, MT1039,MT1060,PET3001,TIS47,TIS84,TIS108,TIS125,*PTC132,*GE-86,*TR-95, *HEPS0016,*SK3018,*ECG108,*WEP56,*BC107B,*276-2011,*RT-113
2N3995	*TR-17,*HEPG0002,*ECG160,*WEP637,*276-2005
2N3996	KSP1035,SDT6413
2N3997	SDT6416
2N3998	2N2880,2N4115,HST6313,HST6413,SDT6313
2N3999	2N4116,HST6316,HST6416,SDT6316
2N4000	2N4001,*PTC144,*GE-32,*HEPS3019,*276-2008
2N4001	*PTC144,*GE-32,*HEPS3019,*276-2008
2N4002	2N4003,HST8003,HST8016,HST8302,SDT8302
2N4003	2N5331
2N4004	2N4005,HST8106,HST8116,SDT8106,SDT8116
2N4005	
2N4006	*GE-22,*TR-20,*BC327,*276-2023
2N4007	*GE-22,*TR-20,*BC327,*276-2023
2N4008	*GE-82,*TR-28,*BC327,*276-2023
2N4009	M.P.2N4006
2N4010	M.P.2N4007
2N4011	M.P.2N4008
2N4012	*GE-28
2N4013	2N5582,TIS133,*GE-20,*TR-21,*HEPS0015,*SK3122,*SN80,*ECG123A,*WEP736, *276-2009,*RT-102
2N4014	2N5582,TIS135,*GE-20,*TR-21,*HEPS0005,*SK3122,*SN80,*ECG123A,*WEP736, *RT-102
2N4017	*HEP715(2)
2N4018	2N4017

To Replace	Substitute This Type
2N4019	*HEP715(2)
2N4026	2N4027,*PTC141,*GE-67,*TR-88,*HEPS5022,*SK3025,*ECG129,*WEP242,*276-2021, *RT-115
2N4027	*PTC141,*TR-88,*HEPS3003,*SK3025,*ECG129,*WEP242,*276-2009,*RT-115
2N4028	2N4029,*GE-67,*TR-88,*HEPS5022,*SK3025,*ECG129,*WEP242,*276-2021,*RT-115
2N4029	*TR-88,*HEPS3032,*SK3025,*ECG129,*WEP242,*276-2009,*RT-115
2N4030	2N4031,*GE-67,*TR-88,*HEPS3001,*SK3025,*ECG129,*WEP242,*276-2014,*RT-115
2N4031	*TR-88,*HEPS3003,*SK3025,*ECG129,*WEP242,*RT-115
2N4032	2N4033,*GE-67,*TR-88,*HEPS3032,*SK3025,*EP25,*ECG129,*WEP242,*RT-115
2N4033	*TR-88,*HEPS3032,*SK3025,*ECG129,*WEP242,*RT-115
2N4034	2N4035,2N5244,NPS404A,*GE-21,*TR-20,*HEPS0013,*EP25,*ECG106,*WEP52, *276-2022,*RT-126
2N4035	2N5244,NPS404A,*GE-21,*TR-20,*HEPS0019,*EP25,*ECG106,*WEP52,*276-2034, *RT-126
2N4036	2N4314,*PTC141,*GE-29,*TR-88,*HEPS3032,*SK3025,*ECG129,*WEP242
2N4037	*GE-67,*TR-88,*HEPS3012,*SK3025,*ECG129,*WEP242,*276-2025
2N4040	2SC599N,2SC690N,2SC692,XB476,*GE-28,*HEPS3005
2N4041	2N4040,2SC599N,2SC690N,2SC692,XB475,*GE-28,*HEPS3005
2N4044	
2N4045	2N4044,2N4100
2N4046	2N3724,2N3724A,2N4960,2N5145,*GE-47,*HEPS3008,*276-2038
2N4047	2N3725,2N3725A,2N4961,*HEPS0005
2N4048	2N4049,2N4051,2N4052,*HEPG6002
2N4049	2N4050,2N4052,2N4053
2N4050	2N4053
2N4051	2N4052,*HEPG6002
2N4052	2N4053
2N4053	
2N4054	*GE-27,*TR-60,*HEPS5015,*SK3103,*ECG157,*WEP244,*RT-135
2N4055	*GE-27,*TR-60,*HEPS5015,*SK3103,*ECG157,*WEP244,*BF338,*RT-135
2N4056	*GE-27,*TR-60,*HEPS5015,*SK3103,*ECG157,*WEP244,*BF338,*RT-135
2N4057	*GE-27,*TR-60,*HEPS5015,*SK3103,*ECG157,*WEP244,*BF338,*RT-135
2N4058	NPS404A,*GE-65,*TR-20,*HEPS0013,*SK3118,*EP35,*ECG106,*WEP52,*BC177, *276-2022,*RT-126
2N4059	2N4062,*GE-65,*TR-20,*HEPS0013,*SK3118,*EP35,*ECG106,*WEP52,*BC177, *276-2022,*RT-126
2N4060	TIS37,*PTC103,*GE-65,*TR-20,*HEPS0013,*SK3118,*EP35,*ECG106,*WEP52, *BC177,*276-2022,*RT-126
2N4061	2N4058,NPS404A,*GE-65,*TR-20,*HEPS0013,*SK3118,*PN66,*ECG106,*WEP52, *BC177,*276-2022,*RT-126
2N4062	*GE-65,*TR-20,*HEPS0013,*SK3118,*EP35,*ECG106,*WEP52,*BC327,*276-2022, *RT-126
2N4063	*TR-21,*SK3045,*ECG123A,*WEP736
2N4064	*GE-32,*TR-21,*SK3045,*ECG123A,*WEP736
2N4068	*PTC125,*GE-18,*TR-78,*HEPS0005,*SK3045,*ECG154,*WEP712,*276-2008, *RT-110
2N4069	2N3500,2N4925,MM2259,MM2260,*PTC125,*GE-32,*HEPS3019,*276-2008
2N4070	KSP1176,KSP1276,SDT7604,*HEPS5004,*SK3561,*276-2020
2N4071	*HEPS5020
2N4072	MPSH34,*PTC136,*GE-17,*TR-95,*HEPS0020,*SK3018,*ECG108,*WEP56,*276-2016, *RT-113
2N4073	*GE-215,*TR-95,*HEPS3013,*SK3018,*ECG108,*WEP56,*276-2009,*RT-113
2N4074	2N760A,2N3242A,2N6426,40458,*PTC125,*GE-210,*TR-21,*HEPS0015,*SK3122, *ECG123A,*WEP736,*BC337,*276-2009,*RT-102
2N4075	2N2880,2N2892,2N3998,2N4076,2N4115,2N5002,2N5284,HST6313,HST6413, SDT6313,STT6313,STT6413
2N4076	2N2880,2N2893,2N3998,2N4115,2N5002,2N5004,2N5284,2N5285,HST6313,HST6413, SDT6313,STT6313,STT6413
2N4077	*ECG155
2N4078	*PTC120A,*GE-30,*TR-94,*HEPG6016,*SK3052,*ECG131,*WEP642,*276-2006, *RT-127

To Replace	Substitute This Type
2N4080	2N4957,2N4958,2N4959,2N5829,*GE-21,*SK3118,*SP70,*RT-126
2N4081	2SC1747,BF168,BF173,BFS62,BFX60,TC3114,*PTC132,*GE-60,*HEPS0025
2N4086	2N2924,2N2925,2N2926,2N2926-GRN,2N2926-YEL,2N3391,2N3391A,2N3392, 2N3395-WHT,2N3395-YEL,2N3396-WHT,2N3396-YEL,2N3397-WHT,2N3397-YEL, 2N3398-WHT,2N3398-YEL,2N3721,2N3900,2N3900A,2N4087,2N4087A,2N5172, 2S503,2SC538,2SC539,91T6,92T6,98T2,40397,40399,A158B,A159B,BC168B, BC169B,BC172B,BC173B,GI-2924,GI-2925,GI-3391,GI-3391A,GI-3392,GI-3395, GI-3398,GI-3721,GI-3900,GI-3900A,MPS2924,MPS2925,MPS2926,MPS3391, MPS3391A,MPS3392,MPS3395,MPS3721,MPS5172,PBC108B,PBC109B,*PTC121, *GE-212,*HEPS0015,*SK3124,*BC107B,*276-2011,*RT-102
2N4087	2N3390,2N3391,2N3391A,2N3395-WHT,2N3396-WHT,2N3397-WHT,2N3398-BLU, 2N3398-WHT,2N3900,2N3900A,2N3901,2N4087A,91T6,A158C,A159C,BC168C, BC169C,BC170C,BC172C,BC173C,GI-3391,GI-3391A,GI-3900,GI-3900A,MPS3390, MPS3391,MPS3391A,PBC108C,PBC109C,*PTC139,*GE-212,*HEPS0015,*SK3124, *BC107B,*276-2011,*RT-102
2N4087A	2N3390,2N3391,2N3391A,2N3395-WHT,2N3396-WHT,2N3397-WHT,2N3398-BLU, 2N3398-WHT,2N3900,2N3900A,2N3901,2N4087,91T6,A158C,A159C,BC168C, BC170C,BC172C,BC173C,GI-3391,GI-3391A,GI-3900,GI-3900A,MPS3390,MPS3391, MPS3391A,PBC108C,PBC109C,*PTC139,*GE-212,*HEPS0015,*SK3124,*BC107B, *276-2011,*RT-102
2N4100	2N4044
2N4104	*GE-20,*SK3124,*276-2010
2N4106	*GE-2,*TR-05,*HEPG6011,*SK3004,*AT20M,*276-2006
2N4111	2N4113,KSP1025,KSP1045,KSP1165,*PTC906,*GE-19,*HEPS7004,*HN100
2N4112	*PTC906,*GE-19,*HN100
2N4113	KSP1026,KSP1046,KSP1166,*PTC906,*GE-19,*HEPS7004,*HN100
2N4114	*PTC906,*GE-19,*HN100
2N4115	
2N4116	
2N4121	2N4034,2N4035,2N4122,2N5244,2SA710,2SA711,NPS404A,*PTC103,*TR-30, *HEPS0013,*SK3025,*SP70,*ECG159,*276-2034,*RT-126
2N4122	2N4035,2N5244,NPS404A,*GE-21,*HEPS0019,*SK3118,*276-2021,*RT-126
2N4123	2N916,2N916A,2N2501,2N2539,2N3903,2N3947,2N4013,2N4227,2N4951,2N4962, 2N5368,2N5380,2SC639,2SC933,2SC943,2SC1216,A5T3903,A5T4123,TN-3903, ZT84,*GE-210,*TR-53,*HEPS0015,*SK3124,*ECG123A,*WEP736,*276-2016, *RT-102
2N4124	2N4419,2N4954,2N5371,2N6012,2SC735-GR,A5T4124,BCY58B,BCY58C,BCY59B, BCY59C,*GE-20,*TR-53,*HEPS0014,*SK3124,*ECG123A,*WEP736,*276-2016, *RT-102
2N4125	2N3136,2N3250,2N3251,2N3504,2N3905,2N3906,2N4403,2N5366,2N5382,2N5383, 2N6001,2N6005,2SA467,2SA719,2SA890,A5T3504,A5T3644,A5T3905,A5T3906, A5T4125,A5T4403,BC213L,BCY78A,BCY79A,GI-3644,MM3906,MPS6534,TN-3905, TN-3906,*PTC103,*GE-21,*TR-54,*HEPS0013,*SK3025,*ECG159,*WEP242, *276-2034,*RT-115
2N4126	2N5356,2N5367,2N6003,A5T4126,*GE-48,*HEPS0019,*SK3114,*ECG159,*WEP242, *276-2034,*RT-115
2N4127	2SC599,2SC599N,2SC688,2SC690-M,2SC690N,2SC737,*GE-66
2N4128	2SC599,2SC599N,2SC690,2SC690N,*GE-66
2N4130	*GE-14,*HEPS7002,*SK3036
2N4131	*HEPS5004,*276-2020
2N4132	
2N4133	2N749,2N751,2N1388,2N1390,2N1491,2N2781,2N2782,2N2783,2N3553,2SC302-M, 2SC307-M,2SC481,2SC590N,2SC845,40605,*HEPS3019,*276-2020
2N4134	2N2808,2N2808A,2N2809,2N2809A,2N3570,2N3600,2N4135,2N4252,2N4253,2N5053, 2N5054,2N6304,2N6305,2SC567,2SC568,2SC1747,40295,40413,BF168,BF173, BFS62,BFX60,MT1061A,*PTC132,*GE-60,*TR-95,*HEPS0017,*SK3039,*ECG108, *WEP56,*BF173,*276-2016
2N4135	2N2808,2N2808A,2N2809,2N2809A,2N3570,2N3600,2N4252,2N4253,2N5053,2N5054, 2N6304,2N6305,2SC567,2SC568,2SC1747,40295,40413,BF168,BF173,BFS62, BFX60,MT1061A,*PTC132,*GE-60,*TR-64,*HEPS0017,*ECG108,*WEP56,*BF173, *276-2016

To Replace	Substitute This Type
2N4137	2N2369,2N2369A,2N3511,2N3862,2N5769,2SC577,2SC639,2SC764,ZT2369, *PTC133,*HEPS0025,*ECG123A,*WEP56,*276-2038
2N4138	*GE-61,*TR-51,*HEPS0020,*BF173,*276-2013
2N4140	2N2221,2N2221A,2N2539,2N2845,2N2847,2N3115,2N3301,2N3903,2N3946,2N4014, 2N4227,2N4962,2N4963,2N5027,2N5380,2N5581,A5T3903,BSW84,GET2221, GET2221A,GI-3641,GI-3642,MM3903,MPS6530,TN-3903,*GE-210,*TR-53, *HEPS0011,*SK3122,*BC107B,*276-2016,*RT-102
2N4141	2N2222,2N2222A,2N2540,2N3116,2N3302,2N3904,2N4401,2N4409,2N4952,2N5028, 2N5107,2N5369,2N5381,2N5582,A5T2222,A5T3904,A5T4409,BSW85,GET2222, GET2222A,GI-3643,MM3904,MPS6531,TIS109,TIS111,TN-3904,*GE-210,*TR-53, *HEPS0015,*SK3122,*ECG199,*WEP736,*BC107B,*276-2016,*RT-102
2N4142	2N2906,2N2906A,2N3250A,2N3485,2N3485A,2N4228,GET2904,GET2905,*PTC103, *GE-21,*HEPS0013,*SK3114,*276-2023,*RT-115
2N4143	2N2907,2N2907A,2N3251A,2N3486,2N3486A,2N3505,2N3672,2N3673,2SA603, 2SA718,2SA720,2SA891,A5T2907,A5T3505,A5T3645,BC212A,BC212L,GET2906, GET2907,TIS112,*PTC103,*GE-48,*HEPS0019,*SK3114,*276-2023,*RT-115
2N4150	2N5327,BFX34,*HEPS3002
2N4207	2N3546,2N4208,2N4209,2N4258A,2N5057,2N5771,*HEPS0019,*ECG159,*WEP242, *276-2021,*RT-126
2N4208	2N3546,2N4209,2N4258A,2N4872,2N5056,2N5057,2N5292,2N5771,*HEPS0019, *SK3118,*ECG159,*WEP242,*276-2021,*RT-126
2N4209	2N5057,2N5771,*GE-21,*HEPS0019,*SK3118,*ECG159,*WEP242,*276-2032,*RT-126
2N4210	2N4211,2N5957,2N5959,2N5966,2N5968
2N4211	2N5957,2N5959,2N5966,2N5968,ST15043,ST17060
2N4225	*GE-28,*HEPS3002
2N4226	*GE-28,*TR-76,*HEPS3002
2N4227	2N915,2N3947,2N4141,2N4951,2N5028,2N5107,2N5368,2N5582,2SC943,GET2222, GET2222A,GI-3643,MM3904,MPS6531,ZT84,ZT89,*GE-210,*TR-53,*HEPS0015, *BC107B,*276-2016
2N4228	2N3250A,2N3672,2N3673,2N4143,2SA603,2SA718,2SA720,2SA891,BC212A,BC212L, GET2906,GET2907,*PTC103,*GE-21,*HEPS0019,*276-2023
2N4231	2N4232,2N6373,2N6374,*PTC169,*GE-66,*TR-81,*HEPS5012,*SK3131,*ECG175, *WEP241,*276-2020
2N4231A	2N4232A,*PTC169
2N4232	2N4233,2N5427,2N5429,2N5616,2N6372,2N6373,2SC1444,2SC1445,KSP1161, KSP1162,KSP1164,KSP1165,*PTC169,*GE-66,*TR-81,*HEPS5012,*SK3131, *ECG175,*WEP241,*276-2020
2N4232A	2N4233A,*PTC169
2N4233	2N3879,2N5429,2N5616,2N5620,2N6372,2SC1113,2SC1445,KSP1162,KSP1165, *PTC148,*HEPS5012,*SK3131,*ECG175,*WEP241,*276-2017
2N4233A	*PTC169,*GE-233
2N4234	2N4235,*PTC142,*GE-29,*TR-88,*HEPS5013,*SK3025,*ECG129,*WEP242,*276-2025, *RT-115
2N4235	2N4236,*TR-88,*HEPS5013,*SK3025,*ECG129,*WEP242,*276-2025,*RT-115
2N4236	*HEPS3032,*ECG129,*WEP242
2N4237	2N4238,*PTC144,*GE-47,*TR-87,*HEPS5014,*SK3024,*ECG128,*WEP243,*276-2012, *RT-114
2N4238	2N4239,*PTC144,*GE-32,*HEPS5014,*276-2012
2N4239	*PTC144,*GE-32,*HEPS3019,*276-2012
2N4240	*TR-81,*SK3021,*ECG124,*WEP240
2N4241	*PTC901,*GE-16,*TR-01,*HEPG6003,*SK3014,*ECG121,*WEP232,*276-2006, *RT-127
2N4242	2N4245,*GE-76,*HEPG6018
2N4243	2N4242,2N4245,2N4246,*PTC901,*GE-76,*HEPG6018
2N4244	2N4243,2N4246,2N4247,*PTC901,*GE-76,*HEPG6005
2N4245	*GE-76,*HEPG6018
2N4246	2N4245,*PTC901,*GE-76,*HEPG6018
2N4247	2N4246,*PTC901,*GE-76,*HEPG6005
2N4248	2N3250,2N3250A,2N3251,2N3251A,2N3581,2N3582,2N3702,2N3905,2N3906,2N4034, 2N4035,2N4121,2N4122,2N4143,2N4228,2N4965,2N5382,2N5383,2N5447,2N6067, 2SA467,2SA467G-O,2SA467G-Y,2SA493-Y,2SA493G-Y,2SA499-Y,2SA548,

To Replace	Substitute This Type
(2N4248)	2SA561-O,2SA561-Y,2SA603,2SA659,2SA661-O,2SA661-Y,2SA710,2SA711,2SA718, 2SA733,2SA767,A5T3905,A5T3906,A5T4249,A5T5447,A8T3702,A177,BC177, BC177A,BC203,BC212A,BC212L,BC213L,BC251A,BC257A,BC261A,BC307A, BC415-6,BC415A,BC416-6,BC416A,BC557A,BC560A,BCY79A,EN3250,GET2906, GET2907,GI-3702,MM3906,MPS3702,TN-3905,TN-3906,TP3644,TP3645,TP3702, TZ552,*PTC103,*GE-65,*TR-20,*HEPS0019,*SK3114,*ECG159,*WEP242,*BC177, *276-2022,*RT-115
2N4249	2N2605A,2N3547,2N3548,2N3549,2N3798,2N3962,2N3963,2N3965,2N4289, 2SA661-GR,2SA776,2SA776A,2SA841-GR,BC256B,BC266B,BC326,BC556B,EN3962, TZ553,*GE-67,*TR-20,*HEPS5022,*SK3118,*ECG159,*WEP242,*276-2021,*RT-115
2N4250	2N3548,2N3549,2N3550,2N3799,2N3962,2N3964,2N3965,2SA493-GR,2SA493G-GR, 2SA561-GR,2SA564A,2SA578,2SA579,2SA640,2SA641,2SA661-GR,2SA750,2SA776, 2SA776A,2SA841-BL,2SA841-GR,2SA842-BL,2SA842-GR,BC177B,BC251B,BC251C, BC257B,BC261B,BC261C,BC307B,BC307C,BC415-10,BC415-16,BC415B,BC415C, BC416-10,BC416-16,BC416B,BC416C,BC557B,BC560B,BC560C,BCY79C,TZ553, *GE-65,*TR-20,*HEPS0019,*SK3118,*ECG159,*WEP242,*BC327,*276-2022, *RT-115
2N4251	*PTC133,*TR-95,*HEPS0016,*SK3039,*ECG108,*WEP56,*276-2011,*RT-113
2N4252	2N4253,*PTC132,*GE-86,*HEPS0017,*276-2015
2N4253	2N4252,MT1061A,*PTC132,*GE-86,*HEPS0017,*276-2015
2N4254	2N4873,2N4996,2SC639,2SC1216,PET3002,TIS86,ZT2369A,*PTC132,*GE-86,*TR-70, *HEPS0016,*SK3039,*EN40,*ECG107,*WEP56,*276-2015,*RT-108
2N4255	2N4873,2N4996,2N4997,2SC639,2SC1216,PET3002,TIS86,TIS87,ZT2369A,*PTC132, *GE-86,*TR-70,*HEPS0016,*SK3018,*EN40,*ECG107,*WEP56,*276-2015,*RT-108
2N4256	2N4954,2N5371,2N5827,BC547B,BC548B,BC549B,BC550B,BCY58C,BCY58D, BCY59C,BCY59D,*PTC121,*GE-212,*TR-21,*HEPS0015,*SK3122,*ECG123A, *WEP736,*BC107B,*276-2016,*RT-102
2N4257	2N3546,2N4207,2N4208,2N4209,2N4257A,2N4258,2N4258A,2N4872,2N5056,2N5057, 2N5292,2N5771,*PTC131,*TR-20,*HEPS0013,*SK3118,*ECG159,*WEP242, *276-2021,*RT-115
2N4257A	2N4258A,2N5771,A5T4260,A5T4261,*HEPS0013,*SK3118,*ECG159,*WEP242, *276-2021
2N4258	2N3546,2N4208,2N4209,2N4258A,2N4872,2N5056,2N5057,2N5292,2N5771,*TR-20, *HEPS0013,*SK3114,*ECG159,*WEP242,*276-2021,*RT-115
2N4258A	2N5771,A5T4260,A5T4261,*HEPS0013,*SK3118,*ECG159,*WEP242,*276-2021
2N4259	40235,40236,40237,40238,40240,40472,40474,*PTC132,*GE-39,*TR-21,*HEPS0011, *ECG123A,*WEP736,*276-2009,*RT-102
2N4260	2N4261
2N4261	
2N4264	2N5845A,*GE-20,*TR-53,*HEPS0011,*276-2016
2N4265	BCY58,BCY66,*GE-20,*TR-53,*HEPS0011,*276-2016
2N4269	*GE-27,*TR-78,*HEPS5025,*SK3045,*ECG154,*WEP712,*BF338,*276-2012,*RT-110
2N4270	*GE-27,*TR-78,*HEPS5025,*SK3045,*ECG154,*WEP712,*BF338,*276-2012,*RT-110
2N4271	*GE-32,*HEPS3019
2N4272	
2N4273	*HEPS5012,*276-2017
2N4274	2N5769,*PTC131,*GE-20,*TR-95,*HEPS0011,*SK3039,*ECG108,*WEP56,*276-2034, *RT-113
2N4275	2N5769,2N5772,*PTC101,*GE-20,*TR-95,*HEPS0011,*SK3039,*ECG108,*WEP56, *276-2016,*RT-113
2N4276	2N4277,2N4278,2N4279,*PTC901
2N4277	2N4279,*PTC901
2N4278	2N4279,2N4280,2N4281,*PTC901
2N4279	2N4281,*PTC901
2N4280	2N4281,2N4282,2N4283,*PTC901
2N4281	2N4283,*PTC901
2N4282	2N4283,*PTC901
2N4283	*PTC901
2N4284	2N2945A,2N4285,2N5230,*GE-65,*TR-30,*HEPS0032,*BC177,*276-2022
2N4285	2N3840,2N5231,*GE-65,*TR-52,*HEPS0032,*BC177,*276-2022
2N4286	2N4287,BC547B,BC550B,BCY58C,BCY58D,BCY59C,BCY59D,MPSA16,PET4002,

To Replace	Substitute This Type
(2N4286)	*PTC139,*GE-212,*TR-21,*HEPS0015,*SK3124,*ECG123A,*WEP736,*BC107B, *276-2013,*RT-102
2N4287	BCY59C,BCY59D,*GE-212,*TR-21,*HEPS0015,*SK3124,*ECG123A,*WEP736, *BC107B,*276-2013,*RT-102
2N4288	2N3964,*GE-65,*TR-20,*HEPS0019,*SK3118,*ECG159,*WEP242,*BC327,*276-2022, *RT-115
2N4289	*GE-65,*TR-20,*SK3114,*ECG159,*WEP242,*BC327,*RT-115
2N4290	2N2838,2N3504,2N4291,2N4402,2N4403,2N5811,2N5817,A5T3504,A5T3644, A5T4402,A5T4403,BC327-16,D29E1,D29E5,TN-3905,TN-3906,TQ60,TQ62,*PTC103, *GE-21,*TR-20,*HEPS0019,*SK3114,*ECG159,*WEP242,*BC327,*276-2034, *RT-115
2N4291	BCW35,BCW37,*PTC103,*GE-48,*TR-20,*HEPS0019,*SK3114,*ECG159,*WEP242, *BC327,*276-2034,*RT-115
2N4292	2N834A,2N2368,2N2729,2N4293,2SC387AG,2SC576,2SC689,2SC1126,EN918, MM1941,MPS918,MPS6507,MPS6542,MPS6543,MPS6546,MPS6547,MPS6548, MT1038,MT1038A,MT1039,MT1060,PET3001,TIS84,ZT2368,*PTC132,*GE-86, *HEPS0025,*276-2011
2N4293	2N834A,2N2368,2N2729,2N4292,2SC387AG,2SC576,2SC689,2SC1126,EN918, MM1941,MPS918,MPS6507,MPS6542,MPS6543,MPS6546,MPS6547,MPS6548, MT1038,MT1038A,MT1039,MT1060,PET3001,TIS84,ZT2368,*PTC132,*GE-86, *HEPS0025,*276-2011
2N4294	2N2369,2N2369A,2N2501,2N2710,2N3011,2N3013,2N3014,2N3211,2N3510,2N3511, 2N3646,2N3647,2N3648,2N3862,2N4137,2N4295,2N4420,2N4421,2N4422,2N5029, 2SC67,2SC169,2SC335,2SC468,2SC577,2SC595,2SC639,2SC752G-R,2SC764, 2SC1274,EN2369A,EN3011,EN3013,EN3014,GET2369,GET3013,GET3014,GET3646, MPS2369,MPS3646,TIS48,TIS49,TIS51,TIS52,TIS55,ZT2369,*GE-20,*TR-21, *HEPS0025,*276-2011
2N4295	2SC169,2SC335,2SC336,2SC601N,2SC639,2SC764,*PTC132,*GE-60,*HEPS0025, *BF173,*276-2038
2N4296	*PTC104,*GE-32,*TR-81,*HEPS5011,*SK3021,*ECG124,*WEP240,*RT-128
2N4297	*PTC104,*GE-32,*TR-81,*HEPS5011,*SK3021,*ECG124,*WEP240,*RT-128
2N4298	*GE-32,*TR-81,*HEPS5011,*SK3021,*ECG124,*WEP240,*RT-128
2N4299	*GE-32,*TR-81,*HEPS5011,*SK3021,*ECG124,*WEP240,*RT-128
2N4300	*HEPS3002
2N4301	HST7140,KSP1155
2N4305	
2N4306	2N4310
2N4307	2N4309,*GE-66,*HEPS3002
2N4308	2N4306,2N4310,2N4312,*GE-66
2N4309	
2N4310	2N4306
2N4311	2N4305,*GE-66,*HEPS3002
2N4312	2N4306,2N4308,2N4310,*GE-66
2N4313	2N2894A,2N3546,2N4258A,2N5056,2N5057,2N5292,EN2894A,*TR-20,*HEPS0013, *SK3118,*ECG106,*WEP52,*276-2021,*RT-126
2N4314	*TR-88,*HEPS3032,*SK3025,*ECG129,*WEP242
2N4346	*PTC122,*GE-25,*TR-27,*HEPG6008,*SK3035,*ECG127,*WEP235
2N4347	2SD110-R,2SD118-R,2SD165,2SD214,HST9204,HST9209,*PTC118,*GE-73,*TR-61, *HEPS5020,*SK3079,*ECG280,*WEP247
2N4348	2N5631,2N5634,1561-1205,1561-1404,1561-1405,1561-1604,1561-1605,1582-1203, 1582-1204,1582-1205,1582-1403,1582-1404,1582-1405,1582-1603,1582-1604, 1582-1605,HST9204,HST9209,IR3773,IR6302,SDT9204,SDT9209,STC2222, *PTC118,*TR-61,*HEPS5020,*SK3079,*ECG280
2N4350	2SC302-M,2SC307-M,2SC590N,*GE-28,*HEPS3001,*276-2035
2N4354	2N2905,2N2905A,2N2907,2N2907A,2N3486,2N3486A,2N3503,2N3505,2N3671, 2N3672,2N3673,2N4028,2N4029,2N4032,2N4033,2N4355,2N5373,2N5823, 2SA503-GR,2SA684,2SA720,2SA777,2SA817,2SA891,A5T2907,A5T3505,A5T3645, BCW35,BSW75,D29E10,FT3644,TIS112,TZ552,*PTC103,*GE-82,*TR-20,*HEPS5022, *SK3025,*ECG159,*WEP242,*276-2021,*RT-115
2N4355	2N2905,2N2905A,2N2907,2N2907A,2N3486,2N3486A,2N3503,2N3505,2N4028, 2N4029,2N4032,2N4033,2N4354,2N5373,2N5374,2SA503-GR,2SA684,2SA720,

To Replace	Substitute This Type
(2N4355)	2SA777,2SA817,2SA891,A5T2907,A5T3505,A5T3645,BCW35,BSW75,TIS112,TZ552, TZ553,*PTC103,*GE-82,*TR-20,*HEPS5022,*SK3114,*ECG159,*WEP242, *276-2021,*RT-115
2N4356	2SA581,2SA606,2SA625,2SA777,2SA850,*PTC171,*GE-82,*TR-20,*HEPS5022, *SK3114,*ECG159,*WEP242,*276-2021,*RT-115
2N4357	*276-2022
2N4358	*276-2021
2N4359	2N3251,2N3251A,2N3547,2N3548,2N3798,2N3962,2N4354,2N4355,2N5373,2N5374, 2N6005,2SA659,2SA661-GR,2SA661-Y,2SA720,2SA889,2SA891,BC326,BC557A, BC557B,BC560A,BC560B,BCY79,BCY79A,BCY79B,GET2906,GET2907,MM4048, TP3644,TP3645,TZ552,TZ553,TZ582,*PTC127,*GE-65,*HEPS0019,*ECG159, *WEP242,*BC327
2N4383	*GE-47,*HEPS3024,*276-2012
2N4384	2N6012,BC337-40,D33D27,*GE-47,*BC337,*276-2009
2N4385	2N4383,*GE-47,*HEPS3024,*276-2012
2N4386	2N4384,2N6012,A5T5451,A8T3706,BC337-25,BC337-40,D33D27,TIS109,*PTC136, *GE-47,*BC337,*276-2009
2N4387	*PTC113,*GE-69,*TR-77,*HEPS5018,*SK3083,*ECG153,*WEP246,*276-2025, *RT-133
2N4388	*PTC113,*GE-69,*TR-77,*HEPS5018,*SK3083,*ECG153,*WEP246,*276-2025, *RT-133
2N4389	2N869A,2N2894,2N3012,2N3209,2N3546,2N3576,2N3640,2N4258A,2N4423,2N5056, 2N5057,2N5292,2SA741,2SA781,MM2894,TIS50,*PTC131,*TR-20,*HEPS0019, *SK3118,*ECG106,*WEP52,*276-2021,*RT-126
2N4390	2N739A,*PTC117,*GE-27,*HEPS5026,*BF338,*276-2008
2N4395	2N5622,2SD369,2SD379,B170003-YEL,B170004-YEL,BDY23C,*PTC140,*GE-19, *TR-59,*HEPS7002,*SK3027,*ECG130,*WEP247,*276-2016
2N4396	2N5622,2N5626,2SC793-BL,2SC793-Y,2SD88A,2SD111,2SD111-O,2SD116, 2SD119-BL,2SD119-Y,2SD334,2SD369,2SD379,2SD428-O,B170003-YEL, B170004-YEL,B170006-YEL,B170007-YEL,BDY24C,*PTC140,*GE-19,*TR-59, *HEPS7002,*SK3027,*ECG130,*WEP247,*276-2012,*RT-131
2N4397	2SC1747,BF168,BF173,BFS62,BFX60,TC3114,*PTC132,*GE-60
2N4398	2N4399,*PTC174,*HEPS7001,*RT-148
2N4399	*PTC174,*HEPS7001,*RT-148
2N4400	2N4962,2N4963,TIS110,TN-3903,*GE-20,*TR-21,*HEPS0015,*SK3122,*ECG123A, *WEP736,*BC337,*276-2016,*RT-102
2N4401	2N2222A,TIS111,TN-3904,*GE-20,*TR-21,*HEPS0015,*SK3122,*ECG123A,*WEP736, *276-2016,*RT-102
2N4402	2N2906,2N2906A,2N3485,2N3485A,2N3672,2N3673,2N4026,A5T4402,BCW35, BCW37,TN-3905,TZ551,*PTC103,*GE-21,*TR-20,*HEPS0019,*SK3114,*ECG159, *WEP242,*276-2034,*RT-115
2N4403	2N2907,2N2907A,2N3486,2N3486A,2N3504,2N3505,2N4028,A5T2907,A5T3504, A5T3505,A5T3644,A5T3645,A5T4403,TIS112,TN-3906,*PTC103,*GE-21,*TR-20, *HEPS0019,*SK3025,*ECG159,*WEP242,*276-2021,*RT-115
2N4404	*TR-88,*HEPS3032,*SK3025,*ECG129,*WEP242,*RT-115
2N4405	*TR-88,*HEPS3032,*SK3025,*ECG129,*WEP242,*RT-115
2N4406	*TR-88,*HEPS3012,*SK3025,*ECG129,*WEP242,*RT-115
2N4407	*TR-88,*HEPS3032,*SK3025,*ECG129,*WEP242,*RT-115
2N4409	A5T4409,*PTC170,*GE-81,*TR-53,*HEPS0005,*ECG194,*WEP712,*276-2008
2N4410	*PTC170,*GE-18,*TR-78,*HEPS0005,*SK3045,*ECG154,*WEP712,*276-2008, *RT-110
2N4411	*TR-20,*HEPS0013,*SK3114,*ECG159,*WEP242,*276-2034,*RT-115
2N4412	2N4412A,2SA773,BC327-25,BC327-40,MPS4354,MPS4355,*PTC127,*GE-48,*TR-88, *HEPS0012,*SK3025,*ECG129,*WEP242,*BC327,*276-2021,*RT-115
2N4412A	MPS4354,MPS4355,*PTC127,*GE-67,*TR-88,*HEPS5013,*SK3025,*ECG129, *WEP242,*276-2025,*RT-115
2N4413	2N4413A,2N6013,BC327-25,BC327-40,D29E7,*PTC127,*GE-48,*TR-20,*HEPS0019, *SK3114,*ECG159,*WEP242,*BC327,*276-2023,*RT-115
2N4413A	2N6017,*PTC127,*GE-67,*TR-20,*HEPS5022,*SK3114,*ECG159,*WEP242,*276-2021, *RT-115
2N4414	2N2905,2N2905A,2N3502,2N3503,2N4032,2N4412,2N4412A,2N4414A,2SA486-BL,

To Replace	Substitute This Type
(2N4414)	2SA498-Y,2SA503-GR,2SA504-GR,2SA684,2SA773,A5T2907,A5T3504,A5T3505, A5T3644,A5T3645,A5T4403,BC160-16,BC161-16,BC327-16,BC327-25,MPS4354, MPS4355,TIS112,TQ59,*PTC127,*GE-48,*TR-88,*HEPS0012,*SK3025,*ECG129, *WEP242,*BC327,*276-2021,*RT-115
2N4414A	2N2905,2N2905A,2N3503,2N4032,2N4033,2N4405,2N4412A,2SA485-BL,2SA497-Y, 2SA503-GR,2SA684,2SA817,A5T2907,A5T3505,A5T3645,BC161-16,MPS4354, MPS4355,TIS112,*PTC127,*GE-67,*TR-88,*HEPS5013,*SK3025,*ECG129, *WEP242,*276-2025,*RT-115
2N4415	2N2907,2N2907A,2N3486,2N3486A,2N3504,2N3505,2N4028,2N4413,2N4413A, 2N4415A,2N5819,2N6011,A5T2907,A5T3504,A5T3505,A5T3644,A5T3645,A5T4403, BC327-16,BC327-25,D29E6,TIS112,TN-3906,TQ60,*PTC127,*GE-48,*TR-20, *HEPS0019,*SK3114,*ECG159,*WEP242,*BC327,*276-2023,*RT-115
2N4415A	2N2907,2N2907A,2N3486,2N3486A,2N3505,2N4028,2N4029,2N4413A,2N6015, 2SA817,A5T2907,A5T3505,A5T3645,TIS112,*PTC127,*GE-67,*TR-20,*HEPS5022, *SK3114,*ECG159,*WEP242,*276-2021,*RT-115
2N4418	2N2501,2N4013,*GE-20,*TR-95,*HEPS0025,*SK3039,*ECG108,*WEP56,*276-2038, *RT-113
2N4419	BCY58B,BCY58C,BCY59B,BCY59C,*GE-20,*TR-95,*HEPS0011,*SK3039,*ECG108, *WEP56,*276-2015,*RT-113
2N4420	2N708,2N914,2N916,2N916A,2N2501,2N3013,2N3014,2N3210,2N3211,2N3646, 2N4013,2N4421,2N4422,2N5144,2SC67,2SC169,2SC335,2SC336,2SC577,2SC639, 2SC752G-O,2SC764,EN708,EN916,MPS3646,MPS6530,ZT84,ZT708,*GE-20,*TR-64, *HEPS0016,*276-2016
2N4421	2N708,2N914,2N916,2N916A,2N2501,2N3013,2N3014,2N3210,2N3211,2N3646, 2N4013,2N4420,2N4422,2N5144,2SC67,2SC169,2SC335,2SC336,2SC577,2SC639, 2SC752G-O,2SC764,EN708,EN916,MPS3646,MPS6530,ZT84,ZT708,*GE-20,*TR-64, *HEPS0011,*SK3118,*BC107B,*276-2015
2N4422	2N708,2N914,2N916,2N916A,2N2501,2N3013,2N3014,2N3210,2N3211,2N3646, 2N4013,2N4420,2N4421,2N5144,2SC67,2SC169,2SC335,2SC336,2SC577,2SC639, 2SC752G-O,2SC764,EN708,EN916,MPS3646,MPS6530,ZT84,ZT708,*GE-20,*TR-64, *HEPS0016,*276-2016
2N4423	2N2894,2N3012,2N3209,2N3546,2N3640,2N4258A,*PTC103,*GE-21,*HEPS0013, *276-2021
2N4424	2N4953,2N5370,2N5376,2N6016,*GE-210,*TR-21,*HEPS0015,*SK3124,*ECG123A, *WEP736,*BC337,*276-2010,*RT-102
2N4425	*GE-47,*TR-21,*HEPS3024,*SK3124,*ECG192,*WEP736,*BC337,*276-2012,*RT-102
2N4427	2N4874,2N5109,2N5421,2SC302-M,2SC307-M,2SC590N,2SC730,2SC731,2SC844, 2SC872,2SC890,2SC908,2SC991,2SC992,2SC1001,2SC1002,2SC1083,2SC1165, 2SC1256,2SC1464,2SC1479,2SC1820,2SC1947,40280,40975,*PTC143,*GE-18, *TR-87,*HEPS3008,*SK3048,*ECG128,*WEP243,*276-2038,*RT-114
2N4428	2N2485,2N2486,2N3866,2N3866A,2SC555,*GE-28,*HEPS3001,*276-2014
2N4429	*GE-28
2N4430	2N4431,2SC977,2SC978,*GE-28
2N4431	2SC978
2N4433	*PTC132,*GE-210,*TR-70,*HEPS0016,*SK3018,*EN10,*ECG107,*WEP56,*BC107B, *276-2011,*RT-108
2N4436	2N2221,2N2221A,2N2539,2N2845,2N2847,2N3115,2N3301,2N4014,2N4400,2N4962, 2N4963,2N5581,BSW84,EN915,GI-3641,GI-3642,MPS6530,TN-3903,*GE-210,*TR-21, *HEPS0015,*SK3122,*EN10,*ECG123A,*WEP736,*BC337,*276-2016,*RT-102
2N4437	2N2222,2N2222A,2N2540,2N3116,2N3302,2N4401,2N4952,2N5107,2N5369,2N5582, BSW85,EN2222,GI-3643,MPS6531,SE6022,SE6023,TN-3904,*GE-210,*TR-21, *HEPS0015,*SK3122,*EN10,*ECG311,*WEP736,*276-2016,*RT-102
2N4438	*276-2008
2N4439	*276-2008
2N4440	2N3733,2N5016,MSA7505,MSA8505,SRF12101,*GE-28
2N4854	D2T2218A,D2T2219A
2N4854+	D2T2905,D2T2905A
2N4855	2N6502,D2T2218,D2T2219
2N4855+	D2T2904,D2T2904A
2N4862	HST5004,HST5005,HST5051,SDT5004,SDT5005,SDT5051
2N4863	

To Replace	Substitute This Type
2N4864	HST5904,HST5905,HST5951,KSP1396,*HEPS5012,*276-2017
2N4865	2N5250,KSP1002,KSP1003
2N4866	2N5251
2N4872	2N2894A,2N4209,2N5056,2N5057,2N5292,2N5771,*TR-20,*HEPS0013,*SK3118, *ECG106,*WEP52,*276-2021,*RT-126
2N4873	ZT2369A,*PTC133,*TR-95,*HEPS0025,*SK3039,*ECG108,*WEP56,*276-2038, *RT-113
2N4874	2SC302-M,*HEPS3008,*276-2038
2N4875	2N4874,2SC302-M,*HEPS3008,*276-2038
2N4876	2N4874,2N4875,2SC302-M,*HEPS3008,*276-2038
2N4877	*GE-66,*HEPS3010,*276-2022
2N4878	
2N4879	2N4878
2N4880	2N4878,2N4879
2N4888	*PTC171
2N4889	2N3930,2N3931,*PTC171
2N4890	2N4405,2N5865,40537,MPS4356,*GE-67,*TR-88,*HEPS5022,*ECG129,*WEP242, *276-2021,*RT-115
2N4895	2N4897,BFX34,HST5004,HST5504,*276-2008
2N4896	
2N4897	2N4895,BFX34,HST5004,HST5504,*276-2008
2N4898	2N3740A,2N4899,2N5956,2SA764,KSP2024,*PTC113,*GE-69,*TR-58,*HEPS5018, *SK3083,*276-2026,*RT-133
2N4899	2N3740A,2N3741A,2N4900,2N5955,2SA764,2SA765,KSP2024,KSP2025,*PTC113, *GE-69,*TR-58,*HEPS5018,*SK3083,*276-2025,*RT-133
2N4900	2N3741A,2N5607,2N5615,2N5954,2N6467,2SA765,KSP2025,*PTC113,*TR-58, *HEPS5018,*ECG218,*WEPS5007,*276-2025
2N4901	2N3789,2N4902,2N4904,2N4905,2N4907,2N4908,2N5867,2N5871,2N5875,*PTC149, *TR-29,*HEPS7003,*ECG218,*WEPS5007,*276-2027
2N4902	2N3789,2N3790,2N4903,2N4905,2N4906,2N4908,2N4909,2N5867,2N5868,2N5871, 2N5872,2N5875,2N5876,2SA746,2SA877,KSP2171,KSP2174,*PTC149,*TR-29, *HEPS7003,*ECG218,*WEPS5007
2N4903	2N3790,2N4906,2N4909,2N5623,2N5868,2N5872,2N5876,2N6248,2SA746,2SA877, KSP2171,KSP2172,KSP2174,KSP2175,*PTC172,*HEPS5005,*ECG218,*WEPS5007
2N4904	2N3789,2N3791,2N4901,2N4902,2N4905,2N4907,2N4908,2N5871,2N5875,*PTC149, *TR-29,*HEPS7003,*ECG218,*WEPS5007,*276-2027
2N4905	2N3789,2N3790,2N3791,2N3792,2N4902,2N4903,2N4906,2N4908,2N4909,2N5871, 2N5872,2N5875,2N5876,2SA746,2SA877,KSP2174,*PTC149,*TR-29,*HEPS7003, *ECG218,*WEPS5007
2N4906	2N3790,2N3792,2N4903,2N4909,2N5623,2N5872,2N5876,2N6248,2SA746,2SA877, KSP2174,KSP2175,*PTC172,*HEPS7003,*ECG218,*WEPS5007
2N4907	2N3789,2N4908,2N5875,2N5879,*PTC149,*HEPS7003,*ECG218,*WEPS5007
2N4908	2N3789,2N3790,2N4909,2N5875,2N5876,2N5879,2N5880,*PTC149,*HEPS7003, *ECG218,*WEPS5007
2N4909	2N3790,2N5876,2N5880,*PTC172,*HEPS5005,*ECG218,*WEPS5007
2N4910	2N4911,*PTC112,*GE-216,*TR-81,*HEPS5019,*SK3131,*ECG175,*WEP241, *276-2018
2N4911	2N4912,*PTC112,*GE-216,*TR-81,*HEPS5019,*SK3131,*ECG175,*WEP241, *276-2020
2N4912	*PTC140,*GE-32,*TR-81,*HEPS5019,*SK3131,*ECG175,*WEP241,*276-2017
2N4913	2N4914,2N5067,2N5068,2N5873,2N5877,180T2,BDY23B,*PTC140,*GE-73,*TR-59, *HEPS7002,*SK3027,*HN100,*ECG130,*WEP247,*RT-131
2N4914	2N3447,2N3713,2N3715,2N4915,2N5068,2N5069,2N5873,2N5874,2N5877,2N5878, 180T2,BDY23B,KSP1174,*PTC140,*GE-73,*TR-59,*HEPS7002,*SK3027,*HN100, *ECG130,*WEP247,*RT-131
2N4915	2N3447,2N3448,2N3713,2N3714,2N3715,2N3716,2N5069,2N5624,2N5874,2N5878, 181T2,BDY24B,KSP1174,KSP1175,*PTC140,*GE-73,*TR-59,*HEPS7002,*SK3027, *ECG130,*WEP247,*RT-131
2N4916	2N4034,2N4035,2N4121,2N4122,2N4917,2N5244,2SA710,2SA711,NPS404A, *PTC103,*GE-22,*TR-20,*HEPS0013,*SK3114,*ECG159,*WEP242,*276-2034, *RT-115

To Replace	Substitute This Type
2N4917	2N4035,2N4122,2N5244,NPS404A,*GE-21,*TR-20,*HEPS0013,*SK3114,*ECG159, *WEP242,*276-2023,*RT-115
2N4918	2N4919,*PTC162,*GE-29,*HEPS5006,*SK3191,*ECG185,*WEPS5007,*276-2026, *RT-153
2N4919	2N4920,*PTC162,*GE-29,*TR-77,*HEPS5006,*SK3191,*ECG185,*WEPS5007, *276-2026,*RT-153
2N4920	*PTC164,*GE-69,*HEPS5006,*SK3191,*ECG185,*WEPS5007,*RT-153
2N4921	2N4922,*PTC163,*GE-28,*HEPS5000,*SK3190,*ECG184,*WEPS5003,*276-2018, *RT-152
2N4922	2N4923,*PTC163,*GE-28,*HEPS5000,*SK3190,*ECG184,*WEPS5003,*276-2020, *RT-152
2N4923	*PTC165,*GE-28,*HEPS5000,*SK3190,*ECG184,*WEPS5003,*276-2018,*RT-152
2N4924	MM2258,*PTC144,*GE-27,*TR-78,*HEPS5026,*SK3045,*ECG154,*WEP712, *276-2012,*RT-110
2N4925	MM2260,*GE-32,*TR-78,*HEPS5025,*SK3045,*ECG154,*WEP712,*RT-110
2N4926	2N4927,2N5059,*GE-40,*TR-78,*HEPS5025,*SK3045,*ECG154,*WEP712,*BF338, *RT-110
2N4927	2N5058,2N5059,*GE-40,*TR-78,*HEPS5025,*SK3045,*ECG154,*WEP712,*BF338, *RT-110
2N4928	*TR-88,*SK3025,*ECG129,*WEP242,*RT-115
2N4929	
2N4930	2N4931
2N4931	
2N4932	2N3927,2N5424,2N5424A,2SC831,40282,SRF32214,SRF52214,*PTC128,*TR-66
2N4933	2N5071
2N4934	2SC1747,*PTC132,*TR-95,*HEPS0016,*SK3039,*ECG108,*WEP56,*276-2011, *RT-113
2N4935	TC3114,*PTC132,*TR-95,*HEPS0016,*SK3039,*ECG108,*WEP56,*276-2011,*RT-113
2N4936	*PTC132,*TR-95,*HEPS0030,*SK3039,*ECG108,*WEP56,*276-2038,*RT-113
2N4943	BSY86,*TR-87,*HEPS3019,*SK3024,*ECG128,*WEP243,*RT-114
2N4944	2N718A,2N736,2N736A,2N736B,2N870,2N910,2N2461,2N2465,2N2516,2N3077, 2N4945,2N4963,2SC979A-O,2SC980AG-O,BC110,*PTC170,*GE-20,*TR-87,*SK3024, *ECG128,*WEP243,*BF338,*276-2009,*RT-114
2N4945	2N718A,2N736,2N736A,2N736B,2N870,2N910,2N2461,2N2465,2N2516,2N3077, 2N4944,2N4963,2SC979A-O,2SC980AG-O,BC110,*PTC170,*GE-81,*TR-87,*SK3024, *ECG194,*WEP243,*BF338,*276-2008,*RT-114
2N4946	2N871,2N956,2N2462,2N2466,2N2645,BC546A,EN871,EN956,GET2222A,PET1075, *PTC170,*GE-20,*TR-87,*HEPS3019,*SK3024,*ECG128,*WEP243,*276-2009, *RT-114
2N4950	2N3149,2N3150,*GE-20
2N4951	2N2222,2N2222A,2N2540,2N3116,2N4952,2N5107,2N5368,2N5369,2N5582,A5T2222, BSW85,GI-3643,TIS111,TN-3904,*GE-210,*TR-21,*HEPS0025,*SK3124,*EN40, *ECG123A,*WEP736,*276-2009,*RT-102
2N4952	2N2222,2N2222A,2N2540,2N3116,2N3302,2N5369,2N5377,A5T2222,BSW85,TIS109, TIS111,TN-3904,*GE-210,*TR-21,*HEPS0025,*SK3124,*EN40,*ECG123A,*WEP736, *276-2009,*RT-102
2N4953	2N5370,*GE-210,*TR-21,*HEPS0015,*SK3124,*EN40,*ECG123A,*WEP736,*276-2009, *RT-102
2N4954	2N4953,2N5370,2N5371,2N5376,2N6012,*GE-210,*TR-21,*HEPS0025,*SK3124, *EN40,*ECG123A,*WEP736,*276-2009,*RT-102
2N4957	2N4958,2N4959
2N4958	2N4957,2N4959
2N4959	2N4957,2N4958
2N4960	2N4961,*TR-87,*HEPS3001,*SK3024,*ECG128,*WEP243,*276-2014,*RT-114
2N4961	*TR-87,*HEPS3002,*SK3024,*ECG128,*WEP243,*RT-114
2N4962	2N4963,*TR-87,*HEPS3001,*SK3024,*ECG128,*WEP243,*276-2014,*RT-114
2N4963	*TR-87,*HEPS3019,*SK3024,*ECG128,*WEP243,*RT-114
2N4964	2N722,2N722A,2N2394,2N2906,2N2906A,2N3039,2N3250,2N3250A,2N3485, 2N3485A,2N3703,2N4142,2N4971,2N5372,2N5448,2N5763,2N6067,2SA499-O, 2SA522A,2SA539,2SA548,2SA549,2SA555,2SA661-R,A5T5448,A8T3703,BSW74, EN722,EN3250,GET2904,GET2905,GI-3703,MPS3703,TZ551,*PTC103,*GE-65,

To Replace	Substitute This Type
(2N4964)	*TR-87,*HEPS0019,*SK3024,*ECG128,*WEP243,*BC177,*276-2024,*RT-114
2N4965	2N2907,2N2907A,2N3251,2N3251A,2N3486,2N3486A,2N3505,2N3548,2N3672, 2N3673,2N3962,2N4143,2N4972,2N5373,2N5374,2N6005,2SA499-Y,2SA561-GR, 2SA561-Y,2SA603,2SA659,2SA661-GR,2SA661-Y,2SA678,2SA705,2SA718,2SA720, 2SA733,2SA891,A5T2907,A5T3505,A5T3645,A177,BC212A,BC256A,BC256B, BC266A,BC266B,BC307A,BC307B,BC416-10,BC416-6,BC416A,BC416B,BC557A, BC557B,BC560A,BC560B,BCW35,BCW37,BSW75,EN2907,GET2906,GET2907, TIS112,TP3645,TZ552,TZ553,*PTC103,*GE-65,*TR-20,*HEPS0019,*SK3114, *ECG159,*WEP242,*BC177,*276-2024,*RT-115
2N4966	2N760A,2N3858A,2N3859A,2N3903,2N4962,2N5380,2S104,2SC943,2SC944, 2SC1999,TN-3903,*PTC121,*GE-61,*TR-21,*HEPS0015,*SK3124,*ECG123A, *WEP736,*BC107B,*276-2013,*RT-102
2N4967	2N929A,2N930A,2N2484,2N2523,2N2586,2N3117,BC107B,BC182B,BC237B,BC547B, BC550B,FT107C,GET2484,SE4020,*PTC121,*GE-212,*TR-21,*HEPS0024,*SK3124, *ECG123A,*WEP736,*BC107B,*276-2013,*RT-102
2N4968	2N760,2N2501,2N3241A,2N4074,2N4418,2N4966,2SC1999,PET4001,ZT44,*PTC101, *GE-60,*TR-21,*HEPS0024,*SK3124,*ECG123A,*WEP736,*BF365,*276-2031, *RT-102
2N4969	2N708A,2N2221,2N2221A,2N2539,2N2845,2N2847,2N3115,2N3301,2N4013,2N4400, 2N4436,2N4962,2N5144,2N5581,BSW84,EN915,GI-3641,GI-3642,MPS6530,TN-3903, *GE-210,*TR-21,*HEPS0025,*SK3122,*EN10,*ECG123A,*WEP736,*BC337, *276-2016,*RT-102
2N4970	2N2222,2N2222A,2N2540,2N3116,2N3302,2N4401,2N4952,2N5369,2N5377,BSW85, EN2222,TN-3904,*GE-210,*TR-21,*HEPS0015,*SK3122,*ECG123A,*WEP736, *BC337,*276-2016,*RT-102
2N4971	2N2906,2N2906A,2N3485,2N3485A,2N4026,*PTC103,*GE-21,*TR-20,*HEPS0019, *SK3114,*ECG159,*WEP242,*276-2023,*RT-115
2N4972	2N2907,2N2907A,2N3486,2N3486A,2N3505,2N3672,2N3673,2N4028,2N6005,2SA684, 2SA720,2SA891,A5T2907,A5T3505,A5T3645,TIS112,*PTC103,*GE-48,*TR-20, *SK3114,*ECG159,*WEP242,*276-2023,*RT-115
2N4974	
2N4975	
2N4980	*GE-82,*BC327,*276-2023
2N4981	2N3840,*GE-82,*BC327
2N4982	*GE-82,*TR-20,*SK3114,*ECG159,*WEP242,*RT-115
2N4994	2N915,2N2847,2N3946,2N3974,2N3976,2N4962,2N4963,2N5380,2N6538,A5T3903, MM3903,TIS96,TIS99,TN-3903,*PTC121,*GE-210,*TR-21,*SK3122,*ECG123A, *WEP736,*BC337,*RT-102
2N4995	2N3302,2N4409,2N5381,2N6540,A5T3904,A5T4409,TIS95,TIS98,TN-3904,*PTC121, *GE-210,*TR-21,*HEPS0015,*SK3122,*ECG123A,*WEP736,*BC337,*276-2013, *RT-102
2N4996	2N4873,2SC639,2SC1216,PET3002,TIS86,ZT2369A,*PTC132,*GE-61,*TR-95, *HEPS0016,*SK3018,*ECG108,*WEP56,*276-2015,*RT-113
2N4997	2N4873,2N4996,2SC639,2SC1216,PET3002,TIS86,TIS87,ZT2369A,*PTC132,*GE-61, *TR-95,*HEPS0016,*SK3018,*ECG108,*WEP56,*276-2015,*RT-113
2N4998	2N2151,2N4115,HST6313,HST6413
2N4999	2N5286
2N5000	2N4116,HST6316,HST6416
2N5001	2N5287
2N5002	2N5284
2N5003	
2N5004	2N5285
2N5005	
2N5006	2N5288
2N5007	2N5290
2N5008	2N5289
2N5009	2N5291
2N5010	2N5011,2N5095,2N5097,*GE-32
2N5011	2N5012,2N5097,2N5098
2N5012	2N5013,2N5098,2N5099
2N5013	2N5014,2N5015,2N5099

To Replace	Substitute This Type
2N5014	2N5015
2N5015	
2N5016	XB408
2N5017	
2N5022	2N3245,2N3468,2N3762A,2N3763,MM3726,*GE-67,*TR-88,*SK3025,*ECG129, *WEP242,*RT-115
2N5023	2N3244,2N3467,2N3762,MM3726,*GE-67,*TR-88,*HEPS3014,*SK3025,*ECG129, *WEP242,*276-2021,*RT-115
2N5025	2N4933,2N5071,40340
2N5026	2SC1037,40341
2N5027	2N915,2N2539,2N2847,2N4014,2N4951,2N4962,2N4963,2N5368,TN-3903,ZT84, *GE-210,*TR-64,*HEPS0014,*276-2016
2N5028	2N2222,2N2222A,2N2540,2N3116,2N3302,2N4401,2N4952,2N5369,A5T2222,BSW85, MPS6531,TIS109,TIS111,TN-3904,*GE-210,*HEPS0014,*276-2016
2N5029	2N2369,2N2369A,2N3862,2N5769,2SC577,2SC639,2SC764,MPS2369,ZT2369, *PTC133,*HEPS0011,*276-2038
2N5030	2N2369,2N2369A,2N2501,2N2710,2N3009,2N3011,2N3013,2N3014,2N3211,2N3510, 2N3511,2N3646,2N3647,2N3648,2N3862,2N4137,2N5029,2N5769,2N5772,2SC67, 2SC577,2SC595,2SC639,2SC764,2SC1274,GET2369,GET3013,GET3014,GET3646, MPS2369,ZT2369,*GE-20,*TR-21,*HEPS0011,*276-2016
2N5031	2N3570,2N3571,2N3572,2N3600,2N4252,2N4253,2N5032,2N6304,2N6305,2SC567, 2SC568,2SC611,2SC653,*PTC132,*GE-11,*TR-70,*HEPS0017,*SK3039,*ECG107, *WEP56,*276-2011,*RT-108
2N5032	2N3570,2N3571,2N3572,2N3600,2N4252,2N4253,2N5031,2N6304,2N6305,2SC567, 2SC568,2SC611,2SC653,*PTC132,*GE-11,*TR-70,*HEPS0017,*SK3039,*ECG107, *WEP56,*276-2011,*RT-108
2N5034	2N5036,40542,40543,40633,*GE-14,*TR-59,*HEPS7002,*SK3036,*ECG223,*WEP247, *276-2019
2N5035	2N5037,*TR-59,*HEPS7002,*ECG223,*WEP247
2N5036	40633,*GE-14,*TR-59,*HEPS7002,*SK3027,*ECG223,*WEP247
2N5037	*TR-59,*HEPS7002,*ECG223,*WEP247
2N5038	*SK3535
2N5039	2N5038,*HEPS7000,*SK3511
2N5040	2N727,2N869,2N869A,2N2392,2N2412,2N3829,2N3905,2N4060,2N4125,2N5041, 2N5042,2N5354,2N5365,2N5382,2SA467,2SA467G-O,2SA523,2SA562-O,2SA732, A5T3905,A5T4125,BF379,BF414,BF441,BF509,FT5040,FT5041,MM3905,MPS6516, MPSA70-RED,MPSA70-WHT,TIS37,TN-3905,*PTC131,*GE-48,*TR-30,*HEPS0012, *ECG129,*WEP242,*BC177,*276-2034
2N5041	2N3250,2N3250A,2N3581,2N3703,2N3905,2N4142,2N5042,2N5382,2N5448,2N6067, 2SA467,2SA467G-O,2SA523,2SA523A,2SA532,2SA544,2SA552,2SA558,2SA594-O, 2SA594-R,2SA594N,2SA661-O,2SA661-R,A5T3905,A5T5448,A8T3703,BC177, FT5041,GET2904,GET2905,GI-3703,MM3905,MPS3703,TN-3905,TP3703,TZ551, *PTC103,*GE-48,*TR-52,*HEPS0019,*ECG129,*WEP242,*BC177,*276-2034
2N5042	2SA544,2SA552,2SA594-O,2SA594-R,2SA594N,*PTC127,*GE-48,*ECG129,*WEP242
2N5043	2N5044,TIXM101
2N5044	2N5043,TIXM101
2N5048	
2N5049	
2N5050	2N5051,*TR-81,*HEPS5012,*SK3131,*ECG286,*WEP241,*276-2017
2N5051	2N5052,*TR-81,*HEPS5012,*SK3131,*ECG291,*WEP241,*276-2017
2N5052	BLY49A,*HEPS5012,*ECG286
2N5053	2N2808,2N2808A,2N2809A,2N3570,2N4252,2N4253,2N5054,2N6304,2N6305, 2SC1747,MT1061A,*PTC133,*GE-86,*TR-95,*HEPS0017,*SK3018,*ECG108, *WEP56,*276-2011,*RT-113
2N5054	2N2808,2N2808A,2N2809A,2N3570,2N4252,2N4253,2N5053,2N6304,2SC1747, MT1061A,*PTC133,*GE-20,*TR-95,*SK3018,*ECG108,*WEP56,*RT-113
2N5055	2N2894A,2N3546,2N4258A,2N4313,2N5056,2N5057,2N5292,EN2894A,*PTC131, *GE-89,*SK3118,*276-2021
2N5056	2N3546,2N5057,*HEPS0013,*276-2023
2N5057	2N3546,2N5056,*HEPS0013,*276-2023
2N5058	*PTC117,*GE-32,*TR-78,*HEPS5024,*SK3045,*ECG154,*WEP712,*276-2008,

To Replace	Substitute This Type
(2N5058)	*RT-110
2N5059	2N5058,*PTC107,*GE-32,*TR-78,*HEPS5024,*SK3045,*ECG154,*WEP712,*276-2008, *RT-110
2N5065	*HEPS0014,*276-2038
2N5066	*GE-210,*TR-22,*HEPS0020,*BC337,*276-2009
2N5067	2N4913,2N4914,2N5068,2N5869,2N5873,2N5877,2N6315,2N6317,180T2,BDY23B, *PTC129A,*GE-73,*TR-59,*HEPS7002,*SK3027,*ECG280,*WEP247,*RT-131
2N5068	2N3445,2N3447,2N3713,2N4914,2N4915,2N5069,2N5869,2N5870,2N5873,2N5874, 2N5877,2N5878,2N6315,2N6316,2N6317,2N6318,180T2,BDY23B,KSP1171,KSP1174, *PTC129A,*GE-73,*TR-59,*HEPS7002,*SK3027,*ECG280,*WEP247,*RT-131
2N5069	2N3445,2N3446,2N3447,2N3448,2N3713,2N3714,2N4915,2N5624,2N5870,2N5874, 2N5878,2N6316,2N6318,181T2,BDY24B,KSP1171,KSP1172,KSP1174,KSP1175, *PTC129A,*GE-73,*HEPS7004,*ECG280,*WEP247
2N5070	2SC1037
2N5071	2N5026,2SC1037,40341
2N5074	2N5076
2N5075	2N5077
2N5076	
2N5077	
2N5083	2N4115,2N5085
2N5084	2N4116
2N5085	
2N5086	2N3798,2N3962,2N3965,2N5087,2N5374,2N6007,2SA661-GR,2SA678,2SA705, A5T5086,A5T5087,BC256B,BC266B,BC307B,BC416-10,BC416B,BC557B,BC560B, TZ553,*GE-65,*TR-20,*HEPS0019,*SK3025,*ECG159,*WEP242,*BC327,*276-2022, *RT-305
2N5087	2N3965,2N6007,2SA844,A5T5087,BC416-16,BC416C,BC560C,*GE-65,*TR-20, *HEPS0019,*SK3025,*ECG159,*WEP242,*BC327,*276-2022,*RT-126A
2N5088	*GE-212,*TR-21,*HEPS0024,*SK3124,*ECG123A,*WEP736,*BC337,*276-2016, *RT-102
2N5089	*GE-85,*TR-21,*HEPS0024,*SK3124,*ECG123A,*WEP736,*276-2016,*RT-102
2N5090	*PTC128,*TR-66
2N5091	
2N5092	2N5095,*GE-32
2N5093	
2N5094	
2N5095	2N5097,*GE-32
2N5097	2N5098
2N5098	2N5099
2N5099	
2N5100	
2N5101	
2N5102	2N5016,XB408
2N5106	2N2219,2N2219A,2N2538,2N3300,*GE-47,*HEPS3001,*276-2038
2N5107	2N2222,2N2222A,2N2540,2N3116,2N3302,2N4952,2N5369,2N5582,A5T2222,BSW85, EN2219,FT3643,GI-3643,SE6021,SE6021A,TIS109,TIS111,TN-3904,*PTC136, *GE-210,*HEPS0015,*ECG123A,*WEP736,*276-2009
2N5108	*GE-18,*TR-87,*HEPS3001,*SK3024,*ECG278,*WEP243,*276-2014,*RT-114
2N5108A	2N2485,2N2486,2N5108,41024,*HEPS3001,*276-2014
2N5109	2SC302-M,2SC307-M,2SC590N,2SC872,2SC890,2SC991,2SC992,2SC1002,2SC1256, 2SC1820,40975,*TR-87,*HEPS3013,*SK3218,*ECG278,*WEP243,*276-2009, *RT-114
2N5110	*GE-84,*TR-88,*HEPS5013,*SK3025,*ECG129,*WEP242,*276-2025,*RT-115
2N5111	*TR-88,*HEPS3032,*SK3025,*ECG129,*WEP242,*RT-115
2N5112	*GE-69,*276-2027
2N5113	
2N5126	2N744,2N2615,2N2616,2N3011,2N3564,2N3959,2N3960,2N4418,2N4996,2N4997, 2N5200,2N5201,2S095A,2S512,2SC99,2SC168,2SC300,2SC301,2SC380-O, 2SC380A-O,2SC388A,2SC400-O,2SC619,2SC735-O,2SC836,2SC922,2SC924, 2SC1123,2SC1128,2SC1129,2SC1215,2SC1372,2SC1393,2SC1394,2SC1395, 2SC1396,2SC1789,2SC1988,A5T3571,BF121,BF123,BF125,BF127,BF254,BF311,

To Replace	Substitute This Type
(2N5126)	EN744,MPS6568,MPS6568A,MPS6569,MPS6570,MPSH02,MPSH10,MPSH11, MPSH30,MPSH31,MT1060A,PT720,SE3005,TIS64A,ZT2205,*PTC132,*GE-86,*TR-83, *HEPS0025,*SK3124,*ECG161,*WEP719,*BF173,*276-2015,*RT-113
2N5127	2N744,2N753,2N2206,2N2331,2N2615,2N2616,2N2922,2N3011,2N3298,2N3564, 2N3691,2N3855A,2N3959,2N3960,2N4418,2N4996,2N4997,2N5126,2N5200,2N5201, 2S095A,2S512,2SC99,2SC168,2SC300,2SC301,2SC372-O,2SC372G-O,2SC380-O, 2SC380A-O,2SC388A,2SC394-Y,2SC400-O,2SC460,2SC461,2SC619,2SC710, 2SC712,2SC713,2SC735-O,2SC836,2SC922,2SC924,2SC963,2SC964,2SC1123, 2SC1128,2SC1129,2SC1215,2SC1372,2SC1393,2SC1394,2SC1395,2SC1396, 2SC1789,2SC1988,A5T3571,BF121,BF123,BF125,BF127,BF254,BF311,BSY61, EN744,MPS6568,MPS6568A,MPS6569,MPS6570,MPS6574-YEL,MPSH02,MPSH10, MPSH11,MPSH30,MPSH31,MT1060A,PET2001,PT720,SE2001,SE3005,TIS64A, ZT2205,*PTC132,*GE-212,*TR-21,*HEPS0025,*SK3124,*ECG123A,*WEP736, *BF365,*276-2011,*RT-102
2N5128	2N703,2N744,2N753,2N1992,2N2206,2N2331,2N2475,2N2615,2N2616,2N2784, 2N3011,2N3298,2N3633,2N3855,2N3855A,2N3959,2N3960,2N4274,2N4418,2N5134, 2N5200,2N5201,2S095A,2S512,2SC99,2SC388A,2SC1068,2SC1387,A5T3571, BF123,BF125,BF127,MPS2926-ORG,MPS6568A,MPS6569,MPS6570,MPSH10, MPSH11,PT720,TIS64A,TIS86,ZT2205,*PTC101,*GE-210,*TR-21,*HEPS0014, *SK3124,*ECG123A,*WEP736,*BF173,*276-2016,*RT-102
2N5129	2N744,2N753,2N1992,2N2206,2N2331,2N2475,2N2615,2N2616,2N2784,2N2922, 2N3011,2N3298,2N3633,2N3855,2N3855A,2N3959,2N3960,2N4418,2N4996,2N4997, 2N5200,2N5201,2S095A,2S512,2SC99,2SC168,2SC300,2SC301,2SC388A, 2SC400-O,2SC460,2SC461,2SC619,2SC710,2SC712,2SC713,2SC922,2SC924, 2SC1215,2SC1372,2SC1393,2SC1394,2SC1395,2SC1396,2SC1789,A5T3571, BF123,BF125,BF127,BF254,BSY61,GI-2712,MPS2926-ORG,MPS6568A,MPS6569, MPS6570,MPSH10,MPSH11,PET3002,PT720,SE3005,TIS64A,TIS86,ZT2205, *PTC132,*GE-210,*TR-87,*HEPS0014,*SK3024,*ECG128,*WEP243,*BF365, *276-2011,*RT-114
2N5130	2N2369,2N2369A,2N2729,2N3011,2N3563,2N3564,2N3688,2N3689,2N3690,2N4137, 2N4294,2N4295,2N5029,2N5030,2SC67,2SC302,2SC335,2SC380-R,2SC380A-R, 2SC387AG,2SC468,2SC595,2SC601,2SC601N,2SC689,2SC752G-R,2SC836, 2SC922,2SC1123,2SC1126,2SC1128,2SC1129,2SC1240,BF125,BF198,BF241, BF311,EN2369A,EN3011,MM1941,MPS6507,MT1060A,SE3001,SE3002,SE5001, SE5002,SE5003,SE5006,TIS48,TIS49,TIS51,TIS129,*PTC132,*GE-86,*TR-83, *HEPS0016,*SK3018,*EN10,*ECG161,*WEP719,*BF173,*276-2011,*RT-113
2N5131	2N2924,2N3241A,2N3392,2N3692,2N3859,2N3860,2N4124,2N4419,2N5224, 2SC372-Y,2SC372G-Y,2SC380-Y,2SC380A-Y,2SC394-GR,2SC400,2SC400-Y, 2SC454,2SC711,2SC735-Y,2SC900,2SC934,2SC941-Y,2SC1085,2SC1281, 2SC1363,2SC1631,2SC1633,2SC1684,2SC1727,2SC1849,2SC1853,2SC1854, 2SD392,A5T3572,BC108,BC108A,BC168A,BC170B,BC172A,BC238A,BC548A, BCY58A,BCY58B,BSW88,BSW89,BSX38,MPS6573,MPS6574-GREEN,MPS6574-SIL, PBC108A,PET2002,SE2002,*PTC132,*GE-64,*TR-95,*HEPS0011,*SK3124,*ECG108, *WEP56,*BF365,*276-2011,*RT-113
2N5132	2N3241A,2N3692,2SC372-Y,2SC372G-Y,2SC380-Y,2SC380A-Y,2SC394-GR,2SC400, 2SC400-Y,2SC454,2SC458,2SC459,2SC735-Y,2SC925,2SC934,2SC1085,2SC1359, 2SC1727,A5T3572,BC108A,BC168A,BC172A,BC238A,BC548A,BCY58A,BSW88, BSW89,BSX38,MPS6573,MPS6574-GREEN,MPS6574-SIL,PET2002,SE2002, *PTC132,*GE-60,*TR-70,*HEPS0033,*SK3124,*ECG107,*WEP56,*BF365, *276-2015,*RT-108
2N5133	2N2924,2N2925,2N3392,2N3565,2N3860,2N4124,2N4256,2N4286,2N4419,2N5172, 2N5219,2N5223,2N5224,2SC373,2SC373G,2SC400-GR,2SC538,2SC539,2SC711, 2SC732-GR,2SC733-GR,2SC735-GR,2SC829,2SC900,2SC1084,2SC1684,2SC1849, 2SC1853,2SC1993,40397,A158B,A159B,BC108,BC108B,BC109B,BC122,BC168B, BC169B,BC172B,BC173B,BC238B,BC239B,BCY58B,BCY58C,MPS5172,PBC108B, PBC109B,PET4002,*PTC121,*GE-212,*TR-83,*HEPS0011,*SK3018,*ECG161, *WEP719,*BC107B,*276-2031,*RT-113
2N5134	*PTC132,*GE-20,*TR-21,*HEPS0015,*SK3124,*EN10,*ECG123A,*WEP736, *276-2011
2N5135	2N5828,2SC733-BL,2SC1006,2SC1010,BC108C,BC109,BC109C,BC238C,BC239C, BC413C,BC414C,BCY58D,BCY59D,MPS6521,MPSA18,PET4003,*PTC139,*GE-64,

To Replace	Substitute This Type
(2N5135)	*TR-87,*HEPS0014,*SK3124,*ECG128,*WEP243,*BC107B,*276-2013,*RT-114
2N5136	2N729,2N760,2N916,2N916A,2N1986,2N2331,2N2501,2N2615,2N2616,2N3241A, 2N3856A,2N3859,2N4074,2N4123,2N4418,2N4873,2N5825,2SC367G,2SC367G-O, 2SC639,2SC733-O,2SC735-O,2SC847,2SC848,2SC852,2SC933,2SC966,2SC1216, 2SC1687,2SC1688,2SC1999,A5T4123,BF240,BFY39-2,MPS3693,MPS6513, MPS6574-BLUE,MPS6574-GREEN,MPS6574-YEL,MPS6576-BLUE,MPS6576-GREEN, MPS6576-YEL,MPSA10-GRN,MPSA10-WHT,MPSA20-GRN,MPSA20-WHT,PET4001, TIS62A,TIS63A,TIS64A,TIS86,TP3705,ZT24,ZT44,*PTC101,*GE-210,*TR-87, *HEPS0014,*SK3124,*EN10,*ECG128,*WEP243,*BF173,*276-2013,*RT-114
2N5137	2N729,2N760,2N916,2N916A,2N2331,2N2501,2N2615,2N2616,2N3241A,2N3856A, 2N3859,2N4074,2N4123,2N4418,2N4873,2N4996,2N5825,2SC367G,2SC367G-O, 2SC372-O,2SC372G-O,2SC380-O,2SC380A-O,2SC394-Y,2SC400-Y,2SC454, 2SC458,2SC459,2SC536,2SC619,2SC639,2SC733-O,2SC735-O,2SC752G-O, 2SC847,2SC848,2SC899,2SC924,2SC925,2SC933,2SC941-O,2SC964,2SC965, 2SC966,2SC1085,2SC1188,2SC1189,2SC1215,2SC1216,2SC1276,2SC1320, 2SC1359,2SC1372,2SC1393,2SC1394,2SC1395,2SC1396,2SC1682-BL,2SC1682-GR, 2SC1687,2SC1688,2SC1999,BF240,BF254,BFY39-2,EN916,MPS3693,MPS6513, MPS6574-BLUE,MPS6574-GREEN,MPS6574-YEL,MPS6576-BLUE,MPS6576-GREEN, MPS6576-YEL,MPSA10-GRN,MPSA10-WHT,MPSA20-GRN,MPSA20-WHT, PET3002,PET4001,SE2001,SE3005,TIS62A,TIS63A,TIS64A,TIS86,TP3705,ZT44, *PTC101,*GE-210,*TR-21,*HEPS0014,*SK3124,*ECG123A,*WEP736,*BF365, *276-2015,*RT-102
2N5138	2N3250,2N3251,2N3581,2N3582,2N3702,2N3905,2N3906,2N4034,2N4035,2N4060, 2N4121,2N4122,2N4248,2N4916,2N4917,2N4965,2N5382,2N5383,2N5447,2N6067, 2SA467,2SA467G-O,2SA467G-Y,2SA493-Y,2SA493G-Y,2SA494-O,2SA494-Y, 2SA495-O,2SA495-Y,2SA495G-O,2SA495G-Y,2SA499-Y,2SA500-Y,2SA548, 2SA561-O,2SA561-Y,2SA562-O,2SA562-Y,2SA642,2SA659,2SA710,2SA711, 2SA733,2SA838,A5T3905,A5T3906,A5T5447,A8T3702,A177,BC177,BC177A,BC178, BC202,BC203,BC213L,BC251A,BC257A,BC261A,BC307A,BC308A,BC415-6,BC415A, BC416-6,BC416A,BC557A,BC558A,BC559A,BC560A,BCY78A,BCY79A,BSW19, BSW20,EN3250,GI-3702,MM3906,MPS3702,TN-3905,TN-3906,TP3644,TP3702, *PTC103,*GE-65,*TR-20,*HEPS0013,*SK3114,*ECG159,*WEP242,*BC177, *276-2022,*RT-115
2N5139	2N4916,2N4917,2SA500-Y,NPS404,*PTC103,*GE-21,*TR-20,*HEPS0013,*SK3114, *ECG159,*WEP242,*276-2034,*RT-115
2N5140	2N2894A,2N3012,2N3304,2N3546,2N3639,2N3640,2N4208,2N4257,2N4257A,2N4258, 2N4258A,2N4313,2N4872,2N5055,2N5056,2N5057,2N5292,EN2894A,MM2894, MPS3639,MPS3640,MPSL07,MPSL08,TIS50,TIS53,*PTC131,*GE-89,*TR-20, *SK3118,*ECG106,*WEP52,*276-2021,*RT-126
2N5141	*PTC131,*GE-22,*TR-20,*SK3118,*ECG106,*WEP52,*276-2021,*RT-126
2N5142	2N869A,2N995,2N2695,2N2696,2N2927,2N3545,2N3638,2N5023,2N5242,2N5243, FT1746,GI-3638,MPS3638,MPS6535,SE8540,SE8541,SE8542,*PTC103,*GE-22, *TR-20,*HEPS0012,*SK3114,*ECG159,*WEP242,*BC327,*276-2034,*RT-115
2N5143	2N869A,2N995,2N2695,2N2696,2N3545,FT1746,GI-3638,MPS3638,MPS6535, *PTC103,*GE-22,*TR-20,*HEPS0012,*SK3114,*ECG159,*WEP242,*BC327, *276-2034,*RT-115
2N5144	2N4013,2N4962,TIS133,TIS134,TN-3903,*PTC136,*GE-20,*HEPS0015,*276-2009
2N5145	2N3724,2N3724A,2N4046,2N4960,*HEPS3001,*276-2038
2N5147	2N5322,2N5675
2N5148	*276-2008
2N5149	
2N5150	*276-2008
2N5151	
2N5152	*276-2008
2N5153	
2N5154	*276-2008
2N5155	2N1908,2N3146,*GE-25
2N5156	*GE-25,*TR-27,*HEPG6018,*OC28,*276-2006
2N5157	IR424,IR425,*TR-88,*HEPS5021,*RT-115
2N5160	*GE-29,*TR-88,*HEPS3014,*SK3025,*ECG129,*WEP242,*276-2021,*RT-115
2N5161	*GE-69

To Replace	Substitute This Type
2N5162	
2N5172	2N2925,2N3391,2N3391A,2N4256,2N4954,2N5371,2N6002,2N6008,BC338,BC548B, BC549B,BCY58C,BCY58D,MPSA16,*PTC121,*GE-212,*TR-70,*HEPS0016,*SK3124, *ECG123A,*WEP56,*BC107B,*276-2016,*RT-108
2N5174	2N2510,*PTC170,*GE-32,*TR-78,*HEPS5026,*SK3045,*ECG154,*WEP712,*276-2012, *RT-110
2N5175	2N740,2N740A,2N2519,2SC1438,2SC1439,PET1075A,*PTC170,*GE-27,*TR-78, *HEPS5025,*SK3045,*ECG154,*WEP712,*BF338,*276-2012,*RT-110
2N5176	2N4410,*GE-32,*TR-78,*HEPS5025,*SK3045,*ECG154,*WEP712,*276-2012,*RT-110
2N5177	2SC1019
2N5178	
2N5179	2SC988,2SC988A,2SC988B,*PTC132,*GE-86,*TR-70,*HEPS0017,*SK3039,*ECG316, *WEP56,*276-2015,*RT-108
2N5180	2SC988,2SC988A,2SC988B,2SC1426,40894,40895,40896,*PTC132,*GE-60,*TR-95, *HEPS0016,*SK3018,*ECG108,*WEP56,*276-2011,*RT-213
2N5181	40235,40236,40237,40238,40240,40472,40474,40477,40478,40480,*PTC132,*GE-39, *TR-83,*HEPS0025,*ECG161,*WEP719,*276-2038,*RT-113
2N5182	2N5181,40235,40236,40237,40238,40240,40472,40474,40477,40478,40480,*PTC132, *GE-39,*TR-83,*HEPS0016,*ECG161,*WEP719,*276-2038,*RT-113
2N5183	2N3241A,*GE-47,*TR-86,*HEPS3001,*SK3124,*ECG123,*WEP53,*BC337,*276-2009, *RT-100
2N5184	2N720A,2N739,2N739A,2N4068,40354,TIS101,*PTC117,*GE-27,*TR-78,*HEPS5025, *SK3040,*ECG154,*WEP712,*BF338,*276-2012,*RT-110
2N5185	2N1613B,2N3500,2N4069,2N4925,40355,MM3008,*PTC117,*GE-27,*HEPS5025, *SK3040,*BF338,*276-2008
2N5186	2N709A,2N2784,2N2967,2N5187,2SC395A-R,MPS6568,MPSH02,*PTC136,*GE-20, *TR-21,*HEPS0011,*SK3122,*ECG123A,*WEP736,*276-2016,*RT-102
2N5187	2N2710,2N3510,2N3511,2N3646,2N3647,2N3648,2SC577,*PTC136,*GE-20,*TR-21, *HEPS0014,*SK3122,*ECG123A,*WEP736,*276-2016,*RT-102
2N5188	*GE-18,*TR-87,*HEPS3001,*SK3024,*ECG128,*WEP243,*276-2038,*RT-114
2N5189	2N5188,*GE-28,*TR-64,*HEPS3010,*SK3529,*ECG282
2N5190	*PTC163,*GE-57,*HEPS5000,*SK3054,*ECG184,*WEPS5003,*276-2019,*RT-152
2N5191	*PTC163,*GE-57,*HEPS5000,*SK3054,*ECG184,*WEPS5003,*276-2020,*RT-152
2N5192	*GE-55,*HEPS5000,*SK3054,*ECG184,*WEPS5003,*276-2019,*RT-152
2N5193	*PTC162,*GE-58,*HEPS5006,*SK3083,*ECG185,*WEPS5007,*276-2027,*RT-153
2N5194	*PTC162,*GE-58,*HEPS5006,*SK3083,*ECG185,*WEPS5007,*276-2027,*RT-153
2N5195	*GE-56,*HEPS5006,*SK3083,*ECG185,*WEPS5007
2N5200	2N5201,*PTC133,*GE-47,*TR-95,*HEPS0020,*SK3124,*ECG108,*WEP56,*276-2015, *RT-113
2N5201	2N5200,*PTC133,*GE-210,*TR-05,*HEPS0020,*SK3020,*ECG100,*WEP254, *276-2015,*RT-118
2N5202	2N5616,2N5620,*GE-12,*SK3562,*ECG175
2N5208	*GE-21,*TR-20,*HEPS0013,*SK3118,*ECG106,*WEP52,*276-2034,*RT-126
2N5209	2N956,2N2390,2N3302,2N3904,2N4141,2N4952,2N5028,2N5107,2N5369,2N5381, 2N5826,2N6004,2SC366G-Y,2SC734-Y,2SC979-Y,2SC1166-Y,2SC1175,2SC1364, 2SC1813,2SC1850,2SC1992,A5T3904,BC107B,BC174B,BC182B,BC182L,BC190B, BC237B,BC414B,BC547A,BSW85,BSX79,GET930,GET2222,GET2222A,GI-3643, MM3904,TN-3904,*PTC121,*GE-212,*TR-21,*HEPS0025,*SK3122,*ECG123A, *WEP736,*BC337,*276-2013,*RT-304
2N5210	2N4953,2N5370,2N5376,2N5827,2N6006,2N6428,2N6428A,2N6539,BC414C,BC547B, BC550B,*GE-212,*TR-21,*SK3122,*ECG123A,*WEP736,*BC337,*276-2013,*RT-102
2N5214	
2N5215	2N3733,2N5016,2N5102,2SC551,40307,MSA7505,MSA8505,SRF52101,XB404
2N5216	
2N5217	
2N5218	HST7141,HST7152,HST7156,HST7801,HST7802,HST7803,KSP1101,KSP1102, KSP1103,SDT7801,SDT7802,SDT7803
2N5219	2N2925,2N4256,2N5223,2N5420,2N6002,2N6008,2SC735-GR,2SC1993,A5T5219, A158B,A159B,BC108B,BC109B,BC168B,BC172B,BC173B,BC238B,BC239B,BC548B, BC549B,BCY58C,*PTC139,*GE-212,*TR-21,*HEPS0015,*SK3122,*ECG123A, *WEP736,*BC337,*276-2016,*RT-102

To Replace	Substitute This Type
2N5220	2N5420,A5T5220,BC338,BC338-40,*GE-210,*TR-21,*HEPS0015,*SK3122,*ECG289, *WEP736,*BC337,*276-2016,*RT-102
2N5221	2SB598,A5T5221,BC328,BC328-25,BC328-40,*GE-82,*TR-20,*HEPS0012,*SK3114, *ECG290,*WEP242,*BC327,*276-2034,*RT-115
2N5222	A5T5222,*GE-11,*TR-70,*HEPS0020,*SK3039,*ECG107,*WEP56,*276-2016,*RT-108
2N5223	2N2925,2N4256,2N4954,2N5371,2N5420,2N6002,2N6008,BC548B,BC549B,BCY58C, *PTC139,*GE-212,*TR-21,*HEPS0025,*SK3122,*ECG123A,*WEP736,*BC337, *276-2016,*RT-102
2N5224	2N4954,2N5371,A5T4124,BCY58B,BCY58C,*PTC121,*GE-10,*TR-21,*HEPS0025, *SK3122,*ECG123A,*WEP736,*276-2016,*RT-102
2N5225	2N3415,2N3706,2N4954,2N5371,2N5420,2N5451,2N6002,2N6008,A5T5225,A5T5451, A8T3706,BC338,BC338-40,MPS3706,*GE-210,*TR-21,*HEPS0015,*SK3122, *ECG123A,*WEP736,*BC337,*276-2016,*RT-102
2N5226	2N5378,2N5379,2N5813,2N6003,2N6009,2SB598,A5T5226,BC328,BC328-25, BC328-40,D29E2,*GE-82,*TR-20,*HEPS0019,*SK3114,*ECG159,*WEP242,*BC327, *276-2022,*RT-115
2N5227	2N3964,2N5087,2N5367,2N5378,2N5379,2N6003,2N6007,2N6009,A5T5087,A5T5227, BC177B,BC251B,BC251C,BC257B,BC261B,BC261C,BC307B,BC307C,BC308B, BC308C,BC327,BC415-10,BC415-16,BC415B,BC415C,BC416-10,BC416-16,BC416B, BC416C,BC557B,BC558B,BC558C,BC559B,BC559C,BC560B,BC560C,BCY78C, BCY78D,BCY79C,*GE-212,*TR-20,*HEPS0019,*SK3114,*ECG159,*WEP242, *BC337,*276-2016,*RT-115
2N5228	2N2894A,2N3012,2N3546,2N3639,2N3640,2N4257A,2N4258A,2N5056,2N5057, 2N5292,MM2894,MPS3639,*PTC103,*GE-65,*TR-20,*ECG106,*WEP52,*BC327, *276-2021,*RT-126
2N5229	2N5230,*GE-22,*TR-19,*BC327,*276-2032
2N5230	2N5231,*GE-82,*TR-19,*BC327
2N5231	*GE-82,*BC327
2N5232	2N5232A,2N5249,2N5249A,2N5311,*GE-212,*TR-21,*HEPS0015,*SK3124,*ECG123A, *WEP736,*276-2012,*RT-102
2N5232A	2N5232,2N5249,2N5249A,2N5311,*GE-212,*TR-21,*HEPS0015,*SK3122,*ECG123A, *WEP736,*276-2012,*RT-102
2N5233	2N871,2N956,2N2390,2N2645,2N6540,2N6541,BC546A,TIS95,TIS98,*PTC170, *GE-81,*TR-87,*SK3024,*ECG128,*WEP243,*276-2012,*RT-114
2N5234	2N5235,*GE-18,*TR-87,*HEPS0005,*SK3024,*ECG128,*WEP243,*276-2012,*RT-114
2N5235	*GE-18,*TR-87,*SK3024,*ECG128,*WEP243,*276-2012,*RT-114
2N5236	2N3137,2N3326,2N5769,2SC651,2SC1360,*PTC136,*HEPS3013,*276-2038
2N5237	2N5238
2N5238	
2N5239	2N5240,2N5804,2N5805,*PTC129A,*GE-73,*HEPS5020,*SK3217,*ECG283
2N5240	2N5805,2N6307M,*PTC129A,*GE-73,*TR-81,*HEPS5020,*ECG283,*WEP240
2N5241	2N6308M,2N6513,2N6560,2N6573,2N6579,2N6580,2N6582,2N6583,DTS-430, DTS-431,FT430,FT431,IR430,IR431,MJ413,MJ431,SDT13303,SDT13304,SDT13305, STS1134,*PTC129A,*GE-73,*HEPS5020,*ECG283,*BU108
2N5242	2N5023,2N5243,*GE-48,*TR-88,*HEPS0012,*ECG129,*WEP242,*276-2021,*RT-115
2N5243	2N3245,2N3468,2N3762,2N3762A,2N3764,2N3764A,2N5022,2N5023,MM3726,*GE-48, *TR-88,*HEPS0019,*ECG129,*WEP242,*276-2021,*RT-115
2N5244	2N4035,NPS404A,*HEPS0013,*276-2023
2N5249	2N5249A,*GE-212,*TR-21,*SK3024,*276-2012,*RT-114
2N5249A	2N5249,*GE-212,*TR-87,*HEPS0015,*SK3024,*ECG199,*WEP736,*276-2012, *RT-114
2N5250	
2N5251	
2N5262	2N4961,*GE-66,*HEPS3010,*SK3529,*ECG282
2N5263	
2N5264	*TR-67,*HEPS5020,*ECG162,*WEP707
2N5279	*GE-32
2N5280	
2N5281	
2N5282	
2N5284	2N5002

To Replace	Substitute This Type
2N5285	2N5004
2N5286	2N4999
2N5287	2N5001
2N5288	
2N5289	
2N5290	2N5007
2N5291	
2N5292	2N2894A,2N3546,2N4258A,2N5056,2N5057,*TR-95,*HEPS0020,*ECG108,*WEP56, *276-2015,*RT-113
2N5293	*PTC169,*GE-66,*TR-76,*HEPS5000,*SK3054,*ECG152,*WEP245,*276-2020
2N5294	*PTC169,*GE-66,*TR-76,*HEPS5000,*SK3054,*ECG152,*WEP245,*276-2020
2N5295	2N5293,2N5297,2N6291,2N6293,*PTC167,*GE-226,*TR-76,*HEPS5000,*SK3054, *ECG152,*WEP245,*276-2020,*RT-197
2N5296	2N5294,2N5298,2N6122,2N6130,2N6131,2N6290,2N6292,NSP5191,*PTC167,*GE-28, *TR-76,*HEPS5000,*SK3054,*ECG152,*WEP245,*276-2020,*RT-197
2N5297	2N5293,2N6293,*PTC167,*GE-66,*TR-76,*HEPS5000,*SK3054,*ECG152,*WEP245, *276-2020
2N5298	2N5294,2N6123,2N6131,2N6292,2N6473,NSP5192,*PTC167,*GE-66,*TR-76, *HEPS5000,*SK3054,*ECG152,*WEP245,*276-2020
2N5301	2N5302,2N5685,IR3771,*PTC175,*GE-75,*HEPS7000,*ECG181
2N5302	2N5685,2N5686,*PTC175,*GE-75,*HEPS7000,*ECG181
2N5303	2N5886,IR3772,*PTC175,*GE-75,*HEPS7000,*ECG181
2N5304	*276-2011
2N5305	2N5307,MPS5305,MPS5307,*TR-69,*ECG172,*WEPS9100
2N5306	2N5306A,2N5308,2N5308A,MPS5306,MPS5306A,MPS5308,MPS5308A,*PTC153, *TR-69,*HEPS9100,*SK3156,*ECG172,*WEPS9100
2N5306A	2N5306,2N5308,2N5308A,MPS5306,MPS5306A,MPS5308,MPS5308A,*GE-64
2N5307	MPS5307,*TR-69,*ECG172,*WEPS9100
2N5308	2N5308A,MPS5308,MPS5308A,*PTC153,*TR-69,*HEPS9100,*SK3156,*ECG172, *WEPS9100,*276-2009
2N5308A	2N5308,MPS5308,MPS5308A
2N5309	2N736A,2N736B,2N910,2N915,2N2516,2N4014,2N4963,2SC1998,GET929,TIS96, TIS99,TIS135,TIS136,*PTC121,*GE-62,*TR-87,*HEPS5026,*SK3024,*ECG128, *WEP243,*BF338,*276-2012,*RT-114
2N5310	2N871,2N956,2N2390,2N2645,2N4409,2N6540,2N6541,A5T4409,BC546A,BSW85, GET930,GET2222A,TIS95,TIS98,*PTC121,*GE-62,*TR-87,*HEPS5026,*SK3024, *ECG128,*WEP243,*276-2012,*RT-114
2N5311	2N5232,2N5232A,2N5234,2N5235,2N5249,2N5249A,*GE-212,*TR-87,*HEPS0005, *SK3024,*ECG128,*WEP243,*276-2012,*RT-114
2N5312	2N5314,2N5316,2N5318,2N5386,2N6182,2N6184,2N6186,2N6188,KSP2151,KSP2152, KSP2154,KSP2155,SDT3103,SDT3104,SDT3107,SDT3108
2N5313	2N5315
2N5314	2N5318,2N5386,2N6184,2N6188,KSP2152,KSP2155,SDT3104,SDT3108,SDT3109
2N5315	HST7140
2N5316	2N5312,2N5314,2N5318,2N5386,2N6182,2N6184,2N6186,2N6188,KSP2151,KSP2152, KSP2154,KSP2155,SDT3103,SDT3104,SDT3107,SDT3108
2N5317	2N4301,2N5006,2N5313,2N5315,2N5319,2N5412,2N6128,KSP1151,KSP1152, KSP1154,KSP1155,SDT3203,SDT3204,SDT3207,SDT3208
2N5318	2N5314,2N5386,2N6184,2N6188,KSP2152,KSP2155,SDT3104,SDT3108,SDT3109
2N5319	2N4301,2N5006,2N5288,2N5315,2N6128,HST7140,KSP1152,KSP1155,SDT3201, SDT3204,SDT3208,SDT3209
2N5320	*TR-25,*HEPS3002,*SK3512,*276-2009
2N5321	*GE-28,*HEPS3010,*SK3047,*ECG195A,*RT-306
2N5322	*GE-28,*TR-20,*HEPS3003,*SK3513,*ECG129,*WEP52,*RT-126
2N5323	*HEPS3003,*SK3513
2N5324	2N5325,*GE-25,*HEPG6007,*276-2006
2N5325	*GE-25,*HEPG6008
2N5326	2N4116,HST6313,HST6413,SDT6313
2N5327	
2N5328	
2N5329	2N5330

To Replace	Substitute This Type
2N5330	
2N5331	
2N5332	MPSH81,*HEPS0013,*276-2023
2N5333	
2N5334	2N5335,*GE-66,*HEPS3010
2N5335	*HEPS3002
2N5336	2N5338,*HEPS3002
2N5337	2N5339,*HEPS3002
2N5338	
2N5339	
2N5344	2N5345
2N5345	
2N5346	2N5347,2N5477,2N5478
2N5347	2N5346,2N5477,2N5478
2N5348	2N5349,2N5479,2N5480
2N5349	2N5348,2N5479,2N5480
2N5354	2N869A,2N5365,*PTC103,*GE-22,*TR-20,*HEPS0012,*SK3114,*ECG159,*WEP242, *276-2023,*RT-115
2N5355	2N5366,2N6001,TN-3906,*PTC103,*GE-48,*TR-20,*HEPS0019,*SK3114,*ECG159, *WEP242,*276-2023,*RT-115
2N5356	2N5367,2N6003,*GE-89,*TR-20,*HEPS0019,*SK3114,*ECG159,*WEP242,*276-2023, *RT-115
2N5357	
2N5365	2N4026,*GE-21,*TR-20,*HEPS0019,*SK3114,*ECG159,*WEP242,*276-2023,*RT-115
2N5366	2N4028,2N6005,TN-3906,*PTC103,*GE-48,*TR-20,*HEPS0019,*SK3114,*ECG159, *WEP242,*276-2023,*RT-115
2N5367	2N6007,*GE-89,*TR-20,*HEPS0019,*SK3114,*ECG159,*WEP242,*276-2023,*RT-115
2N5368	2N2540,2N5369,2N5582,A5T2222,BSW85,TIS111,TN-3904,*GE-210,*TR-21, *HEPS0015,*SK3122,*ECG123A,*WEP736,*276-2009,*RT-102
2N5369	2N2540,2N5377,A5T2222,BSW85,TIS109,TIS111,TN-3904,*GE-210,*TR-21, *HEPS0015,*SK3122,*ECG123A,*WEP736,*276-2009,*RT-102
2N5370	*GE-210,*TR-21,*HEPS0015,*SK3122,*ECG123A,*WEP736,*276-2009,*RT-102
2N5371	2N5370,2N5376,2N6012,*GE-210,*TR-21,*HEPS0014,*SK3122,*ECG123A,*WEP736, *276-2009,*RT-102
2N5372	2N4026,2N4027,*PTC103,*GE-82,*TR-20,*HEPS5022,*SK3114,*ECG159,*WEP242, *276-2023,*RT-115
2N5373	2N4028,2N4029,2N5374,2SA817,2SA891,A5T2907,A5T3505,A5T3645,TIS112, *PTC103,*GE-82,*TR-20,*HEPS5022,*SK3114,*ECG159,*WEP242,*276-2023, *RT-115
2N5374	*GE-48,*TR-20,*HEPS5022,*SK3114,*ECG159,*WEP242,*276-2023,*RT-115
2N5375	2N4028,2N5373,2N5374,2N5378,2N5379,2N5819,A5T2907,A5T3504,A5T3505, A5T3644,A5T3645,A5T4403,D29E6,TIS112,*GE-82,*TR-20,*HEPS0019,*SK3114, *ECG159,*WEP242,*276-2023,*RT-115
2N5376	2N5370,*HEPS0015,*276-2009
2N5377	2N5370,2N5376,*HEPS0015,*276-2009
2N5378	2N5379,*GE-48,*TR-20,*HEPS0019,*SK3114,*ECG159,*WEP242,*276-2023,*RT-115
2N5379	2N5378,*GE-48,*TR-20,*HEPS0019,*SK3114,*ECG159,*WEP242,*276-2023,*RT-115
2N5380	2N4962,2N4963,A5T3903,TIS135,TIS136,TN-3903,*GE-210,*TR-21,*HEPS0015, *SK3122,*ECG123A,*WEP736,*276-2009,*RT-102
2N5381	A5T3904,TIS111,TN-3904,*GE-20,*HEPS0015,*276-2009
2N5382	2SA891,A5T3905,*PTC103,*GE-21,*TR-20,*HEPS0019,*SK3114,*ECG159,*WEP242, *276-2023,*RT-115
2N5383	A5T3906,*GE-48,*TR-20,*HEPS0019,*SK3114,*ECG159,*WEP242,*276-2023, *RT-115
2N5384	2N5409,KSP2033,SDT3128,SDT3129,SDT3308,SDT3309
2N5385	2N6184,2N6188
2N5386	
2N5387	2N5388
2N5388	2N5389
2N5389	
2N5399	2N5769,*PTC133,*GE-20,*TR-95,*HEPS0020,*SK3039,*ECG108,*WEP56,*276-2038,

To Replace	Substitute This Type
(2N5399)	*RT-113
2N5400	A5T5400,*ECG288
2N5401	A5T5401,*GE-221,*SK3114,*ECG288
2N5404	2N5405,*HEPS3003
2N5405	
2N5406	2N5407,*HEPS3003
2N5407	
2N5408	2N5409,KSP2032,SDT3307,SDT3308
2N5409	KSP2033,SDT3308,SDT3309
2N5410	2N5411,KSP2035,SDT3303,SDT3304
2N5411	KSP2036,SDT3304
2N5412	
2N5413	2N5414,*HEPS3001,*276-2014
2N5414	*HEPS3001,*276-2014
2N5415	*SK3053
2N5416	*SK3528
2N5417	2N2219,2N2538,2N2540,2N2959,2N2960,2N2961,2N3300,2N3643,2N5106,2N5369, 2N6010,A5T2222,BSW83,TIS111,TN-3904,*PTC136,*GE-20,*HEPS3008,*276-2009
2N5418	BSW82,*GE-210,*TR-21,*HEPS0025,*SK3122,*ECG123A,*WEP736,*276-2009, *RT-102
2N5419	2N5371,BSW83,*GE-210,*TR-21,*HEPS0025,*SK3122,*ECG123A,*WEP736, *276-2009,*RT-102
2N5420	*GE-88,*TR-21,*HEPS0024,*SK3122,*ECG123A,*WEP736,*276-2009,*RT-102
2N5421	2N4427,2N4874,2N5109,2N5422,2SC302-M,2SC307-M,2SC590N,2SC730,2SC731, 2SC844,2SC872,2SC890,2SC908,2SC991,2SC992,2SC1001,2SC1002,2SC1083, 2SC1165,2SC1256,2SC1464,2SC1479,2SC1820,2SC1947,40280,40975,SRF1001, *GE-28,*TR-87,*HEPS3001,*SK3024,*ECG128,*WEP243,*276-2014,*RT-114
2N5422	2SC302-M,2SC307-M,2SC590N,2SC992,2SC1002,2SC1256,2SC1947,40975,*GE-28, *HEPS3001,*276-2014
2N5423	2N3926,2N4932,2N5424,2N5424A,2SC831,2SC1003,SRF12213,SRF32214,SRF52214, *GE-28
2N5424	2N5102,SRF32214,SRF52214,*GE-66
2N5424A	2N5102,2N5424,SRF32214,SRF52214
2N5427	2N5429,KSP1164,KSP1165,*HEPS5012,*276-2017
2N5428	2N5430,*HEPS5012,*276-2017
2N5429	KSP1165,*HEPS5012,*276-2017
2N5430	*HEPS5012,*276-2017
2N5435	2N5436,*PTC901
2N5436	2N5437
2N5437	
2N5438	2N5439,*PTC901
2N5439	2N5440
2N5440	
2N5447	2N2838,2N2907,2N2907A,2N3251,2N3251A,2N3486,2N3486A,2N3504,2N3505, 2N3672,2N3673,2N3702,2N3962,2N5373,2N5374,2N5375,2N5379,2N5383,2N5817, 2N5819,2N6005,2SA659,2SA661-GR,2SA661-Y,2SA720,2SA891,A5T2907,A5T3504, A5T3505,A5T3644,A5T3645,A5T3906,A5T4403,A5T5447,A8T3702,BC327-16, BC327-25,BCW35,BCY79,BCY79A,BCY79B,BSW73,D29E5,D29E6,GET2906, GET2907,MM3906,TIS112,TN-3906,TQ60,TZ552,TZ553,*PTC103,*GE-67,*TR-20, *HEPS0019,*SK3114,*ECG159,*WEP242,*276-2024,*RT-115
2N5448	2N2838,2N2906,2N2906A,2N3250,2N3250A,2N3485,2N3485A,2N3672,2N3673, 2N3703,2N5372,2N5815,2N5817,2N5821,2N5823,2SA661-O,A5T5448,A8T3703, BSW74,D29E10,D29E5,GET2904,GET2905,TZ551,*PTC103,*GE-67,*TR-20, *HEPS0019,*SK3114,*ECG159,*WEP242,*276-2034,*RT-115
2N5449	2N2222,2N2222A,2N2540,2N3302,2N3704,2N5818,2N6112,A5T2222,A5T5449, A8T3704,D33D26,TIS109,TIS111,TN-3904,*GE-20,*TR-87,*HEPS3001,*SK3024, *ECG128,*WEP243,*BC337,*276-2014,*RT-114
2N5450	2N915,2N2539,2N2897,2N2900,2N3705,2N4962,2N5822,40458,A5T5450,A8T3705, D33D25,D33D30,TIS110,TN54,TN-3903,*GE-20,*TR-87,*HEPS3001,*SK3024, *ECG128,*WEP243,*BC337,*276-2014,*RT-114
2N5451	2N3706,2N6012,2N6112,A5T5451,A8T3706,BC337-40,D33D27,*GE-20,*TR-87,

To Replace	Substitute This Type
(2N5451)	*HEPS3001,*SK3024,*ECG128,*WEP243,*BC337,*276-2014,*RT-114
2N5466	BDY28A,*TR-67,*HEPS5021,*ECG163,*WEP740
2N5467	*TR-93,*HEPS5021,*ECG165,*WEP740B
2N5468	
2N5469	
2N5470	40837,40909†
2N5477	2N5346,2N5347,2N5478
2N5478	2N5346,2N5347,2N5477
2N5479	2N5348,2N5349,2N5480
2N5480	2N5348,2N5349,2N5479
2N5481	2N5482,2N5766,2N5767,2N5768,*GE-28,*HEPS3005,*276-2020
2N5482	2N5483,2N5768,*GE-28,*HEPS3005,*276-2020
2N5483	2N5768,*GE-66,*HEPS3006,*276-2020
2N5489	*GE-28
2N5490	2N5492,2N5494,*PTC169,*GE-66,*TR-92,*HEPS5004,*SK3054,*ECG196,*WEP245, *276-2020
2N5491	2N5493,2N5495,*PTC169,*GE-66,*TR-92,*HEPS5004,*SK3054,*ECG196,*WEP245, *276-2020
2N5492	2N5496,*PTC169,*GE-28,*TR-76,*HEPS5004,*SK3054,*ECG152,*WEP245,*276-2020
2N5493	2N5497,*PTC169,*GE-66,*TR-92,*HEPS5004,*SK3054,*ECG196,*WEP245,*276-2020
2N5494	2N5490,2N5492,*PTC169,*GE-66,*TR-92,*HEPS5004,*SK3054,*ECG196,*WEP245, *276-2020
2N5495	2N5491,2N5493,*PTC169,*GE-66,*TR-92,*HEPS5004,*SK3054,*ECG196,*WEP245, *276-2020
2N5496	*PTC169,*GE-28,*TR-92,*HEPS5004,*SK3054,*ECG196,*WEP245,*276-2020
2N5497	*PTC169,*GE-28,*TR-92,*HEPS5004,*SK3054,*ECG196,*WEP245,*276-2020
2N5498	
2N5525	*TR-69,*HEPS9100,*SK3156,*ECG172,*WEPS9100,*276-2009
2N5526	*276-2009
2N5527	*HEPS3002
2N5528	
2N5529	2N3621,2N3622,2N3625,2N3626,2N5528,2N5530
2N5530	2N3621,2N3622,2N3625,2N3626,2N5528,2N5529
2N5531	*HEPS3002
2N5532	
2N5533	2N5532,2N5534
2N5534	2N5532,2N5533
2N5535	2N5536
2N5536	2N5535
2N5537	2N5538
2N5538	2N5537
2N5539	1768-1610,1768-1620,1768-1630,1768-1810,HST8801,HST8802,KSP1201,SDT8753, SDT8754,SDT8801,SDT8802
2N5540	HST7803,HST7804,HST7805,KSP1103,KSP1104,KSP1105,SDT7803,SDT7804, SDT7805
2N5541	STT2402
2N5542	2N5288,2N5329,HST7140,HST7154,HST7155,KSP1156,SDT3201,SDT3209
2N5550	2N5551,A5T5550,A5T5551,*GE-212,*TR-78,*HEPS0005,*SK3024,*ECG287, *276-2008
2N5551	A5T5551,*GE-220,*HEPS0005,*ECG287,*276-2008
2N5559	2N5631,2N5634,2SD165,2SD166,1561-1404,1561-1405,1561-1604,1561-1605, 1561-1803,1561-1804,1561-1805,1582-1403,1582-1404,1582-1405,1582-1603, 1582-1604,1582-1605,1582-1803,1582-1804,1582-1805,HST9204,HST9209,IR3773, IR6302,KSP1273,SDT9204,SDT9209,STC2222,STC2223,*PTC118,*HEPS5020
2N5575	2N5576,2N5577
2N5576	2N5575,2N5577
2N5577	2N5575,2N5576
2N5578	2N5579,2N5580
2N5579	2N5578,2N5580
2N5580	2N5578,2N5579
2N5581	*GE-20,*HEPS3001,*276-2009

To Replace	Substitute This Type
2N5582	*GE-20,*HEPS3001,*276-2009
2N5583	*HEPS3001,*276-2014
2N5584	
2N5587	2N5588
2N5588	
2N5589	2N5945,2N5995†,2SC891,2SC892,2SC1011,2SC1040,2SC1257,2SC1405,2SC1495, 2SC1606,3TX601,41010†,MSA8508,*TR-76,*HEPS3005,*276-2009
2N5590	2N5646,2N5713,2N5946,2N5994,2SC1081,2SC1118,2SC1258,2SC1528,3TX602, 40893†,MSA8507,SRF54215,*PTC128,*GE-66,*HEPS3006,*SK3176
2N5591	2N5705,2N6082,2N6083,2N6136,MM1603,MSA8506,ZT5591,*HEPS3007,*SK3177
2N5595	2N5596,2N5925,2N6207,41026†
2N5596	2N6208,2SC1334
2N5597	2N5601,*PTC113,*GE-69,*HEPS5018,*276-2025
2N5598	2N5602,*PTC112,*GE-66,*TR-81,*HEPS5012,*SK3131,*ECG175,*WEP241,*276-2017
2N5599	2N5603,*HEPS5018,*276-2025
2N5600	2N5202,2N5604,*PTC104,*TR-81,*HEPS5012,*SK3131,*ECG175,*WEP241, *276-2017
2N5601	*HEPS5018,*276-2025
2N5602	*TR-81,*HEPS5012,*SK3131,*ECG175,*WEP241,*276-2017
2N5603	
2N5604	2N3441,HST6905,*PTC104,*TR-81,*HEPS5012,*SK3131,*ECG175,*WEP241, *276-2017
2N5605	2N5609,2N5613,2N5617,*HEPS5005
2N5606	2N5610,2N5614,2N5618,*GE-66,*HEPS5012
2N5607	2N5611,2N5615,2N5619,KSP2023,KSP2026,KSP2162,KSP2165,*HEPS5005
2N5608	2N5612,2N5616,2N5620,KSP1022,KSP1025,KSP1162,KSP1165,*HEPS5012, *276-2017
2N5609	2N5617
2N5610	2N5618,*HEPS5012,*276-2017
2N5611	2N5619,KSP2023,KSP2026,KSP2163,KSP2166
2N5612	2N5620,KSP1023,KSP1026,KSP1163,KSP1166,*HEPS5012,*276-2017
2N5613	2N5617,*HEPS7003
2N5614	2N5618,*GE-66,*TR-59,*HEPS7002,*ECG130,*WEP247,*RT-131
2N5615	2N5619
2N5616	2N5620,*HEPS7004,*SK3561
2N5617	
2N5618	*HEPS7004,*SK3561
2N5619	
2N5620	*SK3561
2N5621	2N5625,*HEPS7003,*ECG180,*WEPS7001,*RT-148
2N5622	2N5626,*TR-59,*HEPS7002,*ECG130,*WEP247,*RT-131
2N5623	2N5627,KSP2272,KSP2275,*ECG180,*WEPS7001,*RT-148
2N5624	2N5628,KSP1275,*HEPS7004,*SK3561
2N5625	*ECG180,*WEPS7001,*RT-148
2N5626	2N5039,*HEPS7004
2N5627	KSP2273,KSP2276
2N5628	KSP1276
2N5629	2N5630,2SD113-O,*PTC173,*GE-75
2N5630	*PTC173,*ECG284
2N5631	IR3773,IR6302
2N5632	2N3714,2N3716,2N5629,2N5630,2N5633,*PTC173,*GE-75,*TR-67,*HEPS5020, *ECG181,*WEP707
2N5633	2N5630,*PTC173,*TR-67,*HEPS5020,*ECG162,*WEP707
2N5634	2N5631,IR3773,IR6302,*TR-67,*HEPS5020,*ECG284,*WEP707
2N5635	2N5636,2N5918†,2N6203,3TX632,40940†,*HEPS3006
2N5636	2N5919,2N5919A†,2N6203,*HEPS3006,*276-2009
2N5637	2N6105†,2N6204,*GE-66,*HEPS3007
2N5641	2N5918†,*HEPS3005,*276-2009
2N5642	2N6199,*PTC128,*GE-66
2N5643	2N6200,MSA8503
2N5644	2N5645,2N5698,2N5914,2N5944,2N5945,2N5946,2SC891,2SC892,2SC911,2SC911A,

To Replace	Substitute This Type
(2N5644)	2SC1016,2SC1040,2SC1495,2SC1608,2SC1620,40934†,41009A†,41010†,*GE-28, *HEPS3005,*276-2009
2N5645	2N5945,2N5946,2SC892,2SC1040,2SC1081,41010†,*PTC128,*GE-28,*HEPS3006
2N5646	2SC1081,2SC1118,40893†,*PTC128,*GE-66,*HEPS3006
2N5650	2N5651,2N5652,A406,*GE-211,*TR-83,*HEPS0017,*SK3117,*ECG161,*WEP719, *276-2011,*RT-113
2N5651	2N5650,2N5652,A406,*GE-211,*TR-83,*HEPS0017,*SK3117,*ECG161,*WEP719, *276-2011,*RT-113
2N5652	2N5650,2N5651,A406,*GE-86,*TR-83,*HEPS0017,*SK3039,*ECG161,*WEP719, *276-2011,*RT-113
2N5655	2N5656,*PTC124,*GE-32,*TR-60,*HEPS5015,*SK3103,*ECG157,*WEP244,*RT-135
2N5656	2N5657,*PTC124,*GE-232,*TR-60,*HEPS5015,*SK3103,*ECG157,*WEP244,*RT-135
2N5657	*PTC124
2N5660	*PTC104,*GE-32,*HEPS3021
2N5661	*GE-32
2N5662	*GE-32,*HEPS3021
2N5663	*GE-32
2N5664	2N3583
2N5665	2N3584,2N3585
2N5666	
2N5667	
2N5671	2N5672,2N6271,2N6273,HST9203,HST9204,HST9804
2N5672	HST9204
2N5675	
2N5679	2N5322,2N5675,2N5680,*HEPS3032
2N5680	2N5675
2N5681	2N5154,2N5682,*GE-32,*HEPS3019,*ECG282
2N5682	*GE-32,*HEPS3019,*SK3024,*ECG282
2N5683	2N5684
2N5684	
2N5685	2N5686
2N5686	
2N5687	2N749,2N751,2N1388,2N1390,2N1491,2N5913,2SC302-M,2SC307-M,2SC590N, 2SC730,2SC773,2SC774,2SC798,2SC803,2SC865,2SC992,2SC1001,2SC1002, 2SC1223,2SC1256,2SC1820,2SC1947,40953,40972,*PTC158,*GE-28,*HEPS3001, *276-2017
2N5688	2SC1176,*GE-28,*HEPS3005,*276-2017
2N5689	2SC1177,2SC1605,2SC1605A,*GE-66,*276-2020
2N5690	2SC1495,MM1619,MM8012,ZT5591,*GE-66,*276-2020
2N5691	2SC1744
2N5692	*PTC901
2N5693	2N5694,*PTC901
2N5694	2N5695
2N5695	2N5696
2N5696	
2N5697	2N3948,2N4876,2SC731,2SC872,2SC890,2SC908,2SC1001,2SC1002,2SC1165, 40964,40965,BLX89,*PTC158,*GE-28,*HEPS3013,*276-2017
2N5698	2N5644,2N5645,2N5914,2N5945,2N5946,2SC891,2SC892,2SC911,2SC911A, 2SC1016,2SC1040,2SC1608,2SC1620,3TX820,40934†,41009A†,41010†,*GE-28, *276-2017
2N5699	2N5700,2SC973,2SC973A,2SC974,2SC975,2SC975A,2SC1074,2SC1075,2SC1337, 2SC1488,*GE-28,*HEPS3006,*276-2017
2N5700	2N5701,2SC990,2SC1015,2SC1076,2SC1338,2SC1338A,2SC1404,2SC1488, 2SC1489,*GE-66,*HEPS3007,*276-2019
2N5701	2SC1339,2SC1488,2SC1489,*GE-66,*HEPS3007,*276-2019
2N5702	2N3924,2N5422,2SC302-M,2SC307-M,2SC590N,2SC730,2SC992,2SC1001,2SC1002, 2SC1256,2SC1820,2SC1947,40290,*GE-18,*HEPS3013,*276-2008
2N5703	2SC599,2SC688,2SC1176,BLY62,*GE-28,*HEPS3006,*276-2017
2N5704	2SC1177,2SC1605,2SC1605A,BLY63,*GE-66,*HEPS3006,*276-2018
2N5705	2N5591,2N6083,2N6136,MM1603,MSA8506,ZT5591,*GE-66,*HEPS3007,*276-2019
2N5706	2SC1744,40971†

To Replace	Substitute This Type
2N5707	2SC1043
2N5708	2SC1043
2N5709	2N5941,2SC1763
2N5710	2N751,2N1491,2N1493,2N4427,2N4874,2N5109,2N5687,2N5913,2SC302-M, 2SC307-M,2SC590N,2SC730,2SC731,2SC741,2SC844,2SC854,2SC855,2SC865, 2SC872,2SC890,2SC908,2SC991,2SC992,2SC1001,2SC1002,2SC1083,2SC1165, 2SC1256,2SC1303,2SC1315,2SC1464,2SC1479,2SC1820,2SC1947,40953,40972, 40975,40976,*PTC158,*GE-28,*HEPS3013,*276-2017
2N5711	*276-2017
2N5712	*GE-66,*276-2020
2N5713	2N5646,2N5994,2N5996,2SC1081,2SC1118,2SC1258,2SC1528,3TX602,40893†, MSA8507,SRF54215,*GE-66,*276-2020
2N5714	2SC1191
2N5715	2N5766,2N5767,2SC1336,2SC1559,*PTC128
2N5729	*HEPS3002
2N5730	
2N5731	
2N5732	2N5039,2N5734,*HEPS7000
2N5733	
2N5734	
2N5737	2N5741,2SA746,2SA877,KSP2171,KSP2271,*PTC149,*HEPS7003
2N5738	2N5623,2N5627,2N5742,2SA747,2SA878,KSP2172,KSP2272,*PTC172,*HEPS5005, *ECG180,*WEPS7001,*RT-148
2N5739	2N5743,KSP2161,*PTC168,*HEPS5005
2N5740	2N5744,KSP2162,*PTC168,*HEPS5005
2N5741	KSP2271,*PTC172,*ECG180,*WEPS7001,*RT-148
2N5742	KSP2272,*PTC172,*ECG180,*WEPS7001,*RT-148
2N5743	
2N5744	
2N5745	2N5884,2N6436,*PTC172,*ECG180,*WEPS7001,*RT-148
2N5758	2N3714,2N3716,2N5632,2N5633,2N5759,*PTC173,*WEP707
2N5759	2N5633,*WEP707
2N5760	2N5634,*WEP707
2N5761	2SC987,2SC1272,2SC1273
2N5762	2SC985,2SC1038,2SC1039,2SC1041,2SC1042
2N5763	2N2906,2N2906A,2N3485,2N3485A,2N4026,2N4027,*PTC127,*GE-21,*HEPS5022, *276-2021
2N5764	2N5595,2N5765,2N5923,2N5924,2N5925,2N6206,2N6207,41025†,41026†,MM4430, MT5764,MT5765,*PTC128,*GE-28,*HEPS3006,*276-2020
2N5765	2N5595,2N5924,2N5925,2N6207,41026†,MT5765,*GE-66,*HEPS3006,*276-2020
2N5766	2N5767,2SC1559,*PTC128,*GE-28,*276-2020
2N5767	2SC1559,*PTC128,*GE-28,*276-2020
2N5768	*PTC128,*GE-66,*276-2020
2N5769	*TR-21,*HEPS0033,*SK3040,*ECG123A,*WEP736,*276-2011
2N5770	*TR-21,*276-2011
2N5771	*276-2021
2N5772	2N1708A,*276-2011
2N5781	2N5151,2N6190,2N6192,*HEPS3003,*SK3025
2N5782	2N5781,2N6190,MJ8100,MJ8101,*GE-29,*HEPS3003,*SK3025
2N5783	2N5782,MJ8100,MJ8101,*GE-29,*HEPS5013,*SK3025,*276-2025
2N5784	2SC756,*GE-18,*HEPS3002,*SK3024
2N5785	2N5784,*GE-216,*HEPS3010,*SK3024,*276-2020
2N5786	2N5785,*GE-46,*HEPS5013,*SK3024,*276-2020
2N5804	2N5805,*PTC129A,*HEPS5020,*ECG283
2N5805	*PTC129A,*HEPS5020,*ECG283
2N5810	2N6010,BC337-16,D33D25,*GE-20,*HEPS0014,*BC337,*276-2009
2N5811	2N2838,2N5815,2N5817,BC327-16,D29E1,D29E5,*GE-48,*HEPS0012,*BC327, *276-2021
2N5812	2N6012,D33D27,*GE-47,*HEPS0015,*BC337,*276-2009
2N5813	2N5819,2N6013,D29E2,D29E6,D29E7,*GE-48,*HEPS0019,*276-2021
2N5814	2N5816,*GE-20,*HEPS0015,*BC337,*276-2009

To Replace	Substitute This Type
2N5815	2N2838,2N4026,2N5817,2N5821,2N5823,D29E10,D29E5,*GE-48,*TR-54,*HEPS0019, *BC327,*276-2021
2N5816	*GE-20,*HEPS0015,*BC337,*276-2009
2N5817	2N2838,2N4028,2N5819,2N5823,2SA817,D29E10,D29E5,D29E6,*GE-48,*HEPS0019, *276-2021
2N5818	2N6012,2N6016,D33D26,D33D27,TIS109,*GE-47,*HEPS0015,*ECG123A,*BC337, *276-2009
2N5819	2N4028,2N6013,2N6017,D29E6,D29E7,*GE-48,*HEPS0019,*276-2021
2N5820	2N4963,2N5822,D33D30,TN54,*GE-63,*TR-87,*HEPS5014,*SK3024,*ECG128, *WEP243,*276-2018
2N5821	2N4027,2N5823,D29E10,*GE-67,*TR-28,*HEPS5013,*SK3025,*ECG129,*WEP242, *276-2025
2N5822	2N5582,2N6014,D33D30,*GE-63,*HEPS5014,*276-2018
2N5823	2N4029,2SA817,2SB560,D29E10,*GE-67,*HEPS5013,*276-2025
2N5824	2N915,2N2847,2N3858A,2N3859A,2N3974,2N3976,2N4013,2N4951,2N4962,2N5144, 2N5368,2N5380,2N5825,2N6538,2SC1166-O,2SC1999,A5T3903,GET929,TIS133, TIS134,TN-3903,TP3705,*PTC121,*GE-62,*HEPS0025,*BC337,*276-2009
2N5825	2N3116,2N3302,2N3859A,2N3947,2N4952,2N5107,2N5369,2N5381,2N5826,2N6004, 2SC1166-Y,2SC1175,2SC1318,2SC1781,2SC1813,2SC1850,2SC1852,2SC1991, A5T3904,BC547A,BCW34,BSW85,BSX79,GET930,GET2222,GET2222A,GI-3643, MM3904,TN-3904,*PTC121,*GE-62,*HEPS0025,*BC337,*276-2009
2N5826	2N5376,2N5377,2N5827,2N6006,2N6112,2SC1166-GR,2SC1850,BC547B,BC550B, TIS94,TIS97,TP3704,*PTC121,*GE-62,*HEPS0015,*BC337,*276-2009
2N5827	2N6428,2N6428A,*PTC121,*HEPS0015,*276-2009
2N5828	2N6429,2N6429A,BC550C,*GE-62,*HEPS0030,*BC337,*276-2009
2N5829	2N4957,2N4958,2N4959,*PTC127,*HEPS0013
2N5834	*HEPS3032,*276-2021
2N5838	2N5239,2N5839,*PTC129A,*GE-73,*HEPS5020,*ECG163
2N5839	2N5239,2N5240,2N5838,2N5840,*PTC129A,*GE-73,*HEPS5020,*ECG163
2N5840	2N5240,*PTC129A,*GE-73,*HEPS5020,*ECG163
2N5845A	*GE-20,*HEPS0025,*276-2009
2N5846	*276-2017
2N5847	2SC1040,2SC1081,2SC1118,2SC1257,2SC1258,2SC1495,2SC1528,MM8012, *HEPS3006,*276-2017
2N5848	2N5690,2SC1495,MM1619,ZT5591,*HEPS3007,*276-2020
2N5851	2N5852,*PTC136,*HEPS0020,*276-2038
2N5852	*HEPS0020,*276-2038
2N5862	
2N5865	2N4404,2N4407,MPS4356,*PTC141,*HEPS3012,*ECG129
2N5867	2N3789,2N3790,2N4902,2N4903,2N4905,2N4906,2N4908,2N4909,2N5868,2N5871, 2N5872,2N5875,2N5876,2SA746,2SA877,KSP2171,*PTC149,*HEPS5006
2N5868	2N3790,2N4903,2N4906,2N4909,2N5623,2N5872,2N5876,2N6248,2SA746,2SA877, KSP2171,KSP2172,*PTC172,*HEPS5005
2N5869	2N3232,2N3445,2N3447,2N3713,2N4914,2N4915,2N5068,2N5069,2N5870,2N5873, 2N5874,2N5877,2N5878,2N6315,2N6316,2N6317,2N6318,2SD211,180T2,BDY23B, HST9201,HST9801,HST9802,KSP1171,*PTC140,*HEPS5000
2N5870	2N3232,2N3233,2N3445,2N3446,2N3447,2N3448,2N3713,2N3714,2N4915,2N5069, 2N5624,2N5874,2N5878,2N6316,2N6318,2SD212,2SD213,181T2,BDY24B,HST9202, HST9802,HST9803,KSP1171,KSP1172,*PTC140,*HEPS5004,*276-2020
2N5871	2N3789,2N3790,2N3791,2N3792,2N4908,2N4909,2N5872,2N5875,2N5876,*PTC149, *HEPS5005
2N5872	2N3790,2N3792,2N4909,2N5876,2N6248,*PTC172,*HEPS5005
2N5873	2N3447,2N3713,2N3715,2N5874,2N5877,2N5878,*PTC140,*HEPS5004
2N5874	2N3447,2N3448,2N3713,2N3714,2N3715,2N3716,2N5878,*PTC140,*HEPS5004, *SK3563,*276-2020
2N5875	2N3789,2N3790,2N3791,2N3792,2N4908,2N4909,2N5876,2N5879,2N5880,*PTC149, *HEPS7003
2N5876	2N3790,2N3792,2N4909,2N5880,*PTC172,*HEPS7003,*ECG285
2N5877	2N3713,2N3715,2N5878,2N5881,2N5882,*HEPS7002
2N5878	2N3713,2N3714,2N3715,2N3716,2N5882,*PTC173,*HEPS7002,*SK3563,*ECG284
2N5879	2N5880,2N5883,2N5884,*PTC149,*HEPS7003,*ECG180

To Replace	Substitute This Type
2N5880	2N5884,2N6436,*HEPS7003,*ECG180
2N5881	2N5882,2N5885,2N5886,BUY51A,BUY52A,*GE-18,*TR-87,*HEPS7002,*SK3024,*ECG128,*WEP243,*RT-114
2N5882	2N5886,2N6270,2N6272,BUY53A,BUY54A,*PTC173,*GE-18,*TR-87,*HEPS7004,*SK3024,*ECG128,*WEP243,*RT-114
2N5883	2N5884,*PTC174,*HEPS7001,*ECG180,*WEPS7001,*RT-148
2N5884	2N6436,*PTC174,*HEPS7001,*ECG180,*WEPS7001,*RT-148
2N5885	2N5886,*PTC175,*HEPS7000,*ECG181
2N5886	*PTC175,*HEPS7000,*ECG181
2N5887	2N5888,2N5897,2N5901,*PTC120A,*HEPG6016
2N5888	2N5897,2N5898,2N5901,*PTC120A,*HEPG6016
2N5889	2N5890,*PTC120A,*HEPG6016
2N5890	2N5891,*PTC120A
2N5891	2N5892,*PTC120A
2N5892	*PTC120A
2N5893	2N5894,2N5897,2N5898,*PTC120A,*GE-49,*TR-94,*HEPG6016,*SK3082,*ECG131,*WEP642,*276-2006
2N5894	2N5895,2N5898,2N5899,*PTC120A
2N5895	2N5896,2N5899,2N5900,*PTC120A
2N5896	2N5900,*PTC120A
2N5897	2N5888,2N5898,*PTC120A,*GE-30,*TR-94,*HEPG6016,*SK3052,*ECG131,*WEP642,*276-2006
2N5898	2N5899,*PTC120A
2N5899	2N5900,*PTC120A
2N5900	*PTC120A
2N5901	*HEPG6016
2N5910	MPSH81,*HEPS0013
2N5913	2SC302-M,2SC307-M,2SC590N,2SC992,2SC1002,2SC1223,2SC1256,2SC1947,40953,41039,*PTC158,*GE-45,*HEPS3008,*276-2017
2N5914	2N5645,2N5944,2N5945,2N5946,2SC891,2SC892,2SC1040,2SC1495,2SC1620,40934†,41009A†,41010†,*HEPS3005,*276-2017
2N5915	2N5646,2N5946,2SC892,2SC1040,2SC1081,2SC1118,40893†,*HEPS3006
2N5916†	2N5917†,2N6202,*HEPS3005,*276-2020
2N5917†	2N5916†,2N6202,*HEPS3005,*276-2020
2N5918†	*276-2020
2N5919	2N5919A†,2N6204,*276-2020
2N5919A†	2N5919,2N6204,*276-2020
2N5920	*276-2020
2N5921†	*276-2020
2N5922	2N5764,2N5765,2N5923,2N5924,2N5925,2N6206,2N6207,2SC1334,41025†,41026†,MM4429,MM4430,MT5764,*HEPS3005,*276-2020
2N5923	2N5595,2N5764,2N5765,2N5924,2N5925,2N6206,2N6207,2SC1334,41025†,41026†,MM4430,MT5764,*HEPS3005,*276-2020
2N5924	2N5595,2N5765,2N5925,2N6207,2SC1334,41026†,*HEPS3006,*276-2020
2N5925	2N5595,2N5596,2N6207,2N6208,2SC1334,41026†,*HEPS3006,*276-2020
2N5927	2N5250,KSP1003
2N5929	2N5933,2N5935,2N6270,2N6271,2N6272,2N6273,1743-0820,1743-1010,1743-1030,HST9202,HST9203
2N5930	2N5931,2N5934,2N5936,2N5937,2N6271,2N6273,1743-1220,1743-1410,1743-1430,1743-1620,HST9203,HST9204,*SK3563
2N5931	2N5937,1743-1620,1743-1810,1743-1820,1743-1830
2N5932	2N5929,2N5935,2N6270,2N6272,1743-0610,1743-0630,1743-0820,HST9202
2N5933	2N5930,2N5934,2N5936,2N6270,2N6271,2N6272,2N6273,1743-1010,1743-1030,1743-1220,1743-1410,1743-1430,HST9202,HST9203,HST9204,*SK3563
2N5934	2N5931,2N5937,1743-1410,1743-1430,1743-1620,1743-1810,1743-1820,1743-1830,HST9204
2N5935	2N5929,2N5933,2N6270,2N6271,2N6272,2N6273,1743-0820,1743-1010,1743-1030,HST9202,HST9203
2N5936	2N5930,2N5931,2N5934,2N5937,2N6271,2N6273,1743-1220,1743-1410,1743-1430,1743-1620,HST9203,HST9204
2N5937	2N5931,1743-1620,1743-1810,1743-1820,1743-1830

To Replace	Substitute This Type
2N5941	2N5709,2SC1763
2N5942	2SC1764
2N5943	*HEPS3008
2N5944	2N5945,2SC891,2SC892,2SC1040,2SC1495,41009A†,41010†,*276-2017
2N5945	2SC892,2SC1040,2SC1081,2SC1495,41010†,*276-2017
2N5946	2SC1081,2SC1118,40893†,*276-2019
2N5947	*HEPS3005,*276-2009
2N5954	KSP2165,*PTC168,*GE-56,*HEPS5012,*SK3085,*276-2017
2N5955	2N5954,2SA765,KSP2164,KSP2165,*PTC168,*GE-56,*HEPS5012,*SK3085, *276-2017
2N5956	2N5955,2SA764,*PTC168,*GE-56,*HEPS5012,*SK3085,*276-2017
2N5957	2N5959
2N5958	2N5960
2N5959	2N5957
2N5960	2N5958
2N5966	2N5968,2N6060,2N6062,ST15043,ST17060
2N5967	2N5969,2N6061,2N6063
2N5968	2N5966,2N6060,2N6062,ST15043,ST17060
2N5969	2N5967,2N6061,2N6063
2N5974	2N5975,2N5980,2N5981,*PTC164,*HEPS5005,*276-2027
2N5975	2N5976,2N5981,2N5982,*PTC164,*HEPS5005
2N5976	2N5982,*PTC164,*HEPS5005
2N5977	2N5978,2N5983,2N5984,*PTC165,*HEPS5004,*276-2019
2N5978	2N5979,2N5984,2N5985,*PTC165,*HEPS5004,*276-2020
2N5979	2N5985,*PTC165,*HEPS5004,*276-2020
2N5980	2N5981,MJE1291,*PTC164,*HEPS5005,*276-2027
2N5981	2N5982,*PTC164,*HEPS5005,*SK3189
2N5982	*PTC164,*HEPS5005,*SK3189
2N5983	2N5984,*PTC165,*HEPS5004,*276-2019
2N5984	2N5985,*PTC165,*HEPS5004,*276-2020
2N5985	*PTC165,*HEPS5004,*SK3188,*276-2020
2N5986	2N5987
2N5987	2N5988
2N5988	
2N5989	2N5990
2N5990	2N5991
2N5991	
2N5992	2SC892,2SC1040,2SC1081,2SC1118,2SC1257,2SC1258,2SC1528,MM8012, *276-2020
2N5993	2SC1323,ZT5591,*HEPS3007,*276-2019
2N5994	2N5848,2N6082,2SC1495,2SC1528,SRF54215,ZT5591,*276-2020
2N5995†	2SC892,2SC1040,2SC1081,2SC1118,2SC1257,2SC1258,2SC1495,2SC1528,MM8012, *HEPS3006,*276-2017
2N5996	2N5591,2N5705,2N6082,2N6136,2SC1528,40893†,MM1603,MSA8506,SRF54215, ZT5591,*HEPS3007,*276-2018
2N5998	2N5371,2N5812,2N5818,2N6002,2N6006,2N6008,2N6012,D33D22,D33D26,D33D27, *GE-210,*HEPS0024,*276-2009
2N5999	2N3504,2N5375,2N5378,2N5379,2N5813,2N5819,2N6001,2N6003,2N6005,2N6007, 2N6009,2N6013,2SA509-Y,2SA509G-Y,A5T3504,A5T3644,A5T4403,BSW73,D29E2, D29E6,D29E7,TN-3906,*GE-82,*HEPS3028,*276-2023
2N6000	2N3242A,2N5812,2N5818,2N5998,2N6004,2N6010,BSW83,D33D22,D33D26,*GE-210, *HEPS0025,*BC337,*276-2009
2N6001	2N3504,2N6005,A5T3504,A5T3644,A5T4403,TN-3906,*PTC103,*GE-48,*HEPS0012, *276-2023
2N6002	2N6006,2N6008,2N6012,D33D27,*HEPS0015,*BC337,*276-2009
2N6003	2N6007,*HEPS0019,*ECG159,*276-2023
2N6004	2N2222,2N2222A,2N2540,2N3116,2N5369,2N5582,2N5818,2N6010,A5T2222,BSW85, D33D26,TIS109,TIS111,TN-3904,*GE-210,*HEPS0025,*BC337,*276-2009
2N6005	2N2907,2N2907A,2N3486,2N3486A,2N3505,2N3672,2N4028,2SA720,2SA891, A5T2907,A5T3505,A5T3645,TIS112,*PTC103,*GE-48,*HEPS0019,*276-2023
2N6006	2N5370,2N6012,2N6016,D33D27,*HEPS0015,*BC337,*276-2009

To Replace	Substitute This Type
2N6007	*HEPS0019,*276-2023
2N6008	2N6002,2N6006,2N6012,D33D27,*GE-210,*TR-53,*HEPS0025,*276-2009
2N6009	2N5378,2N5379,2N6003,2N6007,2N6013,D29E2,D29E7,*GE-48,*HEPS3028, *276-2023
2N6010	*HEPS3020
2N6011	2N2838,2N4028,2N5817,2N5819,2N5823,2N6015,2SA817,D29E10,D29E5,D29E6, *GE-48,*HEPS3028,*BC327,*276-2026
2N6012	*HEPS3001,*276-2014
2N6013	2N6017,D29E7,*HEPS3032
2N6014	2N5582,*GE-63,*HEPS3020
2N6015	2N4029,2N5823,2SA817,D29E10,*GE-67,*HEPS3032
2N6016	*HEPS3001,*276-2014
2N6017	*HEPS3032
2N6029	2N6030,*PTC172,*GE-74,*ECG285
2N6030	*PTC172,*ECG285
2N6031	*ECG285
2N6032	HST9203,HST9204,HST9208,HST9209,HST9804
2N6033	1763-1410,1763-1420,1763-1430,1763-1610,1763-1620,1763-1630,1763-1810,HST9204, HST9209
2N6034	2N6035,*SK3181,*ECG254,*276-2022
2N6035	2N6036,*ECG254,*276-2021
2N6036	*ECG197,*276-2021
2N6037	2N6038,*SK3180,*ECG184,*276-2011
2N6038	2N6039,*ECG184,*276-2011
2N6039	*SK3180,*ECG253,*276-2011
2N6040	2N6041,*HEPS9122,*ECG262,*276-2021
2N6041	2N6042,*HEPS9122,*ECG262
2N6042	*276-2021
2N6043	2N6044,*HEPS9102,*SK3180,*ECG261,*276-2011
2N6044	2N6045,*HEPS9102,*SK3180,*ECG261,*276-2011
2N6045	*HEPS9102,*276-2011
2N6046	1748-0610,1748-0630,1748-0810,1748-0820,1748-0830,MJ7000
2N6047	2N6048,1748-1010,1748-1030,1748-1210,1748-1220,1748-1230,1748-1410,1748-1430, MJ7000
2N6048	1748-1410,1748-1430,1748-1610,1748-1620,1748-1630,1748-1810,1748-1820, 1748-1830
2N6049	*PTC168,*HEPS5005,*ECG218
2N6050	2N6051,2N6285,2N6286,PMD11K-60,PMD11K-80,PMD17K-60,*ECG248
2N6051	2N6052,2N6286,2N6287,PMD11K-100,PMD11K-80,PMD17K-100,PMD17K-80,*ECG248
2N6052	2N6287,PMD11K-100,PMD17K-100,PMD17K-80,*ECG248
2N6053	2N6050,2N6051,PMD11K-60,PMD11K-80,PMD13K-60,PMD13K-80,*HEPS9141, *SK3183,*ECG244
2N6054	2N6051,2N6052,PMD11K-100,PMD11K-80,PMD13K-100,PMD13K-80,RCA8350B, *HEPS9141,*SK3183,*ECG244
2N6055	2N6057,2N6058,2N6384,2N6385,PMD10K-60,PMD10K-80,PMD12K-60,PMD12K-80, *HEPS9140,*SK3182,*ECG243
2N6056	2N6058,2N6059,2N6385,PMD10K-100,PMD10K-80,PMD12K-100,PMD12K-80, *HEPS9140,*SK3182,*ECG243
2N6057	2N6058,2N6282,2N6283,PMD10K-60,PMD10K-80,PMD16K-60,PMD16K-80,*ECG247
2N6058	2N6059,2N6283,2N6284,PMD10K-100,PMD10K-80,PMD16K-100,PMD16K-80,*ECG247
2N6059	2N6284,PMD10K-100,PMD16K-100,*ECG247
2N6060	2N6062,ST14030
2N6061	2N6063
2N6062	2N6060,ST14030
2N6063	2N6061
2N6064	
2N6065	2N6066
2N6066	
2N6067	2SA891,TP3645,*PTC103,*GE-82,*HEPS0013,*276-2021
2N6076	2N4035,2N5244,2N5355,2N5356,2N5366,2N5367,2N5378,2N5379,2N5383,2N6001, 2N6003,A5T3906,A5T4126,A5T4403,BCY78A,BCY78B,BCY78C,MM3906,NPS404,

To Replace	Substitute This Type
(2N6076)	NPS404A,TN-3906,*GE-65,*HEPS0015,*ECG159,*WEP242,*BC327,*276-2024
2N6077	2N6078,2N6079,HST7903,HST7904,HST7905,KSP1123,KSP1124,KSP1125,SDT7903, SDT7904,SDT7905
2N6078	2N6077,HST7902,HST7903,HST7904,HST7905,KSP1122,KSP1123,KSP1124, KSP1125,SDT7902,SDT7903,SDT7904,SDT7905
2N6079	
2N6080	2SC892,2SC1040,2SC1081,2SC1257,2SC1258,41010†,MM8012,*HEPS3005, *276-2025
2N6081	2N5589,2N5645,2N5914,2N5945,2N5946,2SC891,2SC892,2SC1011,2SC1040, 2SC1257,2SC1606,2SC1608,2SC1620,40934†,41009A†,41010†,*HEPS3006, *276-2018
2N6082	2N6083,ZT5591,*HEPS3007,*276-2019
2N6083	*HEPS3007
2N6084	*HEPS3009,*ECG320
2N6093†	
2N6094	*276-2025
2N6095	*276-2025
2N6096	*276-2027
2N6097	*276-2027
2N6098	2N6100,*SK3534
2N6099	2N6101,*HEPS5004,*SK3534,*276-2019
2N6100	*SK3534
2N6101	*HEPS5004,*SK3534,*276-2020
2N6102	*SK3534
2N6103	*SK3534
2N6104†	*276-2020
2N6105†	2N6205,*HEPS3007,*276-2020
2N6106	*PTC168,*HEPS5005,*SK3083,*ECG197,*WEP246,*276-2020
2N6107	*PTC168,*HEPS5005,*SK3083,*ECG197,*WEP246
2N6108	2N6106,*PTC168,*HEPS5005,*SK3083,*ECG197,*WEP246,*276-2027
2N6109	2N6107,*PTC168,*HEPS5005,*SK3083,*ECG197,*WEP246,*276-2027
2N6110	2N6108,*PTC168,*HEPS5005,*SK3084,*ECG197,*WEP246,*276-2027
2N6111	2N6109,RCA1C11,*PTC168,*HEPS5005,*SK3084,*ECG197,*WEP246,*276-2027
2N6112	2N6428,2N6428A,2N6539,BC547B,BC550B,TIS94,TIS97,*PTC121,*GE-62, *HEPS0015,*BC337,*276-2009
2N6121	2N6122,2N6130,2N6290,*PTC167,*HEPS5000,*ECG152,*WEP245
2N6122	2N6123,2N6130,2N6131,2N6290,2N6292,*PTC167,*TR-76,*HEPS5000,*ECG152, *WEP245,*276-2020
2N6123	2N6131,2N6473,NSP41B,NSP41C,*PTC167,*HEPS5000
2N6124	2N6109,2N6125,2N6133,*PTC166,*TR-77,*HEPS5006,*ECG153,*WEP246
2N6125	2N6107,2N6109,2N6126,2N6133,2N6134,*PTC166,*TR-77,*HEPS5006,*ECG153, *WEP246
2N6126	2N6134,2N6475,NSP42B,NSP42C,*PTC166,*TR-77,*HEPS5006,*ECG153,*WEP246
2N6127	
2N6128	
2N6129	2N6130,*PTC169,*HEPS5004,*276-2019
2N6130	2N6131,*PTC169,*HEPS5004,*276-2020
2N6131	*PTC169,*HEPS5004,*276-2020
2N6132	2N6133,*PTC168,*HEPS5005,*ECG197,*WEP246,*276-2027
2N6133	2N6134,*PTC168,*HEPS5005,*ECG197,*WEP246,*276-2019
2N6134	*PTC168,*HEPS5005,*ECG197,*WEP246
2N6135	
2N6136	ZT5591,*276-2019
2N6166	
2N6175	2N6176,*GE-32,*HEPS5015,*SK3104,*ECG228
2N6176	2N6177,*GE-32,*TR-60,*HEPS5015,*SK3103,*ECG157
2N6177	*GE-32,*SK3103,*ECG228
2N6178	*HEPS5000,*SK3024
2N6179	*PTC110,*HEPS5000,*SK3024
2N6180	*HEPS5006,*SK3025
2N6181	*PTC111,*HEPS5006,*SK3025

To Replace	Substitute This Type
2N6182	2N6184,2N6186,2N6188,*HEPS5005
2N6183	2N6185,2N6187,2N6189,*HEPS5005
2N6184	2N6188,*HEPS5005
2N6185	2N6189,*HEPS5005
2N6186	2N6182,2N6184,2N6188,*HEPS5005
2N6187	2N6183,2N6185,2N6189,*HEPS5005
2N6188	2N6184,*HEPS5005
2N6189	2N6185,*HEPS5005
2N6190	2N6192,*HEPS3003
2N6191	2N6193,*HEPS3003
2N6192	
2N6193	
2N6197	2SC1043,*276-2020
2N6198	*276-2020
2N6199	*276-2020
2N6200	2N5643,MSA8503
2N6201	
2N6202	*276-2020
2N6203	2N5919,2N5919A†,*276-2020
2N6204	*276-2020
2N6205	
2N6206	2N5595,2N5765,41025†,41026†,MT5764,MT5765,*HEPS3005,*276-2020
2N6207	2N5595,2N5596,2N5925,41026†,*HEPS3007,*276-2020
2N6208	2N5596,2SC1334,*HEPS3007,*276-2020
2N6211	2N6212,*ECG162
2N6212	2N6213,2N6214,*ECG162
2N6213	2N6214,*ECG162
2N6214	*ECG167,*276-2021
2N6215	
2N6218	*PTC170
2N6219	2N6218,*PTC170
2N6220	2N6219,*PTC170
2N6221	2N6220,*PTC170
2N6222	2N760A,2N909,2N930B,2N956,2N2390,2N2484A,2N2645,2N3116,2N3302,2N3859A, 2N3947,2N4409,2N4952,2N5107,2N5310,2N5369,2N5381,2N6224,2N6540,2N6541, 2SC1166-Y,2SC1318,2SC1781,2SC1852,2SC1991,2SC1998,A5T3904,A5T4409, BC546A,BCW34,BSW85,GET930,GET2222,GET2222A,GI-3643,MM3904,TIS95, TIS98,TN-3904,TZ82,*PTC170,*GE-81,*276-2011
2N6223	2N2907,2N2907A,2N3061,2N3251A,2N3486,2N3486A,2N3505,2N3547,2N3672, 2N3673,2N3911,2N3912,2N3914,2N3915,2N4415A,2N5373,2N6225,2SA661-Y, 2SA720,2SA742,2SA891,A5T2907,A5T3505,A5T3645,BC556A,BCW35,BSW75, GET2906,GET2907,TIS112,TP3645,TZ552,TZ582,*PTC103,*GE-82,*276-2021
2N6224	2N2484A,2N4409,2N4424,2N5232,2N5232A,2N5311,2N5376,2N5377,2SC1166-GR, 2SC1850,A5T4409,BC546B,TIS94,TIS97,*GE-81,*276-2011
2N6225	2N2907,2N2907A,2N3251A,2N3486,2N3486A,2N3505,2N3547,2N3548,2N3912, 2N3915,2N3962,2N3963,2N4413A,2N4415A,2N5373,2N5374,2SA661-GR,2SA661-Y, A5T2907,A5T3505,A5T3645,BC556A,BC556B,BSW75,GET2906,GET2907,TIS112, TZ552,TZ553,*GE-82,*276-2021
2N6226	2N6227,2N6229,2N6230,*PTC172,*GE-74,*HEPS5005,*ECG281
2N6227	2N6230,*PTC172,*ECG281
2N6228	2N6231,*ECG281
2N6229	2N6029,2N6030,2N6230,*PTC172,*GE-74,*HEPS5005,*ECG281
2N6230	2N6030,*PTC172
2N6231	2N6031,*ECG281
2N6232	
2N6233	2N6234
2N6234	2N6235
2N6235	
2N6246	2N6247,2N6436,*PTC149,*HEPS7001,*SK3173,*ECG219,*WEPS7001
2N6247	2N6436,2N6437,*PTC172,*HEPS7001,*ECG219,*WEPS7001
2N6248	*PTC172,*HEPS7001

To Replace	Substitute This Type
2N6249	2N6250,KSP1223,KSP1224,KSP1225,XDAR10025,XDAR10030,XGSR10025, XGSR15025,*PTC118,*HEPS5020,*SK3559
2N6250	2N6251,XDAR10030,XDAR10035,*HEPS5020,*SK3559
2N6251	*HEPS5020,*SK3559
2N6253	2N6470,2N6471,2SD114-O,B170002-ORG,B170002-RED,B170002-YEL,B170005-ORG, B170005-RED,B170005-YEL,*PTC140,*GE-77,*TR-59,*HEPS7004,*SK3027, *ECG130,*WEP247,*BDY20
2N6254	2N3236,2N5629,2N5630,2N6270,2N6271,2N6272,2N6273,108T2,BUY53A,BUY54A, IR3772,RCS258,*PTC173,*GE-75,*HEPS7004,*SK3511,*ECG181
2N6255	2SC302-M,2SC307-M,2SC590N,*PTC144,*HEPS5014,*276-2017
2N6256	2N5644,2N5645,2N5698,2N5914,2N5944,2N5945,2N5946,2SC891,2SC892,2SC911, 2SC911A,2SC1016,2SC1040,2SC1495,2SC1608,2SC1620,40934†,41008A†, 41009A†,41010†,XB433,*276-2017
2N6257	2N3771,2N5302,1561-0408,1561-0410,1561-0608,1561-0610,1561-0615,IR3771, MJ3771,*PTC173,*GE-75,*TR-36,*HEPS7000,*SK3036,*ECG181,*WEPS7000
2N6258	2N6570,2N6571,*PTC175,*GE-14,*TR-36,*HEPS7000,*SK3036,*ECG181,*WEPS7000
2N6259	*SK3535,*ECG284
2N6260	2N4231,2N4232,2N6373,2N6374,40250,BD148-6,*PTC112,*GE-66,*TR-81, *HEPS5019,*SK3026,*ECG175,*WEP241,*276-2020
2N6261	*PTC169,*GE-66,*TR-81,*HEPS5019,*SK3026,*ECG175,*WEP241,*276-2017
2N6262	2N3773,2N6259,43104,IR3773,*HEPS5020,*ECG284
2N6263	2N3441,2SD244,2SD259,2SD297,HST6905,HST6906,KSP1026,SDT6905,SDT6906, *GE-12,*TR-81,*HEPS5012,*SK3538,*ECG175,*WEP241,*276-2017
2N6264	*HEPS5012,*SK3538,*ECG286,*276-2017
2N6265†	*276-2020
2N6266†	2N6267†,*276-2020
2N6267†	*276-2020
2N6268†	*276-2020
2N6269†	*276-2020
2N6270	2N6271,2N6272,2N6273,*HEPS7000
2N6271	2N6273,*HEPS7000
2N6272	2N6270,2N6271,2N6273
2N6273	2N6271
2N6274	2N6275,2N6276
2N6275	2N6276,2N6277
2N6276	2N6277
2N6277	
2N6278	2N6279,2N6280
2N6279	2N6280,2N6281
2N6280	2N6281
2N6281	
2N6282	2N6283,PMD16K-60,PMD16K-80,*ECG251
2N6283	2N6284,PMD16K-100,PMD16K-80,*ECG251
2N6284	PMD16K-100,*ECG251
2N6285	2N6286,PMD17K-60,*ECG250
2N6286	2N6287,PMD17K-100,PMD17K-80,*ECG250
2N6287	PMD17K-100,PMD17K-80,*ECG250
2N6288	2N6290,RCA1C10,*PTC169,*GE-28,*HEPS5004,*SK3054,*276-2017
2N6289	2N6291,*PTC169,*GE-28,*HEPS5004,*SK3054,*276-2017
2N6290	2N6292,*PTC169,*GE-28,*HEPS5004,*SK3054,*276-2020
2N6291	2N6293,*PTC169,*GE-28,*HEPS5004,*SK3054,*276-2020
2N6292	*PTC169,*GE-28,*HEPS5004,*SK3054,*ECG196,*276-2020
2N6293	*PTC169,*GE-28,*HEPS5004,*SK3054,*276-2020
2N6294	2N6295,2N6300,2N6301,*HEPS9101,*SK3182,*ECG274,*276-2020
2N6295	2N6301,*HEPS9101,*SK3182,*ECG274
2N6296	2N6297,2N6298,2N6299,*TR-58,*HEPS9121,*SK3183,*ECG275,*WEPS5007
2N6297	2N6299,*TR-58,*HEPS9121,*SK3183,*ECG275,*WEPS5007
2N6298	2N6299,*HEPS9121,*ECG244
2N6299	*HEPS9121,*ECG244
2N6300	2N6301,*HEPS9101,*ECG243
2N6301	*ECG243

To Replace	Substitute This Type
2N6302	2N3055C,2N3772C,2N3773,2N5631,2N6259,109T2,1561-1208,1561-1210,1561-1215, 1561-1410,1561-1415,1561-1608,1561-1610,1561-1615,43104,IR3773,IR6302, SDT9703,SDT9706,STC2222,STC2226,STC2230,*HEPS5020
2N6303	
2N6304	2N4252,2N4253,2N6305,MT1061A,*PTC133,*HEPS0017,*ECG316
2N6305	2N4252,2N4253,2N6304,MT1061A,*PTC133,*GE-86,*HEPS0017,*ECG316
2N6306	2N6307,2N6308M,2SC807,*HEPS5020,*ECG283
2N6307	2N6308,*HEPS5020,*ECG283
2N6307M	2N6306,2N6308M,2SC807
2N6308	2SC806,*HEPS5020,*ECG283
2N6308M	2N6306,2N6307,2SC807
2N6309	2N6310,2N6311
2N6310	2N6311
2N6311	
2N6312	2N6313,*PTC168,*HEPS5005,*276-2021
2N6313	2N6314,*PTC168,*HEPS5005,*276-2021
2N6314	*PTC168,*GE-234,*HEPS5005,*276-2021
2N6315	2N6316,2N6317,2N6318,*PTC169,*HEPS5004
2N6316	2N6318,*PTC169,*HEPS5004
2N6317	2N6315,2N6316,2N6318,*PTC169,*HEPS5005
2N6318	2N6316,*PTC169,*HEPS5005
2N6319	2N6320,2N6321
2N6320	2N6321
2N6321	
2N6322	
2N6323	
2N6324	
2N6325	
2N6326	2N6327,*PTC175,*GE-75,*HEPS7000
2N6327	2N6328,*PTC175,*GE-75,*HEPS7000
2N6328	*PTC175,*GE-75,*HEPS7000
2N6329	2N6330,*PTC174,*GE-74,*HEPS7001,*ECG180,*WEPS7001
2N6330	2N6331,*PTC174,*GE-74,*HEPS7001
2N6331	*PTC174,*GE-74,*HEPS7001
2N6338	2N6339,2N6340,*HEPS7000
2N6339	2N6340,2N6341
2N6340	2N6341
2N6341	
2N6350	
2N6351	
2N6352	
2N6353	
2N6354	KSP1176,KSP1276,*ECG284
2N6355	2N6356,*ECG251
2N6356	2N6355,*ECG251,*276-2011
2N6357	2N6358,*ECG247
2N6358	2N6357,*ECG251
2N6359	2N3236,2N5630,2N6254,2N6258,2SD55,108T2,1561-1008,1561-1010,1561-1015, 1561-1208,1561-1210,1561-1215,IR3772,MJ3772,RCS258,SDT9701,SDT9702, SDT9704,SDT9705,STC2221,STC2225,STC2229,*PTC173,*HEPS7000
2N6360	2N3773,2N5630,2N5631,2N6302,1561-1008,1561-1010,1561-1208,1561-1210, 1561-1410,43104,IR3773,IR6302,SDT9702,SDT9703,SDT9705,SDT9706,STC2222, *PTC173,*HEPS7000
2N6361	*276-2020
2N6362	
2N6363	
2N6364	
2N6365	*276-2002
2N6365A	2N6365,*276-2002
2N6366	
2N6367	*276-2011

To Replace	Substitute This Type
2N6368	
2N6369	
2N6370	*276-2011
2N6371	2N3055A,2N3235,2N3771,2N5302,2N5881,2N5885,1561-0403,1561-0404,1582-0403, 1582-0404,1582-0405,1582-0603,1582-0604,1582-0605,B170002-BLK,B170002-BRN, BUY51A,BUY52A,HST9201,HST9205,IR3771,MJ2801,MJ3771,SDT9201,SDT9205, *PTC140,*HEPS7004,*SK3511,*ECG130,*276-2011
2N6372	2N5429,2SC1113,2SC1445,KSP1165,*PTC169,*HEPS5012,*276-2011
2N6373	2N5427,2N5429,2N6372,2SC1444,2SC1445,KSP1164,KSP1165,*PTC169,*HEPS5000, *276-2011
2N6374	2N6373,*PTC169,*HEPS5000
2N6375	
2N6376	2N3722,2N3723,2N4961,2N6375
2N6377	2N6378
2N6378	2N6379
2N6379	
2N6380	2N6381
2N6381	2N6382
2N6382	
2N6383	2N6384,*SK3182,*ECG245
2N6384	2N6385,*SK3182,*ECG245
2N6385	*SK3182,*ECG245
2N6386	2N6387,*SK3180,*ECG263
2N6386+	2N6387+,*SK3180
2N6387	2N6388,*SK3180,*ECG263
2N6387+	2N6388+,*SK3180
2N6388	*SK3180,*ECG263
2N6388+	*SK3180
2N6389	2SC1547,*GE-212,*ECG316,*276-2011
2N6390	2N6390†,2N6391†,2N6393†,RCA2005†,RCA2310†,*276-2020
2N6390†	2N6390,2N6391†,2N6393†,RCA2005†,RCA2310†,*276-2020
2N6391†	2N6393†,RCA2005†,RCA2310†,*276-2020
2N6392†	2N6393†,RCA2010†,*276-2020
2N6393†	RCA2310†,*276-2020
2N6406	2N6407,2N6415
2N6407	
2N6408	2N6409
2N6409	
2N6410	2N6412
2N6411	2N6414
2N6412	2N6413
2N6413	
2N6414	2N6415
2N6415	
2N6416	2N6413,2N6417
2N6417	
2N6418	2N6415,2N6419
2N6419	
2N6420	
2N6421	
2N6422	2N6423
2N6423	
2N6424	2N6425
2N6425	
2N6426	
2N6427	2N6426
2N6428	2N6428A
2N6428A	2N6428
2N6429	2N6429A
2N6429A	2N6429
2N6430	

To Replace	Substitute This Type
2N6431	
2N6432	
2N6433	
2N6436	2N6437
2N6437	2N6438
2N6438	
2N6439	
2N6455	2SC1476,2SC1523,2SC1548,2SC1668,41042†
2N6456	2SC1503
2N6457	2SC1503
2N6458	
2N6459	
2N6460	
2N6461	
2N6462	
2N6463	2N6461
2N6464	2N6462
2N6465	2N5430,2N5618,2N6466,2SC1445,*PTC148
2N6466	
2N6467	2N5617,2N6468
2N6468	
2N6469	*PTC149
2N6470	2N6471,*PTC140,*SK3563
2N6471	2N5734,2N6472,*PTC140,*SK3563
2N6472	2N5039,2N5734,2N6338,*PTC173,*SK3563
2N6473	2N6474,*ECG291
2N6474	*ECG291
2N6475	2N6476,*ECG291
2N6476	*HEPS5005,*ECG292
2N6477	2N6478,*ECG292
2N6478	*ECG292
2N6479	2N6480,2N6481,2N6482
2N6480	2N6479,2N6481,2N6482
2N6481	2N6482
2N6482	2N6481
2N6486	2N6487,*PTC169
2N6487	2N6488,*PTC169
2N6488	*PTC169
2N6489	2N6490,*PTC168
2N6490	2N6491,*PTC168
2N6491	*PTC168
2N6492	2N6576,*ECG243
2N6493	2N6494,2N6578,*ECG243
2N6494	2N6493,2N6578,*ECG243
2N6495	
2N6496	2N5672,KSP1276,*ECG284
2N6497	2N6498,2N6499
2N6498	2N6499
2N6499	
2N6500	HST6901,KSP1023,*SK3562,*ECG175
2N6501	
2N6502	
2N6503	
2N6510	2N6511,SDT14304,*SK3560,*ECG283
2N6511	2N6512,2N6514,SDT14304,*SK3560,*ECG283
2N6512	2N6307M,2N6513,2N6514,*SK3560,*ECG283
2N6513	2N6306,2N6307M,2N6308M,2SC807,*SK3560,*ECG283
2N6514	2N6307M,2N6512,2N6513,*SK3560,*ECG283
2N6515	
2N6516	
2N6517	

To Replace	Substitute This Type
2N6518	2N6519
2N6519	
2N6520	
2N6521	
2N6522	
2N6523	2N6524,2N6525
2N6524	2N6525,RCA2310†
2N6525	RCA2310†
2N6526	2N6527,RCA3001†
2N6527	RCA3001†
2N6528	2N6529,RCA3003†,RCA3005†
2N6529	RCA3005†
2N6530	2N6531,2N6532
2N6531	2N6532,2N6533
2N6532	2N6531,2N6533
2N6533	
2N6534	2N6535,2N6536
2N6535	2N6536,2N6537
2N6536	2N6535,2N6537
2N6537	
2N6538	A5T3903,TIS96,TIS99,TIS110,TIS135,TIS136
2N6539	
2N6540	TIS95,TIS98
2N6541	2N6540,TIS95,TIS98
2N6542	2N6544
2N6543	2N6545
2N6544	2N6546
2N6545	2N6547
2N6546	
2N6547	
2N6548	2N6549
2N6549	2N6548
2N6551	2N6552
2N6552	2N6553
2N6553	
2N6554	
2N6555	
2N6556	
2N6557	2N6558
2N6558	2N6559
2N6559	
2N6560	2N6579,2N6580,2N6581,2N6582,2N6583,2N6584
2N6561	
2N6562	2N6585,2N6586,2N6587,2N6588,2N6589,2N6590
2N6563	
2N6566	
2N6567	2N2946A,2N4008
2N6569	*PTC140
2N6570	2N6571
2N6571	
2N6572	
2N6573	2N6574
2N6574	2N6575
2N6575	
2N6576	
2N6577	
2N6578	
2N6579	2N6580,2N6581,2N6582,2N6583,2N6584
2N6580	2N6581,2N6583,2N6584
2N6581	2N6580,2N6583,2N6584
2N6582	2N6579,2N6580,2N6581,2N6583,2N6584

To Replace	Substitute This Type
2N6583	2N6580,2N6581,2N6584
2N6584	2N6580,2N6581,2N6583
2N6585	2N6586,2N6587,2N6588,2N6589,2N6590
2N6586	2N6587,2N6589,2N6590
2N6587	2N6586,2N6589,2N6590
2N6588	2N6585,2N6586,2N6587,2N6589,2N6590
2N6589	2N6586,2N6587,2N6590
2N6590	2N6586,2N6587,2N6589
2N6591	2N6592
2N6592	2N6593
2N6593	
2N6594	
2N6595	2N6596,2N6599,2N6600
2N6596	2N6595,2N6600
2N6597	2N6598,2N6599,2N6600
2N6598	2N6597,2N6600
2N6599	2N6600
2N6600	2N6599
2N6601	2N6599,2N6600,MT1061
2N6602	
2N6603	2N6604
2N6604	
2N6609	
2N6617	
2N6618	
2N6622	
2N6648	2N6649
2N6649	2N6650
2N6650	
2N6653	2N6654,2N6655
2N6654	2N6655
2N6655	
2N6665	
2N6666	2N6667
2N6667	2N6668
2N6668	
2N6669	
2S024	*HEPS5004
2S025	2S026
2S026	
2S57	SEE 2SA57
2S095A	2N4418,2S512,2SC395A,2SC395A-O,2SC395A-Y,2SC735-O,2SC934,ZT2938, *PTC136,*BC107B,*276-2016
2S102	*PTC123,*GE-210,*TR-21,*HEPS0011,*SK3122,*ECG123A,*WEP736,*BC337, *276-2009,*RT-102
2S103	2N4962,2N4963,2N5380,2N6538,A5T3903,FT3722,TIS135,TIS136,TN-3903,*PTC123, *GE-210,*TR-87,*HEPS0011,*SK3024,*ECG128,*WEP243,*BC337,*276-2009, *RT-114
2S104	2N5381,2N6540,2N6541,2SC1998,A5T3904,BC546A,TIS95,TIS98,TN-3904,*PTC123, *GE-210,*TR-87,*HEPS0011,*SK3024,*ECG128,*WEP243,*BC337,*276-2009, *RT-114
2S131	2N743,2N784,2N5187,2SC68,2SC395A-R,40519,MPS706A,*PTC136,*GE-210,*TR-21, *HEPS0013,*SK3122,*ECG123A,*WEP736,*BC107B,*276-2016,*RT-102
2S189	SEE 2SB189
2S301	HA7535,HA7539,SHA7535,SHA7539,*GE-82,*TR-88,*SK3025,*ECG129,*WEP242, *RT-115
2S302A	2S303,HA7537,SHA7537,*GE-22,*TR-88,*HEPS0013,*SK3025,*ECG129,*WEP242, *BC177,*276-2022,*RT-115
2S303	2N1469A,HA7537,SHA7537,*PTC127,*GE-22,*TR-88,*HEPS0012,*SK3025,*ECG129, *WEP242,*BC177,*276-2022,*RT-115
2S304	ZT152,*PTC127,*GE-22,*TR-88,*HEPS0012,*SK3025,*ECG129,*WEP242,*BC177,

To Replace	Substitute This Type
(2S304)	*276-2022,*RT-115
2S305	2N2551,*SK3025
2S306	2N863,2N864,2N1027,2N1229,2N1429,2N2163,2N2164,2N2276,2N2278,2N2280, 2N3342,2N4006,2N4007,2N5141,2N5229,2S307,2S326,2S327,*PTC127,*GE-65, *TR-20,*SK3114,*ECG159,*WEP242,*BC177,*276-2021,*RT-115
2S307	2N861,2N1027,2N1229,2N1917,2N1918,2N2162,2N2163,2N2274,2N2276,2N2278, 2N2945A,2N3342,2N3344,2N4007,2N4284,2N5229,2N5230,2S327,TW135,*PTC127, *GE-65,*TR-20,*HEPS0012,*SK3118,*ECG159,*WEP242,*BC177,*276-2022, *RT-115
2S321	2N1234,2N3064,2S301,HA7535,HA7539,SHA7535,SHA7539,*GE-82,*TR-20,*SK3114, *ECG159,*WEP242,*RT-115
2S322	2N328B,2N1232,2N1441,2N3219,2N3979,HA7534,SHA7534,*GE-65,*TR-20, *HEPS0013,*SK3114,*ECG159,*WEP242,*BC177,*276-2022,*RT-115
2S322A	2N1231,2N2946A,2N3219,2N3979,2N4008,2N6567,2S302A,2S303,2S322,2S323, HA7537,SHA7537,TW135,*GE-65,*TR-20,*HEPS0013,*SK3114,*ECG159,*WEP242, *BC177,*276-2022,*RT-115
2S323	2N1231,2N1469A,2N2945A,2N2946A,2N3218,2N3219,2N3978,2N3979,2S303, HA7537,SHA7537,*GE-65,*TR-20,*HEPS0013,*SK3114,*ECG159,*WEP242,*BC177, *276-2022,*RT-115
2S324	2N2944A,2N2945A,2N3977,2N4980,2S304,ZT152,*GE-65,*TR-20,*HEPS0013, *SK3114,*ECG159,*WEP242,*BC177,*276-2022,*RT-115
2S325	2N2551,2N3842,2S305
2S326	2N863,2N864,2N1027,2N1229,2N1429,2N2163,2N2164,2N2276,2N2278,2N2280, 2N3342,2N4006,2N4007,2N5141,2N5229,2S306,2S307,2S327,*GE-65,*TR-20, *HEPS0013,*SK3114,*ECG159,*WEP242,*BC177,*276-2021,*RT-115
2S327	2N861,2N1027,2N1229,2N1917,2N1918,2N2162,2N2163,2N2274,2N2276,2N2278, 2N2945A,2N3342,2N3344,2N4007,2N4284,2N5229,2N5230,2S307,TW135,*GE-65, *TR-20,*HEPS0013,*SK3114,*ECG159,*WEP242,*BC177,*276-2022,*RT-115
2S501	2N703,2N708,2N753,2N784A,2N2242,2N2331,2N2501,2N3011,2N3013,2N3014, 2N3394,2N3646,2N4123,2N4418,2N5772,2S512,2SC67,2SC367G-O,2SC601N, 2SC639,2SC733-O,2SC735-O,2SC764,2SC1274,A5T4123,A321,GET2369,PT720, ZT708,ZT2205,*PTC126,*GE-61,*TR-21,*HEPS0011,*SK3122,*ECG123A,*WEP736, *BC107B,*276-2013,*RT-102
2S502	2N2924,2N3242A,2N3392,2N4124,2N4419,2N5224,2S503,2SC367G-Y,2SC538, 2SC539,2SC733-Y,2SC735-Y,2SC1849,2SC1993,40397,A5T4124,BC108,BC108B, BC109B,BC172B,BC173B,BC238B,BC239B,BC548A,BCY58B,BSW88,BSW89, BSX38,MPSA17,TN80,TP3706,TZ81,*PTC123,*GE-61,*TR-21,*HEPS0011, *ECG123A,*WEP736,*BC107B,*276-2013,*RT-102
2S503	2N2925,2N3391,2N3391A,2N3566,2N4256,2N5172,2SC733-GR,2SC735-GR,2SC1993, A5T3565,A322,BC108B,BC109B,BC172B,BC173B,BC238B,BC239B,BC548B,BC549B, BCY58C,MPSA16,MPSA17,PET4002,TP3566,TP3706,*PTC123,*GE-212,*TR-21, *HEPS0011,*SK3122,*ECG123A,*WEP736,*BC107B,*276-2013,*RT-102
2S512	2N2501,2N4123,2N4418,2SC639,2SC735-O,2SC1216,A5T4123,ZT2938,*PTC136, *GE-210,*TR-95,*HEPS0013,*SK3039,*ECG108,*WEP56,*BC107B,*276-2016, *RT-113
2S3010	2N328B,2S3020,2S3210,2S3220,HA7534,SHA7534,*GE-82,*TR-88,*HEPS0019, *SK3025,*ECG129,*WEP242,*BC177,*276-2022,*RT-115
2S3020	2N328B,2S3220,HA7538,SHA7538,*GE-82,*TR-20,*HEPS0019,*SK3114,*ECG159, *WEP242,*BC177,*276-2022,*RT-115
2S3021	2S302A,2S3221,MM4052,SHA7532,*GE-22,*TR-20,*HEPS0013,*SK3114,*ECG159, *WEP242,*BC177,*276-2022,*RT-115
2S3030	2N1469A,2S303,2S3230,HA7537,SHA7537,*GE-22,*TR-20,*HEPS0015,*SK3114, *ECG159,*WEP242,*BC177,*276-2022,*RT-115
2S3210	2N328B,2S3010,2S3020,2S3220,HA7534,SHA7534,*GE-82,*TR-20,*HEPS0013, *SK3114,*ECG159,*WEP242,*BC177,*276-2022,*RT-115
2S3220	2N328B,2S3020,HA7538,SHA7538,*GE-82,*TR-20,*HEPS0013,*SK3114,*ECG159, *WEP242,*BC177,*276-2022
2S3221	2S302A,2S3021,MM4052,SHA7532,*GE-22,*TR-20,*HEPS0013,*SK3114,*ECG159, *WEP242,*BC177,*276-2022,*RT-115
2S3230	2N1469A,2S303,2S3030,HA7537,SHA7537,*GE-22,*TR-20,*HEPS0013,*SK3114, *ECG159,*WEP242,*BC177,*276-2022,*RT-115

To Replace	Substitute This Type
2S3240	2S304,ZT152,*GE-22,*TR-20,*HEPS0013,*SK3114,*ECG159,*WEP242,*BC177, *276-2022,*RT-115
2SA12	2N711A,2N711B,2N829,2N838,2N972,2N985,2N2022,2N2273,2N2402,2N2587, 2N3320,2N6365,2SA15,2SA138,2SA139,2SA202,AFY15,SF.T307,*PTC107,*GE-50, *TR-85,*HEPG0005,*SK3008,*ECG102A,*WEP250,*276-2004,*RT-121
2SA13	2N711,2N711A,2N711B,2N794,2N795,2N829,2N972,2N973,2N974,2N985,2N2022, 2N2273,2N2402,2N3320,2SA12,2SA15,2SA16,2SA138,2SA139,2SA202,2SA451, AFY15,SF.T307,*PTC107,*GE-50,*TR-17,*HEPG0005,*SK3005,*ECG160,*WEP637, *276-2004
2SA14	*PTC107,*HEPG0005,*SK3005,*ECG160,*WEP637
2SA15	2N711A,2N711B,2N829,2N838,2N964A,2N972,2N985,2N2022,2N2273,2N2402, 2N2953,2N3320,2N6365,2SA138,40359,AFY15,SF.T308,*PTC107,*GE-50,*TR-85, *HEPG0005,*SK3005,*ECG102A,*WEP250,*276-2004,*RT-121
2SA16	2N711,2N711A,2N711B,2N794,2N795,2N796,2N829,2N964A,2N972,2N973,2N974, 2N985,2N2022,2N2273,2N2402,2N3320,2SA15,2SA138,2SA451,40359,AFY15, SF.T308,*PTC107,*GE-50,*TR-85,*HEPG0005,*SK3005,*ECG102A,*WEP250, *276-2004,*RT-121
2SA17	*DS56,*GE-50,*TR-85,*HEPG0006,*SK3005,*ECG102A,*WEP250,*276-2021,*RT-121
2SA18	2N827,2N2635,*GE-50,*TR-85,*HEPG0006,*SK3005,*ECG102A,*WEP250,*276-2021, *RT-121
2SA19	*PTC107,*HEPG0008,*SK3008,*ECG126,*WEP635
2SA20	*PTC107,*HEPG0008,*SK3008,*ECG126,*WEP635
2SA21	*PTC107,*HEPG0008,*SK3008,*ECG126,*WEP635
2SA22	
2SA23	
2SA24	*HEPG0002
2SA25	*HEPG0002
2SA26	*HEPG0005,*SK3005,*ECG100,*WEP254
2SA27	
2SA28	SF.T357P,*PTC107,*GE-51,*TR-17,*HEPG0008,*SK3005,*ECG126,*WEP635
2SA29	2N990,2N991,2N993,2N2496,2N2966,2N3127,2N3281,2N3282,2SA341,2SA342, AF137,AFY12,AFZ12,*PTC107,*GE-51,*TR-17,*HEPG0008,*SK3008,*ECG126, *WEP635
2SA30	*PTC107,*HEPG0003,*SK3005,*ECG126,*WEP635
2SA31	*PTC107,*HEPG0005,*SK3005,*ECG100,*WEP254
2SA32	*PTC107,*HEPG0005,*SK3005,*ECG102A,*WEP250
2SA33	*PTC107,*HEPG0005,*ECG102A,*WEP250
2SA35	*PTC107,*HEPG0003,*SK3005,*ECG126,*WEP635
2SA36	*PTC107,*HEPG0005,*SK3005,*ECG100,*WEP254
2SA37	*HEPG0005,*SK3005,*ECG126,*WEP635
2SA38	*HEPG0005,*SK3005,*ECG126,*WEP635
2SA39	*HEPG0005,*SK3005,*ECG126,*WEP635
2SA40	*PTC107,*HEPG0005,*SK3005,*ECG100,*WEP254
2SA41	*DS56,*GE-51,*TR-17,*HEPG0005,*SK3005,*HF35,*ECG160,*WEP637,*276-2007
2SA42	*GE-51,*TR-17,*HEPG0005,*SK3123,*HF35,*ECG160,*WEP637,*276-2007
2SA43	*PTC107,*HEPG0009,*SK3007,*ECG126,*WEP635
2SA44	*PTC102,*HEPG0005,*SK3005,*ECG100,*WEP254
2SA45	*HEPG0001,*SK3006,*ECG126,*WEP635
2SA46	
2SA47	
2SA48	
2SA49	*PTC109,*GE-51,*TR-17,*HEPG0008,*SK3005,*ECG126,*WEP635,*AF125,*276-2007, *RT-117
2SA49G	2N2953,2SA49,SF.T308,*GE-51,*SK3005
2SA50	SF.T308,*PTC109,*GE-53,*TR-85,*HEPG0005,*SK3005,*ECG102A,*WEP250, *RT-121
2SA50G	SF.T308,*GE-51,*SK3005
2SA51	*PTC107,*HEPG0005,*SK3005,*ECG160,*WEP637
2SA52	2SA49,*PTC109,*GE-51,*TR-17,*HEPG0008,*SK3005,*ECG126,*WEP635,*AF125, *276-2007
2SA52G	2SA49,2SA52,*GE-51,*SK3005

To Replace	Substitute This Type
2SA53	2N2953,2SA49,2SA52,SF.T307,SF.T308,*PTC109,*GE-51,*TR-17,*HEPG0008, *SK3007,*ECG126,*WEP635,*AF125,*276-2007
2SA53G	*GE-51,*SK3005
2SA54	*HEPG0003,*SK3005,*ECG160,*WEP637
2SA55	*PTC102,*HEPG0005,*SK3005,*ECG100,*WEP254
2SA56	2N705,2SA405,*PTC107,*HEPG0008,*ECG160,*WEP637
2SA57	*PTC107,*GE-51,*TR-17,*HEPG0008,*SK3007,*ECG126,*WEP635
2SA58	*PTC107,*GE-51,*TR-17,*HEPG0008,*SK3007,*ECG126,*WEP635
2SA58G	SEE 2SA518G
2SA59	*PTC107,*HEPG0008,*SK3007,*ECG126,*WEP635
2SA60	SF.T357P,*PTC107,*GE-51,*TR-17,*HEPG0008,*SK3007,*ECG126,*WEP635
2SA60G	SEE 2SA470G
2SA61	*PTC102,*HEPG0003,*SK3007,*ECG160,*WEP637
2SA64	*PTC107,*HEPG0005,*SK3008,*ECG102A,*WEP250
2SA65	*PTC102,*HEPG0005,*SK3005,*ECG102A,*WEP250
2SA66	*PTC102,*HEPG0005,*SK3005,*ECG102A,*WEP250
2SA67	*PTC102,*HEPG0005,*SK3006,*ECG126,*WEP635
2SA68	
2SA69	2N2496,2N3127,2N3281,2N3282,*PTC107,*GE-50,*TR-17,*HEPG0008,*SK3006, *HF12M,*ECG126,*WEP635
2SA70	2N2496,2N3127,2N3281,2N3282,*PTC107,*GE-50,*TR-17,*HEPG0008,*SK3007, *AT20M,*ECG160,*WEP637
2SA71	2N2496,2N3127,2N3281,2N3282,*PTC107,*GE-50,*TR-17,*HEPG0008,*SK3006, *AT20M,*ECG126,*WEP635
2SA72	SF.T357P,*PTC107,*GE-51,*TR-17,*HEPG0008,*SK3006,*JR100,*ECG126,*WEP635
2SA73	2N3783,2N3784,2SA234,2SA235,2SA246,2SA435,2SA436,2SA437,2SA438,AF134, AF135,AF136,AF138,AFY16,SF.T316,SF.T354,SF.T357,SF.T358,*PTC107,*GE-51, *TR-17,*HEPG0008,*SK3007,*JR100,*ECG160,*WEP637
2SA74	*GE-2,*TR-17,*HEPG0009,*SK3007,*AT20M,*ECG160,*WEP637,*276-2005
2SA75	*PTC145,*GE-2,*TR-17,*HEPG0009,*SK3007,*AT20M,*ECG126,*WEP635,*276-2005
2SA75N	SEE 2SA522N
2SA76	*PTC145,*GE-50,*TR-17,*HEPG0008,*SK3006,*ECG126,*WEP635
2SA77	*PTC145,*GE-51,*TR-17,*HEPG0008,*SK3007,*ECG126,*WEP635,*AF125,*RT-188
2SA78	*GE-51,*TR-85,*HEPG0005,*SK3006,*ECG126,*WEP635
2SA79	*PTC102,*HEPG0005,*SK3123,*ECG160,*WEP637
2SA80	2N3783,2N3784,2SA234,2SA235,2SA246,SF.T316,SF.T354,SF.T357,SF.T358, *PTC107,*GE-50,*TR-17,*HEPG0003,*SK3006,*HF12M,*ECG126,*WEP635,*AF125
2SA81	2N3783,2N3784,2SA234,2SA235,2SA246,SF.T316,SF.T354,SF.T357,SF.T358, *PTC107,*GE-50,*TR-17,*HEPG0009,*SK3008,*JR30X,*ECG160,*WEP637,*AF125, *276-2005
2SA82	2N3783,2N3784,2SA234,2SA235,2SA246,SF.T316,SF.T354,SF.T357,SF.T358, *PTC107,*GE-50,*TR-17,*HEPG0009,*SK3008,*HF12M,*ECG126,*WEP635,*AF125, *276-2005
2SA83	2N384,2N2588,2N3783,2N3784,2SA246,*PTC107,*GE-50,*TR-17,*HEPG0009, *SK3007,*JR30X,*ECG126,*WEP635,*AF125,*276-2005
2SA84	2N384,2N2588,2N3783,2N3784,2SA246,*PTC107,*GE-50,*TR-17,*HEPG0009, *SK3007,*JR30X,*ECG160,*WEP637,*AF125,*276-2005
2SA85	2N2588,2N3783,2N3784,*PTC107,*GE-50,*TR-17,*HEPG0009,*SK3007,*ECG126, *WEP635,*AF125,*276-2005
2SA86	*GE-51,*TR-17,*HEPG0001,*SK3123,*ECG160,*WEP637
2SA87	*HEPG0009,*SK3006,*ECG126,*WEP635
2SA88	*HEPG0009,*ECG160,*WEP637
2SA89	*HEPG0009,*ECG160,*WEP637
2SA90	*HEPG0008,*SK3006,*ECG126,*WEP635
2SA92	SF.T357P,*PTC107,*GE-51,*TR-85,*HEPG0008,*SK3007,*JR30,*ECG126,*WEP635
2SA92G	SEE 2SA468G
2SA93	SF.T357P,*PTC107,*GE-51,*TR-17,*HEPG0008,*SK3007,*JR30,*ECG126,*WEP635
2SA93G	SEE 2SA469G
2SA94	*HEPG0008,*SK3008,*ECG126,*WEP635
2SA95	
2SA96	

To Replace	Substitute This Type
2SA97	
2SA98	
2SA99	
2SA100	2N2188,2N2189,2N2190,2N2191,2N2956,2N2957,2SA104,*PTC107,*GE-51,*TR-17, *HEPG0008,*SK3005,*JR100,*ECG126,*WEP635,*AF125
2SA101	*PTC107,*GE-51,*TR-17,*HEPG0009,*SK3007,*JR100,*ECG126,*WEP635,*AF125, *276-2005
2SA102	2N2955,2SA103,ASY24,*PTC107,*GE-51,*TR-17,*HEPG0009,*SK3007,*JR100, *ECG126,*WEP635,*AF125,*276-2005,*RT-115
2SA103	2N2956,2SA353A,2SA354A,2SA355A,*PTC107,*GE-51,*TR-17,*HEPG0009,*SK3007, *JR100,*ECG126,*WEP635,*AF125,*276-2005
2SA104	2N2188,2N2189,2N2190,2N2191,2N2957,*PTC107,*GE-1,*TR-17,*HEPG0009, *SK3007,*HF12M,*ECG126,*WEP635,*AF125,*276-2005
2SA105	*PTC107,*HEPG0008,*SK3008,*ECG160,*WEP637
2SA106	*PTC107,*HEPG0008,*SK3007,*ECG160,*WEP637
2SA107	*PTC107,*HEPG0008,*SK3007,*ECG160,*WEP635
2SA108	*PTC107,*HEPG0008,*SK3007,*ECG126,*WEP635
2SA109	*PTC107,*HEPG0003,*SK3007,*ECG126,*WEP635
2SA110	*PTC107,*HEPG0003,*SK3007,*ECG126,*WEP635
2SA111	*PTC107,*HEPG0003,*SK3007,*ECG126,*WEP635
2SA112	*PTC107,*HEPG0003,*SK3007,*ECG126,*WEP635
2SA113	*PTC107,*HEPG0005,*SK3007,*ECG126,*WEP635
2SA114	*PTC107,*HEPG0005,*SK3007,*ECG102A,*WEP250
2SA115	*HEPG0005,*SK3007,*ECG102A,*WEP250
2SA116	*HEPG0003,*SK3006,*ECG102A,*WEP250
2SA117	*HEPG0003,*SK3006,*ECG160,*WEP637
2SA118	*HEPG0003,*SK3006,*ECG160,*WEP637
2SA119	*PTC127
2SA120	*PTC127
2SA121	2N499,2N499A,2N501A,2N559,2N711A,2N779,2N779A,2N828,2N964,2N968,2N980, 2N984,2N2401,2N2402,2N2487,2N2488,2N2489,2N2587,2N2717,2N2795,2N2860, 2N6365,2N6365A,2SA56,2SA405,2SA411,ASZ21,MM380,*GE-51,*TR-17, *HEPG0008,*SK3006,*JR100,*ECG126,*WEP635,*AF125
2SA122	2N499,2N499A,2N501A,2N559,2N711A,2N779,2N779A,2N828,2N964,2N968,2N980, 2N984,2N2401,2N2402,2N2487,2N2488,2N2489,2N2587,2N2717,2N2795,2N2860, 2N6365,2N6365A,2SA56,2SA405,2SA411,ASZ21,MM380,*GE-51,*TR-17, *HEPG0008,*SK3006,*JR100,*ECG126,*WEP635,*AF125
2SA123	2N499,2N499A,2N501A,2N559,2N711A,2N779,2N779A,2N828,2N964,2N968,2N980, 2N984,2N2401,2N2402,2N2487,2N2488,2N2489,2N2587,2N2717,2N2795,2N2860, 2N6365,2N6365A,2SA56,2SA405,2SA411,ASZ21,MM380,*GE-51,*TR-17, *HEPG0008,*SK3006,*ECG126,*WEP635,*AF125
2SA124	2N501A,2N559,2N711A,2N711B,2N779,2N779A,2N838,2N980,2N984,2N2401,2N2402, 2N2587,2N2717,2N2795,2N6365,2N6365A,ASZ21,MM380,*GE-51,*TR-17, *HEPG0008,*SK3006,*JR100,*ECG126,*WEP635
2SA125	2N559,2N711A,2N711B,2N779,2N779A,2N829,2N838,2N972,2N980,2N983,2N984, 2N2401,2N2402,2N2587,2N2717,2N2795,2N3320,2N6365,2N6365A,ASZ21,MM380, *GE-50,*TR-17,*HEPG0008,*SK3006,*ECG126,*WEP635
2SA126	2N705,2N710,2N711,2N711A,2N794,2N795,2N828,2N837,2N934,2N964,2N965, 2N966,2N967,2N968,2N969,2N970,2N2401,2N2402,2N2860,2SA56,2SA405,2SA410, 2SA411,2SA450,2SB185,*PTC107,*HEPG0008,*ECG160,*WEP637
2SA127	*GE-51,*TR-17,*ECG160,*WEP637
2SA127G	SEE 2SA429G
2SA128	*DS-25,*GE-51,*TR-84,*HEPG0005,*SK3008,*ECG158,*WEP238
2SA129	*DS25,*GE-51,*TR-84,*HEPG0005,*SK3008,*ECG158,*WEP238
2SA130	*HEPG0003,*SK3005,*ECG160,*WEP637
2SA131	*HEPG0003,*SK3005,*ECG126,*WEP635
2SA132	*HEPG0003,*SK3005,*ECG126,*WEP635
2SA133	*HEPG0003,*SK3006,*ECG126,*WEP635
2SA134	*PTC109,*HEPG0003,*SK3006,*ECG126,*WEP635
2SA135	*PTC109,*HEPG0003,*SK3005,*ECG160,*WEP637
2SA136	*PTC107,*HEPG0008,*SK3006,*ECG126,*WEP635

To Replace	Substitute This Type
2SA137	*PTC107,*HEPG0008,*SK3005,*ECG126,*WEP635
2SA138	*PTC107,*HEPG0005,*ECG102A,*WEP150
2SA139	2N2953,*PTC107,*HEPG0005,*SK3006,*ECG126,*WEP635
2SA141	*PTC107,*HEPG0005,*SK3007,*ECG126,*WEP635
2SA142	*PTC107,*HEPG0005,*SK3008,*ECG126,*WEP635
2SA142A	*PTC145,*HEPG0005,*SK3006,*ECG126,*WEP635
2SA143	*PTC107,*HEPG0005,*SK3006,*ECG126,*WEP635
2SA144	*HEPG0009,*SK3005,*ECG126,*WEP635
2SA145	*PTC109,*HEPG0009,*SK3005,*ECG126,*WEP635
2SA146	*PTC107,*HEPG0008,*SK3005,*ECG100,*WEP254
2SA147	*PTC107,*HEPG0008,*SK3005,*ECG100,*WEP254
2SA148	*PTC107,*HEPG0008,*SK3005,*ECG100,*WEP254
2SA149	*PTC107,*HEPG0008,*SK3005,*ECG100,*WEP254
2SA151	*HEPG0005,*SK3005,*ECG160,*WEP637
2SA152	*HEPG0005,*SK3005,*ECG160,*WEP637
2SA153	*HEPG0008,*SK3006,*ECG126,*WEP635
2SA154	*HEPG0008,*SK3006,*ECG126,*WEP635
2SA155	*HEPG0008,*SK3006,*ECG126,*WEP635
2SA156	*HEPG0008,*SK3006,*ECG126,*WEP635
2SA157	*HEPG0008,*SK3006,*ECG126,*WEP635
2SA159	*HEPG0008,*SK3006,*ECG126,*WEP635
2SA160	*HEPG0008,*SK3005,*ECG100,*WEP254
2SA161	2N700,*GE-51,*TR-17,*HEPG0008,*SK3006,*HF75,*ECG126,*WEP635,*276-2005
2SA162	2N700,2N3285,2N3286,*GE-51,*TR-17,*HEPG0008,*SK3006,*JR100,*ECG126, *WEP635,*276-2005
2SA163	2N3279,2N3280,2N3283,2N3284,AFY37,GM290A,GM378A,MM1139,*PTC107,*GE-51, *TR-17,*HEPG0008,*SK3006,*JR100,*ECG126,*WEP635
2SA164	*PTC109,*HEPG0008,*SK3006,*ECG126,*WEP635
2SA165	2N3279,2N3280,2N3283,2N3284,AFY37,GM290A,GM378A,MM1139,*GE-50,*TR-17, *HEPG0008,*SK3006,*ECG126,*WEP635,*276-2005
2SA166	2N700,2N3285,2N3286,*GE-51,*TR-17,*HEPG0008,*SK3006,*HF75,*ECG126, *WEP635,*276-2005
2SA167	*HEPG0005,*SK3005,*ECG100,*WEP254
2SA168	*HEPG0001,*SK3005,*ECG100,*WEP254
2SA168A	*HEPG0001,*SK3005,*ECG100,*WEP254
2SA169	2N1309A,*HEPG0005,*SK3005,*ECG100,*WEP254
2SA170	*HEPG0005,*SK3005,*ECG100,*WEP254
2SA171	2N414,2N582,2N584,2N1171,2N1307,2N1309A,2SA169,2SA207,2SA210,2SA407, *PTC109,*HEPG0005,*SK3005,*ECG100,*WEP254
2SA172	*HEPG0005,*SK3005,*ECG100,*WEP254
2SA173	2N123,2N414,2N414B,2N414C,2N415,2N415A,2N416,2N450,2N582,2N584,2N1115, 2N1171,2N1305,2N1307,2N1309A,2N1356,2N1681,2SA169,2SA171,2SA204,2SA207, 2SA209,2SA210,2SA212,2SA407,40269,ASY27,SF.T223,SF.T228,SYL792,*PTC109, *HEPG0005,*SK3005,*ECG102A,*WEP254
2SA174	2SA170,*HEPG0005,*SK3005,*ECG102,*WEP631
2SA175	*PTC145,*GE-51,*TR-17,*HEPG0008,*SK3006,*JR30,*ECG126,*WEP635
2SA176	*PTC102,*HEPG0008,*SK3008,*ECG126,*WEP635
2SA178	
2SA180	*PTC109,*HEPG0005,*SK3005,*ECG160,*WEP254
2SA181	*PTC109,*HEPG0005,*SK3005,*ECG102A,*WEP250
2SA182	2SA251,*PTC109,*GE-1,*TR-05,*HEPG0005,*SK3005,*JR10,*ECG100,*WEP254, *276-2007,*RT-118
2SA183	*PTC109,*HEPG0003,*SK3005,*ECG102A,*WEP250
2SA184	
2SA188	*PTC107,*HEPG0009,*SK3005,*ECG126,*WEP635
2SA189	*PTC107,*HEPG0008,*SK3005,*ECG126,*WEP635
2SA190	
2SA191	
2SA192	
2SA193	
2SA194	

To Replace	Substitute This Type
2SA195	
2SA196	
2SA197	*PTC102,*HEPG0008,*SK3006,*ECG102,*WEP631
2SA198	*PTC107,*HEPG0008,*SK3005,*ECG100,*WEP254
2SA199	
2SA200	
2SA201	2SA202,AFY15,SF.T307,*PTC109,*GE-50,*TR-17,*HEPG0005,*SK3005,*JR10,*ECG126,*WEP635,*276-2007
2SA202	AFY15,*PTC109,*GE-50,*TR-17,*HEPG0005,*SK3007,*JR30X,*ECG126,*WEP635,*276-2007
2SA203	2SA201,*PTC109,*GE-50,*TR-85,*HEPG0005,*SK3007,*JR10,*ECG126,*WEP250,*276-2007,*RT-121
2SA204	2N414,2N415,2N415A,2N416,2N1171,2N1305,2N1307,2N1309A,2N1348,2N1349,2N1350,2N1351,2N1356,2N1681,2SA207,2SA407,*PTC109,*HEPG0005,*SK3005,*ECG100,*WEP254
2SA205	2N1348,2N1350,2N1351,2SA206,2SA207,2SA406,2SA407,*PTC102,*HEPG0005,*SK3005,*ECG100,*WEP254
2SA206	2N1349,2N1350,2N1351,2SA207,2SA407,*PTC102,*HEPG0005,*SK3005,*ECG100,*WEP254
2SA207	2N1349,*HEPG0005,*SK3005,*ECG100,*WEP254
2SA208	2N315A,2N316A,2N317A,2N428,2N579,2N579+,2N580,2N580+,2N1017,2N1018,2N1313,2N1942,*PTC135,*GE-1,*TR-05,*HEPG0005,*SK3005,*HF6M,*ECG100,*WEP254,*AC188/01,*276-2004,*RT-118
2SA209	2N660,2N662,2N1171,2SA210,*PTC135,*GE-1,*TR-05,*HEPG0005,*SK3005,*HF6M,*ECG100,*WEP254,*AC188/01,*276-2004,*RT-118
2SA210	2N660,2N661,*DS-25,*GE-1,*TR-05,*HEPG0005,*SK3005,*HF12M,*ECG100,*WEP254,*276-2004,*RT-118
2SA211	2N123,2N414,2N414B,2N414C,2N415,2N415A,2N416,2N450,2N582,2N584,2N1115,2N1171,2N1305,2N1307,2N1309A,2N1356,2N1681,2SA204,2SA207,2SA209,2SA210,2SA212,2SA407,40269,ASY27,SF.T223,SF.T228,SF.T229,SYL792,*PTC109,*GE-1,*TR-05,*HEPG0005,*SK3005,*HF6M,*ECG100,*WEP254,*AC188/01,*276-2004,*RT-118
2SA212	2N414,2N414B,2N414C,2N415,2N415A,2N416,2N505,2N582,2N584,2N1171,2N1305,2N1307,2N1309A,2N1348,2N1349,2N1350,2N1351,2N1356,2N1681,2SA204,2SA207,2SA407,40269,ASY27,SF.T223,SF.T228,*PTC109,*GE-1,*TR-05,*HEPG0005,*SK3005,*HF6M,*ECG100,*WEP254,*AC188/01,*276-2007,*RT-118
2SA213	*PTC107,*HEPG0008,*SK3006,*ECG126,*WEP635
2SA214	*PTC107,*HEPG0008,*SK3006,*ECG126,*WEP635
2SA215	*PTC107,*HEPG0008,*SK3006,*ECG160,*WEP637
2SA216	*PTC107,*HEPG0008,*SK3006,*ECG126,*WEP635
2SA217	*GE-51,*TR-05,*HEPG0005,*SK3005,*HF35,*ECG100,*WEP254,*276-2004,*RT-118
2SA218	*PTC109,*HEPG0008,*SK3008,*ECG160,*WEP637
2SA219	2N827,*PTC107,*GE-50,*TR-17,*HEPG0003,*SK3006,*HF35,*ECG160,*WEP637
2SA220	*PTC145,*HEPG0003,*SK3008,*ECG160,*WEP637
2SA221	2N2273,2N3323,2N3324,2N3325,2N3371,*PTC145,*GE-50,*TR-17,*HEPG0003,*SK3007,*HF35,*ECG160,*WEP637
2SA222	2N827,2N2273,2N3371,*PTC145,*GE-50,*TR-17,*HEPG0003,*SK3007,*HF12M,*ECG160,*WEP637,*RT-183
2SA223	2N827,2N2635,*PTC145,*GE-50,*TR-17,*HEPG0003,*SK3006,*HF12M,*ECG160,*WEP637
2SA224	*PTC109,*HEPG0003,*SK3008,*ECG160,*WEP637
2SA225	*PTC107,*HEPG0003,*SK3006,*ECG160,*WEP637
2SA226	*PTC145,*HEPG0003,*ECG160,*WEP637
2SA227	*PTC109,*HEPG0003,*SK3006,*ECG160,*WEP637
2SA228	*ECG160,*WEP637
2SA229	2N700A,2N3127,2N3783,2N3784,AFY39,AFY42,MM5000,MM5001,MM5002,*GE-50,*TR-17,*HEPG0003,*SK3006,*ECG160,*WEP637
2SA229G	SEE 2SC390
2SA230	2N700A,2N3127,2N3783,2N3784,AFY39,AFY42,MM5000,MM5001,MM5002,*GE-50,*TR-17,*HEPG0003,*SK3006,*ECG160,*WEP637
2SA230G	SEE 2SC387AG

To Replace	Substitute This Type
2SA231	*PTC135,*HEPG0005,*SK3123,*ECG158,*WEP238
2SA232	*PTC135,*HEPG0006,*ECG158,*WEP238
2SA233	2N2496,2N3127,2N3281,2N3282,2N3783,2N3784,2SA234,2SA246,AFZ12,*PTC107, *GE-50,*TR-17,*HEPG0003,*SK3007,*HF12M,*ECG160,*WEP637
2SA234	2N2496,2N3783,2N3784,2SA235,2SA246,AFZ12,*PTC107,*GE-50,*TR-17, *HEPG0003,*SK3007,*HF12M,*ECG160,*WEP637
2SA235	2N3783,2N3784,2SA234,2SA246,*PTC107,*GE-50,*TR-17,*HEPG0003,*SK3006, *HF12M,*ECG160,*WEP637
2SA236	2N3783,2N3784,2SA234,2SA235,2SA246,2SA435,2SA436,2SA437,2SA438,AF134, AF135,AF136,AF138,AFY16,SF.T316,SF.T354,SF.T357,SF.T358,*PTC107,*GE-51, *TR-17,*HEPG0008,*SK3007,*JR10,*ECG126,*WEP635
2SA237	SF.T357P,*PTC107,*GE-51,*TR-17,*HEPG0008,*SK3008,*JR10,*ECG126,*WEP635
2SA238	*HEPG0003,*SK3006,*ECG126,*WEP635
2SA239	2N700,*GE-50,*TR-17,*HEPG0003,*SK3006,*JR100,*ECG160,*WEP637
2SA240	2N700A,2N3127,2N3281,2N3282,2N3399,2N3783,2N3784,AFY37,AFY42,AFZ12, MM5000,MM5001,MM5002,*GE-50,*TR-17,*HEPG0003,*SK3006,*HF12M,*ECG160, *WEP637
2SA241	*PTC107,*GE-50,*TR-17,*HEPG0008,*SK3006,*HF75,*ECG160,*WEP637
2SA242	*PTC107,*HEPG0008,*ECG160,*WEP637
2SA243	*PTC107,*HEPG0008,*ECG160,*WEP637
2SA243A	
2SA244	*HEPG0003,*ECG160,*WEP637
2SA244A	
2SA245	*HEPG0003,*ECG160,*WEP637
2SA246	*PTC107,*GE-51,*TR-17,*HEPG0003,*SK3006,*JR200,*ECG126,*WEP635
2SA247	*PTC145,*GE-51,*TR-17,*HEPG0008,*SK3006,*JR100,*ECG126,*WEP635
2SA248	*GE-51,*TR-05,*HEPG0008,*SK3005,*HF20M,*ECG100,*WEP254,*276-2005,*RT-118
2SA249	*SK3008
2SA250	*DS-25,*GE-2,*TR-17,*HEPG0003,*SK3008,*AT20M,*ECG126,*WEP635
2SA251	*PTC107,*HEPG0001,*SK3006,*ECG126,*WEP635
2SA252	*PTC109,*HEPG0001,*SK3006,*ECG126,*WEP635
2SA253	*PTC145,*HEPG0002,*SK3006,*ECG126,*WEP635
2SA254	*PTC107,*HEPG0003,*SK3005,*ECG126,*WEP635
2SA255	*PTC107,*HEPG0008,*SK3005,*ECG126,*WEP635
2SA256	*PTC107,*HEPG0008,*SK3008,*ECG126,*WEP635
2SA257	*PTC107,*HEPG0008,*SK3007,*ECG126,*WEP635
2SA258	*PTC107,*HEPG0008,*SK3007,*ECG126,*WEP635
2SA259	*PTC107,*HEPG0003,*SK3008,*ECG126,*WEP635
2SA260	*HEPG0003,*SK3006,*ECG126,*WEP635
2SA261	*HEPG0003,*SK3006,*ECG126,*WEP635
2SA262	*HEPG0003,*SK3006,*ECG126,*WEP635
2SA263	*HEPG0003,*SK3006,*ECG126,*WEP635
2SA264	*HEPG0003,*SK3006,*ECG126,*WEP635
2SA265	*HEPG0003,*SK3006,*ECG126,*WEP635
2SA266	*PTC107,*HEPG0008,*SK3008,*ECG126,*WEP635
2SA267	*PTC107,*HEPG0008,*SK3006,*ECG126,*WEP635
2SA268	*PTC107,*HEPG0008,*SK3006,*ECG126,*WEP635
2SA269	*PTC107,*HEPG0008,*SK3008,*ECG126,*WEP635
2SA270	*PTC107,*HEPG0008,*SK3007,*ECG126,*WEP635
2SA271	*PTC107,*HEPG0008,*SK3008,*ECG126,*WEP635
2SA272	*PTC107,*HEPG0008,*SK3008,*ECG126,*WEP635
2SA273	*PTC107,*HEPG0009,*SK3008,*ECG126,*WEP635
2SA274	*PTC107,*HEPG0009,*SK3008,*ECG126,*WEP635
2SA275	*PTC107,*HEPG0009,*SK3008,*ECG126,*WEP635
2SA276	2N3783,2N3784,2N3785,*PTC107,*GE-51,*TR-17,*HEPG0003,*JR200,*ECG160, *WEP637
2SA277	2N414,2N414B,2N414C,2N415,2N415A,2N416,2N582,2N584,2N1171,2N1303,2N1305, 2N1309A,2N1313,2N1681,2SA204,2SA206,2SA207,2SA209,2SA212,2SA406, 2SA407,ASY26,SF.T228,*PTC109,*GE-1,*TR-05,*HEPG0005,*SK3005,*HF6M, *ECG100,*WEP254,*AC188/01,*276-2007,*RT-118
2SA278	2N417,2N1307,2N1309,2N1357,*GE-1,*TR-05,*HEPG0005,*SK3005,*HF12H,

To Replace	Substitute This Type
(2SA278)	*ECG100,*WEP254,*276-2004,*RT-118
2SA278N	SEE 2SA522N
2SA279	*PTC145,*GE-50,*TR-05,*HEPG0009,*SK3005,*ECG100,*WEP254,*276-2005, *RT-118
2SA280	*PTC145,*HEPG0009,*SK3006,*ECG126,*WEP635
2SA281	*HEPG0009,*SK3006,*ECG126,*WEP635
2SA282	2N415,2N415A,2N416,2N1307,2N1309A,2N1356,2N1892,ASY27,SF.T229,*PTC109, *GE-1,*TR-05,*HEPG0005,*SK3005,*HF6M,*ECG100,*WEP254,*AC188/01, *276-2007,*RT-118
2SA283	2N1307,2N1309A,SF.T229,*PTC109,*GE-1,*TR-05,*HEPG0005,*SK3005,*HF12M, *ECG100,*WEP254,*276-2007,*RT-118
2SA284	2N1309A,*PTC109,*GE-1,*TR-05,*HEPG0005,*SK3005,*HF12M,*ECG100,*WEP254, *276-2005,*RT-118
2SA285	*PTC107,*HEPG0008,*SK3008,*ECG160,*WEP637
2SA286	*PTC107,*HEPG0008,*SK3008,*ECG160,*WEP637
2SA287	*HEPG0008,*SK3008,*ECG160,*WEP637
2SA288	2N3279,2N3280,2N3283,2N3284,*PTC145,*GE-51,*TR-17,*HEPG0003,*SK3008, *JR100,*ECG126,*WEP635
2SA289	2N3279,2N3280,2N3283,2N3284,*PTC145,*GE-51,*TR-17,*HEPG0003,*SK3006, *ECG126,*WEP635
2SA290	2N3279,2N3280,2N3283,2N3284,*PTC145,*GE-51,*TR-17,*HEPG0003,*SK3006, *ECG126,*WEP635
2SA291	*PTC107,*HEPG0009,*SK3006,*ECG126,*WEP635
2SA293	*PTC107,*HEPG0009,*SK3006,*ECG126,*WEP635
2SA296	*HEPG0009,*SK3005,*ECG160,*WEP637
2SA297	*HEPG0009,*SK3005,*ECG160,*WEP637
2SA298	*HEPG0009,*SK3008,*ECG160,*WEP637
2SA299	
2SA300	
2SA301	2N2588,A1383,*PTC145,*HEPG0009,*SK3006,*ECG126,*WEP635
2SA302	*PTC109,*HEPG0005,*ECG102A,*WEP250
2SA303	*PTC109,*HEPG0005,*SK3004,*ECG102A,*WEP250
2SA304-GREEN	2N1316,2N1317,2N2613
2SA304-RED	2N269,2N404,2N414,2N414B,2N414C,2N580,2N580+,2N582,2N584,2N1017,2N1171, 2N1313,2SA204,2SA205,2SA206,2SA209,2SA212,2SA406,2SA407,ASY26,SF.T227, SF.T228
2SA304-YELLOW	2N415,2N415A,2N416,2N1307,2N1309A,2N1356,2N1681,2N1808,2N1892,2SA210, 40269,ASY27,SF.T223,SF.T229
2SA305-GREEN	2N1316,2N1317,2N2613
2SA305-RED	2N404,2N580,2N580+,2N582,2N584,2N1017,2N1171,2SA406,2SA407
2SA305-YELLOW	2N1307,2N1309A,2SA210,SF.T229
2SA306	*HEPG0009,*ECG160,*WEP637
2SA307	*HEPG0009,*SK3008,*ECG160,*WEP637
2SA308	2N3279,2N3280,2N3283,2N3284,MM1139,*HEPG0003,*SK3006,*ECG126,*WEP635
2SA309	*HEPG0003,*SK3006,*ECG126,*WEP635
2SA310	*HEPG0003,*SK3006,*ECG126,*WEP635
2SA311	*DS25,*GE-51,*TR-05,*HEPG0005,*SK3008,*ECG100,*WEP254,*276-2007,*RT-118
2SA312	*PTC145,*GE-51,*TR-05,*HEPG0005,*SK3008,*ECG100,*WEP254,*276-2004, *RT-118
2SA313-BLUE	2N982,2N1726,2N1727,2N1748A,2N2273,2N3323,2N3324,2N3325,2N3412
2SA313-GREEN	2N982,2N1726,2N1727,2N1748A,2N2273,2N3323,2N3324,2N3325,2N3412
2SA313-RED	2N502B,2N980,2N1499B,2N1728,2N1746,2N1747,2N1790,2N1865,2N1866,2N1867, 2N2048,2N2168,2N2402,2N2489,2N2587,2N2717,2N2795,2N2860,2N3883,2N6365, 2N6365A,ASZ21
2SA313-YELLOW	2N502B,2N980,2N1499B,2N1728,2N1746,2N1747,2N1790,2N1865,2N1866,2N1867, 2N2048,2N2168,2N2402,2N2489,2N2587,2N2717,2N2795,2N2860,2N3883,2N6365, 2N6365A,ASZ21
2SA314-GREEN	2N2635,2N3995
2SA314-RED	2N502B,2N980,2N1499B,2N1728,2N1746,2N1747,2N1790,2N1865,2N1866,2N1867, 2N2048,2N2168,2N2402,2N2489,2N2587,2N2717,2N2795,2N2860,2N3883,2N6365, 2N6365A,ASZ21

To Replace	Substitute This Type
2SA314-YELLOW	2N982,2N1726,2N1727,2N1748A,2N1788,2N2273,2N3323,2N3324,2N3325,2N3412
2SA315-GREEN	2N2635,2N3995
2SA315-RED	2N502B,2N980,2N1499B,2N1728,2N1746,2N1747,2N1790,2N1865,2N1866,2N1867, 2N2048,2N2168,2N2402,2N2489,2N2587,2N2717,2N2795,2N2860,2N3883,2N6365, 2N6365A,ASZ21
2SA315-YELLOW	2N982,2N1726,2N1727,2N1748A,2N1788,2N2273,2N3323,2N3324,2N3325,2N3412
2SA316-GREEN	2N2635,2N3995
2SA316-RED	2N502B,2N980,2N1499B,2N1728,2N1746,2N1747,2N1790,2N1865,2N1866,2N1867, 2N2048,2N2168,2N2402,2N2489,2N2587,2N2717,2N2795,2N2860,2N3883,2N6365, 2N6365A,ASZ21
2SA316-YELLOW	2N982,2N1726,2N1727,2N1748A,2N1788,2N2273,2N3323,2N3324,2N3325,2N3412
2SA321	2N838,2N3323,2N3324,2N3325,*PTC107,*GE-50,*TR-17,*HEPG0008,*SK3006,*HF35, *ECG160,*WEP637
2SA322	2N838,2N3323,2N3324,2N3325,*PTC107,*GE-50,*TR-17,*HEPG0008,*SK3008, *HF12M,*ECG160,*WEP637
2SA323	2N827,2N2635,*PTC107,*GE-50,*TR-17,*HEPG0003,*SK3008,*HF12M,*ECG160, *WEP637
2SA324	2N838,2N3323,2N3324,2N3325,*PTC107,*GE-50,*TR-17,*HEPG0003,*SK3006,*HF35, *ECG160,*WEP637
2SA325	2N2953,2SA326,ACY23,ACY32,*HEPG0005,*SK3006,*ECG126,*WEP635
2SA326	2N2953,2SA325,ACY23,ACY32,*HEPG0005,*SK3006,*ECG126,*WEP635
2SA327	*HEPG0009,*ECG160,*WEP637
2SA328	
2SA329	40490,*PTC102,*HEPG0001,*SK3006,*ECG126,*WEP635
2SA330	40490,*PTC107,*HEPG0001,*SK3005,*ECG100,*WEP254
2SA331	2N2957,*PTC107,*GE-50,*TR-17,*HEPG0003,*SK3007,*ECG126,*WEP635
2SA332	*HEPG0005,*SK3008,*ECG100,*WEP254
2SA333	
2SA334	
2SA336	
2SA337	*HEPG0001,*SK3006,*ECG126,*WEP635
2SA338	2N979,2N2489,2N2796,*PTC107,*GE-51,*TR-17,*HEPG0008,*SK3008,*HF35, *ECG126,*WEP635,*AF125,*276-2002
2SA339	2N838,2N2273,2N2587,2N3323,2N3324,2N3325,2N6365,*PTC107,*GE-51,*TR-17, *HEPG0008,*SK3008,*HF35,*ECG126,*WEP635,*AF125,*276-2002
2SA340	2N2496,2N2966,2N3127,2N3281,2N3282,2SA341,2SA342,AFY12,AFZ12,*PTC107, *GE-51,*TR-17,*HEPG0003,*ECG160,*WEP637,*AF125
2SA341	2N2496,2N2966,2N3127,2N3281,2N3282,2SA342,AFY12,AFZ12,*PTC107,*GE-51, *TR-17,*HEPG0003,*SK3007,*ECG160,*WEP637,*AF125
2SA342	2N2496,2N2966,2N3127,2N3281,2N3282,AFY12,AFZ12,*PTC107,*GE-50,*TR-17, *HEPG0003,*SK3008,*JR100,*ECG160,*WEP637
2SA343	*HEPG0008,*SK3006,*ECG126,*WEP635
2SA344	*GE-51,*TR-17,*HEPG0009,*SK3006,*ECG126,*WEP635,*276-2005
2SA345	*PTC107,*HEPG0003,*ECG160,*WEP637
2SA346	*PTC107,*HEPG0003,*ECG160,*WEP637
2SA347	*PTC107,*HEPG0003,*ECG160,*WEP637
2SA348	*PTC107,*HEPG0003,*SK3006,*ECG160,*WEP637
2SA349	*PTC107,*HEPG0003,*ECG160,*WEP637
2SA350	2N827,2N2273,SF.T317,*PTC107,*GE-50,*TR-17,*HEPG0008,*SK3007,*JR100, *ECG126,*WEP635,*AF125
2SA351	2N838,2N2273,2N3323,2N3324,2N3325,2SA350,2SA352,SF.T317,SF.T320,*PTC107, *GE-50,*TR-17,*HEPG0008,*SK3007,*HF12M,*ECG126,*WEP635,*AF125
2SA352	2N838,2N2273,2N3323,2N3324,2N3325,2SA350,2SA351,SF.T317,SF.T320,*PTC107, *GE-50,*TR-17,*HEPG0008,*SK3007,*JR100,*ECG126,*WEP635,*AF125
2SA353	2N838,2N2188,2N2273,2N2956,2N2957,2N3323,2N3324,2N3325,2SA354,2SA355, *PTC107,*GE-50,*TR-17,*HEPG0008,*SK3007,*JR100,*ECG126,*WEP635,*AF125
2SA353A	2N2190,2SA354A,2SA355A,*GE-51,*TR-17,*HEPG0005,*SK3006,*ECG126,*WEP635, *276-2007
2SA354	2N838,2N2188,2N2273,2N2956,2N2957,2N3323,2N3324,2N3325,2SA353,2SA355, *PTC107,*GE-50,*TR-17,*HEPG0008,*SK3006,*JR100,*ECG126,*WEP635,*AF125
2SA354A	2N2190,2SA353A,2SA355A,*GE-51,*TR-17,*HEPG0008,*SK3006,*ECG126,*WEP635

To Replace	Substitute This Type
2SA355	2N1526,2N2188,2N2189,2N2273,2N2957,*PTC107,*GE-50,*TR-17,*HEPG0008, *SK3006,*HF35,*ECG126,*WEP635,*AF125
2SA355A	2N2190,2SA353A,2SA354A,*GE-51,*TR-17,*HEPG0009,*SK3006,*ECG126,*WEP635, *276-2005
2SA356	2N796,2N964A,2N972,2N973,2N974,2N985,2N1526,2N2022,2N2699,2N3320,2SA350, 40488,SF.T317,SF.T320,*PTC107,*GE-50,*TR-17,*HEPG0008,*SK3008,*HF35, *ECG126,*WEP635,*AF125
2SA357	2N796,2N964A,2N972,2N973,2N974,2N985,2N1526,2N2022,2N2699,2N3320,2SA350, 2SA356,40488,SF.T317,SF.T320,*PTC107,*GE-50,*TR-17,*HEPG0008,*SK3008, *ECG126,*WEP635
2SA358	*DS-41,*GE-51,*TR-17,*HEPG0006,*SK3007,*ECG126,*WEP635,*276-2021
2SA359	*HEPG0002,*SK3006,*ECG126,*WEP635
2SA360	*PTC107,*HEPG0003,*SK3007,*ECG160,*WEP637
2SA361	*PTC107,*HEPG0003,*SK3007,*ECG160,*WEP637
2SA362	*PTC107,*HEPG0003,*ECG160,*WEP637
2SA363	*HEPG0003,*ECG160,*WEP637
2SA364	*PTC107,*HEPG0003,*SK3006,*ECG126,*WEP635
2SA365	*PTC107,*HEPG0003,*SK3006,*ECG126,*WEP635
2SA366	*PTC107,*SK3006,*ECG126,*WEP635
2SA367	*PTC107,*HEPG0008,*SK3007,*ECG126,*WEP635
2SA368	*PTC107,*HEPG0008,*SK3007,*ECG126,*WEP635
2SA369	*PTC107,*HEPG0008,*SK3008,*ECG126,*WEP635
2SA370	
2SA371	2N2953,*HEPG0005
2SA372	*PTC145,*GE-51,*TR-17,*HEPG0008,*HF35,*ECG160,*WEP637,*276-2005
2SA373	*PTC145,*HEPG0002,*ECG160,*WEP637
2SA373A	
2SA374	*ECG102A
2SA375	2N2955,*PTC145,*HEPG0001,*SK3123,*ECG160,*WEP637
2SA376	2N700A,2N2496,2N2966,2N3127,2N3281,2N3282,2N3588,2N3783,2N3784,2SA341, 2SA342,2SA436,2SA437,2SA438,AF106,AF109R,AF139,AF239,AFY12,AFY37, AFY42,AFZ12,MM5000,MM5001,MM5002,*PTC107,*HEPG0008,*SK3006,*ECG126, *WEP635
2SA377	2N3279,2N3280,2N3283,2N3284,*GE-50,*TR-17,*HEPG0008,*SK3006,*HF75, *ECG160,*WEP637
2SA378	2N3279,2N3280,2N3283,2N3284,*GE-50,*TR-17,*HEPG0008,*SK3006,*HF75, *ECG160,*WEP637
2SA379	2N3279,2N3280,2N3283,2N3284,*GE-50,*TR-17,*HEPG0008,*SK3006,*HF75, *ECG160,*WEP637
2SA380	*HEPG0008,*SK3007,*ECG126,*WEP635
2SA381	*HEPG0008,*SK3008,*ECG126,*WEP635
2SA382	*HEPG0008,*SK3008,*ECG126,*WEP635
2SA383	*HEPG0008,*SK3008,*ECG126,*WEP635
2SA384	*HEPG0008,*SK3008,*ECG126,*WEP635
2SA385	*PTC135,*GE-50,*TR-17,*HEPG0005,*SK3006,*HF35,*ECG126,*WEP635,*AF125, *276-2007
2SA386	
2SA387	
2SA389	
2SA390	
2SA391	2N1307,2N1309A,ASY27,SF.T229,*PTC109,*GE-1,*TR-17,*HEPG0008,*SK3006, *HF35,*ECG126,*WEP635,*AC188/01,*276-2007
2SA392	2N417,2N599,2N1307,2N1309,2N1357,*PTC109,*GE-1,*TR-17,*HEPG0008,*SK3006, *HF35,*ECG126,*WEP635,*276-2007
2SA393	2N1309,*DS56,*GE-51,*TR-17,*HEPG0008,*SK3006,*HF35,*ECG126,*WEP635, *276-2005
2SA393A	2N1309,*DS56,*GE-51,*TR-17,*HEPG0008,*SK3006,*HF35,*ECG126,*WEP635, *276-2005
2SA394	*DS56,*GE-51,*TR-17,*HEPG0008,*SK3006,*HF35,*ECG126,*WEP635,*276-2005
2SA395	*DS56,*GE-51,*TR-17,*HEPG0008,*SK3006,*HF35,*ECG126,*WEP635,*276-2005
2SA396	2N397,2N1356,*PTC135,*GE-1,*TR-05,*HEPG0008,*HF12M,*ECG102,*WEP631,

To Replace	Substitute This Type
(2SA396)	*AC188/01,*276-2005,*RT-120
2SA397	2N599,2N1355,2N1357,*PTC135,*GE-1,*TR-05,*HEPG0008,*HF12M,*ECG102, *WEP631,*AC188/01,*276-2005,*RT-120
2SA398	2N599,2N1316,2N1357,*DS56,*GE-51,*TR-17,*HEPG0008,*SK3006,*HF35,*ECG126, *WEP635,*276-2005
2SA399	2N1349,*DS56,*GE-51,*TR-17,*HEPG0008,*SK3006,*HF35,*ECG126,*WEP635, *276-2005
2SA400	*PTC107,*HEPG0008,*SK3008,*ECG126,*WEP635
2SA401	*PTC107,*GE-50,*TR-17,*HEPG0003,*ECG160,*WEP637
2SA402	2N3136,2N3251,2N3504,2N3906,2N4035,2N4403,2N5244,2N5366,2N5379,2N5383, 2N6001,2N6005,2SA733,A5T3504,A5T3644,A5T3906,A5T4403,BC213L,BC214L, BC257A,BC257B,BC415-10,BC415-6,BC415A,BC415B,BC416-10,BC416-6,BC416A, BC416B,BCY78,BCY78A,BCY78B,BCY79,BCY79A,BCY79B,GI-3644,MM3906, MPS6518,MPS6534,NPS404A,TN-3906,TP3644,*PTC127,*HEPS0013,*ECG159, *WEP242
2SA403	*HEPG0003,*SK3006,*ECG160,*WEP637
2SA404	*HEPG0003,*SK3006,*ECG160,*WEP637
2SA405	2N705,2SA56,*PTC107,*HEPG0003,*ECG160,*WEP637
2SA406	2N1349,2N1350,2N1351,2SA207,2SA407,*PTC109,*HEPG0005,*SK3005,*ECG102A, *WEP250
2SA407	2SA207,*HEPG0005,*SK3005,*ECG102A,*WEP250
2SA408	2SA252,2SA409,*PTC107,*HEPG0001,*SK3123,*ECG160,*WEP637
2SA409	2SA252,*PTC107,*HEPG0001,*SK3123,*ECG160,*WEP637
2SA410	2N828,2N968,2N969,2N970,2N2860,2SA411,*HEPG0003,*ECG160,*WEP637
2SA411	2N828,2N968,*PTC145,*HEPG0008,*ECG160,*WEP637
2SA412	2N711,2N711A,2N711B,2N795,2N829,2N972,2N973,2N974,2N2401,2N2402,2SA451, *PTC145,*GE-51,*TR-17,*HEPG0005,*SK3006,*ECG126,*WEP635,*276-2004
2SA413	2N827,*PTC145,*GE-51,*TR-17,*HEPG0003,*ECG160,*WEP637
2SA414	2N414,2N1017,2N1171,2N1313,2N1350,2N1351,2SA204,2SA206,2SA207,2SA406, 2SA407,*PTC109,*GE-1,*TR-05;*HEPG0005,*SK3005,*ECG100,*WEP254, *AC188/01,*276-2007,*RT-118
2SA415	2N1171,2N1307,2N1309A,2N1349,2SA207,2SA407,*PTC109,*GE-1,*TR-05, *HEPG0005,*SK3005,*ECG100,*WEP254,*276-2007,*RT-118
2SA416	*GE-16,*TR-01,*HEPG6005,*SK3014,*HO300,*ECG121,*WEP232,*276-2006,*RT-127
2SA417	*HEPG0003,*ECG160,*WEP637
2SA419	*HEPG0003,*ECG160,*WEP637
2SA420	*PTC107,*HEPG0003,*SK3006,*ECG160,*WEP637
2SA421	*HEPG0003,*ECG160,*WEP637
2SA422	GM290A,*GE-51,*TR-17,*HEPG0003,*ECG160,*WEP637,*AF125,*276-2002
2SA423	
2SA424	
2SA425	*PTC145,*HEPG0002,*ECG160,*WEP637
2SA426	*PTC145,*HEPG0002,*ECG160,*WEP637
2SA427	*HEPG0009,*SK3006,*ECG126,*WEP635
2SA428	*HEPG0009,*SK3006,*ECG126,*WEP635
2SA429G-O	2SA605,2SA637,2SA638,2SA639,2SA685,2SA857,2SA858
2SA429G-R	2SA605,2SA638,2SA639
2SA429G-Y	2SA637,2SA685,2SA857,2SA858
2SA430	*HEPG0003,*ECG160,*WEP637
2SA431	*PTC107,*GE-50,*TR-17,*HEPG0003,*ECG160,*WEP637
2SA431A	*HEPG0003,*ECG160,*WEP637
2SA431AG	SEE 2SC397
2SA431G	SEE 2SC387G
2SA432	2N700A,2N3127,2N3783,2N3784,AFY39,AFY42,MM5000,MM5001,MM5002,*DS-41, *GE-51,*TR-17,*HEPG0003,*SK3006,*ECG160,*WEP637
2SA432A	*HEPG0003,*SK3006,*ECG160,*WEP637
2SA433	2N2496,2N3783,2N3784,2SA234,2SA235,2SA246,2SA435,2SA436,2SA437,2SA438, AF134,AF135,AF136,AF137,AF138,AFY16,AFZ12,SF.T316,SF.T354,SF.T357, SF.T358,*PTC107,*GE-51,*TR-17,*HEPG0008,*SK3006,*HF35,*ECG160,*WEP637, *AF125
2SA434	*PTC107,*HEPG0003,*SK3006,*ECG160,*WEP637

To Replace	Substitute This Type
2SA435	AFY16,*PTC145,*GE-50,*TR-17,*HEPG0003,*SK3008,*ECG160,*WEP637
2SA436	2N3783,2N3784,2SA437,2SA438,AFY16,*PTC107,*GE-50,*TR-17,*HEPG0003, *SK3006,*HF75,*ECG126,*WEP635
2SA437	2N3783,2N3784,2SA436,2SA438,AFY16,*PTC107,*GE-50,*TR-17,*HEPG0003, *SK3006,*HF75,*ECG126,*WEP635
2SA438	2N3783,2N3784,AFY16,*PTC107,*GE-50,*TR-17,*HEPG0003,*SK3006,*HF75, *ECG126,*WEP635
2SA440	2N3783,2N3784,2SA435,2SA436,2SA437,2SA438,AFY16,*GE-51,*TR-17,*HEPG0003, *SK3006,*HF75,*ECG160,*WEP637,*AF125
2SA440A	*PTC107,*HEPG0003,*SK3007,*ECG160,*WEP637
2SA441	
2SA443	
2SA445	
2SA446	2N985,*PTC145,*GE-53,*TR-17,*HEPG0002,*SK3006,*ECG126,*WEP635,*276-2005
2SA447	*PTC109,*HEPG0009,*SK3006,*ECG126,*WEP635
2SA448	*GE-51,*TR-17,*HEPG0003,*HF75,*ECG160,*WEP637,*276-2003
2SA449	2N828,2N828A,2N964,2N965,2N966,2N967,2N968,2N969,2N970,2N2860,2SA411, 2SA450
2SA450	*PTC145,*GE-51,*TR-17,*HEPG0008,*ECG160,*WEP637
2SA451	*PTC145,*GE-51,*TR-17,*HEPG0008,*ECG160,*WEP637
2SA452	*GE-51,*TR-17,*HEPG0008,*ECG160,*WEP637
2SA453	*HEPG0003,*SK3006,*ECG126,*WEP635
2SA454	*HEPG0003,*SK3006,*ECG126,*WEP635
2SA455	AFY37,GM290A,*HEPG0003,*SK3006,*ECG126,*WEP635
2SA456	2N700A,2N3783,2N3784,AF139,AF239,AFY37,AFY42,MM5000,MM5001,MM5002, *HEPG0003,*SK3006,*ECG126,*WEP635
2SA457	*PTC145,*HEPG0003,*SK3006,*ECG126,*WEP635
2SA458	2N414,2N414A,2N414B,2N414C,2N415,2N415A,2N416,2N654,2N654+,2N1008A, 2N1124,2N1171,2N1192,2N1305,2N1307,2N1309A,2N1348,2N1349,2N1350,2N1351, 2N1352,2N1356,2N1372,2N1374,2N1376,2N1381,2N1382,2N1383,2N1478,2N1681, 2N2375,2SA204,2SA207,2SA407,2SB326,ACY23,ACY32,ASY27,SF.T223, *HEPG0005
2SA459	2N417,2N598,2N599,2N655,2N655+,2N1193,2N1309,2N1316,2N1354,2N1355, 2N1357,2N1379,2N1808,2N1892,2N1998,2N2374,MA1703,*HEPG0005
2SA460	*HEPG0003,*ECG160,*WEP637
2SA461	*HEPG0003,*ECG160,*WEP637
2SA462	*HEPG0003,*ECG160,*WEP637
2SA463	*PTC107,*HEPG0003,*ECG160,*WEP637
2SA464	*HEPG0003,*ECG160,*WEP637
2SA465	*GE-21,*HEPS0019,*SK3114,*RT-115
2SA466	*PTC107,*SK3006,*ECG126,*WEP635
2SA467	2N2906,2N2906A,2N3485,2N3485A,2N3644,2N3645,2N3672,2N3673,2N4026,2SA720, 2SA891,CS9012,FT3644,FT3645,TN-3905,*PTC103,*HEPS0012,*SK3114,*ECG159, *WEP242
2SA467G-O	2N2838,2N3644,2N3645,2N3672,2N3673,2N4354,2N4402,2N5815,2N5817,2N6005, 2SA467,2SA720,2SA891,A5T4402,BC327-16,BCW35,BCW37,CS9012,D29E5, FT3644,FT3645,GI-3644,TN-3905,TQ62,*TR-20,*ECG159,*WEP242
2SA467G-Y	2N2838,2N2907,2N2907A,2N3486,2N3486A,2N3504,2N3505,2N3644,2N3645,2N3672, 2N3673,2N4028,2N4354,2N4355,2N4403,2N5373,2N5374,2N5375,2N5379,2N5817, 2N5819,2N6005,2SA720,2SA891,A5T2907,A5T3504,A5T3505,A5T3644,A5T3645, A5T4403,BC327-16,BC327-25,BCW35,BCW37,BSW73,D29E5,D29E6,EN2905, EN3502,FT3644,FT3645,GI-3644,TIS112,TN-3906,TQ60,TZ552,TZ553,*TR-20, *ECG159,*WEP242
2SA468	*PTC145,*GE-51,*TR-17,*HEPG0008,*SK3007,*ECG126,*WEP635,*AF125
2SA468G	*GE-50,*SK3007
2SA469	2SA472,*PTC145,*GE-51,*TR-17,*HEPG0008,*SK3007,*ECG126,*WEP635,*AF125
2SA469G	2SA472,*GE-50,*SK3007
2SA470	*PTC145,*GE-51,*TR-17,*HEPG0008,*SK3007,*ECG126,*WEP635,*AF125
2SA470G	*GE-51,*SK3006
2SA471-1	2N499A,2N980,2N1524,2N2402,2N2489,2N2587,2N2717,2N2795,2N2860,2N6365, 2N6365A,ASZ21,*PTC107,*GE-51,*TR-17,*HEPG0008,*SK3006,*ECG126,*WEP635,

To Replace	Substitute This Type
(2SA471-1)	*AF125
2SA471-2	2N827,2N982,2N1526,2N2273,2SA350,2SA355,SF.T317,*PTC107,*GE-51,*TR-17, *HEPG0008,*SK3006,*ECG126,*WEP635,*AF125
2SA471-3	*GE-51,*TR-17,*HEPG0006,*SK3006,*ECG126,*WEP635,*276-2021
2SA472	*GE-51,*TR-17,*HEPG0008,*SK3007,*ECG126,*WEP635,*AF125,*RT-183
2SA472-1	2N499A,2N980,2N1524,2N2402,2N2489,2N2587,2N2717,2N2795,2N2860,2N6365, 2N6365A,ASZ21,*PTC107,*GE-50,*TR-17,*HEPG0009,*276-2005
2SA472-2	2N827,2N982,2N1526,2N2273,2SA350,2SA355,SF.T317,*PTC107,*GE-50, *HEPG0008
2SA472-3	*PTC145,*GE-50
2SA472-4	2N499A,2N980,2N1524,2N2402,2N2489,2N2587,2N2717,2N2795,2N2860,2N6365, 2N6365A,ASY24B,ASZ21,*PTC107,*GE-50,*TR-17,*HEPG0009,*276-2005
2SA472-5	2N827,2N982,2N1526,2N2273,2SA350,2SA355,SF.T317,*PTC107,*GE-50, *HEPG0008,*276-2005
2SA472-6	*PTC145,*GE-50,*HEPG0005
2SA473-GR	*TR-77,*ECG153,*WEP246
2SA473-O	*TR-77,*ECG153,*WEP246
2SA473-R	*TR-77,*ECG153,*WEP246
2SA473-Y	2SA473-GR,*TR-77,*ECG153,*WEP246
2SA474	*GE-50,*TR-17,*HEPG0009,*SK3006,*ECG126,*WEP635,*276-2005
2SA474G	*GE-51,*SK3006,*276-2005
2SA475	*HEPG0009,*SK3007
2SA476	*PTC107,*HEPG0008,*SK3008,*ECG126,*WEP635
2SA476G	SEE 2SA495G
2SA477	*PTC107,*HEPG0008,*SK3006,*ECG126,*WEP635
2SA477G	SEE 2SA495G
2SA478	*PTC145,*GE-51,*TR-17,*HEPG0005,*SK3006,*ECG126,*WEP635,*276-2007
2SA478G	*PTC145,*GE-51,*SK3006
2SA479	*PTC145,*GE-51,*TR-17,*HEPG0005,*SK3006,*ECG126,*WEP635
2SA479G	*PTC145,*GE-51,*SK3006
2SA480	2N722,2N722A,2N924,2N940,2N2394,2N2946,2N2946A,2N3039,2N3040,2N3250, 2N3703,2N3829,2N3840,2N3905,2N4402,2N4964,2N4971,2N4981,2N5041,2N5382, 2N5448,2N6067,2SA467,2SA499-O,2SA500-O,2SA522A,2SA548,2SA555,2SA558, 2SA559A,A5T3905,A5T4248,A5T4402,A5T5448,A8T3703,BSW72,C9082,C9084, EN722,EN1132,EN3250,GI-3703,MM3905,MPS3703,TN-3905,TQ62,TW135, *PTC131,*HEPS0013,*SK3118,*ECG159,*WEP242
2SA482	2N1132,2N1132A,2N2800,2N2904,2N2904A,2N3245,2N3467,2N3468,2N3762, 2N3762A,2N3763,2N4030,2N5022,2N5042,2SA498-R,2SA503-O,2SA504-O,2SA717, MM3726,*PTC127,*HEPS0012,*SK3114,*ECG159,*WEP242
2SA483-O	
2SA483-R	
2SA483-Y	
2SA484-BL	
2SA484-R	2N5333,2SA484-Y,2SA510-R,2SA516A
2SA484-Y	2SA510-O
2SA485-BL	2SA484-BL
2SA485-R	2N4406,2N5333,2SA484-R,2SA484-Y,2SA485-Y,2SA512-R,2SA516
2SA485-Y	2N4407,2SA484-Y,2SA512-O
2SA486-BL	
2SA486-R	2N3762A,2N3763,2SA486-Y,92PU51A,MM3726
2SA486-Y	
2SA489-O	2N6107,2SB595,*GE-69,*TR-77,*ECG153,*WEP246
2SA489-R	2N6107,2N6126,2N6134,D45C10,D45C11,D45C12,D45C7,D45C8,D45C9,*GE-69, *TR-77,*ECG153,*WEP246
2SA489-Y	2SB595,*TR-77,*ECG153,*WEP246
2SA490-O	2SA489-O,*ECG292
2SA490-R	2N6125,2SA489-R,D45C4,D45C5,D45C6,D45C7,D45C8,D45C9
2SA490-Y	2SA489-Y,2SB507
2SA493-GR	2N6007,2SA776,2SA776A,BC416-10,BC416-16,BC416B,BC416C,*276-2032
2SA493-O	2N3250,2N3250A,2N3645,2N4143,2N4228,2N4972,2N6005,2SA499-O,2SA499-Y, 2SA548,2SA603,2SA710,2SA711,2SA718,2SA720,2SA891,BC212A,BC212L,

To Replace	Substitute This Type
(2SA493-O)	BC416-6,BC416A,EN3250,FT3644,GET2906,GET2907,TP3645,*ECG159
2SA493-Y	2N3251,2N3251A,2N3645,2N4143,2N4972,2N6005,2SA493-GR,2SA499-Y,2SA603, 2SA718,2SA720,2SA733,2SA776,2SA776A,2SA891,BC212A,BC416-10,BC416-6, BC416A,BC416B,FT3644,GET2906,GET2907,TP3645,*276-2021
2SA493G-GR	2N3798,2N3962,2N3965,2N5374,2N6007,2SA493-GR,2SA561-GR,2SA661-GR, 2SA678,2SA705,2SA776,2SA776A,2SA841-GR,BC256B,BC266B,BC307B,BC416-10, BC416-16,BC416B,BC416C,BC557B,BC560B,BC560C,TZ553
2SA493G-Y	2N3251,2N3251A,2N3645,2N3798,2N3962,2N4143,2N4354,2N4355,2N4972,2N5373, 2N5374,2N6005,2SA493-GR,2SA493-Y,2SA493G-GR,2SA499-Y,2SA561-GR, 2SA561-Y,2SA603,2SA659,2SA661-GR,2SA661-Y,2SA678,2SA705,2SA718,2SA720, 2SA733,2SA776,2SA776A,2SA841-GR,2SA891,A177,BC212A,BC256A,BC256B, BC266A,BC266B,BC307A,BC307B,BC416-10,BC416-6,BC416A,BC416B,BC557A, BC557B,BC560A,BC560B,BSW75,FT3644,GET2906,GET2907,TP3645,TZ552, TZ553
2SA494-GR	2N3964,2SA493-GR,2SA550A,2SA578,2SA579,2SA640,2SA641,2SA678,2SA705, 2SA750,2SA776,2SA776A,2SA842-GR,2SA889,BC177B,BC214L,BC251B,BC251C, BC257B,BC261B,BC261C,BC307B,BC307C,BC415-10,BC415-16,BC415B,BC415C, BC416-10,BC416-16,BC416B,BC416C,BC557B,BC560B,BC560C,BCY78,BCY78B, BCY78C,BCY78D,BCY79,BCY79B,BCY79C,MM4048,*PTC127,*GE-89,*TR-20, *SK3114,*ECG159,*WEP242,*276-2023,*RT-115
2SA494-O	2N3250,2N3581,2N3703,2N3905,2N4034,2N4121,2N5382,2N5448,2N6067,2SA467, 2SA467G-O,2SA495-O,2SA495G-O,2SA499-O,2SA499-Y,2SA548,2SA710,2SA711, 2SA767,A5T3905,A5T5448,A8T3703,A177,BC177,BC177A,BC203,BC251A,BC257A, BC261A,BC307A,BC415-6,BC415A,BC416-6,BC416A,EN3250,FT3645,GI-3702, MM3905,MM3906,MPS3703,TIS37,TN-3905,TP3644,TP3703,*PTC103,*GE-89,*TR-20, *SK3114,*ECG159,*WEP242,*276-2023,*RT-115
2SA494-Y	2N3251,2N3582,2N3702,2N3906,2N4034,2N4035,2N4121,2N4122,2N5244,2N5383, 2N5447,2SA467G-Y,2SA493-GR,2SA493-Y,2SA494-GR,2SA495-Y,2SA495G-Y, 2SA499-Y,2SA550A,2SA640,2SA641,2SA659,2SA678,2SA705,2SA733,2SA750, 2SA776,2SA776A,2SA842-GR,2SA889,A5T3906,A5T5447,A8T3702,A177,BC177A, BC177B,BC213L,BC214L,BC251A,BC251B,BC257A,BC257B,BC261A,BC261B, BC307A,BC307B,BC415-10,BC415-6,BC415A,BC415B,BC416-10,BC416-6,BC416A, BC416B,BC557A,BC557B,BC560A,BC560B,BCY78,BCY78A,BCY78B,BCY79, BCY79A,BCY79B,BSW19,BSW20,FT3645,MM3906,MM4048,MPS3702,NPS404A, TN-3906,TP3644,TP3702,*PTC103,*TR-20,*SK3114,*ECG159,*WEP242,*276-2023, *RT-115
2SA494G	SEE 2SA493G
2SA495-O	2N3250,2N3644,2N3905,2N4034,2N4121,2N4972,2N5382,2N6001,2N6005,2SA467, 2SA495G-O,2SA499-O,2SA499-Y,2SA548,2SA710,2SA711,A5T3905,BC257A, BC415-6,BC415A,BC416-6,BC416A,CS9012,EN3250,FT3645,GI-3644,MM3905, MM3906,TN-3905,TP3644,*PTC103,*GE-65,*TR-20,*HEPS0013,*SK3025,*ECG159, *WEP242,*276-2034,*RT-115
2SA495-R	2N3250,2N3829,2N3905,2N4971,2N5382,2SA467,2SA499-O,2SA522A,2SA548, 2SA555,2SA558,A5T3905,EN3250,MM3905,TN-3905,*PTC103,*GE-65,*TR-20, *HEPS0012,*SK3114,*ECG159,*WEP242,*276-2034,*RT-115
2SA495-Y	2N3251,2N3504,2N3644,2N3906,2N4034,2N4035,2N4121,2N4122,2N4403,2N4972, 2N5244,2N5379,2N5383,2N6001,2N6005,2SA495G-Y,2SA499-Y,2SA733,A5T3504, A5T3644,A5T3906,A5T4403,BC213L,BC214L,BC257A,BC257B,BC415-10,BC415-6, BC415A,BC415B,BC416-10,BC416-6,BC416A,BC416B,BCY78,BCY78A,BCY78B, BCY79,BCY79A,BCY79B,FT3645,GI-3644,MM3906,NPS404A,TN-3906,TP3644, *PTC103,*GE-65,*TR-20,*HEPS5013,*SK3114,*ECG159,*WEP242,*276-2034, *RT-115
2SA495G-O	2N3250,2N3644,2N3905,2N4034,2N4121,2N4972,2N5382,2N6001,2N6005,2SA467, 2SA495-O,2SA499-O,2SA499-Y,2SA548,2SA710,2SA711,A5T3905,BC257A, BC415-6,BC415A,BC416-6,BC416A,CS9012,EN3250,FT3645,GI-3644,MM3905, MM3906,TN-3905,TP3644,*PTC103,*GE-65,*TR-20,*ECG159,*WEP242,*276-2034
2SA495G-R	2N3250,2N3829,2N3905,2N4971,2N5382,2SA467,2SA499-O,2SA522A,2SA548, 2SA555,2SA558,A5T3905,EN3250,MM3905,TN-3905,*PTC131,*GE-65,*TR-20, *ECG159,*WEP242,*276-2034
2SA495G-Y	2N3251,2N3504,2N3644,2N3906,2N4034,2N4035,2N4121,2N4122,2N4403,2N4972, 2N5244,2N5379,2N5383,2N6001,2N6005,2SA495-Y,2SA499-Y,2SA733,A5T3504,

To Replace	Substitute This Type
(2SA495G-Y)	A5T3644,A5T3906,A5T4403,BC213L,BC214L,BC257A,BC257B,BC415-10,BC415-6, BC415A,BC415B,BC416-10,BC416-6,BC416A,BC416B,BCY78,BCY78A,BCY78B, BCY79,BCY79A,BCY79B,FT3645,GI-3644,MM3906,NPS404A,TN-3906,TP3644, *PTC103,*GE-65,*TR-20,*ECG159,*WEP242,*276-2034
2SA496-O	2N2801,2N3244,2SA505-O,2SA571,2SA684,2SA885,2SA886,BC327-16,MPSU52, *PTC905,*GE-48,*TR-77,*ECG185,*WEPS5007,*276-2026,*RT-153
2SA496-R	2N2800,2N3244,2N3245,2N3467,2N3468,2N3762,2N3762A,2N3763,2N4030,2N5042, 2SA505-R,2SA571,2SA717,MM3726,*PTC905,*GE-48,*TR-77,*SK3191,*ECG185, *WEPS5007,*276-2026,*RT-153
2SA496-Y	2N2801,2N4032,2SA505-Y,2SA684,2SA885,BC327-16,BC327-25,MPS4354,MPS4355, *PTC905,*GE-48,*TR-77,*SK3191,*ECG185,*WEPS5007,*276-2021,*RT-153
2SA497-O	2N4314,2N4405,2N4407,2SA512-O,MPS4356
2SA497-R	2N4031,2N4036,2N4404,2N4406,2SA512-O,2SA512-R,2SA546A,2SA560
2SA497-Y	2N4033,2N4405,MPS4356
2SA498-O	2N2801,2N4037,2N5865,2SA571,2SA684,2SA817,*GE-48,*276-2025
2SA498-R	2N2800,2N3245,2N3468,2N3762A,2N3763,2N4030,2N5865,2SA546,2SA571,2SA717, MM3726,*GE-48,*276-2025
2SA498-Y	2N2801,2N4032,2SA684,2SA773,2SA817,MPS4354,MPS4355,*GE-48,*276-2025
2SA499-O	2N3250,2N3250A,2SA499-Y,2SA603,2SA710,2SA711,2SA718,*PTC127,*GE-65, *TR-20,*ECG159,*WEP242,*276-2023
2SA499-R	*PTC127,*GE-65,*TR-20,*ECG159,*WEP242,*276-2023
2SA499-Y	2N3251,2N3251A,2N6005,2SA603,2SA710,2SA711,2SA718,*PTC127,*GE-65,*TR-20, *ECG159,*WEP242,*276-2023
2SA500-O	2N3250,2N3829,2N4034,2SA499-O,2SA499-Y,2SA500-Y,2SA710,2SA711,CS9012, *PTC127,*GE-65,*TR-20,*ECG159,*WEP242,*276-2034
2SA500-R	2N3829,*PTC131,*GE-65,*TR-20,*ECG159,*WEP242,*276-2023
2SA500-Y	2N3251,2N3906,2N4034,2N4035,2N5244,2N5383,2N6001,2N6005,2SA499-Y,2SA710, 2SA711,A5T3906,CS9012,MM3906,NPS404A,TN-3906,*PTC127,*GE-65,*TR-20, *ECG159,*WEP242,*276-2034
2SA501	2N2904,2N2904A,2SA594-O,2SA594-R,2SA594N,BC360-10,BC360-6,BC361-10, BC361-6,*PTC127,*HEPS0012,*SK3025,*ECG129,*WEP242
2SA502-O	
2SA502-R	
2SA502-Y	2SA850
2SA503-GR	2N2905,2N2905A,2N4032,2N4033,2N4405,2SA684,*PTC127,*GE-48,*WEPS3031
2SA503-O	2N2904,2N2904A,2N3763,2N4030,2N4031,2N4404,2SA503-Y,2SA560,2SA571, 2SA717,*PTC127,*GE-48,*TR-88,*SK3025,*ECG129,*WEP242,*RT-115
2SA503-Y	2N2904,2N2904A,2N4030,2N4031,2SA571,2SA684,2SA717,*PTC127,*GE-48,*TR-88, *SK3025,*ECG129,*WEP242,*RT-115
2SA504-GR	2N2905,2N2905A,2N4032,2SA503-GR,2SA684,*PTC127,*GE-48,*WEPS3031
2SA504-O	2N2800,2N2904,2N2904A,2N3244,2N3245,2N3467,2N3468,2N3762,2N3762A,2N3763, 2N4030,2SA503-O,2SA503-Y,2SA504-Y,2SA571,2SA717,MM3726,*PTC127,*GE-48, *TR-88,*SK3025,*ECG129,*WEP242,*RT-115
2SA504-Y	2N2801,2N2904,2N2904A,2N3244,2N3467,2N4030,2SA503-Y,2SA571,2SA684, 2SA717,*PTC127,*GE-48,*TR-88,*SK3025,*ECG129,*WEP242,*RT-115
2SA505-O	2N4405,2N4407,2N5865,2SA571,2SA682-O,2SA684,2SA817,MPS4356,MPSU52, *PTC905,*GE-48,*TR-77,*SK3191,*ECG185,*WEPS5007,*276-2026,*RT-153
2SA505-R	2N3763,2N4030,2N4031,2N4404,2N4406,2N5865,2SA560,2SA571,2SA717,*PTC905, *GE-48,*TR-77,*SK3191,*ECG185,*WEPS5007,*276-2026,*RT-153
2SA505-Y	2N4032,2N4033,2N4405,2SA682-Y,2SA684,2SA817,MPS4354,MPS4355,MPS4356, *PTC905,*GE-48,*TR-77,*SK3191,*ECG185,*WEPS5007,*276-2026,*RT-153
2SA506	*HEPG0003,*ECG160,*WEP637
2SA507	*HEPG0003,*SK3006,*ECG160,*WEP637
2SA508	*HEPG0003,*ECG160,*WEP637
2SA509-O	2SA509G-O,A5T4402,*GE-89,*TR-20,*SK3114,*ECG290,*WEP242,*276-2032, *RT-115
2SA509-Y	2SA509G-Y,A5T3504,A5T3644,A5T4403,*GE-89,*TR-52,*SK3114,*WEP242,*276-2032, *RT-115
2SA509G-O	2SA509-O,A5T4402
2SA509G-Y	2SA509-Y,A5T3504,A5T3644,A5T4403
2SA510-O	*GE-89,*TR-20,*SK3025,*ECG159,*WEP242,*RT-115

To Replace	Substitute This Type
2SA510-R	2SA510-O,2SA516A,*GE-89,*TR-52,*ECG159,*WEP242,*RT-115
2SA511-O	2N4314,2SA510-O,2SA546A,*GE-89,*TR-20,*SK3114,*ECG159,*WEP242,*RT-115
2SA511-R	2N4036,2SA510-O,2SA510-R,2SA516A,2SA546A,*GE-89,*TR-52,*SK3114,*ECG159, *WEP242,*RT-115
2SA512-O	2N4407,*GE-89,*TR-20,*SK3025,*ECG159,*WEP242,*RT-115
2SA512-R	2N4406,2SA512-O,2SA516,*GE-89,*TR-52,*ECG159,*WEP242,*RT-115
2SA513-O	2N4030,2N4037,2N5865,2SA546,2SA571,2SA684,2SA717,BC161-10,BC161-16, *GE-89,*TR-20,*SK3114,*ECG159,*WEP242,*RT-115
2SA513-R	2N3245,2N3468,2N3762A,2N3763,2N4030,2N5865,2SA546,2SA571,2SA717, 92PU51A,BC161-10,BC161-6,MM3726,*GE-89,*TR-52,*SK3114,*ECG159,*WEP242, *RT-115
2SA516	2N4406,2SA512-O,2SA512-R,*PTC127,*HEPS3032,*SK3025,*ECG129,*WEP242
2SA516A	2SA510-O,2SA510-R,*SK3025,*ECG129,*WEP242
2SA517	*PTC107,*HEPG0008,*SK3007,*ECG126,*WEP635
2SA517G	SEE 2SA495G
2SA518	*PTC145,*GE-51,*TR-17,*HEPG0008,*SK3007,*ECG126,*WEP635,*AF125
2SA518G	*PTC145,*GE-51,*SK3006,*AF125
2SA522	2N869A,2N3829,2N3905,2N5382,2SA467,2SA500-O,2SA642,A5T3905,MM3905, TN-3905,*PTC131,*HEPS0013,*SK3114,*ECG159,*WEP242
2SA522A	2N2906,2N2906A,2N3250,2N3250A,2N3485,2N3485A,2N4142,2N4228,2SA499-O, 2SA539,2SA710,2SA711,GET2904,GET2905,*PTC127,*HEPS0013,*SK3114, *ECG159,*WEP242
2SA523	2N2904,2N2904A,2N3244,2N3467,2N4030,2SA523A,2SA571,2SA717,BC360-10, BC360-6,BC361-10,BC361-6,*PTC127
2SA523A	2N2904,2N2904A,2N4030,2N4031,2N4404,2SA571,2SA717,BC361-10,BC361-6, *PTC127
2SA524	*PTC127
2SA525	*HEPS0013,*SK3118,*ECG160,*WEP637
2SA525G	SEE 2SC387AG
2SA526	2N3910,2N4026,2N4030,2N4037,2N5865,2SA486-R,2SA486-Y,2SA546,2SA571, 2SA717,92PU51A,A5T4026,A8T4026,BC161-10,BC161-6,MM3726
2SA527	2SA528,*PTC109,*HEPS5013,*SK3025,*ECG129,*WEP242
2SA528	*PTC107,*HEPS5013,*SK3025,*ECG129,*WEP242
2SA529	
2SA530	2N3250,2N3250A,2SA499-O,2SA548,EN3250,*PTC127,*GE-21,*TR-20,*HEPS0019, *SK3114,*ECG159,*WEP242,*276-2023,*RT-115
2SA532	2N2801,2N2838,2N2904,2N2904A,2N3671,2N4890,2N5372,2N5815,2N5817,2N5821, 2N5823,2SA503-Y,2SA544,2SA552,2SA580,2SA594,2SA594-O,2SA594N,2SA661-O, 2SA817,2SA891,40537,A5T5448,A8T3703,BC361-10,D29E10,D29E4,D29E5,D29E9, *TR-88,*HEPS0012,*SK3025,*ECG129,*WEP242,*276-2021,*RT-115
2SA535	2N4027,2N4031,2N4404,2N5821,2N5865,2SA497-R,2SA537A,2SA546,2SA546A, 2SA560,2SA606,40362,D29E9
2SA536	2N2800,2N2837,2N3245,2N3468,2N3762A,2N3763,2N3764A,2N3765,2N4026,2N4030, 2N5815,2N5821,2N5865,2SA498-R,2SA537,2SA546,2SA571,2SA717,40362,40406, D29E4,D29E9,MM3726,*PTC141,*HEPS3032
2SA537	2N4030,2N4031,2N4404,2SA571,2SA717,*GE-48,*TR-88,*HEPS5022,*SK3025, *ECG129,*WEP242,*276-2021,*RT-115
2SA537A	*TR-88,*HEPS3032,*SK3025,*ECG129,*WEP242,*RT-115
2SA538	*GE-53,*TR-85,*HEPG0005,*ECG102A,*WEP250,*RT-121
2SA538G	
2SA539	2N2906,2N2906A,2N3250A,2N3485,2N3485A,2N4142,2N4228,GET2904,GET2905, *PTC103,*GE-82,*TR-54,*HEPS0019,*SK3025,*ECG159,*WEP242,*BC177, *276-2021,*RT-303
2SA542	2SA543,FX3502,FX4034,MMT3905,*GE-65,*TR-20,*HEPS0019,*SK3114,*ECG159, *WEP242,*BC177,*276-2022,*RT-115
2SA543	2SA542,FX3502,FX4034,MMT3905,*GE-65,*TR-52,*HEPS0019,*SK3025,*WEP243, *BC177,*276-2022,*RT-115
2SA544	2N2904,2N2904A,2SA552,2SA594,2SA594-O,2SA594N,BC361-10,*PTC127,*GE-21, *TR-20,*HEPS5022,*SK3114,*ECG159,*WEP242,*276-2021,*RT-115
2SA545	*PTC177,*GE-82,*TR-52,*HEPS5022,*SK3114,*ECG193,*WEP242,*BC327,*276-2021, *RT-115

To Replace	Substitute This Type
2SA546	2N4031,2N4404,2N4407,2N5865,2SA546A,*PTC141,*GE-67,*TR-88,*HEPS3032, *SK3025,*ECG129,*WEP242,*RT-101
2SA546A	*PTC141,*TR-88,*HEPS3032,*SK3025,*ECG129,*WEP242,*RT-115
2SA547	2SA547A,*GE-69,*HEPS3032,*ECG193,*WEP242
2SA547A	*HEPS3032,*ECG193,*WEP242
2SA548	2SA710,2SA711,*PTC103,*HEPS0019,*SK3025,*ECG129,*WEP242
2SA549	
2SA549A	2N5400,2SA850,2SA912,A5T5400,*ECG288
2SA550	2N3906,2N4035,2N4403,2N5244,2N5375,2N5378,2N5379,2N5383,2N5999,2N6003, 2N6009,A5T3906,A5T4403,BC178,BC178B,BC179,BC252B,BC253B,BC258B, BC262B,BC263B,BC308B,BC309B,BC328,BC558B,BC559B,BCY78,BCY78B, BCY78C,BSW19,BSW20,BSW73,NPS404,NPS404A,TN-3906,TQ60,*PTC127,*GE-21, *TR-20,*HEPS0031,*SK3025,*ECG159,*WEP242,*276-2024,*RT-115
2SA550A	2N2907,2N2907A,2N3251,2N3251A,2N3486,2N3486A,2N3504,2N3505,2N3962, 2N4355,2N5373,2N5374,2N6007,2SA661-GR,2SA678,2SA705,A5T2907,A5T3504, A5T3505,A5T3644,A5T3645,BC177B,BC214L,BC251B,BC256B,BC257B,BC261B, BC266B,BC307B,BC327,BC415-10,BC415B,BC416-10,BC416B,BC557B,BC560B, BCY79,BCY79B,BCY79C,EN2905,EN3502,TIS112,TZ552,TZ553,*PTC127,*GE-65, *TR-20,*HEPS0019,*SK3114,*ECG159,*WEP242,*276-2021,*RT-115
2SA552	2N2904,2N2904A,2SA503-Y,2SA544,2SA594,2SA594-O,2SA594N,BC361-10,*PTC127, *GE-21,*TR-88,*HEPS5022,*SK3025,*ECG129,*WEP242,*276-2021,*RT-115
2SA553	*PTC127
2SA554	*PTC127
2SA555	2N2906,2N2906A,2N3250,2N3250A,2N3485,2N3485A,2N4142,2N4228,2N4971, 2SA539,GET2904,GET2905,*PTC103
2SA556	
2SA557	
2SA558	2N2906,2N2906A,2N3485,2N3485A,GET2904,GET2905,*PTC127
2SA559	2N869A,2N3545,2N3576,2N3829,*PTC127
2SA559A	2N722,2N722A,2N2394,2N2837,2N2906,2N2906A,2N3039,2N3485,2N3485A,2N3910, 2N3913,2N5372,2SA558,2SA661-R,BSW72,GET2904,GET2905,TZ551,*PTC127
2SA560	2N4031,2N4404,*HEPS3032,*SK3025,*ECG129,*WEP242
2SA561-GR	2N3962,2N3965,2N5374,2N6007,2SA661-GR,2SA678,2SA705,TZ553,*GE-48,*TR-30, *WEP242,*RT-115
2SA561-O	2N3250,2N3250A,2N3645,2N3672,2N3673,2N3703,2N4143,2N4228,2N4354,2N5448, 2N6005,2SA603,2SA661-O,2SA718,2SA720,2SA891,A5T5448,A8T3703,A177, BC212A,BC212L,BCW35,BCW37,FT3644,GET2906,GET2907,MPS3703,TP3645, *PTC103,*GE-89,*TR-20,*SK3114,*ECG159,*WEP242,*RT-115
2SA561-R	2N722,2N722A,2N2394,2N2906,2N2906A,2N3250,2N3250A,2N3485,2N3485A, 2N3703,2N4142,2N5372,2N5448,2N5763,2SA661-R,A5T5448,A8T3703,BSW74, EN1132,GET2904,GET2905,GI-3703,MPS3703,TZ551,*PTC103,*GE-89,*TR-52, *SK3114,*ECG159,*WEP242,*RT-115
2SA561-Y	2N2907,2N2907A,2N3251,2N3251A,2N3486,2N3486A,2N3505,2N3645,2N3672, 2N3673,2N3962,2N4143,2N4354,2N4355,2N5373,2N5374,2N6005,2SA561-GR, 2SA603,2SA659,2SA661-GR,2SA661-Y,2SA678,2SA705,2SA718,2SA720,2SA891, A5T2907,A5T3505,A5T3645,A177,BC212A,BCW35,BCW37,BSW75,EN2905,FT3644, GET2906,GET2907,TIS112,TP3645,TZ552,TZ553,*PTC103,*GE-89,*TR-20,*SK3114, *ECG159,*WEP242,*RT-115
2SA562-GR	2N5378,2N5379,2N5813,2N6003,2N6007,2N6009,2N6013,BC327,BC327-25,BC327-40, D29E2,D29E7,*GE-48,*TR-52,*WEP242,*RT-115
2SA562-O	2N2838,2N3644,2N4402,2N5811,2N5815,2N5817,2N6001,2N6005,2N6011,2SA467, 2SA467G-O,2SA509-O,2SA509G-O,2SA719,2SA890,A5T4402,BC327-16,CS9012, D29E1,D29E4,D29E5,FT3645,GI-3644,TN-3905,TQ62,*PTC103,*GE-21,*TR-20, *HEPS0012,*SK3114,*ECG290,*WEP242,*RT-115
2SA562-R	2N722,2N722A,2N2837,2N4402,2N5041,2N5815,2SA467,A5T4402,BSW72,D29E4, EN1132,SE8540,TN-3905,TQ62,*PTC103,*GE-89,*TR-20,*SK3114,*WEP242, *RT-115
2SA562-Y	2N2838,2N3504,2N3644,2N4403,2N5375,2N5379,2N5813,2N5817,2N5819,2N5999, 2N6001,2N6005,2N6011,2SA467G-Y,2SA509-Y,2SA509G-Y,2SA562-GR,2SA719, 2SA890,A5T3504,A5T3644,A5T4403,BC327-16,BC327-25,BSW73,D29E5,D29E6, EN3502,FT3645,GI-3644,TN-3906,TQ60,*PTC103,*GE-89,*TR-20,*HEPS0012,

To Replace	Substitute This Type
(2SA562-Y)	*SK3114,*WEP242,*RT-115
2SA564	2N5378,2N5379,2N6003,2N6009,2SA562-GR,BC178C,BC252C,BC253C,BC258C, BC262C,BC263C,BC308C,BC309C,BC328,BC558C,BC559C,BCY78C,BCY78D, *GE-65,*TR-20,*HEPS0019,*SK3114,*ECG234,*WEP242,*276-2034,*RT-303
2SA564A	2N3964,2N3965,2N5374,2N6007,2SA561-GR,2SA661-GR,BC251C,BC261C,BC307C, BC327,BC415-16,BC415C,BC416-16,BC416C,BC560C,BCY79C,TZ553,*GE-65, *IR2SA564,*HEPS0019,*SK3114,*ECG234,*WEP242,*276-2022,*RT-126
2SA565	2N2906,2N2906A,2N3073,2N3135,2N3485,2N3485A,2N4026,2N5372,2N5815,2N5821, 2SA505-O,2SA673A,2SA697,BSW74,D29E4,D29E9,MPS-A55,MPSA55,TZ551, *PTC103,*HEPS5022,*SK3114,*ECG159,*WEP242
2SA566	*SK3085,*ECG218,*WEPS5007
2SA567	2N3964,2N5367,2N5378,2N6003,2N6007,2N6009,2SA564A,BC251C,BC261C, BC307C,BC308C,BC327,BC415-16,BC415C,BC416-16,BC416C,BC558C,BC559C, BC560C,BCY78C,BCY78D,BCY79C,*PTC127,*HEPS0019,*SK3114,*ECG159, *WEP242
2SA568	2N3135,2N4402,2N5365,2SA467,2SA509-O,2SA509G-O,2SA696,A5T4402,BSW72, MPS6533,TN-3905,ZT181,ZT182,ZT184,ZT281,ZT282,ZT284,*PTC103,*HEPS0013, *SK3118,*ECG159,*WEP242
2SA569	2N2906,2N2906A,2N3073,2N3135,2N3485,2N3485A,2N3672,2N3673,2N5372,2SA697, 2SA720,2SA817,2SA891,BCW35,BCW37,BSW74,TZ551,ZT189,*PTC103, *HEPS0019,*SK3114,*ECG159,*WEP242
2SA570	2N2906,2N2906A,2N3073,2N3485,2N3485A,2N5372,2SA697,BSW74,GET2904, GET2905,TZ551,*PTC103,*HEPS5022,*SK3114,*ECG159,*WEP242
2SA571	2N4030,2N4031,2SA717,*TR-88,*HEPS3032,*SK3025,*ECG129,*WEP242,*RT-115
2SA578	2N3799,2N3965,2SA579,2SA661-GR,BC256B,BC266B,BC307B,BC416-10,BC416-16, BC416B,BC416C,BC557B,BC560B,BC560C,TZ553,*HEPS0019,*SK3247,*ECG234
2SA579	2N3799,2N3965,2SA578,2SA661-GR,BC256B,BC266B,BC307B,BC416-10,BC416-16, BC416B,BC416C,BC557B,BC560B,BC560C,TZ553,*HEPS0019,*SK3247,*ECG234
2SA580	*PTC127
2SA581	
2SA594	2N2905,2N2905A,2N3503,2SA594-Y,BC361-10,*PTC127,*GE-21,*TR-88,*HEPS5022, *SK3025,*ECG129,*WEP242,*RT-115
2SA594-O	2SA594,2SA594N,BC361-10,*TR-88,*ECG129,*WEP242
2SA594-R	2N2904,2N2904A,2SA594N,BC361-6,*TR-88,*SK3025,*ECG129,*WEP242
2SA594-Y	2N2905,2N2905A,2N3503,2SA594,*TR-88,*SK3025,*ECG129,*WEP242
2SA594N	2N2904,2N2904A,2SA594,2SA594-O,BC361-10,BC361-6
2SA597	*PTC142,*HEPS5013,*SK3025,*ECG129,*WEP242
2SA603	2SA718,*GE-82,*TR-20,*HEPS5022,*SK3118,*ECG159,*WEP242,*BC177,*276-2023, *RT-113
2SA604	*PTC117,*GE-27,*TR-88,*SK3025,*ECG129,*WEP242,*BF338,*276-2012,*RT-115
2SA605	2N6432,2SA860,*PTC117,*GE-27,*TR-78,*BF338,*276-2012
2SA606	*PTC141,*TR-88,*SK3025,*ECG129,*WEP242,*RT-115
2SA607	*GE-89,*SK3025,*ECG129
2SA608	2N3962,2N3964,2N3965,2N5374,2N5378,2N5379,2N6007,BC214L,BC257B,BC415-10, BC415-16,BC415B,BC415C,BC416-10,BC416-16,BC416B,BC416C,BC557B,BC560B, BC560C,BCY79B,BCY79C,TZ553,*PTC103,*HEPS0013,*SK3114,*ECG159, *WEP242
2SA609	2N3964,2N5378,2N5379,2N6003,2N6007,2N6009,2SA550A,2SA561-GR,2SA562-GR, 2SA564A,2SA608,2SA678,2SA705,2SA725,2SA726,BC177B,BC214L,BC251B, BC251C,BC257B,BC261B,BC261C,BC307B,BC307C,BC308B,BC308C,BC327, BC415-10,BC415-16,BC415B,BC415C,BC416-10,BC416-16,BC416B,BC416C, BC557B,BC558B,BC558C,BC559B,BC559C,BC560B,BC560C,BCY78,BCY78B, BCY78C,BCY78D,BCY79B,BCY79C,*PTC103,*HEPS0013,*SK3114,*ECG159, *WEP242
2SA613	2SA614,*PTC113,*GE-26,*TR-58,*HEPS5018,*ECG218,*WEP700,*276-2025,*RT-133
2SA614	*PTC113,*GE-69,*TR-58,*HEPS5018,*ECG218,*WEPS5007,*276-2025,*RT-133
2SA616	KSP2025,*PTC113,*HEPS5018,*SK3085,*ECG218,*WEPS5007
2SA617	2N6432,2SA605,2SA618,2SA638,2SA639,2SA778A,2SA860
2SA618	2N6432,2SA605,2SA639,2SA778A,2SA860
2SA623	2SA624,*TR-56,*HEPS3028,*SK3193,*ECG307,*WEPS3031
2SA624	*TR-56,*HEPS3032,*SK3083,*ECG307

To Replace	Substitute This Type
2SA625	2SA581
2SA626	2N5623,2SA627,2SA877,KSP2042,KSP2045,KSP2171,KSP2172,KSP2174,KSP2175, *PTC149,*HEPS7003,*ECG219,*WEPS7001
2SA627	2N5623,2N5627,2SA745A,2SA878,KSP2043,KSP2046,KSP2172,KSP2175,*PTC149, *HEPS5005,*ECG219,*WEPS7001
2SA628	2N3251,2N3504,2N3702,2N3906,2N4035,2N4122,2N4291,2N4403,2N4917,2N4972, 2N5244,2N5375,2N5379,2N5383,2N5447,2N5999,2N6001,2N6005,2SA467G-Y, 2SA495-Y,2SA495G-Y,2SA509-Y,2SA509G-Y,2SA550A,2SA659,2SA678,2SA705, 2SA719,2SA733,2SA890,A5T3504,A5T3644,A5T3906,A5T4403,A5T5447,A8T3702, BC177A,BC177B,BC178,BC213L,BC214L,BC251A,BC251B,BC257A,BC257B, BC261A,BC261B,BC307A,BC307B,BC308A,BC308B,BC415-10,BC415-6,BC415A, BC415B,BC416-10,BC416-6,BC416A,BC416B,BC557A,BC557B,BC558A,BC558B, BC559A,BC559B,BC560A,BC560B,BCY78,BCY78A,BCY78B,BCY79A,BCY79B, BSW19,BSW20,BSW73,EN3504,GI-3644,MM3906,MPS3702,NPS404A,TN-3906, TP3644,TQ60,*PTC103,*TR-20,*HEPS0013,*SK3118,*ECG159,*WEP242, *276-2034,*RT-303
2SA628A	2N2907,2N2907A,2N3251A,2N3486,2N3486A,2N3505,2N3962,2N3963,2N4143, 2N5373,2N5374,2SA661-GR,2SA661-Y,2SA720,2SA891,A5T2907,A5T3505,A5T3645, BC212A,BC256A,BC256B,BC266A,BC266B,BC556A,BC556B,BCW35,BCW37, BSW75,EN2907,GET2906,GET2907,TIS112,TP3645,TZ552,TZ553,ZT189,*PTC103, *TR-20,*HEPS5022,*SK3025,*ECG159,*WEP242,*276-2021
2SA629	2N3251,2N3582,2N3702,2N3906,2N4035,2N4058,2N4061,2N4122,2N4917,2N5244, 2N5383,2N5447,2SA467G-Y,2SA493-GR,2SA493-Y,2SA494-GR,2SA494-Y,2SA495-Y, 2SA495G-Y,2SA550A,2SA628,2SA640,2SA641,2SA659,2SA678,2SA705,2SA733, 2SA750,2SA842-GR,2SA889,A5T3906,A5T5447,A8T3702,BC177A,BC177B,BC178, BC213L,BC214L,BC251A,BC251B,BC257A,BC257B,BC261A,BC261B,BC307A, BC307B,BC308A,BC308B,BC415-10,BC415-6,BC415A,BC415B,BC416-10,BC416-6, BC416A,BC416B,BC557A,BC557B,BC558A,BC558B,BC559A,BC559B,BC560A, BC560B,BCY78,BCY78A,BCY78B,BCY79A,BCY79B,BSW19,BSW20,MM3906, MM4048,MPS3702,NPS404A,TN-3906,TP3644,TP3702,*PTC103,*HEPS0019, *SK3118,*ECG159,*WEP242
2SA633	2SA634,2SA699,2SA699A,*HEPS0015
2SA634	2SA699,2SA699A,*PTC111,*HEPS0015,*SK3203,*ECG187A,*WEP246,*RT-167
2SA635	
2SA636	*PTC111,*HEPS5006,*SK3083,*ECG187A,*WEP246
2SA637	*TR-20,*ECG159,*WEP242
2SA638	2N6432,2SA605,2SA637,2SA639,2SA685,2SA857,2SA858,2SA860,2SA912,*ECG288
2SA639	2N6432,2SA605,2SA860,*PTC171,*ECG288
2SA640	2N3798,2N3962,2N3965,2SA578,2SA579,2SA641,2SA661-GR,2SA678,2SA705, 2SA750,BC256B,BC266B,BC307B,BC416-10,BC416-16,BC416B,BC416C,BC557B, BC560B,BC560C,TZ553,*PTC103,*HEPS0019,*SK3118,*ECG159,*WEP242, *RT-303
2SA641	2N3798,2N3962,2N3965,2SA578,2SA579,2SA640,2SA661-GR,2SA678,2SA705, 2SA750,BC256B,BC266B,BC307B,BC416-10,BC416-16,BC416B,BC416C,BC557B, BC560B,BC560C,TZ553,*PTC103,*HEPS0019,*SK3118,*ECG234,*WEP242, *RT-305
2SA642	2N3504,2N4403,2N6001,2N6005,2SA467,2SA683,2SA719,2SA890,A5T3504,A5T3644, A5T4403,GI-3644,TN-3905,TN-3906,*PTC103,*HEPS3014,*SK3083,*ECG159, *WEP242
2SA643	*HEPS3028,*SK3138,*ECG193,*WEP243,*RT-311
2SA645	2SA636,*HEPS3032,*SK3203,*ECG211,*WEPS3031
2SA646	2SA636,*HEPS3032,*SK3193,*ECG187A,*WEP246
2SA647	2SA698
2SA648	2N5627,2SA649,2SA650,2SA651,2SA745A,2SA747,2SA747A,2SA878,KSP2173, KSP2176,*PTC172,*HEPS5005,*ECG219,*WEPS7001
2SA649	2SA650,2SA651,2SA747A,KSP2173,KSP2176
2SA650	2SA651,2SA747A,2SA753,KSP2276
2SA651	2SA650,2SA747A,2SA753,KSP2276
2SA652	2SA653
2SA653	2SA652
2SA656	2SA656A,2SA714,2SA882,*TR-56,*HEPS5005,*ECG187A

To Replace	Substitute This Type
2SA656A	2SA656,2SA714,2SA882
2SA657	2SA656,2SA656A,2SA657A,2SA663-BL,2SA714L,2SA882,*PTC172,*HEPS5005,*SK3173
2SA657A	2SA656,2SA656A,2SA657,2SA663-BL,2SA714L,2SA882,*SK3173
2SA658	2SA657,2SA657A,2SA658A,2SA663-BL,2SA714L,*PTC149,*HEPS5005,*SK3173,*ECG219,*WEPS7001
2SA658A	2SA657,2SA657A,2SA658,2SA663-BL,2SA714L,*PTC149,*SK3173
2SA659	2N2838,2N2907,2N2907A,2N3486,2N3486A,2N3505,2N3672,2N5373,2N5374,2N5817,2N5819,2N5823,2N6005,2N6011,2N6015,2SA661-GR,2SA661-Y,2SA720,2SA817,2SA891,A5T2907,A5T3505,A5T3645,BSW75,D29E10,D29E5,D29E6,TIS112,*PTC103,*HEPS0019,*SK3025,*ECG159,*WEP242
2SA661-GR	
2SA661-O	2SA777,2SA817,2SA891,*ECG298
2SA661-R	
2SA661-Y	2SA661-GR,2SA777,2SA817,2SA891,A5T2907,A5T3505,A5T3645,TIS112
2SA663-BL	2N5625,2SA714L,2SA882
2SA663-R	2N5623,2N5627,2SA648,2SA679-R,2SA680-R,2SA745,2SA745A,2SA747,2SA878,2SB556-R,2SB557-R,2SB558-R,KSP2172,KSP2175
2SA663-Y	2N5625,2SA663-BL,2SA679-Y,2SA680-Y,2SA757,2SA758,2SB556-O,2SB557-O,2SB558-O,KSP2175
2SA666	2N3906,2N4035,2N4122,2N4917,2N5244,2N5375,2N5378,2N5379,2N5383,2N5999,2N6003,2N6009,2SA550,2SA677,2SA842-GR,A5T3906,BC178,BC178B,BC179,BC252B,BC253B,BC258B,BC262B,BC263B,BC308B,BC309B,BC558B,BC559B,BCY78,BCY78B,BCY78C,BSW19,BSW20,BSW73,NPS404,NPS404A,TN-3906,*PTC103,*GE-65,*TR-30,*HEPS0019,*SK3114,*ECG234,*WEP242,*276-2022
2SA666A	2N3251,2N3251A,2N3547,2N3548,2N3549,2N3798,2N3962,2N4249,2N4289,2N4359,2N5373,2N5374,2N6007,2N6225,2SA493-GR,2SA493G-GR,2SA550A,2SA561-GR,2SA564A,2SA661-GR,2SA678,2SA705,2SA733,2SA776,2SA776A,2SA841-GR,2SA889,BC177B,BC214L,BC251B,BC256B,BC257B,BC261B,BC266B,BC307B,BC326,BC327,BC415-10,BC415B,BC416-10,BC416B,BC557B,BC560B,BCY79B,BCY79C,EN3962,MM4048,TZ552,TZ553,*PTC103,*GE-65,*TR-20,*HEPS0019,*SK3114,*ECG234,*WEP242,*276-2022,*RT-115
2SA668	2N5400,2SA549A,A5T5400
2SA669	2N6432
2SA670	*HEPS5006,*SK3083,*ECG153,*WEP246
2SA671	D45C4,D45C5,D45C6,D45C7,D45C8,D45C9,*HEPS5006,*SK3083,*ECG292,*WEP246
2SA672	2N2838,2N3250,2N3250A,2N3672,2N3673,2N3703,2N4143,2N4228,2N4972,2N5448,2N5815,2N5817,2N5821,2N5823,2N6005,2N6011,2N6015,2SA603,2SA659,2SA661-O,2SA661-Y,2SA697,2SA718,2SA720,2SA817,2SA891,A5T5448,A8T3703,A177,BC212A,BC212L,BCW35,BCW37,D29E10,D29E4,D29E5,D29E9,GET2906,GET2907,MPS3703,MPS-A55,ZT189,*PTC103,*HEPS0019,*SK3114,*ECG159,*WEP242
2SA673	2N2838,2N3136,2N3504,2N5811,2N5817,2N6001,2N6005,2N6011,2SA509-O,2SA509-Y,2SA509G-O,2SA509G-Y,2SA673A,2SA696,A5T3504,A5T3644,A5T4402,A5T4403,BC327-16,BSW73,D29E1,D29E5,TN-3905,TN-3906,TQ60,TQ62,*PTC103,*HEPS5013,*SK3114,*ECG290,*WEP242,*RT-115
2SA673A	2N2838,2N2907,2N2907A,2N3136,2N3486,2N3486A,2N3505,2N3672,2N4028,2N5373,2N5817,2N5823,2N6005,2N6011,2N6015,2SA684,2SA697,2SA720,2SA817,2SA891,A5T2907,A5T3505,A5T3645,BSW75,D29E10,D29E5,MPS-A55,TIS112,*PTC103,*HEPS5022,*SK3114,*ECG290,*WEP242
2SA675	2SA777,2SA850,BC556A,BSW75,*PTC171,*HEPS5023,*ECG159
2SA677	2N3906,2N4403,2N5375,2N5378,2N5379,2N5383,2N5813,2N5999,2N6003,2N6009,A5T3906,A5T4403,BCY78,BCY78B,BCY78C,BSW73,D29E2,TN-3906,*HEPS0019,*SK3118,*ECG159,*WEP242
2SA678	2N2907,2N2907A,2N3251,2N3251A,2N3486,2N3486A,2N3505,2N3962,2N5373,2N5374,2N5819,2N6007,2N6013,2N6017,2SA705,A5T2907,A5T3505,A5T3645,BSW75,D29E6,D29E7,TIS112,TZ552,TZ553,*HEPS0019,*SK3118,*ECG159,*WEP242
2SA679-R	2SB554,2SB555-R,2SB556-R,KSP2276
2SA679-Y	2SB554,2SB555-O,2SB556-O

To Replace	Substitute This Type
2SA680-R	2SA679-R,2SB556-R,KSP2275
2SA680-Y	2SA679-Y,2SB556-O
2SA681	2SA510-O
2SA682-O	2N4405,MPS4356
2SA682-Y	2N4033,2N4405,2SB560,MPS4356
2SA683	*GE-67,*TR-88,*HEPS3028,*SK3114,*ECG294,*WEP242,*RT-115
2SA684	*GE-58,*TR-88,*HEPS3032,*SK3025,*ECG294,*WEP242
2SA685	2SA637,2SA857,2SA858,2SA860,2SA912,*PTC171,*GE-21,*HEPS5023,*SK3114,*ECG159,*WEP242
2SA695	2SA683,*PTC103,*TR-28,*HEPS5013,*ECG298,*WEP242,*276-2032
2SA696	2N4026,2N5372,2SA684,2SA697,2SA891,*PTC103,*SK3025,*ECG288,*WEP242,*RT-303
2SA697	2N4026,2N4027,2N5372,2SA684,2SA817,2SA891,*ECG294,*WEP242
2SA698	
2SA699	2SA699A,*GE-69,*TR-56,*HEPS5006,*SK3203,*ECG187A,*WEP246,*276-2025,*RT-167
2SA699A	*GE-69,*TR-56,*HEPS5006,*SK3203,*ECG187A,*WEP246,*276-2025,*RT-151
2SA700	*HEPS5006,*SK3083,*ECG187A,*WEP246
2SA701	2N3964,2N5378,2N5379,2N6003,2N6007,2N6009,2SA493-GR,2SA493G-GR,2SA561-GR,2SA562-GR,2SA564A,2SA608,2SA609,2SA702,2SA721,2SA725,2SA726,2SA774,2SA842-BL,2SA842-GR,2SA880,BC177B,BC251B,BC251C,BC257B,BC261B,BC261C,BC307B,BC307C,BC308B,BC308C,BC327,BC415-10,BC415-16,BC415B,BC415C,BC416-10,BC416-16,BC416B,BC416C,BC557B,BC558B,BC558C,BC559B,BC559C,BC560B,BC560C,BCY78B,BCY78C,BCY78D,BCY79B,BCY79C,MM4048,*PTC103,*HEPS0019,*SK3118,*ECG159,*WEP242
2SA702	2N3798,2N3962,2N3965,2N5374,2N6007,2SA493-GR,2SA493G-GR,2SA561-GR,2SA661-GR,2SA722,2SA726,2SA774A,2SA776,2SA776A,2SA841-BL,2SA841-GR,BC256B,BC266B,BC307B,BC416-10,BC416-16,BC416B,BC416C,BC557B,BC560B,BC560C,TZ553,*PTC103,*HEPS0019
2SA703	2SA623,2SA633,2SA634,2SA699,*ECG307
2SA704	2N3906,2N4403,2N4413,2N4415,2N5375,2N5378,2N5379,2N5383,2N5813,2N5999,2N6003,2N6009,2SA562-GR,2SA677,A5T3906,A5T4403,BC328,BC328-25,BC328-40,BCY78,BCY78B,BCY78C,BSW73,D29E2,TIS91-GRY,TIS91-VIO,TIS91M-GRY,TIS91M-VIO,TIS93-GRY,TIS93-VIO,TIS93M-GRY,TIS93M-VIO,TN-3906,TQ60,TZ581,*HEPS0019,*SK3118
2SA705	2N2907,2N2907A,2N3251,2N3251A,2N3486,2N3486A,2N3505,2N3962,2N5373,2N5374,2N5819,2N6007,2N6013,2N6017,2SA678,A5T2907,A5T3505,A5T3645,BSW75,D29E6,D29E7,TIS112,TZ552,TZ553,*HEPS0019,*SK3118
2SA706	*HEPS5006,*SK3200,*ECG211,*WEP246
2SA707	2N2905,2N2905A,2N3502,2N3503,2N4032,2SA571,2SA684,2SA885,BC360-10,BC360-16,BC361-10,*HEPS3032,*SK3114,*ECG193,*WEP242
2SA708	2SA708A,*PTC141,*SK3203,*ECG129,*WEPS3031
2SA708A	*PTC141,*SK3203,*ECG129,*WEPS3031
2SA710	2SA711
2SA711	
2SA712	
2SA713	
2SA713A	
2SA714	
2SA714L	
2SA715	*PTC905,*SK3191,*ECG185,*WEPS5007
2SA717	2N4030,2N4031,2SA571,*SK3203,*ECG211,*WEPS3031
2SA718	2SA603,*ECG159,*WEP242
2SA719	2N3504,2N6001,2N6005,2SA890,A5T3504,A5T3644,A5T4403,TN-3906,*GE-21,*TR-30,*HEPS0019,*SK3114,*ECG290,*WEP242,*276-2023,*RT-305
2SA720	2N2907,2N2907A,2N3486,2N3486A,2N3505,2N3672,2N4028,2N4029,2SA891,A5T2907,A5T3505,A5T3645,TIS112,*GE-21,*TR-20,*HEPS0019,*SK3114,*ECG298,*WEP242,*276-2032,*RT-115
2SA721	2SA722,*SK3247,*ECG234,*276-2022
2SA722	*SK3247,*ECG234,*276-2022
2SA725	2N3964,2SA725Y,2SA726,2SA726Y,2SA836,2SA844,BC251C,BC261C,BC307C,

To Replace	Substitute This Type
(2SA725)	BC415-16,BC415C,BC416-16,BC416C,BC560C,BCY78D,*SK3247,*ECG234,*276-2034
2SA725Y	2SA726Y,2SA836,2SA844
2SA726	2N3965,2SA726Y,2SA836,2SA844,BC416-16,BC416C,BC560C,*SK3247,*ECG234
2SA726Y	2SA836,2SA844
2SA728	ZTX213,ZTX213A,ZTX213B,ZTX214B,ZTX501,ZTX502,ZTX503,ZTX3702,ZTX3906, ZTX4403
2SA728A	ZTX212,ZTX212A,ZTX212B,ZTX504,ZTX3904
2SA730	2SA751,*PTC177,*GE-21,*TR-20,*HEPS3028,*ECG159,*WEP242,*276-2026
2SA731	2SA752,*PTC177,*GE-21,*TR-20,*HEPS3032,*SK3138,*ECG159,*WEP242
2SA732	2N2904,2N2904A,2N3133,2N3244,2N3467,2N4030,2SA523,2SA523A,2SA537, 2SA571,2SA717,BC360-6,BC361-6,*PTC127
2SA733	2N2907,2N2907A,2N3251,2N3251A,2N3486,2N3486A,2N3505,2N3962,2N4143, 2N5373,2N5374,2N6005,2SA720,2SA891,A5T2907,A5T3505,A5T3645,BC212A, BC416-10,BC416-6,BC416A,BC416B,BC557A,BC557B,BC560A,BC560B,BCW35, BCW37,BSW75,GET2906,GET2907,TIS112,TP3645,TZ552,TZ553,*HEPS5022, *SK3138,*ECG294,*WEP242,*RT-303
2SA738	2SA715,*PTC905,*HEPS5006,*SK3084,*ECG185,*WEPS5007
2SA739	
2SA740	*SK3083,*ECG292
2SA740A	2SA740,2SA839-O,2SA839-R,*SK3083,*ECG292
2SA741	2N869A,2N3209,2N3576,2N3829,*SK3114
2SA742	MPS4356,*SK3025,*ECG129,*WEP242
2SA743	*PTC905,*HEPS5006,*SK3513,*ECG185
2SA743A	*PTC905,*HEPS5006,*SK3513,*ECG185
2SA744	2SA745,2SA746,2SA877,KSP2174,KSP2175,*ECG219,*WEPS7001
2SA745	2SA745A,2SA747,2SA878,KSP2175
2SA745A	2SA747A,2SA878,KSP2176
2SA746	2SA877,KSP2274,KSP2275,*ECG219,*WEPS7001
2SA747	2SA747A,2SA878,KSP2276
2SA747A	KSP2276
2SA748	
2SA749	2N2592,2N2599,2N2600A,*PTC171,*HEPS3002,*SK3513
2SA750	2N3798,2N3962,2N3965,2SA578,2SA579,2SA640,2SA641,2SA661-GR,2SA678, 2SA705,BC256B,BC266B,BC307B,BC416-10,BC416-16,BC416B,BC416C,BC557B, BC560B,BC560C,TZ553
2SA751	*SK3025,*ECG193,*WEP242
2SA752	*ECG193,*WEP242
2SA753	KSP2276,*ECG281
2SA754	2SA670,2SB513,2SB565,*SK3083,*ECG153
2SA755	2N6125,2SA489-O,2SA489-R,2SA490-O,2SA490-R,2SA671,2SB512,2SB566,2SB604, BD224,D45C4,D45C5,D45C6,D45C7,D45C8,D45C9,*SK3083,*ECG153
2SA756	2SA757,KSP2175,*SK3173,*ECG281
2SA757	KSP2176,*ECG281
2SA758	2SA753,KSP2176,*ECG281
2SA759	
2SA760	
2SA761	
2SA762	
2SA764	2SA765,KSP2164,*ECG218,*WEPS5007
2SA765	KSP2164,KSP2165,*ECG218,*WEPS5007
2SA766	*SK3085
2SA767	2N2907,2N2907A,2N3250,2N3250A,2N3251,2N3251A,2N3486,2N3486A,2N3505, 2N3645,2N3672,2N3673,2N4143,2N4228,2N4972,2N5373,2N6005,2N6067,2SA499-Y, 2SA548,2SA603,2SA710,2SA711,2SA718,2SA720,2SA733,2SA891,A5T2907, A5T3505,A5T3645,A177,BC177,BC212A,BC212L,BC256A,BC266A,BC307A, BC416-6,BC416A,BC556A,BC557A,BC560A,BCW35,BCW37,BSW75,EN2905, EN2907,EN3250,FT3644,GET2906,GET2907,TIS112,TP3645,TZ552
2SA772	*SK3083,*ECG294
2SA773	*HEPS0019,*SK3114,*ECG129,*WEP242
2SA774	2N3964,2SA721,2SA722,2SA725,2SA725Y,2SA726,2SA726Y,2SA774A,2SA836, 2SA842-BL,2SA844,2SA880,BC251C,BC261C,BC307C,BC415-16,BC415C,BC416-16,

To Replace	Substitute This Type
(2SA774)	BC416C,BC560C,BCY78D
2SA774A	2N3799,2N3965,2SA722,2SA726,2SA726Y,2SA836,2SA841-BL,2SA844,BC416-16, BC416C,BC560C
2SA775	2SA775A,*ECG292
2SA775A	*ECG292
2SA776	2N3962,2N3963,2N3965,2N5374,2N6007,2SA493-GR,2SA776A,BC416-10,BC416-16, BC416B,BC416C,BC556B,BC557B,BC560B,BC560C,TZ553
2SA776A	2N3962,2N3963,2N3965,2N5374,2N6007,2SA493-GR,2SA776,BC416-10,BC416-16, BC416B,BC416C,BC556B,BC557B,BC560B,BC560C,TZ553
2SA777	2SA850,2SB560,MPS4356,*ECG298
2SA778	2N6432,2SA605,2SA617,2SA618,2SA638,2SA639,2SA778A,2SA860,*ECG288
2SA778A	2N6432,2SA605,2SA618,2SA639,2SA860,*SK3114,*ECG159
2SA779	*SK3084,*ECG187A
2SA780	*SK3083
2SA780A	*SK3083
2SA781	
2SA782	2SA785
2SA783	2SA786,2SA788
2SA784	2SA789
2SA785	
2SA786	
2SA787	
2SA788	2SA783,2SA786
2SA789	2SA784
2SA790	
2SA791	2SA790
2SA794	
2SA794A	
2SA795	
2SA798	
2SA807	2SA744,2SA746,2SA808,2SA877,KSP2174,*PTC149
2SA808	2SA744,2SA745,2SA746,2SA756,2SA877,KSP2174,KSP2175,*PTC172
2SA809	2SA810,2SA912
2SA810	2SA809,2SA912
2SA811	
2SA812	
2SA813	
2SA814-O	
2SA814-Y	
2SA815-O	2SA814-O
2SA815-Y	2SA814-Y
2SA816-O	
2SA816-Y	
2SA817	*PTC171
2SA818-O	
2SA818-Y	
2SA835	
2SA836	2SA844,*SK3114,*ECG129
2SA837	2N5613,2N5617,2SB532,*PTC172,*ECG281
2SA838	2N3250,2N4034,2SA499-O,2SA499-Y,2SA500-O,2SA500-Y,2SA710,2SA711,*PTC103
2SA839-O	2SA740
2SA839-R	2SA740,2SA740A
2SA839-Y	
2SA840	*SK3083,*ECG288
2SA841-BL	2N3965,2SA836,2SA844
2SA841-GR	2N3962,2N3963,2N3965,2N5374,2SA776,2SA776A,BC256B,BC266B,BC556B,TZ553
2SA842-BL	2N3964,2N3965,2SA836,2SA841-BL,2SA844,BC251C,BC261C,BC307C,BC415-16, BC415C,BC416-16,BC416C,BC560C
2SA842-GR	2N3962,2N3964,2N3965,2N5374,2N5378,2N5379,2N6007,2SA493-GR,2SA550A, 2SA678,2SA705,2SA776,2SA776A,2SA841-GR,BC177B,BC214L,BC251B,BC251C, BC257B,BC261B,BC261C,BC307B,BC307C,BC415-10,BC415-16,BC415B,BC415C,

To Replace	Substitute This Type
(2SA842-GR)	BC416-10,BC416-16,BC416B,BC416C,BC557B,BC560B,BC560C,BCY79,BCY79B, BCY79C,TZ553
2SA843	
2SA844	*ECG129
2SA847	
2SA848	2SA818-Y,2SA849
2SA849	2SA818-Y,2SA848
2SA850	
2SA857	2SA858,2SA860,2SA912,*PTC171
2SA858	2SA857,2SA860,2SA912
2SA859	2N6519,2N6520
2SA860	2N6518
2SA861	
2SA873	*PTC103
2SA877	KSP2274,KSP2275
2SA878	2SA747A,KSP2276
2SA879	2N6519,2SA859,*PTC171
2SA880	2N3964,2SA721,2SA722,2SA725,2SA725Y,2SA726,2SA726Y,2SA774,2SA774A, 2SA836,2SA842-BL,2SA844,BC251C,BC261C,BC307C,BC415-16,BC415C,BC416-16, BC416C,BC560C,BCY78D
2SA882	
2SA884	
2SA885	*PTC905
2SA886	
2SA887	
2SA888	2N4035,2N5244,2N5375,2N5378,2N5379,2N5383,2N5999,2N6003,2N6009,A5T3906, BC328,BC558B,BC559B,BCY78B,BCY78C,BSW73,NPS404,NPS404A,TN-3906
2SA889	2N3251,2N3251A,2N3798,2N3962,2N5373,2N5374,2N6007,2SA661-GR,2SA678, 2SA705,BC327,BC557B,BC560B,BCY79,BCY79B,BCY79C,MM4048,TZ552,TZ553
2SA890	A5T3504,A5T3644,A5T4403
2SA891	A5T2907,A5T3505,A5T3645,TIS112
2SA896	2SA917,*PTC171
2SA900	
2SA912	
2SA913	
2SA917	2SA896,*PTC171
2SB12	
2SB13	
2SB14	
2SB15	
2SB16	
2SB16A	*PTC106,*HEPG6011,*SK3012,*ECG105,*WEP233
2SB17	
2SB17A	*PTC106,*HEPG6011,*SK3012,*ECG105,*WEP233
2SB18	
2SB18A	*PTC106,*HEPG6012,*SK3012,*ECG105,*WEP233
2SB19	*HEPG6011,*SK3009,*ECG104,*WEP230
2SB20	*HEPG6011,*SK3009,*ECG104,*WEP230
2SB21	*HEPG6011,*SK3009,*ECG104,*WEP230
2SB22	2N1008A,*PTC135,*GE-53,*TR-84,*HEPG0005,*SK3004,*JR15,*ECG102A,*WEP250, *AC188/01
2SB23	*PTC135,*HEPG0008,*SK3004,*ECG102A,*WEP250
2SB24	*HEPG0008,*SK3004,*ECG102A,*WEP250
2SB25	*PTC109,*GE-16,*TR-01,*HEPG6005,*SK3009,*PT40,*ECG121,*WEP232,*276-2006, *RT-127
2SB25-N	*PTC138
2SB25G	SEE 2SB425G
2SB26	*PTC114,*GE-16,*TR-01,*HEPG6003,*SK3009,*PT40,*ECG104,*WEP230,*276-2006, *RT-124
2SB26A	*PTC138,*GE-16,*TR-01,*HEPG6003,*SK3009,*ECG104,*WEP230,*RT-124
2SB26G	SEE 2SB426G

To Replace	Substitute This Type
2SB27	OC30,*PTC105A,*GE-49,*TR-01,*HEPG6003,*SK3009,*PT40,*ECG104,*WEP230,*OC28,*276-2006
2SB28	2SB367,*PTC105A,*GE-49,*TR-01,*HEPG6003,*SK3009,*PT40,*ECG104,*WEP230,*AD149,*276-2006
2SB29	2N4078,2SB367,AD155,AD162,AD164,*PTC105A,*GE-49,*TR-01,*HEPG6003,*SK3009,*PT40,*ECG104,*WEP230,*276-2006,*RT-124
2SB30	2SB367,*PTC105A,*GE-49,*TR-01,*HEPG6003,*SK3009,*PT40,*ECG104,*WEP230,*AD149,*276-2006,*RT-124
2SB31	2N4078,2SB367,AD155,AD162,AD164,*PTC105A,*GE-49,*TR-01,*HEPG6003,*SK3009,*PT40,*ECG104,*WEP230,*276-2006,*RT-124
2SB32	*PTC109,*HEPG0005,*SK3004,*ECG102A,*WEP250
2SB33	2N838,2N3323,2N3324,2N3325,2SB37,2SB56,2SB77,2SB187,2SB188,2SB328,2SB329,2SB364,2SB496,ACY23,ACY32,NKT211,NKT213,NKT216,SF.T323,*PTC109,*HEPG0005,*SK3004,*ECG102A,*WEP250
2SB34	2N396,2N1008,2N1374,2N1376,2N1381,2N1382,2N2271,2N2706,2SB22,2SB38,2SB79,2SB496,*PTC135,*HEPG0005,*SK3003,*ECG102,*WEP631
2SB37	2SB56,2SB77A,ACY23,ACY32,*PTC109,*HEPG0005,*SK3004,*ECG102A,*WEP250
2SB38	2N597,2N1008A,2N1125,2N1375,2N1377,*PTC135,*HEPG0005,*SK3004,*ECG102,*WEP631
2SB39	*PTC107,*HEPG0005,*SK3003,*ECG102A,*WEP250
2SB40	2SB136A,*PTC109,*GE-2,*TR-85,*HEPG0005,*SK3004,*JR15,*ECG102A,*WEP250,*AC126,*276-2005,*RT-121
2SB41	2N2143,2N2143A,2N2144,2N2144A,2SB42,2SB131,2SB131A,2SB132,*HEPG6003,*SK3009,*ECG104,*WEP230
2SB42	2N2144,2N2144A,2N2145,2N2145A,2SB132,*HEPG6003,*SK3009,*ECG104,*WEP230
2SB43	2SB56,2SB77,2SB186,2SB187,2SB188,2SB431,ACY23,ACY32,SF.T323,*PTC109,*GE-52,*TR-85,*HEPG0005,*SK3003,*JR15,*ECG102A,*WEP250,*AC126,*276-2007
2SB43A	*PTC109,*GE-2,*TR-85,*HEPG0005,*SK3004,*ECG102A,*WEP250
2SB43G	SEE 2SA495G
2SB44	2N217+,2N2614,2N2953,2SB56,2SB77A,2SB431,ACY23,ACY32,*PTC135,*GE-52,*TR-85,*HEPG0005,*SK3004,*AT20M,*ECG102A,*WEP250,*AC126,*276-2007
2SB44G	SEE 2SA495G
2SB46	*PTC135,*HEPG0006,*SK3003,*ECG102A,*WEP250
2SB47	2N2613+,*PTC105A,*GE-52,*TR-85,*HEPG0006,*SK3003,*JR15,*ECG102A,*WEP250,*276-2007,*RT-121
2SB48	2N302,2N320,2N361,2N414,2N414B,2N414C,2N450,2N711A,2N711B,2N838,2N1115,2N1171,2N1204A,2N1281,2N1282,2N1303,2N1305,2N1313,2N1344,2N1478,2N2048,2N2401,2N2402,2N2587,2N3883,2N6365,2N6365A,2SA204,2SA206,2SA207,2SA406,2SA407,2SB221A,2SB326,ASY26,MM380,SF.T227,SF.T228,*HEPG0005,*SK3004,*ECG102,*WEP631
2SB49	2N324,2N360,2N508,2N520A,2N571,2N654+,2N964A,2N985,2N1307,2N1356,2N1705,2N1706,2N1707,2N1808,2N1892,2N2699,2N2928,2N3371,2SB222,2SB222A,2SB327,40269,ASY27,SF.T223,SF.T229,SF.T237,*GE-52,*TR-05,*HEPG0005,*SK3004,*HF6M,*ECG102,*WEP631,*AC126,*276-2004
2SB50	2N359,2N417,2N521A,2N522A,2N523,2N599,2N655+,2N827,2N1284,2N1309,2N1316,2N1317,2N1353,2N1354,2N1355,2N1357,2N2635,2SB223,2SB223A,MA1703,MA1706,*GE-52,*TR-05,*HEPG0006,*SK3003,*HF6M,*ECG102,*WEP631,*AC126,*276-2005
2SB51	2N381,2N414B,2N414C,2N525,2N525A,2N526,2N526A,2N1348,2N1350,2N1351,2N1446,2N1447,2N1478,2N1495,2SA206,2SA207,2SA406,2SA407,2SB57A,2SB220,2SB220A,2SB221,2SB221A,2SB225,2SB226,2SB326,SF.T222,*GE-53,*TR-05,*HEPG0005,*SK3003,*JR5,*ECG102,*WEP631,*AC188/01,*276-2004,*RT-120
2SB52	2N382,2N383,2N527,2N527A,2N652,2N654+,2N1356,2N1448,2N1449,2N1707,2N1997,2SB222,2SB222A,2SB227,2SB327,SF.T223,*GE-53,*TR-05,*HEPG0005,*SK3003,*HF3H,*ECG102,*WEP631,*AC188/01,*276-2004,*RT-120
2SB53	2N382,2N526,2N526A,2N527A,2N597,2N654+,2N1349,2N1448,2SB226,2SB227,2SB327,SF.T223,*PTC103,*GE-53,*TR-05,*HEPG0005,*SK3004,*JR15,*ECG102,*WEP631,*276-2004,*RT-120
2SB54	2SB439,2SB440,*PTC109,*GE-52,*TR-85,*HEPG0007,*SK3004,*AT20M,*ECG102A,

To Replace	Substitute This Type
(2SB54)	*WEP250,*AC188/01,*276-2007,*RT-123
2SB54G	2SB54,2SB439,2SB440,*GE-53,*SK3004
2SB55	*GE-2,*TR-85,*HEPG0006,*SK3004,*JR15,*ECG102A,*WEP250,*RT-121
2SB55G	*GE-53,*SK3004
2SB56	2SB431,ACY23,ACY32,*PTC109,*TR-84,*HEPG0005,*SK3004,*JR15,*ECG102A, *WEP250,*AC188/01,*276-2007,*RT-123
2SB56A	*GE-2,*TR-05,*HEPG0005,*SK3004,*ECG102A,*WEP631,*276-2007,*RT-120
2SB56G	2SB56,2SB431,ACY23,ACY32,*GE-53,*SK3004
2SB57	*PTC109,*HEPG0005,*SK3004,*ECG102A
2SB57A	2N526,2N526A,2N527A,2N597,2SB226,2SB227,*SK3004
2SB58	*HEPG0005,*ECG158
2SB59	2N109+,2N217+,2SB56,2SB77A,2SB345,2SB347,ACY23,ACY32,NKT211,NKT213, NKT216,NKT226,*PTC109,*HEPG0005,*SK3004,*ECG102A,*WEP250
2SB60	*PTC109,*HEPG0005,*SK3004,*ECG102A,*WEP250
2SB60A	2N109+,2N217+,2N838,2N3323,2N3324,2N3325,2SB33,2SB37,2SB56,2SB59, 2SB77,2SB156A,2SB187,2SB188,2SB328,2SB329,2SB345,2SB347,2SB364,2SB496, ACY23,ACY32,NKT211,NKT213,NKT216,NKT226,SF.T323,*PTC109,*HEPG0005, *SK3004,*ECG102A,*WEP250
2SB61	*PTC109,*HEPG0005,*SK3004,*ECG102A,*WEP250
2SB62	2SB356,*GE-16,*TR-01,*SK3009,*ECG104,*WEP230,*RT-124
2SB62G	SEE 2SB462G
2SB63	*GE-30,*TR-94,*HEPG6016,*SK3082,*ECG131,*WEP642,*276-2006,*RT-127
2SB63G	SEE 2SB463G
2SB64	2N1905,*PTC122,*GE-25,*TR-01,*HEPG6017,*SK3009,*AT20M,*ECG121,*WEP232, *276-2006,*RT-127
2SB65	*PTC109,*HEPG0005,*SK3004,*ECG102A,*WEP250
2SB66	2N217+,2N2614,2N2953,2SB37,2SB56,2SB77A,ACY23,ACY32,*PTC109,*GE-52, *TR-85,*HEPG0005,*SK3004,*JR15,*ECG102A,*WEP250,*AC126,*276-2007, *RT-121
2SB67	*GE-2,*TR-85,*HEPG0005,*SK3004,*AT20M,*ECG102A,*WEP250,*276-2007,*RT-121
2SB67A	*DS26,*GE-2,*TR-85,*HEPG0005,*SK3004,*ECG102A,*WEP250,*276-2007,*RT-121
2SB68	MA202,MA203,*DS-26,*GE-2,*SK3004
2SB69	*PTC122,*GE-25,*TR-01,*HEPG6005,*SK3009,*HO300,*ECG121,*WEP232, *276-2006,*RT-127
2SB69N	SEE 2SD52N
2SB70	*HEPG0005
2SB71	*PTC102,*HEPG0005,*SK3004,*ECG102,*WEP631
2SB73	2SA49,2SA52,2SA53,2SA408,2SA409,2SB77,2SB329,SF.T307,SF.T308,SF.T323, *PTC109,*GE-52,*TR-85,*HEPG0008,*SK3004,*JR15,*ECG102A,*WEP250,*AC126, *276-2007,*RT-121
2SB74	*PTC107,*HEPG0009,*SK3004,*ECG160,*WEP637
2SB75	2SB77,SF.T323,*GE-52,*TR-85,*HEPG0005,*SK3004,*JR15,*ECG102A,*WEP250, *AC126,*276-2007,*RT-121
2SB75A	2SB77A,*GE-2,*TR-85,*HEPG0005,*SK3004,*ECG102A,*WEP250,*276-2007,*RT-121
2SB76	2N796,2N829,2N964A,2N972,2N985,2SB75,2SB77,2SB328,2SB329,2SB496,SF.T307, SF.T308,SF.T323,*PTC109,*GE-52,*TR-85,*HEPG0005,*SK3003,*JR10,*ECG102A, *WEP250,*AC126,*276-2007,*RT-121
2SB77	SF.T323,*PTC135,*GE-53,*TR-85,*HEPG0005,*SK3004,*JR15,*ECG102A,*WEP250, *AC126,*276-2007,*RT-123
2SB77A	*PTC135,*GE-2,*TR-85,*HEPG0005,*SK3004,*ECG102A,*WEP250,*276-2007, *RT-121
2SB78	2N796,2N964A,2N985,2N2699,2SB77,2SB136,2SB186,2SB187,2SB188,2SB329, 2SB364,2SB496,NKT231,SF.T308,SF.T323,SF.T337,*PTC107,*GE-52,*TR-85, *HEPG0005,*SK3004,*JR5,*ECG102A,*WEP250,*AC126,*276-2007,*RT-121
2SB79	2N270,2N1008A,2N1124,2N1125,2N1374,2N1376,2N1381,2N1382,2N2375,2SB22, 2SB38,2SB89,SF.T125,SF.T125P,*PTC109,*HEPG0005,*SK3004,*ECG102A, *WEP250
2SB80	2SB367,*PTC194,*GE-3,*TR-01,*HEPG6016,*SK3009,*PT40,*ECG121,*WEP232, *276-2006,*RT-127
2SB81	2SB356,2SB357,*GE-25,*TR-01,*SK3009,*ECG121,*WEP232,*RT-127
2SB82	2SB357,*GE-25,*TR-01,*SK3009,*ECG121,*WEP232,*RT-127

To Replace	Substitute This Type
2SB83	2N351,2SB84,*PTC138,*GE-3,*TR-01,*HEPG6005,*SK3009,*PT40,*ECG104, *WEP230,*AD149,*276-2006,*RT-124
2SB84	2N1295,*PTC105A,*GE-25,*TR-01,*HEPG6005,*SK3014,*PT40,*ECG121,*WEP232, *276-2006,*RT-127
2SB85	2N1553,2N1554,2SB86,*SK3004,*ECG102,*WEP631
2SB86	2N1554,2N1555,2SB87
2SB87	2SB88,*SK3004,*ECG102,*WEP631
2SB88	
2SB89	2N270,2N1008A,2N1125,2N1374,2N1376,2N1381,2N1382,2SB22,2SB38,2SB79, *PTC135,*GE-53,*TR-85,*HEPG0005,*SK3003,*JR15,*ECG102A,*WEP250, *276-2007,*RT-121
2SB89A	2N597,2N1008B,2N1375,2N1377,*DS-26,*GE-2,*TR-85,*HEPG0005,*SK3004, *ECG102A,*WEP250,*276-2007,*RT-121
2SB90	*PTC135,*HEPG0006,*SK3004,*ECG102A,*WEP250
2SB91	2SB187,2SB188,*PTC135,*HEPG0005,*SK3004,*ECG102A,*WEP250
2SB92	2SB56,2SB96,2SB172,2SB288,NKT213,NKT214,NKT216,NKT221,NKT224,NKT228, *PTC102,*HEPG0005,*SK3003,*ECG102,*WEP631
2SB93	
2SB94	2SB56,2SB77,2SB186,2SB187,2SB188,2SB431,ACY23,ACY32,SF.T323,*PTC109, *GE-52,*TR-85,*HEPG0006,*SK3004,*AT20M,*ECG102A,*WEP250,*AC126, *276-2007,*RT-121
2SB95	2N109+,2N217+,2SB37,2SB56,2SB75,2SB77,2SB187,2SB188,ACY23,ACY32, SF.T323,*PTC102,*HEPG0008,*SK3004,*ECG102,*WEP631
2SB96	2N109+,2N217+,2N2953,2SB56,2SB176,2SB431,AC122-GRN,AC122/30-GRN, ACY23,ACY32,NKT211,NKT213,NKT216,NKT222,NKT223,NKT226,OC76, *HEPG0005
2SB97	*PTC135,*HEPG0005,*SK3004,*ECG102A,*WEP250
2SB98	2N43,2N43A,2N43A+,2N61A,2N61B,2N361,2N363,2N413,2N525,2N525A,2N650, 2N653,2N1017,2N1191,2N1303,2N1313,2N1414,2N1446,2N1495,2N2587,2N2955, 2N6365A,2SA205,2SA206,2SA406,2SB220,2SB220A,2SB224,2SB225,ASY26, SF.T222,SF.T227,*HEPG0005,*SK3004,*ECG102,*WEP631
2SB99	2N417,2N508A,2N598,2N599,2N655,2N655+,2N1185,2N1190,2N1193,2N1309, 2N1316,2N1354,2N1355,2N1357,2N1998,2N2171,2N2613,2N2635,2N3427,2SB223, 2SB223A,MA888,*HEPG0006,*SK3004
2SB100	*PTC109,*HEPG0005,*SK3004,*ECG102,*WEP631
2SB101	*PTC109,*HEPG0005,*SK3004,*ECG102,*WEP631
2SB102	*PTC109,*HEPG0005,*SK3004,*ECG102,*WEP631
2SB103	2N59A,2N59B,2N60A,2N60B,2N382,2N415,2N415A,2N416,2N422A,2N461,2N467, 2N526,2N526A,2N527A,2N569,2N570,2N571,2N572,2N597,2N651,2N651A,2N652A, 2N654,2N654+,2N1008A,2N1175,2N1175A,2N1192,2N1305,2N1307,2N1309A, 2N1349,2N1352,2N1356,2N1371,2N1375,2N1377,2N1415,2N1448,2N1452,2N1681, 2N1707,2SB57A,2SB221,2SB221A,2SB222,2SB222A,2SB226,2SB227,2SB327, 2SB428,MA887,SF.T223,*PTC109,*HEPG0005,*SK3004,*ECG100,*WEP254
2SB104	*PTC102,*HEPG0005,*SK3004,*ECG100,*WEP254
2SB105	2SB108,2SB273,*PTC135,*HEPG0005,*SK3123,*ECG158,*WEP238
2SB106	2SB109,*PTC135,*HEPG6011,*ECG158,*WEP238
2SB107	2N250,2SB474,AD148,NKT451,NKT452,*HEPG6011,*SK3009,*ECG121,*WEP631
2SB107A	2N420,2N665,2N1295,2N2836,2SB138,2SB216A,*PTC138,*HEPG6003,*SK3009, *ECG121,*WEP232
2SB108	2SB108A,*HEPG0005,*SK3123,*ECG158,*WEP238
2SB108A	2SB108B,*HEPG0005,*SK3123,*ECG158,*WEP238
2SB108B	*SK3123,*ECG158,*WEP238
2SB109	2SB109A,*PTC135,*HEPG6011,*ECG158,*WEP238
2SB109A	2SB109B,*HEPG6011
2SB109B	
2SB110	*PTC109,*HEPG0005,*SK3003,*ECG102A,*WEP250
2SB111	*PTC109,*HEPG0005,*SK3003,*ECG102A,*WEP250
2SB112	*PTC109,*HEPG0005,*SK3004,*ECG102A,*WEP250
2SB113	*PTC109,*HEPG0005,*SK3004,*ECG102A,*WEP250
2SB114	2N2953,2SB75,2SB77,2SB115,2SB328,2SB329,ACY23,ACY32,SF.T323,*PTC109, *HEPG0005,*SK3003,*ECG102A,*WEP250

TRANSISTOR SUBSTITUTES

To Replace	Substitute This Type
2SB115	2N2614,2N2953,2SB77,2SB116,2SB117,2SB329,ACY23,ACY32,SF.T323,*PTC109, *HEPG0005,*SK3003,*ECG102A,*WEP250
2SB116	2N2614,2SB117,2SB439,2SB440,*PTC135,*HEPG0005,*SK3004,*ECG102A, *WEP250
2SB117	2N2613+,2SB439,2SB440,*PTC135,*HEPG0005,*SK3004,*ECG102A,*WEP250
2SB118	2N155,2N2142,2N2142A,2SB119
2SB119	2N155,2N2142,2N2142A,2N2143,2N2143A,2SB131,2SB131A,*HEPG6003,*SK3009, *ECG121,*WEP232
2SB120	*PTC109,*HEPG0005,*SK3004,*ECG102A,*WEP250
2SB121	
2SB122	*HEPG6005,*SK3009,*ECG121,*WEP232
2SB122G	SEE 2SB424G
2SB123	*PTC138,*HEPG6005,*SK3009,*ECG104,*WEP230
2SB124	*SK3009,*ECG121,*WEP232
2SB125	
2SB126	2N5889,2N5890,*PTC138,*GE-25,*TR-01,*HEPG6003,*SK3009,*PT50,*ECG121, *WEP232,*OC28,*276-2006,*RT-127A
2SB126A	2N5891,2N5892,2SB128,*PTC138,*GE-25,*TR-01,*SK3009,*ECG121,*WEP232
2SB127	2N5893,2N5894,2SB137,2SB337,*PTC138,*GE-25,*TR-01,*HEPG6003,*SK3009, *PT50,*ECG121,*WEP230,*OC28,*276-2006,*RT-124
2SB127A	2N5895,2N5896,2SB138,*PTC138,*GE-25,*TR-01,*SK3009,*ECG121,*WEP232, *RT-124
2SB128	*PTC105A,*GE-25,*TR-27,*HEPG6013,*SK3035,*PT150,*ECG121,*WEP235,*OC28, *276-2004
2SB128A	*GE-25,*TR-01,*HEPG6018,*SK3009,*ECG121,*WEP232,*OC28,*276-2006,*RT-127
2SB129	*PTC105A,*GE-25,*TR-01,*HEPG6013,*SK3009,*PT150,*ECG121,*WEP232,*OC28, *276-2004,*RT-127
2SB129A	*HEPG6018,*SK3009
2SB130	*PTC120A,*GE-30,*TR-94,*HEPG6016,*SK3035,*ECG131,*WEP642,*276-2006, *RT-127
2SB131	2N2143,2N2143A,2N2144,2N2144A,2SB131A,2SB132,*HEPG6003,*SK3009, *ECG121,*WEP232
2SB131A	2N2143,2N2143A,2N2144,2N2144A,2SB131,2SB132,*HEPG6005,*SK3009,*ECG121, *WEP232
2SB132	2N2144,2N2144A,2N2145,2N2145A,*HEPG6005,*SK3009,*ECG121,*WEP232
2SB134	*PTC135,*HEPG0005,*SK3004,*ECG102A,*WEP250
2SB135	*PTC135,*HEPG0005,*SK3004,*ECG102A,*WEP250,*RT-123
2SB136	2SB54,2SB186,2SB370,2SB370A,2SB408,2SB431,2SB439,2SB440,*PTC135, *HEPG0005,*SK3004,*ECG102A,*WEP250
2SB136A	*HEPG0006,*SK3123,*ECG102A,*WEP250
2SB137	2N5893,2N5894,*PTC138,*HEPG6003,*SK3009,*ECG121,*WEP232
2SB138	2SB339,KR6501,*PTC105A,*HEPG6005,*SK3009,*ECG121,*WEP232
2SB140	2N251,2SB147,*PTC138,*GE-3,*TR-01,*HEPG6003,*SK3009,*PT40,*ECG104, *WEP230,*AD149,*276-2006,*RT-124
2SB141	2N251,2SB147,*PTC138,*GE-25,*TR-01,*HEPG6005,*SK3009,*PT40,*ECG104, *WEP230,*AD149,*276-2006,*RT-124
2SB142	*PTC105A,*GE-49,*TR-01,*HEPG6003,*SK3009,*PT40,*ECG104,*WEP230,*OC28, *276-2006,*RT-124
2SB143	2N250,*PTC105A,*GE-49,*TR-01,*HEPG6003,*SK3009,*PT40,*ECG104,*WEP230, *OC28,*276-2006,*RT-124
2SB144	2N250,2SB254,*PTC105A,*GE-49,*TR-01,*HEPG6003,*SK3009,*PT40,*ECG104, *WEP230,*AD149,*276-2006,*RT-124
2SB145	2N250,*PTC105A,*GE-49,*TR-01,*HEPG6003,*SK3009,*PT40,*ECG104,*WEP230, *OC28,*276-2006,*RT-124
2SB146	2N250,2SB254,*PTC105A,*GE-49,*TR-01,*HEPG6003,*SK3009,*PT40,*ECG104, *WEP230,*AD149,*276-2006,*RT-124
2SB147	2N251,*PTC138,*HEPG6005,*SK3009,*ECG104,*WEP230
2SB148	
2SB149	2SB149-N,KR6500,*GE-25,*TR-01,*HEPG6003,*SK3009,*PT150,*ECG104,*WEP230, *276-2006,*RT-124
2SB149-N	KR6500,*PTC105A

To Replace	Substitute This Type
2SB149N	SEE 2SD51N
2SB150	*DS-26,*GE-2,*SK3004
2SB151	2N561,2N5156,2SB152,*HEPG6018,*SK3009,*ECG104,*WEP230
2SB152	2N1538,2N5156,*HEPG6018,*SK3009,*ECG104,*WEP230
2SB152A	2N1543,2N1548,2SB129A,*SK3009
2SB153	*HEPG0005,*SK3004,*ECG102,*WEP631
2SB154	2N796,2N829,2N964A,2N972,2N985,2SA417,2SB77,2SB156,2SB156A,2SB187, 2SB188,2SB328,2SB329,2SB364,2SB496,NKT231,SF.T307,SF.T308,SF.T323, *HEPG0005,*SK3004,*ECG102,*WEP631
2SB155	*PTC102,*GE-53,*TR-85,*HEPG0005,*SK3003,*JR15,*ECG102A,*WEP250, *AC188/01,*276-2007
2SB156	2SB156A,2SB364,2SB415,2SB495A,NKT211,NKT231,*PTC135,*GE-53,*TR-84, *HEPG0005,*SK3004,*JR15,*ECG102A,*WEP250,*AC188/01,*276-2006,*RT-121
2SB156A	2SB364,2SB415,2SB495A,NKT211,*PTC135,*GE-53,*TR-84,*HEPG0005,*SK3004, *ECG102A,*WEP250,*AC188/01,*276-2004,*RT-121
2SB157	OC57,OC58,*PTC109,*GE-51,*TR-85,*HEPG0008,*SK3004,*JR10,*ECG102A, *WEP250,*AF125,*276-2002,*RT-121
2SB158	2SB91,OC58,OC59,OC60,*PTC109,*GE-51,*TR-85,*HEPG0008,*SK3004,*JR10, *ECG102A,*WEP250,*AF125,*276-2002,*RT-121
2SB159	OC59,*PTC109,*GE-51,*TR-85,*HEPG0008,*SK3004,*JR10,*ECG102A,*WEP250, *AF125,*276-2002,*RT-121
2SB160	OC59,*PTC109,*GE-51,*TR-85,*HEPG0008,*SK3004,*JR10,*ECG102A,*WEP250, *AF125,*276-2002,*RT-121
2SB161	2N60A,2N60B,2N61A,2N61B,2N363,2N381,2N414,2N414A,2N414B,2N414C,2N415, 2N415A,2N416,2N422A,2N466,2N505,2N518,2N525,2N525A,2N526,2N526A,2N567, 2N568,2N569,2N570,2N572,2N597,2N650,2N650A,2N651,2N651A,2N652A,2N654, 2N838,2N1171,2N1175,2N1192,2N1303,2N1305,2N1309A,2N1313,2N1348,2N1349, 2N1350,2N1351,2N1373,2N1414,2N1415,2N1447,2N1451,2N1478,2N1495,2N1681, 2N2587,2N2956,2N3323,2N3324,2N3325,2N6365,2N6365A,2SA204,2SA206, 2SA207,2SA406,2SA407,2SB57A,2SB103,2SB220,2SB220A,2SB221,2SB221A, 2SB225,2SB226,2SB299,2SB304,2SB304A,2SB326,2SB427,ASY26,MA887,NKT221, NKT224,NKT228,SF.T222,*PTC109,*HEPG0005,*SK3004,*ECG102,*WEP631
2SB162	2SB104,*HEPG0005,*SK3004,*ECG102,*WEP631
2SB163	2N59A,2N59B,2N60A,2N60B,2N360,2N382,2N415,2N415A,2N416,2N422A,2N461, 2N467,2N526,2N526A,2N527A,2N569,2N570,2N571,2N572,2N597,2N651,2N651A, 2N652A,2N654,2N654+,2N838,2N1008A,2N1125,2N1130,2N1175,2N1175A, 2N1192,2N1305,2N1307,2N1309A,2N1349,2N1352,2N1356,2N1371,2N1375,2N1377, 2N1415,2N1448,2N1452,2N1681,2N1707,2N2956,2N2957,2N3323,2N3324, 2N3325,2SB57A,2SB103,2SB221,2SB221A,2SB222,2SB222A,2SB226,2SB227, 2SB304,2SB304A,2SB327,2SB428,MA887,MM404A,NKT226,SF.T223,*PTC109, *HEPG0005,*SK3004,*ECG102,*WEP631
2SB164	2SB104,*HEPG0005,*SK3004,*ECG102,*WEP631
2SB165	2N382,2N383,2N417,2N461,2N467,2N508A,2N527,2N571,2N598,2N599,2N652, 2N655,2N655+,2N1008A,2N1189,2N1193,2N1307,2N1309,2N1352,2N1354,2N1355, 2N1356,2N1357,2N1371,2N1377,2N1449,2N1892,2N1997,2N1998,2N2957,2SB223, 2SB223A,2SB327,MA888,SF.T223,*PTC109,*HEPG0006,*SK3004,*ECG102, *WEP631
2SB166	*HEPG0001,*SK3004,*ECG102,*WEP631
2SB168	*PTC109,*HEPG0005,*SK3003,*ECG102A,*WEP250
2SB169	2N796,2N964A,2N972,2N985,2N2699,2SB364,NKT231,SF.T307,SF.T308,SF.T323, SF.T337,*PTC109,*HEPG0005,*SK3003,*ECG102A,*WEP250
2SB170	2SB172,2SB288,NKT214,NKT215,NKT225,*GE-2,*TR-84,*HEPG0005,*SK3004,*JR15, *ECG102A,*WEP250,*276-2004,*RT-121
2SB171	2N109+,2N217+,2N2953,2SB56,2SB59,2SB75A,2SB77A,2SB173,2SB175,2SB176, 2SB345,2SB347,AC122-YEL,AC122/30-YEL,ACY23,ACY32,NKT211,NKT213, NKT216,NKT221,NKT224,NKT226,NKT228,*DS26,*GE-2,*TR-85,*HEPG0005, *SK3004,*JR15,*ECG158,*WEP250,*276-2005,*RT-121
2SB172	2SB56,2SB288,NKT213,NKT214,NKT216,NKT221,NKT224,NKT228,*PTC135,*GE-2, *TR-84,*HEPG0005,*SK3004,*JR15,*ECG158,*WEP250,*276-2005,*RT-121
2SB173	2N2953,2SB54,2SB175,2SB346,2SB348,2SB370A,2SB431,AC122-GRN,AC122-VIO, AC122/30-GRN,AC122/30-VIO,NKT211,NKT218,NKT222,NKT223,NKT226,*GE-50,

TRANSISTOR SUBSTITUTES

To Replace	Substitute This Type
(2SB173)	*TR-85,*HEPG0005,*SK3004,*AT20M,*ECG102A,*WEP250,*276-2007,*RT-121
2SB175	2N2953,2SB54,2SB173,2SB346,2SB348,2SB370A,2SB431,AC122-GRN,AC122-VIO, AC122/30-GRN,AC122/30-VIO,NKT211,NKT218,NKT222,NKT223,NKT226,*GE-2, *TR-85,*HEPG0005,*SK3004,*JR15,*ECG158,*WEP250,*276-2005,*RT-121
2SB176	2N109+,2N217+,2N2953,2SB54,2SB370A,2SB431,AC122-GRN,AC122/30-GRN, NKT211,NKT222,NKT223,NKT226,*PTC135,*GE-2,*TR-84,*HEPG0005,*SK3004, *AT20M,*ECG102A,*WEP250,*RT-121
2SB177	2SB136A,NKT217,NKT227,*GE-2,*TR-85,*HEPG0005,*SK3004,*ECG102A,*WEP250, *276-2007,*RT-121
2SB178	AC117,NKT221,NKT228,*PTC135,*GE-2,*TR-85,*HEPG0005,*SK3004,*AT20M, *ECG102A,*WEP250,*276-2005,*RT-121
2SB178A	*PTC135,*GE-2,*TR-85,*HEPG0005,*SK3004,*ECG102A,*WEP250,*276-2005, *RT-121
2SB178B	*PTC155,*GE-53,*SK3004
2SB179	*PTC107,*HEPG0005
2SB180	2SB181,*HEPG6011,*ECG158,*WEP238
2SB180A	2SB181A,*HEPG6011,*ECG158,*WEP238
2SB181	*HEPG6011,*ECG158,*WEP238
2SB181A	2SB456,*HEPG6011,*ECG158,*WEP238
2SB183	*PTC109,*HEPG0005,*SK3004,*ECG102A,*WEP250
2SB184	2N796,2N827,2N2699,2SA252,2SA408,2SA409,2SB116,2SB117,2SB186,2SB187, 2SB188,2SB303,2SB329,40359,*PTC107,*HEPG0006,*SK3004,*ECG102A,*WEP250
2SB185	*PTC109,*GE-2,*TR-84,*HEPG0005,*SK3004,*JR15,*ECG102A,*WEP250,*276-2007, *RT-121
2SB186	*PTC109,*GE-53,*IR2SB186,*HEPG0007,*SK3004,*JR15,*ECG102A,*WEP250, *RT-117
2SB187	2SB186,2SB188,*PTC109,*GE-53,*IR2SB187,*HEPG0005,*SK3004,*JR15,*ECG102A, *WEP250,*AC188/01,*276-2006,*RT-123
2SB188	2SB186,2SB187,*DS-26,*GE-2,*TR-85,*HEPG0005,*SK3004,*JR15,*ECG102A, *WEP250,*RT-121
2SB189	*PTC135,*GE-53,*TR-05,*HEPG0005,*SK3004,*JR15,*ECG102A,*WEP631, *AC188/01,*276-2007,*RT-123
2SB189G	*GE-53,*SK3004
2SB190	*HEPG0005
2SB191	
2SB192	*HEPG0005
2SB193	*HEPG0005
2SB194	*HEPG0005
2SB195	*HEPG0005
2SB196	2SB197,*HEPG0005
2SB197	2SB198,*HEPG0005
2SB198	*PTC109,*HEPG0005
2SB199	2N1008,NKT231,*PTC135,*HEPG0005,*SK3004,*ECG102,*WEP631
2SB200	2N382,2N383,2N527,2N527A,2N597,2N1997,2SB227,2SB327,2SB428,2SB461, *PTC135,*GE-53,*TR-05,*HEPG0005,*SK3004,*JR15,*ECG102,*WEP631, *AC188/01,*276-2005,*RT-120
2SB200A	2N1997,2N2000,*GE-2,*TR-05,*HEPG0005,*SK3004,*ECG102,*WEP631,*276-2007, *RT-120
2SB200G	SEE 2SA467G
2SB201	2N2000,*PTC135,*GE-53,*TR-85,*HEPG0005,*AT20M,*ECG102A,*WEP250,*AC128, *276-2005,*RT-121
2SB202	2N598,2N599,2N1998,2N1999,2N2000,2N2001,2N2541,*PTC135,*GE-53,*TR-05, *HEPG0005,*SK3004,*JR15,*ECG102,*WEP631,*AC188/01,*276-2005,*RT-120
2SB202G	SEE 2SA467G
2SB203	2SB209
2SB204	2SB210,*TR-35,*ECG179,*WEPG6001,*RT-147
2SB205	2N1165,2N1165A,2N1167,2N1167A,2SB207,2SB211,2SB213,*TR-35,*ECG179, *WEPG6001,*RT-147
2SB206	2SB208,2SB212,2SB214,*TR-35,*ECG179,*WEPG6001,*RT-147
2SB207	2N1167,2N1167A,2SB213
2SB207A	2SB213A

To Replace	Substitute This Type
2SB208	2SB214
2SB208A	2SB214A
2SB209	2SB203
2SB210	2SB204
2SB211	2N1165,2N1165A,2N1167,2N1167A,2SB205,2SB207,2SB213
2SB212	2SB206,2SB208,2SB214
2SB213	2N1167,2N1167A,2SB207
2SB213A	
2SB214	2SB208
2SB214A	
2SB215	*HEPG6005,*SK3009,*ECG121,*WEP232
2SB216	2N420,2SB138,2SB216A,CDT1311,CDT1312,*PTC138,*HEPG6005,*SK3034, *ECG121,*WEP232
2SB216A	2SB138,CDT1311,CDT1312,*SK3034,*ECG121,*WEP232
2SB217	2N257-BLK,2N285B,2N4241,2SB137,CDT1310,CDT1319,*PTC138,*HEPG6003, *SK3009,*ECG121,*WEP232
2SB218	*HEPG6012,*SK3004,*ECG102,*WEP631
2SB219	2N524,2N524A,2N525,2N525A,2SB219A,2SB220,2SB220A,2SB224,2SB225,SF.T222, *PTC135,*HEPG0005,*SK3004,*ECG102,*WEP631
2SB219A	2N524,2N524A,2N525,2N525A,2SB220A,2SB224,2SB225,*SK3004
2SB220	2N381,2N525,2N525A,2N526,2N526A,2N597,2N1478,2SB57A,2SB220A,2SB221, 2SB221A,2SB225,2SB226,2SB326,SF.T222,*PTC135,*HEPG0005,*SK3004, *ECG102,*WEP631
2SB220A	2N525,2N525A,2N526,2N526A,2N597,2N1478,2SB57A,2SB221A,2SB225,2SB226, 2SB326,*PTC102,*SK3004,*ECG102,*WEP631
2SB221	2N382,2N526,2N526A,2N527A,2N597,2N1008A,2N1997,2SB221A,2SB222,2SB222A, 2SB226,2SB227,2SB327,SF.T223,*PTC135,*HEPG0005,*SK3004,*ECG102, *WEP631
2SB221A	2N526,2N526A,2N527A,2N597,2N1997,2SB222A,2SB226,2SB227,2SB327,*PTC102, *HEPG0005,*SK3004,*ECG102,*WEP631
2SB222	2N382,2N383,2N527,2N598,2N599,2N1008A,2N1997,2N1998,2SB222A,2SB223, 2SB223A,2SB327,SF.T223,*PTC135,*HEPG0005,*SK3004,*ECG102,*WEP631
2SB222A	2N527,2N598,2N599,2N1997,2N1998,2N2541,2SB223A,2SB327,*SK3004
2SB223	2N599,2N1998,2N1999,2SB223A,*PTC135,*HEPG0005,*SK3004,*ECG102,*WEP631
2SB223A	2N599,2N1998,2N1999,2N2001,2N2541,*SK3004
2SB224	2N525,2N525A,2SB225,*HEPG0005,*SK3004,*ECG102,*WEP631
2SB225	2N526,2N526A,2N597,2SB226,*HEPG0005,*SK3004,*ECG102,*WEP631
2SB226	2N527,2N527A,2N597,2N1997,2SB227,*HEPG0006,*SK3004,*ECG102,*WEP631
2SB227	2N527,2N527A,2N1997,*HEPG0006,*SK3004,*ECG102,*WEP631
2SB228	2N1532,2SB87,2SB88,2SB340,*PTC105A,*GE-25,*TR-01,*HEPG6018,*SK3009, *PT40,*ECG104,*WEP230,*OC28,*276-2006,*RT-124
2SB229	2N1532,2N1533,2SB88,2SB340,2SB341,*PTC105A,*GE-25,*TR-01,*HEPG6018, *SK3009,*AT20M,*ECG121,*WEP232,*OC28,*276-2006,*RT-127
2SB230	2N1533,2SB341,*GE-25,*TR-01,*HEPG6018,*SK3009,*PT40,*ECG104,*WEP230, *OC28,*276-2006,*RT-124
2SB231	2N1046A,2N1046B,2SB375A,*GE-25,*TR-01,*HEPG6017,*SK3035,*PT150,*ECG127, *WEP235,*276-2006
2SB231G	SEE 2SC520A
2SB232	2N1022,2N2293,2SB129A,*PTC122,*HEPG6007,*SK3034,*ECG127,*WEP235
2SB233	*HEPG6007,*SK3009,*ECG104,*WEP230
2SB234	2SB360,*HEPG6007,*SK3035,*ECG127,*WEP235
2SB235	2SB333,2SB353,2SB354,SF.T267,*GE-4,*TR-03,*HEPG6006,*SK3012,*PT515, *ECG105,*WEP233,*276-2006
2SB235G	*GE-4,*SK3012
2SB236	2SB332,2SB333,2SB352,2SB353,SF.T266,SF.T267,*GE-4,*TR-03,*HEPG6006, *SK3012,*PT515,*ECG105,*WEP233,*276-2006
2SB236G	*GE-4,*SK3012
2SB236N	SEE 2SD55N
2SB237	*GE-4,*TR-03,*HEPG6004,*SK3012,*PT515,*ECG105,*WEP231,*276-2006
2SB237G	*GE-4,*SK3012
2SB238	2SB240,2SB240A,2SB243A,*HEPG6011,*ECG158,*WEP238

To Replace	Substitute This Type
2SB239	2SB239A,2SB241,2SB241A,2SB245,2SB456,*HEPG6012,*SK3009,*ECG121, *WEP232
2SB239A	2SB241A,*HEPG6012,*SK3009,*ECG121,*WEP232
2SB240	2SB238,2SB240A,2SB243A,*HEPG6011,*SK3009,*ECG160,*WEP637
2SB240A	2SB239,2SB241,2SB245,2SB456,*HEPG6011,*SK3009,*ECG160,*WEP637
2SB241	2SB239,2SB239A,2SB241A,2SB245,2SB456,*HEPG6012,*SK3004,*ECG102, *WEP631
2SB241A	2SB239A,*HEPG6012,*SK3004
2SB242	2SB238,2SB240,2SB243,*HEPG6011,*SK3009
2SB242A	2SB239,2SB240A,2SB241,2SB243A,2SB244,2SB245,*HEPG6011,*SK3009
2SB243	2SB238,2SB240,*HEPG6011,*SK3009
2SB243A	2SB239,2SB240A,2SB241,2SB245,2SB456,*HEPG6011,*SK3009
2SB244	2SB239,2SB239A,2SB241,2SB241A,2SB245,*HEPG6012,*SK3009
2SB245	2SB239,2SB239A,2SB241,2SB241A,2SB456,*HEPG6012,*SK3009
2SB246	2N380,2N456A,2N457A,2N1168,2N1534,2N1535,2N1539,2N1540,2N2143,2N2143A, 2N2144,2N2144A,2N3611,2N3612,2N3613,2N3614,2N4243,2N4244,2N4246,2N4247, 2N5890,2N5891,2N5894,2N5895,2SB247,2SB248,2SB248A,2SB251A,MP2060, MP2061,*PTC105A,*HEPG6003,*SK3009,*ECG104,*WEP230
2SB247	2N380,2N457A,2N458A,2N1535,2N1536,2N1540,2N1541,2N2144,2N2144A,2N2145, 2N2145A,2N3612,2N3614,2N3617,2N4242,2N4243,2N4245,2N4246,2N5891,2N5892, 2N5895,2N5896,2SB248A,2SB249,2SB251A,2SB253,2SB339,MP2061,MP2062, *PTC114,*HEPG6005,*SK3009,*ECG121,*WEP232
2SB248	2N380,2N456A,2N457A,2N1168,2N1534,2N1535,2N1539,2N1540,2N2143,2N2143A, 2N2144,2N2144A,2N3611,2N3612,2N3613,2N3614,2N4243,2N4244,2N4246,2N4247, 2N5890,2N5891,2N5894,2N5895,2SB246,2SB247,2SB248A,2SB251A,MP2060, MP2061,*PTC114,*HEPG6003,*SK3009,*ECG121,*WEP232
2SB248A	2N380,2N457A,2N458A,2N1535,2N1536,2N1540,2N1541,2N2144,2N2144A,2N2145, 2N2145A,2N3612,2N3614,2N3617,2N4242,2N4243,2N4245,2N4246,2N5891,2N5892, 2N5895,2N5896,2SB247,2SB249,2SB251A,2SB253,2SB339,MP2061,MP2062, *PTC114,*HEPG6005,*SK3009,*ECG121,*WEP232
2SB249	2N1536,2N1537,2N1541,2N1542,2N2145,2N2145A,2N2146,2N2146A,2N3617,2N3618, 2N4242,2N4245,2SB253,2SB339,*PTC114,*HEPG6005,*SK3009,*ECG104, *WEP230
2SB250	2N2138,2N2138A,2N3611,2N4244,2SB251,*PTC114,*HEPG6003,*SK3009,*ECG104, *WEP230
2SB250A	2N380,2N561,2N1160,2N1535,2N1536,2N2139,2N2139A,2N2140,2N2140A,2N3612, 2N3615,2N3617,2N4242,2N4243,2SB251A,2SB252,2SB253,*PTC105A,*HEPG6003, *SK3009,*ECG104,*WEP230
2SB251	2N1168,2N1534,2N2143,2N2143A,2N3611,2N3613,2N4244,2N4247,MP2060,*PTC114, *HEPG6003,*SK3035,*ECG127,*WEP235
2SB251A	2N380,2N1535,2N1536,2N2144,2N2144A,2N2145,2N2145A,2N3612,2N3614,2N3617, 2N4242,2N4243,2N4245,2N4246,2SB253,MP2061,MP2062,*PTC114,*HEPG6003, *SK3035,*ECG127,*WEP235
2SB252	2N561,2N1536,2N1537,2N2140,2N2140A,2N2141,2N2141A,2N3615,2N3616,2N3617, 2N3618,2N4242,2SB253,*PTC114,*HEPG6005,*SK3009,*ECG127,*WEP235
2SB252A	2N1538,2SB253A,*HEPG6018,*SK3035,*ECG127,*WEP235
2SB253	2N1536,2N1537,2N2145,2N2145A,2N2146,2N2146A,2N3617,2N3618,2N4242,2N4245, *PTC105A,*HEPG6005,*SK3035,*ECG127,*WEP235
2SB253A	2N1538,2SB129A,*HEPG6018,*SK3009,*ECG127,*WEP235
2SB254	*PTC138,*SK3009,*ECG104,*WEP230
2SB255	*PTC194,*HEPG6016,*SK3009,*ECG104,*WEP230
2SB256	2SB254,*PTC194,*HEPG6016,*SK3052,*ECG104,*WEP230
2SB257	*PTC109,*GE-1,*TR-05,*HEPG0008,*SK3004,*JR15,*ECG102,*WEP631,*AC188/01, *276-2007
2SB258	*PTC122,*GE-4,*TR-03,*SK3012,*PT501,*ECG105,*WEP233
2SB258G	SEE 2SD114
2SB259	*PTC122,*GE-4,*TR-03,*HEPG6006,*SK3012,*PT515,*ECG105,*WEP233,*276-2006
2SB259G	SEE 2SD114
2SB260	*PTC122,*GE-4,*TR-03,*HEPG6004,*SK3012,*PT515,*ECG105,*WEP233,*276-2006
2SB260G	SEE 2SD114
2SB261	*PTC107,*HEPG0005,*SK3004,*ECG102A,*WEP250

To Replace	Substitute This Type
2SB262	2N838,2N3323,2N3324,2N3325,40359,*PTC107,*HEPG0005,*SK3004,*ECG102A, *WEP250
2SB263	2N2706,2SB496,SF.T323,*PTC102,*HEPG0005,*SK3004,*ECG102A,*WEP250
2SB264	*HEPG0005,*SK3004,*ECG102A,*WEP250
2SB265	2N382,2N526,2N526A,2N527A,2N597,2N651,2N652A,2N1008A,2N1008B,2N1192, 2N1348,2N1349,2N1350,2N1351,2N1447,2N1448,2N1925,2N1926,2SB57A,2SB226, 2SB227,*PTC109,*GE-80,*TR-85,*HEPG0005,*SK3004,*JR15,*ECG102A,*WEP250, *AC126,*RT-121
2SB266	*HEPG0005,*SK3004,*ECG102A,*WEP250
2SB267	2N109+,2N217+,2SB56,2SB187,2SB188,2SB328,2SB329,2SB496,ACY23,ACY32, MM404,MM404A,NKT211,SF.T323,SF.T353,*HEPG0005,*SK3004,*ECG102A, *WEP250
2SB268	2N597,2N1008A,2N1009,2N1124,2N1125,2N1375,2N1377,2N1997,2N2706,*PTC135, *HEPG0005,*SK3004,*ECG102A,*WEP250
2SB269	*HEPG0005,*SK3004,*ECG102A,*WEP250
2SB270	*PTC109,*HEPG0005,*SK3004,*ECG102A,*WEP250
2SB271	2SB272,2SB273,*HEPG0005,*ECG102A,*WEP250
2SB272	2SB273,ACY33,*HEPG0005,*SK3004,*ECG102A,*WEP250
2SB273	ACY33,*HEPG0005,*SK3004,*ECG102A,*WEP250
2SB274	2N1046,2SB358,*PTC122,*GE-25,*TR-27,*HEPG6005,*SK3035,*HO300,*ECG127, *WEP235,*276-2006
2SB275	2N1046A,2N1046B,*PTC122,*GE-25,*TR-27,*HEPG6017,*SK3035,*HO300,*ECG127, *WEP235,*276-2006
2SB276	2N1046A,2N1046B,*PTC122,*GE-25,*TR-27,*HEPG6017,*SK3035,*HO300,*ECG127, *WEP235,*276-2006
2SB277	
2SB278	
2SB279	
2SB280	BCY10,BCY12,MM4052
2SB281	BCY11
2SB282	2SB87,2SB88,2SB340,*PTC105A,*GE-25,*TR-01,*HEPG6005,*SK3009,*AT20M, *ECG121,*WEP232,*OC28,*276-2006,*RT-127
2SB283	2N5891,2N5892,2SB339,KR6500,*PTC105A,*GE-25,*TR-01,*HEPG6005,*SK3009, *ECG121,*WEP232,*276-2006,*RT-127
2SB284	2N5891,2N5892,2SB86,2SB87,2SB128,*PTC105A,*GE-25,*TR-01,*HEPG6005, *SK3009,*ECG121,*WEP232,*OC28,*276-2006,*RT-127
2SB285	2SB87,2SB88,2SB128,2SB340,*PTC105A,*GE-25,*TR-01,*HEPG6005,*SK3009, *ECG121,*WEP232,*OC28,*RT-127
2SB286	2N458
2SB287	2N5892,2SB87,2SB88,2SB340,2SB358,KR6502
2SB288	2SB56,2SB172,AC122-RED,AC122-YEL,AC122/30-RED,AC122/30-YEL,NKT213, NKT214,NKT216,NKT221,NKT224,NKT228
2SB289	
2SB290-GREEN	2N417,2N1309,2N1316,2N1317,2N1354,2N1355,2N1357,2N2613
2SB290-YELLOW	2N417,2N1307,2N1356,2N1892,2SA210,ASY27,SF.T229
2SB290N	SEE 2SC587AN
2SB291-GREEN	2N599,2N655+,2N1185,2N1190,2N1316,2N1999,2N2171,2N3427,MA888
2SB291-RED	2N381,2N413,2N414,2N414A,2N414B,2N414C,2N525,2N525A,2N650,2N653,2N1017, 2N1171,2N1191,2N1303,2N1313,2N1446,2N1447,2N1478,2SA204,2SA205,2SA206, 2SA207,2SA406,2SA407,2SB225,2SB326,2SB427,ASY26,SF.T222,SF.T227
2SB291-YELLOW	2N382,2N383,2N415,2N415A,2N416,2N422A,2N527,2N527A,2N597,2N651,2N651A, 2N652,2N652A,2N654,2N654+,2N1008A,2N1189,2N1307,2N1309A,2N1349,2N1352, 2N1356,2N1375,2N1377,2N1448,2N1449,2N1681,2N1892,2N1997,2SB227,2SB327, 2SB428,MA887,SF.T223
2SB292-BLUE	2N381,2N414,2N414A,2N414B,2N414C,2N415,2N415A,2N416,2N422A,2N525, 2N525A,2N526,2N526A,2N597,2N650,2N651,2N651A,2N652A,2N654,2N1171, 2N1192,2N1303,2N1305,2N1309A,2N1313,2N1348,2N1349,2N1350,2N1351, 2N1447,2N1478,2N1681,2SA204,2SA206,2SA207,2SA406,2SA407,2SB57A,2SB225, 2SB226,2SB326,2SB427,ASY26,MA887,SF.T222
2SB292-GREEN	2N381,2N414,2N414A,2N414B,2N414C,2N415,2N415A,2N416,2N422A,2N526, 2N526A,2N527A,2N597,2N651,2N651A,2N652A,2N654,2N654+,2N1008A,2N1171,

To Replace	Substitute This Type
(2SB292-GREEN)	2N1192,2N1305,2N1307,2N1309A,2N1348,2N1349,2N1350,2N1351,2N1352,2N1356, 2N1375,2N1377,2N1447,2N1448,2N1478,2N1681,2SA204,2SA207,2SA407,2SB57A, 2SB226,2SB227,2SB326,2SB427,2SB428,MA887,SF.T223
2SB292-ORANGE	2N382,2N383,2N417,2N527,2N527A,2N598,2N652,2N654+,2N655,2N1008A,2N1189, 2N1193,2N1307,2N1352,2N1356,2N1375,2N1377,2N1448,2N1449,2N1892,2N1997, 2SB227,2SB327,2SB428,SF.T223
2SB292-RED	2N383,2N417,2N527,2N598,2N599,2N652,2N655,2N655+,2N1185,2N1189,2N1193, 2N1309,2N1354,2N1355,2N1357,2N1449,2N1892,2N1997,2N1998,2N2171,2SB327, MA888
2SB292-YELLOW	2N382,2N415,2N415A,2N416,2N422A,2N526,2N526A,2N527A,2N597,2N651,2N651A, 2N652,2N652A,2N654,2N654+,2N1008A,2N1192,2N1307,2N1309A,2N1349,2N1352, 2N1356,2N1375,2N1377,2N1448,2N1681,2N1892,2N1997,2SB226,2SB227,2SB327, 2SB428,MA887,SF.T223
2SB292A-BLUE	2N1187,2N1924,2N1925,MA100,MA882,SF.T243
2SB292A-GREEN	2N1008B,2N1187,2N1925,2N1926,MA100,MA882,SF.T243
2SB292A-ORANG	2N1008B,2N1188,2N1926
2SB292A-RED	2N1188,MA883
2SB292A-YELLO	2N1008B,2N1187,2N1188,2N1926,MA882
2SB292AN	SEE 2SA560N
2SB292N	SEE 2SA560N
2SB293	2N369,2N520A,2N569,2N571,2N654,2N654+,2N1008,2N1009,2N1128,2N1130, 2N1305,2N1307,2N1309A,2N1705,2N1706,2N1707,2N2271,2N2375,2SB221A, 2SB222A,2SB304,2SB327,SF.T223,SF.T229,SYL792,*HEPG0005,*SK3004, *ECG102A,*WEP250
2SB294	*HEPG0005,*SK3123,*ECG102A,*WEP250
2SB295	2N1542,2N1543,2SB129A,2SB152A,2SB253A,*HEPG6018,*SK3009,*ECG121, *WEP232
2SB296	2SB410,*PTC122,*GE-25,*TR-27,*HEPG6017,*SK3035,*ECG127,*WEP235, *276-2006
2SB296G	SEE 2SA656
2SB299	2N526A,2N527A,2N597,2N651,2N651A,2N652A,2N654,2N1124,2N1125,2N1192, 2N1305,2N1307,2N1309A,2N1348,2N1349,2N1350,2N1351,2N1375,2N1377,2N2375, *HEPG0005,*SK3004,*ECG102A,*WEP250
2SB300	2N2293,2SB410,2SB484,KR6502,*PTC122,*HEPG6015,*SK3035,*ECG127,*WEP235
2SB300G	SEE 2SC520A
2SB301	2N2144,2N2144A,2N2145,2N2145A,2N2292,2SB339,2SB433,KR6500,*PTC122, *HEPG6005,*SK3035,*ECG127,*WEP235
2SB301G	SEE 2SC521A
2SB302	2SA49,2SA52,2SA53,2SA408,2SA409,2SB77,2SB115,2SB116,2SB136,2SB186, 2SB187,2SB188,2SB303,2SB329,2SB364,AFY15,SF.T307,SF.T308,SF.T323,SF.T337, *PTC109,*GE-51,*TR-85,*HEPG0005,*SK3004,*AT20M,*ECG102A,*WEP250, *AF125,*276-2002
2SB303	2SB186,2SB187,2SB188,*PTC109,*GE-52,*TR-84,*HEPG0005,*SK3004,*AT20M, *ECG102A,*WEP250,*AC126,*276-2007,*RT-121
2SB304	2N526,2N526A,2N527A,2N597,2N2375,2SB57A,2SB221A,2SB222A,2SB226,2SB227, 2SB304A,2SB327,2SB428,2SB461,*PTC135,*HEPG0005,*SK3123,*ECG158, *WEP238
2SB304A	2N526,2N526A,2N527A,2N597,2N1925,2N1926,2SB57A,2SB226,2SB227,2SB428, *PTC135,*HEPG0005,*SK3123,*ECG158,*WEP238
2SB306	2N2043,2N2043A
2SB309	2N2145,2N2145A,2N2146,2N2146A,2N2292,2SB339,2SB433,2SB484,ASZ18,KR6502, *PTC122,*GE-25,*TR-27,*HEPG6005,*SK3035,*ECG127,*WEP235,*276-2006
2SB309A	*PTC122,*GE-25,*SK3035
2SB310	2SB410,*PTC122,*HEPG6017,*SK3035,*ECG127,*WEP235
2SB311	*PTC122,*HEPG6017,*SK3035,*ECG127,*WEP235
2SB312	2SB313,2SB360,*PTC122,*HEPG6007,*SK3035,*ECG127,*WEP235
2SB313	2SB360,*PTC122,*HEPG6017,*SK3035,*ECG127,*WEP235
2SB314	2SB106,2SB109,*PTC102,*SK3004,*ECG100,*WEP254
2SB315	*PTC102,*HEPG0005,*SK3004,*ECG102A,*WEP250
2SB316	2SB156,2SB156A,2SB365,2SB415,2SB475,NKT211,NKT221,NKT228,*PTC102, *HEPG0005,*SK3004,*ECG102A,*WEP250

To Replace	Substitute This Type
2SB317	2N1008,2N1478,2N2271,2N3883,NKT221,NKT228,*PTC135,*HEPG0005,*ECG102A, *WEP250
2SB318	2N2444,*HEPG6005,*SK3014,*ECG127,*WEP235
2SB319	*HEPG6007,*SK3035,*ECG127,*WEP235
2SB320	*HEPG6017,*SK3035,*ECG127,*WEP235
2SB321	*PTC109,*HEPG6006,*ECG102A,*WEP250
2SB322	*PTC109,*HEPG0005,*ECG102A,*WEP250
2SB323	*PTC109,*HEPG0006,*ECG102A,*WEP250
2SB324	*PTC156,*GE-53,*TR-84,*HEPG0005,*SK3004,*AT20M,*ECG158,*WEP238, *AC188/01,*276-2006,*RT-123
2SB325	
2SB326	2N526,2N526A,2N527A,2N597,2SB226,2SB227,*PTC135,*HEPG0005,*SK3004, *ECG102,*WEP631
2SB327	2N527,2N598,2N599,2N1997,2N1998,*PTC135,*HEPG0005,*SK3004,*ECG102, *WEP631
2SB328	2SB329,SF.T323,*HEPG0005,*SK3004,*ECG102,*WEP631
2SB329	SF.T323,*HEPG0005,*SK3004,*ECG102A,*WEP250
2SB330	*HEPG6012
2SB331	2SB332,*GE-4,*TR-03,*SK3012,*PT501,*ECG213,*WEP233
2SB332	2SB333,*GE-4,*TR-03,*HEPG6006,*SK3012,*PT501,*ECG213,*WEP233,*276-2006
2SB333	*GE-4,*TR-03,*HEPG6006,*SK3012,*PT501,*ECG213,*WEP233,*276-2006
2SB334	2N1100,2N1412,*GE-4,*TR-03,*HEPG6006,*SK3012,*PT501,*WEP233,*276-2006
2SB335	2SB186,2SB187,2SB188,*PTC109,*GE-52,*TR-85,*HEPG0005,*ECG102,*AC126, *276-2007
2SB336	2SB186,2SB187,2SB188,*PTC109,*GE-52,*TR-85,*HEPG0005,*ECG102A,*WEP250, *AC126,*276-2007,*RT-121
2SB337	2N5894,2N5895,2N5898,2N5899,2SB338,2SB471,KR6501,*PTC105A,*GE-25,*TR-02, *HEPG6005,*SK3009,*PT40,*ECG121,*WEP232,*276-2006,*RT-127A
2SB338	2N5895,2N5896,2SB339,*PTC105A,*GE-25,*TR-01,*HEPG6005,*SK3009,*PT40, *ECG121,*WEP232,*276-2006,*RT-127
2SB339	2N1560,*PTC105A,*GE-25,*TR-35,*HEPG6018,*PT40,*ECG179,*WEPG6001, *276-2006,*RT-147
2SB340	2SB88,2SB341,*PTC105A,*GE-25,*TR-35,*HEPG6018,*PT40,*ECG179,*WEPG6001, *OC28,*276-2006,*RT-147
2SB341	*DS503,*GE-25,*TR-27,*HEPG6018,*SK3034,*HO300,*ECG127,*WEPG6001,*OC28, *276-2006,*RT-147
2SB342	2SB375A,*PTC122,*GE-25,*TR-27,*HEPG6017,*SK3035,*HO300,*ECG127,*WEP235
2SB343	2SB375A,2SB411,*PTC122,*GE-25,*TR-27,*HEPG6017,*SK3035,*HO300,*ECG127, *WEP235
2SB345	2N109+,2N217+,2SB54,2SB346,2SB347,2SB348,2SB370A,2SB431,NKT211, NKT222,NKT223,NKT226,*PTC135,*GE-52,*TR-85,*HEPG0005,*SK3004,*AT20M, *ECG102A,*WEP250,*AC126,*276-2007,*RT-121
2SB346	2SB54,2SB348,2SB370A,2SB431,2SB439,2SB440,NKT218,NKT222,NKT223,*PTC135, *TR-85,*HEPG0007,*SK3004,*AT20M,*ECG102A,*WEP250,*AC188/01,*RT-121
2SB347	2N109+,2N217+,2SB54,2SB345,2SB346,2SB348,2SB370A,2SB431,NKT211, NKT222,NKT223,NKT226,*PTC135,*GE-52,*TR-05,*HEPG0007,*SK3004,*ECG102, *WEP631,*AC126,*276-2005,*RT-120
2SB348	2SB54,2SB346,2SB370A,2SB431,2SB439,2SB440,NKT218,NKT222,NKT223,*PTC135, *TR-85,*HEPG0006,*SK3004,*ECG102A,*WEP250,*AC188/01,*276-2005,*RT-121
2SB349	2N827,2N2613+,2N2635,2SA18,2SB400,2SB408,2SB439,2SB440,40263,AC122-VIO, AC122-WHT,AC150-VIO,AC150-WHT,AC171,*PTC107,*HEPG0006,*ECG102A, *WEP250
2SB350	2N2614,2N2953,2N2957,2N3371,2SB54,2SB116,2SB117,2SB136,2SB173,2SB175, 2SB186,2SB187,2SB188,2SB329,2SB346,2SB348,2SB370,2SB370A,2SB431, 2SB459,2SB460,AC122-GRN,AC122-VIO,AC150-GRN,AC150-VIO,AC170,NKT211, NKT218,NKT222,NKT223,NKT226,SF.T323,*HEPG0005,*SK3004,*ECG102, *WEP631
2SB351	2SB331,2SB332,2SB352,*PTC106,*SK3012,*ECG105,*WEP233
2SB352	2SB332,2SB333,2SB353,*PTC106,*HEPG6006,*SK3012,*ECG294,*WEP233
2SB353	2SB333,2SB354,*PTC106,*HEPG6006,*SK3012,*ECG105,*WEP233
2SB354	*PTC106,*HEPG6006,*SK3012,*ECG105,*WEP233

To Replace	Substitute This Type
2SB355	2SB255,2SB367,*HEPG6016,*SK3009,*ECG104,*WEP230
2SB356	*SK3009,*ECG104,*WEP230
2SB357	*SK3035,*ECG127,*WEP235
2SB358	*HEPG6005,*SK3035,*ECG127,*WEP235
2SB359	2SB341,*HEPG6017,*SK3035,*ECG127,*WEP235
2SB360	*HEPG6017,*SK3035,*ECG127,*WEP235
2SB361	2N5896,2N5900,KR6503,*PTC138,*GE-3,*TR-27,*HEPG6005,*SK3035,*HO300, *ECG127,*WEP235,*OC28,*276-2006
2SB362	*GE-25,*TR-27,*HEPG6005,*SK3014,*HO300,*ECG127,*WEP235,*OC28,*276-2006
2SB363	2N1046A,2N1046B,2SB366,2SB410
2SB364	2SB431,*PTC135,*GE-53,*TR-85,*HEPG0005,*SK3004,*AT20M,*ECG102A,*WEP238, *AC188/01,*276-2007,*RT-123
2SB365	2SB364,*PTC135,*GE-53,*TR-84,*HEPG0005,*SK3004,*ECG158,*WEP238, *AC188/01,*276-2007
2SB366	2N1046A,2N1046B,2SB363,2SB410,*PTC102,*SK3123,*ECG102A,*WEP250
2SB367	2SB254,2SB256,*PTC120A,*GE-30,*TR-94,*HEPG6016,*SK3052,*PT40,*ECG131, *WEP642,*276-2006,*RT-132
2SB368	*PTC120A,*TR-50,*HEPG6016,*SK3052,*ECG131,*WEP642,*276-2006,*RT-132
2SB370	2SB370A,*PTC135,*GE-53,*TR-85,*HEPG0006,*SK3004,*AT20M,*ECG102A, *WEP250,*AC188/01,*276-2006,*RT-123
2SB370A	*PTC135,*GE-53,*TR-85,*HEPG0006,*SK3004,*ECG102A,*WEP250,*276-2005, *RT-121
2SB371	*PTC135,*GE-1,*TR-85,*HEPG0005,*SK3004,*ECG102A,*WEP250,*AC188/01, *RT-121
2SB372	*PTC156,*HEPG6011,*ECG158,*WEP238
2SB373	*PTC156,*HEPG6011,*SK3009,*ECG158,*WEP238
2SB374	*HEPG6011
_2SB375_	2SB375A,2SB410,*PTC122,*GE-25,*TR-27,*HEPG6007,*SK3035,*PT40,*ECG127, *WEP235,*276-2006
2SB375A	*HEPG6007,*SK3035,*ECG127,*WEP235
_2SB376_	2SB178,AC117,NKT221,NKT228,*PTC135,*GE-53,*TR-85,*HEPG0005,*SK3004, *ECG102A,*WEP250,*AC188/01,*276-2007,*RT-121
_2SB377_	2N598,2N599,2N1997,2N1998,*GE-53,*TR-85,*HEPG0005,*AT20M,*ECG102A, *WEP250,*276-2005,*RT-121
_2SB378_	2N320,2N361,2N414B,2N414C,2N1204A,2N1313,2N1478,2N3883,2SA206,2SA207, 2SA406,2SA407,2SB326,MM380,*GE-2,*TR-05,*HEPG0005,*SK3004,*AT20M, *ECG102A,*WEP254,*AC188/01,*276-2007,*RT-118
2SB378A	2N1478,2N1495,2N3883,MM380,*SK3004,*ECG102,*WEP631
2SB378B	2N1478,2N1495,MM380,*SK3004
_2SB379_	2N360,2N598,2N654+,2N655,2N1356,2N1705,2N1706,2N1707,2SB222,2SB222A, 2SB327,SF.T223,*GE-2,*TR-05,*HEPG0005,*SK3004,*AT20M,*ECG102,*WEP631, *AC188/01,*276-2007,*RT-120
2SB379A	2N598,*PTC102,*SK3004,*ECG102,*WEP631
2SB379B	2N598,*PTC102,*SK3004,*ECG102,*WEP631
_2SB380_	2N359,2N1316,2N1317,2N1999,MA1703,MA1704,*GE-53,*TR-05,*HEPG0006, *SK3004,*ECG102,*WEP631,*AC188/01,*276-2007,*RT-120
2SB380A	2N1999,2N2001,*PTC102,*HEPG0005,*SK3004,*ECG102,*WEP631
_2SB381_	2N1478,2N1495,*GE-53,*TR-85,*HEPG0005,*SK3004,*AT20M,*ECG102A,*WEP250, *276-2004,*RT-121
_2SB382_	2N598,2N1997,*GE-53,*TR-85,*HEPG0005,*SK3004,*AT20M,*ECG102A,*WEP250, *276-2007,*RT-121
_2SB383_	2N598,2N1997,*GE-53,*TR-85,*HEPG0005,*SK3004,*AT20M,*ECG102A,*WEP250, *276-2005,*RT-121
2SB384	*HEPG0005,*SK3004,*ECG126,*WEP635
2SB385	2N196,2N407+,2N408+,2N838,2N2587,2N2717,2N2795,2N3323,2N3324,2N3325, 2N6365,2N6365A,2SA139,2SA325,2SA326,2SB33,2SB37,2SB56,2SB59,2SB60A, 2SB66,2SB75,2SB95,2SB114,2SB156A,2SB171,2SB172,2SB263,2SB288,2SB328, 2SB365,2SB475,AC122-RED,AC122-YEL,AFY15,ASZ21,MM380,NKT213,NKT214, NKT216,NKT224,SF.T322,SF.T352,*HEPG0005,*SK3004,*ECG126, *WEP635
_2SB386_	*PTC135,*HEPG0005,*SK3123,*ECG102A,*WEP250
2SB387	*PTC102,*HEPG0008,*SK3004,*ECG100,*WEP254

To Replace	Substitute This Type
2SB389	*PTC107,*HEPG0003,*SK3004,*ECG102A,*WEP250
2SB390	*PTC122,*GE-25,*TR-27,*HEPG6005,*SK3035,*HO300,*ECG127,*WEP235, *276-2006
2SB391	*PTC122,*GE-25,*TR-01,*HEPG6003,*SK3009,*HO300,*ECG121,*WEP232, *276-2006,*RT-127
2SB392	2N654,2N654+,2N1008,2N1356,2N1374,2N1376,2N1381,2N1382,2SB327,SF.T223, *PTC135,*GE-2,*TR-05,*HEPG0005,*SK3004,*AT20M,*ECG100,*WEP254, *AC188/01,*276-2006,*RT-118
2SB393	2N414B,2N414C,2N525,2N525A,2N650,2N650A,2N653,2N1191,2N1373,2N1446, 2N1478,2SA205,2SA206,2SA406,2SA407,2SB225,2SB427,SF.T222,*PTC135,*GE-53, *TR-05,*HEPG0005,*SK3004,*AT20M,*ECG100,*WEP254,*AC188/01,*276-2006, *RT-118
2SB394	2N527,2N527A,2N597,2N651,2N651A,2N652,2N652A,2N654,2N654+,2N1008A, 2N1124,2N1189,2N1349,2N1356,2N1375,2N1377,2N1448,2N1449,2N1997,2SB227, 2SB327,2SB428,SF.T223,*PTC135,*GE-53,*TR-05,*HEPG0005,*SK3004,*AT20M, *ECG100,*WEP254,*AC188/01,*276-2006,*RT-118
2SB395	2N599,2N655+,2N1185,2N1190,2N1316,2N1354,2N1355,2N1357,2N1998,2N1999, 2N2374,2N3427,*PTC135,*GE-53,*TR-05,*HEPG0005,*SK3004,*AT20M,*ECG100, *WEP254,*AC188/01,*276-2006,*RT-118
2SB396	2N382,2N383,2N527,2N527A,2N597,2N651,2N651A,2N652,2N652A,2N1008A, 2N1008B,2N1124,2N1187,2N1188,2N1189,2N1349,2N1375,2N1377,2N1448,2N1449, 2N1926,2N1954,2N1955,2N1956,2N1957,2N1997,2SB227,2SB428,MA882,MA887, *PTC135,*GE-80,*TR-05,*HEPG0005,*SK3004,*AT20M,*ECG100,*WEP254, *276-2004,*RT-118
2SB397	2N381,2N525,2N525A,2N526,2N526A,2N650,2N650A,2N651A,2N1187,2N1373, 2N1446,2N1447,2N1924,2N1925,2N1954,2N1956,2N1957,2SB57A,2SB225,2SB226, 2SB304A,2SB427,MA100,MA881,MA886,SF.T243
2SB398	
2SB399	2N2042,2N2042A,2SB330
2SB400	*PTC135,*GE-52,*TR-85,*HEPG0005,*SK3004,*ECG102A,*WEP250,*AC126, *276-2007,*RT-121
2SB401	2N381,2N526,2N526A,2N527A,2N597,2N1008A,2N1008B,2N1925,2N1926,2SB57A, 2SB226,2SB227,2SB304A,2SB427,2SB428,SF.T243,*PTC135,*GE-80,*TR-05, *HEPG0005,*SK3004,*ECG100,*WEP254,*AC128,*276-2007,*RT-118
2SB402	2N1008B,2N1925,2N1926,SF.T243,*GE-53,*TR-05,*HEPG0005,*SK3004,*ECG100, *WEP254,*276-2007,*RT-118
2SB403	2N382,2N383,2N1997,2N2000,*GE-80,*TR-05,*HEPG0005,*SK3004,*ECG100, *WEP254,*AC128,*276-2007,*RT-118
2SB404	
2SB405	*GE-53,*IR2SB405,*HEPG6011,*SK3004,*AT20M,*ECG158,*WEP238,*AC188/01, *RT-123
2SB406	2N5324,2SB411
2SB407	*PTC105A,*GE-25,*TR-02,*HEPG6005,*SK3009,*PT40,*ECG121,*WEP232,*276-2006, *RT-124
2SB408	*PTC102,*HEPG0005,*SK3004,*ECG100,*WEP254
2SB409	
2SB410	*TR-27,*HEPG6017,*ECG127,*WEP235,*276-2006
2SB411	*GE-76,*TR-27,*HEPG6007,*ECG127,*WEP235,*276-2006
2SB412	
2SB413	*GE-16,*TR-01,*SK3014,*ECG104,*WEP230,*RT-124
2SB413G	SEE 2SC524
2SB414	*GE-16,*TR-01,*HEPG6016,*SK3014,*ECG104,*WEP230,*276-2006,*RT-124
2SB414G	SEE 2SC525
2SB415	*PTC156,*GE-53,*TR-84,*HEPG6011,*SK3004,*AT20M,*ECG158,*WEP238, *AC188/01,*276-2006,*RT-123
2SB415G	*GE-53,*SK3004
2SB416	2N414,2N414A,2N414B,2N414C,2N415,2N415A,2N416,2N422,2N654,2N654+, 2N1008A,2N1124,2N1125,2N1171,2N1192,2N1305,2N1307,2N1309A,2N1348, 2N1349,2N1350,2N1351,2N1352,2N1356,2N1372,2N1374,2N1376,2N1381,2N1382, 2N1383,2N1478,2N1681,2N2375,2SA204,2SA207,2SA407,2SB299,2SB326,ASY27, SF.T223,*PTC135,*GE-53,*TR-05,*HEPG0005,*SK3004,*AT20M,*ECG100,

TRANSISTOR SUBSTITUTES

To Replace	Substitute This Type
(2SB416)	*WEP254,*276-2006,*RT-118
2SB417	2N381,2N518,2N526,2N526A,2N527A,2N597,2N651,2N651A,2N652A,2N1008B, 2N1187,2N1373,2N1375,2N1377,2N1447,2N1448,2N1925,2N1926,2N1954,2N1956, 2N1957,2SB57A,2SB226,2SB227,2SB304A,2SB427,2SB428,MA100,MA882,MA887, SF.T243,*GE-2,*TR-05,*HEPG0005,*SK3004,*AT20M,*ECG100,*WEP254, *276-2004,*RT-118
2SB418	
2SB419	*PTC120A
2SB420	
2SB421	*PTC127,*TR-85,*HEPG6012,*ECG102A,*WEP250,*RT-121
2SB421G	SEE 2SA512
2SB422	*PTC109,*SK3004,*ECG102
2SB423	2N109+,2N217+,2SB56,ACY23,ACY32,*PTC109,*HEPG0005,*SK3004,*ECG102
2SB424	2N665,2SB253,*PTC105A,*GE-25,*TR-01,*HEPG6005,*SK3014,*PT40,*ECG121, *WEP232,*276-2006,*RT-127
2SB424G	2N665,2SB253,*GE-25,*SK3034
2SB425-BL	
2SB425-R	2N665,2N2836,2SB128,2SB250A,2SB251A,2SB252,2SB253,AD131,AD132,AUY19, AUY20,CDT1311,CDT1312
2SB425-Y	2N665,2N2836,2SB138,2SB251A,2SB253,CDT1311,CDT1312
2SB425G-BL	
2SB425G-R	2N665,2N2836,2SB128,2SB250A,2SB251A,2SB252,2SB253,AD131,AD132,AUY19, AUY20,CDT1311,CDT1312
2SB425G-Y	2N665,2N2836,2SB138,2SB251A,2SB253,CDT1311,CDT1312
2SB426-BL	2N257-WHT,2N4241,*ECG104,*RT-127A
2SB426-R	2N257-GRN,2SB250,2SB251,2SB426A,AD149,CDT1310,*PTC105A,*ECG121, *RT-127A
2SB426-Y	2N257-BLK,2N257-GRN,2SB137,2SB251,2SB426A,CDT1310,CDT1319,*PTC105A, *ECG121,*RT-127A
2SB426A	2N2836,2SB127A,2SB138,2SB247,2SB248A,2SB251A,CDT1311,*SK3009
2SB426G-BL	2N257-WHT,2N4241
2SB426G-R	2N257-GRN,2SB250,2SB251,2SB426A,AD149,CDT1310
2SB426G-Y	2N257-BLK,2N257-GRN,2SB137,2SB251,2SB426A,CDT1310,CDT1319
2SB427	2N526,2N526A,2N527A,2N597,2N1925,2N1926,2SB57A,2SB226,2SB227,2SB428, SF.T243,*PTC135,*HEPG0005,*SK3123,*ECG102A,*WEP250
2SB428	2N527,2N527A,2N1926,2N1997,2N2000,2SB227,*PTC135,*HEPG0005,*SK3123, *ECG102A,*WEP250
2SB429	2N2613+,2N2635,2SB370A,2SB439,2SB440,AC122-VIO,AC122-WHT,AC122/30-VIO, AC122/30-WHT,NKT218
2SB430	*HEPG6010,*ECG105,*WEP233
2SB431	*PTC135,*HEPG0005,*ECG158,*WEP238
2SB432	*HEPG6007,*SK3035,*ECG127,*WEP235
2SB433	2N1164,2N1164A,2N1166,2N1166A,2N1555,2N1555A,2N1556,2N1556A,2N1559, 2N1559A,2N1560,2N1560A,2N2140,2N2140A,2N2141,2N2141A,2N2145,2N2145A, 2N2146,2N2146A,CQT940B,CQT940BA,*PTC106,*HEPG6006,*ECG105,*WEP233
2SB434-O	2SB434G-O,2SB435G-O,*PTC166,*TR-77,*ECG153,*WEP246
2SB434-R	2SB434G-R,2SB435G-R,*PTC166,*TR-77,*ECG153,*WEP246
2SB434-Y	2SB434G-Y,2SB435G-Y,*TR-77,*ECG153,*WEP246
2SB434G-O	
2SB434G-R	
2SB434G-Y	
2SB435-O	2SB434-O,2SB434G-O,2SB435G-O,*PTC166,*TR-77,*ECG153,*WEP246
2SB435-R	2SB434-R,2SB434G-R,2SB435G-R,*PTC166,*TR-77,*ECG153,*WEP246
2SB435-Y	2SB434-Y,2SB434G-Y,2SB435G-Y,*TR-77,*ECG153,*WEP246
2SB435G-O	2SB434-O,2SB434G-O
2SB435G-R	2SB434-R,2SB434G-R
2SB435G-Y	2SB434-Y,2SB434G-Y
2SB436	2N109+,2N217+,2N2953,2SB56,2SB176,2SB187,2SB188,2SB328,2SB329,2SB475, 2SB496,AC122-YEL,ACY23,ACY32,MM404,MM404A,NKT211,NKT213,NKT216, NKT221,NKT224,NKT226,NKT228,SF.T323,SF.T353
2SB437	2SB177,AC122/30-YEL,NKT217

To Replace	Substitute This Type
2SB438	
2SB439	2SB440,*PTC135,*GE-52,*TR-05,*HEPG0005,*SK3004,*ECG102A,*WEP250, *AC188/01,*276-2007,*RT-133
_2SB439G_	2SB439,2SB440,*GE-53,*SK3004
2SB440	2SB439,*PTC135,*GE-52,*TR-85,*HEPG0005,*SK3004,*AT20M,*ECG102A,*WEP250, *AC188/01,*276-2007,*RT-121
_2SB440G_	2SB439,2SB440,*GE-53,*SK3004
2SB443	2N2613+,2SB444,*PTC135,*GE-50,*TR-85,*HEPG0008,*SK3004,*ECG102A, *WEP250,*AC188/01,*276-2007,*RT-121
2SB444	2N2613+,*PTC135,*GE-50,*TR-17,*HEPG0008,*SK3004,*ECG126,*WEP635, *AC188/01,*276-2007
2SB445	2SB446,*HEPG6016,*SK3014,*ECG121,*WEP232
2SB446	*SK3014,*ECG121,*WEP232
2SB447	2N5324,*HEPG6007,*SK3035,*ECG127,*WEP235
_2SB448_	*PTC120A,*GE-3,*TR-94,*HEPG6016,*SK3034,*ECG131,*WEP642,*276-2006, *RT-127
2SB449	2SB127A,2SB138,*PTC105A,*GE-25,*TR-85,*HEPG6003,*SK3009,*ECG121, *WEP232,*276-2006,*RT-127
2SB450	2SB370,2SB370A,2SB431,AC131,NKT218,*HEPG0005,*ECG158,*WEP238
_2SB451_	2N2431,2SB619,*HEPG6011,*ECG158,*WEP238
_2SB452_	2N2001,*PTC156,*HEPG6011,*SK3004,*ECG158,*WEP238
2SB452A	2N2000,*PTC156,*HEPG6011,*SK3004,*ECG158,*WEP238
_2SB453_	2N598,2N599,2N1997,2N1998,2N2000,2N2001,*HEPG0005,*ECG158,*WEP238
_2SB454_	*HEPG0005,*ECG158,*WEP238
_2SB455_	*HEPG0005,*ECG158,*WEP238
2SB456	
_2SB457_	2SB370,2SB370A,2SB431,2SB495A,2SB619,AC131,NKT218,*PTC135,*HEPG0005, *SK3004,*ECG158,*WEP238
_2SB457A_	2SB370A,2SB431,2SB452A,2SB495A,2SB619,AC131,AC131/30,NKT218,*PTC109, *HEPG0005,*SK3004,*ECG158,*WEP238
_2SB458_	*HEPG6016,*SK3052,*ECG131,*WEP642,*RT-132
_2SB458A_	*HEPG6016,*SK3082,*ECG131,*WEP642
2SB458B	*SK3052
2SB459	2N2613+,2N2635,2SB54,2SB370A,2SB439,2SB440,2SB460,AC122-VIO, AC122/30-VIO,NKT218,*PTC145,*GE-52,*TR-85,*HEPG0006,*SK3004,*ECG102A, *WEP250,*AC188/01,*RT-121
2SB460	AC122/30-VIO,*PTC145,*GE-80,*TR-85,*HEPG0001,*SK3123,*ECG102A,*WEP250, *RT-121
2SB461	2N2000,*GE-53,*TR-82,*HEPG0005,*SK3004,*ECG176,*WEP238,*276-2007,*RT-136
_2SB462-BL_	*PTC120A,*GE-30
_2SB462-R_	*PTC120A,*GE-30
_2SB462-Y_	*PTC120A,*GE-30
_2SB462G-BL_	*PTC120A,*GE-30
_2SB462G-R_	*PTC120A,*GE-30
_2SB462G-Y_	*PTC120A,*GE-30
_2SB463-BL_	*PTC120A,*GE-30,*TR-50,*ECG131,*WEP642
_2SB463-R_	*PTC120A,*GE-30,*ECG131,*WEP642,*RT-127
_2SB463-Y_	*PTC120A,*GE-30,*TR-50,*ECG131,*WEP642,*RT-127
_2SB463G-BL_	*PTC120A,*GE-30
_2SB463G-R_	*PTC120A,*GE-30
_2SB463G-Y_	*PTC120A,*GE-30
_2SB464_	2N1022,2N1905,2SB129A,2SB362,KR6502,KR6503,*PTC122,*HEPG6007,*SK3035, *ECG127,*WEP235
2SB464G	SEE 2SA657
_2SB465_	2N457A,2N458A,2N2869,2N2869/2N301,2N2870,2N2870/2N301A,2N5895,2N5896, 2SB301,2SB309A,2SB338,2SB339,2SB471,2SB472,ASZ16,ASZ18,KR6500,KR6501, *PTC122,*HEPG6005,*SK3035,*ECG127,*WEP235
2SB465G	SEE 2SA658
2SB466	2SB467,*HEPG6016,*SK3009,*ECG131,*WEP642
2SB467	*HEPG6016,*SK3009,*ECG131,*WEP642
2SB468	2SB411,*PTC122,*GE-25,*TR-27,*HEPG6008,*SK3035,*HO300,*ECG127,*WEP235

To Replace	Substitute This Type
2SB468A	*PTC122,*GE-25,*TR-27,*HEPG6008,*SK3035,*ECG127,*WEP235
2SB470	*PTC102,*HEPG0006,*SK3004,*ECG102,*WEP631
2SB471	2SB339,2SB472,KR6501,*PTC105A,*GE-25,*TR-01,*HEPG6005,*SK3009,*ECG121, *WEP232,*276-2006,*RT-127A
2SB472	2SB339,KR6503,*PTC105A,*GE-25,*TR-01,*HEPG6005,*SK3034,*ECG121,*WEP232, *276-2006,*RT-127
2SB473	*PTC120A,*GE-25,*TR-50,*HEPG6016,*SK3198,*ECG131,*WEP642,*276-2006, *RT-127
2SB474	*PTC194,*GE-49,*IR2SB474,*HEPG6016,*SK3082,*ECG226,*WEP642,*276-2006, *RT-132
2SB475	2SB415,NKT211,NKT221,NKT228,*PTC135,*GE-53,*TR-84,*HEPG0005,*SK3004, *ECG158,*WEP238,*AC188/01,*276-2007
2SB476	*TR-84,*HEPG6011,*ECG158,*WEP238,*276-2006
2SB477	*HEPG6010
2SB478	2SB479,*HEPG6010
2SB479	2SB480,*HEPG6010
2SB480	*HEPG6010
2SB481	*PTC120A,*GE-25,*IR2SB481,*HEPG6016,*SK3052,*ECG131,*WEP642,*276-2006, *RT-124
2SB482	*GE-2,*TR-85,*HEPG0005,*SK3004,*ECG102A,*WEP250,*RT-121
2SB483	2SB484,*PTC901,*HEPG6018,*ECG179,*WEPG6001
2SB484	*HEPG6018,*ECG179,*WEPG6001
2SB485	*ECG179,*WEPG6001
2SB486	*GE-53,*TR-05,*HEPG0007,*SK3004,*ECG102,*WEP631,*276-2007,*RT-120
2SB487	
2SB488	
2SB492	*PTC155,*GE-80,*TR-82,*HEPG6011,*SK3004,*ECG176,*WEP238,*276-2006, *RT-127
2SB493	*HEPG6011,*276-2006
2SB494	2SB415,*PTC156,*HEPG6011,*ECG158,*WEP238
2SB495	2SB495A,2SB619,AC131,*PTC156,*HEPG6011,*SK3004,*ECG158,*WEP238
2SB495A	2SB452A,2SB619,AC131,AC131/30,*PTC156,*HEPG6011,*SK3004,*ECG158, *WEP238
2SB496	*PTC145,*1/2 GE-54,*TR-85,*HEPG0005,*SK3004,*ECG102A,*WEP250,*AC188/01, *276-2005,*RT-121
2SB497	*PTC107,*HEPG0005,*SK3004,*ECG102,*WEP631
2SB498	*PTC109,*HEPG0005,*SK3004,*ECG102A,*WEP250
2SB502-O	2SB502A-O
2SB502-R	2SB502A-R
2SB502-Y	2SB502A-Y
2SB502A-O	
2SB502A-R	
2SB502A-Y	
2SB503-O	2SB503A-O
2SB503-R	2SB503A-R
2SB503-Y	2SB503A-Y
2SB503A-O	2SB502A-O
2SB503A-R	2SB502A-R
2SB503A-Y	2SB502A-Y
2SB504	2SB504A,*HEPS3003
2SB504A	
2SB505	
2SB506	2SB506A,KSP2176,*ECG281
2SB506A	KSP2176,*ECG281
2SB507	*PTC166,*HEPS5006,*SK3083,*ECG153,*WEP246
2SB508	*PTC166,*HEPS5006,*SK3083,*ECG153,*WEP246
2SB509	*PTC168,*HEPS5018,*ECG153,*WEP246
2SB510	*ECG129,*WEP242
2SB511	2SB514,*PTC111,*HEPS3028,*SK3084,*ECG153,*WEP246
2SB512	2N6125,2N6126,2SA489-R,2SB512A,BD223,BD224,BD225,D45C4,D45C5,D45C6, D45C7,D45C8,D45C9,*PTC166,*TR-77,*ECG153,*WEP246

To Replace	Substitute This Type
2SB512A	2N6126,2N6475,BD223,BD225,D45C10,D45C11,D45C12,*PTC166,*TR-77,*ECG153, *WEP246
2SB513	2SB513A,*PTC166,*TR-77,*ECG153,*WEP246
2SB513A	*PTC166,*TR-77,*ECG153,*WEP246
2SB514	2SB507,*PTC166,*HEPS3032,*ECG153,*WEP246
2SB515	2SB508,*PTC166,*HEPS3032,*ECG153,*WEP246
2SB516	2N827,2N2635,2SB54,2SB136,2SB186,2SB346,2SB348,2SB370,2SB370A,2SB408, 2SB431,2SB439,2SB440,NKT218,NKT222,NKT223,*ECG102A,*WEP250
2SB518	*PTC172
2SB519	KSP2172,KSP2173,KSP2175,KSP2176,KSP2272,KSP2273,KSP2275,KSP2276, *PTC172
2SB520	2SB555-R,KSP2273,KSP2276
2SB521	2N6133,2N6134,NSP42A,NSP42B
2SB522	
2SB523	2SB529,NSE170,*PTC111
2SB524	NSD202,NSD6181,NSE170,NSE171,*PTC111,*HEPS5006
2SB525	2SA683
2SB526	2SB527,NSD204,NSD6180,*ECG292,*WEP635
2SB527	2SB528,NSD204,NSD206,NSD6180
2SB528	NSD206
2SB529	2SB523,NSE170,*PTC111
2SB530	2N5625,2N6229,2SA650,2SA651,2SA679-Y,2SA680-Y,2SA753,2SB539,2SB539A, 2SB539B,2SB541,2SB555-O,2SB556-O,2SB557-O,KSP2175,KSP2176
2SB531-O	2N5625,2SA757,2SB557-O,2SB558-O
2SB531-R	2N5623,2N5627,2SA648,2SA745,2SA745A,2SA747,2SA756,2SA757,2SA878, 2SB557-R,2SB558-R,KSP2175
2SB531-Y	2SA714L
2SB532	2N5621,2N5625,KSP2045,KSP2174,KSP2175,*ECG285
2SB533	
2SB534	*ECG102A
2SB535	2SA529,*SK3004,*ECG158
2SB536	2SA740,2SA740A,2SA839-O,*ECG292
2SB537	*SK3083,*ECG153
2SB538	2N1146A,2N1146B,2N2612
2SB539	2SA747,2SA747A,2SA878,2SB539A,2SB539B,KSP2276,*ECG285
2SB539A	2SB539,2SB539B,KSP2276
2SB539B	2SB539,2SB539A,KSP2276
2SB540	*SK3123
2SB541	2N6229,2SB539,2SB539A,2SB539B,KSP2175,KSP2176,*PTC172,*ECG281
2SB542	2N869,2N5999,2N6001,2SA509-Y,2SA509G-Y,2SA683,2SA719,2SA890,2SB544, *PTC103
2SB544	2SA683,*ECG294
2SB544P1	2SB544P2,*ECG294
2SB544P2	
2SB546	2SB568,*SK3083,*ECG292
2SB547	2SB567
2SB548	2N5322,2N5675,2SA761,2SB549
2SB549	2N5322,2N5675,2SA761,2SB548
2SB552	
2SB554	*ECG285
2SB555-O	2SB554,*ECG281
2SB555-R	2SB554,KSP2276
2SB556-O	2SA679-Y,2SB554,2SB555-O
2SB556-R	2SA679-R,2SB554,2SB555-R,KSP2276
2SB557-O	2SA679-Y,2SA753,2SB554,2SB555-O,2SB556-O
2SB557-R	2N5627,2N6248,2SA650,2SA651,2SA679-R,2SA747,2SA747A,2SA753,2SA878, 2SB554,2SB555-R,2SB556-R,KSP2176
2SB558-O	2N5625,2SA663-BL,2SA663-Y,2SA679-Y,2SA680-Y,2SA757,2SA758,2SA882, 2SB556-O,2SB557-O,*SK3173
2SB558-R	2N5623,2N5627,2SA648,2SA663-R,2SA663-Y,2SA679-R,2SA680-R,2SA745,2SA745A, 2SA747,2SA757,2SA758,2SA878,2SB556-R,2SB557-R,KSP2175

To Replace	Substitute This Type
2SB560	*ECG102A,*WEP250
2SB561	2N5811,2SA683,2SB544,2SB562,BC328-16,D29E1,TIS91,TIS91M,*PTC103,*ECG159, *WEP250
2SB562	*ECG288
2SB564	2SA486-BL,2SA683,2SA773,2SA885,BC160-16
2SB565	2N6108,*PTC162,*SK3004
2SB566	2N6109,*PTC162
2SB567	
2SB568	*SK3083,*ECG292
2SB569	2N6034,2N6035,2SB570
2SB570	2N6035,2N6036,2SB571
2SB571	2N6036
2SB572	2SB573,2SB575,2SB576
2SB573	2SB574,2SB576,2SB577
2SB574	2SB577
2SB575	2SB576
2SB576	2SB577
2SB577	
2SB578	2N5987,2N5988,MJE2955
2SB579	2N5974,2N5975,2N5980,2N5981,2SB578,2SB580,MJE2955
2SB580	2N5975,2N5976,2N5981,2N5982,2SB581
2SB581	2N5976,2N5982
2SB582	2SB583
2SB583	2SB584
2SB584	
2SB585	2SB586,2SB587,2SB588
2SB586	2SB588,2SB589
2SB587	2SB588
2SB588	2SB589
2SB589	
2SB595	*PTC168
2SB596	2SB595,*PTC166
2SB598	
2SB604	2SA489-O,2SA489-Y,2SB595,2SB596,*PTC168
2SB605	2N4032,2N4033,2N4405,2SA485-BL,2SA684,BC161-16,MPS4354,MPS4355
2SB606	
2SB607	
2SB615	
2SB619	
2SC11	2SC180,2SC181,*PTC108,*HEPG0011,*SK3011,*ECG101,*WEP641
2SC12	2N2310,*PTC125,*HEPS0014,*SK3024,*ECG128,*WEP243
2SC13	2N440,2N440A,2N635,2N636,2N1091,2N1996,2SC129,ASY28,*PTC108,*GE-5, *TR-08,*HEPG0011,*SK3011,*NR5,*ECG101,*WEP641,*AC187/01,*276-2001, *RT-119
2SC14	2N636,ASY28,*PTC108,*GE-6,*TR-08,*HEPG0011,*SK3011,*NR10,*ECG101, *WEP641,*RT-119
2SC15-0	*GE-47,*BF338,*276-2030
2SC15-1	2N3295,2N5188,2SC31,*GE-47,*TR-21,*HEPS0005,*SK3122,*ECG123A,*WEP736, *BF338,*276-2010,*RT-102
2SC15-2	2SC59,2SC69,2SC149,2SC154,MM3000,*GE-27,*TR-21,*HEPS0005,*ECG123A, *WEP736,*BF338,*276-2010,*RT-102
2SC15-3	2SC59,2SC69,2SC149,2SC154,2SC154A,MM3001,*GE-27,*TR-21,*HEPS0005, *SK3122,*ECG123A,*WEP736,*BF338,*276-2010,*RT-102
2SC16	2N783,2N834,2N834A,2N959,2N2214,2SC68,2SC172,2SC321,2SC322,BSX25, GET706,MPS706A,MPS834,MPS6567,MPSH37,*PTC136,*HEPS0011,*SK3122, *ECG123A,*WEP736
2SC16A	2N783,2N784,2N834A,2N959,2N2432,2N3261,2N3793,2N5187,2SC54,2SC68, 2SC137,2SC172,2SC321,2SC322,2SC400-R,2SC576,2SC689,40405,BSX25,GET706, GI-3793,MPS706A,MPS834,MPS6567,MPSH37,*PTC136,*HEPS0011,*SK3122, *ECG123A,*WEP736
2SC17	*PTC136,*HEPS0011,*SK3018,*ECG108,*WEP56

To Replace	Substitute This Type
2SC17A	*PTC136,*HEPS0011,*SK3039,*ECG108,*WEP56
2SC18	2N835,2N958,MPSH07,MPSH08,*PTC108,*HEPS0011,*SK3122,*ECG123A,*WEP736
2SC19	2N780,2N1708A,2N1959,2N1959A,2N2218,2N2237,2N2410,2N2846,2N2958,2N3015, 2N3122,2N3252,2N3299,2N3326,2SC97,2SC190,2SC254,2SC344,2SC346,2SC347, 2SC348,2SC479,2SC503-O,2SC504-O,2SC580,BC340-6,BC341-6,BSY51,ZT697, *PTC144,*HEPS0014,*SK3024,*ECG128,*WEP243
2SC20	*HEPS0014,*SK3024,*ECG128,*WEP243
2SC21	HST9201,HST9205,HST9206,HST9801,HST9802,*PTC118,*GE-72,*TR-59, *HEPS7002,*SK3027,*ECG130,*WEP247,*276-2020,*RT-131
2SC22	*GE-66,*TR-87,*HEPS3020,*SK3024,*ECG128,*WEP243,*RT-114
2SC23	2SC22,2SC24,2SC801,*PTC144,*GE-66,*TR-87,*HEPS3020,*SK3024,*ECG128, *WEP243,*RT-114
2SC24	*PTC144,*GE-66,*TR-87,*HEPS3021,*SK3024,*ECG128,*WEP243,*RT-114
2SC25	
2SC26	*HEPS0025,*SK3122,*ECG123A,*WEP736
2SC27	*PTC101,*HEPS0014,*SK3024,*ECG128,*WEP243
2SC28	*PTC115,*HEPS0014,*SK3124,*ECG123,*WEP53
2SC29	*PTC132,*HEPS0014,*SK3019,*ECG123,*WEP53
2SC30	2N2224,2N2846,2N3015,2N3299,2N3326,2SC32,2SC344,2SC345,FT3722,*PTC136, *GE-17,*TR-87,*HEPS0014,*SK3024,*EN10,*ECG128,*WEP243,*276-2009,*RT-114
2SC31	2N2217,2N3295,2N5188,BSX49,*PTC125,*GE-47,*TR-87,*HEPS0014,*SK3024,*EN10, *ECG128,*WEP243,*276-2030,*RT-114
2SC32	2N2218,2N2218A,2N2224,2N2537,2N2846,2N2848,2N3015,2N3299,2N3326,2N3678, 2N4047,2N4960,2N4961,2SC307,2SC1218,TIS136,*PTC144,*GE-47,*TR-87, *HEPS0014,*SK3024,*EN10,*ECG128,*WEP243,*276-2030,*RT-114
2SC32A	*HEPS5026,*SK3024,*ECG128,*WEP243
2SC33	2SC334,*PTC132,*GE-39,*TR-21,*HEPS0011,*SK3122,*EN10,*ECG123A,*WEP736, *BC107B,*276-2016,*RT-102
2SC34	OC139,*PTC108,*GE-8,*TR-08,*HEPG0011,*SK3011,*ECG103A,*WEP641B, *AC187/01,*276-2001,*RT-122
2SC35	OC140,*DS-72,*GE-8,*TR-08,*HEPG0011,*SK3011,*ECG103A,*WEP641B, *AC187/01,*276-2001,*RT-122
2SC36	OC141,*DS-72,*GE-8,*TR-08,*HEPG0011,*SK3011,*ECG101,*WEP641,*276-2001, *RT-119
2SC37	*PTC136,*HEPS0014,*SK3022,*ECG123A,*WEP371,*RT-100A
2SC38	*HEPS0014,*SK3047,*ECG123A,*WEP56,*RT-100A
2SC39	*PTC132,*HEPS0016,*SK3019,*ECG108,*WEP56
2SC39A	*PTC132,*HEPS0016,*SK3019,*ECG108,*WEP56
2SC40	*HEPS0016,*SK3019,*ECG108,*WEP56
2SC41	2N6495,2SC42A,*TR-67,*HEPS5020,*SK3111,*ECG163,*WEP740
2SC42	2N6495,2SC42A,*GE-35,*TR-67,*HEPS5020,*SK3079,*PN66,*ECG162,*WEP707
2SC42A	2N6233,*PTC129A,*GE-35,*TR-67,*HEPS5020,*SK3079,*ECG162,*WEP707
2SC43	*PTC129A,*GE-35,*TR-67,*HEPS5020,*ECG162,*WEP707
2SC44	*PTC129A,*GE-35,*TR-67,*HEPS5020,*NA20,*ECG162,*WEP707
2SC45	2N2224,2N2846,2N2958,2N3015,2N3299,2N3326,2N3641,2N3642,2SC32,2SC46, 2SC190,2SC344,2SC479,2SC503-O,2SC594-R,2SC651,2SC1164-O,2SC1199, 2SC1412,*PTC123,*HEPS0014,*ECG123A,*WEP736
2SC46	2N2218,2N2218A,2N2224,2N2410,2N2479,2N2846,2N2958,2N3015,2N3252,2N3253, 2N3299,2N3326,2N3678,2N3722,2N6376,2SC97,2SC97A,2SC109A-R,2SC190, 2SC344,2SC345,2SC347,2SC348,2SC479,2SC503-O,2SC560,2SC580,2SC1218, BSX60,*HEPS0015,*SK3024,*ECG128,*WEP243
2SC47	2N1708A,2N2218,2N2410,2N2846,2N2958,2N3015,2N3252,2N3299,2N3326,2SC46, 2SC97,2SC190,2SC254,2SC344,2SC346,2SC347,2SC348,2SC479,2SC503-O, 2SC504-O,2SC580,*HEPS0014,*SK3122,*ECG123A,*WEP736
2SC48	2N3500,2SC49,2SC69,2SC590,2SC1217,*HEPS5026,*SK3124,*ECG123,*WEP53
2SC49	2N3500,*PTC125,*GE-20,*TR-87,*HEPS5026,*SK3024,*SN80,*ECG128,*WEP243, *RT-114
2SC50	*PTC108,*HEPG0011,*SK3124,*ECG101,*WEP641
2SC51	2N3252,2N3253,*PTC144,*HEPS5014,*SK3024,*ECG128,*WEP243
2SC52	2N3326,2N3646,2N5772,2SC344,*PTC136,*HEPS0014,*SK3122,*ECG123A, *WEP736

To Replace	Substitute This Type
2SC53	2N1708A,2N2320,2N3137,2N4264,2N4274,2N4275,2N5236,2N5769,2N5772,2SC87, 2SC343,2SC1290,BSY63,*HEPS0014,*SK3122,*ECG123A,*WEP736
2SC54	2N708,2N708A,2N784A,2N914,2N2481,2N2710,2N3013,2N3014,2N3210,2N3261, 2N3510,2N3511,2N3605A,2N3606A,2N3646,2N3647,2N3648,2N5772,2SC67,2SC302, 2SC577,2SC601,2SC601N,2SC689,2SC1274,GET708,GET914,GET2369,GET3013, GET3014,GET3646,ZT83,ZT708,*PTC121,*HEPS0014,*SK3122,*ECG123A, *WEP736
2SC55	2N708,2N784A,2N914,2N2242,2N2318,2N2369,2N2369A,2N2481,2N2656,2N2710, 2N3009,2N3011,2N3013,2N3014,2N3210,2N3510,2N3511,2N3605A,2N3606A, 2N3646,2N3647,2N3648,2N4137,2N4274,2N4275,2N5418,2N5769,2N5772,2SC67, 2SC204,2SC577,BSW82,GET708,GET914,GET2369,GET3013,GET3014,GET3646, PT720,PT2760,ZT708,ZT2369,*PTC121,*HEPS0011,*SK3122,*ECG123A,*WEP736
2SC56	*PTC126,*HEPS0011,*SK3018,*ECG107,*WEP56,*RT-107A
2SC57	2SC22,2SC23,2SC24,2SC117,2SC801,*PTC144,*HEPS3001
2SC58	2N2224,2N2846,2N3015,2N3299,2N3326,2SC32,2SC344,2SC345,2SC1218,*PTC136, *SK3040,*ECG154,*WEP712
2SC58A	2SC154B,2SC507-R,2SC1217,*PTC125,*GE-27,*TR-78,*HEPS5025,*SK3040, *ECG154,*WEP712,*BF338,*276-2012,*RT-110
2SC59	2SC69,2SC149,*PTC125,*GE-18,*TR-87,*HEPS5026,*SK3024,*ECG128,*WEP243, *RT-114
2SC60	*PTC108,*HEPG0011,*SK3011,*ECG101,*WEP641
2SC61	2N744A,*HEPS3013,*SK3024,*ECG128,*WEP243
2SC62	2N916,2N916A,2N2501,2N3947,2N4013,2N4951,2N4962,2N5368,2N5380,2SC639, 2SC1999,A5T3903,A5T4123,TIS133,TIS134,TN-3903,*PTC136,*GE-17,*TR-21, *HEPS0011,*SK3122,*EN10,*ECG123A,*WEP736,*276-2009,*RT-102
2SC63	2N2368,2N2710,2N2729,2N3510,2N3511,2N3544,2N3648,2N4137,2N5030,2N5187, 2SC54,2SC68,2SC321,2SC322,2SC576,2SC595,2SC601,2SC689,2SC1293, GET3013,GET3014,GET3646,MM1941,MPS6507,MT1038,MT1038A,MT1039, MT1060,TIS84,ZT2368,*PTC126,*HEPS0011,*SK3039,*ECG108,*WEP56
2SC64	2SC345,PT1558,*PTC123,*GE-27,*TR-78,*HEPS5026,*SK3045,*ECG154,*WEP712, *BF338,*RT-110
2SC65	2SC154B,2SC589,*PTC125,*GE-27,*TR-78,*HEPS5025,*SK3024,*SN80,*ECG154, *WEP712,*BF338,*276-2012,*RT-110
2SC65Y	2SC507-O,*PTC144,*HEPS5025,*SK3044,*ECG154,*WEP712
2SC66	2N3500,2SC154C,2SC310,2SC470-5,2SC470-6,2SC507-R,2SC686,2SC1048, 2SC1217,MM2259,*GE-27,*TR-78,*HEPS5025,*SK3045,*ECG154,*WEP712,*BF338, *RT-110
2SC67	2N2501,2N3013,2N3014,2N3211,2SC577,2SC639,2SC764,*PTC136,*GE-20,*TR-21, *HEPS0014,*SK3122,*PN66,*ECG123A,*WEP736,*276-2009,*RT-102
2SC68	2N2710,2N3510,2N3511,2N3648,2N5187,2SC321,2SC322,2SC576,2SC595,GET3013, GET3014,GET3646,*PTC136,*GE-20,*TR-21,*HEPS0014,*SK3122,*PN66, *ECG123A,*WEP736,*276-2009,*RT-102
2SC69	2N3500,2SC49,*GE-18,*TR-87,*HEPS5026,*SK3024,*ECG128,*WEP243,*RT-114
2SC70	2SC154C,2SC507-O,MM2260,*PTC117,*GE-27,*TR-78,*HEPS5025,*SK3018, *ECG154,*WEP712,*BF338,*276-2008,*RT-110
2SC71	2N1306,2N1308,ASY29,*PTC108,*GE-8,*TR-08,*HEPG0011,*SK3011,*ECG101, *WEP641,*AC187/01,*276-2001,*RT-119
2SC72	2N1308,ASY29,*PTC108,*GE-8,*TR-08,*HEPG0011,*SK3011,*ECG101,*WEP641, *276-2001,*RT-119
2SC73	2N797,*GE-7,*TR-08,*HEPG0011,*SK3011,*NR10,*ECG101,*WEP641,*RT-119
2SC74-GR	BCY58,BCY66
2SC74-O	2N3724,2N4046,2N5145,A5T4123,TIS133,TIS134
2SC74-R	2N1708A,2N2320,2N4264,2N5772,2SC343,BSY63
2SC74-Y	2N4265,A5T4124
2SC74N	SEE 2SC594N
2SC75	2SC181,*GE-7,*TR-08,*HEPG0011,*SK3011,*NR10,*ECG101,*WEP641,*AC187/01, *276-2001,*RT-119
2SC76	2SC181,*GE-7,*TR-08,*HEPG0011,*SK3011,*NR10,*ECG101,*WEP641,*AC187/01, *276-2001,*RT-119
2SC77	2SC181,*GE-7,*TR-08,*HEPG0011,*SK3011,*NR10,*ECG101,*WEP641,*AC187/01, *276-2001,*RT-119

To Replace	Substitute This Type
2SC78	2N797,*GE-6,*TR-08,*HEPG0011,*SK3011,*NR10,*ECG101,*WEP641,*276-2001, *RT-119
2SC79	*PTC121,*HEPS0020,*SK3039,*ECG108,*WEP56
2SC80	2N3337,2N3338,2N3339,2N3600,2N4134,2N4135,2N5851,2N5852,2SC1747,40413, MT1061A,*PTC132,*GE-61,*TR-95,*HEPS0011,*SK3019,*EN10,*ECG108,*WEP56, *BC107B,*276-2011,*RT-113
2SC81	*HEPS5004,*SK3124,*WEP736
2SC82	*HEPS5004
2SC83	2N1901
2SC84	2SC86,2SC179,2SC180,2SC181,*HEPG0011
2SC85	2SC180,*HEPG0011
2SC86	2SC181,*HEPG0011
2SC87	2N1708A,2N2320,2N3137,2N4264,2N4274,2N4275,2N5236,2N5769,2N5772,2SC343, 2SC651,2SC1164-O,2SC1199,2SC1290,2SC1412,BSY63,*HEPS0014,*SK3122, *ECG123,*WEP736
2SC88	2SC1217,*HEPS5026,*SK3045,*ECG154,*WEP712
2SC89	2N357A,2N358A,2N440,2N440A,2N1012,2N1090,2N1091,2SC90,SF.T298,*DS-72, *GE-7,*TR-08,*HEPG0011,*SK3011,*NR5,*ECG101,*WEP641,*AC187/01,*RT-119
2SC90	2N440,2N440A,2N1091,2SC91,SF.T298,*DS-72,*GE-54,*TR-08,*HEPG0011,*SK3011, *ECG101,*WEP641,*AC187/01,*RT-119
2SC91	*GE-54,*TR-08,*HEPG0011,*SK3011,*ECG101,*WEP641,*RT-119
2SC92	2SC94,*GE-66,*HEPS3002
2SC93	2SC92,2SC94,*GE-66,*HEPS3010
2SC94	2SC92,*GE-66,*HEPS3002
2SC95	*HEPS0005,*SK3045,*WEP712
2SC96	*ECG154
2SC97	2N2846,2N2848,2N3015,2N3299,2N3326,2N3722,2N4960,2N4961,2N5262,2SC97A, 2SC580,BSX60,*PTC144,*GE-47,*TR-87,*HEPS3001,*SK3024,*SN80,*ECG128, *WEP243,*276-2030,*RT-114
2SC97A	2N3725A,*HEPS3001,*SK3024,*ECG128,*WEP243
2SC98	2N2318,2N2319,2N3011,2N5187,2SC395A-R,2SC595,*GE-17,*TR-95,*HEPS0011, *SK3039,*EN10,*ECG108,*WEP56,*276-2009,*RT-113
2SC99	2N744,2N4418,2S095A,2SC395A,2SC395A-O,2SC735-O,ZT2205,ZT2938,*GE-17, *TR-95,*HEPS0011,*SK3039,*ECG108,*WEP56,*276-2009,*RT-113
2SC100	FX3014,MMT3014,*HEPS0014,*SK3122,*ECG123A,*WEP736
2SC101	*GE-12,*TR-81,*HEPS5019,*SK3021,*PN66,*ECG124,*WEP240,*276-2017,*RT-128
2SC101A	2N5614,2N5618,*GE-12,*TR-81,*HEPS5019,*SK3021,*ECG124,*WEP240,*276-2017, *RT-128
2SC102	*TR-03,*HEPS5004,*ECG105,*WEP233,*276-2014
2SC103	2N783,2N784,2N834A,2N959,2N2432,2N3261,2N3793,2N5187,2SC54,2SC68, 2SC137,2SC321,2SC322,2SC400-R,2SC576,2SC689,40405,GI-3793,MPS706A, MPS834,MPS6567,MPSH37,*HEPS0011,*SK3039,*ECG123A,*WEP736
2SC103A	2N916,2N916A,2N2501,2N4013,2N4123,2N4418,2SC400-Y,2SC619,2SC639, 2SC735-O,2SC933,2SC1216,2SC1999,*GE-212,*TR-21,*HEPS0011,*SK3019,*EN10, *ECG123A,*WEP736,*276-2009,*RT-102
2SC104	*PTC109,*HEPS0011,*SK3122,*ECG123A,*WEP736
2SC104A	2N708,2N708A,2N784A,2N914,2N2242,2N2318,2N2319,2N2481,2N2501,2N2710, 2N3011,2N3013,2N3014,2N3210,2N3211,2N3510,2N3511,2N3605A,2N3606A, 2N3646,2N3647,2N3648,2N4123,2N4418,2N5144,2SC67,2SC169,2SC204,2SC335, 2SC400-O,2SC577,2SC595,2SC601,2SC601N,2SC620,2SC639,2SC764,2SC1274, BSW82,GET708,GET914,GET2369,GET3013,GET3014,GET3646,TIS45,TIS46,TIS51, TIS52,TIS55,ZT708,*PTC139,*GE-61,*TR-21,*HEPS0011,*SK3122,*EN10, *ECG123A,*WEP736,*276-2015,*RT-102
2SC105	2N916,2N916A,2N2242,2N2501,2N3013,2N3014,2N3211,2N4013,2N4123,2N4418, 2N5144,2SC67,2SC169,2SC400-O,2SC577,2SC601N,2SC619,2SC620,2SC639, 2SC735-O,2SC764,BSW82,*PTC132,*GE-61,*TR-21,*HEPS0020,*SK3245,*EN10, *ECG123A,*WEP736,*BC107B,*276-2015,*RT-102
2SC106	2N749,2N751,2N1388,2N1390,2SC302-M,2SC307-M,2SC590N,*DS-81,*GE-11, *HEPS3001,*SK3048,*EN10,*ECG282,*276-2014,*RT-182
2SC107	*GE-66,*HEPS3001,*276-2014
2SC108	2SC256,2SC1065,2SC1069,BSX23,*PTC123,*GE-20,*TR-87,*HEPS3019,*SK3024,

To Replace	Substitute This Type
(2SC108)	*SN80,*ECG128,*WEP243,*RT-114
2SC108A-O	2N2405,2N4943
2SC108A-R	2N3723
2SC109	2N1958A,2N2217,2N2477,2N3122,2N3253,2N3735,2N3831,2N5188,2SC254,40539, BSX48,BSX49,*PTC123,*GE-47,*TR-87,*HEPS0014,*SK3024,*SN80,*ECG128, *WEP243,*BC337,*276-2008,*RT-114
2SC109A-O	2N3725A,2SC108A-O
2SC109A-R	2N2218A,2N3678,2N3722,2N3723,2N4961,2N5262,2SC97A,2SC108A-R,2SC560, BSX60
2SC109A-Y	2N2219A,BSY54
2SC110	*HEPS0014,*SK3122,*ECG123A,*WEP736
2SC111	*HEPS0014,*SK3048,*ECG123A,*WEP736
2SC112	2N1708A,2N2218,2N2410,2N2846,2N3015,2N3299,2N3326,2SC32,2SC503-O, 2SC504-O,2SC594-R,BSY63,*GE-47,*TR-87,*HEPS3013,*SK3024,*ECG128, *WEP243,*276-2038,*RT-114
2SC113	2N2218,2N2218A,2N2224,2N2410,2N2846,2N3015,2N3299,2N3326,2N3678,2SC32, 2SC109A-R,2SC503-O,2SC594-R,*GE-47,*TR-87,*HEPS3013,*SK3024,*ECG128, *WEP243,*276-2038,*RT-114
2SC114	2N656,2N656A,2N1613,2N1613A,2N2106,2N2107,2N2218,2N2218A,2N2224,2N2410, 2N2846,2N3015,2N3122,2N3299,2N3326,2N3678,2SC32,2SC109A-R,2SC480, 2SC503-O,2SC594-R,2SC1053,2SC1072,2SC1072A,2SC1385,2SC1386,7A31, BC341-6,BSY51,BSY53,BSY92,*PTC125,*HEPS3013,*SK3024,*ECG128,*WEP243
2SC115-1	2N1708A,2N4264,BSY63,*PTC125,*GE-47,*276-2030
2SC115-2	2N2224,2N2846,2N3015,2N3299,2N3326,2SC32,2SC594-R,*PTC125,*GE-47
2SC115-3	2N2224,2SC1218,SE8010,*PTC125
2SC116	2N2217,2N5188,2SC32,2SC109A-R,2SC503-O,2SC594-R,BSX48,BSX49,*PTC125, *GE-47,*TR-87,*HEPS3013,*SK3047,*ECG195A,*WEP243,*276-2008,*RT-114
2SC117	2SC22,*PTC144,*GE-18,*TR-87,*HEPS3001,*SK3024,*ECG128,*WEP243,*276-2014, *RT-114
2SC118	*PTC144,*GE-18,*TR-87,*HEPS3001,*SK3024,*ECG128,*WEP243,*276-2014, *RT-114
2SC119	*PTC144,*GE-18,*TR-87,*HEPS3001,*SK3024,*ECG128,*WEP243,*276-2014, *RT-114
2SC120	2N717,2N758,2N758A,2N759,2N759A,2N760B,2N783,2N840,2N844,2N2368,2N2387, 2N2395,2N2432A,2N2522,2N3605A,2N3606A,2N3973,2N3975,2N4137,2S103,2SC30, 2SC52,2SC54,2SC121,2SC124,2SC302,2SC321,2SC322,2SC601,2SC689,BF224J, BF225J,BFY39-1,BSY93,GET708,GET914,GET3013,GET3014,GET3646,KT218, MPSA10-RED,MPSA20-RED,TIS45,TIS46,TIS49,TIS52,TIS55,TIS84,TIS129,ZT23, ZT43,*PTC123,*GE-61,*TR-87,*HEPS0014,*SK3024,*EN10,*ECG128,*WEP243, *BF365,*276-2016,*RT-114
2SC121	2N783,2N2368,2N3605A,2N3606A,2N3973,2N3975,2N4137,2S103,2SC30,2SC52, 2SC54,2SC124,2SC302,2SC321,2SC322,2SC601,2SC689,BF224J,BF225J,BFY39-1, GET708,GET914,GET3013,GET3014,GET3646,KT218,MPSA20-RED,TIS45,TIS46, TIS49,TIS52,TIS55,TIS84,TIS129,ZT23,ZT43,*PTC123,*GE-61,*TR-87,*HEPS0014, *SK3024,*EN10,*ECG128,*WEP243,*BF365,*276-2016,*RT-114
2SC122	2N916,2N916A,2N2501,2N2847,2N3643,2N3903,2N3974,2N3976,2N4123,2N4227, 2N4873,2N4962,2N5380,2S104,2SC123,2SC318,2SC366G,2SC366G-O,2SC367G, 2SC367G-O,2SC620,2SC639,2SC852,2SC899,2SC933,2SC943,2SC944,2SC965, 2SC1187,2SC1188,2SC1189,2SC1216,2SC1319,2SC1320,2SC1687,2SC1688, 2SC1999,BF240,BFY39-2,MPS3693,MPS6513,MPS6565,MPS6576-BLUE,MPS6576-GREEN,MPS6576-YEL,TIS87,TN-3903,*PTC123,*GE-61,*TR-87,*HEPS0014, *SK3024,*EN10,*ECG128,*WEP243,*BF365,*276-2016,*RT-114
2SC123	2N3242A,2N3302,2N3643,2N3904,2N3947,2N4141,2N4995,2N5028,2N5381, 2SC366G,2SC366G-Y,2SC367G,2SC367G-Y,2SC876,2SC933,2SC943,2SC1175, 2SC1216,2SC1999,BC107A,BC123,BC167A,BC171A,BC182A,BC182L,BC237A, BC547A,BCY59A,BF240,BFY39-2,BSX79,FT3643,GET2222,MM3904,MPS6514, MPS6575,MPS6576-GREEN,MPS6576-SIL,PET1002,TN-3904,*PTC123,*GE-61, *TR-87,*HEPS0014,*SK3024,*EN10,*ECG128,*WEP243,*BF365,*276-2016, *RT-114
2SC124	2N783,2N2368,2N3605A,2N3606A,2N3973,2N3975,2N4137,2SC30,2SC52,2SC54, 2SC302,2SC321,2SC322,2SC601,2SC689,BF224J,BF225J,GET708,GET914,

To Replace	Substitute This Type
(2SC124)	GET3013,GET3014,GET3646,KT218,TIS45,TIS46,TIS49,TIS52,TIS55,TIS84,TIS129, *PTC123,*GE-61,*TR-87,*HEPS0014,*SK3024,*EN10,*ECG128,*WEP243,*BF365, *276-2016,*RT-114
2SC125	2SC49,SE8010,*PTC117,*GE-27,*TR-17,*HEPG0003,*ECG160,*WEP637,*BF338, *276-2003
2SC126	2SC154B,2SC507-R,2SC1217,*PTC117,*GE-27,*TR-78,*BF338
2SC127	*PTC132,*HEPS0014,*SK3019,*ECG108,*WEP56
2SC128	2N357A,2N358A,2N1000,ASY28,*PTC108,*GE-54,*TR-08,*HEPG0011,*SK3011, *ECG101,*WEP641,*AC187/01,*276-2001,*RT-119
2SC129	2N358A,2N440,2N440A,2N1306,ASY28,ASY29,*PTC108,*GE-5,*TR-08,*HEPG0011, *SK3011,*ECG101,*WEP641,*276-2001,*RT-119
2SC130	2SC413,*HEPS3001,*SK3024,*ECG128,*WEP243
2SC131	2N708A,2N718,2N731,2N760,2N914,2N2221,2N2312,2N2315,2N2396,2N2481, 2N2501,2N2539,2N2710,2N2845,2N2847,2N3115,2N3210,2N3211,2N3301,2N3508, 2N3510,2N3511,2N3641,2N3642,2N3646,2N3647,2N3648,2N3705,2N3973,2N3974, 2N3975,2N3976,2N4962,2N5027,2N5144,2N5450,2SC134,2SC239,2SC577, A5T5450,A8T3705,BSW82,BSY93,D33D24,FT3641,FT3642,GET2221,GI-3641,GI-3705, PET3705,PET6001,TIS110,TN62,TN-3903,ZT708,*PTC136,*HEPS0014,*SK3122, *ECG123A,*WEP736
2SC132	2N2318,2N2331,2SC133,2SC135,SE8040,*PTC136,*HEPS0011,*SK3122,*ECG123A, *WEP736
2SC133	2N2318,2N2331,2SC132,2SC135,SE8040,*PTC136,*HEPS0011,*SK3122,*ECG123A, *WEP736
2SC134	2N708A,2N718,2N731,2N760,2N914,2N2221,2N2312,2N2315,2N2396,2N2481, 2N2501,2N2539,2N2710,2N2845,2N2847,2N3115,2N3210,2N3211,2N3301,2N3508, 2N3510,2N3511,2N3641,2N3642,2N3646,2N3647,2N3648,2N3705,2N3973,2N3974, 2N3975,2N3976,2N4962,2N5027,2N5144,2N5450,2SC131,2SC239,2SC577, A5T5450,A8T3705,BSW82,BSY93,D33D24,FT3641,FT3642,GET2221,GI-3641,GI-3705, PET3705,PET6001,TIS110,TN62,TN-3903,ZT708,*PTC136,*HEPS0014,*SK3018, *ECG123A,*WEP736
2SC135	2N2318,2N2331,2SC132,2SC133,SE8040,*PTC136,*HEPS0011,*SK3018,*ECG123A, *WEP736
2SC136	2N718A,2N870,2N911,2N2221A,2N2313,2N2317,2N2351,2N2351A,2N2389,2N3056, 2N4963,2N5581,A5T2193,BSW84,FT3567,FT3568,FT3722,GET2221A,SE8012,TN54, *HEPS5026,*ECG123A,*WEP736
2SC137	2N914,2N2318,2N2481,2N2710,2N3210,2N3508,2N3510,2N3511,2N3646,2N3647, 2N3648,2SC131,2SC134,2SC577,BSW82,PET6001,SE8040,ZT708,*PTC136, *HEPS0011,*SK3018,*ECG123A,*WEP736
2SC138	2SC138A,2SC139,2SC596,*GE-20,*TR-21,*HEPS3001,*SK3122,*SN80,*ECG123A, *WEP736,*276-2014,*RT-102
2SC138A	2SC138,2SC139,2SC596,*GE-20,*TR-21,*HEPS3001,*SK3122,*SN80,*ECG123A, *WEP736,*276-2038,*RT-102
2SC139	2SC138,2SC138A,2SC596,*GE-47,*TR-21,*HEPS3001,*SK3122,*SN80,*ECG123A, *WEP736,*276-2038,*RT-102
2SC140	*HEPS5014,*SK3024,*ECG128,*WEP243
2SC141	2N3011,*PTC123,*GE-210,*TR-21,*BC337,*276-2014
2SC142	2N3011,2SC67,2SC577,2SC639,2SC764,*PTC136,*GE-210,*TR-21,*BC337, *276-2009,*RT-102A
2SC143	2SC67,2SC577,2SC639,2SC764,*PTC136,*GE-210,*TR-64,*276-2009
2SC144	*PTC136,*GE-210,*TR-64,*276-2009
2SC144A	*PTC136,*GE-210,*TR-64,*276-2009
2SC145	
2SC146	
2SC146A	
2SC147	2N3734,MM3724,MM3725,*PTC144,*HEPS5014,*SK3024,*ECG128,*WEP243
2SC147A	2N2330,2N3724A,2N3734,BSY81,*PTC144,*GE-47,*TR-65,*SK3024,*276-2030
2SC147B	2N2846,2N2848,2N3015,2N3299,2N3326,2N3722,2N3725A,2N4960,2N4961,2SC97A, BSY83,*PTC144,*GE-47,*SK3024,*276-2030
2SC149	2N3500,2SC49,2SC69,*HEPS5026
2SC150	*PTC125,*GE-47,*TR-21,*HEPS0014,*SK3018,*EN10,*ECG128,*WEP736,*BF338, *276-2030,*RT-102

To Replace	Substitute This Type
2SC151	2N1708A,2N2846,2N3015,2N3299,2N3326,2SC32,2SC152,2SC503-O,2SC504-O, 2SC594-R,2SC651,BC340-6,BC341-6,BSY51,BSY63,*PTC144,*GE-47,*TR-21, *HEPS3013,*SK3122,*EN10,*ECG123A,*WEP736,*BF338,*276-2030,*RT-102
2SC152	2N1837A,2N2224,2N2846,2N3015,2N3299,2N3326,2SC32,2SC503-O,2SC594-R, 2SC1218,BC341-6,BSY51,PT1558,*PTC144,*GE-47,*TR-87,*HEPS3008,*SK3024, *ECG128,*WEP243,*276-2038,*RT-114
2SC153	2N3500,2SC309,2SC310,2SC590,2SC1217,MM2258,*HEPS5026
2SC154	2N3500,2SC49,2SC69,2SC154A,2SC470-4,2SC470-5,2SC590,2SC1217,*GE-27, *TR-78,*HEPS5026,*SK3054,*ECG152,*WEP712,*276-2012,*RT-110
2SC154A	2N3500,2SC154B,2SC470-5,2SC470-6,2SC507-R,2SC1217,MM2259,*GE-32,*TR-78, *SK3040,*276-2012
2SC154B	2SC507-R,*GE-27,*TR-78,*SK3040,*ECG152,*WEP712,*276-2012,*RT-110
2SC154C	*GE-20,*TR-78,*HEPS5024,*SK3040,*ECG152,*WEP712,*BF338,*276-2012,*RT-110
2SC155	*PTC132,*HEPS0011,*SK3019,*ECG108,*WEP56
2SC156	*PTC136,*HEPS0033,*SK3018,*ECG108,*WEP56
2SC157	*PTC126,*HEPS0033,*SK3124,*ECG123,*WEP53
2SC158	*PTC126,*HEPS0033,*SK3124,*ECG123,*WEP53
2SC159	*PTC126,*HEPS0033,*SK3018,*ECG123A,*WEP736
2SC160	*PTC132,*HEPS0016,*SK3018,*ECG123A,*WEP736
2SC161	HST9801,HST9802,KSP1041,*PTC118
2SC162	*PTC123,*GE-81,*TR-63,*SN80,*BC337,*276-2030
2SC163	2N3326,*PTC136,*GE-47,*TR-87,*HEPS3021,*SK3047,*ECG195A,*WEP774, *276-2038,*RT-306
2SC164	2N3326,*PTC136,*GE-45,*TR-64,*276-2038
2SC165	2N3326,*GE-88,*TR-64,*276-2038
2SC166	2N760,2N916,2N916A,2N929,2N957,2N2242,2N2331,2N2501,2N3013,2N3014, 2N4123,2N4138,2N4418,2N4966,2N4968,2N5824,2SC67,2SC169,2SC318, 2SC367G-O,2SC372G-O,2SC400-O,2SC459,2SC460,2SC461,2SC536,2SC601N, 2SC619,2SC620,2SC639,2SC733-O,2SC735-O,2SC752G-O,2SC764,2SC899, 2SC924,2SC925,2SC1276,2SC1372,2SC1682-BL,2SC1682-GR,A321,BF254,EN916, TP3705,ZT44,*PTC126,*GE-60,*TR-21,*HEPS0011,*SK3122,*ECG123A,*WEP736, *BC107B,*276-2013,*RT-102
2SC167	2N718,2N718A,2N736,2N736A,2N736B,2N915,2N2396,2N2516,2N2847,2N3077, 2N3858A,2N3903,2N3946,2N3974,2N3976,2N4227,2N4944,2N4945,2N4962,2N4963, 2N4966,2N5027,2N5380,2N5824,2S104,2SC318,2SC366G-O,2SC473,2SC620, 2SC734-O,2SC815,2SC899,2SC938,2SC944,2SC979-O,2SC980G-O,2SC1453, BC110,EN915,GET929,MM3903,TN-3903,TP3705,*PTC123,*GE-212,*TR-21, *HEPS0011,*SK3122,*ECG123A,*WEP736,*BC107B,*276-2013,*RT-102
2SC168	2SC395A,2SC395A-O,ZT2938
2SC169	2SC639,2SC764,2SC1216,MPS6530
2SC170	2N2318,2N2319,2N2368,2N2656,2N2710,2N2729,2N3261,2N3510,2N3511,2N3544, 2N3605A,2N3606A,2N4137,2N4294,2N5030,2N5187,2SC54,2SC55,2SC63,2SC68, 2SC204,2SC321,2SC322,2SC387AG,2SC400-R,2SC455,2SC468,2SC576,2SC595, 2SC601,2SC689,2SC724,2SC752G-R,2SC1126,2SC1293,40405,EN706,EN914, EN918,EN2369A,EN3009,EN3011,EN3013,EN3014,GET708,GET914,GET3013, GET3014,GET3646,KT218,MM1941,MPS6507,MPSH19,MT1038,MT1038A,MT1039, MT1060,PET3001,PT2760,TIS45,TIS46,TIS49,TIS51,TIS52,TIS55,ZT2368,*PTC132, *HEPS0011,*SK3019,*ECG108,*WEP56
2SC171	2N834A,2N2368,2N2729,2N3261,2N3544,2N4292,2N4293,2N5187,2SC54,2SC55, 2SC63,2SC68,2SC172,2SC321,2SC322,2SC387AG,2SC400-R,2SC455,2SC576, 2SC689,2SC1126,40405,EN918,KT218,MM1941,MPS834,MPS918,MPS6507, MPS6540,MPS6542,MPS6543,MPS6546,MPS6547,MPS6548,MPSH19,MPSH20, MPSH32,MPSH37,MT1038,MT1038A,MT1039,MT1060,PET3001,SE5025,TIS47, TIS84,TIS108,TIS125,ZT2368,*PTC121,*HEPS0011,*SK3019,*ECG107,*WEP56
2SC172	2N834A,2N3261,2N5187,2SC54,2SC68,2SC321,2SC322,2SC576,2SC689,40405, MPS834,MPSH37,*PTC121,*HEPS0016,*SK3039,*ECG108,*WEP56
2SC172A	*PTC121,*HEPS0016,*SK3039,*ECG108,*WEP56
2SC173	*HEPG0011,*SK3011,*ECG101,*WEP641
2SC174	*PTC121,*HEPS0017,*SK3019,*ECG107,*WEP56
2SC174A	*PTC136,*HEPS0025,*SK3019,*ECG107,*WEP56
2SC175	*HEPG0011,*SK3011,*ECG101,*WEP641

To Replace	Substitute This Type
2SC176	*HEPG0011,*SK3011,*ECG101,*WEP641
2SC177	*HEPG0011,*SK3011,*ECG101,*WEP641
2SC178	*HEPG0011,*SK3011,*ECG101,*WEP641
2SC179	2SC181,*PTC108,*GE-8,*TR-08,*HEPG0011,*SK3124,*SN80,*ECG103A,*WEP641B, *AC187/01,*RT-122
2SC180	*DS-72,*GE-8,*TR-08,*HEPG0011,*SK3011,*ECG103A,*WEP641B,*AC187/01, *RT-122
2SC181	*GE-8,*TR-08,*HEPG0011,*SK3048,*ECG103A,*WEP641B,*RT-129
2SC182	2SC267,2SC269,MMT3014,*GE-212,*TR-21,*HEPS0016,*SK3020,*SN80,*ECG313, *WEP736,*BC107B,*276-2011,*RT-102
2SC183	2SC183A,2SC184,2SC185,2SC266,2SC269,2SC469,A3T3011,*GE-39,*TR-95, *HEPS0016,*SK3122,*SN80,*ECG313,*WEP56,*BF365,*276-2031,*RT-308
2SC183A	2SC100,2SC266,2SC920,A3T3011,FX3013,FX3014,MMT3014,*SK3124
2SC184	2SC185,2SC266,2SC269,2SC469,A3T3011,*GE-39,*TR-95,*HEPS0016,*SK3018, *SN80,*ECG313,*WEP56,*BF365,*276-2015,*RT-308
2SC185	2SC266,2SC469,A3T3011,*GE-39,*TR-95,*HEPS0016,*SK3018,*EN10,*ECG313, *WEP56,*BF365,*276-2015,*RT-308
2SC186	*PTC132,*HEPS0033,*SK3019,*ECG107,*WEP56
2SC187	*PTC132,*HEPS0033,*SK3019,*ECG107,*WEP56
2SC188	2N2217,2N2477,2N3252,2N5188,2N5236,2SC97,2SC189,2SC254,2SC344,2SC503-O, 2SC504-O,2SC580,BSX48,BSX49,*PTC123,*HEPS0014,*SK3024,*ECG128, *WEP243
2SC189	2N2217,2N2477,2N3252,2N3253,2N3444,2N5188,2N6376,2SC97,2SC109A-R,2SC255, 2SC344,2SC345,2SC503-O,2SC580,BSX49,BSX60,*PTC123,*HEPS0014,*SK3024, *ECG128,*WEP243
2SC190	2N2218,2N2218A,2N2224,2N2410,2N2479,2N2537,2N2846,2N2848,2N2958,2N3015, 2N3299,2N3725,2N3725A,2N4047,2N4960,2N4961,2SC97A,2SC109A-O,2SC307, 2SC479,2SC503-Y,2SC560,2SC1218,TIS110,TIS135,TIS136,*HEPS0014,*SK3024, *ECG128,*WEP243
2SC191	2N718,2N718A,2N736,2N736A,2N736B,2N915,2N2396,2N2516,2N2847,2N3077, 2N3858A,2N3903,2N3946,2N3974,2N3976,2N4227,2N4962,2N4963,2N4994,2N5027, 2N5380,2S104,2SC366G-O,2SC734-O,2SC815,2SC938,2SC944,2SC979-O,BC110, EN697,EN1613,FT3567,FT3568,GET929,MM3903,MPS6565,TN-3903,*PTC123, *GE-212,*TR-21,*HEPS0014,*SK3122,*SN80,*ECG123A,*WEP736,*BC107B, *276-2010,*RT-102
2SC192	2N718,2N718A,2N736,2N736A,2N736B,2N915,2N2396,2N2516,2N2847,2N3077, 2N3858A,2N3903,2N3946,2N3974,2N3976,2N4227,2N4962,2N4963,2N4994,2N5027, 2N5380,2S104,2SC366G-O,2SC734-O,2SC815,2SC938,2SC944,2SC979-O,BC110, EN697,EN1613,FT3567,FT3568,GET929,MM3903,MPS6565,TN-3903,*PTC123, *GE-212,*TR-21,*HEPS0014,*SK3122,*SN80,*ECG123A,*WEP736,*BC107B, *276-2010,*RT-102
2SC193	2N718,2N718A,2N736,2N736A,2N736B,2N915,2N2396,2N2516,2N2847,2N3077, 2N3858A,2N3903,2N3946,2N3974,2N3976,2N4227,2N4962,2N4963,2N4994,2N5027, 2N5380,2S104,2SC366G-O,2SC734-O,2SC815,2SC938,2SC944,2SC979-O,BC110, EN697,EN1613,FT3567,FT3568,GET929,MM3903,MPS6565,TN-3903,*PTC123, *GE-212,*TR-21,*HEPS0014,*SK3122,*SN80,*ECG123A,*WEP736,*BC107B, *276-2010,*RT-102
2SC193A	*SK3122
2SC194	2N718,2N718A,2N736,2N736A,2N736B,2N915,2N2396,2N2516,2N2847,2N3077, 2N3858A,2N3903,2N3946,2N3974,2N3976,2N4227,2N4962,2N4963,2N4994,2N5027, 2N5380,2S104,2SC366G-O,2SC734-O,2SC815,2SC938,2SC944,2SC979-O,BC110, EN697,EN1613,FT3567,FT3568,GET929,MM3903,MPS6565,TN-3903,*PTC123, *GE-212,*TR-21,*HEPS0014,*SK3122,*SN80,*ECG123A,*WEP736,*BC107B, *276-2012,*RT-102
2SC195	2N338A,2N729,2N760,2N843,2N916,2N916A,2N929,2N957,2N1704,2N2242,2N2331, 2N2369,2N2369A,2N2501,2N2615,2N2616,2N3009,2N3013,2N3014,2N3855A, 2N3858,2N3862,2N4123,2N4138,2N4418,2N4873,2N4996,2N4997,2N5136,2N5824, 2SC67,2SC122,2SC169,2SC318,2SC367G-O,2SC400-O,2SC601N,2SC619,2SC620, 2SC639,2SC733-O,2SC735-O,2SC764,2SC852,2SC899,2SC922,2SC924,2SC925, 2SC964,2SC965,2SC1187,2SC1188,2SC1189,2SC1319,2SC1320,2SC1393, 2SC1394,2SC1395,2SC1396,2SC1686,2SC1687,2SC1688,A321,BF123,BF125,

To Replace	Substitute This Type
(2SC195)	BF127,BF198,BF199,BF241,BF377,BF378,MPS3563,MPS3693,MPS6574-BLUE, MPS6574-YEL,MPS6576-BLUE,MPS6576-YEL,MPSA10-WHT,MPSA20-WHT, MPSH10,MPSH11,PET1001,PET2001,PET3002,TIS62A,TIS63A,TIS64A,TIS86, TIS87,TP3705,ZT24,ZT44,*PTC126,*GE-61,*TR-21,*HEPS0014,*SK3018,*EN10, *ECG123A,*WEP736,*BF173,*276-2013,*RT-102
2SC196	2N338A,2N729,2N760,2N843,2N916,2N916A,2N929,2N957,2N2242,2N2331,2N2369, 2N2369A,2N2501,2N2615,2N2616,2N3009,2N3013,2N3014,2N3855A,2N3858, 2N3862,2N4123,2N4418,2N4873,2N4996,2N4997,2N5136,2N5824,2SC67,2SC122, 2SC169,2SC318,2SC367G-O,2SC400-O,2SC601N,2SC619,2SC620,2SC639, 2SC733-O,2SC735-O,2SC764,2SC852,2SC899,2SC922,2SC924,2SC925,2SC964, 2SC965,2SC1187,2SC1188,2SC1189,2SC1319,2SC1320,2SC1393,2SC1394, 2SC1395,2SC1396,2SC1686,2SC1687,2SC1688,A321,BF123,BF125,BF127,BF198, BF199,BF241,BF377,BF378,MPS3563,MPS3693,MPS6574-BLUE,MPS6574-YEL, MPS6576-BLUE,MPS6576-YEL,MPSA10-WHT,MPSA20-WHT,MPSH10,MPSH11, PET1001,PET2001,PET3002,TIS62A,TIS63A,TIS64A,TIS86,TIS87,TP3705,ZT24,ZT44, *PTC126,*GE-61,*TR-21,*HEPS0014,*SK3018,*EN10,*ECG123A,*WEP736,*BF173, *276-2013,*RT-102
2SC197	2N729,2N760,2N916,2N916A,2N957,2N2242,2N2331,2N2369,2N2369A,2N2501, 2N2615,2N2616,2N3009,2N3013,2N3014,2N3855A,2N3858,2N3862,2N4123,2N4418, 2N4873,2N4996,2N4997,2N5824,2SC67,2SC122,2SC169,2SC318,2SC367G-O, 2SC400-O,2SC601N,2SC619,2SC620,2SC639,2SC733-O,2SC735-O,2SC764, 2SC852,2SC899,2SC922,2SC924,2SC925,2SC964,2SC965,2SC1187,2SC1188, 2SC1189,2SC1319,2SC1320,2SC1393,2SC1394,2SC1395,2SC1396,2SC1686, 2SC1687,2SC1688,BF123,BF125,BF127,BF198,BF199,BF241,BF377,BF378, MPS3563,MPS3693,MPS6574-BLUE,MPS6574-YEL,MPS6576-BLUE,MPS6576-YEL, MPSA10-WHT,MPSA20-WHT,MPSH10,MPSH11,PET1001,PET2001,PET3002, TIS62A,TIS63A,TIS64A,TIS86,TIS87,TP3705,ZT24,ZT44,*PTC126,*GE-61,*TR-21, *HEPS0014,*SK3018,*EN10,*ECG123A,*WEP736,*BF173,*276-2013,*RT-102
2SC198	
2SC199	*GE-81,*TR-95,*HEPS5026,*SK3039,*EN10,*ECG108,*WEP56,*BF338,*276-2008, *RT-113
2SC200	*HEPS0014,*SK3122,*ECG123A,*WEP736
2SC201	*HEPS0014,*SK3122,*ECG123A,*WEP736
2SC202	*HEPS5026,*SK3122,*ECG123A,*WEP736
2SC203	*PTC136,*HEPS0011,*SK3122,*ECG123A,*WEP736
2SC204	2N708,2N708A,2N784A,2N914,2N2242,2N2318,2N2481,2N2501,2N2710,2N3011, 2N3013,2N3014,2N3210,2N3211,2N3510,2N3511,2N3605A,2N3606A,2N3646, 2N3647,2N3648,2N5144,2N5772,2SC67,2SC577,2SC639,2SC764,A5T4123,BSW82, GET708,GET914,GET2369,GET3013,GET3014,GET3646,ZT708,*PTC136, *HEPS0011,*SK3122,*ECG123A,*WEP736
2SC205	*HEPS5026,*SK3122,*ECG123A,*WEP736
2SC206	*PTC115,*HEPS0017,*SK3018,*ECG108,*WEP56
2SC207	
2SC208	*PTC136,*SK3024,*ECG128,*WEP243
2SC209	
2SC210	*PTC144,*HEPS0014,*SK3024,*ECG128,*WEP243
2SC211	*PTC144,*HEPS0014,*SK3024,*ECG128,*WEP243
2SC212	*PTC144,*HEPS0005,*SK3024,*ECG194
2SC213	*PTC144,*HEPS3001,*SK3024,*ECG128,*WEP243
2SC214	*PTC144,*HEPS3013,*SK3024,*ECG128,*WEP243
2SC215	*PTC144,*HEPS3001,*SK3024,*ECG128,*WEP243
2SC216	*PTC125,*HEPS0014,*SK3024,*ECG128,*WEP243
2SC217	2N546,2N548,2N550,2N1708A,2N1840,2N2237,2N2309,2N2320,2SC47,2SC147, 2SC227,2SC343,2SC482-O,2SC504-O,2SD215,BC340-6,SE8041,SE8042,TI-484, *PTC125,*HEPS0014,*SK3024,*ECG128,*WEP243
2SC218	2N657A,2N1613,2N1613A,2N1716,2N1837,2N1837A,2N1889,2N1974,2N1988,2N2193, 2N2193A,2N2193B,2N2218A,2N2380,2N2380A,2N2479,2N3108,2N3110,2N3253, 2N3498,2N3678,2N3722,2N3723,2N4000,2SC97A,2SC108A-R,2SC228,2SC308, 2SC345,2SC349,2SC497-R,2SC498-R,2SC560,2SC1053,2SC1218,A5T2193, BC140-6,BC141-6,BSY53,BSY87,KT600G,SE8002,SE8010,ZT1613,*PTC125, *HEPS5026,*SK3024,*ECG123A,*WEP243

To Replace	Substitute This Type
2SC219	*HEPS0014
2SC220	*PTC144,*HEPS0014,*SK3024,*ECG128,*WEP243
2SC221	*PTC144,*HEPS0014,*SK3024,*ECG128,*WEP243
2SC222	*PTC144,*HEPS3002,*SK3024,*ECG128,*WEP243
2SC223	*PTC144,*HEPS3001,*SK3024,*ECG128,*WEP243
2SC224	*PTC144,*HEPS3001,*SK3024,*ECG128,*WEP243
2SC225	*PTC144,*HEPS3001,*SK3024,*ECG128,*WEP243
2SC226	2N545,2N547,2N549,2N1613A,2N1959A,2N1984,2N1987,2N2106,2N2107,2N2218, 2N2218A,2N2410,2N2846,2N2868,2N3015,2N3122,2N3252,2N3253,2N3299,2N3326, 2N3678,2N3734,2N3735,2SC97,2SC109A-R,2SC254,2SC347,2SC348,2SC580, 2SC879,2SC880,2SC1053,2SC1072,2SC1072A,2SC1150,2SC1386,2SD216,7A31, 40347,BSX60,BSY53,KT600,KT600F,KT600T,MM3725,ZT1613,*PTC144,*HEPS0014, *SK3024,*ECG128,*WEP243
2SC227	2N546,2N548,2N550,2N2320,2SC147,2SD215,MM3724,SE8041,SE8042,*PTC144, *HEPS0014,*SK3024,*ECG128,*WEP243
2SC228	2N1613A,2N1716,2N1889,2N1974,2N1988,2N2193,2N2193A,2N2193B,2N2218A, 2N3108,2N3110,2N3253,2N3678,2N3722,2N3723,2N3735,2N4000,2SC97A, 2SC108A-R,2SC349,2SC485-R,2SC497-R,2SC498-R,2SC512-R,2SC516,2SC560, 2SC1053,40348,40367,A5T2193,BC140-6,BC141-6,BSY53,KT600G,SE8002,ZT1613, *PTC144,*HEPS3002,*SK3024,*ECG128,*WEP243
2SC229	2SC416,2SC916,2SC1266,*PTC144,*HEPS3002,*SK3024,*ECG128,*WEP243
2SC230	*PTC126,*HEPS5026,*SK3122,*ECG123A,*WEP736
2SC231	2N545,2N547,2N549,2N1958A,2N1987,2N2107,2N2194,2N2194A,2N2194B,2N2217, 2N2477,2N3122,2N3252,2N3253,2N3735,2N5188,2SC97,2SC109A-R,2SC226, 2SC254,2SC580,2SC1052,2SC1053,2SC1064,2SC1072,2SC1072A,2SC1150, 2SD149,2SD329,7A31,40347,BSX60,KT600,KT600F,KT600T,MM3725,*PTC144, *HEPS0014,*SK3024,*ECG128,*WEP243
2SC232	2N546,2N548,2N550,2N2320,2SC147,2SC227,MM3724,*PTC144,*HEPS0014, *SK3024,*ECG128,*WEP243
2SC233	2N1714,2N1989,2N3108,2N3110,2N3253,2N3444,2N3735,2N4000,2SC108A-R, 2SC228,2SC255,2SC256,2SC485-R,2SC497-R,2SC498-R,2SC512-R,2SC516, 2SC1052,2SC1053,2SC1065,2SC1069,40367,BC140-6,BC141-6,KT600G,TI-486, *HEPS3002,*SK3024,*ECG128,*WEP243
2SC234	*HEPS3002,*SK3024,*ECG128,*WEP243
2SC235	*HEPS3002,*SK3024,*ECG293,*WEP243
2SC236	*PTC144,*HEPS3019,*SK3024,*ECG128,*WEP243
2SC237	*PTC136,*HEPS0011,*SK3122,*ECG123A,*WEP736
2SC238	*PTC132,*HEPS0014,*SK3024,*ECG128,*WEP243
2SC239	2N708A,2N718,2N718A,2N731,2N2221,2N2221A,2N2312,2N2315,2N2317,2N2389, 2N2396,2N2539,2N2845,2N2847,2N3115,2N3301,2N3641,2N3642,2N3705,2N3973, 2N3974,2N3975,2N3976,2N4962,2N5027,2N5144,2N5450,2N5581,2N5820,A5T5450, A8T3705,BSW84,BSY93,D33D24,D33D29,FT3641,FT3642,GET2221,GET2221A, GI-3641,GI-3705,PET3705,TIS110,TN54,TN-3903,*PTC136,*HEPS0011,*SK3122, *ECG123A,*WEP736
2SC240	2N3233,2SC242,2SC245,HST9202,HST9203,HST9207,HST9208,HST9803,HST9804, KSP1172,*PTC129A,*GE-35,*TR-67,*HEPS7004,*SK3027,*ECG162,*WEP707, *BDY20
2SC241	2N3232,2SC244,HST9201,HST9205,HST9206,HST9801,HST9802,KSP1171, *PTC129A,*GE-35,*TR-67,*HEPS7002,*SK3027,*HN100,*ECG162,*WEP707, *BDY20
2SC242	2N3233,2SC240,2SC245,HST9202,HST9203,HST9207,HST9208,HST9803,HST9804, KSP1172,*PTC129A,*GE-35,*TR-67,*HEPS7004,*SK3027,*ECG162,*WEP707, *BDY20
2SC243	2N3234,2SC246,HST9204,HST9209,KSP1173,*PTC129A,*GE-35,*TR-67,*HEPS7004, *ECG162,*WEP707
2SC244	2N3232,2SC241,HST9201,HST9205,HST9206,HST9801,HST9802,KSP1171, *HEPS7002,*SK3027,*ECG130,*WEP247
2SC245	2N3233,2N3234,2SC243,HST9203,HST9204,HST9208,HST9209,HST9804,KSP1173, *HEPS7004,*SK3561,*ECG162,*WEP707
2SC246	KSP1141,*HEPS5020,*ECG128,*WEP707
2SC247	*PTC144,*HEPS5026,*SK3024,*ECG128,*WEP243

To Replace	Substitute This Type
2SC248	*PTC121,*HEPS5026,*SK3024,*ECG128,*WEP243
2SC249	*PTC121,*HEPS5026,*SK3024,*ECG128,*WEP243
2SC250	*PTC132,*HEPS0033,*SK3019,*ECG161,*WEP719
2SC251	2N3600,2SC251A,2SC252,2SC253,2SC602,2SC1747,MT1061,MT1061A,*PTC132, *GE-86,*TR-83,*HEPS0014,*SK3019,*EN10,*ECG161,*WEP719,*276-2011,*RT-113
2SC251A	2N3600,2SC251,2SC252,2SC253,2SC602,2SC1747,MT1061,MT1061A,*PTC132, *GE-86,*TR-83,*HEPS0011,*SK3039,*EN10,*ECG161,*WEP719,*276-2011,*RT-113
2SC252	2N3600,2SC251,2SC251A,2SC253,2SC602,2SC1747,MT1061,MT1061A,*PTC132, *GE-86,*TR-83,*HEPS0016,*SK3019,*EN10,*ECG161,*WEP719,*276-2011,*RT-113
2SC253	2N3600,2SC251,2SC251A,2SC252,2SC602,2SC1747,MT1061,MT1061A,*PTC132, *GE-86,*TR-83,*HEPS0016,*SK3019,*EN10,*ECG161,*WEP719,*276-2011,*RT-113
2SC254	2N2218,2N2218A,2N2846,2N3015,2N3299,2N3326,2N3678,2N3734,2N3735,2N5262, 2SC97,2SC580,BSX60,*HEPS5012,*ECG175,*WEP241
2SC255	2N3678,2N3722,2N3723,2N3735,2N5262,2SC256
2SC256	2N3723
2SC257	
2SC258	2SC259
2SC259	
2SC260	*PTC144,*TR-65
2SC261	*PTC144,*ECG192,*WEP243
2SC262	*PTC144
2SC263	A3T3011,FX709,*GE-212,*TR-95,*SK3039,*EN10,*ECG108,*WEP56,*BC107B, *276-2011,*RT-113
2SC264	2SC100,A3T3011,FX3014,MMT3014,*GE-210,*TR-53,*BC107B,*276-2011
2SC265	2SC100,FX3014,MMT3014,*GE-210,*TR-53,*BC107B,*276-2016
2SC266	2SC100,2SC920,A3T3011,FX3013,FX3014,MMT3014,*GE-39,*TR-70,*HEPS0016, *SK3019,*EN10,*ECG107,*WEP56,*BF365,*276-2015,*RT-108
2SC267	2SC100,FX914,FX3013,FX3014,FX4046,MMT3014,*GE-64,*TR-21,*HEPS0016, *SK3122,*EN10,*ECG123A,*WEP736,*BC107B,*276-2013,*RT-102
2SC268	2SC268A,*GE-81,*TR-87,*HEPS0005,*SK3019,*EN10,*ECG128,*WEP243,*BF338, *276-2008,*RT-114
2SC268A	*GE-81,*TR-87,*HEPS0005,*SK3024,*EN10,*ECG128,*WEP243,*BF338,*276-2008, *RT-114
2SC268B	*SK3039
2SC269	MMT3014,*GE-210,*TR-95,*HEPS0016,*SK3019,*EN10,*ECG108,*WEP56,*BC107B, *276-2015,*RT-113
2SC270	2N6233,2N6234,2N6235,HST7202,HST7203,HST7204,HST7205,KSP1092,KSP1093, KSP1094,KSP1095,KSP1142,KSP1143,KSP1144,KSP1145,SDT7202,SDT7203, SDT7204,SDT7205,*PTC129A,*TR-67,*HEPS5020,*ECG162,*WEP707
2SC271	2SC272,2SC287A,2SC1158,2SC1159,*GE-39,*TR-95,*HEPS0016,*SK3019,*EN10, *ECG108,*WEP56,*276-2011,*RT-113
2SC272	2SC271,2SC287A,2SC1158,2SC1159,*GE-214,*TR-95,*HEPS0016,*SK3039,*EN10, *ECG108,*WEP56,*276-2011,*RT-113
2SC273	2SC48,2SC49,2SC69,2SC154,2SC154A,2SC470-4,2SC470-5,2SC590,2SC1217, *TR-78,*HEPS3019,*ECG154,*WEP712
2SC280	2N2639,2N2640,2N2641,2N2642,2N2643,2N2644,2N2722,2N2910
2SC281	2N3227,2N3242A,2N3794,2N4124,2N4419,2N4970,2N5826,2N5998,2N6000,2N6004, 2SC367G-Y,2SC372G-Y,2SC400,2SC400-GR,2SC733-Y,2SC735-Y,2SC752G-Y, 2SC907,2SC1175,2SC1285,2SC1364,2SC1632,2SC1634,2SC1684,2SC1813, 2SC1849,2SC1992,BC107,BC107A,BC108,BC108A,BC167A,BC171A,BC237A, BC238A,BC547A,BC548A,BCY58A,BCY58B,BCY59A,BCY59B,BFY39-3,BSW83, BSW88,BSW89,BSX38,BSX79,GI-3794,MPS3704,MPSA17,PBC107A,*PTC123, *GE-212,*TR-51,*HEPS0020,*SK3020,*EN10,*ECG123A,*WEP736,*BC107B, *276-2015,*RT-102
2SC282	2N916,2N916A,2N2331,2N2501,2N3241A,2N4013,2N6426,2SC639,2SC1317, 2SC1851,2SC1999,A5T4123,*PTC123,*GE-62,*TR-95,*HEPS0016,*SK3039, *ECG123A,*WEP56,*BC337,*276-2016,*RT-113
2SC283	2N708A,2N2845,2N2847,2N3115,2N3301,2N3641,2N3642,2N3946,2N3974,2N3976, 2N4013,2N4962,2N5027,2N5144,2N5380,2N6538,2SC317,A5T3903,BSW84,FT3641, FT3642,GET2221,GET2221A,GI-3641,MM3903,TN-3903,*PTC123,*GE-62,*TR-21, *HEPS0015,*SK3122,*EN10,*ECG123A,*WEP736,*BC337,*276-2016,*RT-102

To Replace	Substitute This Type
2SC284	2N4963,2SC317,2SC317A,BSW84,FT3722,GET2221A,SE8012,*PTC123,*GE-62, *TR-21,*HEPS5026,*SK3122,*EN10,*ECG123A,*WEP736,*BF338,*276-2009, *RT-102
2SC285	*PTC136,*HEPS0014,*SK3122,*ECG123A,*WEP736
2SC285A	*PTC136,*HEPS0014,*SK3122,*ECG123A,*WEP736
2SC286	*HEPS0016,*SK3019,*ECG108,*WEP56
2SC287	*HEPS0016,*SK3019,*ECG313,*WEP56
2SC287A	2SC1158,2SC1159,*GE-60,*TR-95,*HEPS0016,*SK3018,*ECG313,*WEP56,*BF200, *276-2038,*RT-108
2SC288	*HEPS0016,*SK3019,*ECG313,*WEP56
2SC288A	2SC287A,2SC1158,2SC1159,*GE-60,*TR-70,*HEPS0016,*SK3018,*ECG313,*WEP56, *BF200,*276-2038,*RT-108
2SC289	2SC271,2SC272,2SC287A,2SC288A,2SC1158,2SC1159,*GE-214,*TR-95,*HEPS0016, *SK3039,*EN10,*ECG108,*WEP56,*BF183,*276-2011,*RT-113
2SC290	2SC92,2SC93,2SC94,*SK3048,*ECG195A
2SC291	2N4311,2SC292,*GE-28,*TR-87,*HEPS5014,*SK3024,*ECG128,*WEP243,*276-2008, *RT-114
2SC292	2N4305,2N4311,2SC293,*GE-28,*TR-87,*HEPS3002,*SK3024,*ECG128,*WEP243, *276-2008,*RT-114
2SC293	2N4305,*GE-18,*TR-87,*SK3048,*ECG195A,*WEP243,*276-2008,*RT-114
2SC294	2SC457
2SC295	
2SC296	*ECG108,*WEP56
2SC297	2SC298,2SC697,2SC698,*PTC195,*GE-28,*TR-76,*HEPS3010
2SC298	2SC299,2SC697,2SC697A,2SC698,*PTC195,*GE-10,*HEPS3002,*EN30
2SC299	2SC697A,*PTC195,*GE-20,*TR-86,*SK3124,*SN80,*ECG225,*WEP53,*276-2030, *RT-100
2SC300	2N2501,2N3211,2SC169,2SC301,2SC639,2SC764,2SC1216,ZT2938,*PTC136,*TR-21, *HEPS0016,*SK3122,*ECG123A,*WEP736,*276-2009
2SC301	2N2501,2N3211,2SC169,2SC300,2SC639,2SC764,2SC1216,ZT2938,*PTC136,*TR-21, *HEPS0016,*SK3122,*ECG123A,*WEP736,*276-2009
2SC302	*PTC136,*TR-21,*HEPS0025,*SK3122,*ECG123A,*WEP736,*276-2009
2SC302-M	
2SC303	2N707A,2N3512,BSX59,BSX61,PT2540,*HEPS3001
2SC304	2N2217,2N3830,2N3831,2N5188,BC142,BSX49,BSX59,BSX61,PT2540,*HEPS3001
2SC305	2N3830,2SC1065,2SC1069,BC142,*HEPS3019,*SK3104
2SC306	2N2537,2N2848,2N3724,2N3724A,2N4046,2N4960,2N5145,*PTC136,*TR-87, *HEPS0014,*SK3024,*ECG128,*WEP243,*276-2030
2SC307	2N3725,2N3725A,2N4047,2N4961,*TR-87,*HEPS3001,*SK3024,*ECG128,*WEP243, *276-2014
2SC307-M	2N751,2SC302-M,2SC590N
2SC308	2N3498,2N3723,2SC309,BSY55,BSY85,BSY87,MM2258,SE8010,*PTC125, *HEPS3019,*SK3024,*ECG128,*WEP243
2SC309	2N1893A,2N3020,2SC310,BSY55,BSY85,MM2258,*PTC144,*TR-87,*HEPS3019, *SK3024,*ECG128,*WEP243
2SC310	2N1893A,2N3020,*TR-87,*HEPS0005,*SK3024,*ECG128,*WEP243
2SC313	2SC464,2SC465,2SC466,40414,MT1061,*PTC132,*GE-11,*TR-83,*HEPS0016, *SK3019,*EN10,*ECG161,*WEP719,*BF173,*276-2011,*RT-112
2SC314	
2SC315	*SK3047,*ECG123A,*WEP736
2SC316	*PTC123,*GE-212,*TR-95,*HEPS0015,*SK3039,*EN10,*ECG108,*WEP56,*BC107B, *276-2016,*RT-113
2SC316N	SEE 2SC587AN
2SC317	2N915,2N4014,2N4963,2SC1998,TIS96,TIS99,*PTC123,*GE-62,*TR-21,*SK3122, *ECG123A,*WEP736,*BF338,*276-2009,*RT-102
2SC317A	*SK3122
2SC318	2N915,2N2847,2N3903,2N3974,2N3976,2N4013,2N4227,2N4400,2N4951,2N4962, 2N5144,2N5368,2N5380,2N6538,2SC366G,2SC366G-O,2SC734-O,2SC933,2SC943, 2SC979-O,2SC1999,A5T3903,TN-3903,ZT84,ZT114,*PTC123,*GE-212,*TR-21, *HEPS0025,*SK3046,*EN10,*ECG123A,*WEP736,*BC107B,*276-2016,*RT-102
2SC318A	2N760A,2N915,2N2847,2N3416,2N3643,2N3858A,2N3859A,2N3903,2N3974,2N3976,

To Replace	Substitute This Type
(2SC318A)	2N4013,2N4227,2N4400,2N4951,2N4962,2N5144,2N5309,2N5368,2N5380,2N5824, 2N5825,2N6222,2N6538,2SC318,2SC366G,2SC366G-O,2SC734-O,2SC933,2SC943, 2SC979-O,2SC1166-O,2SC1999,A5T3903,EN1613,GET929,GI-3416,MPS3705, TN-3903,TP3705,ZT84,ZT114,*PTC123,*HEPS0025,*SK3122,*ECG123A,*WEP736
2SC319	*GE-20,*TR-21,*HEPS3013,*SK3122,*SN80,*ECG123A,*WEP736,*276-2038,*RT-102
2SC320	2SC628,*GE-20,*TR-21,*HEPS3013,*SK3122,*SN80,*ECG123A,*WEP736,*276-2009, *RT-102
2SC321	2N2710,2N3511,2N3648,2SC322,2SC576,*PTC136,*GE-20,*TR-21,*SK3122, *ECG123A,*WEP736,*276-2009,*RT-102
2SC322	2N2710,2N3511,2N3648,2SC321,2SC576
2SC323	2N708,2N708A,2N784A,2N914,2N2242,2N2481,2N2710,2N2845,2N3013,2N3014, 2N3115,2N3210,2N3261,2N3301,2N3510,2N3511,2N3605A,2N3606A,2N3646, 2N3647,2N3648,2N3946,2N4140,2SC54,2SC67,2SC302,2SC335,2SC336,2SC577, 2SC601,2SC601N,2SC689,2SC1274,BSW82,FT3641,FT3642,GET708,GET914, GET2221,GET2369,GET3013,GET3014,GET3646,GI-3641,GI-3642,TIS45,TIS46, TIS52,TIS55,ZT83,ZT708,*PTC139,*GE-212,*TR-21,*HEPS0011,*SK3122,*SN80, *ECG123A,*WEP736,*276-2016,*RT-102
2SC324	*PTC136,*HEPS0015,*SK3004,*ECG123A,*WEP736
2SC325	*SK3054,*ECG152
2SC326	
2SC327	
2SC328	
2SC329	*ECG107,*WEP56
2SC330	2SC331,2SC989
2SC331	2SC330,2SC989
2SC332	2SC333
2SC333	2SC334
2SC334	
2SC335	2SC336,MPS6530,*SK3018,*ECG199,*WEP736
2SC336	MPS6530
2SC337	2N703,2N2331,2N3011,2N3394,PT720
2SC338	2N736A,2N2461,2N2465,2N4963,GET929,TIS96,TIS99
2SC339	2N2461,2N2465
2SC340	2SC695
2SC341	
2SC342	
2SC343	
2SC344	2N3326,2SC345,2SC1218
2SC345	2SC1218
2SC346	2N3326,2N3724A,2N3734,2SC347,2SC348
2SC347	2N3326,2N3722,2N3725A,2SC97A,2SC348
2SC348	2N3326,2N3722,2N3725A,2SC97A,2SC347,*SK3117,*ECG161,*WEP719
2SC349	2N3723
2SC350	2N916,2N916A,2N3227,2N3241A,2N3242A,2N4970,2N5825,2N6000,2N6004, 2SC367G,2SC367G-Y,2SC372G-Y,2SC400,2SC400-Y,2SC454,2SC458,2SC459, 2SC536,2SC735-Y,2SC752G-Y,2SC907,2SC933,2SC1085,2SC1175,2SC1216, 2SC1276,2SC1285,2SC1317,2SC1364,2SC1632,2SC1634,2SC1813,2SC1999, BC107A,BC108A,BC167A,BC171A,BC237A,BC238A,BC547A,BC548A,BCY58A, BCY59A,BFY39-2,BSW83,BSW88,BSW89,BSX38,BSX79,EN916,MPS3704, PBC107A,*PTC132,*GE-212,*TR-21,*HEPS0015,*SK3122,*ECG123A,*WEP736, *BC107B,*276-2015,*RT-102
2SC351	*PTC132,*HEPS0016,*SK3039,*ECG107,*WEP56
2SC352	2N2848,2N3724,2N4046,2N4960,2N5145,2SC503-Y,2SC594,2SC594-O,BC341-10, TIS133,TIS134,*PTC125,*GE-47,*TR-87,*HEPS5026,*SK3024,*EN10,*ECG128, *WEP243,*BF338,*276-2030,*RT-114
2SC352A	2N2017,2N2108,2N2848,2N3724,2N4046,2N4960,2N5145,2SC352,2SC503-Y,2SC594, 2SC594-O,7A32,BC341-10,TIS133,TIS134,*HEPS0014,*SK3024,*ECG128,*WEP243
2SC353	2N4924,MM2258,*PTC125,*GE-27,*TR-87,*SK3024,*ECG128,*WEP243,*BF338
2SC353A	2N1973,2N2443,2N3107,2N4924,2SC353,2SC826,2SC827,MM2258,*HEPS5026, *SK3024,*ECG128,*WEP243
2SC354	*HEPS3001,*SK3197

To Replace	Substitute This Type
2SC355	
2SC356	2N2319,2N3013,2N3014,2N3510,2N3511,2N3647,2N3648,2N4123,2N4418,2N5772, A5T4123,*PTC136,*GE-210,*TR-21,*HEPS3013,*SK3039,*PN66,*ECG123A, *WEPS3021,*BC107B,*276-2016,*RT-113
2SC360	*PTC136,*HEPS0011,*SK3019,*ECG123A,*WEP736
2SC361	2N2331,2N2501,2N2923,2N3241A,2N4123,2N4418,2SC367G,2SC367G-O, 2SC372G-O,2SC400-Y,2SC454,2SC458,2SC459,2SC536,2SC619,2SC639, 2SC735-O,2SC752G-O,2SC1085,2SC1216,2SC1276,2SC1317,2SC1372,BSY61, ZT2938,*PTC136,*HEPS0016,*SK3018,*ECG107,*WEP56
2SC361G	SEE 2SC371G
2SC362	2N2924,2N3227,2N3242A,2N4124,2N4419,2N5224,2N5998,2N6000,2SC367G-Y, 2SC372G-Y,2SC400,2SC400-GR,2SC735-Y,2SC752G-Y,2SC1285,2SC1363, 2SC1684,2SC1849,2SC1993,BC108,BC108B,BC109B,BC172B,BC173B,BC238B, BC239B,BC548A,BCY58B,BSW83,BSW88,BSW89,BSX38,*PTC136,*HEPS0016, *SK3124,*ECG107,*WEP56
2SC362G	SEE 2SC372G
2SC363	BC548C,BC549C,*PTC136,*HEPS0015,*SK3019,*ECG123A,*WEP736
2SC363G	SEE 2SC373G
2SC364	
2SC366	*HEPS0015,*SK3122,*ECG123A,*WEP736
2SC366G	2N2222,2N2222A,2N2540,2N3116,2N3302,2N3643,2N4401,2N4952,2N5107,2N5369, 2N5582,2SC366G-Y,2SC1318,2SC1781,2SC1852,40458,A5T2222,BCW34,BCW36, BSW85,EN2219,FT3643,GET2222,GET2222A,GI-3643,MPS6531,SE6020,SE6020A, SE6021,SE6021A,TIS111,TN-3904,ZT89,ZT119,*PTC123,*GE-210,*HEPS0015, *SK3122
2SC366G-O	2N915,2N3643,2N4014,2N4951,2N5107,2N5368,2N5582,2SC366G,2SC1318, 2SC1781,2SC1852,40458,BCW34,BCW36,GET2222,GET2222A,GI-3643,ZT84,ZT89, ZT114,ZT119,*PTC170,*GE-210,*276-2016
2SC366G-R	2N2221,2N2221A,2N2539,2N2845,2N2847,2N3115,2N3301,2N3641,2N3642,2N3973, 2N3974,2N3975,2N3976,2N4400,2N4962,2N4963,2N5581,BSW84,FT3641,FT3642, FT3722,GET2221,GET2221A,GI-3641,GI-3642,MPS6530,TIS110,TN-3903,ZT83, ZT113,*PTC170,*GE-210,*276-2016
2SC366G-Y	2N2222,2N2222A,2N2540,2N3116,2N3302,2N4401,2N4952,2N5369,2N5377,A5T2222, BSW85,EN2219,FT3643,MPS6531,SE6020,SE6020A,SE6021,SE6021A,TIS109, TIS111,TN-3904
2SC367-O	2N916,2N916A,2N2897,2N2900,2N3643,2N4013,2N4951,2N5107,2N5368,2N6004, 2N6010,2N6426,2N6427,2SC366G,2SC366G-O,2SC367G,2SC367G-O,2SC1318, 2SC1852,40458,BC337-16,BCW34,BCW36,BSX75,D33D25,GET2222,GI-3643, GI-3704,PET3704,ZT84,ZT114,*PTC170
2SC367-R	2N708A,2N914,2N2221,2N2481,2N2501,2N2539,2N2710,2N2845,2N2847,2N3115, 2N3210,2N3211,2N3301,2N3510,2N3511,2N3641,2N3642,2N3646,2N3647,2N3648, 2N3705,2N3973,2N3974,2N3975,2N3976,2N4400,2N4962,2N5144,2N5450,2SC577, A5T5450,A8T3705,BSW82,FT3641,FT3642,GET2221,GI-3641,GI-3642,GI-3705, MPS3705,MPS6530,PET3705,TIS110,TN62,TN-3903,ZT83,ZT113,ZT708,*PTC170
2SC367-Y	2N2222,2N2540,2N3116,2N3227,2N3242A,2N3302,2N3704,2N4401,2N4952,2N5369, 2N5377,2N5449,2N5818,2SC366G-Y,2SC367G-Y,2SC1813,A5T2222,A5T5449, A8T3704,BC337-25,BSW83,D33D26,EN2219,FT3643,GI-3706,MPS3704,MPS6531, PET3706,SE6020,SE6020A,TIS109,TIS111,TN60,TN-3904,*PTC170
2SC367G	2N2222,2N2540,2N3116,2N3227,2N3242A,2N3302,2N3643,2N4401,2N4952,2N5107, 2N5369,2N5818,2N6004,2N6010,2N6426,2SC366G,2SC366G-Y,2SC367G-Y, 2SC1318,2SC1852,40458,A5T2222,BCW34,BCW36,BSW83,D33D26,EN2219, FT3643,GET2222,GI-3643,MPS6531,SE6020,SE6020A,TIS111,TN-3904,*PTC123, *GE-210,*HEPS0014,*SK3124
2SC367G-O	2N916,2N916A,2N3643,2N4013,2N4951,2N5107,2N5368,2N6004,2N6010,2N6426, 2N6427,2SC366G,2SC366G-O,2SC367G,2SC1318,2SC1852,40458,BCW34,BCW36, GET2222,GI-3643,ZT84,ZT114,*PTC170,*GE-210,*276-2016
2SC367G-R	2N708A,2N914,2N2221,2N2481,2N2501,2N2539,2N2710,2N2845,2N2847,2N3115, 2N3210,2N3211,2N3301,2N3510,2N3511,2N3641,2N3642,2N3646,2N3647,2N3648, 2N3973,2N3974,2N3975,2N3976,2N4400,2N4962,2N5144,2SC577,BSW82,FT3641, FT3642,GET2221,GI-3641,GI-3642,MPS6530,TIS110,TN-3903,ZT83,ZT113,ZT708, *PTC170,*GE-210,*276-2016

To Replace	Substitute This Type
2SC367G-Y	2N2222,2N2540,2N3116,2N3227,2N3242A,2N3302,2N4401,2N4952,2N5369,2N5377, 2N5818,2SC366G-Y,A5T2222,BSW83,D33D26,EN2219,FT3643,MPS6531,SE6020, SE6020A,TIS109,TIS111,TN-3904,*PTC170,*GE-210,*276-2016
2SC368-BL	BC548C,BC549C,*TR-21,*WEP736
2SC368-GR	2N2925,2N4256,2N4954,2N5371,2N6002,2N6008,2SC735-GR,2SC1993,BC108B, BC109B,BC172B,BC173B,BC238B,BC239B,BC548B,BC549B,BCY58C,*TR-21, *WEP736
2SC369-BL	BC548C,BC549C,*TR-21,*WEP736
2SC369-GR	2N2925,2N4256,2N4954,2N5371,2N6002,2N6008,2SC373G,2SC735-GR,2SC1335, 2SC1344,BC108C,BC109,BC109C,BC172C,BC173C,BC238C,BC239C,BC548B, BC549B,BCY58C,BCY58D,*TR-21,*WEP736,*RT-102
2SC369-V	2N5526,*TR-21,*WEP736
2SC369G-BL	BC548C,BC549C,*PTC139,*GE-212,*TR-51,*ECG123A,*WEP736,*276-2015,*RT-102
2SC369G-GR	2N2925,2N4256,2N4954,2N5371,2N6002,2N6008,2SC373G,2SC735-GR,2SC1993, BC108B,BC109B,BC172B,BC173B,BC238B,BC239B,BC548B,BC549B,BCY58C, *PTC139,*GE-212,*TR-21,*HEPS0020,*ECG123A,*WEP736,*276-2011,*RT-102
2SC370	2N783,2N2368,2N2710,2N3261,2N3510,2N3605A,2N3606A,2N3648,2SC54,2SC302, 2SC321,2SC322,2SC394-O,2SC468,2SC576,2SC601,2SC689,2SC752G-R,40405, EN914,EN2369A,EN3009,EN3013,EN3014,GET708,GET914,GET3013,GET3014, GET3646,KT218,MPS6532,MPSH20,MPSH37,TIS45,TIS46,TIS49,TIS52,TIS55, ZT2368,*PTC101,*GE-212,*TR-21,*HEPS0011,*SK3122,*EN10,*ECG108,*WEP736, *BC107B,*276-2016,*RT-302
2SC370G	2N783,2N2710,2N3261,2N3510,2N3605A,2N3606A,2N3648,2SC54,2SC302,2SC321, 2SC322,2SC468,2SC576,2SC601,2SC689,2SC752G-R,40405,EN914,EN3013, EN3014,GET708,GET914,GET3013,GET3014,GET3646,MPS6532,MPSH37,TIS45, TIS46,TIS52,TIS55,*PTC101,*GE-212,*TR-21,*HEPS0025,*SK3018,*ECG108, *WEP736,*276-2013,*RT-102
2SC370T	2SC302-M,2SC307-M,2SC590N,2SC731,2SC844,2SC854,2SC855,2SC872,2SC890, 2SC991,2SC992,2SC1002,2SC1083,2SC1256,2SC1303,2SC1315,2SC1464, 2SC1479,2SC1820,40637A,*PTC121,*GE-212,*SK3018
2SC371-O	2N916,2N916A,2N4013,2N4873,2SC372-O,2SC372G-O,2SC394-Y,2SC536,2SC735-O, 2SC752G-O,2SC933,2SC965,2SC1216,2SC1276,2SC1999,BC107A,BC167A, BC171A,BC237A,BCY58A,BCY59A,EN916,MPS6513,MPS6574-BLUE, MPS6574-GREEN,MPS6576-BLUE,MPS6576-GREEN,ZT82,ZT112,ZT2369A, *PTC101,*GE-20,*TR-21,*HEPS0011,*SK3122,*ECG123A,*WEP371,*276-2009
2SC371-R	2N708,2N708A,2N784A,2N914,2N957,2N2242,2N2369,2N2369A,2N2481,2N2501, 2N2710,2N3009,2N3013,2N3014,2N3210,2N3211,2N3510,2N3511,2N3605A, 2N3606A,2N3646,2N3647,2N3648,2N3688,2N3689,2N3690,2N3691,2N3693, 2N3862,2N4123,2N4137,2N4295,2N4420,2N4421,2N4422,2N5029,2N5144,2SC67, 2SC169,2SC335,2SC394-O,2SC468,2SC577,2SC601,2SC601N,2SC620,2SC639, 2SC752G-R,2SC764,2SC1274,BSW82,EN708,EN914,EN2369A,EN3009,EN3013, EN3014,GET708,GET914,GET2369,GET3013,GET3014,GET3646,MPS2369, MPS3646,MPS3693,MPS6512,MPS6574-YEL,MPS6576-YEL,PET1001,PET2001, PT2760,SE1001,SE2001,SE5001,SE5002,SE5003,SE5006,TIS45,TIS46,TIS48,TIS49, TIS52,TIS55,ZT81,ZT111,ZT708,ZT2369,*PTC101,*TR-86,*ECG123A,*WEP371
2SC371G-O	2N916,2N916A,2N4013,2SC372G-O,2SC536,2SC735-O,2SC752G-O,2SC933, 2SC1216,2SC1276,2SC1999,BC107A,BC167A,BC171A,BC237A,BCY58A,BCY59A, EN916,*PTC101,*GE-212,*276-2016
2SC371G-R	2N708,2N708A,2N784A,2N914,2N957,2N2242,2N2481,2N2501,2N2710,2N3013, 2N3014,2N3210,2N3211,2N3510,2N3511,2N3605A,2N3606A,2N3646,2N3647, 2N3648,2N4123,2N4295,2N4420,2N4421,2N4422,2N5144,2SC67,2SC169,2SC335, 2SC468,2SC577,2SC601,2SC601N,2SC620,2SC639,2SC752G-R,2SC764,2SC1274, BSW82,EN708,EN914,EN3013,EN3014,GET708,GET914,GET2369,GET3013, GET3014,GET3646,MPS3646,TIS45,TIS46,TIS52,TIS55,ZT708,*PTC101,*GE-212, *276-2016
2SC371T	2SC302-M,2SC307-M,2SC590N,2SC731,2SC844,2SC854,2SC855,2SC872,2SC890, 2SC991,2SC992,2SC1002,2SC1083,2SC1256,2SC1303,2SC1315,2SC1464, 2SC1479,2SC1820,40637A,*PTC121,*GE-212,*SK3018
2SC372-O	2N916,2N916A,2N4013,2N4873,2SC372G-O,2SC394-Y,2SC536,2SC735-O, 2SC752G-O,2SC933,2SC965,2SC1216,2SC1276,2SC1999,BC107A,BC167A, BC171A,BC237A,BCY58A,BCY59A,EN916,MPS6513,MPS6574-BLUE,

To Replace	Substitute This Type
(2SC372-O)	MPS6574-GREEN,MPS6576-BLUE,MPS6576-GREEN,ZT82,ZT112,ZT2369A,*PTC101, *HEPS0015,*SK3018,*ECG123A,*WEP372,*276-2009,*RT-102
2SC372-Y	2N3227,2N3242A,2N3692,2N3694,2N4970,2SC372G-Y,2SC373,2SC373G,2SC394-GR, 2SC735-GR,2SC735-Y,2SC752G-Y,2SC1175,2SC1992,A157B,BC107,BC107B, BC167B,BC171B,BC237B,BC413B,BC414B,BC547A,BCY58B,BCY59B,BSW83, BSW88,BSW89,BSX38,BSX79,MPS3694,MPS6514,MPS6520,MPS6573, MPS6574-SIL,MPS6575,MPS6576-SIL,PET1002,PET2002,SE1002,SE2002,*PTC101, *TR-24,*HEPS0015,*ECG123A,*WEP372,*276-2031,*RT-102
2SC372G-O	2N916,2N916A,2N4013,2SC536,2SC735-O,2SC752G-O,2SC933,2SC1216,2SC1276, 2SC1999,BC107A,BC167A,BC171A,BC237A,BCY58A,BCY59A,EN916,*PTC101, *GE-212,*276-2016
2SC372G-Y	2N3227,2N3242A,2N4970,2SC373G,2SC735-GR,2SC735-Y,2SC752G-Y,2SC1175, 2SC1992,A157B,BC107,BC107B,BC167B,BC171B,BC237B,BC413B,BC414B, BC547A,BCY58B,BCY59B,BSW83,BSW88,BSW89,BSX38,BSX79,*PTC101, *GE-212,*276-2016
2SC373	2N4954,2N5371,2N5827,2SC373G,2SC735-GR,2SC1345,A157B,BC413C,BC414C, BC547B,BC550B,BCY58C,BCY58D,BCY59C,BCY59D,MPS6515,MPS6520,MPS6521, *PTC121,*GE-212,*IR2SC373,*HEPS0015,*SK3122,*EN10,*ECG199,*WEP736, *BC107B,*276-2016,*RT-105
2SC373G	2N4954,2N5371,2N5827,2SC735-GR,2SC1345,A157B,BC413C,BC414C,BC547B, BC550B,BCY58C,BCY58D,BCY59C,BCY59D,*PTC121,*GE-212,*TR-21,*HEPS0014, *SK3122,*ECG199,*WEP736,*276-2016,*RT-102
2SC374	2N5526,2SC1312Y,2SC1313Y,BC550C,*GE-212,*TR-33,*HEPS0015,*SK3018,*EN30, *ECG199,*WEP736,*BC107B,*276-2016,*RT-102
2SC375	*PTC132,*HEPS0016,*SK3039,*ECG311,*WEP56
2SC376	2N915,2N4014,2N4963,2SC734-O,2SC979-O,2SC979A-O,2SC980AG-O,2SC980G-O, 2SC1211,2SC1998,EN915,ZT88,ZT118,*PTC101,*GE-212,*TR-87,*HEPS5026, *SK3024,*EN10,*ECG128,*WEP243,*BF338,*276-2009,*RT-114
2SC376G	SEE 2SC780AG
2SC377-BN	2N783,2N784,2N2318,2N2319,2N2368,2N3605A,2N3606A,2N4294,2N5030,2SC54, 2SC204,2SC302,2SC321,2SC322,2SC356,2SC380-R,2SC380A-R,2SC394-O, 2SC400-R,2SC455,2SC468,2SC595,2SC601,2SC689,2SC724,2SC752G-R,BF224J, BF225J,BFY39-1,EN914,EN2369A,EN3009,EN3011,EN3013,EN3014,GET708, GET914,GET3013,GET3014,GET3646,KT218,MPS6540,MPSA20-RED,MPSH20, MPSH37,MT1038,MT1038A,MT1039,MT1060,PET3001,TIS45,TIS46,TIS49,TIS51, TIS52,TIS55,TIS84,TIS129,*PTC101,*GE-86,*276-2011
2SC377-O	2N916,2N916A,2N3241A,2N3856A,2N4873,2SC367G,2SC367G-O,2SC372-O, 2SC372G-O,2SC380-O,2SC380A-O,2SC394-Y,2SC400-Y,2SC454,2SC458,2SC459, 2SC536,2SC619,2SC735-O,2SC752G-O,2SC899,2SC925,2SC933,2SC965, 2SC1085,2SC1216,2SC1276,2SC1359,2SC1682-BL,2SC1682-GR,2SC1999,BC107A, BC108A,BC167A,BC171A,BC237A,BC238A,BCY58A,BCY59A,BF254,BFY39-2, EN916,MPS6513,MPS6574-BLUE,MPS6574-GREEN,MPS6576-BLUE, MPS6576-GREEN,MPSA20-GRN,MPSA20-WHT,PET3002,TIS86,*PTC101,*GE-60, *276-2011
2SC377-R	2N708,2N708A,2N784A,2N957,2N2242,2N2318,2N2319,2N2331,2N2369,2N2369A, 2N2501,2N3009,2N3011,2N3013,2N3014,2N3564,2N3605A,2N3606A,2N3646, 2N3688,2N3689,2N3690,2N3691,2N3693,2N3855A,2N3862,2N4123,2N4137, 2N4294,2N4295,2N4418,2N4420,2N4421,2N4422,2N4996,2N4997,2N5029,2N5030, 2SC67,2SC169,2SC204,2SC318,2SC335,2SC356,2SC380-R,2SC380A-R,2SC394-O, 2SC400-O,2SC460,2SC461,2SC468,2SC595,2SC601,2SC601N,2SC620,2SC639, 2SC710,2SC712,2SC713,2SC724,2SC752G-R,2SC764,2SC838,2SC839,2SC924, 2SC963,2SC964,2SC1188,2SC1189,2SC1274,2SC1320,2SC1372,2SC1687, 2SC1688,BF123,BF125,BF255,BF311,BFY39-1,EN708,EN914,EN2369A,EN3009, EN3011,EN3013,EN3014,GET708,GET914,GET2369,GET3013,GET3014,GET3646, MPS3646,MPS3693,MPS6512,MPS6574-YEL,MPS6576-YEL,MPSA20-RED,MT1060A, PET1001,PET2001,PT2760,SE1001,SE2001,SE3005,SE5001,SE5002,SE5003, SE5006,TIS45,TIS46,TIS48,TIS49,TIS51,TIS52,TIS55,TIS87,TIS129,ZT708,*PTC101, *GE-86,*276-2011
2SC378-O	2N916,2N916A,2N4873,2SC367G,2SC367G-O,2SC372-O,2SC372G-O,2SC380-O, 2SC380A-O,2SC394-Y,2SC536,2SC735-O,2SC752G-O,2SC899,2SC933,2SC965, 2SC1216,2SC1276,2SC1453,2SC1682-BL,2SC1682-GR,2SC1999,BC107A,BC167A,

To Replace	Substitute This Type
(2SC378-O)	BC171A,BC237A,BCY58A,BCY59A,BFY39-2,EN916,MPS6513,MPS6574-BLUE, MPS6574-GREEN,MPS6576-BLUE,MPS6576-GREEN,MPSA20-GRN,MPSA20-WHT, *PTC101,*GE-60,*276-2011
2SC378-R	2N708,2N708A,2N784A,2N957,2N2242,2N2369,2N2369A,2N2501,2N3009,2N3013, 2N3014,2N3605A,2N3606A,2N3646,2N3688,2N3689,2N3690,2N3691,2N3693, 2N3862,2N4123,2N4137,2N4295,2N4420,2N4421,2N4422,2N5029,2SC67,2SC169, 2SC318,2SC335,2SC380-R,2SC380A-R,2SC394-O,2SC468,2SC601,2SC601N, 2SC620,2SC639,2SC752G-R,2SC764,2SC838,2SC839,2SC963,2SC964,2SC1188, 2SC1189,2SC1274,2SC1320,2SC1687,2SC1688,BF311,BFY39-1,EN708,EN914, EN2369A,EN3009,EN3013,EN3014,GET708,GET914,GET2369,GET3013,GET3014, GET3646,MPS3646,MPS3693,MPS6512,MPS6574-YEL,MPS6576-YEL,MPSA20-RED, PET1001,PET2001,PT2760,SE1001,SE2001,SE5001,SE5002,SE5003,SE5006,TIS45, TIS46,TIS48,TIS49,TIS52,TIS55,TIS87,TIS129,ZT708,*PTC101,*GE-86,*276-2011
2SC378-Y	2N3242A,2N3692,2N3694,2SC367G-Y,2SC372-Y,2SC372G-Y,2SC373,2SC373G, 2SC380-Y,2SC380A-Y,2SC394-GR,2SC711A,2SC735-GR,2SC735-Y,2SC752G-Y, 2SC1084,2SC1175,2SC1285,2SC1364,2SC1992,A157B,BC107,BC107B,BC123, BC167B,BC171B,BC237B,BC413B,BC414B,BC547A,BCY58B,BCY59B,BFY39-3, BSW88,BSW89,BSX38,BSX79,MPS3694,MPS6514,MPS6520,MPS6573, MPS6574-SIL,MPS6575,MPS6576-SIL,MPS-A20,MPS-K20,MPS-K21,MPS-K22, MPSA20-BLU,MPSA20-YEL,NPSA20,PBC107A,PBC107B,PET1002,PET2002, SE1002,SE2002,*PTC101,*GE-211,*276-2011
2SC379	*PTC132,*HEPS0016,*SK3122,*ECG123A,*WEP736
2SC380-O	2N4873,2SC380A-O,2SC752G-O,2SC1216,2SC1276,*PTC132,*GE-60,*TR-70, *HEPS0016,*SK3018,*ECG229,*276-2011,*RT-108
2SC380-R	2N2369,2N2369A,2N2501,2N3009,2N3013,2N3014,2N3646,2N3688,2N3689,2N3690, 2N3862,2N4137,2N4295,2N4420,2N4421,2N4422,2N5029,2SC67,2SC169,2SC335, 2SC380A-R,2SC468,2SC601,2SC601N,2SC639,2SC752G-R,2SC764,2SC1188, 2SC1189,2SC1274,2SC1320,2SC1687,2SC1688,BF311,EN2369A,EN3009,EN3013, EN3014,GET2369,GET3013,GET3014,GET3646,MPS3646,SE5001,SE5002, SE5003,SE5006,TIS48,TIS49,TIS52,TIS55,TIS87,TIS129,*PTC132,*GE-86,*SK3018, *276-2011,*RT-108
2SC380-Y	2SC380A-Y,2SC752G-Y,MPS6514,MPS6520,*PTC132,*GE-211,*TR-70,*WEP56, *276-2011,*RT-108
2SC380A-O	2N4873,2SC380-O,2SC752G-O,2SC1216,2SC1276,*GE-61,*TR-24,*HEPS0016, *SK3018,*ECG229,*WEP380,*276-2015
2SC380A-R	2N2369,2N2369A,2N2501,2N3009,2N3013,2N3014,2N3646,2N3688,2N3689,2N3690, 2N3862,2N4137,2N4295,2N4420,2N4421,2N4422,2N5029,2SC67,2SC169,2SC335, 2SC380-R,2SC468,2SC601,2SC601N,2SC639,2SC752G-R,2SC764,2SC1188, 2SC1189,2SC1274,2SC1320,2SC1687,2SC1688,BF311,EN2369A,EN3009,EN3013, EN3014,GET2369,GET3013,GET3014,GET3646,MPS3646,SE5001,SE5002, SE5003,SE5006,TIS48,TIS49,TIS52,TIS55,TIS87,TIS129,*GE-20,*TR-24,*HEPS0016, *SK3039,*ECG107,*WEP56,*276-2015,*RT-108
2SC380A-Y	2SC380-Y,2SC752G-Y,MPS6514,MPS6520,*TR-21
2SC381-BN	2N2368,2N3605A,2N3606A,2SC54,2SC302,2SC321,2SC322,2SC381-R,2SC468, 2SC601,2SC689,2SC725,2SC752G-R,2SC784-BN,2SC784-R,2SC785-BN,2SC785-R, BF224J,BF225J,EN914,EN2369A,EN3009,EN3013,EN3014,GET708,GET914, GET3013,GET3014,GET3646,KT218,MPSH20,MPSH37,TIS45,TIS46,TIS49,TIS52, TIS55,TIS129,*PTC132,*GE-60,*TR-33,*WEP56,*276-2011,*RT-108
2SC381-O	2N916,2N916A,2N3947,2N4141,2N4227,2N4873,2N5028,2SC752G-O,2SC784-O, 2SC785-O,2SC899,2SC933,2SC943,2SC944,2SC1007,2SC1216,2SC1276,BC107A, BC167A,BC171A,BC237A,BF240,EN916,GET2222,MM3904,MPS6513,*PTC132, *GE-60,*TR-70,*SK3018,*ECG107,*WEP56,*276-2011,*RT-108
2SC381-R	2N708,2N708A,2N784A,2N2242,2N2369,2N2369A,2N2501,2N2845,2N2847,2N2885, 2N3009,2N3013,2N3014,2N3301,2N3605A,2N3606A,2N3688,2N3689,2N3690, 2N3862,2N3903,2N3946,2N4123,2N4137,2N4140,2N4295,2N4420,2N4421,2N4422, 2N5027,2N5029,2SC67,2SC169,2SC335,2SC336,2SC468,2SC601,2SC601N, 2SC620,2SC639,2SC725,2SC752G-R,2SC764,2SC784-R,2SC785-R,2SC838, 2SC839,2SC1187,2SC1188,2SC1189,2SC1274,2SC1319,2SC1320,2SC1675, 2SC1686,2SC1923,BF198,BF199,BF241,EN708,EN914,EN2369A,EN3009,EN3013, EN3014,GET708,GET914,GET2221,GET2369,GET3013,GET3014,GET3646,MM3903, MPS3646,MPS6512,SE5001,SE5002,SE5003,SE5006,TIS45,TIS46,TIS48,TIS49,

To Replace	Substitute This Type
(2SC381-R)	TIS52,TIS55,TIS129,ZT708,*PTC132,*GE-60,*TR-70,*ECG107,*WEP56,*276-2011, *RT-108
2SC382	2N834A,2N2368,2SC576,2SC689,TIS84,ZT2368,*PTC132,*GE-39,*TR-71,*HEPS0016, *SK3018,*EN10,*ECG319,*WEP56,*BC107B,*276-2016,*RT-107A
2SC382-G	2N834A,2N2368,2SC382,2SC576,2SC689,TIS84,ZT2368,*PTC132,*TR-33,*EN10, *WEP56,*276-2011,*RT-108
2SC382-R	2N834A,2N2368,2SC382,2SC576,2SC689,TIS84,ZT2368,*PTC132,*TR-33,*EN10, *WEP56,*276-2011,*RT-108
2SC383	2N4963,2SC979-R,2SC979A-R,BSW84,FT3722,GET2221A,SE8012,ZT86,*PTC115, *GE-61,*TR-24,*HEPS5026,*SK3018,*ECG319,*WEP736,*BC107B,*276-2016, *RT-107A
2SC384	*PTC132,*HEPS0016,*SK3018,*ECG108,*WEP56
2SC385	2N3663,2N4292,2N4293,2SC717,2SC837,D16G6,EN918,MPS918,MPS6542,MPS6543, MPS6546,MPS6547,MPS6548,MT1038,MT1038A,MT1039,MT1060,TIS18,*PTC132, *GE-86,*TR-70,*HEPS0016,*SK3039,*ECG107,*WEP56,*BF173,*276-2015,*RT-108
2SC385A	2N4292,2N4293,2SC387A,D16G6,MPS918,MPS6546,MPS6547,MPSH08,*GE-11, *TR-70,*HEPS0016,*SK3039,*ECG107,*WEP56,*276-2015,*RT-108
2SC386	2N3663,2N4292,2N4293,2SC717,2SC837,D16G6,EN918,MPS918,MPS6542,MPS6543, MPS6546,MPS6547,MPS6548,MT1038,MT1038A,MT1039,MT1060,TIS18,*PTC132, *GE-86,*TR-95,*HEPS0016,*SK3039,*ECG108,*WEP56,*BF173,*276-2015,*RT-113
2SC386A	2N706,2N706A,2N706B,2N743,2N747,2N784,2N947,2N959,2N988,2N989,2N2205, 2N2318,2N2319,2N2432,2N2656,2N2729,2N2921,2N3340,2N3544,2N3663,2N3844, 2N3844A,2N3854A,2N3983,2N4294,2N5030,2SC55,2SC63,2SC68,2SC98,2SC132, 2SC133,2SC135,2SC137,2SC204,2SC356,2SC380-R,2SC380A-R,2SC387AG, 2SC394-O,2SC400-R,2SC455,2SC467,2SC595,2SC718,2SC719,2SC724,2SC726, 2SC812,2SC837,2SC941-R,2SC1126,2SC1231,2SC1293,BC170A,BF255,EN706, EN918,EN3011,GI-2921,MM1941,MPS6507,MPS6511,MPSH19,MT1038,MT1038A, MT1039,MT1060,PET3001,PT2760,TIS18,TIS44,TIS51,ZT41,*PTC132,*HEPS0033, *SK3039
2SC387	2SC387AG,2SC717,2SC718,2SC719,2SC1231,MT1038,MT1038A,MT1039,MT1060, SE3001,*PTC132,*GE-61,*TR-70,*HEPS0016,*SK3039,*ECG316,*WEP56, *276-2011,*RT-134
2SC387A	*PTC101,*GE-11,*TR-70,*HEPS0016,*SK3122,*ECG316,*WEP56,*276-2011,*RT-108
2SC387AG	MT1060A
2SC387G	2SC387A,*PTC132,*GE-86,*TR-33,*HEPS0016,*SK3039,*ECG316,*WEP56, *276-2011,*RT-108
2SC388	*PTC132,*SK3018,*ECG233,*WEP56,*RT-308
2SC388A	2N2615,2N2616,2N3227,2N4873,2SC735-Y,2SC1216,A5T3571,A5T3572,BC108A, BC172A,BC238A,BCY58A,MPS6513,TIS86,ZT87,ZT2369A,ZT2938,*PTC115,*GE-61, *TR-70,*HEPS0016,*SK3018,*ECG233,*WEP56,*BC107B,*276-2016,*RT-108
2SC389	2N2865,2N3571,2N3572,2N4252,2SC390,2SC392,2SC568,2SC653,2SC927,2SC1035, 2SC1036,2SC1180,2SC1331,2SC1553A,*PTC132,*GE-60,*TR-83,*HEPS0017, *SK3117,*ECG161,*WEP719,*BF200,*276-2015,*RT-113
2SC390	2N4252,2SC568,2SC1180,40235,40236,40237,40238,40240,40472,40474,*GE-211, *TR-83,*SK3039,*ECG161,*WEP719,*RT-113
2SC391	2SC1180,*PTC132,*GE-214,*HEPS0017,*SK3039,*276-2011
2SC392	2SC1180,40236,40237,40240,40474,40477,40480,TC3114,*PTC126,*GE-39,*TR-95, *HEPS0017,*SK3039,*ECG108,*WEP56,*276-2011,*RT-113
2SC392A	*GE-86,*SK3039
2SC393	
2SC394-GR	2N3227,2N3242A,2N3692,2N3694,2N4970,2SC372-Y,2SC372G-Y,2SC373,2SC373G, 2SC735-GR,2SC735-Y,2SC752G-Y,2SC1175,2SC1992,A157B,BC107,BC107B, BC167B,BC171B,BC237B,BC413B,BC414B,BC547A,BCY58B,BCY59B,BSW83, BSW88,BSW89,BSX38,BSX79,MPS3694,MPS6514,MPS6520,MPS6573, MPS6574-SIL,MPS6575,MPS6576-SIL,PET1002,PET2002,SE1002,SE2002,*PTC132, *GE-212,*WEP394,*276-2011
2SC394-O	2N708,2N708A,2N784A,2N914,2N957,2N2242,2N2369,2N2369A,2N2481,2N2501, 2N2710,2N3009,2N3013,2N3014,2N3210,2N3211,2N3510,2N3511,2N3605A, 2N3606A,2N3646,2N3647,2N3648,2N3688,2N3689,2N3690,2N3691,2N3693, 2N3862,2N4123,2N4137,2N4295,2N4420,2N4421,2N4422,2N5029,2N5144,2SC67, 2SC169,2SC335,2SC468,2SC577,2SC601,2SC601N,2SC620,2SC639,2SC752G-R,

To Replace	Substitute This Type
(2SC394-O)	2SC764,2SC1274,BSW82,EN708,EN914,EN2369A,EN3009,EN3013,EN3014,GET708, GET914,GET2369,GET3013,GET3014,GET3646,MPS2369,MPS3646,MPS3693, MPS6512,MPS6574-YEL,MPS6576-YEL,PET1001,PET2001,PT2760,SE1001,SE2001, SE5001,SE5002,SE5003,SE5006,TIS45,TIS46,TIS48,TIS49,TIS52,TIS55,ZT81, ZT111,ZT708,ZT2369,*PTC132,*GE-212,*TR-33,*SK3018,*ECG229,*WEP394, *276-2011,*RT-108
2SC394-R	2N783,2N784,2N2318,2N2319,2N2368,2N2710,2N3261,2N3510,2N3605A,2N3606A, 2N3648,2N4294,2N5030,2SC54,2SC204,2SC302,2SC321,2SC322,2SC356, 2SC394-O,2SC400-R,2SC455,2SC468,2SC576,2SC595,2SC601,2SC689,2SC724, 2SC752G-R,40405,EN914,EN2369A,EN3009,EN3011,EN3013,EN3014,GET708, GET914,GET3013,GET3014,GET3646,KT218,MPS6532,MPS6540,MPSH20,MPSH37, MT1038,MT1038A,MT1039,PET3001,TIS45,TIS46,TIS49,TIS51,TIS52,TIS55,ZT2368, *PTC132,*GE-212,*WEP394,*276-2011
2SC394-Y	2N916,2N916A,2N4013,2N4873,2SC372-O,2SC372G-O,2SC536,2SC735-O, 2SC752G-O,2SC933,2SC965,2SC1216,2SC1276,2SC1999,BC107A,BC167A, BC171A,BC237A,BCY58A,BCY59A,EN916,MPS6513,MPS6574-BLUE, MPS6574-GREEN,MPS6576-BLUE,MPS6576-GREEN,ZT82,ZT112,ZT2369A, *PTC132,*GE-212,*SK3122,*WEP394,*276-2011
2SC395	2N744,2N784,2N988,2N989,2N2206,2N2318,2N2319,2N2571,2N2572,2N2656, 2N3011,2N5030,2N5187,2N5418,2SC132,2SC133,2SC135,2SC137,2SC204,2SC356, 2SC395A-R,2SC579,2SC595,2SC718,2SC915,2SC915A,2SC1071,2SC1231, MM1941,MPS6568,MPSH02,MPSH30,MPSH31,PT720,PT2760,SE8040,TIS51,TN64, ZT404P,ZT2205,*PTC133,*HEPS0011,*SK3122,*ECG123A,*WEP736
2SC395A	2SC395A-Y,ZT2938,*PTC123,*GE-210,*TR-51,*HEPS0011,*SK3122,*ECG123A, *WEP736,*RT-102
2SC395A-O	2SC395A,2SC395A-Y,ZT2938,*PTC136,*GE-210,*276-2016
2SC395A-R	2N5187,*PTC136,*GE-20,*276-2016
2SC395A-Y	*PTC136,*GE-210,*276-2016
2SC396	*PTC136,*HEPS0015,*SK3122,*ECG123A,*WEP736
2SC397	2N6599,2N6600,2N6601,2SC313,2SC464,2SC465,2SC466,A406,MT1061,*PTC132, *GE-86,*TR-95,*HEPS0017,*SK3039,*ECG108,*276-2011,*RT-113
2SC398	2N2865,2N3570,2N3571,2N3572,2N4252,2N4253,2N6304,2N6305,2SC399,2SC568, 2SC653,*PTC132,*GE-11,*TR-70,*HEPS0017,*SK3018,*ECG161,*WEP719,*BF173, *276-2011,*RT-113
2SC399	2N2865,2N3570,2N3571,2N3572,2N4252,2N4253,2N6304,2N6305,2SC398,2SC568, 2SC653,*PTC132,*GE-11,*TR-83,*HEPS0017,*SK3018,*ECG161,*WEP719,*BF173, *276-2011,*RT-113
2SC400	2N4124,2N4419,2SC400-GR,2SC735-GR,2SC1992,2SC1993,A157B,BC107,BC107B, BC108,BC108B,BC109B,BC167B,BC171B,BC237B,BC238B,BC239B,BC413B, BC414B,BC547A,BC548A,BCY58B,BCY59B,BSW83,*PTC132,*GE-17,*TR-51, *HEPS0011,*SK3122,*EN10,*ECG123A,*WEP736,*276-2015,*RT-102
2SC400-GR	2N4954,2N5371,2N5827,2SC735-GR,2SC1993,A157B,BC107B,BC108B,BC109B, BC167B,BC171B,BC237B,BC238B,BC239B,BC413B,BC414B,BC547B,BC548B, BC549B,BC550B,BCY58C,BCY59C,*TR-21,*ECG199,*WEP736
2SC400-O	2N916,2N916A,2N2501,2N4013,2N4123,2N4418,2N5144,2SC400-Y,2SC619,2SC620, 2SC639,2SC735-O,2SC933,2SC1216,2SC1999,*TR-21,*WEP736
2SC400-R	2N708,2N708A,2N784A,2N914,2N2242,2N2318,2N2319,2N2481,2N2710,2N3011, 2N3013,2N3014,2N3210,2N3261,2N3510,2N3511,2N3605A,2N3606A,2N3646, 2N3647,2N3648,2SC54,2SC67,2SC204,2SC302,2SC335,2SC577,2SC595,2SC601, 2SC601N,2SC689,2SC1274,BSW82,GET708,GET914,GET2369,GET3013,GET3014, GET3646,TIS45,TIS46,TIS51,TIS52,TIS55,ZT708,*TR-21,*ECG123A,*WEP736
2SC400-Y	2N3227,2N4124,2N4419,2SC400,2SC735-Y,2SC933,2SC1992,BC107,BC107A,BC108, BC108A,BC167A,BC171A,BC237A,BC238A,BC547A,BC548A,BCY58A,BCY58B, BCY59A,BCY59B,BSW83,*TR-21,*ECG123A,*WEP736
2SC401	2N915,2N2847,2N3903,2N3974,2N3976,2N4013,2N4227,2N4400,2N4951,2N5144, 2SC318,2SC366G,2SC366G-O,2SC402,2SC620,2SC734-O,2SC933,2SC943, 2SC944,2SC979-O,2SC980G-O,2SC1453,EN915,MPS6565,ZT84,ZT114,*PTC132, *GE-212,*TR-21,*HEPS0015,*SK3018,*EN30,*ECG123A,*WEP736,*BC107B, *276-2016,*RT-102
2SC402	2N915,2N2847,2N3903,2N3974,2N3976,2N4013,2N4227,2N4400,2N4951,2N5144, 2SC318,2SC366G,2SC366G-O,2SC401,2SC620,2SC734-O,2SC933,2SC943,

To Replace	Substitute This Type
(2SC402)	2SC944,2SC979-O,2SC980G-O,2SC1453,EN915,MPS6565,ZT84,ZT114,*PTC132, *GE-212,*TR-21,*HEPS0015,*SK3124,*EN10,*ECG123A,*WEP736,*BC107B, *276-2016,*RT-102
2SC402A	2N915,2N2847,2N3903,2N3974,2N3976,2N4013,2N4227,2N4400,2N4951,2N5144, 2SC318,2SC366G,2SC366G-O,2SC620,2SC734-O,2SC933,2SC943,2SC944, 2SC979-O,2SC980G-O,2SC1453,EN915,MPS6565,ZT84,ZT114,*PTC132, *HEPS0015,*SK3122,*ECG123A,*WEP736
2SC403	2N708A,2N2845,2N2847,2N3301,2N3862,2N3903,2N3946,2N3973,2N3974,2N3975, 2N3976,2N4140,2N4400,2N4436,2N4969,2N5027,2N5144,2SC318,2SC335,2SC336, 2SC401,2SC402,2SC473,2SC620,2SC714,2SC725,2SC815,2SC896,2SC979-R, 2SC980,2SC980G-R,EN915,GET2221,GET2221A,GI-3641,GI-3642,MM3903, MPS6530,MPS6565,ZT83,ZT113,*PTC132,*GE-212,*IR2SC403A,*HEPS0015, *SK3126,*SN80,*ECG123A,*WEP736,*BC107B,*276-2016,*RT-302
2SC403A	2N708A,2N2845,2N2847,2N3115,2N3301,2N3862,2N3903,2N3946,2N3973,2N3974, 2N3975,2N3976,2N4140,2N4400,2N4436,2N4969,2N5027,2N5144,2SC318,2SC335, 2SC336,2SC473,2SC620,2SC714,2SC725,2SC815,2SC896,2SC979-R,2SC980, 2SC980G-R,EN915,GET2221,GET2221A,GI-3641,GI-3642,MM3903,MPS6530, MPS6565,ZT83,ZT113,*PTC132,*HEPS0015,*SK3124,*ECG123A,*WEP736
2SC403C	*HEPS0015,*SK3122,*ECG123A,*WEP736
2SC404	2N915,2N2847,2N3877,2N3903,2N3974,2N3976,2N4013,2N4227,2N4951,2N5144, 2SC318,2SC366G,2SC366G-O,2SC401,2SC402,2SC620,2SC734-O,2SC899, 2SC933,2SC943,2SC944,2SC979-O,2SC980G-O,2SC1453,EN915,MPS6565,ZT84, ZT114,*PTC132,*GE-61,*TR-21,*HEPS0015,*SK3018,*EN10,*ECG123A,*WEP736, *BC107B,*276-2016,*RT-102
2SC405	2N357A,2N358A,2N377A,2N385,2N440,2N440A,2N1000,2N1012,2N1114,2N1299, 2N1302,2N1304,2N1306,2N1473,2N2426,2SC129,2SD11,ASY28,ASY29,NKT713, SF.T298,*HEPG0011,*SK3039,*ECG108,*WEP56
2SC406	2N385A,2N388,2N388A,2N634A,2N1306,2N1308,2N1624,2N1993,ASY29,NKT713, *HEPG0011,*SK3039,*ECG108,*WEP56
2SC407	2N3240,2SC409,2SC1440,2SC1441,2SC1784,2SC1785,2SD165,2SD166,1561-1404, 1561-1405,1561-1604,1561-1605,1561-1803,1561-1804,1561-1805,1582-1403, 1582-1404,1582-1405,1582-1603,1582-1604,1582-1605,1582-1803,1582-1804, 1582-1805,HST9209,M10C,SDT9209,STC2222,STC2223,*PTC118,*HEPS7004, *ECG162,*WEP707
2SC408	2N4348,2N5559,2N5631,2N5634,2SC410,2SC410A,2SC1115,2SC1116,2SC1116A, 2SC1782,2SC1783,2SD181,2SD375,2SD425-R,2SD426-R,HST9204,HST9209, IR3773,IR6302,KSP1273,SDT9204,SDT9209,*PTC118,*HEPS5020,*SK3024, *ECG163,*WEP740
2SC409	2SC1436,2SC1441,2SC1785,2SC1786,2SD166,1561-1803,1561-1804,1561-1805, 1561-2003,1561-2004,1582-1803,1582-1804,1582-1805,1582-2003,1582-2004, DTS-310,M10C,STC2223,XDAR10020,XGSR15020,*PTC118,*HEPS5020,*SK3111, *ECG163,*WEP740
2SC410	2SC410A,2SC1116A,2SC1229,2SC1322,2SD181A,2SD376,2SD552,DTS-311, KSP1221,KSP1222,*PTC118,*HEPS5020,*SK3111,*ECG163,*WEP740
2SC410A	2SC410,2SC1116A,2SC1229,2SC1322,2SD166,2SD181A,2SD376,2SD552,1561-1803, 1561-1804,1561-1805,1561-2003,1561-2004,1582-1803,1582-1804,1582-1805, 1582-2003,1582-2004,DTS-311,KSP1221,KSP1222,M10C,STC2223,XDAR10020, *SK3111
2SC411	2N6249,2N6250,2N6561,M10D,XDAR10025,XDAR10030,XGSR10025,XGSR10030, XGSR15025,XGSR15030,*PTC118,*HEPS5020,*SK3111,*ECG163,*WEP740
2SC412	2SC1227,2SD376A,IR515,KSP1223,KSP1224,KSP1225,*PTC118,*HEPS5020, *SK3111,*ECG163,*WEP740
2SC413	
2SC414	2SC415
2SC415	
2SC416	2SC916,2SC1266
2SC420	
2SC423	2SC651,*PTC136,*GE-20,*TR-87,*HEPS0014,*SK3024,*ECG123A,*WEP243, *276-2009,*RT-100A
2SC424	2SC577,MPS2369,ZT2369,ZT2369A,*PTC136,*GE-20,*TR-95,*HEPS0014,*SK3018, *ECG123A,*WEP56,*276-2016,*RT-100A

To Replace	Substitute This Type
2SC425	2N5065,*PTC136,*GE-20,*TR-87,*HEPS0011,*SK3024,*ECG123A,*WEP243, *276-2009,*RT-114
2SC426	2N3959,2N3960,2N5200,2N5201,SE3005,ZT2938,*PTC132,*GE-17,*TR-64, *HEPS0011,*276-2011
2SC427	2N916,2N916A,2N2242,2N2501,2N2847,2N3013,2N3014,2N3115,2N3211,2N3903, 2N3946,2N3974,2N3976,2N4013,2N4123,2N4227,2N4400,2N4951,2N4962,2N5027, 2N5144,2N5368,2N5380,2N6538,2SC67,2SC318,2SC366G-O,2SC367G-O,2SC577, 2SC601N,2SC639,2SC764,A5T3903,A5T4123,BSW82,MM3903,MPS6530,TN-3903, ZT84,ZT114,*PTC136,*GE-17,*TR-64,*276-2016
2SC428	2N744,2N2331,2N4418,2S095A,2S512,2SC99,2SC395A,2SC395A-O,2SC735-O, PT720,ZT2205,ZT2938,*PTC136,*GE-17,*TR-64,*276-2016
2SC429	2SC430,*GE-214,*TR-70,*HEPS0016,*SK3019,*EN10,*ECG108,*WEP56,*BF183, *276-2011,*RT-108
2SC430	2SC429,A3T918,MMT918,*GE-214,*TR-95,*HEPS0016,*SK3018,*EN10,*ECG108, *WEP56,*BF183,*276-2011,*RT-113
2SC431	2SC433,L30C
2SC432	2SC434
2SC433	L30C,*PTC136,*SK3124,*ECG123A,*WEP736
2SC434	2SC1301
2SC434A	2SC434,2SC1301,L30C
2SC435	L30D
2SC436	2SC1299,2SC1532
2SC437	*HEPS3010
2SC438	2SC437,*HEPS3010
2SC439	2N3646,2N5772,2SC52,2SC343
2SC440	2N3646,2N5772,2SC52,2SC343,*PTC136,*HEPS0014,*ECG123,*WEP53
2SC441	2N3646,2N5772,2SC52,2SC343,*PTC136,*HEPS0014,*SK3124,*ECG123,*WEP53
2SC442	2N3326,2SC344,*PTC136,*HEPS0014,*SK3124,*ECG123,*WEP53
2SC443	2N707A,2N3512,BSX59,BSX61,PT2540,*PTC125,*HEPS3008,*SK3024,*ECG128, *WEP243
2SC444	2N2217,2N3830,2N3831,2N5188,BC142,BSX49,BSX59,BSX61,PT2540,*PTC144, *HEPS3001
2SC445	*PTC144,*HEPS3019,*SK3104
2SC446	2N2217,2N3830,2N3831,2N5188,BC142,BSX49,BSX59,BSX61,PT2540
2SC447	
2SC448	
2SC449	
2SC450	
2SC451	
2SC452	
2SC453	
2SC454	2N3227,2N4124,2N4419,2N4970,2SC372G-Y,2SC400,2SC400-Y,2SC735-Y, 2SC752G-Y,2SC933,2SC1085,2SC1317,2SC1992,BC107,BC107A,BC108,BC108A, BC167A,BC171A,BC237A,BC238A,BC547A,BC548A,BCY58A,BCY58B,BCY59A, BCY59B,BSW83,BSW88,BSW89,BSX38,BSX79,*PTC132,*GE-61,*TR-21, *HEPS0014,*SK3122,*EN10,*ECG123A,*WEP736,*BC107B,*276-2016,*RT-187
2SC455	2N708,2N708A,2N784,2N784A,2N914,2N2318,2N2319,2N2481,2N2710,2N3011, 2N3210,2N3261,2N3510,2N3511,2N3605A,2N3606A,2N3646,2N3647,2N3648, 2N4295,2N4420,2N4421,2N4422,2SC54,2SC204,2SC302,2SC335,2SC356, 2SC400-R,2SC468,2SC595,2SC601,2SC689,2SC724,2SC752G-R,2SC1274,EN708, EN914,EN3011,EN3013,EN3014,GET708,GET914,GET2369,GET3013,GET3014, GET3646,MPS3646,TIS45,TIS46,TIS51,TIS52,TIS55,ZT708,*PTC132,*GE-210, *TR-95,*HEPS0025,*SK3138,*ECG289,*WEP736,*BC107B,*276-2016,*RT-187
2SC456	2N707,2N3512,*PTC144,*GE-47,*TR-65,*HEPS3001,*SK3048,*ECG195A,*276-2030
2SC457	2SC294
2SC458	2N3227,2N4970,2SC372G-Y,2SC400,2SC400-Y,2SC454,2SC735-Y,2SC752G-Y, 2SC933,2SC1085,2SC1216,2SC1317,2SC1999,BC107A,BC108A,BC167A,BC171A, BC237A,BC238A,BC547A,BC548A,BCY58A,BCY59A,BSW83,BSW88,BSW89, BSX38,BSX79,*PTC132,*GE-210,*TR-21,*HEPS0015,*SK3122,*ECG289,*WEP736, *BC107B,*276-2016,*RT-187
2SC459	2N3227,2N4970,2SC372G-Y,2SC400,2SC400-Y,2SC454,2SC458,2SC735-Y,

To Replace	Substitute This Type
(2SC459)	2SC752G-Y,2SC933,2SC1085,2SC1216,2SC1317,2SC1999,BC107A,BC108A,BC167A, BC171A,BC237A,BC238A,BC547A,BC548A,BCY58A,BCY59A,BSW83,BSW88, BSW89,BSX38,BSX79,*PTC132,*GE-210,*TR-87,*SK3048,*ECG128,*WEPS5003, *BC107B,*276-2011,*RT-114
2SC460	2N916,2N916A,2N2501,2N4013,2N4123,2N4418,2SC372G-O,2SC400-Y,2SC454, 2SC458,2SC459,2SC619,2SC639,2SC735-O,2SC752G-O,2SC933,2SC1085, 2SC1216,2SC1276,2SC1317,2SC1372,2SC1999,EN916,*PTC132,*GE-210,*TR-70, *HEPS0014,*SK3018,*EN10,*ECG233,*WEP460,*BC107B,*276-2011,*RT-134
2SC461	2N916,2N916A,2N2501,2N3211,2N4013,2N4123,2N4418,2N5144,2SC372G-O, 2SC400-O,2SC458,2SC459,2SC460,2SC619,2SC620,2SC639,2SC735-O, 2SC752G-O,2SC1216,2SC1276,2SC1372,2SC1999,EN916,*PTC132,*GE-210, *TR-24,*HEPS0014,*SK3018,*EN10,*ECG229,*WEP56,*BC107B,*276-2011, *RT-113
2SC462	
2SC463	40235,40238,40239,*PTC132,*GE-60,*TR-70,*SK3019,*EN10,*ECG319,*WEP242, *BF200,*276-2016,*RT-108
2SC464	2SC313,2SC465,2SC466,40414,MT1061,*PTC132,*GE-86,*TR-83,*HEPS0016, *SK3018,*SN80,*ECG161,*WEP719,*276-2015,*RT-113
2SC465	2SC313,2SC464,2SC466,40414,MT1061,*PTC132,*GE-86,*TR-83,*HEPS0016, *SK3018,*EN10,*ECG161,*WEP719,*276-2011,*RT-113
2SC466	2SC313,2SC464,2SC465,40414,MT1061,*PTC132,*GE-86,*TR-83,*HEPS0017, *SK3018,*EN10,*ECG161,*WEP719,*276-2011,*RT-113
2SC467	2N702,2N743,2N784,2N2318,2N2319,2N2432,2N2571,2N2572,2N2656,2N3854A, 2N5030,2N5187,2SC68,2SC98,2SC132,2SC133,2SC135,2SC137,2SC204,2SC356, 2SC395A-R,2SC595,2SC718,2SC1231,BC170A,MT1038,MT1038A,MT1039,MT1060, PT2760,ZT41
2SC468	2N2710,2N3511,2N3648,2N4295,2SC67,2SC169,2SC335,2SC336,2SC577,2SC639, 2SC752G-R,2SC764,MPS6530,*PTC136,*GE-20,*TR-21,*ECG123A,*WEP736, *BC107B,*276-2016,*RT-102
2SC469	2SC185,2SC266,*GE-39,*TR-95,*HEPS0016,*SK3018,*EN10,*ECG313,*WEP56, *BF365,*276-2015,*RT-113
2SC470-3	2N3498,2SC49,2SC470-4,2SC590,MM2258,SE8010,*GE-18,*BF338,*276-2008
2SC470-4	2N3500,2SC49,2SC470-5,2SC590,2SC1217,MM2258,*PTC117,*GE-27,*TR-78, *BF338,*276-2008
2SC470-5	2N3500,2SC154C,2SC470-6,2SC507-R,2SC1217,MM2259,MM2260,*PTC117,*GE-27, *TR-78,*BF338,*276-2008
2SC470-6	2SC154C,2SC507-R,MM2259,MM2260,*PTC117,*GE-27,*TR-78,*BF338,*276-2008
2SC471	
2SC472	*HEPS0015,*SK3018
2SC473	2N915,2N4014,2N4963,2SC734-O,2SC979-O,2SC979A-O,2SC980AG-O,2SC980G-O, BSW84,EN915,ZT88,ZT118
2SC474	2SC472,*ECG123A
2SC475	2SC476,2SC540,2SC640,*GE-212,*TR-21,*HEPS0015,*SK3124,*EN10,*ECG123A, *WEP736,*BC107B,*276-2011,*RT-102
2SC476	2SC475,2SC540,2SC640,*GE-212,*TR-21,*HEPS0015,*SK3122,*EN10,*ECG123A, *WEP736,*BC107B,*276-2011,*RT-102
2SC477	*PTC132,*GE-61,*TR-95,*HEPS0015,*SK3039,*EN10,*ECG108,*WEP56,*BC107B, *276-2016,*RT-113
2SC478	*PTC123,*GE-212,*TR-75,*HEPS0025,*SK3018,*EN10,*ECG195A,*BC107B, *276-2016,*RT-102A
2SC479	2N2218,2N2218A,2N2537,2N2848,2N2958,2N3725A,2N4960,2N4961,2SC97A,TIS110, *PTC125,*GE-47,*TR-87,*SK3024,*ECG128,*WEP243,*RT-114
2SC480	2N2017,2N2106,2N2193,2N2193A,2N2193B,2N2218,2N2218A,2N2270,2N2537, 2N2848,2N2868,2N3109,2N3725A,2N4960,2N4961,2SC97A,2SC109A-O,2SC503-Y, 2SC879,2SC880,2SC1386,40635,BC140-10,BSY53,BSY83,TN53,ZT90,ZT95, ZT1613,ZT2270,*SK3124
2SC481	2N749,2N751,2N1388,2N1390,2SC302-M,2SC307-M,2SC590N,2SC845,2SC865, *PTC158,*GE-47,*TR-65,*HEPS3001,*SK3047,*ECG195A,*276-2030,*RT-306
2SC482-GR	2N1420,2N1507,2N1946,2N1972,2N2192,2N2192A,2N2192B,2N2219,2N2538,2N2959, 2N3300,2SC503-GR,2SC504-GR,A5T2192,A5T2222,A5T5449,A5T5451,A8T3704, A8T3706,BC337-25,TIS109,TIS111,TN59,*PTC123,*GE-47,*TR-65,*SK3048,

To Replace	Substitute This Type
(2SC482-GR)	*ECG195A,*276-2008
2SC482-O	2N780,2N1941,2N1959A,2N2218,2N2410,2N2537,2N2846,2N2848,2N2868,2N2958, 2N3015,2N3252,2N3299,2N3326,2N3734,2N4960,2SC97,2SC346,2SC347,2SC348, 2SC479,2SC482-Y,2SC503-O,2SC503-Y,2SC504-O,2SC504-Y,2SC580,A5T5450, A8T3705,BSY81,MM3724,MM3725,TIS110,TN61,*PTC123,*GE-47,*TR-65,*276-2008
2SC482-Y	2N2270,2N2537,2N3053,2N3724A,2N4960,2SC503-Y,2SC504-Y,2SC1384,A5T5450, A8T3705,BC337-16,BSX72,TIS110,TN61,ZT90,ZT95,ZT2270,*PTC123,*GE-47, *TR-65,*ECG195A,*WEP243,*276-2008
2SC483	2N5602,HST5913
2SC484-BL	*TR-78,*276-2012
2SC484-R	2SC484-Y,2SC510-R,2SC516A,SE7001,*TR-78,*276-2012
2SC484-Y	2SC510-O,*TR-78,*276-2012
2SC485-BL	2N1890,*276-2012
2SC485-R	2N699A,2N699B,2N1613B,2N1889,2N1974,2N2849,2N3107,2N3421,2SC292,2SC293, 2SC485-Y,2SC512-R,2SC516,SE7002,TN624,TN624-1,*276-2012
2SC485-Y	2N1973,2N2443,2SC512-O,*276-2012
2SC486-BL	2N1890,2SC485-BL,ZT1711,*276-2012
2SC486-R	2N1613A,2N1889,2N1974,2N2849,2N3107,2N3109,2N3420,2N5262,2SC291,2SC292, 2SC485-R,2SC485-Y,2SC512-R,2SC516,SE8002,TN624,TN624-1,*276-2012
2SC486-Y	2N1973,2N4961,2SC485-Y,2SC512-O,*HEPS5026,*276-2012
2SC487	2N4864,2N5602,2SC487A,2SC895,HST5913,HST5914,*PTC104,*TR-81,*HEPS5012, *SK3131,*ECG175,*WEP241
2SC487A	2N4864,2SC895,HST5904,HST5905,HST5951,HST5952,KSP1396
2SC488	*PTC104,*HEPS5012,*SK3131,*ECG175,*WEP241
2SC489-BL	2N5610,HST5913,*PTC104
2SC489-R	2N5608,2N5612,2N6500,HST5908,KSP1022,*PTC104
2SC489-Y	2N3767,2N5610,2SC981,HST5903,KSP1025,*PTC104
2SC490-BL	2N5606,HST5911,HST5912,*PTC112,*GE-66
2SC490-R	HST5906,HST5907,KSP1021,*PTC112,*GE-216
2SC490-Y	2N3766,2N5606,HST5901,HST5902,KSP1024,*PTC112,*GE-216
2SC491-BL	HST5911,*PTC112,*GE-23,*TR-81,*WEP241
2SC491-R	HST5906,*PTC112,*GE-23,*TR-81,*WEP241
2SC491-Y	HST5901,*PTC112,*GE-23,*TR-81,*WEP241
2SC492	2N5618,2N5626,2SC519A,2SC520A,*PTC121,*GE-10,*TR-67,*HEPS5004,*SK3124, *ECG162,*WEP707,*276-2009,*RT-102
2SC493-BL	2N5614,2N5618,2N5622,2N5626,2SC520A,*TR-59,*ECG223,*WEP247
2SC493-R	2N5616,2N5624,2SC647,2SD68,2SD180,KSP1041,KSP1042,KSP1044,KSP1045, KSP1171,KSP1172,KSP1174,KSP1175,SDT7602,SDT7603,*TR-59,*ECG223, *WEP247
2SC493-Y	2N5614,2N5618,2N5622,2N5626,2SD68,KSP1044,KSP1045,KSP1174,KSP1175, SDT7602,SDT7603,*TR-59,*ECG223,*WEP247
2SC494	2SC521A,*GE-14,*HEPS7002,*SK3029,*EN10,*ECG223,*WEP247,*RT-164A
2SC494-BL	2SC521A,*TR-59,*ECG223,*WEP247
2SC494-R	2SD68,SDT7601,*TR-59,*ECG223,*WEP247
2SC494-Y	2SD68,SDT7601,*TR-59,*ECG223,*WEP247
2SC495-O	2N3725A,2SC108A-O,2SC109A-O,2SC1382-O,*PTC904,*GE-47,*TR-76,*SK3190, *ECG184,*WEPS5003,*276-2019,*RT-152
2SC495-R	2N2218A,2N3678,2N3722,2N3723,2N4961,2N5262,2N6375,2N6376,2SC97A, 2SC108A-R,2SC109A-R,2SC560,BSX60,BSY53,BSY83,TN53,*PTC904,*GE-47, *TR-76,*SK3190,*ECG184,*WEPS5003,*276-2018,*RT-152
2SC495-Y	2N2219A,2SC109A-Y,2SC1382-Y,BSY54,BSY84,*PTC904,*GE-47,*SK3190,*ECG184, *WEPS5003,*RT-152
2SC496-O	2N2270,2N2960,2N2961,2N3724A,2SC1384,BC337-16,BSX72,MPSU02,*PTC904, *GE-20,*TR-77,*HEPS5000,*SK3104,*ECG295,*WEPS5003,*276-2030,*RT-152
2SC496-R	2N1959A,2N2218,2N2410,2N2537,2N2846,2N2848,2N3015,2N3252,2N3299,2N3326, 2N3734,2N4960,2SC97,2SC580,A5T5450,A8T3705,BSY81,MM3724,MM3725,TIS110, TN61,*PTC904,*GE-20,*HEPS5000,*SK3190,*ECG295,*WEPS5003,*276-2030
2SC496-Y	2N2219,2N2538,2N3300,A5T2222,A5T5449,A8T3704,BC337-25,BSY82,TIS109, TIS111,TN59,*PTC904,*GE-20,*TR-76,*SK3190,*ECG295,*WEPS5003,*276-2030, *RT-152
2SC497-O	2N1574,2N1711B,2N1973,2N2405,2N4943,BC141-16,ZT92,*TR-87,*ECG128,

To Replace	Substitute This Type
(2SC497-O)	*WEP243
2SC497-R	2N699,2N699A,2N699B,2N1613B,2N1889,2N1974,2N2102,2N2102A,2N2243, 2N2243A,2N2443,2N3036,2N3107,2N3723,2SC512-O,2SC512-R,2SC516,A5T2243, BC141-10,BC141-6,BSY85,ZT91,ZT93,ZT2102,*PTC144,*TR-87,*ECG128,*WEP243
2SC497-Y	2N1890,2SC1475,BSY86,*TR-87,*ECG128,*WEP243
2SC498-O	2N1711A,2N1973,2N3725A,2SC108A-O,2SC497-O,BC140-16,BC141-16,*GE-47, *TR-87,*ECG128,*WEP243
2SC498-R	2N1613A,2N1889,2N1974,2N2193,2N2193A,2N2193B,2N2218A,2N3107,2N3109, 2N3678,2N3722,2N3723,2N4961,2N5262,2N6375,2N6376,2SC97A,2SC108A-R, 2SC497-R,2SC512-O,2SC512-R,2SC516,2SC560,A5T2193,BC140-10,BC140-6, BC141-10,BC141-6,BSY53,BSY83,TN53,ZT1613,*PTC144,*GE-47,*TR-87,*ECG128, *WEP243
2SC498-Y	2N1890,2N2219A,2SC497-Y,2SC1475,BSY54,BSY84,ZT1711,*GE-47,*TR-87, *ECG128,*WEP243
2SC499-R	2N2461,2N2465,2SC979A-O,*PTC144,*GE-27,*TR-78,*HEPS5026,*SK3122, *276-2008
2SC499-Y	2N2462,2N2466,2N4410,A5T4410,*PTC170,*GE-18,*TR-78,*HEPS5026,*276-2008
2SC500	2N3501,A5T4410,*PTC123,*GE-18,*TR-87,*HEPS5026,*SK3040,*ECG154,*WEP712, *BF338,*276-2008,*RT-110
2SC501-O	2N3725,2N5106,TIS135,*PTC125,*GE-47,*276-2030
2SC501-R	2N2218,2N2218A,2N2224,2N2537,2N2846,2N2848,2N3015,2N3299,2N3326,2N3678, 2N3722,2N4047,2N4960,2N4961,2SC97,2SC97A,2SC307,2SC580,2SC1218,BSX60, TIS136,*PTC125,*GE-47,*276-2030
2SC501-Y	2N2219,2N2219A,2N2538,2N3300,*PTC125,*GE-47,*276-2030
2SC502	2SC302-M,*PTC144,*GE-47,*TR-87,*HEPS3001,*SK3048,*ECG195A,*WEP243, *276-2030,*RT-114
2SC503-GR	2N2219,2N2219A,2N2538,2N3300,BSY54,*GE-18,*TR-87,*WEP243,*RT-114
2SC503-O	2N2218,2N2218A,2N2410,2N2537,2N2846,2N2848,2N3015,2N3252,2N3299,2N3326, 2N3678,2N3722,2N4960,2N4961,2SC97,2SC97A,2SC109A-R,2SC503-Y,2SC560, 2SC580,BSX60,*GE-18,*TR-87,*SK3024,*ECG128,*WEP243,*RT-114
2SC503-Y	2N2537,2N3725A,2N4960,2N4961,2SC109A-O,2SC1384,*GE-18,*TR-87,*SK3024, *ECG128,*WEP243,*RT-114
2SC504-GR	2N2219,2N2538,2N3300,2SC503-GR,*TR-87,*WEP243,*RT-114
2SC504-O	2N744A,2N2218,2N2410,2N2537,2N2846,2N2848,2N3015,2N3252,2N3299,2N3326, 2N3734,2N4960,2SC97,2SC503-O,2SC503-Y,2SC504-Y,2SC580,MM3724,MM3725, *TR-87,*SK3024,*ECG128,*WEP243,*RT-114
2SC504-Y	2N2537,2N3724A,2N4960,2SC503-Y,2SC1384,*TR-87,*SK3024,*ECG128,*WEP243, *RT-114
2SC505-O	2N6516,2N6517,2SC1752,*PTC117,*GE-27,*TR-78,*ECG154,*WEP712,*276-2012
2SC505-R	*PTC117,*GE-27,*TR-78,*ECG154,*WEP712,*276-2012
2SC506-O	2N6515,2SC1056,2SC1753,*PTC117,*GE-27,*TR-78,*SK3045,*ECG154,*WEP712, *276-2012
2SC506-R	2SC154C,2SC1103,2SC1103A,*PTC117,*GE-27,*TR-78,*SK3044,*ECG154,*WEP712, *276-2012
2SC507-O	*TR-78,*ECG154,*WEP712,*276-2012
2SC507-R	*TR-78,*ECG154,*WEP712
2SC507-Y	*TR-78,*ECG154,*WEP712
2SC508	HST6902,HST6903,HST6904,HST7901,KSP1071,*GE-45,*TR-81,*HEPS5012, *SK3538,*ECG286,*WEP777,*RT-116
2SC509-O	2N6426,2N6427,2SC509G-O,BC337-16,TIS133,*ECG289
2SC509-Y	2SC509G-Y,A5T5449,A8T3704,BC337-25,*SK3122
2SC509G-O	2N6426,2N6427,2SC509-O,BC337-16,TIS133
2SC509G-Y	2SC509-Y,A5T5449,A8T3704,BC337-25
2SC510-O	*TR-78,*276-2012
2SC510-R	2SC293,2SC510-O,2SC516A,*TR-78,*276-2012
2SC511-O	2N2405,2N2443,2N3036,2SC510-O,ZT92,*TR-78,*276-2012
2SC511-R	2N699A,2N699B,2N1613B,2N2102,2N2102A,2N2243,2N2243A,2N2443,2N3020, 2N3036,2SC510-O,2SC510-R,2SC516A,BSY85,ZT91,ZT93,ZT2102,*TR-78, *276-2012
2SC512-O	2N1973,2N2443,*GE-18,*TR-87,*ECG128,*WEP243,*276-2012,*RT-114
2SC512-R	2N699A,2N699B,2N1613B,2N1889,2N1974,2N2443,2N3107,2SC292,2SC293,

To Replace	Substitute This Type
(2SC512-R)	2SC512-O,2SC516,*GE-28,*TR-87,*ECG128,*WEP243,*276-2012,*RT-114
2SC513-O	2N1973,2N3725A,2N4961,2SC512-O,BC140-10,BC140-16,BC141-10,BC141-16,*GE-18,*TR-87,*SK3024,*ECG128,*WEP243,*276-2012,*RT-114
2SC513-R	2N1613A,2N1889,2N1974,2N2193,2N2193A,2N2193B,2N3107,2N3109,2N3722,2N3723,2N4961,2N5262,2SC97A,2SC512-O,2SC512-R,2SC516,BC140-10,BC140-6,BC141-10,BC141-6,BSX60,BSY83,ZT1613,*GE-28,*276-2012,*RT-114
2SC514	*HEPS5015,*SK3021,*ECG124,*WEP240
2SC515	2SC1102,2SC1235,*PTC104,*GE-12,*TR-23,*HEPS5011,*SK3021,*PN66,*ECG124,*WEP240,*RT-128
2SC515A	*PTC104,*GE-32,*TR-81,*HEPS5011,*SK3021,*ECG124,*WEP240,*RT-128
2SC516	2N699A,2N699B,2N1613B,2N1889,2N1974,2N2443,2N3107,2SC292,2SC293,2SC512-O,2SC512-R,*PTC110,*HEPS3019,*SK3024,*ECG128,*WEP243
2SC516A	2SC293,2SC510-O,2SC510-R,*HEPS3019,*SK3104
2SC517	*PTC195,*HEPS5014,*SK3049,*ECG237,*WEP517,*RT-306
2SC518	2N6495,2SC243,2SC246,KSP1043,KSP1173,SDT7610,SDT7611,*PTC129A,*HEPS5020
2SC518A	2SC42A,2SC246,2SC887,2SC1116,2SC1116A,2SC1403A,2SC1783,2SD18,2SD84,182T2,BDY25B,FT410,HST7201,KSP1091,KSP1141,SDT7201,SDT7611,SDT7612,SDT7736,*HEPS5020
2SC519	2N4070,2N5624,2N5628,KSP1172,KSP1173,KSP1175,KSP1176,SDT7603,SDT7604,*HEPS5004,*SK3079,*ECG162,*WEP707
2SC519A	*PTC118,*TR-67,*HEPS5020,*SK3079,*ECG162,*WEP707
2SC520	2N5624,KSP1171,KSP1172,KSP1174,KSP1175,SDT7602,SDT7603,*PTC140,*HEPS7002,*SK3510,*ECG130,*WEP247
2SC520A	2N5626,2SC519A,*TR-59,*HEPS7004,*SK3510,*ECG130,*WEP247,*RT-131
2SC521	SDT7601,*PTC140,*HEPS7002,*SK3027,*ECG130,*WEP247
2SC521A	2N5622,2N5626,2SC520A,*TR-59,*HEPS7002,*SK3027,*ECG130,*WEP247,*RT-131
2SC522-O	*TR-78,*SK3045,*ECG225
2SC522-R	2SC299,2SC522-O,2SC697A,*TR-78,*ECG225
2SC523-O	2SC522-O,*PTC195,*TR-78,*ECG225
2SC523-R	2SC299,2SC522-O,2SC522-R,2SC697A,*PTC195,*TR-78,*ECG225
2SC524-O	2SC698,*PTC195,*TR-78,*ECG225
2SC524-R	2SC298,2SC299,2SC524-O,2SC697,2SC697A,2SC698,*PTC195,*TR-78,*ECG225
2SC525-O	2SC524-O,2SC698,*PTC195,*TR-78,*ECG225
2SC525-R	2SC297,2SC298,2SC524-O,2SC524-R,2SC697,2SC698,*PTC195,*TR-78,*ECG225
2SC526	*DS-72,*GE-27,*TR-78,*HEPS5025,*SK3044,*SN80,*ECG154,*WEP712,*BF338,*RT-110
2SC527	2N3287,2N3288,2N3337,2N3338,2N3339,2SC333,2SC463,2SC1054,2SC1067,2SC1747,40239,40295,40413,BF167,BF232,BFX60,TN-3200,*PTC132,*GE-60,*TR-24,*BF167,*276-2016
2SC528	2N2331,2N3241A,2N4418,2SC395A,2SC395A-Y,2SC619,2SC735-O,2SC934,2SC1317,2SC1372,2SD392,BSY61,ZT2938,*PTC101,*GE-212,*TR-53,*BC107B,*276-2011
2SC529	2N3227,2N4124,2N4419,2N4970,2SC372G-Y,2SC400,2SC400-GR,2SC735-Y,2SC752G-Y,2SC1085,2SC1992,BC107,BC107A,BC108,BC108A,BC129A,BC130A,BC167A,BC171A,BC237A,BC238A,BSW88,BSW89,BSX38,BSX79,*ECG123A,*WEP736
2SC530	
2SC531	
2SC532	
2SC533	
2SC534	
2SC535	2N2369,2N2369A,2N3862,2N4873,2N4996,2N4997,2SC764,2SC922,2SC1187,2SC1188,2SC1189,2SC1276,2SC1319,2SC1320,2SC1393,2SC1394,2SC1395,2SC1396,2SC1674,PET3002,SE3005,*PTC132,*GE-39,*TR-70,*HEPS0016,*SK3018,*ECG229,*WEP535,*BF200,*276-2011,*RT-113
2SC536	2N916,2N916A,2N3116,2N3227,2N3242A,2N3904,2N3947,2N4141,2N4401,2N4437,2N4951,2N4952,2N4970,2N5028,2N5107,2N5368,2N5369,2N5381,2N6004,2SC366G,2SC366G-Y,2SC367G,2SC367G-Y,2SC752G-Y,2SC933,2SC943,2SC944,2SC1175,2SC1216,2SC1276,2SC1318,2SC1453,2SC1999,BC107A,BC167A,BC171A,BC182A,BC237A,BC547A,BCW34,BCW36,BCY59A,BFY39-2,BSW83,BSX79,EN916,GET2222,GI-3643,MM3904,MPS6531,PBC107A,TN-3904,ZT84,ZT114,*PTC132,

To Replace	Substitute This Type
(2SC536)	*GE-62,*IR2SC536,*HEPS0016,*SK3122,*ECG199,*WEP736,*BC107B,*276-2016,*RT-107A
2SC536FP	2N5828,2SC536NP,BC550C,*ECG199
2SC536NP	2N5828,2SC536FP,BC550C
2SC537	2N2923,2N3241A,2N6000,2SC372G-Y,2SC395A-Y,2SC400,2SC400-Y,2SC454,2SC458,2SC459,2SC735-Y,2SC934,2SC1085,2SC1317,2SD392,BC108A,BC168A,BC172A,BC238A,BC548A,BCY58A,BSW88,BSW89,BSX38,PBC108A,ZT2938,*PTC132,*GE-212,*TR-21,*HEPS0016,*SK3122,*EN10,*ECG123A,*WEP736,*BC107B,*276-2011,*RT-105
2SC538	2N2925,2N3391,2N3391A,2N4256,2N4419,2N4954,2N5172,2N5371,2N6002,2N6008,2SC539,2SC733-GR,2SC735-GR,2SC1849,2SC1993,A5T3565,BC108,BC108B,BC109B,BC172B,BC173B,BC238B,BC239B,BC338,BC548B,BC549B,BCY58B,BCY58C,MPSA16,MPSA17,TP3706,*PTC126,*GE-61,*IR2SC538A,*HEPS0015,*SK3122,*EN10,*ECG123A,*WEP736,*BC107B,*276-2016,*RT-102
2SC538A	2N4953,2N5370,2N5376,2N5377,2N5827,2N6006,2N6112,2SC1166-GR,2SC1850,A157B,BC107,BC107B,BC167B,BC171B,BC174B,BC182B,BC183L,BC184L,BC190B,BC237B,BC337,BC413B,BC414B,BC547B,BC550B,BCY59B,BCY59C,BFY39-3,TIS94,TIS97,TP3704,*PTC123,*GE-61,*IR2SC538A,*HEPS0030,*SK3020,*EN10,*ECG123A,*WEP736,*BC107B,*276-2016,*RT-172
2SC539	2N2925,2N3391,2N3391A,2N4256,2N4419,2N4954,2N5172,2N5371,2N6002,2N6008,2SC538,2SC733-GR,2SC735-GR,2SC1849,2SC1993,A5T3565,BC108,BC108B,BC109B,BC172B,BC173B,BC238B,BC239B,BC338,BC548B,BC549B,BCY58B,BCY58C,MPSA16,MPSA17,TP3706,*PTC126,*GE-61,*TR-21,*HEPS0015,*SK3020,*EN10,*ECG123A,*WEP736,*BC107B,*276-2016,*RT-102
2SC540	2SC640,*GE-212,*TR-21,*HEPS0016,*SK3122,*EN10,*ECG123A,*WEP736,*BC107B,*276-2031,*RT-102
2SC541	*HEPS3001
2SC542	2SC600
2SC543	*SK3018
2SC544	2N708,2N708A,2N784A,2N2369,2N2369A,2N2501,2N2885,2N3009,2N3013,2N3014,2N3605A,2N3606A,2N3688,2N3689,2N3690,2N3862,2N4137,2N4295,2N4420,2N4421,2N4422,2N5029,2SC67,2SC169,2SC335,2SC336,2SC468,2SC601,2SC601N,2SC639,2SC752G-R,2SC764,2SC1188,2SC1189,2SC1274,2SC1320,2SC1687,2SC1688,EN708,EN914,EN2369A,EN3009,EN3013,EN3014,GET708,GET914,GET2369,GET3013,GET3014,GET3646,MPS3646,SE5001,SE5002,SE5003,SE5006,TIS45,TIS46,TIS48,TIS49,TIS52,TIS55,TIS87,TIS129,ZT708,*PTC132,*GE-39,*TR-95,*HEPS0016,*SK3018,*EN10,*ECG108,*WEP56,*276-2016,*RT-113
2SC545	2N744,2N2318,2N2319,2N3011,2N3564,2N3959,2N3960,2N4294,2N4418,2N4996,2N4997,2N5030,2N5200,2N5201,2S095A,2SC99,2SC168,2SC300,2SC301,2SC380-R,2SC380A-R,2SC400-O,2SC595,2SC718,2SC719,2SC1231,BF123,BF125,BF311,EN744,EN3011,MT1060A,NT3000,SE3005,TIS51,ZT2205,*PTC132,*GE-39,*TR-70,*HEPS0016,*SK3018,*EN10,*ECG107,*WEP56,*276-2011,*RT-113
2SC546	2N2369,2N2369A,2N4137,2N4295,2N5029,2SC601,2SC601N,2SC689,BF311,EN2369A,MT1060A,TIS48,TIS49,TIS129,*PTC132,*GE-39,*276-2011
2SC547	2N1709,2N2874,2SC302-M,2SC307-M,2SC590N,40305,SRF11101,*GE-28,*TR-76,*HEPS3001,*276-2014
2SC548	2N2485,2N2486,2N4874,2SC731,2SC872,2SC890,2SC908,2SC1001,2SC1002,2SC1165,BLX89,BLY61,*PTC115,*GE-28,*TR-76,*HEPS3008,*276-2038
2SC549	2N3733,2N4440,2N5016,2N5102,2N5215,2SC551,40307,MSA7505,MSA8505,SRF12101,*GE-28,*TR-76
2SC550	2SC1003,*GE-28,*TR-76
2SC551	2N4933,40307,SRF52101,*GE-66
2SC552	2SC831,*GE-66
2SC553	*GE-66
2SC554	*HEPS3013
2SC555	2N3866,2N3866A,2SC302-M,*TR-25,*HEPS3008,*SK3195
2SC556	*HEPS3013,*276-2038
2SC557	
2SC558	2N6233,2N6234,2SC270,2SC1227,2SC1229,HST7202,HST7203,HST7204,KSP1092,KSP1093,KSP1094,KSP1095,KSP1142,KSP1143,KSP1144,KSP1145,SDT7202,SDT7203,SDT7204,*PTC129A,*GE-73,*TR-67,*HEPS5020,*SK3115,*ECG163,

To Replace	Substitute This Type
(2SC558)	*WEP740
2SC559	2N749,2N751,2N1388,2N1390,2SC302-M,2SC307-M,2SC590N,*PTC136,*HEPS0014
2SC560	2N2218A,2N3725A,2N4961,2SC97A,2SC108A-O,*HEPS3019,*SK3104,*ECG128, *WEP243
2SC561	*PTC115,*HEPS0011,*SK3018,*ECG108,*WEP56
2SC562	2SC563,2SC563A,*PTC132,*GE-60,*TR-70,*HEPS0017,*SK3018,*SN80,*ECG161, *WEP719,*BF167,*276-2016,*RT-113
2SC563	2SC563A,40239,BF232,*PTC132,*GE-60,*TR-83,*HEPS0017,*SK3018,*EN10, *ECG161,*WEP719,*BF173,*276-2038,*RT-113
2SC563A	*PTC133,*GE-20,*TR-70,*HEPS0020,*SK3018,*ECG161,*WEP719,*RT-113
2SC564	2N545,2N547,2N549,2N656A,2N1613,2N1613A,2N1959,2N1959A,2N1984,2N1987, 2N2106,2N2107,2N2218,2N2218A,2N2224,2N2410,2N2846,2N2868,2N2958,2N3015, 2N3122,2N3252,2N3253,2N3299,2N3326,2N3678,2SC97,2SC109A-R,2SC190, 2SC226,2SC254,2SC344,2SC347,2SC348,2SC479,2SC480,2SC503-O,2SC578, 2SC580,2SC879,2SC880,2SC1053,2SC1072,2SC1072A,2SC1150,2SC1385, 2SC1386,2SD216,7A31,40347,BC341-6,BSX60,BSY51,BSY53,KT600,KT600F, KT600T,ZT697,ZT1613,*ECG128,*WEP242
2SC565	2N708A,2N2312,2N2845,2N3301,2N3641,2N3642,2N3973,2N3975,2N5581,2SC239, 2SC1166-R,BSY93,FT3641,FT3642,GET2221,GET2221A,GI-3641
2SC566	*GE-88,*TR-21,*HEPS3001,*SK3122,*ECG123A,*WEP736,*276-2014,*RT-102
2SC567	2N2808,2N2808A,2N2809A,2N3570,2N4252,2N4253,2N5053,2N5054,2N6304,2N6305, 2SC568,2SC1747,MT1061A,*PTC132,*GE-86,*TR-83,*HEPS0017,*SK3039, *ECG161,*WEP719,*276-2011,*RT-113
2SC568	2N3570,2N4252,2N4253,2N6304,2N6305,2SC567,*PTC132,*GE-86,*TR-95, *HEPS0017,*SK3039,*EN10,*ECG108,*WEP56,*276-2011,*RT-113
2SC569	2SC623
2SC570	2N708,2N708A,2N718,2N760,2N784A,2N914,2N2221,2N2242,2N2312,2N2315, 2N2369,2N2369A,2N2396,2N2481,2N2501,2N2539,2N2710,2N2845,2N2847,2N3013, 2N3014,2N3115,2N3210,2N3211,2N3301,2N3508,2N3510,2N3511,2N3605A, 2N3606A,2N3646,2N3647,2N3648,2N3705,2N3862,2N3903,2N3946,2N3973,2N3974, 2N3975,2N3976,2N4123,2N4137,2N4140,2N4400,2N4962,2N5027,2N5029,2N5144, 2N5380,2N5450,2N5769,2N5772,2N6538,2SC67,2SC131,2SC134,2SC239,2SC577, 2SC613,2SC624,2SC639,2SC764,2SC877,2SC878,2SC913,2SC914,2SC914A, 2SC1166-R,2SC1274,A5T3903,A5T4123,A5T5450,A8T3705,BSW82,BSY93,D33D24, GET708,GET914,GET2221,GET2369,GET3013,GET3014,GET3646,GI-3641,GI-3642, GI-3705,KT218F,MM3903,MPS2369,MPS3705,MPS6530,PET3705,PET6001,TIS110, TN62,TN-3903,ZT83,ZT113,ZT708,ZT2369
2SC571	*HEPS3001,*SK3048,*ECG195A,*WEP224,*RT-306
2SC572	2SC573
2SC573	
2SC574	
2SC575	
2SC576	
2SC577	
2SC578	2N656A,2N1613,2N1613A,2N1837,2N1837A,2N1959,2N1959A,2N2106,2N2107, 2N2193,2N2193A,2N2193B,2N2218,2N2218A,2N2224,2N2380,2N2380A,2N2410, 2N2479,2N2537,2N2846,2N2848,2N2868,2N2958,2N3015,2N3109,2N3252,2N3299, 2N3326,2N3678,2N3722,2N4047,2N4960,2N4961,2N6376,2SC97,2SC97A, 2SC109A-R,2SC190,2SC307,2SC347,2SC348,2SC479,2SC480,2SC498-R, 2SC503-O,2SC503-Y,2SC560,2SC580,2SC879,2SC880,2SC1218,2SC1385, 2SC1386,2SD216,40347,A5T2193,BC140-10,BC140-6,BC341-6,BSX60,BSY51, BSY53,BSY83,KT600F,KT600T,SE8002,TIS110,TIS136,TN53,ZT697,ZT1613
2SC579	2N760,2N916,2N916A,2N2242,2N2331,2N2501,2N3013,2N3014,2N3211,2N3705, 2N4013,2N5144,2N5450,2N5810,2N6427,2SC67,2SC509-O,2SC509G-O,2SC577, 2SC613,2SC639,2SC764,A5T4123,A5T5450,A8T3705,BSW82,D33D21,D33D24, GI-3416,GI-3705,PET3705,PET6002,PET8000,TIS90-GRN,TIS90-YEL,TIS90M-GRN, TIS90M-YEL,TIS92-GRN,TIS92-YEL,TIS92M-GRN,TIS92M-YEL,TIS133,TIS134,TN62
2SC580	2N2846,2N2848,2N3015,2N3299,2N3326,2N3722,2N4960,2N4961,2N5262,2SC97, 2SC97A,BSX60,*HEPS3019,*SK3048,*ECG195A,*WEP243
2SC581	*PTC132,*GE-39,*TR-24,*BF365,*276-2011
2SC582	2SC1102,2SC1235,*GE-12,*TR-23,*HEPS5011,*SK3021,*PN66,*ECG124,*WEP240,

To Replace	Substitute This Type
(2SC582)	*BF338,*RT-128
2SC583	2N2808,2N2808A,2N2809A,2N2857,2N3570,2N4252,2N4253,2N5053,2N5054,2N6304, 2N6305,2SC567,2SC568,2SC1747,A485,BFX89,MT1061A,ZT2857,*PTC132,*GE-39, *ECG278
2SC584	
2SC585	2SC636,*GE-66
2SC586	KSP1043,KSP1173,SDT7610,SDT7611,*PTC129A,*GE-73,*TR-67,*HEPS5020, *ECG162,*WEP707
2SC587	2N5827,2N6112,2N6428,2N6428A,2N6539,2SC587A,2SC1166-GR,A157B,BC183L, BC184L,BC413C,BC414C,BC547B,BC550B,BCY59C,BCY59D,MPSA18,TIS94,TIS97, TP3704,*PTC123,*GE-212,*TR-21,*HEPS0015,*SK3122,*ECG123A,*WEP736, *BC107B,*276-2016,*RT-102
2SC587A	2N5827,2N6112,2N6428,2N6428A,2N6539,2SC587,2SC1166-GR,A157B,BC183L, BC184L,BC413C,BC414C,BC547B,BC550B,BCY59C,BCY59D,MPSA18,TIS94,TIS97, TP3704,*PTC123,*GE-212,*TR-21,*HEPS0015,*SK3122,*ECG123A,*WEP736, *BC107B,*276-2016,*RT-102
2SC588	*PTC136,*HEPS0011,*SK3122,*ECG123A,*WEP736
2SC588N	SEE 2SC594N
2SC589	2SC154B,2SC507-R,*PTC117,*GE-27,*TR-78,*HEPS5025,*SK3044,*ECG154, *WEP712,*BF338,*RT-110
2SC590	2N3500,2SC1217,MM2258,*PTC125,*GE-18,*TR-87,*HEPS3019,*SK3024,*ECG289, *WEP243,*RT-114
2SC590N	2N749,2N751,2N1388,2N1390,2SC302-M,2SC307-M
2SC591	2SC92,2SC94,ZT1484,*GE-66,*HEPS3019,*SK3104
2SC592	2SC355
2SC593	*PTC132,*GE-61,*TR-51,*HEPS0025,*ECG123A,*BC107B,*276-2016
2SC594	2N2219,2N2219A,2N2538,2N3300,2N5106,2SC594-Y,*GE-47,*TR-87,*HEPS0014, *SK3024,*SN80,*ECG128,*WEP243,*276-2038,*RT-114
2SC594-O	2N3725,2N5106,2SC594,TIS135
2SC594-R	2N2218,2N2218A,2N2224,2N2410,2N2537,2N2846,2N2848,2N3015,2N3299,2N3326, 2N3678,2N4047,2N4960,2N4961,2SC32,2SC307,2SC1218,TIS136
2SC594-Y	2N2219,2N2219A,2N2538,2N3300,BCY65
2SC595	2N2710,2N3011,2N3511,2N3648,2SC67,2SC577,2SC639,2SC764,*PTC136,*GE-210, *TR-21,*HEPS0016,*SK3122,*SN80,*ECG123A,*WEP736,*BC107B,*276-2016, *RT-102
2SC596	2SC138,2SC138A,2SC139,*GE-47,*TR-21,*HEPS3001,*SK3122,*SN80,*ECG123A, *WEP736,*276-2038,*RT-102
2SC597	*GE-215,*TR-76,*HEPS3001,*SK3197,*276-2014
2SC598	2SC542,2SC600,*GE-28,*TR-76
2SC599	2SC688,*HEPS3006
2SC599N	2SC690N
2SC600	*GE-66
2SC601	2N2710,2SC577,2SC601N,2SC639,2SC764,*PTC133,*GE-17,*HEPS0016,*EN10, *ECG311,*276-2038
2SC601N	2SC577,2SC639,2SC764,*PTC133
2SC602	2N3570,2N3600,2N4252,2N4253,2N5852,2N6304,2N6305,2SC251,2SC251A,2SC252, 2SC253,2SC1747,40295,40413,MT1061A,*PTC132,*GE-86,*HEPS0016,*EN10, *276-2015
2SC603	
2SC605	2SC100,2SC287A,2SC606,2SC1158,2SC1159,A3T3011,FX2368,FX2369A,MMT2369, *GE-60,*TR-95,*HEPS0016,*SK3018,*EN10,*ECG313,*WEP56,*BF167,*276-2016, *RT-308
2SC606	2SC287A,2SC605,2SC1158,2SC1159,FX2369A,MMT2369,*GE-60,*TR-95,*HEPS0016, *SK3018,*EN10,*ECG313,*WEP56,*BF173,*276-2016,*RT-108
2SC607	
2SC608	2SC524-R,2SC609,2SC697,*GE-18,*TR-87,*SK3049,*ECG237,*WEP243,*276-2008, *RT-174
2SC609	2SC608,2SC697,*GE-46,*TR-87,*SK3048,*ECG237,*WEP777,*276-2008,*RT-174
2SC610	*SK3024,*ECG159,*WEP242
2SC611	2N2808,2N2808A,2N2809,2N2809A,2N2810,2N2810A,2N3570,2N3571,2N4252, 2N4253,2N5053,2N5054,2N6304,2N6305,2SC567,2SC568,2SC653,MT1061A,

To Replace	Substitute This Type
(2SC611)	*PTC132,*GE-11,*TR-95,*HEPS0016,*SK3039,*EN10,*ECG108,*WEP56,*BF173, *276-2011,*RT-113
2SC612	2SC1747,40235,*PTC132,*GE-39,*TR-95,*HEPS0017,*SK3039,*EN10,*ECG108, *WEP56,*276-2011,*RT-113
2SC613	2N718,2N760,2N916,2N916A,2N2221,2N2242,2N2315,2N2396,2N2501,2N2539, 2N2847,2N3013,2N3014,2N3115,2N3211,2N3705,2N3946,2N3974,2N3976,2N4013, 2N4951,2N4962,2N5144,2N5368,2N5380,2N5450,2N6427,2N6538,2SC67,2SC577, 2SC639,2SC764,2SC1166-O,A5T3903,A5T4123,A5T5450,A8T3705,BSW82,D33D24, GI-3416,GI-3705,MM3903,PET3705,PET6002,PET8000,TIS90-GRN,TIS90-YEL, TIS90M-GRN,TIS90M-YEL,TIS92-GRN,TIS92-YEL,TIS92M-GRN,TIS92M-YEL,TIS110, TIS133,TIS134,TN62,TN-3903,*PTC133,*HEPS0016,*SK3039,*ECG108,*WEP56
2SC614	*GE-20,*TR-87,*HEPS3001,*SK3048,*ECG195A,*WEP243,*276-2014,*RT-306
2SC615	2SC354,*GE-18,*TR-65,*HEPS3001,*SK3024,*ECG195A,*276-2014,*RT-114
2SC616	
2SC617	
2SC618	*PTC132
2SC618A	*PTC132
2SC619	2N916,2N916A,2N3227,2N6010,2SC735-Y,2SC933,2SC1216,BCY58A,BCY59A, *TR-24,*HEPS0014,*SK3124,*ECG123A,*WEP736,*276-2016,*RT-107A
2SC620	2N915,2N2539,2N2847,2N3903,2N4013,2N4227,2N4951,2N4962,2N5144,2N5368, 2N5380,2SC933,2SC943,TN-3903,ZT84,*TR-24,*HEPS0014,*SK3020,*ECG123A, *WEP736,*276-2016,*RT-107A
2SC620M	SEE 2SC620
2SC621	2N708,2N914,2N957,2N2242,2N2369,2N2369A,2N2501,2N2922,2N3009,2N3011, 2N3013,2N3014,2N3210,2N3211,2N3691,2N3855A,2N4123,2N4418,2N4420,2N4421, 2N4422,2N4873,2N5418,2S512,2SC67,2SC169,2SC300,2SC301,2SC367G-O, 2SC372-O,2SC372G-O,2SC394-Y,2SC400-O,2SC460,2SC461,2SC536,2SC577, 2SC601N,2SC619,2SC633,2SC634,2SC639,2SC712,2SC713,2SC735-O,2SC752G-O, 2SC764,2SC1276,2SC1372,BSY61,EN708,MPS2369,MPS3646,MPS6512, MPS6574-BLUE,MPS6574-YEL,MPSA20-WHT,PET2001,PET3002,PT720,SE2001, SE3005,ZT80,ZT110,ZT708,ZT2205,ZT2369,ZT2369A,ZT2938,*HEPS0011,*SK3122, *ECG123A,*WEP736
2SC622	2N5851,2N5852,2SC332,2SC333,TN-3200,*PTC132,*HEPS0011,*SK3122,*ECG123A, *WEP736
2SC623	2SC569
2SC624	2N708,2N708A,2N718,2N760,2N784A,2N914,2N2221,2N2242,2N2312,2N2315, 2N2369,2N2369A,2N2396,2N2481,2N2501,2N2539,2N2710,2N2845,2N2847,2N3013, 2N3014,2N3115,2N3210,2N3211,2N3301,2N3508,2N3510,2N3511,2N3605A, 2N3606A,2N3646,2N3647,2N3648,2N3705,2N3862,2N3903,2N3946,2N3973,2N3974, 2N3975,2N3976,2N4123,2N4137,2N4140,2N4400,2N4962,2N5027,2N5029,2N5144, 2N5380,2N5450,2N5769,2N5772,2N6538,2SC67,2SC131,2SC134,2SC239,2SC570, 2SC577,2SC613,2SC639,2SC764,2SC877,2SC878,2SC913,2SC914,2SC914A, 2SC1166-R,2SC1274,A5T3903,A5T4123,A5T5450,A8T3705,BSW82,BSY93,D33D24, GET708,GET914,GET2221,GET2369,GET3013,GET3014,GET3646,GI-3641,GI-3642, GI-3705,KT218F,MM3903,MPS2369,MPS3705,MPS6530,PET3705,PET6001,TIS110, TN62,TN-3903,ZT83,ZT113,ZT708,ZT2369
2SC626	*HEPS0014,*ECG123A
2SC627	2SC154C,2SC1103,2SC1103A,A5T5059,*HEPS3021,*ECG154,*WEP712
2SC628	2SC320,*TR-87,*ECG128,*WEP52,*276-2015
2SC629	2SC387AG,2SC717,MPS6507,MPS6542,MPS6543,MT1038,MT1038A,MT1039, MT1060,PET8101,*PTC132,*HEPS0016,*SK3018,*ECG107,*WEP56
2SC631	2N2925,2N3391,2N3391A,2N5420,2N6002,2N6008,2SC1312,2SC1335,2SC1344, BC108C,BC109,BC109C,BC172C,BC173C,BC238C,BC239C,BCY58C,BCY58D, MPS6515,MPS6521,*PTC139,*HEPS0015,*SK3124,*ECG123A,*WEP736
2SC631A	*HEPS0011,*SK3124,*ECG123A,*WEP736
2SC632	2N4953,2N5827,2N6006,2SC1313,2SC1345,BC184L,BC413C,BC414C,BCY59C, BCY59D,MPS6515,MPS6521,*HEPS0015,*SK3124,*ECG123A,*WEP736
2SC632A	*HEPS0015,*SK3124,*ECG123A,*WEP736
2SC633	2N2501,2N4418,2SC634,*PTC101,*HEPS0016,*SK3124,*ECG123A,*WEP736
2SC633A	*HEPS0011,*SK3124,*ECG123A,*WEP736
2SC634	2N2501,2N3903,2N4013,2N4400,2N5144,2SC943,2SC944,*PTC121,*HEPS0014,

To Replace	Substitute This Type
(2SC634)	*SK3124,*ECG123A,*WEP736,*RT-302
2SC634A	*HEPS0015,*SK3124,*ECG123A,*WEP736
2SC635	*GE-28
2SC636	2SC585,*GE-66,*TR-56,*SK3083,*WEP246
2SC637	BLY79,*GE-28,*TR-76
2SC638	*GE-66,*TR-66,*276-2019
2SC639	*PTC133,*HEPS0014,*276-2009
2SC640	2SC540,*GE-212,*TR-21,*HEPS0015,*SK3122,*EN10,*ECG123A,*WEP736,*BC107B, *276-2031,*RT-102
2SC641	2N916,2N916A,2N2501,2N2847,2N3211,2N3903,2N3974,2N3976,2N4013,2N4123, 2N4227,2N4400,2N4951,2N5027,2N5144,2SC169,2SC536,2SC620,2SC639, 2SC752G-O,2SC764,2SC944,2SC1216,2SC1276,EN916,MM3903,MPS6530,ZT84, ZT114,*PTC132,*GE-210,*TR-95,*SK3039,*ECG108,*WEP56,*BC107B,*276-2016, *RT-113
2SC642	2SC642A,2SC1171,*TR-93,*SK3133,*ECG164,*WEP740A
2SC642A	*PTC130,*GE-37,*TR-93,*SK3133,*ECG164,*WEP740A
2SC643	2SC1005A,2SC1170,2SC1170A,2SC1174,2SC1413,2SC1413A,*TR-68,*SK3115, *ECG238
2SC643A	*GE-38,*TR-93,*SK3115,*ECG238,*WEP740B
2SC644	2N4256,2N4419,2N4954,2N5827,2N6002,2N6006,2N6008,2N6112,2SC373G, 2SC400-GR,2SC735-GR,2SC828,2SC828A,2SC1684,2SC1849,2SC1993,A157B, BC107,BC107B,BC108,BC108B,BC109B,BC122,BC167B,BC171B,BC183L,BC184L, BC237B,BC238B,BC239B,BC413B,BC414B,BCY58B,BCY58C,BCY59B,BCY59C, BFY39-3,PBC107B,*PTC139,*GE-62,*IR2SC644,*HEPS0015,*SK3124,*ECG199, *WEP736,*276-2016,*RT-105
2SC645	2N4124,2SC372G-Y,2SC400,2SC400-Y,2SC454,2SC735-Y,2SC752G-Y,2SC933, 2SC1085,2SC1175,2SC1992,BC107A,BC108A,BC123,BC129A,BC130A,BC167A, BC171A,BC237A,BC238A,BCY58A,BCY59A,BSW88,BSW89,BSX38,BSX79, *PTC132,*GE-39,*TR-21,*HEPS0016,*SK3018,*SN80,*ECG107,*WEP56,*BF365, *276-2011,*RT-108
2SC646	2SD68,KSP1024,*HEPS7002,*SK3027,*ECG223,*WEP247
2SC647	2N5616,*GE-73,*TR-67,*HEPS7002,*SK3111,*ECG130,*WEP247,*RT-131
2SC648	2N5827,2SC735-GR,2SC1084,2SC1993,A157B,BC109,BC109B,BC129B,BC130B, BC131B,MPS6515,MPS6520,MPS6521,*SK3124,*ECG199,*WEP736
2SC649	2SC1999,BC107A,BC167A,BC171A,BC237A,BCY58A,BCY59A,*PTC126,*GE-60, *TR-95,*HEPS0014,*SK3039,*EN10,*ECG108,*WEP56,*BF365,*276-2011,*RT-113
2SC650	2N4419,BC107B,BC167B,BC171B,BC237B,BC547B,BC550B,BCY58B,BCY58C, BCY59B,BCY59C,*PTC136,*GE-210,*TR-95,*HEPS0011,*SK3122,*ECG199, *WEP56,*BC107B,*276-2011,*RT-113
2SC651	*HEPS3001,*ECG311,*276-2014
2SC652	*HEPS3013,*ECG311,*276-2009
2SC653	2N3571,2N3572,2N4252,*PTC132,*GE-11,*TR-70,*SK3039,*ECG161,*WEP709, *RT-113
2SC654	2SC566,*GE-18,*TR-21,*HEPS0014,*SK3122,*ECG123A,*WEP736,*276-2009, *RT-102
2SC655	*HEPS0017,*SK3122,*ECG123A,*WEP736
2SC656	*HEPS0017,*SK3039,*ECG107,*WEP56
2SC657	2N2729,2N3544,2N3563,2N5130,2N5399,2SC387AG,2SC595,2SC718,2SC719, 2SC1123,2SC1126,2SC1128,2SC1231,2SC1988,BF125,BF311,EN744,MM1941, MPS6507,MT1060A,NT3000,SE3001,SE3002,*PTC132,*HEPS0016,*SK3018, *ECG107,*WEP56
2SC658	2N2369,2N2369A,2N4137,2N4295,2N4996,2N4997,2N5029,2SC601,2SC601N, 2SC639,2SC764,2SC836,2SC922,2SC1187,2SC1188,2SC1319,2SC1393,2SC1394, 2SC1395,2SC1396,2SC1687,BF123,BF199,BF311,EN2369A,MT1060A,SE3005, TIS48,TIS49,TIS129,*PTC132,*HEPS0016,*SK3039,*ECG108,*WEP56
2SC659	2N2369,2N2369A,2N2501,2N3009,2N3011,2N3013,2N3014,2N3564,2N3688,2N3689, 2N3690,2N4137,2N4294,2N4295,2N4420,2N4421,2N4422,2N4996,2N4997,2N5029, 2N5030,2SC67,2SC169,2SC300,2SC301,2SC380-R,2SC380A-R,2SC468,2SC595, 2SC601,2SC601N,2SC639,2SC752G-R,2SC764,2SC836,2SC922,2SC1187, 2SC1188,2SC1274,2SC1319,2SC1393,2SC1394,2SC1395,2SC1396,2SC1687, BF121,BF123,BF125,BF127,BF198,BF199,BF241,BF311,EN2369A,EN3009,EN3011,

To Replace	Substitute This Type
(2SC659)	EN3013,EN3014,GET2369,GET3013,GET3014,GET3646,MPS3646,MT1060A,SE3005, SE5001,SE5002,SE5003,SE5006,TIS48,TIS49,TIS51,TIS52,TIS55,TIS129,*PTC132, *HEPS0011,*SK3039,*ECG108,*WEP56
2SC660	2N4252,2N4253,2N5851,2N5852,2SC463,2SC661,BF173,BFS62,BFX60,MT1061A, *HEPS0020
2SC661	2N4252,2N4253,2N5851,2N5852,2SC463,2SC660,BF173,BFS62,BFX60,MT1061A, *HEPS0020
2SC662	*PTC132,*HEPS0016,*SK3039,*ECG108,*WEP56
2SC663	2N3572,2N4259,2SC390,2SC392,2SC653,2SC1180,TC3114,*HEPS0017,*SK3117, *ECG161,*WEP719
2SC664	2N4070,2SC665,2SD67,KSP1045,KSP1175,SDT7603,*PTC140,*GE-14,*TR-67, *HEPS7004,*SK3027,*ECG130,*WEP707,*276-2020
2SC665	2N4070,2N6354,2SC897,2SC939,2SC1343,2SD67,KSP1046,KSP1176,SDT7604, *TR-67,*HEPS7004,*SK3111,*ECG280,*WEP740
2SC666	2SC1051,2SC1343,2SD117,BDY25C,BUY20
2SC667	2N709,2N709A,2N2475,2N2784,2N3633,2N3959,2N3960,2N4996,2N4997,2N5200, 2N5201,2SC718,2SC719,2SC1231,BF123,EN744,MT1060A,NT3000,SE3005, *PTC132,*GE-39,*276-2011
2SC668	2N709,2N709A,2N2475,2N2784,2N3633,2N3959,2N3960,2N4996,2N4997,2N5200, 2N5201,2SC674,2SC706,2SC718,2SC719,2SC1215,2SC1231,2SC1789,2SC1988, BF123,EN744,MPSH10,MPSH11,MT1060A,NT3000,SE3005,*PTC132,*GE-60, *TR-95,*HEPS0016,*SK3018,*EN10,*ECG229,*WEP56,*276-2011,*RT-187
2SC669A	
2SC674	2N709,2N709A,2N2475,2N2784,2N3633,2N3959,2N3960,2N4996,2N4997,2N5200, 2N5201,2SC668,2SC706,2SC718,2SC719,2SC1215,2SC1231,2SC1789,2SC1988, BF123,EN744,MPSH10,MPSH11,MT1060A,NT3000,SE3005,*PTC132,*GE-20, *TR-83,*HEPS0016,*SK3018,*ECG161,*WEP719,*276-2011,*RT-187
2SC675	2N6249,2SD376,2SD376A,2SD417,HST7202,HST7203,HST7204,IR411,KSP1142, KSP1143,KSP1144,KSP1145,SDT7202,SDT7203,SDT7204,XGSR10025,*ECG283
2SC676	2SC675,2SD166,2SD376,2SD417,HST7201,HST7202,KSP1141,KSP1142,SDT7201, SDT7202,SDT7612
2SC677	2N5559,2N6495,2SC902,2SC1115,2SC1116,2SC1116A,2SC1402,2SC1403, 2SC1403A,2SC1782,2SC1783,2SD218,2SD323,2SD375,KSD2201,KSP1173, SDT7610,SDT7611,SDT7735,SDT7736
2SC678	2N5624,2N5628,2SD110-R,2SD111-R,2SD118-R,2SD119-R,2SD213,2SD214,2SD217, 2SD319,2SD322,2SD340,2SD522,HST9803,HST9804,KSD2202,KSD2203,KSP1172, SDT7609,SDT7733,SDT7734
2SC679	
2SC680	*PTC104,*GE-12,*TR-81,*HEPS5012,*SK3131,*ECG175,*WEP241,*276-2017
2SC680A	2SC680,2SC1025,*PTC104,*GE-66,*TR-81,*HEPS5012,*SK3131,*ECG175,*WEP241, *276-2017
2SC681	2SC675,2SC676,2SC681A,2SC770,2SC771,2SC887,2SC1229,2SD18,2SD166, 2SD321,2SD376,2SD417,BDY25A,HST7201,HST7202,KSP1141,KSP1142,SDT7201, SDT7202,SDT7612,*GE-36,*TR-67,*HEPS5020,*SK3111,*ECG163,*WEP740
2SC681A	2N6249,2SC675,2SC771,2SC886,2SC1227,2SC1229,2SD321,2SD376,2SD376A, 2SD417,BDY26A,HST7202,HST7203,HST7204,IR410,IR411,KSP1142,KSP1143, KSP1144,KSP1145,SDT7202,SDT7203,SDT7204,XGSR10025,*PTC129A,*TR-67, *HEPS5020,*SK3111,*ECG163,*WEP740
2SC682	2N2808,2N2808A,2N2809,2N2809A,2N2810,2N2810A,2N3570,2N3600,2N4252, 2N4253,2N5053,2N5054,2N5851,2N5852,2N6304,2N6305,2SC251,2SC251A,2SC252, 2SC253,2SC567,2SC568,2SC602,2SC611,2SC683,2SC1054,2SC1067,2SC1553A, 40295,40413,A406,MT1061A,S1297,*PTC132,*GE-60,*IR2SC682A,*HEPS0017, *SK3018,*SN80,*ECG161,*WEP719,*276-2011,*RT-113
2SC683	2N2808,2N2808A,2N2809,2N2809A,2N2810,2N2810A,2N3570,2N3600,2N4252, 2N4253,2N5053,2N5054,2N5851,2N5852,2N6304,2N6305,2SC251,2SC251A,2SC252, 2SC253,2SC567,2SC568,2SC602,2SC611,2SC682,2SC1054,2SC1067,2SC1553A, 40295,40413,A406,MT1061A,S1297,*PTC132,*GE-86,*TR-83,*HEPS0017,*SK3018, *SN80,*ECG161,*WEP719,*276-2011,*RT-107
2SC684	*PTC132,*GE-61,*TR-70,*HEPS0016,*SK3039,*ECG108,*WEP56,*276-2015,*RT-108
2SC685	2SC582,2SC1102,2SC1235,2SC1391,*PTC104,*GE-27,*TR-23,*HEPS5011,*SK3021, *PN66,*ECG124,*WEP240,*RT-128

To Replace	Substitute This Type
2SC685A	2SC1059,2SC1102,2SC1235,2SC1391,MJ2252,*PTC104,*GE-12,*TR-81,*HEPS5011, *SK3021,*ECG124,*WEP240,*RT-128
2SC686	2SC154C,*PTC117,*GE-40,*TR-78,*HEPS5025,*SK3045,*ECG154,*WEP712,*BF338, *276-2012,*RT-110
2SC687	2N4347,2N5559,2N6495,2SC42A,2SC676,2SC677,2SC901,2SC902,2SD17,2SD18, 2SD45,2SD46,2SD165,2SD166,BDY25A,KSD2201,KSP1043,KSP1173,SDT7610, SDT7611,SDT7735,SDT7736,*PTC129A,*GE-73,*TR-67,*HEPS5020,*SK3079, *ECG162,*WEP707
2SC688	2SC599,*SK3018,*ECG108,*WEP56
2SC689	2N2710,2SC577,2SC601,2SC601N,*PTC133,*GE-88,*TR-21,*SK3122,*ECG123A, *WEP736,*276-2038,*RT-102
2SC690	2SC599N,2SC690N,*TR-23
2SC690-M	2SC599N,2SC690N
2SC690N	2SC599N
2SC691	2SC599N,2SC690N,2SC692,2SC937-M,*TR-76,*HEPS3005
2SC691-M	2SC599N,2SC690N,2SC937-M
2SC692	2SC599N,2SC690N,*TR-76,*HEPS3005
2SC693	2N4954,2N5827,2N6112,2SC711A,2SC828A,2SC1313,2SC1681-GR,2SC1685, 2SC1850,2SC1992,A157B,BC107,BC107B,BC129B,BC167B,BC171B,BC182B, BC183L,BC184L,BC237B,BC413B,BC414B,BFY39-3,MPS3694,MPS6515,MPS6520, MPS6566,PBC107B,*PTC121,*GE-210,*TR-21,*HEPS0015,*SK3124,*ECG199, *WEP736,*BC107B,*276-2016,*RT-102
2SC693FP	2N4104,2N5828,2SC693NP,2SC1570,2SC1571,2SC1571L
2SC693NP	2N4104,2N5828,2SC693FP,2SC1570,2SC1571,2SC1571L
2SC694	2N4954,2N5827,2N6112,2SC693,2SC711A,2SC828A,2SC1313,2SC1681-GR, 2SC1685,2SC1850,2SC1992,A157B,BC107,BC107B,BC129B,BC167B,BC171B, BC182B,BC183L,BC184L,BC237B,BC413B,BC414B,BFY39-3,MPS3694,MPS6515, MPS6520,MPS6566,PBC107B,*PTC132,*GE-210,*TR-21,*HEPS0015,*SK3124, *ECG199,*WEP736,*BC107B,*276-2016,*RT-102
2SC695	*GE-39,*TR-70,*HEPS0016,*ECG107,*WEP56,*BF365,*276-2011,*RT-108
2SC696	2SC292,2SC293,2SC696A,*GE-32,*TR-87,*HEPS3002,*SK3048,*ECG195A,*WEP243, *276-2012,*RT-114
2SC696A	2SC293,*GE-32,*TR-78,*HEPS5014,*SK3124,*ECG195A,*WEP243,*276-2012
2SC697	2SC298,2SC299,2SC697A,2SC698,*PTC195,*GE-28,*TR-76,*SK3049,*ECG237
2SC697A	2SC299,*PTC195,*GE-46,*SK3049,*ECG237
2SC698	2SC298,2SC299
2SC699	*HEPS3020
2SC700	2N751,2N1491,2N1492,2N1493,2SC302-M,2SC307-M,2SC590N,*HEPS3001
2SC701	2N751,2N1493,2SC302-M,2SC307-M,2SC590N,*HEPS3001,*SK3124,*ECG311, *WEP53
2SC702	2SC691-M,2SC937-M,*HEPS3005,*SK3124,*ECG123,*WEP53
2SC703	2SC1605,*HEPS3006
2SC704	*HEPS3007
2SC705	2N709,2N2475,2N2616,2N2784,2N3633,2N3959,2N3960,2N4996,2N4997,2N5200, 2N5201,2SC706,2SC1215,2SC1789,2SC1988,EN744,MPS6568A,MPSH10,MPSH11, MT1060A,SE3005,*PTC132,*GE-17,*TR-70,*HEPS0016,*SK3039,*ECG107, *WEP56,*276-2011,*RT-108
2SC706	2N2784,2N3633,2N3959,2N3960,2N5200,2N5201,2SC1215,2SC1789,SE3005
2SC707	2N918,2N2808,2N2808A,2N2809,2N2809A,2N2810,2N2810A,2N3600,2N5053,2N5054, 2N5650,2N5651,2N5652,2N5852,2SC251,2SC251A,2SC252,2SC253,2SC567, 2SC602,2SC611,2SC660,2SC661,2SC682,2SC683,2SC761,2SC762,2SC1054, 40295,40413,A406,MT1061,MT1061A,S1297,*PTC132,*GE-214,*TR-70,*HEPS0016, *SK3039,*ECG107,*WEP56,*BF200,*276-2011,*RT-108
2SC708	2N2107,2N2194,2N2194A,2N2194B,2N3110,2N3252,2N3253,2N3444,2N3735,2N5188, 2N5262,2SC97,2SC580,BC140-6,BSX60,*GE-47,*TR-87,*HEPS5014,*SK3024, *HN100,*ECG128,*WEP243,*276-2018,*RT-114
2SC708A	2N698,2N3108,2N4000,2SC485-R,2SC512-R,2SC516,2SC1065,2SC1069,BC141-6, SE7002,*GE-18,*TR-87,*HEPS5014,*SK3024,*HN100,*ECG128,*WEP243, *276-2018,*RT-114
2SC709	2N708,2N914,2N957,2N2242,2N2331,2N2369,2N2369A,2N2501,2N2922,2N3009, 2N3011,2N3013,2N3014,2N3210,2N3211,2N3646,2N3691,2N4123,2N4418,2N4420,

To Replace	Substitute This Type
(2SC709)	2N4421,2N4422,2N4873,2N5418,2S512,2SC67,2SC169,2SC300,2SC301,2SC367G-O, 2SC372-O,2SC372G-O,2SC394-Y,2SC400-O,2SC460,2SC461,2SC536,2SC577, 2SC601N,2SC619,2SC639,2SC712,2SC713,2SC735-O,2SC752G-O,2SC764, 2SC1276,2SC1372,BSW82,BSY61,EN708,MPS2369,MPS3646,MPS6512, MPS6574-BLUE,MPS6574-YEL,PET2001,PET3002,PT720,SE2001,SE3005,ZT80, ZT110,ZT708,ZT2205,ZT2369,ZT2369A,ZT2938,*PTC132,*HEPS0016,*SK3122, *ECG123A
2SC710	2N916,2N916A,2N2501,2N3241A,2N4123,2N4418,2N4873,2N4996,2N4997,2SC372-O, 2SC372G-O,2SC380-O,2SC380A-O,2SC394-Y,2SC400-O,2SC400-Y,2SC454, 2SC458,2SC459,2SC460,2SC536,2SC619,2SC620,2SC639,2SC735-O,2SC752G-O, 2SC899,2SC924,2SC925,2SC933,2SC964,2SC965,2SC1188,2SC1189,2SC1216, 2SC1276,2SC1320,2SC1359,2SC1372,2SC1687,2SC1688,2SC1999,BF123,BF254, EN916,MPS3693,MPS6513,MPS6574-BLUE,MPS6574-GREEN,MPS6574-YEL, MPS6576-BLUE,MPS6576-GREEN,MPS6576-YEL,PET3002,SE2001,SE3005,TIS86, TIS87,*PTC132,*GE-20,*IR2SC710,*HEPS0016,*SK3122,*ECG123A,*WEP736, *276-2016,*RT-308
2SC711	2N4256,2N4419,2N4954,2N5371,2N5827,2N6002,2N6006,2N6008,2N6112,2SC373, 2SC373G,2SC400-GR,2SC711A,2SC735-GR,2SC828,2SC828A,2SC1084,2SC1313, 2SC1684,2SC1849,2SC1993,A157B,BC107,BC107B,BC108,BC108B,BC109B, BC122,BC167B,BC171B,BC183L,BC184L,BC237B,BC238B,BC239B,BC413B, BC414B,BC547B,BC548B,BC549B,BC550B,BCY58B,BCY58C,BCY59B,BCY59C, BFY39-3,MPS3694,MPS6515,MPS6520,PBC107B,*PTC139,*TR-21,*HEPS0020, *SK3122,*ECG199,*WEP736,*276-2016,*RT-304
2SC711A	2N4953,2N5370,2N5376,2N5377,2N5827,2N6006,2N6112,2SC734-GR,2SC1313, 2SC1685,2SC1850,BC107B,BC174B,BC182B,BC190B,BC237B,BC414B,BC547B, BC550B,MPS6566,*PTC132,*TR-21,*HEPS0025,*SK3122,*ECG199,*WEP736, *276-2013
2SC712	2N916,2N916A,2N2331,2N2501,2N3241A,2N3856A,2N4013,2N4123,2N4418,2N4873, 2N5144,2SC318,2SC367G,2SC367G-O,2SC372-O,2SC372G-O,2SC394-Y,2SC400-O, 2SC400-Y,2SC454,2SC458,2SC459,2SC460,2SC536,2SC619,2SC620,2SC639, 2SC713,2SC735-O,2SC752G-O,2SC933,2SC965,2SC1210,2SC1216,2SC1276, 2SC1372,2SC1999,BFY39-2,EN916,MPS3693,MPS6513,MPS6574-BLUE,MPS6574-GREEN,MPS6574-YEL,MPS6576-BLUE,MPS6576-GREEN,MPS6576-YEL, MPSA20-GRN,MPSA20-WHT,PET3002,SE2001,SE3005,ZT81,ZT111,ZT2369A, *PTC132,*TR-24,*HEPS0011,*SK3122,*ECG123A,*WEP736,*276-2016,*RT-302
2SC712A	2N708,2N914,2N916,2N916A,2N957,2N2242,2N2331,2N2369,2N2369A,2N2501, 2N3009,2N3011,2N3013,2N3014,2N3210,2N3211,2N3646,2N3691,2N3693,2N3862, 2N4013,2N4123,2N4418,2N4420,2N4421,2N4422,2N4873,2N5144,2SC67,2SC169, 2SC318,2SC335,2SC367G-O,2SC372-O,2SC372G-O,2SC394-Y,2SC400-O,2SC460, 2SC461,2SC536,2SC577,2SC601N,2SC619,2SC620,2SC639,2SC712,2SC713, 2SC735-O,2SC752G-O,2SC764,2SC1210,2SC1276,2SC1372,BSW82,EN708,EN916, MPS2369,MPS3646,MPS3693,MPS6512,MPS6574-BLUE,MPS6574-YEL, MPS6576-BLUE,MPS6576-YEL,PET1001,PET2001,PET3002,SE1001,SE2001, SE3005,ZT81,ZT111,ZT708,ZT2369,ZT2369A,*PTC132,*HEPS0016,*SK3122, *ECG123A,*WEP736
2SC713	2N916,2N916A,2N2331,2N2501,2N3241A,2N3856A,2N4013,2N4123,2N4418,2N4873, 2N5144,2SC318,2SC367G,2SC367G-O,2SC372-O,2SC372G-O,2SC394-Y,2SC400-O, 2SC400-Y,2SC454,2SC458,2SC459,2SC460,2SC536,2SC619,2SC620,2SC639, 2SC712,2SC735-O,2SC752G-O,2SC933,2SC965,2SC1210,2SC1216,2SC1276, 2SC1372,2SC1999,BFY39-2,EN916,MPS3693,MPS6513,MPS6574-BLUE,MPS6574-GREEN,MPS6574-YEL,MPS6576-BLUE,MPS6576-GREEN,MPS6576-YEL, MPSA20-GRN,MPSA20-WHT,PET3002,SE2001,SE3005,ZT81,ZT111,ZT2369A, *PTC132,*TR-21,*HEPS0014,*SK3122,*ECG123A,*WEP736,*276-2016,*RT-302
2SC713M	SEE 2SC713
2SC714	2N2221A,2N4963,2N5581,BSW84,GET2221A,ZT86,ZT116,*TR-21,*HEPS5026, *SK3122,*ECG123A,*WEP736,*276-2009
2SC715	2N3116,2N3227,2N3302,2N3904,2N4401,2N4952,2N4970,2SC366G-Y,2SC367G-Y, 2SC752G-Y,2SC1175,2SC1285,2SC1364,2SC1685,2SC1850,2SC1992,A157B, BC107,BC107B,BC129A,BC129B,BC167B,BC171B,BC182B,BC182L,BC237B, BC413B,BC414B,BCY59B,BFY39-3,BSX79,MPS6531,PBC107A,PBC107B,*PTC132, *GE-212,*TR-21,*HEPS0016,*SK3018,*ECG123A,*WEP736,*BC107B,*276-2016,

To Replace	Substitute This Type
(2SC715)	*RT-302
2SC716	2N2924,2N3392,2N4124,2N4256,2N4419,2N5172,2N5224,2SC372G-Y,2SC373G, 2SC400,2SC400-GR,2SC735-GR,2SC735-Y,2SC1363,2SC1684,2SC1849,2SC1993, A158B,A159B,BC108,BC108B,BC109B,BC130A,BC130B,BC131B,BC168B,BC172B, BC173B,BC238B,BC239B,BSW88,BSW89,BSX38,MPS5172,PBC108A,PBC108B, PBC109B,*PTC132,*GE-212,*TR-70,*HEPS0016,*SK3039,*ECG123A,*WEP56, *BC107B,*276-2016
2SC717	2SC387AG,MT1038,MT1038A,MT1039,MT1060,*PTC133,*GE-86,*IR2SC717, *HEPS0020,*SK3122,*ECG161,*WEP56,*276-2015,*RT-134
2SC718	2N3959,2N3960,2N5200,2N5201,2SC1231,*PTC123
2SC719	2N5200,2N5201,*PTC123,*SK3124
2SC720	
2SC721	
2SC722	*ECG229
2SC723	
2SC724	2N708,2N708A,2N784A,2N914,2N2242,2N2318,2N2319,2N2481,2N2501,2N2710, 2N3011,2N3013,2N3014,2N3210,2N3211,2N3510,2N3511,2N3605A,2N3606A, 2N3646,2N3647,2N3648,2N4123,2N4295,2N4418,2N4420,2N4421,2N4422,2N5144, 2SC67,2SC169,2SC204,2SC335,2SC468,2SC577,2SC595,2SC620,2SC639, 2SC752G-R,2SC764,2SC1274,2SC1372,BSW82,EN708,EN914,EN3011,EN3013, EN3014,GET708,GET914,GET2369,GET3013,GET3014,GET3646,MPS3646,TIS45, TIS46,TIS51,TIS52,TIS55,ZT708
2SC725	2N2221,2N2221A,2N2539,2N2845,2N2847,2N3115,2N3301,2N3903,2N3946,2N4140, 2N4962,2N4963,2N5027,2N5380,2N5581,2SC336,BSW84,EN915,GET2221, GET2221A,GI-3641,GI-3642,MM3903,MPS6530,TN-3903,ZT83,*SK3122,*ECG123A
2SC726	2N744,2N2318,2N2319,2N2331,2N2656,2N3011,2N3564,2N3691,2N3959,2N3960, 2N4294,2N4418,2N5030,2N5200,2N5201,2N5418,2S095A,2S512,2SC132,2SC133, 2SC135,2SC168,2SC204,2SC356,2SC395A-O,2SC579,2SC595,2SC718,2SC719, 2SC724,2SC915,2SC915A,2SC1071,2SC1209,2SC1231,2SC1372,EN744,EN3011, GI-2922,PT720,PT2760,SE2001,SE3005,TIS51,ZT80,ZT110,ZT404P,ZT2205
2SC727	2N740,2N910,*HEPS5026,*SK3024,*ECG154,*WEP712
2SC728	2SC1753,*HEPS5024,*SK3045,*ECG154,*WEP712
2SC729	
2SC730	2N751,2SC302-M,2SC307-M,2SC590N,2SC865,2SC992,2SC1001,2SC1002,2SC1256, 2SC1820,2SC1947,*TR-76,*HEPS3001,*276-2014
2SC731	2SC872,2SC890,2SC1002,*HEPS3001,*SK3195
2SC732-BL	2N5828,2SC733-BL,2SC1000-BL,2SC1000G-BL,2SC1380-BL,2SC1380A-BL,2SC1570, 2SC1571,2SC1571L,BC550C,*PTC139,*GE-212,*WEP736,*276-2016,*RT-102
2SC732-GR	2N4954,2N5371,2N5827,2N6002,2N6006,2N6008,2N6112,2SC373G,2SC733-GR, 2SC735-GR,2SC1000-GR,2SC1000G-GR,2SC1345,2SC1380-GR,2SC1380A-GR, A157B,BC183L,BC184L,BC337,BC413C,BC414C,BC547B,BC550B,BCY58C, BCY58D,BCY59C,BCY59D,MPS3706,MPSA16,TP3704,*PTC139,*GE-62,*WEP736, *276-2016,*RT-100
2SC732-V	2N5526,*PTC139,*GE-64,*WEP736,*276-2016
2SC733-BL	2N5828,2N6429,2N6429A,BC550C,*PTC139,*GE-212,*TR-21,*WEP736,*276-2016, *RT-102
2SC733-GR	2N4954,2N5371,2N5827,2N6002,2N6006,2N6008,2N6112,2SC735-GR,A157B,BC183L, BC184L,BC337,BC413C,BC414C,BC547B,BC550B,BCY58C,BCY58D,BCY59C, BCY59D,MPS3706,MPSA16,TP3704,*PTC139,*GE-212,*TR-24,*WEP736,*276-2016, *RT-102
2SC733-O	2N916,2N916A,2N4013,2N5825,2N6000,2N6004,2N6426,2N6427,2SC367G, 2SC367G-O,2SC509-O,2SC509G-O,2SC735-O,2SC850,2SC933,2SC968,2SC970, 2SC1216,2SC1999,BC107A,BC167A,BC171A,BC237A,BCY58A,BCY59A,BFY39-2, *PTC101,*GE-212,*TR-21,*SK3122,*ECG199,*WEP736,*276-2016,*RT-102
2SC733-Y	2N3227,2N3242A,2N5826,2N5998,2SC367G-Y,2SC509-Y,2SC509G-Y,2SC733-GR, 2SC735-GR,2SC735-Y,2SC1175,2SC1364,2SC1813,2SC1992,A157B,BC107, BC107B,BC167B,BC171B,BC237B,BC413B,BC414B,BC547A,BCY58B,BCY59B, BFY39-3,BSW83,BSW88,BSW89,BSX38,BSX79,MPS3704,MPSA17,TP3704,TP3706, *PTC101,*GE-212,*WEP736,*276-2016
2SC734-GR	*PTC121,*GE-212,*TR-24,*WEP243
2SC734-O	2N915,2N4014,2SC1781,GET2222A,TIS96,TIS99,ZT88,ZT89,ZT118,ZT119,*PTC121,

To Replace	Substitute This Type
(2SC734-O)	*GE-212,*TR-87,*SK3045,*ECG128,*WEP243,*RT-114
2SC734-R	2N4963,BSW84,FT3722,GET2221A,SE8012,ZT86,ZT116,*PTC121,*GE-212,*TR-87, *SK3045,*ECG128,*WEP243,*RT-114
2SC734-Y	2N4409,2N6540,2N6541,2SC734-GR,A5T4409,BSW85,TIS95,TIS98,*PTC121,*GE-212, *TR-87,*HEPS5026,*ECG128,*WEP243,*RT-114
2SC735-GR	2N4954,2N5371,2N6012,*GE-210,*TR-25,*276-2016
2SC735-O	2N916,2N916A,2N4013,2N6010,*PTC170,*GE-20,*TR-21,*HEPS0014,*SK3122, *ECG289,*276-2016,*RT-102
2SC735-R	2N708A,2N914,2N2481,2N2501,2N2710,2N3210,2N3211,2N3510,2N3511,2N3646, 2N3647,2N3648,2N5144,2SC577,BSW82,ZT708,*PTC170,*GE-210,*TR-21,*SK3122, *WEP735A,*276-2016,*RT-102
2SC735-Y	2N3227,2SC735-GR,BSW83,*PTC170,*GE-210,*TR-21,*HEPS0014,*276-2016, *RT-102
2SC736	2N6354,2SC665,2SC897,2SC898,2SC902,2SC939,2SC1030,2SC1111,2SC1112, 2SC1115,2SC1116,2SC1343,2SC1402,2SC1403,2SC1403A,2SC1782,2SC1783, 2SD17,2SD51A,2SD60,2SD74,2SD83,2SD110-O,2SD110-R,2SD117,2SD118-R, 2SD118-Y,2SD203,2SD218,2SD286,2SD323,2SD375,2SD388,KSP1046,KSP1176, SDT7604,SDT7605,*PTC129A,*HEPS7004,*SK3027,*ECG130,*WEP247
2SC737	2SC599N,2SC690-M,2SC690N,*HEPS3005,*ECG123A,*WEP736
2SC738	2SC380-Y,2SC380A-Y,2SC739,2SC752G-Y,MPS6514,*PTC132,*HEPS0011,*SK3020, *ECG107,*WEP56,*276-2011,*RT-309
2SC739	2N4124,2N4419,2SC380-Y,2SC380A-Y,2SC400,2SC400-GR,2SC735-Y,2SC738, 2SC752G-Y,BC108,MPS6514,*PTC132,*HEPS0011,*SK3018,*ECG107,*WEP56, *276-2011,*RT-302
2SC740	2SC251,2SC251A,2SC252,2SC253,2SC313,2SC464,2SC465,2SC466,2SC602,40414, MT1061,*HEPS0016,*SK3039,*ECG108,*WEP56
2SC741	2N751,2SC302-M,2SC307-M,2SC590N,2SC730,2SC731,2SC844,2SC854,2SC855, 2SC865,2SC872,2SC890,2SC908,2SC991,2SC992,2SC1001,2SC1002,2SC1083, 2SC1165,2SC1256,2SC1303,2SC1315,2SC1464,2SC1479,2SC1820,2SC1947, *PTC125,*TR-21,*HEPS3013,*SK3122,*ECG123A,*WEP736,*276-2009
2SC742	
2SC743	
2SC745	
2SC746	
2SC748	*SK3019,*ECG108,*WEP56
2SC749	
2SC751	
2SC752	*PTC121,*HEPS0011,*SK3122,*ECG123A,*WEP736
2SC752G	2N3227,2N6010,2SC752G-Y,2SC1216,MPS6531,*PTC115,*TR-21,*HEPS0014, *SK3122,*ECG123A,*WEP736,*276-2009,*RT-102
2SC752G-O	2N6010,2SC1216,*GE-210,*276-2016
2SC752G-R	2N2501,2N2710,2N3013,2N3014,2N3211,2N3510,2N3511,2N3646,2N3647,2N3648, 2N4295,2N4420,2N4421,2N4422,2SC67,2SC169,2SC335,2SC336,2SC468,2SC577, 2SC639,2SC764,2SC1274,EN3013,EN3014,GET2369,GET3013,GET3014,GET3646, MPS3646,MPS6530,TIS52,TIS55,*GE-210,*276-2016
2SC752G-Y	2N3227,MPS6531,*GE-210,*276-2016
2SC753	
2SC754	2SC182,2SC269
2SC755	2SC475,2SC476,2SC540,2SC640
2SC756	*HEPS3021,*SK3049,*ECG282,*WEP1239,*RT-306F
2SC757	
2SC758	
2SC759	
2SC760	
2SC761	2N918,2N2808,2N2808A,2N2809,2N2809A,2N5053,2N5054,2N5852,2SC251, 2SC251A,2SC252,2SC253,2SC463,2SC602,2SC762,2SC1054,2SC1747,40239, BF232,MT1061,MT1061A,*PTC132,*GE-39,*TR-83,*HEPS0017,*SK3018,*ECG161, *WEP719,*BF200,*276-2011,*RT-113
2SC762	2N918,2N2808,2N2808A,2N2809,2N2809A,2N5053,2N5054,2N5851,2N5852,2SC251, 2SC251A,2SC252,2SC253,2SC463,2SC563,2SC563A,2SC602,2SC761,2SC1054, 2SC1067,2SC1747,40239,BF232,BFX60,MT1061,MT1061A,TIS56,TIS57,*PTC132,

To Replace	Substitute This Type
(2SC762)	*GE-60,*TR-83,*HEPS0017,*SK3018,*ECG161,*WEP719,*BF200,*276-2011,*RT-113
2SC763	2N2369,2N2369A,2N2501,2N2615,2N2616,2N3009,2N3011,2N3013,2N3014,2N3564, 2N3688,2N3689,2N3690,2N4295,2N4420,2N4421,2N4422,2N4996,2N4997,2N5029, 2SC67,2SC169,2SC300,2SC301,2SC380-O,2SC380A-O,2SC535,2SC601N,2SC639, 2SC752G-O,2SC764,2SC784-O,2SC785-O,2SC836,2SC922,2SC1123,2SC1128, 2SC1129,2SC1187,2SC1188,2SC1215,2SC1240,2SC1274,2SC1319,2SC1393, 2SC1394,2SC1395,2SC1396,2SC1674,2SC1789,2SC1923,2SC1988,BF121,BF123, BF125,BF127,BF198,BF199,BF241,BF311,GET2369,MPS3646,MPSH10,MPSH11, MT1060A,SE3005,SE5001,SE5002,SE5003,SE5006,TIS48,*PTC132,*TR-24, *HEPS0033,*SK3122,*ECG229,*WEP56,*276-2011,*RT-309
2SC764	*PTC133
2SC765	*PTC906,*HEPS7002,*SK3027,*ECG130,*WEP247
2SC766	2SC767,*PTC906,*HEPS7004
2SC767	2SC766,*PTC129A,*HEPS5020
2SC768	2SD163,2SD172,2SD211,1723-0605,1723-0610,B170004-BLK,B170004-BRN,B170022, HST9801,HST9802,KSD1051,KSD1055,KSD1056,KSD9707,KSP1171,KSP1271, SDT7607,SDT7608,SDT7731,SDT7732,SDT9901,SDT9902,*PTC140,*HEPS7002, *SK3027,*ECG130,*WEP247
2SC769	2N5559,2N6495,2SC1440,2SC1784,2SD161,2SD165,2SD177,2SD213,2SD214, 2SD319,1723-1005,1723-1010,1723-1205,1723-1210,1723-1405,1723-1410,HST9804, KSD1053,KSD1054,KSD1058,KSD2201,KSD2202,KSD9702,KSD9703,KSD9705, KSD9706,KSP1173,KSP1273,SDT7610,SDT7734,SDT7735,SDT9904,*PTC118, *HEPS7004
2SC770	2SC771,2SC1436,2SC1441,2SC1785,2SC1786,2SD166,2SD376,1723-1805, 1723-1810,HST1050,HST1055,HST7201,HST7202,HST7207,HST7208,KSP1141, KSP1142,KSP1221,KSP1222,SDT7201,SDT7202,SDT7207,SDT7208,SDT7612, XDAR10020,XGSR15020,*PTC118,*HEPS5020
2SC771	2N6249,2SC1436,2SC1786,2SD376,HST1050,HST1055,HST7202,HST7203,HST7204, HST7208,HST7209,KSP1142,KSP1143,KSP1144,KSP1145,KSP1222,KSP1223, KSP1224,KSP1225,SDT7202,SDT7203,SDT7204,SDT7208,SDT7209,XDAR10020, XDAR10025,XGSR10025,XGSR15020,XGSR15025,*PTC118,*HEPS5020,*SK3018
2SC772	2N709,2N709A,2N744,2N1992,2N2318,2N2319,2N2475,2N2656,2N2784,2N3011, 2N3564,2N3605,2N3606,2N3607,2N3633,2N3959,2N3960,2N4294,2N4418,2N4996, 2N4997,2N5030,2N5200,2N5201,2S095A,2S512,2SC99,2SC168,2SC204,2SC300, 2SC301,2SC400-O,2SC595,2SC718,2SC719,2SC724,2SC924,2SC929,2SC930, 2SC1231,2SC1372,BF123,BF125,EN744,EN3011,MPS2713,MPS2926-RED, MT1060A,NT3000,PT720,SE3005,TIS51,ZT2205,*PTC132,*GE-39,*IR2SC772, *HEPS0016,*SK3018,*ECG229,*WEP56,*BC107B,*276-2011,*RT-187
2SC773	2N749,2N751,2N1388,2N1390,2SC302-M,2SC307-M,2SC590N,2SC731,2SC774, 2SC797,2SC798,2SC802,2SC803,2SC844,2SC854,2SC855,2SC865,2SC872, 2SC890,2SC991,2SC992,2SC1002,2SC1083,2SC1256,2SC1303,2SC1315, 2SC1464,2SC1479,2SC1820,40637A,*GE-210,*TR-62,*HEPS0015,*SK3122, *ECG123A,*WEP773,*276-2016,*RT-300
2SC774	2N749,2N751,2N1388,2N1390,2SC302-M,2SC307-M,2SC590N,2SC731,2SC773, 2SC797,2SC798,2SC802,2SC803,2SC844,2SC854,2SC855,2SC865,2SC872, 2SC890,2SC991,2SC992,2SC1002,2SC1083,2SC1256,2SC1303,2SC1315, 2SC1464,2SC1479,2SC1820,40637A,*GE-219,*TR-62,*HEPS3001,*SK3047, *ECG195A,*WEP774,*276-2014,*RT-306
2SC775	2N749,2N751,2N1388,2N1390,2SC302-M,2SC307-M,2SC590N,2SC731,2SC773, 2SC774,2SC797,2SC798,2SC802,2SC803,2SC844,2SC854,2SC855,2SC865, 2SC872,2SC890,2SC991,2SC992,2SC1002,2SC1083,2SC1256,2SC1464,2SC1479, 2SC1820,*PTC115,*HEPS3001,*SK3047,*ECG195A,*WEP243,*RT-306
2SC776	2N749,2N751,2N1388,2N1390,2SC302-M,2SC307-M,2SC590N,2SC773,2SC774, 2SC798,2SC803,2SC865,2SC992,2SC1002,2SC1256,2SC1820,*HEPS3001, *SK3048,*ECG195A,*WEP243,*RT-306F
2SC777	2SC799,2SC866,*PTC144,*HEPS3002,*SK3047,*ECG237,*WEP777,*RT-174
2SC778	*PTC144,*HEPS3002,*SK3049,*ECG237,*WEP778,*RT-174
2SC779-O	2SC779-Y
2SC779-R	2SC679,*ECG286
2SC779-Y	*ECG286
2SC780	*PTC123,*HEPS5026,*SK3045,*ECG287,*WEP712

To Replace	Substitute This Type
2SC780AG-O	2SC1278,2SC1279,*ECG287
2SC780AG-R	2SC1515,*ECG287
2SC780AG-Y	*ECG287
2SC780G-O	2N3077,2N4946,2SC979A-O,2SC980AG-O,EN871,GET2222A,PET1075,*PTC170, *GE-81,*276-2008
2SC780G-R	2N4944,2N4945,2SC339,2SC979A-R,2SC980AG-R,2SC1283,BC110,EN718A,EN870, EN1613,GET2221A,*PTC170,*GE-81,*276-2008
2SC780G-Y	EN956,EN1711,*PTC170,*GE-81,*276-2008
2SC781	*PTC144,*GE-219,*TR-87,*HEPS3001,*SK3048,*ECG195,*WEP243,*276-2014, *RT-306
2SC782	2SC782A,*PTC104,*GE-12,*TR-23,*HEPS3021,*SK3194,*ECG124,*WEP244
2SC782A	*HEPS3021,*SK3021
2SC783-O	BLY50A
2SC783-R	2N5052,2N5664,BLY49A,HST6907,HST6908
2SC783-Y	
2SC784-BN	2N2368,2SC302,2SC321,2SC322,2SC468,2SC601,2SC689,2SC784-R,2SC785-BN, 2SC785-R,EN2369A,TIS49,TIS129,*PTC132,*GE-60,*WEP784,*276-2015,*RT-108
2SC784-O	2N4873,2SC785-O,2SC1216,2SC1276,BF240,*PTC132,*GE-60,*TR-70,*HEPS0016, *SK3122,*ECG229,*WEP784,*276-2015,*RT-108
2SC784-R	2N2369,2N2369A,2N3862,2N4137,2N4295,2N5029,2SC169,2SC335,2SC336,2SC468, 2SC601,2SC601N,2SC639,2SC764,2SC785-R,2SC1187,2SC1188,2SC1189, 2SC1319,2SC1320,2SC1923,BF199,EN2369A,TIS48,TIS49,TIS129,*PTC132,*GE-60, *TR-21,*HEPS0016,*RT-108
2SC785	2N3862,2N4873,2SC169,2SC639,2SC764,2SC784-O,2SC785-O,2SC1187,2SC1188, 2SC1189,2SC1276,2SC1319,2SC1320,2SC1923,BF199,*PTC132,*GE-60,*TR-71, *HEPS0016,*SK3122,*EN10,*ECG229,*WEP56,*276-2011,*RT-187
2SC785-BN	2N2368,2SC302,2SC321,2SC322,2SC468,2SC601,2SC689,2SC784-BN,2SC784-R, 2SC785-R,EN2369A,TIS49,TIS129,*PTC132,*GE-60,*WEP56,*RT-108
2SC785-O	2N4873,2SC784-O,2SC1216,2SC1276,BF240,*PTC132,*GE-60,*HEPS0016,*SK3122, *ECG229,*WEP56,*276-2015
2SC785-R	2N2369,2N2369A,2N3862,2N4137,2N4295,2N5029,2SC169,2SC335,2SC336,2SC468, 2SC601,2SC601N,2SC639,2SC764,2SC784-R,2SC1187,2SC1188,2SC1189, 2SC1319,2SC1320,2SC1923,BF199,EN2369A,TIS48,TIS49,TIS129,*PTC132,*GE-60, *WEP56,*RT-108
2SC785-Y	MPS6520,*PTC132,*GE-39
2SC786	2N2865,2N3570,2N3571,2N3572,2N4252,2N4253,2N6304,2N6305,2SC398,2SC399, 2SC568,2SC653,*PTC132,*GE-11,*TR-83,*HEPS0017,*SK3039,*ECG229,*WEP719, *276-2011,*RT-113
2SC787	MT1061,*PTC126,*GE-214,*TR-24,*HEPS0020,*SK3039,*ECG161,*WEP719, *276-2015,*RT-113
2SC788	2SC995,2SC1056,*GE-27,*TR-78,*HEPS5025,*SK3044,*ECG154,*WEP712, *276-2012,*RT-110
2SC789-O	2N6292,2SD525,2SD526,*GE-66,*TR-76,*WEP745,*276-2019
2SC789-R	2N6123,2N6131,2N6292,D44C10,D44C11,D44C12,D44C7,D44C8,D44C9,*GE-66, *TR-76,*ECG152,*WEP745
2SC789-Y	2SD525,*TR-76,*ECG152,*WEP745
2SC790-O	2SC789-O,*ECG291,*276-2019
2SC790-R	2N6122,2SC789-R,D44C4,D44C5,D44C6,D44C7,D44C8,D44C9
2SC790-Y	2SC789-Y
2SC791	2N5602,HST5913,*PTC104,*GE-12,*HEPS5012,*SK3026,*ECG175,*WEP241
2SC792	*PTC129A,*GE-73,*TR-86,*HEPS5020,*SK3115,*ECG162,*WEP707
2SC793-BL	2N5626,2SC1051L,*PTC140,*TR-61,*WEP247,*RT-131
2SC793-R	2N3446,2N3448,2N4070,2N5624,2N5628,2SC961,KSP1172,KSP1175,SDT7603, *PTC140,*TR-61,*WEP247,*RT-131
2SC793-Y	2N3448,2N4070,2N5626,2SC793-BL,KSP1175,SDT7603,*PTC140,*TR-61,*WEP247, *RT-131
2SC794	2N3445,2N3446,2N3447,2N3448,2N5624,2SC793-R,KSP1171,KSP1172,KSP1174, KSP1175,SDT7602,SDT7603,*SK3027
2SC795	*HEPS5011,*SK3021,*ECG124,*WEP240
2SC796	2N744A,2N1708A,2N3252,*PTC136,*HEPS0014,*SK3122,*ECG123A,*WEP736
2SC797	2N749,2N751,2N1388,2N1390,2SC302-M,2SC307-M,2SC590N,2SC773,2SC774,

To Replace	Substitute This Type
(2SC797)	2SC798,2SC802,2SC803,2SC865,2SC872,2SC890,2SC991,2SC992,2SC1002, 2SC1256,2SC1820,*PTC123,*HEPS0015,*SK3024,*ECG128,*WEP243
2SC798	2N749,2N751,2N1388,2N1390,2SC302-M,2SC307-M,2SC590N,2SC773,2SC774, 2SC803,2SC865,*HEPS3020,*SK3024,*ECG128,*WEP243
2SC799	2SC866,*PTC176,*GE-45,*TR-65,*SK3049,*ECG237,*WEP799,*RT-306F
2SC800	2SC287A,2SC1158,2SC1159,*GE-60,*TR-70,*HEPS0016,*SK3039,*ECG107,*WEP56, *BF183,*276-2038,*RT-108
2SC801	2SC22,2SC23,2SC24
2SC802	2N749,2N751,2N1388,2N1390,2SC302-M,2SC307-M,2SC590N,2SC773,2SC774, 2SC797,2SC798,2SC803,2SC865,2SC872,2SC890,2SC991,2SC992,2SC1002, 2SC1256,2SC1820,*HEPS3001
2SC803	2N749,2N751,2N1388,2N1390,2SC302-M,2SC307-M,2SC590N,2SC773,2SC774, 2SC798,2SC865,*HEPS3002,*SK3024,*ECG128,*WEP243
2SC804	2SC182,2SC330,2SC331,2SC605,2SC606,2SC989,A3T3011,FX709,FX918,MMT2857, MMT3960A,MT3833,MT3834
2SC805	*HEPS5026,*SK3040,*ECG154,*WEP712
2SC806	*HEPS5021,*SK3115,*ECG283,*WEP707
2SC806A	*HEPS5021,*SK3115,*ECG283,*WEP740
2SC807	*HEPS5021,*SK3111,*ECG163,*WEP740
2SC807A	*HEPS5020,*SK3111,*ECG163,*WEP740
2SC808	BDY26C,BUY21
2SC809	*PTC126
2SC810	
2SC811	
2SC812	2N744,2N784,2N2318,2N2319,2N2432,2N2571,2N2572,2N2656,2N3011,2N3394, 2N3397-RED,2N3398-RED,2N3854A,2N3855A,2N5030,2N5187,2N5418,2SC99, 2SC132,2SC133,2SC135,2SC137,2SC204,2SC356,2SC395A-R,2SC400-R,2SC579, 2SC595,2SC718,2SC915,2SC915A,2SC1071,2SC1231,2SC1374,2SC1375,BC170A, MPS3394,PET2001,PT720,PT2760,TIS51,ZT404P,ZT2205,*PTC123
2SC814	*PTC178,*GE-210,*TR-65,*HEPS0014,*SK3124,*ECG192,*WEP243,*BC337, *276-2009,*RT-102
2SC815	2N915,2N2221,2N2221A,2N2539,2N2847,2N3115,2N3903,2N3946,2N3974,2N3976, 2N4014,2N4227,2N4400,2N4951,2N4962,2N4963,2N5027,2N5368,2N5380,BSW84, MM3903,MPS6530,TN-3903,ZT84,ZT114,*PTC170,*GE-210,*TR-21,*HEPS0015, *SK3122,*ECG123A,*WEP736,*BC107B,*276-2010,*RT-302
2SC816	2N3725A,*PTC144,*HEPS3001,*SK3024,*ECG128,*WEP243
2SC817	2N3478,2N6389,2SC988,2SC988A,2SC988B,2SC1117,2SC1260,2SC1275,2SC1424, 2SC1529,2SC1553,2SC1574,35824A,40915,*HEPS3019
2SC818	2SC507-O,MM2260,*PTC117,*HEPS5025,*SK3045,*ECG154,*WEP712
2SC819	
2SC820	
2SC821	*ECG278
2SC822	*HEPS3001,*ECG278
2SC823	*HEPS0020,*ECG278,*276-2011
2SC824	
2SC825	2SC679,40318,40322,*SK3021
2SC826	2N1973,2N2405,2N2443,2SC827,2SC1008A,BC141-10,BC141-16,ZT92,*PTC125, *HEPS5026,*SK3024,*ECG128,*WEP243
2SC827	2N1973,2N2405,2N2443,2N3036,2SC1008A,BC141-10,BC141-16,ZT92,*PTC125, *HEPS3019,*SK3024,*ECG128,*WEP243
2SC828	2SC828A,BC108C,BC109,BC109C,BC238C,BC239C,BC413C,BC414C,BCY58D, BCY59D,MPSA18,*PTC139,*GE-20,*IR2SC828A,*HEPS0015,*SK3124,*ECG199, *WEP736,*276-2016,*RT-302
2SC828A	2N4104,2N4953,2N5370,BC413C,BC414C,BCY59D,MPSA18,*GE-39,*IR2SC828A, *HEPS0015,*SK3024,*ECG199,*WEP736,*276-2009,*RT-107A
2SC829	2N4256,2N5827,2SC735-GR,2SC1993,A157B,BC107B,BC108B,BC109B,BC122, BC167B,BC171B,BC237B,BC238B,BC239B,BC413B,BC414B,BC547B,BC548B, BC549B,BC550B,BCY58C,BCY59C,*PTC139,*GE-39,*TR-24,*HEPS0011,*SK3122, *ECG123A,*WEP56,*276-2015,*RT-308
2SC830	*PTC112,*GE-66,*TR-81,*HEPS5019,*SK3026,*ECG175,*WEP241,*276-2020
2SC831	*GE-69

To Replace	Substitute This Type
2SC833	2N4240
2SC836	2N2369,2N2369A,2N3862,2N4295,2N4996,2N4997,2N5029,2SC601N,2SC639, 2SC764,2SC922,2SC1187,2SC1188,2SC1189,2SC1319,2SC1320,2SC1393, 2SC1394,2SC1395,2SC1396,2SC1687,2SC1688,BF123,BF199,BF311,MT1060A, SE3005,TIS48,TIS87,*WEP736
2SC837	2N2368,2N4137,2SC601,2SC689,MT1038,MT1038A,MT1039,MT1060,PET3001,TIS49, TIS84,TIS129,*HEPS0020,*SK3018,*ECG107,*WEP56,*RT-107A
2SC838	2N915,2N2845,2N2847,2N3301,2N3903,2N3946,2N4227,2N4962,2N5027,2N5380, 2SC335,2SC336,2SC620,2SC839,2SC899,2SC944,2SC979-O,MM3903,TN-3903, *PTC132,*GE-20,*IR2SC838,*HEPS0015,*SK3122,*ECG123A,*WEP243,*BC107B, *276-2016,*RT-301
2SC839	2N915,2N2845,2N2847,2N3301,2N3903,2N3946,2N4227,2N4962,2N5027,2N5380, 2SC335,2SC336,2SC620,2SC838,2SC899,2SC944,2SC979-O,MM3903,TN-3903, *PTC132,*GE-61,*TR-24,*HEPS0015,*SK3122,*ECG123A,*WEP243,*BC107B, *276-2016,*RT-308
2SC840	2N5202,2N5600,2N5604,*GE-12,*TR-81,*HEPS5012,*SK3021,*ECG175,*WEP241, *276-2017
2SC840A	2N3441,HST6905,HST6906,HST6907,*GE-12,*TR-23,*HEPS5011,*SK3021,*ECG175, *WEP241
2SC841	
2SC842	
2SC843	2SC557,2SC573
2SC844	2SC302-M,2SC307-M,2SC590N,2SC731,2SC872,2SC890,2SC991,2SC992,2SC1002, 2SC1083,2SC1256,2SC1464,2SC1479,2SC1820,*HEPS3013,*ECG123A,*WEP736
2SC845	2N749,2N751,2N1388,2N1390,2SC302-M,2SC307-M,2SC590N,*HEPS3001
2SC847	2N3227,2N3242A,2N3704,2N5449,2N5818,2N5998,2N6000,2N6004,2N6010,2N6426, 2SC509-Y,2SC509G-Y,2SC848,2SC850,2SC966,2SC967,2SC968,2SC970,2SC1175, 2SC1246,2SC1247,2SC1317,2SC1364,2SC1813,2SC1851,A5T4124,A5T5449, A8T3704,BC337-16,BC337-25,BCY58A,BCY58B,BCY59A,BCY59B,BSW83,D33D22, D33D26,GI-3704,GI-3706,PET3704,PET3706,TN60,*PTC123,*HEPS0011,*SK3122, *ECG123A,*WEP736
2SC848	2N3227,2N3242A,2N3509,2N3704,2N5449,2N5818,2N5998,2N6000,2N6004,2N6010, 2N6426,2SC509-Y,2SC509G-Y,2SC850,2SC967,2SC968,2SC970,2SC1246, 2SC1247,2SC1317,2SC1813,2SC1851,A5T5449,A8T3704,BC337-16,BC337-25, BSW83,D33D22,D33D26,GI-3704,GI-3706,PET3704,PET3706,TN60,TN80,*PTC123, *HEPS0011,*SK3122,*ECG123A,*WEP736
2SC849	2N3227,2N3242A,2N3509,2N3704,2N4386,2N5449,2N5818,2N5998,2N6000,2N6004, 2N6010,2N6426,2SC509-Y,2SC509G-Y,2SC848,2SC850,2SC967,2SC968,2SC970, 2SC1246,2SC1247,2SC1317,2SC1813,2SC1851,A5T5449,A8T3704,BC337-16, BC337-25,BSW83,D33D22,D33D26,GI-3704,GI-3706,PET3704,PET3706,PET8001, PET8003,TIS90,TIS90-BLU,TIS90-GRY,TIS90-VIO,TIS90M,TIS90M-BLU,TIS90M-GRY, TIS90M-VIO,TIS92-BLU,TIS92-GRY,TIS92-VIO,TIS92M-BLU,TIS92M-GRY,TIS92M-VI- O,TN60,TN80,TZ81,*PTC123,*HEPS0011,*SK3122,*ECG123A,*WEP736
2SC850	2N956,2N2390,2SC970,*PTC123,*HEPS0015,*SK3122,*ECG123A,*WEP736
2SC851	*HEPS7002,*SK3027,*ECG130,*WEP247
2SC852	*HEPS0020
2SC853	*PTC178,*GE-210,*TR-21,*HEPS0015,*SK3122,*ECG192,*WEP243,*BC337, *276-2012,*RT-102
2SC854	2SC302-M,2SC307-M,2SC590N,2SC731,2SC844,2SC855,2SC872,2SC890,2SC991, 2SC992,2SC1002,2SC1083,2SC1256,2SC1464,2SC1479,2SC1820,*HEPS3008
2SC855	2SC302-M,2SC307-M,2SC590N,2SC731,2SC844,2SC872,2SC890,2SC991,2SC992, 2SC1002,2SC1083,2SC1256,2SC1464,2SC1479,2SC1820,*HEPS3008
2SC856	2SC1596,*GE-27,*TR-78,*HEPS5025,*SK3040,*ECG154,*276-2012,*RT-110
2SC857	2N6430,2SC1515,*HEPS0030,*SK3040,*ECG154,*WEP712
2SC858	2N2925,2N3391,2N3391A,2N4124,2N4256,2N4419,2N5172,2N5219,2N5223,2SC373, 2SC373G,2SC400-GR,2SC644,2SC711,2SC735-GR,2SC828,2SC1084,2SC1684, 2SC1849,2SC1993,A158B,A159B,BC108,BC108B,BC109B,BC122,BC130B,BC131B, BC168B,BC169B,BC172B,BC173B,BC238B,BC239B,MPS5172,PBC108B,PBC109B, *PTC123,*HEPS0015,*SK3124
2SC859	2N2925,2N3391,2N3391A,2N4124,2N4256,2N4419,2N5172,2N5219,2N5223,2SC373, 2SC373G,2SC400-GR,2SC644,2SC711,2SC735-GR,2SC828,2SC1084,2SC1684,

To Replace	Substitute This Type
(2SC859)	2SC1849,2SC1993,A158B,A159B,BC108,BC108B,BC109B,BC122,BC130B,BC131B, BC168B,BC169B,BC172B,BC173B,BC238B,BC239B,MPS5172,PBC108B,PBC109B, *PTC123,*HEPS0015,*SK3124,*ECG199,*WEP736
2SC860	2N2475,2N2616,2N2784,2N3633,2N4996,2N4997,2N5200,2N5201,2SC706,2SC1215, 2SC1789,MPS6568A,MPSH10,MPSH11,PET3002,SE3005,*HEPS0017,*SK3117, *ECG161,*WEP719
2SC861	
2SC862	SDT1063,SDT1064,*PTC129A,*TR-61
2SC863	*GE-211,*TR-83,*SK3117,*ECG161,*WEP719,*RT-113
2SC864	2SC563A,BF232,*PTC132,*GE-60,*TR-83,*HEPS0016,*SK3132,*ECG161,*WEP719, *276-2016,*RT-113
2SC865	2N749,2N751,2N1388,2N1390,2SC302-M,2SC307-M,2SC590N
2SC866	2SC799,*HEPS3001,*SK3048,*RT-174
2SC867	*HEPS5012,*SK3021,*ECG127,*WEP240
2SC868	2N740,2N740A,2N2519,2SC869,2SC1278,2SC1438,2SC1439,*HEPS5025,*SK3045, *ECG154,*WEP712
2SC869	2N4269,2N6430,2SC728,2SC926A,2SC1278,2SC1279,2SC1438,2SC1439,*TR-78, *HEPS5025,*SK3045,*ECG154,*WEP712
2SC870	2N3242A,2N3692,2N3694,2N4124,2N4419,2SC367G-Y,2SC372-Y,2SC372G-Y, 2SC380-Y,2SC380A-Y,2SC394-GR,2SC400,2SC400-GR,2SC711,2SC711A, 2SC735-Y,2SC752G-Y,2SC1085,2SC1175,2SC1285,2SC1364,2SC1684,2SC1849, 2SC1853,2SC1854,2SC1992,BC107,BC107A,BC108,BC108A,BC123,BC167A, BC171A,BC237A,BC238A,BC547A,BC548A,BCY58A,BCY58B,BCY59A,BCY59B, BSW88,BSW89,BSX38,BSX79,MPS3694,MPS6514,MPS6573,MPS6574-SIL, MPS6575,MPS6576-SIL,MPS-A20,MPS-K20,MPS-K21,MPS-K22,MPSA20-BLU, MPSA20-YEL,NPSA20,PBC107A,PET1002,PET2002,SE1002,SE2002,*PTC139, *HEPS0015,*SK3122,*ECG123A,*WEP736,*RT-304
2SC871	2N4256,2N4419,2N5827,2N6112,2SC373,2SC373G,2SC400-GR,2SC711,2SC711A, 2SC735-GR,2SC828,2SC828A,2SC829,2SC1084,2SC1313,2SC1684,2SC1849, 2SC1853,2SC1993,A157B,BC107,BC107B,BC108,BC108B,BC109B,BC122,BC167B, BC171B,BC183L,BC184L,BC237B,BC238B,BC239B,BC413B,BC414B,BC547B, BC548B,BC549B,BC550B,BCY58B,BCY58C,BCY59B,BCY59C,BFY39-3,MPS3694, MPS6515,MPS6520,PBC107B,*PTC139,*HEPS0015,*SK3018,*ECG123A,*WEP736
2SC871M	SEE 2SC1312
2SC872	2SC302-M,2SC890,2SC1002
2SC873	2N3722,2N3723,2N4961,2N4963,2N5262,2N6375,2N6376,2SC97A,BSX60,FT3722
2SC874	2N2846,2N2848,2N3015,2N3299,2N3326,2N3736,2N4960,2N4962,2N5262,2N6375, 2N6376,2SC97,2SC580,2SC873,BSX60,TN-3903
2SC875	2N2219A,2N4409,2N6540,2N6541,2SC109A-Y,A5T4409,BSW85,BSY54,BSY88,TIS95, TIS98,*HEPS3019,*SK3024,*ECG128,*WEP243
2SC876	2N2219,2N2219A,2N2538,2N2540,2N2959,2N3300,2N5369,2N5377,2N5381, 2SC109A-Y,2SC503-GR,2SC594-Y,2SC875,A5T2222,A5T3904,BSW85,BSY54, TIS109,TIS111,TN-3904,*PTC123,*HEPS0015,*SK3024,*ECG128,*WEP243
2SC877	2N708,2N708A,2N718,2N760,2N784A,2N914,2N2221,2N2242,2N2312,2N2315, 2N2369,2N2369A,2N2396,2N2481,2N2501,2N2539,2N2710,2N2845,2N2847,2N3013, 2N3014,2N3115,2N3210,2N3211,2N3301,2N3508,2N3510,2N3511,2N3605A, 2N3606A,2N3641,2N3642,2N3646,2N3647,2N3648,2N3705,2N3862,2N3903,2N3946, 2N3973,2N3974,2N3975,2N3976,2N4123,2N4137,2N4140,2N4400,2N4962,2N5027, 2N5029,2N5144,2N5380,2N5450,2N5769,2N5772,2N6538,2SC67,2SC131,2SC134, 2SC239,2SC570,2SC577,2SC613,2SC624,2SC639,2SC764,2SC878,2SC913, 2SC914,2SC914A,2SC1166-R,2SC1274,A5T3903,A5T4123,A5T5450,A8T3705, BSW82,BSY93,D33D24,EN697,FT3641,FT3642,GET708,GET914,GET2221,GET2369, GET3013,GET3014,GET3646,GI-3641,GI-3642,GI-3705,KT218F,MM3903,MPS2369, MPS3705,MPS6530,PET3705,PET6001,TIS110,TN62,TN-3903,ZT83,ZT113,ZT708, ZT2369
2SC878	2N708,2N708A,2N718,2N760,2N784A,2N914,2N2221,2N2242,2N2315,2N2369, 2N2369A,2N2396,2N2481,2N2501,2N2539,2N2845,2N2847,2N3013,2N3014,2N3115, 2N3210,2N3211,2N3301,2N3508,2N3641,2N3642,2N3646,2N3647,2N3705,2N3862, 2N3903,2N3946,2N3974,2N3976,2N4013,2N4123,2N4140,2N4227,2N4400,2N4962, 2N5027,2N5029,2N5144,2N5380,2N5450,2N5769,2N5772,2N6427,2N6538,2SC67, 2SC366G-O,2SC367G-O,2SC577,2SC613,2SC639,2SC764,2SC913,2SC914,

To Replace	Substitute This Type
(2SC878)	2SC914A,2SC1166-O,2SC1274,A5T3903,A5T4123,A5T5450,A8T3705,BSW82,D33D24, EN697,FT3641,FT3642,GET2221,GET2369,GI-3641,GI-3642,GI-3705,KT218F, MM3903,MPS2369,MPS3705,MPS6530,PET3705,PET6001,TIS90-YEL,TIS90M-YEL, TIS92-YEL,TIS92M-YEL,TIS110,TN62,TN-3903,ZT708,ZT2369
2SC879	2N2017,2N2106,2N2193,2N2193A,2N2193B,2N2270,2N2848,2N2868,2N3109, 2N3725A,2N4960,2N4961,2SC97A,2SC880,BC140-10,BSY83,ZT90,ZT95,ZT1613, ZT2270
2SC880	2N2017,2N2106,2N2193,2N2193A,2N2193B,2N2270,2N2848,2N2868,2N3109, 2N3725A,2N4960,2N4961,2SC97A,2SC879,BC140-10,BSY83,ZT90,ZT95,ZT1613, ZT2270
2SC881	*PTC178,*GE-210,*TR-87,*HEPS5026,*SK3137,*ECG192,*WEP243,*BC337, *276-2012,*RT-301
2SC882	
2SC883	
2SC884	2N5600,2SC840,HST5902,HST5903,KSP1394,KSP1395,MJ2250
2SC885	2N6249,2N6250,2SC1227,2SD376A,FT413,HST7204,HST7205,IR410,IR411,KSP1144, KSP1145,SDT7204,SDT7205
2SC886	2N6249,2SC675,2SC885,2SC1227,2SC1229,2SD376,2SD376A,HST7202,HST7203, HST7204,HST7205,IR410,IR411,KSP1142,KSP1143,KSP1144,KSP1145,SDT7202, SDT7203,SDT7204,SDT7205
2SC887	2SC675,2SC676,2SC886,2SC1229,2SD166,2SD376,HST7201,HST7202,HST7203, KSP1141,KSP1142,KSP1143,SDT7201,SDT7202,SDT7203,SDT7612
2SC888	2N5559,2N6495,2SC676,2SC677,2SC898,2SC902,2SD165,2SD166,KSD2201, KSP1173,SDT7610,SDT7611,SDT7735,SDT7736
2SC889	2N5497,2SC678,2SC793-R,2SD119-R,2SD125A,2SD161,2SD164,2SD212,2SD213, 2SD319,2SD339,2SD340,2SD522,B170007-BLK,B170007-BRN,B170025,HST9803, HST9804,KSD2202,KSD2203,KSP1172,SDT7609,SDT7733,SDT7734,*ECG162, *WEP707
2SC890	2SC872,2SC1002,*GE-28,*TR-76,*HEPS3013,*ECG311,*276-2009
2SC891	2SC892,2SC1040,*GE-28,*TR-76,*HEPS3005
2SC892	2SC1040,2SC1081,2SC1118,*GE-66,*HEPS3006
2SC893	*PTC195,*HEPS3019
2SC894	2N2501,2N4418,*PTC123,*HEPS0011,*SK3122,*ECG123A,*WEP736
2SC895	*SK3111,*ECG175,*WEP241
2SC896	2N3903,2N4400,2N5581,2N6538,A5T3903,MM3903,MPS6530,MPS6565,TIS96,TIS99, TIS110,*PTC123,*GE-210,*TR-21,*HEPS5026,*SK3122,*ECG123A,*WEP736, *BC107B,*276-2016,*RT-102
2SC897	2N4071,2N6354,BUY20,KSP1176,SDT7604,SDT7605,*HEPS5020,*SK3535,*ECG280, *WEP707
2SC898	KSP1173,KSP1176,*PTC129A,*HEPS5020,*SK3535,*ECG280,*WEP707
2SC899	2N915,2N3947,2N4141,2N4951,2N5028,2N5107,2N5368,2SC933,2SC943,2SC944, 2SC979-Y,2SC1999,BC107A,BC174A,BC190A,BC237A,BC547A,GET2222, GET2222A,GI-3643,MM3904,ZT84,ZT89,*PTC101,*GE-61,*TR-21,*HEPS0025, *SK3122,*ECG123A,*WEP736,*BF365,*276-2013,*RT-102
2SC900	2N4256,2N4419,2N5827,2N6112,2SC400-GR,2SC735-GR,2SC828,2SC828A,2SC829, 2SC923,2SC1222,2SC1684,2SC1849,2SC1853,2SC1993,A157B,BC107,BC107B, BC108,BC108B,BC109B,BC122,BC167B,BC171B,BC183L,BC184L,BC237B, BC238B,BC239B,BC413B,BC414B,BC547B,BC548B,BC549B,BC550B,BCY58B, BCY58C,BCY59B,BCY59C,BFY39-3,MPSA16,TP3704,TP3706,*PTC139,*GE-62, *TR-21,*HEPS0015,*SK3124,*ECG199,*WEP736,*BC107B,*276-2013,*RT-304
2SC901	2SC42A,2SC901A,2SD321,*PTC129A,*GE-35,*TR-67,*HEPS5020,*SK3111,*ECG163, *WEP740
2SC901A	2SC270,2SD321,*PTC129A,*GE-35,*TR-67,*HEPS5020,*SK3111,*ECG163,*WEP740
2SC902	2N3442,2N6262,2N6496,2SC1115,2SC1116,2SC1116A,2SC1782,2SC1783,2SD181, 2SD286,2SD287,2SD287A,2SD287B,2SD375,KSD9706,KSP1173,KSP1176,KSP1273, KSP1276,SDT7735,SDT7736,SDT9706,STS1121,STS1122
2SC903	2N916,2N916A,2N2501,2N4013,2SC367G,2SC367G-O,2SC735-O,2SC933,2SC1210, EN916,MPS3693,SE2001,ZT81,ZT82,ZT111,ZT112,ZT2369A,*HEPS0014,*SK3244
2SC904	2N915,2N2539,2N3974,2N3976,2N4013,2N4400,2N4951,2N4962,2N5368,2SC366G, 2SC366G-O,2SC933,2SC1211,2SC1318,2SC1781,40458,BCW34,BCW36,EN915, MPS-A05,TN-3903,ZT84,ZT89,ZT114,ZT119,*HEPS0015,*SK3244

To Replace	Substitute This Type
2SC905	2N915,2N2221,2N2221A,2N2539,2N2847,2N3115,2N3974,2N3976,2N4014,2N4400, 2N4951,2N4962,2N4963,2N5027,2N5368,2SC366G-O,2SC1211,BSW84,EN915, MPS6530,TN-3903,ZT84,ZT114,*HEPS0015,*SK3244
2SC906	2N1420A,2N1711,2N1711A,BSY54,ZT1711,*ECG123A,*WEP242
2SC907	2N3116,2N3302,2N3904,2N4401,2N4952,2N4954,2N5369,2N5371,2N5377,2N5381, 2SC1992,A157B,BC107,BC107B,BC167B,BC171B,BC237B,BC413B,BC414B, BCY59B,BSW83,BSX79,FT3643,TN-3904,*SK3122,*ECG199,*WEP736
2SC907A	2N915,2N3116,2N3643,2N3904,2N3947,2N4141,2N4401,2N4437,2N4951,2N4952, 2N5028,2N5107,2N5368,2N5369,2N5381,2SC943,2SC944,2SC979-Y,2SC980G-Y, 2SC1318,2SC1998,BC174A,BC190A,BC546A,BSW85,GET2222,GET2222A,GI-3643, MM3904,MPS6531,TN-3904,ZT84,ZT89,ZT114,ZT119,*SK3122,*ECG199,*WEP736
2SC908	2N2485,2N2486,2SC731,2SC872,2SC890,2SC1001,2SC1002,BLX89,*HEPS3001, *276-2014
2SC909	
2SC910	
2SC911	2SC891,2SC892,2SC911A,2SC1040,2SC1608,41009A†,41010†,*TR-76
2SC911A	2SC891,2SC892,2SC911,2SC1040,2SC1608,41009A†,41010†
2SC912	2N916,2N916A,2N2501,2N4013,2N4123,2N4418,2N5144,2SC318,2SC367G, 2SC367G-O,2SC372G-O,2SC400-O,2SC400-Y,2SC454,2SC458,2SC459,2SC460, 2SC536,2SC619,2SC620,2SC634,2SC639,2SC735-O,2SC752G-O,2SC933, 2SC1216,2SC1276,2SC1372,BFY39-2,EN916,*PTC132,*HEPS0016,*ECG108, *WEP56
2SC912M	2N3227,2N3692,2N3694,2N3794,2N3860,2N4124,2N4419,2N4970,2N5826,2N5998, 2N6000,2N6004,2SC367G-Y,2SC372-Y,2SC372G-Y,2SC394-GR,2SC400,2SC400-GR, 2SC711A,2SC735-Y,2SC752G-Y,2SC1175,2SC1285,2SC1364,2SC1632,2SC1634, 2SC1684,2SC1849,2SC1992,BC107,BC107A,BC108,BC108A,BC167A,BC171A, BC237A,BC238A,BCY58A,BCY58B,BCY59A,BCY59B,BFY39-3,BSW88,BSW89, BSX38,BSX79,GI-3794,MPS3694,MPS3704,MPS6514,MPS6573,MPS6574-SIL, MPS6575,MPS6576-SIL,MPS-A20,MPS-K20,MPS-K21,MPS-K22,MPSA20-BLU, MPSA20-YEL,NPSA20,PBC107A,PET1002,PET2002,SE1002,SE2002
2SC913	2N708A,2N718,2N731,2N760,2N914,2N2221,2N2315,2N2396,2N2481,2N2501, 2N2539,2N2845,2N2847,2N3115,2N3210,2N3211,2N3301,2N3508,2N3641,2N3642, 2N3646,2N3647,2N3705,2N3974,2N3976,2N4013,2N4400,2N4962,2N5027,2N5144, 2N5450,2N6427,2SC366G-O,2SC367G-O,2SC577,2SC914,2SC914A,2SC1274, A5T5450,A8T3705,BSW82,D33D24,EN697,FT3641,FT3642,GET2221,GI-3641,GI-3642, GI-3705,MPS3705,MPS6530,PET3705,PET6001,TIS90-YEL,TIS90M-YEL,TIS92-YEL, TIS92M-YEL,TIS110,TN62,TN-3903,ZT708,*PTC123,*GE-64,*TR-21,*SK3122, *ECG123A,*WEP736,*BC337,*276-2013,*RT-102
2SC914	2N708A,2N718,2N731,2N760,2N914,2N2221,2N2315,2N2396,2N2481,2N2501, 2N2539,2N2845,2N2847,2N3115,2N3210,2N3211,2N3301,2N3508,2N3641,2N3642, 2N3646,2N3647,2N3705,2N3974,2N3976,2N4013,2N4400,2N4962,2N5027,2N5144, 2N5450,2N6427,2SC366G-O,2SC367G-O,2SC577,2SC913,2SC914A,2SC1274, A5T5450,A8T3705,BSW82,D33D24,EN697,FT3641,FT3642,GET2221,GI-3641,GI-3642, GI-3705,MPS3705,MPS6530,PET3705,PET6001,TIS90-YEL,TIS90M-YEL,TIS92-YEL, TIS92M-YEL,TIS110,TN62,TN-3903,ZT708,*PTC123,*GE-64,*TR-53,*SK3244, *BC337,*276-2013
2SC914A	2N708A,2N718,2N731,2N760,2N914,2N2221,2N2315,2N2396,2N2481,2N2501, 2N2539,2N2845,2N2847,2N3115,2N3210,2N3211,2N3301,2N3508,2N3641,2N3642, 2N3646,2N3647,2N3705,2N3974,2N3976,2N4013,2N4400,2N4962,2N5027,2N5144, 2N5450,2N6427,2SC366G-O,2SC367G-O,2SC577,2SC913,2SC914,2SC1274, A5T5450,A8T3705,BSW82,D33D24,EN697,FT3641,FT3642,GET2221,GI-3641,GI-3642, GI-3705,MPS3705,MPS6530,PET3705,PET6001,TIS90-YEL,TIS90M-YEL,TIS92-YEL, TIS92M-YEL,TIS110,TN62,TN-3903,ZT708
2SC915	2N708A,2N760,2N914,2N2331,2N2481,2N2501,2N3210,2N3211,2N3508,2N3646, 2N3647,2N3705,2N4013,2N5144,2N5450,2N5810,2N6427,2SC367G-O,2SC509-O, 2SC509G-O,2SC577,2SC735-O,2SC913,2SC914,2SC914A,2SC915A,2SC1274, A5T5450,A8T3705,BSW82,D33D21,D33D24,GI-3705,MPS3705,PET3705,PET6001, SE8040,TIS90-YEL,TIS90M-YEL,TIS92-YEL,TIS92M-YEL,TN62,ZT708,*PTC123, *GE-64,*TR-53,*SK3244,*BC337,*276-2013
2SC915A	2N708A,2N760,2N914,2N2331,2N2481,2N2501,2N3210,2N3211,2N3508,2N3646, 2N3647,2N3705,2N4013,2N5144,2N5450,2N5810,2N6427,2SC367G-O,2SC509-O,

To Replace	Substitute This Type
(2SC915A)	2SC509G-O,2SC577,2SC735-O,2SC913,2SC914,2SC914A,2SC915,2SC1274, A5T5450,A8T3705,BSW82,D33D21,D33D24,GI-3705,MPS3705,PET3705,PET6001, SE8040,TIS90-YEL,TIS90M-YEL,TIS92-YEL,TIS92M-YEL,TN62,ZT708
2SC916	*GE-66,*HEPS3002,*276-2020
2SC917	*PTC121,*HEPS0005,*SK3132,*ECG229,*WEP736
2SC918	2N2857,2N2865,2N3570,2N3571,2N3600,2N3839,2N3932,2N4134,2N4135,2N4252, 2N4253,2N6304,2N6305,2SC1553A,40294,40295,40413,MT1061A,S1297,SE5055, *HEPS0020,*SK3018,*ECG161,*WEP719
2SC920	MMT3903,*PTC121,*GE-20,*TR-24,*HEPS0016,*SK3018,*ECG229,*WEP56,*BC107B, *276-2016,*RT-308
2SC921	2SC271,2SC272,2SC287A,2SC288A,2SC289,2SC800,2SC1158,2SC1159,FX918, *GE-214,*TR-70,*HEPS0020,*SK3018,*ECG107,*WEP56,*BF183,*276-2011, *RT-108
2SC922	2SC764,2SC1393,2SC1394,2SC1395,2SC1396,*PTC115,*HEPS0020,*SK3018, *ECG107,*WEP56,*RT-107A
2SC923	2N5828,2SC536FP,2SC536NP,2SC1006,2SC1010,2SC1222,BC108C,BC109,BC109C, BC238C,BC239C,BC413C,BC414C,BCY58D,BCY59D,MPSA18,*PTC139,*GE-20, *TR-21,*HEPS0015,*SK3122,*ECG199,*WEP736,*BC107B,*276-2013,*RT-105
2SC924	2N916,2N916A,2N2501,2N4013,2N4123,2N4418,2SC400-Y,2SC619,2SC639, 2SC735-O,2SC899,2SC925,2SC933,2SC1216,2SC1999,*PTC101,*GE-61,*TR-95, *HEPS0011,*SK3124,*ECG108,*WEP56,*BC107B,*276-2015,*RT-113
2SC925	2N3227,2SC400,2SC400-Y,2SC735-Y,2SC933,2SC1216,2SC1999,BC107A,BC108A, BC123,BC167A,BC171A,BC237A,BC238A,BC547A,BC548A,BCY58A,BCY59A, BSW83,*PTC121,*HEPS0030,*SK3124,*ECG123A,*WEP736
2SC926	*HEPS5026,*SK3045,*ECG154,*WEP712
2SC926A	2N6430,*HEPS5026,*SK3024,*ECG154
2SC927	2N4252,2N4253,2N5181,2N6304,2N6305,2SC390,2SC392,2SC568,2SC1035, 2SC1036,2SC1180,40235,40236,40237,40238,40240,40472,40474,40477,40478, 40480,TC3114,*PTC132,*HEPS0016,*SK3117,*ECG319,*WEP719
2SC928	2N2808,2N2808A,2N2809,2N2809A,2N3570,2N3600,2N4135,2N4252,2N4253,2N5053, 2N5054,2N5181,2N6304,2N6305,2SC332,2SC333,2SC390,2SC463,2SC567,2SC568, 2SC927,2SC1035,2SC1036,2SC1747,40235,40238,40239,40295,40413,40472, 40478,BF168,BF173,BFS62,BFX60,MT1061A,*PTC132,*HEPS0016,*SK3117, *ECG161,*WEP719
2SC929	2N2712,2N4418,2SC400-Y,2SC619,2SC924,2SC925,2SC929NP,2SC934,2SC1359, 2SC1372,BF254,GI-2712,MPS2714,MPS2926-ORG,MPS2926-YEL,*PTC132, *HEPS0016,*SK3039,*ECG229,*WEP736
2SC929NP	2N916,2N916A,2SC400,2SC400-Y,2SC735-Y,2SC752G-Y,2SC925,2SC930NP,2SC933, 2SC1216,2SC1276,2SC1359,BC107A,BC108A,BC123,BC129A,BC130A,BC167A, BC171A,BC237A,BC238A,BCY58A,BCY59A,EN916
2SC930	2N744,2N4418,2S095A,2S512,2SC99,2SC168,2SC300,2SC301,2SC400-O,2SC619, 2SC924,2SC925,2SC929,2SC929NP,2SC1372,BF254,GI-2712,MPS2926-ORG, NT3000,PT720,ZT2205,*PTC132,*HEPS0016,*SK3122,*ECG229,*WEP736,*RT-308
2SC930NP	2N4124,2N4419,2SC400,2SC400-GR,2SC735-GR,2SC735-Y,2SC752G-Y,2SC1992, 2SC1993,A157B,BC107,BC107B,BC108,BC108B,BC109B,BC122,BC123,BC129A, BC129B,BC130A,BC130B,BC131B,BC167B,BC171B,BC237B,BC238B,BC239B, BC413B,BC414B,BCY58B,BCY59B
2SC931	*HEPS5000,*SK3202,*ECG186,*WEP245,*RT-194
2SC932	*HEPS5000,*SK3192,*ECG152,*WEP245
2SC933	2N2222,2N2222A,2N2540,2N3116,2N3302,2N4401,2N4952,2N5028,2N5107,2N5369, 2N5582,2N6010,A5T2222,BSW85,GET2222,GET2222A,GI-3643,MPS6531,TIS109, TIS111,TN-3904,ZT89,*PTC136,*HEPS3013,*SK3122,*ECG123A,*WEP736
2SC933FP	2N3417,2N4424,2N4953,2N5370,2N5376,2N5377,2N6006,2N6012,2N6016,2N6112, D33D27,GI-3417,PET8002,PET8004,*ECG123A
2SC934	2SC395A-Y,2SC735-Y,*PTC136,*HEPS0025,*SK3124,*ECG123A,*WEP243,*RT-102A
2SC935	2N5838,HST7209,*PTC129A,*HEPS5020,*SK3111,*ECG162,*WEP707
2SC936	2SC936A,*PTC130,*HEPS5021,*SK3115,*ECG164
2SC936A	2SC936,*SK3115
2SC937	*PTC129A,*HEPS5021,*SK3115,*ECG238,*WEP740B
2SC937-M	2SC599N,2SC690-M,2SC690N
2SC938	2N915,2N2221,2N2221A,2N2539,2N2847,2N3115,2N3903,2N3946,2N3974,2N3976,

To Replace	Substitute This Type
(2SC938)	2N4014,2N4227,2N4400,2N4951,2N4962,2N4963,2N5027,2N5368,2N5380,2N5820, 2SC366G-O,2SC815,BSW84,D33D29,MM3903,MPS6530,TN54,TN-3903,ZT84, ZT114,*PTC170,*GE-81,*TR-21,*SK3122,*ECG123A,*WEP736,*276-2010,*RT-107A
2SC939	2N4071,2SC940,KSP1046,KSP1176,SDT7604,SDT7605,*GE-14,*TR-61,*HEPS5020, *SK3137,*ECG163,*WEP740,*276-2009,*RT-131
2SC940	2N4071,*PTC118,*GE-73,*TR-61,*HEPS5020,*SK3111,*ECG283,*WEP740B
2SC941-O	2N916,2N916A,2N4873,2SC367G,2SC367G-O,2SC372-O,2SC372G-O,2SC380-O, 2SC380A-O,2SC394-Y,2SC536,2SC735-O,2SC752G-O,2SC899,2SC933,2SC965, 2SC1216,2SC1276,2SC1453,2SC1682-BL,2SC1682-GR,2SC1999,BC107A,BC167A, BC171A,BC237A,BCY58A,BCY59A,BF240,BFY39-2,EN916,MPS6513, MPS6574-BLUE,MPS6574-GREEN,MPS6576-BLUE,MPS6576-GREEN,MPSA20-GRN, MPSA20-WHT,ZT44,*GE-88,*TR-21,*SK3122,*ECG123A,*WEP736,*276-2014, *RT-102
2SC941-R	2N708,2N708A,2N784A,2N957,2N2242,2N2369,2N2369A,2N2501,2N3009,2N3013, 2N3014,2N3605A,2N3606A,2N3646,2N3688,2N3689,2N3690,2N3691,2N3693, 2N3862,2N4123,2N4137,2N4295,2N4420,2N4421,2N4422,2N5029,2SC67,2SC169, 2SC318,2SC335,2SC380-R,2SC380A-R,2SC394-O,2SC468,2SC601,2SC601N, 2SC620,2SC639,2SC752G-R,2SC764,2SC838,2SC839,2SC963,2SC964,2SC1187, 2SC1188,2SC1189,2SC1274,2SC1319,2SC1320,2SC1686,2SC1687,2SC1688, BF198,BF199,BF241,BF311,BFY39-1,EN708,EN914,EN2369A,EN3009,EN3013, EN3014,GET708,GET914,GET2369,GET3013,GET3014,GET3646,MPS3646, MPS3693,MPS6512,MPS6574-YEL,MPS6576-YEL,MPSA20-RED,PET1001,PET2001, PT2760,SE1001,SE2001,SE5001,SE5002,SE5003,SE5006,TIS45,TIS46,TIS48,TIS49, TIS52,TIS55,TIS87,TIS129,TP3705,ZT43,ZT708,*GE-88,*TR-21,*SK3122,*ECG123A, *WEP736,*RT-102
2SC941-Y	2N3242A,2N3692,2N3694,2SC367G-Y,2SC372-Y,2SC372G-Y,2SC373,2SC373G, 2SC380-Y,2SC380A-Y,2SC394-GR,2SC711A,2SC735-GR,2SC735-Y,2SC752G-Y, 2SC1084,2SC1175,2SC1285,2SC1364,2SC1632,2SC1634,2SC1992,A157B,BC107, BC107B,BC123,BC167B,BC171B,BC237B,BC413B,BC414B,BC547A,BCY58B, BCY59B,BFY39-3,BSW88,BSW89,BSX38,BSX79,MPS3694,MPS6514,MPS6520, MPS6573,MPS6574-SIL,MPS6575,MPS6576-SIL,MPS-A20,MPS-K20,MPS-K21, MPS-K22,MPSA20-BLU,MPSA20-YEL,NPSA20,PBC107A,PBC107B,PET1002, PET2002,SE1002,SE2002,TP3704,TP3706,*GE-60,*TR-21,*ECG123A,*WEP736, *RT-102
2SC942	
2SC943	*GE-210,*TR-21,*HEPS0015,*SK3122,*ECG123A,*WEP736,*BC107B,*276-2016, *RT-102A
2SC944	2SC943,*HEPS0025,*SK3122,*ECG123A,*RT-302
2SC945	*PTC132,*HEPS0015,*SK3124,*ECG199,*WEP736,*RT-107A
2SC947	2SC787,2SC948,*PTC132,*GE-214,*TR-83,*HEPS0017,*SK3039,*ECG161,*WEP719, *276-2015,*RT-131
2SC948	2SC787,MT1061,*PTC132,*GE-214,*TR-83,*HEPS0017,*SK3019,*ECG161,*WEP719, *276-2015,*RT-131
2SC957	2N758,2N759,2N834A,2N2368,2N3854A,2SC302,2SC321,2SC322,2SC565,2SC576, BSX25,GET706,MPSH19,MPSH32,ZT2368,*PTC133,*HEPS0017,*SK3039,*ECG161, *WEP719
2SC959	2N2405,*PTC144,*GE-32,*TR-87,*HEPS3019,*SK3104,*ECG128,*WEP243,*276-2008
2SC960	*GE-10,*TR-75,*HEPS3021,*SK3024,*ECG225
2SC961	2N4070,2N5628,2N6354,2SC897,2SC898,2SC1343,KSP1176,SDT7604
2SC962	2N3448,2N3864,2N4070,2N5624,2N5628,2SC793-R,2SC793-Y,2SC961,2SC1079-R, 2SC1080-R,2SD110-O,2SD110-R,2SD111-O,2SD111-R,2SD116,2SD118-R, 2SD118-Y,2SD119-R,2SD119-Y,2SD217,2SD322,2SD340,2SD427-R,2SD428-R, B170007-ORG,B170007-RED,B170007-YEL,KSP1175,SDT7603
2SC963	2N708,2N708A,2N784A,2N914,2N957,2N2242,2N2369,2N2369A,2N2481,2N2501, 2N3009,2N3013,2N3014,2N3210,2N3211,2N3646,2N3647,2N3862,2N4013,2N4123, 2N5029,2N5144,2SC67,2SC169,2SC335,2SC577,2SC601N,2SC620,2SC639, 2SC735-O,2SC764,2SC899,2SC964,2SC1274,BSW82,GET2369,MPS2369, MPS3693,MPS6512,MPS6574-YEL,MPS6576-YEL,PET1001,PET2001,TIS48,TIS87, ZT81,ZT111,ZT708,ZT2369,*PTC132
2SC964	2N916,2N916A,2N2501,2N4013,2N4123,2N4873,2SC639,2SC735-O,2SC899,2SC933, 2SC965,2SC1216,2SC1999,MPS3693,MPS6513,MPS6574-BLUE,MPS6574-GREEN,

To Replace	Substitute This Type
(2SC964)	MPS6574-YEL,MPS6576-BLUE,MPS6576-GREEN,MPS6576-YEL,ZT81,ZT82,ZT111, ZT112,ZT2369A,*PTC123
2SC965	2N3116,2N3643,2N3904,2N3947,2N4141,2N4401,2N4951,2N4952,2N5028,2N5107, 2N5368,2N5369,2N5381,2SC933,2SC943,2SC1175,2SC1318,2SC1999,BC107A, BC167A,BC171A,BC174A,BC190A,BC237A,BC547A,BCY59A,BSX79,GET2222, GI-3643,MM3904,MPS6531,MPS6575,MPS6576-GREEN,MPS6576-SIL,MPS-A05, PET1002,TN-3904,ZT82,ZT112,*PTC132
2SC966	2N3242A,2N5818,2N6010,2N6426,2SC509-Y,2SC509G-Y,2SC967,2SC968,2SC970, 2SC1246,2SC1247,2SC1813,2SC1851,A5T4124,A5T5449,A8T3704,BC337-16, BC337-25,BSW83,D33D22,D33D26,TN60,*PTC123,*HEPS5014,*SK3122, *ECG123A,*WEP736
2SC967	2N3242A,2N5818,2N6010,2N6426,2SC509-Y,2SC509G-Y,2SC968,2SC970,2SC1246, 2SC1247,2SC1813,2SC1851,A5T5449,A8T3704,BC337-16,BC337-25,BSW83, D33D22,D33D26,TN60,*PTC123,*SK3122,*ECG123A,*WEP736
2SC968	2N956,2N2540,2N5369,2N5582,2N5818,2N6010,2N6014,2SC970,2SC1247,2SC1813, 2SC1852,A5T2222,A5T5449,A8T3704,BSW85,D33D26,TIS109,TIS111,TN-3904, *PTC123,*HEPS5014,*SK3024,*ECG123A,*WEP243,*RT-102A
2SC969	2N956,2N2540,2N2645,2N5369,2N5381,2N5582,2N5818,2N6010,2N6014,2SC968, 2SC970,2SC1166-Y,2SC1214,2SC1247,2SC1813,2SC1852,A5T2222,A5T3904, A5T5449,A8T3704,BSW85,D33D26,TIS109,TIS111,TN-3904
2SC970	2N956
2SC971	2N1711A,*PTC125,*HEPS5014,*SK3018,*ECG192,*WEP243
2SC973	2SC973A,2SC974,2SC975,2SC975A,2SC1015,2SC1074,2SC1075,2SC1337, 2SC1338A
2SC973A	2SC973,2SC974,2SC975,2SC975A,2SC1015,2SC1074,2SC1075,2SC1337,2SC1338A
2SC974	2SC990,2SC1015,2SC1075,2SC1076
2SC975	2SC975A,2SC990,2SC1015,2SC1075,2SC1076,2SC1337,2SC1338,2SC1338A, 2SC1404
2SC975A	2SC975,2SC990,2SC1015,2SC1075,2SC1076,2SC1337,2SC1338,2SC1338A, 2SC1404
2SC976	2SC977,2SC978
2SC977	2SC978
2SC978	
2SC979-O	2N915,2N4014,2SC979A-O,2SC1998,BC174A,BC190A,GET2222A,ZT88,ZT89
2SC979-R	2N4963,2SC979A-R,BSW84,FT3722,GET2221A,SE8012,ZT86
2SC979-Y	2N4409,A5T4409,BC174B,BC190B,BC546A,BSW85
2SC979A-O	ZT88
2SC979A-R	SE8012,ZT86
2SC980	*PTC121,*HEPS5026,*SK3122
2SC980AG-O	2SC979A-O,ZT88
2SC980AG-R	2SC979A-R,SE8012,ZT86
2SC980G-O	2N915,2N4014,2SC979-O,2SC979A-O,2SC980AG-O,2SC1998,BC174A,BC190A, GET2222A,ZT88,ZT89
2SC980G-R	2N4963,2SC979-R,2SC979A-R,2SC980,2SC980AG-R,BSW84,EN915,FT3722, GET2221A,SE8012,ZT86
2SC980G-Y	2SC979-Y,BC174B,BC190B,BC546A,BSW85
2SC981	2N5429,2N5616,2N5620,2SC1055,2SC1113,2SC1445,KSP1025,KSP1165
2SC982	MPS-A13,*GE-64,*TR-69,*HEPS9100,*SK3156,*ECG172,*WEPS9100,*276-2016
2SC983-O	2N6515,2N6516,*PTC117,*GE-27,*TR-79,*HEPS5025,*SK3044,*WEPS3021,*RT-110
2SC983-R	*PTC117,*GE-27,*TR-32,*HEPS5025,*SK3244,*WEPS3021,*RT-110
2SC983-Y	*PTC117,*GE-27,*TR-32,*HEPS5025,*SK3244,*ECG198,*WEPS3021,*RT-110
2SC984	2N915,2N2221,2N2221A,2N2539,2N2847,2N3115,2N4013,2N4951,2N4962,2N5144, 2N5368,BSW84,TIS110,TN-3903,*PTC123,*HEPS0015,*SK3122,*ECG123A, *WEP736
2SC985	*PTC115
2SC985A	2SC985,2SC987A,2SC1038,2SC1039,2SC1041,2SC1042,2SC1336,2SC1559,35861E, 35861E,*PTC139
2SC986	
2SC987	*PTC132
2SC987A	2SC985,2SC1038,2SC1039,2SC1041,2SC1042,2SC1336,2SC1559,*PTC115
2SC988	2SC988A,2SC988B,*PTC132

To Replace	Substitute This Type
2SC988A	2SC988,2SC988B,*PTC132
2SC988B	2SC988,2SC988A,*PTC132
2SC989	
2SC990	2SC1076,2SC1207,*GE-66,*HEPS3006
2SC991	2SC302-M,2SC307-M,2SC590N,2SC872,2SC890,2SC992,2SC1002,2SC1256, 2SC1820,*PTC136,*HEPS0014,*SK3122,*ECG123A,*WEP736
2SC992	2SC302-M,2SC307-M,2SC590N,2SC1002,2SC1256,*PTC136,*HEPS0014,*SK3122, *ECG123A,*WEP736
2SC993	*ECG128
2SC994	*PTC136,*HEPS3013,*SK3195,*276-2038
2SC995	*GE-27,*TR-78,*HEPS3021,*SK3044,*ECG154,*WEP712,*RT-110
2SC996	*GE-27,*TR-78,*HEPS5015,*SK3045,*ECG225,*RT-135
2SC997	2SC463,40235,40238,40239,BF168,BF173,BFS62,BFX60,*PTC132,*GE-60,*TR-83, *HEPS0016,*SK3018,*ECG161,*WEP719,*276-2038,*RT-113
2SC998	*PTC136,*HEPS3008,*SK3195
2SC999	*PTC146,*SK3115,*ECG165,*WEP740B
2SC999A	*PTC129A,*SK3115,*WEP740B
2SC1000-BL	2N5828,2SC1000G-BL,2SC1380-BL,2SC1380A-BL,2SC1570,BC550C,*TR-21,*SK3245, *ECG199,*WEP736
2SC1000-GR	2N4953,2N5370,2N5376,2N5377,2N5827,2N6006,2N6112,2SC734-GR,2SC1000G-GR, 2SC1345,2SC1380-GR,2SC1380A-GR,BC414C,BC547B,BC550B,TP3704,*GE-62, *TR-21,*SK3245,*ECG199,*WEP736
2SC1000G-BL	2N5828,2SC1000-BL,2SC1380-BL,2SC1380A-BL,2SC1570,BC550C
2SC1000G-GR	2N4953,2N5370,2N5376,2N5377,2N5827,2N6006,2N6112,2SC734-GR,2SC1000-GR, 2SC1345,2SC1380-GR,2SC1380A-GR,BC414C,BC546B,BC547B,BC550B,TP3704
2SC1001	2SC1002,BLX89,*HEPS3001,*SK3195,*ECG278,*276-2014
2SC1002	
2SC1003	*HEPS5000
2SC1004	2SC1004A,*PTC130,*GE-37,*TR-93,*SK3133,*ECG164,*WEP740A
2SC1004A	*PTC130,*GE-37,*TR-86,*SK3133,*ECG164,*WEP740A
2SC1005	2SC1005A,2SC1172,2SC1172B,2SC1413,2SC1413A,*PTC129A,*GE-38,*TR-93, *SK3115,*ECG165,*WEP740B
2SC1005A	2SC1172,2SC1172B,2SC1413A,*PTC129A,*SK3115,*ECG165,*WEP740B
2SC1006	2N4104,2N6429,2N6429A,BC550C,*HEPS0030,*SK3245,*ECG199,*WEP736
2SC1007	2SC943,*ECG123A,*WEP736
2SC1008	*PTC144,*HEPS3019,*ECG128,*WEP243
2SC1008A	
2SC1009	
2SC1010	2N4104,2N6429,2N6429A,2SC1006,BC550C,*HEPS0030,*SK3122,*ECG199, *WEP736
2SC1011	2SC891,2SC892,2SC1040,2SC1257,2SC1606
2SC1012	*GE-235,*TR-78,*HEPS5024,*SK3044,*ECG154,*WEP712,*276-2012,*RT-110
2SC1012A	*GE-40,*TR-78,*HEPS5024,*SK3045,*ECG154,*WEP712,*276-2012,*RT-110
2SC1013	2SC1014,2SC1096,2SC1226,2SC1226A,D42C2,D42C3,*PTC193,*TR-55,*HEPS3020, *ECG300,*WEPS3020,*RT-137
2SC1014	D42C5,D42C8,*PTC193,*GE-226,*TR-55,*HEPS3020,*ECG300,*WEP1014,*RT-137
2SC1015	2SC990,2SC1076,2SC1404,*HEPS3007
2SC1016	2SC891,2SC892,2SC1040,2SC1608,*HEPS5000
2SC1017	2SC1589,*PTC192,*TR-74,*HEPS3020,*ECG299,*WEPS3020,*RT-143
2SC1018	2SC1589,*PTC192,*GE-215,*TR-55,*HEPS3020,*ECG299,*WEP1018,*RT-143
2SC1019	
2SC1020	
2SC1021	2SC1022,2SC1077A
2SC1022	2SC1021,2SC1077A
2SC1023	2N706,2N706A,2N706B,2N783,2N784,2N959,2N2205,2N2318,2N2319,2N2368, 2N2656,2N2729,2N2921,2N3544,2N3605A,2N3606A,2N3663,2N3983,2N4137, 2N4294,2N5030,2SC54,2SC55,2SC63,2SC68,2SC204,2SC321,2SC322,2SC356, 2SC380-R,2SC380A-R,2SC387AG,2SC394-O,2SC400-R,2SC455,2SC468,2SC595, 2SC601,2SC689,2SC724,2SC752G-R,2SC1126,2SC1293,BF255,EN706,EN914, EN918,EN2369A,EN3009,EN3011,EN3013,EN3014,GET708,GET914,GET3013, GET3014,GET3646,KT218,MM1941,MPS6507,MPS6511,MPSH19,MT1038,MT1038A,

To Replace	Substitute This Type
(2SC1023)	MT1039,MT1060,PET3001,PT2760,TIS18,TIS44,TIS45,TIS46,TIS49,TIS51,TIS52, TIS55,TIS129,*PTC132,*HEPS0017,*SK3018,*ECG107,*WEP56
2SC1025	*PTC104,*HEPS5012,*SK3131,*ECG175,*WEP241
2SC1026	2N708,2N729,2N753,2N784A,2N957,2N2206,2N2242,2N2369,2N2369A,2N2501, 2N2615,2N2616,2N2885,2N2922,2N3009,2N3011,2N3013,2N3014,2N3298,2N3394, 2N3397-RED,2N3398-RED,2N3564,2N3688,2N3689,2N3690,2N3691,2N3709, 2N3845,2N3845A,2N3855A,2N3858,2N4123,2N4138,2N4295,2N4418,2N4420, 2N4421,2N4422,2N4968,2N4996,2N4997,2N5029,2N5137,2N6566,2S501,2S512, 2SC67,2SC169,2SC300,2SC301,2SC367G-O,2SC372-O,2SC372G-O,2SC380-O, 2SC380A-O,2SC388A,2SC394-Y,2SC400-O,2SC460,2SC461,2SC579,2SC601N, 2SC613,2SC619,2SC633,2SC634,2SC639,2SC710,2SC712,2SC713,2SC733-O, 2SC735-O,2SC752G-O,2SC764,2SC878,2SC913,2SC914,2SC914A,2SC915, 2SC915A,2SC924,2SC963,2SC964,2SC1032,2SC1071,2SC1123,2SC1128, 2SC1129,2SC1187,2SC1188,2SC1215,2SC1240,2SC1274,2SC1319,2SC1372, 2SC1373,2SC1374,2SC1375,2SC1376,2SC1686,2SC1687,2SC1789,2SC1988,A321, BF121,BF123,BF125,BF127,BF198,BF199,BF241,BF254,BF311,BSY61,EN708, GET2369,GI-2922,GI-3394,GI-3709,KT218F,MPS3394,MPS3646,MPS3709,MPS6512, MPS6574-YEL,MPSH10,MPSH11,MT1060A,PET2001,PT720,SE2001,SE3005, SE5001,SE5002,SE5003,SE5006,TIS48,ZT708,ZT2205,*PTC132,*HEPS0025, *SK3018,*ECG107,*WEP56,*RT-302
2SC1030	2N4071,2N6354,2SC897,BDY25C,BUY20,KSP1176,SDT7604,SDT7605,*PTC118, *HEPS5020,*SK3027,*ECG280,*WEP247
2SC1031	
2SC1032	2N708,2N753,2N784A,2N957,2N2206,2N2242,2N2369,2N2369A,2N2501,2N2615, 2N2616,2N2885,2N2922,2N3009,2N3011,2N3013,2N3014,2N3298,2N3564,2N3688, 2N3689,2N3690,2N3691,2N4123,2N4295,2N4418,2N4420,2N4421,2N4422,2N4996, 2N4997,2N5029,2S512,2SC67,2SC169,2SC300,2SC301,2SC372-O,2SC372G-O, 2SC380-O,2SC380A-O,2SC388A,2SC394-Y,2SC400-O,2SC460,2SC461,2SC601N, 2SC619,2SC639,2SC710,2SC735-O,2SC752G-O,2SC764,2SC924,2SC963,2SC964, 2SC1123,2SC1128,2SC1129,2SC1187,2SC1188,2SC1215,2SC1240,2SC1274, 2SC1319,2SC1372,2SC1686,2SC1687,2SC1789,2SC1988,BF121,BF123,BF125, BF127,BF198,BF199,BF241,BF254,BF311,BSY61,EN708,GET2369,MPS3646, MPS6512,MPS6574-YEL,MPSH10,MPSH11,MT1060A,PET2001,PT720,SE2001, SE3005,SE5001,SE5002,SE5003,SE5006,TIS48,ZT708,ZT2205,*PTC132, *HEPS0015,*SK3018,*ECG107,*WEP56
2SC1033	*PTC117,*GE-230,*TR-21,*HEPS5025,*SK3044,*ECG123A,*WEP736,*276-2009
2SC1033A	2N6218,2N6219,*PTC117,*GE-32,*TR-21,*HEPS5025,*SK3044,*ECG123A,*WEP736
2SC1034	*SK3021,*ECG277
2SC1035	2N3570,2N3600,2N4252,2N4253,2N5181,2N6304,2N6305,2SC390,2SC567,2SC568, 2SC927,2SC1036,40235,40236,40237,40238,40240,40295,40413,40472,40474, 40477,40478,40480,TC3114,*PTC132,*HEPS0016,*SK3117,*ECG161,*WEP719
2SC1036	2N3570,2N3600,2N4252,2N4253,2N5181,2N6304,2N6305,2SC390,2SC567,2SC568, 2SC927,2SC1035,40235,40236,40237,40238,40240,40295,40413,40472,40474, 40477,40478,40480,TC3114,*PTC132,*HEPS0016,*SK3117,*ECG161,*WEP719
2SC1037	
2SC1038	2SC1039,*HEPS3005
2SC1039	*HEPS3005
2SC1040	2SC1081,2SC1118,*HEPS3005
2SC1041	2SC1039,2SC1042,*HEPS3005
2SC1042	*HEPS3005
2SC1043	
2SC1044	2SC988,2SC988A,2SC988B,2SC1254,*HEPS0017,*ECG161
2SC1045	2SC642,2SC642A,2SC1171,*PTC130,*HEPS5021,*SK3115,*ECG164,*WEP740A
2SC1046	2SC1154,2SC1358,2SC1358A,2SC1413,2SC1413A,*PTC129A,*HEPS5021,*SK3115, *ECG165,*WEP740B
2SC1047	2N4873,2SC1276,2SC1727,PET3002,*PTC121,*GE-60,*TR-21,*HEPS0016,*SK3018, *ECG229,*WEP56,*276-2015
2SC1048	2SC154C,*HEPS5025,*SK3045,*ECG154,*WEP712
2SC1050	*PTC129A,*HEPS5021,*ECG124,*WEP240
2SC1051	*PTC129A,*HEPS5020,*ECG280
2SC1051L	

To Replace	Substitute This Type
2SC1052	2N3108,2N3110,2N3253,2N3444,2N3735,2N4000,2N5262,2SC485-R,2SC512-R, 2SC516,2SC1065,2SC1069,2SD149,2SD329,40367,BC140-6,BC141-6,BSX60, KT600G
2SC1053	2N1613A,2N1716,2N1889,2N1974,2N2193,2N2193A,2N2193B,2N2218A,2N3108, 2N3110,2N3253,2N3678,2N3722,2N3723,2N3735,2N4000,2SC97A,2SC108A-R, 2SC109A-R,2SC485-R,2SC512-R,2SC516,2SC560,2SC1386,40348,40367,BC140-6, BC141-6,BSX60,BSY53,KT600G,SE8002,ZT1613
2SC1054	2N5181,2SC1747,40239,40413
2SC1055	2N3879,KSP1166,*SK3538
2SC1056	*HEPS5024,*SK3044,*ECG154,*WEP712
2SC1057	2SC1058
2SC1058	
2SC1059	2SC1456,MJ2252,*PTC104,*HEPS5011,*SK3021,*ECG124,*WEP240
2SC1060	2SD314,*HEPS5000,*SK3054,*ECG152,*WEP245
2SC1061	D44C4,D44C5,D44C6,D44C7,D44C8,D44C9,*HEPS5000,*SK3054,*ECG152, *WEP245,*RT-197
2SC1062	2N2726,2SC154C,2SC627,2SC1103,2SC1103A,2SD413,2SD576,A5T5059,*PTC117, *HEPS5024
2SC1063	ZT600
2SC1064	2N1613A,2N2107,2N2846,2N3015,2N3110,2N3252,2N3253,2N3299,2N3326,2N3722, 2N3735,2N5262,2SC97,2SC580,2SC1150,2SD216,40347,BC140-6,BSX60,KT600F, KT600T,SE8002
2SC1065	2SC1069
2SC1066	
2SC1067	2N5181,2SC1054,2SC1747,40239,40413
2SC1068	2SC1387,*PTC123
2SC1069	*PTC144
2SC1070	*PTC126,*HEPS0017,*SK3018,*ECG161,*WEP709
2SC1071	2N760,2N916,2N916A,2N2242,2N2331,2N2501,2N3013,2N3014,2N3211,2N3705, 2N4013,2N4123,2N4418,2N5144,2N5450,2N5810,2N6427,2SC67,2SC367G-O, 2SC509-O,2SC509G-O,2SC577,2SC579,2SC613,2SC639,2SC735-O,2SC764, A5T4123,A5T5450,A8T3705,BSW82,D33D21,D33D24,GI-3416,GI-3705,MPS3705, PET3705,PET6002,PET8000,TIS90-GRN,TIS90-YEL,TIS90M-GRN,TIS90M-YEL, TIS92-GRN,TIS92-YEL,TIS92M-GRN,TIS92M-YEL,TN62,*PTC123,*GE-64,*TR-21, *SK3244,*ECG123A,*WEP736,*BC107B,*276-2013,*RT-102
2SC1072	2N1613A,2N2106,2N2107,2N2193,2N2193A,2N2193B,2N2218,2N2218A,2N2410, 2N2846,2N2868,2N3015,2N3110,2N3252,2N3253,2N3299,2N3326,2N3678,2N3722, 2N3735,2SC97,2SC97A,2SC109A-R,2SC560,2SC580,2SC879,2SC880,2SC1053, 2SC1072A,2SC1150,2SC1386,2SD216,40347,BC140-6,BSX60,BSY53,KT600F, KT600T,SE8002,ZT1613,*PTC144,*GE-47,*TR-87,*SK3024,*ECG128,*WEP243, *276-2012,*RT-114
2SC1072A	2N1613A,2N2106,2N2107,2N2193,2N2193A,2N2193B,2N2218,2N2218A,2N2410, 2N2846,2N2868,2N3015,2N3110,2N3252,2N3253,2N3299,2N3326,2N3678,2N3722, 2N3735,2SC97,2SC97A,2SC109A-R,2SC560,2SC580,2SC879,2SC880,2SC1053, 2SC1072,2SC1150,2SC1386,2SD216,40347,BC140-6,BSX60,BSY53,KT600F, KT600T,SE8002,ZT1613,*PTC144,*GE-63,*TR-87,*SK3024,*ECG128,*WEP243, *276-2012,*RT-114
2SC1073	2SC974,2SC1074,2SC1075
2SC1074	2SC974,2SC1015,2SC1075
2SC1075	2SC990,2SC1015,2SC1076,2SC1404
2SC1076	2SC990,2SC1404,*HEPS5000
2SC1077	
2SC1077A	
2SC1078	*SK3131,*ECG124,*WEP240
2SC1079-R	2SD424,2SD425-R,2SD426-R,KSP1276
2SC1079-Y	2N5038,2N5039,2SD424,2SD425-O,2SD426-O
2SC1080-R	2SC1079-R,KSP1275
2SC1080-Y	2N5039,2SC1079-Y
2SC1081	2SC1118,*HEPS3006
2SC1082	2SC1561,2SC1562
2SC1083	2SC302-M,2SC307-M,2SC590N,2SC731,2SC872,2SC890,2SC991,2SC992,2SC1002,

To Replace	Substitute This Type
(2SC1083)	2SC1256,2SC1464,2SC1479,2SC1820
2SC1084	2N4954,2N5371,2N5827,2SC735-GR,A157B,BC413C,BC414C,BC547B,BC550B, BCY58C,BCY58D,BCY59C,BCY59D,MPS6515,MPS6520,MPS6521
2SC1085	2N3227,2N4124,2N4419,2N4970,2SC372G-Y,2SC400,2SC400-GR,2SC735-Y, 2SC752G-Y,2SC1317,2SC1992,BC107,BC107A,BC108,BC108A,BC167A,BC171A, BC237A,BC238A,BC547A,BC548A,BCY58A,BCY58B,BCY59A,BCY59B,BSW83, BSW88,BSW89,BSX38,BSX79
2SC1086	*SK3115,*ECG165,*WEP740B
2SC1088	2SC1089,*HEPS3021
2SC1089	*HEPS3021,*SK3104,*ECG154,*WEPS3021
2SC1090	*TR-65,*ECG154,*WEP712,*276-2013
2SC1096	2SC1226,2SC1226A,D42C2,D42C3,D42C5,*PTC110,*HEPS5000,*SK3248, *ECG186A,*WEP1096,*RT-167
2SC1098	*PTC110,*HEPS5000,*SK3192,*ECG186A,*WEP245
2SC1099	2SC1100,*ECG165
2SC1100	2SC1099,*SK3115,*ECG165,*WEP740B
2SC1101	2SC999,*PTC130,*HEPS5020,*SK3115,*ECG164,*WEP740A
2SC1102	2SC1456,*HEPS5011,*ECG124,*WEP240
2SC1103	2SC1103A,*ECG154,*WEPS3021
2SC1103A	2SC1103,*SK3040,*ECG154,*WEPS3021
2SC1104	2SC779-O,2SC825,40318,40322,*PTC104,*HEPS5011,*SK3079,*ECG124,*WEP240
2SC1105	2SC515A,2SC1456,*PTC104,*HEPS5011,*SK3021,*ECG124,*WEP240
2SC1106	BDY27C,*PTC129A,*HEPS5020,*SK3535,*ECG162,*WEP707
2SC1107	2SD526
2SC1108	
2SC1109	
2SC1110	
2SC1111	2N6354,2SC897,2SC1030,2SC1112,2SC1115,2SC1116,2SC1402,2SC1403, 2SC1403A,2SC1782,2SC1783,KSP1176,SDT7604,SDT7605,*HEPS7004,*ECG280
2SC1112	2N4071,2N6354,2SC897,2SC1030,2SC1116,2SC1116A,2SC1403,2SC1403A, 2SC1783,182T2,BDY25B,BUY20,SDT7605,*HEPS7004
2SC1113	
2SC1114	*PTC118,*HEPS7004,*SK3027,*ECG283
2SC1115	2SC1116,2SC1782,2SC1783,KSP1276,*HEPS7004,*SK3079,*ECG280,*WEP707
2SC1116	2SC1116A,2SC1783,*HEPS7004,*SK3079,*ECG154,*RT-141
2SC1116A	
2SC1117	2N6389,2SC1529,35824A,*PTC132,*HEPS0017,*SK3039,*ECG161,*WEP709
2SC1118	*HEPS3006
2SC1119	
2SC1120	2SC1122,2SC1356,2SC1414,2SC1481,2SC1482,2SC1534,2SC1535,2SC1725, 2SC1726,2SC1821,40968†,*HEPS3006
2SC1121	2SC1122,2SC1122A,2SC1414,2SC1415,2SC1482,2SC1535,2SC1726,*HEPS3007
2SC1122	2SC1415,2SC1536,*HEPS3007
2SC1122A	2SC1415,2SC1536
2SC1123	2N2369,2N2369A,2N3862,2N5029,2N5769,2SC577,2SC601N,2SC639,2SC764, 2SC1128,MPS2369,ZT2369,*PTC133,*HEPS0020,*SK3018,*ECG107,*WEP56
2SC1124	*HEPS3019,*SK3199,*ECG198,*WEPS3021
2SC1126	2N2369,2N2369A,2N2710,2N2729,2N4137,2N5029,2SC577,2SC601,2SC601N, 2SC689,2SC1123,2SC1128,2SC1240,MM1941,MPS2369,MPS6507,MT1060A,TIS48, TIS49,ZT2369,*PTC132,*HEPS0016,*SK3018,*ECG107,*WEP56
2SC1127	*HEPS3021,*SK3104,*ECG190,*WEPS3021
2SC1128	2N2369,2N2369A,2N3862,2N5029,2N5769,2SC577,2SC601N,2SC639,2SC764, 2SC1123,MPS2369,ZT2369,*PTC115,*HEPS0016,*SK3018,*ECG161
2SC1129	2N2369,2N2369A,2N2501,2N3009,2N3013,2N3014,2N3646,2N3862,2N4275,2N5029, 2N5769,2N5772,2SC67,2SC601N,2SC639,2SC764,2SC1123,2SC1128,2SC1240, 2SC1274,2SC1687,2SC1688,BF311,GET2369,TIS87,TIS105,*PTC101,*HEPS0016, *SK3018,*ECG161
2SC1130	*PTC129A
2SC1131	2N6542,SDT1053,SDT1054,SDT1058,SDT1059,SDT1063,SDT1064,*PTC129A
2SC1132	*PTC129A
2SC1136	2SC1437,2SC1443

To Replace	Substitute This Type
2SC1138	
2SC1139	
2SC1140	2N6547,*HEPS5021
2SC1141	2N6546
2SC1142	2N6547,2SC1140
2SC1143	2N6546,2N6581,2N6584,2SC1141,2SC1580,IR4065,IR5063
2SC1150	2N1613A,2N2106,2N2107,2N2193,2N2193A,2N2193B,2N2846,2N2868,2N3015, 2N3110,2N3252,2N3253,2N3299,2N3326,2N3722,2N3735,2N5262,2SC97,2SC97A, 2SC580,2SC879,2SC880,2SD216,40347,BC140-6,BSX60,KT600F,KT600T,SE8002, ZT1613,*PTC144
2SC1151	2SC1151A,*PTC130,*HEPS5021,*SK3133,*ECG164,*WEP740A
2SC1151A	*PTC130,*HEPS5021,*SK3133,*ECG164,*WEP740A
2SC1152	2N6234,2N6235,2SC808,2SC1114,*PTC119,*HEPS7004,*ECG162
2SC1153	2SC1099,2SC1100,2SC1154,2SC1309,2SC1358,*HEPS5021,*SK3115,*ECG165, *WEP740
2SC1153A	2SC999A,2SC1170B,2SC1358
2SC1154	2SC1099,2SC1100,2SC1309,2SC1325,2SC1325A,2SC1358,2SC1358A,*HEPS7004, *SK3111,*ECG163,*WEP740
2SC1155	2SC1098,2SC1156,*HEPS3020,*ECG190,*WEPS3021
2SC1156	2SC1157,*HEPS3019,*SK3104,*ECG190,*WEPS3021
2SC1157	2SC1224,*HEPS3019,*SK3104,*ECG190,*WEPS3021
2SC1158	2SC287A,2SC1159,*HEPS0020,*SK3019,*ECG108,*WEP56
2SC1159	2SC287A,2SC1158,*PTC132,*HEPS0020,*ECG107,*WEP56
2SC1160	2N5660,2SC1161,HST5952,HST5953,HST6907,HST6908,*HEPS3021,*ECG175, *WEP241
2SC1161	2N5660,2SC1160,HST5952,HST5953,HST6907,HST6908,*PTC104,*HEPS3021, *SK3021,*ECG175,*WEP241
2SC1162	*PTC904,*HEPS5000,*SK3041,*ECG184,*WEPS5003,*RT-162
2SC1163	
2SC1164-O	2SC1199
2SC1164-R	
2SC1165	2SC731,2SC872,2SC890,2SC908,2SC1001,2SC1002,BLX89,*HEPS3008,*SK3195
2SC1166-GR	2N6428,2N6428A,2N6539,BCY65,TIS94,TIS97
2SC1166-O	2SC1509,2SC1852,TIS96,TIS99,TIS135,*SK3137
2SC1166-R	2N6538,A5T3903,TIS110,TIS136,*SK3047
2SC1166-Y	2N4409,2N6540,2N6541,2SC1166-GR,A5T2222,A5T3904,A5T4409,BCY65,TIS95, TIS98,TIS109,TIS111,*GE-18
2SC1167	*PTC129A,*GE-38,*TR-93,*HEPS5021,*SK3115,*ECG165
2SC1168	2SC515A,*GE-12,*TR-81,*HEPS5011,*SK3131,*ECG124
2SC1169	
2SC1170	2SC1170A,2SC1413,2SC1413A,*PTC129A,*SK3115,*ECG165
2SC1170A	2SC1413A,*PTC129A,*SK3115,*ECG165
2SC1170B	2SC1172,2SC1172B,*PTC129A,*GE-38,*TR-93,*HEPS5021,*SK3115,*ECG165
2SC1171	2SC642,2SC642A,*GE-38,*TR-93,*SK3133,*ECG165,*WEP740B
2SC1172	2SC1172B,*PTC129A,*GE-38,*TR-93,*SK3115,*ECG238,*WEP740B
2SC1172A	2SC1172,2SC1172B,*PTC129A,*GE-38,*TR-93,*SK3115,*ECG238,*WEP740B
2SC1172B	*PTC129A,*GE-38,*SK3115,*ECG238
2SC1173-GR	*TR-76,*ECG152,*WEP245
2SC1173-O	*TR-76,*HEPS3024,*SK3054,*ECG152,*WEP245
2SC1173-R	*PTC154,*GE-28,*TR-55,*ECG152,*WEP245,*276-2019,*RT-154
2SC1173-Y	2SC1173-GR,*TR-76,*ECG152,*WEP245
2SC1174	2SC1413,2SC1413A,*PTC129A,*SK3115,*ECG165,*WEP740B
2SC1175	2N2222,2N2222A,2N2540,2N3116,2N5369,2N5377,2N5381,A5T2222,A5T3904, BSW85,TIS109,TIS111,TN-3904,*HEPS0025,*SK3122,*ECG123A,*WEP243,*RT-300
2SC1176	2SC1177,2SC1605,2SC1605A
2SC1177	2SC1178A,2SC1605,2SC1605A
2SC1178A	
2SC1180	40236,40237,*PTC132,*HEPS0017
2SC1184	2SC642,2SC642A,2SC999,2SC1101,2SC1151,2SC1151A,2SC1171,*ECG164, *WEP740A
2SC1185	2SC1106,*HEPS7004,*SK3079,*ECG162,*WEP707

To Replace	Substitute This Type
2SC1187	2N4873,2SC639,2SC1188,2SC1189,2SC1216,2SC1319,2SC1320,2SC1687,2SC1688, TIS87,*HEPS0016,*SK3117,*ECG229,*WEP719
2SC1188	2SC1189,2SC1320
2SC1189	2SC1320,*RT-107A
2SC1190	2SC1297,2SC1404,*HEPS3007
2SC1191	2SC1192
2SC1192	
2SC1193	
2SC1195	*PTC129A,*HEPS5020,*SK3535,*ECG162
2SC1196	2SC1082,2SC1196A,2SC1197A,2SC1206B,2SC1510,2SC1524,2SC1525,2SC1561, 2SC1562
2SC1196A	2SC1197A,2SC1206B,2SC1524,2SC1525,2SC1562,2SC1563
2SC1197	2SC1197A,2SC1206B,2SC1525,2SC1562,2SC1563
2SC1197A	2SC1198,2SC1525,2SC1526,2SC1563
2SC1198	2SC1526,2SC1564
2SC1199	*PTC123
2SC1200	40837,40909†
2SC1204	2N4256,2N4419,2N4954,2N5371,2N5827,2SC735-GR,2SC1993,A157B,BC107, BC107B,BC108,BC108B,BC109B,BC167B,BC171B,BC237B,BC238B,BC239B, BC413B,BC414B,BC547B,BC548B,BC549B,BC550B,BCY58B,BCY58C,BCY59B, BCY59C,MPS3694,MPS6515,MPS6520,*PTC139,*HEPS0015,*SK3124,*ECG199, *WEP736
2SC1205	2N916,2N916A,2N3227,2N4873,2SC735-Y,2SC933,2SC1216,2SC1317,2SC1851, 2SC1999,BC107A,BC108A,BC167A,BC171A,BC237A,BC238A,BC547A,BC548A, BCY58A,BCY59A,BSW83,BSW88,BSW89,BSX38,BSX79,MPS6513,MPS6573, MPS6574-BLUE,MPS6574-GREEN,MPS6574-SIL,MPS6575,MPS6576-BLUE, MPS6576-GREEN,MPS6576-SIL,PET1002,PET2002,ZT82,ZT112,ZT2369A,*PTC101, *SK3122,*ECG108,*WEP56
2SC1206	2SC1206A,2SC1207,*HEPS0015,*SK3024
2SC1206A	2SC1207
2SC1206B	2SC1197A,2SC1198,2SC1525,2SC1563
2SC1207	
2SC1207B	
2SC1208	2SC1208A
2SC1208A	
2SC1209	2N2331,2N3241A,2SC1383,2SC1518,*TR-25,*HEPS5014,*SK3124,*ECG297, *WEP243,*276-2009
2SC1210	2N2539,2N4962,2N5368,2SC1211,2SC1384,2SC1852,40458,MPS-A05,TIS110,TIS133, TIS134,TN-3903,*TR-51,*SK3122,*ECG287,*WEP736,*276-2009,*RT-172
2SC1211	2N2539,2N4962,2N4963,2N5368,2SC1384,2SC1852,40458,MPS-A05,TIS110,TIS135, TIS136,TN-3903,*PTC125,*TR-21,*HEPS0015,*SK3020,*ECG293,*WEP736, *276-2009
2SC1212	2N3300,*PTC904,*HEPS5000,*SK3190,*ECG184,*WEPS5003
2SC1212A	2N1890,2SC485-BL,2SC1475,BSY86,*PTC904,*HEPS5000,*SK3190,*ECG184, *WEPS5003
2SC1213	2N3242A,2N3509,2N5812,2N5818,2N5998,2SC509-Y,2SC509G-Y,2SC1213A, 2SC1214,2SC1813,A5T5449,A8T3704,BC337-25,BSW83,D33D22,D33D26,TN60, *PTC170,*HEPS0025,*SK3137,*ECG289,*WEP736,*RT-134
2SC1213A	2N909,2N930B,2N956,2N2222,2N2222A,2N2350,2N2350A,2N2390,2N2540,2N2645, 2N3116,2N5369,2N5377,2N5818,2SC1214,2SC1813,A5T2192,A5T2222,A5T5449, A8T3704,BSW85,D33D26,TIS109,TIS111,TN-3904,*HEPS0025,*SK3122,*ECG289, *WEP736
2SC1214	A5T2192,A5T2222,A5T5449,A8T3704,TIS109,TIS111,*PTC144,*HEPS0025,*SK3122, *ECG289,*WEP243
2SC1215	*PTC133,*GE-86,*TR-21,*HEPS5000,*SK3018,*ECG107,*WEP736,*276-2011
2SC1216	2N3227,2N6010,*PTC126
2SC1217	*HEPS3019,*SK3201,*ECG171,*WEPS3021
2SC1218	2SC97A,*HEPS3011,*ECG128,*WEP243
2SC1219	2N3227,2N3242A,2N3509,2N3704,2N3794,2N4970,2N5449,2N5818,2N5998,2N6000, 2N6004,2N6010,2SC850,2SC967,2SC968,2SC970,2SC1220,2SC1246,2SC1247, 2SC1317,2SC1813,BSW83,D33D22,D33D26,GI-3704,GI-3706,GI-3794,MPS3704,

To Replace	Substitute This Type
(2SC1219)	PET3704,PET3706,TN60,TN80
2SC1220	2N909,2N956,2N2222,2N2222A,2N2350,2N2350A,2N2390,2N2484A,2N2540,2N2645, 2N3116,2N3302,2N3704,2N4401,2N4437,2N4952,2N4970,2N5107,2N5369,2N5449, 2N5582,2N5818,2N6004,2N6010,2N6014,2SC850,2SC968,2SC970,2SC1247, 2SC1318,2SC1781,2SC1813,BSW85,D33D26,EN956,EN2222,GI-3643,GI-3704, MPS3704,MPS6531,PET3704,SE6022,TN-3904,ZT89,ZT119,*PTC170
2SC1222	2N4104,2N5828,2SC1006,2SC1010,BC414C,*HEPS0015,*SK3122,*ECG199, *WEP736
2SC1223	2SC302-M,2SC307-M,2SC731,2SC844,2SC854,2SC855,2SC872,2SC890,2SC991, 2SC992,2SC1002,2SC1083,2SC1256,2SC1303,2SC1315,2SC1464,2SC1479, 2SC1820,40637A,41039
2SC1224	
2SC1226	*GE-28,*TR-55,*HEPS5000,*SK3202,*ECG186A,*WEP245,*276-2018,*RT-167
2SC1226A	*TR-55,*HEPS5000,*SK3197,*ECG186A,*WEP245,*276-2018,*RT-150
2SC1227	KSP1223,KSP1224,KSP1225,*PTC118
2SC1228	2N6580,2N6581,2N6583,2N6584,BDY46
2SC1229	2SC1227,2SC1322,KSP1222,KSP1223,KSP1224,KSP1225
2SC1230	2N6560,2N6579,2N6580,2N6581,2N6582,2N6583,2N6584,2SC1228,XGSR10040, XGSR15040
2SC1231	2N3959,2N3960,2N5200,2N5201,2SC718,*PTC123
2SC1232	2SC1057,2SC1058,2SC1233
2SC1233	2SC1058
2SC1235	2SC1102,2SC1168,*PTC104,*HEPS5011,*ECG124,*WEP240
2SC1236	
2SC1237	2SC1306,2SC1816,*PTC110,*HEPS3024,*SK3197,*ECG236,*WEP1237,*RT-191
2SC1238	2SC1082,2SC1355,2SC1356,2SC1414,2SC1533,2SC1534
2SC1239	2SC799,2SC866,*GE-46,*TR-65,*HEPS3001,*SK3049,*ECG237,*WEP1239, *276-2014,*RT-174
2SC1240	2N2369,2N2369A,2N3862,2N5029,2N5769,2SC577,2SC639,2SC764,TIS87,TIS105, ZT2369
2SC1241	2SC1122,2SC1482,40968†
2SC1241A	2SC1122,2SC1242A,2SC1482,2SC1718
2SC1242	2SC1122,2SC1548,41042†
2SC1242A	2SC1548,2SC1668,41042†
2SC1243	2SC1013,2SC1096,2SC1226,*PTC110,*TR-55,*HEPS5000,*SK3041,*ECG300
2SC1246	2N3242A,2N5818,2N6010,2N6426,2SC509-Y,2SC509G-Y,2SC967,2SC968,2SC970, 2SC1247,2SC1383,2SC1813,2SC1851,A5T5449,A8T3704,BC337-16,BC337-25, BSW83,D33D22,D33D26,TN60,TN80
2SC1247	2N956,2N2540,2N2645,2N5369,2N5582,2N5818,2N6010,2N6014,2SC968,2SC970, 2SC1384,2SC1813,2SC1852,A5T2192,A5T2222,A5T5449,A8T3704,BSW85,D33D26, TIS109,TIS111,TN-3904
2SC1248	2SC1090,2SC1458
2SC1249	2SC1250
2SC1250	2SC1249
2SC1251	
2SC1252	2SC302-M,2SC307-M,2SC590N,2SC1253,2SC1365,2SC1366,*PTC144
2SC1253	2SC302-M,2SC307-M,2SC590N,2SC1252,2SC1365,2SC1366,*PTC144
2SC1254	2SC988,2SC988A,2SC988B,2SC1044,*HEPS0017
2SC1255	
2SC1256	2SC302-M,2SC307-M,2SC590N,2SC1002,*HEPS3001
2SC1257	2SC1040,2SC1081,2SC1118,2SC1258,*HEPS3005
2SC1258	2SC1081,2SC1118,*HEPS3006
2SC1259	2SC1298,*HEPS3007
2SC1260	2N6389,2SC817,2SC988,2SC988A,2SC988B,2SC1275,2SC1424,2SC1529,2SC1547, 2SC1553,2SC1574,40915,*PTC133,*ECG316
2SC1261	2SC1090,2SC1458
2SC1262	2SC1263,2SC1484
2SC1263	2SC1262,2SC1484
2SC1264	
2SC1265	
2SC1266	

To Replace	Substitute This Type
2SC1267	2SC1251
2SC1268	2SC987,2SC1269,2SC1270,2SC1271,HXTR-6101,HXTR-6102
2SC1269	2SC987,2SC1268,2SC1270,2SC1271
2SC1270	2SC987,2SC1268,2SC1269,2SC1271,HXTR-6101,HXTR-6102
2SC1271	2SC987,2SC1268,2SC1269,2SC1270
2SC1272	2N5761,2SC987,2SC1273
2SC1273	2N5761,2SC987,2SC1272
2SC1274	2N708A,2N914,2N2481,2N2501,2N3210,2N3211,2N3646,2N3647,2N4013,2N5144, 2SC577,MPS6530,ZT708
2SC1275	2N6389,2SC988,2SC988A,2SC988B,2SC1260,2SC1424,2SC1547,2SC1553,2SC1574, *PTC133
2SC1276	2N3227,2SC1216
2SC1277	2N760A,2N909,2N956,2N2222,2N2222A,2N2350,2N2350A,2N2390,2N2484A,2N2540, 2N2645,2N3116,2N3302,2N4401,2N4952,2N5107,2N5369,2N5582,2N5822,2N6014, 2SC1318,2SC1781,40458,BCW34,BCW36,BSW85,D33D30,GI-3643,MPS6531, TN-3904,ZT89,ZT119
2SC1278	2SC1279,2SC1438,2SC1439,*ECG194
2SC1279	*SK3244,*ECG194
2SC1280	MPSA14,*ECG172,*WEPS9100
2SC1280A	*HEPS9100,*ECG172
2SC1281	2N2924,2N3392,2N4124,2N4256,2N4419,2N5172,2N5224,2SC735-GR,2SC1849, 2SC1993,A5T4124,A158B,A159B,BC108,BC108B,BC109B,BC168B,BC169B,BC172B, BC173B,BC238B,BC239B,BCY58B,BSW88,BSW89,BSX38
2SC1282	2N736A,2N2461,2N2465,2N4963,2SC338,2SC734-O,2SC979-O,2SC979A-O,BC110, TIS96,TIS99
2SC1283	2N2461,2N2465,2SC339,2SC979A-O
2SC1285	2N3116,2N3242A,2N3302,2N3904,2N4401,2N4952,2N4954,2N5369,2N5371,2N5377, 2N5381,2N6112,2SC1364,2SC1685,2SC1850,2SC1992,A157B,BC107,BC107B, BC167B,BC171B,BC182B,BC182L,BC183L,BC237B,BC413B,BC414B,BCY59B, BFY39-3,BSW83,BSX79,PBC107B,TN-3904,*PTC132,*HEPS0015
2SC1288	
2SC1290	
2SC1292	2N5239,2N5240,2N5838,2N5839,2N5840,STS-411
2SC1293	2N2369,2N2369A,2N2501,2N2615,2N2616,2N3009,2N3011,2N3013,2N3014,2N3211, 2N3511,2N3646,2N3647,2N4274,2N4275,2N5029,2N5769,2N5772,2SC67,2SC577, 2SC601N,2SC639,2SC764,2SC1123,2SC1128,2SC1240,2SC1274,2SC1988, A5T3571,GET2369,MPS2369,MPSH10,MPSH11,MT1060A,ZT2369,*PTC121, *HEPS0020,*SK3132,*ECG161,*WEP709
2SC1295	*HEPS5020,*SK3111,*ECG165
2SC1297	2SC1339,2SC1587,*HEPS3007
2SC1298	*HEPS3007
2SC1299	
2SC1300	
2SC1301	2SC1299
2SC1302	2SC1300,2SC1401
2SC1303	2SC302-M,2SC307-M,2SC590N,2SC731,2SC844,2SC854,2SC855,2SC872,2SC890, 2SC991,2SC992,2SC1002,2SC1083,2SC1256,2SC1464,2SC1479,2SC1820, *HEPS3013,*ECG278
2SC1304	2SD326,*PTC104,*HEPS5011,*SK3021,*ECG124,*WEP240
2SC1306	2SC1307,2SC1679,2SC1816,*PTC186,*HEPS3024,*SK3197,*ECG235,*WEP1306, *RT-146
2SC1307	2SC1679,*PTC186,*HEPS5004,*SK3197,*ECG236,*WEP1307,*RT-158
2SC1308	2SD348,*PTC129A,*ECG238
2SC1309	2SC1325,2SC1325A,*HEPS5020,*SK3115,*ECG165,*WEP740
2SC1310	2SC1311,ZTX114,ZTX237B,ZTX238B,ZTX239B,ZTX302
2SC1311	2SC1310,ZTX114,ZTX237B,ZTX238B,ZTX239B,ZTX302
2SC1312	2SC1312Y,2SC1313Y,2SC1345,BC413C,BC414C,BC550C,BCY58D,BCY59D,*TR-24, *HEPS0015,*SK3122,*ECG199,*WEP736,*276-2016
2SC1312Y	2N5526,2SC1313Y,*ECG199
2SC1313	2N4953,2N5370,2N5376,2N5827,2N6006,2SC1345,BC414C,BC547B,BC550B, *SK3245,*ECG199

To Replace	Substitute This Type
2SC1313Y	*SK3245,*ECG199
2SC1314	2SC1208,2SC1208A,*276-2016
2SC1315	2N751,2SC302-M,2SC307-M,2SC590N,2SC731,2SC844,2SC854,2SC855,2SC865, 2SC872,2SC890,2SC991,2SC992,2SC1002,2SC1083,2SC1256,2SC1303,2SC1464, 2SC1479,2SC1820
2SC1316	*HEPS5012,*ECG277,*WEP241
2SC1317	2N3242A,2N6010,2SC1851,BSW83,*GE-210,*TR-24,*HEPS0003,*SK3122,*ECG289, *WEP243,*276-2009,*RT-107A
2SC1318	2N2222,2N2222A,2N2540,2N3116,2N5369,2N5582,2SC1852,A5T2222,BSW85,TIS109, TIS111,TN-3904,*GE-210,*TR-25,*HEPS0015,*SK3122,*ECG297,*WEP243, *276-2009,*RT-302
2SC1319	2N4873,2SC639,2SC1187,2SC1188,2SC1189,2SC1216,2SC1320,2SC1687,2SC1688, TIS87,*HEPS0015
2SC1320	2SC1189,*SK3018
2SC1321	
2SC1322	KSP1222,KSP1223,KSP1224,KSP1225
2SC1323	2SC891,2SC892,2SC1016,2SC1040,2SC1043,2SC1257,2SC1608
2SC1324	*PTC125
2SC1325	2SC1325A,*PTC129A,*HEPS5011,*SK3115,*ECG165,*WEP740B
2SC1325A	2SC1325,*HEPS5020,*SK3115,*ECG165
2SC1327	2SC1328,*PTC153,*GE-64,*TR-51,*HEPS0024,*SK3245,*ECG199,*WEP736, *276-2013
2SC1328	2N4104,*GE-47,*TR-51,*SK3245,*ECG199,*WEP736,*276-2013
2SC1329	2SC1298
2SC1330	*SK3137,*ECG192,*WEP243,*RT-301
2SC1331	2N3570,2N3571,2N3572,2N3600,2N4252,2N4253,2N6304,2N6305,2SC390,2SC567, 2SC568,2SC611,2SC653,2SC927,2SC1035,2SC1036,2SC1553A,40295,40413, S1297
2SC1333	2SC1334,2SC1425,MM4429,MM4430,*HEPS3005
2SC1334	MM4429,MM4430
2SC1335	2SC1344,BC108C,BC109C,BC238C,BC239C,BC413C,BC414C,BC548C,BC549C, BC550C,BCY58D,BCY59D,*PTC132,*HEPS0025,*SK3122,*ECG199,*WEP736
2SC1336	2SC985,2SC987A,2SC1559
2SC1337	2SC990,2SC1015,2SC1076,2SC1338,2SC1338A,2SC1404
2SC1338	2SC1404
2SC1338A	2SC990,2SC1015,2SC1076,2SC1338,2SC1404
2SC1339	
2SC1340	2SC1082,2SC1355,2SC1356,2SC1414,2SC1533,2SC1534,2SC1821,2SC2040
2SC1342	2N916,2N916A,2N2501,2N4123,2N4418,2N4873,2N4996,2SC380-O,2SC380A-O, 2SC400-Y,2SC454,2SC458,2SC459,2SC529,2SC619,2SC639,2SC735-O, 2SC752G-O,2SC899,2SC924,2SC925,2SC929NP,2SC933,2SC1085,2SC1188, 2SC1189,2SC1216,2SC1276,2SC1320,2SC1359,2SC1372,BF254,EN916,MPS6513, PET3002,SE3005,*PTC132,*HEPS0016,*SK3018,*ECG229,*WEP56
2SC1343	2N5038,KSP1276,*PTC118,*HEPS5020,*SK3535,*ECG280,*WEPS7000
2SC1344	2SC1335,BC108C,BC109C,BC238C,BC239C,BC413C,BC414C,BC548C,BC549C, BC550C,BCY58D,BCY59D,*PTC132,*HEPS0015,*SK3124,*ECG199,*WEP736
2SC1345	BC414C,BC550C,*HEPS0025,*SK3122,*ECG199,*WEP736
2SC1346	2SC1406,*PTC178,*GE-63,*HEPS3013,*ECG192,*WEP243,*276-2038
2SC1347	2SC1407,*PTC178,*GE-47,*TR-21,*HEPS3001,*SK3137,*ECG192,*WEP243, *276-2014
2SC1348	2SC1086,*HEPS5021,*ECG163,*WEP740
2SC1354	2SC1020
2SC1355	2SC1082,2SC1356,2SC1414,2SC1533,2SC1534,2SC1535,2SC1561,2SC1562
2SC1356	2SC1414,2SC1415,2SC1534,2SC1535,2SC1562,2SC1563
2SC1358	2SC1325,2SC1325A,2SC1358A,*PTC129A,*HEPS5011,*SK3115,*ECG165, *WEP740B
2SC1358A	2SC1325,2SC1325A,*PTC129A,*ECG165
2SC1359	2SC400,2SC400-Y,2SC735-Y,2SC933,2SC1216,2SC1999,BC107A,BC108A,BC123, BC167A,BC171A,BC237A,BC238A,BC547A,BC548A,BCY58A,BCY59A,*PTC132, *GE-211,*TR-24,*HEPS0015,*SK3122,*ECG229,*WEP56,*276-2015,*RT-308
2SC1360	*GE-11,*TR-70,*HEPS0025,*SK3132,*ECG107,*WEP56,*276-2015

To Replace	Substitute This Type
2SC1361	*SK3124,*ECG123A
2SC1362	*HEPS0015,*SK3124,*ECG199,*WEP736
2SC1363	2N3242A,2N4419,BCY58B,*HEPS0011,*SK3124,*ECG123A,*WEP736
2SC1364	2N2222A,2N3904,2N4401,2N5381,A5T3904,BC182B,TIS111,TN-3904,*HEPS0015, *SK3122,*ECG123A,*WEP736,*RT-308
2SC1365	2SC302-M,2SC307-M,2SC590N,2SC1252,2SC1253,2SC1366
2SC1366	2SC302-M,2SC307-M,2SC590N,2SC1252,2SC1253,2SC1365
2SC1367	2SC1151A,2SC1367A,*HEPS5021,*SK3115,*ECG164
2SC1367A	2SC1151A,2SC1367,*PTC130,*HEPS5021,*SK3131,*ECG164
2SC1368	2SC1162,*PTC904,*SK3190,*ECG184,*WEPS5003
2SC1372	2N916,2N916A,2N2501,2N4013,2N4123,2N4418,2SC619,2SC639,2SC735-O, 2SC752G-O,2SC933,2SC1216,EN916,*HEPS0016
2SC1373	2N2331,2N2501,2N2923,2N3241A,2N3393,2N3396-ORG,2N3397-ORG,2N3398-ORG, 2N3414,2N3856A,2N3858,2N3859,2N4074,2N4123,2N4138,2N4418,2N4873,2N6427, 2N6566,2S512,2SC367G,2SC367G-O,2SC509-O,2SC509G-O,2SC639,2SC733-O, 2SC735-O,2SC1216,40398,A5T4123,GI-3414,MPS3393,MPS6513,MPS6560, MPS6574-BLUE,MPS6574-GREEN,MPS6574-YEL,MPSA10-GRN,MPSA10-WHT, MPSA20-GRN,MPSA20-WHT,NPS6513,PET4001,TIS90-GRN,TIS90-YEL, TIS90M-GRN,TIS90M-YEL,TIS92-GRN,TIS92-YEL,TIS92M-GRN,TIS92M-YEL,ZT80, ZT110,ZT2369A,ZT2938
2SC1374	2N708,2N784A,2N914,2N2242,2N2331,2N2369,2N2369A,2N2481,2N2501,2N3009, 2N3011,2N3013,2N3014,2N3210,2N3211,2N3394,2N3397-RED,2N3398-RED,2N3508, 2N3646,2N3647,2N3855A,2N3858,2N4123,2N4138,2N4274,2N4275,2N4418, 2N5029,2N5418,2N5769,2N5772,2N6427,2N6566,2S512,2SC67,2SC367G-O, 2SC509-O,2SC509G-O,2SC577,2SC579,2SC601N,2SC613,2SC639,2SC733-O, 2SC735-O,2SC764,2SC878,2SC913,2SC914,2SC914A,2SC915,2SC915A,2SC1071, 2SC1274,2SC1373,2SC1375,2SC1376,A5T4123,BSW82,GET2369,KT218F,MPS2369, MPS3394,MPS6512,MPS6574-YEL,NPS6512,PET2001,PT720,TIS90-YEL, TIS90M-YEL,TIS92-YEL,TIS92M-YEL,ZT80,ZT110,ZT404P,ZT708,ZT2205,ZT2369
2SC1375	2N2242,2N2331,2N2369,2N2369A,2N2501,2N3009,2N3013,2N3014,2N3211,2N3394, 2N3397-RED,2N3398-RED,2N3855A,2N3858,2N4123,2N4138,2N4418,2N4873, 2N5418,2N5769,2N6427,2N6566,2S512,2SC67,2SC367G-O,2SC509-O,2SC509G-O, 2SC577,2SC579,2SC601N,2SC613,2SC639,2SC733-O,2SC735-O,2SC764,2SC1071, 2SC1373,A5T4123,BSW82,GI-3414,MPS2369,MPS3394,MPS6560,MPS6574-BLUE, MPS6574-YEL,MPSA10-WHT,MPSA20-WHT,PET2001,PT720,TIS90-GRN,TIS90-YEL, TIS90M-GRN,TIS90M-YEL,TIS92-GRN,TIS92-YEL,TIS92M-GRN,TIS92M-YEL,ZT80, ZT110,ZT2205,ZT2369,ZT2369A,ZT2938
2SC1376	2N708,2N708A,2N718,2N760,2N784A,2N914,2N2242,2N2315,2N2369,2N2369A, 2N2396,2N2481,2N2501,2N2845,2N2847,2N3009,2N3013,2N3014,2N3115,2N3210, 2N3211,2N3301,2N3508,2N3641,2N3642,2N3646,2N3647,2N3862,2N3903,2N3946, 2N3974,2N3976,2N4013,2N4123,2N4140,2N4227,2N4400,2N4962,2N5027,2N5029, 2N5144,2N5380,2N5769,2N5772,2N6427,2N6538,2SC67,2SC366G-O,2SC367G-O, 2SC577,2SC613,2SC639,2SC764,2SC878,2SC913,2SC914,2SC914A,2SC1166-O, 2SC1210,2SC1274,A5T3903,A5T4123,BSW82,EN697,FT3641,FT3642,GET2221, GET2369,GI-3641,GI-3642,KT218F,MM3903,MPS2369,MPS3693,MPS3705, MPS6530,MPS6565,PET1001,TIS90-YEL,TIS90M-YEL,TIS92-YEL,TIS92M-YEL, TN-3903,ZT81,ZT111,ZT708,ZT2369,*SK3024
2SC1377	2SC1307,*PTC110,*GE-216,*SK3197,*ECG236,*RT-146
2SC1378	2SC1314,2SC1476,2SC1668,*HEPS3006
2SC1379	
2SC1380-BL	2N5828,2SC1000-BL,2SC1000G-BL,2SC1380A-BL,2SC1570,BC550C
2SC1380-GR	2N4953,2N5370,2N5376,2N5377,2N5827,2N6006,2N6112,2SC734-GR,2SC1000-GR, 2SC1000G-GR,2SC1345,2SC1380A-GR,BC414C,BC546B,BC547B,BC550B,TP3704
2SC1380A-BL	2N5828,2SC1000-BL,2SC1000G-BL,2SC1380-BL,2SC1570,BC550C
2SC1380A-GR	2N4953,2N5370,2N5376,2N5377,2N5827,2N6006,2N6112,2SC734-GR,2SC1000-GR, 2SC1000G-GR,2SC1345,2SC1380-GR,BC414C,BC546B,BC547B,BC550B,TP3704
2SC1381	2N1711B,2N1890,2N4943,BC141-16,BSY86
2SC1382-O	2N3725A,2SC108A-O,40635,*ECG157
2SC1382-Y	2N2219A,2SD438,BSY54,BSY84
2SC1383	*PTC144,*GE-63,*TR-25,*HEPS3024,*SK3512,*ECG293,*WEPS3020,*276-2038, *RT-302

To Replace	Substitute This Type
2SC1384	*TR-25,*HEPS3001,*SK3122,*ECG293,*WEPS3020,*276-2014
2SC1385	2N1613,2N1837A,2N2017,2N2106,2N2193,2N2193A,2N2193B,2N2218,2N2218A, 2N2270,2N2537,2N2848,2N2868,2N3109,2N3725,2N3725A,2N4047,2N4960,2N4961, 2SC97A,2SC109A-O,2SC307,2SC480,2SC503-Y,2SC879,2SC880,2SC1218, 2SC1386,40635,BC140-10,BC341-10,BSY51,BSY53,BSY83,TN53,ZT90,ZT95, ZT1613,ZT2270,*SK3024
2SC1386	2N1716,2N1889,2N2193,2N2193A,2N2193B,2N2218A,2N3107,2N3109,2N3725A, 2N4961,2SC97A,2SC108A-O,2SC109A-O,2SC485-Y,2SC512-O,2SD328,40635, BC140-10,BC141-10,BSY53,BSY83,TN53,ZT1613,*SK3024
2SC1387	2SC1068,*PTC123
2SC1391	2SC1102,2SC1235,*HEPS0015,*ECG124
2SC1393	2SC1394,2SC1395,2SC1396,*SK3018,*ECG123A
2SC1394	2SC1393,2SC1395,2SC1396,*HEPS0025,*SK3018,*ECG107
2SC1395	2SC1396,*PTC115,*HEPS0011,*SK3122,*ECG107,*WEP56
2SC1396	2SC1395,*ECG316
2SC1398	*GE-66,*TR-76,*HEPS5000,*ECG152,*WEP245,*276-2019,*RT-152
2SC1400	
2SC1401	2SC1300,2SC1302
2SC1402	2N6354,2N6496,2SC1115,2SC1116,2SC1403,2SC1403A,2SC1782,2SC1783, KSP1176,SDT7604,SDT7605,*PTC118,*HEPS5020,*ECG280
2SC1403	2N4071,2N6354,2N6496,2SC1116,2SC1116A,2SC1403A,2SC1783,BUY20,SDT7605, *PTC118,*HEPS5020
2SC1403A	2N4071,2SC1116,2SC1116A,2SC1783,BUY20,SDT7605
2SC1404	2SC1339
2SC1405	2SC891,2SC892,2SC1040,2SC1257,MM8012
2SC1406	*HEPS3024,*ECG192,*WEP243,*276-2038
2SC1407	*HEPS3020,*SK3137,*ECG192
2SC1409	2SC1409A,*HEPS5000,*SK3021,*ECG291,*WEP245
2SC1409A	2SC1409,*ECG291
2SC1410	2SC1410A,*ECG291
2SC1410A	2SC1410,*ECG291
2SC1411	40400,*PTC126
2SC1412	2SC1199,*PTC123
2SC1413	2SC1413A,*PTC129A,*HEPS5020,*SK3115,*ECG165,*WEP740
2SC1413A	*PTC129A,*SK3115,*ECG165
2SC1414	2SC1415,2SC1535,2SC1563
2SC1415	2SC1536
2SC1416	2N4953,2N5370,2N5376,2N5827,2N6006,2SC1345,2SC1416A,2SC1570,2SC1681-BL, BC414C,BC546B,BC547B,BC550B,*HEPS0030,*ECG199
2SC1416A	2N4953,2N5370,2N5376,2N5827,2N6006,2SC1345,2SC1416,2SC1570,2SC1681-BL, BC414C,BC546B,BC547B,BC550B,*HEPS0030,*SK3244
2SC1417	2N2615,2N2616,2N4418,2N4996,2SC380-O,2SC380A-O,2SC388A,2SC400-Y,2SC619, 2SC735-O,2SC924,2SC925,2SC929NP,2SC934,2SC1215,2SC1342,2SC1359, 2SC1372,2SC1789,BF254,MPS3563,MPS6568A,MPS6569,MPS6570,PET3002, SE3005,*PTC132,*HEPS0030,*SK3018,*ECG107,*WEP56
2SC1418	2N5295,2SC1060,2SD314,2SD318,2SD331,2SD366,2SD390,*PTC167,*HEPS5000, *SK3054,*ECG152
2SC1419	2N5296,2N6122,2SC789-O,2SC789-R,2SC790-O,2SC790-R,2SC1061,2SD234-O, 2SD234-R,2SD234G-O,2SD234G-R,2SD235-O,2SD235-R,2SD235G-O,2SD235G-R, 2SD330,2SD365,2SD389,2SD570,BD221,D44C4,D44C5,D44C6,D44C7,D44C8, D44C9,*HEPS5000,*SK3054,*ECG152,*RT-197
2SC1420	2SC1459,2SC1937
2SC1421	2SC1420,2SC1422,2SC1423,2SC1459,2SC1462,2SC1935,2SC1937
2SC1422	2SC1420,2SC1459,2SC1937
2SC1423	2SC1420,2SC1462,2SC1935,2SC1937
2SC1424	2N6389,2SC988,2SC988A,2SC988B,2SC1260,2SC1275,2SC1547,2SC1553,2SC1574, *PTC133,*ECG316
2SC1425	2SC1333,2SC1334,MM4429,MM4430
2SC1426	2N5180,2SC988,2SC988A,2SC988B,2SC1044,2SC1254,*PTC144
2SC1427	
2SC1428	

To Replace	Substitute This Type
2SC1429	2SC1761,*ECG128
2SC1430	2N740,2N910,2N1574,2N1973,2N2443,2SC510-O,2SC512-O
2SC1431	TIP503,*PTC104
2SC1433	*PTC129A,*HEPS5021,*SK3111
2SC1434	2SD573
2SC1435	
2SC1436	1843-2005,1843-2010,1843-2205,1843-2210,1843-2505,1843-2510,1843-2705, 1843-2710,XDAR10020,XDAR10025,XGSR15020,XGSR15025
2SC1437	
2SC1438	2SC1439
2SC1439	2SC1438
2SC1440	2SC1441,HST9209
2SC1441	2SC1436,1843-2005,1843-2010,1843-2205,1843-2210,XDAR10020,XGSR15020
2SC1442	2SC1443
2SC1443	2SC1437
2SC1444	2N5428,2N5430,2SC1445,KSP1164,KSP1165,*HEPS5019
2SC1445	2N5430,KSP1165,*HEPS5012
2SC1446	2SC1447,2SC1505,2SC1723,*HEPS3019,*SK3220,*ECG198,*WEPS3021
2SC1447	*HEPS3021,*SK3104,*ECG198,*WEP244
2SC1448	2N6477,2N6478,2SC1448A,2SC1669-O,2SC1669-R,*PTC129A,*HEPS3021,*SK3104, *ECG291,*WEPS3021
2SC1448A	2N6477,2N6478,2SC1669-O,*SK3054,*ECG196
2SC1449	2N3506,2N5334,*SK3190,*ECG295,*WEPS5003,*RT-162
2SC1450	HST6905,HST6906,HST6907,*HEPS5012,*ECG152,*WEP240
2SC1451	2N3501,2SC507-Y,2SC1452
2SC1452	2N3501,2SC507-Y,2SC1451,*PTC125
2SC1453	2N915,2N3116,2N3643,2N3904,2N3947,2N4141,2N4401,2N4437,2N4951,2N4952, 2N4970,2N5028,2N5107,2N5368,2N5369,2N5381,2N6004,2SC366G,2SC366G-Y, 2SC734-Y,2SC933,2SC943,2SC944,2SC979-Y,2SC980G-Y,2SC1175,2SC1318, 2SC1364,2SC1781,2SC1998,2SC1999,BC107A,BC174A,BC182A,BC190A,BC237A, BC546A,BC547A,BCW34,BCW36,BSW85,BSX79,GET2222,GET2222A,GI-3643, MM3904,MPS6531,TN-3904,ZT84,ZT89,ZT114,ZT119,*PTC123
2SC1454	*PTC129A,*HEPS5020
2SC1456	*HEPS5011,*ECG124,*WEP240
2SC1457	*PTC125
2SC1458	2SC1090
2SC1459	2SC1420,2SC1937
2SC1460	2SC1119,2SC1420,2SC1421,2SC1422,2SC1423,2SC1459,2SC1461,2SC1462, 2SC1935,2SC1937
2SC1461	2SC1420,2SC1422,2SC1459,2SC1937
2SC1462	2SC1420,2SC1421,2SC1422,2SC1423,2SC1935,2SC1937
2SC1463	SDT1052,SDT1057,SDT1062,*PTC129A
2SC1464	2SC302-M,2SC307-M,2SC590N,2SC731,2SC872,2SC890,2SC991,2SC992,2SC1002, 2SC1083,2SC1256,2SC1479,2SC1820,*PTC144
2SC1465	
2SC1466	SDT1152,SDT1157,SDT1162,*PTC129A
2SC1467	SDT1162,SDT1163,*PTC129A
2SC1468	2N6560,2N6579,2N6580,2N6581,2N6582,2N6583,2N6584,2SC1228,2SC1230, XGSR10040,XGSR15040
2SC1469	2N6580,2N6581,2N6583,2N6584,2SC1141,2SC1228,BDY46
2SC1469A	2N6580,2N6581,2N6583,2N6584,2SC1141,2SC1143,2SC1579,2SC1580,BDY46, DTS-4065,IR4065,IR5063
2SC1470	2SC1300
2SC1471	
2SC1471A	
2SC1472	
2SC1474	*ECG282,*WEP243
2SC1475	*PTC125,*HEPS5014,*SK3047,*ECG315,*WEP243,*RT-114
2SC1476	
2SC1477	2N6249,2SC1436,2SC1786,1561-2003,1561-2004,1582-2003,1582-2004,HST1050, HST1055,XDAR10020,XDAR10025,XGSR10025,XGSR15025,*ECG283

To Replace	Substitute This Type
2SC1478	2SC1327,2SC1328,2SC1478A
2SC1478A	2N4104,2SC1328
2SC1479	2SC302-M,2SC307-M,2SC590N,2SC731,2SC872,2SC890,2SC991,2SC992,2SC1002, 2SC1083,2SC1256,2SC1464,2SC1820
2SC1480	2SC1356,2SC1414,2SC1481,2SC1482,2SC1534,2SC1535,2SC1808,2SC1821,40968†
2SC1481	2SC1122,2SC1356,2SC1414,2SC1482,2SC1534,2SC1535,40968†
2SC1482	2SC1122,2SC1414,2SC1415,2SC1535
2SC1483	
2SC1484	
2SC1485	2SC1752,*PTC117
2SC1486	2SC973,2SC973A,2SC974,2SC975,2SC975A,2SC1073,2SC1074,2SC1075,2SC1337, 2SC1487
2SC1487	2SC973,2SC973A,2SC974,2SC975,2SC975A,2SC1073,2SC1074,2SC1075,2SC1337, 2SC1486
2SC1488	2SC973,2SC973A,2SC974,2SC975,2SC975A,2SC1073,2SC1074,2SC1075,2SC1337, 2SC1487
2SC1489	2N5699,2SC973,2SC973A,2SC974,2SC975,2SC975A,2SC1073,2SC1074,2SC1075, 2SC1337
2SC1490	2SC973,2SC973A,2SC974,2SC975,2SC975A,2SC1073,2SC1074,2SC1075,2SC1251, 2SC1267,2SC1337
2SC1491	2SC973,2SC973A,2SC974,2SC975,2SC975A,2SC1073,2SC1074,2SC1075,2SC1337, 2SC1487,2SC1488,2SC1492,2SC1493
2SC1492	2SC973,2SC973A,2SC974,2SC975,2SC975A,2SC1073,2SC1074,2SC1075,2SC1337, 2SC1487,2SC1488,2SC1491,2SC1493
2SC1493	2N5699,2SC973,2SC973A,2SC974,2SC975,2SC975A,2SC1073,2SC1074,2SC1075, 2SC1337,2SC1488,2SC1491,2SC1492
2SC1494	2N5699,2SC973,2SC973A,2SC974,2SC975,2SC975A,2SC1073,2SC1074,2SC1075, 2SC1337,2SC1488,2SC1489,2SC1491,2SC1492,2SC1493
2SC1495	2N5944,2N5945,2SC891,2SC892,2SC911,2SC911A,2SC1016,2SC1040,2SC1608, 41008A†,41009A†,41010†
2SC1498	2SC1499
2SC1499	2SC1498
2SC1500	2SC1499
2SC1501	*SK3103,*ECG157
2SC1503	2SC1082,2SC1355,2SC1356,2SC1414,2SC1533,2SC1534,2SC1561,2SC1562, 2SC1821,2SC1822
2SC1504	
2SC1505	*SK3220,*ECG198
2SC1506	*SK3103,*ECG198
2SC1507	*HEPS3021,*ECG198
2SC1509	2SD438,*HEPS3020,*ECG297
2SC1510	2SC1196A,2SC1197A,2SC1206B,2SC1524,2SC1525,2SC1562
2SC1511	
2SC1512	
2SC1513	2SC988,2SC988A,2SC988B,2SC1260,2SC1275,2SC1424,2SC1553,2SC1574, *PTC125
2SC1514	*SK3220,*ECG198
2SC1515	2N6430
2SC1516	
2SC1517	
2SC1518	2SC1383,2SD400
2SC1519	2SC1507,2SC1520
2SC1520	2SC1507,2SC1519,*SK3220,*ECG198,*WEP712
2SC1521	
2SC1522	
2SC1523	
2SC1524	2SC1197A,2SC1206B,2SC1525,2SC1562,2SC1563
2SC1525	2SC1198,2SC1526,2SC1563
2SC1526	2SC1564
2SC1528	ZT5591
2SC1529	2N6389,2SC1117,35824A

To Replace	Substitute This Type
2SC1530	
2SC1532	
2SC1533	2SC1082,2SC1356,2SC1414,2SC1534,2SC1535,2SC1561,2SC1562
2SC1534	2SC1414,2SC1415,2SC1535,2SC1562,2SC1563
2SC1535	2SC1415,2SC1563
2SC1536	2SC1564
2SC1537	2SC1538,*SK3245,*ECG199,*WEP736
2SC1538	2SC1537,*SK3122,*ECG199,*WEP736
2SC1539	2SC1540,2SC1541,2SC1542
2SC1540	2SC1539,2SC1541,2SC1542
2SC1541	
2SC1542	2SC1541,*PTC101,*HEPS0024,*SK3124,*ECG123A
2SC1543	2SC1537,2SC1538,2SC1541
2SC1544	2SC1539,2SC1540,2SC1541,2SC1542,2SC1543
2SC1545	
2SC1546	2SC1545,*ECG154
2SC1547	2N6389,*PTC132,*ECG316
2SC1548	2SC1536,2SC1564,2SC1824
2SC1550	*SK3130,*ECG154,*WEP712
2SC1551	
2SC1552	2SC1558,2SC1742
2SC1553	2N6389,2SC988,2SC988A,2SC988B,2SC1260,2SC1275,2SC1424,2SC1547,2SC1574, *PTC126
2SC1553A	*PTC123
2SC1554	
2SC1555	
2SC1556	*PTC117,*SK3024,*ECG195A
2SC1557	
2SC1558	2SC1552,2SC1742
2SC1559	2SC985,2SC985A,2SC987A,2SC1336,35861E,35861E
2SC1561	2SC1082,2SC1562
2SC1562	2SC1563
2SC1563	
2SC1564	
2SC1565	*PTC124
2SC1566	*SK3045,*ECG157
2SC1567	2SC1567A
2SC1567A	
2SC1568	*PTC904,*HEPS5000,*SK3190,*ECG184,*WEPS5003
2SC1569	*SK3104,*ECG198
2SC1570	2N5828,BC550C,*PTC121
2SC1571	2N5828,2SC1570,2SC1571L,BC550C,*PTC132,*SK3124
2SC1571L	2N5828,2SC1570,2SC1571,BC550C,*PTC132
2SC1573	2N6516,2SC1752,*PTC123,*SK3044,*ECG287
2SC1574	2N6389,2SC988,2SC988A,2SC988B,2SC1260,2SC1275,2SC1424,2SC1547,2SC1553
2SC1576	
2SC1577	2SC1578
2SC1577A	
2SC1578	
2SC1579	2SC1580
2SC1580	2N6546
2SC1581	
2SC1582	
2SC1583	
2SC1584	2N6339,2N6340,2N6341
2SC1585	
2SC1586	2N6322
2SC1587	
2SC1589	
2SC1590	2SC1591
2SC1591	

To Replace	Substitute This Type
2SC1596	
2SC1603	2SC1480,2SC1481,2SC1482,2SC1808,40967†,40968†,41009†
2SC1604	
2SC1605	
2SC1605A	2SC1177,2SC1178A,2SC1605
2SC1606	2SC891,2SC892,2SC1011,2SC1040,2SC1257,41010†,*PTC143
2SC1607	2N3478,2SC817,2SC988,2SC988A,2SC988B,2SC1260,40915,*PTC123
2SC1608	2SC891,2SC892,2SC1040,*PTC144
2SC1617	BUY21,*PTC129A,*SK3115
2SC1618	2N5624,2SC1619,181T2,BDY24B,KSP1174,KSP1175,SDT7602,SDT7603,*PTC129A
2SC1619	2N4070,2N5624,2N5628,181T2,BDY24B,KSP1175,SDT7603,*PTC129A
2SC1620	2SC891,2SC892,2SC1040
2SC1621	
2SC1622	
2SC1623	
2SC1624-O	
2SC1624-Y	U2TD410
2SC1625-O	2SC1624-O
2SC1625-Y	2SC1624-Y
2SC1626-O	
2SC1626-Y	
2SC1627	2N6540,2N6541,2SC1735,TIS95,TIS98,*PTC170
2SC1628-O	
2SC1628-Y	
2SC1629	DTS-1020
2SC1630	2SC507-O,MM2260,*PTC117
2SC1631	2N3242A,2N4419,2SC1363,2SC1633,BCY58B
2SC1632	2N2222A,2N3904,2N4401,2N5381,2SC1364,2SC1634,BC182B,TN-3904
2SC1633	2N3242A,2N4419,2SC1363,2SC1631,BCY58B,*SK3124,*ECG199
2SC1634	2N2222A,2N3904,2N4401,2N5381,2SC1364,2SC1632,BC182B,TN-3904,*SK3124
2SC1636	
2SC1637	
2SC1638	
2SC1653	
2SC1654	
2SC1663	
2SC1664	*ECG274
2SC1665	2N3571,2N3572,2N4252,2SC653,2SC1457,2SC1638
2SC1666	2SC566,2SC654,2SC1638
2SC1667	2N5614,2N5618,2SD189,2SD189A,2SD379,*PTC129A,*ECG284
2SC1668	2SC1314,2SC1476
2SC1669-O	2N6477,2N6478,2SC1448A
2SC1669-R	2N6477,2N6478,2SC1448
2SC1669-Y	2SD386,2SD386A
2SC1670	*SK3104,*ECG190
2SC1673	2SC1038,2SC1039,V575,V643
2SC1674	2N4873,2N4996,2N4997,2SC639,2SC1187,2SC1188,2SC1189,2SC1216,2SC1276, 2SC1319,2SC1320,2SC1393,2SC1394,2SC1395,2SC1396,2SC1687,2SC1688, BF123,PET3002,SE3005,TIS86,TIS87,*PTC132,*SK3018,*ECG229,*RT-308
2SC1675	2N915,2N2847,2N3903,2N4227,2SC620,2SC899,2SC933,2SC943,2SC944,2SC979-O, 2SC980G-O,EN915,*PTC132,*SK3018,*ECG229,*RT-308
2SC1676	2SC990,2SC1207
2SC1677	
2SC1678	2SC1237,2SC1306,2SC1816,*PTC186,*SK3197,*RT-146
2SC1679	*PTC110,*SK3197,*ECG235,*RT-146
2SC1680	
2SC1681-BL	2N4104
2SC1681-GR	2N4953,2N5370,2N5376,2N5377,2SC734-GR,2SC1345,BC546B
2SC1682-BL	2N3227,2N3242A,2N3643,2N3904,2N3947,2N4141,2N4437,2N4951,2N4952,2N4970, 2N5028,2N5107,2N5368,2N5369,2N5381,2N6004,2S104,2SC366G,2SC366G-Y, 2SC367G,2SC367G-Y,2SC752G-Y,2SC933,2SC943,2SC1175,2SC1216,2SC1285,

To Replace	Substitute This Type
(2SC1682-BL)	2SC1318,2SC1364,2SC1682-GR,2SC1999,BC107A,BC123,BC167A,BC171A,BC182A, BC237A,BC547A,BCY59A,BFY39-2,BSW83,BSX79,GET2222,GI-3643,MM3904, PBC107A,TN-3904
2SC1682-GR	2N3227,2N3242A,2N3643,2N3904,2N3947,2N4141,2N4437,2N4951,2N4952,2N4970, 2N5028,2N5107,2N5368,2N5369,2N5381,2N6004,2S104,2SC366G,2SC366G-Y, 2SC367G,2SC367G-Y,2SC752G-Y,2SC933,2SC943,2SC1175,2SC1216,2SC1285, 2SC1318,2SC1364,2SC1682-BL,2SC1999,BC107A,BC123,BC167A,BC171A, BC182A,BC237A,BC547A,BCY59A,BFY39-2,BSW83,BSX79,GET2222,GI-3643, MM3904,PBC107A,TN-3904
2SC1683	*PTC132
2SC1684	2N4256,2N4419,2N4954,2N5371,2N5827,2N6002,2N6006,2N6008,2N6112, 2SC400-GR,2SC735-GR,2SC1849,2SC1993,A157B,BC107,BC107B,BC108,BC108B, BC109B,BC167B,BC171B,BC183L,BC184L,BC237B,BC238B,BC239B,BC413B, BC414B,BC547B,BC548B,BC549B,BC550B,BCY58B,BCY58C,BCY59B,BCY59C, BFY39-3,*PTC139,*SK3024,*ECG199
2SC1685	2N4953,2N5370,2N5376,2N5377,2SC734-GR,2SC1850,BC174B,BC182B,BC190B, BC546B,*ECG194
2SC1686	2N916,2N916A,2N2369,2N2369A,2N2501,2N3009,2N3013,2N3014,2N3862,2N4873, 2SC67,2SC169,2SC601N,2SC639,2SC764,2SC1187,2SC1188,2SC1189,2SC1319, 2SC1320,2SC1687,2SC1688,BF198,BF199,BF241,TIS87,*PTC132,*SK3018, *ECG229
2SC1687	2SC1688,*SK3246,*ECG123A
2SC1688	*PTC121,*SK3137,*ECG123A
2SC1689	2SC1587
2SC1708	
2SC1709	2SC216,FX3724,FX4046
2SC1710	
2SC1711	2SC1711A
2SC1711A	2SC1711
2SC1712	
2SC1713	
2SC1714	2SC1715,2SC1716,2SC1932
2SC1715	2SC1714,2SC1716,2SC1932
2SC1716	2SC1714,2SC1715,2SC1932
2SC1717	2SC1169,*SK3049
2SC1718	2SC1122,2SC1242A,2SC1415,2SC1536,2SC1548,41042†
2SC1719	2N3501,2SC1628-Y,2SC1720
2SC1720	2N3501,2SC1628-Y,2SC1719
2SC1721	2N3737
2SC1722	*SK3103
2SC1723	2SC1505,*SK3103
2SC1724	2SC1238,2SC1356,2SC1414,2SC1481,2SC1482,2SC1534,2SC1535,2SC1725, 2SC1808,2SC1821,40968†,*PTC144
2SC1725	2SC1122,2SC1356,2SC1414,2SC1415,2SC1482,2SC1534,2SC1535,40968†
2SC1726	2SC1122,2SC1415,2SC1535
2SC1727	*PTC133
2SC1728	2SC1760,*ECG302,*RT-310
2SC1729	
2SC1734	2N3904,2SC1992,BC107B,BC129B,BC167B,BC171B,BC237B
2SC1735	
2SC1742	2SC1552
2SC1743	35862E,35862E,S2509,S2510
2SC1744	
2SC1747	*PTC133
2SC1749	
2SC1752	2N6516
2SC1753	2N6515
2SC1754	
2SC1755	2SC1756,*SK3045,*ECG198
2SC1756	2SC1755,*SK3220,*ECG198
2SC1757	*SK3220,*ECG198

To Replace	Substitute This Type
2SC1760	2SC1728,*SK3197,*ECG306,*RT-310
2SC1761	
2SC1762	
2SC1763	
2SC1764	
2SC1768	2SC1829
2SC1775	2SC1775A
2SC1775A	
2SC1776	FX4960
2SC1777	2N5624,2SC1618,2SC1619,181T2,BDY24B,KSP1174,KSP1175,SDT7602,SDT7603, *PTC129A
2SC1778	2N2369,2N2369A,2N2729,2N3544,2SC387AG,2SC601,2SC689,2SC836,2SC922, 2SC1123,2SC1126,2SC1128,2SC1240,2SC1779,2SC1988,BF311,MM1941,MPS6507, MT1060A,TIS129,*PTC132
2SC1779	2N2369,2N2369A,2N2729,2SC387AG,2SC601,2SC689,2SC836,2SC922,2SC1240, BF311,MM1941,MPS6507,MT1060A,TIS129,*PTC132
2SC1780	2N708,2N747,2N753,2N784,2N784A,2N957,2N2206,2N2242,2N2318,2N2319,2N2369, 2N2369A,2N2432,2N2656,2N2729,2N2885,2N2921,2N2922,2N3009,2N3011,2N3013, 2N3014,2N3340,2N3394,2N3397-RED,2N3398-RED,2N3544,2N3564,2N3605A, 2N3606A,2N3688,2N3689,2N3690,2N3691,2N3844,2N3844A,2N3854A,2N3855A, 2N4137,2N4294,2N4295,2N4420,2N4421,2N4422,2N5029,2N5030,2S501,2SC54, 2SC55,2SC67,2SC131,2SC134,2SC137,2SC204,2SC356,2SC380-R,2SC380A-R, 2SC387AG,2SC394-O,2SC400-R,2SC468,2SC570,2SC579,2SC595,2SC601, 2SC601N,2SC613,2SC624,2SC689,2SC724,2SC752G-R,2SC836,2SC877,2SC878, 2SC913,2SC914,2SC914A,2SC915,2SC915A,2SC922,2SC941-R,2SC963,2SC1026, 2SC1032,2SC1071,2SC1123,2SC1126,2SC1128,2SC1129,2SC1240,2SC1274, 2SC1293,2SC1374,2SC1375,2SC1376,2SC1686,2SC1778,2SC1779,2SC1988,A321, BF121,BF125,BF127,BF198,BF241,BF255,BF311,EN706,EN708,EN914,EN2369A, EN3009,EN3011,EN3013,EN3014,GET708,GET914,GET2369,GET3013,GET3014, GET3646,GI-2921,GI-2922,GI-3394,KT218F,MM1941,MPS3394,MPS3646,MPS6507, MPS6511,MPS6512,MPSA10-RED,MPSA20-RED,MT1060A,PET2001,PT720,PT2760, SE5001,SE5002,SE5003,SE5006,TIS45,TIS46,TIS48,TIS49,TIS51,TIS52,TIS55, TIS129,ZT708,ZT2205,*PTC101
2SC1781	2N2222A,2N5582,BSW85,SE6021,SE6021A
2SC1782	2SC1115,2SC1116,2SC1783,KSP1276
2SC1783	2SC1116,2SC1116A
2SC1784	2N3240,2N6302,2SC1440,2SC1441,2SC1785,2SD232A,1561-1404,1561-1405, 1561-1410,1561-1415,1561-1604,1561-1605,1561-1608,1561-1610,1561-1615, 1561-1803,1561-1804,1561-1805,1561-1808,1561-1810,1561-1815,1582-1403, 1582-1404,1582-1405,1582-1408,1582-1410,1582-1415,1582-1603,1582-1604, 1582-1605,1582-1608,1582-1610,1582-1615,1582-1803,1582-1804,1582-1805, 1582-1808,1582-1810,1582-1815,HST9209,SDT9209,SDT9703,STC2222,STC2223, STC2226,STC2227,STC2230
2SC1785	2SC1436,2SC1441,2SC1786,1561-1803,1561-1804,1561-1805,1561-1808,1561-1810, 1561-1815,1561-2003,1561-2004,1561-2008,1561-2010,1582-1803,1582-1804, 1582-1805,1582-1808,1582-1810,1582-1815,1582-2003,1582-2004,1582-2008, 1582-2010,1843-2005,1843-2010,1843-2205,1843-2210,STC2223,STC2227, XDAR10020,XGSR15020
2SC1786	2SC1436,1561-2003,1561-2004,1561-2008,1561-2010,1582-2003,1582-2004,1582-2008, 1582-2010,1843-2005,1843-2010,1843-2205,1843-2210,1843-2505,1843-2510, 1843-2705,1843-2710,XDAR10020,XDAR10025,XGSR15020,XGSR15025
2SC1787	2SC1006,2SC1327,2SC1328,2SC1478,2SC1478A,*PTC121
2SC1788	2N6426
2SC1789	2SC1215,A5T3571,A5T3572,*PTC133
2SC1790	2N918,2N2808,2N2808A,2N2809,2N2809A,2N2810,2N2810A,2N3287,2N3288,2N3289, 2N3290,2N3337,2N3338,2N3339,2N3600,2N4134,2N4135,2N5053,2N5054,2N5851, 2N5852,2SC80,2SC251,2SC251A,2SC252,2SC253,2SC332,2SC333,2SC463, 2SC567,2SC602,2SC622,2SC660,2SC661,2SC761,2SC762,2SC928,2SC1054, 2SC1067,2SC1747,40295,40413,BF185,BFX60,MT1061,MT1061A,TIS56,TIS57, TN-3200,*PTC126
2SC1804	2SC1805

To Replace	Substitute This Type
2SC1805	
2SC1806	
2SC1808	2SC1356,2SC1414,2SC1481,2SC1482,2SC1534,2SC1535,2SC1821,40968†
2SC1810	
2SC1811	
2SC1813	2N2540,2N5369,2N5377,2N5818,A5T2222,A5T5449,A8T3704,BSW85,D33D26,TIS109, TIS111,TN-3904
2SC1816	2SC1307,*SK3239,*ECG236,*RT-146
2SC1818	*PTC118
2SC1819	
2SC1820	2SC302-M,2SC307-M,2SC590N,*PTC144
2SC1821	2SC1356,2SC1414,2SC1534,2SC1535,2SC1562,2SC1822
2SC1822	2SC1562,2SC1563,2SC1823
2SC1823	2SC1563
2SC1824	2SC1564
2SC1829	
2SC1846	
2SC1847	
2SC1848	
2SC1849	2N4256,2N4954,2N5371,2N5827,2N6002,2N6006,2N6008,2N6112,BC547B,BC548B, BC549B,BC550B,BCY58B,BCY58C,BCY59B,BCY59C
2SC1850	2N4953,2N5370,2N5376,2N5377,BC546B,TIS94,TIS97
2SC1851	
2SC1852	A5T2222,TIS109,TIS111
2SC1853	2N4256,2N4419,2N5827,2SC400-GR,2SC735-GR,2SC829,2SC1993,A157B,BC107, BC107B,BC108,BC108B,BC109B,BC122,BC167B,BC171B,BC237B,BC238B,BC239B, BC413B,BC414B,BC547B,BC548B,BC549B,BC550B,BCY58B,BCY58C,BCY59B, BCY59C
2SC1854	2N3242A,2N4124,2N4256,2N4419,2N4954,2N5371,2N5998,2N6112,2SC400-GR, 2SC735-GR,2SC1364,2SC1684,2SC1849,2SC1992,2SC1993,A157B,BC107,BC107B, BC108,BC108B,BC109B,BC122,BC123,BC167B,BC171B,BC183L,BC237B,BC238B, BC239B,BC413B,BC414B,BCY58B,BCY59B,BFY39-3,BSW83,BSW88,BSW89, BSX38,BSX79
2SC1859	2N3242A
2SC1885	
2SC1888	2SC1889
2SC1889	
2SC1910	2SC1920
2SC1911	2SC1912
2SC1912	2SC1911
2SC1913	
2SC1920	2SC1910
2SC1923	2N4873,2SC639,2SC784-O,2SC785-O,2SC1188,2SC1189,2SC1216,2SC1276, 2SC1320,BF240,*SK3018
2SC1929	
2SC1930	2SC1423,2SC1930A
2SC1930A	2SC1423,2SC1930
2SC1931	2SC1713
2SC1932	2SC1714,2SC1715,2SC1716
2SC1933	
2SC1934	
2SC1935	2SC1420,2SC1421,2SC1422,2SC1423,2SC1459,2SC1462,2SC1937
2SC1936	2SC1713,2SC1933
2SC1937	2SC1420,2SC1459
2SC1938	2SC1714,2SC1715,2SC1716,2SC1932,2SC1939
2SC1939	2SC1714,2SC1715,2SC1716,2SC1932,2SC1938
2SC1944	2SC1945,2SC1969
2SC1945	2SC1944,2SC1969
2SC1946	
2SC1947	2SC302-M,2SC307-M,2SC590N
2SC1964	2SC1944,2SC1945,*SK3197,*ECG236,*RT-146

To Replace	Substitute This Type
2SC1965	
2SC1966	2SC1967
2SC1967	2SC1968
2SC1968	
2SC1969	
2SC1988	
2SC1991	2N4409,A5T3904,A5T4409
2SC1992	BC107B,BC237B,BC547B,BC550B
2SC1993	2N4954,2N5371,2N5827,2SC735-GR,A157B,BC107B,BC108B,BC109,BC109B, BC167B,BC171B,BC237B,BC238B,BC239B,BC413B,BC414B,BC547B,BC548B, BC549B,BC550B,BCY58C,BCY59C
2SC1998	BC546A
2SC1999	2N5381,A5T3904,BC547A,TN-3904
2SC2017	2N6579,2N6580,2N6581,2N6582,2N6583,2N6584,XGSR10040,XGSR15040
2SC2018	2N6653,1843-2520,1843-2720,1843-3020,1843-3220,XGSR15030
2SC2040	2SC1340,2SC1821
2SD11	2N440,2N440A,2N1012,2N1304,2N1306,SF.T298,*PTC108,*SK3010,*ECG103, *WEP641A
2SD12	2N4233A,2N4915,2N5069,2N5870,2SD73,2SD88,2SD188,KSP1041,KSP1042, KSP1044,KSP1045,*PTC140,*GE-72,*TR-59,*HEPS5012,*SK3027,*HN100,*ECG130, *WEP247,*BDY20,*276-2020,*RT-131
2SD13	*GE-4,*TR-03,*HEPS5004,*SK3012,*ECG105,*WEP233
2SD14	2N2015,2SD54,2SD196,2SD196A,*TR-03,*HEPS5004,*ECG105,*WEP233
2SD15	180T2,BDY23B,*PTC173,*HEPS7002,*SK3027,*ECG130,*WEP247
2SD16	2N3446,2N3448,2SD110-R,2SD111-R,2SD118-R,2SD119-R,181T2,BDY24B,*PTC173, *HEPS7004,*SK3027,*ECG130,*WEP247
2SD17	2SD18,182T2,BDY25B,*HEPS5020,*SK3027,*ECG130,*WEP247
2SD18	182T2,BDY25B,*HEPS5020,*ECG162,*WEP707
2SD19	2N356A,2N357A,2N358A,2N438,2N438A,2N439,2N439A,2N576A,2N1302,2N1473, 2N1891,2N1994,2N1995,2SD20,*PTC108,*HEPG0011,*SK3010,*ECG103, *WEP641A
2SD20	2N357A,2N358A,2N439,2N439A,2N440,2N440A,2N445A,2N1012,2N1302,2N1304, 2N1473,2SD11,2SD21,SF.T298,*PTC108,*HEPG0011,*SK3010,*ECG103,*WEP641A
2SD21	2N440,2N440A,2N445A,2N446A,2N634A,2N1012,2N1304,2N1306,2N1624,2SD11, 2SD22,SF.T298,*PTC108,*HEPG0011,*SK3010,*ECG103,*WEP641A
2SD22	2N445A,2N446A,2N447A,2N634A,2N1012,2N1306,2N1308,2N1624,2N1993,2SD23, *PTC108,*HEPG0011,*SK3010,*ECG103,*WEP641A
2SD23	2N446A,2N447A,2N634A,2N635A,2N1308,2N1624,2N1993,*PTC134,*HEPG0011, *SK3010,*ECG103,*WEP641A
2SD24	*GE-12,*TR-23,*HEPS5011,*SK3021,*ECG124,*WEP240,*RT-128
2SD24Y	2SC1168,MJ400,*HEPS5011,*SK3021,*ECG124,*WEP240
2SD25	2SC179,2SC181,2SD77,2SD186,2SD187,NKT713,*PTC108,*HEPG0011,*SK3010, *ECG103A,*WEP641B
2SD26	2SD26A,*HEPS7002,*SK3027
2SD26A	*PTC129A,*HEPS7002,*SK3027,*ECG130,*WEP247
2SD26B	*PTC129A,*HEPS7004,*SK3027,*ECG130,*WEP247
2SD26C	*PTC129A,*HEPS5020,*SK3027,*ECG130,*WEP247
2SD27	2SD352
2SD28	2N3054,2N3766,2N3767,2N5606,2N5610,2SD103,2SD103-O,2SD129-BL,2SD129-Y, 2SD154,2SD242,2SD257,2SD291,2SD292,BD149-10,HST5902,HST5903,KSP1024, KSP1025,RCS29B,RCS29C,*HEPS5019,*SK3026,*ECG175,*WEP241
2SD29	2N3054,2N3766,2N3767,2N5606,2N5610,2SD103,2SD103-O,2SD129-BL,2SD129-Y, 2SD154,2SD242,2SD257,2SD291,2SD292,BD149-10,HST5902,HST5903,KSP1024, KSP1025,RCS29B,RCS29C,*HEPS5019,*SK3026,*ECG175,*WEP241
2SD30	*PTC134,*GE-59,*TR-08,*HEPG0011,*SK3010,*NA20,*ECG103A,*WEP641B,*AC127, *276-2001,*RT-123
2SD31	2SC84,2SC86,2SC179,2SC181,AC127,*PTC108,*GE-8,*TR-08,*HEPG0011,*SK3010, *ECG103A,*WEP641B,*AC127,*276-2001,*RT-122
2SD32	2N2430,2SD27,2SD170A,2SD186,2SD187,NKT713,*PTC108,*GE-8,*TR-08, *HEPG0011,*SK3010,*NA20,*ECG103A,*WEP641B,*AC127,*276-2001,*RT-122
2SD33	2N797,2SD77,2SD96,2SD195,*PTC108,*HEPG0011,*SK3010,*ECG103A,*WEP641B

To Replace	Substitute This Type
2SD34	2SD96,*PTC134,*HEPG0011,*SK3124,*ECG103A,*WEP641B,*RT-123
2SD35	2SD186,2SD187,*PTC108,*GE-5,*TR-08,*HEPG0011,*ECG103A,*WEP641B,*AC127,*276-2001,*RT-122
2SD36	*PTC108,*GE-5,*TR-08,*HEPG0011,*SK3011,*ECG101,*WEP641,*AC127,*276-2001,*RT-119
2SD37	*PTC108,*HEPG0011,*SK3010,*ECG103A,*WEP641B
2SD38	*PTC134,*HEPG0011,*SK3124,*ECG103A,*WEP641B
2SD41	*PTC173,*HEPS7002,*SK3027,*ECG130,*WEP247
2SD43	*PTC108,*GE-5,*TR-08,*HEPG0011,*SK3010,*NA20,*ECG101,*WEP641,*AC127,*276-2001,*RT-119
2SD43A	*PTC108,*GE-8,*TR-08,*HEPG0011,*SK3010,*ECG101,*WEP641,*276-2001,*RT-119
2SD44	2N647,2SC86,2SC179,2SC181,2SD186,2SD187,NKT713,*PTC108,*GE-59,*TR-08,*HEPG0011,*SK3011,*NA20,*ECG101,*WEP641,*276-2001,*RT-119
2SD45	2N4347,2N5559,2N6495,2SC42A,2SC677,2SC902,2SC1111,2SC1112,2SC1115,2SC1116,2SC1116A,2SC1402,2SC1403,2SC1403A,2SC1782,2SC1783,2SD17,2SD18,2SD46,2SD51A,2SD83,2SD84,2SD218,2SD323,2SD375,182T2,BDY25B,KSD2201,KSP1043,KSP1173,SDT7610,SDT7611,SDT7735,SDT7736,*PTC129A,*GE-35,*TR-67,*HEPS5020,*ECG162,*WEP707
2SD46	2N4347,2N5559,2N6495,2SC42A,2SC677,2SC902,2SC1111,2SC1112,2SC1115,2SC1116,2SC1116A,2SC1402,2SC1403,2SC1403A,2SC1782,2SC1783,2SD17,2SD18,2SD45,2SD51A,2SD83,2SD84,2SD218,2SD323,2SD375,182T2,BDY25B,KSD2201,KSP1043,KSP1173,SDT7610,SDT7611,SDT7735,SDT7736,*PTC118,*GE-35,*TR-67,*HEPS5020,*ECG162,*WEP707
2SD47	2N1490,2N5616,2N5620,2N5624,2N5628,2SC678,2SC1619,2SD16,2SD51,2SD82,2SD110-R,2SD111-R,2SD118-R,2SD119-R,2SD202,2SD203,2SD213,2SD214,2SD217,2SD319,2SD322,2SD522,181T2,40369,BDY24B,HST9803,HST9804,KSD2202,KSD2203,KSP1042,KSP1172,SDT7609,SDT7733,SDT7734,ZT1490,*PTC129A,*GE-14,*TR-67,*HEPS7004,*ECG162,*WEP707
2SD48	2N1486,S2N1486,ZT1486,*HEPS3002
2SD49	2SD92,*HEPS5000,*SK3026,*ECG175,*WEP241
2SD50	2SD16,2SD118-R,2SD119-R,2SD164,BDY24A,HST9803,HST9804,*PTC173,*HEPS7004,*SK3027,*ECG130,*WEP247
2SD51	2N4070,2N5616,2N5620,2N5624,2N5628,2SC1619,181T2,BDY24B,KSP1045,KSP1175,SDT7603,*PTC140,*GE-14,*TR-59,*HEPS7004,*SK3027,*HN100,*ECG130,*WEP247,*BDY20,*276-2020,*RT-131
2SD51A	2N4071,2N6354,2SC897,2SC939,2SC940,2SC1030,2SC1111,2SC1112,2SC1115,2SC1116,2SC1116A,2SC1402,2SC1403,2SC1403A,2SC1782,2SC1783,182T2,BDY25B,BUY20,KSP1046,KSP1176,SDT7604,SDT7605,*PTC129A,*HEPS7004,*SK3027
2SD52	2SD164,BDY24A,HST9207,HST9208,*PTC173
2SD53	2SD161,2SD176,2SD177,HST9202,HST9203,HST9207,HST9208,*HEPS7004,*SK3027,*ECG130,*WEP247
2SD54	*HEPS5004,*WEP233
2SD55	*PTC173,*HEPS7000,*ECG181,*WEPS7000
2SD56	*SK3026,*ECG175,*WEP241
2SD57	*HEPS5000,*SK3131,*ECG175,*WEP241
2SD58	2SD103-O,2SD103-R,2SD130-Y,2SD291,2SD292,*HEPS5000,*SK3131,*ECG175,*WEP241
2SD59	2N1490,2N4070,2N5624,2N5628,2SC793-R,2SC793-Y,2SC961,2SC962,2SC1079-R,2SC1080-R,2SC1619,2SD16,2SD82,2SD110-O,2SD110-R,2SD111-O,2SD111-R,2SD116,2SD118-R,2SD118-Y,2SD119-R,2SD119-Y,2SD202,2SD203,2SD217,2SD322,2SD334,2SD340,2SD427-R,2SD428-R,181T2,B170006-ORG,B170006-RED,B170006-YEL,B170007-ORG,B170007-RED,B170007-YEL,B170024,BDY24B,KSP1175,SDT7603,ZT1490,*PTC129A,*HEPS7004,*ECG162,*WEP707
2SD60	2N4071,2N6354,2SC897,2SC898,2SC902,2SC1030,2SC1111,2SC1112,2SC1115,2SC1116,2SC1116A,2SC1343,2SC1402,2SC1403,2SC1403A,2SC1782,2SC1783,2SD17,2SD18,2SD83,2SD84,2SD117,2SD218,2SD286,2SD287,2SD287A,2SD287B,2SD323,2SD375,2SD388,2SD425-R,2SD426-R,182T2,BDY25B,BUY20,FT410,KSP1176,SDT7604,SDT7605,*HEPS5020,*ECG162,*WEP707
2SD61	2SD62,2SD75A,2SD77A,AC127,*PTC108,*GE-59,*TR-08,*HEPG0011,*SK3010,*NA30,*ECG103A,*WEP641B,*AC127,*RT-122

To Replace	Substitute This Type
2SD62	2SD61,2SD75A,2SD77A,AC127,*PTC108,*GE-59,*TR-08,*HEPG0011,*SK3010, *NA30,*ECG103A,*WEP641B,*AC127,*RT-122
2SD63	2SC179,2SC181,2SD25,2SD61,2SD62,2SD75,2SD77,AC127,*GE-5,*TR-08,*SK3010, *NA30,*ECG103A,*WEP641B,*AC127,*RT-122
2SD64	2SD96,2SD186,2SD187,NKT713,*GE-5,*TR-08,*HEPG0011,*SK3010,*NA30, *ECG103A,*WEP641B,*AC187/01,*276-2001,*RT-122
2SD65	2SC179,2SC181,2SD25,2SD61,2SD62,2SD75,2SD77,AC127,*GE-5,*TR-08, *HEPG0011,*SK3010,*ECG103A,*WEP641B,*AC127,*RT-122
2SD66	2SC84,2SC85,2SC180,2SD75,*GE-5,*TR-08,*HEPG0011,*SK3010,*NA20,*ECG103A, *WEP641B,*AC127,*RT-122
2SD67	KSP1046,KSP1176,*GE-35,*TR-67,*HEPS5020,*ECG162,*WEP707
2SD68	KSP1044,KSP1045,KSP1174,KSP1175,*GE-75,*TR-59,*HEPS7002,*SK3027, *ECG223,*WEP247,*276-2018,*RT-131
2SD69	2SC666,2SD334A,*HEPS5020,*SK3111,*ECG130,*WEP247
2SD70	2SD58,2SD130-Y,2SD142,2SD241,2SD256,BDY12,HST5901,MJ2249,MJ3101,RCS29, RCS29A,*PTC906,*GE-23,*TR-81,*HEPS7002,*SK3027,*ECG175,*WEP241, *276-2017,*RT-131
2SD71	2N3767,2N5602,2SD102-O,2SD144,HST5903,KSP1395,RCS29C,*PTC906,*TR-81, *HEPS5012,*ECG175,*WEP241,*276-2020
2SD72	*GE-59,*TR-08,*HEPG0011,*SK3010,*NA20,*ECG103A,*WEP641B,*AC187/01, *276-2001,*RT-122
2SD73	2N1490,2N3446,2N3448,2N3864,2N4070,2N5624,2N5628,2SC793-R,2SC961,2SC962, 2SD16,2SD88,2SD110-R,2SD111-R,2SD118-R,2SD119-R,2SD188,2SD217,2SD322, 2SD427-R,2SD428-R,2SD522,181T2,40369,B170006-BRN,B170006-ORG, B170006-RED,B170007-BRN,B170007-ORG,B170007-RED,B170024,B170025, BDY24B,KSP1042,KSP1045,KSP1172,KSP1175,SDT7603,SDT7733,SDT7734, ZT1490,*PTC140,*TR-67,*HEPS7004,*ECG280,*WEP707,*BDY20,*276-2020
2SD74	2N3442,2N4071,2SC897,2SC898,2SC902,2SC1115,2SC1116,2SC1116A,2SC1402, 2SC1403,2SC1403A,2SC1782,2SC1783,2SD17,2SD18,2SD218,2SD286,2SD287, 2SD287A,2SD287B,2SD323,2SD375,2SD388,182T2,BDY25B,FT410,KSP1043, KSP1046,KSP1173,KSP1176,SDT7604,SDT7605,SDT7735,SDT7736,STS1121, *PTC118,*GE-73,*TR-67,*HEPS5020,*ECG280,*WEP707
2SD75	2SD77,*PTC108,*GE-5,*TR-08,*HEPG0011,*SK3010,*NA30,*ECG103A,*WEP641B, *RT-122
2SD75A	2SD77A,*GE-10,*TR-08,*HEPG0011,*SK3010,*ECG103A,*WEP641B,*RT-122
2SD77	*PTC108,*GE-5,*TR-08,*HEPG0011,*SK3010,*NA30,*ECG103A,*WEP641B,*AC127, *276-2001,*RT-123
2SD77A	*PTC108,*GE-8,*TR-08,*HEPG0011,*SK3010,*ECG103A,*WEP641B,*RT-122
2SD78	2SD78A,2SD378,*GE-66,*HEPS3002,*ECG282,*276-2008,*RT-119
2SD78A	2SD78,2SD378,*ECG282
2SD79	*GE-66,*ECG225,*276-2008
2SD80	*PTC173,*HEPS7002,*SK3027,*ECG130,*WEP247
2SD81	180T2,BDY23B,*PTC173,*HEPS7002,*SK3027,*ECG130,*WEP247,*RT-131
2SD82	181T2,BDY24B,*PTC173,*HEPS7004,*SK3027,*ECG130,*WEP247
2SD83	2SD84,182T2,BDY25B,*HEPS5020,*ECG162,*WEP707
2SD84	182T2,BDY25B,*HEPS5020,*ECG162,*WEP707
2SD88	2N3448,2SD110-O,2SD111-O,2SD118-Y,2SD119-Y,BDY24C,KSP1175,*HEPS5020, *ECG162,*WEP707
2SD88A	*HEPS5020,*ECG162,*WEP707
2SD90	*PTC906,*HEPS5014,*SK3024,*ECG152,*WEP245
2SD91	*PTC906,*HEPS5019,*SK3026,*ECG152,*WEP245
2SD92	2SC1113,*PTC906,*HEPS5012,*SK3131,*ECG175,*WEP241
2SD93	2SD94,*HEPS5012,*WEP245
2SD94	*HEPS5012,*ECG124,*WEP240
2SD96	*PTC134,*GE-59,*TR-08,*HEPG0011,*SK3010,*ECG103A,*WEP641B,*AC187/01, *276-2001,*RT-122
2SD100	*PTC134,*GE-54R,*TR-08,*HEPG0011,*SK3010,*ECG103A,*WEP641B,*RT-122
2SD100A	*PTC134,*GE-54R,*TR-08,*HEPG0011,*SK3010,*ECG103A,*WEP641B,*RT-122
2SD100AG	SEE 2SC367G
2SD100G	SEE 2SC367G
2SD101	*GE-54,*TR-08,*ECG101,*WEP641,*RT-119

To Replace	Substitute This Type
2SD101G	SEE 2SC503
2SD102	2SD102-Y,*GE-47,*TR-81,*HEPS5012,*SK3538,*ECG175,*WEP241,*276-2017
2SD102-O	2SD102,*TR-81,*SK3538,*ECG175,*WEP241
2SD102-R	2SC1113,2SD102-O,*TR-81,*SK3538,*ECG175,*WEP241
2SD102-Y	2SD102,*TR-81,*SK3538,*ECG175,*WEP241
2SD103	2SD102,2SD102-Y,2SD103-Y,2SD129-BL,*TR-81,*SK3026,*ECG175,*WEP241
2SD103-O	2SD102,2SD102-O,2SD103,2SD129-BL,2SD129-Y,*TR-81,*SK3026,*ECG175, *WEP241
2SD103-R	2SD102-O,2SD102-R,2SD103-O,2SD129-R,2SD129-Y,*TR-81,*SK3131,*ECG175, *WEP241
2SD103-Y	2SD102,2SD102-Y,2SD103,*TR-81,*SK3026,*ECG175,*WEP241
2SD104	*GE-59,*TR-08,*HEPG0011,*SK3010,*NA20,*ECG103A,*WEP641B,*RT-122
2SD105	*GE-8,*TR-08,*HEPG0011,*SK3010,*NA20,*ECG103A,*WEP641B,*RT-122
2SD105G	SEE 2SC367G
2SD107	*HEPS9140
2SD108	2SD107,*HEPS9140
2SD110	2SD110-Y,*PTC118,*GE-35,*TR-67,*HEPS5020,*SK3535,*ECG162,*WEP707
2SD110-O	2SD110,*TR-67,*ECG162,*WEP707
2SD110-R	2SD110-O,*GE-35,*TR-67,*ECG162,*WEP707
2SD110-Y	2SD110,*GE-35,*TR-67,*ECG162,*WEP707
2SD111	2SD110,2SD110-Y,2SD111-Y,*PTC118,*GE-75,*TR-67,*HEPS7004,*SK3510, *ECG162,*WEP707
2SD111-O	2SD110,2SD110-O,2SD111,*GE-75,*TR-67,*ECG162,*WEP707
2SD111-R	2SD110-O,2SD110-R,2SD111-O,*PTC173,*GE-75,*TR-67,*ECG162,*WEP707
2SD111-Y	2SD110,2SD110-Y,2SD111,*GE-75,*TR-67,*ECG162,*WEP707
2SD113	2SD113-Y,*GE-75,*TR-36,*HEPS7000,*SK3535,*ECG181,*WEPS7000,*RT-149
2SD113-O	2SD113,*GE-75,*TR-36,*ECG181,*WEPS7000,*RT-149
2SD113-R	2SD113-O,*GE-75,*TR-36,*ECG181,*WEPS7000,*RT-149
2SD113-Y	2SD113,*GE-75,*TR-36,*ECG181,*WEPS7000,*RT-149
2SD114	2SD113,2SD113-Y,2SD114-Y,*GE-75,*TR-36,*HEPS7000,*SK3535,*ECG181, *WEPS7000,*RT-149
2SD114-O	2SD113,2SD113-O,2SD114,*GE-75,*TR-36,*ECG181,*WEPS7000,*RT-149
2SD114-R	2SD113-O,2SD114-O,*GE-75,*TR-36,*ECG181,*WEPS7000,*RT-149
2SD114-Y	2SD113,2SD113-Y,2SD114,*GE-75,*TR-36,*ECG181,*WEPS7000,*RT-149
2SD116	2N3716,2N5626,2SD88A,2SD110,2SD110-O,2SD111,2SD111-O,2SD118-BL, 2SD118-Y,2SD119-BL,2SD119-Y,KSP1175
2SD117	2N6354,2SD287A,2SD287B,2SD388,BUY20,KSP1176
2SD118-BL	2SD110,2SD110-Y,*GE-75,*TR-67,*WEP707
2SD118-R	2SD110-R,*PTC173,*GE-75,*TR-67,*WEP707
2SD118-Y	2SD110-O,2SD118-BL,*GE-75,*TR-67,*WEP707
2SD119-BL	2SD110,2SD110-Y,2SD111,2SD111-Y,2SD118-BL,*GE-75,*TR-67,*WEP707
2SD119-R	2N3446,2N3448,2SD110-R,2SD111-R,2SD118-R,*PTC173,*GE-75,*TR-67,*WEP707
2SD119-Y	2N3448,2SD110-O,2SD111-O,2SD118-BL,2SD118-Y,2SD119-BL,*GE-75,*TR-67, *WEP707
2SD120	*PTC143,*GE-46,*TR-87,*HEPS5014,*SK3024,*ECG282,*WEP243,*276-2008, *RT-114
2SD121	*PTC110,*GE-18,*TR-87,*HEPS3019,*SK3024,*ECG282,*WEP243,*276-2008, *RT-114
2SD122	*HEPS5014,*276-2020
2SD123	*HEPS3002,*276-2020
2SD124	2N1487,2N1489,2N2305,2SD15,2SD124A,2SD163,BDY23A,HST9201,HST9205, HST9206,HST9801,HST9802,SDT7731,SDT7732,ZT1487,ZT1489,*PTC173,*GE-77, *TR-59,*HEPS7002,*SK3027,*ECG280,*WEP247,*BDY20,*276-2020,*RT-131
2SD124A	2N3446,2SD53,2SD119-R,2SD125A,2SD164,2SD176,HST9202,HST9206,HST9207, HST9802,HST9803,*PTC173,*GE-77,*TR-26,*SK3027,*ECG280,*BDY20,*276-2020
2SD125	2N1488,2N1490,2N3446,2SD16,2SD118-R,2SD119-R,2SD125A,2SD161,2SD164, 2SD177,2SD522,40369,BDY24A,HST9202,HST9203,HST9207,HST9208,HST9803, HST9804,SDT7733,SDT7734,ZT1488,ZT1490,*PTC173,*GE-14,*TR-67,*HEPS7004, *SK3027,*ECG280,*WEP740,*BDY20,*276-2020
2SD125A	2N3446,2SD118-R,2SD119-R,2SD161,2SD164,2SD177,HST9202,HST9203,HST9207, HST9208,HST9803,HST9804,*PTC173,*GE-14,*TR-61,*SK3027,*ECG280,*BDY20,

To Replace	Substitute This Type
(2SD125A)	*276-2020
2SD126	HST7206,SDT7206,*TR-61,*ECG280
2SD127	*GE-54,*TR-21,*HEPG0011,*SK3010,*ECG103A,*WEP641B,*AC187/01
2SD127A	*GE-54,*TR-08,*HEPG0011,*SK3010,*ECG103A,*WEP641B,*AC187/01,*RT-122
2SD128	*GE-59,*TR-08,*HEPG0011,*SK3010,*NA20,*ECG103A,*WEP641B,*RT-122
2SD128A	*GE-59,*TR-08,*HEPG0011,*SK3010,*ECG103A,*WEP641B,*RT-122
2SD129-BL	2SD102,2SD102-Y,*TR-81,*ECG175,*WEP241
2SD129-R	2SC1113,2SD102-R,*TR-81,*ECG175,*WEP241
2SD129-Y	2SD102-O,2SD129-BL,*TR-81,*ECG175,*WEP241
2SD130-BL	2SD103,2SD103-Y,*GE-66,*TR-67,*WEP241,*276-2020
2SD130-R	2SD103-R,*GE-216,*TR-81,*ECG175,*WEP241,*276-2020
2SD130-Y	2SD103-O,2SD130-BL,2SD291,2SD292,*GE-216,*TR-81,*ECG175,*WEP241, *276-2020
2SD131	2N1490,2N4070,2N5616,2N5620,2N5624,2N5628,2SC664,2SC665,2SC736,2SC793-R, 2SC793-Y,2SC961,2SC962,2SC1619,2SD16,2SD51,2SD59,2SD67,2SD73,2SD82, 2SD88,2SD110-O,2SD110-R,2SD111-O,2SD111-R,2SD116,2SD118-R,2SD118-Y, 2SD119-R,2SD119-Y,2SD188,2SD202,2SD203,2SD217,2SD260,2SD322,2SD334, 2SD427-R,2SD428-R,181T2,B170006-ORG,B170006-RED,B170006-YEL, B170007-ORG,B170007-RED,B170007-YEL,B170024,BDY24B,KSP1045,KSP1175, SDT7603,ZT1490
2SD132	*TR-36,*HEPS7004,*ECG181,*WEPS7000
2SD134	*DS66,*GE-20,*TR-86,*SK3124,*SN80,*ECG123,*WEP53,*RT-100
2SD136	*HEPS5024,*SK3021,*ECG124,*WEP240
2SD137	*HEPS5024,*SK3021,*ECG124,*WEP240
2SD138	2SD158,2SD251,RCA1E02
2SD139	2SC679,2SC825,2SD159,40318,40322
2SD141	2SD57,*PTC906,*GE-28,*TR-76,*HEPS5000,*SK3041,*ECG152,*WEP245,*276-2017, *RT-154
2SD142	2SD58,2SD130-Y,2SD241,2SD256,HST5901,RCS29,RCS29A,*PTC906,*GE-28, *TR-81,*HEPS5000,*SK3026,*ECG175,*WEP241,*276-2017,*RT-145
2SD143	2N3054,2N3766,2N3767,2N5598,2N5602,2SD71,2SD102-O,2SD103-O,2SD129-Y, 2SD144,2SD154,2SD155,2SD242,2SD257,BDY13,HST5902,HST5903,KSP1394, KSP1395,MJ2250,RCS29B,RCS29C,*PTC906,*GE-12,*TR-81,*HEPS5019,*SK3026, *PN66,*ECG175,*WEP241,*276-2020,*RT-197
2SD144	2N3767,2N5602,2SD71,2SD102-O,HST5903,KSP1395,RCS29C,*PTC906,*TR-81, *HEPS5012,*SK3131,*ECG175,*WEP241,*276-2020
2SD146	2N4910,2N4911,2SD147,MJ2249,MJ3101,*HEPS5000,*SK3026,*ECG130,*WEP247, *RT-197
2SD147	2N4911,2N4912,*HEPS5000,*SK3027,*ECG130,*WEP247
2SD148	2N5202,2N5600,2SD92,2SD129-R,1571-0601,1571-0602,1571-0801,1571-0802, 1571-0820,1571-0825,1571-1020,1571-1025,BLY47A
2SD149	2N3108,2N3110,2N3253,2N3444,2N3735,2N4000,2N5262,2SC485-R,2SC512-R, 2SC516,2SC1065,2SC1069,40367,BC140-6,BC141-6,BSX60,*PTC144
2SD150	BDY12,HST5901,MJ2249,MJ3101,*PTC906,*GE-12,*TR-81,*HEPS5019,*SK3026, *ECG175,*WEP241,*276-2020,*RT-145
2SD151	2N3236,2N3714,2N5630,2N5633,2N6254,2N6359,2N6360,1561-1008,1561-1010, 1561-1208,1561-1210,IR3772,MJ3772,RCS258,SDT9701,SDT9702,SDT9704, SDT9705,STC2221,*PTC173,*GE-75,*TR-59,*HEPS7004,*SK3036,*ECG130, *WEP247,*RT-131
2SD152	2N4864,2SC1160,2SC1161,HST5904,HST5905,HST5951,HST5952,HST6905, HST6906,HST6907,KSP1396,*PTC118,*TR-93,*HEPS5012,*SK3026,*ECG164, *WEP740A,*276-2017
2SD153	ST18008
2SD154	2N3054,2N3766,2N3767,2N5606,2N5610,2N6465,2SC981,2SC1444,2SC1445, 2SD102-O,2SD103-O,2SD129-Y,2SD155,2SD242,2SD257,BD149-10,BD149-6, BLY48A,KSP1024,KSP1025,RCS29B,RCS29C,*PTC906,*TR-76,*HEPS5019, *SK3054,*ECG152,*WEP245,*276-2020,*RT-197
2SD155	2N3054,2N4233,2N5202,2N5608,2N5616,2N6261,2N6372,2N6465,2SC981,2SC1444, 2SC1445,2SD102-O,2SD102-R,2SD103-O,2SD103-R,2SD129-R,2SD129-Y,2SD226B, 2SD242,2SD243,2SD257,2SD258,BD149-10,BD149-6,BLY47A,KSP1024,KSP1025, RCS29B,RCS29C,*PTC112,*TR-81,*HEPS5019,*SK3131,*ECG175,*WEP241,

To Replace	Substitute This Type
(2SD155)	*276-2020
2SD156	*HEPS5011,*SK3021,*ECG124,*WEP240
2SD157	2SC582,2SC1235,*HEPS5011,*SK3021,*ECG124,*WEP240
2SD158	*HEPS5011,*SK3021,*ECG124,*WEP240
2SD159	2SC679,*HEPS5011,*SK3021,*ECG124,*WEP240
2SD160	2N1486,40368,S2N1486,ZT1486
2SD161	2SD177,HST9203,HST9204,HST9208,HST9209,*SK3011,*ECG101,*WEP641
2SD162	2N797,*PTC108,*HEPG0011,*SK3010,*ECG101,*WEP641
2SD163	HST9201,HST9205,HST9206,*PTC173,*HEPS7002,*SK3027,*ECG130,*WEP247
2SD164	HST9202,HST9203,HST9207,HST9208,*PTC173,*HEPS7004,*SK3027,*ECG130, *WEP247
2SD165	2SD166,HST9204,HST9209,*HEPS5020,*ECG162,*WEP707
2SD166	*HEPS5020,*ECG162,*WEP707
2SD167	2SD27,2SD72,2SD170,2SD170A,2SD352,AC186,*PTC134,*HEPG0011,*SK3010, *ECG101,*WEP641
2SD168	2N6493,2N6494,2SD301
2SD170	2SD72,2SD170A,2SD352,AC186,*HEPG0011,*ECG103A,*WEP641B
2SD170A	*HEPG0011,*SK3010,*ECG103A,*WEP641B
2SD171	*HEPS5020
2SD172	2SD163,2SD211,1582-0603,1582-0604,1582-0605,HST9201,HST9205,HST9206, KSP1271,*PTC140,*HEPS7002,*SK3027
2SD173	2N3236,2SD55,2SD161,2SD164,2SD177,2SD213,2SD214,2SD319,1561-1005, 1561-1205,1582-1003,1582-1004,1582-1005,1582-1203,1582-1204,1582-1205, HST9202,HST9203,HST9207,HST9208,IR3772,KSP1272,MJ3772,*PTC140, *HEPS7004,*SK3027,*ECG130,*WEP247
2SD174	2N1487,2N1702,2SD124A,2SD163,2SD172,2SD211,BDY23A,HST9801,HST9802, KSP1041,KSP1171,SDT7607,SDT7608,ZT1487,*PTC129A,*HEPS7002,*SK3027, *ECG130,*WEP247
2SD175	2N1488,2SC43,2SD50,2SD52,2SD125A,2SD161,2SD164,2SD173,2SD177,2SD213, 2SD214,2SD319,BDY24A,HST9803,HST9804,KSP1042,KSP1172,SDT7609,ZT1488, *PTC129A,*HEPS7004,*SK3027,*ECG130,*WEP247
2SD176	2SD53,2SD161,2SD177,HST9202,HST9203,HST9207,HST9208,*HEPS7004,*SK3027, *ECG130,*WEP247
2SD177	2SD161,HST9203,HST9204,HST9208,HST9209,*HEPS7004,*ECG162,*WEP707
2SD178	2N2430,2SD27,2SD72,2SD352,AC175,AC186,*PTC134,*GE-08,*TR-08,*HEPG0011, *SK3124,*SN80,*ECG103A,*WEP641B,*AC127,*276-2001,*RT-122
2SD178A	2SD178B,*PTC134,*GE-59,*TR-08,*HEPG0011,*SK3124,*SN80,*ECG103A, *WEP641B,*AC127,*276-2001,*RT-122
2SD178B	*GE-47,*SK3124
2SD179	
2SD180	KSP1044,KSP1045,KSP1174,KSP1175,SDT7602,SDT7603,*PTC140,*GE-59,*TR-59, *HEPS7004,*SK3027,*ECG223,*WEP247,*BDY20,*276-2020,*RT-131
2SD181	2N3055C,KSP1276,*HEPS5020,*ECG162,*WEP707
2SD181A	2SD376A
2SD182	*HEPS5014,*SK3024,*ECG128,*WEP243
2SD183	*HEPS3019,*SK3024,*ECG128,*WEP243
2SD184	2N1483,ZT1483,*HEPS5014,*SK3024,*ECG152,*WEP245
2SD185	2N1484,ZT1484,*HEPS3019
2SD186	2SD187,*PTC108,*GE-59,*TR-08,*HEPG0011,*SK3010,*ECG103A,*WEP641B, *AC127,*276-2001
2SD187	2SD186,*PTC108,*GE-59,*TR-08,*HEPG0011,*SK3010,*ECG103A,*WEP641B, *AC127,*RT-129
2SD188	2N3448,2N4070,2N5624,2N5628,2SC793-Y,2SC961,2SC962,2SD88,2SD110-O, 2SD110-R,2SD111-O,2SD111-R,2SD116,2SD118-Y,2SD119-Y,2SD217,2SD322, 2SD334,2SD427-O,2SD427-R,2SD428-O,2SD428-R,181T2,B170006-ORG, B170006-RED,B170006-YEL,B170007-ORG,B170007-RED,B170007-YEL,B170024, BDY24B,KSP1045,KSP1175,SDT7603,*PTC140,*HEPS7004,*SK3027,*ECG130, *WEP247
2SD189	2N5614,2N5618,2SD189A,2SD379,*PTC129A,*GE-75,*TR-59,*HEPS7004,*SK3027, *ECG162,*WEP247,*RT-131
2SD189A	2N5618,*PTC129A,*GE-75,*TR-67,*HEPS7004,*SK3027,*ECG162,*WEP162

To Replace	Substitute This Type
2SD190	2SD137,*PTC104,*SK3021,*ECG124,*WEP240
2SD191	2N385A,2N440,2N440A,2N1012,2N1299,SF.T298,*PTC108,*GE-59,*TR-08, *HEPG0011,*ECG103A,*WEP641B,*AC127,*276-2001,*RT-122
2SD192	2N385A,2N440,2N440A,2N1012,2N1299,SF.T298,*PTC108,*GE-59,*TR-08, *HEPG0011,*ECG103A,*WEP641B,*AC127,*276-2001,*RT-122
2SD193	*PTC134,*GE-59,*TR-08,*HEPG0011,*ECG103A,*WEP641B,*AC127,*276-2001, *RT-122
2SD194	*PTC134,*GE-59,*TR-08,*HEPG0011,*ECG103A,*WEP641B,*AC127,*RT-122
2SD195	*PTC108,*HEPG0011,*SK3010,*ECG101,*WEP641B
2SD196	2SD54,2SD196A,2SD197,2SD197A,*HEPS5004
2SD196A	2SD54,2SD197A,*HEPS5004
2SD197	2SD197A
2SD197A	
2SD198	*GE-35,*TR-67,*HEPS5020,*SK3560,*ECG162,*WEP707,*RT-131
2SD199	*PTC146,*GE-35,*TR-67,*HEPS5021,*SK3111,*ECG162,*WEP707
2SD200	2SD200A,2SD300,2SD517,*PTC146,*GE-38,*TR-93,*SK3115,*ECG165,*WEP740B
2SD200A	2SD200,2SD300,2SD517,*PTC146,*GE-37,*TR-93,*HEPS5021,*SK3115,*ECG165, *WEP740B
2SD201	2N4070,2SD202,181T2,BDY24B,KSP1172,KSP1175,SDT7603,*PTC140,*HEPS7004, *SK3027,*ECG130,*WEP247
2SD202	2N4070,2SD203,181T2,BDY24B,KSP1172,KSP1173,KSP1175,KSP1176,SDT7603, SDT7604,*PTC140,*HEPS7004,*ECG162,*WEP707
2SD203	2N4070,KSP1173,KSP1176,SDT7604,SDT7605,*PTC118,*HEPS5020,*ECG162, *WEP707
2SD204	*PTC123,*GE-47,*TR-87,*HEPS5014,*SK3024,*ECG128,*WEP243,*BC337,*276-2012, *RT-156
2SD205	*PTC144,*GE-63,*TR-87,*HEPS5014,*SK3024,*ECG225,*276-2008,*RT-114
2SD206	2SD206A,A580-0402,A580-0403,A580-0405
2SD206A	2SD206,A580-0402,A580-0403,A580-0405
2SD207	2SD55,2SD207A,A580-1202,A580-1203,A580-1205,STC2221
2SD207A	2SD55,2SD207,A580-1202,A580-1203,A580-1205,STC2221
2SD208	2N3240,2SD208A,A580-1602,A580-1603,A580-1605,A580-1802,A580-1803,A580-1805, STC2222,STC2223
2SD208A	2N3240,2SD208,A580-1602,A580-1603,A580-1605,A580-1802,A580-1803,A580-1805, STC2222,STC2223
2SD211	2N3235,HST9201,HST9205,HST9206,KSP1271,*PTC140,*HEPS7002,*SK3027, *ECG130,*WEP247
2SD212	2N3236,2SD213,HST9202,HST9203,HST9207,HST9208,KSP1272,*PTC140, *HEPS7004,*SK3027,*ECG130,*WEP247
2SD213	2N3236,2SD214,HST9202,HST9203,HST9204,HST9207,HST9208,HST9209,KSP1272, KSP1273,*PTC140,*ECG162,*WEP707
2SD214	HST9203,HST9204,HST9208,HST9209,KSP1273,*PTC118,*HEPS5020
2SD215	2N780,2N2106,2N2846,2N2848,2N2868,2N3015,2N3299,2N3326,2N3724A,2N3734, 2N4960,2SD216,BSY81,ZT95,*PTC144,*ECG103,*WEP641A
2SD216	2N1613A,2N2106,2N2193,2N2193A,2N2193B,2N2846,2N2848,2N2868,2N3015, 2N3109,2N3299,2N3326,2N3722,2N3725A,2N4960,2N4961,2SC97A,BC140-10, BSY83,SE8002,ZT95,ZT1613,*PTC144
2SD217	2N3864,2N4070,2SD110-O,2SD110-R,2SD118-R,2SD118-Y,2SD218,2SD322,2SD323, 2SD340,2SD375,2SD388,KSP1176,SDT7604,*PTC118,*GE-14,*TR-67,*HEPS7004, *ECG162,*WEP707
2SD218	2N4071,2SD286,2SD287,2SD287A,2SD287B,2SD323,2SD375,2SD388,BUY20, KSP1176,SDT7604,SDT7605,*PTC118,*GE-35,*TR-67,*HEPS5020,*ECG162, *WEP707
2SD219	2N760,2N760A,2N2897,2N2900,40458,*PTC144,*HEPS5014,*SK3024,*ECG128, *WEP243
2SD220	2N718A,2N870,2N910,2N1889,2N2193,2N2193A,2N2193B,2N3107,2N3109,A5T2193, BC140-10,BC141-10,ZT1613,*PTC144,*HEPS5014,*SK3024,*ECG128,*WEP243
2SD221	2N699B,2N720A,2N870,2N911,2N1889,2N1974,2N2243,2N2243A,2N2443,2N2895, 2N2898,2N3020,2N3036,2N3107,2N3701,A5T2243,BC141-10,BC141-6,*PTC144, *HEPS3019,*ECG128
2SD222	*HEPS5014,*SK3024

To Replace	Substitute This Type
2SD223	2N2196,*HEPS5014,*SK3049,*ECG224
2SD224	*HEPS3019
2SD226	2SD130-R,2SD226A,*GE-66,*TR-92,*HEPS5019,*SK3026,*ECG175,*WEP241, *276-2018
2SD226A	2SD103-R,2SD130-R,2SD226B,KSP1021,KSP1024,*TR-81,*HEPS5019,*SK3026, *ECG175,*WEP241,*276-2020
2SD226B	2SD102-R,2SD103-R,2SD129-R,BLY47A,KSP1021,KSP1022,KSP1024,KSP1025, *GE-66,*TR-81,*HEPS5019,*SK3026,*ECG175,*WEP241,*276-2017
2SD227	2N916,2N916A,2N3227,2N3241A,2N3242A,2N3416,2N3509,2N3704,2N4074,2N5449, 2N6000,2N6004,2N6010,2SC367G,2SC367G-Y,2SC735-Y,2SC848,2SC849,2SC850, 2SC933,2SC967,2SC968,2SC970,2SC1246,2SC1247,2SC1317,2SC1813,2SD327, BSW83,BSX75,D33D25,GI-3416,GI-3704,GI-3706,MPS3704,PET3704,PET3706, PET6002,PET8000,TIS60,TIS60M,TN60,TN80,*PTC170,*GE-63,*TR-24,*HEPS0015, *SK3124,*ECG123A,*WEP736,*BC337,*276-2015,*RT-187
2SD228	2N3404,2SC814,2SC1346,2SC1406,2SD261,D33D21J1,D33D25J1,*PTC178,*GE-210, *TR-21,*HEPS0011,*SK3122,*ECG192,*WEP243,*BC337,*276-2009,*RT-102
2SD231	2N3771,2N5302,2SD249,1561-0615,1582-0408,1582-0410,1582-0415,1582-0608, 1582-0610,1582-0615,HST9201,HST9205,IR3771,MJ3771,*PTC175
2SD232	2N3772C,2N6571,2SD232A,108T2,109T2,1561-1015,1561-1215,1561-1415,1582-1208, 1582-1210,1582-1215,1582-1408,1582-1410,1582-1415,1582-1608,1582-1610, 1582-1615,HST9203,HST9204,HST9208,HST9209,STC2230,*PTC175
2SD232A	109T2,1561-1415,1561-1615,1561-1815,1582-1408,1582-1608,1582-1610,1582-1615, 1582-1808,1582-1810,1582-1815,1582-2008,1582-2010,STC2230
2SD234-O	2SD234G-O,*GE-66,*TR-76,*SK3054,*ECG152,*WEP245,*276-2020,*RT-154
2SD234-R	2SD234G-R,*GE-66,*TR-76,*ECG152,*WEP245,*276-2020,*RT-154
2SD234-Y	*TR-76,*ECG152,*WEP245,*276-2020,*RT-154
2SD234G-O	2SD234-O
2SD234G-R	2SD234-R
2SD234G-Y	2SC789-Y,2SD234-Y
2SD235-O	2SD234-O,2SD234G-O,2SD235G-O,*GE-66,*TR-76,*SK3054,*ECG152,*WEP245, *276-2018,*RT-154
2SD235-R	2SD234-R,2SD234G-R,2SD235G-R,*GE-66,*TR-76,*ECG152,*WEP245,*276-2018, *RT-154
2SD235-Y	2SD234-Y,*TR-76,*ECG152,*WEP245,*276-2018,*RT-154
2SD235G-O	2SD234-O,2SD234G-O,2SD235-O
2SD235G-R	2SD234-R,2SD234G-R,2SD235-R
2SD235G-Y	2SC789-Y,2SC790-Y,2SD234-Y,2SD234G-Y,2SD235-Y
2SD236	HST5901,*PTC112,*HEPS5019,*SK3026,*ECG175,*WEP241
2SD237	HST5902,HST5903,KSP1394,KSP1395,*PTC112,*HEPS5019
2SD238	HST5908,HST5909,HST6901,KSP1392,KSP1393,*HEPS5012,*SK3131,*ECG175, *WEP241
2SD241	2SD256,KSP1024
2SD242	2N3878,2SD257,KSP1025
2SD243	2N3878,2N3879,2SC1113,2SD244,2SD258,2SD259,KSP1025,KSP1026
2SD244	2N3878,2N3879,2SC1113,2SD259,KSP1026
2SD246	2SD299,*PTC146,*GE-38,*TR-93,*SK3115,*ECG165,*WEP740B
2SD247	2N1490,2N4915,2N5069,2N5616,2N5624,2SC664,2SC793-R,2SC793-Y,2SC962, 2SC1618,2SC1619,2SD16,2SD51,2SD59,2SD68,2SD73,2SD82,2SD88,2SD111-O, 2SD111-R,2SD116,2SD119-R,2SD119-Y,2SD131,2SD180,2SD188,2SD201,2SD202, 2SD260,2SD316,2SD334,2SD339,2SD369,2SD371-R,2SD379,2SD428-R,181T2, B170003-ORG,B170003-RED,B170003-YEL,B170004-ORG,B170004-RED,B170004-Y EL,B170006-ORG,B170006-RED,B170006-YEL,B170007-ORG,B170007-RED, B170007-YEL,B170021,B170024,BDY24B,KSP1044,KSP1045,KSP1174,KSP1175, SDT7602,SDT7603,STT2300,STT3500,ZT1490
2SD249	2N3771,2N5302,2SD231,1582-0408,1582-0410,1582-0415,1582-0608,1582-0610, 1582-0615,HST9201,HST9205,IR3771,MJ3771,*PTC175
2SD250	2N6570,2N6571,2SD232,108T2,1582-1008,1582-1010,1582-1015,1582-1208,1582-1210, 1582-1215,HST9202,HST9203,HST9207,HST9208,STC2229
2SD251	BLY50A,RCA1E02
2SD254	2N3054,2N3766,2N3767,2N4232,2N4233,2N5202,2N5608,2N6372,2N6373,2SC981, 2SC1444,2SC1445,2SD103-O,2SD103-R,2SD129-R,2SD129-Y,2SD154,2SD155,

To Replace	Substitute This Type
(2SD254)	2SD226B,2SD242,2SD255,2SD257,2SD291,2SD292,BD149-10,BD149-6,BLY47A, KSP1024,KSP1025,RCS29B,RCS29C,*PTC112,*GE-23,*TR-57,*HEPS5000,*SK3026, *ECG175,*WEP241,*276-2020
2SD255	2N3054,2N3766,2N3767,2N4232,2N4233,2N5202,2N5608,2N6372,2N6373,2SC981, 2SC1444,2SC1445,2SD103-O,2SD103-R,2SD129-R,2SD129-Y,2SD154,2SD155, 2SD226B,2SD242,2SD254,2SD257,2SD291,2SD292,BD149-10,BD149-6,BLY47A, KSP1024,KSP1025,RCS29B,RCS29C,*PTC112,*GE-216,*TR-81,*HEPS5019, *SK3026,*ECG175,*WEP241,*276-2020
2SD256	2SD241,KSP1024,*PTC112,*HEPS5019,*SK3239
2SD257	2N3878,2SD242,KSP1025,*PTC112,*HEPS5012,*ECG175,*WEP241
2SD258	2N3878,2N3879,2SC1113,2SD243,2SD244,2SD259,KSP1025,KSP1026,*PTC104, *HEPS5012,*ECG124,*WEP240
2SD259	2N3878,2N3879,2SC1113,2SD244,KSP1026,*HEPS5012,*SK3021
2SD260	2N1490,2N4070,2N5616,2N5620,2N5624,2N5628,2SC664,2SC665,2SC736,2SC793-R, 2SC793-Y,2SC961,2SC962,2SC1619,2SD16,2SD51,2SD59,2SD67,2SD73,2SD82, 2SD88,2SD110-O,2SD110-R,2SD111-O,2SD111-R,2SD116,2SD118-R,2SD118-Y, 2SD119-R,2SD119-Y,2SD131,2SD188,2SD202,2SD203,2SD217,2SD322,2SD334, 2SD427-R,2SD428-R,181T2,B170006-ORG,B170006-RED,B170006-YEL, B170007-ORG,B170007-RED,B170007-YEL,B170024,BDY24B,KSP1045,KSP1175, SDT7603,ZT1490
2SD261	*PTC127,*HEPS3024,*SK3137,*ECG192,*WEP243,*RT-301
2SD262	
2SD265	2N6622,2SD293
2SD266	2SD294
2SD271	
2SD272	
2SD273	2N6622,2SD265,2SD293
2SD274	2SD266,2SD294
2SD283	2N3879,2N5612,2N5620,2SC1055,2SC1113,2SD297,KSP1026,KSP1166,SDT7A04, *ECG162,*WEP707
2SD284	2N3879,2N5612,2N5620,2SC1055,2SC1113,2SD283,KSP1023,KSP1163,SDT7744, SDT7745,*ECG162,*WEP707
2SD285	2N3879,2N5429,2N5608,2N5612,2N5616,2N5620,2SC1055,2SC1113,2SD283, 2SD284,KSP1022,KSP1162,SDT7743,SDT7744,*ECG162,*WEP707
2SD286	2SD287,2SD287A,2SD287B
2SD287	2SD181A,2SD287A,2SD287B,*PTC118,*HEPS5020,*ECG284
2SD287A	2SD287,2SD287B
2SD287B	2SD287,2SD287A
2SD288	2N5294,2N6473,2SD365A,2SD389A,2SD526,BD220,D44C10,D44C11,D44C12, *PTC167,*HEPS5000,*ECG152,*WEP245
2SD289	2N5293,2SD366A,2SD390A,*PTC167,*HEPS5000,*ECG152,*WEP245
2SD290	HST5902,HST5903,*HEPS5019,*ECG175,*WEP241
2SD291	2SD292,*HEPS5019,*SK3026,*ECG175,*WEP241
2SD292	2SD291,*HEPS5019,*SK3026,*ECG175,*WEP241
2SD293	2SD310
2SD294	2SD311
2SD295	
2SD296	
2SD297	KSP1026,KSP1166,SDT7A04,SDT7A05,*HEPS5012,*ECG175,*WEP241
2SD299	*PTC146,*GE-38,*TR-93,*HEPS5020,*SK3115,*ECG165,*WEP740B
2SD299B	*SK3115
2SD300	2SD350,2SD350A,*PTC146,*GE-38,*TR-93,*SK3115,*ECG165,*WEP740B
2SD301	2N6493,2N6494,2SD168
2SD310	
2SD311	
2SD312	2SD199,*PTC129A,*GE-37,*TR-93,*HEPS5021,*SK3079,*ECG164,*WEP740A
2SD313	2SD526,D44C4,D44C5,D44C6,D44C7,D44C8,D44C9,*PTC167,*HEPS5000,*SK3054, *ECG196,*WEP245,*RT-197
2SD314	*PTC167,*HEPS5000,*SK3054
2SD315	2N5428,2N5614,2SC1444,*PTC169,*HEPS5019,*ECG175,*WEP241
2SD316	2N3447,2N3448,2N3713,2N3714,2N3715,2N3716,2N5624,2N5632,2N5874,2N5878,

To Replace	Substitute This Type
(2SD316)	2SC1080-R,2SC1080-Y,2SD111-O,2SD111-R,2SD116,2SD119-Y,2SD369, B170004-ORG,B170004-RED,B170004-YEL,B170007-ORG,B170007-RED, B170007-YEL,KSP1174,KSP1175,*PTC118,*HEPS5004
2SD317	2SD234-R,2SD234G-R,2SD317A,2SD389,2SD389A,*TR-76,*HEPS5000,*ECG152, *WEP245,*276-2020
2SD317A	2SD389A,2SD531,*GE-28,*TR-76,*HEPS5000,*SK3054,*ECG152,*WEP245, *276-2019
2SD318	2SD318A,2SD390,2SD390A,*TR-76,*HEPS5000,*ECG152,*WEP245,*276-2020
2SD318A	2SD390A,2SD544,*TR-76,*HEPS5000,*ECG152,*WEP245
2SD319	2N3236,*PTC175,*GE-75,*TR-36,*HEPS7000,*ECG181,*WEPS7000
2SD320	*PTC129A,*HEPS5021,*SK3559,*ECG162,*WEP707
2SD321	*GE-38,*TR-67,*HEPS5020,*SK3115,*ECG283
2SD322	2N3864,2N5633,2SD110-O,2SD110-R,2SD118-R,2SD118-Y,2SD323,2SD340,2SD375, 2SD388,KSP1176,*PTC118,*HEPS7004,*ECG280,*WEP707
2SD323	2SD286,2SD287,2SD287A,2SD287B,2SD375,2SD388,BUY20,KSP1176,*PTC118, *HEPS5020,*ECG280,*WEP707
2SD324	2N3739,2SC515A,2SC1102,2SC1105,2SC1168,2SC1456,MJ2252,MJ3202,*TR-81, *HEPS5015,*ECG124,*WEP240
2SD325	2SD330,*PTC110,*HEPS5000,*SK3054,*ECG152,*WEP245,*RT-197
2SD326	2SC779-O,2SC867,*SK3021,*ECG124,*WEP240
2SD327	2N760A,2N909,2N930B,2N2222,2N2350,2N2350A,2N2484A,2N2540,2N3116,2N3227, 2N3242A,2N3302,2N3416,2N3509,2N3704,2N4401,2N4952,2N5107,2N5369,2N5449, 2N5818,2N6004,2N6010,2SC850,2SC968,2SC970,2SC1247,2SC1277,2SC1318, 2SC1813,40458,BCW34,BCW36,BSW83,BSX75,D33D25,D33D26,GI-3643,GI-3704, GI-3706,MPS3704,MPS6531,PET3704,PET3706,TN60,TN-3904,TZ81,TZ82, *HEPS0015,*SK3024,*ECG123A,*WEP736
2SD328	2N1973,2N2443,*HEPS3019
2SD329	2N3108,2N3110,2N3253,2N3444,2N3735,2N4000,2N5262,2SC485-R,2SC512-R, 2SC516,2SC1052,2SC1065,2SC1069,2SD149,40367,BC140-6,BC141-6,BSX60, KT600G
2SD330	D44C4,D44C5,D44C6,D44C7,D44C8,D44C9,*PTC167,*HEPS3020,*SK3054
2SD331	2SD314,*PTC167,*HEPS3020
2SD332	
2SD334	2SD334A,*PTC140,*TR-67,*HEPS7004,*ECG162,*WEP707
2SD334A	*PTC118,*TR-67,*HEPS5020,*ECG162,*WEP707
2SD335	2N4111,BD149-10,BD149-6,BLY48,BLY48A,KSP1024,KSP1025,KSP1044,KSP1045, STT2300,STT3500,*HEPS5000
2SD336	2N1420A,2N2192,2N2192A,2N2192B,2N2219,2N2538,2N3053,2N3300,2N5106, 2SC503-GR,2SC504-GR,2SC971,2SC1384,2SD571,40451,40459,BC340-16,BSX72, BSY52,BSY82,TN59,*HEPS3020,*SK3122,*ECG192,*WEP243
2SD338	2N3446,2N3448,2SD111-R,2SD119-R,2SD339,2SD522,KSP1171,KSP1172,KSP1174, KSP1175,SDT7602,SDT7603,SDT7732,SDT7733,*PTC173
2SD339	2SD111-R,2SD340,2SD522,KSP1172,KSP1175,KSP1272,KSP1275,SDT7733, SDT7734,*PTC173
2SD340	KSP1272,KSP1273,KSP1275,KSP1276
2SD341	2N3055,2N3055SD,2N3236,2N5629,2N5630,2N6254,2N6270,2N6271,2N6272,2N6273, 2SD492,108T2,40636,BUY53A,BUY54A,HST9202,HST9203,HST9207,HST9208, IR3772,SDT9202,SDT9203,SDT9207,SDT9208,SDT9704,SDT9705,*ECG180
2SD342	2SD525,2SD531
2SD343	2SD544,*SK3054,*ECG152,*WEP245
2SD344	2SD525,2SD531
2SD345	2SD544
2SD346	2N6130,2N6131,NSP41A,NSP41B
2SD347	
2SD348	*HEPS5021,*SK3115,*ECG238
2SD349	
2SD350	2SD350A,*GE-36,*TR-61,*HEPS5020,*SK3111,*ECG238
2SD350A	*PTC146
2SD351	*PTC129A,*GE-36,*ECG283
2SD352	*GE-59,*TR-08,*SK3010,*ECG103A,*WEP736
2SD353	*PTC129A,*HEPS5020,*SK3133,*ECG164,*WEP740A

To Replace	Substitute This Type
2SD355	2SC1383
2SD356	2SD357,NSD104,NSD6178,*ECG291
2SD357	2SD358,NSD104,NSD106,NSD6178,*ECG291
2SD358	NSD106,*ECG291
2SD359	2SD360,NSE180,*PTC110,*GE-66,*SK3054,*ECG152,*276-2019,*RT-158
2SD360	2SD359,NSE180,*PTC110,*GE-66,*SK3041,*ECG152,*RT-158
2SD361	NSD102,NSD6179,NSE180,NSE181,*PTC110
2SD363	2SD373
2SD363A	2SD364,2SD373A,2SD374
2SD364	2SD374
2SD365	2N5294,2N5296,2N5298,2N6122,2SC789-O,2SC789-R,2SD234-O,2SD234-R, 2SD234G-O,2SD234G-R,2SD365A,BD220,BD221,D44C4,D44C5,D44C6,D44C7, D44C8,D44C9,*PTC167,*ECG152,*276-2020
2SD365A	2N5294,2N5298,2N6473,2SD531,BD220,D44C10,D44C11,D44C12,*PTC167
2SD366	2N5293,2N5295,2N5297,2SD314,2SD366A,*PTC167,*ECG152,*276-2020
2SD366A	2N5293,2N5297,2SD544,*PTC167
2SD367	2N2430,2SD170,2SD170A,*PTC134,*GE-47,*SK3124,*ECG103A,*276-2001
2SD368	2SD348
2SD369	2SD111,2SD111-O,*PTC173
2SD371-O	2N5626,2SC520A,2SC793-BL,2SC793-Y,2SD427-O,2SD428-O,BDY24C
2SD371-R	2N4070,2N5624,2N5628,2SC793-R,2SC793-Y,2SC961,2SC1619,2SD201,2SD202, 2SD427-R,2SD428-R,181T2,BDY24B,KSP1175,SDT7603
2SD371-Y	2SC520A,2SC793-BL,2SC1051L
2SD372	
2SD373	2SD363
2SD373A	2SD363A,2SD364,2SD374
2SD374	2SD364
2SD375	2SD181,2SD286,2SD287,2SD287A,2SD287B,KSP1276
2SD376	2SD181A,2SD376A,KSP1222,KSP1223,KSP1224,KSP1225
2SD376A	KSP1223,KSP1224,KSP1225
2SD377	2SC807
2SD378	2SD78,2SD78A
2SD379	2N5622,2N5626,2SC793-BL,2SC793-Y,BDY24C,KSP1044,KSP1045,KSP1174, KSP1175,*PTC140,*ECG284
2SD380	2SC1413A,2SD348,*ECG238
2SD380A	*PTC129A
2SD381	2SC1448,2SC1448A,2SC1669-O,*ECG291
2SD382	*SK3054,*ECG152
2SD383	*ECG283
2SD384	2SD385
2SD385	
2SD386	
2SD386A	2SD386,*SK3104,*ECG198
2SD387	
2SD387A	2SD387
2SD388	2SD286,2SD287,2SD287A,2SD287B,BUY20,KSP1176,*PTC118
2SD389	2SD234-O,2SD234-R,2SD234G-O,2SD234G-R,2SD389A,*SK3054,*ECG152,*WEP245
2SD389A	2SD531,*ECG152
2SD390	2SD318,2SD318A,2SD390A,*HEPS5000,*SK3054,*ECG152,*WEP245
2SD390A	2SD318A,2SD544
2SD392	2N3241A,2N5998,2N6000,2SC395A-Y,2SC735-Y,2SC934,2SC1317,2SC1788, 2SC1851,*SK3244,*276-2016
2SD393	2SC1469A,RCS579
2SD394	2SD395
2SD395	IR5064
2SD396	2SD311
2SD400	*ECG293
2SD400P1	2SD400P2,*ECG293
2SD400P2	
2SD404	
2SD413	

To Replace	Substitute This Type
2SD414	2N3665,2SD415
2SD415	2N3665,2SD414
2SD416	
2SD417	KSP1142,KSP1143,KSP1144,KSP1145,XGSR10025,*PTC129A
2SD418	2SC1309,2SC1325,2SC1325A,*ECG165
2SD419	2SD420
2SD420	2SD421
2SD421	
2SD422	SDT1150,SDT1155
2SD423	2SD422,HST1155,HST1160,SDT1150,SDT1155
2SD424	2N5038,2N6340,2N6341,*ECG284
2SD425-O	2N5038,2SD424,2SD426-O
2SD425-R	2SD424,2SD426-R,KSP1276
2SD426-O	2N5038,2SD424,2SD425-O
2SD426-R	2SD424,2SD425-R,KSP1276
2SD427-O	2N6354,2SC1343,2SD424,2SD425-O,2SD426-O
2SD427-R	2N5628,2N6354,2N6496,2SC1115,2SC1343,2SC1782,2SD424,2SD425-R,2SD426-R, KSP1176
2SD428-O	2N5626,2SC793-BL,2SC793-Y,2SC1818,2SD427-O
2SD428-R	2N3448,2N4070,2N5624,2N5628,2SC793-R,2SC793-Y,2SC961,2SD427-R,KSP1175, SDT7603
2SD429	2N6547,2SC1140,2SD310,BDY47
2SD437	2SC1141,2SC1143,DTS-4060,DTS-4065,HST1063,HST1064,IR4060,IR4065,IR5063, IR5064
2SD438	
2SD458	2N6542,2N6544,2N6581,2N6584,2SC806,2SC1325,2SC1325A,2SC1578,HST1053, HST1054,HST1058,HST1059,MJ3260,SDT1053,SDT1054,SDT1058,SDT1059, *PTC129A,*ECG163
2SD459	2SD460
2SD460	
2SD461	2SC808,BDY26C
2SD463	2SD464,2SD503,2SD505,2SD506,RCA-1001
2SD464	2SD506
2SD465	2SD466
2SD466	
2SD467	2N3241A,2N3242A,2N4386,2SC1383,2SC1518,2SC1859,2SD400,BC338-16,BC338-25, TIS90,TIS90M,TN60,TN80
2SD468	
2SD470	*SK3131,*ECG321
2SD471	TN79,*SK3124,*ECG289,*RT-127A
2SD475	*PTC163
2SD479	2N6037,2N6038,2SD480
2SD480	2N6038,2N6039,2SD481
2SD481	2N6039
2SD482	2N5655,2N5656,2SD483
2SD483	2N5656,2N5657,2SD484
2SD484	2N5657
2SD485	2SD486
2SD486	2SD487
2SD487	
2SD488	2SD489
2SD489	2SD490
2SD490	
2SD491	2N5990,2N5991
2SD492	2N3055,2N3055SD,2N3236,2N5629,2N5630,2N6254,2N6270,2N6271,2N6272,2N6273, 2SD341,108T2,40636,BUY53A,BUY54A,HST9202,HST9203,HST9207,HST9208, IR3772,SDT9202,SDT9203,SDT9207,SDT9208,SDT9704,SDT9705
2SD493	2N5977,2N5978,2N5983,2N5984,2SD491,2SD494,2SD499,2SD500
2SD494	2N5978,2N5979,2N5984,2N5985,2SD495,2SD500,2SD501
2SD495	2N5979,2N5985,2SD501
2SD496	2SD497

To Replace	Substitute This Type
2SD497	2SD498
2SD498	
2SD499	2SD500
2SD500	2SD501
2SD501	
2SD502	2N6492,2N6576,2SD503,2SD504,2SD505
2SD503	2N6493,2N6494,2N6577,2SD505,2SD506
2SD504	2SD505
2SD505	2SD506
2SD506	
2SD507	2SD508,2SD511,2SD512,2SD514,2SD515,1401-0820,1401-1015,1401-1020,1401-1025, 1401-1215
2SD508	2SD509,2SD510,2SD512,2SD513,2SD515,1401-1015,1401-1020,1401-1025,1401-1215, 1401-1220
2SD509	2SD510,2SD513,1401-1215,1401-1220
2SD510	
2SD511	2SD507,2SD508,2SD512,2SD514,2SD515,1401-0820,1401-1015,1401-1020,1401-1025, 1401-1215,*HEPS3028
2SD512	2SD508,2SD509,2SD510,2SD513,2SD515,1401-1015,1401-1020,1401-1025,1401-1215, 1401-1220
2SD513	2SD509,2SD510,1401-1215,1401-1220
2SD514	2SD507,2SD508,2SD511,2SD512,2SD515,1401-0820,1401-1015,1401-1020,1401-1025, 1401-1215
2SD515	2SD508,2SD509,2SD510,2SD512,2SD513,1401-1015,1401-1020,1401-1025,1401-1215, 1401-1220
2SD517	2SD300,2SD350,2SD350A,*PTC146
2SD518	2SD422,2SD423,1571-2001,HST1155,HST1160,SDT1150,SDT1155
2SD519	2N6560,2N6579,2N6580,2N6582,2N6583,2SC1228,2SC1230,2SC1469A,2SC1579, 2SC2017,BDY45,FT431,MJ431,XDAR10035,XGSR10035,XGSR10040,XGSR15035, XGSR15040
2SD521	
2SD522	2SD110-R,2SD111-R,HST9202,HST9203,HST9207,HST9208,SDT9202,SDT9203, SDT9207,SDT9208
2SD523	2N6055,2N6056,2N6384,2N6385,2SD524,PMD12K-60,PMD12K-80,RCA1B07
2SD524	2N6058,2N6059,2N6385,PMD10K-100,PMD10K-80,RCA1B07
2SD525	
2SD526	*PTC167
2SD531	
2SD544	*PTC140
2SD545	
2SD546	
2SD552	
2SD570	2SC789-O,2SC789-Y,2SD525,2SD526,2SD531,*PTC169
2SD571	2N2192,2N2192A,2N2192B,2N2219,2N2219A,2N2538,2N3300,BSY54,ZT1711
2SD572	2SD573
2SD573	
2SD574	2SD574A
2SD574A	2SD574
2SD576	40321,40327
2SD585	
2T15	SEE 2SB50
2T64	SEE 2SD64
2T65	SEE 2SD65
2T66	SEE 2SD66
2T69	SEE 2SD63
2T73	SEE 2SC73
2T75	SEE 2SC75
2T76	SEE 2SC76
2T77	SEE 2SC77
2T78	SEE 2SC78
2T83	SEE 2SB53

To Replace	Substitute This Type
2T201	SEE 2SA121,2,3
2T203	SEE 2SA124
2T205A	SEE 2SA125
2T311	SEE 2SB48
2T312	SEE 2SB48
2T313	SEE 2SB49
2T314	SEE 2SB49
2T321	SEE 2SB51
2T322	SEE 2SB51
2T323	SEE 2SB52
2T324	SEE 2SB52
2T681	SEE 2SD61
2T682	SEE 2SD62
2T3011	SEE 2SB140
2T3021	SEE 2SB141
2T3030	SEE 2SB142
2T3031	SEE 2SB143
2T3032	SEE 2SB143,4
2T3033	SEE 2SB144
2T3041	SEE 2SB145
2T3042	SEE 2SB145,6
2T3043	SEE 2SB146
3N74	*PTC133,*SK3245,*276-2013
3N75	*PTC133,*SK3245,*276-2013
3N76	*PTC133,*SK3245,*276-2013
3N77	*PTC139,*SK3245,*276-2013
3N78	*PTC139,*SK3245,*276-2013
3N79	*PTC139,*SK3245,*276-2013
3N108	
3N109	
3N110	
3N111	
3TE110	SEE 2N4131
3TE120	SEE 2N4130
3TE245	
3TE350	
3TE440	
3TE450	
3TE604	
3TE609	
3TE610	
3TE611	
3TX601	2N5945,2N5995†,2SC892,2SC1040,2SC1081,2SC1257,2SC1258,2SC1495,41010†, *HEPS3006
3TX602	2N5994,2SC1081,2SC1118,2SC1258,2SC1528,SRF54215,*276-2017
3TX632	2N5636,2N5918†,2N6203,40940†,*HEPS3005
3TX820	2N5644,2N5645,2N5698,2N5914,2N5915,2N5946,2SC891,2SC892,2SC911,2SC911A, 2SC1016,2SC1040,2SC1608,2SC1620,3TX821,40934†,41009A†,41010†,*HEPS3005
3TX821	2N5645,2N5646,2N5915,2N5946,2SC891,2SC892,2SC1040,41009A†,41010†, *HEPS3006
4C28	2N472,2N2520,2SC945,*PTC123,*GE-60,*TR-21,*HEPS0014,*SK3124,*SN80, *ECG123A,*WEP736,*BF167,*276-2013,*RT-102
4C29	*PTC123,*GE-20,*TR-21,*HEPS0014,*SK3124,*SN80,*ECG123A,*WEP736,*BC337, *276-2013,*RT-102
4C30	2N480,2N543A,2N708A,2N784A,2N840,2N2387,2N2432A,2N2522,2N3605A,2N3606A, 2N3688,2N3689,2N3690,2N3973,2N3975,2N4137,2N4140,2N4295,2N5029,2S103, 2SC54,2SC302,2SC468,2SC601,2SC689,2SC725,2SC752G-R,2SC896,2SC1240, 2SC1274,BF224J,BF225J,BFY39-1,BSY93,EN914,EN2369A,EN3009,EN3013, EN3014,GET708,GET914,GET2221,GET2369,GET3013,GET3014,GET3646, KT218,MPSA10-RED,MPSA20-RED,SE5001,SE5002,SE5003,SE5006,TIS22,TIS45, TIS46,TIS48,TIS49,TIS52,TIS55,TIS129,ZT23,ZT43,*PTC123,*GE-60,*TR-21,

To Replace	Substitute This Type
(4C30)	*HEPS0014,*SK3124,*SN80,*ECG123A,*WEP736,*BF167,*276-2013,*RT-102
4C31	2N718,2N843,2N916,2N916A,2N929,2N957,2N2161,2N2242,2N2369,2N2369A, 2N2501,2N2847,2N3009,2N3013,2N3014,2N3693,2N3826,2N3858A,2N3862,2N3903, 2N3946,2N3974,2N3976,2N4123,2N4227,2N4873,2N4966,2N4994,2N5027,2N5824, 2S104,2SC67,2SC169,2SC318,2SC366G-O,2SC367G-O,2SC536,2SC601N,2SC620, 2SC634,2SC639,2SC752G-O,2SC764,2SC815,2SC899,2SC938,2SC944,2SC965, 2SC1187,2SC1188,2SC1189,2SC1276,2SC1319,2SC1320,2SC1453,2SC1682-BL, 2SC1682-GR,2SC1686,2SC1687,2SC1688,BF198,BF199,BF241,EN697,EN916, MM3903,MPS3693,MPS6565,MPS6576-BLUE,MPS6576-YEL,MPSA10-WHT, MPSA20-WHT,PET1001,SE1001,TIS87,TP3705,ZT24,ZT44,*PTC123,*GE-60,*TR-21, *HEPS0014,*SK3124,*ECG123A,*WEP736,*BF167,*276-2013,*RT-102
4C43	2N498,2N498A,BFY65,MM3000,*PTC123,*GE-81,*HEPS0014,*BF338,*276-2012
4D20	2N162A,2N163A,2N475,2N717,2N783,2N834A,2N842,2N844,2N2368,2N2432A, 2N2522,2S102,2SC54,2SC302,2SC321,2SC322,2SC565,2SC689,BF224J,BF225J, BF237,KT218,MPS834,MPS6567,MPSH20,MPSH37,TIS47,*PTC123,*GE-60,*TR-21, *HEPS0014,*SK3122,*SN80,*ECG123A,*WEP736,*BF167,*276-2013,*RT-102
4D21	2N543,2N843,2N916,2N916A,2N2161,2N2501,2N2847,2N3858A,2N3862,2N3903, 2N3974,2N3976,2N4123,2N4227,2N4873,2N4966,2N5027,2N5824,2N6222,2S104, 2SC318,2SC318A,2SC366G,2SC366G-O,2SC367G,2SC367G-O,2SC536,2SC620, 2SC634,2SC639,2SC752G-O,2SC899,2SC944,2SC965,2SC1187,2SC1188, 2SC1189,2SC1216,2SC1276,2SC1319,2SC1320,2SC1453,2SC1675,2SC1682-BL, 2SC1682-GR,2SC1687,2SC1688,BF199,BF238,BF240,EN916,MM3903,MPS3693, MPS6513,MPS6565,MPS6576-BLUE,MPS6576-YEL,MPSA10-WHT,MPSA20-WHT, TIS87,TP3705,ZT24,ZT44,*PTC123,*GE-60,*TR-21,*HEPS0014,*SK3122,*SN80, *ECG123A,*WEP736,*BF167,*276-2013,*RT-102
4D22	SEE 2N2349
4D24	2N118A,2N473,2N474,2N542A,2N706,2N706A,2N706B,2N743,2N784,2N947,2N959, 2N988,2N989,2N1589,2N1590,2N2205,2N2432,2N2729,2N2921,2N2926-BRN, 2N3544,2N3662,2N3663,2N3844,2N3844A,2N3854,2N3854A,2N3983,2N5130, 2SC55,2SC63,2SC68,2SC98,2SC137,2SC387AG,2SC400-R,2SC455,2SC467, 2SC717,2SC812,2SC1023,2SC1126,EN918,GI-2711,GI-2713,GI-2715,GI-2921, GI-3605,GI-3606,GI-3607,MM1748,MM1941,MPS706,MPS706A,MPS2926-BRN, MPS6507,MPS6511,MPS6540,MPS6542,MPS6548,MPSH19,MT1038,MT1038A, MT1039,MT1060,PET3001,PET8101,SE1010,SE3002,SE5025,TIS18,TIS44,*PTC126, *GE-60,*TR-21,*HEPS0014,*SK3124,*SN80,*ECG123A,*WEP736,*BF167, *276-2011,*RT-102
4D25	2N541,2N542,2N728,2N729,2N1592,2N1593,2N2475,2N2569,2N2570,2N2615, 2N2616,2N2923,2N2926-ORG,2N3298,2N3393,2N3396-ORG,2N3397-ORG, 2N3398-ORG,2N3633,2N3709,2N3845,2N3845A,2N3856,2N3856A,2N3858,2N3959, 2N3960,2N4138,2N4418,2N4968,2N4996,2N4997,2N5132,2N5136,2N5137,2N5200, 2N5201,2N6566,2S095A,2S512,2SC388A,2SC400-O,2SC458,2SC459,2SC460, 2SC528,2SC619,2SC633,2SC645,2SC710,2SC712,2SC713,2SC924,2SC925, 2SC929,2SC929NP,2SC1215,2SC1359,2SC1372,2SC1373,2SC1789,2SD227,40398, 40400,BF123,BF254,BF377,BF378,BSY61,GI-2712,GI-2714,GI-2716,GI-2923,GI-3393, MPS2923,MPS2926-ORG,MPS3393,MPS3563,MPS3709,MPS6568A,MPS6569, MPS6570,MPSH10,MPSH11,PET3002,PET4001,SE3005,TIS86,*PTC126,*GE-60, *TR-21,*HEPS0014,*SK3124,*SN80,*ECG123A,*WEP736,*BF167,*276-2011, *RT-102
4D26	2N4265,A5T3392,A5T3707,A5T3710,A5T4124,A5T5172,A5T5219,A7T3392,A7T5172, A8T3392,A8T3707,A8T3710,A8T5172,BCY58,TN79,TP5136,*PTC123,*GE-81,*TR-21, *HEPS0014,*SK3124,*SN80,*ECG123A,*WEP736,*BC337,*276-2012,*RT-102
7A30	*PTC125,*GE-63,*TR-87,*HEPS5014,*SK3039,*EN10,*ECG128,*WEP56,*BF338, *276-2008,*RT-108
7A31	2N1613A,2N2106,2N2107,*PTC125,*GE-63,*TR-87,*HEPS3019,*SK3039,*EN10, *ECG128,*WEP56,*BF338,*276-2008,*RT-108
7A32	2N1711A,2N2108,*PTC125,*GE-63,*TR-87,*HEPS0015,*SK3039,*EN10,*ECG128, *WEP56,*BF338,*276-2008,*RT-108
7B1	*GE-12,*TR-81,*HEPS5011,*SK3021,*PN66,*ECG124,*WEP240,*RT-128
7B2	*GE-12,*TR-81,*HEPS5011,*SK3021,*PN66,*ECG124,*WEP240,*RT-128
7C1	*GE-12,*TR-81,*HEPS5011,*SK3021,*PN66,*ECG124,*WEP240,*RT-128
7C2	*GE-12,*TR-81,*HEPS5011,*SK3021,*PN66,*ECG124,*WEP240,*RT-128

TRANSISTOR SUBSTITUTES

To Replace	Substitute This Type
7C3	*GE-12,*TR-81,*HEPS5011,*SK3021,*PN66,*ECG124,*WEP240,*RT-128
7D1	*GE-12,*TR-81,*HEPS5011,*SK3021,*PN66,*ECG124,*WEP240,*RT-128
7D2	*GE-12,*TR-81,*HEPS5011,*SK3021,*ECG124,*WEP240,*RT-128
7D3	*GE-12,*TR-81,*HEPS5011,*SK3021,*PN66,*ECG124,*WEP240,*RT-128
7E1	*GE-12,*TR-81,*HEPS5011,*SK3021,*PN66,*ECG124,*WEP240,*RT-128
7E2	*GE-12,*TR-81,*HEPS5011,*SK3021,*PN66,*ECG124,*WEP240,*RT-128
7E3	*GE-12,*TR-81,*HEPS5011,*SK3021,*ECG124,*WEP240,*RT-128
7F1	*HEPS3020
7F2	*HEPS3020
7F3	*HEPS3019
7F4	*HEPS3019
7G1	*GE-12,*TR-81,*HEPS5011,*SK3021,*PN66,*ECG124,*WEP240,*RT-128
7G2	*GE-12,*TR-81,*HEPS5011,*SK3021,*PN66,*ECG124,*WEP240,*RT-128
7G3	*GE-12,*TR-81,*HEPS5011,*SK3021,*PN66,*ECG124,*WEP240,*RT-128
7G4	*GE-12,*TR-81,*HEPS5011,*SK3021,*PN66,*ECG124,*WEP240,*RT-128
10B551-2	*GE-60,*TR-51,*HEPS0016,*BF200,*276-2016
10B551-3	*GE-60,*TR-51,*HEPS0016,*BF200,*276-2016
10B553-2	*GE-60,*TR-51,*HEPS0016,*BF200,*276-2016
10B553-3	*GE-60,*TR-51,*HEPS0016,*BF200,*276-2016
10B555-2	*GE-214,*TR-51,*HEPS0016,*BF183,*276-2015
10B555-3	*GE-214,*TR-51,*BF183,*276-2015
10B556-2	*GE-214,*TR-51,*HEPS0016,*BF183,*276-2011
10B556-3	*GE-214,*TR-51,*HEPS0016,*BF183,*276-2011
10C573-2	*GE-39,*TR-51,*HEPS0016,*BC107B,*276-2015
10C573-3	*GE-39,*TR-51,*HEPS0016,*BC107B
10C574-2	*GE-39,*TR-51,*HEPS0016,*BC107B,*276-2015
10C574-3	*GE-39,*TR-51,*HEPS0016,*BC107B,*276-2015
11B551-2	*GE-212,*TR-51,*HEPS0016,*SK3245,*BC107B,*276-2016
11B551-3	*GE-212,*TR-51,*HEPS0016,*SK3245,*BC107B,*276-2016
11B552-2	*GE-212,*TR-51,*HEPS0016,*SK3245,*BC107B,*276-2016
11B552-3	*GE-212,*TR-51,*HEPS0016,*SK3245,*BC107B,*276-2016
11B554-2	*GE-212,*TR-51,*HEPS0016,*SK3245,*BC107B,*276-2016
11B554-3	*GE-212,*TR-51,*HEPS0025,*SK3245,*BC107B,*276-2016
11B555-2	*GE-212,*TR-51,*HEPS0016,*SK3245,*BC107B,*276-2016
11B555-3	*GE-212,*TR-51,*HEPS0016,*SK3245,*BC107B,*276-2016
11B556-2	*GE-18,*SK3244,*BF338,*276-2008
11B556-3	*GE-18,*SK3244,*BF338,*276-2008
11B560-2	*GE-18,*TR-51,*HEPS0005,*SK3244,*BF338,*276-2008
11B560-3	*GE-18,*TR-51,*HEPS0005,*SK3244,*BF338,*276-2008
11C1B1	*GE-12,*TR-81,*HEPS3020,*SK3021,*PN66,*ECG124,*WEP240,*RT-128
11C1F1	*HEPS3020
11C3B1	*GE-12,*TR-81,*HEPS3020,*SK3021,*ECG124,*WEP240,*RT-128
11C3F1	*HEPS3020
11C5B1	*GE-12,*TR-81,*HEPS3020,*SK3021,*PN66,*ECG124,*WEP240,*RT-128
11C5F1	*HEPS3020
11C7B1	*PTC195,*GE-12,*TR-81,*SK3021,*PN66,*ECG124,*WEP240,*RT-128
11C7F1	*TR-65
11C10B1	*GE-12,*TR-81,*HEPS3019,*SK3021,*PN66,*ECG124,*WEP240,*RT-128
11C10F1	*HEPS3019
11C11B1	*PTC144,*GE-12,*TR-81,*HEPS3020,*SK3021,*PN66,*ECG124,*WEP240,*RT-128
11C11B20	*PTC144
11C11F1	*HEPS3020
11C201B20	*HEPS3020
11C203B20	*HEPS3020
11C205B20	*HEPS3020
11C207B20	*TR-65
11C210B20	*HEPS3019
11C551-2	*GE-212,*TR-51,*SK3245,*BC107B,*276-2016
11C551-3	*GE-212,*TR-51,*SK3245,*BC107B,*276-2016
11C553-2	*GE-212,*TR-51,*SK3245,*BC107B,*276-2016
11C553-3	*GE-212,*TR-51,*SK3245,*BC107B,*276-2016

To Replace	Substitute This Type
11C557-2	*GE-39,*TR-51,*BF167,*276-2016
11C557-3	*GE-39,*TR-51,*BF167,*276-2016
11C702	2N2350,2N2350A,*GE-47,*HEPS3020,*BC337,*276-2016
11C704	2N2351,2N2351A,*HEPS3020,*276-2008
11C710	2N2364,2N2364A,2N2898,2N3056A,*HEPS3019,*276-2008
11C1536	2N2270,2N2868,2N3724A,2N4960,ZT90,ZT95,ZT1613,ZT2270,*GE-47,*TR-87, *HEPS3011,*SK3024,*ECG128,*WEP243,*276-2008
16A1	SEE 2N2711
16A2	SEE 2N2712
16A667-GRN	2N2925,2N2926-GRN,2N3391,2N3391A,2N3395-WHT,2N3396-WHT,2N3397-WHT, 2N3398-WHT,2N3415,2N3721,2N3900,2N3900A,2N3901,2N6002,2N6008,2SC732-BL, 2SC733-BL,2SC1335,2SC1344,91T6,98T2,BC108C,BC109,BC109C,BC168C, BC170C,BC172C,BC173C,BC238C,BC239C,BC338,BC548B,BC549B,BCY58C, BCY58D,GI-2925,GI-3391,GI-3391A,GI-3900,GI-3900A,MPS2925,MPS3391, MPS3391A,MPS3721,PBC108C,PBC109C,PET4002,PET4003,*PTC139,*GE-212, *TR-51,*SK3245,*BC107B,*276-2031
16A667-ORG	2N2712,2N2714,2N2924,2N2926-YEL,2N3241A,2N3392,2N3395-YEL,2N3396-YEL, 2N3397-YEL,2N3398-YEL,2N3414,2N5224,2N5998,2N6000,2SC372G-Y,2SC395A-Y, 2SC400,2SC400-Y,2SC454,2SC458,2SC733-Y,2SC735-Y,2SC847,2SC848,2SC849, 2SC934,2SC966,2SC967,2SC1085,2SC1219,2SC1246,2SC1317,2SC1363,2SC1631, 2SC1633,2SD392,92T6,93T6,40232,40233,40398,40400,BC108A,BC168A,BC170B, BC172A,BC238A,BC548A,BCY58A,BSW88,BSW89,BSX38,GI-2924,GI-2926,GI-3392, GI-3396,GI-3397,GI-3721,MPS2712,MPS2714,MPS2924,MPS2926-GRN, MPS2926-YEL,MPS3392,PBC108A,SE4001,TN80,*PTC101,*GE-212,*TR-51, *SK3245,*BC107B,*276-2031
16A667-RED	2N2331,2N2569,2N2570,2N2922,2N2923,2N2926-ORG,2N3393,2N3394,2N3396-ORG, 2N3397-ORG,2N3397-RED,2N3398-ORG,2N3398-RED,2N4138,2N4418,2N6566, 2S095A,2S512,2SC168,2SC300,2SC301,2SC372G-O,2SC395A,2SC395A-O, 2SC400-O,2SC459,2SC460,2SC461,2SC528,2SC619,2SC733-O,2SC735-O, 2SC1372,2SD227,BSY61,GI-2712,GI-2714,GI-2922,GI-2923,GI-3393,GI-3394,GI-3402, GI-3414,MPS2923,MPS2926-ORG,MPS3393,MPS3394,MPS6560,MPS6561,PET4001, ZT2938,*PTC101,*GE-212,*TR-51,*SK3245,*BC107B,*276-2031
16A667-YEL	2N2925,2N2926,2N2926-GRN,2N3391,2N3391A,2N3395-WHT,2N3396-WHT, 2N3397-WHT,2N3398-WHT,2N3415,2N3565,2N3721,2N3900,2N3900A,2N4124, 2N4256,2N4286,2N4419,2N5172,2N6002,2N6008,2SC373G,2SC400-GR, 2SC732-GR,2SC733-GR,2SC735-GR,2SC1684,2SC1849,2SC1993,91T6,98T2,40397, 40399,A158B,A159B,BC108,BC108B,BC109B,BC168B,BC172B,BC173B,BC238B, BC239B,BC338,BC548B,BC549B,BCY58B,BCY58C,GI-2925,GI-3391,GI-3391A, GI-3395,GI-3398,GI-3403,GI-3415,GI-3900,GI-3900A,MPS2925,MPS2926,MPS3391, MPS3391A,MPS3395,MPS3721,MPS5172,PBC108B,PBC109B,PET4002,*PTC121, *GE-212,*TR-51,*SK3245,*BC107B,*276-2031
16A668-GRN	2N2925,2N2926-GRN,2N3391,2N3391A,2N3395-WHT,2N3396-WHT,2N3397-WHT, 2N3398-WHT,2N3415,2N3721,2N3900,2N3900A,2N3901,2N6002,2N6008,2SC732-BL, 2SC733-BL,2SC1335,2SC1344,91T6,98T2,BC108C,BC109,BC109C,BC168C, BC170C,BC172C,BC173C,BC238C,BC239C,BC338,BC548B,BC549B,BCY58C, BCY58D,GI-2925,GI-3391,GI-3391A,GI-3900,GI-3900A,MPS2925,MPS3391, MPS3391A,MPS3721,PBC108C,PBC109C,PET4002,PET4003,*PTC139,*GE-212, *TR-51,*SK3245,*BC107B,*276-2031
16A668-ORG	2N2712,2N2714,2N2924,2N2926-YEL,2N3241A,2N3392,2N3395-YEL,2N3396-YEL, 2N3397-YEL,2N3398-YEL,2N3414,2N5224,2N5998,2N6000,2SC372G-Y,2SC395A-Y, 2SC400,2SC400-Y,2SC454,2SC458,2SC733-Y,2SC735-Y,2SC847,2SC848,2SC849, 2SC934,2SC966,2SC967,2SC1085,2SC1219,2SC1246,2SC1317,2SC1363,2SC1631, 2SC1633,2SD392,92T6,93T6,40232,40233,40398,40400,BC108A,BC168A,BC170B, BC172A,BC238A,BC548A,BCY58A,BSW88,BSW89,BSX38,GI-2924,GI-2926,GI-3392, GI-3396,GI-3397,GI-3721,MPS2712,MPS2714,MPS2924,MPS2926-GRN, MPS2926-YEL,MPS3392,PBC108A,SE4001,TN80,*PTC101,*GE-212,*TR-51, *SK3245,*BC107B,*276-2031
16A668-YEL	2N2925,2N2926,2N2926-GRN,2N3391,2N3391A,2N3395-WHT,2N3396-WHT, 2N3397-WHT,2N3398-WHT,2N3415,2N3565,2N3721,2N3900,2N3900A,2N4124, 2N4256,2N4286,2N4419,2N5172,2N6002,2N6008,2SC373G,2SC400-GR, 2SC732-GR,2SC733-GR,2SC735-GR,2SC1684,2SC1849,2SC1993,91T6,98T2,40397,

To Replace	Substitute This Type
(16A668-YEL)	40399,A158B,A159B,BC108,BC108B,BC109B,BC168B,BC172B,BC173B,BC238B, BC239B,BC338,BC548B,BC549B,BCY58B,BCY58C,GI-2925,GI-3391,GI-3391A, GI-3395,GI-3398,GI-3403,GI-3415,GI-3900,GI-3900A,MPS2925,MPS2926,MPS3391, MPS3391A,MPS3395,MPS3721,MPS5172,PBC108B,PBC109B,PET4002,*PTC121, *GE-212,*TR-51,*SK3245,*BC107B,*276-2031
16A669-GRN	2N2925,2N2926-GRN,2N3391,2N3391A,2N3395-WHT,2N3396-WHT,2N3397-WHT, 2N3398-WHT,2N3415,2N3721,2N3900,2N3900A,2N3901,2N6002,2N6008,2SC732-BL, 2SC733-BL,2SC1335,2SC1344,91T6,98T2,BC108C,BC109,BC109C,BC168C, BC170C,BC172C,BC173C,BC238C,BC239C,BC338,BC548B,BC549B,BCY58C, BCY58D,GI-2925,GI-3391,GI-3391A,GI-3900,GI-3900A,MPS2925,MPS3391, MPS3391A,MPS3721,PBC108C,PBC109C,PET4002,PET4003,*PTC139,*GE-212, *TR-51,*SK3245,*BC107B,*276-2031
16A669-YEL	2N2925,2N2926,2N2926-GRN,2N3391,2N3391A,2N3395-WHT,2N3396-WHT, 2N3397-WHT,2N3398-WHT,2N3415,2N3565,2N3721,2N3900,2N3900A,2N4124, 2N4256,2N4286,2N4419,2N5172,2N6002,2N6008,2SC373G,2SC400-GR, 2SC732-GR,2SC733-GR,2SC735-GR,2SC1684,2SC1849,2SC1993,91T6,98T2,40397, 40399,A158B,A159B,BC108,BC108B,BC109B,BC168B,BC172B,BC173B,BC238B, BC239B,BC338,BC548B,BC549B,BCY58B,BCY58C,GI-2925,GI-3391,GI-3391A, GI-3395,GI-3398,GI-3403,GI-3415,GI-3900,GI-3900A,MPS2925,MPS2926,MPS3391, MPS3391A,MPS3395,MPS3721,MPS5172,PBC108B,PBC109B,PET4002,*PTC121, *GE-212,*TR-51,*SK3245,*BC107B,*276-2031
16B1	SEE 2N2713
16B2	SEE 2N2714
16B670-GRN	2N3415,2N6002,2N6008,BCY58D,GI-3391,GI-3391A,MPS3390,MPS3391,MPS3391A, SE4002,SE4010,*GE-64,*TR-53,*SK3156,*BC107B,*276-2031
16B670-RED	2N744,2N2318,2N2319,2N2331,2N2656,2N2713,2N3011,2N3605,2N3606,2N3607, 2N4418,2S095A,2S512,2SC132,2SC133,2SC135,2SC168,2SC204,2SC356, 2SC395A-O,2SC579,2SC595,2SC724,2SC915,2SC915A,2SC1071,2SC1372,EN744, EN3011,GI-2922,MPS2713,PT720,TIS51,ZT2205,*PTC170,*GE-64,*TR-53,*SK3244, *BC107B,*276-2031
16B670-YEL	2N2714,2N3241A,2N3414,2N4124,2N4419,2N5998,2N6000,2SC395A-Y,2SC735-Y, 2SC847,2SC848,2SC849,2SC934,2SC966,2SC967,2SC1219,2SC1246,2SC1317, 2SC1363,2SC1631,2SC1633,2SD392,2SD467,40397,40399,BCY58A,BCY58B, GI-2924,GI-2926,MPS2714,SE4001,TN80,*PTC170,*GE-64,*TR-53,*SK3244, *BC107B,*276-2031
16G2	SEE 2N3663
16J1	SEE 2N3605
16J2	SEE 2N3606
16J3	SEE 2N3607
16K1	2N708,2N708A,2N784A,2N2318,2N2319,2N2369,2N2369A,2N2501,2N3009,2N3011, 2N3013,2N3014,2N3564,2N3605A,2N3606A,2N3646,2N3688,2N3689,2N3690, 2N3862,2N4137,2N4294,2N4295,2N4418,2N4420,2N4421,2N4422,2N4996,2N4997, 2N5029,2N5030,2SC67,2SC169,2SC335,2SC380-R,2SC380A-R,2SC400-O,2SC468, 2SC595,2SC601,2SC601N,2SC639,2SC752G-R,2SC764,2SC1187,2SC1188, 2SC1189,2SC1274,2SC1319,2SC1320,2SC1686,2SC1687,2SC1688,BF121,BF123, BF125,BF127,BF198,BF199,BF241,BF311,EN708,EN914,EN2369A,EN3009,EN3011, EN3013,EN3014,GET708,GET914,GET2369,GET3013,GET3014,GET3646,MPS3646, MT1060A,SE3005,SE5001,SE5002,SE5003,SE5006,TIS45,TIS46,TIS48,TIS49,TIS51, TIS52,TIS55,TIS87,TIS129,ZT708,*PTC132,*GE-60,*TR-95,*HEPS0015,*SK3018, *SN80,*ECG108,*WEP56,*BF173,*276-2016,*RT-113
16K2	2N708,2N708A,2N784A,2N2318,2N2319,2N2369,2N2369A,2N2501,2N3009,2N3011, 2N3013,2N3014,2N3564,2N3605A,2N3606A,2N3646,2N3688,2N3689,2N3690, 2N3862,2N4137,2N4294,2N4295,2N4418,2N4420,2N4421,2N4422,2N4996,2N4997, 2N5029,2N5030,2SC67,2SC169,2SC335,2SC380-R,2SC380A-R,2SC400-O,2SC468, 2SC595,2SC601,2SC601N,2SC639,2SC752G-R,2SC764,2SC1187,2SC1188, 2SC1189,2SC1274,2SC1319,2SC1320,2SC1686,2SC1687,2SC1688,BF121,BF123, BF125,BF127,BF198,BF199,BF241,BF311,EN708,EN914,EN2369A,EN3009,EN3011, EN3013,EN3014,GET708,GET914,GET2369,GET3013,GET3014,GET3646,MPS3646, MT1060A,SE3005,SE5001,SE5002,SE5003,SE5006,TIS45,TIS46,TIS48,TIS49,TIS51, TIS52,TIS55,TIS87,TIS129,ZT708,*PTC132,*GE-60,*TR-95,*HEPS0015,*SK3018, *SN80,*ECG108,*WEP56,*BF173,*276-2016,*RT-113

To Replace	Substitute This Type
16K3	2N708,2N708A,2N784A,2N2318,2N2319,2N2369,2N2369A,2N2501,2N3009,2N3011, 2N3013,2N3014,2N3564,2N3605A,2N3606A,2N3646,2N3688,2N3689,2N3690, 2N3862,2N4137,2N4294,2N4295,2N4418,2N4420,2N4421,2N4422,2N4996,2N4997, 2N5029,2N5030,2SC67,2SC169,2SC335,2SC380-R,2SC380A-R,2SC400-O,2SC468, 2SC595,2SC601,2SC601N,2SC639,2SC752G-R,2SC764,2SC1187,2SC1188, 2SC1189,2SC1274,2SC1319,2SC1320,2SC1686,2SC1687,2SC1688,BF121,BF123, BF125,BF127,BF198,BF199,BF241,BF311,EN708,EN914,EN2369A,EN3009,EN3011, EN3013,EN3014,GET708,GET914,GET2369,GET3013,GET3014,GET3646,MPS3646, MT1060A,SE3005,SE5001,SE5002,SE5003,SE5006,TIS45,TIS46,TIS48,TIS49,TIS51, TIS52,TIS55,TIS87,TIS129,ZT708,*PTC132,*GE-60,*TR-95,*HEPS0015,*SK3039, *EN10,*ECG108,*WEP56,*BF173,*276-2016,*RT-113
16L2	2N783,2N784,2N834A,2N2368,2N3261,2N3793,2N3825,2N3854A,2SC54,2SC302, 2SC321,2SC322,2SC400-R,2SC455,2SC576,2SC689,40405,GET706,GI-3793,KT218, MPS834,MPS6532,MPS6540,MPSH20,MPSH32,MPSH37,MT1038,MT1038A, MT1039,PET3001,SE1010,TIS47,ZT403P,ZT2368,*PTC101,*GE-212,*TR-95, *HEPS0016,*SK3039,*EN10,*ECG108,*WEP56,*BC107B,*276-2016,*RT-113
16L3	2N708,2N708A,2N784A,2N914,2N957,2N2242,2N2318,2N2319,2N2369,2N2369A, 2N2481,2N2710,2N3009,2N3011,2N3013,2N3014,2N3210,2N3211,2N3510,2N3511, 2N3564,2N3605A,2N3606A,2N3646,2N3647,2N3648,2N3688,2N3689,2N3690, 2N3691,2N3693,2N3845,2N3845A,2N3855A,2N3858,2N3862,2N4137,2N4294, 2N4295,2N4420,2N4421,2N4422,2N4969,2N5029,2N5030,2N5144,2N5824,2SC67, 2SC169,2SC204,2SC318,2SC335,2SC356,2SC394-O,2SC400-O,2SC461,2SC468, 2SC577,2SC595,2SC601,2SC601N,2SC620,2SC712,2SC713,2SC724,2SC752G-R, 2SC764,2SC1274,BFY39-1,BSW82,EN708,EN914,EN2369A,EN3009,EN3011, EN3013,EN3014,GET708,GET914,GET2369,GET3013,GET3014,GET3646,MPS2369, MPS3646,MPS6512,MPSA20-RED,PET1001,PET2001,PT2760,SE1001,SE5001, SE5002,SE5003,SE5006,TIS45,TIS46,TIS48,TIS49,TIS51,TIS52,TIS55,ZT404P, ZT708,ZT2369,*PTC101,*GE-212,*TR-95,*HEPS0016,*SK3039,*EN10,*ECG108, *WEP56,*BC107B,*276-2016,*RT-113
16L4	2N916,2N916A,2N2331,2N2501,2N3241A,2N3845,2N3845A,2N3856A,2N4013, 2N4123,2N4418,2N4873,2N5144,2SC318,2SC367G,2SC367G-O,2SC372-O, 2SC372G-O,2SC394-Y,2SC400-O,2SC400-Y,2SC454,2SC458,2SC459,2SC460, 2SC536,2SC619,2SC620,2SC639,2SC712,2SC713,2SC735-O,2SC752G-O,2SC933, 2SC965,2SC1210,2SC1216,2SC1276,2SC1372,2SC1999,BFY39-2,EN916,MPS3693, MPS3705,MPS6513,MPS6574-BLUE,MPS6574-GREEN,MPS6574-YEL, MPS6576-BLUE,MPS6576-GREEN,MPS6576-YEL,MPSA20-GRN,MPSA20-WHT, PET3002,SE2001,SE3005,TP3705,ZT81,ZT111,ZT2369A,*PTC132,*GE-212,*TR-53, *HEPS0020,*BC107B,*276-2016
16L5	MPSH34,*PTC132,*GE-20,*TR-95,*HEPS0016,*SK3039,*EN10,*ECG108,*WEP56, *276-2016,*RT-113
16L22	2N783,2N784,2N834A,2N2368,2N3261,2N3793,2N3825,2N3854A,2SC54,2SC302, 2SC321,2SC322,2SC400-R,2SC455,2SC576,2SC689,40405,GET706,GI-3793,KT218, MPS834,MPS6532,MPS6540,MPSH20,MPSH32,MPSH37,MT1038,MT1038A, MT1039,PET3001,SE1010,TIS47,ZT403P,ZT2368,*PTC101,*GE-212,*TR-95, *HEPS0016,*SK3039,*EN10,*ECG108,*WEP56,*BC107B,*276-2016,*RT-113
16L23	2N708,2N708A,2N784A,2N914,2N957,2N2242,2N2318,2N2319,2N2369,2N2369A, 2N2481,2N2710,2N3009,2N3011,2N3013,2N3014,2N3210,2N3211,2N3510,2N3511, 2N3564,2N3605A,2N3606A,2N3646,2N3647,2N3648,2N3688,2N3689,2N3690, 2N3691,2N3693,2N3845,2N3845A,2N3855A,2N3858,2N3862,2N4137,2N4294, 2N4295,2N4420,2N4421,2N4422,2N4969,2N5029,2N5030,2N5144,2N5824,2SC67, 2SC169,2SC204,2SC318,2SC335,2SC356,2SC394-O,2SC400-O,2SC461,2SC468, 2SC577,2SC595,2SC601,2SC601N,2SC620,2SC712,2SC713,2SC724,2SC752G-R, 2SC764,2SC1274,BFY39-1,BSW82,EN708,EN914,EN2369A,EN3009,EN3011, EN3013,EN3014,GET708,GET914,GET2369,GET3013,GET3014,GET3646,MPS2369, MPS3646,MPS6512,MPSA20-RED,PET1001,PET2001,PT2760,SE1001,SE5001, SE5002,SE5003,SE5006,TIS45,TIS46,TIS48,TIS49,TIS51,TIS52,TIS55,ZT404P, ZT708,ZT2369,*PTC101,*GE-212,*TR-95,*HEPS0016,*SK3039,*EN10,*ECG108, *WEP56,*BC107B,*276-2016,*RT-113
16L24	2N916,2N916A,2N2331,2N2501,2N3241A,2N3845,2N3845A,2N3856A,2N4013, 2N4123,2N4418,2N4873,2N5144,2SC318,2SC367G,2SC367G-O,2SC372-O, 2SC372G-O,2SC394-Y,2SC400-O,2SC400-Y,2SC454,2SC458,2SC459,2SC460,

To Replace	Substitute This Type
(16L24)	2SC536,2SC619,2SC620,2SC639,2SC712,2SC713,2SC735-O,2SC752G-O,2SC933, 2SC965,2SC1210,2SC1216,2SC1276,2SC1372,2SC1999,BFY39-2,EN916,MPS3693, MPS3705,MPS6513,MPS6574-BLUE,MPS6574-GREEN,MPS6574-YEL, MPS6576-BLUE,MPS6576-GREEN,MPS6576-YEL,MPSA20-GRN,MPSA20-WHT, PET3002,SE2001,SE3005,TP3705,ZT81,ZT111,ZT2369A,*PTC132,*GE-212,*TR-64, *SK3245,*BC107B,*276-2016
16L25	2N916,2N916A,2N3241A,2N3856A,2N4013,2N4873,2SC367G,2SC367G-O,2SC372-O, 2SC372G-O,2SC394-Y,2SC400-Y,2SC454,2SC458,2SC459,2SC536,2SC619, 2SC735-O,2SC752G-O,2SC933,2SC965,2SC1085,2SC1216,2SC1276,2SC1317, 2SC1999,BC107A,BC108A,BC167A,BC171A,BC237A,BC238A,BCY58A,BCY59A, BFY39-2,EN916,MPS6513,MPS6574-BLUE,MPS6574-GREEN,MPS6576-BLUE, MPS6576-GREEN,MPSA20-GRN,MPSA20-WHT,PET3002,ZT82,ZT112,ZT2369A, *PTC132,*GE-212,*HEPS0020,*SK3245,*BC107B,*276-2016
16L42	2N743,2N784,2N959,2N2571,2N2572,2N2926-BRN,2N3825,2N3854,2N3854A,2N5187, 2SC68,2SC98,2SC395A-R,2SC400-R,2SC455,GI-3605,GI-3606,GI-3607,MPS706A, MPS2926-BRN,MPS6540,MT1038,MT1038A,MT1039,PET3001,PET8101,SE1010, ZT403P,*PTC101,*GE-212,*TR-21,*HEPS0011,*SK3122,*EN10,*ECG123A, *WEP736,*BC107B,*276-2015,*RT-102
16L43	SEE 2N3854
16L44	SEE 2N3855
16L45	SEE 2N3856
16L62	2N783,2N784,2N834A,2N2368,2N3261,2N3793,2N3825,2N3854A,2SC54,2SC302, 2SC321,2SC322,2SC400-R,2SC455,2SC576,2SC689,40405,GET706,GI-3793,KT218, MPS834,MPS6532,MPS6540,MPSH20,MPSH32,MPSH37,MT1038,MT1038A, MT1039,PET3001,SE1010,TIS47,ZT403P,ZT2368,*PTC101,*GE-212,*TR-21, *HEPS0011,*SK3122,*EN10,*ECG123A,*WEP736,*BC107B,*276-2016,*RT-102
16L63	SEE 2N3854A
16L64	SEE 2N3855A
16L65	SEE 2N3856A
16X1	SEE 2N3877A
16X2	SEE 2N3877
71T2	*GE-27
72T2	
73T2	71T2,*GE-27
74T2	72T2
80T2	*PTC144,*TR-63,*HEPG0005
81T2	2N707A,*PTC125,*GE-63,*TR-87,*HEPS3001,*SK3024,*ECG128,*276-2030,*RT-114
82T2	2SC542,83T2,*HEPS3020
83T2	*HEPS3020
90T2	2N719,2N738,2N738A,2N912,2N2460,2N2464,2N2517,2N4390,BFY80,*PTC170, *GE-18,*TR-70,*SK3039,*ECG107,*WEP56,*BF338,*276-2012,*RT-108
91T6	2N3391,2N3391A,2N3900,2N3900A,2N3901,2N6002,2N6008,2SC732-BL,2SC733-BL, 2SC1335,2SC1344,A158C,A159C,BC108C,BC109,BC109C,BC168C,BC170C, BC172C,BC173C,BC238C,BC239C,BCY58D,PBC108C,PBC109C,PET4003,SE4002, SE4010,*PTC139,*GE-212,*TR-51,*SK3245,*BC107B,*276-2011
92PE37A	92PE37B
92PE37B	92PE37C
92PE37C	
92PE77A	92PE77B
92PE77B	92PE77C
92PE77C	
92PE487	
92PE488	92PE489
92PE489	
92PU01	2N744A,92PU01A
92PU01A	
92PU05	92PU06
92PU06	92PU07
92PU07	
92PU10	
92PU45	92PU45A

To Replace	Substitute This Type
92PU45A	
92PU51	92PU51A
92PU51A	
92PU55	92PU56
92PU56	92PU57
92PU57	
92T6	2N2924,2N2925,2N2926-GRN,2N2926-YEL,2N3391,2N3391A,2N3392,2N3415,2N3565, 2N3900,2N3900A,2N4124,2N4256,2N4286,2N4419,2N5172,2N5998,2N6002,2N6008, 2SC373G,2SC400-GR,2SC732-GR,2SC733-GR,2SC735-GR,2SC1684,2SC1849, 2SC1993,91T6,98T2,40397,40399,A158B,A159B,BC108,BC108B,BC109B,BC168B, BC172B,BC173B,BC238B,BC239B,BC338,BC548B,BC549B,BCY58B,BCY58C, MPS3721,MPS5172,PBC108B,PBC109B,PET4002,*PTC121,*GE-212,*TR-51, *SK3245,*BC107B,*276-2011
93T6	2N2712,2N2714,2N2924,2N2926-YEL,2N3241A,2N3392,2N3414,2N4124,2N4419, 2N5224,2N5998,2N6000,2SC372G-Y,2SC395A-Y,2SC400,2SC400-Y,2SC454, 2SC733-Y,2SC735-Y,2SC847,2SC848,2SC934,2SC966,2SC967,2SC1085,2SC1219, 2SC1246,2SC1317,2SC1363,2SC1631,2SC1633,2SC1684,2SC1849,2SD392,92T6, 40232,40233,40397,40399,BC108,BC108A,BC168A,BC170B,BC172A,BC238A, BC548A,BCY58A,BCY58B,BSW88,BSW89,BSX38,MPS2714,MPS2926-GRN, MPS2926-YEL,PBC108A,SE4001,TN80,*PTC101,*GE-212,*TR-51,*BC107B, *276-2011
98T2	BCY58C,BCY58D,*GE-210,*TR-53,*HEPS0015,*BC107B,*276-2015
100T2	*GE-19,*TR-59,*SK3027,*HN100,*ECG130,*WEP247,*RT-131
108T2	109T2
109T2	
111T2	2N657,2N699A,2N699B,2N1335,2N1336,2N1337,2N1613B,2N1889,2N1974,2N2443, 2N3107,2N4924,2SC49,2SC353,2SC353A,2SC470-3,2SC470-4,2SC590,MM3008, SE7002,*PTC125,*GE-27,*TR-78,*SK3045,*ECG154,*WEP712,*BF338,*276-2012, *RT-110
121-6	SEE 2N216
121-7	SEE 2N35
121-9	SEE 2N112
121-10	SEE 2N111
121-11	SEE CK725
121-12	SEE CK725
121-14	SEE 2N112
121-21	SEE 2N193
121-22	SEE 2N194
121-24	SEE 2N168A
121-25	SEE 2N168A
121-26	SEE 2N168
121-27	SEE 2N190
121-33	SEE 2N169A
121-34	SEE 2N186A
121-44	SEE 2N370
121-45	SEE 2N139
121-46	SEE 2N109
121-47	SEE 2N270
121-48	SEE 2N371
121-49	SEE 2N372
121-50	SEE 2N253
121-51	SEE 2N254
121-54	SEE 2N252
121-60	SEE 2N213
121-61	SEE 2N407
121-62	SEE 2N411
121-63	SEE 2N247
121-64	SEE 2N407
121-65	SEE 2N409
121-66	SEE 2N409
121-67	SEE 2N308

To Replace	Substitute This Type
121-70	SEE 2N515
121-71	SEE 2N516
121-73	SEE 2N409
121-74	SEE 2N409
121-75	SEE 2N139
121-76	SEE 2N139
121-78	SEE 2N544
121-91	SEE 2N483
121-92	SEE 2N485
121-93	SEE 2N483
121-94	SEE 2N482
121-95	SEE 2N362
121-96	SEE 2N632
121-101	SEE 2N544
121-102	SEE 2N409
121-103	SEE 2N411
121-104	SEE 2N409
121-105	SEE 2N409
121-107	SEE 2N407
121-113	SEE 2N309
121-128	SEE 2N1108
121-134	SEE 2N1177
121-135	SEE 2N1178
121-136	SEE 2N1179
121-138	SEE 2N1180
121-139	SEE 2N1180
121-145	SEE 2N1108
121-146	SEE 2N1110
121-147	SEE 2N1111
121-148	SEE 2N407
121-150	SEE 2N1631
121-153	SEE 2N1108
121-154	SEE 2N1110
121-161	SEE 2N410
121-162	SEE 2N410
121-164	SEE 2N408
121-179	SEE 2N1527
121-180	SEE 2N1525
121-181	SEE 2N1525
121-184	SEE 2N1374
121-185	SEE 2N1525
121-205	SEE 2N1374
121-225	SEE 2N407
121-228	SEE 2N1742
121-229	SEE 2N1745
121-230	SEE 2N1745
121-231	SEE 2N1865
121-232	SEE 2N1865
121-233	SEE 2N1747
121-240	SEE 2N2614
121-241	SEE 2N218
121-242	SEE 2N993
121-243	SEE 2N993
121-244	SEE 2N993
121-256	SEE 2N1632
121-257	SEE 2N1526
121-258	SEE 2N1524
121-259	SEE 2N1524
121-260	SEE 2N1524
121-266	SEE 2N406
121-267	SEE 2N408

To Replace	Substitute This Type
121-268	SEE 2N1742
121-269	SEE 2N1745
121-294	SEE 2N2654
121-295	SEE 2N2654
121-296	SEE 2N2654
121-297	SEE 2N2671
121-298	SEE 2N2671
121-299	SEE 2N2671
121-300	SEE 2N2429
121-301	SEE 2N2428
121-302	SEE 2N1302
121-309	SEE 2N2429
121-310	SEE 2N2428
121-311	SEE 2N2706
151-04	151-06,152-04,152-06
151-06	2N1016B,151-08,152-06,152-08
151-08	2N1016B,151-10,152-08,152-10
151-10	2N1016C,151-12,151-14,152-10,152-12,152-14
151-12	2N1016C,151-14,151-16,152-12,152-14,152-16
151-14	2N1016C,2N1016D,151-16,151-18,152-14,152-16,152-18
151-16	2N1016D,151-18,151-20,152-16,152-18,152-20
151-18	2N1016D,151-20,151-22,151-24,152-18,152-20,152-22,152-24
151-20	151-18,151-22,151-24,151-26,152-18,152-20,152-22,152-24,152-26
151-22	151-20,151-24,151-26,151-28,152-20,152-22,152-24,152-26,152-28
151-24	151-22,151-26,151-28,151-30,152-22,152-24,152-26,152-28,152-30
151-26	151-24,151-28,151-30,152-24,152-26,152-28,152-30
151-28	151-26,151-30,152-26,152-28,152-30
151-30	151-28,152-28,152-30
152-04	152-06
152-06	2N1016B,152-08
152-08	2N1016B,152-10
152-10	2N1016C,152-12,152-14
152-12	2N1016C,152-14,152-16
152-14	2N1016C,2N1016D,152-16,152-18
152-16	2N1016D,152-18,152-20
152-18	2N1016D,152-20,152-22,152-24
152-20	152-18,152-22,152-24,152-26
152-22	152-20,152-24,152-26,152-28
152-24	152-22,152-26,152-28,152-30
152-26	152-24,152-28,152-30
152-28	152-26,152-30
152-30	152-28
153-04	153-06,154-04,154-06
153-06	153-08,154-06,154-08
153-08	153-10,154-08,154-10
153-10	153-12,153-14,154-10,154-12,154-14
153-12	153-14,153-16,154-12,154-14,154-16
153-14	153-16,153-18,154-14,154-16,154-18
153-16	153-18,153-20,154-16,154-18,154-20,154-22,154-24
153-18	153-20,153-22,153-24,154-18,154-20,154-22,154-24,154-26
153-20	153-18,153-22,153-24,153-26,154-18,154-20,154-22,154-24,154-26,154-28
153-22	153-20,153-24,153-26,153-28,154-20,154-24,154-26,154-28
153-24	153-22,153-26,153-28,153-30,154-26,154-28,154-30
153-26	153-24,153-28,153-30,154-26,154-28,154-30
153-28	153-26,153-30,154-28,154-30
153-30	154-30
154-04	154-06
154-06	154-08
154-08	154-10
154-10	154-12,154-14
154-12	154-14,154-16

To Replace	Substitute This Type
154-14	154-16,154-18
154-16	154-18,154-20,154-22,154-24
154-18	154-20,154-22,154-24,154-26
154-20	154-18,154-22,154-24,154-26,154-28
154-22	154-18,154-20,154-24,154-26,154-28
154-24	154-20,154-22,154-26,154-28
154-26	154-24,154-28,154-30
154-28	154-26,154-30
154-30	
156-043	2N3055A,1561-0403,1561-0404,1582-0403,1582-0404,1582-0405,1582-0408,1582-0410, 1582-0415,1582-0608,1582-0615,HST9201,HST9205,MJ2801,*PTC140,*GE-77, *TR-59,*HEPS7004,*SK3027,*ECG130,*WEP247,*BDY20,*RT-131
156-044	2N3055A,1561-0403,1561-0404,1582-0403,1582-0404,1582-0405,HST9201,HST9205, MJ2801,*PTC140,*GE-77,*HEPS7004,*BDY20
156-083	2N3055,2N3236,2N3237,2SD55,1561-0803,1561-0804,1561-0805,1561-1005, 1582-0803,1582-0804,1582-0805,1582-0808,1582-0810,1582-0815,1582-1003, 1582-1004,1582-1005,1582-1008,1582-1010,1582-1015,HST9202,HST9206,HST9207, IR3772,MJ3772,*PTC140,*GE-75,*HEPS7004
156-084	2N3055,2N3236,2N3237,2SD55,1561-0803,1561-0804,1561-0805,1561-1005, 1582-0803,1582-0804,1582-0805,1582-1003,1582-1004,1582-1005,HST9202, HST9206,HST9207,IR3772,MJ3772,*PTC140,*GE-75,*HEPS7004
156-104	2N3055,2N3236,2N5630,2SD55,108T2,1561-1005,1561-1205,1582-1003,1582-1004, 1582-1005,1582-1203,1582-1204,1582-1205,HST9202,HST9203,HST9207,HST9208, IR3772,MJ3772,*PTC173,*GE-75,*HEPS7004
156-123	2N5631,108T2,109T2,1561-1005,1561-1205,1561-1404,1561-1405,1582-1003, 1582-1004,1582-1005,1582-1203,1582-1204,1582-1205,1582-1208,1582-1210, 1582-1215,1582-1403,1582-1404,1582-1405,1582-1408,1582-1410,1582-1415, 1582-1608,1582-1610,1582-1615,HST9203,HST9204,HST9208,HST9209,IR3773, IR6302,*PTC173
156-124	2N5630,2N5631,108T2,109T2,1561-1005,1561-1205,1561-1404,1561-1405,1582-1003, 1582-1004,1582-1005,1582-1203,1582-1204,1582-1205,1582-1403,1582-1404, 1582-1405,HST9203,HST9204,HST9208,HST9209,IR3773,IR6302,*PTC173
156-144	2N5631,109T2,1561-1205,1561-1404,1561-1405,1561-1604,1561-1605,1582-1203, 1582-1204,1582-1205,1582-1403,1582-1404,1582-1405,1582-1603,1582-1604, 1582-1605,HST9204,HST9209,IR3773,IR6302
156-164	109T2,1561-1404,1561-1405,1561-1604,1561-1605,1561-1803,1561-1804,1561-1805, 1561-2003,1561-2004,1582-1403,1582-1404,1582-1405,1582-1603,1582-1604, 1582-1605,1582-1803,1582-1804,1582-1805,1582-2003,1582-2004,IR3773
163-04	163-06,164-04,164-06
163-06	163-08,164-06,164-08
163-08	163-10,164-08,164-10
163-10	163-12,163-14,164-10,164-12,164-14
163-12	163-14,163-16,164-12,164-14,164-16
163-14	163-16,163-18,164-14,164-16,164-18
163-16	163-18,163-20,164-16,164-18,164-20
163-18	163-20,163-22,163-24,164-18,164-20,164-22,164-24
163-20	163-22,163-24,163-26,164-20,164-22,164-24,164-26
163-22	163-20,163-24,163-26,163-28,164-20,164-22,164-24,164-26,164-28
163-24	163-22,163-26,163-28,163-30,164-22,164-24,164-26,164-28,164-30
163-26	163-24,163-28,163-30,164-24,164-26,164-28,164-30
163-28	163-26,163-30,164-26,164-28,164-30
163-30	163-28,164-28,164-30
164-04	164-06
164-06	164-08
164-08	164-10
164-10	164-12,164-14
164-12	164-14,164-16
164-14	164-16,164-18
164-16	164-18,164-20
164-18	164-20,164-22,164-24
164-20	164-22,164-24,164-26

To Replace	Substitute This Type
164-22	164-20,164-24,164-26,164-28
164-24	164-22,164-26,164-28,164-30
164-26	164-24,164-28,164-30
164-28	164-26,164-30
164-30	164-28
176-04	164-04
176-06	164-06,164-08,STC1726,STC1728
176-08	164-08,164-10,STC1726,STC1728,STC1731,STC1733
176-10	164-10,164-12,STC1731,STC1733
176-12	164-12,164-14,STC1736,STC1738
176-14	164-12,164-14,164-16,164-18,STC1736,STC1738
176-16	164-14,164-16,164-18,164-20,STC3706
180T2	BDY23B,*PTC173,*GE-19,*TR-59,*HEPS7002,*SK3027,*HN100,*ECG130,*WEP247, *RT-131
181T2	2N3448,BDY24B,*PTC173,*HEPS7004
182T2	BDY25B,*HEPS5020
183T2	BDY26B,*HEPS5020
184T2	185T2,BDY27B,BDY28B,*HEPS5020
185T2	BDY28B,*HEPS5021
276-2001	REFER TO SECTION 2
276-2002	REFER TO SECTION 2
276-2003	REFER TO SECTION 2
276-2004	REFER TO SECTION 2
276-2005	REFER TO SECTION 2
276-2006	REFER TO SECTION 2
276-2007	REFER TO SECTION 2
276-2008	REFER TO SECTION 2
276-2009	REFER TO SECTION 2
276-2010	REFER TO SECTION 2
276-2011	REFER TO SECTION 2
276-2012	REFER TO SECTION 2
276-2013	REFER TO SECTION 2
276-2014	REFER TO SECTION 2
276-2015	REFER TO SECTION 2
276-2016	REFER TO SECTION 2
276-2017	REFER TO SECTION 2
276-2018	REFER TO SECTION 2
276-2019	REFER TO SECTION 2
276-2020	REFER TO SECTION 2
276-2021	REFER TO SECTION 2
276-2022	REFER TO SECTION 2
276-2023	REFER TO SECTION 2
276-2024	REFER TO SECTION 2
276-2025	REFER TO SECTION 2
276-2026	REFER TO SECTION 2
276-2027	REFER TO SECTION 2
276-2030	REFER TO SECTION 2
276-2031	REFER TO SECTION 2
276-2032	REFER TO SECTION 2
276-2033	REFER TO SECTION 2
276-2034	REFER TO SECTION 2
903	SEE 2N1149
904	SEE 2N1150
904A	SEE 2N1151
905	SEE 2N1152
910	SEE 2N1153
951	SEE 2N1154
952	SEE 2N1155
953	SEE 2N1156
1401-14	
1401-0405	1401-0407,1401-0410,1401-0605,1401-0607,1401-0610

To Replace	Substitute This Type
1401-0407	1401-0405,1401-0410,1401-0605,1401-0607,1401-0610
1401-0410	1401-0405,1401-0407,1401-0605,1401-0607,1401-0610
1401-0415	1401-0420,1401-0425,1401-0615,1401-0625
1401-0420	1401-0615,1401-0620,1401-0625
1401-0425	1401-0415,1401-0420,1401-0615,1401-0625
1401-0605	1401-0607,1401-0610,1401-0805,1401-0807,1401-0810
1401-0607	1401-0605,1401-0610,1401-0805,1401-0807,1401-0810
1401-0610	1401-0605,1401-0607,1401-0805,1401-0807,1401-0810
1401-0615	1401-0620,1401-0625,1401-0815,1401-0825
1401-0620	1401-0815,1401-0820,1401-0825,1401-1015,1401-1025
1401-0625	1401-0615,1401-0620,1401-0815,1401-0825
1401-0805	1401-0807,1401-0810,1401-1005,1401-1007,1401-1010
1401-0807	1401-0805,1401-0810,1401-1005,1401-1007,1401-1010
1401-0810	1401-0805,1401-0807,1401-1005,1401-1007,1401-1010
1401-0815	1401-0820,1401-0825,1401-1015,1401-1020,1401-1025
1401-0820	1401-1015,1401-1020,1401-1025,1401-1215
1401-0825	1401-0815,1401-0820,1401-1015,1401-1020,1401-1025
1401-1005	1401-1007,1401-1010,1401-1205,1401-1207,1401-1210
1401-1007	1401-1005,1401-1010,1401-1205,1401-1207,1401-1210
1401-1010	1401-1005,1401-1007,1401-1205,1401-1207,1401-1210
1401-1015	1401-1020,1401-1025,1401-1215,1401-1220
1401-1020	1401-1015,1401-1025,1401-1215,1401-1220
1401-1025	1401-1015,1401-1020,1401-1215,1401-1220
1401-1205	1401-1207,1401-1210
1401-1207	1401-1205,1401-1210
1401-1210	1401-1205,1401-1207
1401-1215	1401-1020,1401-1220
1401-1220	1401-1215
1441-0405	
1441-0407	
1441-0410	
1441-0605	
1441-0607	
1441-0610	
1441-0805	
1441-0807	
1441-0810	
1441-1005	
1441-1007	
1441-1010	
1441-1205	
1441-1207	
1441-1210	
1561-0403	2N3055A,2N3235,1561-0404,1582-0403,1582-0404,1582-0405,1582-0603,1582-0604, 1582-0605,HST9201,HST9205,MJ2801,*PTC140,*GE-77,*BDY20
1561-0404	2N3055A,2N3235,1561-0403,1582-0403,1582-0404,1582-0405,1582-0603,1582-0604, 1582-0605,HST9201,HST9205,MJ2801,*PTC140,*GE-77,*SK3027,*BDY20
1561-0408	1561-0410,1561-0608,1561-0610,1561-0615,*PTC173,*GE-75
1561-0410	1561-0408,1561-0608,1561-0610,1561-0615,*PTC173,*GE-75
1561-0608	2N3237,2N6258,2N6570,2SD55,1561-0610,1561-0615,1561-0808,1561-0810, 1561-0815,IR3772,MJ3772,RCS258,STC2220,STC2221,STC2224,STC2225, STC2228,STC2229,*PTC173,*GE-75
1561-0610	2N3237,2N6258,2N6570,2SD55,1561-0608,1561-0615,1561-0808,1561-0810, 1561-0815,IR3772,MJ3772,RCS258,STC2220,STC2221,STC2224,STC2225, STC2228,STC2229,*PTC173,*GE-75
1561-0615	2N6258,2N6570,1561-0815,STC2228,STC2229,*PTC175,*GE-75
1561-0803	2N3055,2N3236,2N3237,2N5630,2SD55,108T2,1561-0804,1561-0805,1561-1005, 1582-0803,1582-0804,1582-0805,1582-1003,1582-1004,1582-1005,HST9202, HST9203,HST9207,HST9208,IR3772,MJ3772,*PTC140,*GE-75
1561-0804	2N3055,2N3236,2N3237,2N5630,2SD55,108T2,1561-0803,1561-0805,1561-1005, 1582-0803,1582-0804,1582-0805,1582-1003,1582-1004,1582-1005,HST9202,

To Replace	Substitute This Type
(1561-0804)	HST9203,HST9207,HST9208,IR3772,MJ3772,*PTC140,*GE-75
1561-0805	2N3055,2N3236,2N3237,2N5630,2SD55,108T2,1561-0803,1561-0804,1561-1005, 1582-0803,1582-0804,1582-0805,1582-1003,1582-1004,1582-1005,HST9202, HST9203,HST9207,HST9208,IR3772,MJ3772,*PTC140,*GE-75
1561-0808	2N3237,2N6258,2N6570,2N6571,2SD55,108T2,1561-0810,1561-0815,1561-1008, 1561-1010,1561-1015,IR3772,MJ3772,RCS258,STC2221,STC2225,STC2229, *PTC173,*GE-75,*SK3036
1561-0810	2N3237,2N6258,2N6570,2N6571,2SD55,108T2,1561-0808,1561-0815,1561-1008, 1561-1010,1561-1015,IR3772,MJ3772,RCS258,STC2221,STC2225,STC2229, *PTC173,*GE-75,*SK3036
1561-0815	2N6258,2N6570,2N6571,108T2,1561-1015,STC2229,*PTC175,*GE-75
1561-1005	2N3055,2N3236,2N5630,2N5631,2SD55,108T2,1561-1205,1561-1404,1561-1405, 1582-1003,1582-1004,1582-1005,1582-1203,1582-1204,1582-1205,1582-1403, 1582-1404,1582-1405,HST9202,HST9203,HST9204,HST9207,HST9208,HST9209, IR3772,IR6302,MJ3772,*PTC173,*GE-75
1561-1008	2N3055C,2N3772C,2N6258,2N6570,2N6571,2SD55,108T2,1561-1010,1561-1015, 1561-1208,1561-1210,1561-1215,1561-1410,1561-1415,IR3772,MJ3772,RCS258, STC2221,STC2222,STC2225,STC2226,STC2229,STC2230,*PTC173
1561-1010	2N3055C,2N3772C,2N6258,2N6570,2N6571,2SD55,108T2,1561-1008,1561-1015, 1561-1208,1561-1210,1561-1215,1561-1410,1561-1415,IR3772,MJ3772,RCS258, STC2221,STC2222,STC2225,STC2226,STC2229,STC2230,*PTC173,*GE-75, *SK3036
1561-1015	2N3055C,2N3772C,2N6258,2N6570,2N6571,108T2,1561-1215,1561-1415,STC2229, STC2230,*PTC175,*GE-75
1561-1205	2N5630,2N5631,108T2,109T2,1561-1404,1561-1405,1561-1604,1561-1605,1582-1203, 1582-1204,1582-1205,1582-1403,1582-1404,1582-1405,1582-1603,1582-1604, 1582-1605,HST9203,HST9204,HST9208,HST9209,IR3773,IR6302
1561-1208	2N3055C,2N3772C,108T2,109T2,1561-1210,1561-1215,1561-1410,1561-1415, 1561-1608,1561-1610,1561-1615,STC2222,STC2226,STC2230
1561-1210	2N3055C,2N3772C,108T2,109T2,1561-1208,1561-1215,1561-1410,1561-1415, 1561-1608,1561-1610,1561-1615,STC2222,STC2226,STC2230
1561-1215	2N3055C,2N3772C,108T2,109T2,1561-1415,1561-1615,STC2230
1561-1404	2N5631,109T2,1561-1405,1561-1604,1561-1605,1561-1803,1561-1804,1561-1805, 1582-1403,1582-1404,1582-1405,1582-1603,1582-1604,1582-1605,1582-1803, 1582-1804,1582-1805,HST9204,HST9209,IR3773,IR6302
1561-1405	2N5631,109T2,1561-1404,1561-1604,1561-1605,1561-1803,1561-1804,1561-1805, 1582-1403,1582-1404,1582-1405,1582-1603,1582-1604,1582-1605,1582-1803, 1582-1804,1582-1805,HST9204,HST9209,IR3773,IR6302
1561-1410	2N3055C,2N3772C,109T2,1561-1415,1561-1608,1561-1610,1561-1615,1561-1808, 1561-1810,1561-1815,STC2222,STC2223,STC2226,STC2227,STC2230
1561-1415	2N3055C,2N3772C,109T2,1561-1615,1561-1815,STC2230
1561-1604	109T2,1561-1605,1561-1803,1561-1804,1561-1805,1561-2003,1561-2004,1582-1603, 1582-1604,1582-1605,1582-1803,1582-1804,1582-1805,1582-2003,1582-2004, 1843-2005,IR3773
1561-1605	109T2,1561-1604,1561-1803,1561-1804,1561-1805,1561-2003,1561-2004,1582-1603, 1582-1604,1582-1605,1582-1803,1582-1804,1582-1805,1582-2003,1582-2004, 1843-2005,IR3773
1561-1608	109T2,1561-1610,1561-1615,1561-1808,1561-1810,1561-1815,1561-2008,1561-2010, 1843-2005,STC2223,STC2227
1561-1610	109T2,1561-1608,1561-1615,1561-1808,1561-1810,1561-1815,1561-2008,1561-2010, 1843-2005,STC2223,STC2227
1561-1615	109T2,1561-1815,1843-2005
1561-1803	1561-1804,1561-1805,1561-2003,1561-2004,1582-1803,1582-1804,1582-1805, 1582-2003,1582-2004,1843-2005,1843-2205
1561-1804	1561-1803,1561-1805,1561-2003,1561-2004,1582-1803,1582-1804,1582-1805, 1582-2003,1582-2004,1843-2005,1843-2205
1561-1805	1561-1803,1561-1804,1561-2003,1561-2004,1582-1803,1582-1804,1582-1805, 1582-2003,1582-2004,1843-2005,1843-2205
1561-1808	1561-1810,1561-1815,1561-2008,1561-2010,1843-2005,1843-2205,STC2223,STC2227
1561-1810	1561-1808,1561-1815,1561-2008,1561-2010,1843-2005,1843-2205,STC2223,STC2227
1561-1815	1843-2005,1843-2205

To Replace	Substitute This Type
1561-2003	1561-2004,1582-2003,1582-2004,1843-2005,1843-2205,1843-2505
1561-2004	1561-2003,1582-2003,1582-2004,1843-2005,1843-2205,1843-2505
1561-2008	1561-2010,1843-2005,1843-2205,1843-2505,STC2223,STC2227
1561-2010	1561-2008,1843-2005,1843-2205,1843-2505,STC2223,STC2227
1571-0401	1571-0402,1571-0601,1571-0602,1571-0620,*PTC112,*GE-66,*276-2020
1571-0402	1571-0401,1571-0601,1571-0602,1571-0620,*PTC112,*GE-66,*276-2020
1571-0420	1571-0401,1571-0402,1571-0425,1571-0620,*PTC112,*GE-66,*276-2018
1571-0425	1571-0401,1571-0402,1571-0420,1571-0620,BD148-6,*PTC112,*GE-66,*276-2018
1571-0601	1571-0602,1571-0801,1571-0802,1571-0820,1571-0825,1571-1020,1571-1025,KSP1021, KSP1022,*PTC112,*GE-66,*276-2020
1571-0602	1571-0601,1571-0801,1571-0802,1571-0820,1571-0825,1571-1020,1571-1025,KSP1021, KSP1022,*PTC112,*GE-66,*276-2020
1571-0620	1571-0601,1571-0602,1571-0820,1571-0825,KSP1021,*PTC112,*GE-66,*276-2020
1571-0801	2N3879,1571-0802,1571-1001,1571-1002,1571-1020,1571-1025,1571-1220,1571-1225, KSP1022,*PTC112
1571-0802	2N3879,1571-0801,1571-1001,1571-1002,1571-1020,1571-1025,1571-1220,1571-1225, KSP1022,*PTC112
1571-0820	1571-0801,1571-0802,1571-0825,1571-1001,1571-1002,1571-1020,1571-1025,KSP1021, KSP1022,*PTC112
1571-0825	1571-0801,1571-0802,1571-0820,1571-1001,1571-1002,1571-1020,1571-1025,BD149-6, KSP1021,KSP1022,*PTC112
1571-1001	2N3879,1571-1002,1571-1020,1571-1025,1571-1201,1571-1202,1571-1220,1571-1225, 1571-1401,1571-1402,1571-1425,KSP1022,KSP1023
1571-1002	2N3879,1571-1001,1571-1020,1571-1025,1571-1201,1571-1202,1571-1220,1571-1225, 1571-1401,1571-1402,1571-1425,KSP1022,KSP1023
1571-1020	2N3879,1571-1001,1571-1002,1571-1025,1571-1201,1571-1202,1571-1220,1571-1225, KSP1022,*PTC169
1571-1025	2N3879,2SC1113,1571-1001,1571-1002,1571-1020,1571-1201,1571-1202,1571-1220, 1571-1225,KSP1022,*PTC169,*SK3538
1571-1201	2N3879,1571-1202,1571-1220,1571-1225,1571-1401,1571-1402,1571-1425,1571-1601, 1571-1602,1571-1620,1571-1625,KSP1023
1571-1202	2N3879,1571-1201,1571-1220,1571-1225,1571-1401,1571-1402,1571-1425,1571-1601, 1571-1602,1571-1620,1571-1625,KSP1023
1571-1220	2N3879,1571-1001,1571-1002,1571-1201,1571-1202,1571-1225,1571-1401,1571-1402, 1571-1425,1571-1620,1571-1625,KSP1023
1571-1225	2N3879,2SC1113,1571-1001,1571-1002,1571-1201,1571-1202,1571-1220,1571-1401, 1571-1402,1571-1425,1571-1620,1571-1625,KSP1023
1571-1401	1571-1402,1571-1425,1571-1601,1571-1602,1571-1620,1571-1625,1571-1801, 1571-1802,1571-1820,1571-1825,KSP1023
1571-1402	1571-1401,1571-1425,1571-1601,1571-1602,1571-1620,1571-1625,1571-1801, 1571-1802,1571-1820,1571-1825,KSP1023
1571-1425	1571-1201,1571-1202,1571-1401,1571-1402,1571-1601,1571-1602,1571-1620, 1571-1625,KSP1023
1571-1601	1571-1602,1571-1620,1571-1625,1571-1801,1571-1802,1571-1820,1571-1825, 1571-2001,1571-2020,1571-2025,KSP1071
1571-1602	1571-1601,1571-1620,1571-1625,1571-1801,1571-1802,1571-1820,1571-1825, 1571-2001,1571-2020,1571-2025,KSP1071
1571-1620	1571-1401,1571-1402,1571-1601,1571-1602,1571-1625,1571-1801,1571-1802, 1571-1820,1571-1825,1571-2001,1571-2020,1571-2025
1571-1625	1571-1401,1571-1402,1571-1601,1571-1602,1571-1620,1571-1801,1571-1802, 1571-1820,1571-1825,1571-2001,1571-2020,1571-2025
1571-1801	1571-1802,1571-1820,1571-1825,1571-2001,1571-2020,1571-2025,KSP1071,KSP1072
1571-1802	1571-1801,1571-1820,1571-1825,1571-2001,1571-2020,1571-2025,KSP1071,KSP1072
1571-1820	1571-1801,1571-1802,1571-1825,1571-2001,1571-2020,1571-2025,KSP1071,KSP1072
1571-1825	1571-1801,1571-1802,1571-1820,1571-2001,1571-2020,1571-2025,KSP1071,KSP1072
1571-2001	1571-2020,1571-2025,KSP1071,KSP1072,KSP1073
1571-2020	1571-2001,1571-2025,KSP1071,KSP1072,KSP1073
1571-2025	1571-2001,1571-2020,KSP1071,KSP1072,KSP1073
1582-0403	2N3055A,2N3235,1561-0403,1561-0404,1582-0404,1582-0405,1582-0603,1582-0604, 1582-0605,HST9201,HST9205,MJ2801,*GE-77,*BDY20
1582-0404	2N3055A,2N3235,1561-0403,1561-0404,1582-0403,1582-0405,1582-0603,1582-0604,

To Replace	Substitute This Type
(1582-0404)	1582-0605,HST9201,HST9205,MJ2801,*PTC140,*GE-77,*BDY20
1582-0405	2N3055A,2N3235,1561-0403,1561-0404,1582-0403,1582-0404,1582-0603,1582-0604, 1582-0605,HST9201,HST9205,MJ2801,*PTC140,*GE-77,*BDY20
1582-0408	1582-0410,1582-0415,1582-0608,1582-0610,1582-0615,*PTC175,*GE-75
1582-0410	1582-0408,1582-0415,1582-0608,1582-0610,1582-0615,*PTC175,*GE-75
1582-0415	1582-0408,1582-0410,1582-0608,1582-0610,1582-0615,*PTC175,*GE-75
1582-0508	1582-0408,1582-0410,1582-0415,1582-0608,1582-0615,*PTC175,*GE-75
1582-0510	1582-0408,1582-0410,1582-0415,1582-0608,1582-0610,1582-0615,*PTC175,*GE-75
1582-0603	2N3055,2N3235,2N3236,2N3237,2SD55,1561-0803,1561-0804,1561-0805,1582-0604, 1582-0605,1582-0803,1582-0804,1582-0805,HST9202,HST9206,HST9207,IR3772, MJ3772,*PTC140,*GE-77,*BDY20
1582-0604	2N3055,2N3235,2N3236,2N3237,2SD55,1561-0803,1561-0804,1561-0805,1582-0603, 1582-0605,1582-0803,1582-0804,1582-0805,HST9202,HST9206,HST9207,IR3772, MJ3772,*PTC140,*GE-77,*BDY20
1582-0605	2N3055,2N3235,2N3236,2N3237,2SD55,1561-0803,1561-0804,1561-0805,1582-0603, 1582-0604,1582-0803,1582-0804,1582-0805,HST9202,HST9206,HST9207,IR3772, MJ3772,*PTC140,*GE-77,*BDY20
1582-0608	1582-0610,1582-0615,1582-0808,1582-0815,*PTC175,*GE-75
1582-0610	1582-0808,1582-0810,1582-0815,1582-1008,1582-1010,1582-1015,*PTC175,*GE-75
1582-0615	1582-0608,1582-0610,1582-0808,1582-0815,*PTC175,*GE-75
1582-0803	2N3055,2N3236,2N3237,2N5630,2SD55,108T2,1561-0803,1561-0804,1561-0805, 1561-1005,1582-0804,1582-0805,1582-1003,1582-1004,1582-1005,HST9202, HST9203,HST9207,HST9208,IR3772,MJ3772,*PTC140,*GE-75
1582-0804	2N3055,2N3236,2N3237,2N5630,2SD55,108T2,1561-0803,1561-0804,1561-0805, 1561-1005,1582-0803,1582-0805,1582-1003,1582-1004,1582-1005,HST9202, HST9203,HST9207,HST9208,IR3772,MJ3772,*PTC140,*GE-75
1582-0805	2N3055,2N3236,2N3237,2N5630,2SD55,108T2,1561-0803,1561-0804,1561-0805, 1561-1005,1582-0803,1582-0804,1582-1003,1582-1004,1582-1005,HST9202, HST9203,HST9207,HST9208,IR3772,MJ3772,*PTC140,*GE-75
1582-0808	1582-0810,1582-0815,1582-1008,1582-1010,1582-1015,*PTC175,*GE-75
1582-0810	108T2,1582-1008,1582-1010,1582-1015,1582-1208,1582-1210,1582-1215,*PTC175, *GE-75
1582-0815	1582-0808,1582-0810,1582-1008,1582-1010,1582-1015,*PTC175,*GE-75
1582-1003	2N3055,2N3236,2N5630,2N5631,2SD55,108T2,1561-1005,1561-1205,1561-1404, 1561-1405,1582-1004,1582-1005,1582-1203,1582-1204,1582-1205,1582-1403, 1582-1404,1582-1405,HST9202,HST9203,HST9204,HST9207,HST9208,HST9209, IR3772,IR6302,MJ3772,*PTC173,*GE-75
1582-1004	2N3055,2N3236,2N5630,2N5631,2SD55,108T2,1561-1005,1561-1205,1561-1404, 1561-1405,1582-1003,1582-1005,1582-1203,1582-1204,1582-1205,1582-1403, 1582-1404,1582-1405,HST9202,HST9203,HST9204,HST9207,HST9208,HST9209, IR3772,IR6302,MJ3772,*PTC173,*GE-75
1582-1005	2N3055,2N3236,2N5630,2N5631,2SD55,108T2,1561-1005,1561-1205,1561-1404, 1561-1405,1582-1003,1582-1004,1582-1203,1582-1204,1582-1205,1582-1403, 1582-1404,1582-1405,HST9202,HST9203,HST9204,HST9207,HST9208,HST9209, IR3772,IR6302,MJ3772,*PTC173,*GE-75
1582-1008	108T2,1582-1010,1582-1015,1582-1208,1582-1210,1582-1215,*PTC175,*GE-75
1582-1010	108T2,1582-1008,1582-1015,1582-1208,1582-1210,1582-1215,*PTC175,*GE-75
1582-1015	108T2,1582-1008,1582-1010,1582-1208,1582-1210,1582-1215,*PTC175,*GE-75
1582-1203	2N5630,2N5631,108T2,109T2,1561-1205,1561-1404,1561-1405,1561-1604,1561-1605, 1582-1204,1582-1205,1582-1403,1582-1404,1582-1405,1582-1603,1582-1604, 1582-1605,HST9203,HST9204,HST9208,HST9209,IR3773,IR6302
1582-1204	2N5630,2N5631,108T2,109T2,1561-1205,1561-1404,1561-1405,1561-1604,1561-1605, 1582-1203,1582-1205,1582-1403,1582-1404,1582-1405,1582-1603,1582-1604, 1582-1605,HST9203,HST9204,HST9208,HST9209,IR3773,IR6302
1582-1205	2N5630,2N5631,108T2,109T2,1561-1205,1561-1404,1561-1405,1561-1604,1561-1605, 1582-1203,1582-1204,1582-1403,1582-1404,1582-1405,1582-1603,1582-1604, 1582-1605,HST9203,HST9204,HST9208,HST9209,IR3773,IR6302
1582-1208	108T2,109T2,1582-1210,1582-1215,1582-1408,1582-1410,1582-1415,1582-1608, 1582-1610,1582-1615,*PTC175
1582-1210	108T2,109T2,1582-1208,1582-1215,1582-1408,1582-1410,1582-1415,1582-1608, 1582-1610,1582-1615,*PTC175

To Replace	Substitute This Type
1582-1215	108T2,109T2,1582-1208,1582-1210,1582-1408,1582-1410,1582-1415,1582-1608, 1582-1610,1582-1615,*PTC175
1582-1403	2N5631,109T2,1561-1404,1561-1405,1561-1604,1561-1605,1561-1803,1561-1804, 1561-1805,1582-1404,1582-1405,1582-1603,1582-1604,1582-1605,1582-1803, 1582-1804,1582-1805,HST9204,HST9209,IR3773,IR6302
1582-1404	2N5631,109T2,1561-1404,1561-1405,1561-1604,1561-1605,1561-1803,1561-1804, 1561-1805,1582-1403,1582-1405,1582-1603,1582-1604,1582-1605,1582-1803, 1582-1804,1582-1805,HST9204,HST9209,IR3773,IR6302
1582-1405	2N5631,109T2,1561-1404,1561-1405,1561-1604,1561-1605,1561-1803,1561-1804, 1561-1805,1582-1403,1582-1404,1582-1603,1582-1604,1582-1605,1582-1803, 1582-1804,1582-1805,HST9204,HST9209,IR3773,IR6302
1582-1408	109T2,1582-1410,1582-1415,1582-1608,1582-1610,1582-1615,1582-1808,1582-1810, 1582-1815
1582-1410	109T2,1582-1408,1582-1415,1582-1608,1582-1610,1582-1615
1582-1415	109T2,1582-1408,1582-1410,1582-1608,1582-1610,1582-1615
1582-1603	109T2,1561-1604,1561-1605,1561-1803,1561-1804,1561-1805,1561-2003,1561-2004, 1582-1604,1582-1605,1582-1803,1582-1804,1582-1805,1582-2003,1582-2004, 1843-2005,IR3773
1582-1604	109T2,1561-1604,1561-1605,1561-1803,1561-1804,1561-1805,1561-2003,1561-2004, 1582-1603,1582-1605,1582-1803,1582-1804,1582-1805,1582-2003,1582-2004, 1843-2005,IR3773
1582-1605	109T2,1561-1604,1561-1605,1561-1803,1561-1804,1561-1805,1561-2003,1561-2004, 1582-1603,1582-1604,1582-1803,1582-1804,1582-1805,1582-2003,1582-2004, 1843-2005,IR3773
1582-1608	109T2,1582-1408,1582-1610,1582-1615,1582-1808,1582-1810,1582-1815,1582-2008, 1582-2010
1582-1610	109T2,1582-1408,1582-1608,1582-1615,1582-1808,1582-1810,1582-1815,1582-2008, 1582-2010
1582-1615	109T2,1582-1408,1582-1608,1582-1610,1582-1808,1582-1810,1582-1815,1582-2008, 1582-2010
1582-1803	1561-1803,1561-1804,1561-1805,1561-2003,1561-2004,1582-1804,1582-1805, 1582-2003,1582-2004,1843-2005,1843-2205
1582-1804	1561-1803,1561-1804,1561-1805,1561-2003,1561-2004,1582-1803,1582-1805, 1582-2003,1582-2004,1843-2005,1843-2205
1582-1805	1561-1803,1561-1804,1561-1805,1561-2003,1561-2004,1582-1803,1582-1804, 1582-2003,1582-2004,1843-2005,1843-2205
1582-1808	1582-1810,1582-1815,1582-2008,1582-2010,1843-2005,1843-2010,1843-2205, 1843-2210
1582-1810	1582-1808,1582-1815,1582-2008,1582-2010,1843-2005,1843-2010,1843-2205, 1843-2210,*PTC175
1582-1815	1582-1808,1582-1810,1582-2008,1582-2010,1843-2005,1843-2010,1843-2205, 1843-2210
1582-2003	1561-2003,1561-2004,1582-2004,1843-2005,1843-2205,1843-2505
1582-2004	1561-2003,1561-2004,1582-2003,1843-2005,1843-2205,1843-2505
1582-2008	1582-2010,1843-2005,1843-2010,1843-2205,1843-2210,1843-2505,1843-2510
1582-2010	1582-2008,1843-2005,1843-2010,1843-2205,1843-2210,1843-2505,1843-2510
1714-0402	1714-0405,1714-0602,1714-0605,*GE-66,*TR-81,*SK3026,*ECG175,*WEP241
1714-0405	1714-0402,1714-0602,1714-0605,*GE-66,*SK3026
1714-0602	1714-0605,1714-0802,1714-0805,KSP1161,KSP1162,*GE-66,*TR-81,*SK3026, *ECG175,*WEP241
1714-0605	1714-0602,1714-0802,1714-0805,KSP1161,KSP1162,*GE-66,*TR-81,*SK3026, *ECG175,*WEP241
1714-0802	1714-0805,1714-1002,1714-1005,KSP1162,*TR-81,*SK3026,*ECG175,*WEP241
1714-0805	1714-0802,1714-1002,1714-1005,KSP1162,*TR-81,*SK3026,*ECG175,*WEP241
1714-1002	1714-1005,1714-1202,1714-1205,1714-1402,1714-1405,KSP1162,KSP1163,*TR-81, *SK3026,*ECG175,*WEP241
1714-1005	1714-1002,1714-1202,1714-1205,1714-1402,1714-1405,KSP1162,KSP1163,*TR-81, *SK3026,*ECG175,*WEP241
1714-1202	1714-1205,1714-1402,1714-1405,1714-1602,1714-1605,KSP1163
1714-1205	1714-1202,1714-1402,1714-1405,1714-1602,1714-1605,KSP1163
1714-1402	1714-1405,1714-1602,1714-1605,1714-1802,1714-1805,KSP1163

To Replace	Substitute This Type
1714-1405	1714-1402,1714-1602,1714-1605,1714-1802,1714-1805,KSP1163
1714-1602	1714-1605,1714-1802,1714-1805,HST7901,KSP1121,SDT7901
1714-1605	1714-1602,1714-1802,1714-1805,HST7901,KSP1121,SDT7901
1714-1802	1714-1805,HST7901,HST7902,KSP1121,KSP1122,SDT7901,SDT7902
1714-1805	1714-1802,HST7901,HST7902,KSP1121,KSP1122,SDT7901,SDT7902
1716-0602	1716-0605,1716-0802,1716-0805,B148004,HST9901,HST9902,KSP1151,*HEPS5004
1716-0605	1716-0602,1716-0802,1716-0805,B148004,HST9901,HST9902,KSP1151,*HEPS5004
1716-0802	1716-0805,1716-1002,1716-1005,B148000,B148004,HST9902,HST9903,KSP1151, KSP1152,*HEPS5004
1716-0805	1716-0802,1716-1002,1716-1005,B148000,B148004,HST9902,HST9903,KSP1151, KSP1152,*HEPS5004
1716-1002	1716-1005,1716-1202,1716-1205,B148000,B148001,HST9903,HST9904,KSP1152, *HEPS5004
1716-1005	1716-1002,1716-1202,1716-1205,B148000,B148001,HST9903,HST9904,KSP1152, *HEPS5004
1716-1202	1716-1205,1716-1402,1716-1405,1716-1602,1716-1605,B148001,HST7150,HST9904, KSP1153
1716-1205	1716-1202,1716-1402,1716-1405,1716-1602,1716-1605,B148001,HST7150,HST9904, KSP1153
1716-1402	1716-1405,1716-1602,1716-1605,1716-1802,HST7150,HST7151,KSP1153
1716-1405	1716-1402,1716-1602,1716-1605,1716-1802,HST7150,HST7151,KSP1153
1716-1602	1716-1605,1716-1802,HST7151
1716-1605	1716-1602,1716-1802,HST7151
1716-1802	HST7151,HST7152,KSP1101
1718-0602	1718-0605,1718-0802,1718-0805
1718-0605	1718-0602,1718-0802,1718-0805
1718-0802	1718-0805,1718-1002,1718-1005
1718-0805	1718-0802,1718-1002,1718-1005
1718-1002	1718-1005,1718-1202,1718-1205
1718-1005	1718-1002,1718-1202,1718-1205
1718-1202	1718-1205,1718-1402,1718-1405,1718-1602,1718-1605
1718-1205	1718-1202,1718-1402,1718-1405,1718-1602,1718-1605
1718-1402	1718-1405,1718-1602,1718-1605,1718-1802
1718-1405	1718-1402,1718-1602,1718-1605,1718-1802
1718-1602	1718-1605,1718-1802
1718-1605	1718-1602,1718-1802
1718-1802	
1723-0405	2N5932,1723-0410,1723-0605,1723-0610,1743-0610,1743-0630,1763-0610,1763-0620, 1763-0630,HST9201,HST9205,HST9801,*TR-36,*SK3036,*ECG181,*WEPS7000, *RT-149
1723-0410	2N5932,1723-0405,1723-0605,1723-0610,1743-0610,1743-0630,1763-0610,1763-0620, 1763-0630,HST9201,HST9205,HST9801,*TR-36,*SK3036,*ECG181,*WEPS7000, *RT-149
1723-0605	2N5929,2N5932,2N5935,1723-0610,1723-0805,1723-0810,1743-0610,1743-0630, 1743-0820,1763-0810,1763-0820,1763-0830,1763-1010,1763-1020,1763-1030, HST9202,HST9206,HST9207,HST9802,HST9803,KSP1271,KSP1272,*TR-36, *SK3036,*ECG181,*WEPS7000,*RT-149
1723-0610	2N5929,2N5932,2N5935,1723-0605,1723-0805,1723-0810,1743-0610,1743-0630, 1743-0820,1763-0810,1763-0820,1763-0830,1763-1010,1763-1020,1763-1030, HST9202,HST9206,HST9207,HST9802,HST9803,KSP1271,KSP1272,*TR-36, *SK3036,*ECG181,*WEPS7000
1723-0805	2N5671,2N5929,2N5933,2N5935,1723-0810,1723-1005,1723-1010,1743-0820, 1743-1010,1743-1030,1763-1010,1763-1020,1763-1030,1763-1210,1763-1220, 1763-1230,HST9202,HST9203,HST9207,HST9208,HST9803,HST9804,KSP1272, *TR-36,*SK3036,*ECG181,*WEPS7000,*RT-149
1723-0810	2N5671,2N5929,2N5933,2N5935,1723-0805,1723-1005,1723-1010,1743-0820, 1743-1010,1743-1030,1763-1010,1763-1020,1763-1030,1763-1210,1763-1220, 1763-1230,HST9202,HST9203,HST9207,HST9208,HST9803,HST9804,KSP1272, *TR-36,*SK3036,*ECG181,*WEPS7000,*RT-149
1723-1005	2N5671,2N5672,2N5930,2N5933,2N5934,2N5936,2N6033,1723-1010,1723-1205, 1723-1210,1723-1405,1723-1410,1743-1010,1743-1030,1743-1220,1743-1410,

To Replace	Substitute This Type
(1723-1005)	1743-1430,1763-1010,1763-1020,1763-1030,1763-1210,1763-1220,1763-1230,1763-1410,1763-1420,1763-1430,HST9202,HST9203,HST9204,HST9207,HST9208,HST9209,HST9803,HST9804,KSP1272,KSP1273,*TR-36,*SK3036,*ECG181,*WEPS7000,*RT-149
1723-1010	2N5671,2N5672,2N5930,2N5933,2N5934,2N5936,2N6033,1723-1005,1723-1205,1723-1210,1723-1405,1723-1410,1743-1010,1743-1030,1743-1220,1743-1410,1743-1430,1763-1010,1763-1020,1763-1030,1763-1210,1763-1220,1763-1230,1763-1410,1763-1420,1763-1430,HST9202,HST9203,HST9204,HST9207,HST9208,HST9209,HST9803,HST9804,KSP1272,KSP1273,*TR-36,*SK3036,*ECG181,*WEPS7000,*RT-149
1723-1205	2N5671,2N5672,2N5930,2N5931,2N5934,2N5936,2N5937,2N6033,1723-1210,1723-1405,1723-1410,1723-1605,1723-1610,1743-1220,1743-1410,1743-1430,1743-1620,1763-1210,1763-1220,1763-1230,1763-1410,1763-1420,1763-1430,1763-1610,1763-1620,1763-1630,HST9203,HST9204,HST9208,HST9209,HST9804,KSP1273,*TR-36,*SK3036,*ECG181,*WEPS7000,*RT-149
1723-1210	2N5671,2N5672,2N5930,2N5931,2N5934,2N5936,2N5937,2N6033,1723-1205,1723-1405,1723-1410,1723-1605,1723-1610,1743-1220,1743-1410,1743-1430,1743-1620,1763-1210,1763-1220,1763-1230,1763-1410,1763-1420,1763-1430,1763-1610,1763-1620,1763-1630,HST9203,HST9204,HST9208,HST9209,HST9804,KSP1273,*TR-36,*SK3036,*ECG181,*WEPS7000,*RT-149
1723-1405	2N5672,2N5931,2N5934,2N5937,2N6033,1723-1410,1723-1605,1723-1610,1723-1805,1723-1810,1743-1410,1743-1430,1743-1620,1743-1810,1743-1820,1743-1830,1763-1410,1763-1420,1763-1430,1763-1610,1763-1620,1763-1630,1763-1810,HST9204,HST9209,KSP1273,*TR-36,*SK3036,*ECG181,*WEPS7000,*RT-149
1723-1410	2N5672,2N5931,2N5934,2N5937,2N6033,1723-1405,1723-1605,1723-1610,1723-1805,1723-1810,1743-1410,1743-1430,1743-1620,1743-1810,1743-1820,1743-1830,1763-1410,1763-1420,1763-1430,1763-1610,1763-1620,1763-1630,1763-1810,HST9204,HST9209,KSP1273,*TR-36,*SK3036,*ECG181,*WEPS7000,*RT-149
1723-1605	2N5931,2N5937,1723-1610,1723-1805,1723-1810,1743-1620,1743-1810,1743-1820,1743-1830,1763-1610,1763-1620,1763-1630,1763-1810,KSP1221,*TR-36,*SK3036,*ECG181,*WEPS7000,*RT-149
1723-1610	2N5931,2N5937,1723-1605,1723-1805,1723-1810,1743-1620,1743-1810,1743-1820,1743-1830,1763-1610,1763-1620,1763-1630,1763-1810,KSP1221,*TR-36,*SK3036,*ECG181,*WEPS7000,*RT-149
1723-1805	1723-1810,1743-1810,1743-1820,1743-1830,1763-1810,KSP1221,KSP1222,XGSR15020,*TR-36,*SK3036,*ECG181,*WEPS7000,*RT-149
1723-1810	1723-1805,1743-1810,1743-1820,1743-1830,1763-1810,KSP1221,KSP1222,XGSR15020,*TR-36,*SK3036,*ECG181,*WEPS7000,*RT-149
1723-2005	
1743-0610	2N5929,2N5932,2N5935,2N6270,2N6272,1743-0630,1743-0820,HST9202
1743-0630	2N5929,2N5932,2N5935,2N6270,2N6272,1743-0610,1743-0820,HST9202
1743-0820	2N5929,2N5933,2N5935,2N6270,2N6271,2N6272,2N6273,1743-1010,1743-1030,HST9202,HST9203
1743-1010	2N5930,2N5933,2N5934,2N5936,2N6270,2N6271,2N6272,2N6273,1743-1030,1743-1220,1743-1410,1743-1430,HST9202,HST9203,HST9204
1743-1030	2N5930,2N5933,2N5934,2N5936,2N6270,2N6271,2N6272,2N6273,1743-1010,1743-1220,1743-1410,1743-1430,HST9202,HST9203,HST9204
1743-1220	2N5930,2N5931,2N5934,2N5936,2N5937,2N6271,2N6273,1743-1410,1743-1430,1743-1620,HST9203,HST9204
1743-1410	2N5931,2N5934,2N5937,1743-1430,1743-1620,1743-1810,1743-1820,1743-1830,HST9204
1743-1430	2N5931,2N5934,2N5937,1743-1410,1743-1620,1743-1810,1743-1820,1743-1830,HST9204
1743-1620	2N5931,2N5937,1743-1810,1743-1820,1743-1830
1743-1810	1743-1820,1743-1830
1743-1820	1743-1810,1743-1830
1743-1830	1743-1810,1743-1820
1748-0610	1748-0630,1748-0810,1748-0820,1748-0830
1748-0630	1748-0610,1748-0810,1748-0820,1748-0830
1748-0810	1748-0820,1748-0830,1748-1010,1748-1030
1748-0820	1748-0810,1748-0830,1748-1010,1748-1030

To Replace	Substitute This Type
1748-0830	1748-0810,1748-0820,1748-1010,1748-1030
1748-1010	1748-1030,1748-1210,1748-1220,1748-1230,1748-1410,1748-1430
1748-1030	1748-1010,1748-1210,1748-1220,1748-1230,1748-1410,1748-1430
1748-1210	1748-1220,1748-1230,1748-1410,1748-1430,1748-1610,1748-1620,1748-1630
1748-1220	1748-1210,1748-1230,1748-1410,1748-1430,1748-1610,1748-1620,1748-1630
1748-1230	1748-1210,1748-1220,1748-1410,1748-1430,1748-1610,1748-1620,1748-1630
1748-1410	1748-1430,1748-1610,1748-1620,1748-1630,1748-1810,1748-1820,1748-1830
1748-1430	1748-1410,1748-1610,1748-1620,1748-1630,1748-1810,1748-1820,1748-1830
1748-1610	1748-1620,1748-1630,1748-1810,1748-1820,1748-1830
1748-1620	1748-1610,1748-1630,1748-1810,1748-1820,1748-1830
1748-1630	1748-1610,1748-1620,1748-1810,1748-1820,1748-1830
1748-1810	1748-1820,1748-1830
1748-1820	1748-1810,1748-1830
1748-1830	1748-1810,1748-1820
1756-0440	1756-0460,1756-0640,1756-0660,1776-0640,1776-0660
1756-0460	1756-0440,1756-0640,1756-0660,1776-0640,1776-0660
1756-0640	1756-0660,1756-0840,1756-0860,1776-0840,1776-0860,1776-1040,1776-1060
1756-0660	1756-0640,1756-0840,1756-0860,1776-0840,1776-0860,1776-1040,1776-1060
1756-0840	1756-0860,1756-1040,1756-1060,1776-1040,1776-1060,1776-1240,1776-1260
1756-0860	1756-0840,1756-1040,1756-1060,1776-1040,1776-1060,1776-1240,1776-1260
1756-1040	1756-1060,1756-1240,1756-1260,1756-1440,1756-1460,1776-1040,1776-1060, 1776-1240,1776-1260,1776-1440,1776-1460
1756-1060	1756-1040,1756-1240,1756-1260,1756-1440,1756-1460,1776-1040,1776-1060, 1776-1240,1776-1260,1776-1440,1776-1460
1756-1240	1756-1260,1756-1440,1756-1460,1756-1640,1756-1660,1776-1240,1776-1260, 1776-1440,1776-1460,1776-1640
1756-1260	1756-1240,1756-1440,1756-1460,1756-1640,1756-1660,1776-1240,1776-1260, 1776-1440,1776-1460,1776-1640
1756-1440	1756-1460,1756-1640,1756-1660,1776-1440,1776-1460,1776-1640
1756-1460	1756-1440,1756-1640,1756-1660,1776-1440,1776-1460,1776-1640
1756-1640	1756-1660,1776-1640
1756-1660	1756-1640,1776-1640
1761-04	HST9201,HST9205
1761-06	HST9206
1761-08	HST9202,HST9203,HST9206,HST9207,HST9208
1761-10	HST9202,HST9203,HST9204,HST9207,HST9208,HST9209
1761-12	HST9203,HST9204,HST9208,HST9209
1761-14	HST9204,HST9209
1761-16	
1763-0610	1763-0620,1763-0630,1763-0810,1763-0820,1763-0830,HST9201,HST9205,HST9206, HST9801,HST9802
1763-0620	1763-0610,1763-0630,1763-0810,1763-0820,1763-0830,HST9201,HST9205,HST9206, HST9801,HST9802,*TR-36,*SK3036,*ECG181,*WEPS7000,*RT-149
1763-0630	1763-0610,1763-0620,1763-0810,1763-0820,1763-0830,HST9201,HST9205,HST9206, HST9801,HST9802
1763-0810	1763-0820,1763-0830,1763-1010,1763-1020,1763-1030,HST9202,HST9206,HST9207, HST9802,HST9803
1763-0820	1763-0810,1763-0830,1763-1010,1763-1020,1763-1030,HST9202,HST9206,HST9207, HST9802,HST9803,*SK3036,*RT-149
1763-0830	1763-0810,1763-0820,1763-1010,1763-1020,1763-1030,HST9202,HST9206,HST9207, HST9802,HST9803
1763-1010	2N6032,1763-1020,1763-1030,1763-1210,1763-1220,1763-1230,HST9202,HST9203, HST9207,HST9208,HST9803,HST9804
1763-1020	2N6032,1763-1010,1763-1030,1763-1210,1763-1220,1763-1230,HST9202,HST9203, HST9207,HST9208,HST9803,HST9804,*TR-36,*SK3036,*ECG181,*WEPS7000, *RT-149
1763-1030	2N6032,1763-1010,1763-1020,1763-1210,1763-1220,1763-1230,HST9202,HST9203, HST9207,HST9208,HST9803,HST9804
1763-1210	2N6032,2N6033,1763-1220,1763-1230,1763-1410,1763-1420,1763-1430,1763-1610, 1763-1620,1763-1630,HST9203,HST9204,HST9208,HST9209,HST9804
1763-1220	2N6032,2N6033,1763-1210,1763-1230,1763-1410,1763-1420,1763-1430,1763-1610,

To Replace	Substitute This Type
(1763-1220)	1763-1620,1763-1630,HST9203,HST9204,HST9208,HST9209,HST9804,*TR-36, *SK3036,*ECG181,*WEPS7000,*RT-149
1763-1230	2N6032,2N6033,1763-1210,1763-1220,1763-1410,1763-1420,1763-1430,1763-1610, 1763-1620,1763-1630,HST9203,HST9204,HST9208,HST9209,HST9804
1763-1410	2N6033,1763-1420,1763-1430,1763-1610,1763-1620,1763-1630,1763-1810,HST9204, HST9209
1763-1420	2N6033,1763-1410,1763-1430,1763-1610,1763-1620,1763-1630,1763-1810,HST9204, HST9209,*TR-36,*SK3036,*ECG181,*WEPS7000,*RT-149
1763-1430	2N6033,1763-1410,1763-1420,1763-1610,1763-1620,1763-1630,1763-1810,HST9204, HST9209
1763-1610	2N6033,1763-1620,1763-1630,1763-1810
1763-1620	2N6033,1763-1610,1763-1630,1763-1810
1763-1630	2N6033,1763-1610,1763-1620,1763-1810
1763-1810	
1768-0610	1768-0620,1768-0630,1768-0810,1768-0820,1768-0830
1768-0620	1768-0610,1768-0630,1768-0810,1768-0820,1768-0830
1768-0630	1768-0610,1768-0620,1768-0810,1768-0820,1768-0830
1768-0810	1768-0820,1768-0830,1768-1010,1768-1020,1768-1030
1768-0820	1768-0810,1768-0830,1768-1010,1768-1020,1768-1030
1768-0830	1768-0810,1768-0820,1768-1010,1768-1020,1768-1030
1768-1010	1768-1020,1768-1030,1768-1210,1768-1220,1768-1230
1768-1020	1768-1010,1768-1030,1768-1210,1768-1220,1768-1230
1768-1030	1768-1010,1768-1020,1768-1210,1768-1220,1768-1230
1768-1210	1768-1220,1768-1230,1768-1410,1768-1420,1768-1430,1768-1610,1768-1620, 1768-1630
1768-1220	1768-1210,1768-1230,1768-1410,1768-1420,1768-1430,1768-1610,1768-1620, 1768-1630
1768-1230	1768-1210,1768-1220,1768-1410,1768-1420,1768-1430,1768-1610,1768-1620, 1768-1630
1768-1410	1768-1420,1768-1430,1768-1610,1768-1620,1768-1630,1768-1810
1768-1420	1768-1410,1768-1430,1768-1610,1768-1620,1768-1630,1768-1810
1768-1430	1768-1410,1768-1420,1768-1610,1768-1620,1768-1630,1768-1810
1768-1610	1768-1620,1768-1630,1768-1810
1768-1620	1768-1610,1768-1630,1768-1810
1768-1630	1768-1610,1768-1620,1768-1810
1768-1810	
1776-0440	1756-0440,1756-0460,1776-0460,1776-0640,1776-0660
1776-0460	1756-0440,1756-0460,1776-0440,1776-0640,1776-0660
1776-0640	1756-0640,1756-0660,1776-0660,1776-0840,1776-0860
1776-0660	1756-0640,1756-0660,1776-0640,1776-0840,1776-0860
1776-0840	1756-0840,1756-0860,1756-1040,1756-1060,1776-0860,1776-1040,1776-1060
1776-0860	1756-0840,1756-0860,1756-1040,1756-1060,1776-0840,1776-1040,1776-1060
1776-1040	1756-1040,1756-1060,1756-1240,1756-1260,1776-1060,1776-1240,1776-1260
1776-1060	1756-1040,1756-1060,1756-1240,1756-1260,1776-1040,1776-1240,1776-1260
1776-1240	1756-1040,1756-1060,1756-1240,1756-1260,1756-1440,1756-1460,1776-1260, 1776-1440,1776-1460,1776-1640
1776-1260	1756-1040,1756-1060,1756-1240,1756-1260,1756-1440,1756-1460,1776-1240, 1776-1440,1776-1460,1776-1640
1776-1440	1756-1240,1756-1260,1756-1440,1756-1460,1756-1640,1756-1660,1776-1460, 1776-1640
1776-1460	1756-1240,1756-1260,1756-1440,1756-1460,1756-1640,1756-1660,1776-1440, 1776-1640
1776-1640	1756-1440,1756-1460,1756-1640,1756-1660
1781-0450	1781-0470,1781-0650,1781-0670
1781-0470	1781-0450,1781-0650,1781-0670
1781-0490	1781-0450,1781-0470,1781-0650,1781-0670,1781-0690
1781-0650	1781-0670,1781-0850,1781-0870,KSP1001,KSP1002
1781-0670	1781-0650,1781-0850,1781-0870,KSP1001,KSP1002
1781-0690	1781-0650,1781-0670,1781-0850,1781-0870,1781-0890,KSP1001,KSP1002
1781-0850	1781-0870,1781-1050,1781-1070,KSP1002,KSP1003
1781-0870	1781-0850,1781-1050,1781-1070,KSP1002,KSP1003

To Replace	Substitute This Type
1781-0890	1781-0850,1781-0870,1781-1050,1781-1070,1781-1090,KSP1002,KSP1003
1781-1050	1781-1070,1781-1250,1781-1270,1781-1450,1781-1470,KSP1002,KSP1003
1781-1070	1781-1050,1781-1250,1781-1270,1781-1450,1781-1470,KSP1002,KSP1003
1781-1090	1781-1050,1781-1070,1781-1250,1781-1270,1781-1290,1781-1450,1781-1470,KSP1002, KSP1003
1781-1250	1781-1270,1781-1450,1781-1470,1781-1650,KSP1003
1781-1270	1781-1250,1781-1450,1781-1470,1781-1650,KSP1003
1781-1290	1781-1250,1781-1270,1781-1450,1781-1470,1781-1650,KSP1003
1781-1450	1781-1470,1781-1650
1781-1470	1781-1450,1781-1650
1781-1650	
1814-2001	KSP1121,KSP1122,KSP1123
1814-2002	KSP1121,KSP1122,KSP1123
1814-2005	KSP1121,KSP1122,KSP1123
1814-2201	KSP1122,KSP1123,KSP1124,KSP1125
1814-2202	KSP1122,KSP1123,KSP1124,KSP1125
1814-2205	KSP1122,KSP1123,KSP1124,KSP1125
1814-2501	KSP1122,KSP1123,KSP1124,KSP1125
1814-2502	KSP1122,KSP1123,KSP1124,KSP1125
1814-2505	KSP1122,KSP1123,KSP1124,KSP1125
1814-2701	KSP1123,KSP1124,KSP1125
1814-2702	KSP1123,KSP1124,KSP1125
1814-2705	KSP1123,KSP1124,KSP1125
1814-3001	KSP1124,KSP1125
1814-3002	KSP1124,KSP1125
1814-3005	KSP1124,KSP1125
1814-3201	KSP1125
1814-3202	KSP1125
1814-3205	KSP1125
1814-3501	
1814-3502	
1814-3505	
1814-3701	
1814-3702	
1814-3705	
1843-2005	1843-2205,1843-2505
1843-2010	1843-2005,1843-2205,1843-2210,1843-2505,1843-2510
1843-2020	1843-2010,1843-2210,1843-2220,1843-2510,1843-2520
1843-2205	1843-2505,1843-2705,1843-3005
1843-2210	1843-2205,1843-2505,1843-2510,1843-2705,1843-2710,1843-3005,1843-3010
1843-2220	1843-2210,1843-2510,1843-2520,1843-2710,1843-2720,1843-3010,1843-3020
1843-2505	1843-2205,1843-2705,1843-3005,1843-3205
1843-2510	1843-2205,1843-2210,1843-2505,1843-2705,1843-2710,1843-3005,1843-3010, 1843-3205,1843-3210
1843-2520	1843-2210,1843-2220,1843-2510,1843-2710,1843-2720,1843-3010,1843-3020, 1843-3210,1843-3220
1843-2705	1843-2505,1843-3005,1843-3205,1843-3505
1843-2710	1843-2505,1843-2510,1843-2705,1843-3005,1843-3010,1843-3205,1843-3210, 1843-3505,1843-3510
1843-2720	1843-2510,1843-2520,1843-2710,1843-3010,1843-3020,1843-3210,1843-3220, 1843-3510
1843-3005	1843-2705,1843-3205,1843-3505,1843-3705
1843-3010	1843-2705,1843-2710,1843-3005,1843-3205,1843-3210,1843-3505,1843-3510, 1843-3705
1843-3020	1843-2710,1843-2720,1843-3010,1843-3210,1843-3220,1843-3510
1843-3205	1843-3005,1843-3505,1843-3705
1843-3210	1843-3005,1843-3010,1843-3205,1843-3505,1843-3510,1843-3705
1843-3220	1843-3010,1843-3020,1843-3210,1843-3510
1843-3505	1843-3205,1843-3705
1843-3510	1843-3205,1843-3210,1843-3505,1843-3705
1843-3705	1843-3505

To Replace	Substitute This Type
1856-2030	1856-2040,1856-2230,1856-2240,1856-2530,1856-2540
1856-2040	1856-2030,1856-2230,1856-2240,1856-2530,1856-2540
1856-2230	1856-2240,1856-2530,1856-2540,1856-2730,1856-2740,1856-3030,1856-3040
1856-2240	1856-2230,1856-2530,1856-2540,1856-2730,1856-2740,1856-3030,1856-3040
1856-2530	1856-2230,1856-2240,1856-2540,1856-2730,1856-2740,1856-3030,1856-3040, 1856-3230,1856-3240
1856-2540	1856-2230,1856-2240,1856-2530,1856-2730,1856-2740,1856-3030,1856-3040, 1856-3230,1856-3240
1856-2730	1856-2530,1856-2540,1856-2740,1856-3030,1856-3040,1856-3230,1856-3240, 1856-3530
1856-2740	1856-2530,1856-2540,1856-2730,1856-3030,1856-3040,1856-3230,1856-3240, 1856-3530
1856-3030	1856-2730,1856-2740,1856-3040,1856-3230,1856-3240,1856-3530
1856-3040	1856-2730,1856-2740,1856-3030,1856-3230,1856-3240,1856-3530
1856-3230	1856-3030,1856-3040,1856-3240,1856-3530
1856-3240	1856-3030,1856-3040,1856-3230,1856-3530
1856-3530	1856-3230,1856-3240
3907	SEE 2N404A
35821B	2SC987,2SC1268,2SC1269,2SC1270,2SC1271,35821E
35821E	2SC987,2SC1268,2SC1269,2SC1270,2SC1271,35821B
35822E	
35824A	
35826E	
35827B	2SC1200,35827E,40836,40837,40909†
35827E	2SC1200,35827B,40836,40837,40909†
35831B OPT005	35831E
35831E OPT005	35831B
35832B OPT005	2SC976,2SC977,2SC978,35832E
35832E OPT005	2SC976,2SC977,2SC978,35832B
35853B	35853E
35853E	35853B
35854B	35854E
35854E	35854B
35861E	2SC985,2SC985A,2SC987A,2SC1038,2SC1039,2SC1041,2SC1042,2SC1336, 2SC1559,35861E
35861E OPT100	2SC985,2SC985A,2SC987A,2SC1038,2SC1039,2SC1041,2SC1042,2SC1336, 2SC1559,35861E
35862E	2SC1743,35862E,S2509,S2510
35862E OPT100	2SC1743,35862E,S2509,S2510
35866E	35866E
35866E OPT100	35866E
35867B	2SC1200,35867E
35867E	2SC1200,35867B
40022	2SB149-N,40050,40051,40254,40462,40623,*PTC138,*GE-25,*TR-01,*HEPG6013, *SK3009,*PT40,*ECG104,*WEP230,*276-2006,*RT-124
40050	2N1138,40051,40462,40623,40626,NKT402,NKT404,NKT405,*PTC138,*GE-25,*TR-01, *HEPG6003,*SK3009,*PT40,*ECG104,*WEP230,*276-2006,*RT-124
40051	2N1138,40626,NKT402,NKT404,NKT405,*PTC138,*GE-25,*TR-01,*HEPG6005, *SK3009,*PT40,*ECG121,*WEP232,*276-2006,*RT-127
40053	SEE 2N3053
40080	2N749,2N751,2N1388,2N1390,2SC302-M,2SC307-M,2SC590N,2SC773,2SC774, *PTC123,*GE-20,*TR-87,*HEPS0014,*SK3046,*SN80,*ECG128,*WEP243,*BC337, *276-2016,*RT-114
40081	2N749,2N751,2N1388,2N1390,2SC302-M,2SC307-M,2SC590N,2SC773,2SC774, *DS-81,*GE-215,*TR-87,*HEPS5026,*SK3048,*SN80,*ECG195A,*WEP243,*RT-114
40082	2N749,2N751,2N1388,2N1390,2SC302-M,2SC307-M,2SC590N,2SC773,2SC774, 2SC798,2SC803,2SC865,2SC1002,2SC1256,40581,40975,*GE-28,*TR-65, *HEPS3001,*SK3048,*ECG195A,*276-2014
40084	*TR-21,*HEPS0011,*SK3122,*ECG123A,*WEP736,*276-2009,*RT-102
40217	SEE 2N3261
40218	SEE 2N3261

To Replace	Substitute This Type
40219	SEE 2N3261
40220	SEE 2N3261
40221	SEE 2N3261
40222	SEE 2N3261
40231	2N2331,2SC509-O,2SC509G-O,MPS6560,MPS6561,*PTC123,*GE-17,*TR-83, *HEPS0011,*SK3124,*SN80,*ECG161,*WEP719,*BC337,*276-2009,*RT-113
40232	2SC509-Y,2SC509G-Y,40233,A5T4124,BC548A,TN80,*PTC123,*GE-20,*TR-83, *HEPS0011,*SK3122,*EN10,*ECG123A,*WEP719,*BC337,*276-2009,*RT-113
40233	2SC509-Y,2SC509G-Y,40232,A5T4124,BC548A,TN80,*PTC123,*GE-20,*TR-83, *HEPS0011,*SK3124,*EN10,*ECG161,*WEP719,*BC337,*276-2009,*RT-113
40234	2N2331,2SC509-O,2SC509G-O,40231,MPS6560,MPS6561,*PTC123,*GE-210,*TR-83, *HEPS0011,*SK3018,*EN10,*ECG161,*WEP719,*BC337,*276-2009,*RT-113
40235	40236,40237,*PTC133,*GE-17,*TR-83,*HEPS0016,*SK3018,*EN10,*ECG161, *WEP719,*276-2011,*RT-107
40236	40237,*PTC133,*GE-17,*TR-83,*HEPS0016,*SK3018,*EN10,*ECG161,*WEP719, *276-2015,*RT-113
40237	40236,*PTC133,*GE-17,*TR-83,*HEPS0016,*SK3018,*EN10,*ECG161,*WEP719, *276-2015,*RT-113
40238	40235,40236,40237,40240,*PTC132,*GE-39,*TR-83,*HEPS0017,*SK3117,*EN10, *ECG161,*WEP719,*276-2011,*RT-113
40239	40235,40238,*PTC132,*GE-39,*HEPS0016,*SK3117,*EN10,*276-2015,*RT-113
40240	40236,40237,*PTC132,*GE-39,*TR-83,*HEPS0017,*SK3117,*EN10,*ECG161, *WEP719,*276-2011,*RT-113
40242	40235,40236,40237,40238,40240,*PTC132,*GE-39,*TR-83,*HEPS0015,*SK3122, *EN10,*ECG161,*WEP719,*276-2016,*RT-113
40243	40235,40236,40237,40238,40240,40242,*PTC123,*GE-39,*TR-83,*HEPS0016, *SK3122,*EN10,*ECG161,*WEP719,*BC107B,*276-2016,*RT-113
40244	40235,40238,40239,40242,*PTC123,*GE-39,*TR-83,*HEPS0016,*SK3018,*EN10, *ECG161,*WEP719,*BC107B,*276-2016,*RT-113
40245	40236,40237,40240,*PTC123,*GE-39,*TR-83,*HEPS0015,*SK3018,*EN10,*ECG161, *WEP719,*BC107B,*276-2016,*RT-113
40246	40239,40244,*PTC123,*GE-39,*TR-83,*HEPS0015,*SK3122,*EN10,*ECG161, *WEP719,*BC107B,*276-2016,*RT-113
40250	2N4231,2N4232,2N6373,2N6374,BD148-6,*PTC112,*GE-66,*TR-81,*HEPS5012, *SK3026,*PN66,*ECG175,*WEP241,*276-2013
40250V1	*GE-28,*TR-76,*HEPS5000,*SK3026,*ECG175,*276-2013
40251	SEE 2N6371
40253	2SB431,NKT211,*PTC135,*GE-53,*TR-85,*HEPG0005,*SK3004,*ECG102A,*WEP250, *276-2004,*RT-121
40254	2SB149-N,40022,40050,40051,40462,40623,*PTC138,*GE-25,*TR-01,*HEPG6013, *SK3009,*ECG104,*WEP230,*276-2006,*RT-124
40261	2N2190,2N2191,*DS-56,*GE-51,*TR-17,*HEPS0030,*HF35,*ECG126,*WEP635, *276-2013,*RT-113
40262	40489,*DS-56,*GE-51,*TR-17,*HEPG0009,*SK3008,*HF35,*ECG126,*WEP635
40263	40490,*PTC145,*GE-50,*TR-17,*HEPG0005,*SK3004,*ECG160,*WEP635,*276-2007
40264	*GE-27,*TR-81,*HEPS3021,*SK3021,*ECG124,*WEP240,*RT-128
40268	2N838,2N2956,2N6365,2N6365A,*TR-17,*HEPG0003,*ECG160,*WEP637
40269	2N417,2N598,2N599,2N1307,2N1309,2N1354,2N1355,2N1356,2N1357,2N1808, 2N1892,2N1998,ASY27,SF.T223,*PTC109,*GE-1,*TR-05,*HEPG0005,*SK3005, *ECG100,*WEP254,*AC188/01,*276-2007,*RT-118
40279	2N3632,40306,40665,*GE-28,*TR-76
40280	2N4427,2N4874,2N5109,2N5421,2N5422,2SC302-M,2SC307-M,2SC590N,2SC730, 2SC731,2SC844,2SC872,2SC890,2SC908,2SC991,2SC992,2SC1001,2SC1002, 2SC1083,2SC1165,2SC1256,2SC1464,2SC1479,2SC1820,2SC1947,40975, SRF1001,*GE-20,*TR-21,*HEPS3013,*SN80,*276-2009,*RT-102
40281	2N3926,2N4932,2N5423,2N5424,2N5424A,2SC831,2SC1003,SRF12213,SRF52214, *GE-28,*TR-76
40282	2N3927,2N4932,2N5102,2N5424,2N5424A,2SC831,SRF52214,*GE-66,*TR-81, *ECG175,*WEP241
40283	*PTC136,*GE-210,*TR-21,*HEPS0014,*SN80,*ECG123A,*WEP736,*276-2009, *RT-102

To Replace	Substitute This Type
40290	2N751,2N3924,2N5422,2SC302-M,2SC307-M,2SC590N,2SC865,2SC992,2SC1002, 2SC1256,2SC1947,*GE-20,*TR-86,*HEPS3008,*SN80,*ECG123,*WEP53,*RT-100
40291	2N3926,2N4932,2N5423,2SC831,2SC1003,40281,SRF12213
40292	2N3926,2N3927,2N4932,2N5424,2N5424A,2SC831,2SC1003,40282,SRF12213, SRF52214,*GE-66
40294	2N2857,2N3570,2N3600,2N3839,2N4252,2N4253,2N6304,2N6305,2SC1747,40414, MT1061A,*PTC132,*GE-86,*TR-95,*HEPS0016,*EN10,*ECG108,*WEP56,*276-2015, *RT-113
40295	2SC1747,40413,*PTC132,*TR-95,*HEPS0016,*SK3039,*ECG108,*WEP56,*276-2038, *RT-113
40296	2N2857,2N3570,2N3600,2N3839,2N4252,2N4253,2N6304,2N6305,2SC1747,40294, 40414,MT1061A,*PTC132,*GE-86,*TR-83,*HEPS0017,*ECG161,*WEP719
40305	2N1709,2N2874,2SC302-M,2SC307-M,2SC547,2SC590N,SRF11101,*GE-28,*TR-76, *HEPS3001,*276-2014
40306	2N3632,40279,40665,*GE-28,*TR-76
40307	2N4933,2N5016,2SC551,SRF52101,*GE-66
40309	40311,40315,40323,40611,40616,*GE-63,*TR-87,*HEPS0014,*SK3024,*SN80, *ECG128,*WEP243,*276-2008,*RT-114
40310	2N4231,2N6374,40250,40316,40324,40621,40622,40624,40627,*PTC112,*GE-66, *TR-81,*HEPS5012,*SK3026,*ECG175,*WEP241,*276-2018
40311	40314,40315,40616,*GE-63,*TR-72,*HEPS0014,*SK3024,*SN80,*ECG188, *WEPS3020,*276-2008,*RT-100
40312	2N4232,2N5427,2N5491,2N5493,2N5495,2N6373,2SC1444,40364,40627,BD148-6, BD149-6,KSP1021,KSP1024,*PTC112,*GE-66,*TR-81,*HEPS5012,*SK3026, *ECG175,*WEP241,*276-2020
40313	40318,40322,*GE-18,*TR-81,*HEPS5011,*SK3021,*ECG124,*WEP240,*RT-128
40314	*GE-63,*TR-87,*HEPS0015,*SK3024,*SN80,*ECG128,*WEP243,*276-2008,*RT-114
40315	40314,40616,*GE-63,*TR-87,*HEPS0014,*SK3024,*SN80,*ECG128,*WEP243, *276-2008,*RT-114
40316	2N4231,2N5491,2N5495,2N6374,40250,40622,40624,40627,BD148-6,*PTC112, *GE-66,*TR-81,*HEPS5012,*SK3026,*ECG175,*WEP241,*276-2018
40317	2N2270,2N3053,2N3724A,40407,*PTC144,*GE-63,*TR-87,*HEPS0015,*SK3024, *ECG128,*WEP243,*276-2030,*RT-114
40318	40322,*TR-81,*HEPS5011,*SK3021,*ECG124,*WEP240,*RT-128
40319	2N3244,2N3467,2N3762,2N4037,2N4890,40406,40537,*PTC141,*GE-67,*TR-88, *HEPS5022,*SK3025,*ECG129,*WEP242,*276-2021,*RT-114
40320	2N2017,2N2108,2N2270,2N3053,2N3724A,7A32,40317,40326,40407,ZT90,ZT95, ZT2270,*PTC144,*GE-63,*TR-87,*HEPS0015,*SK3024,*SN80,*ECG128,*WEP243, *276-2008,*RT-114
40321	40327,*GE-32,*TR-78,*HEPS3021,*SK3044,*SN80,*ECG198,*WEP244,*276-2008, *RT-114
40322	40318,*GE-12,*TR-81,*HEPS5011,*SK3021,*SN80,*ECG124,*WEP240,*RT-128
40323	40309,40311,40315,40611,40616,*GE-63,*TR-87,*HEPS0014,*SK3024,*SN80, *ECG128,*WEP243,*276-2008,*RT-114
40324	2N4231,2N6374,40250,40310,40316,40621,40622,40624,40627,*PTC112,*GE-66, *TR-81,*HEPS5012,*SK3026,*ECG175,*WEP241,*276-2018
40325	2N3667,2N3771,2N5301,2N6371,2SD231,2SD249,1561-0403,1561-0404,1582-0403, 1582-0404,1582-0405,1582-0408,1582-0410,1582-0415,B170002-BLK,B170002-BRN, B170020,HST9201,HST9205,HST9210,IR3771,MJ2801,MJ3771,SDT9201,SDT9205, SDT9210,*PTC140,*GE-77,*TR-59,*HEPS7004,*SK3027,*ECG130,*WEP247, *RT-131
40326	2N2017,2N2108,2N2270,2N3053,2N3724A,7A32,40317,40320,40407,ZT90,ZT95, ZT2270,*PTC144,*GE-63,*TR-87,*HEPS0014,*SK3024,*SN80,*ECG128,*WEP243, *276-2008,*RT-114
40327	40321,*GE-32,*TR-87,*SK3044,*SN80,*ECG128,*WEP243,*276-2008,*RT-114
40328	*TR-81,*HEPS5011,*SK3021,*ECG124,*WEP240,*RT-128
40329	2N729,2N2501,2N2615,2N2923,2N3298,2N3393,2N3414,2N3845,2N3845A,2N3856A, 2N3858,2N3859,2N4074,2N4123,2N4138,2N4418,2N4873,2N5136,2N5137,2N6566, 2S512,2SC282,2SC350,2SC367G,2SC367G-O,2SC372-O,2SC372G-O,2SC394-Y, 2SC400-O,2SC400-Y,2SC454,2SC458,2SC459,2SC460,2SC536,2SC619,2SC633, 2SC634,2SC639,2SC712,2SC713,2SC733-O,2SC735-O,2SC752G-O,2SC1216,

To Replace	Substitute This Type
(40329)	2SC1276,2SC1372,2SC1373,2SD227,2SD327,40398,BSY61,GI-2923,GI-3393,GI-3402, MPS2923,MPS6513,MPS6574-BLUE,MPS6574-GREEN,MPS6574-YEL,MPSA10-GRN, MPSA10-WHT,MPSA20-GRN,MPSA20-WHT,MPSH10,MPSH11,PET3002,PET4001, SE2001,SE3005,ZT80,ZT110,ZT2369A,ZT2938,*PTC123,*GE-212,*TR-85, *HEPG0005,*SK3004,*ECG102A,*WEP250,*BC107B,*276-2031,*RT-121
40340	
40341	2SC1037
40342	2SC543,*HEPS5000,*276-2020
40343	2SC543,*HEPS5000,*276-2020
40346	*GE-32,*HEPS3021,*SK3044,*ECG198,*WEP244,*276-2008
40346V1	*TR-75,*HEPS3021,*SK3537,*ECG191,*WEPS3021,*276-2012
40346V2	*TR-78,*HEPS3021,*SK3045,*ECG225,*276-2012
40347	2N1613A,2N2106,*PTC144,*GE-63,*TR-87,*HEPS5014,*ECG128,*WEP243, *276-2008,*RT-114
40347V1	*GE-28,*TR-76,*HEPS3020,*SK3536,*276-2020
40347V2	2N2196,*GE-28,*TR-76,*HEPS3020,*ECG225,*276-2020
40348	2N3420,2N3421,2SD78,2SD78A,2SD378,*GE-32,*TR-87,*HEPS5014,*276-2008, *RT-114
40348V1	*HEPS3019,*SK3536,*276-2020
40348V2	2SD79,*HEPS3019,*ECG225,*276-2020
40349	*GE-32,*TR-21,*HEPS3021,*276-2008
40349V1	*HEPS3021,*SK3536
40349V2	*HEPS3021,*ECG225
40354	2N4068,TIS101,*PTC117,*GE-27,*TR-78,*HEPS5025,*SK3040,*ECG154,*WEP712, *BF338,*276-2008,*RT-110
40355	2N3500,2N4069,2N4925,MM2259,MM3009,*PTC117,*GE-27,*TR-78,*HEPS5025, *SK3040,*ECG154,*WEP712,*BF338,*276-2008,*RT-110
40359	2N827,*PTC126,*GE-2,*TR-05,*HEPG0005,*SK3004,*ECG102,*WEP631,*276-2007, *RT-120
40360	2N3725A,40408,40635,*PTC144,*GE-18,*TR-87,*HEPS3020,*SK3024,*ECG128, *WEP243,*RT-114
40361	*GE-18,*TR-87,*HEPS5014,*SK3024,*ECG128,*WEP243,*276-2018,*RT-114
40362	2N4404,2N5865,MPS4356,*PTC141,*TR-88,*HEPS5022,*SK3025,*ECG129,*WEP242, *276-2021,*RT-115
40363	2N3055,2N3055SD,2N3235,2N3236,2N3772,2N5303,2N5629,2N5882,2N5886,2N6254, 2N6270,2N6272,2SD341,2SD492,1561-0803,1561-0804,1561-0805,1582-0603, 1582-0604,1582-0605,1582-0803,1582-0804,1582-0805,40411,B170005-BLK, B170005-BRN,B170008-BLK,B170008-BRN,BUY53A,BUY54A,HST9202,HST9206, HST9207,IR802,IR3772,MJ802,MJ3772,SDT9202,SDT9206,SDT9207,STC2220, STC2221,STC2224,STC2225,*PTC140,*GE-75,*TR-59,*HEPS7004,*SK3027, *ECG130,*WEP247,*BDY20,*RT-131
40364	2N5428,KSP1164,*TR-81,*HEPS5019,*SK3026,*ECG175,*WEP241,*276-2017
40366	2N1613B,2N2102,2N2102A,2N4001,40349,*PTC144,*GE-32,*TR-87,*HEPS3019, *SK3024,*ECG128,*WEP243,*276-2008,*RT-114
40367	*TR-87,*HEPS3019,*SK3044,*ECG128,*WEP243,*276-2008,*RT-114
40368	2N1486,S2N1486,ZT1486,*GE-66,*276-2020
40369	2N1490,2N3446,2SD16,2SD110-R,2SD111-R,2SD118-R,2SD119-R,2SD522,181T2, BDY24B,HST9202,HST9203,HST9803,HST9804,SDT7733,SDT7734,ZT1490, *PTC173,*GE-14,*TR-59,*HEPS7004,*SK3079,*ECG130,*WEP247,*BDY20, *RT-131
40372	*GE-28,*TR-81,*HEPS5019,*SK3026,*ECG175,*WEP241,*276-2020
40373	*TR-81,*HEPS5012,*ECG175,*WEP241
40374	*TR-81,*SK3021,*ECG124,*WEP240
40375	*SK3562
40385	*GE-32,*TR-87,*ECG128,*WEP243,*RT-114
40389	*GE-28,*TR-76,*SK3045,*276-2020
40390	*GE-32,*TR-75,*SK3044,*ECG191,*WEPS3021
40391	*GE-29,*TR-77,*ECG128,*WEP243,*276-2023
40392	*PTC195,*GE-28,*TR-76,*HEPS6001,*SK3045,*ECG225,*276-2020
40394	*GE-29,*TR-77,*HEPS6001,*276-2023
40395	*GE-53,*TR-85,*HEPG0005,*SK3004,*ECG102A,*WEP250,*276-2007,*RT-121

To Replace	Substitute This Type
40397	*PTC125,*GE-210,*TR-21,*HEPS0015,*SK3124,*ECG123A,*WEP736,*BC337, *276-2009,*RT-102
40398	2N3241A,2N3242A,2N4074,2N6426,*PTC125,*GE-210,*TR-21,*HEPS0015,*SK3124, *ECG123A,*WEP736,*BC337,*276-2009,*RT-102
40399	40397,BCY58B,BCY58C,*PTC123,*GE-210,*TR-21,*HEPS0015,*SK3124,*ECG123A, *WEP736,*BC337,*276-2009,*RT-102
40400	2N3241A,2SC1788,40398,BCY58A,*PTC123,*GE-210,*TR-21,*HEPS0015,*SK3124, *ECG123A,*WEP736,*BC337,*276-2009,*RT-102
40403	2N1348,2N1349,2N1350,2N1351,2N1356,*TR-05,*HEPG0007,*SK3005,*ECG100, *WEP254,*AC188/01,*276-2005,*RT-118
40404	2N2710,2N3261,2N3510,2N3511,2N3648,2SC576,40405,MPS6532,ZT83,*PTC136, *GE-20,*TR-21,*HEPS0015,*SK3122,*ECG123A,*WEP736,*276-2016,*RT-102
40405	2N3261,2N3510,2N3511,2N3648,*PTC136,*GE-20,*TR-21,*HEPS0014,*ECG123A, *WEP736,*276-2016,*RT-102
40406	2N4890,2N5865,40362,40537,*PTC141,*GE-67,*TR-88,*HEPS5013,*SK3025, *ECG129,*WEP242,*276-2025,*RT-115
40407	2N2270,2N3053,2N3724A,40360,40635,*PTC144,*GE-63,*TR-88,*HEPS5014, *SK3024,*ECG129,*WEP242,*276-2030,*RT-115
40408	2N2405,*PTC144,*GE-20,*TR-74,*HEPS3019,*SK3024,*ECG190,*WEPS3021, *276-2030,*RT-159
40409	*GE-18,*TR-87,*SK3024,*ECG128,*WEP243,*RT-114
40410	*GE-21,*TR-88,*SK3025,*ECG129,*WEP242,*RT-115
40411	*PTC175,*GE-14,*TR-36,*HEPS7000,*SK3036,*ECG181,*WEPS7000,*RT-149
40412	*GE-32,*TR-87,*HEPS3021,*SK3044,*ECG198,*WEP244,*276-2012,*RT-114
40412V1	*HEPS3021,*SK3537,*276-2012
40412V2	*TR-78,*HEPS3021,*SK3045,*ECG225,*276-2012
40413	2SC1747,40295,*PTC132,*TR-95,*SK3039,*ECG108,*WEP56,*276-2038
40414	2N2808,2N2808A,2N2809,2N2809A,2N2857,2N3570,2N3600,2N3839,2N4253,2N5053, 2N5054,2N6304,2N6305,2SC251,2SC251A,2SC252,2SC253,2SC567,2SC583, 2SC1747,40294,A485,BFX89,BFY90,MT1061A,ZT2857,*PTC132,*GE-86,*TR-95, *SK3039,*ECG108,*WEP56
40421	*PTC138,*GE-16,*TR-59,*HEPG6005,*SK3014,*ECG130,*WEP247,*276-2006, *RT-131
40422	2SC1168,2SC1456,*PTC104,*GE-12,*TR-81,*HEPS5011,*SK3021,*PN66,*ECG124, *WEP240,*RT-128
40423	*GE-12,*TR-81,*HEPS5011,*SK3021,*ECG124,*WEP240,*RT-128
40424	2SC1456,*PTC104,*GE-12,*TR-81,*HEPS5012,*SK3021,*PN66,*ECG175,*WEP241
40425	*GE-12,*TR-81,*HEPS5011,*SK3021,*ECG124,*WEP240,*RT-128
40426	*PTC104,*GE-12,*TR-81,*HEPS5011,*SK3021,*PN66,*ECG124,*WEP240,*RT-128
40427	*GE-12,*TR-81,*HEPS5011,*SK3021,*ECG124,*WEP240,*RT-128
40439	*PTC122,*GE-25,*TR-75,*HEPG6008,*SK3035,*ECG191,*WEPS3021
40440	*PTC122,*GE-25,*TR-27,*HEPG6007,*SK3035,*ECG127,*WEP235
40444	2N6271,2N6273,*TR-36,*HEPS7000,*ECG181,*WEPS7000,*RT-149
40446	2SC699,2SC799,2SC866,40582,*GE-28,*TR-76,*ECG224,*WEP224
40450	40451,*GE-63,*TR-76,*HEPS0014,*276-2038
40451	*GE-63,*TR-76,*HEPS0014,*276-2009
40452	40451,40459,*PTC125,*GE-63,*TR-76,*HEPS5014,*276-2008
40453	*GE-63,*TR-76,*HEPS0015,*276-2008
40454	40450,40451,40452,*PTC125,*GE-63,*TR-76,*HEPS0014,*276-2008
40455	40453,BCY58,*GE-63,*TR-76,*HEPS0015,*276-2008
40456	40450,*PTC125,*GE-63,*TR-76,*HEPS0014,*ECG123A,*276-2008
40458	*GE-47,*TR-65,*HEPS3001,*SK3122,*ECG195A,*BC337,*276-2009
40459	*GE-63,*TR-78,*HEPS5014,*SK3040,*ECG154,*WEP712,*276-2030,*RT-110
40462	40050,40051,40623,40626,NKT402,NKT404,*PTC138,*GE-25,*TR-01,*HEPG6005, *SK3009,*ECG104,*WEP230,*276-2006,*RT-124
40464	2N6374,*PTC140,*GE-73,*TR-59,*HEPS7002,*SK3027,*HN100,*ECG130,*WEP247, *276-2019,*RT-131
40465	SDT7601,*PTC140,*GE-75,*TR-59,*HEPS7002,*SK3027,*HN100,*ECG130,*WEP247, *276-2019,*RT-131
40466	2SD68,SDT7601,*PTC140,*GE-75,*TR-59,*HEPS7002,*HN100,*ECG130,*WEP247, *276-2020,*RT-131

To Replace	Substitute This Type
40469	2N5181,40235,40238,40239,40472,40478,*PTC132,*GE-39,*TR-95,*HEPS0016, *SK3039,*ECG108,*WEP56,*276-2015,*RT-113
40470	2N5181,40235,40238,40239,40472,40478,*PTC132,*GE-39,*TR-95,*HEPS0016, *SK3039,*ECG108,*WEP56,*276-2038,*RT-113
40471	40239,*PTC132,*GE-39,*TR-59,*HEPS0016,*EN10,*ECG130,*WEP247,*276-2038, *RT-131
40472	40235,40236,40237,40238,40240,40474,40477,40478,40480,*PTC132,*GE-39,*TR-95, *HEPS0015,*ECG108,*WEP56,*276-2011,*RT-113
40473	40236,40237,40240,40474,40477,40480,*PTC132,*GE-39,*TR-21,*HEPS0030, *ECG123A,*WEP736
40474	40236,40237,40240,40477,40480,*PTC132,*GE-39,*TR-21,*HEPS0030,*ECG123A, *WEP736
40475	2N5181,40235,40236,40237,40238,40240,40472,40474,40477,40478,40480,*PTC132, *GE-39,*TR-95,*HEPS0015,*ECG108,*WEP56,*276-2011,*RT-113
40476	2N5181,40235,40238,40239,40472,40478,*PTC132,*GE-39,*TR-86,*HEPS0016, *ECG123,*WEP53,*276-2015,*RT-100
40477	40236,40237,40240,40474,40480,*PTC132,*GE-39,*TR-21,*HEPS0015,*SK3122, *ECG123A,*WEP736,*276-2031,*RT-102
40478	2N5181,40235,40236,40237,40238,40240,40472,40474,40477,40480,*PTC132, *GE-39,*TR-95,*HEPS0016,*SK3018,*ECG108,*WEP56,*276-2011,*RT-113
40479	2N5181,40235,40236,40237,40238,40240,40472,40474,40477,40478,40480,*PTC132, *GE-39,*TR-95,*HEPS0016,*SK3018,*ECG108,*WEP56,*276-2015,*RT-113
40480	40236,40237,40240,40474,40477,*PTC132,*GE-39,*TR-95,*HEPS0016,*ECG108, *WEP56,*276-2030,*RT-113
40481	40236,40237,40240,40474,40477,40480,*GE-39,*TR-95,*HEPS0016,*ECG108, *WEP56,*276-2011,*RT-113
40482	40239,*PTC132,*GE-39,*TR-95,*HEPS0016,*ECG108,*WEP56,*276-2030,*RT-113
40487	*GE-1,*TR-17,*HEPG0008,*ECG160,*WEP637,*276-2007
40488	2N796,2N964A,2N972,2N973,2N974,2N985,2N1526,2N2022,2N2273,2N2699,2N3320, 2SA350,2SA355,SF.T317,SF.T320,*PTC107,*GE-50,*TR-17,*HEPG0008,*ECG126, *WEP635,*AF125,*276-2007
40489	*GE-1,*TR-17,*HEPG0008,*ECG160,*WEP637,*276-2007
40490	*GE-2,*TR-85,*HEPG0006,*SK3004,*ECG102A,*WEP250,*276-2007,*RT-121
40491	*GE-12,*TR-81,*HEPS5011,*SK3021,*ECG124,*WEP240,*RT-128
40513	*TR-59,*HEPS5004,*ECG223,*WEP233
40514	*TR-59,*HEPS7002,*SK3036,*ECG223,*WEP233
40517	2N2857,2N3570,2N3600,2N3839,2N4252,2N4253,2N6304,2N6305,2SC1747,40294, 40414,MT1061A,*PTC132,*GE-86,*TR-86,*HEPS0017,*SK3039,*ECG123,*WEP53, *276-2011,*RT-100
40518	2N2857,2N3570,2N3600,2N3839,2N4252,2N4253,2N6304,2N6305,2SC1747,40294, 40414,MT1061A,*PTC132,*GE-86,*TR-86,*HEPS0017,*ECG123,*WEP53,*276-2011, *RT-100
40519	2N2318,2N2319,2N5187,2SC395A-R,*PTC136,*GE-20,*TR-86,*HEPS0014,*ECG123, *WEP53,*276-2016,*RT-100
40537	2N4405,2N4890,MPS4356,*GE-67,*TR-88,*HEPS5022,*SK3025,*ECG129,*WEP242, *276-2021
40538	2N3245,2N3468,2N3762A,2N3763,2N5022,MM3726,*PTC141,*GE-67,*TR-88, *HEPS5022,*SK3025,*ECG129,*WEP242,*276-2021
40539	2N3252,2N3253,2N3444,2N3735,MM3725,MPS3725,*PTC144,*GE-63,*TR-87, *HEPS0015,*SK3024,*ECG128,*WEP243,*276-2009,*RT-114
40542	2N5034,2N5036,40543,40633,*TR-59,*HEPS7002,*ECG223,*WEP247
40543	2N5036,40633,*TR-59,*HEPS7002,*ECG223,*WEP247
40544	*SK3045,*ECG225
40546	2SC1168,2SC1456,*PTC104,*GE-12,*TR-81,*HEPS5011,*SK3021,*ECG124, *WEP240,*RT-128
40547	2SC1168,2SC1456,*PTC104,*GE-12,*TR-81,*HEPS5011,*SK3021,*ECG124, *WEP240,*RT-128
40577	2N749,2N751,2N1388,2N1390,2SC302-M,2SC307-M,2SC590N,2SC845,SRF11101, *PTC158,*TR-21,*HEPS0015,*ECG123A,*WEP736,*276-2009
40578	2N749,2N751,2N1388,2N1390,2SC302-M,2SC307-M,2SC590N,2SC1252,2SC1365, *GE-28,*TR-87,*HEPS3008,*ECG128,*WEP243,*RT-114

To Replace	Substitute This Type
40581	2N749,2N751,2N1388,2N1390,2SC302-M,2SC307-M,2SC590N,2SC773,2SC774, 2SC798,2SC803,2SC865,40975,*PTC158,*GE-28,*TR-65,*HEPS3001,*SK3048, *ECG195A,*276-2014
40582	2SC699,2SC799,2SC866,*PTC176,*GE-28,*TR-76,*SK3049,*ECG224,*WEP224
40594	*GE-66,*TR-72,*HEPS3002,*SK3512,*ECG188,*WEPS3020
40595	*TR-73,*HEPS3032,*SK3513,*ECG189,*WEPS3031
40605	2N3553,2SC302-M,2SC307-M,2SC590N,*GE-28,*TR-76,*HEPS3008
40608	*GE-28,*TR-76,*HEPS3008,*ECG278
40611	40616,*GE-63,*TR-87,*HEPS0014,*SK3024,*ECG128,*WEP243,*276-2008
40612	40022,40050,40254,40462,*PTC138,*GE-25,*TR-01,*HEPG6003,*SK3014,*ECG121, *WEP232,*276-2006
40613	40618,40621,40622,*PTC167,*GE-66,*TR-76,*HEPS5000,*SK3054,*ECG152, *WEP245,*276-2019,*RT-154
40616	*GE-63,*TR-87,*HEPS0014,*SK3024,*ECG128,*WEP243,*276-2008
40618	40621,40622,*PTC167,*GE-66,*TR-76,*HEPS5000,*SK3054,*ECG152,*WEP245, *276-2019,*RT-154
40621	40618,40622,*PTC167,*GE-66,*TR-76,*HEPS5000,*SK3054,*ECG152,*WEP245, *276-2019,*RT-154
40622	2N5491,2N5495,*PTC167,*GE-66,*TR-76,*HEPS5000,*SK3054,*ECG152,*WEP245, *276-2019,*RT-154
40623	2N1138,40051,40626,NKT402,NKT404,NKT405,*PTC138,*GE-25,*TR-01,*HEPG6005, *SK3009,*ECG121MP,*WEP232MP,*276-2006
40624	2N5491,2N5495,40627,40632,*PTC167,*GE-66,*TR-76,*SK3054,*ECG196,*WEP245, *276-2020,*RT-154
40625	40628,*TR-72,*SK3024,*ECG128,*WEPS3020
40626	2N1138,40051,NKT402,NKT403,NKT404,NKT405,*PTC138,*GE-25,*TR-01, *HEPG6005,*SK3014,*ECG121,*WEP232,*276-2006
40627	2N5491,2N5493,2N5495,40632,*PTC167,*GE-66,*TR-76,*SK3054,*ECG196,*WEP245, *276-2020,*RT-154
40628	40625,*TR-72,*SK3024,*ECG128,*WEPS3020
40629	40621,40622,40630,40631,*PTC167,*GE-66,*TR-76,*HEPS5000,*SK3054,*ECG152, *WEP245,*276-2019,*RT-154
40630	2N5491,2N5495,40622,40631,*PTC167,*GE-66,*TR-76,*HEPS5000,*SK3054, *ECG152,*WEP245,*276-2019,*RT-154
40631	2N5491,2N5495,*PTC167,*GE-66,*TR-76,*HEPS5000,*SK3054,*ECG152,*WEP245, *276-2020,*RT-154
40632	2N5491,2N5493,2N5495,40627,*PTC167,*TR-92,*SK3054,*ECG196,*WEP245, *276-2020
40633	2N5036,*TR-59,*HEPS5004,*ECG223,*WEP247,*276-2020
40634	2N4314,*TR-88,*HEPS5013,*SK3025,*ECG129,*WEP242,*276-2025
40635	*TR-87,*HEPS5014,*SK3024,*ECG128,*WEP243,*276-2018
40636	2N3055,2N3055SD,2N3236,2N5629,2N5630,2N6254,2N6270,2N6271,2N6272,2N6273, 2SD341,2SD492,108T2,1561-0803,1561-0804,1561-0805,1561-1005,1561-1205, 1582-0803,1582-0804,1582-0805,1582-1003,1582-1004,1582-1005,1582-1203, 1582-1204,1582-1205,BUY53A,BUY54A,HST9202,HST9203,HST9207,HST9208, IR3772,MJ3772,SDT9202,SDT9203,SDT9207,SDT9208,STC2221,STC2225,*GE-75, *TR-59,*HEPS7004,*SK3027,*ECG284,*WEP247
40637	*GE-212,*TR-21,*HEPS0015,*ECG123A,*WEP736,*BC107B,*276-2016
40637A	2SC302-M,2SC307-M,2SC590N,2SC731,2SC844,2SC854,2SC855,2SC872,2SC890, 2SC991,2SC992,2SC1002,2SC1083,2SC1256,2SC1303,2SC1315,2SC1464, 2SC1479,2SC1820,*PTC125,*GE-47,*276-2012
40665	2N3632,2N4933,*GE-66
40666	*GE-28,*TR-76
40675	SEE 2N6093
40829	
40830	40829
40831	40830
40836	2SC1200,40837,40909†,*276-2020
40837	40909†,*276-2020
40850	2N5468,HST1156,HST1157,HST1161,HST1162
40851	

To Replace	Substitute This Type
40852	2N6579,2N6582,2SC1230,2SC1468,2SC2017,2SD519,IR409,XGSR10030, XGSR10035,XGSR10040
40853	2N6653,2N6654,2N6655,2SC1230,2SC1468,2SC2017,2SD519,XGSR10035, XGSR15035,XGSR15040
40854	
40893†	ZT5591,*HEPS3006
40894	2SC988,2SC988A,2SC988B,2SC1044,2SC1254,2SC1426,40895,40896,*PTC126, *HEPS0017,*SK3039,*276-2011
40895	2SC988,2SC988A,2SC988B,2SC1044,2SC1254,2SC1426,40894,40896,*PTC126, *HEPS0017,*SK3039,*276-2011
40896	2SC988,2SC988A,2SC988B,2SC1044,2SC1254,2SC1426,40894,40895,*PTC126, *HEPS0017,*SK3039,*276-2011
40897	2SC988,2SC988A,2SC988B,*PTC126,*TR-83,*HEPS0017,*SK3039,*ECG161, *WEP719,*276-2011
40898†	40899†,*276-2020
40899†	*ECG231,*276-2020
40909†	*276-2020
40915	2N3478,2N6389,2SC817,2SC988,2SC988A,2SC988B,2SC1117,2SC1260,2SC1275, 2SC1424,2SC1529,2SC1553,2SC1574,*PTC123
40934†	2N5645,2N5914,2N5944,2N5945,2N5946,2SC891,2SC892,2SC1040,2SC1495, 2SC1620,41009A†,41010†
40936†	2N5070,*276-2019
40940†	2N5636,2N6203,3TX632
40941†	2N5916†,2N5917†,2N6202,2SC1465,*276-2020
40953	2N1491,2N5913,2SC302-M,2SC307-M,2SC590N,2SC992,2SC1002,2SC1223, 2SC1256,2SC1947,41039,*GE-45,*HEPS3013
40954†	2SC1122,2SC1718,40973†,41042†,*HEPS3006,*SK3176
40955†	2SC1314,2SC1378,2SC1476,2SC1668,40974†,*HEPS3007,*SK3177
40964	2N2486,*HEPS3001
40965	2N2485,2N2486,40964,BLX89,*HEPS3001
40967†	2SC1355,2SC1356,2SC1414,2SC1480,2SC1481,2SC1482,2SC1533,2SC1534, 2SC1535,2SC1808,2SC1821,40968†,41009†,*HEPS3005
40968†	2SC1122,2SC1356,2SC1414,2SC1415,2SC1482,2SC1534,2SC1535,*HEPS3005
40970†	*276-2019
40971†	
40972	2N5913,2SC302-M,2SC307-M,2SC590N,2SC992,2SC1002,2SC1223,2SC1256, 2SC1947,40953,*GE-45,*HEPS3013
40973†	2SC1122,2SC1718,40954†,41042†,*HEPS3006
40974†	2SC1314,2SC1378,2SC1476,2SC1668,40955†,*HEPS3007
40975	2N749,2N751,2N1493,2N4874,2N4875,2N5109,2SC302-M,2SC307-M,2SC590N, 2SC731,2SC773,2SC774,2SC797,2SC798,2SC802,2SC803,2SC844,2SC854, 2SC855,2SC865,2SC872,2SC890,2SC991,2SC992,2SC1002,2SC1083,2SC1256, 2SC1303,2SC1315,2SC1464,2SC1479,2SC1820,40082,40581,40637A,40976, *GE-45
40976	2N749,2N751,2N1493,2N4874,2N4875,2N5109,2SC302-M,2SC307-M,2SC590N, 2SC731,2SC773,2SC774,2SC797,2SC798,2SC802,2SC803,2SC844,2SC854, 2SC855,2SC865,2SC872,2SC890,2SC991,2SC992,2SC1002,2SC1083,2SC1256, 2SC1303,2SC1315,2SC1464,2SC1479,2SC1820,40082,40581,40975,*GE-45
40977†	2SC1122,2SC1242A,2SC1482,40968†
40989	SEE 2N6389
41008†	2SC1355,2SC1356,2SC1414,2SC1480,2SC1481,2SC1482,2SC1533,2SC1534, 2SC1603,2SC1808,40967†,40968†,41009†,*276-2038
41008A†	2SC891,2SC892,2SC911,2SC911A,2SC1016,2SC1040,2SC1608,41009A†,41010†, *276-2038
41009†	2SC1355,2SC1356,2SC1414,2SC1480,2SC1481,2SC1482,2SC1533,2SC1534, 2SC1535,2SC1808,40967†,40968†
41009A†	2SC891,2SC892,2SC1040,41010†
41010†	2SC892,2SC1040,2SC1081,2SC1118
41024	2N2485,2N2486,2N5108,2N5108A
41025†	2N5595,2N5765,2N6206,41026†,MT5764,MT5765
41026†	2N5595,2N5596,2N5925,2N6207

To Replace	Substitute This Type
41027†	2N5595,2N5765,2N5924,2N5925,2N6206,2N6207,2SC1333,2SC1425,41025†,41026†, 41028†,MT5764,MT5765
41028†	2N5595,2N5596,2N5925,2N6207,2SC1333,2SC1425,41026†
41038	
41039	2SC302-M,2SC307-M,2SC590N,2SC731,2SC845,2SC872,2SC890,2SC1252,2SC1303, 2SC1315,2SC1365,2SC1820
41042†	2SC1476,2SC1548,2SC1668
41044	
41500	*PTC169,*ECG196
41501	*PTC168,*ECG197,*276-2027
41502	2N1409A,7A30,92PE37A,*PTC144,*ECG128
41503	92PE77A,*PTC142,*ECG129
41504	RCA31,*PTC167,*ECG152
41505	*ECG228
41506	*PTC129A,*SK3560,*ECG162
43104	2N3773,2N6259,109T2,1561-1410,1561-1415,1561-1608,1561-1610,1561-1615, 1561-1808,1561-1810,1561-1815,1561-2008,1561-2010,IR3773,STC2222,STC2223, STC2226,STC2227,*ECG284
95101	SEE 2N1526
95102	SEE 2N1524
95103	SEE 2N1524
95107	SEE 2N1526
95108	SEE 2N1745
95109	SEE 2N501
95110	SEE 2N501
95111	SEE 2N1865
95112	SEE 2N2711
95113	SEE 2N2712
95114	SEE 2N193
95115	SEE 2N211
95116	SEE 2N769
95117	SEE 2N769
95201	SEE 2N406
95202	SEE 2N649
95203	SEE 2N408
95204	SEE 2N270
95208	SEE 2N2374
95209	SEE 2N2376
95212	SEE 2N321
95214	SEE 2N406
99101	SEE 2SA84
99102	SEE 2SA83
99103	SEE 2SA83
99104	SEE 2SA15
99201	SEE 2SB75
99203	SEE 2SB77
99204	SEE 2SB89
A3T2221	A3T2221A,FX3725,FX4047,*PTC136,*GE-210,*TR-21,*HEPS0025,*ECG123A, *WEP736,*276-2016
A3T2221A	FX3725,FX4047,*PTC136,*GE-20,*HEPS0005,*SK3122,*ECG123A,*WEP736, *276-2009
A3T2222	A3T2222A,FX3300,*PTC136,*GE-210,*HEPS0015,*SK3122,*ECG123A,*WEP736, *276-2016
A3T2222A	*PTC136,*GE-20,*HEPS0005,*ECG123A,*WEP736,*276-2009
A3T2484	A3T2222A,FX2483,FX4960,*PTC123,*GE-63,*TR-51,*HEPS0005,*SK3137,*ECG192, *WEP243,*276-2008
A3T2894	FX2894,FX2894A,*TR-20,*HEPS0013,*SK3118,*ECG159,*WEP242,*276-2021
A3T2906	A3T2906A,*PTC103,*GE-21,*TR-20,*HEPS0019,*SK3118,*ECG159,*WEP242, *276-2023
A3T2906A	A3T2906,*PTC103,*GE-21,*TR-20,*HEPS0019,*SK3118,*ECG159,*WEP242
A3T2907	A3T2907A,FX3503,*PTC127,*GE-48,*TR-20,*HEPS0019,*SK3118,*ECG159,*WEP242,

To Replace	Substitute This Type
(A3T2907)	*276-2023
A3T2907A	A3T2907,FX3503,*PTC127,*GE-67,*HEPS0019,*SK3138,*ECG193,*WEP242
A3T3011	FX3014,*PTC136,*GE-20,*TR-64,*HEPS0011,*SK3122,*ECG123A,*WEP736, *276-2016
A3T918	FX918,*PTC115,*GE-61,*HEPS0020,*ECG161,*WEP719,*276-2015
A3T929	FX4046,*PTC133,*GE-212,*TR-51,*HEPS0025,*ECG123A,*WEP736,*BC107B, *276-2013
A3T930	A3T2484,FX2483,FX4960,*PTC153,*GE-212,*TR-51,*HEPS0015,*SK3245,*ECG123A, *WEP736,*BC107B,*276-2013
A5T2192	*276-2008
A5T2193	*276-2008
A5T2222	TIS109,TIS111,*PTC136,*SK3122,*ECG123A,*WEP736,*276-2009
A5T2243	*276-2008
A5T2604	*PTC103
A5T2605	*PTC103
A5T2907	A5T3505,A5T3645,TIS112,*PTC103,*HEPS0019,*SK3025,*ECG159,*WEP242, *276-2023
A5T3391	A5T3391A,A7T3391,A7T3391A,A8T3391,A8T3391A,TP3566,*276-2012
A5T3391A	A5T3391,A7T3391,A7T3391A,A8T3391,A8T3391A,TP3566,*276-2012
A5T3392	A5T3391,A5T3391A,A5T4124,A5T5172,A7T3391,A7T3391A,A7T3392,A7T5172, A8T3391,A8T3391A,A8T3392,A8T5172,BCY58,TIS90-GRY,TIS90M-GRY,TIS92-GRY, TIS92M-GRY,TP3566,*PTC170,*276-2012
A5T3496	
A5T3497	
A5T3504	A5T2907,A5T3505,A5T3644,A5T3645,TIS112,*PTC103
A5T3505	A5T2907,A5T3645,TIS112,*PTC103
A5T3565	BCY59,*PTC170
A5T3571	A5T3572
A5T3572	*PTC101
A5T3638	*PTC103
A5T3638A	2SA509-O,2SA509G-O,2SA890,A5T4402,*PTC103
A5T3644	A5T2907,A5T3504,A5T3505,A5T3645,TIS112,*PTC103,*SK3025,*ECG159,*WEP242
A5T3645	A5T2907,A5T3505,TIS112,*PTC103,*SK3025,*ECG159,*WEP242
A5T3707	A5T3565,A5T3708,A8T3707,A8T3708,BCY58,BCY66,*PTC101,*276-2010
A5T3708	A5T3565,A5T3711,A8T3708,A8T3711,BCY59,*PTC101,*276-2010
A5T3709	A8T3709,*PTC101,*276-2010
A5T3710	A5T3707,A8T3707,A8T3710,BCY58,BCY66,*PTC101,*276-2010
A5T3711	A8T3711,BCY59,*PTC101,*276-2010
A5T3903	TIS135,TIS136,*SK3122,*ECG123A,*WEP736
A5T3904	BCY65,TIS111,*HEPS0025,*SK3122,*ECG123A,*WEP736
A5T3905	2SA891,*PTC103,*SK3025,*ECG159,*WEP242,*276-2023
A5T3906	*PTC103,*SK3025,*ECG159,*WEP242,*276-2023
A5T4026	A5T4027,A8T4026,A8T4027
A5T4027	A8T4027
A5T4028	A5T4029,A8T4028,A8T4029,MPS4354,MPS4355,*PTC103
A5T4029	A8T4029
A5T404	A5T404A,A8T404,A8T404A,*276-2025
A5T404A	A8T404A,*276-2025
A5T4058	A5T404A,A5T4059,A5T4061,A8T404A,A8T4058,A8T4059,A8T4061,*PTC103, *276-2023
A5T4059	A5T4062,A8T4059,A8T4062,*PTC103,*276-2023
A5T4060	A8T4060,TIS37,*PTC103,*276-2023
A5T4061	A5T404A,A5T4058,A8T404A,A8T4058,A8T4061,*PTC103,*276-2023
A5T4062	A5T4059,A8T4059,A8T4062,*PTC103,*276-2023
A5T4123	A5T3903,TIS133,TIS134,*PTC136,*276-2016
A5T4124	BCY58,BCY66,*276-2016
A5T4125	2SA890,A5T3504,A5T3644,A5T3905,A5T3906,A5T4403,*PTC103,*SK3025,*ECG159, *WEP242,*276-2034
A5T4126	*PTC103,*SK3025,*ECG159,*WEP242,*276-2034
A5T4248	2SA661-R,*PTC103
A5T4249	2SA661-O,2SA891,TP3645,*PTC103

To Replace	Substitute This Type
A5T4250	2SA661-GR,A5T2907,A5T3504,A5T3505,A5T3644,A5T3645,A5T3906,A5T4403,BCY79, TIS112
A5T4260	A5T4261,*PTC103
A5T4261	
A5T4402	*PTC103
A5T4403	A5T2907,A5T3504,A5T3505,A5T3644,A5T3645,TIS112,*PTC103
A5T4409	2N4409,*PTC170
A5T4410	2N4410
A5T5058	*276-2012
A5T5059	A5T5058,*ECG287,*276-2012
A5T5086	2SA661-GR,A5T5087,*PTC103
A5T5087	
A5T5172	A5T3391,A5T3391A,A7T3391,A7T3391A,A7T5172,A8T3391,A8T3391A,A8T5172, BCY58,TP3566,*276-2012
A5T5209	2SC1166-Y,A5T3904,*PTC170
A5T5210	2N6428,2N6428A,2N6539
A5T5219	BCY58,*276-2030
A5T5220	BC338-40,*PTC170,*276-2030
A5T5221	BC328-25,BC328-40,*PTC103
A5T5222	
A5T5225	A5T5451,A8T3706,BC338-40,*276-2030
A5T5226	BC328-25,BC328-40,*PTC103
A5T5227	A5T5087,*PTC103
A5T5400	*PTC171
A5T5401	*PTC171
A5T5447	2SA661-GR,2SA661-Y,2SA891,A5T2907,A5T3504,A5T3505,A5T3644,A5T3645, A5T3906,A5T4403,A8T3702,BC327-16,BC327-25,BCY79,TIS112
A5T5448	2SA661-O,A8T3703
A5T5449	A5T2222,A8T3704,TIS109,TIS111
A5T5450	A8T3705,TIS110
A5T5451	A8T3706,BC337-40
A5T5550	A5T5551,*PTC170
A5T5551	*PTC170
A7T3391	A5T3391,A5T3391A,A7T3391A,A8T3391,A8T3391A,TP3566,*276-2012
A7T3391A	A5T3391,A5T3391A,A7T3391,A8T3391,A8T3391A,TP3566,*276-2012
A7T3392	A5T3391,A5T3391A,A5T3392,A5T4124,A5T5172,A7T3391,A7T3391A,A7T5172, A8T3391,A8T3391A,A8T3392,A8T5172,BCY58,TIS90-GRY,TIS90M-GRY,TIS92-GRY, TIS92M-GRY,TP3566,*PTC170,*276-2012
A7T5172	A5T3391,A5T3391A,A5T5172,A7T3391,A7T3391A,A8T3391,A8T3391A,A8T5172, BCY58,TP3566,*PTC170,*276-2012
A8T3391	A5T3391,A5T3391A,A7T3391,A7T3391A,A8T3391A,TP3566,*276-2012
A8T3391A	A5T3391,A5T3391A,A7T3391,A7T3391A,A8T3391,TP3566,*276-2012
A8T3392	A5T3391,A5T3391A,A5T3392,A5T4124,A5T5172,A7T3391,A7T3391A,A7T3392, A7T5172,A8T3391,A8T3391A,A8T5172,BCY58,TIS90-GRY,TIS90M-GRY,TIS92-GRY, TIS92M-GRY,TP3566,*PTC170,*276-2012
A8T3702	2SA661-GR,2SA661-Y,2SA891,A5T2907,A5T3504,A5T3505,A5T3644,A5T3645, A5T3906,A5T4403,A5T5447,BC327-16,BC327-25,BCY79,TIS112,*PTC103
A8T3703	2SA661-O,A5T5448
A8T3704	A5T2222,A5T5449,TIS109,TIS111
A8T3705	A5T5450,TIS110
A8T3706	A5T5451,BC337-40
A8T3707	A5T3565,A5T3707,A5T3708,A8T3708,BCY58,BCY66,*PTC170,*276-2012
A8T3708	A5T3565,A5T3708,A5T3711,A8T3711,BCY59,*PTC170,*276-2012
A8T3709	A5T3709,*PTC170,*276-2012
A8T3710	A5T3707,A5T3710,A8T3707,BCY58,BCY66,*PTC170,*276-2012
A8T3711	A5T3711,BCY59,*PTC170,*276-2012
A8T4026	A5T4026,A5T4027,A8T4027
A8T4027	A5T4027
A8T4028	A5T4028,A5T4029,A8T4029,MPS4354,MPS4355
A8T4029	A5T4029
A8T404	A5T404,A5T404A,A8T404A,*276-2025

To Replace	Substitute This Type
A8T404A	A5T404A,*276-2025
A8T4058	A5T404A,A5T4058,A5T4059,A5T4061,A8T404A,A8T4059,A8T4061,*PTC103, *276-2025
A8T4059	A5T4059,A5T4062,A8T4062,*PTC103,*276-2025
A8T4060	A5T4060,TIS37,*PTC103,*276-2025
A8T4061	A5T404A,A5T4058,A5T4061,A8T404A,A8T4058,*PTC103,*276-2025
A8T4062	A5T4059,A5T4062,A8T4059,*PTC103,*276-2025
A8T5172	A5T3391,A5T3391A,A5T5172,A7T3391,A7T3391A,A7T5172,A8T3391,A8T3391A, BCY58,TP3566,*PTC170,*276-2012
A104	2N703,2N2331,2N2569,2N2570,2N2923,2N3241A,2N3393,2N3856A,2N3859,2N4418, 2SC733-O,2SC735-O,2SC847,2SC848,2SC934,2SC966,2SD392,40398,BC170B, MPS6574-BLUE,MPS6574-GREEN,MPS6574-YEL,PET4001,TIS86,*PTC126,*GE-61, *TR-21,*HEPS0015,*SK3007,*ECG126,*WEP736,*BC107B,*276-2013,*RT-102
A106	2N703,2N2331,2N2569,2N2570,2N2923,2N3241A,2N3393,2N3856A,2N3858,2N3859, 2N3959,2N3960,2N4418,2S095A,2S512,2SC733-O,2SC735-O,2SC934,2SD392, 40398,BF123,MPS6574-BLUE,MPS6574-GREEN,MPS6574-YEL,PET4001,TIS86, *PTC126,*GE-61,*TR-21,*HEPS0033,*SK3124,*ECG123A,*WEP736,*BC107B, *276-2013,*RT-102
A108	2N2924,2N3392,2N3860,2N4124,2N4256,2N4419,2N5172,2N5224,2S503,2SC538, 2SC539,2SC733-GR,2SC733-Y,2SC735-GR,2SC735-Y,2SC1849,2SC1993,40397, A5T4124,A158B,A159B,A322,BC108,BC108B,BC109B,BC168B,BC169B,BC172B, BC173B,BC238B,BC239B,BC548A,BCY58B,BSW88,BSW89,BSX38,MPS6573, MPS6574-SIL,PET2002,TN80,*PTC123,*GE-61,*TR-21,*HEPS0015,*SK3124, *EN30,*ECG123A,*WEP736,*BC107B,*276-2013,*RT-102
A110	2N3391,2N3391A,2N5135,2SC733-BL,A5T3565,A158C,A159C,BC108C,BC109, BC109C,BC168C,BC169C,BC170C,BC172C,BC173C,BC238C,BC239C,BC548B, BC549B,BCY58D,PET4003,*GE-212,*TR-21,*HEPS0024,*SK3156,*ECG199, *WEP736,*BC107B,*276-2013
A111	2N930,2N2388,2N3242A,2N3566,2N3860,2N4124,2N4256,2N4419,2N5826, 2SC367G-Y,2SC538A,2SC587,2SC587A,2SC733-GR,2SC733-Y,2SC735-GR, 2SC735-Y,2SC1175,2SC1364,2SC1849,2SC1992,2SC1993,A5T4124,A157B,A322, BC107,BC107B,BC108,BC108B,BC109B,BC167B,BC171B,BC237B,BC238B,BC239B, BC413B,BC414B,BC547A,BC548A,BCY58B,BCY59B,BFY39-3,BSW88,BSW89, BSX38,BSX79,GI-3566,MPS3694,MPS6514,MPS6520,MPS6573,MPS6574-SIL, MPS6575,MPS6576-SIL,MPS-A10,MPS-A20,MPS-K20,MPS-K21,MPS-K22, MPSA10-BLU,MPSA10-YEL,MPSA17,MPSA20-BLU,MPSA20-YEL,NPSA20,PET1002, PET2002,TN80,TP3704,TP3706,TZ81,*PTC123,*GE-61,*TR-21,*HEPS0015,*SK3122, *EN30,*ECG123A,*WEP736,*BC107B,*276-2013,*RT-102
A115	2N708,2N708A,2N759,2N784,2N784A,2N929,2N2242,2N2318,2N2319,2N2369, 2N2369A,2N2387,2N3009,2N3011,2N3013,2N3014,2N3340,2N3605A,2N3606A, 2N3646,2N3854A,2N3855A,2N4137,2N4274,2N4275,2N5029,2N5030,2N5769, 2N5772,2SC54,2SC67,2SC204,2SC302,2SC356,2SC595,2SC601,2SC601N,2SC689, 2SC1274,A321,BF125,BF311,BFY39-1,GET708,GET914,GET2369,GET3013, GET3014,GET3646,MPS6512,MPSA10-RED,MPSA20-RED,MT1060A,PET1001, PET2001,PT2760,ZT43,ZT708,*PTC126,*GE-61,*TR-87,*HEPS0015,*SK3124, *ECG128,*WEP243,*BC107B,*276-2013,*RT-114
A116	2N916,2N916A,2N3241A,2N3856A,2N3859,2N4074,2N4873,2N5209,2N5825, 2SC367G,2SC367G-Y,2SC733-Y,2SC735-Y,2SC847,2SC848,2SC933,2SC966, 2SC1175,2SC1216,2SC1999,A5T5209,BC107A,BC108A,BC167A,BC171A,BC237A, BC238A,BCY58A,BCY59A,BFY39-2,MPS6513,MPS6574-BLUE,MPS6574-GREEN, MPS6576-BLUE,MPS6576-GREEN,MPSA10-GRN,MPSA10-WHT,MPSA20-GRN, MPSA20-WHT,PET4001,TIS86,ZT44,*PTC126,*GE-61,*TR-21,*HEPS0015,*SK3122, *ECG123A,*WEP736,*BC107B,*276-2013,*RT-102
A130	2N738A,2N912,2N1572,2N2459,2N2463,2N2517,A777,BFY65,SF.T186,*PTC123, *GE-18,*TR-78,*HEPS5026,*SK3039,*ECG154,*WEP719,*BF338,*276-2012
A141	BC198A,*GE-39,*TR-51,*HEPS0015,*SK3007,*ECG123A,*WEP736,*BC107B, *276-2011,*RT-102
A142	BC122,BC198B,BC199B,*PTC121,*GE-39,*TR-21,*HEPS0015,*SK3122,*ECG123A, *WEP736,*BC107B,*276-2011,*RT-102
A143	BC198C,BC199C,*PTC139,*GE-212,*TR-21,*HEPS0015,*SK3122,*ECG123A, *WEP736,*BC107B,*276-2011,*RT-102

To Replace	Substitute This Type
A151	2SC529,BC198A,*PTC132,*GE-39,*TR-21,*HEPS0015,*ECG123A,*WEP736,*BC107B,*276-2011
A152	BC122,BC198B,BC199B,*PTC121,*GE-39,*TR-21,*HEPS0015,*ECG123A,*WEP736,*BC107B,*276-2011
A153	BC198C,BC199C,*PTC139,*GE-212,*TR-21,*HEPS0015,*ECG123A,*WEP736,*BC107B,*276-2011
A157B	2N4953,2N5370,2N5376,2N5377,2N5827,2N6428,2N6428A,BC413C,BC414C,BC547B,BC550B,BCY59C,BCY59D,*PTC136,*TR-21,*HEPS0015,*SK3122,*ECG123A,*WEP736,*BC107B,*276-2016,*RT-102
A157C	*TR-21,*HEPS0015,*ECG123A,*WEP736,*276-2016
A158B	2SC735-GR,A159B,BC108C,BC109,BC109C,BC168C,BC172C,BC173C,BC238C,BC239C,BC548B,BC549B,BCY58C,BCY58D,*PTC136,*GE-10,*TR-21,*HEPS0015,*ECG123A,*WEP736,*BC107B,*276-2016,*RT-102
A158C	A159C,BC548C,BC549C,*GE-10,*TR-21,*HEPS0024,*ECG123A,*WEP736,*276-2016,*RT-102
A159B	2SC735-GR,A158B,BC108C,BC109,BC109C,BC168C,BC172C,BC173C,BC238C,BC239C,BC548B,BC549B,BCY58C,BCY58D,*PTC136,*GE-20,*TR-21,*HEPS0015,*SK3124,*ECG123A,*WEP736,*BC107B,*276-2016,*RT-102
A159C	A158C,BC548C,BC549C,*TR-21,*HEPS0024,*SK3122,*ECG123A,*WEP736,*276-2016,*RT-102
A177	2N2838,2N2907,2N2907A,2N3251,2N3251A,2N3486,2N3486A,2N3505,2N3645,2N3672,2N3673,2N4143,2N4228,2N5373,2N5817,2N5819,2N5823,2N6005,2SA603,2SA718,2SA720,2SA817,2SA891,A5T2907,A5T3505,A5T3645,BC212A,BC212L,BCW35,BCW37,BSW75,D29E10,D29E5,D29E6,EN2905,FT3644,GET2906,GET2907,TIS112,TZ552,*PTC127,*GE-82,*TR-20,*HEPS0019,*ECG159,*WEP242,*BC177,*276-2023
A178A	2N3906,2N4034,2N4035,2N4403,2N5244,2N5375,2N5379,2N5383,2N5999,2N6001,2SA509-Y,2SA509G-Y,2SA550,2SA719,2SA890,A5T3906,A5T4403,A178B,A179A,BC178,BC178A,BC178B,BC179,BC252A,BC252B,BC253A,BC253B,BC258A,BC258B,BC262A,BC262B,BC263A,BC263B,BC308A,BC308B,BC309A,BC309B,BC558A,BC558B,BC559A,BC559B,BCY78,BCY78A,BCY78B,BSW19,BSW20,BSW73,MM3906,NPS404,NPS404A,TN-3906,*PTC127,*GE-65,*TR-20,*HEPS0013,*SK3114,*ECG159,*WEP242,*BC177,*276-2034,*RT-115
A178B	2N3906,2N4034,2N4035,2N4403,2N5244,2N5375,2N5379,2N5383,2N5999,2N6001,2SA509-Y,2SA509G-Y,2SA550,2SA719,2SA890,A5T3906,A5T4403,A178A,A179A,BC178,BC178A,BC178B,BC179,BC252A,BC252B,BC253A,BC253B,BC258A,BC258B,BC262A,BC262B,BC263A,BC263B,BC308A,BC308B,BC309A,BC309B,BC558A,BC558B,BC559A,BC559B,BCY78,BCY78A,BCY78B,BSW19,BSW20,BSW73,MM3906,NPS404,NPS404A,TN-3906,*PTC127,*GE-65,*TR-20,*HEPS0031,*SK3114,*ECG159,*WEP242,*BC177,*276-2034,*RT-115
A179A	2N3906,2N4034,2N4035,2N4403,2N5244,2N5375,2N5379,2N5383,2N5999,2N6001,2SA509-Y,2SA509G-Y,2SA550,2SA719,2SA890,A5T3906,A5T4403,A178A,A178B,BC178,BC178A,BC178B,BC179,BC252A,BC252B,BC253A,BC253B,BC258A,BC258B,BC262A,BC262B,BC263A,BC263B,BC308A,BC308B,BC309A,BC309B,BC558A,BC558B,BC559A,BC559B,BCY78,BCY78A,BCY78B,BSW19,BSW20,BSW73,MM3906,NPS404,NPS404A,TN-3906,*PTC127,*GE-65,*TR-20,*HEPS0013,*SK3114,*ECG159,*WEP242,*BC177,*276-2034,*RT-115
A179B	2N5244,2N5999,2N6003,2N6009,2SA550,2SA608,2SA677,BC178B,BC179,BC196B,BC252B,BC253B,BC258B,BC262B,BC263B,BC308B,BC309B,BCY78,BCY78B,BCY78C,*PTC127,*GE-65,*TR-20,*HEPS0031,*SK3114,*ECG159,*WEP242,*276-2034,*RT-115
A201	*GE-28,*TR-76,*HEPS3020
A202	*GE-66,*TR-66,*HEPS3006,*SK3007,*276-2019
A210	A211,BFR95,*GE-28,*TR-76,*HEPS3008
A211	A210,BFR95,*GE-28,*TR-76,*HEPS3008
A306	2N2924,2N3241A,2N3242A,2N3392,2N5224,BC548A,BCY58A,BSX38,*PTC126,*GE-62,*TR-21,*HEPS0011,*SK3122,*EN10,*ECG123A,*WEP736,*BC337,*276-2014,*RT-102
A307	2N2925,2N3391,2N3391A,A5T3565,BC548B,BC549B,BCY58C,BCY58D,*GE-62,*TR-21,*HEPS0011,*SK3122,*EN10,*ECG123A,*WEP736,*BC337,*276-2014,*RT-102

To Replace	Substitute This Type
A310	A778,BF178,*PTC125,*GE-27,*TR-78,*HEPS5025,*SK3045,*ECG154,*WEP712, *BF338,*RT-110
A311	2N734,2N2459,2N2463,A777,BF177,BFY65,*PTC123,*GE-213,*TR-87,*HEPS5026, *SK3024,*ECG128,*WEP243,*BF338,*276-2009,*RT-114
A321	2N760,2N916,2N916A,2N929,2N2242,2N2331,2N2501,2N3013,2N3014,2N4123, 2N4418,2N5824,2SC67,2SC318,2SC367G-O,2SC601N,2SC639,2SC733-O, 2SC735-O,2SC764,A5T4123,TP3705,ZT44,*PTC126,*GE-61,*TR-95,*HEPS0015, *SK3039,*ECG108,*WEP56,*BC107B,*276-2013,*RT-113
A322	2N3566,2N4256,2N5827,2N6112,2SC587,2SC587A,2SC733-GR,2SC735-GR,A5T3565, A157B,BC108C,BC109,BC109C,BC183L,BC184L,BC238C,BC239C,BC413C,BC414C, BC547B,BC548B,BC549B,BC550B,BCY58C,BCY58D,BCY59C,BCY59D,GI-3566, MPSA16,MPSA18,PET4002,TP3566,TP3704,*PTC123,*GE-212,*TR-21,*HEPS0015, *SK3122,*ECG160,*WEP736,*BC107B,*276-2013
A323	2N930A,2N2523,2N2586,2N3117,2N6428,2N6428A,BC182B,BC546B,TIS94,TIS97, *PTC123,*GE-212,*TR-21,*HEPS0015,*SK3122,*ECG123A,*WEP736,*BC107B, *276-2016
A324	2N2511,2N2524,2N4104,2N6428,2N6428A,2N6539,*GE-212,*TR-21,*HEPS0015, *SK3122,*ECG123A,*WEP736,*BC107B,*276-2016
A344	2N743,2N5187,2SC68,2SC395A-R,*PTC136,*GE-20,*TR-21,*HEPS0011,*SK3122, *EN10,*ECG123A,*WEP736,*276-2016,*RT-102
A345	2N743,2SC68,*PTC136,*GE-20,*TR-21,*HEPS0011,*SK3122,*EN10,*ECG123A, *WEP736,*276-2016,*RT-102
A346	2N743,2N2318,2N2319,2N5187,2SC68,2SC395A-R,2SC595,*PTC136,*GE-20,*TR-21, *HEPS0011,*SK3122,*EN10,*ECG123A,*WEP736,*276-2016,*RT-102
A400	A406,*PTC126,*SK3031,*ECG126,*WEP635
A401	A406,*PTC126,*ECG160,*WEP637
A406	*PTC123,*ECG102A,*WEP250
A415	*PTC132,*GE-61,*TR-21,*HEPS0016,*SK3018,*EN10,*ECG123A,*WEP736,*BC107B, *276-2016,*RT-102
A417	TC3200,*PTC132,*GE-39,*TR-95,*HEPS0016,*SK3039,*EN10,*ECG108,*WEP56, *BC107B,*276-2015,*RT-113
A418	A417,TC3200,*PTC132,*GE-39,*TR-95,*HEPS0017,*SK3039,*EN10,*ECG108, *WEP56,*BF365,*276-2015,*RT-113
A419	2SC332,A420,BF184,S1297,*PTC132,*GE-39,*TR-95,*HEPS0017,*SK3039,*EN10, *ECG108,*WEP56,*BF365,*276-2015,*RT-113
A420	A419,BF184,*PTC132,*GE-39,*TR-95,*HEPS0017,*SK3039,*EN10,*ECG108,*WEP56, *BF365,*276-2015,*RT-113
A430	2N6599,2N6600,*PTC133,*TR-83,*HEPS0017,*SK3039,*ECG161,*WEP719, *276-2011
A467	2N3933,2N4259,2N5181,2SC333,2SC334,2SC463,2SC593,2SC1747,40235,40238, 40239,40242,40472,40478,BF115,*PTC132,*GE-39,*TR-95,*HEPS0016,*SK3039, *EN10,*ECG229,*WEP56,*BC107B,*276-2016,*RT-113
A473	TC3114,*PTC132,*GE-61,*TR-95,*HEPS0016,*SK3018,*EN10,*ECG108,*WEP56, *276-2015,*RT-113
A482	*PTC132,*GE-214,*TR-83,*HEPS0017,*SK3117,*EN10,*ECG161,*WEP719,*BF183, *276-2011,*RT-113
A483	*PTC132,*GE-214,*TR-83,*HEPS0017,*SK3039,*EN10,*ECG161,*WEP719,*BF183, *276-2011,*RT-113
A484	*PTC132,*GE-214,*TR-83,*HEPS0017,*SK3117,*EN10,*ECG161,*WEP719,*BF200, *276-2011,*RT-113
A485	2N2808,2N2808A,2N2809A,2N2857,2N3570,2N4252,2N4253,2N5053,2N5054,2N6304, 2N6305,2SC1747,BFX89,MT1061A,*PTC132,*GE-86,*TR-87,*HEPS0017,*SK3024, *WEP243,*276-2011,*RT-114
A486	*TR-83,*HEPS0017,*SK3039,*ECG161,*WEP719,*276-2011
A490	2N2857,2N3570,2N3839,2N4252,2N4253,2N6304,2N6305,A485,BFX89,BFY90, *PTC132,*GE-86,*TR-87,*HEPS0017,*SK3024,*ECG128,*WEP243,*276-2011, *RT-114
A492	MT1061,*PTC132,*GE-86,*TR-83,*HEPS0017,*SK3117,*ECG161,*WEP719, *276-2011,*RT-113
A496	*PTC132,*GE-61,*TR-70,*HEPS0016,*SK3191,*ECG107,*WEP56,*BF173,*276-2016
A497	*PTC132,*GE-61,*TR-95,*HEPS0016,*ECG108,*WEP56,*BF173,*276-2038

To Replace	Substitute This Type
A580-0402	A580-0403,A580-0405,*HEPS7002
A580-0403	A580-0402,A580-0405,*HEPS7002
A580-0405	A580-0402,A580-0403,*HEPS7002
A580-0802	A580-0803,A580-0805,*HEPS7004
A580-0803	A580-0802,A580-0805,*HEPS7004
A580-0805	A580-0802,A580-0803,*HEPS7004
A580-1202	A580-1203,A580-1205,A580-1602,A580-1603,A580-1605,*HEPS5020
A580-1203	A580-1202,A580-1205,A580-1602,A580-1603,A580-1605,*HEPS5020
A580-1205	A580-1202,A580-1203,A580-1602,A580-1603,A580-1605,*HEPS5020
A580-1602	A580-1603,A580-1605,A580-1802,A580-1803,A580-1805,A580-2002,A580-2003, A580-2005,*HEPS5020
A580-1603	A580-1602,A580-1605,A580-1802,A580-1803,A580-1805,A580-2002,A580-2003, A580-2005,*HEPS5020
A580-1605	A580-1602,A580-1603,A580-1802,A580-1803,A580-1805,A580-2002,A580-2003, A580-2005,*HEPS5020
A580-1802	A580-1803,A580-1805,A580-2002,A580-2003,A580-2005,A580-2202,A580-2203, A580-2205,A580-2402,A580-2403,A580-2405,*HEPS5020
A580-1803	A580-1802,A580-1805,A580-2002,A580-2003,A580-2005,A580-2202,A580-2203, A580-2205,A580-2402,A580-2403,A580-2405,*HEPS5020
A580-1805	A580-1802,A580-1803,A580-2002,A580-2003,A580-2005,A580-2202,A580-2203, A580-2205,A580-2402,A580-2403,A580-2405
A580-2002	A580-2003,A580-2005,A580-2202,A580-2203,A580-2205,A580-2402,A580-2403, A580-2405,*HEPS5020
A580-2003	A580-2002,A580-2005,A580-2202,A580-2203,A580-2205,A580-2402,A580-2403, A580-2405,*HEPS5020
A580-2005	A580-2002,A580-2003,A580-2202,A580-2203,A580-2205,A580-2402,A580-2403, A580-2405,*HEPS5020
A580-2202	A580-2002,A580-2003,A580-2005,A580-2203,A580-2205,A580-2402,A580-2403, A580-2405
A580-2203	A580-2002,A580-2003,A580-2005,A580-2202,A580-2205,A580-2402,A580-2403, A580-2405,*HEPS5020
A580-2205	A580-2002,A580-2003,A580-2005,A580-2202,A580-2203,A580-2402,A580-2403, A580-2405,*HEPS5020
A580-2402	A580-2202,A580-2203,A580-2205,A580-2403,A580-2405,*HEPS5020
A580-2403	A580-2202,A580-2203,A580-2205,A580-2402,A580-2405,*HEPS5020
A580-2405	A580-2202,A580-2203,A580-2205,A580-2402,A580-2403,*HEPS5020
A667-GRN	2N2925,2N2926-GRN,2N3391,2N3391A,2N3395-WHT,2N3396-WHT,2N3397-WHT, 2N3398-WHT,2N3415,2N3721,2N3900,2N3900A,2N3901,2N6002,2N6008,2SC732-BL, 2SC733-BL,2SC1335,2SC1344,91T6,98T2,BC108C,BC109,BC109C,BC168C, BC170C,BC172C,BC173C,BC238C,BC239C,BC548B,BC549B,BCY58C,BCY58D, GI-2925,GI-3391,GI-3391A,GI-3900,GI-3900A,MPS2925,MPS3391,MPS3391A, MPS3721,PBC108C,PBC109C,PET4002,PET4003,*PTC139,*GE-212,*TR-51, *SK3245,*ECG199,*WEP736,*BC107B,*276-2031
A667-ORG	2N2712,2N2714,2N2924,2N2926-YEL,2N3241A,2N3392,2N3395-YEL,2N3396-YEL, 2N3397-YEL,2N3398-YEL,2N3414,2N5224,2N5998,2N6000,2SC372G-Y,2SC395A-Y, 2SC400,2SC400-Y,2SC454,2SC458,2SC733-Y,2SC735-Y,2SC847,2SC848,2SC849, 2SC934,2SC966,2SC967,2SC1085,2SC1219,2SC1246,2SC1317,2SC1363,2SC1631, 2SC1633,2SD392,92T6,93T6,40232,40233,40398,40400,BC108A,BC168A,BC170B, BC172A,BC238A,BC548A,BCY58A,BSW88,BSW89,BSX38,GI-2924,GI-2926,GI-3392, GI-3396,GI-3397,GI-3721,MPS2712,MPS2714,MPS2924,MPS2926-GRN, MPS2926-YEL,MPS3392,PBC108A,SE4001,TN80,*PTC101,*GE-212,*TR-51, *SK3245,*ECG199,*WEP736,*BC107B,*276-2031
A667-RED	2N744,2N2331,2N2569,2N2570,2N2922,2N3394,2N3397-RED,2N3398-RED,2N4138, 2N4418,2N6566,2S095A,2S512,2SC99,2SC168,2SC300,2SC301,2SC372G-O, 2SC395A,2SC395A-O,2SC400-O,2SC460,2SC461,2SC528,2SC579,2SC619, 2SC733-O,2SC735-O,2SC1071,2SC1372,2SD227,40231,40234,BSY61,GI-2712, GI-2922,GI-3394,MPS2926-ORG,MPS3394,MPS6560,MPS6561,PT720,ZT2205, ZT2938,*PTC101,*GE-212,*TR-51,*SK3245,*ECG199,*WEP736,*BC107B, *276-2031
A667-YEL	2N2924,2N2925,2N2926,2N2926-GRN,2N2926-YEL,2N3392,2N3395-YEL,2N3396-YEL, 2N3397-YEL,2N3398-YEL,2N3415,2N3565,2N3721,2N4124,2N4256,2N4286,2N4419,

To Replace	Substitute This Type
(A667-YEL)	2N5172,2N5998,2SC373G,2SC400-GR,2SC732-GR,2SC733-GR,2SC735-GR,2SC1684, 2SC1849,2SC1993,92T6,98T2,40397,40399,A158B,A159B,BC108,BC108B,BC109B, BC168B,BC172B,BC173B,BC238B,BC239B,BC338,BC548B,BC549B,BCY58B, BCY58C,GI-2924,GI-2925,GI-3392,GI-3395,GI-3398,GI-3403,GI-3415,GI-3721, MPS2924,MPS2925,MPS2926,MPS3392,MPS3395,MPS3721,MPS5172,PBC108B, PBC109B,PET4002,*PTC121,*GE-212,*TR-51,*SK3245,*ECG199,*WEP736, *BC107B,*276-2031
A668-GRN	2N2925,2N2926-GRN,2N3391,2N3391A,2N3395-WHT,2N3396-WHT,2N3397-WHT, 2N3398-WHT,2N3415,2N3721,2N3900,2N3900A,2N3901,2N6002,2N6008,2SC732-BL, 2SC733-BL,2SC1335,2SC1344,91T6,98T2,BC108C,BC109,BC109C,BC168C, BC170C,BC172C,BC173C,BC238C,BC239C,BC548B,BC549B,BCY58C,BCY58D, GI-2925,GI-3391,GI-3391A,GI-3900,GI-3900A,MPS2925,MPS3391,MPS3391A, MPS3721,PBC108C,PBC109C,PET4002,PET4003,*PTC139,*GE-212,*TR-51, *SK3245,*ECG199,*WEP736,*BC107B,*276-2031
A668-ORG	2N2712,2N2714,2N2924,2N2926-YEL,2N3241A,2N3392,2N3395-YEL,2N3396-YEL, 2N3397-YEL,2N3398-YEL,2N3414,2N5224,2N5998,2N6000,2SC372G-Y,2SC395A-Y, 2SC400,2SC400-Y,2SC454,2SC458,2SC733-Y,2SC735-Y,2SC847,2SC848,2SC849, 2SC934,2SC966,2SC967,2SC1085,2SC1219,2SC1246,2SC1317,2SC1363,2SC1631, 2SC1633,2SD392,92T6,93T6,40232,40233,40398,40400,BC108A,BC168A,BC170B, BC172A,BC238A,BC548A,BCY58A,BSW88,BSW89,BSX38,GI-2924,GI-2926,GI-3392, GI-3396,GI-3397,GI-3721,MPS2712,MPS2714,MPS2924,MPS2926-GRN, MPS2926-YEL,MPS3392,PBC108A,SE4001,TN80,*PTC101,*GE-212,*TR-51, *SK3245,*ECG199,*WEP736,*BC107B,*276-2031
A668-YEL	2N2924,2N2925,2N2926,2N2926-GRN,2N2926-YEL,2N3392,2N3395-YEL,2N3396-YEL, 2N3397-YEL,2N3398-YEL,2N3415,2N3565,2N3721,2N4124,2N4256,2N4286,2N4419, 2N5172,2N5998,2SC373G,2SC400-GR,2SC732-GR,2SC733-GR,2SC735-GR, 2SC1684,2SC1849,2SC1993,92T6,98T2,40397,40399,A158B,A159B,BC108, BC108B,BC109B,BC168B,BC172B,BC173B,BC238B,BC239B,BC338,BC548B, BC549B,BCY58B,BCY58C,GI-2924,GI-2925,GI-3392,GI-3395,GI-3398,GI-3403, GI-3415,GI-3721,MPS2924,MPS2925,MPS2926,MPS3392,MPS3395,MPS3721, MPS5172,PBC108B,PBC109B,PET4002,*PTC121,*GE-212,*TR-51,*SK3245, *ECG199,*WEP736,*BC107B,*276-2031
A669-GRN	2N2925,2N2926-GRN,2N3391,2N3391A,2N3395-WHT,2N3396-WHT,2N3397-WHT, 2N3398-WHT,2N3415,2N3721,2N3900,2N3900A,2N3901,2N6002,2N6008,2SC732-BL, 2SC733-BL,2SC1335,2SC1344,91T6,98T2,BC108C,BC109,BC109C,BC168C, BC170C,BC172C,BC173C,BC238C,BC239C,BC548B,BC549B,BCY58C,BCY58D, GI-2925,GI-3391,GI-3391A,GI-3900,GI-3900A,MPS2925,MPS3391,MPS3391A, MPS3721,PBC108C,PBC109C,PET4002,PET4003,*PTC139,*GE-212,*TR-51, *SK3245,*ECG199,*WEP736,*BC107B,*276-2031
A669-YEL	2N2924,2N2925,2N2926,2N2926-GRN,2N2926-YEL,2N3392,2N3395-YEL,2N3396-YEL, 2N3397-YEL,2N3398-YEL,2N3415,2N3565,2N3721,2N4124,2N4256,2N4286,2N4419, 2N5172,2N5998,2SC373G,2SC400-GR,2SC732-GR,2SC733-GR,2SC735-GR, 2SC1684,2SC1849,2SC1993,92T6,98T2,40397,40399,A158B,A159B,BC108, BC108B,BC109B,BC168B,BC172B,BC173B,BC238B,BC239B,BC338,BC548B, BC549B,BCY58B,BCY58C,GI-2924,GI-2925,GI-3392,GI-3395,GI-3398,GI-3403, GI-3415,GI-3721,MPS2924,MPS2925,MPS2926,MPS3392,MPS3395,MPS3721, MPS5172,PBC108B,PBC109B,PET4002,*PTC121,*GE-212,*SK3245,*ECG199, *WEP736,*BC107B,*276-2031
A747	BC147B,*PTC121,*GE-210,*TR-21,*HEPS0015,*SK3122,*ECG123A,*WEP736, *BC107B,*276-2016
A748	*PTC101,*GE-10,*TR-21,*HEPS0015,*SK3124,*ECG123A,*WEP736,*BC107B, *276-2016,*RT-102
A749	BC148B,BC148C,*PTC139,*GE-212,*TR-21,*HEPS0024,*SK3124,*ECG199,*WEP736, *BC107B,*276-2016,*RT-102
A757	*PTC103,*GE-82,*TR-21,*HEPS0015,*SK3122,*ECG281,*WEP736,*BC177,*276-2023
A758	BC158,BC158A,BC158B,BC159,*PTC103,*GE-65,*TR-52,*SK3025,*ECG281, *WEP242,*BC177,*276-2034
A759	A758,BC158,BC158A,BC158B,BC159,*PTC103,*GE-65,*TR-52,*BC177,*276-2034
A777	BF177,MM3000,*PTC123,*GE-81,*TR-78,*HEPS5026,*SK3044,*ECG298,*WEP712, *BF338
A778	BF178,BF179A,MM3001,*GE-27,*TR-78,*HEPS5025,*SK3044,*ECG288,*WEP712,

To Replace	Substitute This Type
(A778)	*BF338
A779	BF179A,BF179B,BF179C,MM3002,MM3003,*GE-27,*TR-78,*HEPS5024,*SK3201, *ECG187A,*WEPS3021,*BF338
A1238	2N3242A,2N4256,2N5826,2N6112,2SC1849,A5T4124,BCY58B,BCY59B,BSX38, BSX79,MPSA16,MPSA17,TN80,TP3704,TP3706,*PTC123,*GE-62,*SK3245,*ECG199, *WEP736,*BC337,*276-2014
A1243	SEE 2N3399
A1379	2N780,2N4264,2N5772,A5T4123,BSY63,*PTC123,*GE-81,*TR-21,*HEPS0011,*SN80, *ECG123A,*WEP736,*BC337,*276-2012,*RT-102
A1380	2N4265,A5T4124,BCY58,BCY66,*PTC123,*GE-81,*TR-21,*HEPS0015,*ECG123A, *WEP736,*BC337,*276-2012,*RT-102
A1383	*PTC107,*GE-51,*TR-17,*HEPG0001,*SK3006,*HF35,*ECG126,*WEP635
A1384	2N838,2N2956,2N2957,2N3323,2N3324,2N3325,*GE-51,*TR-17,*HEPG0001,*SK3123, *HF75,*ECG126,*WEP635
AA1	REFER TO SECTION 2
AA2	REFER TO SECTION 2
AA3	REFER TO SECTION 2
AA4	REFER TO SECTION 2
AA5	REFER TO SECTION 2
AC105	SEE AC117
AC106	SEE AC117
AC107	2N2953,2SA138,2SA139,2SB75,2SB77,2SB328,2SB329,AFY15,OC44,OC45,OC47, SF.T307,SF.T308,SF.T323,*PTC109,*GE-50,*TR-17,*HEPG0003,*HF35,*ECG160, *WEP637,*AF125
AC116-GRN	2N383,2N417,2N527,2N598,2N599,2N652,2N655,2N655+,2N1185,2N1189,2N1193, 2N1309,2N1354,2N1355,2N1357,2N1449,2N1892,2N1997,2N1998,2N2171,2SB54, 2SB370A,2SB431,2SB439,2SB440,AC123-GRN,MA888,OC80,*PTC135
AC116-YEL	2N109+,2N217+,2N382,2N383,2N415,2N415A,2N416,2N422A,2N467,2N527, 2N527A,2N597,2N651,2N651A,2N652,2N652A,2N654,2N654+,2N1008A,2N1124, 2N1125,2N1189,2N1307,2N1309A,2N1349,2N1352,2N1356,2N1375,2N1377, 2N1448,2N1449,2N1681,2N1892,2N1997,2SB56,2SB77A,2SB227,2SB327,2SB428, 2SB431,AC116-GRN,AC123-GRN,AC123-YEL,ACY23,ACY32,MA887,SF.T223, *PTC135
AC117	2N2431,2SB461,2SB619,AC124,*PTC156,*GE-2,*TR-85,*HEPG0005,*AT20M, *ECG102A,*WEP250,*276-2005,*RT-121
AC121	*PTC135,*GE-53,*TR-85,*HEPG0005,*ECG102A,*WEP250,*AC188/01,*276-2005, *RT-121
AC122-GRN	2SB370A,2SB431,AC122-VIO,AC122/30-GRN,AC122/30-VIO,*PTC135,*GE-53,*TR-05, *SK3123,*WEP250,*AC188/01,*RT-121
AC122-RED	AC122-YEL,AC122/30-RED,AC122/30-YEL,ACY23,ACY32,*PTC135,*GE-53,*TR-05, *SK3123,*WEP250,*AC188/01,*RT-121
AC122-VIO	AC122-WHT,AC122/30-VIO,AC122/30-WHT,*PTC135,*GE-53,*AC188/01
AC122-WHT	AC122/30-WHT,*DS26,*AC188/01
AC122-YEL	2SB431,AC122-GRN,AC122/30-GRN,AC122/30-YEL,ACY23,ACY32,*PTC135,*GE-53, *TR-05,*SK3123,*WEP250,*AC188/01,*RT-121
AC122/30-GRN	2SB136A,AC122/30-VIO
AC122/30-RED	AC122/30-YEL
AC122/30-VIO	AC122/30-WHT
AC122/30-WHT	
AC122/30-YEL	2SB136A,AC122/30-GRN
AC123-GRN	2N383,2N527,2N652,2N1185,2N1188,2N1189,2N1449,2N1955,2N1997,2N2171, 2SB136A,MA883,MA888
AC123-YEL	2N382,2N383,2N527,2N527A,2N597,2N651,2N651A,2N652,2N652A,2N1008B,2N1187, 2N1188,2N1189,2N1375,2N1377,2N1448,2N1449,2N1926,2N1954,2N1955,2N1956, 2N1957,2N1997,2SB77A,2SB136A,2SB227,2SB428,AC123-GRN,MA882,MA887
AC124	2N2000,*DS-26,*GE-53,*TR-85,*HEPS0011,*ECG102A,*WEP250,*276-2009,*RT-121
AC125	*GE-53,*TR-85,*HEPG0005,*AT20M,*ECG102A,*WEP250,*276-2005,*RT-121
AC126	*GE-2,*TR-85,*HEPG0005,*AT20M,*ECG102A,*WEP250,*276-2004,*RT-121
AC127	*PTC108,*GE-59,*TR-08,*HEPG0011,*SK3010,*NA20,*ECG103A,*WEP641B, *276-2001,*RT-122
AC127Z	AC127,*HEPG0011

To Replace	Substitute This Type
AC128	AC153,*GE-53,*TR-84,*HEPG0005,*SK3004,*AT20M,*ECG158,*WEP238,*276-2006
AC129	*GE-51,*TR-17,*HEPG0003,*HF35,*ECG160,*WEP637,*AF125
AC130	2N797,2SD38,2SD75,2SD77,AC127,*GE-5,*TR-08,*HEPG0011,*SK3010,*NR10, *ECG101,*WEP641,*AC127,*276-2001,*RT-119
AC131	AC131/30,*PTC156,*GE-2,*TR-85,*HEPG0005,*AT20M,*ECG102A,*WEP250, *276-2005,*RT-121
AC131/30	*WEP238
AC132	*GE-2,*TR-85,*HEPG0005,*AT20M,*ECG102A,*WEP250,*276-2005,*RT-121
AC150-GRN	2SB54,2SB370A,2SB431,2SB439,2SB440,AC122-GRN,AC122-VIO,AC122/30-GRN, AC122/30-VIO,AC150-VIO,*PTC135,*GE-51,*TR-85,*WEP250,*AF125,*RT-121
AC150-VIO	2N2613+,2SB439,2SB440,AC122-VIO,AC122-WHT,AC122/30-VIO,AC122/30-WHT, AC150-WHT,*PTC135
AC150-WHT	
AC150-YEL	2N109+,2N217+,2N2614,2N2953,2SB56,2SB77A,2SB431,AC122-GRN,AC122-YEL, AC122/30-GRN,AC122/30-YEL,AC150-GRN,ACY23,ACY32,*PTC135,*GE-51,*TR-85, *WEP250,*AF125,*RT-121
AC151	AC151R,AC152,ASY70,*PTC135,*GE-53,*TR-85,*HEPG0005,*SK3004,*ECG102A, *WEP250,*276-2005,*RT-121
AC151R	AC151,AC152,ASY70,*PTC135,*GE-53,*TR-85,*HEPG0005,*SK3004,*ECG102A, *WEP250,*276-2005,*RT-121
AC152	*PTC135,*GE-53,*TR-84,*HEPG0005,*SK3004,*AT20M,*ECG158,*WEP238, *276-2005
AC153	*DS-26,*GE-53,*TR-84,*HEPG6011,*SK3004,*ECG158,*WEP238,*276-2005
AC153K	AC153,*GE-53,*TR-84,*HEPG6011,*SK3004,*ECG158,*WEP238,*276-2005
AC160-GRN	2SA49,2SA252,2SB54,2SB116,2SB117,2SB136,2SB186,2SB439,2SB440,2SB443, 2SB444,AC160-VIO,AC170,AC171,*PTC135,*GE-51,*TR-85,*WEP250,*AF125
AC160-RED	2N140+,2SA53,2SA79,2SA138,2SA139,2SA182,2SA251,2SA325,2SA326,2SB37, 2SB56,2SB59,2SB66,2SB75,2SB91,2SB95,2SB114,2SB171,2SB172,2SB288,2SB328, 2SB365,AC160-YEL,NKT213,NKT214,NKT216,SF.T307,SF.T322,*PTC109,*GE-51, *TR-85,*WEP250,*AF125
AC160-VIO	2N2613+,2SB400,*PTC135,*GE-51,*AF125
AC160-YEL	2N2953,2SA49,2SA52,2SA53,2SA408,2SA409,2SB56,2SB77,2SB115,2SB116, 2SB136,2SB173,2SB175,2SB176,2SB186,2SB187,2SB188,2SB303,2SB329,2SB364, AC160-GRN,AC170,ACY23,ACY32,NKT213,NKT216,SF.T307,SF.T308,SF.T323, *PTC107,*GE-51,*TR-85,*WEP250,*AF125
AC162	AC163,*PTC135,*GE-53,*TR-85,*HEPG0005,*ECG102A,*WEP250,*276-2005,*RT-121
AC163	*PTC135,*GE-53,*TR-85,*HEPG0005,*ECG102A,*WEP250,*276-2005,*RT-121
AC170	AC171,*PTC135,*GE-53,*TR-85,*HEPG0005,*SK3123,*ECG102A,*WEP250, *AC188/01,*276-2005,*RT-121
AC171	*PTC135,*GE-53,*TR-85,*HEPG0005,*ECG102A,*WEP250,*AC188/01,*276-2005, *RT-121
AC175	*GE-20,*TR-08,*HEPG0011,*SK3124,*ECG103A,*WEP641B,*RT-122
AC176	*TR-08,*HEPG0011,*SK3010,*ECG103A,*WEP641B,*AC187/01,*RT-122
AC176K	AC176,*TR-84,*HEPG6011,*ECG158,*WEP238,*AC187/01,*276-2006
AC178	2N2001,*PTC156,*TR-84,*ECG158,*WEP238
AC179	AC175,*GE-54,*TR-08,*HEPG0011,*ECG103A,*WEP641B,*276-2001,*RT-122
AC186	*GE-54,*TR-08,*HEPG0011,*ECG103A,*WEP641B,*RT-122
AC187	*GE-54R,*TR-08,*HEPG0011,*SK3010,*ECG103A,*WEP641B,*276-2001,*RT-122
AC187/01	REFER TO SECTION 2
AC187K	*ECG103A,*WEP238,*AC187/01
AC188	*GE-53,*TR-84,*HEPG6011,*SK3004,*ECG158,*WEP238,*276-2005
AC188/01	REFER TO SECTION 2
AC188K	*TR-84,*HEPG6011,*ECG158,*WEP238,*AC188/01
ACY16	*PTC135,*GE-16,*TR-01,*HEPG0005,*SK3009,*ECG121,*WEP232,*RT-127
ACY23	ACY32,*GE-53,*TR-85,*HEPG0005,*SK3123,*ECG102A,*WEP250,*276-2005,*RT-121
ACY24	*GE-50,*HEPG6011,*ECG160,*WEP637
ACY32	ACY23,*GE-53,*TR-85,*HEPG0005,*SK3123,*ECG102A,*WEP250,*276-2005,*RT-121
ACY33	*PTC156,*GE-53,*TR-85,*HEPG0005,*ECG102A,*WEP250,*276-2005,*RT-121
ACZ10	*HEPG6012
AD130	2N257-GRN,2SB250,2SB251,2SB426A,AD149,AD150,ADY27,CDT1310,*PTC138, *GE-25,*TR-01,*HEPG6003,*SK3009,*ECG121,*WEP232,*OC28,*276-2006,*RT-127

TRANSISTOR SUBSTITUTES

To Replace	Substitute This Type
AD131	2N665,2SB128,2SB250A,2SB251A,2SB252,2SB253,AD132,AUY19,AUY20,CDT1311, CDT1312,*PTC105A,*GE-25,*TR-01,*HEPG6005,*SK3009,*ECG121,*WEP232, *OC28,*276-2006,*RT-127
AD132	2N665,2SB128,2SB252,2SB253,AD163,AUY20,AUY34,CDT1312,CDT1313,*PTC105A, *GE-25,*TR-01,*HEPG6005,*SK3014,*ECG121,*WEP232,*OC28,*276-2006,*RT-127
AD133	AUY29,*PTC901,*GE-76,*TR-35,*HEPG6005,*SK3014,*ECG179,*WEPG6001, *276-2006,*RT-147
AD136	
AD138	*PTC901,*GE-76,*TR-01,*HEPG6003,*SK3009,*PT40,*ECG121,*WEP232,*276-2006, *RT-127
AD139	AD148,*PTC120A,*GE-30,*TR-01,*HEPG6016,*SK3009,*ECG104,*WEP230, *276-2006,*RT-124
AD148	*PTC120A,*GE-3,*TR-94,*HEPG6003,*PT40,*WEP642,*276-2006,*RT-127
AD149	CDT1311,*GE-25,*TR-01,*HEPG6003,*SK3013,*PT40,*ECG104,*WEP230,*OC28, *276-2006,*RT-124
AD150	AD149,ADY27,CDT1310,*PTC138,*GE-25,*TR-01,*HEPG6003,*SK3030,*PT40, *ECG121,*WEP232,*OC28,*276-2006,*RT-127
AD152	2SB368,*PTC120A,*GE-16,*TR-94,*HEPG6016,*SK3198,*ECG131,*WEP642, *276-2006,*RT-127
AD155	2SB254,2SB256,2SB367,*PTC120A,*TR-94,*HEPG6016,*SK3082,*ECG131,*WEP642, *276-2006,*RT-127
AD159	AD136,*DS-520,*GE-16,*TR-01,*ECG121,*WEP232,*RT-127
AD160	*GE-12,*TR-81,*ECG175,*WEP241
AD161	*GE-43,*TR-76,*ECG155
AD162	*PTC120A,*GE-44,*TR-94,*HEPG6016,*SK3052,*ECG131,*WEP642,*276-2006, *RT-127
AD163	2SB128A,2SB252A,2SB253A,AUY34,CDT1313,*PTC105A,*GE-25,*TR-35,*HEPG6007, *WEPG6001,*RT-147
AD164	2N4078,AD162,*GE-44,*TR-94,*HEPG6016,*SK3198,*ECG131,*WEP642,*276-2006, *RT-127
AD165	2N4077,*ECG155
ADY26	2SB480,*GE-4,*TR-05,*HEPG6010,*SK3012,*PT501,*ECG213,*WEP233
ADY27	2SB137,AD149,AD150,CDT1310,CDT1319,*PTC138,*GE-25,*TR-01,*HEPG6003, *ECG121,*WEP232,*OC28,*276-2006,*RT-127
ADZ11	2SB478,*GE-4,*TR-03,*HEPG6004,*SK3012,*PT501,*ECG105,*WEP233,*276-2006
ADZ12	2N1100,2N1358,2N1412,2SB334,2SB480,ADY26,*GE-4,*TR-03,*HEPG6006,*SK3012, *PT501,*ECG105,*WEP233
AF101	2SA138,2SA139,AFY15,*PTC109,*GE-50,*TR-17,*HEPG0005,*HF20M,*ECG160, *WEP637,*AF125,*276-2004
AF102	SEE 2N2494
AF105	2N838,2N2188,2N2273,2N2956,2N2957,2N3323,2N3324,2N3325,2N6365,*PTC107, *GE-50,*TR-17,*HEPG0003,*SK3008,*JR100,*ECG126,*WEP635,*AF125
AF106	2N3783,2N3784,AFY16,AFY39,MM5000,MM5001,MM5002,*GE-50,*TR-17, *HEPG0003,*SK3006,*JR200,*ECG160,*WEP637
AF109R	2N2966,2N3127,2N3783,2N3784,2SA436,2SA437,2SA438,AF106,AF139,AF239, AFY12,AFY16,AFY39,AFY42,MM5000,MM5001,MM5002,*GE-50,*TR-17,*HEPG0003, *ECG160,*WEP637
AF114	SEE 2N2089
AF115	SEE 2N2089
AF116	SEE 2N2092
AF117	SEE 2N2092
AF118	*GE-51,*TR-17,*HEPG0003,*SK3006,*ECG160,*WEP637
AF121	*GE-50,*TR-17,*HEPG0003,*ECG160,*WEP637
AF124	2N987,AF125,AF126,AF127,*PTC145,*GE-51,*TR-17,*HEPG0008,*SK3006,*JR100, *ECG160,*WEP637,*AF125,*276-2005
AF125	2N987,AF124,AF126,AF127,*PTC145,*GE-51,*TR-17,*HEPG0008,*SK3006,*JR100, *ECG160,*WEP637,*AF125,*276-2005
AF126	2N987,AF124,AF125,AF127,*PTC145,*GE-51,*TR-17,*HEPG0008,*SK3006,*JR100, *ECG160,*WEP637,*AF125,*276-2005
AF127	2N987,AF124,AF125,AF126,*PTC145,*GE-51,*TR-17,*HEPG0008,*SK3005,*JR100, *ECG160,*WEP637,*AF125,*276-2005

To Replace	Substitute This Type
AF128	*GE-50,*TR-17,*HEPG0005,*SK3005,*HF12M,*ECG160,*WEP637,*276-2004
AF134	2N987,2N2588,AF124,AF125,AF126,AF127,AF135,*DS-25,*GE-50,*TR-17, *HEPG0008,*SK3008,*JR100,*ECG126,*WEP242,*276-2005
AF135	2N987,2N2588,AF124,AF125,AF126,AF127,AF134,*PTC107,*GE-50,*TR-17, *HEPG0008,*SK3008,*HF50H,*ECG126,*WEP242,*276-2005
AF136	2N2588,2N3783,2N3784,AF134,AF135,AF138,*PTC107,*GE-50,*TR-17,*HEPG0008, *SK3008,*HF50H,*ECG126,*WEP242,*276-2005
AF137	2N2588,2N3783,2N3784,AF135,AF136,AF138,*DS25,*GE-50,*TR-17,*HEPG0008, *SK3008,*HF50H,*ECG126,*WEP242,*276-2005
AF138	2N987,2N2588,AF124,AF125,AF126,AF127,AF134,AF135,AF136,*PTC107,*GE-50, *TR-17,*HEPG0008,*SK3006,*HF50H,*ECG126,*WEP242,*276-2005
AF139	2N2966,2N3783,2N3784,2SA438,AF239,AFY16,AFY39,AFY42,MM5000,MM5001, MM5002,*TR-17,*HEPG0003,*ECG160,*WEP637
AF178	2N3279,2N3280,2N3283,2N3284,MM1139,*GE-50,*TR-17,*HEPG0002,*SK3006, *ECG160,*WEP637,*276-2005
AF179	*GE-50,*TR-17,*HEPG0002,*SK3006,*ECG160,*WEP637,*276-2005
AF180	SEE 2N3074
AF181	SEE 2N3075
AF200	*GE-51,*TR-17,*HEPG0005,*SK3006,*ECG160,*WEP637,*276-2005
AF200U	2N384,2N2588,2N3588,2N3783,2N3784,2SA217,2SA246,AF106,AF124,AF125,AF126, AF127,AF134,AF135,AF136,AF138,AF200,AF201,AF201U,AF202,AF202S,AFY16, AFY39,*GE-53,*ECG102,*276-2006
AF201	AF200,AF202,AF202S,*GE-51,*TR-17,*HEPG0005,*SK3006,*ECG160,*WEP637, *276-2005
AF201U	2N384,2N2588,2N3588,2N3783,2N3784,2SA217,2SA246,AF106,AF124,AF125,AF126, AF127,AF134,AF135,AF136,AF138,AF200,AF200U,AF201,AF202,AF202S,AFY16, AFY39,*GE-53,*ECG102,*276-2006
AF202	AF202S,*GE-51,*TR-17,*HEPG0005,*SK3006,*ECG160,*WEP637,*276-2005
AF202S	*TR-17,*HEPG0005,*SK3006,*ECG160,*WEP637,*276-2005
AF239	2N3783,2N3784,AFY37,AFY42,MM5000,MM5001,MM5002,*TR-17,*HEPG0003, *SK3006,*ECG160,*WEP637
AF239S	2N3783,2N3784,2N3785,AF239,AFY42,MM5000,MM5001,MM5002,TIXM101,*TR-17, *HEPG0003,*SK3006,*ECG160,*WEP637
AF240	AFY37,GM290A,*TR-17,*HEPG0003,*SK3006,*ECG160,*WEP637
AF256	*GE-50,*TR-17,*HEPG0002,*SK3006,*ECG160,*WEP637,*276-2005
AF279	*GE-51,*TR-17,*HEPG0003,*SK3006,*ECG160,*WEP637
AF280	*HEPG0003,*ECG121,*WEP232
AFY11	*GE-50,*TR-17,*HEPG0002,*SK3006,*ECG126,*WEP635,*276-2002
AFY12	2N3127,2N3281,2N3282,2N3783,2N3784,*PTC107,*GE-50,*TR-17,*HEPG0003, *SK3006,*ECG160,*WEP637
AFY14	*PTC145,*GE-50,*TR-17,*HEPG0008,*SK3006,*ECG160,*WEP637,*276-2005
AFY15	*GE-50,*TR-17,*HEPG0008,*SK3006,*ECG160,*WEP637,*276-2005
AFY16	2N3783,2N3784,*PTC145,*TR-17,*HEPG0003,*SK3006,*ECG160,*WEP637
AFY18	*GE-53,*TR-17,*HEPG0002,*SK3006,*ECG126,*WEP635,*276-2005
AFY19	*GE-80,*HEPS0012,*ECG160
AFY37	*GE-50,*TR-17,*HEPG0003,*SK3006,*ECG160,*WEP637
AFY39	*GE-50,*TR-17,*HEPG0003,*SK3006,*ECG160,*WEP637
AFY42	2N3783,2N3784,AFY37,MM5000,MM5001,MM5002,*TR-17,*HEPG0008,*SK3006, *ECG160,*WEP637
AFZ10	
AFZ12	2N3783,2N3784,*PTC107,*GE-50,*TR-17,*HEPG0003,*SK3006,*ECG160,*WEP637
ALZ10	*TR-05,*SK3005,*ECG100,*WEP254,*RT-118
AMF101	2N2032,AMF122,AMF122A,AMF123,AMF123A,*PTC165,*HEPS5000
AMF102	AMF112,AMF113,AMF121,AMF121A,*HEPS5000
AMF103	2S024,TI-1126,TI-1145,TI-1146,*HEPS5000
AMF104	2N5067,40514,AMF118,AMF118A,AMF119,AMF119A,M5A,*PTC129A,*GE-19,*TR-59, *HEPS7002,*SK3027,*ECG130,*WEP247,*BDY20,*RT-131
AMF105	2N1487,2N1489,2N1702,2N3232,2N3445,2N5068,2N5069,2N5869,2N5870,2N6315, 2N6316,2N6317,2N6318,2SC241,2SC244,2SD15,AMF115,AMF116,AMF117, AMF117A,BDY23A,HST9201,HST9205,HST9206,HST9801,HST9802,ZT1487, ZT1489,*PTC129A,*GE-77,*TR-59,*HEPS7002,*SK3027,*ECG130,*WEP247,

To Replace	Substitute This Type
(AMF105)	*BDY20,*RT-131
AMF106	2N1488,2N1490,2N3233,2N3446,2N5759,2SC240,2SC242,2SC245,2SD16,2SD118-R, 2SD119-R,BDY24A,HST9202,HST9203,HST9207,HST9208,HST9803,HST9804,M5B, ZT1488,ZT1490,*PTC129A,*GE-73,*TR-61,*HEPS5004,*ECG163,*WEP740
AMF107	2N1209,ST401,*HEPS5000
AMF108	2N1208,2N1616A,2N1617A,2N3487,AMF110,AMF111,AMF114,ST400,*HEPS5000
AMF109	2N1618A,2N3488,2N3489,TI-1136,TI-1155,*HEPS5004
AMF110	2N1208,2N1616A,2N1617A,2N3487,AMF108,AMF111,AMF114,ST400,*HEPS5000
AMF111	2N1616A,2N1617A,2N3487,AMF114,ST400,*HEPS5004
AMF112	AMF113,*HEPS5004
AMF113	AMF112,*HEPS5004
AMF114	2N1616A,2N1617A,2N3487,AMF111,ST400,*HEPS5004
AMF115	2N3055A,2N3232,2N3235,2N3445,2N6315,2N6316,2N6317,2N6318,2SD163,2SD172, 2SD211,1582-0603,1582-0604,1582-0605,AMF116,B170004-BLK,B170004-BRN, B170005-BLK,B170005-BRN,B170013,B170014,B170022,B170023,HST9201, HST9205,HST9206,HST9801,HST9802,KSP1171,*PTC140,*GE-77,*TR-59, *HEPS7002,*SK3027,*ECG130,*WEP247,*BDY20,*RT-131
AMF116	2N3055A,2N3232,2N3235,2N3445,2N6315,2N6316,2N6317,2N6318,2SD163,2SD172, 2SD211,1582-0603,1582-0604,1582-0605,AMF115,B170004-BLK,B170004-BRN, B170005-BLK,B170005-BRN,B170013,B170014,B170022,B170023,HST9201, HST9205,HST9206,HST9801,HST9802,KSP1171,*PTC140,*GE-77,*TR-59, *HEPS7002,*SK3027,*ECG130,*WEP247,*BDY20,*RT-131
AMF117	2N1487,2N1489,2N1702,2N3232,2N3445,2N5068,2N5069,2N5869,2N5870,2N6315, 2N6316,2N6317,2N6318,2SC241,2SC244,2SD15,AMF105,AMF115,AMF116, AMF117A,BDY23A,HST9201,HST9205,HST9206,HST9801,HST9802,M5A,ZT1487, ZT1489,*PTC129A,*GE-19,*TR-59,*HEPS7002,*SK3027,*ECG130,*WEP247, *BDY20,*RT-131
AMF117A	2N1487,2N1489,2N1702,2N3232,2N3445,2N5068,2N5069,2N5869,2N5870,2N6315, 2N6316,2N6317,2N6318,2SC241,2SC244,2SD15,AMF105,AMF115,AMF116,AMF117, BDY23A,HST9201,HST9205,HST9206,HST9801,HST9802,M5A,ZT1487,ZT1489, *PTC129A,*GE-19,*TR-59,*HEPS7002,*SK3027,*ECG130,*WEP247,*BDY20, *RT-131
AMF118	2N1487,2N1489,2N1702,2N5068,2N5869,2N6315,2N6317,2SC241,2SC244,2SD15, 40514,AMF105,AMF115,AMF116,AMF117,AMF117A,AMF118A,BDY23A,HST9201, HST9205,HST9801,M5A,ZT1487,ZT1489,*PTC129A,*GE-19,*TR-59,*HEPS7002, *SK3027,*ECG130,*WEP247,*BDY20,*RT-131
AMF118A	2N1487,2N1489,2N1702,2N5068,2N5869,2N6315,2N6317,2SC241,2SC244,2SD15, 40514,AMF105,AMF115,AMF116,AMF117,AMF117A,AMF118,BDY23A,HST9201, HST9205,HST9801,M5A,ZT1487,ZT1489,*PTC129A,*GE-19,*TR-59,*HEPS7002, *SK3027,*ECG130,*WEP247,*BDY20,*RT-131
AMF119	2N5067,40514,AMF117,AMF117A,AMF118,AMF118A,AMF119A,HST9201,HST9205, M5A,*PTC129A,*GE-19,*TR-59,*HEPS7002,*SK3027,*ECG130,*WEP247,*BDY20, *RT-131
AMF119A	2N5067,40514,AMF117,AMF117A,AMF118,AMF118A,AMF119,HST9201,HST9205, M5A,*PTC129A,*GE-19,*TR-59,*HEPS7002,*SK3027,*ECG130,*WEP247,*BDY20, *RT-131
AMF120	2N5067,AMF104,AMF119,AMF119A,AMF120A,*PTC129A,*GE-19,*TR-59,*HEPS7002, *SK3027,*ECG130,*WEP247,*BDY20,*RT-131
AMF120A	2N5067,AMF104,AMF119,AMF119A,AMF120,*PTC129A,*GE-19,*TR-59,*HEPS7002, *SK3027,*ECG130,*WEP247,*BDY20,*RT-131
AMF121	AMF102,AMF112,AMF113,AMF121A,*GE-14,*HEPS5000
AMF121A	AMF102,AMF112,AMF113,AMF121,*HEPS5000
AMF122	2N2032,AMF102,AMF112,AMF113,AMF121,AMF121A,AMF122A,AMF123A, *HEPS5000
AMF122A	2N2032,AMF102,AMF112,AMF113,AMF121,AMF121A,AMF122,AMF123A,*HEPS5000
AMF123	2N2032,AMF121,AMF121A,AMF122,AMF122A,AMF123A,*HEPS5000
AMF123A	2N2032,AMF102,AMF112,AMF113,AMF121,AMF121A,AMF122,AMF122A,*HEPS5000
AMF124	AMF101,AMF123,AMF124A,*HEPS5000
AMF124A	AMF101,AMF123,AMF124,*HEPS5000
AMF201	1561-0403,1561-0404,1582-0403,1582-0404,1582-0405,1723-0405,1723-0410, B170002-BLK,B170011,B170020,HST9210,MJ2801,*PTC140,*GE-77,*HEPS5004,

To Replace	Substitute This Type
(AMF201)	*ECG130,*WEP247,*BDY20
AMF201B	2N3238,2N3239,1561-0803,1561-0804,1561-0805,1561-1005,1582-0803,1582-0804, 1582-0805,1582-1003,1582-1004,1582-1005,1723-0805,1723-0810,1723-1005, 1723-1010,AMF201C,B170005-BLK,B170008-BLK,B170014,B170017,B170023, B170026,HST9206,HST9207,MJ3772,*PTC140,*GE-75,*HEPS5004,*ECG181, *WEPS7000
AMF201C	1561-1005,1561-1205,1582-1003,1582-1004,1582-1005,1582-1203,1582-1204, 1582-1205,1723-1005,1723-1010,1723-1205,1723-1210,AMF201D,B170008-BLK, B170017,B170026,HST9207,HST9208,MJ3772,*PTC173,*GE-75,*HEPS5004, *ECG181,*WEPS7000
AMF201D	2N3240,2SC1440,1561-1205,1561-1404,1561-1405,1561-1604,1561-1605,1582-1203, 1582-1204,1582-1205,1582-1403,1582-1404,1582-1405,1582-1603,1582-1604, 1582-1605,1723-1205,1723-1210,1723-1405,1723-1410,1723-1605,1723-1610, AMF201E,HST9208,HST9209
AMF201E	2N3240,2SC1440,2SC1441,1561-1404,1561-1405,1561-1604,1561-1605,1561-1803, 1561-1804,1561-1805,1582-1403,1582-1404,1582-1405,1582-1603,1582-1604, 1582-1605,1582-1803,1582-1804,1582-1805,1723-1405,1723-1410,1723-1605, 1723-1610,1723-1805,1723-1810,HST9209
AMF210	HST9210,M10A,*PTC140,*GE-75,*TR-59,*HEPS5004,*SK3027,*ECG130,*WEP247, *RT-131
AMF210A	2SD163,HST9205,HST9206,*PTC140,*GE-75,*TR-59,*HEPS5004,*SK3027,*ECG130, *WEP247,*RT-131
AMF210B	2SD164,HST9207,HST9208,M10B,*PTC118,*GE-75,*TR-61,*HEPS5004,*ECG181, *WEPS7000
AMF210C	2SC1440,2SC1441,2SD165,2SD166,HST9209,M10C,*PTC118,*TR-61
AMF227	AMF228,*HEPS5004
AMF227A	AMF228A,*HEPS5004
AMF227B	AMF228B,*HEPS5004
AMF227C	AMF228C
AMF228	AMF227,*HEPS5004
AMF228A	AMF227A,*HEPS5004
AMF228B	AMF227B,*HEPS5004
AMF228C	AMF227C
AMF229	AMF227,AMF228,*HEPS5000
AMF229A	AMF227A,AMF228A,*HEPS5000
AMF229B	AMF227B,AMF228B,*HEPS5000
AMF229C	AMF227C,AMF228C
ASY23	
ASY24	*DS25,*GE-51,*TR-85,*HEPG0008,*JR100,*ECG102A,*WEP250,*276-2005,*RT-121
ASY24B	ASY24,*PTC145
ASY26	2N414,2N414B,2N414C,2N415,2N415A,2N416,2N526A,2N1171,2N1305,2N1309A, 2N1348,2N1349,2N1350,2N1351,2N1681,2SA204,2SA206,2SA207,2SA407,*GE-51, *TR-05,*HEPG0005,*SK3123,*HF35,*ECG100,*WEP254,*276-2004,*RT-118
ASY27	2N417,2N598,2N599,2N1307,2N1309,2N1355,2N1356,2N1357,2N1892,2N1998, *GE-51,*TR-05,*HEPG0005,*HF35,*ECG100,*WEP254,*276-2004,*RT-118
ASY28	2N440,2N440A,*PTC108,*GE-51,*TR-08,*HEPG0011,*SK3011,*HF35,*ECG101, *WEP641,*276-2001,*RT-119
ASY29	*PTC108,*GE-5,*TR-08,*HEPG0011,*SK3011,*HF35,*ECG101,*WEP641,*276-2001, *RT-119
ASY30	*TR-17,*SK3006,*ECG126,*WEP635
ASY48	*GE-53,*TR-85,*HEPG0005,*SK3004,*ECG102A,*WEP250,*276-2004,*RT-121
ASY70	*GE-53,*TR-85,*HEPG0005,*SK3004,*ECG102A,*WEP250,*RT-121
ASZ15	2N2140,2N2140A,2N2141,2N2141A,2SB87,2SB88,2SB340,*PTC122,*GE-25,*TR-01, *HEPG6005,*SK3009,*ECG121,*WEP232,*276-2006,*RT-127
ASZ16	2N2144,2N2144A,2N2145,2N2145A,2SB339,2SB538,ASZ18,KR6501,*PTC105A, *GE-25,*TR-01,*HEPG6005,*SK3009,*ECG121,*WEP232,*276-2006,*RT-127
ASZ17	2N2139,2N2139A,2N2140,2N2140A,2N2144,2N2144A,2N2145,2N2145A,2SB339, 2SB433,ASZ18,KR6500,*PTC105A,*GE-25,*TR-01,*HEPG6005,*SK3009,*ECG121, *WEP232,*276-2006,*RT-127
ASZ18	2N2145,2N2145A,2N2146,2N2146A,2SB339,2SB484,*PTC105A,*GE-25,*TR-01, *HEPG6005,*SK3009,*ECG121,*WEP232,*276-2006,*RT-127

To Replace	Substitute This Type
ASZ21	2N838,2N2717,*PTC107,*GE-51,*TR-17,*HEPG0003,*SK3006,*ECG160,*WEP637
AT10H	REFER TO SECTION 2
AT10M	REFER TO SECTION 2
AT10N	REFER TO SECTION 2
AT20H	REFER TO SECTION 2
AT20M	REFER TO SECTION 2
AT20N	REFER TO SECTION 2
AT30H	REFER TO SECTION 2
AT30M	REFER TO SECTION 2
AT30N	REFER TO SECTION 2
AT100H	REFER TO SECTION 2
AT100M	REFER TO SECTION 2
AT100N	REFER TO SECTION 2
AUY10	*GE-16,*TR-01,*HEPG6005,*SK3009,*PT40,*ECG121,*WEP230,*276-2006,*RT-124
AUY18	
AUY19	2N665,2SB128,2SB250A,2SB251A,2SB252,2SB253,AD131,AD132,AUY20,CDT1311, CDT1312,*PTC105A,*GE-25,*TR-01,*HEPG6005,*SK3014,*ECG121,*WEP232, *OC28,*276-2006,*RT-127
AUY20	2N665,2SB128,2SB252,2SB253,AD132,AD163,AUY34,CDT1312,CDT1313,*PTC105A, *GE-25,*TR-01,*HEPG6005,*SK3014,*ECG121,*WEP232,*OC28,*276-2006,*RT-127
AUY21	*PTC105A,*GE-16,*TR-01,*HEPG6005,*PT40,*ECG104,*WEP230,*276-2006,*RT-124
AUY22	*PTC105A,*GE-16,*TR-01,*HEPG6005,*PT40,*ECG104,*WEP230,*276-2006,*RT-124
AUY28	2SB88,2SB128A,2SB340,2SB341,KR6502,*PTC105A,*GE-25,*TR-27,*HEPG6005, *SK3035,*ECG127,*WEP235,*OC28
AUY29	AD133,*PTC901,*TR-35,*HEPG6009,*ECG179,*WEPG6001
AUY34	2SB128A,2SB252A,2SB253A,AD163,*PTC105A,*GE-25,*HEPG6018
B1013	SEE 2N2282
B1013A	SEE 2N2283
B1274	SEE 2N2291
B1274A	SEE 2N2292
B1274B	SEE 2N2293
B3456	
B3458	*HEPS3002
B3459	
B3459A	
B3460	SEE 2N4225
B3461	SEE 2N4226
B3465	*SK3048,*ECG195
B3466	
B3531	HST5007,HST5008,HST5507,HST5508,*GE-28,*SK3192,*ECG186,*WEP245
B3532	HST5008,HST5508
B3533	HST4453,HST4454,HST5002,HST5003,HST5502,HST5503,*GE-28,*TR-76,*SK3192, *ECG186,*WEP245
B3534	HST4454,HST5003,HST5503
B3535	
B3536	
B3537	HST5006,HST5007,HST5506,HST5507,HST9005,*GE-28,*TR-76,*SK3192,*ECG186, *WEP245
B3538	*SK3192,*ECG186,*WEP245
B3539	
B3540	*SK3192,*ECG186,*WEP245
B3541	*SK3192,*ECG186,*WEP245
B3542	*SK3192,*ECG186,*WEP245
B3543	
B3544	
B3545	
B3546	
B3547	2N2877,HST6308,HST6309,HST6408,SDT6308,SDT6309,STT2805,STT6309, STT6409,*SK3054,*ECG152,*WEP245
B3548	2N2877,2N2879,HST6309,HST6310,HST6409,HST6410,SDT6309,SDT6310,STT2804, STT2805,STT6309,STT6310,STT6409,STT6410,*SK3054,*ECG152,*WEP245

To Replace	Substitute This Type
B3549	2N2879,HST6310,HST6409,HST6410,SDT6310,STT2803,STT2804,STT6310,STT6410
B3550	2N2878,HST6311,HST6312,HST6411,HST6412,SDT6311,SDT6312,STT2200, STT3400,STT6312,STT6412,*SK3054,*ECG152,*WEP245
B3551	2N2878,2N2880,2N3998,HST6312,HST6313,HST6412,HST6413,SDT6312,SDT6313, STT2200,STT3400,STT6312,STT6313,STT6412,STT6413,*SK3054,*ECG152, *WEP245
B3552	2N2880,2N3998,2N4115,HST6313,HST6413,SDT6313,ST91057,ST91085,STT6313, STT6413
B3553	HST6314,HST6315,HST6414,HST6415,SDT6314,SDT6315,STT6315,STT6415
B3554	2N3999,HST6315,HST6316,HST6415,HST6416,SDT6315,SDT6316,STT6315, STT6316,STT6415,STT6416
B3555	2N3999,2N4116,HST6316,HST6416,SDT6316,STT6316,STT6416
B3556	2N2880,2N3998,2N4115,HST6313,HST6413,SDT6313,STT6313,STT6413
B3557	2N4115,STT2802
B3558	STT2802
B3559	2N2877,HST6308,HST6309,HST6408,SDT6308,SDT6309,STT2805,STT6309, STT6409
B3560	2N2877,2N2879,HST6309,HST6310,HST6409,HST6410,SDT6309,SDT6310,STT2804, STT2805,STT6309,STT6310,STT6409,STT6410
B3561	2N2879,HST6310,HST6409,HST6410,SDT6310,STT2803,STT2804,STT6310,STT6410
B3562	2N2878,HST6311,HST6312,HST6411,HST6412,SDT6311,SDT6312
B3563	2N2878,2N2880,2N3998,HST6312,HST6313,HST6412,HST6413,SDT6312,SDT6313
B3564	2N2880,2N3998,2N4115,HST6313,HST6413,SDT6313
B3565	HST6314,HST6315,HST6414,HST6415,SDT6314,SDT6315
B3566	HST6315,HST6316,HST6415,HST6416,SDT6315,SDT6316
B3567	2N4116,HST6316,HST6416,SDT6316
B3568	
B3569	
B3570	*SK3192,*ECG186,*WEP245
B3571	
B3572	
B3573	
B3574	
B3575	
B3576	*SK3192,*ECG186,*WEP245
B3577	*SK3054,*ECG152,*WEP245
B3578	STT2804,STT2805,*SK3054,*ECG152,*WEP245
B3579	STT2803,STT2804
B3580	*SK3054,*ECG152,*WEP245
B3581	2N4115
B3582	
B3583	2N4116
B3584	STT2805,*SK3054,*ECG152,*WEP245
B3585	2N2877,HST6011,HST6308,HST6309,HST6408,KSP1151,SDT6011,SDT6308, SDT6309,STT2805,STT6309,STT6409,*SK3054,*ECG152,*WEP245
B3586	2N2877,2N2879,2N4301,HST6011,HST6012,HST6309,HST6310,HST6409,HST6410, KSP1151,KSP1152,SDT6011,SDT6012,SDT6309,SDT6310,STT2804,STT2805, STT6309,STT6310,STT6409,STT6410,*SK3054,*ECG152,*WEP245
B3587	2N2879,2N4301,HST6012,HST6310,HST6409,HST6410,KSP1152,SDT6012,SDT6310, STT2803,STT2804,STT6310,STT6410
B3588	2N2878,HST6013,HST6311,HST6312,HST6411,HST6412,KSP1154,SDT6013, SDT6311,SDT6312,STT6312,STT6412,*SK3054,*ECG152,*WEP245
B3589	2N2878,2N2880,2N3998,2N4301,HST6013,HST6014,HST6312,HST6313,HST6412, HST6413,KSP1154,KSP1155,SDT6013,SDT6014,SDT6312,SDT6313,STT6312, STT6313,STT6412,STT6413,*SK3054,*ECG152,*WEP245
B3590	2N2880,2N3998,2N4115,2N4301,HST6014,HST6313,HST6413,HST7140,KSP1155, SDT6014,SDT6313,STT6313,STT6413
B3591	HST6015,HST6314,HST6315,HST6414,HST6415,SDT6015,SDT6314,SDT6315, STT6315,STT6415
B3592	2N3999,HST6015,HST6016,HST6315,HST6316,HST6415,HST6416,SDT6015, SDT6016,SDT6315,SDT6316,STT6315,STT6316,STT6415,STT6416

To Replace	Substitute This Type
B3593	2N3999,2N4116,HST6016,HST6316,HST6416,SDT6016,SDT6316,STT6316,STT6416
B3594	
B3595	2N5327
B3596	2N5327
B3597	
B3598	
B3599	
B3600	SDT7401,SDT7402,SDT7414,SDT7415
B3601	2N5327,SDT7402,SDT7403,SDT7415,SDT7416
B3602	2N5327,SDT7403,SDT7416
B3603	
B3604	
B3605	
B3606	HST5006,HST5506,HST9004,HST9005,STT9004,STT9005,*GE-216,*TR-76,*SK3192, *ECG186,*WEP245,*276-2020
B3607	HST5007,HST5008,HST5507,HST5508,HST9005,HST9006,STT9005,STT9006, *GE-216,*TR-76,*SK3192,*ECG186,*WEP245,*276-2020
B3608	HST5008,HST5508,HST9006,STT9006,*GE-28,*TR-76,*SK3192,*ECG186,*WEP245, *276-2020
B3609	HST5001,HST5501,HST7401,HST7414,HST9004,HST9005,HST9007,HST9008, SDT3422,SDT7401,SDT7414,STT9004,STT9005,STT9007,STT9008,*GE-216,*TR-76, *SK3192,*ECG186,*WEP245,*276-2020
B3610	2N4150,2N5327,HST4453,HST4454,HST5002,HST5003,HST5502,HST5503,HST7402, HST7403,HST7415,HST7416,HST9005,HST9006,HST9008,HST9009,SDT3423, SDT3424,SDT7402,SDT7403,SDT7415,SDT7416,STT1800,STT3000,STT4453, STT9005,STT9006,STT9008,STT9009,*GE-216,*TR-76,*SK3192,*ECG186,*WEP245, *276-2020
B3611	2N4150,2N4895,2N4897,2N5327,HST4454,HST5003,HST5503,HST7403,HST7416, HST9006,HST9009,SDT3424,SDT7403,SDT7416,STT9006,STT9009,*GE-28,*TR-76, *SK3192,*ECG186,*WEP245,*276-2020
B3612	HST5001,HST5501,HST9007,HST9008,STT9007,STT9008,*GE-28,*TR-76,*SK3192, *ECG186,*WEP245,*276-2020
B3613	2N5327,HST5002,HST5003,HST5502,HST5503,HST9008,HST9009,STT9008, STT9009,*GE-28,*TR-76,*SK3192,*ECG186,*WEP245,*276-2020
B3614	2N5327,BFX34,HST5003,HST5503,HST9009,STT9009,*GE-28,*TR-76,*SK3192, *ECG186,*WEP245,*276-2020
B3615	HST5011,HST5511,*276-2020
B3616	HST5012,HST5013,HST5512,HST5513,*276-2020
B3617	2N4896,HST5013,HST5513,*276-2020
B3618	2N2811,2N5313,2N5317,1716-0602,1716-0605,1716-0802,1716-0805,HST6011, HST6031,HST7011,HST7012,HST9901,HST9902,KSP1151,SDT3202,SDT3203, SDT3206,SDT3207,SDT7011,SDT7012
B3619	2N2811,2N4301,2N5313,2N5315,2N5317,2N5319,2N6128,1716-0802,1716-0805, 1716-1002,1716-1005,HST6011,HST6012,HST7012,HST7013,HST9902,HST9903, KSP1151,KSP1152,SDT3203,SDT3204,SDT3207,SDT3208,SDT7012,SDT7013
B3620	2N2813,2N4301,2N5315,2N5319,2N6128,1716-1002,1716-1005,1716-1202,1716-1205, HST6012,HST7013,HST9903,HST9904,KSP1152,SDT3201,SDT3204,SDT3208, SDT3209,SDT7013
B3621	2N2812,HST6013,HST7014,HST7015,KSP1154,SDT7014,SDT7015
B3622	2N2812,2N4301,2N5730,2N5731,2N6128,HST6013,HST6014,HST7015,HST7016, KSP1154,KSP1155,SDT7015,SDT7016
B3623	2N2814,2N4115,2N4301,2N5730,2N5731,2N6128,HST6014,HST7016,HST7140, KSP1155,SDT7016
B3624	HST6015,HST7017,HST7018,SDT7017,SDT7018
B3625	HST6015,HST6016,HST7018,HST7019,SDT7018,SDT7019
B3626	2N4116,HST6016,HST7019,SDT7019
B3629	
B3630	
B3631	2N2877,2N2879,HST6309,HST6310,HST6409,HST6410,SDT6309,SDT6310,STT2804, STT2805,STT6309,STT6310,STT6409,STT6410
B3632	2N2878,2N2880,HST6312,HST6313,HST6412,HST6413,SDT6312,SDT6313

To Replace	Substitute This Type
B3633	2N2879,HST6310,HST6409,HST6410,SDT6310,STT2803,STT2804,STT6310,STT6410
_B3634_	2N2880,2N4115,HST6313,HST6413,SDT6313
_B3746_	2N5130,BFY69,BFY69A,BFY69B,BFY87,BFY87A,EN706,EN3011,GI-2921,SE3001, SE3002,*PTC136,*GE-63,*TR-87,*SK3024,*ECG128,*WEP243,*BC337,*276-2008, *RT-114
_B3747_	*SK3192,*ECG186,*WEP245
_B3748_	*SK3192,*ECG186,*WEP245
_B3749_	
_B3750_	*SK3192,*ECG186,*WEP245
_B5000_	B5030,B5040,*ECG152,*WEP245
_B5001_	*GEMR-6,*SK3192,*ECG186,*WEP245,*276-2019
_B5002_	*GE-66,*SK3054,*ECG152,*WEP245,*276-2020
B5020	*SK3027,*ECG130,*WEP247
_B5021_	*GE-66,*SK3054,*ECG152,*WEP245,*276-2019
_B5022_	*GE-66,*SK3054,*ECG152,*WEP245,*276-2020
B5030	B5040
_B5031_	*GE-66,*SK3054,*ECG152,*WEP245,*276-2019
_B5032_	*GE-66,*SK3054,*ECG152,*WEP245,*276-2020
B5040	B5050
_B5041_	*276-2019
_B5042_	*276-2020
B5050	
_B5051_	*276-2019
_B5052_	*276-2020
_B5100_	
B5120	
B5130	
B5140	
B5150	
_B10142_	2N5325,MP3731,*GE-25,*TR-27,*HEPG6008,*ECG127,*WEP235
_B10142A_	2N5325,2SB412,MP3731,*GE-25,*TR-27,*HEPG6008,*ECG127,*WEP235
_B10142B_	2N5324,*GE-25,*TR-27,*HEPG6008,*ECG127,*WEP235
_B10474_	*GE-16,*TR-01,*HEPG6005,*SK3014,*ECG121,*WEP232,*276-2006,*RT-127
_B10475_	*GE-25,*TR-01,*HEPG6005,*SK3014,*ECG121,*WEP232,*276-2006,*RT-127
_B10912_	*GE-16,*TR-01,*SK3009,*ECG121,*WEP232,*OC28,*RT-127
_B10913_	*GE-16,*TR-01,*SK3009,*ECG121,*WEP232,*RT-127
_B102000_	2N2291,2N3613,2N3614,2N5894,2N5895,2N5898,2N5899,ASZ16,KR6501,MP2060, MP2061,*PTC105A,*GE-25,*SK3035,*ECG127,*WEP235
_B102001_	2N3614,2N5895,2N5896,2N5899,2N5900,ASZ16,KR6501,MP2061,MP2062,*PTC105A, *GE-25,*SK3035,*ECG127,*WEP235
_B102002_	2N2292,2N2444,2N3614,2N5895,2N5896,2N5899,2N5900,ASZ16,KR6501,MP2061, MP2062,*PTC105A,*GE-25,*SK3035,*ECG127,*WEP235
_B102003_	2N2292,2N2444,2N5896,2N5900,KR6503,MP2062,MP2063,*PTC105A,*GE-25, *SK3035,*ECG127,*WEP235
_B103000_	2N678,2N678A,2N1558,2N1558A,2N2144,2N2144A,2N2145,2N2145A,2SB538, KR6501,*PTC105A,*GE-25,*SK3035,*ECG127,*WEP235
_B103001_	2N678A,2N1558,2N1558A,2N1559,2N1559A,2N2144,2N2144A,2N2145,2N2145A, 2N2292,2N2444,2N2832,2SB339,2SB538,KR6501,*PTC105A,*GE-25,*SK3035, *ECG127,*WEP235
_B103002_	2N678B,2N678C,2N1559,2N1559A,2N1560,2N1560A,2N2145,2N2145A,2N2146, 2N2146A,2N2292,2N2444,2N2832,2SB339,2SB484,2SB538,DTG-603M,KR6503, *PTC105A,*GE-25,*SK3035,*ECG127,*WEP235
_B103003_	2N678B,2N678C,2N1559,2N1559A,2N1560,2N1560A,2N2145,2N2145A,2N2146, 2N2146A,2N2292,2N2444,2N2832,2SB339,2SB484,DTG-603M,KR6503,*PTC105A, *GE-25,*SK3035,*ECG127,*WEP235
_B103004_	2N678B,2N678C,2N1560,2N1560A,2N2146,2N2146A,2N2293,2N2833,2SB484, DTG-603M,KR6503,*PTC105A,*GE-25,*SK3035,*ECG127,*WEP235
B113000-BRN	
B113000-ORG	
B113000-RED	
B113001-BRN	

To Replace	Substitute This Type
B113001-ORG	
B113001-RED	
B113002-BRN	
B113002-ORG	
B113002-RED	
B113003-BRN	
B113003-RED	
B113004-BRN	
B113004-RED	
B113005-BRN	
B113005-RED	
B133000	*276-2020
B133001	B133000,B133002,*PTC169,*276-2020
B133002	B133001,*PTC169,*276-2020
B133003	B133006,*PTC169,*276-2020
B133004	B133003,B133005,B133006,B133007,B133008,*PTC169,*276-2020
B133005	B133004,B133007,B133008,*PTC169,*276-2020
B133006	*PTC169
B133007	B133006,B133008,*PTC169
B133008	B133007,*PTC169
B143000	*GE-28,*TR-76,*SK3192,*ECG186,*WEP245
B143001	*GE-28,*TR-76,*SK3192,*ECG186,*WEP245
B143002	
B143003	*GE-28,*TR-76,*SK3192,*ECG186,*WEP245
B143004	*GE-66,*SK3054,*ECG152,*WEP245
B143005	
B143006	
B143007	
B143008	
B143009	*GE-28,*TR-76,*SK3192,*ECG186,*WEP245
B143010	*GE-28,*SK3192,*ECG186,*WEP245
B143011	*GE-66,*SK3054,*ECG152,*WEP245
B143012	*GE-66,*SK3054,*ECG152,*WEP245
B143013	
B143014	
B143015	*SK3192,*ECG186,*WEP245
B143016	*SK3192,*ECG186,*WEP245
B143017	
B143018	*SK3054,*WEP245
B143019	*SK3054,*ECG152,*WEP245
B143020	*ECG152
B143021	
B143022	
B143023	
B143024	*SK3192,*ECG186,*WEP245
B143025	*SK3192,*ECG186,*WEP245
B143026	*SK3054,*ECG152,*WEP245
B143027	*SK3054,*ECG152,*WEP245
B143028	
B143029	
B144000	B144001,B144003,B144004
B144001	B144000,B144002,B144003,B144004,B144005
B144002	B144005
B144003	B144004,B144006,B144007
B144004	B144003,B144005,B144006,B144007,B144008
B144005	B144008
B144006	B144007
B144007	B144006,B144008
B144008	
B145000	B144000,B144001,B144003,B144004,B145001
B145001	B144000,B144001,B144002,B144003,B144004,B144005,B145000

To Replace	Substitute This Type
B145002	B144002,B144005
B145003	B144003,B144004,B144006,B144007
B145004	B144003,B144004,B144005,B144006,B144007,B144008
B145005	B144005,B144008
B145006	B144006,B144007
B145007	B144006,B144007,B144008
B145008	2N4116,2N5084,B144008,HST6016,HST6316,HST6416,HST7019,SDT7019
B145009	
B145010	
B145011	
B145012	
B145013	
B145014	
B146000	B144000,B144001,B144003,B144004,B145000,B145001
B146001	B144000,B144001,B144002,B144003,B144004,B144005,B145000,B145001
B146002	B144002,B144005
B146003	B144003,B144004,B144006,B144007
B146004	B144003,B144004,B144005,B144006,B144007,B144008
B146005	B144005,B144008
B146006	B144006,B144007
B146007	B144006,B144007,B144008
B146008	B144008
B146009	
B146010	
B146011	
B146012	
B146013	
B146014	
B148000	B148001
B148001	
B148002	B148000,B148001,B148003
B148003	B148001
B148004	B148000
B148005	SEE 2N5412
B149000	
B149001	
B149002	
B149003	
B149004	
B149005	
B155000	
B155001	2N5329,2N5330
B155002	
B155003	
B155004	
B155005	2N5412
B170000-BLK	2N1487,2N1489,2N6315,2N6317,2SC794,2SD15,2SD124A,2SD163,2SD172,2SD211, B170000-BRN,B170001-BLK,B170001-BRN,B170019,BDY23A,HST9201,HST9205, HST9801,SDT7607,ZT1487,ZT1489,*PTC129A,*GE-77,*TR-26,*WEP247,*BDY20, *276-2020,*RT-131
B170000-BRN	2N1489,2N5873,2N6315,2N6317,2SC794,2SD15,2SD211,180T2,B170000-RED, B170001-BRN,B170001-RED,B170018,B170019,BDY23B,HST9201,HST9205, HST9801,SDT7607,ZT1489,*PTC129A,*GE-77,*TR-26,*WEP247,*BDY20,*276-2020, *RT-131
B170000-ORG	2N5873,180T2,B170000-RED,B170000-YEL,B170001-ORG,B170001-RED, B170001-YEL,B170018,BDY23B,BDY23C,SDT7601,*PTC140,*GE-77,*TR-61, *ECG130,*WEP247,*BDY20,*276-2020
B170000-RED	2N5873,180T2,B170000-ORG,B170000-YEL,B170001-ORG,B170001-RED, B170001-YEL,B170018,BDY23B,SDT7601,*PTC140,*GE-77,*TR-61,*ECG130, *WEP247,*BDY20,*276-2020
B170000-YEL	B170001-YEL,BDY23C,*PTC140,*GE-75,*276-2020

To Replace	Substitute This Type
B170001-BLK	2N3055A,2N3235,2SD163,2SD172,2SD211,1561-0403,1561-0404,1582-0403, 1582-0404,1582-0405,1582-0603,1582-0604,1582-0605,1723-0405,1723-0410, 1723-0605,1723-0610,B170001-BRN,B170002-BLK,B170002-BRN,B170019,B170020, HST9201,HST9205,HST9801,MJ2801,*PTC140,*GE-77,*TR-26,*ECG130,*WEP247, *BDY20,*RT-131
B170001-BRN	2N3055A,2N3235,2N5877,2N5881,2SD211,B170001-RED,B170002-BRN,B170002-RED, B170019,HST9201,HST9205,HST9801,*PTC140,*GE-77,*TR-26,*ECG130,*WEP247, *BDY20,*RT-131
B170001-ORG	2N5877,2N5881,2N6470,2N6471,B170001-RED,B170001-YEL,B170002-ORG, B170002-RED,B170002-YEL,*PTC140,*GE-77,*TR-61,*ECG130,*WEP247,*BDY20
B170001-RED	2N5877,2N5881,2N6470,2N6471,B170001-ORG,B170001-YEL,B170002-ORG, B170002-RED,B170002-YEL,*PTC140,*GE-77,*TR-61,*ECG130,*WEP247,*BDY20
B170001-YEL	2N6470,2N6471,B170002-YEL,*PTC140,*GE-75
B170002-BLK	2N3055A,2N3235,2N5302,1561-0403,1561-0404,1582-0403,1582-0404,1582-0405, 1582-0603,1582-0604,1582-0605,B170002-BRN,B170020,HST9201,HST9205,IR3771, MJ2801,MJ3771,*GE-77
B170002-BRN	2N3055A,2N3235,2N5302,2N5881,2N5885,B170002-RED,BUY51A,BUY52A,HST9201, HST9205,IR3771,*GE-77
B170002-ORG	2N5881,2N5885,2N6470,2N6471,2SD114-O,B170002-RED,B170002-YEL,*PTC140, *GE-77,*ECG130,*WEP247
B170002-RED	2N5881,2N5885,2N6470,2N6471,2SD114-O,B170002-ORG,B170002-YEL,BUY51A, BUY52A,*PTC140,*GE-77,*ECG130,*WEP247
B170002-YEL	2N6470,2N6471,2SD114,2SD114-O,*PTC140,*GE-75
B170003-BLK	2N1488,2N1490,2N3232,2N3233,2N3445,2N3446,2N6316,2N6318,2SC793-R,2SD16, 2SD53,2SD119-R,2SD124A,2SD125A,2SD164,2SD173,2SD176,2SD212,2SD213, 2SD319,B170003-BRN,B170004-BLK,B170004-BRN,B170006-BLK,B170006-BRN, B170007-BLK,B170007-BRN,B170022,B170025,BDY24A,HST9202,HST9206, HST9207,HST9802,HST9803,KSP1171,KSP1172,SDT7608,SDT7609,ZT1488,ZT1490, *PTC129A,*GE-75,*TR-61,*ECG181,*WEPS7000,*BDY20,*276-2020
B170003-BRN	2N1490,2N3232,2N3233,2N3445,2N3446,2N5624,2N5874,2N6316,2N6318,2SC793-R, 2SC1080-R,2SD16,2SD111-R,2SD119-R,2SD212,2SD213,2SD319,2SD428-R,181T2, B170003-RED,B170004-BRN,B170004-RED,B170006-BRN,B170006-RED, B170007-BRN,B170007-RED,B170021,B170022,B170024,B170025,BDY24B, HST9202,HST9206,HST9207,HST9802,HST9803,KSP1171,KSP1172,SDT7608, SDT7609,ZT1490,*PTC129A,*GE-75,*TR-61,*ECG181,*WEPS7000,*BDY20, *276-2020
B170003-ORG	2N3447,2N3448,2N5624,2N5874,2SC793-Y,2SC1080-R,2SC1080-Y,2SD111-O, 2SD111-R,2SD119-Y,2SD428-O,2SD428-R,181T2,B170003-RED,B170003-YEL, B170004-ORG,B170004-RED,B170004-YEL,B170006-ORG,B170006-RED, B170006-YEL,B170007-ORG,B170007-RED,B170007-YEL,B170021,B170024, BDY24B,BDY24C,KSP1174,KSP1175,SDT7602,SDT7603,*PTC140,*GE-75,*TR-61, *BDY20,*276-2020
B170003-RED	2N3447,2N3448,2N5624,2N5874,2SC793-Y,2SC1080-R,2SC1080-Y,2SD111-O, 2SD111-R,2SD119-Y,2SD428-O,2SD428-R,181T2,B170003-ORG,B170003-YEL, B170004-ORG,B170004-RED,B170004-YEL,B170006-ORG,B170006-RED, B170006-YEL,B170007-ORG,B170007-RED,B170007-YEL,B170021,B170024, BDY24B,KSP1174,KSP1175,SDT7602,SDT7603,*PTC140,*GE-75,*TR-61,*ECG181, *WEPS7000,*BDY20,*276-2020
B170003-YEL	2N5622,2N5626,2SC793-BL,2SC793-Y,2SC1080-Y,2SD111,2SD111-O,2SD119-BL, 2SD119-Y,2SD428-O,B170004-YEL,B170006-YEL,B170007-YEL,BDY24C,KSP1174, KSP1175,*PTC140,*GE-75,*276-2020
B170004-BLK	2N3055,2N3236,2SD53,2SD164,2SD173,2SD176,2SD212,2SD213,2SD319,1561-0803, 1561-0804,1561-0805,1561-1005,1582-0803,1582-0804,1582-0805,1582-1003, 1582-1004,1582-1005,1723-0805,1723-0810,1723-1005,1723-1010,B170004-BRN, B170005-BLK,B170005-BRN,B170007-BLK,B170007-BRN,B170008-BLK, B170008-BRN,B170022,B170023,B170025,B170026,BD130,HST9202,HST9206, HST9207,HST9802,HST9803,KSP1271,KSP1272,MJ3772,*PTC140,*GE-75,*TR-61, *ECG181,*WEPS7000,*BDY20
B170004-BRN	2N3055,2N3236,2N3713,2N3714,2N5624,2N5632,2N5878,2N5882,2SC1080-R, 2SD111-R,2SD212,2SD213,2SD319,B170004-RED,B170005-BRN,B170005-RED, B170007-BRN,B170007-RED,B170008-BRN,B170008-RED,B170022,B170025,BD130,

To Replace	Substitute This Type
(B170004-BRN)	HST9202,HST9206,HST9207,HST9802,HST9803,KSP1271,KSP1272,*PTC140,*GE-75,*TR-61,*ECG181,*WEPS7000,*BDY20
B170004-ORG	2N3715,2N3716,2N5624,2N5632,2N5878,2N5882,2N6472,2SC1080-R,2SC1080-Y,2SD111-O,2SD111-R,B170004-RED,B170004-YEL,B170005-ORG,B170005-RED,B170005-YEL,B170007-ORG,B170007-RED,B170007-YEL,B170008-ORG,B170008-RED,B170008-YEL,KSP1274,KSP1275,*PTC140,*GE-75,*TR-61,*BDY20
B170004-RED	2N3713,2N3714,2N3715,2N3716,2N5624,2N5632,2N5878,2N5882,2N6472,2SC1080-R,2SC1080-Y,2SD111-O,2SD111-R,B170004-ORG,B170004-YEL,B170005-ORG,B170005-RED,B170005-YEL,B170007-ORG,B170007-RED,B170007-YEL,B170008-ORG,B170008-RED,B170008-YEL,KSP1274,KSP1275,*PTC140,*GE-75,*TR-61,*ECG181,*WEPS7000,*BDY20
B170004-YEL	2N3715,2N3716,2N5622,2N5626,2N6472,2SC1080-Y,2SD111,2SD111-O,B170005-YEL,B170007-YEL,B170008-YEL,KSP1274,KSP1275,*PTC140,*GE-75
B170005-BLK	2N3055,2N3236,2N3237,2N5303,2SD55,1561-0803,1561-0804,1561-0805,1561-1005,1582-0803,1582-0804,1582-0805,1582-1003,1582-1004,1582-1005,B170005-BRN,B170008-BLK,B170008-BRN,B170023,B170026,HST9202,HST9206,HST9207,IR3772,MJ3772,*PTC140,*GE-75,*ECG181,*WEPS7000
B170005-BRN	2N3055,2N3236,2N5303,2N5629,2N5882,2N5886,2N6270,2N6272,B170005-RED,B170008-BRN,B170008-RED,BUY53A,BUY54A,HST9202,HST9206,HST9207,IR3772,*PTC140,*GE-75,*ECG181,*WEPS7000
B170005-ORG	2N5629,2N5734,2N5882,2N5886,2N6270,2N6272,2N6472,2SD113-O,B170005-RED,B170005-YEL,B170008-ORG,B170008-RED,B170008-YEL,*PTC140,*GE-75
B170005-RED	2N5629,2N5734,2N5882,2N5886,2N6270,2N6272,2N6472,2SD113-O,B170005-ORG,B170005-YEL,B170008-ORG,B170008-RED,B170008-YEL,BUY53A,BUY54A,*PTC140,*GE-75,*ECG181,*WEPS7000
B170005-YEL	2N5734,2N6472,2SD113,2SD113-O,B170008-YEL,*PTC140,*GE-75
B170006-BLK	2N1488,2N1490,2N3233,2N3446,2SC793-R,2SD16,2SD118-R,2SD119-R,2SD125A,2SD161,2SD164,2SD173,2SD177,2SD213,2SD214,2SD319,B170006-BRN,B170007-BLK,B170007-BRN,B170025,BDY24A,HST9202,HST9203,HST9207,HST9208,HST9803,HST9804,KSP1172,SDT7609,ZT1488,ZT1490,*PTC129A,*GE-75,*TR-61,*ECG181,*WEPS7000
B170006-BRN	2N1490,2N3233,2N3446,2N5624,2N5628,2SC793-R,2SC961,2SC1079-R,2SC1080-R,2SD16,2SD110-R,2SD111-R,2SD118-R,2SD119-R,2SD213,2SD214,2SD319,2SD427-R,2SD428-R,181T2,B170006-RED,B170007-BRN,B170007-RED,B170024,B170025,BDY24B,HST9202,HST9203,HST9207,HST9208,HST9803,HST9804,KSP1172,SDT7609,ZT1490,*PTC129A,*GE-75,*TR-61,*ECG181,*WEPS7000
B170006-ORG	2N3448,2N4070,2N5624,2N5628,2SC793-Y,2SC961,2SC1079-R,2SC1079-Y,2SC1080-R,2SC1080-Y,2SD110-O,2SD110-R,2SD111-O,2SD111-R,2SD118-Y,2SD119-Y,2SD427-O,2SD427-R,2SD428-O,2SD428-R,181T2,B170006-RED,B170006-YEL,B170007-ORG,B170007-RED,B170007-YEL,B170024,BDY24B,BDY24C,KSP1175,SDT7603,*PTC118,*GE-75,*TR-61
B170006-RED	2N3448,2N4070,2N5624,2N5628,2SC793-Y,2SC961,2SC1079-R,2SC1079-Y,2SC1080-R,2SC1080-Y,2SD110-O,2SD110-R,2SD111-O,2SD111-R,2SD118-Y,2SD119-Y,2SD427-O,2SD427-R,2SD428-O,2SD428-R,181T2,B170006-ORG,B170006-YEL,B170007-ORG,B170007-RED,B170007-YEL,B170024,BDY24B,KSP1175,SDT7603,*PTC118,*GE-75,*TR-61,*ECG181,*WEPS7000
B170006-YEL	2N5626,2SC793-BL,2SC793-Y,2SC1079-Y,2SC1080-Y,2SC1818,2SD110,2SD110-O,2SD111,2SD111-O,2SD118-BL,2SD118-Y,2SD119-BL,2SD119-Y,2SD427-O,2SD428-O,B170007-YEL,BDY24C,KSP1175,*PTC118,*GE-75
B170007-BLK	2N3055,2N3236,2N5633,2SD161,2SD164,2SD173,2SD177,2SD213,2SD214,2SD319,1561-1005,1561-1205,1582-1003,1582-1004,1582-1005,1582-1203,1582-1204,1582-1205,1723-1005,1723-1010,1723-1205,1723-1210,B170007-BRN,B170008-BLK,B170008-BRN,B170025,B170026,BD130,HST9202,HST9203,HST9207,HST9208,HST9803,HST9804,KSP1272,MJ3772,*PTC118,*GE-75,*TR-61,*ECG181,*WEPS7000
B170007-BRN	2N3055,2N3236,2N3714,2N5624,2N5628,2N5632,2N5633,2SC1079-R,2SC1080-R,2SD110-R,2SD111-R,2SD213,2SD214,2SD319,B170007-RED,B170008-BRN,B170008-RED,B170025,BD130,HST9202,HST9203,HST9207,HST9208,HST9803,HST9804,KSP1272,*PTC118,*GE-75,*TR-61,*ECG181,*WEPS7000
B170007-ORG	2N3716,2N5624,2N5628,2N5632,2SC1079-R,2SC1079-Y,2SC1080-R,2SC1080-Y,2SD110-O,2SD110-R,2SD111-O,2SD111-R,B170007-RED,B170007-YEL,

To Replace	Substitute This Type
(B170007-ORG)	B170008-ORG,B170008-RED,B170008-YEL,KSP1275,*PTC118,*GE-75,*TR-61
B170007-RED	2N3714,2N3716,2N5624,2N5628,2N5632,2SC1079-R,2SC1079-Y,2SC1080-R, 2SC1080-Y,2SD110-O,2SD110-R,2SD111-O,2SD111-R,B170007-ORG,B170007-YEL, B170008-ORG,B170008-RED,B170008-YEL,KSP1275,*PTC118,*GE-75,*TR-61, *ECG181,*WEPS7000
B170007-YEL	2N3716,2N5039,2N5626,2SC1079-Y,2SC1080-Y,2SD110,2SD110-O,2SD111, 2SD111-O,B170008-YEL,KSP1275,*PTC118,*GE-75
B170008-BLK	2N3055,2N3236,2N5630,2SD55,108T2,1561-1005,1561-1205,1582-1003,1582-1004, 1582-1005,1582-1203,1582-1204,1582-1205,B170008-BRN,B170026,HST9202, HST9203,HST9207,HST9208,IR3772,MJ3772,*GE-75,*ECG181,*WEPS7000
B170008-BRN	2N3055,2N3236,2N5629,2N5630,2N6270,2N6271,2N6272,2N6273,108T2, B170008-RED,BUY53A,BUY54A,HST9202,HST9203,HST9207,HST9208,IR3772, *PTC173,*GE-75,*ECG181,*WEPS7000
B170008-ORG	2N5629,2N5734,2N6270,2N6271,2N6272,2N6273,2N6338,2SD113-O,B170008-RED, B170008-YEL,*PTC173,*GE-75
B170008-RED	2N5629,2N5734,2N6270,2N6271,2N6272,2N6273,2N6338,2SD113-O,B170008-ORG, B170008-YEL,BUY53A,BUY54A,*PTC173,*GE-75,*ECG181,*WEPS7000
B170008-YEL	2N5039,2N5734,2SD113,2SD113-O,*PTC173,*GE-75
B170009	2N1489,2N5873,2SC794,2SD15,180T2,B170000-ORG,B170000-RED,B170000-YEL, B170001-ORG,B170001-RED,B170001-YEL,B170018,BDY23B,SDT7601,ZT1489, *PTC140,*GE-77,*TR-59,*HEPS7002,*SK3027,*ECG130,*WEP247,*BDY20, *276-2020,*RT-131
B170010	2N3055A,2N3235,2N5877,2N5881,2SD211,B170001-BRN,B170001-RED,B170002-BRN, B170002-RED,B170019,HST9201,HST9205,HST9801,*PTC140,*GE-77,*TR-59, *HEPS7002,*SK3027,*ECG130,*WEP247,*BDY20,*RT-131
B170011	2N5302,1561-0403,1561-0404,1582-0403,1582-0404,1582-0405,1582-0408,1582-0410, 1582-0415,1582-0603,1582-0604,1582-0605,1582-0608,1582-0610,1582-0615, B170002-BLK,B170002-BRN,B170020,HST9201,HST9205,IR3771,MJ2801,MJ3771, *PTC140,*GE-77,*TR-59,*HEPS7002,*SK3027,*ECG130,*WEP247,*RT-131
B170012	2N1490,2N3447,2N3448,2N5624,2N5874,2SC793-R,2SC793-Y,2SC1080-R,2SD16, 2SD111-O,2SD111-R,2SD119-R,2SD119-Y,2SD428-R,181T2,B170003-ORG, B170003-RED,B170003-YEL,B170004-ORG,B170004-RED,B170004-YEL, B170006-ORG,B170006-RED,B170006-YEL,B170007-ORG,B170007-RED, B170007-YEL,B170015,B170021,B170024,BDY24B,KSP1174,KSP1175,SDT7602, SDT7603,ZT1490,*PTC140,*GE-75,*TR-59,*SK3027,*HN100,*ECG130,*WEP247, *BDY20,*276-2020,*RT-131
B170013	2N3055,2N3236,2N3713,2N3714,2N5624,2N5632,2N5878,2N5882,2SC1080-R, 2SD111-R,2SD212,2SD213,2SD319,B170004-BRN,B170004-RED,B170005-BRN, B170005-RED,B170007-BRN,B170007-RED,B170008-BRN,B170008-RED,B170016, B170022,B170025,BD130,HST9202,HST9206,HST9207,HST9802,HST9803, KSP1271,KSP1272,*PTC140,*GE-75,*TR-59,*SK3027,*ECG130,*WEP247, *BDY20,*RT-131
B170014	2N3236,2N3237,2N5303,2SD55,1561-0803,1561-0804,1561-0805,1561-1005, 1582-0803,1582-0804,1582-0805,1582-0808,1582-0810,1582-0815,1582-1003, 1582-1004,1582-1005,1582-1008,1582-1010,1582-1015,B170005-BLK,B170005-BRN, B170008-BLK,B170008-BRN,B170017,B170023,B170026,HST9202,HST9206, HST9207,IR3772,MJ3772,*PTC140,*GE-75,*TR-59,*SK3027,*ECG130,*WEP247, *RT-131
B170015	2N1490,2N3448,2N4070,2N5624,2N5628,2SC793-R,2SC793-Y,2SC961,2SC1079-R, 2SC1080-R,2SD16,2SD110-O,2SD110-R,2SD111-O,2SD111-R,2SD118-R,2SD118-Y, 2SD119-R,2SD119-Y,2SD427-R,2SD428-R,181T2,B170006-ORG,B170006-RED, B170006-YEL,B170007-ORG,B170007-RED,B170007-YEL,B170024,BDY24B, KSP1175,SDT7603,ZT1490,*PTC118,*GE-75,*TR-59,*SK3027,*ECG130,*WEP247, *RT-131
B170016	2N3055,2N3236,2N3714,2N5624,2N5628,2N5632,2N5633,2SC1079-R,2SC1080-R, 2SD110-R,2SD111-R,2SD213,2SD214,2SD319,B170007-BRN,B170007-RED, B170008-BRN,B170008-RED,B170025,BD130,HST9202,HST9203,HST9207, HST9208,HST9803,HST9804,KSP1272,*PTC118,*GE-75,*TR-59,*SK3027,*ECG130, *WEP247,*RT-131
B170017	2N3236,2SD55,108T2,1561-1005,1561-1205,1582-1003,1582-1004,1582-1005, 1582-1008,1582-1010,1582-1015,1582-1203,1582-1204,1582-1205,1582-1208,

To Replace	Substitute This Type
(B170017)	1582-1210,1582-1215,B170008-BLK,B170008-BRN,B170026,HST9202,HST9203, HST9207,HST9208,IR3772,MJ3772,*PTC173,*GE-75,*SK3027,*ECG181, *WEPS7000,*RT-131
B170018	2N1489,2N5873,2SC794,2SD15,180T2,B170000-ORG,B170000-RED,B170000-YEL, B170001-ORG,B170001-RED,B170001-YEL,BDY23B,SDT7601,ZT1489,*PTC140, *GE-77,*TR-59,*HEPS7002,*SK3027,*HN100,*ECG130,*WEP247,*BDY20, *276-2020,*RT-131
B170019	2N3055A,2N3235,2N5877,2N5881,2SD211,B170001-BRN,B170001-RED,B170002-BRN, B170002-RED,HST9201,HST9205,HST9801,*PTC140,*GE-77,*TR-59,*HEPS7002, *SK3027,*ECG130,*WEP247,*BDY20,*RT-131
B170020	2N5302,1561-0403,1561-0404,1582-0403,1582-0404,1582-0405,1582-0408,1582-0410, 1582-0415,1582-0603,1582-0604,1582-0605,1582-0608,1582-0610,1582-0615, B170002-BLK,B170002-BRN,HST9201,HST9205,IR3771,MJ2801,MJ3771,*PTC140, *GE-77,*TR-59,*HEPS7002,*SK3027,*ECG130,*WEP247,*RT-131
B170021	2N1490,2N3447,2N3448,2N5624,2N5874,2SC793-R,2SC793-Y,2SC1080-R,2SD16, 2SD111-O,2SD111-R,2SD119-R,2SD119-Y,2SD428-R,181T2,B170003-ORG, B170003-RED,B170003-YEL,B170004-ORG,B170004-RED,B170004-YEL, B170006-ORG,B170006-RED,B170006-YEL,B170007-ORG,B170007-RED, B170007-YEL,B170024,BDY24B,KSP1174,KSP1175,SDT7602,SDT7603,ZT1490, *PTC140,*GE-75,*TR-59,*SK3027,*ECG130,*WEP247,*BDY20,*276-2020,*RT-131
B170022	2N3055,2N3236,2N3713,2N3714,2N5624,2N5632,2N5878,2N5882,2SC1080-R, 2SD111-R,2SD212,2SD213,2SD319,B170004-BRN,B170004-RED,B170005-BRN, B170005-RED,B170007-BRN,B170007-RED,B170008-BRN,B170008-RED,B170025, BD130,HST9202,HST9206,HST9207,HST9802,HST9803,KSP1271,KSP1272, *PTC140,*GE-75,*TR-59,*SK3027,*ECG130,*WEP247,*BDY20,*RT-131
B170023	2N3236,2N3237,2N5303,2SD55,1561-0803,1561-0804,1561-0805,1561-1005, 1582-0803,1582-0804,1582-0805,1582-0808,1582-0810,1582-0815,1582-1003, 1582-1004,1582-1005,1582-1008,1582-1010,1582-1015,B170005-BLK,B170005-BRN, B170008-BLK,B170008-BRN,B170026,HST9202,HST9206,HST9207,IR3772,MJ3772, *PTC140,*GE-75,*SK3027,*ECG181,*WEPS7000,*RT-131
B170024	2N1490,2N3448,2N4070,2N5624,2N5628,2SC793-R,2SC793-Y,2SC961,2SC1079-R, 2SC1080-R,2SD16,2SD110-O,2SD110-R,2SD111-O,2SD111-R,2SD118-R,2SD118-Y, 2SD119-R,2SD119-Y,2SD427-R,2SD428-R,181T2,B170006-ORG,B170006-RED, B170006-YEL,B170007-ORG,B170007-RED,B170007-YEL,BDY24B,KSP1175, SDT7603,ZT1490,*PTC118,*GE-75,*TR-59,*SK3027,*ECG130,*WEP247,*RT-113
B170025	2N3055,2N3236,2N3714,2N5624,2N5628,2N5632,2N5633,2SC1079-R,2SC1080-R, 2SD110-R,2SD111-R,2SD213,2SD214,2SD319,B170007-BRN,B170007-RED, B170008-BRN,B170008-RED,BD130,HST9202,HST9203,HST9207,HST9208, HST9803,HST9804,KSP1272,*PTC118,*GE-75,*TR-59,*SK3027,*ECG130,*WEP247, *RT-131
B170026	2N3236,2SD55,108T2,1561-1005,1561-1205,1582-1003,1582-1004,1582-1005, 1582-1008,1582-1010,1582-1015,1582-1203,1582-1204,1582-1205,1582-1208, 1582-1210,1582-1215,B170008-BLK,B170008-BRN,HST9202,HST9203,HST9207, HST9208,IR3772,MJ3772,*PTC173,*GE-75,*SK3027,*ECG181,*WEPS7000
B176000	2N6233,2N6234,2SC270,2SC558,2SC1227,2SC1229,2SD181A,2SD321,2SD376, 2SD376A,183T2,B176001,BDY26B,FT411,IR410,IR411,KSP1092,KSP1093,KSP1094, KSP1095,KSP1142,KSP1143,KSP1144,KSP1145,MJ1800,*PTC118,*GE-73,*TR-61, *HEPS5020,*SK3559,*ECG163,*WEP740
B176001	2N6233,2N6234,2SC270,2SC558,2SC886,2SC1227,2SC1229,2SD181A,2SD321, 2SD376,2SD376A,183T2,B176000,BDY26B,FT411,HST7202,HST7203,HST7204, IR410,IR411,KSP1092,KSP1093,KSP1094,KSP1095,KSP1142,KSP1143,KSP1144, KSP1145,SDT7202,SDT7203,SDT7204,*PTC129A,*GE-35,*TR-61,*HEPS5020, *ECG162,*WEP707
B176002	2N6249,2SC270,2SC558,2SC675,2SC681A,2SC771,2SC886,2SC901A,2SC1227, 2SC1229,2SD321,2SD376,2SD376A,2SD417,B176000,B176001,BDY26A,HST7202, HST7203,HST7204,IR410,IR411,KSP1092,KSP1093,KSP1094,KSP1095,KSP1142, KSP1143,KSP1144,KSP1145,SDT7202,SDT7203,SDT7204,XGSR10025,*PTC129A, *GE-35,*TR-61,*HEPS5020,*ECG162,*WEP707
B176003	2N6249,2SC675,2SC681A,2SC771,2SC901A,2SC1172,2SC1172A,2SC1172B, 2SC1325,2SC1325A,2SC1358A,2SC1477,2SD417,B176002,BDY26A,HST1050, HST1055,HST7208,HST7209,SDT1050,SDT1055,SDT7208,SDT7209,XGSR10025,

To Replace	Substitute This Type
(B176003)	*PTC129A,*GE-35,*TR-61,*HEPS5020,*ECG162,*WEP707
B176004	2N5240,184T2,185T2,B176005,BDY27B,BDY28B,FT413,FT423,IR401,IR413,*GE-73, *TR-61,*HEPS5020,*SK3559,*ECG163,*WEP740
B176005	2N5240,184T2,185T2,B176004,BDY27B,BDY28B,FT413,FT423,IR401,IR413, *PTC129A,*GE-36,*TR-61,*HEPS5020,*ECG163,*WEP740
B176006	2N5240,2N6250,2N6251,2SC1577,2SD377,B176004,B176005,B176024,BDY27A, BDY28A,FT413,IR413,*PTC129A,*GE-36,*TR-61,*HEPS5020,*ECG163,*WEP740
B176007	2N6250,2N6251,2SC1172,2SC1172A,2SC1172B,2SC1325,2SC1325A,2SC1358A, 2SC1577,2SD377,B176006,B176024,B176025,BDY27A,BDY28A,BDY42,HST1051, HST1052,HST1056,HST1057,SDT1051,SDT1052,SDT1056,SDT1057,XGSR10040, *PTC129A,*GE-36,*TR-61,*HEPS5020,*ECG163,*WEP740
B176008	185T2,B176009,B176012,B176013,BDY28B,BUY23,BUY23A,*HEPS5021
B176009	185T2,B176008,B176012,B176013,BDY28B,*PTC129A,*GE-36,*HEPS5021,*ECG163, *WEP740
B176010	2SC1577,2SC1578,B176008,B176009,B176012,B176013,B176014,B176026,B176028, BDY28A,*PTC129A,*GE-36,*HEPS5021,*ECG163,*WEP740
B176011	2N6542,2SC1172,2SC1172A,2SC1172B,2SC1325,2SC1325A,2SC1358A,2SC1577, 2SC1578,2SD458,B176010,B176014,B176015,B176026,B176027,B176028,B176029, BDY28A,BDY43,HST1052,HST1053,HST1054,HST1057,HST1058,HST1059, MJ3260,SDT1052,SDT1053,SDT1054,SDT1057,SDT1058,SDT1059,*PTC129A, *GE-36,*HEPS5021,*ECG163,*WEP740
B176012	2SD266,2SD274,B176013,BUY23A,*HEPS5021
B176013	2SD266,2SD274,B176012,*PTC129A,*GE-36,*HEPS5021,*ECG163,*WEP740
B176014	2SD266,2SD274,B176012,B176013,B176028,*PTC129A,*GE-36,*HEPS5021, *ECG163,*WEP740
B176015	2N6542,2N6543,2SC1172,2SC1172A,2SC1172B,2SC1325,2SC1325A,2SC1358A, 2SD265,2SD273,B176014,B176028,B176029,BDY44,HST1054,HST1059,MJ3260, SDT1054,SDT1059,*PTC129A,*GE-36,*HEPS5021,*ECG163,*WEP740
B176024	2N5240,2N6250,2N6251,2SC1577,2SD377,B176004,B176005,B176006,BDY27A, BDY28A,FT413,IR413,*PTC129A,*GE-36,*TR-61,*HEPS5020,*ECG163,*WEP740
B176025	2N6250,2N6251,2SC1172,2SC1172A,2SC1172B,2SC1325,2SC1325A,2SC1358A, 2SC1577,2SD377,B176006,B176007,B176024,BDY27A,BDY28A,BDY42,HST1051, HST1052,HST1056,HST1057,SDT1051,SDT1052,SDT1056,SDT1057,XGSR10040, *PTC129A,*GE-36,*TR-61,*HEPS5020,*ECG163,*WEP740
B176026	2SC1577,2SC1578,B176008,B176009,B176010,B176012,B176013,B176014,B176028, BDY28A,*PTC129A,*GE-36,*HEPS5021,*ECG163,*WEP740
B176027	2N6542,2SC1172,2SC1172A,2SC1172B,2SC1325,2SC1325A,2SC1358A,2SC1577, 2SC1578,2SD458,B176010,B176011,B176014,B176015,B176026,B176028,B176029, BDY28A,BDY43,HST1052,HST1053,HST1054,HST1057,HST1058,HST1059, MJ3260,SDT1052,SDT1053,SDT1054,SDT1057,SDT1058,SDT1059,*PTC129A, *GE-36,*HEPS5021,*ECG163,*WEP740
B176028	2SD266,2SD274,B176012,B176013,B176014,*PTC129A,*GE-36,*HEPS5021, *ECG163,*WEP740
B176029	2N6542,2N6543,2SC1172,2SC1172A,2SC1172B,2SC1325,2SC1325A,2SC1358A, 2SD265,2SD273,B176014,B176015,B176028,BDY44,HST1054,HST1059,MJ3260, SDT1054,SDT1059,*PTC129A,*GE-36,*HEPS5021,*ECG163,*WEP740
B176030	2N6545,2SC1142,2SD293
B177000	2N6570,2N6571,108T2,STC2229,*PTC175,*GE-75,*ECG181,*WEPS7000
BC107	2N4953,2N5370,2N5376,2N5377,2N5827,A157B,BC107B,BC167B,BC171B,BC174B, BC190B,BC237B,BC413B,BC414B,BC547B,BC550B,BCY59B,BCY59C,*GE-20, *TR-21,*HEPS0015,*SK3122,*SN80,*ECG123A,*WEP736,*BC107B,*276-2031, *RT-102
BC107A	2N3904,2N4401,2N5381,2SC1992,A5T3904,BC237A,BC547A,MM3904,TN-3904, *GE-210,*TR-21,*HEPS0015,*SK3122,*ECG123A,*WEP736,*BC107B,*276-2031, *RT-102
BC107B	2N6428,2N6428A,BC237B,BC547B,BC550B,*GE-20,*TR-21,*HEPS0015,*SK3122, *ECG123A,*WEP736,*BC107B,*276-2009,*RT-102
BC108	2N4419,2N4954,2N5371,2N5827,2SC735-GR,2SC1993,A157B,BC107,BC107B, BC108B,BC109B,BC167B,BC171B,BC237B,BC238B,BC239B,BC413B,BC414B, BC547B,BC548B,BC549B,BC550B,BCY58B,BCY58C,BCY59B,BCY59C,*PTC136, *GE-20,*TR-21,*HEPS0011,*SK3124,*SN80,*ECG123A,*WEP736,*BC107B,

To Replace	Substitute This Type
(BC108)	*276-2016,*RT-102
BC108A	2N3227,2N4124,2N4419,2SC735-Y,2SC1992,A5T4124,BC107,BC107A,BC108, BC167A,BC171A,BC237A,BC238A,BC547A,BC548A,BCY58A,BCY58B,BCY59A, BCY59B,BSW83,*PTC136,*GE-210,*TR-21,*HEPS0015,*SK3122,*ECG123A, *WEP736,*BC107B,*276-2016,*RT-102
BC108B	2N4954,2N5371,2N5827,2SC735-GR,2SC1993,A157B,BC107B,BC109,BC109B, BC167B,BC171B,BC237B,BC238B,BC239B,BC413B,BC414B,BC547B,BC548B, BC549B,BC550B,BCY58C,BCY59C,*PTC136,*GE-20,*TR-21,*HEPS0015,*SK3122, *ECG123A,*WEP736,*BC107B,*276-2016,*RT-102
BC108C	BC109C,BC238C,BC239C,BC413C,BC414C,BC548C,BC549C,BC550C,BCY58D, BCY59D,*TR-21,*HEPS0015,*SK3122,*ECG123A,*WEP736,*276-2016,*RT-102
BC109	BC108C,BC109C,BC238C,BC239C,BC413C,BC414C,BC548C,BC549C,BC550C, BCY58D,BCY59D,*GE-62,*TR-21,*HEPS0024,*SK3124,*EN10,*ECG123A,*WEP736, *BC107B,*276-2016,*RT-102
BC109B	2N4954,2N5371,2N5827,2SC735-GR,2SC1993,A157B,BC107B,BC108B,BC109, BC167B,BC171B,BC237B,BC238B,BC239B,BC413B,BC414B,BC547B,BC548B, BC549B,BC550B,BCY58C,BCY59C,*PTC136,*GE-10,*TR-21,*HEPS0015,*SK3124, *ECG123A,*WEP736,*276-2016,*RT-102
BC109C	BC108C,BC238C,BC239C,BC413C,BC414C,BC548C,BC549C,BC550C,BCY58D, BCY59D,*GE-20,*TR-21,*HEPS0015,*SK3122,*ECG123A,*WEP736,*BC107B, *276-2016,*RT-102
BC110	2N736A,2N736B,2N2461,2N2462,2N2465,2N2466,2N2516,*PTC125,*GE-81,*TR-21, *HEPS0005,*SK3122,*ECG123A,*WEP736,*BF338,*276-2010,*RT-102
BC121	*PTC139,*GE-212,*TR-70,*HEPS0025,*SK3245,*SN80,*ECG107,*WEP56,*BC107B, *276-2015,*RT-108
BC122	*PTC139,*GE-212,*TR-70,*HEPS0020,*SK3245,*SN80,*ECG107,*WEP56,*BC107B, *276-2015,*RT-108
BC123	*PTC132,*GE-212,*TR-70,*HEPS0016,*SK3245,*SN80,*ECG107,*WEP56,*BC107B, *276-2016,*RT-108
BC129A	2N3904,2N4401,2SC1992,BC107B,BC129B,BC237B,*HEPS0025
BC129B	BC107B,BC237B,*HEPS0030
BC130A	2N3227,2N4124,2N4419,2SC400,2SC400-GR,2SC735-GR,2SC735-Y,2SC752G-Y, 2SC1992,2SC1993,A157B,BC107,BC107B,BC108,BC108B,BC109B,BC129A,BC129B, BC130B,BC131B,BC167B,BC171B,BC237B,BC238B,BC239B,BC413B,BC414B, BCY58B,BCY59B,*PTC132,*HEPS0015
BC130B	2N4954,2N5827,2SC735-GR,2SC1993,A157B,BC107B,BC108B,BC109,BC109B, BC129B,BC131B,BC167B,BC171B,BC237B,BC238B,BC239B,BC413B,BC414B, BCY58C,BCY59C,*PTC136,*HEPS0015
BC130C	BC131C,*HEPS0024
BC131B	2N4954,2N5827,2SC735-GR,2SC1993,A157B,BC107B,BC108B,BC109,BC109B, BC129B,BC130B,BC167B,BC171B,BC237B,BC238B,BC239B,BC413B,BC414B, BCY58C,BCY59C,*PTC136,*HEPS0015
BC131C	BC130C,*HEPS0024
BC140	HST4455,HST4456,*GE-28,*TR-87,*HEPS3002,*SK3024,*ECG128,*WEP243,*RT-114
BC140-10	2N1973,BC140-16,BC141-10,BC141-16,*PTC144,*TR-87,*HEPS3001,*SK3024, *ECG128,*WEP243,*276-2012,*RT-114
BC140-16	2N1711A,2N1890,BC141-16,BSY84,ZT1711,*TR-87,*HEPS3001,*SK3024,*ECG128, *WEP243,*276-2012,*RT-114
BC140-6	2N1613A,2N1889,2N1974,2N2193,2N2193A,2N2193B,2N3107,2N3109,BC140-10, BC141-10,BC141-6,BSY83,ZT1613,*PTC144,*TR-87,*HEPS3001,*SK3024,*ECG128, *WEP243,*276-2014,*RT-114
BC141	2N1711,2N1711A,2N2049,ZT1711,*PTC125,*GE-18,*TR-87,*HEPS3019,*SK3024, *ECG128,*WEP243,*RT-114
BC141-10	2N1973,2N2405,2N2443,2N3036,BC141-16,*PTC144,*TR-87,*HEPS3019,*SK3024, *ECG128,*WEP243,*276-2012,*RT-114
BC141-16	2N1711B,2N1890,2N4943,BSY86,*TR-87,*HEPS3019,*SK3024,*ECG128,*WEP243, *276-2012,*RT-114
BC141-6	2N699B,2N1613B,2N1889,2N1974,2N2102,2N2102A,2N2243,2N2243A,2N2443, 2N3036,2N3107,BC141-10,BSY85,*PTC144,*TR-87,*HEPS3019,*SK3024,*ECG128, *WEP243,*276-2012,*RT-114
BC142	2N3253,2N3735,2N3830,2SC1065,2SC1069,*GE-18,*TR-87,*HEPS3001,*SK3024,

To Replace	Substitute This Type
(BC142)	*ECG128,*WEP243,*276-2014,*RT-114
BC143	*PTC127,*GE-21,*TR-88,*HEPS0012,*SK3025,*ECG129,*WEP242,*276-2021, *RT-115
BC147	*GE-210,*TR-21,*HEPS0015,*SK3124,*ECG123A,*WEP736,*BC107B,*276-2013, *RT-102
BC147A	*TR-21,*HEPS0015,*SK3124,*ECG123A,*WEP736,*276-2013,*RT-102
BC147B	*PTC121,*GE-10,*TR-21,*HEPS0015,*SK3124,*ECG123A,*WEP736,*276-2014, *RT-102
BC148	BC147B,BC149,*PTC139,*GE-20,*TR-21,*HEPS0015,*SK3124,*ECG123A,*WEP736, *BC107B,*276-2016,*RT-102
BC148A	*PTC101,*GE-20,*TR-21,*HEPS0015,*SK3124,*ECG123A,*WEP736,*276-2015, *RT-102
BC148B	*PTC139,*GE-20,*TR-21,*HEPS0015,*SK3124,*ECG123A,*WEP736,*276-2015, *RT-102
BC148C	*PTC139,*GE-20,*TR-21,*HEPS0024,*SK3124,*ECG123A,*WEP736,*276-2015, *RT-102
BC149	*PTC139,*GE-20,*TR-21,*HEPS0015,*SK3124,*ECG123A,*WEP736,*BC107B, *276-2016,*RT-102
BC149B	BC148B,*PTC139,*GE-20,*TR-21,*HEPS0015,*SK3124,*ECG123A,*WEP736, *276-2015,*RT-102
BC149C	BC148C,*PTC139,*GE-62,*TR-21,*HEPS0024,*SK3124,*ECG123A,*WEP736, *276-2015,*RT-102
BC155A	MPS2926-GRN,MPS2926-YEL,*PTC101,*GE-39,*HEPS0015,*SK3039,*ECG108, *WEP56,*276-2011
BC155B	MPS3721,*PTC139,*GE-39,*HEPS0015,*SK3122,*ECG123A,*WEP736,*276-2011
BC155C	*PTC139,*GE-212,*HEPS0024,*SK3245,*ECG199,*WEP736,*276-2011
BC156A	BC155A,*PTC101,*GE-39,*HEPS0015,*SK3039,*ECG108,*WEP56,*276-2011
BC156B	BC155B,*PTC139,*GE-39,*HEPS0015,*SK3122,*ECG123A,*WEP736,*276-2011
BC156C	BC155C,*PTC139,*GE-212,*HEPS0024,*SK3245,*ECG199,*WEP736,*276-2011
BC157	A757,*PTC103,*GE-65,*TR-20,*HEPS0019,*SK3114,*ECG159,*WEP242,*BC177, *276-2024,*RT-115
BC157A	*PTC103,*GE-65,*TR-20,*ECG159,*WEP242
BC157B	*GE-65,*SK3025,*ECG159,*WEP242
BC158	BC157A,BC157B,BC158A,BC158B,*PTC103,*GE-65,*TR-20,*HEPS0013,*SK3118, *ECG159,*WEP242,*BC177,*276-2034,*RT-115
BC158A	BC157A,*PTC103,*GE-65,*TR-20,*HEPS0013,*SK3118,*ECG159,*WEP242, *276-2034
BC158B	BC157B,*GE-65,*TR-20,*HEPS0031,*SK3114,*ECG159,*WEP242,*276-2034
BC158C	*GE-65,*SK3247,*276-2034
BC159	A758,BC158,BC158B,*GE-65,*TR-20,*HEPS0031,*SK3118,*ECG159,*WEP242, *BC177,*276-2034,*RT-115
BC159B	BC158B,*GE-65,*TR-20,*HEPS0031,*SK3114,*ECG159,*WEP242,*276-2034
BC159C	BC158C,*GE-65,*SK3247,*276-2034
BC160	2N4032,2N4037,2SA684,BC160-16,BC161,BC161-16,*GE-29,*TR-88,*HEPS3028, *SK3025,*ECG129,*WEP242
BC160-10	2N3244,2N3467,2N4030,2N4037,2SA571,2SA684,2SA717,BC160-16,BC161-10, BC161-16,*PTC141,*GE-48,*TR-88,*HEPS3012,*SK3025,*ECG129,*WEP242, *276-2025,*RT-115
BC160-16	2N4032,2SA684,2SA773,BC161-16,MPS4355,*GE-48,*TR-88,*HEPS3012,*SK3025, *ECG129,*WEP242,*276-2025,*RT-115
BC160-6	2N3244,2N3245,2N3467,2N3468,2N3762,2N3762A,2N3763,2N4030,2N5042,2SA571, 2SA717,92PU51,92PU51A,BC160-10,BC161-10,BC161-6,MM3726,*PTC142,*GE-48, *TR-88,*HEPS3012,*SK3025,*ECG129,*WEP242,*276-2025,*RT-115
BC161	2N4032,2N4033,2N4037,2N4405,2N4407,2SA684,BC161-16,MPS4356,*TR-88, *HEPS3012,*SK3025,*ECG129,*WEP242
BC161-10	2N4030,2N4031,2N4037,2N4407,2N5865,2SA512-O,2SA546,2SA571,2SA684,2SA717, BC161-16,MPS4356,*PTC141,*GE-67,*TR-88,*HEPS3012,*SK3025,*ECG129, *WEP242,*RT-115
BC161-16	2N4032,2N4033,2N4405,2N4407,2SA684,MPS4355,MPS4356,*GE-67,*TR-88, *HEPS3012,*SK3025,*ECG129,*WEP242,*RT-115
BC161-6	2N3763,2N4030,2N4031,2N4404,2N4406,2N5865,2SA512-O,2SA512-R,2SA516,

To Replace	Substitute This Type
(BC161-6)	2SA546,2SA571,2SA717,BC161-10,*PTC141,*GE-67,*TR-88,*HEPS3012,*SK3025, *ECG129,*WEP242,*RT-115
BC167A	2N3904,2N4401,2N5381,2SC1992,A5T3904,BC107A,BC171A,BC237A,BC547A, BCY59A,BCY59B,MM3904,TN-3904,*PTC121,*GE-210
BC167B	2N6428,2N6428A,BC107B,BC171B,BC237B,BC547B,BC550B,BCY59C,*PTC121, *GE-210,*SK3122,*ECG123A,*WEP736,*276-2009
BC168A	2N4124,2N4419,2N5224,2SC735-Y,A5T4124,BC108,BC108A,BC172A,BC238A, BC548A,BCY58A,BCY58B,*PTC101,*GE-10,*TR-21,*SK3124,*ECG123A,*WEP736, *276-2016,*RT-102
BC168B	2SC735-GR,2SC1993,A158B,A159B,BC108B,BC109,BC109B,BC172B,BC173B, BC238B,BC239B,BC548B,BC549B,BCY58C,*PTC139,*GE-10,*TR-21,*SK3124, *ECG123A,*WEP736,*276-2016,*RT-102
BC168C	A158C,A159C,BC108C,BC109C,BC172C,BC173C,BC238C,BC239C,BC548C,BC549C, BCY58D,*PTC139,*TR-21,*ECG123A,*WEP736,*276-2016
BC169B	2SC735-GR,2SC1993,A158B,A159B,BC108B,BC109,BC109B,BC168B,BC172B, BC173B,BC238B,BC239B,BC548B,BC549B,BCY58C,*PTC139,*GE-10,*TR-21, *HEPS0014,*SK3124,*ECG123A,*WEP736,*276-2016,*RT-102
BC169C	A158C,A159C,BC108C,BC109C,BC168C,BC172C,BC173C,BC238C,BC239C,BC548C, BC549C,BCY58D,*PTC139,*GE-62,*TR-21,*HEPS0015,*SK3018,*ECG123A, *WEP736,*276-2016,*RT-102
BC170A	2N744,2N2331,2N2569,2N2570,2N3011,2N3394,2N4418,2S095A,2S512,2SC99, 2SC395A-O,2SC509-O,2SC509G-O,2SC735-O,PT720,ZT2205,*PTC101,*GE-212, *TR-21,*HEPS0025,*SK3122,*ECG199,*WEP736,*BC107B,*276-2016,*RT-102
BC170B	2N2924,2N3392,2N4124,2N4419,2N5224,2N5998,2N6000,2SC509-Y,2SC509G-Y, 2SC735-Y,2SC1849,A5T4124,BC108,BC108A,BC168A,BC172A,BC238A,BC548A, BCY58A,BCY58B,BSW88,BSW89,BSX38,*PTC101,*GE-212,*TR-21,*HEPS0015, *SK3122,*ECG199,*WEP736,*BC107B,*276-2016,*RT-102
BC170C	A158C,A159C,BC108C,BC109,BC109C,BC168C,BC172C,BC173C,BC238C,BC239C, BCY58D,*PTC139,*GE-212,*TR-21,*HEPS0015,*SK3122,*ECG199,*WEP736, *BC107B,*276-2016,*RT-102
BC171A	2N3904,2N4401,2N5381,2SC1992,A5T3904,BC107A,BC167A,BC237A,BC547A, BCY59A,BCY59B,MM3904,TN-3904,*PTC121,*GE-210,*TR-21,*HEPS0015,*SK3122, *ECG123A,*WEP736,*BC107B,*276-2011,*RT-102
BC171B	2N6428,2N6428A,BC107B,BC167B,BC237B,BC547B,BC550B,BCY59C,*PTC121, *GE-210,*TR-21,*HEPS0030,*SK3122,*ECG123A,*WEP736,*BC107B,*RT-102
BC172A	2N3227,2N4124,2N4419,2N5224,2SC735-Y,A5T4124,BC108,BC108A,BC238A, BC548A,BCY58A,BCY58B,BSW83,*PTC101,*GE-210,*TR-21,*HEPS0015,*SK3122, *ECG123A,*WEP736,*BC107B,*276-2016,*RT-102
BC172B	2N4954,2N5371,2SC735-GR,2SC1993,BC108B,BC109,BC109B,BC173B,BC238B, BC239B,BC548B,BC549B,BCY58C,*PTC139,*GE-210,*TR-21,*HEPS0015,*SK3122, *ECG123A,*WEP736,*BC107B,*276-2016,*RT-102
BC172C	BC108C,BC109C,BC173C,BC238C,BC239C,BC548C,BC549C,BCY58D,*PTC139, *TR-21,*HEPS0015,*SK3122,*ECG123A,*WEP736,*276-2016,*RT-102
BC173B	2N4954,2N5371,2SC735-GR,2SC1993,BC108B,BC109,BC109B,BC172B,BC238B, BC239B,BC548B,BC549B,BCY58C,*PTC139,*GE-17,*TR-21,*HEPS0015,*SK3245, *ECG199,*WEP736,*BC107B,*276-2016,*RT-102
BC173C	BC108C,BC109C,BC172C,BC238C,BC239C,BC548C,BC549C,BCY58D,*PTC139, *GE-212,*TR-21,*HEPS0024,*SK3122,*ECG199,*WEP736,*BC107B,*276-2015, *RT-102
BC174A	2N3116,2N3302,2N3904,2N4141,2N4401,2N4409,2N4952,2N5028,2N5107,2N5369, 2N5381,2SC979-Y,2SC1991,A5T3904,A5T4409,BC190A,BC546A,BSW85,GET2222, GET2222A,GI-3643,MM3904,MPS6531,TN-3904,*GE-63,*TR-87,*HEPS0025, *SK3024,*ECG128,*WEP243,*276-2013,*RT-114
BC174B	2N4953,2N5370,2N5376,2N5377,2N6428,2N6428A,BC190B,BC546B,*TR-87, *HEPS0015,*SK3024,*ECG128,*WEP243,*276-2011,*RT-114
BC177	2N3250,2N3250A,2N3645,2N3672,2N3673,2N4143,2N4228,2N6005,2N6067,2SA603, 2SA710,2SA711,2SA718,2SA720,2SA891,A177,BC212A,BC212L,BC256A,BC266A, BC307A,BC416-6,BC416A,BCW35,BCW37,FT3644,GET2906,GET2907,TP3645, *PTC127,*GE-65,*TR-20,*HEPS0019,*SK3004,*ECG234,*WEP242,*BC177, *276-2024,*RT-115
BC177A	2N2907,2N2907A,2N3251,2N3251A,2N3486,2N3486A,2N3504,2N3505,2N3644,

To Replace	Substitute This Type
(BC177A)	2N3645,2N3672,2N3673,2N3962,2N4143,2N5373,2N6005,2SA550A,2SA603,2SA678, 2SA705,2SA718,2SA720,2SA891,A5T2907,A5T3504,A5T3505,A5T3644,A5T3645, A177,BC212A,BC212L,BC213L,BC214L,BC251A,BC256A,BC257A,BC261A,BC266A, BC307A,BC415-6,BC415A,BC416-6,BC416A,BC557A,BC560A,BCW35,BCW37, BCY79,BCY79A,BCY79B,EN2905,EN3502,FT3644,FT3645,GET2906,GET2907, GI-3644,TIS112,TP3644,TP3645,TZ552,*PTC127,*GE-48,*TR-20,*HEPS0019, *SK3114,*ECG234,*WEP242,*276-2024
BC177B	2N3962,2N3964,2N3965,2N5374,2N6007,2SA550A,2SA678,2SA705,BC214L,BC251B, BC256B,BC257B,BC261B,BC266B,BC307B,BC415-10,BC415B,BC416-10,BC416B, BC557B,BC560B,BCY79,BCY79B,BCY79C,TZ553,*PTC127,*GE-48,*TR-20, *HEPS0032,*SK3114,*ECG234,*WEP242
BC178	2N3251,2N3504,2N3906,2N4035,2N4403,2N5244,2N5375,2N5379,2N5383,2N5999, 2N6001,2N6005,2SA509-Y,2SA509G-Y,2SA550A,2SA678,2SA705,2SA719,2SA890, A5T3504,A5T3644,A5T3906,A5T4403,BC177A,BC177B,BC213L,BC214L,BC251A, BC251B,BC257A,BC257B,BC261A,BC261B,BC307A,BC307B,BC308A,BC308B, BC415-10,BC415-6,BC415A,BC415B,BC416-10,BC416-6,BC416A,BC416B, BC557A,BC557B,BC558A,BC558B,BC559A,BC559B,BC560A,BC560B,BCY78, BCY78A,BCY78B,BCY79,BCY79A,BCY79B,BSW19,BSW20,BSW73,EN3502,GI-3644, MM3906,NPS404A,TN-3906,TP3644,*PTC127,*GE-65,*TR-20,*HEPS0031,*SK3114, *ECG234,*WEP242,*BC177,*276-2034,*RT-115
BC178A	2N3906,2N4034,2N4035,2N4403,2N5244,2N5375,2N5383,2N5999,2N6001,2SA509-Y, 2SA509G-Y,2SA550,2SA719,2SA890,A5T3906,A5T4403,A178A,A178B,A179A, BC178,BC179,BC252A,BC253A,BC258A,BC262A,BC263A,BC308A,BC309A, BC558A,BC559A,BCY78,BCY78A,BCY78B,BSW19,BSW20,BSW73,CS9012, MM3906,NPS404,NPS404A,TN-3906,*PTC127,*GE-48,*TR-20,*HEPS0013,*SK3114, *ECG234,*WEP242,*276-2034
BC178B	2N5244,2N5378,2N5379,2N5999,2N6003,2N6009,2SA550,BC179,BC252B,BC253B, BC258B,BC262B,BC263B,BC308B,BC309B,BC558B,BC559B,BCY78,BCY78B, BCY78C,*PTC127,*GE-48,*TR-20,*HEPS0019,*SK3053,*ECG234,*WEP242, *276-2034
BC178C	BC252C,BC253C,BC258C,BC262C,BC263C,BC308C,BC309C,BC558C,BC559C, BCY78D,*GE-65,*SK3247,*ECG234,*276-2034
BC179	2N3906,2N4035,2N4403,2N5244,2N5375,2N5378,2N5379,2N5383,2N5999,2N6003, 2N6009,2SA550,A5T3906,A5T4403,BC178,BC178B,BC252B,BC253B,BC258B, BC262B,BC263B,BC308B,BC309B,BC558B,BC559B,BCY78,BCY78B,BCY78C, BSW19,BSW20,BSW73,NPS404,NPS404A,TN-3906,*PTC127,*GE-65,*TR-20, *HEPS0019,*SK3114,*ECG234,*WEP242,*BC177,*276-2034,*RT-115
BC179B	2N5999,2N6003,2N6009,2SA550,BC178B,BC179,BC250C,BC252B,BC253B,BC258B, BC259B,BC260C,BC262B,BC263B,BC308B,BC309B,BC558B,BC559B,BCY78, BCY78B,BCY78C,*PTC127,*GE-48,*TR-20,*HEPS0031,*SK3114,*ECG234, *WEP242,*276-2034
BC179C	BC178C,BC250C,BC252C,BC253C,BC258C,BC259C,BC260C,BC262C,BC263C, BC308C,BC309C,BC558C,BC559C,BCY78D,*GE-65,*SK3247,*ECG234,*276-2034
BC182A	2N2222A,2N3904,2N4401,2N5381,2N5582,2N6540,2N6541,A5T3904,MM3904,TIS95, TIS98,TIS111,TN-3904,*GE-210,*HEPS0015,*ECG123A,*WEP736,*276-2013
BC182B	2N6428,2N6428A,2N6539,TIS94,TIS97
BC182L	2N3302,2N4409,2N5376,2N5377,A5T4409,BC182B,TIS94,TIS97,TIS109,*GE-210, *TR-21,*HEPS0015,*SK3122,*ECG123A,*WEP736,*BC107B,*276-2009,*RT-102
BC183L	2N4953,2N5370,2N5376,2N6006,2N6012,2N6112,2N6428,2N6428A,2N6539,BC184L, BCY59C,BCY59D,D33D27,TIS94,TIS97,*GE-210,*TR-21,*HEPS0015,*SK3122, *ECG123A,*WEP736,*BC107B,*276-2016,*RT-102
BC184L	2N4953,2N5370,2N6006,2N6012,2N6428,2N6428A,2N6539,BCY59D,D33D27,*TR-21, *HEPS0015,*SK3122,*ECG123A,*WEP736,*276-2013,*RT-102
BC186	2N3250,2N3250A,2N3644,2N3645,2N3672,2N3673,2N3905,2N4143,2N4228,2N4402, 2N5382,2N6005,2SA467,2SA603,2SA696,2SA718,2SA720,2SA891,A5T3905, A5T4402,BC212A,BC212L,BCW35,BCW37,CS9012,FT3644,FT3645,GET2906, GET2907,GI-3644,MM3905,MM3906,TN-3905,ZT181,ZT182,ZT184,ZT281,ZT282, ZT284,*PTC127,*GE-65,*TR-21,*HEPS0013,*SK3122,*ECG123A,*WEP736, *276-2034,*RT-102
BC187	2N3136,2N3251,2N3504,2N3906,2N4403,2N5366,2N5379,2N5383,2N6001,2N6005, 2SA719,2SA890,A5T3504,A5T3644,A5T3906,A5T4403,BC213L,BC214L,BCY78,

To Replace	Substitute This Type
(BC187)	BCY78A,BCY78B,BCY79,BCY79A,BCY79B,FT3645,GI-3644,MM3906,MPS6534, TN-3906,*PTC127,*GE-65,*TR-20,*HEPS0013,*SK3114,*ECG159,*WEP242, *276-2034,*RT-115
BC190A	2N3116,2N3302,2N3904,2N4141,2N4401,2N4409,2N4952,2N5028,2N5107,2N5369, 2N5381,2SC979-Y,2SC1991,A5T3904,A5T4409,BC174A,BC546A,BSW85,FT3643, GET2222,GET2222A,GI-3643,MM3904,MPS6531,TN-3904,*GE-63,*HEPS0005, *276-2010
BC190B	2N4953,2N5370,2N5376,2N5377,2N6428,2N6428A,BC174B,BC546B,*GE-63, *ECG192,*WEP243
BC192	2N5811,2N6001,2SA509-O,2SA509-Y,2SA509G-O,2SA509G-Y,2SA719,2SA890, A5T4402,A5T4403,BC328-16,BSW73,D29E1,TN-3905,TN-3906,TQ60,TQ62,*PTC127, *GE-82,*TR-20,*HEPS0019,*SK3114,*ECG159,*WEP242,*BC327,*276-2023, *RT-115
BC194	2N5449,*GE-20,*TR-95,*HEPS0025,*ECG108,*WEP56,*276-2016,*RT-113
BC196A	BC196B,*PTC103,*GE-65,*TR-20,*SK3114,*ECG159,*WEP242,*276-2034,*RT-115
BC196B	*GE-65,*TR-20,*SK3118,*ECG159,*WEP242,*276-2034,*RT-115
BC197A	BC197B,*PTC121,*TR-21,*SK3122,*ECG123A,*WEP736,*RT-102
BC197B	*PTC121,*TR-21,*SK3122,*ECG123A,*WEP736,*RT-102
BC198A	BC197A,BC197B,BC198B,BC199B,*PTC132,*GE-17,*276-2015
BC198B	BC197B,BC199B,*PTC139,*276-2015
BC198C	BC199C,*PTC139,*276-2015
BC199B	BC197B,BC198B,*PTC139,*276-2015
BC199C	BC198C,*PTC139,*276-2015
BC201	*PTC103,*GE-65,*TR-20,*SK3114,*ECG159,*WEP242,*BC177,*276-2021,*RT-115
BC202	BC203,*PTC103,*GE-65,*TR-20,*SK3114,*ECG159,*WEP242,*BC177,*276-2034, *RT-115
BC203	*PTC103,*GE-65,*TR-20,*SK3114,*ECG159,*WEP242,*BC177,*276-2023,*RT-115
BC212A	2N2907,2N2907A,2N3251A,2N3486,2N3486A,2N3505,2N3672,2N3673,2N4143, 2SA603,2SA718,2SA720,2SA891,A5T2907,A5T3505,A5T3645,BC212L,GET2906, GET2907,TIS112,*PTC103,*GE-48,*ECG193,*WEP242
BC212B	*GE-48,*ECG193,*WEP242
BC212L	2N2907,2N2907A,2N3251A,2N3486,2N3486A,2N3505,2N3672,2N3673,2N4143, 2N4228,2SA603,2SA718,2SA720,2SA891,A5T2907,A5T3505,A5T3645,BC212A, GET2906,GET2907,TIS112,*PTC103,*GE-48,*TR-20,*SK3114,*ECG159,*WEP242, *RT-115
BC213L	2N2907,2N2907A,2N3251,2N3251A,2N3486,2N3486A,2N3504,2N3505,2N3672, 2N3673,2N4143,2N6005,2SA603,2SA718,2SA720,2SA891,A5T2907,A5T3504, A5T3505,A5T3644,A5T3645,BC212A,BC214L,BCY79,BCY79A,BCY79B,GET2906, GET2907,GI-3644,TIS112,*PTC103,*GE-48,*TR-20,*SK3114,*ECG159,*WEP242, *276-2023,*RT-115
BC214L	2N2907,2N2907A,2N3251,2N3251A,2N3486,2N3486A,2N3504,2N3505,2N6007, A5T2907,A5T3504,A5T3505,A5T3644,A5T3645,BCY79,BCY79B,BCY79C,TIS112, *GE-48,*TR-20,*SK3114,*ECG159,*WEP242,*276-2023,*RT-115
BC237A	2N3904,2N4401,2N5381,2SC1992,A5T3904,BC107A,BC547A,MM3904,TN-3904, *PTC121,*GE-212,*TR-21,*HEPS0030,*SK3245,*ECG123A,*WEP736,*BC107B, *276-2013
BC237B	2N6428,2N6428A,BC107B,BC547B,BC550B,*PTC121,*GE-212,*TR-21,*HEPS0030, *SK3245,*ECG123A,*WEP736,*BC107B,*276-2013
BC238A	2N3227,2N4124,2N4419,2SC735-Y,2SC1992,A5T4124,BC107,BC107A,BC108, BC108A,BC167A,BC171A,BC237A,BC547A,BC548A,BCY58A,BCY58B,BCY59A, BCY59B,BSW83,*PTC139,*GE-212,*TR-21,*SK3122,*ECG123A,*WEP736,*BC107B, *276-2013,*RT-102
BC238B	2N4954,2N5371,2N5827,2SC735-GR,2SC1993,A157B,BC107B,BC108B,BC109, BC109B,BC167B,BC171B,BC237B,BC239B,BC413B,BC414B,BC547B,BC548B, BC549B,BC550B,BCY58C,BCY59C,*PTC139,*GE-212,*TR-21,*HEPS0015,*SK3122, *ECG123A,*WEP736,*BC107B,*276-2013,*RT-102
BC238C	BC108C,BC109C,BC239C,BC413C,BC414C,BC548C,BC549C,BC550C,BCY58D, BCY59D,*PTC139,*GE-212,*TR-21,*HEPS0024,*SK3156,*ECG199,*WEP736, *BC107B,*276-2013
BC239B	2N4954,2N5371,2N5827,2SC735-GR,2SC1993,A157B,BC107B,BC108B,BC109, BC109B,BC167B,BC171B,BC237B,BC238B,BC413B,BC414B,BC547B,BC548B,

TRANSISTOR SUBSTITUTES

To Replace	Substitute This Type
(BC239B)	BC549B,BC550B,BCY58C,BCY59C,*PTC139,*GE-212,*TR-21,*HEPS0015,*SK3122, *ECG199,*WEP736,*BC107B,*276-2013,*RT-102
BC239C	BC108C,BC109C,BC238C,BC413C,BC414C,BC548C,BC549C,BC550C,BCY58D, BCY59D,*PTC139,*TR-21,*HEPS0024,*SK3245,*ECG199,*WEP736,*276-2016
BC250A	2N869A,2N3545,2N3576,2N3829,BC260A,*PTC131,*GE-65,*TR-20,*HEPS0013, *SK3114,*ECG159,*WEP242,*276-2034,*RT-115
BC250B	2N6001,2SA719,2SA890,2SB542,BC258A,BC260B,BC558A,BC559A,BCY78,BCY78A, BCY78B,BSW19,BSW20,NPS404,*PTC103,*GE-65,*TR-20,*HEPS0019,*SK3114, *ECG159,*WEP242,*276-2034,*RT-115
BC250C	2N6003,BC258C,BC260C,BC558C,BC559C,BCY78C,BCY78D,*GE-65,*TR-20, *HEPS0031,*SK3114,*ECG159,*WEP242,*276-2034,*RT-115
BC251A	2N2907,2N2907A,2N3251,2N3251A,2N3486,2N3486A,2N3504,2N3505,2N3672, 2N3673,2N3962,2N4143,2N5373,2N6005,2SA550A,2SA603,2SA678,2SA705,2SA718, 2SA720,2SA891,A5T2907,A5T3504,A5T3505,A5T3644,A5T3645,A177,BC177A, BC212A,BC212L,BC213L,BC214L,BC256A,BC257A,BC261A,BC266A,BC307A, BC415-6,BC415A,BC416-6,BC416A,BC557A,BC560A,BCW35,BCW37,BCY79, BCY79A,BCY79B,GET2906,GET2907,GI-3644,TIS112,TP3644,TP3645,TZ552, *PTC103,*GE-48,*TR-20,*SK3114,*ECG159,*WEP242,*RT-115
BC251B	2N3962,2N3964,2N3965,2N5374,2N6007,2SA550A,2SA678,2SA705,BC177B,BC214L, BC256B,BC257B,BC261B,BC266B,BC307B,BC415-10,BC415B,BC416-10,BC416B, BC557B,BC560B,BCY79,BCY79B,BCY79C,TZ553,*TR-20,*SK3114,*ECG159, *WEP242,*RT-115
BC251C	2N3964,2N3965,2SA844,BC261C,BC307C,BC415-16,BC415C,BC416-16,BC416C, BC560C,*TR-20,*SK3114,*ECG159,*WEP242,*RT-115
BC252A	2N3906,2N4034,2N4035,2N4403,2N5244,2N5375,2N5383,2N5999,2N6001,2SA509-Y, 2SA509G-Y,2SA550,2SA719,2SA890,A5T3906,A5T4403,A178A,A178B,A179A, BC178,BC178A,BC179,BC253A,BC258A,BC262A,BC263A,BC308A,BC309A, BC558A,BC559A,BCY78,BCY78A,BCY78B,BSW19,BSW20,BSW73,MM3906, NPS404,NPS404A,TN-3906,*PTC103,*GE-48,*TR-20,*SK3114,*ECG159,*WEP242, *276-2034,*RT-115
BC252B	2N5244,2N5378,2N5379,2N5999,2N6003,2N6009,2SA550,BC178B,BC179,BC253B, BC258B,BC262B,BC263B,BC308B,BC309B,BC558B,BC559B,BCY78,BCY78B, BCY78C,*TR-20,*SK3114,*ECG159,*WEP242,*276-2034,*RT-115
BC252C	BC178C,BC253C,BC258C,BC262C,BC263C,BC308C,BC309C,BC558C,BC559C, BCY78D,*TR-20,*SK3114,*ECG159,*WEP242,*276-2034,*RT-115
BC253A	2N3906,2N4034,2N4035,2N4403,2N5244,2N5375,2N5383,2N5999,2N6001,2SA509-Y, 2SA509G-Y,2SA550,2SA719,2SA890,A5T3906,A5T4403,A178A,A178B,A179A, BC178,BC178A,BC179,BC252A,BC258A,BC262A,BC263A,BC308A,BC309A, BC558A,BC559A,BCY78,BCY78A,BCY78B,BSW19,BSW20,BSW73,MM3906, NPS404,NPS404A,TN-3906,*PTC103,*GE-22,*TR-20,*SK3114,*ECG159,*WEP242, *276-2034,*RT-115
BC253B	2N5244,2N5378,2N5379,2N5999,2N6003,2N6009,2SA550,BC178B,BC179,BC252B, BC258B,BC262B,BC263B,BC308B,BC309B,BC558B,BC559B,BCY78,BCY78B, BCY78C,*GE-22,*TR-20,*SK3114,*ECG159,*WEP242,*276-2034,*RT-115
BC253C	BC178C,BC252C,BC258C,BC262C,BC263C,BC308C,BC309C,BC558C,BC559C, BCY78D,*GE-48,*TR-20,*SK3114,*ECG159,*WEP242,*276-2034,*RT-115
BC256A	2N2907,2N2907A,2N3251A,2N3486,2N3486A,2N3505,2N3672,2N3673,2N3962, 2N3963,2N4143,2N5373,2SA603,2SA718,2SA720,2SA891,A5T2907,A5T3505, A5T3645,BC212A,BC212L,BC266A,BC556A,BCW35,BCW37,BSW75,GET2906, GET2907,TIS112,TP3645,TZ552,*PTC103,*GE-67,*TR-20,*SK3114,*ECG159, *WEP242,*RT-115
BC256B	2N3962,2N3963,2N3965,2N5374,BC266B,BC556B,TZ553,*TR-20,*SK3114,*ECG159, *WEP242,*RT-115
BC257A	2N2907,2N2907A,2N3251,2N3251A,2N3486,2N3486A,2N3504,2N3505,2N3672, 2N3673,2N4143,2N6005,2SA603,2SA718,2SA720,2SA891,A5T2907,A5T3504, A5T3505,A5T3644,A5T3645,BC212A,BC212L,BC213L,BC214L,BC415-6,BC415A, BC416-6,BC416A,BCY79,BCY79A,BCY79B,GET2906,GET2907,GI-3644,TIS112, TP3644,TP3645,*PTC103,*GE-65,*SK3138,*ECG193,*WEP242
BC257B	2N6007,BC214L,BC415-10,BC415B,BC416-10,BC416B,BCY79,BCY79B,BCY79C, *GE-65,*SK3138,*ECG193,*WEP242
BC258A	2N3906,2N4034,2N4035,2N4403,2N5244,2N5383,2N6001,2SA719,2SA890,A5T3906,

To Replace	Substitute This Type
(BC258A)	A5T4403,BCY78,BCY78A,BCY78B,MM3906,NPS404,NPS404A,TN-3906,*PTC103, *GE-65,*SK3138,*ECG193,*WEP242,*276-2034
BC258B	2N5244,2N5378,2N5379,2N6003,BCY78,BCY78B,BCY78C,*GE-65,*SK3138,*ECG193, *WEP242,*276-2034
BC258C	BCY78D,*GE-65,*SK3247,*276-2034
BC259B	2N6003,BC250C,BC258B,BC260C,BCY78,BCY78B,BCY78C,*GE-65,*SK3138, *ECG193,*WEP242,*276-2034
BC259C	BC250C,BC258C,BC260C,BCY78D,*276-2034
BC260A	2N869A,2N3545,2N3576,2N3829,BC250A,*PTC131,*GE-65,*TR-20,*HEPS0013, *SK3114,*ECG159,*WEP242,*276-2034,*RT-115
BC260B	2N6001,2SA719,2SA890,2SB542,BC250B,BC258A,BC558A,BC559A,BCY78,BCY78A, BCY78B,BSW19,BSW20,NPS404,*PTC127,*GE-65,*TR-20,*HEPS0013,*SK3114, *ECG159,*WEP242,*276-2034,*RT-115
BC260C	2N6003,BC250C,BC258C,BC558C,BC559C,BCY78C,BCY78D,*GE-65,*TR-20, *HEPS0019,*SK3114,*ECG159,*WEP242,*276-2034,*RT-115
BC261A	2N2907,2N2907A,2N3251,2N3251A,2N3486,2N3486A,2N3504,2N3505,2N3644, 2N3645,2N3672,2N3673,2N3962,2N4143,2N5373,2N6005,2SA550A,2SA603,2SA678, 2SA705,2SA718,2SA720,2SA891,A5T2907,A5T3504,A5T3505,A5T3644,A5T3645, A177,BC177A,BC212A,BC212L,BC213L,BC214L,BC251A,BC256A,BC257A,BC266A, BC307A,BC415-6,BC415A,BC416-6,BC416A,BC557A,BC560A,BCW35,BCW37, BCY79,BCY79A,BCY79B,EN2905,EN3502,FT3644,FT3645,GET2906,GET2907, GI-3644,TIS112,TP3644,TP3645,TZ552,*PTC127,*GE-48,*TR-20,*HEPS0019, *SK3114,*ECG159,*WEP242,*276-2024,*RT-115
BC261B	2N3962,2N3964,2N3965,2N5374,2N6007,2SA550A,2SA678,2SA705,BC177B,BC214L, BC251B,BC256B,BC257B,BC266B,BC307B,BC415-10,BC415B,BC416-10,BC416B, BC557B,BC560B,BCY79,BCY79B,BCY79C,TZ553,*PTC127,*TR-20,*HEPS0019, *SK3025,*ECG159,*WEP242,*276-2024,*RT-115
BC261C	2N3964,2N3965,2SA844,BC251C,BC307C,BC415-16,BC415C,BC416-16,BC416C, BC560C,*TR-20,*HEPS0019,*SK3114,*ECG159,*WEP242,*RT-115
BC262A	2N3906,2N4034,2N4035,2N4403,2N5244,2N5375,2N5383,2N5999,2N6001,2SA509-Y, 2SA509G-Y,2SA550,2SA719,2SA890,A5T3906,A5T4403,A178A,A178B,A179A, BC178,BC178A,BC179,BC252A,BC253A,BC258A,BC263A,BC308A,BC309A, BC558A,BC559A,BCY78,BCY78A,BCY78B,BSW19,BSW20,BSW73,CS9012, MM3906,NPS404,NPS404A,TN-3906,*PTC127,*GE-48,*TR-20,*SK3114,*ECG159, *WEP242,*276-2034,*RT-115
BC262B	2N5244,2N5378,2N5379,2N5999,2N6003,2N6009,2SA550,BC178B,BC179,BC252B, BC253B,BC258B,BC263B,BC308B,BC309B,BC558B,BC559B,BCY78,BCY78B, BCY78C,*PTC127,*TR-20,*SK3114,*ECG159,*WEP242,*276-2034,*RT-115
BC262C	BC178C,BC252C,BC253C,BC258C,BC263C,BC308C,BC309C,BC558C,BC559C, BCY78D,*TR-20,*SK3114,*ECG159,*WEP242,*276-2034,*RT-115
BC263A	2N3906,2N4034,2N4035,2N4403,2N5244,2N5375,2N5383,2N5999,2N6001,2SA509-Y, 2SA509G-Y,2SA550,2SA719,2SA890,A5T3906,A5T4403,A178A,A178B,A179A, BC178,BC178A,BC179,BC252A,BC253A,BC258A,BC262A,BC308A,BC309A, BC558A,BC559A,BCY78,BCY78A,BCY78B,BSW19,BSW20,BSW73,CS9012, MM3906,NPS404,NPS404A,TN-3906,*PTC127,*GE-22,*TR-20,*SK3114,*ECG159, *WEP242,*276-2034,*RT-115
BC263B	2N5244,2N5378,2N5379,2N5999,2N6003,2N6009,2SA550,BC178B,BC179,BC252B, BC253B,BC258B,BC262B,BC308B,BC309B,BC558B,BC559B,BCY78,BCY78B, BCY78C,*PTC127,*GE-22,*TR-20,*SK3114,*ECG159,*WEP242,*276-2034,*RT-115
BC263C	BC178C,BC252C,BC253C,BC258C,BC262C,BC308C,BC309C,BC558C,BC559C, BCY78D,*GE-48,*TR-20,*SK3114,*ECG159,*WEP242,*276-2034,*RT-115
BC266A	2N2907,2N2907A,2N3251A,2N3486,2N3486A,2N3505,2N3645,2N3672,2N3673, 2N3962,2N3963,2N4143,2N5373,2SA603,2SA718,2SA720,2SA891,A5T2907, A5T3505,A5T3645,BC212A,BC212L,BC256A,BC556A,BCW35,BCW37,BSW75, EN2905,FT3644,GET2906,GET2907,TIS112,TP3645,TZ552,*PTC127,*GE-67,*TR-20, *SK3114,*ECG159,*WEP242,*RT-115
BC266B	2N3962,2N3963,2N3965,2N5374,BC256B,BC556B,TZ553,*PTC127,*TR-20,*SK3114, *ECG159,*WEP242,*RT-115
BC307A	2N2907,2N2907A,2N3251,2N3251A,2N3486,2N3486A,2N3505,2N3672,2N3673, 2N3962,2N4143,2N5373,2N6005,2SA603,2SA678,2SA705,2SA718,2SA720,2SA891, A5T2907,A5T3505,A5T3645,A177,BC212A,BC212L,BC256A,BC266A,BC416-6,

To Replace	Substitute This Type
(BC307A)	BC416A,BC557A,BC560A,BCW35,BCW37,BSW75,GET2906,GET2907,TIS112,TP3645, TZ552,*PTC103,*GE-65,*TR-20,*HEPS0019,*SK3114,*ECG159,*WEP242,*BC177, *276-2022
BC307B	2N3962,2N3965,2N5374,2N6007,2SA678,2SA705,BC256B,BC266B,BC416-10, BC416B,BC557B,BC560B,TZ553,*GE-65,*TR-52,*SK3138,*ECG193,*WEP242, *BC327,*276-2022
BC307C	2N3964,2N3965,2SA844,BC251C,BC261C,BC415-16,BC415C,BC416-16,BC416C, BC560C,*GE-65,*SK3247,*BC327,*276-2022
BC308A	2N3251,2N3504,2N3906,2N4034,2N4035,2N4403,2N5244,2N5375,2N5383,2N5999, 2N6001,2N6005,2SA509-Y,2SA509G-Y,2SA550A,2SA678,2SA705,2SA719,2SA890, A5T3504,A5T3644,A5T3906,A5T4403,A177,BC177A,BC178,BC213L,BC214L, BC251A,BC257A,BC261A,BC307A,BC415-6,BC415A,BC416-6,BC416A,BC557A, BC558A,BC559A,BC560A,BCY78,BCY78A,BCY78B,BCY79,BCY79A,BCY79B, BSW19,BSW20,BSW73,GI-3644,MM3906,NPS404A,TN-3906,TP3644,*PTC103, *GE-65,*TR-20,*HEPS0013,*SK3114,*ECG159,*WEP242,*BC177,*276-2022
BC308B	2N3964,2N5244,2N5378,2N5379,2N5999,2N6003,2N6007,2N6009,2SA550A,2SA678, 2SA705,BC177B,BC214L,BC251B,BC257B,BC261B,BC307B,BC415-10,BC415B, BC416-10,BC416B,BC557B,BC558B,BC559B,BC560B,BCY78,BCY78B,BCY78C, BCY79,BCY79B,BCY79C,*GE-65,*TR-20,*HEPS0019,*SK3247,*ECG159,*WEP242, *BC327,*276-2022
BC308C	2N3964,BC251C,BC261C,BC307C,BC415-16,BC415C,BC416-16,BC416C,BC558C, BC559C,BC560C,BCY78D,*GE-65,*SK3247,*BC327,*276-2022
BC309A	2N3906,2N4034,2N4035,2N4403,2N5244,2N5375,2N5383,2N5999,2N6001,2SA509-Y, 2SA509G-Y,2SA550,2SA719,2SA890,A5T3906,A5T4403,A178A,A178B,A179A, BC178,BC178A,BC179,BC252A,BC253A,BC258A,BC262A,BC263A,BC308A, BC558A,BC559A,BCY78,BCY78A,BCY78B,BSW19,BSW20,BSW73,MM3906, NPS404,NPS404A,TN-3906,*PTC103,*GE-65,*TR-20,*HEPS0013,*SK3114, *ECG159,*WEP242,*BC177,*276-2022
BC309B	2N5244,2N5378,2N5379,2N5999,2N6003,2N6009,2SA550,BC178B,BC179,BC252B, BC253B,BC258B,BC262B,BC263B,BC308B,BC558B,BC559B,BCY78,BCY78B, BCY78C,*GE-65,*TR-20,*HEPS0031,*ECG159,*WEP242,*BC177,*276-2022
BC309C	BC178C,BC252C,BC253C,BC258C,BC262C,BC263C,BC308C,BC558C,BC559C, BCY78D,*GE-65,*TR-52,*SK3138,*ECG193,*WEP242,*BC177,*276-2022
BC323	
BC325	*PTC127,*GE-82,*TR-20,*SK3114,*ECG159,*WEP242,*RT-115
BC326	2N3548,2N3962,2N3963,*PTC127,*GE-67,*TR-20,*SK3114,*ECG159,*WEP242, *RT-115
BC327	2N5374,2N6013,BC327-40,D29E7,*GE-87,*TR-73,*HEPS3032,*SK3200,*ECG189, *WEPS3031
BC327-16	BC327-25,MPS4355,*GE-48,*SK3138,*ECG193,*WEP242,*BC327
BC327-25	BC327-40,MPS4354,MPS4355,*BC327
BC327-40	*BC327
BC328	2N5378,2N5379,2N5813,2SB598,BC328-40,D29E2,*GE-22,*TR-73,*HEPS3028, *SK3200,*ECG189,*WEPS3031,*276-2032
BC328-16	BC328-25,*GE-48,*SK3138,*ECG193,*WEP242,*BC327,*276-2023
BC328-25	BC328-40,*BC327,*276-2023
BC328-40	*BC327,*276-2023
BC337	2N5370,2N5376,2N6012,BC337-40,D33D27,*GE-63,*TR-72,*HEPS3020,*SK3124, *ECG188,*WEPS3020
BC337-16	A5T2222,A5T5449,A8T3704,BC337-25,TIS109,TIS111,*GE-47
BC337-25	BC337-40,TIS109
BC337-40	
BC338	BC338-40,*GE-20,*TR-72,*HEPS3024,*SK3124,*ECG188,*WEPS3020,*276-2009
BC338-16	BC338-25,*GE-47,*SK3137,*ECG192,*BC337,*276-2009
BC338-25	A5T5451,A8T3706,BC338-40,*BC337,*276-2009
BC338-40	*BC337,*276-2009
BC340-10	2N2270,2N3053,2N5106,2SC1384,BC340-16,BC341-10,*PTC125,*GE-47,*TR-87, *SK3024,*ECG128,*WEP243,*276-2012,*RT-114
BC340-16	2N2219,2N2538,2N3300,2N5106,2SC503-GR,2SC504-GR,BSY52,BSY82,TN59, *PTC125,*GE-47,*TR-87,*SK3024,*ECG128,*WEP243,*276-2012,*RT-114
BC340-6	2N1708A,2N2218,2N2410,2N2537,2N2846,2N2848,2N3015,2N3299,2N3326,2N3724,

To Replace	Substitute This Type
(BC340-6)	2N3724A,2N4046,2N4960,2N5145,2SC306,2SC503-Y,2SC504-Y,BC340-10,BC341-10, BC341-6,BSY51,BSY81,TN61,*PTC123,*GE-47,*TR-87,*SK3024,*ECG128, *WEP243,*276-2012,*RT-114
BC341-10	2N2270,2N3053,2N5106,2SC109A-Y,2SC1384,2SC1509,40635,*PTC125,*GE-18, *TR-87,*SK3024,*ECG128,*WEP243,*276-2012,*RT-114
BC341-6	2N1837A,2N2218,2N2218A,2N2224,2N2410,2N2537,2N2846,2N2848,2N3015,2N3299, 2N3326,2N3678,2N3722,2N3725,2N3725A,2N4047,2N4960,2N4961,2SC97A, 2SC109A-O,2SC307,2SC503-Y,2SC560,2SC1218,40635,BC341-10,BSY51,BSY53, BSY83,TN53,*PTC125,*GE-18,*TR-87,*SK3024,*ECG128,*WEP243,*276-2012, *RT-114
BC360-10	2SA571,BC360-16,BC361-10,*PTC127,*GE-48,*TR-88,*SK3025,*ECG129,*WEP242, *276-2025,*RT-115
BC360-16	2N4032,*PTC127,*GE-48,*TR-88,*SK3025,*ECG129,*WEP242,*276-2025,*RT-115
BC360-6	2N4030,2SA571,2SA717,BC360-10,BC361-10,BC361-6,*PTC127,*GE-48,*TR-88, *SK3025,*ECG129,*WEP242,*276-2025,*RT-115
BC361-10	2SA571,*PTC127,*GE-67,*TR-88,*SK3025,*ECG129,*WEP242,*RT-115
BC361-6	2N4030,2N4031,2SA571,2SA717,BC361-10,*PTC127,*GE-67,*TR-88,*SK3025, *ECG129,*WEP242,*RT-115
BC413B	2N4953,2N5370,2N5376,2N5377,2N5827,2N6428,2N6428A,A157B,BC107B,BC167B, BC171B,BC174B,BC190B,BC237B,BC414B,BC547B,BC550B,BCY59C,*GE-210, *TR-21,*HEPS0015,*SK3245,*ECG199,*WEP736,*BC107,*276-2016
BC413C	BC414C,BC550C,BCY59D,*TR-21,*HEPS0024,*SK3245,*ECG199,*WEP736, *276-2016
BC414B	2N4953,2N5370,2N5376,2N5377,2N5827,2N6428,2N6428A,BC107B,BC174B,BC190B, BC237B,BC547B,BC550B,*PTC121,*GE-210,*TR-21,*HEPS0015,*SK3124,*ECG199, *WEP736,*BC107
BC414C	BC550C,*TR-21,*HEPS0015,*SK3245,*ECG199,*WEP736,*276-2013
BC415-10	2N6007,BC214L,BC257B,BC415B,BC416-10,BC416B,BCY79,BCY79B,BCY79C
BC415-16	2SA844,BC415C,BC416-16,BC416C
BC415-6	2N2907,2N2907A,2N3251,2N3251A,2N3486,2N3486A,2N3504,2N3505,2N3672, 2N3673,2N4143,2N6005,2SA603,2SA718,2SA720,2SA891,A5T2907,A5T3504, A5T3505,A5T3644,A5T3645,BC212A,BC212L,BC213L,BC214L,BC257A,BC415A, BC416-6,BC416A,BCY79,BCY79A,BCY79B,GET2906,GET2907,GI-3644,TIS112, TP3644,TP3645,*PTC103
BC415A	2N2907,2N2907A,2N3251,2N3251A,2N3486,2N3486A,2N3504,2N3505,2N3672, 2N3673,2N4143,2N6005,2SA603,2SA718,2SA720,2SA891,A5T2907,A5T3504, A5T3505,A5T3644,A5T3645,BC212A,BC212L,BC213L,BC214L,BC257A,BC415-6, BC416-6,BC416A,BCY79,BCY79A,BCY79B,GET2906,GET2907,GI-3644,TIS112, TP3644,TP3645,*PTC103,*GE-48,*TR-20,*HEPS0019,*SK3025,*ECG159,*WEP242, *276-2023
BC415B	2N6007,BC214L,BC257B,BC415-10,BC416-10,BC416B,BCY79,BCY79B,BCY79C, *GE-48,*TR-20,*HEPS0019,*SK3025,*ECG159,*WEP242,*276-2023
BC415C	2SA844,BC415-16,BC416-16,BC416C,*276-2023
BC416-10	2N6007,BC416B
BC416-16	2SA844,BC416C
BC416-6	2N2907,2N2907A,2N3251,2N3251A,2N3486,2N3486A,2N3505,2N3672,2N3673, 2N4143,2N6005,2SA603,2SA718,2SA720,2SA891,A5T2907,A5T3505,A5T3645, BC212A,BC212L,BC416A,GET2906,GET2907,TIS112,TP3645,*PTC103
BC416A	2N2907,2N2907A,2N3251,2N3251A,2N3486,2N3486A,2N3505,2N3672,2N3673, 2N4143,2N6005,2SA603,2SA718,2SA720,2SA891,A5T2907,A5T3505,A5T3645, BC212A,BC212L,BC416-6,GET2906,GET2907,TIS112,TP3645,*GE-48,*TR-20, *HEPS0019,*SK3025,*ECG159,*WEP242,*276-2034
BC416B	2N6007,BC416-10,*GE-48,*TR-20,*HEPS0019,*SK3025,*ECG159,*WEP242, *276-2021
BC416C	2SA844,BC416-16
BC546A	*SK3122
BC546B	*SK3122
BC547A	2N5381,A5T3904,TN-3904,*PTC121
BC547B	2N6428,2N6428A,BC550B
BC548A	A5T4124,BC547A,BSW83,*PTC121,*276-2009
BC548B	BC547B,BC549B,BC550B,*PTC139,*276-2009

To Replace	Substitute This Type
BC548C	BC549C,BC550C,*276-2009
BC549B	BC547B,BC548B,BC550B,*PTC139,*276-2009
BC549C	BC548C,BC550C,*PTC139
BC550B	2N6428,2N6428A,BC547B,*276-2009
BC550C	2N6429,2N6429A,*276-2009
BC556A	2SA850,BC556B
BC556B	
BC557A	2N5373,2N5374,2SA891,A5T2907,A5T3505,A5T3645,BC557B,BC560A,BC560B, TIS112,TP3645,*PTC103
BC557B	2N5374,BC560B
BC558A	2N5375,2N5379,2N5383,2SA509-Y,2SA509G-Y,2SA890,A5T3504,A5T3644,A5T3906, A5T4403,BC557A,BC557B,BC558B,BC559A,BC559B,BC560A,BC560B,BCY79, TP3644,*PTC103
BC558B	2N5378,2N5379,BC557B,BC559B,BC560B,BCY79
BC558C	BC559C,BC560C
BC559A	2N5375,2N5379,2N5383,2SA509-Y,2SA509G-Y,2SA890,A5T3504,A5T3644,A5T3906, A5T4403,BC557A,BC557B,BC558A,BC558B,BC559B,BC560A,BC560B,BCY79, TP3644,*PTC103
BC559B	2N5378,2N5379,BC557B,BC558B,BC560B,BCY79
BC559C	BC558C,BC560C
BC560A	2N5373,2N5374,2SA891,A5T2907,A5T3505,A5T3645,BC557A,BC557B,BC560B, TIS112,TP3645
BC560B	2N5374,BC557B
BC560C	
BC1274	SEE 2N2294
BC1274A	SEE 2N2295
BC1274B	SEE 2N2296
BCW34	2N2222A,2N5582,SE6021,SE6021A,TIS111,TN-3904,*PTC123,*GE-47,*TR-21, *SK3122,*ECG123A,*WEP736,*BC337,*RT-102
BCW35	*PTC127,*GE-48,*TR-20,*SK3114,*ECG159,*WEP242,*RT-115
BCW36	2N2222A,2N4401,2N5582,BCW34,TIS111,TN-3904,*GE-47,*TR-21,*SK3122, *ECG123A,*WEP736,*BC337,*RT-102
BCW37	BCW35,*PTC103,*GE-48,*TR-20,*SK3114,*ECG159,*WEP242,*RT-115
BCY10	BCY12,MM4052,*GE-82,*TR-20,*HEPS0012,*SK3114,*SP70,*ECG159,*WEP242, *BC327,*276-2022,*RT-115
BCY11	*GE-82,*TR-20,*HEPS0012,*SK3114,*SP70,*ECG159,*WEP242,*276-2021,*RT-115
BCY12	*GE-82,*TR-20,*HEPS0013,*SK3114,*ECG159,*WEP242,*BC327,*276-2022,*RT-115
BCY30	2N928,2N1232,2N1474,2N1474A,BCY31,HA7534,HA7539,SHA7534,SHA7539,*GE-82, *TR-88,*HEPS0013,*SK3025,*SP70,*ECG129,*WEP242,*276-2023,*RT-115
BCY31	2N928,2N1233,2N1475,2N4982,HA7538,SHA7538,*GE-82,*TR-88,*HEPS0013, *SK3025,*SP70,*ECG129,*WEP242,*276-2023,*RT-115
BCY33	2N328B,2N923,2N926,2N939,2N1026,2N1219,2N1230,2N1441,2N3219,2N3979, 2N4008,2N6567,BCY34,HA7533,MM4052,SHA7533,*GE-65,*TR-88,*HEPS0013, *SK3025,*SP70,*ECG129,*WEP242,*BC177,*276-2022,*RT-115
BCY34	2N329B,2N924,2N926,2N939,2N940,2N1231,2N1442,2N1443,2N1469,2N1469A, 2N2946A,2N3219,2N3979,HA7537,SHA7537,*GE-65,*TR-88,*HEPS0013,*SK3025, *SP70,*ECG129,*WEP242,*BC177,*276-2022,*RT-115
BCY58	BCY59,*DS-66,*GE-20,*TR-21,*HEPS0011,*SN80,*ECG123A,*WEP736,*276-2038, *RT-102
BCY58A	BCY58B,BCY59A,BCY59B,*PTC136,*TR-51,*HEPS0014,*SK3122,*BC107B, *276-2016,*RT-102
BCY58B	BCY58C,BCY59B,BCY59C,*PTC136,*TR-21,*HEPS0014,*SK3122,*ECG123A, *WEP736,*BC107B,*276-2016,*RT-102
BCY58C	BCY58D,BCY59C,BCY59D,*TR-21,*SK3122,*ECG123A,*WEP736,*BC107B, *276-2016,*RT-102
BCY58D	BCY59D,*TR-21,*SK3122,*ECG123A,*WEP736,*BC107B,*276-2016,*RT-102
BCY59	*DS-66,*GE-17,*TR-21,*HEPS0011,*EN10,*ECG123A,*WEP736,*276-2009,*RT-102
BCY59A	BCY59B,*TR-21,*HEPS0015,*SK3122,*ECG123A,*WEP736,*BC107B,*RT-102
BCY59B	BCY59C,*TR-21,*HEPS0015,*SK3122,*ECG123A,*WEP736,*BC107B,*RT-102
BCY59C	BCY59D,*TR-21,*HEPS0015,*SK3122,*ECG123A,*WEP736,*BC107B,*RT-102
BCY59D	*TR-51,*HEPS0015,*SK3122,*ECG123A,*WEP736,*BC107B,*RT-102

To Replace	Substitute This Type
BCY65	*TR-87,*HEPS3020,*SK3024,*ECG128,*WEP243,*RT-114
BCY66	BCY65,*GE-88,*TR-87,*SK3024,*ECG128,*WEP243,*RT-114
BCY78	2N3251,2N3504,2N3906,2N4403,2N5378,2N5379,2N5383,2N6003,2N6007,A5T3504, A5T3644,A5T3906,A5T4403,BC214L,BCY78B,BCY78C,BCY79,BCY79B,BCY79C, TN-3906,*PTC127,*GE-48,*TR-20,*HEPS0013,*SK3114,*ECG234,*WEP242, *276-2034,*RT-115
BCY78A	2N3251,2N3504,2N5375,2N5379,2N5383,2N6001,2N6005,2SA719,2SA890,A5T3504, A5T3644,A5T3906,A5T4403,BCY78B,BCY79,BCY79A,BCY79B,BSW73,FT3645, MM3906,TN-3906,*PTC127,*GE-82,*SK3114,*ECG193,*WEP242,*BC327,*276-2022
BCY78B	2N3251,2N3504,2N3964,2N5375,2N5378,2N5379,2N5383,2N6003,2N6007,A5T3504, A5T3644,A5T3906,A5T4403,BCY78C,BCY79,BCY79B,BCY79C,BSW73,D29E7, TN-3906,*PTC127,*GE-48,*ECG193,*WEP242,*BC327,*276-2022
BCY78C	2N3964,2N5378,2N5379,2N6003,2N6007,BCY78D,BCY79C,D29E7,*GE-48,*BC327, *276-2022
BCY78D	2N3964,*BC327,*276-2022
BCY79	*PTC127,*GE-48,*TR-20,*HEPS5022,*SK3114,*ECG234,*WEP242,*276-2022, *RT-115
BCY79A	2N2907,2N2907A,2N3251,2N3251A,2N3486,2N3486A,2N3504,2N3505,2N3672, 2N3962,2N5373,2N5374,2N6005,2SA720,2SA891,A5T2907,A5T3504,A5T3505, A5T3644,A5T3645,BCW35,BCY79,BCY79B,FT3644,FT3645,GET2906,GET2907, TIS112,TZ552,TZ553,*PTC127,*GE-82,*SK3114,*ECG193,*WEP242,*BC327, *276-2022
BCY79B	2N2907,2N2907A,2N3251,2N3251A,2N3486,2N3486A,2N3504,2N3505,2N3962, 2N3964,2N3965,2N5373,2N5374,2N6007,A5T2907,A5T3504,A5T3505,A5T3644, A5T3645,BCY79,BCY79C,D29E7,TIS112,TZ552,TZ553,*PTC127,*GE-48,*ECG193, *WEP242,*BC327,*276-2022
BCY79C	2N3964,2N3965,2N5374,2N6007,D29E7,TZ553,*GE-48,*BC327,*276-2022
BCZ10	2N938,2N939,2N4008,2N6567,MM4052,*GE-65,*TR-20,*HEPS0013,*SK3114,*SP70, *ECG159,*WEP242,*BC177,*276-2022,*RT-115
BCZ11	2N2946A,2N4008,2N6567,*GE-65,*TR-20,*HEPS0032,*SK3114,*SP70,*ECG159, *WEP242,*BC177,*276-2022,*RT-115
BCZ12	*GE-82,*TR-20,*HEPS0013,*SK3114,*ECG159,*WEP242,*276-2023,*RT-115
BCZ13	2N860,2N2391,2N2971,MPSH85,TP3638,*PTC127,*GE-65,*TR-20,*HEPS0013, *SK3114,*SP70,*ECG159,*WEP242,*BC177,*276-2022,*RT-115
BCZ14	2N860,2N2391,2N2971,MPSH85,TP3638,*PTC127,*GE-65,*TR-20,*HEPS0013, *SK3114,*SP70,*ECG159,*WEP242,*BC177,*276-2022,*RT-115
BD106A	BDY15A,*GE-28,*TR-76,*HEPS5000,*SK3054,*ECG152,*WEP245,*RT-154
BD106B	BDY15B,*GE-28,*TR-76,*SK3054,*ECG152,*WEP245,*RT-154
BD107A	BDY16A,HST5901,HST5902,*GE-28,*TR-76,*HEPS5000,*SK3054,*ECG152,*WEP245, *RT-154
BD107B	BDY16B,HST5911,HST5912,*GE-28,*TR-76,*SK3054,*ECG152,*WEP245,*RT-154
BD109-10	BD109-16,*PTC112
BD109-16	*PTC112
BD109-6	BD109-10,*PTC112,*GE-66
BD124	2N5608,HST5907,HST5908,KSP1021,KSP1022,*PTC112,*GE-66,*TR-76,*HEPS5000, *SK3054,*ECG152,*WEP245
BD127	BD128,*PTC124,*TR-81,*HEPS5011,*ECG124,*WEP240,*RT-128
BD128	BD129,*PTC124,*TR-81,*HEPS5011,*ECG124,*WEP240,*RT-128
BD129	*PTC124,*TR-81,*HEPS5011,*ECG124,*WEP240,*RT-128
BD130	2N3055,2N3236,2N5629,2N5630,2N5930,2N5933,2N5936,2N6270,2N6271,2N6272, 2N6273,108T2,1743-1010,1743-1030,1743-1220,BUY53A,BUY54A,HST9202, HST9203,IR3772,KSP1272,*PTC140,*GE-14,*TR-59,*HEPS7004,*SK3027,*ECG130, *WEP247,*BDY20,*RT-131
BD131	*GE-57,*HEPS5000,*SK3054,*ECG184,*WEPS5003,*BD230,*RT-152
BD132	*GE-58,*HEPS5006,*SK3083,*ECG242,*WEPS5007,*RT-153
BD133	
BD135	BD135-16,*PTC904,*GE-57,*TR-76,*HEPS3024,*SK3054,*ECG184,*WEPS5003
BD135-10	BD135-16,BD137-10,BD307A,BDY34,*PTC110
BD135-16	BD307B,BDY34
BD135-6	BD135-10,BD137-10,BD137-6,BD307A,*PTC110
BD136	BD136-16,*PTC905,*GE-58,*TR-77,*HEPS3028,*SK3083,*ECG185,*WEPS5007,

To Replace	Substitute This Type
(BD136)	*276-2026,*RT-153
BD136-10	BD136-16,BD138-10,*PTC111
BD136-16	
BD136-6	BD136-10,BD138-10,BD138-6,*PTC111
BD137	BD137-10,BD139,BD139-10,*PTC904,*GE-57,*HEPS5000,*SK3054,*ECG184, *WEPS5003,*RT-152
BD137-10	BD139-10,BD307A,*PTC110
BD137-6	BD137-10,BD139-10,BD139-6,BD307A,*PTC110
BD138	BD138-10,BD140,BD140-10,*PTC905,*GE-69,*HEPS5006,*SK3083,*ECG185, *WEPS5007,*RT-153
BD138-10	BD140-10,*PTC111
BD138-6	BD138-10,BD140-10,BD140-6,*PTC111
BD139	BD139-10,*PTC904,*GE-57,*HEPS5000,*SK3054,*ECG184,*WEPS5003,*RT-152
BD139-10	*PTC110
BD139-6	BD139-10,*PTC110
BD140	BD140-10,*PTC905,*HEPS5006,*SK3083,*ECG185,*WEPS5007,*RT-153
BD140-10	*PTC111
BD140-6	BD140-10,*PTC111
BD148-10	BD148-16,BD149-10,KSP1024,*PTC169
BD148-16	
BD148-6	BD148-10,BD149-10,BD149-6,KSP1024,*PTC169
BD149-10	KSP1024,KSP1025,*PTC169
BD149-6	BD149-10,KSP1024,KSP1025,*PTC169
BD165	BD167,*PTC163,*GE-28,*ECG184,*276-2020,*RT-113
BD166	BD168,*PTC162,*ECG185,*RT-113
BD167	BD169,*PTC163,*ECG184,*276-2020,*RT-113
BD168	BD170,*PTC162,*ECG185,*RT-113
BD169	*PTC165,*ECG182,*RT-113
BD170	*PTC164,*ECG183
BD171	BD172,*PTC124
BD172	BD173,*PTC124
BD173	*PTC124
BD175	BD177,BD437,BD439,*PTC163,*ECG184,*276-2020
BD176	BD178,BD438,BD440,*PTC162,*ECG185
BD177	BD179,BD439,BD441,*PTC163,*ECG184,*276-2020
BD178	BD180,BD440,BD442,*PTC162,*ECG185
BD179	BD441,*PTC165,*ECG182
BD180	BD442,*PTC164,*ECG183
BD181	2N3055A,2N3235,2N6371,B170002-BRN,B170002-ORG,B170002-RED,HST9201, HST9801,KSD1051,KSD1055,SDT9201,SDT9901,*PTC140,*GE-77,*HEPS7002, *ECG130,*BDY20
BD182	2N3055A,2N3235,2N5302,2N5303,2N5881,2N5882,2N5885,2N5886,40363, B170005-BRN,B170005-ORG,B170005-RED,BUY51A,BUY52A,HST9201,SDT9201, *GE-77,*HEPS7002,*ECG130
BD183	2N3055,2N3055SD,2N3236,2N3772,2N5303,2N5629,2N5882,2N5886,2N6254,2N6270, 2N6272,2SD341,2SD492,40411,40636,B170005-BRN,B170005-ORG,B170005-RED, B170008-BRN,B170008-ORG,B170008-RED,BUY53A,BUY54A,HST9202,IR802, IR3772,MJ802,SDT9202,SDT9704,*GE-75,*HEPS7004,*SK3027,*ECG284,*WEP247
BD184	2N3055,2N3055SD,2N3236,2N3772,2N5629,2N5630,2N6254,2N6270,2N6271,2N6272, 2N6273,2N6338,2SD341,2SD492,108T2,40411,40636,B170008-BRN,B170008-ORG, B170008-RED,BUY53A,BUY54A,HST9202,HST9203,IR802,IR3772,MJ802,SDT9202, SDT9203,SDT9704,SDT9705,*PTC173,*ECG280
BD185	BD187,*PTC163,*ECG184
BD186	BD188,*PTC162,*ECG185
BD187	BD189,*PTC163,*ECG184
BD188	BD190,*PTC162,*ECG185
BD189	*PTC163,*ECG184
BD190	*PTC162,*ECG185
BD213/45	BD213/60,*PTC169
BD213/60	*PTC169
BD213/80	*PTC169

To Replace	Substitute This Type
BD214/45	BD214/60,*PTC168
BD214/60	*PTC168
BD214/80	*PTC168
BD220	2N5294,2SD531,*WEP245
BD221	2N5294,BD220,*WEP245
BD222	2N5294,BD220,*WEP245
BD223	*WEP246
BD224	BD223,*WEP246
BD225	BD223,*WEP246
BD230	REFER TO SECTION 2
BD231	REFER TO SECTION 2
BD232	REFER TO SECTION 2
BD233	BD175,BD177,BD235,BD437,BD439,*276-2020
BD234	BD176,BD178,BD236,BD438,BD440
BD235	BD177,BD179,BD439,BD441,*276-2020
BD236	BD178,BD180,BD440,BD442
BD237	REFER TO SECTION 2
BD238	REFER TO SECTION 2
BD306A	*PTC110
BD306B	
BD307A	*PTC110
BD307B	
BD433	BD435,*PTC165,*ECG152,*276-2019
BD434	BD436,*PTC164,*ECG153,*276-2027
BD435	*PTC165,*ECG152,*276-2019
BD436	*PTC164,*ECG153,*276-2027
BD437	BD439,*PTC165,*ECG152,*276-2020
BD438	BD440,*PTC164,*ECG153
BD439	BD441,*PTC165,*ECG152,*276-2020
BD440	BD442,*PTC164,*ECG153
BD441	*PTC165,*ECG196
BD442	*PTC164,*ECG197
BD585	BD587,BD595,BD597,*PTC167,*ECG241,*276-2020
BD586	BD588,BD596,BD598,*PTC166,*ECG242
BD587	BD597,*PTC167,*ECG241,*276-2020
BD588	BD598,*PTC166,*ECG242
BD589	BD599,*PTC167,*ECG241
BD590	BD600,*PTC166,*ECG242
BD595	BD597,*PTC169,*276-2020
BD596	BD598,*PTC168
BD597	*PTC169,*276-2020
BD598	*PTC168
BD599	*PTC169
BD600	*PTC168
BD675	BD677,*ECG257
BD676	BD678,*ECG258
BD677	BD679,*ECG257
BD678	BD680,*ECG258
BD679	BD681,*ECG257
BD680	BD682,*ECG258
BD681	
BD682	
BD695	BD697
BD696	BD698
BD697	BD699
BD698	BD700
BD699	BD701
BD700	BD702
BD701	
BD702	
BDY10	*PTC173,*GE-75,*TR-36,*ECG181,*WEPS7000,*RT-149

To Replace	Substitute This Type
BDY11	*PTC173,*GE-75,*TR-36,*SK3036,*ECG181,*WEPS7000,*RT-149
BDY12	BDY13,*PTC112,*GE-66,*TR-76,*SK3054,*ECG152,*WEP245,*RT-154
BDY13	*PTC112,*GE-66,*TR-76,*SK3054,*ECG152,*WEP245,*RT-154
BDY15A	BD106A,*GE-66,*ECG155
BDY15B	BD106B,*GE-66,*ECG155
BDY15C	*GE-66,*ECG155
BDY16A	BD107A,HST5901,HST5902,*GE-66,*ECG155
BDY16B	BD107B,HST5911,HST5912,*GE-66,*ECG155
BDY20	REFER TO SECTION 2
BDY23A	HST9201,HST9205,HST9206,HST9801,HST9802,*PTC173
BDY23B	180T2,*PTC173
BDY23C	
BDY24A	2N3446,HST9202,HST9203,HST9207,HST9208,HST9803,HST9804,*PTC173
BDY24B	2N3448,181T2,*PTC173
BDY24C	
BDY25A	
BDY25B	182T2
BDY25C	
BDY26A	
BDY26B	183T2
BDY26C	
BDY27A	BDY28A
BDY27B	184T2,185T2,BDY28B
BDY27C	BDY28C
BDY28A	
BDY28B	185T2
BDY28C	
BDY34	*GE-28,*TR-76,*SK3054,*ECG152,*WEP245,*RT-154
BDY42	BDY27A,BDY28A,SDT1051,SDT1052,SDT1056,SDT1057,XGSR10040
BDY43	2N6542,BDY44,SDT1053,SDT1054,SDT1058,SDT1059
BDY44	2N6543,2N6622,SDT1054,SDT1059
BDY45	1843-3505,1843-3510,1843-3705,XDAR10035
BDY46	2N6546,BDY47
BDY47	2N6547
BF110	*GE-27,*TR-78,*HEPS5025,*SK3045,*ECG154,*WEP712,*RT-110
BF111	*GE-27,*TR-78,*SK3045,*ECG154,*WEP712,*BF338,*RT-110
BF114	*GE-27,*TR-78,*HEPS5025,*SK3045,*ECG154,*WEP712,*BF338,*RT-110
BF115	2SC334,2SC593,*PTC132,*GE-61,*TR-21,*HEPS0011,*SK3122,*ECG123A,*WEP736, *BC107B,*276-2016,*RT-102
BF117	*GE-27,*TR-78,*HEPS5025,*SK3045,*ECG154,*WEP712,*BF338,*RT-110
BF118	*GE-27,*TR-78,*HEPS5025,*SK3045,*ECG154,*WEP712,*BF338,*RT-110
BF119	*GE-27,*TR-78,*HEPS5025,*SK3045,*ECG154,*WEP712,*BF338,*RT-110
BF120	*PTC117,*GE-27,*TR-78,*HEPS0005,*BF338,*276-2012
BF121	2N3009,2N3013,2N3014,2N4418,2N5769,2N5772,2SC1687,2SC1688,BF123,BF125, BF127,BF198,BF199,BF241,BF311,TIS86,TIS87,TIS105,*GE-61,*TR-70,*SK3039, *ECG107,*WEP56,*BC337,*276-2016,*RT-108
BF123	2SC1687,2SC1688,TIS86,TIS87,TIS105,*GE-61,*TR-70,*SK3039,*ECG107,*WEP56, *BC337,*276-2013,*RT-108
BF125	2SC1687,2SC1688,BF123,TIS86,TIS87,TIS105,*GE-61,*TR-70,*SK3039,*ECG107, *WEP56,*BC337,*276-2016,*RT-108
BF127	2N3009,2N3013,2N3014,2N4418,2N5769,2SC1687,2SC1688,BF123,BF125,BF198, BF199,BF241,TIS86,TIS87,TIS105,*GE-60,*TR-70,*SK3039,*ECG107,*WEP56, *BF167,*276-2013,*RT-108
BF137	MM3001,MM3002,*PTC117,*GE-27,*TR-78,*HEPS5025,*BF338
BF167	2N3337,2N3338,2N3339,2SC333,2SC334,2SC463,40239,BF173,BF232,BFS62,BFX60, TN-3200,*PTC132,*GE-60,*TR-83,*HEPS0011,*SK3117,*EN10,*ECG161,*WEP719, *BF167,*276-2016,*RT-113
BF168	*PTC133,*GE-61,*TR-83,*HEPS0016,*SK3117,*EN10,*ECG161,*WEP719,*BF173, *276-2038,*RT-113
BF173	BF168,TC3114,*PTC133,*GE-61,*TR-83,*HEPS0016,*SK3018,*EN10,*ECG161, *WEP719,*BF173,*276-2038,*RT-113

To Replace	Substitute This Type
BF177	MM3000,*PTC144,*GE-27,*TR-78,*HEPS5026,*SK3045,*ECG154,*WEP712,*BF338,*RT-110
BF178	2SC154A,2SC154B,2SC589,BF179A,MM3001,*GE-27,*TR-78,*HEPS5025,*SK3045,*ECG154,*WEP712,*BF338,*RT-110
BF179A	2SC154B,BF179B,MM3002,*GE-27,*TR-78,*HEPS5025,*SK3045,*ECG154,*WEP712,*BF338,*RT-110
BF179B	BF179C,MM3002,MM3003,*GE-27,*TR-78,*HEPS5024,*SK3045,*ECG154,*WEP712,*BF338,*276-2008,*RT-110
BF179C	MM3003,*GE-27,*TR-78,*HEPS5024,*SK3045,*ECG154,*WEP712,*BF338,*276-2008,*RT-110
BF180	*PTC132,*GE-39,*TR-83,*HEPS0017,*SK3117,*ECG161,*WEP719,*BF200,*276-2015,*RT-113
BF181	*PTC132,*GE-60,*TR-83,*HEPS0017,*SK3117,*ECG161,*WEP719,*BF200,*276-2015,*RT-113
BF182	*PTC132,*GE-214,*TR-83,*HEPS0017,*SK3117,*ECG161,*WEP719,*BF183,*276-2015,*RT-113
BF183	*PTC132,*GE-214,*TR-83,*HEPS0017,*SK3117,*ECG161,*WEP719,*276-2015,*RT-113
BF184	*PTC132,*GE-39,*TR-83,*HEPS0017,*SK3117,*EN10,*ECG161,*WEP719,*BC107B,*276-2015,*RT-113
BF185	2SC332,2SC333,2SC593,*PTC132,*GE-39,*TR-83,*HEPS0017,*SK3117,*EN10,*ECG161,*WEP719,*BF365,*276-2015,*RT-113
BF194	BC147A,*PTC101,*GE-60,*TR-70,*HEPS0016,*SK3018,*ECG107,*WEP56,*BC107B,*276-2015,*RT-108
BF195	*PTC101,*GE-86,*TR-70,*HEPS0016,*SK3018,*ECG107,*WEP56,*BF365,*276-2015,*RT-108
BF196	BF197,*PTC101,*GE-60,*TR-70,*HEPS0016,*SK3039,*ECG107,*WEP56,*BF173,*276-2016,*RT-108
BF197	*PTC133,*GE-60,*TR-70,*HEPS0016,*SK3039,*ECG107,*WEP56,*BF173,*276-2038
BF198	2N2369,2N2369A,2N2501,2N3009,2N3013,2N3014,2N3862,2N4873,2N5769,2SC67,2SC601N,2SC639,2SC764,2SC1687,2SC1688,BF199,BF241,TIS87,TIS105,*PTC101,*GE-20,*TR-70,*SK3039,*ECG107,*WEP56,*276-2009,*RT-108
BF199	2N3862,2N4873,2SC639,2SC1216,2SC1687,2SC1688,TIS87,TIS105,*PTC133,*SK3039,*ECG161,*276-2038
BF200	*PTC132,*GE-60,*TR-83,*HEPS0017,*SK3018,*ECG161,*WEP719,*276-2015,*RT-108
BF223	*PTC133,*TR-95,*SK3039,*ECG108,*WEP56,*RT-113
BF224J	2N708A,2N2845,2N3301,2SC302,BF225J,GET2221,GI-3641,*PTC101,*GE-17,*SK3122,*ECG123A,*276-2009
BF225J	2SC302,*PTC121,*GE-20,*SK3122,*ECG123A,*276-2009
BF227	*PTC132,*GE-60,*276-2038
BF228	*PTC125,*GE-18,*SK3244,*276-2008
BF229	BC197A,BC198A,*PTC132,*GE-39,*TR-70,*SK3039,*ECG107,*WEP56,*276-2015
BF230	2N2885,BSW12,*PTC132,*GE-39,*TR-70,*SK3039,*ECG107,*WEP56,*276-2015
BF232	*PTC133,*GE-61,*TR-83,*SK3117,*ECG161,*WEP719,*RT-113
BF237	2N708A,2N735,2N735A,2N759A,2N760B,2N2515,2N2522,2N2845,2N3301,2N3641,2N3642,2N3973,2N3975,2S103,2SC302,2SC1166-R,BF225J,BSY93,FT3641,FT3642,FT3722,GET2221,GET2221A,*PTC123,*GE-62,*TR-95,*HEPS0025,*SK3039,*ECG108,*WEP56,*BC337,*276-2010,*RT-113
BF238	2N760A,2N916,2N916A,2N2483,2N3643,2N3859A,2N3947,2N5825,2S104,2SC1166-O,2SC1991,2SC1999,A5T5209,BCY59A,GET2222,MM3904,TZ82,*PTC123,*GE-62,*TR-95,*HEPS0025,*SK3039,*ECG108,*WEP56,*BC337,*276-2010,*RT-113
BF240	MPS6514,*PTC121,*GE-61,*TR-70,*HEPS0020,*SK3039,*ECG107,*WEP56,*276-2016,*RT-108
BF241	2N2501,2N3862,2N4873,2SC639,2SC764,2SC1687,2SC1688,BF199,TIS87,TIS105,*PTC121,*GE-61,*TR-70,*SK3039,*ECG107,*WEP56,*276-2016,*RT-108
BF254	2N916,2N916A,2SC400,2SC400-Y,2SC454,2SC458,2SC459,2SC735-Y,2SC752G-Y,2SC899,2SC925,2SC933,2SC1085,2SC1216,2SC1276,2SC1359,2SC1999,BC107A,BC108A,BC167A,BC171A,BC237A,BC238A,BC547A,BC548A,BCY58A,BCY59A,EN916,*PTC101,*GE-60,*TR-70,*HEPS0016,*SK3039,*ECG107,*WEP720,*276-2015,*RT-108

To Replace	Substitute This Type
BF255	2N708,2N708A,2N784A,2N957,2N2242,2N2501,2N3011,2N3013,2N3014,2N3646, 2N4123,2N4295,2N4418,2N4420,2N4421,2N4422,2SC67,2SC169,2SC335, 2SC372G-O,2SC400-O,2SC460,2SC461,2SC601N,2SC619,2SC620,2SC639, 2SC735-O,2SC752G-O,2SC764,2SC838,2SC839,2SC924,2SC1274,2SC1372, EN708,GET2369,MPS3646,ZT708,*PTC132,*GE-86,*TR-70,*HEPS0016,*SK3039, *ECG107,*WEP720,*276-2015,*RT-108
BF257	BF119,*GE-27,*TR-78,*HEPS5025,*SK3045,*ECG154,*WEP712,*BF338,*RT-110
BF258	BF118,BF259,*GE-27,*TR-78,*HEPS3021,*SK3045,*ECG154,*WEP712,*BF338, *RT-110
BF259	*GE-27,*TR-78,*HEPS3021,*SK3045,*ECG154,*WEP712,*RT-110
BF272A	
BF310	*PTC133,*GE-61,*TR-70,*SK3039,*ECG107,*WEP56,*276-2016,*RT-108
BF311	2N2369,2N2369A,2N4873,2SC764,TIS105,*PTC133,*TR-70,*SK3039,*ECG107, *WEP56,*276-2038,*RT-108
BF314	*PTC133,*GE-61,*SK3039,*ECG311,*WEP56,*276-2016
BF316A	
BF324	2N3829,BF424,BF509,*PTC131
BF333	2N957,2N4966,2N4968,2SC169,2SC372G-O,2SC400-O,2SC459,2SC460,2SC461, 2SC536,2SC619,2SC620,2SC634,2SC641,2SC752G-O,2SC899,2SC922,2SC924, 2SC925,2SC929NP,2SC1276,2SC1372,2SC1393,2SC1394,2SC1395,2SC1396, 2SC1675,2SC1682-BL,2SC1682-GR,2SD227,BF229,BF254,BSW12,EN916,GI-3709, *PTC101,*GE-86,*TR-70,*SK3039,*ECG107,*WEP56,*276-2015,*RT-108
BF336	MM3002,*ECG154,*WEP712,*BF338
BF337	BF338,MM3003,*HEPS5024,*ECG154,*WEP712,*BF338
BF338	*ECG154,*WEP712
BF377	BF378,TIS62A,TIS63A,TIS64A,*PTC133
BF378	BF377,TIS62A,TIS63A,TIS64A,*PTC133
BF379	2SA710,2SA711,BF414,BF509,*PTC103
BF414	2SA710,2SA711,BF379,BF509,*PTC103,*276-2021
BF424	BF509
BF440	2N3251,2N3251A,2N3906,2N4034,2N4035,2N5244,2N5366,2N5383,2SA603,2SA710, 2SA711,2SA718,A5T3906,MM3906,MPS6518,NPS404A,TN-3906,*PTC103,*GE-21, *TR-21,*SK3118,*ECG123A,*WEP736,*276-2034
BF441	2N3250,2N3250A,2N5365,2SA710,2SA711,BF379,BF414,BF509,*PTC103,*GE-21, *TR-21,*SK3118,*ECG123A,*WEP736,*276-2034
BF450	2SA548,BF379,BF414,BF509,*PTC103,*276-2034
BF451	BF506,TIS128,*PTC103,*276-2034
BF457	*GE-27,*SK3201,*ECG171,*WEPS3021,*BF338
BF458	BF459,*GE-27,*SK3201,*ECG171,*WEPS3021,*BF338
BF459	*GE-27,*SK3201,*ECG171,*WEPS3021
BF479	
BF479S	
BF506	
BF509	
BF516	
BF679	
BF679M	
BF679S	
BF680A	
BFR36	
BFR38	
BFR49	
BFR90	BFR96
BFR90A	
BFR91	BFR96
BFR94	
BFR95	
BFR96	
BFR99	
BFS29	*GE-210,*SK3024,*ECG128,*WEP243,*BC107B,*276-2013
BFS30	*GE-210,*SK3137,*ECG192,*WEP243,*BC107B,*276-2013

To Replace	Substitute This Type
BFS31	*GE-31,*SK3118,*ECG159,*WEP242,*BC107B,*276-2013
BFS32	*GE-21,*SK3118,*ECG159,*WEP242
BFS33	*GE-48,*ECG159,*WEP242
BFS34	*GE-48,*ECG193,*WEP242,*276-2023
BFS50	*GE-28,*SK3192,*ECG186,*WEP245
BFS51	*GE-28,*SK3192,*ECG186,*WEP245
BFS57	
BFS58	*GE-86
BFS59	BFS60,*GE-47,*SK3137,*ECG192,*WEP243,*276-2009
BFS60	BFS59,*GE-47,*SK3137,*ECG192,*WEP243,*276-2009
BFS61	
BFS62	BF168,BF173,BFX60,TC3114,*PTC132,*GE-61,*SK3039,*ECG108,*WEP56, *276-2038
BFS69	2N3040,2N3905,2N4290,2N4402,2N5448,MM3905,MPS3703,*PTC103,*GE-65, *SK3025,*ECG159,*WEP242,*276-2022
BFS85	BFS88
BFS86	
BFS88	*RT-115
BFS96	BFS97,*GE-48,*SK3138,*ECG193,*WEP242
BFS97	BFS96,*GE-48,*SK3138,*ECG193,*WEP242
BFS98	
BFT24	BFR90,BFR91
BFT95	
BFT95H	
BFT96	
BFW16A	BFW17A
BFW17A	BFW16A
BFW60	*GE-210,*HEPS0014,*SK3122,*ECG123A,*WEP736,*BC337,*276-2016
BFW92	
BFW93	
BFW94	
BFX29	2N2905,2N2905A,2N3503,2N3671,2N4032,2N4033,2N4405,2N4890,2SA503-GR, 2SA503-Y,2SA571,2SA580,2SA684,2SA817,40537,A5T2907,A5T3505,A5T3645, MPS4356,TIS112,*PTC127,*GE-21,*HEPS5022,*SK3025,*ECG159,*WEP242, *276-2021
BFX33	*HEPS3008,*ECG278
BFX34	
BFX55	*GE-28,*HEPS3008
BFX59	2N2708,TC3114,*PTC132,*276-2015
BFX60	BF168,BF173,*PTC132,*GE-61,*TR-83,*SK3117,*ECG161,*WEP719,*BF173, *276-2038,*RT-113
BFX62	2N5852,2SC563,2SC563A,40239,BF232,MT1061,*PTC132,*GE-60,*TR-83,*SK3117, *ECG161,*WEP719,*BF183,*276-2011,*RT-113
BFX84	2N1711B,2N1973,2N2405,2N4943,40594,BC141-16,ZT92,*PTC144,*GE-32,*TR-87, *HEPS3019,*SK3024,*ECG128,*WEP243,*276-2008
BFX88	2N2801,2N2905,2N2905A,2N3134,2N3502,2N3503,2N3671,2N4032,2N4890, 2SA503-GR,2SA503-Y,2SA504-GR,2SA504-Y,2SA571,2SA580,2SA684,40537, A5T2907,A5T3504,A5T3505,A5T3644,A5T3645,A5T4402,A5T4403,BC327-16, TIS112,TQ59,TQ61,*PTC127,*GE-21,*TR-88,*HEPS0012,*SK3025,*ECG129, *WEP242,*BC327,*276-2021
BFX89	2N2808,2N2808A,2N2809,2N2809A,2N2857,2N3570,2N3839,2N4252,2N4253,2N5053, 2N5054,2N6304,2N6305,2SC1747,40294,A485,A490,BFY90,MT1061A,*PTC132, *GE-86,*ECG161
BFY10	2N708A,2N718,2N759,2N759A,2N760B,2N1410,2N1959,2N2315,2N2396,2N2522, 2N2845,2N3301,2N3641,2N3642,2N3946,2N3973,2N3975,2N4140,2S103,2SC46, 2SC190,2SC302,2SC344,2SC896,2SC1164-O,2SC1166-R,2SC1199,2SC1412, BFY39-1,BSY93,EN697,FT3641,FT3642,GET2221,GI-3641,GI-3642,MPSA05, PET1001,ZT23,ZT43,ZT83,ZT113,ZT697,*PTC123,*GE-212,*TR-87,*HEPS0014, *SK3018,*ECG128,*WEP243,*BC107B,*276-2008,*RT-114
BFY11	2N718,2N760,2N916,2N916A,2N1838,2N2315,2N2396,2N2847,2N3858A,2N3862, 2N3903,2N3946,2N3974,2N3976,2N4013,2N4227,2N4951,2N4962,2N5027,2N5144,

To Replace	Substitute This Type
(BFY11)	2N5368,2N5380,2N5824,2N6538,2S104,2SC318,2SC366G-O,2SC852,2SC1166-O, 2SC1199,2SC1210,2SC1211,A5T3903,EN697,MM3903,MPS3693,MPS6565, MPS6576-BLUE,MPS6576-YEL,PET1001,TIS87,TIS105,TN-3903,TP3705,ZT24, ZT44,ZT81,ZT84,ZT111,ZT114,*PTC123,*GE-212,*TR-87,*HEPS0014,*SK3018, *EN10,*ECG128,*WEP243,*BC107B,*276-2008,*RT-114
BFY27	SEE 2N915
BFY33	*PTC144,*GE-20,*TR-21,*HEPS0014,*SK3122,*SN80,*ECG123A,*WEP736, *276-2009,*RT-102
BFY34	*PTC144,*GE-20,*TR-87,*HEPS3011,*SK3024,*SN80,*ECG128,*WEP243,*RT-114
BFY39-1	2N708A,2N2845,2N2847,2N3115,2N3301,2N3641,2N3642,2N3903,2N3946,2N3974, 2N3976,2N4140,2N4400,2N4962,2N5027,2N5144,2N5380,2N6538,2SC318,A5T3903, FT3641,FT3642,GET2221,GI-3641,GI-3642,MM3903,MPS6530,TN-3903,ZT83,ZT113, *PTC123,*GE-212,*TR-65,*SK3122,*ECG199,*WEP736,*BC107B,*276-2016
BFY39-2	2N3116,2N3302,2N3904,2N3947,2N4141,2N4401,2N4952,2N5028,2N5107,2N5369, 2N5381,2N6004,2SC366G-Y,2SC933,2SC943,2SC1175,2SC1318,2SC1364, 2SC1850,2SC1852,2SC1991,2SC1992,A5T3904,BC107,BC107A,BC167A,BC171A, BC174A,BC182A,BC182L,BC190A,BC237A,BC547A,BCW34,BCW36,BCY59A, BCY59B,BSX79,FT3643,GET2222,GI-3643,MM3904,MPS6531,TN-3904,*PTC123, *GE-212,*SK3245,*ECG199,*WEP736,*BC107B,*276-2016
BFY39-3	2N4953,2N5370,2N5376,2N5377,2N5827,2N6006,2N6112,2N6428,2N6428A,A157B, BC107B,BC167B,BC171B,BC174B,BC182B,BC183L,BC184L,BC190B,BC237B, BC413B,BC414B,BC547B,BC550B,BCY59C,TIS94,TIS97,*PTC123,*GE-212,*TR-51, *SK3245,*ECG199,*WEP736,*BC107B,*276-2016
BFY40	2N2270,40635,BSY53,BSY83,*PTC144,*GE-47,*TR-87,*HEPS3011,*SK3024, *ECG128,*WEP243,*276-2030,*RT-114
BFY41	2N1893A,2N3020,2SC309,2SC310,BSY55,BSY85,*PTC125,*GE-32,*HEPS3019, *SK3245,*ECG198,*WEP244
BFY45	*TR-78,*HEPS5025,*SK3045,*ECG154,*WEP712,*BF338,*RT-110
BFY46	2N2434,*GE-18,*TR-87,*HEPS3011,*SK3024,*ECG128,*WEP243,*RT-114
BFY50	*PTC144,*GE-18,*TR-87,*HEPS3019,*SK3104,*EN10,*ECG128,*WEP243,*RT-114
BFY51	*PTC144,*GE-17,*TR-87,*HEPS3001,*SK3024,*EN10,*ECG128,*WEP243,*276-2014, *RT-114
BFY52	BFY51,*PTC144,*GE-18,*TR-87,*HEPS3001,*SK3024,*ECG128,*WEP243,*276-2014, *RT-114
BFY65	2N3108,*PTC125,*GE-27,*TR-78,*HEPS5026,*SK3045,*ECG154,*WEP712,*BF338, *276-2008,*RT-110
BFY66	SEE 2N918
BFY69	2N2319,2N2885,2N3013,2N3014,2N4295,2N4420,2N4421,2N4422,2SC356,BF255, BFY69A,MPS3646,TIS45,TIS46,TIS51,TIS52,TIS55,*PTC101,*GE-214,*TR-70, *SK3039,*ECG107,*WEP56,*276-2011
BFY69A	2N2319,2N2885,2N3013,2N3014,2N4295,2N4420,2N4421,2N4422,2SC356,BF255, BFY69,MPS3646,TIS45,TIS46,TIS51,TIS52,TIS55,*PTC101,*GE-214,*TR-70,*SK3039, *ECG107,*WEP56,*BF183,*276-2031,*RT-108
BFY69B	BFY69,BFY69A,*PTC101,*GE-214,*TR-24,*SK3039,*ECG161,*WEP709,*BF183, *276-2031,*RT-108
BFY80	*PTC125,*GE-27,*TR-78,*HEPS0005,*SK3045,*ECG154,*WEP712,*BF338,*276-2008, *RT-110
BFY85	
BFY86	
BFY87	2N2885,BF230,BFY69,BFY69A,BFY87A,BSW11,*PTC101,*GE-214,*ECG161, *WEP709,*276-2011
BFY87A	2N2885,BF230,BFY69,BFY69A,BFY87,BSW11,*PTC101,*GE-214,*ECG116,*WEP158, *276-2011
BFY88	*PTC132,*HEPS0017
BFY90	2N2808,2N2808A,2N2809,2N2809A,2N2857,2N3570,2N3600,2N3839,2N4252,2N4253, 2N5053,2N5054,2N6304,2N6305,2SC1747,40294,A485,A490,BFX89,MT1061A, *PTC132,*GE-86,*TR-95,*HEPS0017,*SK3039,*ECG108,*WEP56,*276-2011
BLW17	BLW39
BLW39	BLW17
BLX89	2N2485,2N2486,*ECG278
BLY38	*SK3192,*WEP245

To Replace	Substitute This Type
BLY39	2SC1206,2SC1206A,2SC1207
BLY47	BLY47A,KSP1045,*GE-19,*TR-59,*HEPS7002,*SK3027,*ECG130,*WEP247,*RT-131
BLY47A	*TR-81,*HEPS5012,*SK3131,*ECG175,*WEP241
BLY48	BLY48A,*GE-19,*TR-59,*HEPS7002,*SK3027,*ECG130,*WEP247,*RT-131
BLY48A	*TR-81,*HEPS5012,*SK3131,*ECG175,*WEP241
BLY49	BLY49A,*GE-19,*TR-67,*HEPS5020,*ECG162,*WEP707
BLY49A	*TR-81,*HEPS5012,*SK3131,*ECG175,*WEP241
BLY50	BLY50A,*GE-19,*TR-67,*HEPS5020,*ECG162,*WEP707
BLY50A	*HEPS5012
BLY53B	
BLY61	2N2485,2N2486,BLX89,*PTC144,*GE-28,*TR-87,*HEPS5014,*SK3024,*ECG128, *WEP243,*276-2017,*RT-114
BLY62	*SK3192,*ECG186,*WEP245
BLY63	*ECG152,*WEP245
BLY78	2SC637,*SK3192,*ECG186,*WEP245
BLY79	
BR100A	SEE 2N5527
BR100B	*SK3192,*ECG186,*WEP245
BR100C	SEE 2N5528
BR100D	
BR100E	SEE 2N5529
BR100F	SEE 2N5530
BR101A	SEE 2N5531
BR101B	*SK3054,*ECG152,*WEP245
BR101C	SEE 2N5532
BR101D	
BR101E	SEE 2N5533
BR101F	SEE 2N5534
BR200A	SEE 2N5535
BR200B	SEE 2N5536
BR201A	SEE 2N5537
BR201B	SEE 2N5538
BR300A	HST7014,HST7015,HST7017,HST7018,*276-2020
BR300B	HST7014,HST7015,HST7017,HST7018,*276-2020
BR301A	HST7016,HST7019,HST7140,*276-2020
BR301B	HST7016,HST7019,HST7140,*276-2020
BR400A	HST7014,HST7015,HST7017,HST7018,*276-2020
BR400B	HST7014,HST7015,HST7017,HST7018,*276-2020
BR401A	*276-2020
BR401B	HST7016,HST7019,HST7140,*276-2020
BSC-1015	*HEPS5004,*276-2017
BSC-1015A	*HEPS5004,*276-2020
BSC-1015B	*HEPS5004,*276-2020
BSC-1016	*HEPS5004,*276-2017
BSC-1016A	*HEPS5004,*276-2020
BSC-1016B	*HEPS5004,*276-2020
BSV15	*PTC141,*GE-28,*TR-88,*HEPS3012,*SK3025,*ECG129,*WEP242
BSV16	*PTC141,*TR-88,*HEPS3012,*SK3025,*ECG129,*WEP242,*RT-115
BSV51	*PTC170,*GE-18,*HEPS5026,*SK3024,*ECG128,*WEP243,*276-2008
BSV53	*PTC115,*SK3122,*ECG123A,*WEP736
BSV54	*SK3122,*ECG123A,*WEP736
BSV55	*276-2034
BSV55A	*SK3118,*ECG159,*WEP242
BSV60	2N2848,2N4960,SDT9007,*GE-28,*HEPS3002,*SK3054,*ECG186,*WEP245, *276-2008
BSV69	2N2107,40347,MM3725,*PTC144,*GE-47,*HEPS3011,*SK3137,*ECG192,*WEP243, *276-2012
BSW11	*PTC132,*GE-20,*TR-21,*SK3122,*ECG123A,*WEP736,*276-2011,*RT-102
BSW12	*GE-210,*TR-21,*SK3122,*ECG123A,*WEP736,*276-2016,*RT-102
BSW19	2N3251,2N3504,2N3906,2N4035,2N4403,2N5244,2N5375,2N5379,2N5383,2N5999, 2N6001,2N6005,2SA509-Y,2SA509G-Y,2SA678,2SA705,A5T3504,A5T3644,A5T3906,

To Replace	Substitute This Type
(BSW19)	A5T4403,BC177A,BC177B,BC213L,BC214L,BC251A,BC251B,BC257A,BC257B, BC261A,BC261B,BC307A,BC307B,BC415-10,BC415-6,BC415A,BC415B,BC416-10, BC416-6,BC416A,BC416B,BC557A,BC557B,BC560A,BC560B,BCY78,BCY78A, BCY78B,BCY79,BCY79A,BCY79B,BSW20,BSW73,EN3502,GI-3644,MM3906, NPS404A,TN-3906,TP3644,*PTC127,*GE-17,*TR-21,*HEPS0011,*SK3122, *ECG123A,*WEP736,*276-2016,*RT-102
BSW20	2N3251,2N3504,2N3906,2N4035,2N4403,2N5244,2N5375,2N5379,2N5383,2N5999, 2N6001,2N6005,2SA509-Y,2SA509G-Y,2SA678,2SA705,A5T3504,A5T3644,A5T3906, A5T4403,BC177A,BC177B,BC213L,BC214L,BC251A,BC251B,BC257A,BC257B, BC261A,BC261B,BC307A,BC307B,BC415-10,BC415-6,BC415A,BC415B,BC416-10, BC416-6,BC416A,BC416B,BC557A,BC557B,BC560A,BC560B,BCY78,BCY78A, BCY78B,BCY79,BCY79A,BCY79B,BSW19,BSW73,GI-3644,MM3906,NPS404A, TN-3906,TP3644,*PTC103,*HEPS0019
BSW72	2N2906,2N2906A,2N3485,2N3485A,2N4026,2N5372,A5T4402,TN-3905,*PTC127, *GE-21,*TR-20,*HEPS0012,*SK3114,*ECG159,*WEP242,*276-2023,*RT-115
BSW73	2N2907,2N2907A,2N3486,2N3486A,2N3504,2N3505,2N4028,2N5373,2N5374,2N5375, 2N5379,2N5819,2N6005,2SA720,2SA891,A5T2907,A5T3504,A5T3505,A5T3644, A5T3645,A5T4403,D29E6,TIS112,TN-3906,*PTC127,*GE-48,*TR-20,*HEPS0012, *SK3114,*ECG159,*WEP242,*276-2023,*RT-115
BSW74	2N4027,*PTC127,*GE-67,*TR-20,*HEPS3012,*SK3114,*ECG159,*WEP242,*RT-115
BSW75	2N4029,2SA817,2SA850,*PTC127,*GE-67,*TR-20,*HEPS3012,*SK3114,*ECG159, *WEP242,*RT-115
BSW82	2N2539,2N4962,2N5368,TIS133,TIS134,TN-3903,*PTC136,*GE-20,*TR-21,*SK3122, *ECG123A,*WEP736,*276-2009,*RT-102
BSW83	2N2540,2N5369,2N5371,2N5377,A5T2222,TIS109,TIS111,TN-3904,*PTC136,*GE-20, *TR-21,*SK3122,*ECG123A,*WEP736,*276-2009,*RT-102
BSW84	2N4963,TIS135,TIS136,*GE-20,*TR-21,*SK3122,*ECG123A,*WEP736,*276-2009, *RT-102
BSW85	*GE-20,*TR-21,*SK3122,*ECG123A,*WEP736,*276-2009,*RT-102
BSW88	2N3242A,2N4954,2N5371,2SC735-GR,2SC1992,A157B,BC107,BC107B,BC167B, BC171B,BC237B,BC413B,BC414B,BCY58B,BCY59B,BSW83,BSW89,BSX38,BSX79, *PTC101,*GE-210,*TR-21,*HEPS0025,*SK3122,*ECG123A,*WEP736,*276-2016, *RT-102
BSW89	2N3242A,2N4954,2N5371,2SC735-GR,2SC1992,A157B,BC107,BC107B,BC167B, BC171B,BC237B,BC413B,BC414B,BCY58B,BCY59B,BSW83,BSW88,BSX38,BSX79, *PTC101,*GE-210,*TR-21,*HEPS0025,*SK3122,*ECG123A,*WEP736,*276-2016, *RT-102
BSX22	2N743A,2N5188,*GE-18,*TR-87,*SK3024,*ECG128,*WEP243,*RT-114
BSX23	2SC696,*TR-87,*SK3024,*ECG128,*WEP243,*RT-114
BSX25	2N758,2N758A,2N2220,2N2352,2N2352A,2N2353,2N2353A,2SC576,*PTC123,*GE-61, *TR-21,*HEPS0016,*SK3122,*EN10,*ECG123A,*WEP736,*BC337,*276-2016, *RT-102
BSX38	2N3242A,2N4954,2N5371,BCY58B,BCY59B,BSW83,BSX79,*PTC136,*GE-210,*TR-21, *HEPS0011,*SK3122,*EN10,*ECG123A,*WEP736,*BC107B,*276-2016,*RT-102
BSX40	2N2800,2N3244,2N3467,2N5023,2N5042,2SA504-O,2SA504-Y,BC360-10,BC360-6, SE8541,SE8542,*PTC127,*GE-48,*TR-88,*SK3025,*ECG129,*WEP242,*RT-115
BSX41	2SA504-GR,2SA683,BC360-16,*PTC127,*GE-48,*TR-88,*HEPS3028,*SK3025, *ECG129,*WEP242,*RT-115
BSX45	BSX46,*PTC144,*GE-28,*TR-87,*SK3024,*ECG128,*WEP243,*RT-114
BSX46	2N2297,2N3945,*PTC144,*GE-28,*TR-87,*SK3024,*ECG128,*WEP243,*RT-114
BSX48	BSX49,*GE-28,*TR-21,*SK3122,*ECG123A,*WEP736,*276-2038,*RT-102
BSX49	*GE-28,*TR-21,*SK3122,*ECG123A,*WEP736,*RT-102
BSX59	2SC1065,2SC1069,BSX61,*GE-X18,*HEPS3001,*SK3048,*ECG195A
BSX60	2N3722,2N3723,2N4961,2N5262,2SC97A,*GE-18,*TR-87,*HEPS3001,*SK3024, *ECG128,*WEP243,*276-2014,*RT-114
BSX61	2SC1065,2SC1069,BSX59,*GE-18,*TR-87,*HEPS3001,*SK3024,*ECG128,*WEP243, *276-2014,*RT-114
BSX62	2N3506,HST4455,HST9011,*GE-28,*TR-87,*SK3024,*ECG128,*RT-114
BSX63	2N3507,HST4453,HST4454,HST5002,HST5003,HST5502,HST5503,HST9009, SDT5002,SDT5003,SDT5502,SDT5503,*TR-87,*SK3024,*ECG128,*RT-114
BSX68	*PTC132,*GE-212,*TR-21,*SK3122,*ECG123A,*WEP736,*276-2011,*RT-102

To Replace	Substitute This Type
BSX69	BSX68,*PTC132,*GE-212,*TR-21,*HEPS0015,*SK3122,*ECG123A,*WEP736,*276-2015,*RT-114
BSX72	2N3300,2SC1384,BSY82,*GE-47,*TR-87,*SK3024,*ECG128,*276-2038,*RT-102
BSX75	2N2222,2N2540,2N3242A,2N5818,2N6010,40458,A5T2222,A5T5449,A8T3704,BC337-16,D33D25,D33D26,FT3643,TIS111,TN60,TN-3904,*PTC136,*GE-17,*TR-21,*SK3122,*ECG123A,*WEP736,*276-2009,*RT-102
BSX79	2N3116,2N3302,2N4952,2N5369,2N5377,2N5381,A5T3904,BSW85,FT3643,TN-3904,*GE-10,*TR-21,*SK3122,*ECG123A,*WEP736,*276-2009,*RT-102
BSX80	*TR-21,*HEPS3019,*SK3122,*ECG123A,*WEP736,*RT-102
BSX81	A747,BC147A,BC147B,*PTC101,*GE-210,*TR-21,*HEPS0025,*SK3122,*ECG123A,*WEP736,*276-2016,*RT-102
BSY10	2N736,2N736A,2N736B,2N760A,2N915,2N1566,2N2516,2N2847,2N3077,2N3643,2N3858A,2N3859A,2N3877,2N3877A,2N3903,2N3974,2N3976,2N4014,2N4227,2N4951,2N4962,2N4963,2N5368,2N5380,2N6538,2S104,2SC366G,2SC366G-O,2SC734-O,2SC943,2SC979-O,2SC1166-O,2SC1211,2SC1998,A5T3903,BC110,EN1613,FT3569,GET929,MPS6565,MPS-A05,TIS96,TIS99,TN-3903,ZT84,ZT114,*PTC123,*GE-81,*TR-86,*HEPS0014,*SK3124,*ECG123,*WEP53,*BF338,*276-2008,*RT-100
BSY11	2N736,2N736B,2N760A,2N909,2N915,2N930B,2N956,2N1566,2N2390,2N2483,2N2516,2N2645,2N2960,2N2961,2N3077,2N3569,2N3643,2N3859A,2N3877,2N3877A,2N3904,2N3947,2N4141,2N4951,2N4952,2N5028,2N5107,2N5368,2N5369,2N5381,2N6540,2N6541,2S104,2SC366G,2SC366G-Y,2SC734-Y,2SC875,2SC943,2SC979-Y,2SC1166-Y,2SC1318,2SC1781,2SC1852,2SC1991,2SC1998,A5T3904,BC174A,BC182A,BC190A,BC546A,BSW85,EN1711,FT3569,GET930,GET2222,GET2222A,GI-3643,MM3904,MPS-A05,TIS95,TIS98,TN-3904,ZT84,ZT89,ZT114,ZT119,*PTC123,*GE-81,*TR-86,*HEPS0014,*SK3124,*EN10,*ECG123,*WEP53,*BF338,*276-2008,*RT-100
BSY17	*GE-17,*TR-21,*HEPS0011,*SK3122,*EN10,*ECG123A,*WEP736,*276-2038,*RT-102
BSY18	2N4264,*GE-17,*TR-21,*HEPS0011,*SK3122,*EN10,*ECG123A,*WEP736,*276-2038,*RT-102
BSY19	SEE 2N708
BSY21	SEE 2N914
BSY34	*GE-28,*TR-21,*HEPS0014,*SK3122,*ECG123A,*WEP736,*276-2009,*RT-102
BSY44	SEE 2N1613
BSY45	SEE 2N1893
BSY46	SEE 2N2193
BSY51	2N1837A,2N2218,2N2218A,2N2270,2N2537,2N2848,2N3725,2N3725A,2N4047,2N4960,2N4961,2SC97A,2SC109A-O,2SC307,2SC503-Y,2SC1218,40635,BC341-10,BSY53,BSY83,TN53,*PTC125,*GE-47,*TR-87,*SK3024,*SN80,*ECG128,*WEP243,*276-2030,*RT-114
BSY52	2N2219,2N2219A,2N2538,2N3300,2SC503-GR,BSY54,*GE-47,*TR-87,*SK3024,*SN80,*ECG128,*WEP243,*276-2030,*RT-114
BSY53	40635,BSY83,*PTC144,*GE-20,*TR-87,*SK3024,*SN80,*ECG128,*WEP243,*RT-114
BSY54	*GE-20,*TR-87,*SK3024,*SN80,*ECG128,*WEP243,*RT-114
BSY55	2N1893A,2N2405,2N3020,BSY85,*PTC125,*GE-20,*TR-87,*SK3024,*ECG128,*WEP243,*RT-114
BSY56	*PTC125,*TR-87,*SK3024,*ECG128,*WEP243,*RT-114
BSY58	BSY34,*GE-28,*TR-21,*HEPS0014,*SK3122,*ECG123A,*WEP736,*276-2038,*RT-102
BSY61	2N3241A,2SC619,2SC735-O,2SC752G-O,2SC1216,2SC1317,BCY58A,ZT2938,*GE-210,*TR-21,*HEPS0025,*SK3122,*ECG123A,*WEP736,*BC107B,*276-2011,*RT-102
BSY62	*GE-210,*TR-21,*HEPS0011,*EN10,*ECG123A,*WEP736,*276-2016,*RT-102
BSY63	2N744A,2N1708A,*GE-20,*TR-21,*HEPS0011,*SK3122,*EN10,*ECG123A,*WEP736,*276-2009,*RT-102
BSY70	SEE 2N706
BSY71	SEE 2N1711
BSY72	2N2924,2N3227,2N3241A,2N3242A,2N5224,2N6000,2N6426,2SC367G-Y,2SC735-Y,2SC1317,2SC1788,2SC1851,BC108A,BC172A,BC238A,BC548A,BCY58A,BSW83,BSW88,BSW89,BSX38,*PTC123,*GE-61,*TR-95,*HEPS0011,*SK3245,*EN10,*ECG108,*WEP56,*BC107B,*276-2016,*RT-113
BSY73	2N708,2N784A,2N914,2N2242,2N2318,2N2319,2N2331,2N2481,2N2501,2N2656,

To Replace	Substitute This Type
(BSY73)	2N2710,2N3011,2N3013,2N3014,2N3210,2N3211,2N3510,2N3511,2N3605A,2N3606A, 2N3646,2N3647,2N3648,2N4123,2N4418,2N5772,2S512,2SC67,2SC204,2SC356, 2SC577,2SC595,2SC601,2SC601N,2SC639,2SC764,2SC1274,A5T4123,BSW82, GET708,GET914,GET2369,GET3013,GET3014,GET3646,PT720,ZT708,ZT2205, *PTC123,*GE-212,*TR-21,*HEPS0011,*SK3245,*EN10,*ECG123A,*WEP736, *BC107B,*276-2016,*RT-102
BSY74	2N2924,2N3227,2N3241A,2N3242A,2N5224,2N6000,2N6426,2SC367G-Y,2SC735-Y, 2SC1317,2SC1788,2SC1851,BC108A,BC172A,BC238A,BC548A,BCY58A,BSW83, BSW88,BSW89,BSX38,*PTC123,*GE-212,*TR-21,*HEPS0011,*SK3245,*EN10, *ECG123A,*WEP736,*BC107B,*276-2016,*RT-102
BSY75	2N708,2N708A,2N784A,2N914,2N2242,2N2481,2N2501,2N2710,2N2845,2N2847, 2N3013,2N3014,2N3115,2N3210,2N3211,2N3301,2N3510,2N3511,2N3605A, 2N3606A,2N3641,2N3642,2N3646,2N3647,2N3648,2N3903,2N3946,2N3973, 2N3974,2N3975,2N3976,2N4123,2N4140,2N4400,2N4962,2N5027,2N5144,2N5380, 2N5772,2N6538,2SC67,2SC318,2SC577,2SC601,2SC601N,2SC639,2SC764, 2SC1274,A5T3903,A5T4123,BFY39-1,BSW82,FT3641,FT3642,GET708,GET914, GET2221,GET2369,GET3013,GET3014,GET3646,GI-3641,GI-3642,MM3903,MPS6530, TN-3903,ZT83,ZT113,ZT708,*PTC123,*GE-210,*TR-21,*HEPS0011,*SK3122,*EN10, *ECG123A,*WEP736,*BC107B,*276-2016,*RT-102
BSY76	2N3116,2N3227,2N3242A,2N3302,2N3904,2N3947,2N4141,2N4401,2N4952,2N5028, 2N5107,2N5369,2N5381,2N6004,2N6426,2SC366G-Y,2SC367G-Y,2SC933,2SC943, 2SC1175,2SC1318,2SC1852,2SC1991,2SC1992,A5T3904,BC107A,BC167A, BC171A,BC182A,BC182L,BC237A,BC547A,BCW34,BCW36,BCY59A,BFY39-2, BSW83,BSX79,FT3643,GET2222,GI-3643,MM3904,MPS6531,TN-3904,*PTC123, *GE-210,*TR-21,*HEPS0011,*SK3122,*EN10,*ECG123A,*WEP736,*BC107B, *276-2016,*RT-102
BSY77	2N4963,2SC979A-R,BSW84,FT3722,GET2221A,SE8012,ZT86,ZT116,*PTC123, *GE-81,*TR-87,*SK3024,*EN10,*ECG128,*WEP243,*RT-114
BSY78	2N4409,2N6540,2N6541,A5T4409,BC546A,BSW85,GET2222A,TIS95,TIS98,*PTC123, *GE-81,*TR-87,*SK3024,*ECG128,*WEP243,*RT-114
BSY79	2SC1596,*PTC123,*GE-18,*TR-87,*HEPS5025,*SK3044,*EN10,*ECG128,*WEP243, *BF338,*RT-114
BSY80	2N4256,2N4954,2N5371,2SC735-GR,BC108C,BC109,BC109C,BC172C,BC173C, BC238C,BC239C,BC548B,BC549B,BCY58C,BCY58D,*PTC136,*GE-10,*TR-21, *HEPS0015,*EN10,*ECG123A,*WEP736,*BC107B,*276-2016,*RT-102
BSY81	2N2270,2N3724A,*PTC144,*GE-63,*TR-87,*HEPS3001,*SK3024,*ECG128,*WEP243, *276-2030,*RT-114
BSY82	*GE-63,*TR-87,*HEPS3024,*SK3024,*ECG128,*WEP243,*276-2030,*RT-114
BSY83	*PTC144,*GE-18,*TR-87,*HEPS3001,*SK3024,*ECG128,*WEP243,*276-2014, *RT-114
BSY84	*TR-87,*HEPS3020,*SK3024,*ECG128,*WEP243,*RT-114
BSY85	2N2405,*PTC144,*TR-87,*HEPS3019,*SK3024,*ECG128,*WEP243,*RT-114
BSY86	*TR-87,*HEPS3019,*SK3024,*ECG128,*WEP243,*RT-114
BSY87	2N2405,BSY55,BSY85,*PTC125,*GE-18,*TR-87,*HEPS5026,*SK3024,*ECG128, *WEP243,*RT-114
BSY88	BSY56,*GE-27,*TR-87,*HEPS5026,*SK3024,*ECG128,*WEP243,*RT-114
BSY90	*GE-47,*TR-87,*SK3024,*SN80,*ECG128,*WEP243,*276-2030,*RT-114
BSY91	2N3122,2N3295,2N5188,2SC31,2SC116,*PTC125,*GE-47,*TR-87,*HEPS3013, *SK3024,*EN10,*ECG128,*WEP243,*BF338,*276-2008,*RT-114
BSY92	2N1613,2N1613A,2N1837A,2N2193,2N2193A,2N2193B,2N2218,2N2218A,2N2224, 2N2410,2N2537,2N2846,2N2848,2N2868,2N3015,2N3109,2N3252,2N3299,2N3326, 2N3678,2N3722,2N4047,2N4960,2N4961,2SC97,2SC97A,2SC109A-R,2SC307, 2SC503-O,2SC503-Y,2SC560,2SC580,2SC1218,BC140-10,BC140-6,BC341-6,BSX60, BSY51,BSY53,BSY83,TN53,ZT1613,*PTC125,*GE-47,*TR-87,*HEPS5014,*SK3024, *EN10,*ECG128,*WEP243,*BF338,*276-2008,*RT-114
BSY93	2N718,2N718A,2N731,2N2221,2N2221A,2N2315,2N2317,2N2351,2N2351A,2N2389, 2N2396,2N2539,2N2845,2N2847,2N3115,2N3301,2N3641,2N3642,2N3973,2N3974, 2N3975,2N3976,2N4962,2N4963,2N5581,2N5820,A5T2193,BSW84,D33D29,FT3567, FT3568,FT3641,FT3642,FT3722,GET2221,GET2221A,GI-3641,TIS110,TIS136,TN54, TN-3903,*PTC123,*GE-62,*TR-21,*HEPS0016,*SK3122,*SN80,*ECG123A,*WEP736, *BC337,*276-2009,*RT-102

To Replace	Substitute This Type
BSY95A	2N3241A,2SC395A-Y,2SC735-Y,2SC934,2SC1317,2SC1851,ZT2938,*PTC136,*GE-10,*TR-21,*HEPS0016,*SK3122,*EN10,*ECG123A,*WEP736,*BC337,*276-2016,*RT-102
BU105	*PTC906,*GE-38,*TR-93,*SK3115,*ECG165,*WEP740B,*BU108
BU106	*TR-67,*ECG162,*WEP707
BU107	BU106,*TR-67,*ECG162,*WEP707
BU108	REFER TO SECTION 2
BU110	*PTC129A,*HEPS5020,*ECG162,*WEP707
BU111	*PTC129A,*HEPS5021,*ECG163,*WEP740
BU114	*PTC129A,*SK3115
BU126	HST7209,*PTC129A
BU204	*PTC129A,*SK3115
BU205	*PTC146,*SK3115
BU206	*PTC146
BU207	*PTC129A,*SK3115
BU208	*PTC146,*SK3115
BU209	*PTC146
BU225	
BUY12	*PTC118,*HEPS5020
BUY13	*PTC118,*HEPS5020
BUY14	*PTC169
BUY20	*HEPS5020,*ECG162,*WEP707
BUY21	*HEPS5020,*ECG162,*WEP707
BUY22	*HEPS5020,*ECG162,*WEP707
BUY23	BUY23A
BUY23A	
BUY51	BUY52
BUY51A	BUY52A
BUY52	BUY51
BUY52A	BUY51A
BUY53	BUY54
BUY53A	2N6270,2N6271,2N6272,2N6273,BUY54A
BUY54	BUY53
BUY54A	2N6270,2N6271,2N6272,2N6273,BUY53A
BUY80	
BUY81	BUY82
BUY82	
C106	2N940,2N1231,2N1442,2N1443,2N1469,2N1469A,2N2425,2N2946A,2N3219,2N3840,2N3979,2N4285,2N5230,2N5231,*GE-65,*TR-88,*SK3025,*ECG282,*WEP242,*276-2022,*RT-115
C9080	2N329B,2N722,2N722A,2N1132,2N1442,2N1443,2N1469A,2N2394,2N2946,2N2946A,2N3840,2N3857,2N4981,2N5042,2N5382,2N6067,2SA482,2SA504-O,2SA504-Y,2SA523,2SA532,A5T3905,A5T4248,A5T4402,A5T5448,A8T3703,BC360-6,BSW72,C9082,C9084,TN-3905,TQ61,TQ62,TW135,*PTC127,*GE-82,*TR-88,*HEPS0012,*ECG129,*WEP242,*BC327,*276-2023,*RT-115
C9081	2N2303,2N3502,2N3504,2N3527,2N3857,2N4414,2N4415,2N4980,2N5382,2N5383,2N6001,2N6005,2N6067,2SA504-GR,2SA504-Y,2SA509-O,2SA509-Y,2SA509G-O,2SA509G-Y,2SA532,2SA659,2SA719,2SA890,A5T3504,A5T3644,A5T3905,A5T3906,A5T4402,A5T4403,A5T5447,A8T3702,BC360-10,BC360-16,BC557A,BC558A,BC559A,BC560A,BCY78A,BCY79A,BSW73,C9083,C9085,FT3645,TN-3905,TN-3906,TP3644,TQ59,TQ60,TQ61,TQ62,*PTC127,*GE-82,*TR-88,*HEPS0012,*ECG129,*WEP242,*BC327,*276-2023,*RT-115
C9082	2N722,2N722A,2N2394,2N2946,2N2946A,2N3840,2N4981,2N5382,2N6067,A5T3905,A5T4248,A5T4402,A5T5448,A8T3703,BSW72,C9084,TN-3905,TQ62,TW135,*PTC127,*GE-82,*TR-20,*HEPS0013,*SK3114,*ECG159,*WEP242,*BC327,*276-2023,*RT-115
C9083	2N3504,2N3527,2N4415,2N4980,2N5382,2N5383,2N6001,2N6005,2N6067,2SA509-O,2SA509-Y,2SA509G-O,2SA509G-Y,2SA659,2SA719,2SA890,A5T3504,A5T3644,A5T3905,A5T3906,A5T4402,A5T4403,A5T5447,A8T3702,BC557A,BC558A,BC559A,BC560A,BCY78A,BCY79A,BSW73,C9085,FT3645,TN-3905,TN-3906,TP3644,TQ60,TQ62,*PTC127,*GE-82,*TR-20,*HEPS0013,*SK3114,*ECG159,*WEP242,*BC327,

To Replace	Substitute This Type
(C9083)	*276-2023,*RT-115
C9084	2N2394,2N2946,2N2946A,2N3840,2N4981,2N6067,A5T3905,A5T4248,A5T4402, A5T5448,*PTC127,*GE-82,*TR-20,*SK3114,*ECG159,*WEP242,*BC327,*276-2023, *RT-115
C9085	2N3527,2N4980,2N6067,A5T3504,A5T3644,A5T3905,A5T3906,A5T4402,A5T4403, A5T5447,BC557A,BC558A,BC559A,BC560A,*PTC127,*GE-82,*TR-20,*HEPS0013, *SK3114,*ECG159,*WEP242,*BC327,*276-2023,*RT-115
CA2D2	*TR-85,*SK3123,*ECG102A,*WEP250,*RT-121
CA3724G	CA3725G
CA3725G	
CDT1309	*PTC105A,*GE-25,*TR-01,*HEPG6003,*SK3009,*ECG121,*WEP232,*276-2006, *RT-127
CDT1310	CDT1311,CDT1319,*PTC105A,*GE-25,*TR-01,*HEPG6003,*SK3009,*ECG121, *WEP232,*276-2006,*RT-127
CDT1311	2N2145,2N2145A,CDT1312,*PTC105A,*GE-25,*TR-01,*HEPG6005,*SK3009, *ECG121,*WEP232,*276-2006,*RT-127
CDT1312	2N2145,2N2145A,2N2146,2N2146A,CDT1313,*PTC105A,*GE-25,*TR-01
CDT1313	2SB129A,*PTC105A,*GE-25
CDT1315	2N1146C,2N2293,2SB484,KR6503,*PTC105A,*GE-25
CDT1319	CDT1310,CDT1311,*PTC105A,*GE-25,*TR-01,*HEPG6003,*SK3009,*ECG121, *WEP232,*276-2006,*RT-127
CDT1320	2N561,2N1531,2N1536,2N2140,2N2140A,2N3615,2SB87,2SB252,ASZ15,CDT1321, SF.T240,*PTC105A,*GE-25,*TR-01,*HEPG6005,*SK3014,*ECG121,*WEP232, *276-2006,*RT-127
CDT1321	2N561,2N1531,2N1532,2N1536,2N1537,2N2140,2N2140A,2N2141,2N2141A,2N3615, 2N3616,2SB87,2SB88,2SB252,2SB340,ASZ15,CDT1322,SF.T240,*PTC105A,*GE-25, *TR-01,*HEPG6005,*SK3014,*ECG121,*WEP232,*276-2006,*RT-127
CDT1322	2N1532,2N1537,2N1538,2N3616,2SB88,2SB252A,2SB340,2SB341,*PTC105A,*GE-25, *HEPG6018,*WEPG6001,*276-2006,*RT-147
CK4	*TR-17,*HEPG0005,*SK3006,*HF6M,*ECG126,*WEP635
CK4A	*TR-17,*HEPG0005,*SK3006,*HF6M,*ECG126,*WEP635
CK13	*GE-2,*TR-85,*HEPG0003,*SK3005,*HF20M,*ECG102A,*WEP250,*AC188/01, *RT-121
CK14	2N829,2N972,2N985,*GE-1,*TR-05,*HEPG0005,*SK3005,*HF12M,*ECG100, *WEP254,*AC188/01,*276-2005,*RT-118
CK16	2N972,2N973,2N974,2N985,*GE-1,*TR-05,*HEPG0005,*SK3005,*HF20H,*ECG100, *WEP254,*276-2005,*RT-118
CK17	*DS-25,*GE-1,*TR-05,*HEPG0005,*SK3005,*HF20H,*ECG100,*WEP254,*276-2005, *RT-118
CK22	*PTC135,*GE-53,*TR-85,*HEPG0005,*SK3004,*AT20H,*ECG102A,*WEP250,*AC128, *276-2004,*RT-121
CK22A	*PTC135,*GE-53,*TR-85,*HEPG0005,*SK3004,*AT20H,*ECG102A,*WEP250,*AC128, *276-2005,*RT-121
CK22B	*PTC135,*GE-53,*TR-85,*HEPG0005,*SK3004,*AT20H,*ECG102A,*WEP250,*AC128, *276-2004,*RT-121
CK22C	*GE-53,*TR-85,*HEPG0005,*SK3004,*AT30H,*ECG102A,*WEP250,*AC128, *276-2004,*RT-121
CK25	MM380,*DS25,*GE-1,*TR-05,*HEPG0005,*SK3005,*HF6M,*ECG100,*WEP254, *AC188/01,*276-2005,*RT-118
CK26	MM380,*DS25,*GE-1,*TR-05,*HEPG0005,*SK3005,*HF6M,*ECG100,*WEP254, *276-2005,*RT-118
CK27	*DS-25,*GE-1,*TR-05,*HEPG0005,*SK3005,*HF12M,*ECG100,*WEP254,*276-2005, *RT-118
CK28	*DS25,*GE-51,*TR-17,*HEPG0005,*SK3008,*HF20M,*ECG126,*WEP635
CK28A	*DS25,*GE-1,*TR-17,*HEPG0005,*SK3008,*HF20H,*ECG126,*WEP635
CK64	*GE-2,*TR-85,*HEPG0005,*SK3004,*AT30N,*ECG102A,*WEP250,*276-2004,*RT-121
CK65	*GE-2,*TR-85,*HEPG0005,*SK3004,*AT30M,*ECG102A,*WEP250,*276-2004,*RT-121
CK65A	*GE-2,*TR-85,*HEPG0005,*SK3004,*AT30M,*ECG102A,*WEP250,*276-2004,*RT-121
CK66	*PTC135,*GE-53,*TR-85,*HEPG0005,*SK3004,*AT20H,*ECG102A,*WEP250,*AC128, *276-2004,*RT-121
CK66A	*PTC135,*GE-53,*TR-85,*HEPG0005,*SK3004,*AT30H,*ECG102A,*WEP250,*AC128,

To Replace	Substitute This Type
(CK66A)	*276-2004,*RT-121
CK67	*PTC135,*GE-53,*TR-85,*HEPG0005,*SK3004,*AT20H,*ECG102A,*WEP250,*AC128, *276-2004,*RT-121
CK256	*HEPG6011
CK258	*HEPG6011
CK311	*HEPG6012
CK312	*HEPG6012
CK313	
CK314	
CK315	
CK398	2N739A,2N2518,2N4390,*PTC117,*GE-27,*TR-78,*BF338,*276-2012
CK419	2N742,2N757A,2N759B,2N2310,2N2521,2SC945,*PTC123,*GE-61,*TR-87, *HEPS0025,*SK3024,*ECG128,*WEP243,*BC107B,*276-2013,*RT-114
CK420	*GE-212,*TR-51,*HEPS0014,*SK3245,*BC107B,*276-2013
CK421	*GE-212,*TR-51,*HEPS0014,*SK3156,*BC107B,*276-2013
CK422	2N717,2N742,2N757A,2N758A,2N759B,2N2314,2N2395,2N2521,2S102,*PTC123, *GE-212,*TR-87,*HEPS0014,*SK3024,*ECG128,*WEP243,*BC107B,*276-2013, *RT-114
CK474	2N742,2N756A,2N759B,2N2310,2N2521,2SC945,*PTC123,*GE-61,*TR-87, *HEPS0014,*SK3024,*ECG128,*WEP243,*BC107B,*276-2013,*RT-114
CK475	2N717,2N757A,2N758A,2N2314,2N2395,2S102,2SC30,*PTC123,*GE-61,*TR-87, *HEPS0014,*SK3024,*ECG128,*WEP243,*BC107B,*276-2013,*RT-114
CK476	2N708A,2N718,2N718A,2N2312,2N2315,2N2317,2N2389,2N2396,2N2845,2N2847, 2N3301,2N3641,2N3642,2N3858A,2N3903,2N3946,2N3973,2N3974,2N3975,2N3976, 2N4140,2N4962,2N5027,2N5144,2N5380,2N5824,2S103,2SC318,2SC335,2SC336, 2SC620,2SC714,2SC815,2SC938,2SC979-R,BSW84,BSY93,EN697,EN1613,FT3641, FT3642,GET929,GET2221,GET2221A,GI-3641,GI-3642,MM3903,TN-3903,TP3705, ZT83,ZT113,*PTC123,*GE-61,*TR-95,*HEPS0014,*SK3039,*ECG108,*WEP56, *BC107B,*276-2013,*RT-113
CK477	2N717,2N742,2N757A,2N758A,2N759B,2N2314,2N2395,2N2521,2S102,*PTC123, *GE-61,*TR-95,*HEPS0014,*SK3039,*ECG108,*WEP56,*BC107B,*276-2013, *RT-113
CK721	2N186A,2N187A,2N319,2N396A,2N741A,2N1018,2N1195,2N2630,2N2718,2SB219, NKT215,NKT225,SF.T221,*GE-2,*TR-05,*HEPG0005,*SK3004,*AT20M,*ECG102, *WEP631,*AC126,*276-2004,*RT-120
CK722	2N60,2N60A,2N61,2N61A,2N188A,2N270,2N320,2N363,2N414B,2N414C,2N422A, 2N610,2N611,2N1175,2N1192,2N1313,2N1348,2N1349,2N1350,2N1351,2N1372, 2N1383,2N1414,2N1415,2N1681,2N3883,2SA206,2SA207,2SA406,2SA407,2SB34, 2SB38,2SB89,2SB185,2SB220,2SB221,2SB263,MM380,NKT214,NKT224,SF.T222, SF.T322,SF.T352,*GE-2,*TR-05,*HEPG0005,*SK3004,*AT20M,*ECG102,*WEP631, *276-2004,*RT-120
CK725	2N467,2N508A,2N654+,2N655,2N1008A,2N1175A,2N1193,2N1354,2N1355,2N1356, 2N1357,2N1374,2N1376,2N1381,2N1382,2N1706,2N1707,2N2706,2SB186,2SB187, 2SB188,2SB222,2SB223,2SB496,SF.T223,SF.T323,SF.T353,*GE-53,*TR-05, *HEPG0005,*SK3004,*AT30H,*ECG102,*WEP631,*AC126,*276-2004,*RT-120
CK727	2N140,2N219,2N302,2N303,2N322,2N323,2N407,2N411,2N412,2N450,2N481,2N482, 2N483,2N485,2N486,2N521,2N559,2N711,2N711A,2N711B,2N794,2N795,2N829, 2N972,2N973,2N974,2N975,2N1097,2N1098,2N1115,2N1265/5,2N1281,2N1282, 2N1300,2N1301,2N1344,2N1683,2N2048,2N2401,2N2402,2N2717,2N3320,2N3322, 2SA12,2SA13,2SA15,2SA16,2SA138,2SA139,2SA169,2SA171,2SA173,2SA201, 2SA202,2SA209,2SA303,2SA325,2SA326,2SA351,2SA352,2SA356,2SA412, 2SA417,2SA451,2SB33,2SB34,2SB60A,2SB154,2SB169,2SB199,2SB263,2SB365, 40488,ASZ21,MA113,MA115,SF.T307,SF.T319,SF.T320,SYL792,*GE-50,*TR-05, *HEPG0005,*SK3004,*AT30H,*ECG102,*WEP631,*AF125,*276-2007,*RT-120
CK751	*GE-53,*TR-05,*HEPG0005,*SK3004,*AT30H,*ECG102,*WEP631,*AC126,*276-2004, *RT-120
CK754	*DS-25,*GE-1,*TR-05,*HEPG0005,*SK3004,*ECG102,*WEP631,*AC126,*276-2005, *RT-120
CK759	SEE 2N111
CK760	SEE 2N112
CK761	SEE 2N113

To Replace	Substitute This Type
CK762	SEE 2N114
CK766	SEE 2N211
CK768	2N111A,2N112A,2N123A,2N315A,2N316A,2N396A,2N425,2N426,2N427,2N428, 2N741,2N741A,2N960,2N1018,2N1347,2N2381,2N2630,2N2718,*GE-52,*TR-05, *HEPG0005,*SK3005,*HF12M,*ECG100,*WEP254,*AC126,*276-2005,*RT-118
CK790	2N925,2N1232,2N1439,2N1440,2N1441,2N1474A,2N3345,BCY30,HA7534,SHA7534, *GE-65,*TR-05,*HEPG0005,*SK3004,*AT20M,*ECG102,*WEP631,*BC177, *276-2022,*RT-120
CK791	2N923,2N926,2N939,2N1026,2N1230,2N1441,2N1642,2N3219,2N3344,2N3345, 2N3346,2N3979,2N4008,2N6567,HA7533,MM4052,SHA7533,*GE-65,*TR-05, *HEPG0005,*SK3004,*AT20M,*ECG102,*WEP631,*BC177,*276-2022,*RT-120
CK793	2N923,2N925,2N938,2N939,2N1026,2N1230,2N1439,2N1441,2N1642,2N3345, 2N4008,2N6567,HA7533,MM4052,SHA7533,*GE-65,*TR-05,*HEPG0005,*SK3004, *AT20M,*ECG102,*WEP631,*BC177,*276-2022,*RT-120
CK870	*GE-52,*TR-05,*HEPG0005,*SK3004,*AT20M,*ECG102,*WEP631,*AC188/01, *276-2005,*RT-120
CK871	2SB155,*GE-52,*TR-05,*HEPG0003,*SK3004,*HF35,*ECG102,*WEP631,*AC188/01, *RT-120
CK882	2N59,2N59A,2N362,2N417,2N467,2N508A,2N520A,2N571,2N598,2N654+,2N655, 2N1008A,2N1128,2N1175A,2N1193,2N1274,2N1307,2N1309,2N1352,2N1354, 2N1355,2N1356,2N1357,2N1370,2N1374,2N1376,2N1381,2N1382,2N1706,2N1707, 2N1808,2N1892,2N1998,2N2957,2N3371,2SB22,2SB54,2SB186,2SB187,2SB188, 2SB222,2SB222A,2SB223,2SB223A,2SB327,2SB329,2SB431,2SB496,40269, ASY27,NKT211,SF.T125,SF.T125P,SF.T223,SF.T323,SF.T353,*GE-53,*TR-05, *HEPG0005,*SK3004,*AT30H,*ECG102,*WEP631,*276-2004,*RT-120
CK888	2N359,2N521A,2N522A,2N523,2N827,2N1284,2N1317,2N1353,2N1378,2N1379, 2N1380,2N1471,2SA452,MA1703,MA1706,*GE-52,*TR-05,*HEPG0005,*SK3004, *AT30H,*ECG102,*WEP631,*AC126,*276-2004,*RT-120
CK942	SEE 2N1623
CQT940A	2N1166,2N1166A,2N1556,2N1556A,CTP1500,*GE-16,*TR-01,*HEPG6005,*SK3009, *ECG121,*WEP232,*276-2006,*RT-127
CQT940B	CQT940BA,*GE-76,*TR-01,*HEPG6005,*SK3009,*ECG121,*WEP232,*276-2006, *RT-127
CQT940BA	CQT940B,*GE-76,*TR-01,*HEPG6005,*SK3009,*ECG121,*WEP232,*276-2006, *RT-127
CQT1075	*GE-16,*TR-01,*SK3009,*ECG121,*WEP232,*RT-127
CQT1076	*GE-16,*TR-01,*SK3009,*ECG121,*WEP232,*RT-127
CQT1077	CQT1076,*GE-16,*TR-01,*SK3009,*ECG121,*WEP232,*RT-127
CQT1110	2N669,2N1168,2N1544,2N1545,2N3613,2N3614,2N4246,2N4247,CQT1110A, CQT1111,CQT1111A,*PTC105A,*GE-76,*TR-01,*HEPG6003,*SK3014,*ECG121, *WEP232,*276-2006,*RT-127
CQT1110A	2N669,2N1168,2N1544,2N1545,2N3613,2N3614,2N4246,2N4247,CQT1110,CQT1111, CQT1111A,*PTC105A,*GE-76,*TR-01,*HEPG6003,*SK3009,*ECG121,*WEP232, *276-2006,*RT-127
CQT1111	2N669,2N1168,2N1544,2N1545,2N3613,2N3614,2N4246,2N4247,CQT1110, CQT1110A,CQT1111A,*PTC105A,*GE-76,*TR-01,*HEPG6005,*SK3009,*ECG121, *WEP232,*276-2006,*RT-127
CQT1111A	2N669,2N1168,2N1544,2N1545,2N3613,2N3614,2N4246,2N4247,CQT1110, CQT1110A,CQT1111,*PTC105A,*GE-76,*TR-01,*HEPG6005,*SK3009,*ECG121, *WEP232,*276-2006,*RT-127
CQT1112	*PTC901,*GE-76,*TR-01,*HEPG6005,*SK3014,*ECG121,*WEP232,*276-2006, *RT-127
CRT1544	CRT1545,*PTC901,*GE-76,*TR-01,*SK3009,*ECG121,*WEP232,*RT-127
CRT1545	*PTC901,*GE-76,*TR-01,*SK3009,*ECG121,*WEP232,*RT-127
CRT1552	2N511,2N511A,2N512,2N512A,2N513,2N513A,CRT1544,*PTC901,*GE-76,*TR-01, *SK3009,*ECG121,*WEP232,*RT-127
CRT1553	2N1167,2N1167A,*PTC901,*GE-16,*TR-01,*SK3009,*ECG121,*WEP232,*RT-127
CS9010	2N5131,2N5132,2SC372-Y,2SC372G-Y,2SC380-Y,2SC380A-Y,2SC394-GR,2SC400-Y, 2SC454,2SC458,2SC459,2SC529,2SC619,2SC645,2SC738,2SC739,2SC912M, 2SC925,2SC929NP,2SC930NP,2SC941-Y,2SC1047,2SC1085,2SC1359,2SD227, BC130A,BC198A,BF229,BF254,GI-2923,GI-3393,GI-3397,GI-3710,MPS2923,PBC108A,

To Replace	Substitute This Type
(CS9010)	PET3002,*PTC101
CS9011	2N3565,2N4286,2SC373,2SC373G,2SC400-GR,2SC732-GR,2SC828,2SC829,2SC923, 2SC1084,91T6,98T2,BC122,BC130B,BC131B,BC198B,BC199B,GI-2925,GI-3391, GI-3391A,GI-3395,GI-3398,GI-3711,GI-3900,GI-3900A,MPS2925,MPS5172,PBC108B, PBC109B,*PTC139,*GE-20,*TR-21,*HEPS0015,*SK3122,*ECG123A,*WEP736, *276-2016
CS9012	*GE-21,*TR-20,*HEPS5013,*SK3025,*ECG159,*WEP242,*276-2025
CS9013	*TR-83,*HEPS0015,*SK3132,*ECG161,*WEP719,*276-2016
CS9016	2N3845,2N3845A,2N4968,2N4996,2N4997,2N5132,2N5137,2SC372-O,2SC372G-O, 2SC380-O,2SC380A-O,2SC394-Y,2SC400-O,2SC458,2SC459,2SC460,2SC528, 2SC619,2SC633,2SC645,2SC710,2SC712,2SC713,2SC894,2SC924,2SC925, 2SC929NP,2SC941-O,2SC964,2SC1047,2SC1215,2SC1342,2SC1359,2SC1372, 2SC1393,2SC1394,2SC1395,2SC1396,2SC1674,2SC1789,2SD227,BF229,BF254, BSY61,GI-2923,GI-3393,MPS2923,PET3002,SE2001,SE3005,*PTC101,*GE-20, *TR-83,*HEPS0015,*SK3018,*ECG161,*WEP719,*276-2011
CS9017	2N5131,2N5132,2SC372-Y,2SC372G-Y,2SC380-Y,2SC380A-Y,2SC394-GR,2SC400, 2SC400-Y,2SC454,2SC458,2SC459,2SC529,2SC645,2SC738,2SC739,2SC912M, 2SC925,2SC929NP,2SC930NP,2SC941-Y,2SC1047,2SC1085,2SC1359,2SD227, 93T6,BC130A,BC198A,GI-2714,GI-2716,GI-2923,GI-2926,GI-3393,GI-3397,GI-3710, MPS2923,PBC108A,PET3002,SE4001,*PTC101
CS9018	2N541,2N4251,2N5131,2N5132,2SC738,2SC739,93T6,GI-2714,GI-2716,GI-2923, GI-2926,GI-3393,GI-3397,MPS2923,PBC108A,*PTC101,*GE-11,*TR-95,*HEPS0011, *SK3018,*ECG108,*WEP56,*276-2011
CS9019	2N3692,2N5131,2N5133,2SC372-Y,2SC372G-Y,2SC380-Y,2SC380A-Y,2SC394-GR, 2SC400,2SC400-Y,2SC454,2SC529,2SC738,2SC739,2SC912M,2SC930NP, 2SC941-Y,2SC1085,2SC1363,2SC1631,2SC1633,2SC1854,BC130A,BC198A, GI-2924,GI-3392,GI-3396,GI-3397,GI-3707,GI-3708,GI-3710,MPS2924,PBC108A, SE2002,SE4001,*PTC126
CTP1111	2N2141,2N2141A,2N2146,2N2146A,*PTC105A,*GE-25,*TR-01,*HEPG6005,*PT40, *ECG121,*WEP232,*276-2006,*RT-127
CTP1117	*PTC114,*GE-25,*TR-01,*HEPG6005,*PT40,*ECG104,*WEP230,*276-2006,*RT-124
CTP1133	2N2143,2N2143A,2N2144,2N2144A,*PTC105A,*GE-25,*TR-01,*HEPG6005,*SK3014, *PT40,*ECG121,*WEP232,*276-2006,*RT-127
CTP1135	2N2143,2N2143A,2N2144,2N2144A,*PTC105A,*GE-25,*TR-01,*HEPG6005,*SK3014, *PT40,*ECG121,*WEP232,*276-2006,*RT-127
CTP1136	2N2144,2N2144A,2N2145,2N2145A,*PTC105A,*GE-25,*TR-01,*HEPG6005,*PT50, *ECG121,*WEP232,*276-2006,*RT-127
CTP1500	2N1166,2N1166A,2N1556,2N1556A,2N1560,2N1560A,*DS503,*GE-3,*TR-01, *HEPG6005,*PT150,*ECG121,*WEP232,*276-2006,*RT-127
CTP1503	2N1164,2N1164A,2N1166,2N1166A,2N1555,2N1555A,2N1556,2N1556A,2N1559, 2N1559A,2N1560,2N1560A,CQT940B,CQT940BA,CTP1500,*DS503,*GE-76,*TR-01, *HEPG6005,*PT150,*ECG121,*WEP232,*276-2006,*RT-127
CTP1504	2N1164,2N1164A,2N1554,2N1554A,2N1555,2N1555A,2N1558A,2N1559,2N1559A, CQT940B,CQT940BA,CTP1503,*PTC901,*GE-76,*TR-01,*HEPG6005,*SK3014, *PT150,*ECG121,*WEP232,*276-2006,*RT-127
CTP1508	2N1554,2N1554A,2N1558A,CTP1504,*PTC901,*GE-76,*TR-01,*HEPG6005,*SK3009, *PT150,*ECG121,*WEP232,*276-2006,*RT-127
CTP1544	2N2139A,2N2140A,2N2144A,2N2145A,CTP1545,*PTC901,*GE-76,*TR-35, *HEPG6009,*PT250,*ECG179,*WEPG6001,*RT-147
CTP1545	2N2140A,2N2141A,2N2145A,2N2146A,CTP1553,*PTC901,*GE-76,*TR-35, *HEPG6009,*PT250,*ECG179,*WEPG6001,*RT-147
CTP1552	2N2139A,2N2144A,CTP1544,*PTC901,*GE-76,*TR-35,*HEPG6009,*PT250,*ECG179, *WEPG6001,*RT-147
CTP1553	*DS503,*GE-16,*TR-35,*HEPG6009,*PT40,*ECG179,*WEPG6001,*RT-147
CTP3500	2N1167,2N1167A,2N2445,CRT1553,MP1556,MP1556A,MP1560,MP1560A,*DS503, *GE-16,*TR-01,*HEPG6005,*PT40,*ECG121,*WEP232,*276-2006,*RT-127
CTP3503	2N511B,2N512B,2N513B,2N1165,2N1165A,2N1167,2N1167A,2N2445,CRT1545, CRT1553,CTP3500,MP1555,MP1555A,MP1556,MP1556A,MP1559,MP1559A,MP1560, MP1560A,*DS503,*GE-76,*TR-01,*HEPG6005,*PT40,*ECG121,*WEP232,*276-2006, *RT-127
CTP3504	CTP3544,CTP3545,*DS503,*GE-16,*TR-01,*HEPG6005,*PT40,*ECG121,*WEP232,

To Replace	Substitute This Type
(CTP3504)	*276-2006,*RT-127
CTP3508	CTP3504,CTP3544,CTP3552,*DS503,*GE-16,*TR-01,*HEPG6005,*PT40,*ECG121,*WEP232,*276-2006,*RT-127
CTP3544	CTP3545,*DS503,*GE-16,*TR-35,*HEPG6009,*PT40,*ECG179,*WEPG6001,*RT-147
CTP3545	CTP3553,*DS503,*GE-16,*TR-35,*HEPG6009,*PT40,*ECG179,*WEPG6001,*RT-147
CTP3552	CTP3544,*DS503,*GE-16,*TR-35,*HEPG6009,*PT40,*ECG179,*WEPG6001,*RT-147
CTP3553	*DS503,*GE-16,*TR-35,*HEPG6009,*PT40,*ECG179,*WEPG6001,*RT-147
D2T2218	2N6502,D2T2219
D2T2218A	D2T2219A
D2T2219	2N6502,D2T2218
D2T2219A	D2T2218A
D2T2904	D2T2904A
D2T2904A	D2T2904
D2T2905	D2T2905A
D2T2905A	D2T2905
D2T918	
D6C	*HEPS0005
D16G6	2N834A,2N2368,2N2729,2N3663,2N4292,2N4293,2SC302,2SC321,2SC322,2SC387AG,2SC689,2SC1126,EN918,MM1941,MPS918,MPS6507,MPS6542,MPS6543,MPS6546,MPS6547,MPS6548,MT1038,MT1038A,MT1039,MT1060,PET3001,TIS84,TIS125,*PTC132,*GE-11,*TR-70,*SK3124,*ECG107,*WEP56,*BF173,*276-2011,*RT-108
D16K4	2N4873,2SC1216,2SC1276,PET3002,TIS86,*PTC132,*GE-60,*TR-70,*SK3039,*ECG107,*WEP56,*BF173,*276-2038,*RT-108
D16P1	
D24A3391	*GE-212,*TR-51,*BC107B,*276-2016
D24A3391A	*GE-212,*TR-51,*BC107B,*276-2016
D24A3392	*GE-62,*TR-51,*BC107B,*276-2016
D24A3393	*GE-212,*TR-51,*BC107B,*276-2016
D24A3394	MMT3014,*GE-212,*TR-95,*SK3039,*ECG108,*WEP56,*BC107B,*276-2016,*RT-113
D24A3900	*GE-212,*TR-51,*BC107B,*276-2015
D24A3900A	*GE-212,*TR-51,*BC107B,*276-2015
D26B1	*PTC132,*GE-60,*TR-70,*SK3039,*ECG107,*WEP56,*BF200,*276-2016,*RT-108
D26B2	*PTC132,*GE-60,*TR-70,*SK3039,*ECG107,*WEP56,*BF200,*276-2038,*RT-108
D26C1	2N2885,*PTC101,*GE-60,*TR-70,*SK3039,*ECG107,*WEP56,*BF183,*276-2031,*RT-108
D26C2	2SC536,*PTC101,*GE-60,*TR-70,*SK3039,*ECG107,*WEP56,*BF183,*276-2031,*RT-108
D26C3	BC122,*PTC101,*GE-39,*TR-70,*SK3039,*ECG107,*WEP56,*BC107B,*276-2031,*RT-108
D26C4	*PTC139,*GE-212,*TR-51,*SK3156,*BC107B,*276-2031
D26C5	*PTC139,*GE-212,*SK3156,*BC337,*276-2031
D26E-1	*PTC121,*GE-212,*TR-51,*BC107B
D26E-2	*PTC101,*GE-214,*WEP56,*BF183,*276-2015
D26E-3	*PTC101,*GE-214,*TR-51,*BF183,*276-2015
D26E-4	*PTC101,*GE-39,*TR-51,*BC107B,*276-2015
D26E-5	BC122,*PTC121,*GE-212,*TR-51,*SK3245,*BC107B,*276-2015
D26E-6	*PTC101,*GE-214,*TR-51,*BF365,*276-2015
D26G-1	*PTC132,*GE-60,*WEP56,*BF183,*276-2011
D27C1	*GE-28,*TR-76,*SK3054,*ECG152,*WEP245,*RT-154
D27C2	*GE-28,*TR-76,*SK3041,*ECG152,*WEP245,*RT-154
D27C3	*GE-28,*TR-76,*SK3041,*ECG152,*WEP245,*RT-154
D27C4	*GE-28,*TR-76,*SK3041,*ECG152,*WEP245,*RT-154
D27D1	*GE-29,*TR-76
D27D2	*GE-29,*TR-76
D27D3	*GE-29,*TR-76
D27D4	*GE-29,*TR-76
D28A12	*PTC193,*GE-63,*TR-76,*HEPS5015,*SK3054,*ECG152,*WEP245,*276-2008,*RT-154
D28A13	*GE-88,*TR-76,*HEPS5015,*SK3054,*ECG152,*WEP245,*276-2008,*RT-154
D28A5	*PTC104,*GE-63,*TR-76,*HEPS5015,*SK3041,*ECG152,*WEP245,*276-2008,

To Replace	Substitute This Type
(D28A5)	*RT-154
D28A6	*GE-88,*TR-76,*HEPS5015,*SK3041,*ECG152,*WEP245,*276-2008,*RT-154
D28D1	*PTC144,*GE-63,*TR-76,*HEPS5015,*SK3054,*ECG152,*WEP245,*276-2008, *RT-154
D28D10	2SC1848,*TR-76,*SK3054,*ECG152,*WEP245,*RT-154
D28D2	*PTC144,*GE-63,*TR-76,*HEPS5015,*SK3054,*ECG152,*WEP245,*276-2008, *RT-154
D28D3	*GE-28,*TR-76,*HEPS5015,*SK3054,*ECG152,*WEP245,*276-2008,*RT-154
D28D4	*PTC144,*GE-63,*TR-76,*HEPS5015,*SK3054,*ECG152,*WEP245,*276-2008, *RT-154
D28D5	*PTC144,*GE-63,*TR-76,*HEPS5015,*SK3054,*ECG152,*WEP245,*276-2008, *RT-154
D28D7	2SC1848,*GE-28,*TR-76,*SK3054,*ECG152,*WEP245,*RT-154
D28E	SEE D40N
D29A10	2N2906,2N2906A,2N3485,2N3485A,2N3910,2N3911,2N3913,2N3914,2N4026,2N4027, 2N5372,2N5821,2SA742,A5T4026,A5T4027,A8T4026,A8T4027,BSW74,D29E9, TZ551,*GE-81,*SK3244
D29A11	2N2907,2N2907A,2N3486,2N3486A,2N3505,2N3912,2N3915,2N4028,2N4029, 2N4413A,2N4415A,2N5373,2N5374,2N6015,2SA684,2SA720,2SA777,2SA817, 2SA891,2SB605,A5T2907,A5T3505,A5T3645,A5T4028,A5T4029,A8T4028,A8T4029, BCW35,BCW37,BSW75,TIS112,TZ552,TZ553,TZ582,*GE-81,*SK3244
D29A12	2N4413,2N5378,2N5379,2N6003,2N6007,2N6009,2N6013,BC327,BC327-40,D29E2, D29E7,*BC337,*276-2016
D29A4	2N3040,2N3121,2N3135,2N4402,2N5811,2N5815,2SA509-O,2SA509G-O,2SA673, 2SA673A,2SA696,2SA742,A5T4402,BSW72,D29E1,D29E4,MPS6533,TN-3905,TQ62, ZT181,ZT183,ZT184,ZT281,ZT283,ZT284,*PTC123,*GE-210,*TR-20,*HEPS0012, *SK3025,*ECG159,*WEP242,*276-2016,*RT-115
D29A5	2N3136,2N3504,2N4403,2N4413,2N4415,2N5375,2N5379,2N5813,2N5819,2N5999, 2N6001,2N6005,2N6011,2SA509-Y,2SA509G-Y,2SA773,A5T3504,A5T3644,A5T4403, BC327-16,BC327-25,BSW73,D29E2,D29E6,GI-3644,MPS6534,TIS91,TIS91M, TN-3906,TQ60,TZ581,ZT182,ZT282,*PTC123,*GE-210,*TR-20,*HEPS0013,*SK3114, *ECG159,*WEP242,*276-2016,*RT-115
D29A6	2N4413,2N5378,2N5379,2N6003,2N6007,2N6009,2N6013,BC327,BC327-40,D29E2, D29E7,*TR-20,*HEPS0019,*SK3118,*ECG159,*WEP242,*276-2016,*RT-115
D29A7	2N2906,2N2906A,2N3073,2N3135,2N3485,2N3485A,2N3910,2N3911,2N3913,2N3914, 2N4026,2N5372,2N5815,2N5821,2SA673A,2SA697,2SA742,A5T4026,A8T4026, BSW74,D29E4,D29E9,MPS-A55,MPSA55,TZ551,*PTC123,*GE-210,*SK3118, *276-2016
D29A8	2N2907,2N2907A,2N3136,2N3486,2N3486A,2N3505,2N3912,2N3915,2N4028, 2N4413A,2N4415A,2N5373,2N5374,2N5819,2N6005,2N6011,2N6015,2SA684, 2SA720,2SA773,2SA817,2SA891,2SB605,A5T2907,A5T3505,A5T3645,A5T4028, A8T4028,BCW35,BCW37,BSW75,D29E6,TIS112,TZ552,TZ553,TZ582,ZT189, *PTC123,*GE-210,*SK3118,*276-2016
D29A9	2N4413A,2N5374,2N6007,2N6013,2N6017,D29E7,TZ553,*TR-20,*SK3114,*ECG159, *WEP242,*276-2016,*RT-115
D29E1	2N2838,2N5811,2N5817,BC327-16,D29E5,*GE-48,*TR-20,*SK3114,*ECG159, *WEP242,*BC327
D29E1J1	D29E5J1,*PTC177,*ECG193,*WEP242
D29E10	2N4029,2N5823,2SA817,2SB560,*GE-67,*SK3114,*ECG193,*WEP242
D29E10J1	*PTC177,*GE-67,*ECG193,*WEP242
D29E2	2N5813,2N6013,D29E7,*GE-48,*TR-20,*SK3114,*ECG159,*WEP242
D29E2J1	D29E7J1,*PTC177,*GE-48,*ECG193,*WEP242
D29E4	2N2838,2N4026,2N5815,2N5817,2N5821,2N5823,D29E10,D29E5,D29E9,*GE-48, *TR-20,*SK3114,*ECG159,*WEP242,*BC327
D29E4J1	D29E10J1,D29E5J1,D29E9J1,*PTC177,*GE-48,*ECG193,*WEP242
D29E5	2N2838,2N4028,2N5817,2N5819,2N5823,2SA684,2SA817,D29E10,D29E6,*GE-48, *TR-20,*SK3114,*ECG159,*WEP242
D29E5J1	D29E10J1,D29E6J1,*PTC177,*GE-48,*ECG193,*WEP242
D29E6	2N4028,2N5819,2N6013,2N6017,D29E7,*GE-48,*TR-20,*SK3114,*ECG159,*WEP242
D29E6J1	D29E7J1,*PTC177,*GE-48,*ECG193,*WEP242
D29E7	*TR-20,*SK3114,*ECG159,*WEP242

To Replace	Substitute This Type
D29E7J1	*ECG193,*WEP242
D29E9	2N4027,2N5821,2N5823,D29E10,*GE-67,*SK3114,*ECG193,*WEP242
D29E9J1	D29E10J1,*PTC177,*GE-67,*ECG193,*WEP242
D30A1	*PTC103,*GE-65,*TR-20,*SK3114,*ECG159,*WEP242,*BC177,*276-2022,*RT-115
D30A2	BC202,BFS69,*PTC103,*GE-65,*TR-20,*SK3118,*ECG159,*WEP242,*BC177,*276-2022,*RT-115
D30A3	2SA608,2SA609,2SA701,*PTC103,*GE-65,*TR-20,*SK3118,*ECG159,*WEP242,*BC177,*276-2022,*RT-115
D30A4	2SA608,2SA609,2SA701,*GE-65,*TR-20,*SK3114,*ECG159,*WEP242,*BC327,*276-2022,*RT-115
D30A5	*GE-65,*TR-20,*SK3114,*ECG159,*WEP242,*BC327,*276-2022,*RT-115
D31B	SEE D41D
D33D21	2N5810,2N6010,BC337-16,D33D25,*GE-20,*TR-21,*SK3122,*ECG123A,*WEP736,*BC337,*276-2009,*RT-102
D33D21J1	D33D25J1,*PTC178,*GE-47,*ECG192,*WEP243,*276-2030
D33D22	2N5812,2N6012,D33D27,*GE-47,*TR-21,*SK3122,*ECG123A,*WEP736,*BC337,*276-2009,*RT-102
D33D22J1	D33D26J1,D33D27J1,*PTC178,*GE-47,*ECG192,*WEP243,*276-2030
D33D24	2N2539,2N2897,2N2900,2N4962,2N5820,2N5822,40458,A5T5450,A8T3705,D33D25,D33D29,D33D30,TIS110,TN54,TN-3903,*GE-20,*TR-21,*SK3122,*ECG123A,*WEP736,*BC337,*276-2009,*RT-102
D33D24J1	D33D25J1,D33D29J1,D33D30J1,*PTC178,*GE-47,*ECG192,*WEP243,*276-2030
D33D25	2N2540,2N5582,2N5818,2N5822,2N6010,2N6014,2SC1384,40458,A5T2222,A5T5449,A8T3704,D33D26,D33D30,TIS109,TIS111,TN-3904,*GE-20,*TR-21,*SK3122,*ECG123A,*WEP736,*BC337,*276-2009,*RT-102
D33D25J1	D33D26J1,D33D30J1,*PTC178,*GE-47,*ECG192,*WEP243,*276-2030
D33D26	2N5818,2N6012,2N6016,D33D27,TIS109,*GE-47,*TR-21,*SK3122,*ECG123A,*WEP736,*BC337,*276-2009,*RT-102
D33D26J1	D33D27J1,*PTC178,*GE-47,*ECG192,*WEP243,*276-2030
D33D27	2N6012,2N6016,*TR-21,*SK3122,*ECG123A,*WEP736,*BC337,*276-2009,*RT-102
D33D27J1	*ECG192,*WEP243,*276-2030
D33D29	2N4963,2N5820,2N5822,D33D30,TN54,*GE-63,*SK3122,*ECG192,*WEP243,*RT-102
D33D29J1	D33D30J1,*PTC178,*GE-63,*ECG192,*WEP243
D33D30	2N5582,2N5822,2N6014,2SD438,*SK3122,*ECG192,*WEP243,*RT-102
D33D30J1	*PTC178,*GE-63,*ECG192,*WEP243
D40D1	D40D4,*GE-X18,*TR-76,*SK3054,*ECG210,*WEPS3020,*RT-154
D40D2	D40D5,*GE-X18,*TR-76,*SK3192,*ECG210,*WEPS3020,*RT-154
D40D3	*GE-28,*TR-76,*SK3192,*ECG210,*WEPS3020,*RT-154
D40D4	D40D7,*GE-X18,*TR-76,*SK3192,*ECG210,*WEPS3020,*RT-154
D40D5	D40D8,*GE-X18,*TR-76,*SK3192,*ECG210,*WEPS3020,*RT-154
D40D7	*GE-28,*TR-76,*SK3192,*ECG210,*WEPS3020,*RT-154
D40D8	*GE-28,*TR-76,*SK3192,*ECG210,*WEPS3020,*RT-154
D40N1	D40N3,RCP113B,RCP113C,RCP115B,*GE-27,*TR-79,*HEPS3021,*SK3201,*ECG171,*WEPS3021,*RT-111
D40N3	RCP113C,RCP113D,*GE-27,*TR-79,*HEPS3021,*SK3201,*ECG171,*WEPS3021,*276-2012,*RT-111
D41D1	2SA699A,D41D4,*GE-29,*TR-77,*HEPS3028,*SK3203,*ECG211,*WEPS3031,*276-2026,*RT-155
D41D2	D41D5,*GE-29,*TR-77,*HEPS3028,*SK3083,*ECG211,*WEPS3031,*RT-133
D41D4	D41D7,*GE-29,*TR-77,*HEPS3028,*SK3203,*ECG211,*WEPS3031,*276-2026,*RT-155
D41D5	D41D8,*GE-29,*TR-77,*HEPS3028,*SK3203,*ECG211,*WEPS3031,*RT-155
D41D7	*GE-29,*TR-77,*HEPS3032,*SK3203,*ECG211,*WEPS3031,*RT-155
D41D8	*GE-29,*TR-77,*HEPS3032,*SK3203,*ECG211,*WEPS3031,*RT-155
D42C1	D42C4,*GE-28,*TR-55,*HEPS5000,*SK3202,*ECG186,*WEP245,*RT-154
D42C2	D42C3,D42C5,*GE-28,*TR-55,*HEPS5000,*SK3083,*ECG186,*WEP245,*RT-154
D42C3	D42C2,D42C5,*GE-28,*TR-55,*HEPS5000,*SK3083,*ECG186,*WEP245,*RT-154
D42C4	D42C7,*GE-28,*TR-55,*HEPS5000,*SK3083,*ECG186,*WEP245,*RT-154
D42C5	D42C8,*GE-28,*TR-55,*HEPS5000,*SK3083,*ECG186,*WEP245,*RT-154
D42C7	*GE-28,*TR-55,*HEPS5000,*SK3054,*ECG186,*WEP245,*RT-154
D42C8	*GE-28,*TR-55,*HEPS5000,*SK3083,*ECG186,*WEP245,*RT-154

To Replace	Substitute This Type
D43C1	D43C4,*GE-29,*TR-56,*HEPS5006,*ECG187,*WEP246,*RT-155
D43C2	D43C3,D43C5,*GE-29,*TR-56,*HEPS5006,*ECG187,*WEP246,*RT-155
D43C3	D43C2,D43C5,*GE-29,*TR-56,*HEPS5006,*ECG187,*WEP246,*RT-155
D43C4	D43C7,*GE-29,*TR-56,*HEPS5006,*ECG187,*WEP246,*RT-155
D43C5	D43C8,*GE-29,*TR-56,*HEPS5006,*ECG187,*WEP246,*RT-155
D43C7	*GE-29,*TR-77,*HEPS5006,*ECG153,*WEP246,*RT-133
D43C8	*GE-29,*TR-56,*HEPS5006,*ECG187,*WEP246,*RT-155
D44C1	D44C2,D44C3,D44C4,D44C5,D44C6,*GE-66,*TR-76,*SK3054,*ECG186A,*WEP245, *RT-154
D44C10	D44C11,D44C12
D44C11	D44C10,D44C12
D44C12	D44C10,D44C11
D44C2	D44C1,D44C3,D44C4,D44C5,D44C6,*GE-66,*TR-76,*SK3054,*ECG152,*WEP245, *RT-154
D44C3	D44C1,D44C2,D44C4,D44C5,D44C6,*GE-66,*TR-76,*SK3054,*ECG152,*WEP245, *RT-154
D44C4	D44C5,D44C6,D44C7,D44C8,D44C9,*GE-66,*TR-76,*SK3054,*ECG152,*WEP245, *RT-154
D44C5	D44C4,D44C6,D44C7,D44C8,D44C9,*GE-66,*TR-76,*SK3054,*ECG152,*WEP245, *RT-154
D44C6	D44C4,D44C5,D44C7,D44C8,D44C9,*GE-66,*TR-76,*SK3054,*ECG152,*WEP245, *RT-154
D44C7	D44C10,D44C11,D44C12,D44C8,D44C9,*GE-66,*TR-76,*SK3054,*ECG152,*WEP245, *RT-154
D44C8	D44C10,D44C11,D44C12,D44C7,D44C9,*GE-66,*TR-76,*SK3054,*ECG152,*WEP245, *RT-154
D44C9	D44C10,D44C11,D44C12,D44C7,D44C8
D45C1	D45C2,D45C3,D45C4,D45C5,D45C6,*GE-69,*TR-77,*SK3083,*ECG153,*WEP246, *RT-133
D45C10	D45C11,D45C12
D45C11	D45C10,D45C12
D45C12	D45C10,D45C11
D45C2	D45C1,D45C3,D45C4,D45C5,D45C6,*GE-69,*TR-77,*SK3083,*ECG153,*WEP246, *RT-133
D45C3	D45C1,D45C2,D45C4,D45C5,D45C6,*GE-69,*TR-77,*SK3083,*ECG153,*WEP246, *RT-133
D45C4	D45C5,D45C6,D45C7,D45C8,D45C9,*GE-69,*TR-77,*SK3083,*ECG153,*WEP246, *RT-133
D45C5	D45C4,D45C6,D45C7,D45C8,D45C9,*GE-69,*TR-77,*SK3083,*ECG153,*WEP246, *RT-133
D45C6	D45C4,D45C5,D45C7,D45C8,D45C9,*GE-69,*TR-77,*SK3083,*ECG153,*WEP246, *RT-133
D45C7	D45C10,D45C11,D45C12,D45C8,D45C9,*GE-69,*TR-77,*SK3083,*ECG153,*WEP246, *RT-133
D45C8	D45C10,D45C11,D45C12,D45C7,D45C9,*GE-69,*TR-77,*SK3083,*ECG153,*WEP246, *RT-133
D45C9	D45C10,D45C11,D45C12,D45C7,D45C8
DA3F3	
DPT657	SEE 2N2887
DS25	REFER TO SECTION 2
DS26	REFER TO SECTION 2
DS41	REFER TO SECTION 2
DS56	REFER TO SECTION 2
DS66	REFER TO SECTION 2
DS71	REFER TO SECTION 2
DS72	REFER TO SECTION 2
DS74	REFER TO SECTION 2
DS76	REFER TO SECTION 2
DS81	REFER TO SECTION 2
DS83	REFER TO SECTION 2
DS501	REFER TO SECTION 2

TRANSISTOR SUBSTITUTES

To Replace	Substitute This Type
DS503	REFER TO SECTION 2
DS509	REFER TO SECTION 2
DS520	REFER TO SECTION 2
DS525	REFER TO SECTION 2
DTG-110	*GE-16,*TR-01,*HEPG6005,*SK3009,*PT40,*ECG121,*WEP232,*RT-127
DTG-110A	*GE-76,*TR-35,*ECG179,*WEPG6001,*RT-147
DTG-110B	*GE-76,*TR-01,*HEPG6009,*SK3009,*PT40,*ECG179,*WEPG6001,*RT-147
DTG-600	2N1751,2N2145,2N2145A,2N2146,2N2146A,DTG-601,DTG-602,DTG-603,MP600, MP601,MP602,MP603,*GE-76,*TR-35,*HEPG6009,*ECG179,*WEPG6001,*RT-147
DTG-601	2N1751,2N2145,2N2145A,2N2146,2N2146A,DTG-600,DTG-602,DTG-603,MP600, MP601,MP602,MP603,*GE-76,*TR-35,*ECG179,*WEPG6001,*RT-147
DTG-602	2N2146,2N2146A,DTG-603,MP602,MP603,*GE-76,*TR-35,*HEPG6009,*ECG179, *WEPG6001,*RT-147
DTG-603	2N2146,2N2146A,DTG-602,MP602,MP603,*PTC147,*GE-76,*TR-35,*HEPG6009, *ECG179,*WEPG6001,*RT-147
DTG-603M	DTG-110B,*GE-76,*TR-35,*ECG179,*WEPG6001,*RT-147
DTG-1010	*GE-25,*TR-27,*HEPG6008,*ECG127,*WEP235
DTG-1110	*GE-25,*TR-27,*HEPG6008,*ECG127,*WEP235
DTG-1200	*GE-76,*TR-35,*HEPG6009,*ECG179,*WEPG6001,*RT-127
DTG-2000	2N1751,2N2139,2N2139A,2N2140,2N2140A,DTG-2100,SDG600,SDG601,SDG602, *PTC147,*GE-76,*TR-35,*HEPG6009,*ECG179,*WEPG6001,*RT-127
DTG-2100	2N1751,2N2140,2N2140A,2N2141,2N2141A,DTG-2200,SDG600,SDG601,SDG602, SDG603,*PTC147,*GE-76,*TR-35,*HEPG6009,*ECG179,*WEPG6001,*RT-127
DTG-2200	DTG-2300,*GE-76,*TR-35,*HEPG6009,*ECG179,*WEPG6001,*RT-127
DTG-2300	DTG-2400
DTG-2400	
DTG-2400M	
DTS-103	2N3055A,2N3235,2N5881,2N5882,2N5885,2N5886,BUY51A,BUY52A,HST9201, HST9205,HST9206,*PTC140,*GE-35,*TR-67,*HEPS7004,*ECG162,*WEP707
DTS-104	2N6471,*PTC140,*GE-35,*TR-67,*HEPS7004,*ECG162,*WEP707
DTS-105	2N3236,2N5882,2N5886,2N6270,2N6272,BUY53A,BUY54A,HST9202,HST9206, HST9207,*PTC173,*HEPS7004
DTS-106	2N3236,2N6270,2N6271,2N6272,2N6273,108T2,BUY53A,BUY54A,HST9202,HST9203, HST9207,HST9208,*PTC173,*HEPS7004
DTS-107	2N3236,2N6270,2N6271,2N6272,2N6273,108T2,BUY53A,BUY54A,HST9202,HST9203, HST9207,HST9208,*PTC173,*HEPS7004
DTS-310	
DTS-311	
DTS-401	DTS-413,DTS-423,DTS-423M,DTS-424,DTS-425,*GE-35,*TR-67,*ECG162,*WEP707
DTS-402	*PTC129A,*GE-73,*TR-67,*HEPS5021,*SK3111,*ECG163,*WEP740,*BU108
DTS-403	*PTC129A,*GE-73,*SK3559
DTS-409	DTS-431,*PTC129A,*GE-73,*SK3559
DTS-410	*PTC118,*GE-72,*TR-67,*HEPS5020,*SK3560,*ECG162,*WEP707
DTS-411	*PTC118,*GE-73,*TR-67,*HEPS5020,*SK3560,*ECG162,*WEP707
DTS-413	DTS-423,*GE-73,*TR-67,*HEPS5020,*SK3560,*ECG162,*WEP707
DTS-423	*GE-35,*TR-67,*HEPS5020,*SK3560,*ECG162,*WEP707
DTS-423M	DTS-423,DTS-424,DTS-425,*GE-73,*TR-67,*HEPS5020,*SK3559,*ECG162,*WEP707
DTS-424	DTS-425,*GE-36,*TR-67,*SK3111,*ECG163,*WEP740
DTS-425	DTS-424,*GE-36,*TR-67,*SK3111,*ECG163,*WEP740
DTS-430	DTS-431,*GE-73,*TR-67,*HEPS5020,*SK3559,*ECG162,*WEP707,*BU108
DTS-431	DTS-430,*PTC129A,*GE-73,*TR-67,*HEPS5020,*SK3560,*ECG162,*WEP707
DTS-431M	*PTC129A,*GE-73,*TR-67,*HEPS5020,*SK3559,*ECG162,*WEP707,*BU108
DTS-515	
DTS-516	
DTS-517	
DTS-518	
DTS-519	
DTS-660	IR660
DTS-663	IR663
DTS-665	
DTS-701	*TR-93,*SK3115,*ECG164,*WEP740A

To Replace	Substitute This Type
DTS-702	DTS-704,DTS-802,DTS-804,*PTC146,*GE-38,*TR-93,*SK3115,*ECG165,*WEP740B
DTS-704	DTS-702,DTS-802,DTS-804,*PTC146,*GE-38,*TR-93,*SK3115,*ECG165,*WEP740B
DTS-708	*PTC146
DTS-709	*PTC146
DTS-710	*PTC146
DTS-712	DTS-702,DTS-704,DTS-714,DTS-802,DTS-804,DTS-812,DTS-814,*PTC146
DTS-714	DTS-702,DTS-704,DTS-712,DTS-802,DTS-804,DTS-812,DTS-814,*PTC146
DTS-720	
DTS-721	
DTS-723	*PTC146
DTS-801	*GE-37,*TR-93,*SK3115,*ECG164,*WEP740A
DTS-802	DTS-804,*GE-38,*TR-93,*SK3115,*ECG165,*WEP740B
DTS-804	DTS-802,*GE-38,*TR-93,*SK3115,*ECG165,*WEP740B
DTS-812	DTS-802,DTS-804,DTS-814
DTS-814	DTS-802,DTS-804,DTS-812
DTS-1010	
DTS-1020	*PTC118
DTS-4010	
DTS-4025	
DTS-4026	
DTS-4039	DTS-4040,DTS-4041
DTS-4040	DTS-4039,DTS-4041
DTS-4041	DTS-4039,DTS-4040
DTS-4045	
DTS-4059	DTS-4061
DTS-4060	
DTS-4061	DTS-4059
DTS-4065	
ECG100	REFER TO SECTION 2
ECG101	REFER TO SECTION 2
ECG102	REFER TO SECTION 2
ECG102A	REFER TO SECTION 2
ECG103	REFER TO SECTION 2
ECG103A	REFER TO SECTION 2
ECG104	REFER TO SECTION 2
ECG104MP	M.P.ECG104
ECG105	REFER TO SECTION 2
ECG106	REFER TO SECTION 2
ECG107	REFER TO SECTION 2
ECG108	REFER TO SECTION 2
ECG121	REFER TO SECTION 2
ECG121MP	M.P.ECG121
ECG123	REFER TO SECTION 2
ECG123A	REFER TO SECTION 2
ECG124	REFER TO SECTION 2
ECG126	REFER TO SECTION 2
ECG127	REFER TO SECTION 2
ECG128	REFER TO SECTION 2
ECG129	REFER TO SECTION 2
ECG130	REFER TO SECTION 2
ECG130MP	M.P.ECG130
ECG131	REFER TO SECTION 2
ECG131MP	M.P.ECG131
ECG152	REFER TO SECTION 2
ECG153	REFER TO SECTION 2
ECG154	REFER TO SECTION 2
ECG155	REFER TO SECTION 2
ECG157	REFER TO SECTION 2
ECG158	REFER TO SECTION 2
ECG159	REFER TO SECTION 2
ECG159+	REFER TO SECTION 2

To Replace	Substitute This Type
ECG160	REFER TO SECTION 2
ECG161	REFER TO SECTION 2
ECG162	REFER TO SECTION 2
ECG163	REFER TO SECTION 2
ECG164	REFER TO SECTION 2
ECG165	REFER TO SECTION 2
ECG171	REFER TO SECTION 2
ECG172	REFER TO SECTION 2
ECG175	REFER TO SECTION 2
ECG176	REFER TO SECTION 2
ECG179	REFER TO SECTION 2
ECG180	REFER TO SECTION 2
ECG181	REFER TO SECTION 2
ECG182	REFER TO SECTION 2
ECG183	REFER TO SECTION 2
ECG184	REFER TO SECTION 2
ECG185	REFER TO SECTION 2
ECG186	REFER TO SECTION 2
ECG186A	REFER TO SECTION 2
ECG187	REFER TO SECTION 2
ECG187A	REFER TO SECTION 2
ECG188	REFER TO SECTION 2
ECG189	REFER TO SECTION 2
ECG190	REFER TO SECTION 2
ECG191	REFER TO SECTION 2
ECG192	REFER TO SECTION 2
ECG192+	REFER TO SECTION 2
ECG193	REFER TO SECTION 2
ECG193+	REFER TO SECTION 2
ECG194	REFER TO SECTION 2
ECG195	REFER TO SECTION 2
ECG195A	REFER TO SECTION 2
ECG196	REFER TO SECTION 2
ECG197	REFER TO SECTION 2
ECG198	REFER TO SECTION 2
ECG199	REFER TO SECTION 2
ECG210	REFER TO SECTION 2
ECG211	REFER TO SECTION 2
ECG213	REFER TO SECTION 2
ECG218	REFER TO SECTION 2
ECG219	REFER TO SECTION 2
ECG223	REFER TO SECTION 2
ECG224	REFER TO SECTION 2
ECG225	REFER TO SECTION 2
ECG226	REFER TO SECTION 2
ECG226MP	M.P.ECG226
ECG228	REFER TO SECTION 2
ECG229	REFER TO SECTION 2
ECG232	REFER TO SECTION 2
ECG233	REFER TO SECTION 2
ECG234	REFER TO SECTION 2
ECG235	REFER TO SECTION 2
ECG236	REFER TO SECTION 2
ECG237	REFER TO SECTION 2
ECG238	REFER TO SECTION 2
ECG241	REFER TO SECTION 2
ECG242	REFER TO SECTION 2
ECG243	REFER TO SECTION 2
ECG244	REFER TO SECTION 2
ECG245	REFER TO SECTION 2
ECG246	REFER TO SECTION 2

To Replace	Substitute This Type
ECG247	REFER TO SECTION 2
ECG248	REFER TO SECTION 2
ECG249	REFER TO SECTION 2
ECG250	REFER TO SECTION 2
ECG251	REFER TO SECTION 2
ECG252	REFER TO SECTION 2
ECG253	REFER TO SECTION 2
ECG254	REFER TO SECTION 2
ECG257	REFER TO SECTION 2
ECG258	REFER TO SECTION 2
ECG259	REFER TO SECTION 2
ECG260	REFER TO SECTION 2
ECG261	REFER TO SECTION 2
ECG262	REFER TO SECTION 2
ECG263	REFER TO SECTION 2
ECG264	REFER TO SECTION 2
ECG265	REFER TO SECTION 2
ECG266	REFER TO SECTION 2
ECG267	REFER TO SECTION 2
ECG268	REFER TO SECTION 2
ECG269	REFER TO SECTION 2
ECG270	REFER TO SECTION 2
ECG271	REFER TO SECTION 2
ECG272	REFER TO SECTION 2
ECG273	REFER TO SECTION 2
ECG274	REFER TO SECTION 2
ECG275	REFER TO SECTION 2
ECG277	REFER TO SECTION 2
ECG278	REFER TO SECTION 2
ECG280	REFER TO SECTION 2
ECG280MP	M.P. ECG280
ECG281	REFER TO SECTION 2
ECG282	REFER TO SECTION 2
ECG283	REFER TO SECTION 2
ECG284	REFER TO SECTION 2
ECG284MP	M.P. ECG284
ECG285	REFER TO SECTION 2
ECG286	REFER TO SECTION 2
ECG287	REFER TO SECTION 2
ECG288	REFER TO SECTION 2
ECG289	REFER TO SECTION 2
ECG289MP	M.P. ECG289
ECG290	REFER TO SECTION 2
ECG291	REFER TO SECTION 2
ECG292	REFER TO SECTION 2
ECG293	REFER TO SECTION 2
ECG293MP	M.P. ECG293
ECG294	REFER TO SECTION 2
ECG295	REFER TO SECTION 2
ECG297	REFER TO SECTION 2
ECG297MP	M.P. ECG297
ECG298	REFER TO SECTION 2
ECG299	REFER TO SECTION 2
ECG300	REFER TO SECTION 2
ECG300MP	M.P. ECG300
ECG302	REFER TO SECTION 2
ECG306	REFER TO SECTION 2
ECG307	REFER TO SECTION 2
ECG311	REFER TO SECTION 2
ECG313	REFER TO SECTION 2
ECG315	REFER TO SECTION 2

To Replace	Substitute This Type
ECG316	REFER TO SECTION 2
ECG317	REFER TO SECTION 2
ECG318	REFER TO SECTION 2
ECG319	REFER TO SECTION 2
ECG320	REFER TO SECTION 2
ECG321	REFER TO SECTION 2
EN10	REFER TO SECTION 2
EN30	REFER TO SECTION 2
EN40	REFER TO SECTION 2
EN697	2N718,2N718A,2N736,2N736A,2N736B,2N915,2N1566,2N2396,2N2516,2N2847, 2N3077,2N3858A,2N3903,2N3946,2N3974,2N3976,2N4227,2N4962,2N4963,2N5027, 2N5380,2N6538,2S104,2SC366G-O,2SC734-O,2SC979-O,2SC1166-O,A5T3903, BC110,EN1613,FT3567,FT3568,GET929,MM3903,TIS96,TIS99,TN-3903,*PTC101, *GE-212,*TR-21,*HEPS0015,*SK3122,*ECG123A,*WEP736,*BC107B,*276-2016, *RT-102
EN706	2N708,2N784A,2N2242,2N2318,2N2319,2N2369,2N2369A,2N2501,2N2656,2N3009, 2N3011,2N3013,2N3014,2N3564,2N3605A,2N3606A,2N3646,2N3688,2N3689, 2N3690,2N4123,2N4137,2N4294,2N4295,2N4418,2N4420,2N4421,2N4422,2N4996, 2N4997,2N5029,2N5030,2S512,2SC67,2SC169,2SC204,2SC300,2SC301,2SC380-R, 2SC380A-R,2SC400-O,2SC468,2SC595,2SC601,2SC601N,2SC639,2SC724, 2SC752G-R,2SC764,2SC836,2SC922,2SC924,2SC1123,2SC1128,2SC1129, 2SC1187,2SC1188,2SC1215,2SC1240,2SC1274,2SC1293,2SC1319,2SC1372, 2SC1393,2SC1394,2SC1395,2SC1396,2SC1686,2SC1687,2SC1789,2SC1988, BF121,BF123,BF125,BF127,BF198,BF199,BF241,BF311,EN708,EN914,EN2369A, EN3009,EN3011,EN3013,EN3014,GET708,GET914,GET2369,GET3013,GET3014, GET3646,MPS3646,MPS6512,MPSH10,MPSH11,MT1060A,PT720,PT2760,SE3005, SE5001,SE5002,SE5003,SE5006,TIS45,TIS46,TIS48,TIS49,TIS51,TIS52,TIS55, TIS64A,TIS129,ZT708,ZT2205,*PTC132,*GE-86,*TR-21,*HEPS0033,*SK3122, *ECG123A,*WEP736,*BF173,*276-2011,*RT-102
EN708	2N708,2N916,2N916A,2N2242,2N2501,2N2845,2N2847,2N3013,2N3014,2N3301, 2N3646,2N3903,2N3946,2N4123,2N4227,2N4420,2N4421,2N4422,2N4962,2N5027, 2N5380,2SC67,2SC169,2SC335,2SC336,2SC601N,2SC620,2SC639,2SC752G-O, 2SC764,2SC838,2SC839,2SC899,2SC944,2SC1276,EN916,MM3903,MPS3646, TN-3903,ZT708,*PTC132,*GE-60,*TR-21,*HEPS0025,*SK3122,*ECG123A, *WEP736,*BF173,*276-2016,*RT-102
EN718A	2N718A,2N736A,2N870,2N911,2N2461,2N2465,BC110,EN870,*PTC170,*GE-212, *TR-51,*SK3245,*ECG108,*WEP56,*BF338,*276-2009
EN722	2N722,2N722A,2N2394,2N2603,2N3250,2N3250A,2N3581,2N3703,2N4142,2N4964, 2N5448,2N6067,2SA499-O,2SA522A,2SA539,2SA548,2SA549,2SA555,2SA661-R, A5T5448,A8T3703,BC325,EN3250,GET2904,GET2905,GI-3703,MPS3703,TP3703, TZ551,*PTC103,*GE-65,*TR-52,*SK3118,*ECG159,*WEP242,*BC177,*276-2022
EN744	*PTC133,*ECG108,*WEP56,*276-2011
EN870	2N720A,2N870,2N911,2N2461,2N2465,2N2509,*PTC170,*GE-18,*TR-51,*SK3244, *BF338,*276-2008
EN871	2N871,*PTC170,*TR-51,*SK3244,*276-2008
EN914	2N708,2N708A,2N784A,2N2242,2N2501,2N2845,2N2847,2N3013,2N3014,2N3301, 2N3605A,2N3606A,2N3646,2N3903,2N3946,2N4123,2N4140,2N4295,2N4420, 2N4421,2N4422,2N4962,2N5027,2N5380,2SC67,2SC169,2SC335,2SC336,2SC468, 2SC601,2SC601N,2SC620,2SC639,2SC725,2SC752G-R,2SC764,2SC838,2SC839, 2SC1274,EN708,EN3013,EN3014,GET708,GET914,GET2221,GET2369,GET3013, GET3014,GET3646,MM3903,MPS3646,TIS45,TIS46,TIS52,TIS55,TN-3903,ZT708, *PTC132,*GE-60,*TR-95,*ECG108,*WEP56,*BF173,*276-2016,*RT-113
EN915	2N915,2N4963,2SC979-O,2SC979A-O,2SC980AG-O,2SC980G-O,2SC1998,*PTC121, *TR-51
EN916	2N916,2N916A,2N3904,2N3947,2N4141,2N5028,2N5381,2SC933,2SC1999,BC107A, BC123,BC167A,BC171A,BC174A,BC190A,BC237A,BC547A,BCY59A,GET2222, MM3904,TN-3904,*PTC132,*GE-61,*TR-51,*ECG108,*WEP56,*BF173,*276-2016
EN918	2N2368,2N2710,2N2729,2N4137,2SC387AG,2SC576,2SC601,2SC689,2SC1126, EN2369A,MM1941,MPS6507,MT1038,MT1038A,MT1039,MT1060,PET3001,TIS49, TIS84,ZT2368,*PTC132,*GE-86,*TR-95,*ECG108,*WEP56,*276-2011,*RT-113
EN930	2N909,2N929A,2N930,2N930B,2N2388,2N2483,2N2484,2N2484A,2N3302,2N3904,

To Replace	Substitute This Type
(EN930)	2N4141,2N4967,2N5028,2N5381,2N5826,2SC366G-Y,2SC538A,2SC1175,2SC1364, 2SC1632,2SC1634,2SC1685,2SC1850,2SC1992,BC107,BC107B,BC123,BC167B, BC171B,BC174B,BC182B,BC182L,BC190B,BC237B,BC413B,BC414B,BC547A, BCY59B,BFY39-3,BSX79,EN2484,GET2222,MM3904,PBC107A,PBC107B,TN-3904, *PTC121,*GE-212,*TR-51,*SK3245,*ECG123A,*WEP736,*BC107B,*276-2013
EN956	2N871,2N956,*PTC170,*GE-212,*TR-51,*SK3245,*276-2009
EN1132	2N722,2N722A,2N2394,2N2603,2N3250,2N3250A,2N3581,2N3703,2N4142,2N5448, 2N6067,2SA523A,2SA532,2SA544,2SA552,2SA594-R,2SA594N,2SA661-R,A5T5448, A8T3703,BC325,GET2904,GET2905,GI-3703,MPS3703,TP3703,TZ551,*PTC103, *GE-65,*TR-52,*ECG159,*WEP242,*BC177,*276-2022
EN1613	2N736A,2N736B,2N910,2N2461,2N2462,2N2465,2N2466,2N2516,2N3077,*PTC170, *GE-212,*TR-53,*SK3245,*ECG128,*WEP243,*BF338,*276-2009
EN1711	2N871,2N956,*PTC170,*GE-212,*TR-53,*SK3245,*ECG128,*WEP243,*BF338, *276-2009
EN2219	2N2222,2N2222A,2N2540,2N3302,A5T2222,FT3643,TIS109,TIS111,TN-3904,*GE-20, *SK3122,*276-2016
EN2222	2N2222,2N2222A,2N2540,2N3302,TN-3904,*GE-20,*ECG123A,*WEP736,*276-2016
EN2369A	2N2369,2N2369A,2N2710,2N3511,2N3648,2N3862,2N4137,2N4295,2N5029,2SC169, 2SC335,2SC336,2SC468,2SC577,2SC639,2SC764,MPS2369,TIS48,TIS49,ZT2369, *PTC133,*276-2038
EN2484	2N2484,2N2484A,BC182B,BC546B,FT107C,GET2484,SE4020,*GE-63,*TR-51, *ECG123A,*WEP736,*276-2012
EN2894A	2N2894A,2N3546,2N4208,2N4209,2N4258,2N4258A,2N4313,2N4872,2N5056,2N5057, 2N5292,2N5771,A5T4260,A5T4261,MPSL08,*TR-20,*SK3114,*ECG159,*WEP242, *RT-115
EN2905	2N2905,2N2905A,2N2907,2N2907A,2N3486,2N3486A,2N3503,2N3505,2N4028, 2N4029,2N4032,2N4033,2SA503-GR,2SA684,2SA817,A5T2907,A5T3505,A5T3645, BCW35,BCW37,TIS112,TZ552,TZ553,*PTC103,*GE-48,*ECG129,*WEP242, *276-2023
EN2907	2N2907,2N2907A,2N3486,2N3486A,2N3505,2N4028,2N4029,2SA684,2SA817, A5T2907,A5T3505,A5T3645,BCW35,BCW37,TIS112,TZ552,TZ553,*PTC103,*GE-48, *SK3118,*ECG159,*WEP242,*276-2023
EN3009	2N708,2N708A,2N784A,2N2369,2N2369A,2N2501,2N3009,2N3013,2N3014,2N3605A, 2N3606A,2N3646,2N3688,2N3689,2N3690,2N3862,2N4137,2N4295,2N4420,2N4421, 2N4422,2N5029,2SC67,2SC169,2SC335,2SC336,2SC468,2SC601,2SC601N, 2SC639,2SC752G-R,2SC764,2SC1187,2SC1188,2SC1189,2SC1274,2SC1319, 2SC1320,2SC1686,2SC1687,2SC1688,BF198,BF199,BF241,EN708,EN914,EN2369A, EN3013,EN3014,GET708,GET914,GET2369,GET3013,GET3014,GET3646,MPS3646, SE5001,SE5002,SE5003,SE5006,TIS45,TIS46,TIS48,TIS49,TIS52,TIS55,TIS87, TIS129,ZT708,*PTC132,*GE-60,*TR-21,*SK3122,*ECG123A,*WEP736,*BF173, *276-2016,*RT-102
EN3011	2N2501,2N3011,2N3013,2N3014,2N3646,2N4295,2N4420,2N4421,2N4422,2SC67, 2SC169,2SC335,2SC468,2SC595,2SC601,2SC601N,2SC639,2SC752G-R,2SC764, 2SC922,2SC1274,2SC1393,2SC1394,2SC1395,2SC1396,EN3013,EN3014, GET2369,GET3013,GET3014,GET3646,MPS3646,TIS51,TIS52,TIS55,*PTC132, *GE-11,*TR-64,*ECG108,*WEP56,*BF173,*276-2011
EN3013	2N708,2N708A,2N784A,2N2501,2N3013,2N3014,2N3605A,2N3606A,2N3646,2N4295, 2N4420,2N4421,2N4422,2SC67,2SC169,2SC335,2SC336,2SC468,2SC601,2SC601N, 2SC639,2SC752G-R,2SC764,2SC1274,EN708,EN914,EN3014,GET708,GET914, GET2369,GET3013,GET3014,GET3646,MPS3646,TIS45,TIS46,TIS52,TIS55,ZT708, *PTC132,*GE-60,*TR-21,*SK3122,*ECG123A,*WEP736,*BF173,*276-2016,*RT-102
EN3014	2N708,2N708A,2N784A,2N2501,2N3013,2N3014,2N3605A,2N3606A,2N3646,2N4295, 2N4420,2N4421,2N4422,2SC67,2SC169,2SC335,2SC336,2SC468,2SC601,2SC601N, 2SC639,2SC752G-R,2SC764,2SC1274,EN708,EN914,EN3013,GET708,GET914, GET2369,GET3013,GET3014,GET3646,MPS3646,TIS45,TIS46,TIS52,TIS55,ZT708, *PTC132,*GE-60,*TR-21,*SK3122,*ECG123A,*WEP736,*BF173,*276-2016,*RT-102
EN3250	2N3250,2N3250A,2SA499-O,2SA499-Y,2SA548,2SA603,2SA710,2SA711,2SA718, *PTC103,*GE-21,*ECG159,*WEP242,*276-2034
EN3502	2N2905,2N2905A,2N2907,2N2907A,2N3486,2N3486A,2N3502,2N3503,2N3504, 2N3505,2N4028,2N4032,2N5819,2SA503-GR,2SA684,A5T2907,A5T3504,A5T3505, A5T3644,A5T3645,BCW35,BCW37,D29E6,EN2905,TIS112,TZ552,TZ553,*PTC103,

To Replace	Substitute This Type
(EN3502)	*GE-48,*ECG129,*WEP242
EN3504	2N2907,2N2907A,2N3486,2N3486A,2N3504,2N3505,2N4028,2N5819,2SA684, A5T2907,A5T3504,A5T3505,A5T3644,A5T3645,BCW35,BCW37,D29E6,EN2907, TIS112,TZ552,TZ553,*PTC103,*GE-48,*SK3118,*ECG159,*WEP242
EN3962	2N3547,2N3548,2N3549,2N3962,2N3963,2N3965,2N4289,*GE-67,*TR-52,*ECG159, *WEP242
EP20	REFER TO SECTION 2
EP25	REFER TO SECTION 2
EP35	REFER TO SECTION 2
ES3110	*GE-51,*TR-17,*HEPG0005,*SK3006,*AT20M,*ECG126,*WEP635,*AF125,*276-2004
ES3111	*GE-51,*TR-17,*HEPG0005,*SK3006,*AT20M,*ECG126,*WEP635,*AF125,*276-2004
ES3112	2N2955,*PTC107,*GE-51,*TR-17,*HEPG0005,*SK3006,*HF35,*ECG126,*WEP635, *AF125,*276-2004
ES3113	2N2587,2N2955,2N6365,2N6365A,*GE-51,*TR-17,*HEPG0005,*SK3006,*HF35, *ECG126,*WEP635,*AF125,*276-2004
ES3114	2N838,2N2587,2N2956,2N3323,2N3324,2N3325,2N6365,*GE-51,*TR-17,*HEPG0005, *SK3006,*HF35,*ECG126,*WEP635,*AF125,*276-2005
ES3115	2N2957,*GE-51,*TR-17,*HEPG0005,*SK3006,*HF35,*ECG126,*WEP635,*AF125, *276-2005
ES3116	2N2635,*GE-51,*TR-17,*HEPG0005,*SK3006,*HF35,*ECG126,*WEP635,*AF125, *276-2005
ES3120	2N45,2N199,2N204,2N205,2N367,2N425,2N563,2N564,2N1408,*GE-52,*TR-85, *HEPG0005,*SK3004,*AT20M,*ECG102A,*WEP250,*AF125,*276-2004,*RT-121
ES3121	2N44,2N44+,2N44A,2N45A,2N104,2N111A,2N112A,2N198,2N200,2N396A,2N404A, 2N413A,2N425,2N426,2N427,2N428,2N460,2N464,2N465,2N563,2N564,2N573, 2N1018,2N1056,2N1195,2N1408,2SB219,NKT225,SF.T221,*GE-52,*TR-85, *HEPG0005,*SK3123,*AT20M,*ECG102A,*WEP250,*AF125,*276-2004,*RT-121
ES3122	2N43,2N43A,2N43A+,2N44,2N44+,2N44A,2N104,2N111A,2N112A,2N197,2N198, 2N215,2N237,2N283,2N361,2N396A,2N404A,2N413,2N428,2N465,2N565,2N566, 2N573,2N1017,2N1018,2N1056,2N1191,2N1413,2N1446,2N1450,2N1499B,2N2955, 2SA205,2SB219,2SB397,NKT225,SF.T221,SF.T226,SF.T227,*GE-52,*TR-85, *HEPG0005,*SK3004,*HF35,*ECG102A,*WEP250,*AF125,*276-2004,*RT-121
ES3123	2N43,2N43A,2N43A+,2N61A,2N61B,2N196,2N197,2N206,2N215,2N361,2N363, 2N381,2N413,2N414,2N414A,2N414B,2N414C,2N466,2N502A,2N502B,2N505,2N518, 2N567,2N568,2N1017,2N1171,2N1191,2N1303,2N1313,2N1348,2N1350,2N1351, 2N1373,2N1414,2N1446,2N1447,2N1450,2N1451,2N1499B,2N1749,2N2587,2N2955, 2N6365,2N6365A,2SA204,2SA205,2SA206,2SA207,2SA406,2SA407,2SB220, 2SB299,2SB397,ASY26,NKT224,SF.T222,SF.T227,*GE-52,*TR-85,*HEPG0005, *SK3004,*HF35,*ECG102A,*WEP250,*AF125,*276-2005,*RT-121
ES3124	2N59A,2N59B,2N60A,2N60B,2N360,2N381,2N414,2N414A,2N414B,2N414C,2N415, 2N415A,2N416,2N422A,2N466,2N502A,2N505,2N518,2N569,2N570,2N572,2N654, 2N654+,2N838,2N1171,2N1175,2N1175A,2N1192,2N1305,2N1309A,2N1348, 2N1349,2N1350,2N1351,2N1375,2N1415,2N1447,2N1448,2N1451,2N1452,2N1681, 2N1707,2N1749,2N2587,2N2956,2N3323,2N3324,2N3325,2N6365,2SA204,2SA207, 2SA407,2SB103,2SB221,*GE-52,*TR-85,*HEPG0005,*SK3004,*HF35,*ECG102A, *WEP250,*AF125,*276-2005,*RT-121
ES3125	2N59A,2N59B,2N360,2N382,2N383,2N415,2N415A,2N416,2N461,2N571,2N654, 2N654+,2N1175A,2N1307,2N1309A,2N1352,2N1356,2N1448,2N1449,2N1452, 2N1707,2N1892,2N2188,2N2189,2N2956,2N2957,2N3323,2N3324,2N3325,2SB222, SF.T223,*GE-52,*TR-85,*HEPG0005,*SK3004,*HF35,*ECG102A,*WEP250,*AF125, *276-2005,*RT-121
ES3126	2N383,2N417,2N508A,2N655,2N655+,2N1193,2N1309,2N1316,2N1354,2N1355, 2N1357,2N1449,2N1892,2N2171,2N2189,2N2613,2N2635,2SB223,*GE-52,*TR-85, *HEPG0005,*SK3004,*HF35,*ECG102A,*WEP250,*AF125,*276-2005,*RT-121
ET110	
ET670	*PTC155,*GE-80
FK914	FX914,FX3013,FX3014,*GE-60,*TR-95,*SK3039,*EN10,*ECG108,*WEP56,*BF173, *276-2016,*RT-113
FK918	FX918,FX2369A,*GE-39,*TR-95,*SK3039,*EN10,*ECG108,*WEP56,*276-2011, *RT-113
FK2369A	FX2369A,*GE-17,*TR-95,*SK3039,*EN10,*ECG108,*WEP56,*276-2038,*RT-113

To Replace	Substitute This Type
FK2484	A3T2484,FX2483,FX2484,*GE-63,*TR-95,*SK3039,*ECG108,*WEP56,*276-2008, *RT-113
FK2894	A3T2894,FX2894,*TR-20,*HEPS0013,*ECG106,*WEP52,*276-2021,*RT-126
FK3014	FX914,FX3013,FX3014,*GE-60,*TR-95,*SK3039,*EN10,*ECG108,*WEP56,*BF173, *276-2016,*RT-113
FK3299	FX3299,*GE-213,*TR-95,*SK3039,*EN10,*ECG108,*WEP56,*BC107B,*276-2016, *RT-113
FK3300	FX3300,FX4960,*GE-210,*TR-95,*SK3039,*EN10,*ECG108,*WEP56,*BC107B, *276-2016,*RT-113
FK3502	A3T2907,A3T2907A,FX3502,FX3503,*GE-82,*HEPS0019,*276-2021
FK3503	A3T2907,A3T2907A,FX3503,*GE-82
FM870	2N2461,2N2519,*GE-18,*TR-95,*SK3039,*EN10,*ECG108,*WEP56,*BF338, *276-2008,*RT-113
FM871	*PTC123,*GE-18,*TR-95,*SK3039,*EN10,*ECG108,*WEP56,*276-2008,*RT-113
FT19H	SEE 2N3963
FT19M	SEE 2N3963
FT34A	
FT34B	
FT34C	*GE-17,*TR-78,*SK3045,*EN10,*ECG154,*WEP712,*RT-110
FT34D	*GE-17,*TR-78,*SK3045,*EN10,*ECG154,*WEP712,*RT-110
FT40	SEE 2N4251
FT45	SEE 2N4134
FT47	FT48,FT49
FT48	FT49,FT50
FT49	FT50
FT50	
FT107A	*276-2016
FT107B	*GE-85
FT107C	*PTC123,*TR-51
FT207A	SEE 2N4115
FT207B	SEE 2N4116
FT317	FT317A
FT317A	FT317B
FT317B	
FT359	IR4040
FT400A	
FT400B	
FT401	2N3902,DTS-423,DTS-423M,DTS-424,DTS-425,FT402,*PTC129A
FT402	2N3902,DTS-423,DTS-423M,DTS-424,DTS-425,FT401,*PTC129A
FT410	2SC1116A,2SC1229,2SC1322,*PTC118
FT411	2SC1227,IR515,*PTC118
FT413	2N6306,2N6307M,2SC807,FT423,IR516,IR517,MJ413,MJ423
FT417	FT417A
FT417A	FT417B
FT417B	
FT423	2SC807,2SC1576,FT413,IR516,IR517,IR4050,IR5000,IR5001,IR5002,IR5251,IR5252, IR5253,MJ423
FT430	2N6560,2N6573,2SC807,FT431,IR516,IR517,MJ413,SDT13303,SDT13304,SDT13305, SDT14305
FT431	2N6560,2N6573,2N6579,2N6580,2N6582,2N6583,2SC1579,FT430,MJ413,SDT13303, SDT13304,SDT13305
FT709	2N709,2N2475,2N2784,2N3633,2N3959,2N3960,2N5200,2N5201,BF123,MT1060A, *PTC126,*GE-61,*TR-95,*HEPS0016,*SK3039,*PN66,*ECG108,*WEP56,*BF173, *276-2038,*RT-113
FT1341	2N708,2N784A,2N2242,2N2318,2N2501,2N2656,2N3011,2N3013,2N3014,2N3605A, 2N3606A,2N3646,2N5772,2SC67,2SC204,2SC639,2SC764,A5T4123,GET708, GET914,GET2369,GET3013,GET3014,GET3646,PT720,ZT708,*PTC126,*GE-17, *TR-20,*HEPS0013,*SK3114,*EN10,*ECG159,*WEP242,*276-2009,*RT-115
FT1702	SEE 2N4208
FT1746	2N3121,2SA558,FT5041,TIS38,TIS137,*PTC127,*GE-65,*TR-20,*HEPS0013,*SK3114, *SP70,*ECG159,*WEP242,*BC327,*276-2023,*RT-115

To Replace	Substitute This Type
FT2955	
FT3055	
FT3567	2N718A,2N736,2N736A,2N736B,2N870,2N910,2N1566,2N1889,2N2465,2N3107, 2N3109,2N4961,2N4963,2SC353,2SC805,FT3568,TIS96,TIS99,*PTC170,*GE-20, *BF338,*276-2009
FT3568	2N718A,2N736,2N736A,2N736B,2N870,2N910,2N1566,2N1889,2N2465,2N3107, 2N3109,2N4961,2N4963,2SC353,2SC805,FT3567,TIS96,TIS99,*PTC170,*GE-81, *BF338,*276-2008
FT3569	2N871,2N956,2N1890,2N2466,2N2645,2N4409,2N6540,2N6541,2SC875,A5T4409, BC546A,TIS95,TIS98,ZT1711,*PTC170,*GE-20,*276-2009
FT3641	2N2846,2N2848,2N3015,2N3299,2N4960,2N4961,2N4962,2N4963,2N5380,A5T3903, FT3642,TN-3903,*PTC101,*GE-17,*TR-78,*ECG154,*WEP712,*276-2009
FT3642	2N2846,2N2848,2N3015,2N3299,2N4960,2N4961,2N4962,2N4963,2N5380,A5T3903, FT3641,TN-3903,*PTC121,*GE-47
FT3643	2N3300,2N4409,A5T4409,*PTC121,*GE-47,*TR-21,*ECG123A,*WEP736,*276-2009
FT3644	2SA594,2SA594-Y,TP3645,*PTC103,*GE-67
FT3645	2SA594,2SA594-Y,BCY79,FT3644,TP3644,TP3645,*PTC103,*GE-48
FT3722	2N4961,2N4963
FT5040	2N5042,2N5382,2SA523,2SA732,A5T3905,A5T4125,FT5041,TIS37,*PTC103,*GE-22, *TR-20,*HEPS0019,*BC327,*276-2032
FT5041	2N5042,2N5372,2N5382,2N6067,2SA523,2SA523A,2SA532,2SA544,2SA552, 2SA594-O,2SA594-R,2SA594N,2SA661-O,2SA661-R,A5T3905,A5T5448,A8T3703, BC360-6,BC361-6,*PTC103,*GE-82,*HEPS0019,*BC327
FT7202A	2N4115
FT7202B	2N4115,HST6313,HST6413
FV914	2N3013,2N3014,2N3510,2N3511,2N3647,2N3648,MM3903,*GE-39,*TR-95, *HEPS0025,*SK3039,*EN10,*ECG108,*WEP56,*BC107B,*276-2016,*RT-113
FV918	*GE-17,*TR-95,*HEPS0015,*SK3039,*EN10,*ECG108,*WEP56,*276-2011,*RT-113
FV2369A	2N3511,*GE-39,*TR-95,*HEPS0025,*SK3039,*EN10,*ECG108,*WEP56,*276-2038, *RT-113
FV2484	*GE-212,*TR-95,*HEPS0015,*SK3039,*EN10,*ECG108,*WEP56,*BC107B,*276-2016, *RT-113
FV2894	*GE-21,*TR-20,*HEPS0013,*SP70,*ECG106,*WEP52,*276-2021,*RT-126
FV3014	2N3013,2N3014,MM3903,*GE-60,*TR-95,*HEPS0016,*SK3039,*EN10,*ECG108, *WEP56,*BF173,*276-2016,*RT-113
FV3299	MM3903,*GE-213,*TR-95,*HEPS0016,*SK3039,*EN10,*ECG108,*WEP56,*BC107B, *276-2016,*RT-113
FV3300	*GE-210,*TR-95,*HEPS0030,*SK3039,*EN10,*ECG108,*WEP56,*BC107B,*276-2016, *RT-113
FV3502	2N3486,2N3486A,2N3673,*GE-82,*HEPS0019,*276-2023
FV3503	2N3486,2N3486A,2N3673,*GE-82,*HEPS0019,*276-2023
FX709	*PTC126,*GE-11,*BF173,*276-2015
FX914	FX3013,FX3014,*GE-61,*TR-51,*BF173,*276-2016
FX918	FX2369A,*GE-61,*276-2015
FX2368	FX2369A,FX3014,*GE-20,*TR-64,*276-2016
FX2369A	*276-2038
FX2483	FX4960,*GE-63,*TR-51,*276-2008
FX2484	*276-2008
FX2894	A3T2894,*TR-20,*276-2021
FX2894A	*276-2021
FX3013	FX914,FX3014,*GE-61,*TR-51,*BF173,*276-2016
FX3014	FX914,FX3013,*GE-61,*TR-64,*BF173,*276-2016
FX3299	*GE-213,*TR-64,*BC107B,*276-2016
FX3300	*GE-210,*BC107B,*276-2016
FX3502	FX3503,*GE-82
FX3503	
FX3724	*GE-47,*276-2016
FX3725	
FX3962	FX3963,FX3965,*GE-67,*TR-52
FX3963	*TR-52
FX3964	FX3962,FX3965,*GE-65,*TR-52,*SK3247,*BC177,*276-2022

To Replace	Substitute This Type
FX3965	FX3962,FX3963,*TR-52
FX4034	*276-2034
FX4046	FX3724,*GE-210,*276-2016
FX4047	FX3725
FX4207	*276-2021
FX4960	*GE-63,*TR-51
GE-1	REFER TO SECTION 2
GE-2	REFER TO SECTION 2
GE-3	REFER TO SECTION 2
GE-4	REFER TO SECTION 2
GE-5	REFER TO SECTION 2
GE-6	REFER TO SECTION 2
GE-7	REFER TO SECTION 2
GE-8	REFER TO SECTION 2
GE-9	REFER TO SECTION 2
GE-10	REFER TO SECTION 2
GE-11	REFER TO SECTION 2
GE-12	REFER TO SECTION 2
GE-13MP	M.P.GE-3
GE-14	REFER TO SECTION 2
GE-15MP	M.P.GE-14
GE-16	REFER TO SECTION 2
GE-17	REFER TO SECTION 2
GE-18	REFER TO SECTION 2
GE-19	REFER TO SECTION 2
GE-20	REFER TO SECTION 2
GE-21	REFER TO SECTION 2
GE-22	REFER TO SECTION 2
GE-23	REFER TO SECTION 2
GE-24MP	M.P.GE-23
GE-25	REFER TO SECTION 2
GE-26	REFER TO SECTION 2
GE-27	REFER TO SECTION 2
GE-28	REFER TO SECTION 2
GE-29	REFER TO SECTION 2
GE-30	REFER TO SECTION 2
GE-31MP	M.P.GE-30
GE-32	REFER TO SECTION 2
GE-35	REFER TO SECTION 2
GE-36	REFER TO SECTION 2
GE-37	REFER TO SECTION 2
GE-38	REFER TO SECTION 2
GE-39	REFER TO SECTION 2
GE-40	REFER TO SECTION 2
GE-43	REFER TO SECTION 2
GE-44	REFER TO SECTION 2
GE-45	REFER TO SECTION 2
GE-46	REFER TO SECTION 2
GE-47	REFER TO SECTION 2
GE-48	REFER TO SECTION 2
GE-49	REFER TO SECTION 2
GE-50	REFER TO SECTION 2
GE-51	REFER TO SECTION 2
GE-52	REFER TO SECTION 2
GE-53	REFER TO SECTION 2
GE-54G	REFER TO SECTION 2
GE-54R	REFER TO SECTION 2
GE-55	REFER TO SECTION 2
GE-56	REFER TO SECTION 2
GE-57	REFER TO SECTION 2
GE-58	REFER TO SECTION 2

To Replace	Substitute This Type
GE-59	REFER TO SECTION 2
GE-60	REFER TO SECTION 2
GE-61	REFER TO SECTION 2
GE-62	REFER TO SECTION 2
GE-63	REFER TO SECTION 2
GE-64	REFER TO SECTION 2
GE-65	REFER TO SECTION 2
GE-66	REFER TO SECTION 2
GE-67	REFER TO SECTION 2
GE-69	REFER TO SECTION 2
GE-72	REFER TO SECTION 2
GE-73	REFER TO SECTION 2
GE-74	REFER TO SECTION 2
GE-75	REFER TO SECTION 2
GE-76	REFER TO SECTION 2
GE-80	REFER TO SECTION 2
GE-81	REFER TO SECTION 2
GE-82	REFER TO SECTION 2
GE-83	REFER TO SECTION 2
GE-84	REFER TO SECTION 2
GE-85	REFER TO SECTION 2
GE-86	REFER TO SECTION 2
GE-88	REFER TO SECTION 2
GE-89	REFER TO SECTION 2
GE-210	REFER TO SECTION 2
GE-211	REFER TO SECTION 2
GE-212	REFER TO SECTION 2
GE-213	REFER TO SECTION 2
GE-214	REFER TO SECTION 2
GE-215	REFER TO SECTION 2
GE-216	REFER TO SECTION 2
GE-217	REFER TO SECTION 2
GE-218	REFER TO SECTION 2
GE-219	REFER TO SECTION 2
GE-220	REFER TO SECTION 2
GE-221	REFER TO SECTION 2
GE-222	REFER TO SECTION 2
GE-223	REFER TO SECTION 2
GE-224	REFER TO SECTION 2
GE-225	REFER TO SECTION 2
GE-226	REFER TO SECTION 2
GE-227	REFER TO SECTION 2
GE-228	REFER TO SECTION 2
GE-229	REFER TO SECTION 2
GE-230	REFER TO SECTION 2
GE-231	REFER TO SECTION 2
GE-232	REFER TO SECTION 2
GE-233	REFER TO SECTION 2
GE-234	REFER TO SECTION 2
GE-235	REFER TO SECTION 2
GEMR-6	REFER TO SECTION 2
GET706	2N834A,2N2220,2SC321,2SC322,2SC576,*GE-210,*TR-21,*HEPS0011,*SK3122, *ECG123A,*WEP736,*BC337,*276-2009,*RT-102
GET708	2N708,2N708A,2N784A,2N914,2N2221,2N2242,2N2481,2N2501,2N2539,2N2710, 2N2845,2N2847,2N3013,2N3014,2N3115,2N3210,2N3211,2N3301,2N3510,2N3511, 2N3605A,2N3606A,2N3646,2N3647,2N3648,2N3946,2N4962,2N5144,2N5380, 2N5772,2SC67,2SC577,2SC639,2SC764,A5T3903,A5T4123,BSW82,GET914, GET2221,GET2369,GET3013,GET3014,GET3646,GI-3641,MM3903,TIS134,TN-3903, ZT708,*GE-20,*TR-21,*HEPS0025,*SK3122,*ECG123A,*WEP736,*276-2009, *RT-102
GET914	2N708,2N708A,2N784A,2N914,2N2221,2N2242,2N2481,2N2501,2N2539,2N2710,

To Replace	Substitute This Type
(GET914)	2N2845,2N2847,2N3013,2N3014,2N3115,2N3210,2N3211,2N3301,2N3510,2N3511, 2N3605A,2N3606A,2N3646,2N3647,2N3648,2N3946,2N4962,2N5144,2N5380, 2N5772,2SC67,2SC577,2SC639,2SC764,A5T3903,A5T4123,BSW82,GET708, GET2221,GET2369,GET3013,GET3014,GET3646,GI-3641,MM3903,TIS134,TN-3903, ZT708,*GE-20,*TR-21,*HEPS0025,*SK3122,*ECG123A,*WEP736,*276-2009, *RT-102
GET929	2N736A,2N736B,2N915,2N2516,2N4014,2N4963,2SC1998,TIS96,TIS99,TIS135, TIS136,*PTC121,*GE-62,*TR-87,*HEPS0015,*SK3122,*ECG128,*WEP243,*BF338, *276-2009,*RT-114
GET930	2N4409,2N6540,2N6541,A5T4409,BC546A,BSW85,GET2222A,TIS95,TIS98,*PTC121, *GE-62,*TR-87,*HEPS0015,*SK3122,*ECG128,*WEP243,*276-2009,*RT-114
GET2221	2N2221,2N2221A,2N2539,2N2845,2N2847,2N3115,2N3301,2N4014,2N4962,2N4963, 2N5581,BSW84,GET2221A,GI-3641,TIS135,TIS136,TN-3903,*GE-210,*TR-21, *HEPS0015,*SK3122,*ECG123A,*WEP736,*276-2009,*RT-102
GET2221A	2N2221A,2N4014,2N4963,2N5581,BSW84,TIS135,TIS136,*GE-20,*HEPS0015, *WEP736,*276-2009
GET2222	2N2222,2N2222A,2N2540,2N3116,2N3302,2N4952,2N5107,2N5369,2N5582,A5T2222, BSW85,GET2222A,GI-3643,TIS109,TIS111,TN-3904,*GE-210,*TR-21,*HEPS0015, *SK3122,*ECG123A,*WEP736,*276-2009,*RT-102
GET2222A	2N2222A,2N5582,BSW85,*GE-20,*HEPS0015,*276-2009
GET2369	2N708,2N708A,2N784A,2N914,2N2481,2N2501,2N3013,2N3014,2N3210,2N3211, 2N3646,2N3647,2N4013,2N5144,2N5772,2SC67,2SC577,2SC639,2SC764,ZT708, *GE-20,*TR-21,*HEPS0011,*SK3122,*ECG123A,*WEP736,*276-2009,*RT-102
GET2484	*SK3122,*276-2008
GET2904	2N2906,2N2906A,2N3485,2N3485A,2N4026,2N4027,GET2905,*PTC103,*GE-21, *HEPS5022,*SK3025,*276-2023
GET2905	2N2906,2N2906A,2N3485,2N3485A,2N4026,2N4027,GET2904,*PTC103,*GE-21, *HEPS5022,*SK3025,*276-2023
GET2906	2N2907,2N2907A,2N3486,2N3486A,2N3505,2N3672,2N3673,2N4028,2N4029,2SA720, 2SA891,A5T2907,A5T3505,A5T3645,GET2907,TIS112,*PTC103,*GE-48,*HEPS5022, *SK3114,*276-2023
GET2907	2N2907,2N2907A,2N3486,2N3486A,2N3505,2N3672,2N3673,2N4028,2N4029,2SA720, 2SA891,A5T2907,A5T3505,A5T3645,GET2906,TIS112,*PTC103,*GE-48,*HEPS5022, *SK3114,*276-2023
GET3013	2N708,2N708A,2N784A,2N914,2N2481,2N2501,2N2710,2N3013,2N3014,2N3210, 2N3211,2N3510,2N3511,2N3605A,2N3606A,2N3646,2N3647,2N3648,2N5144, 2N5772,2SC67,2SC577,2SC639,2SC764,GET708,GET914,GET2369,GET3014, GET3646,ZT708,*GE-20,*TR-21,*HEPS0011,*SK3122,*ECG123A,*WEP736, *276-2009,*RT-102
GET3014	2N708,2N708A,2N784A,2N914,2N2481,2N2501,2N2710,2N3013,2N3014,2N3210, 2N3211,2N3510,2N3511,2N3605A,2N3606A,2N3646,2N3647,2N3648,2N5144, 2N5772,2SC67,2SC577,2SC639,2SC764,GET708,GET914,GET2369,GET3013, GET3646,ZT708,*GE-20,*TR-21,*HEPS0011,*SK3122,*ECG123A,*WEP736, *276-2009,*RT-102
GET3638	FT1746,GET3638A,*PTC103,*GE-22,*TR-20,*HEPS0012,*SK3114,*ECG159, *WEP242,*BC327,*276-2024,*RT-115
GET3638A	2N869A,2N2695,2N2696,BSW72,FT1746,GET3638,*PTC103,*GE-22,*TR-20, *HEPS0013,*SK3114,*ECG159,*WEP242,*BC327,*276-2024,*RT-115
GET3646	2N708,2N708A,2N784A,2N914,2N2481,2N2501,2N2710,2N3013,2N3014,2N3210, 2N3211,2N3510,2N3511,2N3605A,2N3606A,2N3646,2N3647,2N3648,2N5144, 2N5772,2SC67,2SC577,2SC639,2SC764,GET708,GET914,GET2369,GET3013, GET3014,ZT708,*GE-20,*TR-21,*HEPS0011,*SK3122,*ECG123A,*WEP736, *276-2009,*RT-102
GI-2711	2N744,2N2318,2N2319,2N2656,2N2711,2N2921,2N2922,2N3011,2N3605,2N3606, 2N3607,2SC99,2SC168,2SC204,2SC300,2SC301,2SC356,2SC461,2SC595,2SC724, EN744,EN3011,MPS2713,MPS2926-RED,PT720,TIS51,ZT2205,*PTC132,*GE-10, *TR-24,*SK3122,*ECG123A,*WEP736,*BC107B,*276-2015
GI-2712	2N2712,2SC395A,2SC395A-Y,2SC400-Y,2SC619,2SC934,BC108A,BC168A,BC172A, BC238A,BCY58A,MPS2714,MPS2926-ORG,MPS2926-YEL,ZT2938,*PTC132,*GE-17, *TR-51,*SK3122,*ECG123A,*WEP736,*BC107B,*276-2015
GI-2713	2N744,2N2318,2N2319,2N2656,2N2713,2N3011,2N3605,2N3606,2N3607,2SC168,

To Replace	Substitute This Type
(GI-2713)	2SC204,2SC356,2SC595,2SC724,EN744,EN3011,MPS2713,PT720,TIS51,ZT2205, *GE-210,*TR-53,*SK3122,*ECG123A,*WEP736,*BC107B,*276-2015
GI-2714	2N2714,2N3241A,2SC395A-Y,2SC735-Y,2SC934,2SC1317,BCY58A,MPS2714, *GE-210,*TR-53,*SK3122,*ECG123A,*WEP736,*BC107B,*276-2015
GI-2715	2N744,2N753,2N2318,2N2319,2N2656,2N2711,2N2713,2N2921,2N2922,2N2926-RED, 2N3011,2N3394,2N3605,2N3606,2N3607,2SC99,2SC168,2SC204,2SC300,2SC301, 2SC356,2SC461,2SC595,2SC724,BC170A,BF255,EN744,EN3011,GI-2711,GI-2713, MPS2713,MPS2926-RED,PT720,TIS51,ZT41,ZT2205,*PTC132,*GE-86,*TR-53, *SK3122,*ECG123A,*WEP736,*BF365,*276-2015
GI-2716	2N2712,2N2714,2N2923,2N2926-ORG,2N3241A,2SC372G-Y,2SC400,2SC400-Y, 2SC454,2SC458,2SC735-Y,2SC934,2SC1085,2SC1359,2SC1363,2SC1854,2SD392, BC108A,BC168A,BC172A,BC238A,BC548A,BCY58A,BSW88,BSW89,BSX38, GI-2714,MPS2714,MPS2926-GRN,MPS2926-YEL,PBC108A,*PTC132,*GE-60,*TR-53, *SK3122,*ECG123A,*WEP736,*BF365,*276-2015
GI-2921	2N708,2N753,2N784A,2N957,2N2242,2N2318,2N2319,2N2656,2N2921,2N2922, 2N3011,2N3013,2N3014,2N3340,2N3394,2N3397-RED,2N3398-RED,2N3605A, 2N3606A,2N3646,2N4295,2N4420,2N4421,2N4422,2S501,2SC67,2SC131,2SC134, 2SC169,2SC204,2SC300,2SC301,2SC356,2SC461,2SC468,2SC579,2SC595, 2SC601,2SC601N,2SC613,2SC724,2SC752G-R,2SC764,2SC913,2SC914,2SC914A, 2SC915,2SC915A,2SC922,2SC1071,2SC1274,A321,BF255,EN708,EN914,EN3011, EN3013,EN3014,GET708,GET914,GET2369,GET3013,GET3014,GET3646,GI-2922, GI-3394,GI-3709,MPS3394,MPS3646,PT720,TIS45,TIS46,TIS51,TIS52,TIS55,ZT708, ZT2205,*PTC101,*GE-60,*TR-53,*SK3122,*ECG123A,*WEP736,*BF365,*276-2031
GI-2922	2N2331,2N2501,2N2922,2N2923,2N3393,2N3396-ORG,2N3397-ORG,2N3398-ORG, 2N3709,2N4123,2N4138,2N4418,2N4968,2N6566,2S512,2SC169,2SC300,2SC301, 2SC367G-O,2SC372G-O,2SC400-O,2SC459,2SC460,2SC461,2SC536,2SC619, 2SC639,2SC733-O,2SC735-O,2SC752G-O,2SC764,2SC924,2SC925,2SC1276, 2SC1359,2SC1372,2SC1393,2SC1394,2SC1395,2SC1396,2SC1682-BL,2SC1682-GR, 2SD227,BF254,BSY61,GI-2923,GI-3393,GI-3394,GI-3709,MPS2923,MPS3393, MPS3394,MPS3709,PET4001,TP3709,*PTC101,*GE-60,*TR-53,*SK3122,*ECG123A, *WEP736,*BF365,*276-2031
GI-2923	2N2923,2N2924,2N3241A,2N3242A,2N3392,2N3393,2N3395-YEL,2N3396-ORG, 2N3396-YEL,2N3397-ORG,2N3397-YEL,2N3398-ORG,2N3398-YEL,2N3710,2N4074, 2N5224,2S502,2SC367G,2SC367G-Y,2SC372G-Y,2SC400,2SC400-Y,2SC454, 2SC458,2SC733-Y,2SC735-Y,2SC752G-Y,2SC847,2SC848,2SC849,2SC966, 2SC1085,2SC1216,2SC1285,2SC1359,2SC1363,2SC1631,2SC1633,2SC1854, 40398,BC108A,BC172A,BC238A,BC548A,BCY58A,BSW88,BSW89,BSX38,GI-2924, GI-3392,GI-3393,GI-3396,GI-3397,GI-3707,GI-3708,GI-3710,MPS2923,MPS2924, MPS3392,MPS3393,MPS3707,MPS3710,PET4001,SE4001,TIS60,TIS60M,TN80, TP3710,TZ81,*PTC101,*GE-60,*TR-53,*SK3122,*ECG123A,*WEP736,*BF365, *276-2031
GI-2924	2N2924,2N2925,2N3391,2N3391A,2N3392,2N3395-WHT,2N3395-YEL,2N3396-WHT, 2N3396-YEL,2N3397-WHT,2N3397-YEL,2N3398-WHT,2N3398-YEL,2N3565,2N3707, 2N3708,2N4124,2N4256,2N4286,2N4419,2N5172,2S503,2SC373G,2SC400-GR, 2SC538,2SC539,2SC732-GR,2SC733-GR,2SC735-GR,2SC829,2SC900,2SC1684, 2SC1849,2SC1853,2SC1993,40397,A322,BC108,BC108B,BC109B,BC122,BC172B, BC173B,BC238B,BC239B,BC548B,BC549B,BCY58B,BCY58C,GI-2925,GI-3391, GI-3391A,GI-3392,GI-3395,GI-3398,GI-3711,MPS2924,MPS2925,MPS3391, MPS3391A,MPS3392,MPS3395,MPS3708,MPS5172,MPSA16,MPSA17,PET4002, TP3706,TP3707,TP3708,TZ81,*PTC121,*GE-212,*TR-53,*SK3122,*ECG123A, *WEP736,*BC107B,*276-2031
GI-2925	2N2925,2N3391,2N3391A,2N3395-WHT,2N3396-WHT,2N3397-WHT,2N3398-WHT, 2N3708,2N3711,2SC536FP,2SC536NP,2SC732-BL,2SC733-BL,2SC828,2SC923, 2SC1335,2SC1344,2SC1571,2SC1571L,BC108C,BC109,BC109C,BC172C,BC173C, BC238C,BC239C,BC548B,BC549B,BCY58D,GI-3391,GI-3391A,MPS2925,MPS3391, MPS3391A,MPS3711,PET4003,TP3708,TP3711,*PTC139,*GE-212,*TR-53, *BC107B,*276-2031
GI-2926	2N2924,2N2926,2N2926-YEL,2N3392,2N3395-YEL,2N3396-YEL,2N3397-YEL, 2N3398-YEL,2N3707,2N3710,2N4124,2N4419,2N5224,2S502,2SC372G-Y,2SC400, 2SC538,2SC539,2SC733-Y,2SC735-Y,2SC847,2SC848,2SC849,2SC900,2SC966, 2SC1085,2SC1281,2SC1363,2SC1631,2SC1633,2SC1684,2SC1849,2SC1853,

To Replace	Substitute This Type
(GI-2926)	2SC1854,92T6,40232,40233,40397,40399,BC108,BC108A,BC168A,BC170B,BC172A, BC238A,BC548A,BCY58A,BCY58B,BSW88,BSW89,BSX38,GI-2924,GI-3392,GI-3396, GI-3397,GI-3398,GI-3707,GI-3708,GI-3710,GI-3721,MPS2924,MPS2926, MPS2926-GRN,MPS2926-YEL,MPS3392,MPS3707,MPS3710,PBC108A,SE4001, TN80,TP3707,TP3710,*PTC101,*GE-211,*TR-53,*BC107B,*276-2031
GI-3391	2N3391,2N3391A,2N3395-WHT,2N3396-WHT,2N3397-WHT,2N3398-BLU,2N3398-WHT, 2N3711,2SC536FP,2SC536NP,2SC732-BL,2SC733-BL,2SC828,2SC923,2SC1335, 2SC1344,2SC1571,2SC1571L,BC108C,BC109,BC109C,BC172C,BC173C,BC238C, BC239C,BCY58D,GI-3391A,MPS3391,MPS3391A,MPS3711,PET4003,SE4002, SE4010,TP3711,*PTC139,*GE-212,*TR-53,*BC107B,*276-2031
GI-3391A	2N3391,2N3391A,2N3395-WHT,2N3396-WHT,2N3397-WHT,2N3398-BLU,2N3398-WHT, 2N3711,2SC536FP,2SC536NP,2SC732-BL,2SC733-BL,2SC828,2SC923,2SC1335, 2SC1344,2SC1571,2SC1571L,BC108C,BC109,BC109C,BC172C,BC173C,BC238C, BC239C,BCY58D,GI-3391,MPS3391,MPS3391A,MPS3711,PET4003,SE4002,SE4010, TP3711,*PTC139,*GE-212,*TR-53,*BC107B,*276-2031
GI-3392	2N2924,2N2925,2N3391,2N3391A,2N3392,2N3395-WHT,2N3395-YEL,2N3396-WHT, 2N3396-YEL,2N3397-WHT,2N3397-YEL,2N3398-WHT,2N3398-YEL,2N3415,2N3565, 2N3794,2N4124,2N4256,2N4286,2N4419,2N4954,2N5172,2N5371,2N5998,2N6002, 2N6008,2SC373G,2SC400-GR,2SC732-GR,2SC733-GR,2SC735-GR,2SC1684, 2SC1849,2SC1993,40397,BC108,BC108B,BC109B,BC172B,BC173B,BC238B, BC239B,BC338,BC548B,BC549B,BCY58B,BCY58C,GI-2924,GI-2925,GI-3391, GI-3391A,GI-3395,GI-3398,GI-3403,GI-3415,GI-3794,MPS2924,MPS2925,MPS3391, MPS3391A,MPS3392,MPS3395,MPS3706,MPS5172,MPSA16,MPSA17,PET4002, TP3706,TZ81,*PTC121,*GE-212,*TR-53,*BC107B,*276-2031
GI-3393	2N2923,2N2924,2N3227,2N3241A,2N3242A,2N3392,2N3393,2N3395-YEL, 2N3396-ORG,2N3396-YEL,2N3397-ORG,2N3397-YEL,2N3398-ORG,2N3398-YEL, 2N3414,2N3509,2N4074,2N5224,2N5998,2N6000,2SC367G,2SC367G-Y, 2SC372G-Y,2SC400,2SC400-Y,2SC454,2SC458,2SC733-Y,2SC735-Y,2SC752G-Y, 2SC847,2SC848,2SC849,2SC966,2SC967,2SC1085,2SC1216,2SC1219,2SC1246, 2SC1285,2SC1317,2SC1363,2SC1631,2SC1633,2SD327,40398,BC108A,BC172A, BC238A,BC548A,BCY58A,BSW83,BSW88,BSW89,BSX38,GI-2923,GI-2924,GI-3392, GI-3396,GI-3397,MPS2923,MPS2924,MPS3392,MPS3393,PET4001,SE4001,TIS60, TIS60M,TN80,TZ81,*PTC101,*GE-212,*TR-53,*BC107B,*276-2031
GI-3394	2N2331,2N2501,2N2922,2N2923,2N3211,2N3393,2N3396-ORG,2N3397-ORG, 2N3398-ORG,2N4123,2N4138,2N4418,2N6566,2S512,2SC169,2SC300,2SC301, 2SC367G-O,2SC372G-O,2SC400-O,2SC459,2SC460,2SC461,2SC536,2SC619, 2SC639,2SC733-O,2SC735-O,2SC752G-O,2SC764,2SC1276,2SC1372,2SD227, BSY61,GI-2922,GI-2923,GI-3393,GI-3402,GI-3414,MPS2923,MPS3393,MPS3394, MPS6560,PET4001,ZT2938,*PTC101,*GE-212,*TR-53,*BC107B,*276-2031
GI-3395	2N2925,2N3391,2N3391A,2N3395-WHT,2N3396-WHT,2N3397-WHT,2N3398-WHT, 2N3415,2N3565,2N4256,2N4286,2N4954,2N5172,2N5371,2N6002,2N6008,2SC373G, 2SC732-GR,2SC733-GR,2SC735-GR,2SC1993,BC108B,BC109B,BC172B,BC173B, BC238B,BC239B,BC338,BC548B,BC549B,BCY58C,GI-2925,GI-3391,GI-3391A, GI-3403,GI-3415,MPS2925,MPS3391,MPS3391A,MPS3395,MPS3706,MPS5172, MPSA16,MPSA17,PET4002,TP3706,*PTC139,*GE-212,*TR-53,*BC107B,*276-2031
GI-3396	2N2924,2N2925,2N3392,2N3395-YEL,2N3396-YEL,2N3397-YEL,2N3398-YEL,2N3565, 2N3794,2N4124,2N4256,2N4286,2N4419,2N4954,2N5172,2N5224,2N5371,2N5998, 2SC373G,2SC400-GR,2SC732-GR,2SC733-GR,2SC735-GR,2SC1684,2SC1849, 2SC1993,40397,BC108,BC108B,BC109B,BC172B,BC173B,BC238B,BC239B,BC338, BCY58B,BCY58C,GI-2924,GI-2925,GI-3392,GI-3395,GI-3398,GI-3403,GI-3415,GI-3794, MPS2924,MPS2925,MPS3392,MPS3395,MPS3706,MPS5172,MPSA16,MPSA17, PET4002,TP3706,TZ81,*PTC121,*GE-212,*TR-53,*BC107B,*276-2031
GI-3397	2N2924,2N3227,2N3242A,2N3392,2N3395-YEL,2N3396-YEL,2N3397-YEL,2N3398-YEL, 2N3509,2N3565,2N3794,2N4124,2N4256,2N4286,2N4419,2N5172,2N5224,2N5998, 2SC367G-Y,2SC372G-Y,2SC373G,2SC400,2SC400-GR,2SC732-GR,2SC733-GR, 2SC733-Y,2SC735-GR,2SC735-Y,2SC752G-Y,2SC1285,2SC1363,2SC1631, 2SC1633,2SC1684,2SC1849,2SC1993,40397,BC108,BC108B,BC109B,BC172B, BC173B,BC238B,BC239B,BC548A,BCY58B,BSW83,BSW88,BSW89,BSX38,GI-2924, GI-3392,GI-3395,GI-3396,GI-3398,GI-3794,MPS2924,MPS3392,MPS5172,MPSA17, SE4001,TN80,TP3706,TZ81,*PTC101,*GE-212,*TR-53,*BC107B,*276-2031
GI-3398	2N2925,2N3391,2N3391A,2N3395-WHT,2N3396-WHT,2N3397-WHT,2N3398-WHT,

To Replace	Substitute This Type
(GI-3398)	2N3415,2N3565,2N4256,2N4286,2N4954,2N5172,2N5371,2N6002,2N6008,2SC373G, 2SC400-GR,2SC732-GR,2SC733-GR,2SC735-GR,2SC1993,BC108B,BC109B, BC172B,BC173B,BC238B,BC239B,BC338,BC548B,BC549B,BCY58C,GI-2925, GI-3391,GI-3391A,GI-3395,GI-3403,GI-3415,MPS2925,MPS3391,MPS3391A, MPS3395,MPS3706,MPS5172,MPSA16,MPSA17,PET4002,TP3706,*PTC139, *GE-212,*TR-53,*BC107B,*276-2031
GI-3402	2N3227,2N3241A,2N3242A,2N3414,2N3509,2N6000,2SC967,2SC1219,2SC1246, 2SC1317,2SD327,2SD467,BSW83,BSX75,GI-2923,GI-3414,GI-3706,PET3706,SE4001, TN60,TN80,*PTC170,*GE-210,*BC337,*276-2031
GI-3403	2N3415,2N3706,2N4384,2N4954,2N5371,2N5451,2N6002,2N6008,BC338,GI-2925, GI-3391,GI-3391A,GI-3415,MPS3391,MPS3391A,MPS3706,*GE-210,*BC337, *276-2031
GI-3404	2N760A,2N909,2N930B,2N956,2N2222,2N2222A,2N2350,2N2350A,2N2390,2N2540, 2N2645,2N3116,2N3416,2N3704,2N4401,2N4437,2N4952,2N4970,2N5107,2N5369, 2N5449,2N5582,2N5822,2N6004,2N6010,2N6014,2SC850,2SC968,2SC970, 2SC1220,2SC1247,2SC1277,2SC1318,2SC1781,2SC1813,40458,BCW34,BCW36, BSW85,D33D25,D33D30,EN956,EN2222,GI-3416,GI-3643,GI-3704,MPS3704, MPS6531,PET3704,SE6022,TN-3904,TZ82,ZT89,ZT119,*PTC170,*GE-210,*BC337, *276-2013
GI-3405	2N3417,2N4424,2N4953,2N5370,2N5376,2N6006,2N6012,2N6016,2N6112,D33D27, GI-3417,PET8002,PET8004,*GE-210,*BC337,*276-2013
GI-3414	2N3227,2N3241A,2N3242A,2N3414,2N3509,2N6000,2N6426,2SC509-Y,2SC509G-Y, 2SC967,2SC1246,2SC1317,2SC1788,2SC1851,2SD467,BC338-16,BSW83,BSX75, GI-3706,PET3706,TIS90,TIS90M,TN60,TN80,*PTC170,*GE-210,*BC337,*276-2013
GI-3415	2N3415,2N3706,2N4384,2N4954,2N5371,2N5451,2N6002,2N6008,A5T5451,A8T3706, BC338,BC338-40,MPS3391,MPS3391A,MPS3706,*GE-210,*BC337,*276-2013
GI-3416	2N760A,2N909,2N930B,2N956,2N2222,2N2222A,2N2350,2N2350A,2N2390,2N2540, 2N2645,2N3116,2N3416,2N3704,2N4952,2N5107,2N5369,2N5449,2N5582,2N5822, 2N6004,2N6010,2N6014,2SC850,2SC968,2SC970,2SC1247,2SC1318,2SC1781, 2SC1813,2SC1852,40458,A5T2192,A5T2222,A5T5449,A8T3704,BCW34,BSW85, D33D25,D33D30,GI-3643,GI-3704,PET3704,TIS111,TN-3904,TZ82,*PTC170,*GE-210, *BC337,*276-2010
GI-3417	2N3417,2N4424,2N4953,2N5370,2N5376,2N6006,2N6012,2N6016,2N6112,D33D27, PET8002,PET8004,*GE-210,*BC337,*276-2010
GI-3566	2N3417,2N4953,2N5370,2N5376,2N5377,2N6006,2N6012,2N6016,2N6112,2N6428, 2N6428A,2N6539,2SC1166-GR,D33D27,PET8002,PET8004,TIS94,TIS97,*GE-210, *TR-53,*BC107B,*276-2013
GI-3605	2N2318,2N2319,2N2656,2N3011,2N3605,2N3606,2N3607,2SC98,2SC204,2SC400-R, 2SC595,2SC724,EN744,EN3011,GI-3606,GI-3607,MPS2713,MPS2926-RED,TIS51, *PTC132,*GE-86,*TR-53,*BF173,*276-2011
GI-3606	2N2318,2N2319,2N2656,2N3011,2N3605,2N3606,2N3607,2SC98,2SC204,2SC400-R, 2SC595,2SC724,EN744,EN3011,GI-3605,GI-3607,MPS2713,MPS2926-RED,TIS51, *PTC132,*GE-86,*TR-53,*BF173,*276-2011
GI-3607	2N2318,2N2319,2N2656,2N3011,2N3605,2N3606,2N3607,2SC98,2SC204,2SC400-R, 2SC595,2SC724,EN744,EN3011,GI-3605,GI-3606,MPS2713,MPS2926-RED,TIS51, *PTC132,*GE-86,*TR-53,*BF173,*276-2011
GI-3638	2N727,2N869A,2N2392,2N2411,2N3829,2SA558,BF509,FT1746,GET3638,GET3638A, MPS6516,MPSA70-RED,TIS38,TIS137,*PTC131,*GE-65,*TR-54,*SK3114,*ECG159, *WEP242,*BC177,*276-2034
GI-3638A	2N869,2N3906,2N4034,2N4035,2N4058,2N4061,2N4125,2N5244,2N5355,2N5366, 2N5383,2N6076,A5T3906,A5T4125,A178A,A178B,A179A,BC178,BC178A,BC179, BC252A,BC253A,BC258A,BC262A,BC263A,BC308A,BC309A,BC558A,BC559A, BCY78,BCY78A,BF440,BSW19,BSW20,MM3906,MPS6517,MPS6518,MPS-A70, MPSA70-BLU,MPSA70-GRN,MPSA70-WHT,MPSA70-YEL,NPS404,NPS404A, TN-3906,*PTC103,*GE-65,*SK3114,*ECG159,*WEP243,*BC177,*276-2034
GI-3641	2N2221,2N2221A,2N2539,2N2845,2N2847,2N3115,2N3301,2N4014,2N4962,2N4963, 2N5581,BSW84,TN-3903,*GE-210,*SK3122,*ECG123A,*WEP736,*276-2016
GI-3642	2N2221,2N2221A,2N2539,2N2845,2N2847,2N3115,2N3301,2N4014,2N4962,2N4963, 2N5581,BSW84,GI-3641,MPS6530,TN-3903,*GE-210
GI-3643	2N2222,2N2222A,2N2540,2N3116,2N3302,2N4952,2N5107,2N5369,2N5582,A5T2222, BSW85,TIS109,TIS111,TN-3904,*GE-210,*SK3122,*ECG123A,*WEP736,*276-2016

To Replace	Substitute This Type
GI-3644	2N2907,2N2907A,2N3486,2N3486A,2N3504,2N3505,2N3672,2N3673,2N4028,2N6005, 2SA684,2SA720,2SA891,A5T2907,A5T3504,A5T3505,A5T3644,A5T3645,TIS112, *PTC103,*GE-48,*SK3114,*ECG159,*WEP242
GI-3702	2N2838,2N2907,2N2907A,2N3251,2N3251A,2N3486,2N3486A,2N3504,2N3505, 2N3672,2N3673,2N3702,2N3906,2N4143,2N4228,2N4403,2N5373,2N5375,2N5383, 2N5447,2N5817,2N5819,2N6005,2SA467G-O,2SA467G-Y,2SA603,2SA659, 2SA661-O,2SA661-Y,2SA718,2SA720,2SA891,A5T2907,A5T3504,A5T3505, A5T3644,A5T3645,A5T3906,A5T4403,A5T5447,A8T3702,A177,BC212A,BC212L, BC213L,BC327-16,BCW35,BCW37,BCY79A,BSW73,D29E5,D29E6,GET2906, GET2907,GI-3644,MM3906,MPS3702,TIS112,TN-3906,TQ60,TZ552,*PTC103,*GE-82, *TR-54,*SK3114,*ECG159,*WEP242,*BC177,*276-2034
GI-3703	2N2837,2N2906,2N2906A,2N3250,2N3250A,2N3485,2N3485A,2N3703,2N4142, 2N4228,2N5372,2N5448,2N5763,2N5815,2N5821,2SA661-O,2SA661-R,A5T5448, A8T3703,BSW74,GET2904,GET2905,MPS3703,TZ551,*PTC103,*GE-82,*TR-54, *SK3114,*ECG159,*WEP242,*BC177,*276-2023
GI-3704	2N2222,2N2222A,2N2540,2N3302,2N3704,2N5449,2N5582,2N5818,2N6010,2N6014, A5T2222,A5T5449,A8T3704,D33D26,PET3704,TIS109,TIS111,TN-3904,*GE-20, *SK3122,*ECG123A,*WEP736,*BC337,*276-2014
GI-3705	2N915,2N2539,2N2847,2N2897,2N2900,2N3705,2N4962,2N5450,2N5820,A5T5450, A8T3705,PET3705,TIS110,TN54,TN-3903,*GE-20,*TR-65,*SK3122,*ECG123A, *WEP736,*BC337,*276-2014
GI-3706	2N2222,2N2540,2N3242A,2N3302,2N3704,2N5449,2N5818,A5T2222,A5T5449, A8T3704,BC337-25,D33D26,PET3706,TIS109,TIS111,TN60,TN-3904,*GE-20,*SK3122, *ECG123A,*WEP736,*BC337,*276-2014
GI-3707	2N3242A,2N4286,2N4287,2N4419,2SC1364,2SC1632,2SC1634,2SC1992,BC107B, BC167B,BC171B,BC237B,BCY58B,BCY59B,GI-3708,GI-3711,MPS3707,MPS3710, MPSA16,MPSA17,*PTC101,*GE-212,*TR-51,*SK3122,*ECG123A,*WEP736, *BC107B,*276-2013
GI-3708	2N4286,2N4287,2N4419,2SC1992,BC107B,BC167B,BC171B,BC237B,BCY58B, BCY58C,BCY59B,BCY59C,GI-3711,MPS3708,MPS3710,MPSA16,MPSA17,PET4002, *PTC121,*GE-212,*TR-51,*SK3122,*ECG123A,*WEP736,*BC107B,*276-2013
GI-3709	2N760,2N2501,2N4138,2N4418,2N6566,2SC1999,MPS3709,PET4001,ZT44,*PTC101, *GE-61,*TR-51,*SK3122,*ECG123A,*WEP736,*BC107B,*276-2013
GI-3710	2N3242A,2N4419,2SC1364,2SC1632,2SC1634,2SC1992,BC547A,BCY58B,BCY59B, GI-3707,GI-3708,MPS3707,MPS3710,MPSA17,*PTC101,*GE-61,*TR-51,*SK3122, *ECG123A,*WEP736,*BC107B,*276-2013
GI-3711	2N3708,BC547B,BC550B,BCY58C,BCY58D,BCY59C,BCY59D,MPS3708,MPS3711, MPSA18,PET4002,PET4003,*PTC139,*GE-212,*TR-51,*SK3122,*ECG123A, *WEP736,*BC107B,*276-2013
GI-3721	2N2924,2N2925,2N2926,2N2926-GRN,2N2926-YEL,2N3391,2N3391A,2N3392, 2N3395-WHT,2N3395-YEL,2N3396-WHT,2N3396-YEL,2N3397-WHT,2N3397-YEL, 2N3398-WHT,2N3398-YEL,2N3415,2N3565,2N3721,2N3900,2N3900A,2N4124, 2N4256,2N4286,2N4419,2N5172,2N5998,2N6002,2N6008,2SC373G,2SC400-GR, 2SC732-GR,2SC733-GR,2SC735-GR,2SC1684,2SC1849,2SC1993,91T6,92T6, 98T2,40397,40399,A158B,A159B,BC108,BC108B,BC109B,BC168B,BC172B, BC173B,BC238B,BC239B,BC338,BC548B,BC549B,BCY58B,BCY58C,GI-2924, GI-2925,GI-3391,GI-3391A,GI-3392,GI-3395,GI-3398,GI-3403,GI-3415,GI-3900, GI-3900A,MPS2924,MPS2925,MPS2926,MPS3391,MPS3391A,MPS3392,MPS3395, MPS3721,MPS5172,PBC108B,PBC109B,PET4002,*PTC121,*GE-212,*TR-51, *BC107B,*276-2031
GI-3793	2N708A,2N914,2N957,2N2221,2N2481,2N2710,2N2845,2N3115,2N3210,2N3261, 2N3301,2N3510,2N3511,2N3646,2N3647,2N3648,2N3793,2SC577,BSW82,GI-3641, GI-3642,ZT83,ZT113,ZT708,*PTC170,*GE-210,*TR-25,*ECG161,*BC337,*276-2016
GI-3794	2N3706,2N3794,2N4953,2N4954,2N5370,2N5371,2N5376,2N5377,2N5451,2N6006, 2N6012,2N6112,BC337,D33D27,MPS3706,*GE-210,*BC337,*276-2016
GI-3900	2N3390,2N3391,2N3391A,2N3395-WHT,2N3396-WHT,2N3397-WHT,2N3398-BLU, 2N3398-WHT,2N3900,2N3900A,2N3901,2N6002,2N6008,2SC732-BL,2SC733-BL, 2SC1335,2SC1344,91T6,A158C,A159C,BC108C,BC109,BC109C,BC168C,BC170C, BC172C,BC173C,BC238C,BC239C,BCY58D,GI-3391,GI-3391A,GI-3900A,MPS3390, MPS3391,MPS3391A,PBC108C,PBC109C,PET4003,SE4002,SE4010,*PTC139, *GE-212,*TR-51,*BC107B,*276-2031

To Replace	Substitute This Type
GI-3900A	2N3390,2N3391,2N3391A,2N3395-WHT,2N3396-WHT,2N3397-WHT,2N3398-BLU, 2N3398-WHT,2N3900,2N3900A,2N3901,2N6002,2N6008,2SC732-BL,2SC733-BL, 2SC1335,2SC1344,91T6,A158C,A159C,BC108C,BC109,BC109C,BC168C,BC170C, BC172C,BC173C,BC238C,BC239C,BCY58D,GI-3391,GI-3391A,GI-3900,MPS3390, MPS3391,MPS3391A,PBC108C,PBC109C,PET4003,SE4002,SE4010,*PTC139, *GE-212,*TR-51,*BC107B,*276-2031
GM290A	*PTC107,*GE-1,*TR-17,*HEPG0003,*ECG160,*WEP637
GM378A	2N3127,2N3279,2N3280,2N3283,2N3284,GM290A,MM1139,*PTC109,*GE-1,*TR-17, *HEPG0003,*ECG160,*WEP637
GT34HV	SEE 2N1408
GT40	*GE-51,*TR-17,*HEPG0003,*HF35,*ECG160,*WEP637,*AC188/01
GT41	*GE-51,*TR-85,*HEPG0005,*SK3005,*HF35,*ECG102A,*WEP250,*AC188/01, *276-2004,*RT-121
GT42	*PTC109,*GE-51,*TR-85,*HEPG0005,*SK3005,*HF35,*ECG102A,*WEP250, *AC188/01,*276-2005,*RT-121
GT43	*PTC109,*GE-51,*TR-17,*HEPG0001,*SK3005,*HF35,*ECG126,*WEP635
GT44	*PTC135,*GE-51,*TR-85,*HEPG0005,*HF35,*ECG102A,*WEP250,*276-2004,*RT-121
GT45	*PTC135,*GE-51,*TR-85,*HEPG0005,*HF35,*ECG102A,*WEP250,*276-2004,*RT-121
GT46	*PTC135,*GE-51,*TR-17,*HEPG0001,*SK3006,*HF35,*ECG126,*WEP635
GT47	*DS-56,*GE-51,*TR-17,*HEPG0001,*SK3006,*ECG126,*WEP635
GT81	2N59,2N59A,2N60,2N60A,2N109,2N217,2N270,2N369,2N415,2N415A,2N416,2N422A, 2N467,2N520A,2N569,2N571,2N609,2N654,2N654+,2N1008A,2N1125,2N1175, 2N1175A,2N1192,2N1274,2N1305,2N1307,2N1309A,2N1349,2N1352,2N1356, 2N1370,2N1374,2N1376,2N1381,2N1382,2N1383,2N1415,2N1681,2N1707,2SB187, 2SB188,2SB221,2SB222,2SB328,2SB329,40269,ASY27,NKT226,SF.T223,*PTC135, *GE-50,*TR-85,*HEPG0005,*SK3004,*AT30M,*ECG102A,*WEP250,*AC126, *276-2005,*RT-121
GT1200	SEE 2N1310
GT1658	SEE 2N1605
H5	SEE 2N538A
H6	SEE 2N539A
H7	SEE 2N540A
H12	SEE 2N1157
H12A	SEE 2N1157A
HA7520	SHA7520,*GE-67,*TR-28,*HEPS5022,*276-2025
HA7521	SHA7521,*GE-67,*TR-28,*HEPS5022
HA7522	2N1238,SHA7522,*GE-67,*TR-28,*HEPS5022,*276-2025
HA7523	2N1240,SHA7523,*GE-67,*TR-28,*HEPS5022,*276-2025
HA7524	2N1242,SHA7524,*GE-67,*TR-28,*HEPS5022
HA7525	SHA7525
HA7526	2N1239,SHA7526,*GE-67,*TR-28,*HEPS5022,*276-2025
HA7527	2N1241,SHA7527,*GE-67,*TR-28,*HEPS5022,*276-2025
HA7528	2N1243,SHA7528,*GE-67,*TR-28,*HEPS5022
HA7529	2N1244,HA7525,SHA7525,SHA7529
HA7530	2N1439,SHA7530,*GE-82,*TR-88,*HEPS0032,*SK3025,*ECG129,*WEP242,*BC327, *276-2023,*RT-115
HA7531	2N1440,SHA7531,*GE-82,*TR-88,*SK3025,*ECG129,*WEP242,*RT-115
HA7532	2N2944,*GE-48,*HEPS0032,*BC327,*276-2023
HA7533	2N1230,2N1441,2N4008,SHA7533,*GE-82,*TR-28,*HEPS0032,*BC327,*276-2023
HA7534	2N1232,SHA7534,*GE-82,*TR-28
HA7535	2N2551,SHA7535
HA7536	2N1229,2N2945A,2N3217,2N3218,2N3978,2N4007,SHA7536,TW135,*GE-22,*TR-20, *HEPS0032,*BC327,*276-2023
HA7537	2N1231,2N1442,2N1443,2N1469A,2N2946A,2N3219,2N3979,SHA7537,*GE-82, *TR-28,*HEPS0032,*BC327,*276-2023
HA7538	2N1233,2N4982,SHA7538,*GE-82,*TR-28
HA7539	2N1234,HA7535,SHA7535,SHA7539
HA7597	2N1242,HA7524,SHA7524,SHA7597,*GE-82,*TR-28,*HEPS0032,*BC327,*276-2023
HA7598	2N1241,HA7527,SHA7527,SHA7598,*GE-82,*TR-28,*HEPS0032,*BC327,*276-2023
HA7599	2N1241,SHA7599,*GE-82,*TR-28,*HEPS0032,*BC327,*276-2023
HEP1	REFER TO SECTION 2

To Replace	Substitute This Type
HEP2	REFER TO SECTION 2
HEP3	REFER TO SECTION 2
HEP50	REFER TO SECTION 2
HEP51	REFER TO SECTION 2
HEP52	REFER TO SECTION 2
HEP53	REFER TO SECTION 2
HEP54	REFER TO SECTION 2
HEP55	REFER TO SECTION 2
HEP56	REFER TO SECTION 2
HEP57	REFER TO SECTION 2
HEP75	REFER TO SECTION 2
HEP76	REFER TO SECTION 2
HEP200	REFER TO SECTION 2
HEP230	REFER TO SECTION 2
HEP231	REFER TO SECTION 2
HEP232	REFER TO SECTION 2
HEP233	REFER TO SECTION 2
HEP234	REFER TO SECTION 2
HEP235	REFER TO SECTION 2
HEP236	REFER TO SECTION 2
HEP237	REFER TO SECTION 2
HEP238	REFER TO SECTION 2
HEP239	REFER TO SECTION 2
HEP240	REFER TO SECTION 2
HEP241	REFER TO SECTION 2
HEP242	REFER TO SECTION 2
HEP243	REFER TO SECTION 2
HEP244	REFER TO SECTION 2
HEP245	REFER TO SECTION 2
HEP246	REFER TO SECTION 2
HEP247	REFER TO SECTION 2
HEP248	REFER TO SECTION 2
HEP250	REFER TO SECTION 2
HEP251	REFER TO SECTION 2
HEP252	REFER TO SECTION 2
HEP253	REFER TO SECTION 2
HEP254	REFER TO SECTION 2
HEP623	REFER TO SECTION 2
HEP624	REFER TO SECTION 2
HEP625	REFER TO SECTION 2
HEP626	REFER TO SECTION 2
HEP627	REFER TO SECTION 2
HEP628	REFER TO SECTION 2
HEP629	REFER TO SECTION 2
HEP630	REFER TO SECTION 2
HEP631	REFER TO SECTION 2
HEP632	REFER TO SECTION 2
HEP633	REFER TO SECTION 2
HEP634	REFER TO SECTION 2
HEP635	REFER TO SECTION 2
HEP636	REFER TO SECTION 2
HEP637	REFER TO SECTION 2
HEP638	REFER TO SECTION 2
HEP639	REFER TO SECTION 2
HEP640	REFER TO SECTION 2
HEP641	REFER TO SECTION 2
HEP642	REFER TO SECTION 2
HEP643	REFER TO SECTION 2
HEP644	REFER TO SECTION 2
HEP700	REFER TO SECTION 2
HEP701	REFER TO SECTION 2

TRANSISTOR SUBSTITUTES

To Replace	Substitute This Type
HEP702	REFER TO SECTION 2
HEP703	REFER TO SECTION 2
HEP704	REFER TO SECTION 2
HEP705	REFER TO SECTION 2
HEP706	REFER TO SECTION 2
HEP707	REFER TO SECTION 2
HEP708	REFER TO SECTION 2
HEP709	REFER TO SECTION 2
HEP710	REFER TO SECTION 2
HEP712	REFER TO SECTION 2
HEP713	REFER TO SECTION 2
HEP714	REFER TO SECTION 2
HEP715	REFER TO SECTION 2
HEP716	REFER TO SECTION 2
HEP717	REFER TO SECTION 2
HEP718	REFER TO SECTION 2
HEP719	REFER TO SECTION 2
HEP720	REFER TO SECTION 2
HEP721	REFER TO SECTION 2
HEP722	REFER TO SECTION 2
HEP723	REFER TO SECTION 2
HEP724	REFER TO SECTION 2
HEP725	REFER TO SECTION 2
HEP726	REFER TO SECTION 2
HEP727	REFER TO SECTION 2
HEP728	REFER TO SECTION 2
HEP729	REFER TO SECTION 2
HEP730	REFER TO SECTION 2
HEP731	REFER TO SECTION 2
HEP732	REFER TO SECTION 2
HEP733	REFER TO SECTION 2
HEP734	REFER TO SECTION 2
HEP735	REFER TO SECTION 2
HEP736	REFER TO SECTION 2
HEP737	REFER TO SECTION 2
HEP738	REFER TO SECTION 2
HEP739	REFER TO SECTION 2
HEP740	REFER TO SECTION 2
HEPG0001	REFER TO SECTION 2
HEPG0002	REFER TO SECTION 2
HEPG0003	REFER TO SECTION 2
HEPG0005	REFER TO SECTION 2
HEPG0006	REFER TO SECTION 2
HEPG0007	REFER TO SECTION 2
HEPG0008	REFER TO SECTION 2
HEPG0009	REFER TO SECTION 2
HEPG0011	REFER TO SECTION 2
HEPG6000	REFER TO SECTION 2
HEPG6001	REFER TO SECTION 2
HEPG6002	REFER TO SECTION 2
HEPG6003	REFER TO SECTION 2
HEPG6004	REFER TO SECTION 2
HEPG6005	REFER TO SECTION 2
HEPG6006	REFER TO SECTION 2
HEPG6007	REFER TO SECTION 2
HEPG6008	REFER TO SECTION 2
HEPG6009	REFER TO SECTION 2
HEPG6010	REFER TO SECTION 2
HEPG6011	REFER TO SECTION 2
HEPG6012	REFER TO SECTION 2
HEPG6013	REFER TO SECTION 2

To Replace	Substitute This Type
HEPG6014	REFER TO SECTION 2
HEPG6015	REFER TO SECTION 2
HEPG6016	REFER TO SECTION 2
HEPG6017	REFER TO SECTION 2
HEPG6018	REFER TO SECTION 2
HEPS0001	REFER TO SECTION 2
HEPS0002	REFER TO SECTION 2
HEPS0003	REFER TO SECTION 2
HEPS0004	REFER TO SECTION 2
HEPS0005	REFER TO SECTION 2
HEPS0006	REFER TO SECTION 2
HEPS0007	REFER TO SECTION 2
HEPS0008	REFER TO SECTION 2
HEPS0009	REFER TO SECTION 2
HEPS0010	REFER TO SECTION 2
HEPS0011	REFER TO SECTION 2
HEPS0012	REFER TO SECTION 2
HEPS0013	REFER TO SECTION 2
HEPS0014	REFER TO SECTION 2
HEPS0015	REFER TO SECTION 2
HEPS0016	REFER TO SECTION 2
HEPS0017	REFER TO SECTION 2
HEPS0019	REFER TO SECTION 2
HEPS0020	REFER TO SECTION 2
HEPS0024	REFER TO SECTION 2
HEPS0025	REFER TO SECTION 2
HEPS0026	REFER TO SECTION 2
HEPS0027	REFER TO SECTION 2
HEPS0028	REFER TO SECTION 2
HEPS0029	REFER TO SECTION 2
HEPS0030	REFER TO SECTION 2
HEPS0031	REFER TO SECTION 2
HEPS0032	REFER TO SECTION 2
HEPS0033	REFER TO SECTION 2
HEPS3001	REFER TO SECTION 2
HEPS3002	REFER TO SECTION 2
HEPS3003	REFER TO SECTION 2
HEPS3004	REFER TO SECTION 2
HEPS3005	REFER TO SECTION 2
HEPS3006	REFER TO SECTION 2
HEPS3007	REFER TO SECTION 2
HEPS3008	REFER TO SECTION 2
HEPS3009	REFER TO SECTION 2
HEPS3010	REFER TO SECTION 2
HEPS3011	REFER TO SECTION 2
HEPS3012	REFER TO SECTION 2
HEPS3013	REFER TO SECTION 2
HEPS3014	REFER TO SECTION 2
HEPS3019	REFER TO SECTION 2
HEPS3020	REFER TO SECTION 2
HEPS3021	REFER TO SECTION 2
HEPS3022	REFER TO SECTION 2
HEPS3023	REFER TO SECTION 2
HEPS3024	REFER TO SECTION 2
HEPS3025	REFER TO SECTION 2
HEPS3026	REFER TO SECTION 2
HEPS3027	REFER TO SECTION 2
HEPS3028	REFER TO SECTION 2
HEPS3029	REFER TO SECTION 2
HEPS3030	REFER TO SECTION 2
HEPS3031	REFER TO SECTION 2

TRANSISTOR SUBSTITUTES

To Replace	Substitute This Type
HEPS3032	REFER TO SECTION 2
HEPS3033	REFER TO SECTION 2
HEPS3034	REFER TO SECTION 2
HEPS3035	REFER TO SECTION 2
HEPS3036	REFER TO SECTION 2
HEPS3037	REFER TO SECTION 2
HEPS3038	REFER TO SECTION 2
HEPS3039	REFER TO SECTION 2
HEPS3040	REFER TO SECTION 2
HEPS3041	REFER TO SECTION 2
HEPS5000	REFER TO SECTION 2
HEPS5001	REFER TO SECTION 2
HEPS5002	REFER TO SECTION 2
HEPS5003	REFER TO SECTION 2
HEPS5004	REFER TO SECTION 2
HEPS5005	REFER TO SECTION 2
HEPS5006	REFER TO SECTION 2
HEPS5007	REFER TO SECTION 2
HEPS5008	REFER TO SECTION 2
HEPS5009	REFER TO SECTION 2
HEPS5010	REFER TO SECTION 2
HEPS5011	REFER TO SECTION 2
HEPS5012	REFER TO SECTION 2
HEPS5013	REFER TO SECTION 2
HEPS5014	REFER TO SECTION 2
HEPS5015	REFER TO SECTION 2
HEPS5018	REFER TO SECTION 2
HEPS5019	REFER TO SECTION 2
HEPS5020	REFER TO SECTION 2
HEPS5021	REFER TO SECTION 2
HEPS5022	REFER TO SECTION 2
HEPS5023	REFER TO SECTION 2
HEPS5024	REFER TO SECTION 2
HEPS5025	REFER TO SECTION 2
HEPS5026	REFER TO SECTION 2
HEPS7000	REFER TO SECTION 2
HEPS7001	REFER TO SECTION 2
HEPS7002	REFER TO SECTION 2
HEPS7003	REFER TO SECTION 2
HEPS7004	REFER TO SECTION 2
HEPS7005	REFER TO SECTION 2
HEPS9100	REFER TO SECTION 2
HEPS9101	REFER TO SECTION 2
HEPS9102	REFER TO SECTION 2
HEPS9103	REFER TO SECTION 2
HEPS9120	REFER TO SECTION 2
HEPS9121	REFER TO SECTION 2
HEPS9122	REFER TO SECTION 2
HEPS9123	REFER TO SECTION 2
HEPS9140	REFER TO SECTION 2
HEPS9141	REFER TO SECTION 2
HEPS9142	REFER TO SECTION 2
HEPS9143	REFER TO SECTION 2
HF3H	REFER TO SECTION 2
HF3M	REFER TO SECTION 2
HF6H	REFER TO SECTION 2
HF6M	REFER TO SECTION 2
HF12H	REFER TO SECTION 2
HF12M	REFER TO SECTION 2
HF20H	REFER TO SECTION 2
HF20M	REFER TO SECTION 2

To Replace	Substitute This Type
HF35	REFER TO SECTION 2
HF50	REFER TO SECTION 2
HF75	REFER TO SECTION 2
HJ22D	SEE 2SA12
HJ23D	SEE 2SA15
HJ56	SEE 2SA13
HJ57	SEE 2SA16
HJ60	SEE 2SA17
HJ70	SEE 2SA80
HJ71	SEE 2SA81
HJ72	SEE 2SA82
HJ73	SEE 2SA83
HJ74	SEE 2SA84
HJ75	SEE 2SA85
HN100	REFER TO SECTION 2
HN150	REFER TO SECTION 2
HO300	REFER TO SECTION 2
HST1050	HST1055,*PTC118,*GE-35,*TR-61
HST1051	HST1052,HST1056,HST1057,*PTC118,*GE-36,*TR-61
HST1052	HST1053,HST1057,HST1058,*PTC118,*GE-36,*TR-61
HST1053	HST1054,HST1058,HST1059,*GE-36
HST1054	HST1059,*GE-36
HST1055	HST1050,*PTC118,*GE-35,*TR-61
HST1056	HST1051,HST1052,HST1057,*PTC118,*GE-36,*TR-61
HST1057	HST1052,HST1053,HST1058,*PTC118,*GE-36,*TR-61
HST1058	HST1053,HST1054,HST1059,*GE-36
HST1059	HST1054,*GE-36
HST1060	HST1050,HST1055,*PTC118,*GE-35,*TR-61
HST1061	HST1051,HST1052,HST1056,HST1057,HST1062,*PTC118,*GE-36,*TR-61
HST1062	HST1052,HST1053,HST1057,HST1058,HST1063,*PTC118,*GE-36,*TR-61
HST1063	HST1053,HST1054,HST1058,HST1059,HST1064,*GE-36
HST1064	HST1054,HST1059,*GE-36
HST1150	HST1155,HST1160
HST1151	HST1152,HST1156,HST1157,HST1161,HST1162
HST1152	HST1153,HST1157,HST1158,HST1162,HST1163
HST1153	HST1154,HST1158,HST1159,HST1163,HST1164
HST1154	HST1159,HST1164
HST1155	HST1160
HST1156	HST1157,HST1161,HST1162
HST1157	HST1158,HST1162,HST1163
HST1158	HST1159,HST1163,HST1164
HST1159	HST1164
HST1160	HST1155
HST1161	HST1156,HST1157,HST1162
HST1162	HST1157,HST1158,HST1163
HST1163	HST1158,HST1159,HST1164
HST1164	HST1159
HST1250	HST1255
HST1251	HST1252,HST1256,HST1257
HST1252	HST1253,HST1257,HST1258
HST1253	HST1254,HST1258,HST1259
HST1254	HST1259
HST1255	HST1250
HST1256	HST1251,HST1252,HST1257
HST1257	HST1252,HST1253,HST1258
HST1258	HST1253,HST1254,HST1259
HST1259	HST1254
HST1260	HST1250,HST1255
HST1261	HST1251,HST1252,HST1256,HST1257,HST1262
HST1262	HST1252,HST1253,HST1257,HST1258,HST1263
HST1263	HST1253,HST1254,HST1258,HST1259,HST1264

TRANSISTOR SUBSTITUTES

To Replace	Substitute This Type
HST1264	HST1254,HST1259
HST1808	
HST1809	
HST1810	
HST1860	
HST1861	
HST1862	
HST1908	
HST1909	
HST1910	
HST1960	
HST1961	
HST1962	
HST2008	
HST2009	
HST2010	
HST2101	
HST2110	
HST2111	
HST2112	
HST2150	
HST2151	
HST2152	
HST4451	HST4452,HST5007,HST5008,HST5507,HST5508,*PTC144,*GE-27,*TR-78,*BF338
HST4452	HST5008,HST5508,*PTC144,*GE-27,*TR-78,*BF338
HST4453	HST4454,HST5002,HST5003,HST5502,HST5503,*PTC144,*GE-27,*TR-78,*BF338
HST4454	HST5003,HST5503,*PTC144,*GE-27,*TR-78,*BF338
HST4455	HST4456,HST5012,HST5013,HST5512,HST5513
HST4456	HST5013,HST5513
HST4483	HST4451,HST5006,HST5007,HST5506,HST5507,*PTC144,*GE-27,*TR-78,*BF338
HST4551	HST4552
HST4552	
HST4553	HST4554
HST4554	
HST4555	HST4556
HST4556	
HST4583	HST4551,*GE-27
HST5001	HST5002,*PTC144,*GE-27,*TR-78,*BF338,*276-2016
HST5002	HST5003,*PTC144,*GE-27,*TR-78,*BF338
HST5003	*PTC144,*GE-27,*TR-78,*BF338
HST5004	HST5005,HST5051,*PTC144,*GE-27,*TR-78,*BF338
HST5005	HST5051,HST5052,HST5053,*GE-27,*TR-78,*BF338
HST5006	HST5007,*PTC144,*GE-27,*TR-78,*BF338
HST5007	HST5008,*PTC144,*GE-27,*TR-78,*BF338
HST5008	*PTC144,*GE-27,*TR-78,*BF338
HST5009	HST5010,HST5054,*PTC144,*GE-27,*TR-78,*BF338
HST5010	HST5054,HST5055,HST5056,*GE-27,*TR-78,*BF338
HST5011	HST5012
HST5012	HST5013
HST5013	
HST5014	HST5015
HST5015	
HST5051	HST5005,HST5052,HST5053,*GE-27,*TR-78,*BF338
HST5052	HST5053,*GE-27,*TR-78,*BF338
HST5053	*GE-27,*TR-78,*BF338
HST5054	HST5010,HST5055,HST5056,*GE-27,*TR-78,*BF338
HST5055	HST5056,*GE-27,*TR-78,*BF338
HST5056	*GE-27,*TR-78,*BF338
HST5501	HST5001,HST5002,HST5502,*PTC144,*GE-27,*TR-78,*BF338
HST5502	HST5002,HST5003,HST5503,*PTC144,*GE-27,*TR-78,*BF338
HST5503	HST5003,*PTC144,*GE-27,*TR-78,*BF338

To Replace	Substitute This Type
HST5504	HST5004,HST5005,HST5051,HST5505,HST5551,*PTC144,*GE-27,*TR-78,*BF338
HST5505	HST5005,HST5051,HST5052,HST5053,HST5551,HST5552,HST5553,*GE-27,*TR-78, *BF338
HST5506	HST5006,HST5007,HST5507,*PTC144,*GE-27,*TR-78,*BF338
HST5507	HST5007,HST5008,HST5508,*PTC144,*GE-27,*TR-78,*BF338
HST5508	HST5008,*PTC144,*GE-27,*TR-78,*BF338
HST5509	HST5009,HST5010,HST5054,HST5510,HST5554,*PTC144,*GE-27,*TR-78,*BF338
HST5510	HST5010,HST5054,HST5055,HST5056,HST5554,HST5555,HST5556,*GE-27,*TR-78, *BF338
HST5511	HST5011,HST5012,HST5512,*GE-46
HST5512	HST5012,HST5013,HST5513
HST5513	HST5013
HST5514	HST5014,HST5015,HST5515
HST5515	HST5015
HST5551	HST5005,HST5051,HST5052,HST5053,HST5505,HST5552,HST5553,*GE-27,*TR-78, *BF338
HST5552	HST5052,HST5053,HST5553,*GE-27,*TR-78,*BF338
HST5553	HST5053,*GE-27,*TR-78,*BF338
HST5554	HST5010,HST5054,HST5055,HST5056,HST5510,HST5555,HST5556,*GE-27,*TR-78, *BF338
HST5555	HST5055,HST5056,HST5556,*GE-27,*TR-78,*BF338
HST5556	HST5056,*GE-27,*TR-78,*BF338
HST5901	HST5902
HST5902	HST5903
HST5903	
HST5904	HST5905,HST5951
HST5905	HST5951,HST5952,HST5953
HST5906	HST5907,*GE-216
HST5907	HST5908
HST5908	
HST5909	HST5910,HST5954
HST5910	HST5954,HST5955,HST5956
HST5911	HST5912
HST5912	HST5913
HST5913	
HST5914	HST5915
HST5915	
HST5951	HST5905,HST5952,HST5953
HST5952	HST5953
HST5953	
HST5954	HST5910,HST5955,HST5956
HST5955	HST5956
HST5956	
HST6001	
HST6011	HST6012
HST6012	
HST6013	HST6014
HST6014	2N4115
HST6015	HST6016
HST6016	2N4116
HST6031	HST6011
HST6308	HST6309,HST6408
HST6309	HST6310,HST6409,HST6410
HST6310	HST6409,HST6410
HST6311	HST6312,HST6411,HST6412
HST6312	HST6313,HST6412,HST6413
HST6313	2N4115,HST6413
HST6314	HST6315,HST6414,HST6415
HST6315	HST6316,HST6415,HST6416
HST6316	2N4116,HST6416
HST6408	HST6308,HST6309

To Replace	Substitute This Type
HST6409	HST6310,HST6410
HST6410	HST6310,HST6409
HST6411	HST6311,HST6312,HST6412
HST6412	HST6312,HST6313,HST6413
HST6413	2N4115,HST6313
HST6414	HST6314,HST6315,HST6415
HST6415	HST6315,HST6316,HST6416
HST6416	2N4116,HST6316
HST6901	HST6902
HST6902	HST6903,HST6904
HST6903	HST6904
HST6904	
HST6905	HST6906
HST6906	HST6907,HST6908
HST6907	HST6908
HST6908	
HST7011	HST7012,HST9901,HST9902
HST7012	HST7013,HST9902,HST9903
HST7013	HST9903,HST9904
HST7014	HST7015
HST7015	HST7016
HST7016	HST7140
HST7017	HST7018
HST7018	HST7019
HST7019	
HST7140	HST7154
HST7141	HST7156
HST7150	HST7151
HST7151	HST7152,HST7801
HST7152	HST7801,HST7802,HST7803
HST7154	HST7155
HST7155	HST7141,HST7156
HST7156	HST7141
HST7201	HST7202,HST7203
HST7202	HST7203,HST7204
HST7203	HST7202,HST7204,HST7205
HST7204	HST7205
HST7205	HST7204
HST7206	HST7207
HST7207	HST7201,HST7202,HST7208
HST7208	HST7202,HST7203,HST7204,HST7209
HST7209	HST7203,HST7204,HST7205
HST7401	HST7402,HST7414,HST7415,*PTC144,*GE-27,*TR-78,*BF338,*276-2020
HST7402	HST7403,HST7415,HST7416,*PTC144,*GE-27,*TR-78,*BF338,*276-2020
HST7403	HST7416,*PTC144,*GE-27,*TR-78,*BF338
HST7411	HST7412,*PTC144,*GE-27,*TR-78,*BF338,*276-2020
HST7412	HST7413,*PTC144,*GE-27,*TR-78,*BF338,*276-2020
HST7413	*PTC144,*GE-27,*TR-78,*BF338
HST7414	HST7401,HST7402,HST7415,*PTC144,*GE-27,*TR-78,*BF338,*276-2020
HST7415	HST7402,HST7403,HST7416,*PTC144,*GE-27,*TR-78,*BF338,*276-2020
HST7416	HST7403,*PTC144,*GE-27,*TR-78,*BF338
HST7417	HST7418,*GE-46,*276-2020
HST7418	HST7419,*GE-32,*276-2020
HST7419	*GE-32
HST7601	HST7602
HST7602	HST7603
HST7603	
HST7604	HST7605
HST7605	HST7606
HST7606	
HST7607	HST7608,*PTC116

To Replace	Substitute This Type
HST7608	HST7609,*PTC116
HST7609	
HST7610	HST7611
HST7611	HST7612
HST7612	
HST7801	HST7152,HST7802,HST7803
HST7802	HST7803,HST7804
HST7803	HST7802,HST7804,HST7805
HST7804	HST7805
HST7805	HST7804
HST7806	HST7807
HST7807	HST7152,HST7801,HST7802,HST7808
HST7808	HST7802,HST7803,HST7804,HST7809
HST7809	HST7803,HST7804,HST7805
HST7901	HST7902,HST7903,*GE-32
HST7902	HST7903,HST7904,*GE-32
HST7903	HST7902,HST7904,HST7905,*GE-32
HST7904	HST7905,*GE-32
HST7905	HST7904,*GE-32
HST7907	HST7901,HST7902,HST7908,*GE-32
HST7908	HST7902,HST7903,HST7904,HST7909,*GE-32
HST7909	HST7903,HST7904,HST7905,*GE-32
HST7910	HST7907,*GE-32
HST8002	HST8003,HST8015,HST8016,HST8301,HST8302
HST8003	HST8016,HST8302
HST8012	HST8013
HST8013	
HST8015	HST8002,HST8003,HST8016,HST8301,HST8302
HST8016	HST8003,HST8302
HST8045	
HST8070	HST8071,HST8303,HST8304
HST8071	HST8304
HST8105	HST8106,HST8115,HST8116
HST8106	HST8116
HST8110	HST8111
HST8111	
HST8112	HST8113
HST8113	
HST8114	
HST8115	HST8105,HST8106,HST8116
HST8116	HST8106
HST8301	HST8002,HST8003,HST8015,HST8016,HST8302
HST8302	HST8003,HST8016
HST8303	HST8070,HST8071,HST8304
HST8304	HST8071
HST8601	HST8602
HST8602	HST8603
HST8603	HST8604
HST8604	
HST8651	HST8652,HST8653
HST8652	HST8653,HST8654
HST8653	HST8654,HST8655
HST8654	HST8653,HST8655
HST8655	HST8654
HST8801	HST8802,HST8803
HST8802	HST8803,HST8804
HST8803	HST8804,HST8805
HST8804	HST8803,HST8805
HST8805	HST8804
HST8920	HST8921
HST8921	HST8922

To Replace	Substitute This Type
HST8922	HST8923
HST8923	
HST8951	HST8952,HST8953
HST8952	HST8953,HST8954
HST8953	HST8954,HST8955
HST8954	HST8953,HST8955
HST8955	HST8954
HST9001	HST4483,HST5006,HST5506,HST9002,*PTC144,*GE-27,*TR-78,*BF338
HST9002	HST4451,HST4452,HST5007,HST5008,HST5507,HST5508,HST9003,*PTC144, *GE-27,*TR-78,*BF338
HST9003	HST4452,HST5008,HST5508,*PTC144,*GE-27,*TR-78,*BF338
HST9004	HST5001,HST5501,HST9005,HST9007,HST9008,*PTC144,*GE-27,*TR-78,*BF338
HST9005	HST4453,HST4454,HST5002,HST5003,HST5502,HST5503,HST9006,HST9008, HST9009,*PTC144,*GE-27,*TR-78,*BF338
HST9006	HST4454,HST5003,HST5503,HST9009,*PTC144,*GE-27,*TR-78,*BF338
HST9007	HST5001,HST5501,HST9008,*PTC144,*GE-27,*TR-78,*BF338
HST9008	HST5002,HST5003,HST5502,HST5503,HST9009,*PTC144,*GE-27,*TR-78,*BF338
HST9009	HST5003,HST5503,*PTC144,*GE-27,*TR-78,*BF338
HST9010	HST5011,HST5511,HST9011,*GE-46
HST9011	HST4455,HST4456,HST5012,HST5013,HST5512,HST5513,HST9012
HST9012	HST4456,HST5013,HST5513
HST9201	HST9205,HST9206,*GE-14
HST9202	HST9203,HST9207,HST9208
HST9203	HST9204,HST9208,HST9209
HST9204	HST9209
HST9205	HST9201,HST9206,*GE-14
HST9206	HST9202,HST9207,*GE-14
HST9207	HST9202,HST9203,HST9208
HST9208	HST9203,HST9204,HST9209
HST9209	HST9204
HST9210	HST9201,HST9205,*GE-14
HST9801	HST9201,HST9205,HST9206,HST9802
HST9802	HST9202,HST9206,HST9207,HST9803
HST9803	HST9202,HST9203,HST9207,HST9208,HST9804
HST9804	HST9203,HST9204,HST9208,HST9209
HST9901	HST9902
HST9902	HST9903
HST9903	HST9904
HST9904	
HXTR-6101	2SC987,HXTR-6102
HXTR-6102	2SC987,HXTR-6101
IR401	2SC807,IR423,IR5251,IR5252,IR5253
IR402	
IR403	
IR409	2N6579,2N6580,2N6582,2N6583,IR431
IR410	2SC1227,FT411,IR411,IR515,STS1132,STS1133
IR411	STS1132,STS1133
IR413	2N5240,2N5805,2N6306,2N6307M,2SC807,FT413,FT423,IR516,IR517,MJ413,MJ423, STS1134
IR423	2N6306,2N6307M,2SC807,STS1134
IR424	IR425
IR425	IR424
IR430	2N6306,2N6307M,2N6308M,2SC807,IR423,IR431,STS1134
IR431	2N6308M,2N6579,2N6580,2N6582,2N6583,IR430,STS1134
IR515	SDT14304
IR516	2SC807,IR517,MJ413,MJ423,SDT14305
IR517	2N6574,2SC807,IR518
IR518	2N6574,2N6575,IR519
IR519	2N6575
IR640	IR641
IR641	IR642

To Replace	Substitute This Type
IR642	
IR645	IR646,TIP645,TIP646
IR646	IR647,TIP646,TIP647
IR647	TIP647
IR660	DTS-660
IR663	DTS-663
IR665	DTS-665
IR701	DTS-801
IR710	DTS-710
IR721	
IR801	2N6543
IR802	2N6274,MJ802
IR900	IR645,IR646,IR901,IR2500,IR2501,SE9403,SE9404,TIP645,TIP646
IR901	IR646,IR647,IR2501,SE9404,SE9405,TIP646,TIP647
IR1000	IR640,IR641,IR1001,IR3000,IR3001,SE9303,SE9304
IR1001	IR641,IR642,IR3001,MJ3521,SE9304,SE9305
IR1010	2SD110,2SD110-Y
IR1020	DTS-1010
IR2500	IR645,IR646,IR2501,TIP645,TIP646
IR2501	IR646,IR647,TIP646,TIP647
IR3000	2N6357,2N6358,IR640,IR641,IR3001
IR3001	2N6357,2N6358,IR641,IR642,MJ3521
IR3771	2N5302,2N5685
IR3772	
IR3773	
IR4039	DTS-4039,DTS-4040,DTS-4041,IR4040,IR4041,IR4055
IR4040	IR4055
IR4041	DTS-4039,DTS-4040,DTS-4041,IR4039,IR4040,IR4055
IR4045	
IR4050	IR5002
IR4055	
IR4059	DTS-4059,DTS-4061,IR4061
IR4060	IR5063,IR5064
IR4061	DTS-4059,DTS-4061,IR4059
IR4065	IR5063
IR4502	MJ4502
IR5000	IR4050,IR5001,IR5002,IR5061,IR5062
IR5001	IR4050,IR5002,IR5062
IR5002	IR4050
IR5060	IR5061,IR5062
IR5061	IR5062
IR5062	
IR5063	IR4065
IR5064	
IR5065	IR5066
IR5066	
IR5251	IR4050,IR5000,IR5001,IR5002,IR5061,IR5062,IR5252,IR5253
IR5252	IR4050,IR5001,IR5002,IR5062,IR5253
IR5253	IR4050,IR5002
IR6000	IR4040,IR4055
IR6001	IR4055
IR6002	IR4055
IR6060	IR5060,IR5061,IR5062
IR6061	IR5061,IR5062
IR6062	IR5062
IR6251	IR4050,IR5000,IR5001,IR5002,IR5061,IR5062,IR5251,IR5252,IR5253
IR6252	IR4050,IR5001,IR5002,IR5062,IR5252,IR5253
IR6253	IR4050,IR5002,IR5253
IR6302	2N5631,IR3773
IRTR62	REFER TO SECTION 2
IRTR63	REFER TO SECTION 2

To Replace	Substitute This Type
IRTR64	REFER TO SECTION 2
IRTR65	REFER TO SECTION 2
IRTR66	REFER TO SECTION 2
IRTR67	REFER TO SECTION 2
IRTR68	REFER TO SECTION 2
IRTR69	REFER TO SECTION 2
IRTR70	REFER TO SECTION 2
IRTR71	REFER TO SECTION 2
IRTR72	REFER TO SECTION 2
IRTR73	REFER TO SECTION 2
IRTR74	REFER TO SECTION 2
IRTR75	REFER TO SECTION 2
IRTR76	REFER TO SECTION 2
IRTR77	REFER TO SECTION 2
IRTR78	REFER TO SECTION 2
IRTR79	REFER TO SECTION 2
IRTR80	REFER TO SECTION 2
IRTR81	REFER TO SECTION 2
IRTR82	REFER TO SECTION 2
IRTR83	REFER TO SECTION 2
IRTR84	REFER TO SECTION 2
IRTR85	REFER TO SECTION 2
IRTR86	REFER TO SECTION 2
IRTR87	REFER TO SECTION 2
IRTR88	REFER TO SECTION 2
IRTR91	REFER TO SECTION 2
IRTR92	REFER TO SECTION 2
IRTR93	REFER TO SECTION 2
IRTR94MP	REFER TO SECTION 2
IRTR95	REFER TO SECTION 2
IRTR-50	REFER TO SECTION 2
IRTR-51	REFER TO SECTION 2
IRTR-52	REFER TO SECTION 2
IRTR-53	REFER TO SECTION 2
IRTR-54	REFER TO SECTION 2
IRTR-55	REFER TO SECTION 2
IRTR-56	REFER TO SECTION 2
IRTR-57	REFER TO SECTION 2
IRTR-58	REFER TO SECTION 2
IRTR-59	REFER TO SECTION 2
IRTR-60	REFER TO SECTION 2
IRTR-61	REFER TO SECTION 2
JR5	REFER TO SECTION 2
JR10	REFER TO SECTION 2
JR15	REFER TO SECTION 2
JR30	REFER TO SECTION 2
JR30X	REFER TO SECTION 2
JR100	REFER TO SECTION 2
JR200	REFER TO SECTION 2
KF2000	KF2002
KF2001	
KF2002	KF2000
KF2003	
KL8010	KL8011
KL8011	KL8010
KL8012	KL8013
KL8013	KL8012
KL8503	KL8504
KL8504	
KL8505	KL8506
KL8506	

To Replace	Substitute This Type
KP3446	
KR6003	KR6004
KR6004	
KR6005	KR6006
KR6006	
KR6500	
KR6501	
KR6502	
KR6503	
KS6101	
KS6102	KS6101,KS6105,KS6106
KS6103	KS6101
KS6104	KS6101,KS6102,KS6103
KS6105	
KS6106	KS6105
KS6107	
KS6108	KS6107
KS6109	KS6111,KS6112
KS6110	
KS6111	KS6109,KS6112
KS6112	KS6109,KS6111
KS6113	KS6115
KS6114	KS6113,KS6115,KS6116
KS6115	
KS6116	KS6115
KS6117	KS6120
KS6118	KS6121,KS6123,KS6125
KS6119	KS6118,KS6121,KS6122,KS6124,KS6126
KS6120	
KS6121	KS6125
KS6122	KS6121,KS6126
KS6123	KS6117,KS6120,KS6125
KS6124	KS6118,KS6119,KS6121,KS6122,KS6126
KS6125	KS6120
KS6126	KS6121,KS6122
KSD1051	2N3055A,2N3235,2N5932,1561-0403,1561-0404,1582-0403,1582-0404,1582-0405, 1582-0603,1582-0604,1582-0605,1723-0405,1723-0410,1723-0605,1723-0610, 1743-0610,1743-0630,HST9201,HST9205,HST9206,HST9801,HST9802,KSD1055, KSD1056,KSP1271,MJ2801,SDT9201,SDT9205,SDT9206,SDT9901,SDT9902, *PTC140,*GE-77,*BDY20
KSD1052	2N3055,2N3055SD,2N5671,2N5930,2N5933,2N5936,2SD341,2SD492,1561-1005, 1561-1205,1582-1003,1582-1004,1582-1005,1582-1203,1582-1204,1582-1205, 1723-1005,1723-1010,1723-1205,1723-1210,1743-1010,1743-1030,1743-1220,BD130, HST9202,HST9203,HST9207,HST9208,HST9803,HST9804,KSD1053,KSD1057, KSD1058,KSD3055,KSP1272,SDT9202,SDT9203,SDT9207,SDT9208,SDT9903, SDT9904,*PTC140,*GE-75
KSD1053	2N5671,2N5672,2N5930,2N5933,2N5934,2N5936,2N6496,2SD181,1561-1005, 1561-1205,1561-1404,1561-1405,1582-1003,1582-1004,1582-1005,1582-1203, 1582-1204,1582-1205,1582-1403,1582-1404,1582-1405,1723-1005,1723-1010, 1723-1205,1723-1210,1723-1405,1723-1410,1743-1010,1743-1030,1743-1220, 1743-1410,1743-1430,HST9203,HST9204,HST9208,HST9209,HST9804,KSD1054, KSD1058,KSP1273,SDT9203,SDT9204,SDT9208,SDT9209,SDT9904,*PTC173
KSD1054	2N5672,2N5930,2N5931,2N5934,2N5936,2N5937,2N6496,2SD181,1561-1205, 1561-1404,1561-1405,1561-1604,1561-1605,1582-1203,1582-1204,1582-1205, 1582-1403,1582-1404,1582-1405,1582-1603,1582-1604,1582-1605,1723-1205, 1723-1210,1723-1405,1723-1410,1723-1605,1723-1610,1743-1220,1743-1410, 1743-1430,1743-1620,HST9204,HST9209,KSP1273,SDT9204,SDT9209
KSD1055	2N3055A,2N3235,2N5932,1561-0403,1561-0404,1582-0403,1582-0404,1582-0405, 1582-0603,1582-0604,1582-0605,1723-0405,1723-0410,1723-0605,1723-0610, 1743-0610,1743-0630,HST9201,HST9205,HST9206,HST9801,HST9802,KSD1051, KSD1056,KSP1271,MJ2801,SDT9201,SDT9205,SDT9206,SDT9901,SDT9902,

To Replace	Substitute This Type
(KSD1055)	*PTC140,*GE-77,*BDY20
KSD1056	2N3055,2N3055SD,2N5929,2N5933,2N5935,2SD341,2SD492,1561-0803,1561-0804,1561-0805,1561-1005,1582-0803,1582-0804,1582-0805,1582-1003,1582-1004,1582-1005,1723-0805,1723-0810,1723-1005,1723-1010,1743-0820,1743-1010,1743-1030,BD130,HST9202,HST9206,HST9207,HST9802,HST9803,KSD1052,KSD1057,KSD3055,KSP1271,KSP1272,SDT9202,SDT9206,SDT9207,SDT9902,SDT9903,*PTC140,*GE-77,*BDY20
KSD1057	2N3055,2N3055SD,2N5671,2N5930,2N5933,2N5936,2SD341,2SD492,1561-1005,1561-1205,1582-1003,1582-1004,1582-1005,1582-1203,1582-1204,1582-1205,1723-1005,1723-1010,1723-1205,1723-1210,1743-1010,1743-1030,1743-1220,BD130,HST9202,HST9203,HST9207,HST9208,HST9803,HST9804,KSD1052,KSD1053,KSD1058,KSD3055,KSP1272,SDT9202,SDT9203,SDT9207,SDT9208,SDT9903,SDT9904,*PTC140,*GE-75
KSD1058	2N5671,2N5672,2N5930,2N5933,2N5934,2N5936,2N6496,2SD181,1561-1005,1561-1205,1561-1404,1561-1405,1582-1003,1582-1004,1582-1005,1582-1203,1582-1204,1582-1205,1582-1403,1582-1404,1582-1405,1723-1005,1723-1010,1723-1205,1723-1210,1723-1405,1723-1410,1743-1010,1743-1030,1743-1220,1743-1410,1743-1430,HST9203,HST9204,HST9208,HST9209,HST9804,KSD1053,KSD1054,KSP1273,SDT9203,SDT9204,SDT9208,SDT9209,SDT9904,*PTC173
KSD2101	2SD242,2SD243,2SD257,2SD258,BD149-6,HST5907,HST5908,KSP1021,KSP1022,KSP1024,KSP1025,*PTC112,*GE-68,*TR-57,*276-2020
KSD2102	2SD242,2SD257,BD149-6,HST5907,HST5908,KSD2101,KSP1021,KSP1022,KSP1024,KSP1025,*PTC112,*GE-216,*TR-57,*276-2020
KSD2103	2SD241,2SD256,BD148-6,BD149-6,HST5906,HST5907,KSD2101,KSD2102,KSP1021,KSP1024,*PTC112,*GE-216,*TR-57,*276-2020
KSD2201	2N4348,2N5559,2N6496,2SD110-R,2SD165,2SD181,2SD214,2SD375,1561-1205,1561-1404,1561-1405,1561-1604,1561-1605,1582-1203,1582-1204,1582-1205,1582-1403,1582-1404,1582-1405,1582-1603,1582-1604,1582-1605,1723-1205,1723-1210,1723-1405,1723-1410,1723-1605,1723-1610,HST9204,HST9209,KSD1054,KSD9703,KSD9706,KSP1173,KSP1273,SDT7610,SDT7611,SDT7735,SDT7736,SDT9204,SDT9209,*PTC118,*TR-61
KSD2202	2N4348,2N5559,2N6496,2SD110-R,2SD161,2SD165,2SD181,2SD213,2SD214,2SD319,2SD340,2SD375,1561-1005,1561-1205,1561-1404,1561-1405,1582-1003,1582-1004,1582-1005,1582-1203,1582-1204,1582-1205,1582-1403,1582-1404,1582-1405,1723-1005,1723-1010,1723-1205,1723-1210,1723-1405,1723-1410,HST9203,HST9204,HST9208,HST9209,HST9804,KSD1053,KSD1054,KSD1058,KSD2201,KSD9702,KSD9703,KSD9705,KSD9706,KSP1173,KSP1273,SDT7610,SDT7734,SDT7735,SDT9203,SDT9204,SDT9208,SDT9209,SDT9904,*PTC118,*TR-61
KSD2203	2N3055,2N3055SD,2SD110-R,2SD111-R,2SD161,2SD164,2SD213,2SD214,2SD319,2SD340,2SD341,2SD492,2SD522,1561-1005,1561-1205,1582-1003,1582-1004,1582-1005,1582-1203,1582-1204,1582-1205,1723-1005,1723-1010,1723-1205,1723-1210,40636,BD130,HST9202,HST9203,HST9207,HST9208,HST9803,HST9804,KSD1052,KSD1053,KSD1057,KSD1058,KSD2202,KSD3055,KSD9701,KSD9702,KSD9704,KSD9705,KSP1172,KSP1272,SDT7609,SDT7733,SDT7734,SDT9202,SDT9203,SDT9207,SDT9208,SDT9903,SDT9904,*PTC140,*GE-75,*TR-61
KSD3055	2N3055,2N3055SD,2N5671,2N5930,2N5933,2N5936,2SD232,2SD341,2SD492,1561-1005,1561-1205,1582-1003,1582-1004,1582-1005,1582-1203,1582-1204,1582-1205,1723-1005,1723-1010,1723-1205,1723-1210,1743-1010,1743-1030,1743-1220,BD130,HST9202,HST9203,HST9207,HST9208,HST9803,HST9804,KSD1052,KSD1053,KSD1057,KSD1058,KSP1272,SDT9202,SDT9203,SDT9207,SDT9208,SDT9903,SDT9904,*PTC140,*GE-14,*BDY20
KSD3771	2SD231,2SD249,1582-0408,1582-0410,1582-0415,1582-0608,1582-0610,1582-0615,*PTC175,*GE-75
KSD3772	2SD232,2SD250,1582-1008,1582-1010,1582-1015,1582-1208,1582-1210,1582-1215,*PTC173,*GE-75
KSD3773	2N3240,2SD232A,1582-1408,1582-1608,1582-1610,1582-1615,1582-1808,1582-1810,1582-1815,1582-2008,1582-2010
KSD9701	2N3055,2N3055SD,2N3236,2N5671,2N5930,2N5933,2N5936,2N6254,2SD232,2SD250,2SD341,2SD492,1561-1005,1561-1205,1582-1003,1582-1004,1582-1005,1582-1203,1582-1204,1582-1205,1723-1005,1723-1010,1723-1205,1723-1210,

To Replace	Substitute This Type
(KSD9701)	1743-1010,1743-1030,1743-1220,40636,BD130,HST9202,HST9203,HST9207, HST9208,HST9803,HST9804,KSD9702,KSD9704,KSD9705,KSP1272,MJ3772, SDT9202,SDT9203,SDT9207,SDT9208,SDT9903,SDT9904,*PTC140,*GE-75
KSD9701A	2SD232,2SD250,1582-1008,1582-1010,1582-1015,1582-1208,1582-1210,1582-1215, KSD3772,KSD9702A,*PTC140,*GE-75
KSD9702	2N3055C,2N3772C,2N5671,2N5672,2N5930,2N5933,2N5934,2N5936,2SD232, 2SD232A,1561-1005,1561-1205,1561-1404,1561-1405,1582-1003,1582-1004, 1582-1005,1582-1203,1582-1204,1582-1205,1582-1403,1582-1404,1582-1405, 1723-1005,1723-1010,1723-1205,1723-1210,1723-1405,1723-1410,1743-1010, 1743-1030,1743-1220,1743-1410,1743-1430,HST9203,HST9204,HST9208,HST9209, HST9804,KSD9703,KSD9705,KSD9706,KSP1273,SDT9203,SDT9204,SDT9208, SDT9209,SDT9904,*PTC173
KSD9702A	2N3240,2SD232,2SD232A,1582-1208,1582-1210,1582-1215,1582-1408,1582-1410, 1582-1415,1582-1608,1582-1610,1582-1615,KSD3773,KSD9703A,*PTC173
KSD9703	2N3055C,2N3772C,2N5672,2N5930,2N5931,2N5934,2N5936,2N5937,2SD232A, 1561-1205,1561-1404,1561-1405,1561-1604,1561-1605,1582-1203,1582-1204, 1582-1205,1582-1403,1582-1404,1582-1405,1582-1603,1582-1604,1582-1605, 1723-1205,1723-1210,1723-1405,1723-1410,1723-1605,1723-1610,1743-1220, 1743-1410,1743-1430,1743-1620,HST9204,HST9209,KSD9706,KSP1273,SDT9204, SDT9209
KSD9703A	2N3240,2SD232A,1582-1408,1582-1410,1582-1415,1582-1608,1582-1610,1582-1615, KSD3773
KSD9704	2N3055,2N3055SD,2N3236,2N5671,2N5930,2N5933,2N5936,2N6254,2N6270,2N6271, 2N6272,2N6273,2SD341,2SD492,1743-1010,1743-1030,1743-1220,40636,BD130, BUY53A,BUY54A,HST9202,HST9203,HST9207,HST9208,HST9803,HST9804, KSD9705,KSP1272,SDT9202,SDT9203,SDT9207,SDT9208,SDT9903,SDT9904, *PTC140,*GE-75
KSD9705	2N3055C,2N3772C,2N5671,2N5672,2N5930,2N5933,2N5934,2N5936,2N6271,2N6273, 2N6496,2SD181,1743-1010,1743-1030,1743-1220,1743-1410,1743-1430,HST9203, HST9204,HST9208,HST9209,HST9804,KSD9706,KSP1273,SDT9203,SDT9204, SDT9208,SDT9209,SDT9904,*PTC173
KSD9706	2N3055C,2N3772C,2N5672,2N5930,2N5931,2N5934,2N5936,2N5937,2N6496, 2SD181,1743-1220,1743-1410,1743-1430,1743-1620,HST9204,HST9209,KSP1273, SDT9204,SDT9209
KSD9707	2SD250,1561-0803,1561-0804,1561-0805,1561-1005,1582-0803,1582-0804,1582-0805, 1582-0808,1582-0810,1582-0815,1582-1003,1582-1004,1582-1005,1582-1008, 1582-1010,1582-1015,1723-0805,1723-0810,1723-1005,1723-1010,40636,HST9202, HST9206,HST9207,HST9802,HST9803,KSD9701,MJ3772,SDT9202,SDT9206, SDT9207,SDT9902,SDT9903,*PTC140,*GE-77
KSP1001	KSP1002
KSP1002	KSP1003
KSP1003	
KSP1004	KSP1005
KSP1005	KSP1006
KSP1006	
KSP1021	KSP1022,KSP1161,KSP1162
KSP1022	KSP1162
KSP1023	KSP1163
KSP1024	KSP1025,KSP1164,KSP1165
KSP1025	KSP1165
KSP1026	KSP1166
KSP1031	KSP1032
KSP1032	
KSP1033	
KSP1034	KSP1035
KSP1035	
KSP1036	
KSP1041	KSP1042,KSP1171,KSP1172
KSP1042	KSP1172
KSP1043	KSP1173
KSP1044	KSP1045,KSP1174,KSP1175

To Replace	Substitute This Type
KSP1045	KSP1175
KSP1046	KSP1176
KSP1051	HST5056,HST5556,KSP1052,KSP1053
KSP1052	KSP1053,KSP1054,KSP1055
KSP1053	KSP1052,KSP1054,KSP1055
KSP1054	KSP1053,KSP1055
KSP1055	KSP1054
KSP1071	KSP1072,KSP1073,KSP1121,KSP1122,KSP1123
KSP1072	KSP1073,KSP1074,KSP1075,KSP1122,KSP1123,KSP1124,KSP1125
KSP1073	KSP1072,KSP1074,KSP1075,KSP1122,KSP1123,KSP1124,KSP1125
KSP1074	KSP1073,KSP1075,KSP1123,KSP1124,KSP1125
KSP1075	KSP1074,KSP1124,KSP1125
KSP1091	KSP1092,KSP1093,KSP1141,KSP1142,KSP1143
KSP1092	KSP1093,KSP1094,KSP1095,KSP1142,KSP1143,KSP1144,KSP1145
KSP1093	KSP1092,KSP1094,KSP1095,KSP1142,KSP1143,KSP1144,KSP1145
KSP1094	KSP1093,KSP1095,KSP1143,KSP1144,KSP1145
KSP1095	KSP1094,KSP1144,KSP1145
KSP1101	KSP1102,KSP1103
KSP1102	KSP1103,KSP1104,KSP1105
KSP1103	KSP1102,KSP1104,KSP1105
KSP1104	KSP1103,KSP1105
KSP1105	KSP1104
KSP1121	KSP1122,KSP1123,*276-2008
KSP1122	KSP1123,KSP1124,KSP1125,*276-2008
KSP1123	KSP1122,KSP1124,KSP1125,*276-2008
KSP1124	KSP1123,KSP1125,*276-2008
KSP1125	KSP1124,*276-2012
KSP1141	KSP1142,KSP1143,KSP1221,KSP1222,KSP1223
KSP1142	KSP1143,KSP1144,KSP1145,KSP1222,KSP1223,KSP1224,KSP1225
KSP1143	KSP1142,KSP1144,KSP1145,KSP1222,KSP1223,KSP1224,KSP1225
KSP1144	KSP1143,KSP1145,KSP1223,KSP1224,KSP1225
KSP1145	KSP1144,KSP1224,KSP1225
KSP1151	KSP1152
KSP1152	
KSP1153	
KSP1154	KSP1155
KSP1155	
KSP1156	
KSP1161	KSP1162
KSP1162	
KSP1163	
KSP1164	KSP1165
KSP1165	
KSP1166	
KSP1171	KSP1172,KSP1271,KSP1272
KSP1172	KSP1272
KSP1173	KSP1273
KSP1174	KSP1175,KSP1274,KSP1275
KSP1175	KSP1275
KSP1176	KSP1276
KSP1201	
KSP1202	KSP1203,KSP1204,KSP1205
KSP1203	KSP1202,KSP1204,KSP1205
KSP1204	KSP1203,KSP1205
KSP1205	KSP1204
KSP1221	KSP1222,KSP1223
KSP1222	KSP1223,KSP1224,KSP1225
KSP1223	KSP1222,KSP1224,KSP1225
KSP1224	HST1223,KSP1225
KSP1225	KSP1224
KSP1251	KSP1252

To Replace	Substitute This Type
KSP1252	
KSP1253	
KSP1254	KSP1255
KSP1255	
KSP1256	2N5331
KSP1271	2N6270,2N6272,KSP1272
KSP1272	2N6270,2N6271,2N6272,2N6273
KSP1273	
KSP1274	KSP1275
KSP1275	
KSP1276	
KSP1341	HST5956,KSP1342,KSP1343
KSP1342	KSP1343,KSP1344,KSP1345
KSP1343	KSP1342,KSP1344,KSP1345
KSP1344	KSP1343,KSP1345
KSP1345	KSP1344
KSP1391	KSP1392
KSP1392	
KSP1393	
KSP1394	KSP1395
KSP1395	
KSP1396	
KSP1601	KSP1602,KSP1603
KSP1602	KSP1603,KSP1604
KSP1603	KSP1604,KSP1605
KSP1604	KSP1603,KSP1605
KSP1605	KSP1604
KSP1611	KSP1612,KSP1613
KSP1612	KSP1613,KSP1614
KSP1613	KSP1614,KSP1615
KSP1614	KSP1613,KSP1615
KSP1615	KSP1614
KSP1651	KSP1652
KSP1652	
KSP1653	
KSP1654	KSP1655
KSP1655	
KSP1656	
KSP1671	KSP1672
KSP1672	
KSP1673	
KSP1674	KSP1675
KSP1675	
KSP1676	
KSP2001	KSP2002
KSP2002	KSP2003
KSP2003	
KSP2004	KSP2005
KSP2005	KSP2006
KSP2006	
KSP2021	KSP2022
KSP2022	
KSP2023	
KSP2024	KSP2025
KSP2025	
KSP2026	
KSP2031	KSP2032
KSP2032	
KSP2033	
KSP2034	KSP2035
KSP2035	

TRANSISTOR SUBSTITUTES

To Replace	Substitute This Type
KSP2036	
KSP2041	KSP2042
KSP2042	
KSP2043	
KSP2044	KSP2045
KSP2045	
KSP2046	
KSP2151	KSP2152
KSP2152	
KSP2153	
KSP2154	KSP2155
KSP2155	
KSP2156	
KSP2161	KSP2162
KSP2162	
KSP2163	
KSP2164	KSP2165
KSP2165	
KSP2166	
KSP2171	KSP2172,KSP2271,KSP2272
KSP2172	KSP2272
KSP2173	KSP2273
KSP2174	KSP2175,KSP2274,KSP2275
KSP2175	KSP2275
KSP2176	KSP2276
KSP2251	KSP2252
KSP2252	
KSP2253	
KSP2254	KSP2255
KSP2255	
KSP2256	
KSP2271	KSP2272
KSP2272	
KSP2273	
KSP2274	KSP2275
KSP2275	
KSP2276	
KSP2391	KSP2392
KSP2392	
KSP2393	
KSP2394	KSP2395
KSP2395	
KSP2396	
KSP2601	KSP2602
KSP2602	
KSP2603	
KSP2604	KSP2605
KSP2605	
KSP2606	
KSP2611	KSP2612
KSP2612	
KSP2613	
KSP2614	KSP2615
KSP2615	
KSP2616	
KT218	2N708,2N708A,2N784A,2N914,2N2481,2N2710,2N2845,2N3210,2N3261,2N3301, 2N3510,2N3511,2N3605A,2N3606A,2N3646,2N3647,2N3648,2N4137,2N4140, 2N5029,2N5772,2SC1274,FT3641,FT3642,GET708,GET914,GET2221,GET2369, GET3013,GET3014,GET3646,GI-3641,GI-3642,MPS6532,ZT83,ZT708,*PTC136, *GE-20,*TR-53,*BC107B,*276-2016
KT218F	2N708,2N708A,2N718,2N760,2N784A,2N914,2N2221,2N2242,2N2315,2N2369,

To Replace	Substitute This Type
(KT218F)	2N2369A,2N2396,2N2481,2N2501,2N2539,2N2845,2N2847,2N3013,2N3014,2N3115, 2N3210,2N3211,2N3301,2N3508,2N3641,2N3642,2N3646,2N3647,2N3705,2N3862, 2N3903,2N3946,2N3974,2N3976,2N4013,2N4123,2N4140,2N4227,2N4400,2N4962, 2N5027,2N5029,2N5144,2N5380,2N5450,2N5769,2N5772,2N6427,2N6538,2SC67, 2SC366G-O,2SC367G-O,2SC577,2SC613,2SC639,2SC764,2SC878,2SC913, 2SC914,2SC914A,2SC1166-O,2SC1274,A5T3903,A5T4123,A5T5450,A8T3705, BSW82,D33D24,EN697,FT3641,FT3642,GET2221,GET2369,GI-3641,GI-3642, GI-3705,MM3903,MPS2369,MPS3705,MPS6530,PET3705,PET6001,TIS90-YEL, TIS90M-YEL,TIS92-YEL,TIS92M-YEL,TIS110,TN62,TN-3903,ZT708,ZT2369,*PTC123, *GE-64,*TR-53,*BC107B,*276-2013
KT600	2N1613A,2N2106,2N2107,2N2846,2N2848,2N2868,2N3015,2N3252,2N3299,2N3326, 2N3734,2N4960,2N5262,2SC97,2SC580,2SC879,2SC880,2SD216,7A31,40347, BSX60,KT600F,KT600T,MM3725,ZT1613,*PTC144,*GE-47,*TR-65,*276-2012
KT600F	2N1613A,2N2106,2N2107,2N2193,2N2193A,2N2193B,2N2846,2N2848,2N2868, 2N3015,2N3109,2N3252,2N3299,2N3326,2N3722,2N4960,2N4961,2N5262,2SC97, 2SC97A,2SC580,2SC879,2SC880,2SD216,40347,BC140-10,BC140-6,BSX60,BSY83, KT600T,SE8002,ZT1613,*PTC144,*GE-47,*276-2012
KT600G	2N699A,2N699B,2N1613B,2N1889,2N1974,2N2102,2N2102A,2N2243,2N2243A, 2N2443,2N3036,2N3107,2N3723,2N4000,2N4001,2SC485-Y,2SC512-O,2SC512-R, 2SC516,2SD78,2SD78A,2SD328,2SD378,40348,40366,40367,BC141-10,BC141-6, BSY85,SE7002,ZT91,ZT93,ZT2102,*PTC144,*GE-32,*276-2012
KT600T	2N1613A,2N2106,2N2107,2N2193,2N2193A,2N2193B,2N2846,2N2848,2N2868, 2N3015,2N3109,2N3252,2N3299,2N3326,2N3722,2N4960,2N4961,2N5262,2SC97, 2SC97A,2SC580,2SC879,2SC880,2SD216,40347,BC140-10,BC140-6,BSX60,BSY83, KT600F,SE8002,ZT1613,*PTC144,*GE-47,*276-2012
L10A	L20A,*HEPS7002
L10B	L20B,*HEPS7004
L10C	L20C,*HEPS5020
L10D	L20D,*HEPS5020
L20A	L30A,*HEPS7000
L20B	L30B
L20C	2SC1301,L30C
L20D	2SC1299,L30D
L30A	*HEPS7000
L30B	*HEPS7000
L30C	2SC1301
L30D	2SC1299
LDA400	*GE-210,*TR-64,*276-2009
LDA401	*GE-210,*276-2009
LDA402	*GE-210,*276-2009
LDA403	*276-2009
LDA404	*GE-20,*TR-87,*SK3024,*ECG128,*WEP243,*BC337,*276-2016,*RT-114
LDA405	*GE-20,*TR-87,*SK3024,*ECG128,*WEP243,*BC337,*276-2016,*RT-114
LDA406	*GE-61,*TR-87,*SK3024,*ECG128,*WEP243,*BF173,*276-2038,*RT-114
LDA407	
LDA410	*GE-62,*TR-63,*BC337,*276-2010
LDA420	*GE-64,*TR-63,*BC337,*276-2010
LDA450	*GE-21,*TR-88,*SK3025,*ECG129,*WEP242,*276-2023,*RT-115
LDA451	*GE-48,*276-2023
LDA452	*GE-21,*TR-20,*276-2023
LDA453	*GE-48,*276-2023
LDS200	*TR-87,*SK3024,*ECG128,*WEP243,*276-2038,*RT-114
LDS201	*TR-87,*SK3024,*ECG128,*WEP243,*276-2038,*RT-114
LDS205	*276-2038
LDS206	*GE-10,*276-2009
LDS208	*GE-64,*TR-64,*BC337,*276-2010
M5A	2SD211,HST9201,HST9205,M10A,*PTC129A,*GE-73,*TR-59,*HEPS5000
M5B	2N3233,2N3446,2SD213,2SD214,HST9202,HST9203,HST9207,HST9208,M10B, *PTC129A,*GE-73,*HEPS5000
M5C	2N6510,2SC1229,M10C,*PTC129A,*GE-73,*HEPS5020
M5D	2N5239,2N5240,2N6511,2N6512,2N6514,2N6561,2SC1227,IR515,M10D,XGSR10025,

TRANSISTOR SUBSTITUTES

To Replace	Substitute This Type
(M5D)	*PTC129A,*GE-73,*HEPS5020
M10A	2N3055A,2N3235,2SD211,HST9201,HST9205,*PTC140,*TR-59,*HEPS5004
M10B	2N3236,2SD213,2SD214,HST9202,HST9203,HST9207,HST9208,KSP1272,*PTC118, *HEPS5004
M10C	2SC1229,2SC1322,KSP1221,KSP1222,XDAR10020,XGSR15020,*HEPS5020
M10D	2N6561,2SC1227,IR515,KSP1223,KSP1224,KSP1225,XDAR10025,XDAR10030, XGSR10025,XGSR15025,*PTC118,*HEPS5020
M8073B	*PTC109
M8073C	SF.T307,*PTC135,*TR-05,*SK3123,*AC188/01
M9010	2SC387A,*PTC133,*GE-86,*TR-95,*SK3019,*EN10,*ECG108,*WEP56,*RT-113
MA100	2N1187,2N1925,2N1926,MA882,SF.T243,*TR-05,*SK3051,*ECG100,*WEP254, *RT-118
MA112	2N414B,2N414C,2N654,2N1313,2N1372,2N1383,2N1478,2N1681,2SA206,2SA207, 2SA406,2SA407,2SB326,MA113,MA115,SF.T222,*PTC109,*GE-2,*TR-84, *HEPG0005,*SK3123,*ECG158,*WEP238,*AC188/01,*276-2006
MA113	2N654+,2N1008,2N1356,2N1374,2N1376,2N1381,2N1382,2SB327,MA116,MA117, SF.T223,*PTC109,*GE-2,*TR-84,*HEPG0005,*SK3025,*ECG158,*WEP238, *AC188/01,*276-2006
MA114	2N1316,2N1317,2N1379,2N1999,MA1703,MA1704,*PTC109,*GE-53,*TR-84, *HEPG0005,*ECG158,*WEP238,*AC188/01,*276-2006
MA115	2N654,2N654+,2N1008,2N1356,2N1374,2N1376,2N1381,2N1382,2N1383,2N1681, 2SB327,MA113,MA116,MA117,SF.T223,*PTC109,*GE-2,*TR-84,*HEPG0005, *SK3123,*ECG158,*WEP238,*AC188/01,*276-2006
MA116	2N599,2N655,2N655+,2N1317,2N1354,2N1355,2N1357,MA114,*PTC109,*GE-2, *TR-84,*HEPG0005,*SK3123,*ECG158,*WEP238,*AC188/01,*276-2006
MA117	2N655,2N1354,2N1355,2N1356,2N1357,2N1376,2N1381,2SB327,MA116,SF.T223, *PTC109,*GE-2,*TR-84,*HEPG0005,*SK3123,*ECG158,*WEP238,*AC188/01, *276-2006
MA200	MA201
MA201	MA200
MA202	MA203
MA203	MA202,*SK3016,*ECG116,*WEP158,*RT-215
MA204	
MA205	MA204
MA206	*TR-85,*SK3100,*ECG102A,*WEP250,*RT-121
MA286	2N1942,*PTC109,*GE-2,*TR-05,*HEPG0005,*SK3005,*ECG100,*WEP254,*AC188/01, *276-2006,*RT-118
MA287	2N1376,2N1381,MA116,MA117,*PTC109,*GE-2,*TR-05,*HEPG0005,*SK3005, *ECG100,*WEP254,*AC188/01,*276-2006,*RT-118
MA288	MA1705,*TR-05,*HEPG0005,*SK3005,*ECG100,*WEP254,*AC188/01,*276-2006, *RT-118
MA881	2N1187,2N1924,2N1925,MA100,MA882,SF.T243,*TR-85,*HEPG0005,*ECG102A, *WEP250,*276-2005,*RT-121
MA882	2N1188,2N1926,*TR-85,*HEPG0005,*ECG102A,*WEP250,*276-2005,*RT-121
MA883	*TR-85,*HEPG0006,*ECG102A,*WEP250,*276-2005,*RT-121
MA884	*TR-85,*HEPG0006,*ECG102A,*WEP250,*276-2005,*RT-121
MA885	2N1186,*TR-85,*HEPG0005,*ECG102A,*WEP250,*276-2004,*RT-121
MA886	2N1187,2N1924,2N1925,MA100,MA881,MA882,MA887,SF.T243,*TR-85,*HEPG0005, *ECG102A,*WEP250,*276-2004,*RT-121
MA887	2N1188,2N1926,MA882,*TR-85,*HEPG0005,*ECG102A,*WEP250,*276-2005,*RT-121
MA888	MA883,*TR-85,*HEPG0006,*ECG102A,*WEP250,*276-2005,*RT-121
MA889	MA884,*TR-85,*HEPG0006,*ECG102A,*WEP250,*276-2005,*RT-121
MA909	2N2042,2N2042A,MA910,*TR-85,*HEPG0005,*ECG102A,*WEP250,*276-2004, *RT-121
MA910	2N2042,2N2042A,*TR-85,*HEPG0005,*ECG102A,*WEP250,*276-2004,*RT-121
MA1702	*DS25,*GE-1,*TR-84,*HEPG0007,*SK3005,*ECG158,*WEP238,*276-2005
MA1703	MA1704,*DS26,*GE-2,*TR-84,*HEPG0007,*SK3004,*ECG158,*WEP238,*276-2005
MA1704	MA1705,*DS25,*GE-1,*TR-84,*HEPG0007,*SK3005,*ECG158,*WEP238,*276-2005
MA1705	*DS25,*GE-1,*TR-17,*HEPG0007,*SK3005,*ECG126,*WEP635
MA1706	2N1999,2N2001,MA1703,MA1704,MA1707,*PTC135,*GE-2,*TR-84,*HEPG0007, *SK3004,*ECG158,*WEP238,*AC188/01,*276-2005

To Replace	Substitute This Type
MA1707	MA1704,MA1705,MA1708,*DS25,*GE-1,*TR-84,*HEPG0007,*SK3005,*ECG158, *WEP238,*AC188/01,*276-2005
MA1708	MA1705,*TR-84,*HEPG0007,*ECG158,*WEP238,*AC188/01,*276-2005
MCS2135	*GE-81,*TR-51,*HEPS0005
MCS2136	*TR-51,*HEPS0005
MCS2137	*GE-82,*TR-52,*HEPS0005
MCS2138	*TR-52,*HEPS0030
MD420	2N700,2N2360,2N2361,2N2362,2N2398,2N2399,2N3285,2N3286,*TR-17,*HEPG0003, *ECG160,*WEP637
ME213	2N708A,2N2845,2N3301,2N3641,2N3642,2N3946,2N3973,2N3975,2N4140,2S103, 2SC302,2SC1166-R,BFY39-1,FT3641,FT3642,GET2221,ZT43,*PTC123,*GE-61, *TR-21,*HEPS0019,*SK3122,*ECG123A,*WEP736,*BF173,*276-2016,*RT-102
ME213A	2N708A,2N2845,2N3301,2N3641,2N3642,2N3946,2N3973,2N3975,2N4140,2S103, 2SC302,2SC1166-R,BFY39-1,FT3641,FT3642,GET2221,ZT43,*PTC123,*GE-61, *TR-21,*HEPS0013,*SK3122,*ECG123A,*WEP736,*BF173,*276-2016,*RT-102
ME900	2N3242A,2N3302,2N3643,2N3859A,2N3947,2N5381,2N5825,2N5826,2SC1166-Y, 2SC1175,2SC1991,2SC1999,A5T3904,BC547A,BCY59A,BSX79,FT3643,GET2222, MM3904,TN-3904,*PTC123,*GE-62,*TR-21,*HEPS0013,*SK3122,*ECG123A, *WEP736,*BC337,*276-2014,*RT-102
ME900A	2N3242A,2N3302,2N3643,2N3859A,2N3947,2N5381,2N5825,2N5826,2SC1166-Y, 2SC1175,2SC1991,2SC1999,A5T3904,BC547A,BCY59A,BSX79,FT3643,GET2222, MM3904,TN-3904,*PTC123,*GE-62,*TR-21,*HEPS0013,*SK3122,*ECG123A, *WEP736,*BC337,*276-2014,*RT-102
ME901	2N5827,2N6428,2N6428A,2N6539,BC547B,BC550B,BCY59C,BCY59D,*GE-62,*TR-21, *HEPS0019,*SK3122,*ECG123A,*WEP736,*BC337,*276-2014,*RT-102
ME901A	2N5827,2N6428,2N6428A,2N6539,BC547B,BC550B,BCY59C,BCY59D,*GE-62,*TR-21, *HEPS0019,*SK3122,*ECG123A,*WEP736,*BC337,*276-2014,*RT-102
MF3304	*GE-21,*TR-20,*HEPS0013,*SK3114,*ECG159,*WEP242,*276-2025,*RT-115
MHT1802	*GE-4,*TR-03,*HEPG6006,*SK3012,*PT501,*ECG105,*WEP233,*276-2006
MHT1803	*GE-4,*TR-03,*HEPG6006,*SK3012,*PT501,*ECG105,*WEP233,*276-2006
MHT1807	MP504,MP504A,MP505,MP505A,*GE-4,*TR-03,*HEPG6006,*SK3012,*PT501, *ECG105,*WEP233,*276-2006
MHT1808-1810	SEE SDT1808-1810
MHT1902	
MHT1903	
MHT1904	
MHT1908-1910	SEE SDT1908-1910
MHT2002	*TR-08,*HEPG0011,*SK3011,*ECG101,*WEP641,*RT-119
MHT2003	*TR-08,*HEPG0011,*SK3011,*ECG101,*WEP641,*276-2002,*RT-119
MHT2004	*TR-08,*HEPG0011,*SK3011,*ECG101,*WEP641,*RT-119
MHT2008-2010	SEE SDT2008-2010
MHT2101	
MHT2110	
MHT2111	
MHT2112	
MHT2150	
MHT2151	
MHT2152	
MHT2205	
MHT2305	*DS-501,*GE-4,*TR-03,*HEPG6002,*SK3012,*PT501,*ECG105,*WEP233
MHT4401	2N656,2N1613A,2N2846,2N2848,2N3015,2N3109,2N3299,2N3326,2N3554,2N4960, 2N4961,2SC32,2SC594-R,BSY92,SE8002,*PTC125,*GE-18,*TR-87,*HEPS3011, *SK3024,*ECG128,*WEP243,*BF338,*276-2012,*RT-114
MHT4402	2N699A,2N699B,2N1342,2N1613B,2N2443,2N3114,2N3500,2N3712,2N3923,2N4925, 2SC49,2SC470-4,2SC470-5,2SC590,2SC686,2SC1217,MM3008,SE7001,SE7002, *PTC117,*GE-27,*TR-78,*HEPS3019,*BF338,*276-2012
MHT4411	2N2194,2N2194A,2N2194B,2N2217,2N3110,2N3252,2N3253,2N3444,2N3735,2N5188, 2SC97,2SC109A-R,2SC503-O,2SC580,BC140-6,BSX49,BSX60,*PTC125,*GE-47, *TR-87,*HEPS0015,*SK3024,*ECG128,*WEP243,*276-2008,*RT-114
MHT4412	2N2193,2N2193A,2N2193B,2N2218,2N2218A,2N2270,2N2537,2N2848,2N2868, 2N3109,2N3725A,2N4960,2N4961,2SC97A,2SC109A-O,2SC503-Y,40635,BC140-10,

To Replace	Substitute This Type
(MHT4412)	BSY53,BSY83,TN53,ZT90,ZT95,ZT1613,ZT2270,*PTC125,*GE-47,*TR-87,*HEPS0015,*SK3024,*ECG128,*WEP243,*276-2008,*RT-114
MHT4413	2N1420A,2N1711A,2N2192,2N2192A,2N2192B,2N2219,2N2219A,2N2538,2N3300,2SC109A-Y,2SC503-GR,2SC1384,BC140-16,BSY54,BSY84,ZT1711,*PTC125,*GE-47,*TR-87,*HEPS0015,*SK3024,*ECG128,*WEP243,*276-2008,*RT-114
MHT4414	2N3108,2N3110,2N3253,2N3444,2N3735,2SC108A-R,2SC512-R,2SC516,2SC1065,2SC1069,BC140-6,BC141-6,*PTC125,*GE-32,*HEPS3011,*276-2008
MHT4415	2N1889,2N2193,2N2193A,2N2193B,2N2218A,2N3107,2N3109,2N3725A,2N4961,2SC97A,2SC108A-O,2SC512-O,40635,BC140-10,BC141-10,BSY53,BSY83,TN53,ZT1613,*PTC125,*GE-32,*HEPS3011,*276-2008
MHT4416	2N1711A,2N1890,2N2219A,2SD438,BC140-16,BC141-16,BSY54,BSY84,ZT1711,*PTC125,*GE-32,*HEPS3011,*276-2008
MHT4417	2SC510-R,2SC516A,*PTC125,*GE-32,*HEPS3019,*276-2008
MHT4418	2N699A,2N699B,2N2102,2N2102A,2N2243,2N2243A,2N2405,2N2443,2N3020,2N3036,2SC510-O,BSY85,ZT91,ZT92,ZT93,ZT2102,*PTC125,*GE-32,*HEPS3019,*276-2008
MHT4419	2N1711B,2N3019,2N4943,BSY86,*PTC125,*GE-32,*HEPS3019,*276-2008
MHT4451-4483	SEE SDT4451-4483
MHT4501	HST4551,HST4583,*HEPS3019
MHT4502	*HEPS3019
MHT4503	HST4555
MHT4514	HST4551,HST4552,*HEPS3019
MHT4515	HST4553,HST4554,*HEPS3019
MHT4516	HST4555,HST4556,*HEPS3019
MHT4517	*HEPS3019
MHT4518	*HEPS3019
MHT4519	*HEPS3019
MHT4551-4583	SEE SDT4551-4583
MHT4611	*HEPS3002
MHT4612	*HEPS3002
MHT4613	*HEPS3002
MHT4614	*HEPS3002
MHT4615	*HEPS3002
MHT4616	*HEPS3019
MHT4617	*HEPS3002
MHT4618	*HEPS3002
MHT4619	*HEPS3019
MHT5001-5015	SEE SDT5001-5015
MHT5051-5056	SEE SDT5051-5056
MHT5501-5515	SEE SDT5501-5515
MHT5551-5556	SEE SDT5551-5556
MHT5901-5915	SEE SDT5901-5915
MHT5951-5956	SEE SDT5951-5956
MHT6001-6031	SEE SDT6001-6031
MHT6308-6316	SEE SDT6308-6316
MHT6408-6416	SEE SDT6408-6416
MHT7011-7019	SEE SDT7011-7019
MHT7201-7209	SEE SDT7201-7209
MHT7401-7419	SEE SDT7401-7419
MHT7511	*HEPS5004
MHT7512	*HEPS5004
MHT7513	*HEPS5004
MHT7514	*HEPS5004
MHT7515	*HEPS5004
MHT7516	*HEPS5004
MHT7517	*HEPS5004
MHT7518	*HEPS5004
MHT7519	*HEPS5004
MHT7601-7612	SEE SDT7601-7612
MHT7801-7809	SEE SDT7801-7809
MHT7901-7910	SEE SDT7901-7910

To Replace	Substitute This Type
MHT8002-8071	SEE SDT8002-8071
MHT8301-8304	SEE SDT8301-8304
MHT9001-9012	SEE SDT9001-9012
MJ400	*PTC104,*GE-32,*TR-81,*HEPS5011,*SK3021,*ECG124,*WEP240,*RT-128
MJ413	2N6573,2SC807,IR516,IR517,MJ423,SDT14305,*HEPS5020,*ECG283
MJ420	A5T5058,A5T5059,MJ421,*PTC117,*GE-27,*TR-78,*HEPS5024,*SK3045,*ECG154, *WEP712,*BF338,*276-2012,*RT-110
MJ421	*PTC117,*GE-32,*TR-78,*HEPS5024,*SK3045,*ECG154,*WEP712,*276-2008, *RT-110
MJ423	2SC807,IR516,IR517,MJ413,SDT14305,*HEPS5020,*ECG283
MJ431	2N6560,2N6573,2N6579,2N6580,2N6582,2N6583,2SC1579,FT430,FT431,MJ413, SDT13303,SDT13304,SDT13305,*HEPS5020,*ECG283
MJ432	*GE-73,*TR-61
MJ440	*PTC143,*GE-83,*HEPS5014,*276-2017
MJ450	*PTC149,*GE-74,*HEPS7001,*ECG180,*WEPS7001,*RT-148
MJ480	*PTC140,*GE-19,*TR-59,*HEPS7002,*SK3027,*HN100,*ECG130,*WEP247,*RT-131
MJ481	*PTC140,*GE-19,*TR-59,*HEPS7002,*SK3027,*HN100,*ECG130,*WEP247,*RT-131
MJ490	*PTC149,*HEPS7003,*ECG180,*WEPS7001,*RT-148
MJ491	*PTC149,*HEPS7003,*ECG180,*WEPS7001,*RT-148
MJ500	2N6182,2N6186,MJ501,MJ6700,MJ6701
MJ501	2N6182,2N6184,2N6186,2N6188,MJ6701
MJ802	2N6274,IR802,*PTC173,*GE-75,*TR-36,*HEPS7000,*ECG181,*WEPS7000,*RT-149
MJ900	
MJ901	
MJ920	MJ921
MJ921	
MJ1000	
MJ1001	
MJ1200	MJ1201
MJ1201	
MJ1800	*PTC140,*GE-35,*TR-67,*HEPS5021,*ECG162,*WEP707
MJ2249	2N5598,MJ2250,*PTC112,*GE-66,*TR-81,*HEPS5012,*SK3131,*ECG175,*WEP241
MJ2250	2N5598,2N5602,*PTC104,*GE-23,*TR-81,*HEPS5012,*SK3131,*ECG175,*WEP241
MJ2251	2N5660,HST5953,HST6908,*PTC104,*GE-32,*TR-81,*HEPS5011,*SK3021,*ECG124, *WEP240,*RT-128
MJ2252	*PTC104,*GE-32,*TR-81,*HEPS5011,*SK3021,*ECG124,*WEP240,*RT-128
MJ2253	2N3741A,KSP2022,*PTC113,*GE-69,*TR-58,*HEPS5018,*ECG218,*WEPS5007, *276-2025
MJ2254	KSP2023,*PTC113,*TR-58,*HEPS5018,*ECG218,*WEPS5007,*276-2025
MJ2267	MJ2268,*PTC149,*GE-74,*HEPS7003,*ECG180,*WEPS7001,*RT-148
MJ2268	*PTC149,*GE-74,*HEPS7003,*ECG180,*WEPS7001,*RT-148
MJ2500	
MJ2501	
MJ2801	2N3055A,2N3235,1561-0403,1561-0404,1582-0403,1582-0404,1582-0405,1582-0603, 1582-0604,1582-0605,HST9201,HST9205,*PTC140,*GE-77,*TR-59,*HEPS7002, *SK3027,*ECG130,*WEP247,*BDY20,*RT-131
MJ2840	2N3713,2N5303,2N5877,2N5878,2N5881,2N5882,2SD41,MJ2841,*PTC173,*GE-75, *TR-59,*HEPS7002,*SK3027,*ECG130,*WEP247,*RT-131
MJ2841	2N3236,2N3713,2N3714,2N3772,2N5303,2N5629,2N5632,2N5878,2N5882,2N6254, 2SD41,2SD151,IR3772,RCS258,SDT9704,*PTC173,*GE-75,*TR-59,*HEPS7002, *SK3027,*ECG130,*WEP247,*RT-131
MJ2901	*PTC149,*GE-74,*HEPS7003,*ECG180,*WEPS7001,*RT-148
MJ2940	MJ2941,*PTC149,*GE-74,*HEPS7003
MJ2941	*PTC172,*GE-74,*HEPS5005
MJ3000	
MJ3001	
MJ3029	*GE-73,*TR-67,*HEPS5020,*ECG162,*WEP707
MJ3030	IR423,IR515,IR516,MJ413,MJ423,SDT14304,SDT14305,STS1133,STS1134,*GE-73, *TR-67,*HEPS5021,*SK3111,*ECG163,*WEP740
MJ3040	2SC1576,IR5251,IR5252,IR5253
MJ3041	FT359,MJ3042

To Replace	Substitute This Type
MJ3042	
MJ3101	MJ2249,*PTC112,*GE-66,*TR-81,*HEPS5012,*SK3131,*ECG175,*WEP241
MJ3201	*PTC104,*GE-32,*TR-81,*HEPS5011,*SK3021,*ECG124,*WEP240,*RT-128
MJ3202	2SC515A,*PTC104,*GE-32,*TR-81,*HEPS5011,*SK3021,*ECG124,*WEP240,*RT-128
MJ3260	2N6544,2N6545,2N6622,*PTC129A
MJ3520	*ECG245
MJ3521	*ECG245
MJ3701	2N3740A,2N4899,2N5955,2N5956,KSP2021,*PTC113,*GE-69,*TR-58,*HEPS5018, *SK3083,*ECG218,*WEPS5007,*276-2025
MJ3771	2N5302,2N5685,IR3771,*PTC175,*GE-75,*HEPS7000
MJ3772	108T2,IR3772,*PTC173,*GE-75,*HEPS7000
MJ4000	
MJ4001	
MJ4010	
MJ4011	
MJ4030	
MJ4031	
MJ4032	
MJ4033	
MJ4034	
MJ4035	
MJ4200	MJ4201
MJ4201	
MJ4210	MJ4211
MJ4211	
MJ4502	IR4502,*PTC174,*GE-74,*HEPS7001,*ECG180,*WEPS7001,*RT-148
MJ6700	2N6182,2N6186,MJ500,MJ501,MJ6701
MJ6701	2N6182,2N6184,2N6186,2N6188,MJ501
MJ7000	
MJ7200	MJ7201
MJ7201	
MJ8100	2N6190,MJ8101,*GE-29,*HEPS3003
MJ8101	2N6190,MJ8100,*HEPS3003
MJ9000	
MJE105	2N5974,2N5980,2SB578,2SB579,MJE2901,*PTC164,*GE-56,*HEPS5005,*SK3189, *ECG183,*WEPS5005,*RT-151
MJE170	MJE171,*HEPS5006,*276-2025
MJE171	MJE172,*HEPS5006
MJE172	*HEPS5006
MJE180	MJE181,*HEPS5000,*276-2017
MJE181	MJE182,*GE-57,*TR-60,*HEPS5000,*SK3054,*ECG182,*WEPS5004
MJE182	*HEPS5000,*ECG182
MJE200	*HEPS5000,*276-2017
MJE205	2N5977,*PTC165,*GE-55,*HEPS5004,*ECG182,*WEPS5004,*276-2019,*RT-150
MJE210	*HEPS5006,*276-2025
MJE220	MJE221,MJE223,MJE224
MJE221	MJE220,MJE223,MJE224
MJE222	MJE225
MJE223	MJE224
MJE224	MJE223
MJE225	
MJE230	MJE231,MJE233,MJE234
MJE231	MJE230,MJE233,MJE234
MJE232	MJE235
MJE233	MJE234
MJE234	MJE233
MJE235	
MJE240	*HEPS5000
MJE241	*HEPS5000
MJE242	MJE223,MJE224,*HEPS5000
MJE243	

To Replace	Substitute This Type
MJE244	
MJE250	*HEPS5006
MJE251	*HEPS5006
MJE252	MJE233,MJE234,*HEPS5006
MJE253	
MJE254	
MJE340	2N5655,2N5656,2N5657,*PTC124,*TR-60,*HEPS5015,*SK3104,*ECG157,*WEP244, *RT-135
MJE341	MJE344,*PTC124,*GE-32,*TR-60,*HEPS5015,*ECG157,*WEP244,*RT-135
MJE344	*PTC124,*GE-66,*TR-60,*HEPS5015,*ECG157,*WEP244,*RT-135
MJE350	
MJE370	2SB572,2SB575,MJE371,*PTC162,*GE-58,*TR-77,*HEPS5006,*SK3083,*ECG185, *WEPS5007,*276-2026,*RT-153
MJE371	2SB575,2SB576,*PTC162,*GE-58,*TR-76,*HEPS5006,*SK3083,*ECG241, *WEPS5007,*276-2027,*RT-153
MJE520	2SD488,*PTC163,*GE-57,*TR-76,*HEPS5000,*SK3054,*ECG184,*WEPS5003, *276-2018,*RT-152
MJE521	2SD475,*PTC163,*GE-57,*TR-76,*HEPS5000,*SK3190,*ECG184,*WEPS5003, *276-2019,*RT-152
MJE700	MJE701,MJE702,MJE703,*HEPS9121,*ECG254
MJE701	MJE700,MJE702,MJE703,*HEPS9121,*ECG254
MJE702	MJE703,*HEPS9121,*ECG254
MJE703	MJE702,*HEPS9121,*ECG254
MJE800	MJE801,MJE802,MJE803,*HEPS9101,*SK3180,*ECG253,*276-2020
MJE801	MJE800,MJE802,MJE803,*HEPS9101,*SK3180,*ECG253,*276-2020
MJE802	MJE803,*HEPS9101,*SK3180,*ECG253
MJE803	MJE802,*HEPS9101,*SK3180,*ECG253
MJE1090	MJE1091,MJE1092,MJE1093,MJE2090,MJE2091,MJE2092,MJE2093,*HEPS9122, *SK3181,*ECG258
MJE1091	MJE1090,MJE1092,MJE1093,MJE2090,MJE2091,MJE2092,MJE2093,*HEPS9122, *ECG258
MJE1092	MJE1093,MJE2092,MJE2093,*HEPS9122,*ECG258
MJE1093	MJE1092,MJE2092,MJE2093,*HEPS9122,*ECG258
MJE1100	MJE1101,MJE1102,MJE1103,MJE2100,MJE2101,MJE2102,MJE2103,*HEPS9102, *SK3180,*ECG257
MJE1101	MJE1100,MJE1102,MJE1103,MJE2100,MJE2101,MJE2102,MJE2103,*HEPS9102, *SK3180,*ECG257
MJE1102	MJE1103,MJE2102,MJE2103,*HEPS9102,*SK3180,*ECG257
MJE1103	MJE1102,MJE2102,MJE2103,*HEPS9102,*SK3180,*ECG257
MJE1290	MJE1291,*PTC164,*HEPS5005
MJE1291	*PTC164,*HEPS5005
MJE1660	MJE1661,*PTC165,*HEPS5004
MJE1661	*PTC165,*HEPS5004
MJE2010	MJE2011,*PTC166,*HEPS5005
MJE2011	*PTC166,*HEPS5005
MJE2020	MJE2021,*PTC167,*HEPS5004,*276-2019
MJE2021	*PTC167,*HEPS5004,*276-2019
MJE2090	MJE1090,MJE1091,MJE1092,MJE1093,MJE2091,MJE2092,MJE2093,*SK3181
MJE2091	MJE1090,MJE1091,MJE1092,MJE1093,MJE2090,MJE2092,MJE2093
MJE2092	MJE1092,MJE1093,MJE2093
MJE2093	MJE1092,MJE1093,MJE2092
MJE2100	MJE1100,MJE1101,MJE1102,MJE1103,MJE2101,MJE2102,MJE2103,*SK3180
MJE2101	MJE1100,MJE1101,MJE1102,MJE1103,MJE2100,MJE2102,MJE2103,*SK3180
MJE2102	MJE1102,MJE1103,MJE2103,*SK3180
MJE2103	MJE1102,MJE1103,MJE2102,*SK3180
MJE2160	
MJE2360	MJE2361,*SK3220,*ECG198
MJE2361	
MJE2370	MJE2371,*PTC166,*HEPS5005,*ECG242,*276-2027
MJE2371	*PTC166,*HEPS5005,*ECG242,*276-2027
MJE2480	MJE2020,MJE2021,MJE2481,MJE2482,MJE2483,*GE-69,*HEPS5004,*SK3188,

To Replace	Substitute This Type
(MJE2480)	*ECG241,*276-2019
MJE2481	MJE2021,MJE2483,*GE-66,*HEPS5004,*SK3190,*ECG241,*276-2020
MJE2482	MJE2020,MJE2021,MJE2480,MJE2481,MJE2483,*GE-66,*HEPS5004,*SK3190, *ECG241,*276-2019
MJE2483	MJE2021,MJE2481,*GE-66,*HEPS5004,*SK3190,*ECG241,*276-2020
MJE2490	MJE2010,MJE2011,MJE2491,*PTC166,*HEPS5005,*ECG242,*276-2027
MJE2491	MJE2011,*PTC166,*HEPS5005,*ECG242
MJE2520	MJE2521,*PTC167,*GE-66,*HEPS5004,*SK3054,*ECG241,*276-2019
MJE2521	*PTC167,*GE-66,*HEPS5004,*SK3054,*ECG241,*276-2020
MJE2522	MJE2020,MJE2021,MJE2523,*PTC167,*GE-66,*HEPS5004,*SK3054,*ECG241, *276-2019
MJE2523	MJE2021,*PTC167,*GE-66,*HEPS5004,*SK3054,*ECG241,*276-2020
MJE2801	2N5989,2N5990,*PTC165,*GE-55,*HEPS5004,*ECG182,*WEPS5004,*276-2020, *RT-150
MJE2901	2N5986,2N5987,2SB578,MJE1291,*PTC164,*GE-56,*HEPS5005,*SK3189,*ECG183, *WEPS5005,*RT-151
MJE2955	2N5987,2N5988,*PTC164,*GE-56,*HEPS5005,*SK3189,*ECG183,*WEPS5005, *276-2027,*RT-151
MJE3054	*PTC167,*GE-66,*HEPS5004,*SK3054,*ECG241,*276-2020
MJE3055	2N5990,2N5991,*PTC165,*GE-55,*HEPS5004,*ECG182,*WEPS5004,*276-2019, *RT-150
MJE3055F	FT3055
MJE3738	
MJE3739	MJE3738
MJE3740	MJE3741,*PTC166,*HEPS5005,*ECG242
MJE3741	*PTC166,*HEPS5005,*ECG242
MM380	*TR-17,*ECG160,*WEP637
MM404	2SB186,2SB187,2SB188,MM404A,*PTC135
MM404A	*PTC145
MM709	*PTC136,*GE-17,*TR-95,*HEPS0016,*SK3039,*ECG108,*WEP56,*276-2009,*RT-113
MM1139	*GE-51,*TR-17,*HEPG0003,*ECG160,*WEP637
MM1500	
MM1501	
MM1549	2N5635,2N5636,2N5918†,2N6203,3TX632,40940†
MM1550	2N5636,2N5919,2N5919A†,2N6203
MM1551	2N5637,2N6105†,2N6204
MM1557	2N5641,2N5918†
MM1558	2N5642,2N6199
MM1559	2N5643,2N6200,MSA8503
MM1601	SEE 2N5589
MM1602	SEE 2N5590
MM1603	2N5591,2N5705,2N6082,2N6083,2N6136,MSA8506,ZT5591,*HEPS3007
MM1619	2N5690,2SC1495,ZT5591,*GE-66
MM1620	2SC1495
MM1748	2N709,2N709A,2N2784,2SC718,2SC1231,*PTC133,*276-2038
MM1803	*GE-17,*TR-95,*SK3039,*EN10,*ECG108,*WEP56,*RT-113
MM1812	*GE-27,*TR-78,*BF338
MM1893	MM3019,*PTC144,*276-2008
MM1941	2N2369,2N2369A,2N2710,2N4137,2N5029,2N5769,2SC577,MPS2369,ZT2369, *PTC133,*GE-20,*TR-95,*HEPS0016,*SK3039,*SN80,*ECG108,*WEP56,*276-2038, *RT-113
MM1943	2N2369,2N2369A,2N2710,2N3511,2N3648,2N4137,2N5029,2N5769,2SC577, MPS2369,ZT2369,*PTC133,*TR-87,*HEPS0025,*SK3024,*ECG128,*WEP243, *RT-114
MM2193A	*PTC144,*GE-63
MM2258	2N2405,*GE-32,*HEPS3019
MM2259	MM2260,*GE-32,*HEPS3019
MM2260	*GE-32,*TR-78,*HEPS3019,*ECG154,*WEP712
MM2264	*GE-46,*HEPS3001
MM2270	*PTC144,*GE-63
MM2483	

To Replace	Substitute This Type
MM2484	
MM2894	2N869A,2N3546,2N3576,2N5056,2N5057,2N5771,A5T4260,A5T4261,*GE-21,*TR-17, *HEPS0013,*SP70,*ECG160,*WEP637
MM2894A	
MM3000	*PTC144,*TR-78,*HEPS5026,*SK3045,*ECG154,*WEP712,*RT-110
MM3001	MM3002,*GE-32,*HEPS5026
MM3002	MM3003,*GE-27,*TR-78,*HEPS5025,*SK3045,*ECG154,*WEP712,*RT-110
MM3003	*GE-27,*TR-78
MM3005	2SC756,MM3006,*TR-78,*HEPS5014,*SK3045,*ECG154,*WEP712,*276-2008, *RT-110
MM3006	2SC756,MM3007,*HEPS5014,*276-2008
MM3007	*HEPS5014,*276-2008
MM3008	2N1613B,2N2102,2N2102A,ZT91,ZT93,ZT2102,*GE-32,*276-2008
MM3009	*GE-32,*TR-78,*SK3045,*ECG154,*WEP712,*276-2008,*RT-110
MM3019	
MM3020	MM3019
MM3053	2N5414,*PTC144,*HEPS3011,*276-2030
MM3724	MM3725,*GE-215,*TR-65,*276-2038
MM3725	*GE-215
MM3726	*GE-21,*TR-20,*SK3114,*ECG159,*WEP242,*RT-115
MM3736	2N3737,2N5188,2N5189
MM3737	
MM3903	2N4014,2N4962,2N4963,2N5380,A5T3903,TIS135,TIS136,TN-3903,*PTC136,*GE-210, *HEPS0025,*276-2009
MM3904	2N2222A,2N5381,2N5582,A5T3904,TIS111,TN-3904,*PTC136,*GE-20,*HEPS0025, *276-2009
MM3905	2N2906,2N2906A,2N3250,2N3250A,2N3485,2N3485A,2N5382,A5T3905,GET2904, GET2905,TN-3905,*PTC127,*GE-21,*TR-20,*HEPS0013,*SK3114,*ECG159, *WEP242,*276-2023,*RT-115
MM3906	2N3251,2N3251A,2N5383,2N6005,A5T3906,TN-3906,*PTC127,*GE-48,*TR-20, *HEPS0013,*SK3114,*ECG159,*WEP242,*276-2023,*RT-115
MM4000	A5T3497,*PTC141
MM4001	MM4002,MPSA93
MM4002	MM4003,MPSA93
MM4003	MPSA92
MM4018	*276-2025
MM4019	*GE-29,*TR-88,*SK3025,*ECG129,*WEP242,*276-2025,*RT-115
MM4020	*GE-69,*276-2026
MM4021	
MM4022	
MM4023	
MM4048	2N3251,2N3251A,2N3798,2N3962,2N3964,2N3965,2N4355,2N5373,2N5374,2N6007, 2SA661-GR,2SA889,BC327,BC557B,BC560B,BCY79,BCY79B,BCY79C,TZ552, TZ553,*GE-65,*TR-20,*SK3114,*ECG159,*WEP242,*BC327,*RT-115
MM4052	*GE-82,*BC327
MM4261H	2N4261
MM4429	MM4430,*GE-28,*TR-76
MM4430	2SC1334,*GE-28,*TR-76
MM4545	
MM4546	
MM4547	
MM4645	
MM4646	
MM4647	
MM5000	2N3783,2N3784,MM5001,MM5002,*TR-17,*HEPG0003,*ECG160,*WEP637
MM5001	2N3783,2N3784,MM5000,MM5002,*TR-17,*HEPG0003,*ECG160,*WEP637
MM5002	2N3783,2N3784,MM5000,MM5001,*TR-17,*HEPG0003,*ECG160,*WEP637
MM8000	MM8001,MM8002
MM8001	MM8000,MM8002
MM8002	MM8000,MM8001,*ECG278
MM8003	*HEPS3005,*276-2009

To Replace	Substitute This Type
MM8006	2SC251,2SC251A,2SC252,2SC253,A406,MM8007,MT1061,*PTC132,*GE-11
MM8007	2SC251,2SC251A,2SC252,2SC253,A406,MM8006,MT1061,*PTC132,*GE-11
MM8008	
MM8009	
MM8010	
MM8011	
MM8012	2SC891,2SC892,2SC1016,2SC1040,2SC1257,2SC1608,*276-2017
MMT73	*TR-20,*SK3118,*276-2021
MMT918	*GE-61,*276-2015
MMT930	
MMT2222	*GE-210,*TR-51,*BC107B,*276-2016
MMT2369	*276-2038
MMT2484	
MMT2857	
MMT2907	*GE-48,*276-2023
MMT3014	*GE-20,*TR-64,*276-2016
MMT3546	*SK3118
MMT3798	*GE-67,*TR-52,*SK3114
MMT3799	*SK3114
MMT3903	*GE-210,*TR-53,*BC107B,*276-2016
MMT3904	*GE-20,*TR-53,*BC107B,*276-2016
MMT3905	*GE-21,*TR-54,*SK3118,*276-2034
MMT3906	*GE-48,*SK3118,*276-2034
MMT3960A	
MMT8015	*GE-11
MN21	SEE 2N375
MN24	SEE 2N350
MN25	SEE 2N351
MN26	SEE 2N376
MN29	SEE 2N555
MN48	SEE 2N669
MN49	SEE 2N618
MN61A	SEE 2N627
MN62A	SEE 2N628
MN63A	SEE 2N629
MN64A	SEE 2N630
MP110	*PTC901,*GE-76,*HEPG6005,*276-2006
MP110B-BLU	
MP110B-GRN	MP110B-BLU,*GE-76
MP110B-RED	MP110B-GRN,*PTC147,*GE-76
MP500	MP500A,MP501,MP501A,MP504,MP504A,MP505,MP505A,*GE-4,*TR-03,*HEPG6002, *SK3012,*PT515,*ECG213,*WEP233,*RT-215
MP500A	MP500,MP501,MP501A,MP504,MP504A,MP505,MP505A,*GE-4,*TR-03,*SK3012, *PT515,*ECG105,*WEP233
MP501	MP501A,MP502,MP502A,MP505,MP505A,MP506,MP506A,*GE-4,*TR-03,*SK3012, *PT515,*ECG105,*WEP233
MP501A	MP501,MP502,MP502A,MP505,MP505A,MP506,MP506A,*GE-4,*TR-03,*SK3012, *PT515,*ECG105,*WEP233
MP502	MP502A,MP506,MP506A,*GE-4,*TR-03,*SK3012,*PT515,*ECG105,*WEP233
MP502A	MP502,MP506,MP506A,*GE-4,*TR-03,*SK3012,*PT515,*ECG105,*WEP233
MP504	MP504A,MP505,MP505A,*GE-4,*TR-03,*HEPG6002,*SK3012,*PT515,*ECG105, *WEP233
MP504A	MP504,MP505,MP505A,*GE-4,*TR-03,*SK3012,*PT515,*ECG105,*WEP233
MP505	MP505A,MP506,MP506A,*GE-4,*TR-03,*SK3012,*PT515,*ECG105,*WEP233
MP505A	MP505,MP506,MP506A,*GE-4,*TR-03,*SK3012,*PT515,*ECG105,*WEP233
MP506	MP506A,*GE-4,*TR-03,*SK3012,*PT515,*ECG105,*WEP233
MP506A	MP506,*GE-4,*TR-03,*SK3012,*PT515,*ECG105,*WEP233
MP525-1	*PTC901,*GE-76
MP525-2	*PTC901,*GE-76
MP525-3	*PTC901,*GE-76
MP525-4	*PTC901,*GE-76

To Replace	Substitute This Type
MP525-5	*PTC901,*GE-76
MP525-6	*PTC901,*GE-76
MP600	MP601,MP602,MP603,*TR-35,*ECG179,*WEPG6001,*RT-147
MP601	MP600,MP602,MP603,*TR-35,*ECG179,*WEPG6001,*RT-147
MP602	MP603,*PTC147,*TR-35,*ECG179,*WEPG6001,*RT-147
MP603	MP602,*PTC147,*TR-35,*ECG179,*WEPG6001,*RT-147
MP1001	
MP1002	
MP1003	
MP1004	
MP1005	MP1001,MP1002
MP1529	MP1530,MP3611,MP3612,*PTC105A,*GE-76,*HEPG6003
MP1529A	MP1530A,*PTC901,*GE-76,*HEPG6003
MP1530	MP1531,MP3612,MP3615,*PTC105A,*GE-76,*HEPG6005
MP1530A	MP1531A,*PTC901,*GE-76,*HEPG6005
MP1531	MP1532,MP3615,MP3616,*PTC105A,*GE-76
MP1531A	MP1532A,*PTC901,*GE-76,*HEPG6005
MP1532	MP1533,MP3616,*PTC105A,*HEPG6018
MP1532A	*HEPG6018
MP1533	*HEPG6018
MP1534	MP1535,*PTC105A,*GE-76,*HEPG6003
MP1534A	MP1535A,*PTC901,*GE-76,*HEPG6003
MP1535	MP1536,*PTC105A,*GE-76,*HEPG6005
MP1535A	MP1536A,*PTC901,*GE-76,*HEPG6005
MP1536	MP1537,*PTC105A,*GE-76,*HEPG6005
MP1536A	MP1537A,*PTC901,*GE-76,*HEPG6005
MP1537	MP1538,*PTC105A,*HEPG6018
MP1537A	*HEPG6018
MP1538	*HEPG6018
MP1539A	MP1540A,MP1544A,MP1545A,MP3613,MP3614,*PTC105A,*GE-76
MP1540A	MP1541A,MP1545A,MP1546A,MP3614,MP3617,*PTC105A,*GE-76
MP1541A	MP1542A,MP1546A,MP1547A,MP3617,MP3618,*PTC105A,*GE-76
MP1542A	MP1547A,MP3618,*PTC105A
MP1543	MP1548
MP1544A	MP1545A,*PTC105A,*GE-76
MP1545A	MP1546A,*PTC105A,*GE-76
MP1546A	MP1547A,*PTC105A,*GE-76
MP1547A	*PTC105A
MP1548	
MP1549	MP1549A,MP1550,MP1550A,*PTC901,*GE-76
MP1549A	MP1550A,*PTC901
MP1550	MP1550A,MP1551,MP1551A,*PTC901,*GE-76
MP1550A	MP1551A,*PTC901
MP1551	MP1551A,MP1552,MP1552A,*PTC901,*GE-76
MP1551A	MP1552A,*PTC901
MP1552	MP1552A,*HEPG6018,*276-2006
MP1552A	
MP1553	2N511,2N511A,2N512,2N512A,2N513,2N513A,CRT1544,MP1553A,MP1554,MP1554A, MP1557,MP1557A,MP1558,MP1558A,*PTC901,*GE-76
MP1553A	2N511,2N511A,2N512,2N512A,2N513,2N513A,CRT1544,MP1553,MP1554,MP1554A, MP1557,MP1557A,MP1558,MP1558A,*PTC901,*GE-76
MP1554	2N511A,2N511B,2N512A,2N512B,2N513A,2N513B,CRT1544,CRT1545,CTP3503, MP1554A,MP1555,MP1555A,MP1558,MP1558A,MP1559,MP1559A,*PTC901,*GE-76
MP1554A	2N511A,2N511B,2N512A,2N512B,2N513A,2N513B,CRT1544,CRT1545,CTP3503, MP1554,MP1555,MP1555A,MP1558,MP1558A,MP1559,MP1559A,*PTC901,*GE-76
MP1555	2N1167,MP1555A,MP1556,MP1556A,MP1559,MP1559A,MP1560,MP1560A,*PTC901, *GE-76
MP1555A	2N1167,MP1555,MP1556,MP1556A,MP1559,MP1559A,MP1560,MP1560A,*PTC901, *GE-76
MP1556	2N1167,MP1556A,MP1560,MP1560A
MP1556A	2N1167,MP1556,MP1560,MP1560A

To Replace	Substitute This Type
MP1557	CRT1544,MP1557A,MP1558,MP1558A,*PTC901,*GE-76
MP1557A	MP1557,MP1558,MP1558A,*PTC901,*GE-76
MP1558	CRT1544,CRT1545,MP1558A,MP1559,MP1559A,*GE-76
MP1558A	MP1558,MP1559,MP1559A,*PTC901,*GE-76
MP1559	MP1559A,MP1560,MP1560A,*PTC901,*GE-76
MP1559A	MP1559,MP1560,MP1560A,*PTC901,*GE-76
MP1560	MP1560A
MP1560A	MP1560
MP1612	*TR-27,*HEPG6009,*ECG127,*WEP235
MP1612A	MP1612B,*TR-27,*ECG127,*WEP235
MP1612B	*TR-27,*ECG127,*WEP235
MP1613	2N637B,2N1021,2N1021A,2N1022A,2N3616,2N3618,2N5156,*PTC105A,*TR-27, *HEPG6018,*ECG127,*WEP235,*276-2006
MP2000A	2N5692,*PTC147,*GE-76
MP2060	2N3613,2N3614,2N4246,2N4247,MP2061,*PTC105A,*GE-76,*TR-01,*HEPG6003, *SK3009,*ECG121,*WEP232,*276-2006,*RT-127
MP2061	2N3614,2N3617,2N4245,2N4246,MP2062,*PTC105A,*GE-76,*TR-01,*HEPG6005, *SK3014,*ECG121,*WEP232,*276-2006,*RT-127
MP2062	2N3617,2N3618,2N4245,MP2063,*PTC105A,*GE-76,*TR-01,*HEPG6005,*SK3009, *ECG121,*WEP232,*276-2006,*RT-127
MP2063	2N3618,*PTC105A,*GE-76,*HEPG6005
MP2100A	2N1751,2N5693,MP2200A,*PTC147,*GE-76
MP2137A	MP2138A,MP2142A,MP2143A,*PTC114,*GE-16,*TR-01,*HEPG6003,*SK3009, *ECG121,*WEP232,*RT-127
MP2138A	MP2139A,MP2143A,MP2144A,*PTC105A,*GE-16,*TR-01,*HEPG6005,*SK3009, *ECG121,*WEP232,*RT-127
MP2139A	MP2140A,MP2144A,MP2145A,*PTC105A,*GE-16,*TR-01,*HEPG6005,*SK3009, *ECG121,*WEP232,*RT-127
MP2140A	MP2141A,MP2145A,MP2146A,*PTC105A,*GE-76,*TR-01,*HEPG6005
MP2141A	MP2146A,*PTC105A,*GE-76,*HEPG6018
MP2142A	MP2143A,*PTC114,*GE-16,*TR-01,*HEPG6003,*SK3009,*ECG121,*WEP232,*RT-127
MP2143A	MP2144A,*PTC105A,*GE-16,*TR-01,*HEPG6005,*SK3009,*ECG121,*WEP232, *RT-127
MP2144A	MP2145A,*PTC105A,*GE-16,*TR-01,*HEPG6005,*SK3009,*ECG121,*WEP232, *RT-127
MP2145A	MP2146A,*PTC105A,*HEPG6005
MP2146A	*PTC105A
MP2200A	2N1751,2N5693,2N5694,MP2300A,*PTC147,*GE-76,*TR-35,*HEPG6009,*ECG179, *WEPG6001,*RT-147
MP2300A	2N5694,2N5695,MP2400A,*PTC147,*TR-35,*HEPG6009,*ECG179,*WEPG6001, *RT-147
MP2400A	2N5695,2N5696
MP2526	CTP3500,CTP3503,*PTC105A,*GE-76
MP2527	MP2528
MP2528	
MP2832	
MP2833	MP2834
MP2834	
MP3611	MP3612,MP3613,MP3614,*PTC105A,*GE-76,*TR-01,*SK3009,*ECG121,*WEP232, *RT-127
MP3612	MP3614,MP3615,MP3617,*PTC105A,*GE-76,*TR-01,*SK3009,*ECG121,*WEP232, *RT-127
MP3613	MP3614,*PTC105A,*GE-76,*TR-01,*SK3009,*ECG121,*WEP232,*RT-127
MP3614	*PTC105A,*GE-76,*TR-01,*SK3009,*ECG121,*WEP232,*RT-127
MP3615	MP3616,MP3617,MP3618,*PTC105A,*GE-76,*TR-01,*SK3009,*ECG121,*WEP232, *RT-127
MP3616	MP3618,*PTC105A
MP3617	MP3618,*PTC105A,*GE-76,*TR-01,*SK3009,*ECG121,*WEP232,*RT-127
MP3618	*PTC105A
MP3730	*PTC122,*GE-25,*TR-27,*HEPG6007,*ECG127,*WEP235
MP3731	*PTC122,*GE-25,*TR-27,*HEPG6008,*ECG127,*WEP235

To Replace	Substitute This Type
MP5435	MP5436,*PTC901
MP5436	MP5437
MP5437	
MP5438	MP5439,*PTC901
MP5439	MP5440
MP5440	
MP5692	*PTC901
MP5693	MP5435,MP5436,MP5694,*PTC901
MP5694	MP5436,MP5695
MP5695	MP5436,MP5437,MP5696,*HEPG6009
MP5696	MP5437
MPM5006	2N2369,2N2369A,2N2710,2N3009,2N3013,2N3014,2N3510,2N3511,2N3646,2N3647, 2N3648,2N4137,2N4275,2N5029,2N5769,2N5772,2SC54,2SC67,2SC302,2SC577, 2SC601,2SC601N,2SC689,2SC1274,GET2369,GET3013,GET3014,GET3646, MPS2369,ZT2369,*PTC133,*GE-20,*HEPS0033,*276-2016
MPN350	M.P.PN350
MPS404	NPS404,NPS404A,*PTC103,*GE-82,*TR-20,*HEPS0032,*SK3118,*ECG159,*WEP242, *BC327,*276-2022,*RT-115
MPS404A	*PTC103,*GE-82,*TR-20,*HEPS0032,*SK3118,*ECG159,*WEP242,*BC327,*276-2022, *RT-115
MPS706	2N706,2N706A,2N706B,2N783,2N784,2N834A,2N2205,2N2368,2N2729,2N3544, 2SC54,2SC55,2SC63,2SC68,2SC321,2SC322,2SC689,KT218,MM1941,MPS706A, MPS834,MPS6507,MPS6511,MPS6540,MPS6542,MPS6543,MPS6548,MPSH19, MPSH20,MPSH32,MPSH37,MT1038,MT1038A,MT1039,MT1060,TIS84,TIS108, *PTC101,*GE-61,*TR-95,*HEPS0025,*SK3018,*EN10,*ECG108,*WEP56,*BC337, *276-2016,*RT-113
MPS706A	2N783,2N784,2N834A,2SC54,2SC68,2SC321,2SC322,2SC689,MPS834,MPSH37, *PTC101,*GE-61,*TR-95,*HEPS0025,*SK3039,*ECG108,*WEP56,*BC337, *276-2016,*RT-113
MPS834	2N834A,2N2710,2N3261,2N3510,2N3605A,2N3606A,2N3648,2SC321,2SC322, 2SC576,40405,GET708,GET914,GET3013,GET3014,GET3646,MPS6532,MPSH37, ZT83,*GE-20,*TR-70,*HEPS0025,*SK3039,*SN80,*ECG107,*WEP56,*276-2016, *RT-108
MPS918	2N834A,2N2368,2SC576,2SC689,MM1941,MPS6507,MPS6542,MPS6543,MPS6548, MT1038,MT1038A,MT1039,ZT2368,*PTC133,*GE-17,*TR-70,*HEPS0020,*SK3039, *EN10,*ECG107,*WEP56,*276-2038,*RT-108
MPS2369	2SC577,ZT2369,ZT2369A,*GE-18,*TR-70,*HEPS0025,*SK3039,*ECG107,*WEP56, *276-2038,*RT-108
MPS2711	2N744,2N2318,2N2319,2N2331,2N2569,2N2570,2N2656,2N2711,2N2713, 2N2926-RED,2N3011,2N3394,2N3397-RED,2N3398-RED,2N4138,2N4418,2N6566, 2S095A,2S512,2SC99,2SC132,2SC133,2SC135,2SC204,2SC356,2SC395A-O, 2SC579,2SC595,2SC915,2SC915A,2SC1071,40231,40234,BC170A,MPS2713, MPS2926-RED,MPS3394,PT720,ZT2205,*PTC101,*GE-212,*TR-21,*HEPS0015, *SK3122,*ECG123A,*WEP736,*BC337,*276-2013,*RT-102
MPS2712	2N2712,2N2714,2N2924,2N2926-YEL,2N3241A,2N3392,2N3395-YEL,2N3396-YEL, 2N3397-YEL,2N3398-YEL,2N3414,2N4124,2N4419,2N5224,2N5998,2N6000, 2SC395A-Y,2SC509-Y,2SC509G-Y,2SC733-Y,2SC735-Y,2SC847,2SC848,2SC849, 2SC934,2SC966,2SC967,2SC1246,2SC1317,2SC1788,2SC1849,2SC1851,2SD392, 40232,40233,40397,40399,A5T3392,A5T4124,A7T3392,A8T3392,BC108,BC108A, BC168A,BC170B,BC172A,BC238A,BC548A,BCY58A,BCY58B,BSW88,BSW89, BSX38,MPS2714,MPS2926-GRN,MPS2926-YEL,MPS3392,TN80,*PTC101,*GE-212, *TR-21,*HEPS0015,*SK3122,*ECG123A,*WEP736,*BC337,*276-2013,*RT-102
MPS2713	2N744,2N2318,2N2319,2N2656,2N3011,2N4418,2S095A,2S512,2SC204,2SC395A-O, 2SC595,PT720,ZT2205,*GE-210,*TR-21,*HEPS0025,*SK3122,*ECG123A,*WEP736, *276-2016,*RT-102
MPS2714	2N4124,2N4419,2SC395A-Y,2SC735-Y,2SC934,A5T4124,BCY58A,BCY58B,*GE-210, *TR-21,*HEPS0015,*SK3122,*ECG123A,*WEP736,*276-2016,*RT-102
MPS2716	2SC372G-Y,2SC400,2SC400-Y,2SC454,2SC529,2SC644,2SC752G-Y,2SC900, 2SC930NP,2SC1085,2SC1285,2SC1363,2SC1631,2SC1633,2SC1684,2SC1734, 2SC1853,2SC1854,BC130A,BC198A,GI-2924,GI-3392,GI-3396,GI-3397,GI-3707, GI-3708,GI-3710,MPS2924,SE4001,*PTC101,*GE-61,*TR-21,*HEPS0015,*SK3020,

To Replace	Substitute This Type
(MPS2716)	*ECG123A,*WEP736,*276-2013
MPS2923	2N2923,2N2924,2N3227,2N3241A,2N3242A,2N3392,2N3393,2N3395-YEL, 2N3396-ORG,2N3396-YEL,2N3397-ORG,2N3397-YEL,2N3398-ORG,2N3398-YEL, 2N3414,2N3509,2N4074,2N5224,2N5998,2N6000,2SC367G,2SC367G-Y, 2SC372G-Y,2SC400,2SC400-Y,2SC454,2SC458,2SC733-Y,2SC735-Y,2SC752G-Y, 2SC847,2SC848,2SC849,2SC966,2SC967,2SC1085,2SC1216,2SC1219,2SC1246, 2SC1285,2SC1317,2SC1363,2SC1631,2SC1633,2SD327,40398,BC108A,BC172A, BC238A,BC548A,BCY58A,BSW83,BSW88,BSW89,BSX38,GI-2923,GI-2924,GI-3392, GI-3393,GI-3396,GI-3397,MPS2924,MPS3392,MPS3393,PET4001,SE4001,TIS60, TIS60M,TN80,TZ81,*PTC101,*GE-212,*TR-21,*HEPS0015,*SK3122,*EN10, *ECG123A,*WEP736,*BC107B,*276-2031,*RT-102
MPS2924	2N2924,2N2925,2N3391,2N3391A,2N3392,2N3395-WHT,2N3395-YEL,2N3396-WHT, 2N3396-YEL,2N3397-WHT,2N3397-YEL,2N3398-WHT,2N3398-YEL,2N3415,2N3565, 2N3794,2N4124,2N4256,2N4286,2N4419,2N4954,2N5172,2N5371,2N5998,2N6002, 2N6008,2SC373G,2SC400-GR,2SC732-GR,2SC733-GR,2SC735-GR,2SC1684, 2SC1849,2SC1993,40397,BC108,BC108B,BC109B,BC172B,BC173B,BC238B, BC239B,BC338,BC548B,BC549B,BCY58B,BCY58C,GI-2924,GI-2925,GI-3391, GI-3391A,GI-3392,GI-3395,GI-3398,GI-3403,GI-3415,GI-3794,MPS2925,MPS3391, MPS3391A,MPS3392,MPS3395,MPS3706,MPS5172,MPSA16,MPSA17,PET4002, TP3706,TZ81,*PTC121,*GE-212,*TR-21,*HEPS0015,*SK3122,*EN10,*ECG123A, *WEP736,*BC107B,*276-2031,*RT-102
MPS2925	2N2925,2N3391,2N3391A,2N3395-WHT,2N3396-WHT,2N3397-WHT,2N3398-WHT, 2N3415,2N6002,2N6008,2SC536FP,2SC536NP,2SC732-BL,2SC733-BL,2SC1335, 2SC1344,2SC1571,2SC1571L,BC108C,BC109,BC109C,BC172C,BC173C,BC238C, BC239C,BC548B,BC549B,BCY58D,GI-2925,GI-3391,GI-3391A,MPS3391,MPS3391A, PET4003,*PTC139,*GE-212,*TR-21,*HEPS0015,*SK3122,*EN10,*ECG123A, *WEP736,*BC107B,*276-2031,*RT-102
MPS2926	2N2925,2N2926-GRN,2N3391,2N3391A,2N3395-WHT,2N3396-WHT,2N3397-WHT, 2N3398-WHT,2N3708,2N3711,2N3721,2N3900,2N3900A,2N4256,2N5172,2S503, 2SC733-GR,2SC735-GR,2SC1993,A5T3391,A5T3391A,A5T3565,A5T3708,A5T3711, A5T5172,A7T3391,A7T3391A,A7T5172,A8T3391,A8T3391A,A8T3708,A8T3711, A8T5172,A158B,A159B,A322,BC108B,BC109B,BC168B,BC169B,BC170C,BC172B, BC173B,BC238B,BC239B,BC548B,BC549B,BCY58C,MPS3391,MPS3391A, MPS3395,MPS3708,MPS3711,MPS3721,PET4002,TP3708,TP3711,*PTC139,*GE-20, *TR-21,*HEPS0015,*SK3122,*ECG123A,*WEP736,*276-2016
MPS2926-BRN	2N743,2N2318,2N2319,2N5187,2SC68,2SC98,2SC204,2SC395A-R,2SC595,MPS2713, *PTC101,*GE-17,*TR-53,*HEPS0015,*SK3122,*ECG123A,*WEP736,*276-2016, *RT-102
MPS2926-GRN	2N4124,2N4419,2N5224,2SC735-GR,2SC1993,A5T4124,A158B,A159B,BC108, BC108B,BC109B,BC168B,BC172B,BC173B,BC238B,BC239B,BCY58B,BCY58C, *PTC121,*TR-53,*HEPS0015,*SK3122,*ECG123A,*WEP736,*276-2016,*RT-102
MPS2926-ORG	2N2712,2SC395A,2SC395A-Y,2SC735-Y,2SC934,BC108A,BC168A,BC172A,BC238A, BC548A,BCY58A,MPS2714,MPS2926-YEL,ZT2938,*PTC101,*GE-17,*TR-53, *HEPS0015,*SK3122,*ECG123A,*WEP736,*276-2016,*RT-102
MPS2926-RED	2N744,2N2656,2N3011,2N4418,2S095A,2S512,2SC99,2SC395A-O,PT720,ZT2205, *PTC101,*GE-17,*TR-53,*HEPS0015,*SK3122,*ECG123A,*WEP736,*276-2016, *RT-102
MPS2926-YEL	2N4124,2N4419,2N5224,2SC735-Y,A5T4124,BC108,BC108A,BC168A,BC172A, BC238A,BC548A,BCY58A,BCY58B,MPS2926-GRN,*PTC101,*GE-17,*TR-53, *HEPS0015,*SK3122,*ECG123A,*WEP736,*276-2016,*RT-102
MPS3390	2N3390,2N3398-BLU,BC548C,BC549C
MPS3391	2N3391,2N3391A,2N3395-WHT,2N3396-WHT,2N3397-WHT,2N3398-BLU,2N3398-WHT, 2N3711,A5T3391,A5T3391A,A5T3565,A5T3711,A7T3391,A7T3391A,A8T3391, A8T3391A,A8T3711,BCY58D,MPS3391A,PET4003,TP3566,TP3711
MPS3391A	2N3391,2N3391A,2N3395-WHT,2N3396-WHT,2N3397-WHT,2N3398-BLU,2N3398-WHT, 2N3711,A5T3391,A5T3391A,A5T3565,A5T3711,A7T3391,A7T3391A,A8T3391, A8T3391A,A8T3711,BCY58D,MPS3391,PET4003,TP3566,TP3711,*PTC139
MPS3392	2N2924,2N2925,2N3391,2N3391A,2N3392,2N3395-WHT,2N3395-YEL,2N3396-WHT, 2N3396-YEL,2N3397-WHT,2N3397-YEL,2N3398-WHT,2N3398-YEL,2N3415,2N4124, 2N4256,2N4419,2N4954,2N5172,2N5371,2N5998,2N6002,2N6008,2SC733-GR, 2SC735-GR,2SC1849,2SC1993,40397,A5T3391,A5T3391A,A5T3392,A5T4124,

To Replace	Substitute This Type
(MPS3392)	A5T5172,A7T3391,A7T3391A,A7T3392,A7T5172,A8T3391,A8T3391A,A8T3392, A8T5172,BC108,BC108B,BC109B,BC172B,BC173B,BC238B,BC239B,BC338, BC548B,BC549B,BCY58B,BCY58C,GI-3415,MPS3391,MPS3391A,MPS3395, MPS3706,MPSA16,MPSA17,PET4002,TIS90-GRY,TIS90M-GRY,TIS92-GRY, TIS92M-GRY,TP3566,TP3706,TZ81,*PTC121,*GE-212,*TR-21,*HEPS0015,*SK3124, *ECG123A,*WEP736,*BC337,*276-2013,*RT-102
MPS3393	2N2923,2N2924,2N3227,2N3241A,2N3242A,2N3392,2N3393,2N3395-YEL, 2N3396-ORG,2N3396-YEL,2N3397-ORG,2N3397-YEL,2N3398-ORG,2N3398-YEL, 2N3414,2N3509,2N4074,2N5224,2N5998,2N6000,2N6426,2SC367G,2SC367G-Y, 2SC509-Y,2SC509G-Y,2SC733-Y,2SC735-Y,2SC847,2SC848,2SC849,2SC966, 2SC967,2SC1216,2SC1246,2SC1317,2SC1788,2SC1851,40398,A5T3392,A7T3392, A8T3392,BC108A,BC172A,BC238A,BC548A,BCY58A,BSW83,BSW88,BSW89, BSX38,MPS3392,PET4001,TIS60,TIS60M,TIS90-BLU,TIS90-VIO,TIS90M-BLU, TIS90M-VIO,TIS92-BLU,TIS92-VIO,TIS92M-BLU,TIS92M-VIO,TN80,TZ81,*PTC101, *GE-212,*TR-21,*HEPS0015,*SK3122,*ECG123A,*WEP736,*BC337,*276-2013, *RT-102
MPS3394	2N2331,2N2501,2N2923,2N3211,2N3393,2N3396-ORG,2N3397-ORG,2N3398-ORG, 2N4123,2N4138,2N4418,2N6427,2N6566,2S512,2SC367G-O,2SC509-O,2SC509G-O, 2SC639,2SC733-O,2SC735-O,2SC764,A5T4123,GI-3414,MPS3393,MPS6560, PET4001,TIS90-GRN,TIS90-YEL,TIS90M-GRN,TIS90M-YEL,TIS92-GRN,TIS92-YEL, TIS92M-GRN,TIS92M-YEL,ZT2938,*PTC101,*GE-212,*TR-21,*HEPS0015,*SK3122, *ECG123A,*WEP736,*BC337,*276-2013,*RT-102
MPS3395	2N2925,2N3391,2N3391A,2N3395-WHT,2N3396-WHT,2N3397-WHT,2N3398-WHT, 2N3415,2N4954,2N5371,2N6002,2N6008,2SC733-BL,A5T3391,A5T3391A,A7T3391, A7T3391A,A8T3391,A8T3391A,BC108C,BC109,BC109C,BC172C,BC173C,BC238C, BC239C,BC338,BC548B,BC549B,BCY58C,BCY58D,MPS3391,MPS3391A,PET4002, PET4003,TP3566,*PTC139,*GE-212,*TR-21,*HEPS0015,*SK3122,*ECG123A, *WEP736,*BC337,*276-2013,*RT-102
MPS3396	2N3565,2N4286,2SC373G,2SC732-GR,2SC828,2SC923,BC122,GI-2925,GI-3391, GI-3391A,GI-3711,MPS2925,MPS5172,*PTC139,*GE-61,*TR-21,*HEPS0015, *SK3122,*ECG123A,*WEP736,*276-2013,*RT-102
MPS3397	2N3565,2N4286,2SC373G,2SC732-GR,2SC828,2SC923,BC122,BC130B,BC131B, BC198B,BC199B,GI-2925,GI-3391,GI-3391A,GI-3711,MPS2925,MPS5172,*PTC139, *GE-212,*TR-21,*HEPS0015,*SK3122,*ECG123A,*WEP736,*276-2013,*RT-102
MPS3398	2SC536FP,2SC536NP,2SC693FP,2SC693NP,2SC732-BL,2SC1327,2SC1335, 2SC1344,2SC1478,2SC1571,2SC1571L,2SC1787,BC130C,BC131C,BC198C, BC199C,SE4002,SE4010,*PTC139,*GE-85,*TR-21,*HEPS0015,*SK3122,*ECG123A, *WEP736,*276-2013,*RT-102
MPS3563	TIS62A,TIS63A,*PTC133,*GE-61,*TR-70,*HEPS0020,*SK3018,*EN10,*ECG229, *WEP56,*276-2038,*RT-108
MPS3638	2N869A,2N2695,2N2696,2N4402,2SA695,A5T3638A,A5T4402,BSW72,MPS6533, TN-3905,TQ62,ZT180,ZT280,*PTC103,*GE-22,*TR-20,*HEPS0019,*SK3025,*EN10, *ECG159,*WEP242,*BC327,*276-2034,*RT-115
MPS3638A	2N4403,2N5375,2N5379,2N5813,2N5999,2N6001,2SA509-Y,2SA509G-Y,2SA719, 2SA890,2SB598,A5T4403,BSW73,D29E2,MPS6534,TN-3906,ZT187,ZT287,*PTC103, *GE-82,*TR-20,*HEPS0019,*SK3025,*EN10,*ECG159,*WEP242,*276-2034,*RT-115
MPS3639	2N3639,2N3640,2N4257A,2N4258A,*PTC131,*GE-21,*TR-20,*SK3114,*SP70, *ECG159,*WEP242,*276-2021,*RT-115
MPS3640	2N3012,2N3546,2N3640,2N4258A,2N5056,2N5057,2N5292,*GE-21,*TR-20, *HEPS0015,*SK3114,*SP70,*ECG159,*WEP242,*276-2021,*RT-115
MPS3642	2SC944,2SC980G-O,EN915,*PTC121,*GE-210,*TR-87,*HEPS0015,*SK3024, *ECG128,*WEP243,*276-2009,*RT-114
MPS3644	2N3251,2N3251A,2N4143,2SA493-GR,2SA493-Y,2SA733,2SA776,2SA776A,BC212A, BC213L,BC214L,BC257A,BC257B,BC415-10,BC415-6,BC415A,BC415B,BC416-10, BC416-6,BC416A,BC416B,BCY79A,BCY79B,GET2906,GET2907,*PTC103,*GE-21
MPS3645	2N3251A,2N4143,2SA776,2SA776A,BC212A,GET2906,GET2907,*PTC103,*GE-21
MPS3646	2N708,2N914,2N916,2N916A,2N2501,2N3013,2N3014,2N3210,2N3211,2N3646, 2N4013,2N4420,2N4421,2N4422,2N5144,2SC67,2SC169,2SC335,2SC336,2SC577, 2SC639,2SC752G-O,2SC764,EN708,EN916,MPS6530,ZT84,ZT708,*GE-20,*TR-95, *HEPS0014,*SK3047,*EN10,*ECG123A,*WEP56,*276-2016,*RT-113
MPS3693	2N916,2N916A,2N3903,2N3947,2N3974,2N3976,2N4227,2N4962,2N5380,2N6538,

To Replace	Substitute This Type
(MPS3693)	2SC933,2SC943,2SC1007,2SC1688,2SC1999,A5T3903,MPS6565,MPS6576-BLUE, MPS6576-GREEN,MPS6576-YEL,TN-3903,*PTC121,*GE-210,*TR-21,*HEPS0030, *SK3018,*ECG229,*WEP736,*BC337,*276-2013,*RT-102
MPS3694	2N5827,A157B,BC107,BC107B,BC167B,BC171B,BC174B,BC190B,BC237B,BC413B, BC414B,BC547B,BC550B,BCY59B,BCY59C,MPS6566,TIS94,TIS97,*PTC121, *GE-210,*TR-21,*HEPS0030,*SK3124,*ECG229,*WEP736,*BC337,*276-2009, *RT-102
MPS3702	2N2838,2N2907,2N2907A,2N3251,2N3251A,2N3486,2N3486A,2N3504,2N3505, 2N3672,2N3673,2N3702,2N3906,2N3962,2N4143,2N4403,2N5373,2N5374,2N5375, 2N5379,2N5383,2N5447,2N5817,2N5819,2N6005,2SA467G-Y,2SA603,2SA659, 2SA661-GR,2SA661-Y,2SA678,2SA705,2SA718,2SA720,2SA891,A5T2907,A5T3504, A5T3505,A5T3644,A5T3645,A5T3906,A5T4403,A5T5447,A8T3702,A177,BC212A, BC213L,BC214L,BC327-16,BC327-25,BCW35,BCW37,BCY79,BCY79A,BCY79B, BSW73,D29E5,D29E6,GET2906,GET2907,GI-3644,MM3906,TIS112,TN-3906,TQ60, TZ552,TZ553,*PTC103,*GE-82,*TR-20,*HEPS0019,*SK3025,*ECG159,*WEP242, *BC327,*276-2034,*RT-115
MPS3703	2N2838,2N2906,2N2906A,2N3250,2N3250A,2N3485,2N3485A,2N3672,2N3673, 2N3703,2N4142,2N4228,2N5372,2N5448,2N5815,2N5817,2N5821,2N5823,2SA603, 2SA661-O,2SA718,A5T5448,A8T3703,A177,BC212L,BSW74,D29E10,D29E5, GET2904,GET2905,TZ551,*PTC103,*GE-82,*TR-20,*HEPS0019,*SK3114,*ECG159, *WEP242,*BC327,*276-2023,*RT-115
MPS3704	2N2222,2N2222A,2N2540,2N3116,2N3302,2N3704,2N4401,2N5449,2N5818,2N6112, A5T2222,A5T5449,A8T3704,D33D26,TIS109,TIS111,TN-3904,*PTC170,*GE-20, *TR-21,*HEPS0015,*SK3122,*ECG123A,*WEP736,*BC337,*276-2016,*RT-102
MPS3705	2N915,2N2539,2N2897,2N2900,2N3705,2N4400,2N4962,2N5450,2N5822,40458, A5T5450,A8T3705,BCW34,BCW36,D33D25,D33D30,TIS110,TN54,TN-3903,*PTC170, *GE-20,*TR-21,*HEPS0015,*SK3122,*ECG123A,*WEP736,*BC337,*276-2016, *RT-102
MPS3706	2N3706,2N5451,2N6012,2N6112,A5T5451,A8T3706,BC337-40,D33D27,*GE-20, *TR-21,*HEPS0015,*SK3122,*ECG123A,*WEP736,*BC337,*276-2016,*RT-102
MPS3707	2N3242A,2N4419,2SC1364,2SC1992,A5T3707,A5T3710,A8T3707,A8T3710,BC107B, BC167B,BC171B,BC237B,BCY58B,BCY59B,MPS3710,MPSA16,MPSA17,*PTC101, *GE-212,*TR-21,*HEPS0015,*SK3122,*ECG123A,*WEP736,*BC337,*276-2013, *RT-102
MPS3708	2N3708,A5T3565,A5T3708,A5T3711,A8T3708,A8T3711,BC547B,BC550B,BCY58C, BCY58D,BCY59C,BCY59D,MPS3711,MPSA18,PET4002,PET4003,*PTC139,*GE-212, *TR-21,*HEPS0025,*SK3122,*ECG123A,*WEP736,*BC337,*276-2013,*RT-102
MPS3709	2N3241A,2N4074,2SC1999,A5T3709,A8T3709,BC107A,BC167A,BC171A,BC237A, BCY58A,BCY59A,PET4001,ZT44,*PTC101,*GE-61,*TR-21,*HEPS0025,*SK3122, *ECG123A,*WEP736,*BC337,*276-2013,*RT-102
MPS3710	2N4419,2SC1992,A5T3707,A5T3710,A8T3707,A8T3710,BC107B,BC167B,BC171B, BC237B,BCY58B,BCY58C,BCY59B,BCY59C,MPS3708,MPSA16,MPSA17,PET4002, *PTC121,*GE-61,*TR-21,*HEPS0025,*SK3122,*ECG199,*WEP736,*BC337, *276-2013,*RT-102
MPS3711	A5T3711,A8T3711,BC550C,BCY58D,BCY59D,MPSA18,PET4003,*PTC139,*GE-212, *TR-21,*HEPS0015,*SK3122,*ECG123A,*WEP736,*BC337,*276-2013,*RT-102
MPS3721	A158C,A159C,BC108C,BC109,BC109C,BC168C,BC172C,BC173C,BC238C,BC239C, BC548B,BC549B,BCY58D,*PTC139,*GE-212,*TR-21,*HEPS0015,*SK3122, *ECG123A,*WEP736,*BC337,*276-2013,*RT-102
MPS3725	
MPS3826	*PTC121,*GE-210,*TR-21,*HEPS0011,*SK3122,*ECG123A,*WEP736,*BC337, *276-2009,*RT-102
MPS3827	*PTC121,*GE-210,*TR-21,*HEPS0011,*SK3122,*ECG123A,*WEP736,*BC337, *276-2009,*RT-102
MPS4354	MPS4355,*GE-67
MPS4355	MPS4354,*GE-67,*SK3025
MPS4356	
MPS5172	2N2925,2N3391,2N3391A,2N4256,2N4954,2N5172,2N5371,2N6002,2N6008, 2SC373G,2SC735-GR,2SC1335,2SC1344,BC108C,BC109,BC109C,BC172C,BC173C, BC238C,BC239C,BC338,BC548B,BC549B,BCY58C,BCY58D,MPS3706,MPSA16, *PTC139,*GE-212,*TR-51,*HEPS0015,*SK3114,*ECG123A,*WEP736,*BC107B,

To Replace	Substitute This Type
(MPS5172)	*276-2016,*RT-115
MPS5305	2N5305,2N5307,MPS5307
MPS5306	2N5306,2N5306A,2N5308,2N5308A,MPS5306A,MPS5308,MPS5308A
MPS5306A	2N5306,2N5306A,2N5308,2N5308A,MPS5306,MPS5308,MPS5308A
MPS5307	2N5307
MPS5308	2N5308,2N5308A,MPS5308A
MPS5308A	2N5308,2N5308A,MPS5308
MPS6507	2N2369,2N2369A,2SC577,2SC689,MM1941,ZT2369,*PTC133,*GE-20,*TR-95, *HEPS0020,*SK3122,*ECG229,*WEP56,*276-2038,*RT-113
MPS6511	2N708,2N708A,2N784,2N784A,2N914,2N2242,2N2318,2N2319,2N2369,2N2369A, 2N2481,2N2710,2N3009,2N3011,2N3013,2N3014,2N3210,2N3261,2N3510,2N3511, 2N3605A,2N3606A,2N3646,2N3647,2N3648,2N4137,2N4274,2N4275,2N5029, 2N5030,2N5769,2N5772,2SC54,2SC67,2SC204,2SC302,2SC356,2SC577,2SC595, 2SC601,2SC601N,2SC689,2SC1123,2SC1128,2SC1274,BSW82,GET708,GET914, GET2369,GET3013,GET3014,GET3646,MM1941,MPS2369,MPS6507,MPS6512, PET1001,PET2001,PT2760,ZT708,ZT2369,*PTC101,*GE-210,*TR-95,*HEPS0020, *SK3018,*ECG108,*WEP56,*BC337,*276-2016,*RT-113
MPS6512	2N708,2N914,2N916,2N916A,2N2242,2N2369,2N2369A,2N2501,2N2845,2N2847, 2N3009,2N3013,2N3014,2N3115,2N3210,2N3211,2N3301,2N3646,2N3862,2N3903, 2N3946,2N4013,2N4123,2N4227,2N4275,2N4873,2N4962,2N5027,2N5144,2N5380, 2N5769,2N5772,2SC67,2SC577,2SC601N,2SC639,2SC764,A5T3903,A5T4123, BSW82,MM3903,MPS2369,MPS6530,TN-3903,ZT81,ZT84,ZT708,ZT2369,ZT2369A, *PTC101,*GE-210,*TR-21,*HEPS0025,*SK3018,*ECG229,*WEP736,*276-2016, *RT-102
MPS6513	2N3116,2N3227,2N3302,2N3904,2N3947,2N4141,2N4401,2N4952,2N5028,2N5107, 2N5369,2N5381,2SC933,2SC943,2SC1216,2SC1991,2SC1999,A5T3904,BC107A, BC167A,BC171A,BC237A,BC547A,BCY59A,BSW83,GET2222,GI-3643,MM3904, MPS6514,MPS6531,TN-3904,ZT82,*PTC101,*GE-210,*TR-21,*HEPS0015,*SK3018, *ECG229,*WEP736,*276-2016,*RT-102
MPS6514	2N5827,MPS6515,MPS6520,*PTC121,*GE-17,*TR-21,*HEPS0015,*SK3122,*EN10, *ECG123A,*WEP736,*276-2016,*RT-102
MPS6515	2N5827,2N6428,2N6428A,MPS6521,*GE-62,*TR-21,*HEPS0015,*ECG289,*WEP736, *276-2016,*RT-102
MPS6516	2N2906,2N2906A,2N3135,2N3250,2N3250A,2N3485,2N3485A,2N3905,2N4142, 2N4228,2N5365,2N5382,2SA467,2SA710,2SA711,A5T3905,GET2904,GET2905, MM3905,MPS6533,TN-3905,*PTC103,*GE-21,*TR-21,*HEPS0019,*SK3025, *ECG159,*WEP242,*276-2034,*RT-115
MPS6517	2N2907,2N2907A,2N3136,2N3251,2N3251A,2N3486,2N3486A,2N3504,2N3505, 2N3672,2N3673,2N3906,2N4034,2N4035,2N4143,2N4228,2N4403,2N5244,2N5366, 2N5383,2N6005,2SA603,2SA710,2SA711,2SA718,2SA720,2SA891,A5T2907, A5T3504,A5T3505,A5T3644,A5T3645,A5T3906,A5T4403,BC212A,BC212L,BC213L, BC257A,BC415-6,BC415A,BC416-6,BC416A,BCY79A,GET2906,GET2907,GI-3644, MM3906,MPS6518,MPS6534,NPS404A,TIS112,TN-3906,TP3644,TP3645,*PTC103, *GE-48,*TR-20,*HEPS0019,*SK3025,*SP70,*ECG159,*WEP242,*276-2034,*RT-115
MPS6518	2N3251,2N3251A,2N4035,2N5244,NPS404A,*GE-67,*TR-20,*HEPS0019,*SK3114, *ECG159,*WEP242,*276-2034,*RT-115
MPS6519	MPS6522,MPS6523,*GE-67,*TR-20,*HEPS0019,*SK3114,*ECG159,*WEP242, *276-2034,*RT-115
MPS6520	2N6428,2N6428A,MPS6521,*GE-62,*TR-21,*HEPS0015,*SK3122,*ECG123A, *WEP736,*276-2038,*RT-102
MPS6521	2N6428,2N6428A,*GE-62,*TR-21,*HEPS0015,*SK3122,*ECG123A,*WEP736, *276-2038,*RT-102
MPS6522	MPS6523,*GE-67,*TR-20,*HEPS0019,*SK3114,*ECG159,*WEP242,*276-2021, *RT-115
MPS6523	*GE-67,*TR-20,*HEPS0019,*SK3114,*ECG159,*WEP242,*276-2021,*RT-115
MPS6530	*GE-20,*TR-21,*HEPS0015,*SK3024,*ECG123A,*WEP736,*276-2016,*RT-102
MPS6531	*GE-20,*TR-70,*HEPS0015,*SK3124,*ECG107,*WEP56,*276-2016,*RT-108
MPS6532	*GE-20,*TR-70,*HEPS0025,*SK3039,*ECG107,*WEP56,*276-2016,*RT-108
MPS6533	2N4026,*PTC103,*GE-21,*TR-20,*HEPS0019,*SK3114,*ECG159,*WEP242,*276-2034, *RT-115
MPS6534	2N4028,TN-3906,*PTC103,*GE-21,*TR-20,*HEPS0019,*SK3138,*ECG159,*WEP242,

To Replace	Substitute This Type
(MPS6534)	*276-2034,*RT-115
MPS6535	*PTC103,*GE-21,*TR-20,*HEPS0019,*SK3114,*ECG159,*WEP242,*276-2034, *RT-115
MPS6539	2N706,2N706A,2N706B,2N743,2N784,2N947,2N2205,2N2432,2N2719,2N2729, 2N3544,2N3854A,2SC55,2SC63,2SC68,2SC98,2SC137,2SC172,2SC467,MM1941, MPS706,MPS706A,MPS918,MPS6507,MPS6511,MPS6540,MPS6542,MPS6543, MPS6546,MPS6547,MPS6548,MPSH19,MT1038,MT1038A,MT1039,MT1060,ZT40, *PTC101,*GE-61,*TR-95,*HEPS0016,*SK3039,*ECG108,*WEP56,*BC337, *276-2013,*RT-113
MPS6540	2N2318,2N2319,2N2368,2N3605A,2N3606A,2N5030,2SC54,2SC302,2SC321,2SC322, 2SC595,2SC601,2SC689,BF224J,BF225J,GET708,GET914,GET3013,GET3014, GET3646,MPSH20,MPSH37,MT1038,MT1038A,MT1039,MT1060,TIS84,*PTC101, *GE-61,*TR-70,*HEPS0016,*SK3039,*ECG107,*WEP56,*276-2016,*RT-108
MPS6542	2N2368,2SC576,2SC689,MM1941,MPS6507,MPS6543,MPS6548,MT1038,MT1038A, MT1039,*PTC133,*TR-95,*HEPS0016,*SK3039,*ECG108,*WEP56,*276-2038, *RT-113
MPS6543	2N2368,TIS84,*PTC133,*GE-11,*TR-95,*HEPS0016,*SK3039,*ECG108,*WEP56, *276-2038,*RT-108
MPS6544	2N734,2N758A,2N759A,2N760B,2N2522,2N3078,2S102,*PTC121,*GE-212,*TR-21, *HEPS0016,*SK3018,*ECG123A,*WEP736,*BC337,*276-2008,*RT-102
MPS6545	2N758,2N758A,2N759,2N759A,2N760B,2N2522,2S102,2SC302,BF224J,BF225J, MPS6544,ZT42,*PTC178,*GE-212,*TR-21,*HEPS0016,*SK3122,*ECG123A, *WEP736,*BC337,*276-2008,*RT-102
MPS6546	2N834A,2N2368,2SC576,2SC689,MPS6543,MPS6547,TIS84,ZT2368,*PTC133, *GE-61,*TR-95,*HEPS0016,*SK3039,*ECG108,*WEP56,*276-2015,*RT-113
MPS6547	2N834A,2N2368,2SC576,2SC689,MPS6543,MPS6546,TIS84,ZT2368,*PTC133, *GE-61,*TR-95,*HEPS0016,*SK3039,*ECG108,*WEP56,*276-2015,*RT-113
MPS6548	2N2368,2N2729,2SC601,2SC689,MM1941,MPS6507,MPS6542,MPS6543,MT1038, MT1038A,MT1039,MT1060,TIS84,*PTC133,*GE-61,*TR-95,*HEPS0016,*SK3039, *ECG108,*WEP56,*276-2038,*RT-113
MPS6560	2N3241A,2N3242A,2SC1859,BC338-16,TN60,TN80,*PTC170,*GE-20,*TR-21, *HEPS0024,*SK3122,*ECG287,*WEP736,*BC337,*276-2009,*RT-102
MPS6561	2N3241A,2SC1859,BC338-16,MPS6560,TN80,*PTC170,*GE-20,*TR-21,*HEPS0024, *SK3124,*ECG123A,*WEP736,*BC337,*276-2009,*RT-102
MPS6562	2N5811,A5T4402,A5T4403,BC328-16,D29E1,*PTC103,*GE-21,*TR-20,*HEPS0019, *SK3114,*ECG159,*WEP242,*BC327,*276-2032,*RT-115
MPS6563	2N5811,BC328-16,D29E1,MPS6562,*PTC103,*GE-21,*TR-20,*HEPS0019,*SK3114, *ECG159,*WEP242,*BC327,*276-2032,*RT-115
MPS6564	*PTC121,*GE-212,*SK3245,*276-2013
MPS6565	2N915,2N2539,2N3903,2N3947,2N3974,2N3976,2N4014,2N4227,2N4400,2N4951, 2N4962,2N4963,2N5368,2N5380,2N6538,2SC943,2SC1318,2SC1852,A5T3903, MPS-A05,TIS96,TIS99,TIS110,TN-3903,ZT84,ZT89,ZT114,ZT119,*GE-210,*TR-21, *HEPS0015,*SK3122,*ECG123A,*WEP736,*BC337,*276-2009,*RT-102
MPS6566	2N4953,2N5370,2N5376,2N5377,TIS94,TIS97,TIS109,*GE-210,*TR-21,*HEPS0015, *SK3124,*ECG123A,*WEP736,*BC337,*276-2009,*RT-108
MPS6567	2N717,2N758,2N758A,2N759,2N759A,2N760B,2N783,2N2387,2N2395,2N2432A, 2N2522,2N3605A,2N3606A,2N3973,2N3975,2SC54,2SC131,2SC134,2SC239, 2SC302,2SC321,2SC322,2SC565,2SC601,2SC689,2SC1166-R,BFY39-1,BSY93, GET708,GET914,GET3013,GET3014,GET3646,MPSH37,ZT43,*PTC121,*GE-61, *TR-21,*HEPS0025,*SK3122,*ECG123A,*WEP736,*BC337,*276-2013,*RT-102
MPS6568	2N2615,2N2616,2N3011,2N3959,2N3960,2N4274,2N5134,2N5200,2N5201,2SC99, 2SC1123,2SC1128,2SC1129,2SC1988,A5T3571,BF121,BF123,BF125,BF127,BF311, MPS6568A,MPS6569,MPS6570,MPSH02,MPSH10,MPSH11,MT1060A,TIS64A, *GE-61,*TR-21,*HEPS0016,*SK3122,*ECG123A,*WEP736,*276-2016,*RT-102
MPS6568A	2N2615,2N2616,A5T3571,A5T3572,MPS6569,MPS6570,TIS62A,TIS63A,TIS86, *PTC133,*GE-61,*TR-21,*HEPS0016,*276-2016
MPS6569	2N2615,2N2616,A5T3571,A5T3572,MPS6568A,MPS6570,TIS62A,TIS63A,TIS86, *GE-61,*TR-70,*HEPS0015,*SK3039,*ECG107,*WEP56,*276-2016,*RT-108
MPS6570	2N2615,2N2616,A5T3571,A5T3572,MPS6568A,MPS6569,TIS62A,TIS63A,TIS86, *GE-61,*TR-70,*HEPS0024,*SK3039,*ECG107,*WEP56,*276-2016,*RT-108
MPS6571	A158C,A159C,BC108C,BC109C,BC168C,BC169C,BC172C,BC173C,BC238C,BC239C,

To Replace	Substitute This Type
(MPS6571)	BC548C,BC549C,BCY58D,*PTC139,*GE-212,*TR-21,*HEPS0024,*SK3124,*ECG123A, *WEP736,*BC337,*276-2016,*RT-102
MPS6573	2N3242A,2N4954,2N5371,2SC735-GR,2SC1992,A157B,BC107,BC107B,BC167B, BC171B,BC237B,BC413B,BC414B,BCY58B,BCY59B,BSW83,BSW88,BSW89,BSX38, BSX79,MPS3694,MPS6514,MPS6520,MPS6574-SIL,MPS6575,MPS6576-SIL, PET1002,PET2002,*PTC101,*GE-20,*TR-21,*HEPS0015,*SK3124,*ECG123A, *WEP736,*276-2030,*RT-102
MPS6574-BLUE	2N916,2N916A,2N3227,2N3242A,2SC735-Y,2SC933,2SC1175,2SC1216,2SC1999, BC107A,BC167A,BC171A,BC237A,BC547A,BCY58A,BCY59A,BSW83,BSW88, BSW89,BSX38,BSX79,MPS6513,MPS6573,MPS6574-GREEN,MPS6574-SIL, MPS6575,MPS6576-BLUE,MPS6576-GREEN,MPS6576-SIL,PET1002,PET2002,ZT82, ZT112,*PTC101
MPS6574-GREEN	2N3227,2N3242A,2SC735-Y,2SC933,2SC1175,2SC1992,BC107,BC107A,BC167A, BC171A,BC237A,BC547A,BCY58A,BCY58B,BCY59A,BCY59B,BSW83,BSW88, BSW89,BSX38,BSX79,MPS3694,MPS6514,MPS6573,MPS6574-SIL,MPS6575, MPS6576-GREEN,MPS6576-SIL,PET1002,PET2002,ZT82,ZT112,*PTC101
MPS6574-SIL	2N3242A,2N4954,2N5371,2SC735-GR,2SC1992,A157B,BC107,BC107B,BC167B, BC171B,BC237B,BC413B,BC414B,BCY58B,BCY59B,BSW83,BSW88,BSW89,BSX38, BSX79,MPS3694,MPS6514,MPS6520,MPS6573,MPS6575,MPS6576-SIL,PET1002, PET2002,*PTC101
MPS6574-YEL	2N916,2N916A,2N2501,2N4013,2N4123,2N4873,2SC639,2SC735-O,2SC933, 2SC1216,2SC1999,A5T4123,MPS3693,MPS6513,MPS6574-BLUE,MPS6574-GREEN, MPS6576-BLUE,MPS6576-GREEN,MPS6576-YEL,ZT81,ZT82,ZT111,ZT112,ZT2369A, *PTC101
MPS6575	2N3116,2N3302,2N3904,2N4401,2N4952,2N5369,2N5377,2N5381,2SC1992,A5T3904, A157B,BC107,BC107B,BC167B,BC171B,BC174B,BC190B,BC237B,BC413B,BC414B, BCY59B,BSX79,MPS3694,MPS6566,MPS6576-SIL,PET1002,TN-3904,*GE-210, *TR-21,*HEPS0015,*SK3122,*ECG123A,*WEP736,*276-2016,*RT-102
MPS6576-BLUE	2N916,2N916A,2N3116,2N3904,2N3947,2N4141,2N4401,2N4951,2N4952,2N5028, 2N5107,2N5368,2N5369,2N5381,2SC933,2SC943,2SC1175,2SC1318,2SC1852, 2SC1991,2SC1999,A5T3904,BC107A,BC167A,BC171A,BC174A,BC190A,BC237A, BC547A,BCY59A,BSX79,GET2222,GI-3643,MM3904,MPS6531,MPS6575, MPS6576-GREEN,MPS6576-SIL,MPS-A05,PET1002,TN-3904,ZT82,ZT112,*PTC121
MPS6576-GREEN	2N3116,2N3302,2N3904,2N3947,2N4141,2N4401,2N4952,2N5028,2N5107,2N5369, 2N5381,2SC933,2SC943,2SC1175,2SC1318,2SC1852,2SC1991,2SC1992,A5T3904, BC107,BC107A,BC167A,BC171A,BC174A,BC190A,BC237A,BC547A,BCY59A, BCY59B,BSX79,GET2222,GI-3643,MM3904,MPS3694,MPS6531,MPS6566,MPS6575, MPS6576-SIL,MPS-A05,PET1002,TN-3904,ZT82,ZT112,*PTC121
MPS6576-SIL	2N3116,2N3302,2N3904,2N4401,2N4952,2N5369,2N5377,2N5381,2SC1992,A5T3904, A157B,BC107,BC107B,BC167B,BC171B,BC174B,BC190B,BC237B,BC413B,BC414B, BCY59B,BSX79,MPS3694,MPS6566,MPS6575,PET1002,TN-3904,*PTC121
MPS6576-YEL	2N916,2N916A,2N3903,2N3947,2N3974,2N3976,2N4013,2N4227,2N4400,2N4951, 2N4962,2N5368,2N5380,2N6538,2SC933,2SC943,2SC1318,2SC1852,2SC1999, A5T3903,MPS3693,MPS6565,MPS6576-BLUE,MPS6576-GREEN,MPS-A05,TN-3903, ZT81,ZT82,ZT84,ZT111,ZT112,ZT114,*PTC121
MPS6580	BF506,*PTC103,*GE-22,*TR-20,*HEPS0019,*SK3118,*ECG106,*WEP52,*276-2034, *RT-126
MPS6590	2N720,2N720A,2N739,2N870,2N911,2N2316,2N3037,ZT86,ZT116,*PTC170,*GE-18, *TR-21,*ECG123A,*WEP736,*276-2008
MPS6591	2N718,2N718A,2N2221,2N2221A,2N2315,2N2317,2N2389,2N2396,2N2539,2N2845, 2N2847,2N3115,2N3301,2N3973,2N3974,2N3975,2N3976,2N4400,2N4962,2N4963, 2N5027,2N5581,2N5820,2SC1211,BSW84,BSY93,D33D29,GET2221,GET2221A, GI-3641,GI-3642,MPS6530,TIS110,TN54,TN-3903,ZT83,ZT113,*PTC170,*GE-210, *TR-53,*HEPS0025,*BC337,*276-2008
MPS-A05	2N2540,2N5369,2N5582,2SC1852,A5T2222,BSW85,TIS109,TIS111,TN-3904,*GE-63, *HEPS0015,*SK3122,*276-2009
MPS-A09	2N2523,2N3117,2N5210,2N5827,2N6428,2N6428A,2N6539,2SC1010,A5T5210, BC414C,BC547B,BC550B,*HEPS0015,*276-2011
MPS-A10	2N2484A,2N3302,2N3417,2N4954,2N5371,2N5376,2N5377,2N5826,2N6112, 2SC1166-GR,2SC1850,2SC1992,A157B,BC107,BC107B,BC167B,BC171B,BC182B, BC182L,BC183L,BC237B,BC337,BC413B,BC414B,BCY59B,BCY59C,BFY39-3,

To Replace	Substitute This Type
(MPS-A10)	MPS3694,MPS3706,MPS6514,MPS6520,MPS6566,MPS-A20,MPS-K20,MPS-K21, MPS-K22,MPSA10-YEL,MPSA16,MPSA17,MPSA20-YEL,NPSA20,TIS94,TIS97, TP3704,TP3706,*PTC121,*GE-20,*HEPS0015,*276-2011
MPS-A12	MPSA12,MPSA14,*HEPS9100
MPS-A13	*HEPS9100
MPS-A14	MPSA13,*HEPS9100
MPS-A20	2N3302,2N4954,2N5371,2N5376,2N5377,2N6112,2SC1166-GR,2SC1850,2SC1992, A157B,BC107,BC107B,BC167B,BC171B,BC182B,BC182L,BC183L,BC237B,BC413B, BC414B,BCY59B,BCY59C,BFY39-3,MPS3694,MPS6514,MPS6520,MPS6566, MPS-K20,MPS-K21,MPS-K22,MPSA20-YEL,NPSA20,TIS94,TIS97,*PTC121, *HEPS0015,*SK3124,*276-2011
MPS-A55	2N4026,2N4027,2N5372,2N5821,2N5823,2SA697,2SA817,2SA891,D29E10,*PTC103, *GE-67,*HEPS5022,*SK3114,*276-2021
MPS-A70	2N2907,2N2907A,2N3136,2N3251,2N3251A,2N3486,2N3486A,2N3504,2N3505, 2N3906,2N3962,2N4035,2N4143,2N4403,2N5244,2N5366,2N5373,2N5374,2N5375, 2N5378,2N5379,2N5383,2N6005,2SA550A,2SA678,2SA705,A5T2907,A5T3504, A5T3505,A5T3644,A5T3645,A5T3906,A5T4403,BC177A,BC177B,BC212A,BC213L, BC214L,BC251A,BC251B,BC257A,BC257B,BC261A,BC261B,BC307A,BC307B, BC415-10,BC415-6,BC415A,BC415B,BC416-10,BC416-6,BC416A,BC416B,BC557A, BC557B,BC560A,BC560B,BCY79,BCY79A,BCY79B,BCY79C,BSW73,GET2906, GET2907,GI-3644,MM3906,MPS6518,MPS6534,MPSA70-BLU,MPSA70-YEL, NPS404A,TIS112,TN-3906,TZ552,TZ553,*HEPS0019,*276-2034
MPS-K20	2N3302,2N4954,2N5371,2N5376,2N5377,2N6112,2SC1166-GR,2SC1850,2SC1992, A157B,BC107,BC107B,BC167B,BC171B,BC182B,BC182L,BC183L,BC237B,BC413B, BC414B,BCY59B,BCY59C,BFY39-3,MPS3694,MPS6514,MPS6520,MPS6566, MPS-A20,MPS-K21,MPS-K22,MPSA20-YEL,NPSA20,TIS94,TIS97,*PTC121,*GE-212
MPS-K21	2N3302,2N4954,2N5371,2N5376,2N5377,2N6112,2SC1166-GR,2SC1850,2SC1992, A157B,BC107,BC107B,BC167B,BC171B,BC182B,BC182L,BC183L,BC237B,BC413B, BC414B,BCY59B,BCY59C,BFY39-3,MPS3694,MPS6514,MPS6520,MPS6566, MPS-A20,MPS-K20,MPS-K22,MPSA20-YEL,NPSA20,TIS94,TIS97,*PTC121,*GE-212
MPS-K22	2N3302,2N4954,2N5371,2N5376,2N5377,2N6112,2SC1166-GR,2SC1850,2SC1992, A157B,BC107,BC107B,BC167B,BC171B,BC182B,BC182L,BC183L,BC237B,BC413B, BC414B,BCY59B,BCY59C,BFY39-3,MPS3694,MPS6514,MPS6520,MPS6566, MPS-A20,MPS-K20,MPS-K21,MPSA20-YEL,NPSA20,TIS94,TIS97,*PTC121,*GE-212
MPSA05	2N718A,2N731,2N5581,A5T2193,BSW84,MPSA06,*PTC170,*GE-81,*TR-21,*SK3122, *ECG123A,*WEP736,*276-2008,*RT-102
MPSA06	2N718A,2N870,2N911,2N5581,A5T2193,BSW84,*PTC170,*GE-81,*TR-87,*HEPS5014, *SK3024,*ECG128,*WEP243,*276-2008
MPSA09	2N2484,2N3117,2N3417,2N4953,2N5210,2N5370,2N5376,2N5377,2N5827,2N6006, 2N6112,2SC734-GR,2SC1166-GR,2SC1850,A5T5210,BC107B,BC174B,BC182B, BC190B,BC237B,BC414B,BC547B,BC550B,GI-3566,MPS6566,MPS-A09,TIS94, TIS97,TP3704,*PTC170,*GE-212,*TR-51,*SK3245,*BC337,*276-2013
MPSA10	2SC693,2SC711A,2SC1000-GR,2SC1000G-GR,2SC1313,2SC1380-GR,2SC1380A-GR, 2SC1416,2SC1416A,2SC1681-GR,2SC1685,BC129B,BC197B,EN2484,PBC107B, SE4020,TIS24,*PTC121,*GE-212,*TR-51,*SK3245,*276-2013
MPSA10-BLU	2N909,2N930B,2N2484A,2N3116,2N3242A,2N3302,2N3904,2N4401,2N4952,2N5369, 2N5377,2N5381,2N5826,2SC1166-GR,2SC1364,2SC1813,2SC1850,2SC1992, A5T3904,A157B,BC107,BC107B,BC167B,BC171B,BC182B,BC182L,BC183L, BC237B,BC413B,BC414B,BCY59B,BFY39-3,BSW83,BSX79,MPS3694,MPS3704, MPS3706,MPS6514,MPS6520,MPS6566,MPS6575,MPS6576-SIL,MPS-A10,MPS-A20, MPS-K20,MPS-K21,MPS-K22,MPSA10-YEL,MPSA16,MPSA17,MPSA20-BLU, MPSA20-YEL,NPSA20,PET1002,TN-3904,TP3704,TP3706,*PTC121,*GE-212,*TR-51, *SK3245,*BC107B,*276-2016
MPSA10-GRN	2N760A,2N909,2N930B,2N2484A,2N3116,2N3227,2N3242A,2N3302,2N3416,2N3509, 2N3859A,2N3904,2N3947,2N4074,2N4141,2N4401,2N4952,2N5028,2N5107,2N5369, 2N5381,2N5825,2N5826,2N6004,2N6426,2SC366G-Y,2SC367G-Y,2SC850,2SC933, 2SC943,2SC968,2SC970,2SC1166-Y,2SC1175,2SC1213A,2SC1214,2SC1247, 2SC1318,2SC1364,2SC1813,2SC1850,2SC1852,2SC1991,2SC1992,A5T3904,BC107, BC107A,BC167A,BC171A,BC182A,BC182L,BC237A,BC547A,BCW34,BCW36, BCY59A,BCY59B,BFY39-2,BSW83,BSX79,GET2222,GI-3643,MM3904,MPS3694, MPS3704,MPS6514,MPS6531,MPS6566,MPS6575,MPS6576-GREEN,MPS6576-SIL,

To Replace	Substitute This Type
(MPSA10-GRN)	MPS-A05,MPS-A10,MPS-A20,MPS-K20,MPS-K21,MPS-K22,MPSA10-BLU,MPSA10-YEL, MPSA20-BLU,MPSA20-GRN,MPSA20-YEL,NPSA20,PET1002,TN-3904,ZT82,ZT112, *PTC121,*GE-212,*TR-51,*SK3245,*BC107B,*276-2016
MPSA10-RED	2N708,2N708A,2N718,2N760,2N784A,2N914,2N2242,2N2315,2N2369,2N2369A, 2N2396,2N2481,2N2501,2N2710,2N2845,2N2847,2N3009,2N3013,2N3014,2N3115, 2N3210,2N3211,2N3301,2N3508,2N3510,2N3511,2N3605A,2N3606A,2N3646, 2N3647,2N3648,2N3858A,2N3862,2N3903,2N3946,2N3973,2N3974,2N3975,2N3976, 2N4123,2N4137,2N4140,2N4275,2N4400,2N4962,2N5027,2N5029,2N5144,2N5380, 2N5769,2N5772,2N5824,2N6538,2SC67,2SC318,2SC577,2SC601,2SC601N, 2SC639,2SC764,2SC1166-R,2SC1210,2SC1274,A5T3903,A5T4123,BFY39-1,BSW82, BSY93,GET708,GET914,GET2221,GET2369,GET3013,GET3014,GET3646,GI-3641, GI-3642,MM3903,MPS2369,MPS3693,MPS3705,MPS6512,MPS6530,MPS6565, MPS6576-YEL,MPSA20-RED,PET1001,TN-3903,TP3705,ZT81,ZT83,ZT111,ZT113, ZT708,ZT2369,*PTC121,*GE-212,*TR-51,*SK3245,*BC107B,*276-2016
MPSA10-WHT	2N760A,2N909,2N916,2N916A,2N930B,2N3116,2N3227,2N3242A,2N3416,2N3509, 2N3859A,2N3904,2N3947,2N4074,2N4141,2N4401,2N4873,2N4951,2N4952,2N5028, 2N5107,2N5368,2N5369,2N5381,2N5825,2N6004,2N6426,2SC366G,2SC366G-Y, 2SC367G,2SC367G-Y,2SC850,2SC933,2SC943,2SC968,2SC970,2SC1166-Y, 2SC1175,2SC1213A,2SC1214,2SC1216,2SC1247,2SC1318,2SC1364,2SC1813, 2SC1852,2SC1991,2SC1999,A5T3904,BC107A,BC167A,BC171A,BC182A,BC237A, BC547A,BCW34,BCW36,BCY59A,BFY39-2,BSW83,BSX79,GET2222,GI-3643, MM3904,MPS3704,MPS6513,MPS6531,MPS6575,MPS6576-BLUE,MPS6576-GREEN, MPS6576-SIL,MPS-A05,MPSA10-BLU,MPSA10-GRN,MPSA20-BLU,MPSA20-GRN, MPSA20-WHT,PET1002,TN-3904,ZT82,ZT84,ZT112,ZT114,ZT2369A,*PTC121, *GE-212,*TR-51,*SK3245,*BC107B,*276-2016
MPSA10-YEL	2N2484A,2N3417,2N4954,2N5371,2N5376,2N5377,2N5826,2N5827,2N6006,2N6112, 2SC1166-GR,2SC1850,2SC1992,A157B,BC107,BC107B,BC167B,BC171B,BC182B, BC183L,BC184L,BC237B,BC337,BC413B,BC414B,BC547B,BC550B,BCY59B, BCY59C,BFY39-3,MPS3694,MPS3706,MPS6514,MPS6515,MPS6520,MPS6566, MPSA16,MPSA17,MPSA20-YEL,TIS94,TIS97,TP3704,TP3706,*PTC121,*GE-212, *TR-51,*SK3245,*BC107B,*276-2016
MPSA12	*TR-69,*ECG172,*WEPS9100,*276-2009
MPSA13	*GE-64,*TR-69,*HEPS9100,*ECG172,*WEPS9100,*276-2009
MPSA14	*TR-69,*ECG172,*WEPS9100,*276-2009
MPSA16	*GE-62,*SK3122,*BC337,*276-2016
MPSA17	*GE-62,*SK3122,*BC337,*276-2016
MPSA18	2N4104,BCY59D,*GE-85
MPSA20	2SC693,2SC711A,2SC1313,2SC1681-GR,2SC1685,BC129B,BC197B,PBC107B, *PTC121,*GE-20,*TR-51,*HEPS0015,*SK3122,*ECG123A,*WEP736,*276-2009
MPSA20-BLU	2N3116,2N3242A,2N3302,2N3904,2N4401,2N4952,2N5369,2N5377,2N5381, 2SC1166-GR,2SC1364,2SC1850,2SC1992,A5T3904,A157B,BC107,BC107B,BC167B, BC171B,BC182B,BC182L,BC183L,BC237B,BC413B,BC414B,BCY59B,BFY39-3, BSW83,BSX79,MPS3694,MPS6514,MPS6520,MPS6566,MPS6575,MPS6576-SIL, MPS-A20,MPS-K20,MPS-K21,MPS-K22,MPSA20-YEL,NPSA20,PET1002,TN-3904, *PTC121,*GE-212,*TR-51,*SK3245,*BC107B,*276-2016
MPSA20-GRN	2N3116,2N3227,2N3242A,2N3302,2N3904,2N3947,2N4141,2N4401,2N4952,2N5028, 2N5107,2N5369,2N5381,2N6004,2N6426,2SC366G-Y,2SC367G-Y,2SC933,2SC943, 2SC1166-Y,2SC1175,2SC1318,2SC1364,2SC1850,2SC1852,2SC1991,2SC1992, A5T3904,BC107,BC107A,BC167A,BC171A,BC182A,BC182L,BC237A,BC547A, BCW34,BCW36,BCY59A,BCY59B,BFY39-2,BSW83,BSX79,GET2222,GI-3643, MM3904,MPS3694,MPS6514,MPS6531,MPS6566,MPS6575,MPS6576-GREEN, MPS6576-SIL,MPS-A05,MPS-A20,MPS-K20,MPS-K21,MPS-K22,MPSA20-BLU, MPSA20-YEL,NPSA20,PET1002,TN-3904,ZT82,ZT112,*PTC121,*GE-212,*TR-51, *SK3245,*BC107B,*276-2016
MPSA20-RED	2N708,2N708A,2N784A,2N914,2N2242,2N2369,2N2369A,2N2481,2N2501,2N2710, 2N2845,2N2847,2N3009,2N3013,2N3014,2N3115,2N3210,2N3211,2N3301,2N3510, 2N3511,2N3605A,2N3606A,2N3646,2N3647,2N3648,2N3862,2N3903,2N3946, 2N3973,2N3974,2N3975,2N3976,2N4123,2N4137,2N4140,2N4275,2N4400,2N4962, 2N5027,2N5029,2N5144,2N5380,2N5769,2N5772,2N6538,2SC67,2SC318,2SC577, 2SC601,2SC601N,2SC639,2SC764,2SC1166-R,2SC1210,2SC1274,A5T3903, A5T4123,BFY39-1,BSW82,GET708,GET914,GET2221,GET2369,GET3013,GET3014,

To Replace	Substitute This Type
(MPSA20-RED)	GET3646,GI-3641,GI-3642,MM3903,MPS2369,MPS3693,MPS6512,MPS6530,MPS6565, MPS6576-YEL,PET1001,TN-3903,ZT81,ZT83,ZT111,ZT113,ZT708,ZT2369,*PTC121, *GE-212,*TR-51,*SK3245,*BC107B,*276-2016
MPSA20-WHT	2N916,2N916A,2N3116,2N3227,2N3242A,2N3904,2N3947,2N4141,2N4401,2N4873, 2N4951,2N4952,2N5028,2N5107,2N5368,2N5369,2N5381,2N6004,2N6426,2SC366G, 2SC366G-Y,2SC367G,2SC367G-Y,2SC933,2SC943,2SC1166-Y,2SC1175,2SC1216, 2SC1318,2SC1364,2SC1852,2SC1991,2SC1999,A5T3904,BC107A,BC167A, BC171A,BC182A,BC237A,BC547A,BCW34,BCW36,BCY59A,BFY39-2,BSW83, BSX79,GET2222,GI-3643,MM3904,MPS6513,MPS6531,MPS6575,MPS6576-BLUE, MPS6576-GREEN,MPS6576-SIL,MPS-A05,MPSA20-BLU,MPSA20-GRN,PET1002, TN-3904,ZT82,ZT84,ZT112,ZT114,ZT2369A,*PTC121,*GE-212,*TR-51,*SK3245, *BC107B,*276-2016
MPSA20-YEL	2N4954,2N5371,2N5376,2N5377,2N5827,2N6006,2N6112,2SC1166-GR,2SC1850, 2SC1992,A157B,BC107,BC107B,BC167B,BC171B,BC182B,BC183L,BC184L,BC237B, BC413B,BC414B,BC547B,BC550B,BCY59B,BCY59C,BFY39-3,MPS3694,MPS6514, MPS6515,MPS6520,MPS6566,TIS94,TIS97,*PTC121,*GE-212,*TR-51,*SK3245, *BC107B,*276-2016
MPSA42	*GE-222,*SK3044,*276-2008
MPSA43	*PTC170,*276-2008
MPSA55	2N4026,2N4027,2N5372,2N5821,2SA697,D29E9,MPS-A55,MPSA56,*PTC103,*GE-82, *TR-20,*SK3114,*ECG159,*WEP242,*BC327,*RT-115
MPSA56	2N4027,*PTC171,*GE-82,*TR-20,*HEPS5022,*SK3025,*ECG159,*WEP242,*RT-115
MPSA65	
MPSA66	
MPSA70	2N3251,2N3251A,2N3906,2N3962,2N4035,2N4122,2N4143,2N5244,2N5366, 2SA493-GR,2SA493-Y,2SA550A,2SA608,2SA678,2SA705,2SA733,2SA776,2SA776A, 2SA841-GR,2SA842-GR,BC177A,BC177B,BC212A,BC213L,BC214L,BC251A, BC251B,BC257A,BC257B,BC261A,BC261B,BC307A,BC307B,BC415-10,BC415-6, BC415A,BC415B,BC416-10,BC416-6,BC416A,BC416B,BCY79A,BCY79B,BCY79C, GET2906,GET2907,MM3906,MPS6518,MPS-A70,MPSA70-BLU,MPSA70-YEL, NPS404A,TN-3906,TZ552,TZ553,*GE-65,*TR-20,*HEPS0019,*SK3114,*ECG159, *WEP242,*276-2034,*RT-115
MPSA70-BLU	2N2907,2N2907A,2N3136,2N3251,2N3251A,2N3486,2N3486A,2N3504,2N3505, 2N3672,2N3673,2N3906,2N3962,2N4034,2N4035,2N4143,2N4403,2N5244,2N5366, 2N5373,2N5374,2N5375,2N5379,2N5383,2N6005,2SA550A,2SA603,2SA678, 2SA705,2SA720,2SA891,A5T2907,A5T3504,A5T3505,A5T3644,A5T3645,A5T3906, A5T4403,BC177A,BC177B,BC212A,BC213L,BC214L,BC251A,BC251B,BC257A, BC257B,BC261A,BC261B,BC307A,BC307B,BC415-10,BC415-6,BC415A,BC415B, BC416-10,BC416-6,BC416A,BC416B,BC557A,BC557B,BC560A,BC560B,BCW35, BCW37,BCY79,BCY79A,BCY79B,BSW73,GET2906,GET2907,GI-3644,MM3906, MPS6518,MPS6534,MPS-A70,MPSA70-GRN,MPSA70-YEL,NPS404A,TIS112, TN-3906,TP3644,TP3645,TZ552,TZ553,ZT182,ZT282,*PTC103,*GE-65,*TR-52, *ECG159,*BC177,*276-2034
MPSA70-GRN	2N2907,2N2907A,2N3136,2N3251,2N3251A,2N3486,2N3486A,2N3504,2N3505, 2N3672,2N3673,2N3906,2N4034,2N4035,2N4143,2N4403,2N5244,2N5366,2N5373, 2N5375,2N5383,2N6005,2SA550A,2SA603,2SA678,2SA705,2SA710,2SA711, 2SA718,2SA720,2SA891,A5T2907,A5T3504,A5T3505,A5T3644,A5T3645,A5T3906, A5T4403,A177,BC177A,BC212A,BC212L,BC213L,BC214L,BC251A,BC257A, BC261A,BC307A,BC415-6,BC415A,BC416-6,BC416A,BC557A,BC560A,BCW35, BCW37,BCY79,BCY79A,BSW73,GET2906,GET2907,GI-3644,MM3906,MPS6517, MPS6518,MPS6534,MPS-A70,MPSA70-BLU,MPSA70-WHT,MPSA70-YEL,NPS404A, TIS112,TN-3906,TP3644,TP3645,TZ552,ZT182,ZT184,ZT282,ZT284,*PTC103, *GE-65,*TR-52,*ECG159,*BC177,*276-2034
MPSA70-RED	2N2906,2N2906A,2N3073,2N3121,2N3135,2N3250,2N3250A,2N3485,2N3485A, 2N3905,2N4142,2N4402,2N5365,2N5372,2N5382,2N6067,2SA467,2SA558,2SA696, A5T3905,A5T4402,BSW72,GET2904,GET2905,MM3905,MPS6516,MPS6533, TN-3905,TZ551,ZT181,ZT183,ZT281,ZT283,*PTC103,*GE-65,*TR-52,*ECG159, *BC177,*276-2034
MPSA70-WHT	2N2907,2N2907A,2N3136,2N3250,2N3250A,2N3251,2N3251A,2N3486,2N3486A, 2N3504,2N3505,2N3672,2N3673,2N3905,2N3906,2N4034,2N4035,2N4143,2N4228, 2N4402,2N4403,2N5366,2N5373,2N5382,2N5383,2N6005,2N6067,2SA467,2SA603,

To Replace	Substitute This Type
(MPSA70-WHT)	2SA696,2SA710,2SA711,2SA718,2SA720,2SA891,A5T2907,A5T3504,A5T3505, A5T3644,A5T3645,A5T3905,A5T3906,A5T4402,A5T4403,A177,BC177,BC177A, BC212A,BC212L,BC213L,BC251A,BC257A,BC261A,BC307A,BC415-6,BC415A, BC416-6,BC416A,BC557A,BC560A,BCW35,BCW37,BCY79A,BSW73,GET2906, GET2907,GI-3644,MM3906,MPS6517,MPS6534,MPSA70-BLU,MPSA70-GRN,TIS112, TN-3905,TN-3906,TP3644,TP3645,TZ552,ZT181,ZT182,ZT184,ZT281,ZT282,ZT284, *PTC103,*GE-65,*TR-52,*ECG159,*BC177,*276-2034
MPSA70-YEL	2N2907,2N2907A,2N3136,2N3251,2N3251A,2N3486,2N3486A,2N3504,2N3505, 2N3906,2N3962,2N4035,2N4143,2N4403,2N5244,2N5366,2N5367,2N5373,2N5374, 2N5375,2N5378,2N5379,2N5383,2N6005,2N6007,2SA550A,2SA678,2SA705, A5T2907,A5T3504,A5T3505,A5T3644,A5T3645,A5T3906,A5T4403,BC177B,BC213L, BC214L,BC251B,BC257B,BC261B,BC307B,BC415-10,BC415B,BC416-10,BC416B, BC557A,BC557B,BC560A,BC560B,BCY79,BCY79A,BCY79B,BCY79C,BSW73, GET2906,GET2907,GI-3644,MM3906,MPS6518,MPS6534,MPS-A70,MPSA70-BLU, NPS404A,TIS112,TN-3906,TZ552,TZ553,*GE-65,*TR-52,*ECG159,*BC177, *276-2034
MPSA92	*PTC171,*GE-223
MPSA93	*PTC171
MPSH02	2N4274,2N5134,A5T3571,TIS64A,*GE-17,*TR-22,*276-2009
MPSH04	2N736A,2N4014,2N4963,2SC979A-O,BSW84,GET2221A,MPSH05,TIS96,TIS99, *GE-81,*TR-78,*BF338
MPSH05	2N736A,2N4014,2N4963,2SC979A-O,BSW84,GET2221A,MPSH04,TIS96,TIS99, *GE-81,*TR-78,*BF338
MPSH07	MPSH08,*PTC101,*GE-20,*276-2009
MPSH08	*PTC133,*276-2038
MPSH10	2N2616,2N4873,2SC1687,2SC1688,BF123,MPSH11,TIS62A,TIS63A,TIS64A,TIS105, *GE-61,*SK3039,*276-2038
MPSH11	2N2616,2N4873,2SC1687,2SC1688,BF123,MPSH10,TIS62A,TIS63A,TIS64A,TIS105, *GE-61,*276-2038
MPSH19	2N708,2N708A,2N784A,2N2318,2N3011,2N3605A,2N3606A,2N3646,2N4137,2N4275, 2N5029,2N5030,2N5772,2SC204,2SC302,2SC1240,BF121,BF224J,BF225J,GET708, GET914,GET2369,GET3013,GET3014,GET3646,PT2760,TIS84,ZT708,*PTC101, *GE-61,*276-2016
MPSH20	2N2368,2N2710,2N3510,2N3648,2SC54,2SC302,2SC321,2SC322,2SC576,2SC601, 2SC689,GET3013,GET3014,GET3646,MPS6532,ZT2368,*GE-20,*TR-64,*276-2016
MPSH24	*276-2038
MPSH30	2N744,2N2615,2N2616,2N3011,2N3959,2N3960,2N4274,2N4418,2N5134,2N5200, 2N5201,2S095A,2S512,2SC99,2SC388A,2SC735-O,2SC1123,2SC1128,2SC1129, 2SC1988,A5T3571,BF121,BF123,BF125,BF127,BF311,MPS6568,MPS6568A, MPS6569,MPS6570,MPSH02,MPSH10,MPSH11,MPSH31,MT1060A,PT720,TIS64A, ZT2205,*GE-61,*TR-64,*276-2016
MPSH31	2N744,2N2615,2N2616,2N3011,2N3959,2N3960,2N4274,2N4418,2N5134,2N5200, 2N5201,2S095A,2S512,2SC99,2SC388A,2SC735-O,2SC1123,2SC1128,2SC1129, 2SC1988,A5T3571,BF121,BF123,BF125,BF127,BF311,MPS6568,MPS6568A, MPS6569,MPS6570,MPSH02,MPSH10,MPSH11,MPSH30,MT1060A,PT720,TIS64A, ZT2205,*GE-61,*TR-64,*276-2016
MPSH32	TIS84,TIS108,*PTC101,*GE-17,*TR-64,*276-2009
MPSH34	
MPSH37	2N3605A,2N3606A,2SC54,2SC302,2SC321,2SC322,2SC601,2SC689,GET708, GET914,GET3013,GET3014,GET3646,*GE-61,*276-2016
MPSH81	
MPSH85	BF506,*PTC103,*276-2034
MPSL01	2N2896,2N2899,2N5550,2N5551,A5T5550,A5T5551,*PTC170,*SK3024,*ECG194, *276-2008
MPSL5	2N2593,2N2600,2N2600A,2SA549A,*PTC171
MPSL07	2N2894A,2N3546,2N4257A,2N4258A,2N5056,2N5057,2N5292,MPSL08,*276-2021
MPSL08	2N2894A,2N3546,2N4258A,2N5056,2N5057,2N5292,*276-2021
MPSL51	2N5400,A5T5400,*SK3025
MPSU01	2SC1847,*PTC197,*GE-83,*TR-72,*SK3199,*ECG188,*WEPS3020,*RT-156
MPSU02	*PTC197,*GE-63,*TR-72,*SK3199,*ECG188,*WEPS3020,*RT-156
MPSU03	*PTC197,*GE-32,*TR-74,*ECG190,*WEPS3021,*RT-159

To Replace	Substitute This Type
MPSU04	*GE-32,*TR-74,*ECG190,*WEPS3021,*RT-159
MPSU05	MPSU06,*PTC197,*GE-18,*TR-72,*SK3199,*ECG188,*WEPS3020,*RT-156
MPSU06	2N5150,RCA1A17,*PTC197,*TR-72,*HEPS3019,*SK3054,*ECG188,*WEPS3020, *RT-156
MPSU10	*GE-27,*TR-75,*SK3232,*ECG191,*WEPS3021,*276-2008,*RT-159
MPSU51	2N3867,2SA886,*GE-84,*TR-73,*SK3200,*ECG189,*WEPS3031,*RT-157
MPSU52	2N4407,*PTC198,*GE-29,*TR-73,*SK3025,*ECG189,*WEPS3031,*RT-157
MPSU55	2SA743A,MPSU56,*PTC198,*TR-73,*SK3025,*ECG189,*WEPS3031,*RT-157
MPSU56	2N5149,2N5322,2SA743A,2SA761,*PTC198,*TR-73,*SK3025,*ECG189,*WEPS3031, *RT-157
MPSU60	
MPSU95	*HEPS9123
MPT40	M.P.PT40
MPT501	M.P.PT501
MRF8004	2N749,2N751,2N1388,2N1390,2SC302-M,2SC307-M,2SC590N,2SC773,2SC774, 2SC798,2SC803,2SC865,2SC1947,40581,40975,*WEP243
MSA7505	2N5016,MSA8505,XB404,*GE-66
MSA8503	*276-2016
MSA8505	2N3733,2N5016,MSA7505,*GE-66
MSA8506	2N5591,2N5705,2N6082,2N6083,2N6136,MM1603,ZT5591,*TR-76
MSA8507	2N5646,2N5994,2N5996,2SC1081,2SC1118,2SC1258,2SC1528,3TX602,40893†, SRF54215,*276-2013
MSA8508	2N5589,2N5945,2N5995†,2SC891,2SC892,2SC1011,2SC1040,2SC1257,2SC1405, 2SC1495,2SC1606,3TX601,41010†,*GE-28,*TR-76
MT1038	MT1038A,MT1039,*PTC133
MT1038A	MT1038,MT1039,*PTC133
MT1039	MT1038,MT1038A,*PTC133
MT1039A	
MT1050	
MT1060	MT1038,MT1038A,MT1039,*PTC133,*HEPS0016,*276-2015
MT1060A	*PTC133
MT1061	*PTC132
MT1061A	*PTC133
MT1062	
MT1063	
MT1070	*GE-28,*TR-76
MT1115	MT1116
MT1116	MT1115
MT3833	MT3834
MT3834	MT3833
MT5763	2N5947,2N6206,41025†,MT5764,MT5765,*276-2017
MT5764	2N5595,2N5765,41026†,MT5765,*276-2017
MT5765	
NA20	REFER TO SECTION 2
NA30	REFER TO SECTION 2
ND5700	
ND5701	
ND5702	
NKT121	2N660,2N661,NKT124,NKT127,*GE-51,*TR-17,*HEPG0002,*SK3006,*HF35,*ECG160, *WEP637,*276-2005
NKT122	2N660,2N661,NKT121,NKT124,NKT125,NKT127,NKT128,*GE-51,*TR-17,*HEPG0002, *SK3006,*HF35,*ECG160,*WEP637,*276-2005
NKT123	2N520A,2N660,2N661,2SB222A,2SB327,NKT121,NKT122,NKT124,NKT125,NKT126, NKT127,NKT128,NKT129,*PTC135,*GE-51,*TR-85,*HEPG0005,*SK3004,*HF35, *ECG102A,*WEP250,*AC188/01,*276-2005,*RT-121
NKT124	2N522A,2N661,*GE-51,*TR-17,*HEPG0002,*SK3006,*HF35,*ECG160,*WEP637, *276-2005
NKT125	2N521A,2N522A,2N599,2N661,2N2001,2N2541,NKT124,*GE-51,*TR-17,*HEPG0002, *SK3006,*HF35,*ECG160,*WEP637,*276-2005
NKT126	2N521A,2N522A,2N598,2N599,2N661,2N1998,2N2001,2N2541,2SB223A,MA1703, NKT124,NKT125,*PTC135,*GE-51,*TR-85,*HEPG0005,*SK3004,*HF35,*ECG102A,

To Replace	Substitute This Type
(NKT126)	*WEP250,*AC188/01,*276-2005,*RT-121
NKT127	2N660,2N661,NKT121,NKT124,*DS-56,*GE-51,*TR-17,*HEPG0002,*SK3006,*HF35, *ECG126,*WEP635,*276-2005
NKT128	2N660,2N661,NKT121,NKT122,NKT124,NKT125,NKT127,*GE-51,*TR-05,*HEPG0002, *SK3005,*HF35,*ECG100,*WEP254,*276-2005,*RT-118
NKT129	2N520A,2N660,2N661,2SB222A,2SB327,NKT121,NKT122,NKT123,NKT124,NKT125, NKT126,NKT127,NKT128,*PTC135,*GE-51,*TR-05,*HEPG6011,*SK3005,*HF35, *ECG100,*WEP254,*AC188/01,*RT-118
NKT211	2SB370A,2SB431,NKT218,*PTC135,*GE-53,*TR-84,*HEPG0005,*SK3004,*AT20M, *ECG158,*WEP238,*276-2005
NKT213	2N2706,2SB431,NKT211,NKT216,*PTC135,*GE-53,*TR-84,*HEPG0005,*SK3004, *AT20M,*ECG158,*WEP238,*276-2005
NKT214	NKT213,NKT216,NKT221,NKT228,*PTC135,*GE-53,*TR-84,*HEPG0005,*SK3004, *AT20M,*ECG158,*WEP238,*276-2005
NKT215	NKT214,*PTC135,*GE-53,*TR-84,*HEPG0005,*SK3004,*AT20M,*ECG158,*WEP238, *276-2004
NKT216	2N2706,2SB431,NKT211,NKT213,*PTC135,*GE-53,*TR-84,*HEPG0005,*SK3004, *AT20M,*ECG158,*WEP238,*276-2005
NKT217	*DS-26,*GE-2,*TR-84,*HEPG0005,*SK3004,*AT20M,*ECG158,*WEP238,*276-2005
NKT218	*PTC135,*GE-53,*TR-84,*HEPG0005,*SK3004,*AT20M,*ECG158,*WEP238, *276-2005
NKT221	NKT228,*PTC135,*GE-53,*TR-05,*HEPG0005,*SK3005,*AT20M,*ECG100,*WEP254, *276-2004,*RT-118
NKT222	2N383,2N508A,2N527,2N598,2N599,2N652,2N655,2N655+,2N1185,2N1189,2N1193, 2N1316,2N1354,2N1355,2N1357,2N1449,2N1997,2N1998,2N2171,2SB223,2SB223A, MA888,NKT223,*PTC135,*GE-53,*TR-85,*HEPG0005,*SK3004,*AT20M,*ECG102A, *WEP250,*276-2005,*RT-121
NKT223	2N383,2N508A,2N527,2N598,2N599,2N652,2N655,2N655+,2N1185,2N1189,2N1193, 2N1316,2N1354,2N1355,2N1357,2N1449,2N1997,2N1998,2N2171,2SB223,2SB223A, MA888,NKT222,*PTC135,*GE-53,*TR-05,*HEPG0005,*SK3004,*AT20M,*ECG102, *WEP631,*276-2005,*RT-120
NKT224	2N59A,2N59B,2N60A,2N60B,2N381,2N414B,2N414C,2N422A,2N461,2N526,2N526A, 2N527A,2N597,2N651,2N651A,2N652A,2N654,2N654+,2N1008A,2N1125,2N1175, 2N1175A,2N1192,2N1348,2N1349,2N1350,2N1351,2N1356,2N1373,2N1375, 2N1377,2N1415,2N1447,2N1448,2N1451,2N1452,2N1478,2N1681,2N1707,2SA207, 2SA407,2SB57A,2SB221,2SB221A,2SB222,2SB222A,2SB226,2SB227,2SB304, 2SB304A,2SB326,2SB427,2SB428,MA887,NKT221,NKT226,NKT228,SF.T223, *PTC135,*GE-53,*TR-05,*HEPG0005,*SK3004,*AT20M,*ECG102,*WEP631, *276-2004,*RT-120
NKT225	2N61A,2N61B,2N396A,2N524,2N524A,2N525,2N525A,2N573,2N650,2N650A,2N653, 2N1018,2N1191,2N1313,2N1413,2N1446,2SA205,2SA406,2SB219,2SB219A,2SB220, 2SB220A,2SB224,MA886,SF.T221,SF.T222,*PTC135,*GE-53,*TR-05,*HEPG0005, *SK3004,*AT20M,*ECG102,*WEP631,*276-2004,*RT-120
NKT226	2N382,2N383,2N461,2N508A,2N527,2N598,2N599,2N652,2N655,2N655+,2N1008A, 2N1125,2N1189,2N1193,2N1354,2N1355,2N1356,2N1357,2N1377,2N1449,2N1997, 2N1998,2SB223,2SB223A,2SB327,MA888,NKT222,NKT223,SF.T223,*PTC135, *GE-53,*TR-05,*HEPG0005,*SK3004,*AT20M,*ECG102,*WEP631,*276-2005, *RT-120
NKT227	2N1955,MA883,*GE-2,*TR-05,*HEPG0005,*SK3004,*AT20M,*ECG102,*WEP631, *276-2005,*RT-120
NKT228	NKT221,*PTC135,*GE-53,*TR-05,*HEPG0005,*SK3004,*AT20M,*ECG102,*WEP631, *276-2004,*RT-120
NKT231	NKT232,*PTC135,*GE-53,*TR-05,*HEPG0005,*SK3004,*AT20M,*ECG102,*WEP631, *AC188/01,*276-2005,*RT-120
NKT232	2N2001,*PTC135,*GE-53,*TR-05,*HEPG0005,*SK3004,*AT20M,*ECG102,*WEP631, *AC188/01,*276-2005,*RT-120
NKT301	NKT302,*HEPG6011
NKT302	*HEPG6011
NKT303	*GE-2,*TR-85,*HEPG6011,*ECG102A,*WEP250,*RT-121
NKT304	*HEPG6011
NKT401	*PTC105A,*GE-25,*TR-01,*HEPG6005,*PT40,*ECG121,*WEP232,*OC28,*276-2006,

To Replace	Substitute This Type
(NKT401)	*RT-127
NKT402	NKT403,NKT404,*PTC105A,*GE-25,*TR-01,*HEPG6005,*PT40,*ECG121,*WEP232, *276-2006,*RT-127
NKT403	*PTC105A,*GE-25,*TR-01,*HEPG6005,*PT40,*ECG121,*WEP232,*276-2006,*RT-127
NKT404	NKT402,NKT403,*PTC105A,*GE-25,*TR-01,*HEPG6005,*PT40,*ECG121,*WEP232, *276-2006,*RT-127
NKT405	NKT402,NKT403,NKT404,*PTC105A,*GE-25,*TR-01,*HEPG6005,*PT40,*ECG121, *WEP232,*276-2006,*RT-127
NKT451	2N285B,*PTC138,*GE-49,*TR-94,*HEPG6003,*PT40,*ECG131,*WEP642,*AD149, *276-2006,*RT-127
NKT452	2N236A,2N285B,NKT451,*PTC138,*GE-49,*TR-94,*HEPG6003,*PT40,*ECG131, *WEP642,*OC28,*RT-127
NKT453	2N235A,2N285A,2N399,2N1291,*PTC138,*GE-49,*TR-94,*HEPG6003,*PT40, *ECG131,*WEP642,*OC28,*276-2006,*RT-127
NKT713	*PTC134,*GE-8,*TR-08,*HEPG0011,*SK3010,*NA20,*ECG103A,*WEP641B, *276-2001,*RT-122
NKT773	AC127,*PTC108,*GE-8,*TR-08,*HEPG0011,*NA20,*ECG103A,*WEP641B,*AC127, *276-2001,*RT-122
NN7000	2N4264,2N4274,2N4275,2N5769,2N5770,2N5772,A5T3709,A5T4123,A8T3709, TIS90-YEL,TIS90M-YEL,TIS92-YEL,TIS92M-YEL,TIS105,*ECG192,*WEP243, *276-2038
NN7001	2N5770,A5T5209,TIS90-BLU,TIS90-GRN,TIS90M-BLU,TIS90M-GRN,TIS92-BLU, TIS92-GRN,TIS92M-BLU,TIS92M-GRN,*PTC110,*TR-55,*HEPS3024,*ECG192, *WEP243,*276-2038
NN7002	2N4265,A5T3707,A5T3710,A5T4124,A8T3707,A8T3710,BCY58,BCY66,TIS90-GRY, TIS90-VIO,TIS90M-GRY,TIS90M-VIO,TIS92-GRY,TIS92-VIO,TIS92M-GRY, TIS92M-VIO,TP5136,*ECG192,*WEP243,*276-2038
NN7500	2N6067,A5T3905,A5T4060,A5T4125,A5T5448,A8T3703,A8T4060,TIS37,TIS91-YEL, TIS91M-YEL,TIS93-YEL,TIS93M-YEL,*ECG193,*WEP242
NN7501	2N6067,A5T3905,A5T4060,A5T4125,A5T5447,A5T5448,A8T3702,A8T3703,A8T4060, TIS37,TIS91-BLU,TIS91-GRN,TIS91-YEL,TIS91M-BLU,TIS91M-GRN,TIS91M-YEL, TIS93-BLU,TIS93-GRN,TIS93-YEL,TIS93M-BLU,TIS93M-GRN,TIS93M-YEL,TP3644, *ECG193,*WEP242
NN7502	A5T3906,A5T404A,A5T4058,A5T4061,A5T4250,A5T5086,A5T5447,A8T3702,A8T404A, A8T4058,A8T4061,BCY79,TIS91-BLU,TIS91-GRY,TIS91-VIO,TIS91M-BLU, TIS91M-GRY,TIS91M-VIO,TIS93-BLU,TIS93-GRY,TIS93-VIO,TIS93M-BLU, TIS93M-GRY,TIS93M-VIO,TP3644,*ECG193,*WEP242
NO400	REFER TO SECTION 2
NPS404	NPS404A,*GE-22,*276-2024
NPS404A	*GE-82,*276-2023
NPS6512	2N760,2N3646,2N4275,2N4962,2N5380,2N5769,2N5772,2N6427,2N6538,2SC1166-O, 2SC1210,A5T3903,A5T4123,BSW82,TIS90-YEL,TIS90M-YEL,TIS92-YEL,TIS92M-YEL, TIS133,TIS134,TN-3903,*PTC101,*GE-210,*276-2009
NPS6513	2N760A,2N930B,2N3242A,2N5369,2N5381,2N6426,2SC968,2SC970,2SC1166-Y, 2SC1214,2SC1247,2SC1813,2SC1852,2SC1991,2SC1999,A5T3904,BC547A,BSW83, MPS-A05,NPS6514,TIS90-BLU,TIS90-VIO,TIS90M-BLU,TIS90M-VIO,TIS92-BLU, TIS92-VIO,TIS92M-BLU,TIS92M-VIO,TN-3904,*PTC101,*GE-210,*276-2009
NPS6514	2N5371,2N5376,2N5377,2SC1166-GR,BC337,BC547B,BC550B,NPS6515,NPS6520, TIS90-GRY,TIS90M-GRY,TIS92-GRY,TIS92M-GRY,TIS94,TIS97,TP3566,*PTC121, *GE-20,*276-2009
NPS6515	2N5370,2N6428,2N6428A,2N6539,NPS6521,TP3566,*276-2009
NPS6516	2N3910,2N3911,2N5372,2N5382,2N6067,2SA661-O,2SA661-R,2SA696,A5T3905, A5T4249,A5T4402,A5T5448,A8T3703,MPS-A55,MPSA55,TIS91-YEL,TIS91M-YEL, TIS93-YEL,TIS93M-YEL,*PTC103,*GE-21,*276-2023
NPS6517	2N3911,2N3912,2N5373,2N5375,2N5383,2SA661-O,2SA661-Y,2SA742,2SA891, A5T2907,A5T3504,A5T3505,A5T3644,A5T3645,A5T3906,A5T404A,A5T4403, A5T5447,A8T3702,A8T404A,BC557A,BC560A,NPS6518,TIS91-BLU,TIS91-GRN, TIS91-VIO,TIS91-YEL,TIS91M-BLU,TIS91M-GRN,TIS91M-VIO,TIS91M-YEL,TIS93-BL- U,TIS93-GRN,TIS93-VIO,TIS93-YEL,TIS93M-BLU,TIS93M-GRN,TIS93M-VIO, TIS93M-YEL,TIS112,TP3644,TP3645,*PTC103,*GE-21,*276-2023
NPS6518	2N3912,2N5373,2N5374,2N5375,2N5378,2N5379,2N5383,2SA661-GR,2SA661-Y,

To Replace	Substitute This Type
(NPS6518)	A5T2907,A5T3504,A5T3505,A5T3644,A5T3645,A5T3906,A5T404A,A5T4250,A5T4403, A5T5447,A8T3702,A8T404A,BC327,BC557A,BC557B,BC560A,BC560B,BCY79, TIS91-GRY,TIS91-VIO,TIS91M-GRY,TIS91M-VIO,TIS93-GRY,TIS93-VIO,TIS93M-GRY, TIS93M-VIO,TIS112,*TR-77,*HEPS5006,*SK3083,*ECG153,*WEP246,*276-2026, *RT-133
NPS6519	2N5378,2N5379,A5T4126,A5T5226,BC328,BC558B,BC558C,BC559B,BC559C, NPS6522,NPS6523,TP3638A
NPS6520	2N5370,2N5371,2N5376,2N5377,2N6428,2N6428A,2N6539,2SC1166-GR,BC337, BC547B,BC550B,NPS6515,NPS6521,TIS94,TIS97,TP3566,*PTC121,*GE-20, *276-2009
NPS6521	2N6428,2N6428A,2N6539,BC550C,*276-2009
NPS6522	2N5378,2N5379,A5T4126,A5T4250,A5T5226,BC328,BC558B,BC558C,BC559B, BC559C,NPS6519,NPS6523,TIS91-GRY,TIS91M-GRY,TIS93-GRY,TIS93M-GRY, TP3638A
NPS6523	2N5378,BC328,BC558C,BC559C,NPS6519
NPSA20	2N3302,2N4954,2N5371,2N5376,2N5377,2N6112,2SC1166-GR,2SC1850,BCY59B, BCY59C,TIS94,TIS97,*PTC121,*GE-62,*276-2009
NR5	REFER TO SECTION 2
NR10	REFER TO SECTION 2
NR700	REFER TO SECTION 2
NS381	2SC54,2SC68,2SC321,2SC322,2SC595,2SC601,2SC689,GET3013,GET3014, GET3646,*PTC126,*GE-61,*TR-95,*HEPS0011,*SK3039,*EN10,*ECG108,*WEP56, *BF173,*276-2016,*RT-113
NS475	2N3648,*PTC123,*GE-210,*TR-21,*HEPS0011,*SK3122,*EN10,*ECG123A,*WEP736, *BC337,*276-2009,*RT-102
NS476	2N3647,*PTC123,*GE-210,*TR-21,*HEPS0011,*SK3122,*EN10,*ECG123A,*WEP736, *BC337,*276-2009,*RT-102
NS477	BC547A,*PTC123,*GE-210,*TR-21,*HEPS0015,*SK3122,*EN10,*ECG123A,*WEP736, *BC337,*276-2009,*RT-102
NS478	*PTC125,*GE-81,*TR-21,*HEPS0011,*SK3122,*ECG123A,*WEP736,*BF338, *276-2009,*RT-102
NS479	2N2520,*PTC125,*GE-81,*TR-21,*HEPS0011,*SK3122,*ECG123A,*WEP736,*BF338, *276-2012,*RT-102
NS480	*PTC125,*GE-81,*TR-21,*HEPS0015,*SK3122,*ECG123A,*WEP736,*276-2012, *RT-102
NS731	2N702,2N3340,*PTC123,*GE-210,*TR-21,*HEPS0011,*SK3122,*EN10,*ECG123A, *WEP736,*BC337,*276-2009,*RT-102
NS732	2N703,2N2331,2N3959,2N3960,2N5770,TIS86,*PTC123,*GE-210,*TR-20,*HEPS0013, *SK3114,*EN10,*ECG159,*WEP242,*BC337,*276-2009,*RT-115
NS733	2N3340,TIS84,TIS108,*PTC123,*GE-210,*TR-21,*HEPS0011,*SK3122,*EN10, *ECG123A,*WEP736,*BC337,*276-2009,*RT-102
NS734	2N2331,2N5769,2N5770,2SC1687,2SC1688,A5T4123,TIS86,TIS87,TIS105,*PTC123, *GE-210,*TR-21,*HEPS0011,*SK3122,*ECG123A,*WEP736,*BC337,*276-2009, *RT-102
NS792	*GE-215
NS793	*GE-215
NS3762	2N3250,2N3250A,2N3905,2N4142,2N4228,2SA467,2SA499-O,2SA522A,2SA539, 2SA548,2SA555,2SA710,2SA711,2SA767,BC203,EN3250,GET2904,GET2905, MM3905,TN-3905,TZ551
NS3763	2N3250A,2N4142,2N4228,2SA539,2SA549,2SA767,GET2904,GET2905,TZ551, *PTC127
NS3903	2SC944,*PTC136,*GE-210,*276-2009
NS3904	*PTC136,*GE-20,*276-2009
NS3905	2N3250,2N3250A,2N3905,2N4034,2N4121,2N4228,2SA467,2SA499-O,2SA499-Y, 2SA539,2SA548,2SA555,2SA603,2SA710,2SA711,2SA718,BC203,BC212L,EN3250, MM3905,TN-3905,*PTC127,*GE-21,*276-2023
NS3906	2N3251,2N3251A,2N3906,2N4035,2N4122,2N5244,MM3906,NPS404A,TN-3906, *PTC127,*GE-21,*276-2023
NSD102	NSD6179,NSE180,NSE181
NSD103	
NSD104	

TRANSISTOR SUBSTITUTES

To Replace	Substitute This Type
NSD105	
NSD106	
NSD126	
NSD127	
NSD128	
NSD129	
NSD131	NSD133
NSD132	NSD134
NSD133	NSD135
NSD134	
NSD135	
NSD151	
NSD152	
NSD202	NSD6181,NSE170,NSE171
NSD203	
NSD204	
NSD205	
NSD206	
NSD3439	
NSD3440	
NSD6178	
NSD6179	NSD6178,NSE181
NSD6180	
NSD6181	NSD6180,NSE171
NSDU01	NSDU01A
NSDU01A	
NSDU05	NSDU06
NSDU06	NSDU07
NSDU07	
NSDU45	NSDU45A
NSDU45A	
NSDU51	NSDU51A
NSDU51A	
NSDU55	NSDU56
NSDU56	NSDU57
NSDU57	
NSE170	NSE171
NSE171	
NSE180	NSE181
NSE181	
NSP41	2N6129,2N6130,NSP41A
NSP41A	2N6130,2N6131,NSP41B
NSP41B	2N6131,FT3055,NSP41C
NSP41C	FT3055
NSP42	2N6132,2N6133,NSP42A
NSP42A	2N6133,2N6134,NSP42B
NSP42B	2N6134,NSP42C
NSP42C	
NSP370	NSP371,TIP32
NSP371	2N6124,2N6125,2N6132,2N6133,NSP42,NSP42A,NSP5193,NSP5194
NSP520	NSP521,TIP31
NSP521	2N6121,2N6122,2N6129,2N6130,NSP41,NSP41A,NSP5190,NSP5191
NSP4918	NSP4919
NSP4919	NSP4920
NSP4920	
NSP4921	NSP4922
NSP4922	NSP4923
NSP4923	
NSP5190	NSP5191
NSP5191	NSP5192
NSP5192	

To Replace	Substitute This Type
NSP5193	NSP5194
NSP5194	NSP5195
NSP5195	
NT3000	
OC16	*DS503,*GE-3,*TR-01,*HEPG6003,*PT40,*ECG104,*WEP230,*276-2006,*RT-124
OC22	OC23,OC24,*PTC138,*GE-16,*TR-01,*HEPG6003,*PT50,*ECG104,*WEP230, *276-2006,*RT-124
OC23	*PTC138,*GE-16,*TR-01,*HEPG6003,*PT50,*ECG104,*WEP230,*276-2006,*RT-124
OC24	OC22,OC23,*PTC138,*GE-16,*TR-01,*HEPG6003,*PT40,*ECG104,*WEP230, *276-2006,*RT-124
OC26	SEE 2N1314
OC27	2N242,2N1168,2N1534,2N2142,2N2142A,2N2143,2N2143A,2N3613,2N5893,2N5894, 2SB251,CQT1110,CQT1110A,CQT1111,CQT1111A,MP2060,*PTC114,*GE-25, *TR-01,*HEPG6003,*SK3009,*PT40,*ECG104,*WEP230,*276-2006,*RT-124
OC28	SEE 2N1666
OC29	SEE 2N1667
OC30	*PTC120A,*GE-30,*TR-94,*HEPG6016,*SK3009,*PT40,*ECG131,*WEP642,*276-2006, *RT-127
OC35	SEE 2N1668
OC36	SEE 2N1669
OC44	2N2953,SF.T308,*PTC109,*GE-50,*TR-17,*HEPG0008,*JR30X,*ECG126,*WEP635, *AF125
OC45	2N2953,OC44,OC47,SF.T307,SF.T308,SF.T323,*PTC109,*GE-50,*TR-85,*HEPG0008, *SK3123,*JR10,*ECG102A,*WEP250,*AF125,*RT-121
OC46	*PTC135,*GE-2,*TR-85,*HEPG0005,*SK3123,*JR10,*ECG102A,*WEP250,*AC188/01, *276-2005,*RT-121
OC47	2N2953,*PTC135,*GE-51,*TR-85,*HEPG0005,*JR10,*ECG102A,*WEP250,*AC188/01, *276-2005,*RT-121
OC57	OC58,*PTC109,*GE-51,*TR-85,*HEPG0005,*HF35,*ECG102A,*WEP250,*AF125, *276-2004,*RT-121
OC58	OC59,OC60,*PTC109,*GE-51,*TR-85,*HEPG0005,*SK3003,*HF35,*ECG102A, *WEP250,*AF125,*276-2005,*RT-121
OC59	*PTC109,*GE-51,*TR-85,*HEPG0005,*SK3123,*HF35,*ECG102A,*WEP250,*AF125, *276-2005,*RT-121
OC60	OC59,*PTC109,*GE-51,*TR-85,*HEPG0005,*SK3003,*HF35,*ECG102A,*WEP250, *AF125,*276-2005,*RT-121
OC65	2SB185,OC66,*PTC109,*GE-51,*TR-85,*HEPG0005,*SK3123,*JR15,*ECG102A, *WEP250,*AF125,*276-2004,*RT-121
OC66	2SB114,2SB328,*PTC109,*GE-51,*TR-85,*HEPG0005,*SK3123,*JR15,*ECG102A, *WEP250,*AF125,*276-2004,*RT-121
OC70	2N215+,2SB92,2SB170,2SB172,2SB288,NKT214,NKT215,OC73,*PTC109,*GE-53, *TR-85,*HEPG0005,*SK3123,*JR5,*ECG102A,*WEP250,*AC126,*276-2004,*RT-121
OC71	2N838,2N2587,2N2956,2N3323,2N3324,2N3325,2N6365,2N6365A,2SB37,2SB56, 2SB75A,ASY30,NKT213,NKT214,NKT216,*GE-53,*TR-85,*HEPG0005,*SK3123,*JR5, *ECG102A,*WEP250,*276-2004,*RT-121
OC71N	2N217+,2N838,2N2188,2N2953,2N2956,2N2957,2N3323,2N3324,2N3325,2N6365, 2SB37,2SB56,2SB59,2SB66,2SB75A,2SB77A,2SB171,2SB173,2SB175,2SB176, ACY23,ACY32,MM404A,NKT213,NKT216,NKT224,NKT226,SF.T315,*GE-53,*TR-85, *HEPG0005,*SK3123,*ECG102A,*WEP250,*276-2005,*RT-121
OC72	NKT211,*GE-53,*TR-85,*HEPG0005,*SK3003,*JR15,*ECG102A,*WEP250,*276-2005, *RT-121
OC73	*GE-53,*TR-85,*HEPG0005,*AT20M,*ECG102A,*WEP250,*276-2005,*RT-121
OC74	*DS26,*GE-53,*TR-85,*HEPG0005,*SK3123,*AT100H,*ECG102A,*WEP250, *AC188/01,*276-2005,*RT-121
OC75	2N2614,2N2953,2N2957,2SB54,2SB431,NKT211,*GE-53,*TR-85,*HEPG0005,*AT30H, *ECG102A,*WEP250,*276-2005,*RT-121
OC75N	2N2189,2N2613+,2N2635,2SB54,2SB439,2SB440,*GE-53,*TR-85,*HEPG0005, *ECG102A,*WEP250,*276-2005,*RT-121
OC76	2N109+,2N217+,2SB370A,2SB431,NKT211,NKT222,NKT223,NKT226,*PTC135, *GE-53,*TR-85,*HEPG0005,*SK3004,*JR15,*ECG102A,*WEP250,*276-2004, *RT-121

To Replace	Substitute This Type
OC77	NKT217,NKT227,*DS26,*GE-2,*TR-85,*HEPG0005,*JR15,*ECG102A,*WEP250, *276-2004,*RT-121
OC79	AC152,ASY70,*PTC135,*GE-53,*TR-85,*HEPG0005,*AT100M,*ECG102A,*WEP250, *276-2004,*RT-121
OC80	*TR-85,*HEPG0005,*ECG102A,*WEP250,*276-2005,*RT-121
OC139	*PTC108,*GE-8,*TR-08,*HEPG0011,*SK3011,*NR5,*ECG101,*WEP641,*276-2001, *RT-119
OC140	*DS25,*GE-7,*TR-08,*HEPG0011,*NR5,*ECG101,*WEP641,*RT-119
OC141	*GE-54,*TR-08,*HEPG0011,*SK3011,*ECG101,*WEP641,*276-2001,*RT-119
OC170	2N3783,2N3784,2SA246,A1383,SF.T316,SF.T354,SF.T357,SF.T358,*PTC107,*GE-50, *TR-17,*HEPG0003,*JR100,*ECG160,*WEP637
OC602	2SB75,2SB185,2SB365,NKT214,SF.T322,*PTC109,*GE-52,*TR-85,*HEPG0005,*JR15, *ECG102A,*WEP250,*AC126,*276-2005,*RT-121
OC602-SPEZ	*PTC135,*TR-85,*HEPG0005,*ECG102A,*WEP250,*AC128,*276-2004
OC603	*PTC135,*GE-50,*TR-85,*HEPG0005,*JR15,*ECG102A,*WEP250,*AC126,*276-2005, *RT-121
OC604	SEE AC122
OC604-SPEZ	*PTC135,*TR-85,*HEPG0005,*ECG102A,*WEP250,*AC128,*276-2004
OC613	SEE AF101
OC614	2N2189,2N2635,OC615,*PTC145,*GE-51,*TR-17,*HEPG0008,*JR100,*ECG126, *WEP635,*AF125
OC615	2N2635,*PTC145,*GE-51,*TR-17,*HEPG0008,*JR100,*ECG126,*WEP635,*AF125
OD603	
OD603/50	
PADT50	AUY10,*GE-16,*TR-01,*HEPG6005,*SK3009,*ECG104,*WEP230,*RT-124
PAR12	REFER TO SECTION 2
PBC107A	2N3116,2N3302,2N3904,2N4401,2N4952,2N4970,2N5369,2N5377,2N5381, 2SC366G-Y,2SC1175,2SC1364,2SC1685,2SC1850,2SC1992,A157B,BC107,BC107B, BC167B,BC171B,BC174B,BC182B,BC182L,BC190B,BC237B,BC413B,BC414B, BC547A,BCY59B,BFY39-3,BSX79,MPS6531,PBC107B,TN-3904,*PTC132,*GE-212, *TR-51,*SK3245,*ECG199,*WEP736,*BC107B
PBC107B	2N4953,2N5370,2N5376,2N5377,2N5827,2N6006,2N6112,A157B,BC107B,BC167B, BC171B,BC174B,BC182B,BC183L,BC184L,BC190B,BC237B,BC413B,BC414B, BC547B,BC550B,BCY59C,*PTC121,*GE-212,*TR-51,*SK3245,*ECG199,*WEP736, *BC107B
PBC108A	2N2924,2N4124,2N4256,2N4419,2N5224,2N5998,2SC372G-Y,2SC373G,2SC400, 2SC400-GR,2SC735-GR,2SC735-Y,2SC1363,2SC1684,2SC1849,2SC1993,A158B, A159B,BC108,BC108B,BC109B,BC168B,BC172B,BC173B,BC238B,BC239B, BC548A,BCY58B,BSW88,BSW89,BSX38,PBC108B,PBC109B,*PTC132,*GE-212, *TR-51,*SK3245,*ECG199,*WEP736,*BC107B,*276-2015
PBC108B	2N2925,2N4256,2N6002,2N6008,2SC373G,2SC735-GR,2SC1993,A158B,A159B, BC108B,BC109,BC109B,BC168B,BC172B,BC173B,BC238B,BC239B,BC548B, BC549B,BCY58C,PBC109B,*PTC139,*GE-212,*TR-51,*SK3245,*ECG199,*WEP736, *BC107B,*276-2015
PBC108C	A158C,A159C,BC548C,BC549C,PBC109C,*PTC139,*GE-212,*SK3245,*ECG199, *WEP736,*BC107B,*276-2015
PBC109B	2N2925,2N4256,2N6002,2N6008,2SC373G,2SC735-GR,2SC1993,A158B,A159B, BC108B,BC109,BC109B,BC168B,BC172B,BC173B,BC238B,BC239B,BC548B, BC549B,BCY58C,PBC108B,*PTC139,*GE-212,*TR-51,*SK3245,*ECG199,*WEP736, *BC107B,*276-2015
PBC109C	A158C,A159C,BC548C,BC549C,PBC108C,*PTC139,*GE-212,*SK3245,*ECG199, *WEP736,*BC107B,*276-2015
PEP9	2N708,2N708A,2N784A,2N914,2N2242,2N2481,2N2501,2N2845,2N2847,2N3013, 2N3014,2N3115,2N3210,2N3211,2N3301,2N3646,2N3647,2N3903,2N3946,2N3974, 2N3976,2N4013,2N4123,2N4140,2N4227,2N4400,2N4962,2N5027,2N5144,2N5380, 2N5772,2N6538,2SC67,2SC577,2SC601N,2SC639,2SC764,2SC1274,A5T3903, A5T4123,BSW82,FT3641,FT3642,GET2221,GET2369,GI-3641,GI-3642,MM3903, MPS6530,TN-3903,ZT708,*PTC136,*GE-210,*TR-21,*HEPS0011,*SK3122, *ECG123A,*WEP736,*BC107B,*276-2016,*RT-102
PEP95	2N2331,2N2501,2N3241A,2N4123,2N4418,2N5810,2N6426,2N6427,2SC367G, 2SC367G-O,2SC509-O,2SC509G-O,2SC639,2SC735-O,2SC1216,2SC1317,

To Replace	Substitute This Type
(PEP95)	2SC1788,2SC1851,A5T4123,BC338-16,BSX75,D33D21,TN62,ZT2938,*PTC123, *GE-64,*TR-53,*ECG172,*WEPS9100,*BC107B,*276-2016
PET1001	2N916,2N916A,2N2847,2N3115,2N3862,2N3946,2N3974,2N3976,2N4013,2N4951, 2N4962,2N5144,2N5368,2N5380,2N6538,A5T3903,MM3903,TIS133,TIS134,TN-3903, *PTC121,*GE-210,*SK3137,*ECG192,*WEP243,*BC337
PET1002	2N3116,2N3302,2N4952,2N5369,2N5377,2N5381,A5T3904,BCY59B,TN-3904, *PTC121,*GE-210,*HEPS0025,*SK3122,*ECG123A,*WEP736,*BC337,*276-2013
PET1075	2N871,2N3038,*PTC170,*GE-18,*TR-95,*SK3039,*ECG108,*WEP56,*276-2008, *RT-113
PET1075A	A5T5551,*PTC170,*TR-78,*ECG154,*WEP712,*276-2008,*RT-110
PET2001	2N916,2N916A,2N2242,2N2369,2N2369A,2N2501,2N3009,2N3013,2N3014,2N3211, 2N3862,2N4013,2N4873,2N5144,2N5769,2SC67,2SC577,2SC639,2SC764,A5T4123, BSW82,PET1001,TIS133,TIS134,ZT2369,ZT2369A,*PTC101,*GE-210,*TR-63, *SK3122,*ECG123A,*WEP736,*BC337,*276-2009
PET2002	2N3242A,2N4954,2N5371,BCY58B,BCY59B,BSW83,PET1002,*PTC121,*GE-210, *SK3122,*ECG123A,*WEP736,*BC337,*276-2009
PET3001	2N2710,2N4137,2N5029,2SC601,2SC689,TIS48,TIS49,*PTC132,*GE-61,*TR-95, *HEPS0020,*SK3018,*ECG108,*WEP56,*276-2015,*RT-107
PET3002	2N3227,2N4873,2SC1216,ZT2369A,*PTC132,*276-2015,*RT-113
PET3704	2N2222,2N2222A,2N2540,2N3302,2N3704,2N5449,2N5582,2N5818,2N6010,2N6014, A5T2222,A5T5449,A8T3704,D33D26,GI-3704,TIS109,TIS111,TN-3904,*GE-20, *SK3122,*ECG123A,*BC337,*276-2014
PET3705	2N915,2N2539,2N2847,2N2897,2N2900,2N3705,2N4962,2N5450,2N5820,A5T5450, A8T3705,GI-3705,TIS110,TN54,TN-3903,*GE-20,*TR-65,*SK3122,*ECG123A,*BC337, *276-2014
PET3706	2N2222,2N2540,2N3242A,2N3302,2N3704,2N5449,2N5818,A5T2222,A5T5449, A8T3704,BC337-25,D33D26,GI-3706,TIS109,TIS111,TN60,TN-3904,*GE-20,*SK3122, *ECG123A,*WEP736,*BC337,*276-2014
PET4001	2N3242A,2N4074,2N6426,*GE-62,*TR-87,*HEPS0015,*SK3024,*EN10,*ECG128, *WEP243,*BC337,*276-2014,*RT-114
PET4002	PET4003,*GE-62,*HEPS0015,*SK3122,*ECG123A,*WEP736,*BC337,*276-2014
PET4003	*GE-62,*HEPS0015,*SK3156,*ECG199,*WEP736,*BC337,*276-2014
PET6001	2N718,2N731,2N760,2N916,2N916A,2N2221,2N2501,2N2539,2N2845,2N2847, 2N2897,2N2900,2N3301,2N3646,2N3705,2N3736,2N4962,2N5450,A5T5450,A8T3705, D33D24,GI-3705,PET3705,TIS110,TN62,TN-3903,ZT708,*GE-20,*TR-65,*HEPS0025, *SK3122,*ECG123A,*WEP736,*BC337,*276-2014
PET6002	2N760A,2N909,2N930B,2N2222,2N2350,2N2350A,2N2540,2N3242A,2N3704,2N5449, 2N6010,40458,A5T2192,A5T2222,A5T5449,A8T3704,BC337-16,BSX75,D33D25, GI-3704,GI-3706,PET3704,PET3706,PET8000,TIS111,TN60,TN-3904,*GE-20, *HEPS0025,*SK3122,*ECG123A,*WEP736,*BC337,*276-2014
PET8000	2N760A,40458,*GE-20,*HEPS0015,*SK3122,*ECG123A,*WEP736,*BC337,*276-2009
PET8001	PET8002,PET8003,PET8004,*GE-47,*HEPS0030,*SK3122,*ECG123A,*WEP736, *BC337,*276-2009
PET8002	PET8004,*HEPS0015,*SK3122,*ECG123A,*WEP736,*BC337,*276-2009
PET8003	PET8001,PET8002,PET8004,*GE-47,*HEPS0030,*SK3122,*ECG123A,*WEP736, *BC337,*276-2009
PET8004	PET8002,*HEPS0030,*SK3122,*ECG123A,*WEP736,*BC337,*276-2009
PET8101	2SC718,2SC1231,*PTC132,*TR-24,*HEPS0025,*276-2015
PMD10K-100	2N6059,2N6284,PMD16K-100
PMD10K-40	2N6057,2N6282,PMD10K-60,PMD16K-40,PMD16K-60
PMD10K-60	2N6057,2N6058,2N6282,2N6283,PMD10K-80,PMD16K-60,PMD16K-80
PMD10K-80	2N6058,2N6059,2N6283,2N6284,PMD10K-100,PMD16K-100,PMD16K-80
PMD11K-100	2N6052,2N6287,PMD17K-100,PMD17K-80
PMD11K-40	2N6050,2N6285,PMD11K-60,PMD17K-40,PMD17K-60
PMD11K-60	2N6050,2N6051,2N6285,2N6286,PMD11K-80,PMD17K-60
PMD11K-80	2N6051,2N6052,2N6286,2N6287,PMD11K-100,PMD17K-100,PMD17K-80
PMD12K-100	2N6059,PMD10K-100
PMD12K-40	2N6055,2N6057,2N6383,2N6384,PMD10K-40,PMD10K-60,PMD12K-60
PMD12K-60	2N6055,2N6057,2N6058,2N6384,2N6385,PMD10K-60,PMD10K-80,PMD12K-80
PMD12K-80	2N6058,2N6059,2N6385,PMD10K-100,PMD10K-80,PMD12K-100
PMD13K-100	2N6052,PMD11K-100

To Replace	Substitute This Type
PMD13K-40	2N6050,2N6053,PMD11K-40,PMD11K-60,PMD13K-60
PMD13K-60	2N6050,2N6051,2N6053,PMD11K-60,PMD11K-80,PMD13K-80
PMD13K-80	2N6051,2N6052,PMD11K-100,PMD11K-80,PMD13K-100
PMD16K-100	
PMD16K-40	PMD16K-60
PMD16K-60	PMD16K-80
PMD16K-80	PMD16K-100
PMD17K-100	PMD17K-80
PMD17K-40	PMD17K-60
PMD17K-60	
PMD17K-80	PMD17K-100
PN26	REFER TO SECTION 2
PN66	REFER TO SECTION 2
PN350	REFER TO SECTION 2
PP7535	
PP7536	
PT2A	*GE-51,*TR-17,*HEPG0003,*SK3006,*HF35,*ECG160,*WEP637,*AC126
PT2S	*GE-51,*TR-17,*HEPG0003,*SK3006,*HF35,*ECG160,*WEP637,*AC126
PT6	REFER TO SECTION 2
PT12	REFER TO SECTION 2
PT25	REFER TO SECTION 2
PT32	REFER TO SECTION 2
PT40	REFER TO SECTION 2
PT50	REFER TO SECTION 2
PT150	REFER TO SECTION 2
PT201	REFER TO SECTION 2
PT250	REFER TO SECTION 2
PT501	REFER TO SECTION 2
PT515	REFER TO SECTION 2
PT530	SEE 2N1709
PT530A	SEE 2N2783
PT531	SEE 2N2874
PT600	*GE-28,*TR-76,*SK3192,*ECG186,*WEP245
PT601	*GE-28,*TR-76,*SK3192,*ECG186,*WEP245
PT612	*PTC144,*GE-66,*TR-87,*SK3024,*ECG128,*WEP243,*RT-114
PT613	SEE 2N2782
PT665	HST6011,HST6012,HST6309,HST6310,HST6409,HST6410,PT665A,*GE-66,*ECG152, *WEP245
PT665A	HST6012,HST6310,HST6409,HST6410
PT692	SEE 2N2781
PT703	*PTC123,*GE-62,*TR-21,*HEPS0011,*SK3122,*SN80,*ECG123A,*WEP736,*BC337, *276-2010,*RT-102
PT720	2N2242,2N2501,2N3013,2N3014,2N3211,2SC67,2SC577,2SC639,2SC764,A5T4123, BSW82,*PTC136,*GE-210,*TR-21,*SK3122,*SN80,*ECG123A,*WEP736,*276-2009, *RT-102
PT886	*PTC123,*GE-81,*TR-21,*HEPS0014,*SK3046,*SN80,*ECG123A,*WEP736,*BC337, *276-2012,*RT-102
PT887	*PTC123,*GE-81,*TR-21,*HEPS0014,*SK3024,*ECG123A,*WEP736,*BC337, *276-2012,*RT-102
PT888	*PTC123,*GE-81,*TR-21,*HEPS0014,*SK3047,*ECG128,*WEP736,*BC337,*276-2012, *RT-102
PT896	*PTC144,*GE-27,*TR-87,*HEPS5014,*SK3024,*ECG128,*WEP243,*BF338,*276-2020, *RT-114
PT897	*PTC123,*GE-81,*TR-21,*HEPS0014,*SK3122,*EN10,*ECG123A,*WEP736,*BC337, *276-2012,*RT-102
PT898	*PTC144,*GE-27,*TR-21,*HEPS0014,*SK3122,*EN10,*ECG123A,*WEP736,*BF338, *276-2020,*RT-102
PT1515	*GE-32
PT1544	*PTC125,*GE-215,*TR-25,*ECG128,*WEP243
PT1545	*PTC125,*GE-215,*TR-25,*ECG128,*WEP243
PT1558	2N1837A,2N3498,2N4047,2SC307,2SC470-3,2SC1218,SE8010,*PTC144,*TR-21,

To Replace	Substitute This Type
(PT1558)	*SK3122,*ECG123A,*WEP736,*RT-102
PT1937	
PT1941	2N1904,*GE-19,*TR-59,*HN100,*ECG130,*WEP247,*RT-131
PT1949	
PT1963	HST7806
PT2523	PT2524,PT2525,*GE-27,*TR-78,*HEPS3021,*SK3201,*ECG171,*WEPS3021,*BF338
PT2524	PT2525,*GE-27,*TR-78,*HEPS3021,*SK3201,*ECG171,*WEPS3021,*BF338
PT2525	PT2524,*GE-27,*TR-78,*SK3201,*ECG171,*WEPS3021,*BF338
PT2540	2N3295,2N5188,*GE-63,*TR-63,*HEPS3001,*ECG192,*WEP243,*276-2038
PT2610	
PT2620	PT2620A,*GE-28,*TR-76,*ECG186,*WEP245
PT2620A	PT2620
PT2630	*GE-32
PT2635	HST6012,HST6310,HST6409,HST6410,HST7013,*GE-66,*ECG152,*WEP245, *276-2020
PT2640	*PTC110,*GE-32,*TR-76,*SK3192,*ECG186,*WEP245
PT2660	*GE-215,*TR-76,*SK3192,*ECG186,*WEP245
PT2670	PT2610,PT2630,*GE-32,*276-2020
PT2760	2N708,2N708A,2N784A,2N914,2N2242,2N2369,2N2369A,2N2481,2N2501,2N3009, 2N3013,2N3014,2N3210,2N3211,2N3511,2N3646,2N3647,2N3862,2N4013,2N4137, 2N5144,2N5769,2N5772,2SC67,2SC577,2SC639,2SC764,A5T4123,BSW82, GET2369,TIS133,TIS134,ZT708,ZT2369,*PTC136,*GE-210,*TR-21,*HEPS0011, *SK3122,*SN80,*ECG123A,*WEP736,*276-2009,*RT-102
PT4800	*PTC144,*GE-83,*TR-21,*HEPS0025,*SK3024,*SN80,*ECG123A,*WEP736,*RT-102
PT4816	HST5001,HST5002,HST5501,HST5502,*PTC144,*GE-27,*TR-70,*SK3039,*EN10, *ECG107,*WEP56,*BF338,*RT-108
PT4830	*PTC144,*GE-27,*TR-70,*HEPS5014,*SK3039,*EN10,*ECG107,*WEP56,*BF338, *RT-108
PTC101	REFER TO SECTION 2
PTC102	REFER TO SECTION 2
PTC103	REFER TO SECTION 2
PTC104	REFER TO SECTION 2
PTC105	REFER TO SECTION 2
PTC105A	REFER TO SECTION 2
PTC106	REFER TO SECTION 2
PTC107	REFER TO SECTION 2
PTC108	REFER TO SECTION 2
PTC109	REFER TO SECTION 2
PTC110	REFER TO SECTION 2
PTC111	REFER TO SECTION 2
PTC112	REFER TO SECTION 2
PTC113	REFER TO SECTION 2
PTC114	REFER TO SECTION 2
PTC115	REFER TO SECTION 2
PTC116	REFER TO SECTION 2
PTC117	REFER TO SECTION 2
PTC118	REFER TO SECTION 2
PTC119	REFER TO SECTION 2
PTC120	REFER TO SECTION 2
PTC120A	REFER TO SECTION 2
PTC121	REFER TO SECTION 2
PTC122	REFER TO SECTION 2
PTC123	REFER TO SECTION 2
PTC124	REFER TO SECTION 2
PTC125	REFER TO SECTION 2
PTC126	REFER TO SECTION 2
PTC127	REFER TO SECTION 2
PTC128	REFER TO SECTION 2
PTC129	REFER TO SECTION 2
PTC129A	REFER TO SECTION 2
PTC130	REFER TO SECTION 2

TRANSISTOR SUBSTITUTES

To Replace	Substitute This Type
PTC131	REFER TO SECTION 2
PTC132	REFER TO SECTION 2
PTC133	REFER TO SECTION 2
PTC134	REFER TO SECTION 2
PTC135	REFER TO SECTION 2
PTC136	REFER TO SECTION 2
PTC137	REFER TO SECTION 2
PTC138	REFER TO SECTION 2
PTC139	REFER TO SECTION 2
PTC140	REFER TO SECTION 2
PTC141	REFER TO SECTION 2
PTC142	REFER TO SECTION 2
PTC143	REFER TO SECTION 2
PTC144	REFER TO SECTION 2
PTC145	REFER TO SECTION 2
PTC146	REFER TO SECTION 2
PTC147	REFER TO SECTION 2
PTC148	REFER TO SECTION 2
PTC149	REFER TO SECTION 2
PTC150	REFER TO SECTION 2
PTC153	REFER TO SECTION 2
PTC154	REFER TO SECTION 2
PTC155	REFER TO SECTION 2
PTC156	REFER TO SECTION 2
PTC157	REFER TO SECTION 2
PTC158	REFER TO SECTION 2
PTC160	REFER TO SECTION 2
PTC162	REFER TO SECTION 2
PTC163	REFER TO SECTION 2
PTC164	REFER TO SECTION 2
PTC165	REFER TO SECTION 2
PTC166	REFER TO SECTION 2
PTC167	REFER TO SECTION 2
PTC168	REFER TO SECTION 2
PTC169	REFER TO SECTION 2
PTC170	REFER TO SECTION 2
PTC171	REFER TO SECTION 2
PTC172	REFER TO SECTION 2
PTC173	REFER TO SECTION 2
PTC174	REFER TO SECTION 2
PTC175	REFER TO SECTION 2
PTC176	REFER TO SECTION 2
PTC177	REFER TO SECTION 2
PTC178	REFER TO SECTION 2
PTC179	REFER TO SECTION 2
PTC180	REFER TO SECTION 2
PTC186	REFER TO SECTION 2
PTC191	REFER TO SECTION 2
PTC192	REFER TO SECTION 2
PTC193	REFER TO SECTION 2
PTC194	REFER TO SECTION 2
PTC195	REFER TO SECTION 2
PTC196	REFER TO SECTION 2
PTC197	REFER TO SECTION 2
PTC198	REFER TO SECTION 2
PTC901	REFER TO SECTION 2
PTC904	REFER TO SECTION 2
PTC905	REFER TO SECTION 2
PTC906	REFER TO SECTION 2
PTC1101	C.P.PTC110/PTC111
PTC1123	C.P.PTC112/PTC113

To Replace	Substitute This Type
PTC1423	C.P.PTC142/PTC143
PTC1623	C.P. PTC162/PTC163
PTC1645	C.P.PTC164/PTC165
PTC1667	C.P. PTC166/PTC167
PTC1689	C.P. PTC168/PTC169
PTC1701	C.P.PTC170/PTC171
PTC1723	C.P. PTC172/PTC173
PTC1745	C.P. PTC174/PTC175
PTC1778	C.P. PTC177/PTC178
PTC1956	C.P.PTC195/PTC196
PTC1978	C.P.PTC197/PTC198
PTC9045	C.P.PTC904/PTC905
Q2T2222	
Q2T2905	
Q2T3244	
Q2T3725	
RCA1A15	2N5152,2N5320,ST74049,*ECG128
RCA1A16	*ECG129
RCA1A17	*PTC144,*ECG128
RCA1A18	*ECG128
RCA1A19	*PTC141,*ECG129
RCA1B07	2N6058,2N6059,2N6283,2N6284,2N6385,PMD10K-100,PMD10K-80,*PTC169,*SK3182
RCA1B08	2N6051,2N6052,2N6286,2N6287,PMD11K-100,PMD11K-80,*PTC168
RCA1C10	*PTC169,*ECG196,*276-2019
RCA1C11	*PTC168,*ECG197
RCA1E02	BLY50A
RCA1E03	
RCA29	MJE2522,MJE2523,RCA29A,*PTC167,*SK3054,*ECG152,*276-2018
RCA29A	MJE2523,RCA29B,*PTC167,*SK3054,*ECG152,*276-2020
RCA29B	RCA29C,*PTC167,*SK3054,*ECG291
RCA29C	*PTC169,*ECG291
RCA30	MJE2490,MJE2491,MJE3740,RCA30A,*PTC166,*SK3083,*ECG153,*276-2026
RCA30A	MJE2491,MJE3740,MJE3741,RCA30B,*PTC166,*SK3083,*ECG153
RCA30B	MJE3741,RCA30C,*PTC166,*SK3083,*ECG292
RCA30C	*PTC168,*ECG292
RCA31	RCA31A,*PTC167,*SK3054,*276-2019
RCA31A	RCA31B,*PTC167,*SK3054,*276-2020
RCA31B	RCA31C,*PTC167,*SK3054
RCA31C	*PTC169
RCA32	RCA32A,*PTC166,*SK3083,*276-2027
RCA32A	RCA32B,*PTC166,*SK3083
RCA32B	RCA32C,*PTC166,*SK3083
RCA32C	*PTC168
RCA41	RCA41A,*PTC169
RCA41A	RCA41B,*PTC169
RCA41B	RCA41C,*PTC169
RCA41C	*PTC169
RCA42	RCA42A,*PTC168
RCA42A	RCA42B,*PTC168
RCA42B	RCA42C,*PTC168
RCA42C	*PTC168
RCA120	RCA121,*SK3180,*ECG261
RCA121	RCA122,*SK3180,*ECG261
RCA122	
RCA125	RCA126,*ECG262
RCA126	*ECG262
RCA0610-30†	
RCA2001†	2N6390,2N6390†,2N6391†,2N6393†,RCA2003†,RCA2005†,RCA2310†
RCA2003†	2N6390,2N6390†,2N6391†,2N6393†,RCA2005†,RCA2310†
RCA2005†	2N6391†,2N6393†,RCA2310†,*276-2020
RCA2010†	2N6392†,2N6393†,*276-2020

To Replace	Substitute This Type
RCA2023-12†	
RCA2310†	
RCA3001†	2N6527
RCA3003†	2N6528,2N6529,RCA3005†
RCA3005†	2N6529
RCA3054	*PTC169,*276-2020
RCA3055	*PTC169
RCA3441	
RCA6263	RCA3441
RCA8203	RCA8203A,*SK3181,*ECG262
RCA8203+	RCA8203A+,*SK3181,*ECG262
RCA8203A	RCA8203B,*SK3221,*ECG264
RCA8203A+	RCA8203B+,*SK3221,*ECG264
RCA8203B	*ECG264
RCA8203B+	*ECG264
RCA8350	RCA8350A,*SK3183,*ECG246
RCA8350A	RCA8350B,*ECG246
RCA8350B	*ECG246
RCA-1000	2SD502,2SD503,2SD504,2SD505,RCA-1001,*ECG243
RCA-1001	2SD503,2SD505,2SD506,*ECG243
RCP111A	RCP111B,*SK3201,*ECG171,*WEPS3021
RCP111B	RCP111C,*SK3201,*ECG171,*WEPS3021
RCP111C	RCP111D,*SK3201,*ECG171,*WEPS3021
RCP111D	*SK3201,*ECG171,*WEPS3021
RCP113A	RCP113B,*SK3201,*ECG171,*WEPS3021
RCP113B	RCP113C,*SK3201,*ECG171,*WEPS3021
RCP113C	RCP113D,*SK3201,*ECG171,*WEPS3021
RCP113D	*SK3201,*ECG171,*WEPS3021
RCP115	*SK3201,*ECG171,*WEPS3021
RCP115B	*SK3201,*ECG171,*WEPS3021
RCP117	*SK3201,*ECG171,*WEPS3021
RCP117B	*SK3201,*ECG171,*WEPS3021
RCP700A	
RCP700B	RCP700C
RCP700C	RCP700D
RCP700D	
RCP701A	
RCP701B	RCP701C
RCP701C	RCP701D
RCP701D	
RCP702A	RCP700A
RCP702B	RCP700B,RCP700C,RCP702C
RCP702C	RCP700C,RCP700D,RCP702D
RCP702D	RCP700D
RCP703A	RCP701A
RCP703B	RCP701B,RCP701C,RCP703C
RCP703C	RCP701C,RCP701D,RCP703D
RCP703D	RCP701D
RCP704	
RCP704B	
RCP705	
RCP705B	
RCP706	
RCP706B	
RCP707	
RCP707B	
RCS29	2SD315,RCS29A,*PTC112
RCS29A	2N5614,2SC1444,2SD315,KSP1024,RCS29B,*PTC112
RCS29B	2N5614,2N5618,2N6465,2SC1444,2SC1445,BLY48A,KSP1024,KSP1025,RCS29C, *PTC112
RCS29C	2N3878,2N5618,2N6465,2N6466,2SC1445,BLY48A,KSP1025,*PTC148

To Replace	Substitute This Type
RCS30	2SA764,KSP2024,RCS30A,*PTC113
RCS30A	2N5613,2SA764,2SA765,KSP2024,KSP2025,RCS30B,*PTC113
RCS30B	2N5613,2N5617,2N6467,2SA765,KSP2025,RCS30C,*PTC113
RCS30C	2N5617,2N6467,2N6468,KSP2026,*PTC168
RCS31	RCS31A,*PTC169
RCS31A	KSP1161,RCS31B,*PTC169
RCS31B	KSP1161,KSP1162,RCS31C,*PTC148
RCS31C	KSP1162,*PTC148
RCS32	RCS32A
RCS32A	KSP2161,RCS32B,*PTC168
RCS32B	KSP2161,KSP2162,RCS32C,*PTC168
RCS32C	KSP2162,*PTC168
RCS242	2SD231,2SD249,1561-0403,1561-0404,1582-0403,1582-0404,1582-0405,1582-0408,1582-0410,1582-0415,1582-0603,1582-0604,1582-0605,1582-0608,1582-0610,1582-0615,B170002-BLK,B170020,HST9205,MJ2801,MJ3771,SDT9205,*PTC140,*ECG130,*WEP247
RCS258	2N6570,2N6571,*PTC175
RCS579	2N6580,2N6581,2N6583,2N6584,2SC1141,2SC1579,2SC1580,IR4065,IR5063
RE 1	REFER TO SECTION 2
RE 2	REFER TO SECTION 2
RE 3	REFER TO SECTION 2
RE 4	REFER TO SECTION 2
RE 5	REFER TO SECTION 2
RE 6	REFER TO SECTION 2
RE 7	REFER TO SECTION 2
RE 7MP	M.P. RE 7
RE 8	REFER TO SECTION 2
RE 9	REFER TO SECTION 2
RE 10	REFER TO SECTION 2
RE 11	REFER TO SECTION 2
RE 11MP	M.P. RE 11
RE 12	REFER TO SECTION 2
RE 13	REFER TO SECTION 2
RE 14	REFER TO SECTION 2
RE 15	REFER TO SECTION 2
RE 16	REFER TO SECTION 2
RE 17	REFER TO SECTION 2
RE 18	REFER TO SECTION 2
RE 19	REFER TO SECTION 2
RE 20	REFER TO SECTION 2
RE 20MP	M.P. RE 20
RE 21	REFER TO SECTION 2
RE 22	REFER TO SECTION 2
RE 23	REFER TO SECTION 2
RE 24	REFER TO SECTION 2
RE 25	REFER TO SECTION 2
RE 26	REFER TO SECTION 2
RE 27	REFER TO SECTION 2
RE 28	REFER TO SECTION 2
RE 29	REFER TO SECTION 2
RE 30	REFER TO SECTION 2
RE 31	REFER TO SECTION 2
RE 32	REFER TO SECTION 2
RE 33	REFER TO SECTION 2
RE 34	REFER TO SECTION 2
RE 35	REFER TO SECTION 2
RE 36	REFER TO SECTION 2
RE 37	REFER TO SECTION 2
RE 38	REFER TO SECTION 2
RE 39	REFER TO SECTION 2
RE 40	REFER TO SECTION 2

TRANSISTOR SUBSTITUTES

To Replace	Substitute This Type
RE 41	REFER TO SECTION 2
RE 42	REFER TO SECTION 2
RE 43	REFER TO SECTION 2
RE 44	REFER TO SECTION 2
RE 53	REFER TO SECTION 2
RE 54	REFER TO SECTION 2
RE 56	REFER TO SECTION 2
RE 57	REFER TO SECTION 2
RE 58	REFER TO SECTION 2
RE 59	REFER TO SECTION 2
RE 60	REFER TO SECTION 2
RE 61	REFER TO SECTION 2
RE 62	REFER TO SECTION 2
RE 63	REFER TO SECTION 2
RE 64	REFER TO SECTION 2
RE 66	REFER TO SECTION 2
RE 67	REFER TO SECTION 2
RE 68	REFER TO SECTION 2
RE 69	REFER TO SECTION 2
RE 70	REFER TO SECTION 2
RE 71	REFER TO SECTION 2
RE 72	REFER TO SECTION 2
RE 73	REFER TO SECTION 2
RE 74	REFER TO SECTION 2
RE 75	REFER TO SECTION 2
RE 76	REFER TO SECTION 2
RE 77	REFER TO SECTION 2
RE 78	REFER TO SECTION 2
RE 79	REFER TO SECTION 2
RE 80	REFER TO SECTION 2
RE 81	REFER TO SECTION 2
RE 82	REFER TO SECTION 2
RE 83	REFER TO SECTION 2
RE 83MP	M.P. RE 83
RE 191	REFER TO SECTION 2
RE 192	REFER TO SECTION 2
RE 193	REFER TO SECTION 2
RE 194	REFER TO SECTION 2
RS-2001	REFER TO SECTION 2
RS-2002	REFER TO SECTION 2
RS-2003	REFER TO SECTION 2
RS-2004	REFER TO SECTION 2
RS-2005	REFER TO SECTION 2
RS-2006	REFER TO SECTION 2
RS-2007	REFER TO SECTION 2
RS-2008	REFER TO SECTION 2
RS-2009	REFER TO SECTION 2
RS-2010	REFER TO SECTION 2
RS-2011	REFER TO SECTION 2
RS-2012	REFER TO SECTION 2
RS-2013	REFER TO SECTION 2
RS-2014	REFER TO SECTION 2
RS-2015	REFER TO SECTION 2
RS-2016	REFER TO SECTION 2
RS-2017	REFER TO SECTION 2
RS-2018	REFER TO SECTION 2
RS-2019	REFER TO SECTION 2
RS-2020	REFER TO SECTION 2
RS-2021	REFER TO SECTION 2
RS-2022	REFER TO SECTION 2
RS-2023	REFER TO SECTION 2

To Replace	Substitute This Type
RS-2024	REFER TO SECTION 2
RS-2025	REFER TO SECTION 2
RS-2026	REFER TO SECTION 2
RS-2027	REFER TO SECTION 2
RS-2030	REFER TO SECTION 2
RS-2031	REFER TO SECTION 2
RS-2032	REFER TO SECTION 2
RS-2033	REFER TO SECTION 2
RS-2034	REFER TO SECTION 2
RT482	2N702,2N2320,TN81,*PTC123,*GE-81,*TR-87,*HEPS0014,*SK3024,*ECG128, *WEP243,*BC337,*276-2012,*RT-114
RT483	2N497,2N3512,BSY91,SE8001,*PTC123,*GE-81,*TR-87,*HEPS0014,*SK3024, *ECG128,*WEP243,*BC337,*276-2012,*RT-114
RT484	2N656,2N1987,2N3122,2N3554,2N5188,2SC32,2SC46,2SC47,2SC116,2SC594-R, 2SC1166-R,BSY63,BSY92,*PTC123,*GE-81,*TR-87,*HEPS0014,*SK3024,*ECG128, *WEP243,*BC337,*276-2012,*RT-114
RT697M	2N2396,*GE-210,*ECG123A,*WEP736,*BC337,*276-2009
RT699M	2N2518,*GE-18,*ECG128,*WEP243,*BF338,*276-2008
RT1890M	HST5013,*HEPS3019
RT1893	HST5004,*PTC144,*GE-27,*TR-78,*SK3201,*ECG171,*WEPS3021,*BF338
RT5151	2N656,2N780,2N2106,2N2107,2N2846,2N2848,2N3015,2N3299,2N3326,2N3554, 2N4960,2N6538,2SC32,2SC352,2SC352A,2SC594-R,2SC651,2SC1166-R,2SC1360, 7A31,A5T3903,BSY92,TIS105,*PTC123,*GE-81,*TR-87,*HEPS3024,*SK3024, *ECG128,*WEP243,*BC337,*276-2012,*RT-114
RT5152	2N656,2N780,2N2106,2N2107,2N2846,2N2848,2N3015,2N3299,2N3326,2N3554, 2N4960,2N6538,2SC32,2SC352,2SC352A,2SC594-R,2SC651,2SC1166-R,2SC1360, 7A31,A5T3903,BSY92,TIS105,*PTC123,*GE-81,*TR-87,*HEPS3024,*SK3024, *ECG128,*WEP243,*BC337,*276-2012,*RT-114
RT5203	*PTC123,*GE-81,*TR-63,*HEPS3024,*ECG128,*WEP243,*BC337,*276-2012
RT5204	2N780,2N2330,2N4264,2N5772,2SC352,A5T4123,*PTC123,*GE-81,*TR-63, *HEPS0016,*SK3039,*ECG128,*WEP243,*BC337,*276-2012
RT5212	2N1613A,2N2846,2N2848,2N3015,2N3109,2N3299,2N3326,2N4960,2N4961,2N6538, 2SC594-O,2SC1166-O,A5T3903,SE8002,TIS96,TIS99,*PTC123.*GE-81,*SK3088, *ECG110,*BF338,*276-2012
RT5230	SEE 2N2309
RT5401	*PTC144,*GE-47,*TR-87,*HEPS3024,*SK3024,*ECG128,*WEP243,*276-2030, *RT-114
RT5402	*GE-47,*TR-87,*HEPS3024,*SK3024,*ECG128,*WEP243,*276-2030,*RT-114
RT5403	2N2270,40635,*PTC144,*GE-47,*TR-87,*HEPS3021,*SK3024,*ECG128,*WEP243, *276-2030,*RT-114
RT5404	2N2270,40635,*PTC144,*GE-47,*TR-87,*HEPS3021,*SK3024,*ECG128,*WEP243, *276-2030,*RT-114
RT-100	REFER TO SECTION 2
RT-100A	REFER TO SECTION 2
RT-101	REFER TO SECTION 2
RT-101A	REFER TO SECTION 2
RT-102	REFER TO SECTION 2
RT-102A	REFER TO SECTION 2
RT-103	REFER TO SECTION 2
RT-103A	REFER TO SECTION 2
RT-104	REFER TO SECTION 2
RT-105	REFER TO SECTION 2
RT-106	REFER TO SECTION 2
RT-107	REFER TO SECTION 2
RT-107A	REFER TO SECTION 2
RT-108	REFER TO SECTION 2
RT-108A	REFER TO SECTION 2
RT-109	REFER TO SECTION 2
RT-110	REFER TO SECTION 2
RT-111	REFER TO SECTION 2
RT-112	REFER TO SECTION 2

TRANSISTOR SUBSTITUTES

To Replace	Substitute This Type
RT-113	REFER TO SECTION 2
RT-114	REFER TO SECTION 2
RT-115	REFER TO SECTION 2
RT-116	REFER TO SECTION 2
RT-117	REFER TO SECTION 2
RT-118	REFER TO SECTION 2
RT-119	REFER TO SECTION 2
RT-120	REFER TO SECTION 2
RT-121	REFER TO SECTION 2
RT-122	REFER TO SECTION 2
RT-123	REFER TO SECTION 2
RT-124	REFER TO SECTION 2
RT-125	REFER TO SECTION 2
RT-126	REFER TO SECTION 2
RT-126A	REFER TO SECTION 2
RT-127	REFER TO SECTION 2
RT-127A	REFER TO SECTION 2
RT-128	REFER TO SECTION 2
RT-129	REFER TO SECTION 2
RT-130	REFER TO SECTION 2
RT-131	REFER TO SECTION 2
RT-132	REFER TO SECTION 2
RT-133	REFER TO SECTION 2
RT-134	REFER TO SECTION 2
RT-135	REFER TO SECTION 2
RT-136	REFER TO SECTION 2
RT-137	REFER TO SECTION 2
RT-139	REFER TO SECTION 2
RT-140	REFER TO SECTION 2
RT-141	REFER TO SECTION 2
RT-143	REFER TO SECTION 2
RT-144	REFER TO SECTION 2
RT-145	REFER TO SECTION 2
RT-146	REFER TO SECTION 2
RT-147	REFER TO SECTION 2
RT-148	REFER TO SECTION 2
RT-149	REFER TO SECTION 2
RT-150	REFER TO SECTION 2
RT-151	REFER TO SECTION 2
RT-152	REFER TO SECTION 2
RT-153	REFER TO SECTION 2
RT-154	REFER TO SECTION 2
RT-155	REFER TO SECTION 2
RT-156	REFER TO SECTION 2
RT-157	REFER TO SECTION 2
RT-158	REFER TO SECTION 2
RT-159	REFER TO SECTION 2
RT-159A	REFER TO SECTION 2
RT-160	REFER TO SECTION 2
RT-162	REFER TO SECTION 2
RT-163	REFER TO SECTION 2
RT-164	REFER TO SECTION 2
RT-164A	REFER TO SECTION 2
RT-165	REFER TO SECTION 2
RT-165A	REFER TO SECTION 2
RT-166	REFER TO SECTION 2
RT-167	REFER TO SECTION 2
RT-172	REFER TO SECTION 2
RT-174	REFER TO SECTION 2
RT-182	REFER TO SECTION 2
RT-183	REFER TO SECTION 2

To Replace	Substitute This Type
RT-185	REFER TO SECTION 2
RT-186	REFER TO SECTION 2
RT-187	REFER TO SECTION 2
RT-188	REFER TO SECTION 2
RT-189	REFER TO SECTION 2
RT-190	REFER TO SECTION 2
RT-191	REFER TO SECTION 2
RT-194	REFER TO SECTION 2
RT-195	REFER TO SECTION 2
RT-196	REFER TO SECTION 2
RT-197	REFER TO SECTION 2
RT-300	REFER TO SECTION 2
RT-301	REFER TO SECTION 2
RT-302	REFER TO SECTION 2
RT-303	REFER TO SECTION 2
RT-304	REFER TO SECTION 2
RT-305	REFER TO SECTION 2
RT-306	REFER TO SECTION 2
RT-306F	REFER TO SECTION 2
RT-307	REFER TO SECTION 2
RT-308	REFER TO SECTION 2
RT-309	REFER TO SECTION 2
RT-310	REFER TO SECTION 2
RT-311	REFER TO SECTION 2
RT-312	REFER TO SECTION 2
RT-313	REFER TO SECTION 2
RT-314	REFER TO SECTION 2
RT-315	REFER TO SECTION 2
S2N1486	2N1486,ZT1486,*276-2020
S2N2034A	2N2033,2N2034,2N2858,*276-2020
S1003A	
S1010	A406
S1071	
S1151	
S1166	
S1182	
S1200	S1201,S1229
S1201	
S1229	
S1230	
S1253	S1254
S1254	S1253
S1297	
S2508	
S2509	2SC1743,35862E,35862E,S2510
S2510	2SC1743,35862E,35862E,S2509
S15649	BC547B,BC550B,BCY58C,BCY58D,BCY59C,BCY59D,MPSA18,PET4002,PET4003, *PTC139,*GE-212,*TR-21,*SK3122,*EN10,*ECG123A,*WEP736,*BC107B,*276-2031, *RT-102
S15650	2N3565,2N4286,2N4287,BC547B,BC550B,BCY58C,BCY58D,BCY59C,BCY59D, MPSA16,MPSA18,PET4002,*PTC139,*GE-212,*TR-83,*SK3117,*EN10,*ECG161, *WEP719,*BC107B,*276-2031,*RT-113
S15657	2N2369,2N2369A,2N2501,2N3009,2N3013,2N3014,2N3646,2N3688,2N3689,2N3690, 2N3862,2N4295,2N4420,2N4421,2N4422,2N5029,2SC67,2SC169,2SC335,2SC336, 2SC601N,2SC639,2SC752G-O,2SC764,2SC1187,2SC1188,2SC1189,2SC1274, 2SC1319,2SC1320,2SC1687,2SC1688,BF198,BF199,BF241,GET2369,MPS3646, SE5001,SE5002,SE5003,SE5006,TIS48,TIS87,*PTC132,*GE-60,*TR-83,*HEPS0016, *SK3117,*EN10,*ECG161,*WEP719,*BF173,*276-2016,*RT-113
S15658	2N2369,2N2369A,2N2501,2N3009,2N3013,2N3014,2N3646,2N3688,2N3689,2N3690, 2N3862,2N4295,2N4420,2N4421,2N4422,2N5029,2SC67,2SC169,2SC335,2SC336, 2SC601N,2SC639,2SC752G-O,2SC764,2SC1187,2SC1188,2SC1189,2SC1274,

To Replace	Substitute This Type
(S15658)	2SC1319,2SC1320,2SC1687,2SC1688,BF198,BF199,BF241,GET2369,MPS3646, SE5001,SE5002,SE5003,SE5006,TIS48,TIS87,*PTC132,*GE-60,*TR-83,*HEPS0016, *SK3117,*EN10,*ECG161,*WEP719,*BF173,*276-2016,*RT-113
S15659	2N3137,2N3326,2N5769,2SC651,TIS105,*PTC136,*GE-17,*TR-83,*SK3117,*EN10, *ECG161,*WEP719,*276-2038,*RT-113
S15660	*TR-87,*SK3024,*ECG128,*WEP243,*276-2038,*RT-114
S18100	*GE-82,*HEPS5022,*ECG159,*WEP242
S18200	2N2461,2N2462,2N2519,*GE-18,*ECG128,*WEP243,*BF338,*276-2008
SA310	2N721,2N721A,2N722A,2N978,2N1254,2N1256,2N1991,2N2162,2N2393,2N2968, 2N2969,2N3039,2N4008,2SA558,A5T4248,FT1746,TIS38,TIS137,TIS138,TW135, *PTC127,*GE-65,*TR-20,*HEPS0012,*SK3114,*SP70,*ECG159,*WEP242,*BC177, *276-2022,*RT-115
SA311	2N721,2N721A,2N722A,2N978,2N1254,2N1256,2N1991,2N2162,2N2393,2N2968, 2N2969,2N3039,2N4008,2SA558,A5T4248,FT1746,TIS38,TIS137,TIS138,TW135, *PTC127,*GE-65,*TR-20,*HEPS0012,*SK3114,*SP70,*ECG159,*WEP242,*BC177, *276-2022,*RT-115
SA312	2N721,2N721A,2N722A,2N978,2N1254,2N1256,2N1991,2N2393,2N3039,2N4008, 2SA558,A5T4248,FT1746,TIS38,TIS137,TIS138,*PTC127,*GE-65,*TR-20,*HEPS0012, *SK3114,*SP70,*ECG159,*WEP242,*BC177,*276-2022,*RT-115
SA313	2N721,2N721A,2N722A,2N978,2N1254,2N1256,2N1991,2N2393,2N3039,2N4008, 2SA558,A5T4248,FT1746,TIS38,TIS137,TIS138,*PTC127,*GE-65,*TR-20,*HEPS0012, *SK3114,*SP70,*ECG159,*WEP242,*BC177,*276-2022,*RT-115
SA314	2N721,2N721A,2N722A,2N978,2N1254,2N1256,2N1991,2N2393,2N3039,2N4008, 2SA558,A5T4248,FT1746,TIS38,TIS137,TIS138,*PTC127,*GE-65,*TR-20,*HEPS0012, *SK3114,*SP70,*ECG159,*WEP242,*BC177,*276-2022,*RT-115
SA315	2N721,2N721A,2N722A,2N978,2N1254,2N1256,2N1991,2N2393,2N3039,2N4008, 2SA558,A5T4248,FT1746,TIS38,TIS137,TIS138,*PTC127,*GE-65,*TR-20,*HEPS0012, *SK3114,*SP70,*ECG159,*WEP242,*BC177,*276-2022,*RT-115
SA316	2N721,2N721A,2N2393,2N4008,MPSH85,TIS138,*PTC127,*GE-65,*TR-20, *HEPS0012,*SK3114,*SP70,*ECG159,*WEP242,*BC177,*276-2022,*RT-115
SA410	2N721,2N721A,2N722A,2N978,2N2393,2N2969,2N3039,2N4008,2SA558,A5T4248, FT1746,TIS38,TIS137,TIS138,TW135,*PTC127,*GE-65,*TR-20,*HEPS0013,*SK3114, *SP70,*ECG159,*WEP242,*BC177,*276-2022,*RT-115
SA411	2N721,2N721A,2N722A,2N978,2N2393,2N2969,2N3039,2N4008,2SA558,A5T4248, FT1746,TIS38,TIS137,TIS138,TW135,*PTC127,*GE-65,*TR-20,*HEPS0013,*SK3114, *SP70,*ECG159,*WEP242,*BC177,*276-2022,*RT-115
SA412	2N721,2N721A,2N722A,2N978,2N2393,2N2969,2N3039,2N4008,2SA558,A5T4248, FT1746,TIS38,TIS137,TIS138,TW135,*PTC127,*GE-65,*TR-20,*HEPS0013,*SK3114, *SP70,*ECG159,*WEP242,*BC177,*276-2022,*RT-115
SA413	2N721,2N721A,2N722A,2N978,2N2393,2N2969,2N3039,2N4008,2SA558,A5T4248, FT1746,TIS38,TIS137,TIS138,TW135,*PTC127,*GE-65,*TR-20,*HEPS0013,*SK3114, *SP70,*ECG159,*WEP242,*BC177,*276-2022,*RT-115
SA414	2N721,2N721A,2N722A,2N978,2N2393,2N2969,2N3039,2N4008,2SA558,A5T4248, FT1746,TIS38,TIS137,TIS138,TW135,*PTC127,*GE-65,*TR-20,*HEPS0013,*SK3114, *SP70,*ECG159,*WEP242,*BC177,*276-2022,*RT-115
SA415	2N721,2N721A,2N722A,2N978,2N2393,2N2969,2N3039,2N4008,2SA558,A5T4248, FT1746,TIS38,TIS137,TIS138,TW135,*PTC127,*GE-65,*TR-20,*HEPS0013,*SK3114, *SP70,*ECG159,*WEP242,*BC177,*276-2022,*RT-115
SA416	2N721,2N721A,2N2393,2N2969,2N2971,2N4008,MM4052,MPSH85,TIS138,*PTC127, *GE-65,*TR-20,*HEPS0013,*SK3114,*SP70,*ECG159,*WEP242,*BC177,*276-2022, *RT-115
SA537	2N858,2N860,2N923,2N938,2N2971,2S322,2S3210,MM4052,MPSH85,TIS138, *PTC127,*GE-65,*TR-20,*HEPS0013,*SP70,*ECG106,*WEP52,*BC177,*276-2022, *RT-126
SA538	2N860,2N862,2N2372,2N2373,2N2378,2S3221,*PTC127,*GE-65,*TR-20,*HEPS0013, *SP70,*ECG106,*WEP52,*BC177,*276-2021,*RT-126
SA539	2N858,2N860,2N923,2N938,2N2971,MM4052,MPSH85,TIS138,*PTC127,*GE-65, *TR-20,*HEPS0013,*SP70,*ECG106,*WEP52,*BC177,*276-2022,*RT-126
SA540	2N860,2N862,2N2372,2N2373,2N2378,*PTC127,*GE-65,*TR-20,*HEPS0013,*SP70, *ECG106,*WEP52,*BC177,*276-2021,*RT-126
SB100	2N346,2N501,2N741,2N846A,2N960,2N961,2N962,2N963,2N2400,2SA449,*GE-51,

To Replace	Substitute This Type
(SB100)	*TR-17,*HEPG0002,*JR30X,*ECG126,*WEP635,*AF125,*276-2002
SB101	SEE 2N344
SB102	SEE 2N345
SB103	SEE 2N346
SB200	2N344,2N346,2N393,2N741,2N960,2N961,2N962,2N963,2N2258,*GE-51,*TR-17, *HEPG0002,*HF35,*ECG126,*WEP635,*AF125,*276-2002
SDG600	2N1164,2N1164A,2N1166,2N1166A,2N1751,MP600,MP601,MP602,MP603,SDG601, SDG602,SDG603,SDG607,*PTC138
SDG601	2N1164,2N1164A,2N1166,2N1166A,2N1751,MP600,MP601,MP602,MP603,SDG600, SDG602,SDG603,SDG607,*PTC138
SDG602	2N1164,2N1164A,2N1166,2N1166A,2N1751,MP600,MP601,MP602,MP603,SDG600, SDG601,SDG603,SDG607,*PTC138
SDG603	2N1166,2N1166A,MP602,MP603,*PTC122
SDG604	2N4276,2N4278,SDG605,*PTC901
SDG605	2N4278,2N4280,SDG606,*PTC901
SDG606	2N4280,2N4282,SDG607,*PTC901
SDG607	2N4282,*PTC901
SDN101	
SDN105	
SDN201	
SDN205	
SDN301	
SDN305	
SDN401	
SDN405	
SDN501	
SDN505	
SDN601	
SDN605	
SDN722	
SDN723	
SDN822	
SDN823	
SDN832	
SDN833	
SDN1010	DTS-1010
SDN1020	DTS-1020
SDN4040	DTS-4039,DTS-4040,DTS-4041,IR4039,IR4040,IR4041,IR4055
SDN4045	DTS-4045,IR4045
SDN4060	DTS-4059,DTS-4061,IR4059,IR4061
SDN4061	DTS-4059,DTS-4061,IR4059,IR4061
SDN4065	
SDN6000	2SD572,DTS-4039,DTS-4040,DTS-4041,IR4039,IR4040,IR4041,IR4055
SDN6001	2SD572,IR4055
SDN6002	2SC1434,2SD572,2SD573,DTS-4059,DTS-4061,IR4055,IR4059,IR4061
SDN6060	IR5060,IR5061,IR5062
SDN6061	IR5061,IR5062
SDN6062	IR5062
SDN6251	2SD572,DTS-4039,DTS-4041,IR4039,IR4055
SDN6252	2SD572,IR4055
SDN6253	2SC1434,2SD572,2SD573,IR4055,IR4059
SDN22301	
SDN22302	
SDN22303	
SDN22311	
SDN22312	
SDN22313	
SDP151	
SDP155	
SDP251	
SDP255	

To Replace	Substitute This Type
SDP351	
SDP355	
SDP451	
SDP455	
SDT7A01	KSP1164,SDT7A02,*276-2020
SDT7A02	KSP1164,KSP1165,SDT7A03,*276-2020
SDT7A03	KSP1165
SDT7A04	KSP1166,SDT7A05
SDT7A05	SDT7A06
SDT7A06	
SDT7B01	SDT7B02,*276-2020
SDT7B02	SDT7B03
SDT7B03	SDT7B04
SDT7B04	
SDT7B05	1718-0802,1718-0805,1718-1002,1718-1005,SDT7B06,SDT3227,SDT3228,SDT7712, SDT7713,*276-2020
SDT7B06	1718-1002,1718-1005,1718-1202,1718-1205,SDT7B07,SDT3228,SDT3229,SDT7713, SDT7714
SDT7B07	1718-1202,1718-1205,1718-1402,1718-1405,1718-1602,1718-1605,SDT7B08,SDT3229, SDT7714,SDT7715
SDT7B08	1718-1402,1718-1405,1718-1602,1718-1605,1718-1802,SDT7715,SDT7716
SDT1001	2N6510,*PTC129A,*GE-73,*HEPS5020,*SK3559
SDT1002	2N6510,2N6511,*PTC129A,*GE-73,*HEPS5020,*SK3559
SDT1003	2N5239,2N5240,2N5804,2N5805,2SC1114,*PTC129A,*GE-73,*HEPS5020,*SK3559
SDT1004	2N5239,2N5240,2N5805,*PTC129A,*GE-73,*HEPS5021,*SK3559
SDT1005	2N5240,2N5805,*PTC129A,*GE-73,*HEPS5021,*SK3559
SDT1006	*PTC129A,*HEPS5021
SDT1007	*PTC129A,*HEPS5021
SDT1011	2N6510,*PTC129A,*GE-73,*HEPS5020,*SK3559
SDT1012	2N6510,2N6511,2N6561,XGSR10025,*PTC129A,*GE-73,*HEPS5020,*SK3559
SDT1013	2N5239,2N5240,2N5804,2N5805,2N6511,2N6512,2N6514,2N6561,2SC1227,FT411, IR515,*PTC129A,*GE-73,*HEPS5020,*SK3559
SDT1014	2N5239,2N5240,2N5805,2N6307M,2N6512,2N6513,2N6514,FT413,FT423,FT430, IR516,MJ413,MJ423,*PTC129A,*GE-73,*HEPS5021,*SK3559
SDT1015	2N5240,2N5805,2N6306,2N6307M,2N6308M,2N6513,2N6573,2SC807,FT413,FT423, FT430,IR516,IR517,MJ413,MJ423,*PTC129A,*GE-73,*HEPS5021,*SK3559
SDT1016	2N6306,2N6307M,2N6308M,2N6573,2SC807,IR517,*PTC129A,*HEPS5021
SDT1017	2N6306,2N6307,2N6308M,2N6573,2N6574,2SC807,IR517,IR518,*PTC129A, *HEPS5021
SDT1050	SDT1055
SDT1051	2N6579,2N6580,2N6582,2N6583,SDT1052,SDT1056,SDT1057
SDT1052	2N6580,2N6581,2N6583,2N6584,SDT1053,SDT1057,SDT1058
SDT1053	2N6581,2N6584,SDT1054,SDT1058,SDT1059
SDT1054	SDT1059
SDT1055	SDT1050
SDT1056	2N6579,2N6580,2N6582,2N6583,SDT1051,SDT1052,SDT1057
SDT1057	2N6580,2N6581,2N6583,2N6584,SDT1052,SDT1053,SDT1058
SDT1058	2N6581,2N6584,SDT1053,SDT1054,SDT1059
SDT1059	SDT1054
SDT1060	SDT1050,SDT1055
SDT1061	2N6579,2N6580,2N6582,2N6583,SDT1051,SDT1052,SDT1056,SDT1057,SDT1062
SDT1062	2N6580,2N6581,2N6583,2N6584,SDT1052,SDT1053,SDT1057,SDT1058,SDT1063
SDT1063	2N6581,2N6584,SDT1053,SDT1054,SDT1058,SDT1059,SDT1064
SDT1064	SDT1054,SDT1059
SDT1150	SDT1155
SDT1151	SDT1152,SDT1156,SDT1157
SDT1152	SDT1153,SDT1157,SDT1158
SDT1153	SDT1154,SDT1158,SDT1159
SDT1154	SDT1159
SDT1155	SDT1150
SDT1156	SDT1151,SDT1152,SDT1157

To Replace	Substitute This Type
SDT1157	SDT1152,SDT1153,SDT1158
SDT1158	SDT1153,SDT1154,SDT1159
SDT1159	SDT1154
SDT1160	SDT1150,SDT1155
SDT1161	SDT1151,SDT1152,SDT1156,SDT1157,SDT1162
SDT1162	SDT1152,SDT1153,SDT1157,SDT1158,SDT1163
SDT1163	SDT1153,SDT1154,SDT1158,SDT1159,SDT1164
SDT1164	SDT1154,SDT1159
SDT1250	SDT1255
SDT1251	2N6585,2N6586,2N6588,2N6589,SDT1252,SDT1256,SDT1257
SDT1252	2N6586,2N6587,2N6589,2N6590,SDT1253,SDT1257,SDT1258
SDT1253	2N6587,2N6590,SDT1254,SDT1258,SDT1259
SDT1254	SDT1259
SDT1255	SDT1250
SDT1256	2N6585,2N6586,2N6588,2N6589,SDT1251,SDT1252,SDT1257
SDT1257	2N6586,2N6587,2N6589,2N6590,SDT1252,SDT1253,SDT1258
SDT1258	2N6587,2N6590,SDT1253,SDT1254,SDT1259
SDT1259	SDT1254
SDT1260	SDT1250,SDT1255
SDT1261	2N6585,2N6586,2N6588,2N6589,SDT1251,SDT1252,SDT1256,SDT1257,SDT1262
SDT1262	2N6586,2N6587,2N6589,2N6590,SDT1252,SDT1253,SDT1257,SDT1258,SDT1263
SDT1263	2N6587,2N6590,SDT1253,SDT1254,SDT1258,SDT1259,SDT1264
SDT1264	SDT1254,SDT1259
SDT1808	SDT1860
SDT1809	SDT1860,SDT1861
SDT1810	SDT1861,SDT1862
SDT1860	
SDT1861	SDT1860
SDT1862	SDT1861
SDT1908	SDT1960
SDT1909	SDT1960,SDT1961
SDT1910	SDT1961,SDT1962
SDT1960	
SDT1961	SDT1960
SDT1962	SDT1961
SDT2008	
SDT2009	
SDT2010	
SDT2101	
SDT2110	
SDT2111	
SDT2112	
SDT2150	
SDT2151	
SDT2152	
SDT2205	
SDT2305	
SDT3101	SDT3102,SDT3105,SDT3106
SDT3102	KSP2151,KSP2154,SDT3106
SDT3103	2N5312,2N5314,2N5316,2N5318,2N5386,2N6182,2N6184,2N6186,2N6188,KSP2151, KSP2152,KSP2154,KSP2155,SDT3104,SDT3107,SDT3108
SDT3104	2N5314,2N5318,2N5386,2N6184,2N6188,KSP2152,KSP2155,SDT3108,SDT3109
SDT3105	SDT3101,SDT3102,SDT3106
SDT3106	KSP2151,KSP2154,SDT3102
SDT3107	2N5312,2N5314,2N5316,2N5318,2N5386,2N6182,2N6184,2N6186,2N6188,KSP2151, KSP2152,KSP2154,KSP2155,SDT3103,SDT3104,SDT3108
SDT3108	2N5314,2N5318,2N5386,2N6184,2N6188,KSP2152,KSP2155,SDT3104,SDT3109
SDT3109	KSP2153,KSP2156
SDT3125	SDT3126
SDT3126	SDT3127
SDT3127	SDT3128

To Replace	Substitute This Type
SDT3128	SDT3129
SDT3129	
SDT3201	2N5288,2N5329,HST7140,HST7154,KSP1153,KSP1156,SDT3209
SDT3202	2N5313,2N5317,2N5412,KSP1151,KSP1154,SDT3203,SDT3206,SDT3207
SDT3203	2N4301,2N5006,2N5313,2N5315,2N5317,2N5319,2N5412,2N6128,KSP1151,KSP1152, KSP1154,KSP1155,SDT3204,SDT3207,SDT3208
SDT3204	2N4301,2N5006,2N5288,2N5315,2N5319,2N6128,HST7140,KSP1152,KSP1155, SDT3201,SDT3208,SDT3209
SDT3205	SDT3202,SDT3206
SDT3206	2N5313,2N5317,2N5412,KSP1151,KSP1154,SDT3202,SDT3203,SDT3207
SDT3207	2N4301,2N5006,2N5313,2N5315,2N5317,2N5319,2N5412,2N6128,KSP1151,KSP1152, KSP1154,KSP1155,SDT3203,SDT3204,SDT3208
SDT3208	2N4301,2N5006,2N5288,2N5315,2N5319,2N6128,HST7140,KSP1152,KSP1155, SDT3201,SDT3204,SDT3209
SDT3209	2N5288,2N5329,HST7140,HST7154,KSP1153,KSP1156,SDT3201
SDT3225	SDT3226
SDT3226	SDT3227
SDT3227	SDT3228
SDT3228	SDT3229
SDT3229	
SDT3301	KSP2034,SDT3302
SDT3302	2N5410,KSP2034,KSP2035,SDT3303
SDT3303	2N5410,2N5411,KSP2035,SDT3304
SDT3304	2N5411,KSP2036
SDT3305	KSP2031,SDT3306
SDT3306	2N5408,KSP2031,KSP2032,SDT3307
SDT3307	2N5408,2N5409,KSP2032,SDT3308
SDT3308	2N5409,KSP2033,SDT3309
SDT3309	KSP2033
SDT3321	SDT3322,*GE-28,*TR-88,*SK3025,*ECG129,*WEP242,*RT-115
SDT3322	2N5406,SDT3323,*GE-29,*TR-88,*SK3025,*ECG129,*WEP242,*RT-115
SDT3323	2N5406,2N5407,SDT3324
SDT3324	2N5407
SDT3325	SDT3326,*GE-29,*SK3193,*ECG187,*WEP246
SDT3326	2N5404,SDT3327,*GE-28,*SK3192,*ECG186,*WEP245
SDT3327	2N5404,2N5405,SDT3328
SDT3328	2N5405,SDT3329
SDT3329	
SDT3401	2N3747,SDT3402,SDT6411
SDT3402	2N3747,2N3748,KSP1034,SDT3403,SDT6411,SDT6412
SDT3403	2N3748,2N3749,2N3996,KSP1034,KSP1035,SDT3404,SDT6412,SDT6413
SDT3404	2N3749,2N3996,KSP1035,SDT6413
SDT3405	1718-0602,1718-0605,SDT3406,SDT6408
SDT3406	1718-0602,1718-0605,1718-0802,1718-0805,KSP1031,SDT3407,SDT6408,SDT6409
SDT3407	1718-0802,1718-0805,1718-1002,1718-1005,KSP1031,KSP1032,SDT3408,SDT6409, SDT6410
SDT3408	1718-1002,1718-1005,1718-1202,1718-1205,KSP1032,SDT3409,SDT6410
SDT3409	1718-1202,1718-1205,1718-1402,1718-1405,1718-1602,1718-1605,KSP1033
SDT3421	HST5001,HST5501,SDT3422,*GE-216,*SK3192,*ECG186,*WEP245
SDT3422	HST4453,HST5001,HST5002,HST5501,HST5502,SDT3423,*GE-216,*ECG152, *WEP245
SDT3423	HST4453,HST4454,HST5002,HST5003,HST5502,HST5503,SDT3424
SDT3424	2N4895,2N4897,BFX34,HST4454,HST5003,HST5503
SDT3425	HST4483,HST5006,HST5506,SDT3426,*GE-216,*SK3192,*ECG186,*WEP245
SDT3426	HST4451,HST4483,HST5006,HST5007,HST5506,HST5507,SDT3427,*GE-216, *SK3192,*ECG186,*WEP245
SDT3427	HST4451,HST4452,HST5007,HST5008,HST5507,HST5508,SDT3428
SDT3428	HST4452,HST5008,HST5508,SDT3429
SDT3429	HST5009,HST5509
SDT3501	SDT3502,*GE-84,*TR-88,*SK3025,*ECG129,*WEP242,*RT-115
SDT3502	SDT3503,*GE-29,*TR-88,*SK3025,*ECG129,*WEP242,*RT-115

To Replace	Substitute This Type
SDT3503	SDT3504,*TR-88,*SK3025,*ECG129,*WEP242,*RT-115
SDT3504	*SK3513
SDT3505	SDT3506,*GE-84,*SK3513,*ECG187,*WEP246
SDT3506	SDT3507,*GE-29,*SK3513,*ECG187,*WEP246
SDT3507	SDT3508,*SK3513
SDT3508	*SK3513
SDT3509	SDT3510,*PTC113,*GE-69,*SK3085,*ECG153,*WEP246
SDT3510	SDT3511,*PTC113,*GE-69,*ECG153,*WEP246
SDT3511	SDT3512,*PTC113
SDT3512	
SDT3513	SDT3514,*PTC113,*GE-69,*SK3085,*ECG153,*WEP246
SDT3514	SDT3515,*PTC113,*GE-69,*ECG153,*WEP246
SDT3515	SDT3516,*PTC113
SDT3516	
SDT3550	2N4235,2N4236,SDT3551,*GE-29,*SK3193,*ECG187,*WEP246
SDT3551	2N4236,2N5322
SDT3552	2N2881,2N3202,2N3203,2N3208,2N3782,SDT3550,SDT3553,*PTC142,*GE-84, *SK3193,*ECG187,*WEP246
SDT3553	2N2881,2N3203,2N3204,2N5781,2N5782,SDT3550,SDT3551,SDT3554,*GE-29, *SK3193,*ECG187,*WEP246
SDT3554	2N2882,2N3204,2N5147,2N5781,SDT3551
SDT3575	SDT3576,*PTC168,*GE-26,*TR-58,*ECG218,*WEPS5007
SDT3576	SDT3577,*PTC168,*GE-69,*TR-58,*ECG218,*WEPS5007
SDT3577	*PTC168,*TR-58,*ECG218,*WEPS5007
SDT3578	SDT3575,SDT3576,SDT3579,*PTC168,*GE-69,*TR-58,*ECG218,*WEPS5007
SDT3579	SDT3576,SDT3577,*PTC168,*GE-69,*TR-58,*ECG218,*WEPS5007
SDT3601	SDT3602
SDT3602	SDT3603
SDT3603	SDT3604
SDT3604	
SDT3620	SDT3621
SDT3621	SDT3622
SDT3622	SDT3623
SDT3623	
SDT3701	SDT3702,SDT3720,SDT3725,SDT3733,*PTC168,*GE-69,*SK3085,*ECG153, *WEP246
SDT3702	KSP2161,*PTC168,*GE-69,*ECG153,*WEP246
SDT3703	2N5956,SDT3701,SDT3702,SDT3704,SDT3712,SDT3713,SDT3715,SDT3720, SDT3725,SDT3726,SDT3727,SDT3733,SDT3805,SDT3807,SDT3850,SDT3852, *PTC168,*GE-69,*SK3085,*ECG153,*WEP246
SDT3704	2N5955,KSP2161,SDT3702,SDT3705,SDT3713,SDT3714,SDT3727,SDT3728, SDT3806,SDT3851,*PTC168,*GE-69,*ECG153,*WEP246
SDT3705	2N5615,2N5954,KSP2161,KSP2162,SDT3714,SDT3728,SDT3806,SDT3851,*PTC168
SDT3706	RCS32,SDT3701,SDT3703,SDT3707,SDT3712,SDT3715,SDT3726,SDT3805, SDT3850,*PTC168,*GE-69,*SK3085,*ECG153,*WEP246
SDT3707	RCS32A,SDT3702,SDT3704,SDT3708,SDT3713,SDT3727,*PTC168,*GE-69,*SK3085, *ECG153,*WEP246
SDT3708	KSP2161,RCS32A,RCS32B,SDT3702,SDT3704,SDT3705,SDT3713,SDT3714, SDT3727,SDT3728,SDT3806,SDT3851,*PTC168,*GE-69,*ECG153,*WEP246
SDT3709	SDT3710,*PTC168,*GE-69,*SK3085,*ECG153,*WEP246
SDT3710	2SA764,SDT3711,*PTC168,*GE-69,*SK3085,*ECG153,*WEP246
SDT3711	2N5613,2SA764,2SA765,KSP2164,*PTC168,*GE-69,*ECG153,*WEP246
SDT3712	SDT3701,SDT3702,SDT3713,SDT3715,*PTC168,*GE-69,*SK3085,*ECG153, *WEP246
SDT3713	KSP2161,SDT3702,SDT3714,*PTC168,*GE-69,*ECG153,*WEP246
SDT3714	KSP2161,KSP2162,*PTC168
SDT3715	SDT3701,SDT3702,SDT3712,SDT3713,*PTC168,*GE-69,*SK3085,*ECG153, *WEP246
SDT3716	SDT3712,SDT3713,SDT3715,SDT3717,SDT3721,SDT3722,SDT3729,SDT3730, *PTC168,*GE-69,*SK3085,*ECG153,*WEP246
SDT3717	SDT3713,SDT3714,SDT3718,SDT3722,SDT3723,SDT3730,SDT3731,*PTC168,*GE-69,

To Replace	Substitute This Type
(SDT3717)	*ECG153,*WEP246
SDT3718	SDT3714,SDT3719,SDT3723,SDT3724,SDT3731,SDT3732,*PTC168
SDT3719	SDT3724,SDT3732,*PTC168
SDT3720	2SA764,SDT3701,SDT3702,SDT3725,SDT3733,*PTC168,*GE-69,*SK3085,*ECG153, *WEP246
SDT3721	SDT3716,SDT3717,SDT3722,SDT3729,SDT3730,*PTC168,*GE-69,*SK3085,*ECG153, *WEP246
SDT3722	SDT3717,SDT3718,SDT3723,SDT3730,SDT3731,*PTC168,*GE-69,*ECG153, *WEP246
SDT3723	SDT3718,SDT3719,SDT3724,SDT3731,SDT3732,*PTC168
SDT3724	SDT3719,SDT3732,*PTC168
SDT3725	2SA764,SDT3701,SDT3702,SDT3720,SDT3733,*PTC168,*GE-69,*SK3085,*ECG153, *WEP246
SDT3726	2N5956,2SA764,SDT3701,SDT3702,SDT3703,SDT3704,SDT3720,SDT3725,SDT3727, SDT3733,SDT3805,SDT3807,SDT3850,SDT3852,*PTC168,*GE-69,*SK3085, *ECG153,*WEP246
SDT3727	2N5955,2SA764,2SA765,KSP2161,KSP2164,SDT3702,SDT3704,SDT3705,SDT3728, SDT3806,SDT3851,*PTC168,*GE-69,*ECG153,*WEP246
SDT3728	2N5615,2N5954,2SA765,KSP2161,KSP2162,KSP2164,KSP2165,SDT3705,SDT3806, SDT3851,*PTC168
SDT3729	SDT3716,SDT3717,SDT3721,SDT3722,SDT3730,*PTC168,*GE-69,*SK3085,*ECG153, *WEP246
SDT3730	SDT3717,SDT3718,SDT3722,SDT3723,SDT3731,*PTC168,*GE-69,*ECG153, *WEP246
SDT3731	SDT3718,SDT3719,SDT3723,SDT3724,SDT3732,*PTC168
SDT3732	SDT3719,SDT3724,*PTC168
SDT3733	2SA764,SDT3701,SDT3702,SDT3720,SDT3725,*PTC168,*GE-69,*SK3085,*ECG153, *WEP246
SDT3750	2SA807,KSP2044,SDT3751,SDT3756,SDT3761,SDT3766,*PTC149
SDT3751	2SA744,2SA746,2SA807,2SA808,2SA877,KSP2044,KSP2045,KSP2174,*PTC149
SDT3752	SDT3753,SDT3757,SDT3758,SDT3762,SDT3763,*PTC149
SDT3753	SDT3754,SDT3758,SDT3759,SDT3763,SDT3764,*PTC149
SDT3754	SDT3755,SDT3759,SDT3760,SDT3764,SDT3765,*PTC172
SDT3755	SDT3760,SDT3765,*PTC172
SDT3756	2SA807,KSP2044,SDT3750,SDT3751,SDT3761,SDT3766,*PTC149
SDT3757	2N3183,2N3184,2N3195,2N3196,SDT3752,SDT3753,SDT3758,SDT3762,SDT3763, *PTC149
SDT3758	2N3184,2N3185,2N3196,2N3197,SDT3753,SDT3754,SDT3759,SDT3763,SDT3764, *PTC149
SDT3759	2N3185,2N3186,2N3197,2N3198,SDT3754,SDT3755,SDT3760,SDT3764,SDT3765, *PTC172
SDT3760	2N3186,2N3198,SDT3755,SDT3765,*PTC172,*GE-74,*ECG180,*WEPS7001
SDT3761	2SA807,KSP2044,SDT3750,SDT3751,SDT3756,SDT3766,*PTC149
SDT3762	2N3183,2N3184,2N3195,2N3196,SDT3752,SDT3753,SDT3757,SDT3758,SDT3763, *PTC149
SDT3763	2N3184,2N3185,2N3196,2N3197,SDT3753,SDT3754,SDT3758,SDT3759,SDT3764, *PTC149
SDT3764	2N3185,2N3186,2N3197,2N3198,SDT3754,SDT3755,SDT3759,SDT3760,SDT3765, *PTC172,*GE-74,*ECG180,*WEPS7001
SDT3765	2N3186,2N3198,SDT3755,SDT3760,*PTC172,*GE-74,*ECG180,*WEPS7001
SDT3766	2SA807,KSP2044,SDT3750,SDT3751,SDT3756,SDT3761,*PTC149,*GE-74,*ECG180, *WEPS7001,*276-2027
SDT3775	SDT3325,SDT3326,SDT3776,SDT3778,*GE-29,*SK3193,*ECG187,*WEP246
SDT3776	2N5404,SDT3326,SDT3327,SDT3777,*GE-29,*SK3193,*ECG187,*WEP246
SDT3777	2N5404,2N5405,SDT3327,SDT3328
SDT3778	SDT3325,SDT3326,SDT3775,SDT3776,*GE-29,*SK3193,*ECG187,*WEP246
SDT3801	SDT3802,*PTC168
SDT3802	*PTC168
SDT3803	SDT3804,*PTC168
SDT3804	*PTC168
SDT3805	SDT3801,SDT3807,SDT3850,SDT3852,*PTC168

To Replace	Substitute This Type
SDT3806	KSP2161,KSP2162,SDT3802,SDT3851,*PTC168
SDT3807	SDT3801,SDT3803,SDT3852,*PTC168
SDT3825	2N5737,2N5741,SDT3827,SDT3875,SDT3877,*PTC149
SDT3826	2N5623,2N5738,2N5742,2SA680-R,2SA746,2SA877,2SB519,KSP2171,KSP2172, KSP2271,KSP2272,SDT3876,*PTC172,*GE-74,*ECG180,*WEPS7001
SDT3827	SDT3877,*PTC149,*GE-74,*ECG180,*WEPS7001,*276-2027
SDT3850	SDT3801,SDT3805,SDT3807,SDT3852,*PTC168,*276-2027
SDT3851	KSP2161,KSP2162,SDT3802,SDT3806,*PTC168
SDT3852	SDT3801,SDT3803,SDT3807,*PTC168,*276-2027
SDT3875	2N5737,2N5741,SDT3825,SDT3827,SDT3877,*PTC149,*GE-74,*ECG180, *WEPS7001,*276-2027
SDT3876	2N5623,2N5738,2N5742,2SA680-R,2SA746,2SA877,2SB519,KSP2171,KSP2172, KSP2271,KSP2272,SDT3826,*PTC172,*GE-74,*ECG180,*WEPS7001
SDT3877	SDT3827,*PTC149,*GE-74,*ECG180,*WEPS7001,*276-2027
SDT3901	SDT3902
SDT3902	KSP2001,KSP2601,SDT3903
SDT3903	KSP2001,KSP2002,KSP2601,KSP2602,SDT3904
SDT3904	KSP2002,KSP2003,KSP2602
SDT3920	KSP2001,KSP2002,SDT3921
SDT3921	KSP2002,KSP2003,SDT3922
SDT3922	KSP2003,SDT3923
SDT3923	
SDT4301	SDT4302,SDT4307,SDT4308,*GE-83,*SK3192,*ECG186,*WEP245,*276-2017
SDT4302	SDT4303,SDT4308,*GE-215,*SK3192,*ECG186,*WEP245,*276-2020
SDT4303	
SDT4304	SDT4305,SDT4310,SDT4311,*GE-83,*SK3192,*ECG186,*WEP245,*276-2017
SDT4305	SDT4306,SDT4311,SDT4312,*GE-215,*SK3192,*ECG186,*WEP245,*276-2020
SDT4306	SDT4312
SDT4307	SDT4301,SDT4302,SDT4308,*GE-83,*SK3192,*ECG186,*WEP245,*276-2017
SDT4308	SDT4302,SDT4303,*GE-215,*SK3192,*ECG186,*WEP245,*276-2020
SDT4309	
SDT4310	SDT4304,SDT4305,SDT4311,*GE-83,*SK3192,*ECG186,*WEP245,*276-2017
SDT4311	SDT4305,SDT4306,SDT4312,*GE-215,*SK3192,*ECG186,*WEP245,*276-2020
SDT4312	SDT4306
SDT4451	SDT4452
SDT4452	
SDT4453	SDT4454
SDT4454	
SDT4455	SDT4456,*GE-28,*SK3192,*ECG186,*WEP245
SDT4456	
SDT4483	SDT4451,SDT9005,*GE-216,*SK3192,*ECG186,*WEP245
SDT4551	HST4551,HST4552,SDT4552,*GE-28,*SK3192,*ECG186,*WEP245
SDT4552	HST4552
SDT4553	HST4553,HST4554,SDT4554,*GE-28,*SK3192,*ECG186,*WEP245
SDT4554	HST4554
SDT4555	HST4555,HST4556,SDT4556
SDT4556	HST4556
SDT4583	HST4551,HST4583,SDT4551,*GE-28,*SK3192,*ECG186,*WEP245
SDT4901	HST5056,HST5556,KSP1051,KSP1052,KSP1053,SDT4902,SDT4903
SDT4902	KSP1052,KSP1053,KSP1054,KSP1055,SDT4903,SDT4904,SDT4905
SDT4903	KSP1052,KSP1053,KSP1054,KSP1055,SDT4902,SDT4904,SDT4905
SDT4904	KSP1053,KSP1054,KSP1055,SDT4903,SDT4905
SDT4905	KSP1054,KSP1055,SDT4904
SDT4921	HST5056,HST5556,KSP1051,KSP1052,KSP1053,SDT4901,SDT4902,SDT4903, SDT4922,SDT4923
SDT4922	KSP1052,KSP1053,KSP1054,KSP1055,SDT4902,SDT4903,SDT4904,SDT4905, SDT4923,SDT4924,SDT4925
SDT4923	KSP1052,KSP1053,KSP1054,KSP1055,SDT4902,SDT4903,SDT4904,SDT4905, SDT4922,SDT4924,SDT4925
SDT4924	KSP1053,KSP1054,KSP1055,SDT4903,SDT4904,SDT4905,SDT4923,SDT4925
SDT4925	KSP1054,KSP1055,SDT4904,SDT4905,SDT4924

TRANSISTOR SUBSTITUTES

To Replace	Substitute This Type
SDT5001	HST5001,HST5002,SDT5002,*GE-28,*SK3512,*ECG186,*WEP245
SDT5002	HST5002,HST5003,SDT5003,*SK3512
SDT5003	HST5003,*SK3512
SDT5004	HST5004,HST5005,HST5051,SDT5005,SDT5051
SDT5005	HST5005,HST5051,HST5052,HST5053,SDT5051,SDT5052,SDT5053
SDT5006	HST5006,HST5007,SDT5007,*GE-28,*SK3512,*ECG186,*WEP245
SDT5007	HST5007,HST5008,SDT5008,*SK3512
SDT5008	HST5008,*SK3512
SDT5009	HST5009,HST5010,HST5054,SDT5010,SDT5054
SDT5010	HST5010,HST5054,HST5055,HST5056,SDT5054,SDT5055,SDT5056,*WEP245
SDT5011	HST5011,HST5012,SDT5012,*GE-28,*SK3512,*ECG186
SDT5012	HST5012,HST5013,SDT5013,*SK3512
SDT5013	HST5013,*SK3512
SDT5014	HST5014,HST5015,SDT5015
SDT5015	HST5015
SDT5051	HST5005,HST5051,HST5052,HST5053,SDT5005,SDT5052,SDT5053
SDT5052	HST5052,HST5053,SDT5053
SDT5053	HST5053
SDT5054	HST5010,HST5054,HST5055,HST5056,SDT5010,SDT5055,SDT5056
SDT5055	HST5055,HST5056,SDT5056
SDT5056	HST5056
SDT5501	HST5001,HST5002,HST5501,HST5502,SDT5001,SDT5002,SDT5502,*GE-46,*SK3512, *ECG186,*WEP245
SDT5502	HST5002,HST5003,HST5502,HST5503,SDT5002,SDT5003,SDT5503,*SK3512
SDT5503	HST5003,HST5503,SDT5003,*SK3512
SDT5504	HST5004,HST5005,HST5051,HST5504,HST5505,HST5551,SDT5004,SDT5005, SDT5051,SDT5505,SDT5551
SDT5505	HST5005,HST5051,HST5052,HST5053,HST5505,HST5551,HST5552,HST5553, SDT5005,SDT5051,SDT5052,SDT5053,SDT5551,SDT5552,SDT5553
SDT5506	HST5006,HST5007,HST5506,HST5507,SDT5006,SDT5007,SDT5507,*GE-46,*SK3512, *ECG186,*WEP245
SDT5507	HST5007,HST5008,HST5507,HST5508,SDT5007,SDT5008,SDT5508,*SK3512
SDT5508	HST5008,HST5508,SDT5008,*SK3512
SDT5509	HST5009,HST5010,HST5054,HST5509,HST5510,HST5554,SDT5009,SDT5010, SDT5054,SDT5510,SDT5554
SDT5510	HST5010,HST5054,HST5055,HST5056,HST5510,HST5554,HST5555,HST5556, SDT5010,SDT5054,SDT5055,SDT5056,SDT5554,SDT5555,SDT5556
SDT5511	HST5011,HST5012,HST5511,HST5512,SDT5011,SDT5012,SDT5512,*GE-46,*SK3512, *ECG186,*WEP245
SDT5512	HST5012,HST5013,HST5512,HST5513,SDT5012,SDT5013,SDT5513,*SK3512
SDT5513	HST5013,HST5513,SDT5013,*SK3512
SDT5514	HST5014,HST5015,HST5514,HST5515,SDT5014,SDT5015,SDT5515
SDT5515	HST5015,HST5515,SDT5015
SDT5551	HST5005,HST5051,HST5052,HST5053,HST5505,HST5551,HST5552,HST5553, SDT5005,SDT5051,SDT5052,SDT5053,SDT5505,SDT5552,SDT5553
SDT5552	HST5052,HST5053,HST5552,HST5553,SDT5052,SDT5053,SDT5553
SDT5553	HST5053,HST5553,SDT5053
SDT5554	HST5010,HST5054,HST5055,HST5056,HST5510,HST5554,HST5555,HST5556, SDT5010,SDT5054,SDT5055,SDT5056,SDT5510,SDT5555,SDT5556
SDT5555	HST5055,HST5056,HST5555,HST5556,SDT5055,SDT5056,SDT5556
SDT5556	HST5056,HST5556,SDT5056
SDT5901	HST5901,HST5902,KSP1394,SDT5902,*PTC906,*GE-28,*SK3026,*ECG186, *WEP245
SDT5902	HST5902,HST5903,KSP1394,KSP1395,SDT5903,*PTC906,*GE-28,*SK3026,*ECG186, *WEP245
SDT5903	HST5903,KSP1395,*PTC906
SDT5904	HST5904,HST5905,HST5951,KSP1396,SDT5905,SDT5951
SDT5905	HST5905,HST5951,HST5952,HST5953,SDT5951,SDT5952,SDT5953
SDT5906	HST5906,HST5907,KSP1391,SDT5907,*PTC906,*GE-28,*SK3192,*ECG186, *WEP245
SDT5907	HST5907,HST5908,KSP1391,KSP1392,SDT5908,*PTC906,*GE-66,*SK3026,*ECG152,

To Replace	Substitute This Type
(SDT5907)	*WEP245
SDT5908	HST5908,KSP1392,*PTC906
SDT5909	HST5909,HST5910,HST5954,KSP1393,SDT5910,SDT5954
SDT5910	HST5910,HST5954,HST5955,HST5956,KSP1341,SDT5954,SDT5955,SDT5956
SDT5911	HST5911,HST5912,SDT5912,*PTC906,*SK3026
SDT5912	HST5912,HST5913,SDT5913,*PTC906,*SK3026
SDT5913	HST5913,*PTC906
SDT5914	HST5914,HST5915,SDT5915
SDT5915	HST5915
SDT5951	HST5905,HST5951,HST5952,HST5953,SDT5905,SDT5952,SDT5953
SDT5952	HST5952,HST5953,SDT5953
SDT5953	HST5953
SDT5954	HST5910,HST5954,HST5955,HST5956,KSP1341,SDT5910,SDT5955,SDT5956
SDT5955	HST5955,HST5956,KSP1341,KSP1342,SDT5956
SDT5956	HST5956,KSP1341,KSP1342,KSP1343
SDT6001	HST6001,*GE-66,*ECG152,*WEP245
SDT6011	HST6011,HST6012,SDT6012,*GE-66,*ECG152,*WEP245
SDT6012	HST6012
SDT6013	HST6013,HST6014,SDT6014,*GE-66,*ECG152,*WEP245
SDT6014	2N4115,HST6014
SDT6015	HST6015,HST6016,SDT6016
SDT6016	2N4116,HST6016
SDT6031	HST6011,HST6031,SDT6011,*GE-66,*ECG152,*WEP245
SDT6101	SDT6102,SDT6103,*GE-28,*SK3192,*ECG186,*WEP245
SDT6102	SDT6101,SDT6103,*GE-28,*SK3192,*ECG186,*WEP245
SDT6103	SDT6101,SDT6102,*GE-66,*ECG152,*WEP245
SDT6104	SDT6105,SDT6106,*GE-28,*SK3192,*ECG186,*WEP245
SDT6105	SDT6104,SDT6106,*GE-28,*SK3192,*ECG186,*WEP245
SDT6106	SDT6104,SDT6105,*GE-28,*SK3192,*ECG186,*WEP245
SDT6110	SDT6101,SDT6102,SDT6103,SDT6111,SDT6112
SDT6111	SDT6101,SDT6102,SDT6103,SDT6110,SDT6112
SDT6112	SDT6101,SDT6102,SDT6103,SDT6110,SDT6111
SDT6113	SDT6114,SDT6115
SDT6114	SDT6113,SDT6115
SDT6115	SDT6113,SDT6114
SDT6308	HST6308,HST6309,HST6408,SDT6309
SDT6309	HST6309,HST6310,HST6409,HST6410,SDT6310
SDT6310	HST6310,HST6409,HST6410
SDT6311	HST6311,HST6312,HST6411,HST6412,SDT6312
SDT6312	HST6312,HST6313,HST6412,HST6413,SDT6313
SDT6313	2N4115,HST6313,HST6413
SDT6314	HST6314,HST6315,HST6414,HST6415,SDT6315
SDT6315	HST6315,HST6316,HST6415,HST6416,SDT6316
SDT6316	2N4116,HST6316,HST6416
SDT6408	KSP1031,SDT6409
SDT6409	KSP1031,KSP1032,SDT6410
SDT6410	KSP1032
SDT6411	KSP1034,SDT6412
SDT6412	KSP1034,KSP1035,SDT6413
SDT6413	KSP1035
SDT6414	SDT6415
SDT6415	SDT6416
SDT6416	
SDT6901	HST6901,HST6902,KSP1023,KSP1163,SDT6902,*TR-81,*HEPS5012,*SK3131, *ECG175,*WEP241
SDT6902	HST6902,HST6903,HST6904,HST7901,KSP1071,KSP1121,SDT6903,SDT6904, SDT7901
SDT6903	HST6903,HST6904,HST7901,HST7902,KSP1071,KSP1072,KSP1121,KSP1122, SDT6904,SDT7901,SDT7902
SDT6904	HST6904,HST7901,HST7902,HST7903,KSP1071,KSP1072,KSP1073,KSP1121, KSP1122,KSP1123,SDT7901,SDT7902,SDT7903

To Replace	Substitute This Type
SDT6905	HST6905,HST6906,KSP1026,KSP1166,SDT6906,*TR-81,*HEPS5012,*SK3131, *ECG175,*WEP241
SDT6906	HST6906,HST6907,HST6908,SDT6907,SDT6908
SDT6907	HST6907,HST6908,SDT6908
SDT6908	HST6908
SDT7011	1716-0602,1716-0605,1716-0802,1716-0805,B148004,HST7011,HST7012,HST9901, HST9902,KSP1151,SDT7012
SDT7012	1716-0802,1716-0805,1716-1002,1716-1005,B148000,B148004,HST7012,HST7013, HST9902,HST9903,KSP1151,KSP1152,SDT7013
SDT7013	1716-1002,1716-1005,1716-1202,1716-1205,B148000,B148001,HST7013,HST9903, HST9904,KSP1152
SDT7014	2N5412,HST7014,HST7015,KSP1154,SDT7015
SDT7015	2N5412,2N6128,HST7015,HST7016,KSP1154,KSP1155,SDT7016
SDT7016	2N6128,HST7016,HST7140,KSP1155
SDT7017	HST7017,HST7018,SDT7018
SDT7018	HST7018,HST7019,SDT7019
SDT7019	HST7019
SDT7140	2N5329,HST7140,HST7154,KSP1156,SDT7154,SDT8151,ST91057,ST91058
SDT7141	HST7141,HST7156,SDT7156,ST18013
SDT7150	1716-1402,1716-1405,1716-1602,1716-1605,1716-1802,HST7150,HST7151,KSP1153, SDT7151,SDT7765,SDT7766,SDT8156
SDT7151	2N5218,1716-1602,1716-1605,1716-1802,HST7151,HST7152,HST7801,KSP1101, SDT7152,SDT7766,SDT7801,SDT12201
SDT7152	2N5218,HST7152,HST7801,HST7802,HST7803,KSP1101,KSP1102,KSP1103, SDT7801,SDT7802,SDT7803,SDT12201,SDT12202
SDT7154	2N5329,HST7154,HST7155,KSP1156,SDT7155,SDT8151,ST91058,ST91059
SDT7155	HST7141,HST7155,HST7156,SDT7141,SDT7156,ST18013,ST91059
SDT7156	HST7141,HST7156,SDT7141,ST18013
SDT7201	HST7201,HST7202,HST7203,KSP1141,KSP1142,KSP1143,KSP1221,KSP1222, KSP1223,SDT7202,SDT7203,SDT7612
SDT7202	HST7202,HST7203,HST7204,KSP1142,KSP1143,KSP1144,KSP1145,KSP1222, KSP1223,KSP1224,KSP1225,SDT7203,SDT7204
SDT7203	HST7202,HST7203,HST7204,HST7205,KSP1142,KSP1143,KSP1144,KSP1145, KSP1222,KSP1223,KSP1224,KSP1225,SDT7202,SDT7204,SDT7205
SDT7204	HST7204,HST7205,KSP1144,KSP1145,KSP1224,KSP1225,SDT7205
SDT7205	HST7204,HST7205,KSP1145,KSP1225,SDT7204
SDT7206	HST7206,HST7207,SDT7207,*SK3079
SDT7207	HST7201,HST7202,HST7207,HST7208,SDT7201,SDT7202,SDT7208,SDT7612
SDT7208	HST7202,HST7203,HST7204,HST7208,HST7209,SDT7202,SDT7203,SDT7204, SDT7209
SDT7209	HST7203,HST7204,HST7205,HST7209,SDT7203,SDT7204,SDT7205
SDT7401	SDT7402,SDT7414,SDT7415,*GE-28,*SK3192,*ECG186,*WEP245
SDT7402	2N5327,SDT7403,SDT7415,SDT7416,*GE-66,*ECG152,*WEP245
SDT7403	2N5327,SDT7416
SDT7411	SDT7412,*GE-28,*SK3192,*ECG186,*WEP245
SDT7412	SDT7413,*GE-66,*ECG152,*WEP245
SDT7413	
SDT7414	SDT7401,SDT7402,SDT7415,*GE-28,*ECG186,*WEP245
SDT7415	2N5327,SDT7402,SDT7403,SDT7416,*GE-66,*ECG152,*WEP245
SDT7416	2N5327,SDT7403
SDT7417	SDT7418
SDT7418	SDT7419
SDT7419	
SDT7601	KSP1174,KSP1274,SDT7602,*SK3027
SDT7602	KSP1174,KSP1175,KSP1274,KSP1275,SDT7603,*SK3027
SDT7603	2N4070,KSP1175,KSP1275,*SK3027
SDT7604	KSP1176,KSP1276,SDT7605,*SK3079
SDT7605	2N4071
SDT7607	HST9201,HST9205,HST9206,HST9801,HST9802,KSP1171,KSP1271,SDT7608, *SK3027
SDT7608	HST9202,HST9206,HST9207,HST9802,HST9803,KSP1171,KSP1172,KSP1271,

To Replace	Substitute This Type
(SDT7608)	KSP1272,SDT7609,*SK3027
SDT7609	HST9202,HST9203,HST9207,HST9208,HST9803,HST9804,KSP1172,KSP1272, *SK3027
SDT7610	HST9204,HST9209,KSP1173,KSP1273,SDT7611,*SK3079
SDT7611	HST7201,KSP1141,KSP1221,SDT7201,SDT7612
SDT7612	HST7201,HST7202,HST7203,KSP1141,KSP1142,KSP1143,KSP1221,KSP1222, KSP1223,SDT7201,SDT7202,SDT7203
SDT7711	SDT7712,*276-2020
SDT7712	SDT7713,*276-2020
SDT7713	SDT7714
SDT7714	SDT7715
SDT7715	SDT7716
SDT7716	
SDT7731	SDT7732,*SK3027
SDT7732	SDT7733,*SK3027
SDT7733	SDT7734,*SK3027
SDT7734	SDT7735,*SK3079
SDT7735	SDT7736,*SK3079
SDT7736	
SDT7741	SDT7742,*276-2020
SDT7742	SDT7743,*276-2020
SDT7743	SDT7744
SDT7744	SDT7745
SDT7745	SDT7746
SDT7746	
SDT7761	SDT7762
SDT7762	SDT7763
SDT7763	SDT7764
SDT7764	SDT7765
SDT7765	SDT7766
SDT7766	
SDT7801	HST7152,HST7801,HST7802,HST7803,KSP1101,KSP1102,KSP1103,SDT7802, SDT7803
SDT7802	HST7802,HST7803,HST7804,KSP1102,KSP1103,KSP1104,KSP1105,SDT7803, SDT7804
SDT7803	HST7802,HST7803,HST7804,HST7805,KSP1102,KSP1103,KSP1104,KSP1105, SDT7802,SDT7804,SDT7805
SDT7804	HST7804,HST7805,KSP1104,KSP1105,SDT7805
SDT7805	HST7804,HST7805,KSP1105,SDT7804
SDT7806	HST7806,HST7807,SDT7807
SDT7807	HST7152,HST7801,HST7802,HST7807,HST7808,SDT7801,SDT7802,SDT7808
SDT7808	HST7802,HST7803,HST7804,HST7808,HST7809,SDT7802,SDT7803,SDT7804, SDT7809
SDT7809	HST7803,HST7804,HST7805,HST7809,SDT7803,SDT7804,SDT7805
SDT7901	HST7901,HST7902,HST7903,KSP1121,KSP1122,KSP1123,SDT7902,SDT7903
SDT7902	HST7902,HST7903,HST7904,KSP1122,KSP1123,KSP1124,KSP1125,SDT7903, SDT7904
SDT7903	HST7902,HST7903,HST7904,HST7905,KSP1122,KSP1123,KSP1124,KSP1125, SDT7902,SDT7904,SDT7905
SDT7904	HST7904,HST7905,KSP1124,KSP1125,SDT7905
SDT7905	HST7904,HST7905,KSP1125,SDT7904
SDT7907	HST7901,HST7902,HST7907,HST7908,SDT7901,SDT7902,SDT7908
SDT7908	HST7902,HST7903,HST7904,HST7908,HST7909,SDT7902,SDT7903,SDT7904, SDT7909
SDT7909	HST7903,HST7904,HST7905,HST7909,SDT7903,SDT7904,SDT7905
SDT7910	HST7907,HST7910,SDT7907
SDT8002	2N3598,2N3599,2N5733,HST8002,HST8003,HST8015,HST8016,HST8301,HST8302, KSP1254,KSP1255,SDT8003,SDT8015,SDT8016,SDT8301,SDT8302
SDT8003	2N3599,2N5733,HST8003,HST8016,HST8302,KSP1255,SDT8016,SDT8302
SDT8012	2N4002,HST8012,HST8013,KSP1251,KSP1252,SDT8013
SDT8013	2N4002,2N4003,HST8013,KSP1252,SDT8751,SDT8755

To Replace	Substitute This Type
SDT8015	2N3598,2N3599,2N5733,HST8002,HST8003,HST8015,HST8016,HST8301,HST8302, KSP1254,KSP1255,SDT8002,SDT8003,SDT8016,SDT8301,SDT8302
SDT8016	2N3599,2N5733,HST8003,HST8016,HST8302,KSP1255,SDT8003,SDT8302
SDT8045	2N3597,HST8045
SDT8070	HST8070,HST8071,HST8303,HST8304,SDT8071,SDT8303,SDT8304
SDT8071	HST8071,HST8304,SDT8304
SDT8105	HST8105,HST8106,HST8115,HST8116,SDT8106,SDT8115,SDT8116
SDT8106	HST8106,HST8116,SDT8116
SDT8110	HST8110,HST8111,SDT8111
SDT8111	HST8111
SDT8112	HST8112,HST8113,SDT8113
SDT8113	HST8113
SDT8114	HST8114
SDT8115	HST8105,HST8106,HST8115,HST8116,SDT8105,SDT8106,SDT8116
SDT8116	HST8106,HST8116,SDT8106
SDT8151	2N5329,2N5330
SDT8152	2N5957,2N5959
SDT8153	2N5957,2N5959,SDT8152
SDT8154	
SDT8155	SDT8154
SDT8156	
SDT8157	HST9903,HST9904
SDT8158	HST9902,HST9903,SDT8157
SDT8159	HST9901,ST400
SDT8301	2N5733,HST8002,HST8003,HST8015,HST8016,HST8301,HST8302,SDT8302
SDT8302	2N5733,HST8003,HST8016,HST8302
SDT8303	HST8070,HST8071,HST8303,HST8304,SDT8304
SDT8304	HST8071,HST8304
SDT8601	
SDT8602	
SDT8603	
SDT8604	
SDT8651	HST8651,HST8652,HST8653,SDT8652,SDT8653
SDT8652	HST8652,HST8653,HST8654,SDT8653,SDT8654
SDT8653	HST8653,HST8654,HST8655,SDT8654,SDT8655
SDT8654	HST8653,HST8654,HST8655,SDT8653,SDT8655
SDT8655	HST8654,HST8655,SDT8654
SDT8751	2N4003,KSP1253,SDT8752
SDT8752	KSP1253,SDT8753
SDT8753	HST8801,HST8802,KSP1201,SDT8754,SDT8801,SDT8802
SDT8754	HST8801,HST8802,HST8803,KSP1201,SDT8801,SDT8802,SDT8803
SDT8755	2N4003,2N5331,KSP1253,KSP1256,SDT8756
SDT8756	2N5331,KSP1253,KSP1256,SDT8757
SDT8757	2N5584,KSP1201,SDT8758
SDT8758	2N5584,KSP1201
SDT8801	HST8801,HST8802,HST8803,KSP1201,SDT8754,SDT8802,SDT8803
SDT8802	HST8802,HST8803,HST8804,KSP1201,SDT8803,SDT8804
SDT8803	HST8803,HST8804,HST8805,SDT8804,SDT8805
SDT8804	HST8803,HST8804,HST8805,SDT8803,SDT8805
SDT8805	HST8804,HST8805,SDT8804
SDT8920	HST8920,HST8921,KSP1001,KSP1002,SDT8921
SDT8921	HST8921,HST8922,KSP1002,KSP1003,SDT8922
SDT8922	HST8922,HST8923,KSP1003,SDT8923
SDT8923	HST8923
SDT8951	KSP1601,KSP1602,KSP1603,SDT8952,SDT8953
SDT8952	KSP1602,KSP1603,KSP1604,SDT8953,SDT8954
SDT8953	KSP1603,KSP1604,KSP1605,SDT8954,SDT8955
SDT8954	KSP1603,KSP1604,KSP1605,SDT8953,SDT8955
SDT8955	KSP1604,KSP1605,SDT8954
SDT9001	SDT4483,SDT9002,*GE-216,*ECG186,*WEP245
SDT9002	SDT4451,SDT4452,SDT9003,*GE-216,*SK3192,*ECG186,*WEP245

To Replace	Substitute This Type
SDT9003	SDT4452,*GE-28,*SK3192,*ECG186,*WEP245
SDT9004	SDT9005,SDT9007,SDT9008,*GE-216,*SK3193,*ECG187,*WEP246
SDT9005	2N4311,SDT4453,SDT4454,SDT9006,SDT9008,SDT9009,*GE-216,*SK3192,*ECG186, *WEP245
SDT9006	2N4305,2N4311,SDT4454,SDT9009,*GE-28,*SK3192,*ECG186,*WEP245
SDT9007	SDT9008,*GE-28,*SK3192,*ECG186,*WEP245
SDT9008	SDT9009,*GE-28,*SK3192,*ECG186,*WEP245
SDT9009	2N4305,*GE-66,*ECG152,*WEP245
SDT9010	SDT9011
SDT9011	SDT4455,SDT4456,SDT9012
SDT9012	SDT4456
SDT9201	HST9201,HST9205,HST9206,SDT9205,SDT9206,*GE-77,*TR-59,*SK3027,*HN100, *ECG130,*WEP247,*BDY20,*RT-131
SDT9202	HST9202,HST9203,HST9207,HST9208,SDT9203,SDT9207,SDT9208,*GE-75,*SK3027, *ECG180,*WEPS7001
SDT9203	HST9203,HST9204,HST9208,HST9209,SDT9204,SDT9208,SDT9209,*SK3079
SDT9204	HST9204,HST9209,SDT9209,*SK3079
SDT9205	HST9201,HST9205,HST9206,SDT9201,SDT9206,*GE-77,*SK3027,*ECG130, *WEP247,*BDY20
SDT9206	HST9202,HST9206,HST9207,SDT9202,SDT9207,*GE-77,*SK3027,*ECG130, *WEP247,*BDY20
SDT9207	HST9202,HST9203,HST9207,HST9208,SDT9202,SDT9203,SDT9208,*GE-75,*SK3027, *ECG180,*WEPS7001
SDT9208	HST9203,HST9204,HST9208,HST9209,SDT9203,SDT9204,SDT9209,*SK3079
SDT9209	HST9204,HST9209,SDT9204,*SK3079
SDT9210	2N3055A,2N3667,2N3771,2N5301,2N5302,2N6257,2N6371,2SD231,2SD249, 1561-0403,1561-0404,1561-0408,1561-0410,1582-0403,1582-0404,1582-0405, B170002-BLK,B170002-BRN,B170020,HST9201,HST9205,IR3771,MJ2801,MJ3771, SDT9201,SDT9205,*PTC140,*GE-77,*SK3027,*ECG130,*WEP247,*BDY20
SDT9301	2SD172,SDT9302,SDT9304,SDT9305,SDT9307,SDT9308,*PTC140,*GE-77
SDT9302	2SD172,SDT9303,SDT9305,SDT9306,SDT9308,SDT9309,*PTC140,*GE-77
SDT9303	2SD53,2SD173,2SD176,SDT9306,SDT9309,*PTC140,*GE-75
SDT9304	2SD172,SDT9301,SDT9302,SDT9305,SDT9307,SDT9308,*PTC140,*GE-77
SDT9305	2SD172,SDT9302,SDT9303,SDT9306,SDT9308,SDT9309,*PTC140,*GE-77
SDT9306	2SD53,2SD173,2SD176,SDT9303,SDT9309,*GE-75
SDT9307	2SD172,SDT9301,SDT9302,SDT9304,SDT9305,SDT9308,*GE-77
SDT9308	2SD172,SDT9302,SDT9303,SDT9305,SDT9306,SDT9309,*GE-77
SDT9309	2SD53,2SD173,2SD176,SDT9303,SDT9306,*GE-75
SDT9701	2N3236,2N5630,2N6254,2N6258,2N6359,2SD55,108T2,1561-1008,1561-1010, 1561-1015,1561-1208,1561-1210,1561-1215,IR3772,MJ3772,RCS258,SDT9702, SDT9704,SDT9705,STC2221,STC2225,STC2229,*PTC173,*GE-75,*ECG180, *WEPS7001
SDT9702	2N3055C,2N3772C,2N3773,2N5630,2N5631,2N6302,108T2,109T2,1561-1008, 1561-1010,1561-1015,1561-1208,1561-1210,1561-1215,1561-1410,1561-1415,43104, IR3773,IR6302,SDT9703,SDT9705,SDT9706,STC2222,STC2226,STC2230,*PTC173
SDT9703	2N3055C,2N3772C,2N3773,2N5631,2N6259,2N6302,109T2,1561-1208,1561-1210, 1561-1215,1561-1410,1561-1415,1561-1608,1561-1610,1561-1615,43104,IR3773, IR6302,SDT9706,STC2222,STC2226,STC2230
SDT9704	2N3236,2N5629,2N5630,2N6254,2N6258,2N6270,2N6271,2N6272,2N6273,108T2, BUY53A,BUY54A,IR3772,RCS258,SDT9705,*PTC173,*GE-75,*ECG180,*WEPS7001
SDT9705	2N3055C,2N3772C,2N3773,2N5630,2N5631,2N6271,2N6273,108T2,109T2,43104, IR3773,IR6302,SDT9706,*PTC173
SDT9706	2N3055C,2N3772C,2N3773,2N5631,2N6259,109T2,43104,IR3773,IR6302
SDT9707	2N3237,2N6359,2SD55,1561-0808,1561-0810,1561-0815,1561-1008,1561-1010, 1561-1015,1582-0808,1582-0810,1582-0815,1582-1008,1582-1010,1582-1015,IR3772, MJ3772,RCS258,SDT9701,STC2220,STC2221,STC2224,STC2225,STC2228, STC2229,*PTC173,*GE-75,*ECG180,*WEPS7001
SDT9801	HST9901,HST9902,SDT9802,*HEPS7004
SDT9802	HST9902,HST9903,SDT9803,*HEPS7004
SDT9803	HST9903,HST9904,SDT9804,*HEPS7004
SDT9804	HST9904

To Replace	Substitute This Type
SDT9901	HST9201,HST9205,HST9206,HST9801,HST9802,SDT9201,SDT9205,SDT9206, SDT9902,*GE-35,*ECG162,*WEP707,*BDY20
SDT9902	HST9202,HST9206,HST9207,HST9802,HST9803,SDT9202,SDT9206,SDT9207, SDT9903,*GE-35,*ECG162,*WEP707,*BDY20
SDT9903	HST9202,HST9203,HST9207,HST9208,HST9803,HST9804,SDT9202,SDT9203, SDT9207,SDT9208,SDT9904,*GE-35,*ECG162,*WEP707
SDT9904	HST9203,HST9204,HST9208,HST9209,HST9804,SDT9203,SDT9204,SDT9208, SDT9209,*GE-35,*ECG162,*WEP707
SDT12201	SDT12202
SDT12202	2N6563,SDT12203
SDT12203	2N6563
SDT12301	SDT12302,*PTC118
SDT12301+	SDT12302+
SDT12302	SDT12303,*PTC118
SDT12302+	SDT12303+
SDT12303	*PTC118
SDT12303+	
SDT13301	SDT12303,SDT13302,*PTC118
SDT13301+	SDT12303+,SDT13302+
SDT13302	SDT13303,SDT13304
SDT13302+	SDT13303+,SDT13304+
SDT13303	SDT13304,SDT13305
SDT13303+	SDT13304+,SDT13305+
SDT13304	SDT13305
SDT13304+	SDT13305+
SDT13305	
SDT13305+	
SDT14304	
SDT14304+	
SDT14305	
SDT14305+	
SE1001	2N916,2N916A,2N2847,2N3693,2N3826,2N3862,2N3903,2N3946,2N3974,2N3976, 2N4227,2N4962,2N4994,2N5027,2N5380,2SC620,2SC815,2SC899,2SC944,2SC965, 2SC1189,2SC1320,2SC1688,EN916,MM3903,MPS3693,MPS6565,MPS6576-BLUE, MPS6576-YEL,PET1001,TIS87,TN-3903,*PTC132,*GE-210,*TR-83,*HEPS0030, *SK3018,*EN10,*ECG229,*WEP719,*BC107B,*276-2009,*RT-113
SE1002	2N3302,2N3694,2N3827,2N3904,2N4995,2N5381,2SC1992,A157B,BC107,BC107B, BC123,BC167B,BC171B,BC174B,BC190B,BC237B,BC413B,BC414B,BCY59B,BSX79, MPS3694,MPS6566,MPS6575,MPS6576-SIL,PET1002,TN-3904,*PTC132,*GE-210, *TR-83,*HEPS0030,*SK3132,*SN80,*ECG161,*WEP719,*BC107B,*276-2009, *RT-113
SE1010	2N783,2N784,2N834A,2N2368,2SC54,2SC302,2SC321,2SC322,2SC400-R,2SC455, 2SC689,2SC837,BF224J,BF225J,KT218,MPS834,MPS6540,MPSH20,MPSH32, MPSH37,MT1038,MT1038A,MT1039,MT1060,PET3001,TIS47,TIS84,TIS108, *PTC132,*GE-86,*TR-83,*HEPS0020,*SK3019,*SN80,*ECG161,*WEP719,*BF365, *276-2011,*RT-113
SE2001	2N916,2N916A,2N2501,2N4123,2N4873,2SC372-O,2SC372G-O,2SC380-O, 2SC380A-O,2SC394-Y,2SC536,2SC639,2SC735-O,2SC752G-O,2SC899,2SC933, 2SC964,2SC965,2SC1188,2SC1189,2SC1216,2SC1276,2SC1320,2SC1687, 2SC1688,2SC1999,BF240,EN916,MPS3693,MPS6513,MPS6574-BLUE, MPS6574-GREEN,MPS6574-YEL,MPS6576-BLUE,MPS6576-GREEN,MPS6576-YEL, *PTC132,*GE-60,*TR-83,*HEPS0025,*SK3018,*SN80,*ECG161,*WEP719,*BF365, *276-2016,*RT-113
SE2002	2N3242A,2N3692,2N3694,2SC373,2SC373G,2SC735-GR,2SC1084,2SC1992,A157B, BC107,BC107B,BC123,BC167B,BC171B,BC237B,BC413B,BC414B,BCY58B,BCY59B, BSW88,BSW89,BSX38,BSX79,MPS3694,MPS6514,MPS6520,MPS6573, MPS6574-SIL,MPS6575,MPS6576-SIL,PET1002,PET2002,SE1002,*PTC132,*GE-210, *TR-83,*HEPS0015,*SK3039,*EN10,*ECG161,*WEP719,*BC107B,*276-2016, *RT-113
SE3001	2SC764,2SC1188,2SC1189,2SC1215,2SC1320,2SC1395,2SC1396,BF311,BF377, BF378,MT1060A,SE3005,TIS64A,TIS129,*PTC132,*GE-11,*TR-83,*HEPS0020,

To Replace	Substitute This Type
(SE3001)	*SK3018,*EN10,*ECG161,*WEP719,*276-2011,*RT-113
SE3002	2N2369,2N2369A,2N2729,2N3563,2N4137,2N4295,2N5029,2SC387AG,2SC601, 2SC601N,2SC689,2SC836,2SC922,2SC1123,2SC1126,2SC1128,2SC1240,BF311, EN2369A,MM1941,MPS6507,MT1060A,SE3001,TIS48,TIS49,TIS129,*PTC132, *GE-11,*TR-95,*HEPS0020,*SK3018,*EN10,*ECG108,*WEP56,*BF173,*276-2011, *RT-113
SE3005	2N4873,2SC1188,2SC1189,2SC1320,2SC1393,2SC1394,2SC1395,2SC1396, *PTC132,*GE-86,*TR-83,*HEPS0020,*SK3117,*ECG161,*WEP719,*276-2011, *RT-113
SE3031	HST5904,HST5905,HST5951,HST5952,HST6905,HST6906,HST6907,KSP1026, *GE-19,*TR-67,*HEPS5020,*HN100,*ECG162,*WEP707
SE3032	2N3920,HST5914,*PTC906,*GE-19,*TR-59,*HEPS7002,*SK3027,*HN100,*ECG130, *WEP247,*RT-131
SE3033	2N3919,2N4113,HST5904,HST6905,KSP1026,*PTC906,*GE-19,*TR-59,*HEPS7002, *SK3027,*HN100,*WEP247,*RT-131
SE3035	KSP1024,KSP1025,*GE-19,*TR-59,*HEPS7002,*SK3027,*HN100,*ECG130,*WEP247, *RT-131
SE3036	*GE-19,*TR-59,*HEPS7002,*SK3027,*HN100,*ECG130,*WEP247,*RT-131
SE3040	2N5606,2N5610,HST5902,HST5903,KSP1024,KSP1025,*TR-81,*HEPS5019,*ECG175, *WEP241,*276-2017
SE3041	HST5904,HST6905,KSP1026,*TR-81,*HEPS5012,*ECG175,*WEP241
SE4001	2N3242A,2N3565,2N4286,2N4287,2N4419,2N4967,2SC1364,2SC1632,2SC1634, 2SC1992,BC107B,BC167B,BC171B,BC237B,BC547A,BCY58B,BCY59B,MPSA17, *PTC101,*GE-60,*TR-21,*HEPS0025,*SK3018,*SN80,*ECG229,*WEP736,*BF365, *276-2016,*RT-102
SE4002	2N5526,BC550C,FT107B,SE4010,SE4021,*PTC139,*GE-212,*TR-21,*HEPS0015, *SK3122,*SN80,*ECG123A,*WEP736,*BC107B,*276-2016,*RT-102
SE4010	2N5526,BC550C,FT107B,SE4002,SE4021,*PTC139,*GE-212,*TR-21,*HEPS0015, *SK3124,*SN80,*ECG123A,*WEP736,*BC107B,*276-2016,*RT-102
SE4020	FT107C,*GE-39,*TR-83,*HEPS0005,*SK3117,*ECG161,*WEP719,*276-2009,*RT-113
SE4021	FT107B,*GE-85,*TR-83,*SK3117,*ECG161,*WEP719,*RT-113
SE4022	FT107A,*GE-39,*TR-83,*SK3117,*ECG161,*276-2016,*RT-113
SE5001	2N2369,2N2369A,2N2501,2N3009,2N3013,2N3014,2N3646,2N3688,2N3689,2N3690, 2N3862,2N4295,2N4420,2N4421,2N4422,2N5029,2SC67,2SC169,2SC335,2SC336, 2SC601N,2SC639,2SC752G-O,2SC764,2SC1187,2SC1188,2SC1189,2SC1274, 2SC1319,2SC1320,2SC1687,2SC1688,BF198,BF199,BF241,GET2369,MPS3646, SE5002,SE5003,SE5006,TIS48,TIS87,*PTC132,*GE-60,*TR-83,*HEPS0020,*SK3018, *SN80,*ECG161,*WEP719,*276-2016,*RT-113
SE5002	2N2369,2N2369A,2N2501,2N3009,2N3013,2N3014,2N3646,2N3688,2N3689,2N3690, 2N3862,2N4295,2N4420,2N4421,2N4422,2N5029,2SC67,2SC169,2SC335,2SC336, 2SC601N,2SC639,2SC752G-O,2SC764,2SC1187,2SC1188,2SC1189,2SC1274, 2SC1319,2SC1320,2SC1687,2SC1688,BF198,BF199,BF241,GET2369,MPS3646, SE5001,SE5003,SE5006,TIS48,TIS87,*PTC132,*GE-60,*TR-83,*HEPS0020,*SK3018, *EN10,*ECG161,*WEP719,*276-2016,*RT-113
SE5003	2N2369,2N2369A,2N2501,2N3009,2N3013,2N3014,2N3646,2N3688,2N3689,2N3690, 2N3862,2N4295,2N4420,2N4421,2N4422,2N5029,2SC67,2SC169,2SC335,2SC336, 2SC601N,2SC639,2SC752G-O,2SC764,2SC1187,2SC1188,2SC1189,2SC1274, 2SC1319,2SC1320,2SC1687,2SC1688,BF198,BF199,BF241,GET2369,MPS3646, SE5001,SE5002,SE5006,TIS48,TIS87,*PTC132,*GE-60,*TR-83,*HEPS0020,*SK3132, *EN10,*ECG161,*WEP719,*276-2016,*RT-113
SE5006	2N2369,2N2369A,2N2501,2N3009,2N3013,2N3014,2N3646,2N3688,2N3689,2N3690, 2N3862,2N4295,2N4420,2N4421,2N4422,2N5029,2SC67,2SC169,2SC335,2SC336, 2SC601N,2SC639,2SC752G-O,2SC764,2SC1187,2SC1188,2SC1189,2SC1274, 2SC1319,2SC1320,2SC1687,2SC1688,BF198,BF199,BF241,GET2369,MPS3646, SE5001,SE5002,SE5003,TIS48,TIS87,*PTC132,*GE-60,*TR-95,*HEPS0016,*SK3018, *EN10,*ECG108,*WEP56,*276-2016,*RT-113
SE5020	2N918,2N5851,2N5852,2SC251,2SC251A,2SC252,2SC253,2SC602,2SC682,2SC683, 2SC1054,2SC1067,A406,MT1061,SE5021,TIS56,TIS57,*PTC132,*GE-60,*TR-95, *HEPS0017,*SK3018,*EN10,*ECG108,*WEP56,*BF173,*276-2015,*RT-113
SE5021	2N918,2N5851,2N5852,2SC251,2SC251A,2SC252,2SC253,2SC602,2SC682,2SC683, 2SC1054,2SC1067,A406,MT1061,SE5020,TIS56,TIS57,*PTC132,*GE-60,*TR-83,

To Replace	Substitute This Type
(SE5021)	*HEPS0017,*SK3018,*EN10,*ECG161,*WEP719,*BF173,*276-2015,*RT-113
SE5022	2N918,2N3289,2N3290,2N5851,2N5852,2SC251,2SC251A,2SC252,2SC253,2SC602, 2SC682,2SC683,2SC1054,2SC1067,A406,MT1061,SE5020,SE5021,SE5023,SE5024, SE5050,SE5051,TIS56,TIS57,*PTC132,*GE-60,*TR-83,*HEPS0017,*SK3117,*EN10, *ECG161,*WEP719,*BF173,*276-2015,*RT-113
SE5023	2N918,2N3289,2N3290,2N5851,2N5852,2SC251,2SC251A,2SC252,2SC253,2SC602, 2SC682,2SC683,2SC1054,2SC1067,A406,MT1061,SE5020,SE5021,SE5022,SE5024, SE5050,SE5051,TIS56,TIS57,*PTC132,*GE-60,*TR-83,*HEPS0017,*SK3018,*EN10, *ECG161,*WEP709,*BF173,*276-2015,*RT-113
SE5024	2N918,2N3289,2N3290,2N5851,2N5852,2SC251,2SC251A,2SC252,2SC253,2SC602, 2SC682,2SC683,2SC1054,2SC1067,A406,MT1061,SE5020,SE5021,SE5022,SE5023, SE5050,SE5051,TIS56,TIS57,*PTC132,*GE-60,*TR-83,*HEPS0017,*SK3018,*EN10, *ECG161,*WEP719,*BF173,*276-2015,*RT-113
SE5025	2N834A,2N2368,2N2729,2SC52,2SC54,2SC302,2SC321,2SC322,2SC400-R,2SC689, 2SC837,2SC1126,BF224J,BF225J,KT218,MM1941,MPS834,MPS6507,MPS6540, MPS6542,MPS6543,MPS6548,MPSH19,MPSH20,MPSH32,MPSH37,MT1038, MT1038A,MT1039,MT1060,PET3001,TIS47,TIS84,TIS108,*PTC132,*GE-61,*TR-83, *HEPS0020,*SK3018,*EN10,*ECG161,*WEP719,*BF173,*276-2016,*RT-113
SE5050	2N918,2N3289,2N3290,2N5851,2N5852,2SC251,2SC251A,2SC252,2SC253,2SC602, 2SC682,2SC683,2SC1054,2SC1067,A406,MT1061,SE5020,SE5021,SE5022,SE5023, SE5024,SE5051,TIS56,TIS57,*PTC132,*GE-60,*TR-83,*HEPS0017,*SK3018,*EN10, *ECG161,*WEP719,*BF173,*276-2015,*RT-113
SE5051	2N918,2N3289,2N3290,2N5851,2N5852,2SC251,2SC251A,2SC252,2SC253,2SC602, 2SC682,2SC683,2SC1054,2SC1067,A406,MT1061,SE5020,SE5021,SE5022,SE5023, SE5024,SE5050,TIS56,TIS57,*PTC132,*GE-60,*TR-83,*HEPS0017,*SK3117,*EN10, *ECG161,*WEP719,*BF173,*276-2015,*RT-113
SE5052	*PTC132,*GE-60,*TR-83,*HEPS0017,*SK3117,*ECG161,*WEP719,*BF173,*276-2015, *RT-113
SE5055	2N2808,2N2808A,2N2809,2N2809A,2N2810,2N2810A,2N2865,2N3570,2N3571, 2N3600,2N4134,2N4135,2N4252,2N4253,2N5053,2N5054,2N6304,2N6305,2SC398, 2SC399,2SC567,2SC568,2SC611,2SC653,2SC1553A,40295,40413,MT1061A, S1297,*PTC132,*GE-60,*TR-21,*HEPS0011,*SK3117,*ECG319,*WEP736,*BF173, *276-2015,*RT-102
SE6001	2N760A,2N909,2N916,2N916A,2N930B,2N1420,2N1507,2N1983,2N1986,2N2350, 2N2350A,2N2897,2N2900,2N2960,2N2961,2N3242A,2N3643,40458,A5T2192, SE6020,SE6020A,TN-3904,*GE-47,*TR-83,*HEPS0015,*SK3124,*SN80,*ECG128, *WEP719,*BC337,*276-2013,*RT-113
SE6002	2N1952,*DS-66,*GE-47,*TR-83,*HEPS0015,*SK3122,*SN80,*ECG161,*WEP719, *BC337,*276-2013,*RT-113
SE6020	SE6020A,SE6021,SE6021A,*GE-18,*TR-87,*HEPS3001,*SK3024,*ECG128,*WEP243, *RT-114
SE6020A	SE6020,SE6021,SE6021A,*TR-87,*HEPS3011,*SK3024,*ECG128,*WEP243,*RT-114
SE6021	SE6021A,*GE-18,*TR-87,*HEPS3019,*SK3024,*ECG128,*WEP243,*RT-114
SE6021A	SE6021,*TR-87,*HEPS3011,*SK3024,*ECG128,*WEP243,*RT-114
SE6022	SE6023,*GE-18,*TR-87,*HEPS3001,*SK3024,*ECG128,*WEP243,*276-2014,*RT-114
SE6023	*GE-18,*TR-87,*HEPS3019,*SK3024,*ECG128,*WEP243,*RT-114
SE7001	2N1342,2N3114,2N3500,2N3712,2N3923,2N4925,2SC154C,2SC470-5,2SC470-6, 2SC507-R,2SC686,2SC1217,MM2259,MM2260,MM3009,*PTC117,*GE-40,*TR-78, *HEPS5026,*SN80,*ECG154,*WEP712,*BF338,*276-2012,*RT-110
SE7002	2N699A,2N699B,2N1342,2N1613B,2N2443,2N3114,2N3500,2N3712,2N3923,2N4925, 2SC49,2SC470-4,2SC470-5,2SC590,2SC686,2SC1217,MM3008,SE7001,*PTC117, *GE-40,*TR-78,*HEPS5026,*SK3045,*ECG154,*WEP712,*BF338,*276-2012, *RT-110
SE7015	2N740,2N740A,2N910,2N1574,2N1973,2N2443,2N2465,2N2466,2N3107,2SC353, 2SC805,*PTC170,*GE-27,*TR-87,*HEPS0005,*SK3024,*ECG128,*WEP243,*BF338, *276-2008,*RT-114
SE7016	2SC686,SE7017,*PTC170,*GE-27,*TR-78,*HEPS0005,*SK3045,*ECG154,*WEP712, *BF338,*276-2008,*RT-110
SE7017	2N6430,2SC154C,2SC1048,*PTC170,*GE-27,*TR-78,*HEPS0005,*SK3045,*ECG154, *WEP712,*BF338,*276-2008,*RT-110
SE7030	*GE-12,*TR-81,*HEPS3021,*SK3021,*ECG124,*WEP240,*RT-128

To Replace	Substitute This Type
SE7055	*GE-40,*TR-78,*SK3045,*ECG154,*WEP712,*276-2008,*RT-110
SE7056	*PTC117,*GE-27,*TR-78,*SK3045,*ECG154,*WEP712,*276-2008,*RT-110
SE7057	
SE8001	2N3110,2N3295,2N5188,*PTC125,*GE-18,*TR-87,*HEPS0015,*SK3024,*SN80, *ECG128,*WEP243,*BF338,*276-2008,*RT-114
SE8002	2N1613A,2N1889,2N1974,2N3107,2N3109,2N4924,2N4961,*PTC125,*GE-18,*TR-87, *HEPS0015,*SK3024,*EN10,*ECG128,*WEP243,*BF338,*276-2008,*RT-114
SE8010	*GE-18,*TR-78,*HEPS0005,*SK3024,*ECG128,*WEP712,*276-2012,*RT-110
SE8012	SE8010,*GE-18,*TR-87,*HEPS0005,*SK3024,*ECG128,*WEP243,*RT-114
SE8040	SE8041,*GE-20,*TR-21,*HEPS3001,*SK3122,*ECG123A,*WEP736,*BC337, *276-2009,*RT-102
SE8041	2N3724A,SE8042,*PTC144,*GE-47,*TR-87,*HEPS3001,*SK3024,*ECG128,*WEP243, *276-2030,*RT-114
SE8042	2N3724A,*PTC144,*GE-63,*TR-87,*HEPS3001,*SK3024,*ECG128,*WEP243, *276-2030,*RT-114
SE8540	2N2800,2N2837,2N3244,2N3467,2N3762,2N3764,2N5023,2N5042,2N5811,2N5815, MM3726,SE8541,SE8542,*GE-48,*TR-88,*HEPS0019,*SK3025,*ECG129,*WEP242, *BC327,*276-2032,*RT-115
SE8541	2N2800,2N3244,2N3467,2N3762,2N5023,2N5042,MM3726,SE8542,*PTC141,*GE-48, *TR-88,*SK3025,*ECG129,*WEP242,*RT-115
SE8542	2N3244,2N3467,2N3762,2N5023,MM3726,*PTC141,*GE-67,*TR-88,*SK3025, *ECG129,*WEP242,*RT-115
SE9060	2N5202,2N5600,*PTC104,*HEPS5012,*ECG175,*WEP241
SE9061	2N5598,2N5602,*PTC104,*HEPS5012,*ECG175,*WEP241
SE9062	2N5202,2N5600,2N5604,*PTC104,*HEPS5012,*ECG175,*WEP241
SE9063	2N5602,*PTC104,*HEPS5012,*ECG175,*WEP241
SE9070	2N5202,*PTC906
SE9071	*PTC906
SE9072	2N5202
SE9073	
SE9080	2N4111,2N5616,KSP1021,KSP1022,KSP1024,KSP1025,KSP1041,KSP1042,KSP1044, KSP1045,KSP1161,KSP1162,KSP1164,KSP1165,SDT7602,SDT7603,*PTC906, *GE-14,*TR-59,*SK3027,*ECG130,*WEP247,*RT-131
SE9081	2N4112,2N5614,2N5618,*PTC906
SE9082	2N4070,2N4111,2N4113,2N5616,2N5620,KSP1022,KSP1025,KSP1042,KSP1045, KSP1162,KSP1165,SDT7603
SE9083	2N4112,2N5618
SE9300	SE9301
SE9301	SE9302
SE9302	
SE9303	SE9304
SE9304	SE9305
SE9305	DTS-1020
SE9331	
SE9400	SE9401
SE9401	SE9402
SE9402	
SE9403	SE9404
SE9404	SE9405
SE9405	
SE9570	*SK3191,*ECG185,*WEPS5007
SE9571	*SK3191,*ECG185,*WEPS5007
SE9572	*HEPS5006,*SK3191,*ECG185,*WEPS5007
SE9573	*HEPS5006,*SK3191,*ECG185,*WEPS5007
SE9580	2N5615,KSP2022,KSP2025,KSP2042,KSP2045,KSP2161,KSP2162,KSP2164, KSP2165
SE9581	2N5613,2N5617
SE9582	2N5615,2N5619,KSP2023,KSP2026,KSP2043,KSP2046,KSP2162,KSP2165
SE9583	2N5617
SEN-T-173	
SF.T124	*PTC135,*GE-53,*ECG158,*WEP238

To Replace	Substitute This Type
SF.T125	SF.T125P,*PTC135,*AC188/01
SF.T125P	*PTC135,*AC188/01
SF.T130	*GE-53,*ECG158,*WEP238
SF.T131	*AC188/01
SF.T131P	*GE-53,*ECG158,*WEP238,*AC188/01
SF.T143	
SF.T144	*GE-2,*AT20M
SF.T145	
SF.T146	
SF.T163	SF.T358,*PTC145,*GE-50,*ECG160,*WEP637
SF.T186	2SC48,2SC59,2SC69,2SC149,2SC154,2SC154A,BF178,MM3001,*PTC123,*BF338
SF.T212	2N257-BLK,2N4241,2SB137,CDT1310,CDT1319,OC27,*PTC138,*GE-25,*TR-16, *PT40,*ECG127,*WEP235,*AD149
SF.T213	2N257-GRN,2N1535,2N2138,2N2138A,2N2139,2N2139A,2SB246,2SB247,2SB248, 2SB248A,2SB250A,2SB251A,CDT1320,SF.T214,SF.T238,SF.T239,*PTC105A,*GE-25, *TR-16,*ECG127,*WEP235
SF.T214	2N561,2N1535,2N1536,2N2139,2N2139A,2N2140,2N2140A,2SB249,2SB252,2SB253, CDT1320,CDT1321,SF.T239,SF.T240,*PTC105A,*GE-25,*TR-16,*ECG127,*WEP235
SF.T221	2N524,2N524A,2N525,2N525A,2SB224,SF.T222,*PTC135,*GE-53,*HF12M,*ECG158, *WEP238
SF.T222	2N381,2N525,2N525A,2N526,2N526A,2N597,2N1478,2SB57A,2SB225,2SB226, 2SB326,*PTC135,*GE-2,*ECG102,*WEP631
SF.T223	2N382,2N383,2N527,2N598,2N599,2N1998,*PTC135,*GE-53,*HF12M,*ECG158, *WEP238,*AC188/01
SF.T226	2N525A,*HF35
SF.T227	2N526A,2N1171,2N1305,2N1313,2N1348,2N1349,2N1350,2N1351,*GE-2,*HF35, *ECG102,*WEP631
SF.T228	2N1305,2N1307,2N1309A,2N1348,2N1349,2N1350,2N1351,*PTC135,*HF35, *AC188/01
SF.T229	2N1307,2N1309,*PTC109,*HF35
SF.T232	SF.T233
SF.T233	SF.T234
SF.T234	
SF.T237	2N417,2N508A,2N521A,2N523,2N571,2N599,2N655,2N655+,2N1284,2N1307, 2N1309,2N1353,2N1354,2N1355,2N1356,2N1357,2N1808,2N1892,2SB223,2SB223A, 2SB327,40269,ASY27,SF.T223,*PTC109,*GE-1,*TR-05,*SK3123,*HF35,*ECG102A, *WEP250,*AC188/01,*276-2007
SF.T238	2N456A,2N457A,2N2138,2N2138A,2N2139,2N2139A,2N3611,2N3612,2N5890, 2N5891,ASZ17,KR6500,SF.T239,*PTC105A,*GE-25,*TR-01,*ECG127,*WEP235
SF.T239	2N2139,2N2139A,2N2140,2N2140A,2N3612,2N3615,2N3617,ASZ15,KR6500,SF.T240, *PTC105A,*GE-25,*TR-01,*ECG127,*WEP235
SF.T240	2N2140,2N2140A,2N2141,2N2141A,2N3615,2N3616,2N3617,2N3618,2SB340,ASZ15, *PTC105A,*GE-25,*TR-01,*ECG127,*WEP235
SF.T243	
SF.T250	2N1541,2N1542,2N1546,2N1546A,2N1547,2N1547A,2N2145,2N2145A,2N2146, 2N2146A,*PTC105A,*GE-25,*ECG127,*WEP235
SF.T264	2SB331,2SB351,SF.T265,*GE-4,*TR-03,*PT501,*ECG105,*WEP233
SF.T265	2SB331,2SB332,2SB351,2SB352,2SB478,SF.T266,*GE-4,*TR-03,*PT501
SF.T266	2SB332,2SB333,2SB352,2SB353,2SB479,SF.T267,*GE-4,*TR-03,*PT501
SF.T267	2SB334,2SB354,*GE-4,*PT501
SF.T288	*HF35,*276-2007
SF.T298	*DS-72,*GE-5
SF.T306	*PTC109,*GE-2,*TR-05,*SK3123,*HF35,*ECG102A,*WEP250,*AC188/01
SF.T307	SF.T308,*PTC109,*HF35
SF.T308	*DS-56,*HF35
SF.T315	2N2188,2N2189,2N2190,2N2191,2N2957,*PTC107,*HF35
SF.T316	SF.T354,SF.T357,SF.T357P,SF.T358,*PTC145,*GE-50,*HF35,*ECG160,*WEP637
SF.T317	2N827,*PTC107,*GE-50,*HF35,*ECG160,*WEP637
SF.T319	2N838,2N2587,2N3323,2N3324,2N3325,2N6365,2N6365A,SF.T320,*PTC107,*GE-50, *HF35,*ECG160,*WEP637
SF.T320	2N3323,2N3324,2N3325,SF.T317,*PTC107,*GE-50,*HF35,*ECG160,*WEP637

To Replace	Substitute This Type
SF.T322	*PTC135,*GE-53,*AT20M,*ECG158,*WEP238,*AC188/01
SF.T323	*PTC135,*AT20M,*AC188/01
SF.T337	*PTC135,*GE-2,*TR-05,*SK3123,*HF35,*ECG102A,*WEP250,*AC188/01
SF.T352	2N2706,2SB496,SF.T353,*PTC102,*GE-53,*TR-05,*SK3123,*AT20M,*ECG158, *WEP238,*AC188/01
SF.T353	2N2706,SF.T323,*PTC102,*TR-05,*SK3123,*AT20M,*ECG176,*WEP238,*AC188/01
SF.T354	SF.T316,SF.T357,SF.T357P,SF.T358,*PTC145,*GE-50,*JR100,*ECG160,*WEP637
SF.T357	SF.T358,*PTC145,*GE-50,*JR100,*ECG160,*WEP637
SF.T357P	*PTC145
SF.T358	*PTC145,*GE-50,*JR100,*ECG160,*WEP637
SHA7520	HA7520,*GE-67,*TR-28,*ECG193,*WEP242,*276-2025
SHA7521	HA7521,*GE-67,*TR-28,*ECG193,*WEP242
SHA7522	2N1238,HA7522,*GE-67,*TR-28,*ECG193,*WEP242,*276-2025
SHA7523	2N1240,HA7523,*GE-67,*TR-28,*ECG193,*WEP242,*276-2025
SHA7524	2N1242,HA7524,*GE-67,*TR-28,*ECG193,*WEP242
SHA7525	HA7525
SHA7526	2N1239,HA7526,*GE-67,*TR-28,*ECG193,*WEP242,*276-2025
SHA7527	2N1241,HA7527,*GE-67,*TR-28,*ECG193,*WEP242,*276-2025
SHA7528	2N1243,HA7528,*GE-67,*TR-28,*ECG193,*WEP242
SHA7529	2N1244,HA7525,HA7529,SHA7525
SHA7530	HA7530,*GE-82,*TR-28,*SK3114,*ECG159,*WEP242,*BC327,*276-2023
SHA7531	2N1440,HA7531,*GE-82,*TR-28,*SK3114,*ECG159,*WEP242
SHA7532	2N1228,2N6567,*GE-22,*TR-28,*SK3114,*ECG159,*WEP242,*BC327,*276-2023
SHA7533	2N1230,2N1441,2N4008,HA7533,*GE-82,*TR-28,*SK3114,*ECG159,*WEP242, *BC327,*276-2023
SHA7534	2N1232,HA7534,*GE-82,*TR-28,*SK3114,*ECG159,*WEP242
SHA7535	2N2551,HA7535
SHA7536	2N1229,2N2945A,2N3217,2N3218,2N3978,2N4007,HA7536,TW135,*GE-22,*TR-20, *SK3114,*ECG159,*WEP242,*BC327,*276-2023
SHA7537	2N1231,2N1442,2N1443,2N1469A,2N2946A,2N3219,2N3979,HA7537,*GE-82,*TR-28, *SK3114,*ECG159,*WEP242,*BC327,*276-2023
SHA7538	2N1233,2N4982,HA7538,*GE-82,*TR-28,*SK3114,*ECG159,*WEP242
SHA7539	2N1234,HA7535,HA7539,SHA7535
SHA7597	2N1242,HA7524,SHA7524,*GE-67,*TR-28,*ECG193,*WEP242
SHA7598	2N1243,HA7528,SHA7528,*GE-67,*TR-28,*ECG193,*WEP242
SHA7599	2N1243,*GE-67,*TR-28,*ECG193,*WEP242
SK3003	REFER TO SECTION 2
SK3004	REFER TO SECTION 2
SK3005	REFER TO SECTION 2
SK3006	REFER TO SECTION 2
SK3007	REFER TO SECTION 2
SK3008	REFER TO SECTION 2
SK3009	REFER TO SECTION 2
SK3010	REFER TO SECTION 2
SK3011	REFER TO SECTION 2
SK3012	REFER TO SECTION 2
SK3013	M.P.SK3009
SK3014	REFER TO SECTION 2
SK3015	M.P.SK3014
SK3018	REFER TO SECTION 2
SK3019	REFER TO SECTION 2
SK3020	REFER TO SECTION 2
SK3021	REFER TO SECTION 2
SK3024	REFER TO SECTION 2
SK3024+	REFER TO SECTION 2
SK3025	REFER TO SECTION 2
SK3025+	REFER TO SECTION 2
SK3026	REFER TO SECTION 2
SK3027	REFER TO SECTION 2
SK3028	M.P.SK3026
SK3029	M.P.SK3027

TRANSISTOR SUBSTITUTES

To Replace	Substitute This Type
SK3034	REFER TO SECTION 2
SK3035	REFER TO SECTION 2
SK3036	REFER TO SECTION 2
SK3037	M.P.SK3036
SK3038	REFER TO SECTION 2
SK3039	REFER TO SECTION 2
SK3040	REFER TO SECTION 2
SK3041	REFER TO SECTION 2
SK3044	REFER TO SECTION 2
SK3045	REFER TO SECTION 2
SK3046	REFER TO SECTION 2
SK3047	REFER TO SECTION 2
SK3048	REFER TO SECTION 2
SK3049	REFER TO SECTION 2
SK3052	REFER TO SECTION 2
SK3053	REFER TO SECTION 2
SK3054	REFER TO SECTION 2
SK3079	REFER TO SECTION 2
SK3082	REFER TO SECTION 2
SK3083	REFER TO SECTION 2
SK3084	REFER TO SECTION 2
SK3085	REFER TO SECTION 2
SK3086	M.P.SK3082
SK3103	REFER TO SECTION 2
SK3104	REFER TO SECTION 2
SK3111	REFER TO SECTION 2
SK3114	REFER TO SECTION 2
SK3115	REFER TO SECTION 2
SK3117	REFER TO SECTION 2
SK3118	REFER TO SECTION 2
SK3122	REFER TO SECTION 2
SK3123	REFER TO SECTION 2
SK3124	REFER TO SECTION 2
SK3131	REFER TO SECTION 2
SK3132	REFER TO SECTION 2
SK3132+	REFER TO SECTION 2
SK3133	REFER TO SECTION 2
SK3137	REFER TO SECTION 2
SK3138	REFER TO SECTION 2
SK3156	REFER TO SECTION 2
SK3173	REFER TO SECTION 2
SK3176	REFER TO SECTION 2
SK3177	REFER TO SECTION 2
SK3178	REFER TO SECTION 2
SK3179	REFER TO SECTION 2
SK3180	REFER TO SECTION 2
SK3181	REFER TO SECTION 2
SK3182	REFER TO SECTION 2
SK3183	REFER TO SECTION 2
SK3188	REFER TO SECTION 2
SK3189	REFER TO SECTION 2
SK3190	REFER TO SECTION 2
SK3191	REFER TO SECTION 2
SK3192	REFER TO SECTION 2
SK3193	REFER TO SECTION 2
SK3194	REFER TO SECTION 2
SK3195	REFER TO SECTION 2
SK3197	REFER TO SECTION 2
SK3198	REFER TO SECTION 2
SK3199	REFER TO SECTION 2
SK3200	REFER TO SECTION 2

To Replace	Substitute This Type
SK3201	REFER TO SECTION 2
SK3202	REFER TO SECTION 2
SK3203	REFER TO SECTION 2
SK3217	REFER TO SECTION 2
SK3218	REFER TO SECTION 2
SK3219	REFER TO SECTION 2
SK3220	REFER TO SECTION 2
SK3221	REFER TO SECTION 2
SK3222	REFER TO SECTION 2
SK3232	REFER TO SECTION 2
SK3239	REFER TO SECTION 2
SK3241	REFER TO SECTION 2
SK3244	REFER TO SECTION 2
SK3245	REFER TO SECTION 2
SK3246	REFER TO SECTION 2
SK3247	REFER TO SECTION 2
SK3248	REFER TO SECTION 2
SK3510	REFER TO SECTION 2
SK3511	REFER TO SECTION 2
SK3512	REFER TO SECTION 2
SK3513	REFER TO SECTION 2
SK3528	REFER TO SECTION 2
SK3529	REFER TO SECTION 2
SK3530	REFER TO SECTION 2
SK3534	REFER TO SECTION 2
SK3535	REFER TO SECTION 2
SK3536	REFER TO SECTION 2
SK3537	REFER TO SECTION 2
SK3538	REFER TO SECTION 2
SK3559	REFER TO SECTION 2
SK3560	REFER TO SECTION 2
SK3561	REFER TO SECTION 2
SK3562	REFER TO SECTION 2
SK3563	REFER TO SECTION 2
SN60	REFER TO SECTION 2
SN80	REFER TO SECTION 2
SO-1	2N2258,2N2999,*GE-51,*TR-17,*HF35,*ECG160,*WEP637,*AF125
SO-2	2N2258,2N2999,*GE-51,*TR-17,*HEPG0008,*HF35,*ECG160,*WEP637,*AF125
SO-3	2N2258,2N2999,*GE-51,*TR-17,*HEPG0008,*HF35,*ECG160,*WEP637,*AF125
SP70	REFER TO SECTION 2
SP90	REFER TO SECTION 2
SRF1001	2SC302-M,2SC307-M,2SC590N,2SC992,2SC1002,2SC1256,2SC1820,40975
SRF1002	2SC302-M
SRF11101	2SC302-M,2SC307-M,2SC590N
SRF12101	
SRF12212	2SC831
SRF12213	2SC831,SRF32214,SRF52214
SRF13113	SRF53104
SRF13122	
SRF13123	
SRF13212	
SRF13213	SRF13212
SRF21140	
SRF23121	
SRF23211	
SRF32214	
SRF52101	
SRF52214	SRF32214
SRF53104	SRF13113
SRF53114	
SRF53215	

To Replace	Substitute This Type
SRF54215	
SRSP4296	2N5344,2N5345,2N6420
SRSP4297	
SRSP4298	2N6214,2N6421
SRSP4299	
ST400	
ST401	ST400
ST440	*HEPS5000,*276-2020
ST450	*HEPS5000
ST4150	*PTC123,*GE-81,*TR-25,*ECG128,*WEP243,*BF338,*276-2012
ST4201	2N560,2N3295,2N5188,2SC31,2SC116,7A30,PT2540,SE8001,*PTC123,*GE-81, *TR-25,*ECG128,*WEP243,*BC337,*276-2012
ST4202	2N1564,2N1615,2N1975,BF177,BFY65,MM3000,*PTC125,*GE-81,*ECG128,*WEP243, *BF338,*276-2012
ST4203	2N656,2N780,2N2106,2N2107,2N2846,2N2848,2N3015,2N3299,2N3326,2N3554, 2N4960,2N6538,2SC32,2SC352,2SC352A,2SC594-R,2SC651,2SC1166-R,2SC1360, 7A31,A5T3903,BSY92,TIS105,*PTC123,*GE-81,*TR-25,*ECG128,*WEP243,*BC337, *276-2012
ST4204	2N657,2N1565,2N1613A,2N1889,2N1974,2N3107,2N3109,2N4924,2N4961,2SC353, 2SC353A,2SC470-3,2SC826,111T2,SE8002,*PTC125,*GE-81,*ECG128,*WEP243, *BF338,*276-2012
ST5641	2N957,2N2885,2N3691,2N3693,2N3845,2N3845A,2N4420,2N4421,2N4422,2N4966, 2N4968,2N4996,2N4997,2N5137,2SC169,2SC335,2SC372-O,2SC372G-O,2SC380-O, 2SC380A-O,2SC381-O,2SC394-Y,2SC400-O,2SC401,2SC402,2SC404,2SC460, 2SC461,2SC535,2SC536,2SC619,2SC620,2SC634,2SC641,2SC710,2SC712, 2SC713,2SC752G-O,2SC784-O,2SC785-O,2SC838,2SC839,2SC899,2SC922, 2SC924,2SC929NP,2SC941-O,2SC964,2SC1047,2SC1187,2SC1188,2SC1189, 2SC1215,2SC1276,2SC1319,2SC1320,2SC1342,2SC1372,2SC1393,2SC1394, 2SC1395,2SC1396,2SC1674,2SC1675,2SC1686,2SC1923,2SD227,BF227,BF229, BF254,BSW12,EN708,EN916,GI-3709,MPS3646,PET3002,SE1001,SE2001,SE3005, *PTC126,*GE-60,*TR-24,*ECG161,*WEP709,*BF183,*276-2031
ST7100	2N6429,2N6429A,*GE-62,*HEPS0015,*SK3245,*ECG199,*WEP736,*BC337, *276-2009
ST10000	2N5386,KSP2151,KSP2152
ST10001	2N5386,KSP2152
ST10007	ST10008,ST40002,ST40003,ST54004,ST54005
ST10008	ST10009,ST40003,ST40004,ST54005,ST54006
ST10009	ST40004,ST54006
ST14000	
ST14000A	
ST14011	ST15043
ST14012	ST15043,ST15044
ST14013	ST15044,ST15045
ST14030	ST14031
ST14031	ST14032
ST14032	
ST15013	
ST15014	
ST15015	
ST15043	ST14030,ST14031,ST15044
ST15044	ST14031,ST14032,ST15045
ST15045	ST14032
ST17003	2N4002,2N4210,2N4211,1768-0810,1768-0820,1768-0830,1768-1010,1768-1020, 1768-1030,B148000,B148004,HST7012,HST7013,HST8012,HST8013,HST9902, HST9903,KSP1251,KSP1252,SDT8012,SDT8013,SDT8157,SDT8158
ST17004	2N4002,2N4003,2N4211,1768-1010,1768-1020,1768-1030,1768-1210,1768-1220, 1768-1230,B148000,B148001,HST7013,HST8013,HST9903,HST9904,KSP1252, SDT8013,SDT8157,SDT8751
ST17060	ST14030,ST14031,ST15043,ST15044,ST17061
ST17061	ST14031,ST14032,ST15044,ST15045,ST17062
ST17062	ST14032,ST15045

To Replace	Substitute This Type
ST18007	
ST18008	
ST18009	
ST18010	HST8801,HST8802,HST8803
ST18011	
ST18012	
ST18013	
ST18014	2N5218,HST7152,HST7801,HST7802,KSP1101,KSP1102,SDT7152,SDT7801, SDT7802,SDT12201,SDT12202
ST18015	
ST18016	
ST18017	
ST18018	
ST40000	
ST40001	
ST40002	ST40003
ST40003	ST40004
ST40004	
ST54000	KSP2251,KSP2252
ST54001	KSP2252
ST54004	ST40002,ST40003,ST54005
ST54005	ST40003,ST40004,ST54006
ST54006	ST40004
ST61000	2N2605,2N2605A,2N3251,2N3251A,2N3547,2N3582,2N3702,2N3906,2N3914,2N3915, 2N4034,2N4035,2N4121,2N4122,2N4143,2N4248,2N4359,2N4965,2N5244,2N5366, 2N5447,2N6223,2N6225,2SA467G-Y,2SA493-Y,2SA493G-Y,2SA499-Y,2SA550A, 2SA561-Y,2SA603,2SA628A,2SA659,2SA666A,2SA678,2SA705,2SA710,2SA711, 2SA718,2SA733,2SA767,2SA889,A177,BC177A,BC212A,BC212L,BC213L,BC214L, BC251A,BC257A,BC261A,BC307A,BC326,BC415-6,BC415A,BC416-6,BC416A, BCY79A,BF440,GET2906,GET2907,GI-3702,MM3906,MPS3702,MPS6517,MPS6518, MPS-A70,MPSA70-BLU,MPSA70-GRN,MPSA70-WHT,MPSA70-YEL,NPS404A,TIS61, TIS61M,TN-3906,TP3702,TZ552,TZ581,TZ582,*PTC127,*GE-65,*TR-52,*HEPS0032, *ECG193,*WEP242,*BC177,*276-2022
ST72000	2N5005
ST72000A	2N5005
ST72011	
ST72012	
ST72013	
ST72014	
ST72015	
ST72016	
ST72017	
ST72018	
ST72019	
ST72020	
ST72021	
ST72036	ST72037,ST76018,ST76019
ST72037	ST72038,ST76019,ST76020
ST72038	ST76020
ST72039	ST72040,*TR-88,*SK3025,*ECG129,*WEP242
ST72040	ST72041,*TR-88,*SK3025,*ECG129,*WEP242
ST72041	
ST74000	*GE-32
ST74049	ST74050
ST74050	ST74051
ST74051	
ST75000	2N5322,2N5679,ST75004,ST75005
ST75001	2N5322,2N5675,2N5679,2N5680,ST75005,ST75006
ST75004	ST75005
ST75005	ST75006
ST75006	

TRANSISTOR SUBSTITUTES

To Replace	Substitute This Type
ST76006	2N4999,2N5286,2N5385
ST76007	2N4999,2N5286,2N5385
ST76018	ST76019
ST76019	ST76020
ST76020	
ST84000	2N3665,*PTC144,*GE-32
ST84027	ST84028,ST84029,*GE-32,*SK3220,*ECG198,*WEP244
ST84028	ST84029,*GE-32,*SK3220,*ECG198,*WEP244
ST84029	*GE-32,*SK3220,*ECG198,*WEP244
ST86020	2N3492,ST86021,STT2650,STT2651,STT2653
ST86021	ST86022,STT2650,STT2651
ST86022	
ST90000	2N2893,2N3999,2N4116,2N5004,2N5084,2N5285,2N5348,2N5349,2N5479,2N5480, 2N5730,HST6316,HST6416,SDT6316,STT6316,STT6416
ST91000	2N4116,2N5000,HST6316,HST6416
ST91054	ST91055
ST91055	ST91056
ST91056	
ST91057	ST91058
ST91058	ST91059
ST91059	
ST91085	ST91057,ST91058,ST91086
ST91086	ST91058,ST91059,ST91087
ST91087	ST18017,ST91059
ST92006	ST92007
ST92007	ST92008
ST92008	
STA8309	
STA8341	SDT14305
STA9364	2N6308M,2N6513,2N6560,2N6573,2N6579,2N6580,2N6582,2N6583,DTS-430, DTS-431,FT430,FT431,MJ413,MJ431,SDT13303,SDT13304,SDT13305,STS1134
STA9370	
STA9371	
STA9372	
STC1024	
STC1080	2N1487,2SC241,2SC244,AMF105,AMF117,AMF117A,AMF118,AMF118A,BDY23A, HST9205,HST9210,HST9801,M5A,ZT1487,*PTC129A,*GE-72,*TR-59,*HEPS7002, *SK3027,*HN100,*ECG130,*WEP247,*BDY20,*RT-131
STC1081	2N1487,2SC241,2SC244,AMF105,AMF117,AMF117A,BDY23A,HST9205,HST9206, HST9801,HST9802,ZT1487,*PTC129A,*GE-72,*TR-59,*HEPS7002,*SK3027,*HN100, *ECG130,*WEP247,*BDY20,*RT-131
STC1082	2N1488,2SC240,2SC242,2SD52,AMF106,BDY24A,HST9206,HST9207,HST9802, HST9803,M5B,ZT1488,*PTC129A,*GE-72,*TR-59,*SK3027,*ECG130,*WEP247, *RT-131
STC1083	2N1487,2SD163,2SD172,2SD206,2SD206A,AMF115,AMF116,AMF210A,B170001-BLK, BDY23A,HST9210,M5A,M10A,ZT1487,*PTC129A,*GE-35,*TR-59,*HEPS7002, *SK3027,*HN100,*ECG130,*WEP247,*BDY20,*RT-131
STC1084	2N1487,2SD163,2SD172,AMF115,AMF116,AMF210A,B170004-BLK,BDY23A,ZT1487, *PTC129A,*GE-35,*TR-59,*HEPS7002,*SK3027,*HN100,*ECG130,*WEP247, *BDY20,*RT-131
STC1085	2N1488,2SD52,2SD53,2SD164,2SD173,2SD176,2SD207,2SD207A,AMF210B, B170004-BLK,B170007-BLK,BDY24A,M5B,M10B,ZT1488,*PTC129A,*GE-35,*TR-61, *SK3027,*WEP247,*RT-131
STC1300	2N1768,2N2339,*GE-66,*ECG152,*WEP245,*276-2019
STC1550	2N1208,2N1212,AMF108,AMF110
STC1551	2N1208,2N1212,AMF108,AMF110
STC1552	AMF109
STC1553	2N1212,2N3429
STC1726	STC1731
STC1728	2N2823,2N2824,STC1733
STC1731	ST15043

To Replace	Substitute This Type
STC1733	2N2824
STC1736	ST15044,ST15045
STC1738	2N2825,STC3706
STC1800	*GE-28,*ECG186,*WEP245,*276-2017
STC1850	*GE-66,*ECG152,*WEP245,*276-2020
STC2220	2SD55,STC2221,STC2224,STC2225,STC2228,STC2229,*PTC173,*GE-75,*ECG181, *WEPS7000
STC2221	2SD55,STC2225,STC2229,*PTC173,*GE-75,*ECG181,*WEPS7000
STC2222	STC2223,STC2226,STC2227,STC2230
STC2223	STC2227
STC2224	STC2225,STC2228,STC2229,*GE-75,*ECG181,*WEPS7000
STC2225	STC2229,*GE-75,*ECG181,*WEPS7000
STC2226	STC2227,STC2230
STC2227	
STC2228	STC2229,*GE-75,*ECG181,*WEPS7000
STC2229	*GE-75,*ECG181,*WEPS7000
STC2230	
STC2231	
STC3706	
STC4401	2N4231,2N4232,2N5491,2N5493,2N5495,2N6373,2N6374,2SD241,2SD256,40250, 40627,BD148-6,*PTC112,*GE-216,*TR-57,*HEPS5012,*ECG175,*WEP241, *276-2020
STC5109/I	
STC5110/I	
STC5111/I	
STC5112/I	
STC5113/I	
STC5114/I	
STC5202	STC5203,*GE-69,*ECG153,*WEP246,*276-2020
STC5203	STC5204,*GE-69,*ECG153,*WEP246
STC5204	
STC5205	STC5202,STC5203,STC5206,*GE-69,*ECG153,*WEP246,*276-2026
STC5206	STC5203,STC5204,STC5207,*GE-69,*ECG153,*WEP246
STC5207	STC5204
STC5519/I	
STC5520/I	
STC5521/I	
STC5522/I	
STC5523/I	
STC5524/I	
STC5610	2N3203,2N3204,*PTC111,*GE-69,*TR-88,*HEPS5013,*SK3025,*ECG129,*WEP242, *276-2025,*RT-115
STC5611	STC5612,*GE-21,*TR-88,*HEPS5013,*SK3025,*ECG129,*WEP242,*276-2025, *RT-115
STC5612	*TR-88,*HEPS5013,*SK3025,*ECG129,*WEP242,*276-2025,*RT-115
STC5624	STC5612
STC5802	STC5803,*GE-69,*276-2025
STC5803	STC5804,*GE-69
STC5804	
STC5805	STC5802,STC5803,STC5806,*GE-69,*ECG153,*WEP246,*276-2025
STC5806	STC5803,STC5804,STC5807,*GE-69,*ECG153,*WEP246
STC5807	STC5804
STC7644	2N2911
STC7645	
STC9142	
STC9160	
STI-10	MM3000,*PTC117
STI-20	MM3002,MM3003,*PTC117
STI-30	*PTC117
STI-40	
STI-50	

TRANSISTOR SUBSTITUTES

To Replace	Substitute This Type
STI-60	
STI-70	
STI-105	
STI-205	
STI-305	
STI-405	
STI-505	
STI-605	
STI-705	
STIP-10	*PTC141
STIP-20	
STIP-30	
STIP-40	
STIP-105	
STIP-205	
STIP-305	
STIP-405	
STS103	2N3055A,2N3235,2N5881,2N5882,2N5885,2N5886,BUY51A,BUY52A,HST9201, HST9205,HST9206
STS104	2N6471
STS105	2N3236,2N5882,2N5886,2N6270,2N6272,40411,BUY53A,BUY54A,HST9202,HST9206, HST9207
STS106	2N3236,2N5882,2N5886,2N6270,2N6272,40411,BUY53A,BUY54A,HST9202,HST9206, HST9207
STS107	2N3236,2N5882,2N5886,2N6270,2N6272,40411,BUY53A,BUY54A,HST9202,HST9206, HST9207
STS401	2SC1106,2SC1114,DTS-411
STS402	
STS403	
STS409	
STS410	
STS411	
STS413	
STS423	
STS424	
STS425	
STS430	
STS431	
STS660	
STS663	
STS665	
STS1121	*PTC118,*TR-61,*HEPS5020
STS1122	*HEPS5020
STS1131	STS1132,*PTC118
STS1132	SDT14304,STS1133,*PTC118
STS1133	2N6307M,STS1134
STS1134	2N6306,2N6307M,2N6308M,2SC807
STS-410	2N6510,2SD417,BDY25A
STS-411	2N5239,2N5240,2N6511,2N6512,2N6514,IR410,IR411
STT1800	STT3000
STT1900	STT3100
STT2000	STT3200
STT2050	STT3250
STT2200	STT3400
STT2300	2SD82,2SD111-O,2SD111-R,2SD119-Y,181T2,BDY24B,STT3500,*PTC173,*GE-75, *TR-61
STT2400	STT2401,STT2402
STT2401	STT2400,STT2402
STT2402	
STT2403	STT2400,STT2401,STT2402
STT2404	STT2403

To Replace	Substitute This Type
STT2405	STT2404,*GE-216,*SK3239
STT2406	*GE-216,*SK3239
STT2650	STT2651,STT2652
STT2651	STT2650,STT2652
STT2652	
STT2653	STT2650,STT2651,STT2652
STT2654	STT2653
STT2655	STT2654
STT2656	
STT2800	STT2801,STT2802
STT2801	STT2800,STT2802
STT2802	
STT2803	STT2800,STT2801,STT2802
STT2804	STT2803
STT2805	STT2804
STT2806	
STT3000	STT1800
STT3100	STT1900
STT3200	STT2000
STT3250	STT2050
STT3400	STT2200
STT3500	2SD82,2SD111-O,2SD111-R,2SD119-Y,181T2,BDY24B,STT2300,*PTC173,*GE-75, *TR-61
STT4451	HST4451,HST4452,HST5007,HST5008,HST5507,HST5508,STT4452,STT4454, STT4456,*GE-28,*TR-76,*ECG186,*WEP245
STT4452	HST4452,HST5008,HST5508,STT4454,STT4456
STT4453	HST4453,HST4454,HST5002,HST5003,HST5502,HST5503
STT4454	HST4452,HST5008,HST5508,STT4452,STT4456
STT4455	HST4455,HST4456,HST5012,HST5013,HST5512,HST5513
STT4456	HST4452,HST5008,HST5508,STT4452,STT4454
STT6309	2N2877,2N2879,HST6309,HST6310,HST6409,HST6410,SDT6309,SDT6310,STT2804, STT2805,STT6310,STT6409,STT6410
STT6310	2N2879,HST6310,HST6409,HST6410,SDT6310,STT2803,STT2804,STT6410
STT6312	2N2878,2N2880,2N3998,HST6312,HST6313,HST6412,HST6413,SDT6312,SDT6313, STT6313,STT6412,STT6413
STT6313	2N2880,2N3998,2N4115,HST6313,HST6413,SDT6313,STT6413
STT6315	2N3999,HST6315,HST6316,HST6415,HST6416,SDT6315,SDT6316,STT6316, STT6415,STT6416
STT6316	2N3999,2N4116,HST6316,HST6416,SDT6316,STT6416
STT6409	2N2877,2N2879,HST6309,HST6310,HST6409,HST6410,SDT6309,SDT6310,STT2804, STT2805,STT6309,STT6310,STT6410
STT6410	2N2879,HST6310,HST6409,HST6410,SDT6310,STT2803,STT2804,STT6310
STT6412	2N2878,2N2880,2N3998,HST6312,HST6313,HST6412,HST6413,SDT6312,SDT6313, STT6312,STT6313,STT6413
STT6413	2N2880,2N3998,2N4115,HST6313,HST6413,SDT6313,STT6313
STT6415	2N3999,HST6315,HST6316,HST6415,HST6416,SDT6315,SDT6316,STT6315, STT6316,STT6416
STT6416	2N3999,2N4116,HST6316,HST6416,SDT6316,STT6316
STT9001	HST4483,HST5006,HST5506,HST9001,HST9002,SDT3426,SDT7411,STT9002, *GE-216,*SK3239
STT9002	HST4451,HST4452,HST5007,HST5008,HST5507,HST5508,HST9002,HST9003, SDT3427,SDT3428,SDT7412,SDT7413,STT4451,STT4452,STT4454,STT4456, STT9003,*GE-216,*SK3239
STT9003	HST4452,HST5008,HST5508,HST9003,SDT3428,SDT3429,SDT7413,STT4452, STT4454,STT4456
STT9004	HST5001,HST5501,HST9004,HST9005,HST9007,HST9008,SDT3422,SDT7401, SDT7414,STT9005,STT9007,STT9008,*GE-216,*SK3239
STT9005	2N4150,2N5327,HST4453,HST4454,HST5002,HST5003,HST5502,HST5503,HST9005, HST9006,HST9008,HST9009,SDT3423,SDT3424,SDT7402,SDT7403,SDT7415, SDT7416,STT4453,STT9006,STT9008,STT9009,*GE-216,*SK3239
STT9006	2N4150,2N4895,2N4897,2N5327,HST4454,HST5003,HST5503,HST9006,HST9009,

To Replace	Substitute This Type
(STT9006)	SDT3424,SDT7403,SDT7416,STT9009
STT9007	HST5001,HST5501,HST9007,HST9008,STT9008
STT9008	2N5327,HST5002,HST5003,HST5502,HST5503,HST9008,HST9009,STT9009
STT9009	2N5327,BFX34,HST5003,HST5503,HST9009
STT9010	HST5011,HST5511,HST9010,HST9011,SDT7417,STT9011
STT9011	HST4455,HST4456,HST5012,HST5013,HST5512,HST5513,HST9011,HST9012, SDT7418,SDT7419,STT4455,STT9012
STT9012	2N4896,HST4456,HST5013,HST5513,HST9012,SDT7419
SYL792	2N415,2N415A,2N416,2N1307,2N1309A,2N1356,2N1681,2N1808,2N1892,40269, ASY27,SF.T223,*PTC109,*GE-50,*TR-08,*SK3005,*ECG101,*WEP254,*AC188/01, *276-2007,*RT-118
SYL1182	SEE 2N2354
SYL4443	
T0003	SEE 2N207
T0004	SEE 2N207A
T0005	SEE 2N207B
T0014	SEE 2N536
T0015	SEE 2N535A
T0033	SEE 2N535
T1013	SEE 2N223
T1028	SEE 2N128
T1032	SEE 2N346
T1038	SEE 2N240
T1042	SEE 2N226
T1046	SEE 2N224
T1166	SEE 2N393
T1224	SEE 2N344
T1225	SEE 2N345
T1250	SEE 2N588
T1251	SEE 2N499
T1275	SEE 2N495
T1276	SEE 2N496
T1289	SEE 2N300
T1291	SEE 2N299
T1312	SEE 2N501
T1314	SEE 2N504
T1322	SEE 2N503
T1326	SEE 2N598
T1327	SEE 2N1122
T1328	SEE 2N1122A
T1334	SEE 2N597
T1342	SEE 2N502
T1346	SEE 2N599
T1347	SEE 2N670
T1392	SEE 2N1126
T1393	SEE 2N671
T1395	SEE 2N600
T1396	SEE 2N1124
T1397	SEE 2N1125
T1398	SEE 2N1127
T1431	SEE 2N672
T1472	SEE 2N1495
T1473	SEE 2N1496
T1474	SEE 2N1500
T1475	SEE 2N673
T1510	SEE 2N501A
T1512	SEE 2N601
T1537	SEE 2N1123
T1546	SEE 2N1129
T1573	SEE 2N1130
T1574	SEE 2N1128

To Replace	Substitute This Type
T1808	SEE 2N1494
T1930	SEE 2N768
TA2363	SEE 2N3839
TA2509	SEE 2N3878
TA2509A	SEE 2N3879
TA2658	SEE 2N3866
TA2675	SEE 2N5016
TA2714	SEE 2N4012
TA2761	SEE 40608
TA2791	SEE 2N5102
TA2792	SEE 2N4933
TA2793	SEE 2N5070
TA2800	SEE 2N5109
TA2827	SEE 2N5071
TA2828	SEE 2N4932
TA2875	SEE 2N4440
TA7003	SEE 2N5470
TA7146	SEE 2N5090
TA7205	SEE 2N5921
TA7238	SEE 2N5262
TA7285	SEE 2N5202
TA7303	SEE 2N5180
TA7319	SEE 2N5179
TA7322	SEE 2N5189
TA7367	SEE 2N5918
TA7403	SEE 40836
TA7408	SEE 2N5914
TA7409	SEE 2N5915
TA7411	SEE 2N5916
TA7420	SEE 2N5840
TA7477	SEE 2N5913
TA7487	SEE 2N5920
TA7513	SEE 2N5838
TA7514	SEE 40964
TA7530	SEE 2N5839
TA7532	SEE 2N5919A
TA7534	SEE 2N6354
TA7554	SEE 2N6178
TA7555	SEE 2N6179
TA7556	SEE 2N6180
TA7557	SEE 2N6181
TA7588	SEE 40965
TA7679	SEE 40837
TA7680	SEE 40941
TA7686	SEE 40893
TA7706	SEE 2N6105
TA7707	SEE 2N6104
TA7852	SEE 2N5917
TA7922	SEE 2N5995
TA7941	SEE 40934
TA7943	SEE 40909
TA7982	SEE 40940
TA7993	SEE 2N6265
TA7994	SEE 2N6266
TA7995	SEE 2N6267
TA7995A	SEE 2N6269
TA8007	SEE 2N6479
TA8007B	SEE 2N6480
TA8100	SEE 2N6481
TA8100B	SEE 2N6482
TA8104	SEE 40915

TRANSISTOR SUBSTITUTES

To Replace	Substitute This Type
TA8172	SEE 40970
TA8201	SEE 2N6388
TA8202	SEE 2N6386
TA8203	SEE RCA8203B
TA8204	SEE RCA8203
TA8236	SEE 40936
TA8340	SEE 41038
TA8343	SEE 2N6478
TA8348	SEE 2N6385
TA8349	SEE 2N6383
TA8350	SEE RCA8350B
TA8351	SEE RCA8350
TA8352	SEE 2N6372
TA8353	SEE 2N6373
TA8354	SEE 2N6374
TA8405	SEE 2N6477
TA8407	SEE 2N6268
TA8439	SEE 40898
TA8440	SEE 40899
TA8485	SEE 2N6387
TA8486	SEE 2N6384
TA8487	SEE RCA8203A
TA8488	SEE RCA8350A
TA8493	SEE 40971
TA8559	SEE 40954
TA8561	SEE 40955
TA8562	SEE 40967
TA8563	SEE 40968
TA8647	SEE 41025
TA8648	SEE 41026
TA8649	SEE 41027
TA8650	SEE 41028
TA8706	SEE 2N6466
TA8707	SEE 2N6465
TA8709	SEE 2N6468
TA8710	SEE 2N6467
TA8719	SEE 41008
TA8720	SEE 41009
TA8721	SEE 41010
TA8746	SEE 2N6393
TA8747	SEE 2N6390
TA8748	SEE RCA2003
TA8749	SEE 2N6391
TA8750	SEE RCA2005
TA8751	SEE 2N6392
TA8752	SEE RCA2010
TA8803	SEE RCA2310
TA8865	SEE 41039
TA8894	SEE 41042
TA8900A	SEE 2N6308
TA8900B	SEE RCA579
TA8900C	SEE 2N6307
TA8900D	SEE 2N6306
TA8923	SEE RCA0610-30
TA8932	SEE 2N6500
TA8955	SEE 41044
TC3114	*PTC133
TC3200	
TCVM-553	REFER TO SECTION 2
TF78/30	*TR-01,*HEPG6003,*SK3009,*ECG104,*WEP230,*276-2006,*RT-124
TF78/60	*TR-01,*SK3009,*ECG121,*WEP232,*RT-127

To Replace	Substitute This Type
TI-156	
TI-156L	
TI-158	
TI-158A	
TI-158AL	
TI-158L	
TI-159	
TI-160	TI-161,TI-162
TI-161	TI-160,TI-162
TI-162	
TI-363	2N382,2N415,2N415A,2N416,2N526,2N526A,2N527A,2N838,2N1305,2N1307, 2N1309A,2N1349,2N1356,2N1448,2N1681,2N2956,2N2957,2N3323,2N3324,2N3325, SF.T223,*GE-1,*TR-85,*HEPG0003,*SK3005,*HF35,*ECG100,*WEP254, *AC188/01,*276-2007
TI-364	2N414,2N414B,2N414C,2N518,2N586,2N1017,2N1171,2N1313,2N1348,2N1350, 2N1351,2N1446,2N1447,2N1495,2N2587,2N2955,2N6365,2N6365A,2SA204,2SA205, 2SA206,2SA207,2SA406,2SA407,ASY26,SF.T227,*GE-1,*TR-85,*HEPG0003,*HF35, *ECG100,*WEP254,*AC188/01,*276-2007
TI-365	2N2635,*GE-51,*HEPG0003
TI-388	2N838,2N2956,2N2957,2N3323,2N3324,2N3325,*DS41,*GE-51,*HEPG0003,*SK3006, *HF35,*ECG160,*WEP637
TI-389	2N1495,2N2587,2N2955,2N6365,2N6365A,*GE-51,*HEPG0003,*HF35
TI-390	*GE-51,*HEPG0003
TI-391	2N700A,*GE-51,*HEPG0003
TI-392	2N700A
TI-393	2N700A
TI-395	2N2635,*GE-51,*HEPG0003
TI-396	
TI-397	2N2956,2N2957,2N3323,2N3324,2N3325,*GE-51,*HEPG0003
TI-398	2N586,2N2956,2N2957,2N3323,2N3324,2N3325,*GE-51,*HEPG0003
TI-399	2N518,2N586,2N1348,2N1350,2N1351,2N1446,2N1447,2N1495,2N2955,*GE-51, *HEPG0003
TI-400	*GE-51,*HEPG0003,*SK3006,*ECG160,*WEP637
TI-401	2N3127,2N3281,2N3282,TI-402,*GE-51,*TR-17,*HEPG0003,*SK3006,*HF35,*ECG160, *WEP637
TI-402	2N3127,2N3281,2N3282,TI-401,*GE-51,*TR-17,*HEPG0003
TI-403	TI-400,*GE-51,*TR-17,*HEPG0003,*SK3006,*ECG160,*WEP637
TI-407	2N4292,2N4293,2SC387A,MPS918,MPS6546,MPS6547,MPSH07,MPSH08,TIS125, *PTC132,*GE-86,*HEPS0016,*SK3019,*EN10,*ECG161,*WEP709,*276-2011
TI-408	2N834,2N835,2SC387A,MPSH07,MPSH08,*PTC132,*GE-86,*TR-51,*HEPS0016, *SK3019,*ECG161,*WEP709,*BC107B,*276-2011
TI-409	2N834,2N835,2SC387A,MPSH07,MPSH08,*PTC132,*GE-86,*TR-51,*HEPS0016, *SK3019,*EN10,*ECG161,*WEP709,*BC107B,*276-2011
TI-411	*GE-20,*HEPS0015,*BC337,*276-2016
TI-412	SEE 2N3704
TI-413	SEE 2N3705
TI-414	SEE 2N3706
TI-415	SEE 2N3707
TI-416	SEE 2N3708
TI-417	SEE 2N3710
TI-418	SEE 2N3711
TI-419	BC547B,BC550B,BCY58D,BCY59D,MPSA18,PET4003,*PTC139,*GE-212,*HEPS0015, *SK3019,*ECG199,*WEP736,*BC107B,*276-2013
TI-420	SEE 2N2387
TI-421	SEE 2N2388
TI-422	SEE 2N851
TI-423	SEE 2N852
TI-424	SEE 2N2389
TI-425	SEE 2N2390
TI-428	SEE 2N2393
TI-429	SEE 2N2394

TRANSISTOR SUBSTITUTES

To Replace	Substitute This Type
TI-430	SEE 2N849
TI-431	SEE 2N850
TI-432	SEE 2N2395
TI-433	SEE 2N2396
TI-474	SEE 2N929
TI-475	SEE 2N930
TI-480	2N3295,2SC31,2SC1164-R,7A30,PT2540,SE8001,*PTC123,*GE-81,*TR-25, *HEPS0014,*EN10,*ECG128,*WEP243,*BC337,*276-2012
TI-481	2N1615,2N1975,BFY65,MM3000,*PTC125,*GE-81,*SN80,*BF338,*276-2012
TI-482	2N1840,2N2320,2SC147,2SC343,*PTC123,*GE-81,*TR-25,*HEPS0014,*SN80, *ECG128,*WEP243,*BC337,*276-2012
TI-483	2N1958,2N1958A,2N1987,2N2194,2N2194A,2N2194B,2N2217,2N2477,2N3122, 2N3252,2N5188,2SC97,2SC189,2SC254,2SC344,2SC482-O,2SC503-O,2SC504-O, 2SC580,BSX48,BSX49,ZT696,*PTC123,*GE-81,*TR-25,*HEPS0014,*SN80, *ECG128,*WEP243,*BC337,*276-2012
TI-484	2N1838,2N1941,2N1959A,2N2218,2N2237,2N2270,2N2537,2N2848,2N2868,2N2958, 2N3724,2N3724A,2N4046,2N4960,2N5145,2N6427,2SC306,2SC479,2SC482-Y, 2SC503-Y,2SC504-Y,A5T5450,A8T3705,BC340-10,BC341-10,BSY51,BSY81,TIS110, TIS133,TIS134,TN61,ZT90,ZT95,ZT2270,*PTC123,*GE-81,*TR-65,*HEPS0014, *SN80,*ECG128,*WEP243,*BC337,*276-2012
TI-485	2N1708,2N2318,2N2319,MPS6540,MPS6543,MPS6548,MPSH19,*PTC144,*GE-47, *TR-65,*HEPS0011,*EN10,*ECG192,*WEP243,*BC337,*276-2016
TI-486	2N1714,2N3108,2N3110,2N4000,2SC1069,40367,BC140-6,BC141-6,*PTC144,*GE-32, *276-2012
TI-487	2N2991,*GE-66
TI-490	SEE 2N780
TI-492	2N162,2N162A,2N163,2N334,2N475,2N480,2N543A,2N717,2N783,2N834A,2N840, 2N842,2N844,2N1151,2N1152,2N1278,2N1591,2N2368,2N2387,2N2432A,2N2522, 2N3605A,2N3606A,2N3973,2N3975,2SC54,2SC120,2SC121,2SC124,2SC302, 2SC321,2SC322,2SC468,2SC601,2SC689,2SC725,2SC752G-R,BF224J,BF225J, BSY93,EN914,EN2369A,EN3009,EN3013,EN3014,GET708,GET914,GET3013, GET3014,GET3646,KT218,MPS834,MPSA10-RED,MPSA20-RED,MPSH20,MPSH37, TIS45,TIS46,TIS47,TIS49,TIS52,TIS55,TIS129,ZT23,ZT43,*PTC123,*GE-60,*TR-51, *HEPS0014,*SK3020,*SN80,*ECG161,*WEP709,*BF167,*276-2013
TI-493	2N264,2N337,2N475,2N480A,2N717,2N783,2N834A,2N842,2N844,2N1150,2N1151, 2N2368,2N2432A,2N2522,2S102,2SC54,2SC120,2SC121,2SC124,2SC302,2SC321, 2SC322,2SC382,2SC689,BF224J,BF225J,BSX25,GET706,KT218,MPS834, MPS6544,MPS6545,MPSH20,MPSH37,TIS47,TIS125,ZT22,ZT42,*PTC123,*GE-60, *TR-51,*HEPS0014,*SK3020,*SN80,*ECG161,*WEP709,*BF200,*276-2013
TI-494	2N263,2N338,2N718,2N843,2N916,2N916A,2N929,2N957,2N2161,2N2242,2N2369, 2N2369A,2N2501,2N2847,2N3009,2N3013,2N3014,2N3693,2N3826,2N3858A, 2N3862,2N3903,2N3946,2N3974,2N3976,2N4123,2N4227,2N4873,2N4966,2N4994, 2N5027,2N5824,2S104,2SC67,2SC122,2SC169,2SC318,2SC366G-O,2SC367G-O, 2SC536,2SC601N,2SC620,2SC634,2SC639,2SC752G-O,2SC764,2SC815, 2SC899,2SC938,2SC944,2SC965,2SC1187,2SC1188,2SC1189,2SC1276,2SC1319, 2SC1320,2SC1453,2SC1675,2SC1682-BL,2SC1682-GR,2SC1686,2SC1687, 2SC1688,BF198,BF199,BF241,EN697,EN916,MM3903,MPS3693,MPS6565, MPS6576-BLUE,MPS6576-YEL,MPSA10-WHT,MPSA20-WHT,PET1001,SE1001, TIS87,TP3705,ZT24,ZT44,*PTC123,*GE-60,*TR-51,*HEPS0014,*SK3020,*SN80, *ECG161,*WEP709,*BF200,*276-2013
TI-495	2N909,2N930,2N1153,2N2349,2N2388,2N2484,2N2484A,2N2586,2N3302,2N3566, 2N3694,2N3827,2N3904,2N4287,2N4967,2N4995,2N5826,2SC366G-Y,2SC367G-Y, 2SC538A,2SC587,2SC587A,2SC711A,2SC752G-Y,2SC1000-GR,2SC1000G-GR, 2SC1175,2SC1285,2SC1364,2SC1380-GR,2SC1380A-GR,2SC1632,2SC1634, 2SC1681-GR,2SC1685,2SC1850,2SC1992,A157B,BC107,BC107B,BC123,BC129A, BC129B,BC167B,BC171B,BC182B,BC182L,BC237B,BC413B,BC414B,BCY59B, BFY39-3,BSX79,EN2484,GI-3566,MPS3694,MPS6514,MPS6520,MPS6566,MPS6575, MPS6576-SIL,MPS-A10,MPS-A20,MPS-K20,MPS-K21,MPS-K22,MPSA10-BLU, MPSA10-YEL,MPSA17,MPSA20-BLU,MPSA20-YEL,NPSA20,PBC107A,PBC107B, PET1002,SE1002,TIS23,TIS24,TP3704,TP3706,TZ81,*PTC123,*GE-39,*TR-51, *HEPS0014,*SK3020,*ECG108,*WEP56,*BC107B,*276-2013

To Replace	Substitute This Type
TI-496	2N342B,2N498,2N498A,BFY65,MM3000,TI-481,*PTC123,*GE-81,*TR-25,*HEPS0014, *SN80,*ECG128,*WEP243,*BF338,*276-2012
TI-539	
TI-540	
TI-890	SEE 2N2861
TI-891	SEE 2N2862
TI-1121	
TI-1122	2S026,TI-1141
TI-1123	TI-1121
TI-1124	2S025,2S026,TI-1122,TI-1141,TI-1143
TI-1125	
TI-1126	2S024,TI-1145
TI-1131	ST18013
TI-1132	TI-1151
TI-1133	STT2652,TI-1131
TI-1134	STT2650,STT2651,TI-1132,TI-1151,TI-1153
TI-1135	2N3491,2N3492
TI-1136	2N3488,2N3489,2N3492,STT2653,STT2654,TI-1155
TI-1141	TI-1121
TI-1142	2S026,TI-1122
TI-1143	TI-1121,TI-1123,TI-1141
TI-1144	2S025,2S026,TI-1122,TI-1124,TI-1142
TI-1145	TI-1125
TI-1146	2S024,TI-1126
TI-1151	TI-1131
TI-1152	TI-1132
TI-1153	STT2650,STT2651,TI-1131,TI-1133,TI-1151
TI-1154	TI-1132,TI-1134,TI-1152
TI-1155	2N3488,2N3491,2N3492,STT2653,STT2654,TI-1135
TI-1156	2N3488,2N3489,TI-1136
TI-3015	SEE 2N3570
TI-3027	TI-3028,*PTC901,*GE-16,*HEPG6003
TI-3028	TI-3029,*PTC901,*GE-3,*HEPG6005
TI-3029	TI-3030,*PTC901,*GE-16,*HEPG6005
TI-3030	TI-3031,*TR-35,*ECG179,*WEPG6001,*RT-147
TI-3031	*TR-35,*ECG179,*WEPG6001,*RT-147
TIA01	*TR-85,*ECG102A,*WEP250,*RT-121
TIA02	*GE-50,*TR-17,*SK3123,*ECG160,*WEP637
TIA03	*GE-53,*TR-05,*HEPG0005,*SK3005,*ECG100,*WEP254,*AC188/01,*276-2004, *RT-118
TIA04	*GE-53,*TR-05,*HEPG0005,*SK3004,*ECG102,*WEP631,*AC188/01,*276-2005, *RT-120
TIA05	*PTC109,*GE-53,*TR-05,*HEPG0005,*SK3005,*ECG100,*WEP254,*AC188/01, *276-2004,*RT-118
TIG05	2N4280,2N4281,2N4282,2N4283,*PTC901
TIG06	2N4282,2N4283,*PTC901
TIG07	
TIG08	*PTC901
TIG09	*PTC901
TIG10	
TIP04	2N3902,2N5240,2N5805,DTS-423,DTS-423M,DTS-424,DTS-425,DTS-663,IR401,IR413, IR663,*GE-73,*TR-61
TIP14	*GE-28,*HEPS5000,*ECG184,*WEPS5003,*RT-154
TIP24	*GE-66,*TR-76,*HEPS5000,*ECG152,*WEP245,*276-2018,*RT-154
TIP27	*TR-60,*HEPS5015,*SK3103,*ECG157,*WEP244,*RT-135
TIP29	NSP4921,NSP4922,TIP29A,*PTC167,*GE-23,*TR-76,*HEPS5000,*SK3041,*ECG152, *WEP245,*276-2018,*RT-154
TIP29A	NSP4922,NSP4923,TIP29B,*PTC167,*GE-28,*TR-76,*HEPS5000,*SK3054,*ECG152, *WEP244,*276-2020,*RT-154
TIP29B	NSP4923,TIP29C,*PTC167,*GE-32,*HEPS5000,*ECG198
TIP29C	*PTC169,*GE-32,*ECG198,*WEP244

TRANSISTOR SUBSTITUTES

To Replace	Substitute This Type
TIP30	NSP4918,NSP4919,TIP30A,*PTC166,*GE-69,*TR-77,*HEPS3028,*ECG153,*WEP246, *276-2026,*RT-133
TIP30A	NSP4919,NSP4920,TIP30B,*PTC166,*GE-69,*TR-77,*HEPS3032,*ECG153,*WEP246, *RT-133
TIP30B	NSP4920,TIP30C,*PTC166,*HEPS3032,*SK3200,*ECG189,*WEPS3031
TIP30C	*PTC168
TIP31	NSP41,NSP41A,TIP31A,*PTC167,*GEMR-6,*TR-76,*HEPS5000,*SK3054,*ECG152, *WEP245,*276-2019,*RT-154
TIP31A	2N6123,NSP41A,NSP41B,TIP31B,*PTC167,*GE-66,*TR-76,*HEPS5000,*SK3054, *ECG152,*WEP245,*276-2020,*RT-154
TIP31B	2N6123,FT317,NSP41B,NSP41C,TIP31C,*PTC167,*HEPS5000,*ECG184
TIP31C	FT317,FT317A,NSP41C,*PTC169,*HEPS5004,*276-2020
TIP32	NSP42,NSP42A,TIP32A,*PTC166,*GE-69,*TR-77,*HEPS5006,*ECG153,*WEP246, *276-2025,*RT-133
TIP32A	2N6126,NSP42A,NSP42B,TIP32B,*PTC166,*GE-69,*TR-77,*HEPS5006,*ECG153, *WEP246,*RT-133
TIP32B	2N6126,FT417,NSP42B,NSP42C,TIP32C,*PTC166,*HEPS5006,*ECG197,*WEP246
TIP32C	FT417,FT417A,NSP42C,*PTC168,*HEPS5005
TIP33	2N6486,TIP33A,*PTC169,*HEPS5004,*SK3188,*276-2017
TIP33A	2N6487,TIP33B,*PTC169,*HEPS5004,*SK3188,*276-2020
TIP33B	2N6488,TIP33C,*PTC169,*HEPS5004,*276-2020
TIP33C	*PTC169,*HEPS5004,*276-2020
TIP34	2N6489,TIP34A,*PTC168,*GE-56,*HEPS5005,*SK3189,*ECG183,*WEPS5005, *RT-151
TIP34A	2N6490,TIP34B,*PTC168,*GE-56,*HEPS5005,*SK3189,*ECG183,*WEPS5005, *RT-151
TIP34B	2N6491,TIP34C,*PTC168,*GE-56,*HEPS5005,*SK3189,*ECG183,*WEPS5005, *RT-151
TIP34C	*PTC168,*HEPS5005
TIP35	TIP35A
TIP35A	TIP35B
TIP35B	TIP35C
TIP35C	
TIP36	TIP36A
TIP36A	TIP36B
TIP36B	TIP36C
TIP36C	
TIP41	TIP41A,*PTC167,*GE-55,*TR-92,*HEPS5004,*ECG196,*WEP245,*276-2017
TIP41A	TIP41B,*PTC167,*GE-55,*TR-92,*HEPS5004,*ECG196,*WEP245,*276-2019
TIP41B	TIP41C,*PTC167,*TR-92,*HEPS5004,*ECG196,*WEP245,*276-2020
TIP41C	*PTC169,*HEPS5004,*276-2020
TIP42	TIP42A,*PTC166,*GE-56,*HEPS5005,*ECG197,*WEPS5005,*276-2025
TIP42A	TIP42B,*PTC166,*GE-56,*HEPS5005,*ECG197,*WEPS5005,*276-2027
TIP42B	TIP42C,*PTC166,*GE-56,*HEPS5005,*ECG197,*WEPS5005
TIP42C	*PTC168,*HEPS5005
TIP47	*SK3219,*ECG198,*WEP244
TIP48	TIP49,TIP50,*SK3219,*ECG198,*WEP244
TIP49	TIP50,*SK3220,*ECG198,*WEP244
TIP50	*SK3220
TIP51	TIP52,TIP53
TIP52	TIP53,TIP54
TIP53	TIP54
TIP54	
TIP61	TIP29,TIP29A,TIP61A,*PTC193
TIP61A	TIP29A,TIP29B,TIP61B
TIP61B	TIP29B,TIP29C,TIP61C
TIP61C	TIP29C
TIP62	TIP30,TIP30A,TIP62A,*PTC162
TIP62A	TIP30A,TIP30B,TIP62B,*PTC162
TIP62B	TIP30B,TIP30C,TIP62C,*PTC164
TIP62C	TIP30C,*PTC164

To Replace	Substitute This Type
TIP110	TIP111
TIP111	TIP112
TIP112	
TIP115	TIP116
TIP116	TIP117
TIP117	
TIP120	2N6043,2N6044,RCA120,RCA121,SE9300,SE9301,TIP121,*ECG261
TIP121	2N6044,2N6045,RCA121,RCA122,SE9301,SE9302,TIP122,*ECG261
TIP122	2N6045,RCA122,SE9302
TIP125	2N6040,2N6041,RCA125,RCA126,SE9400,SE9401,TIP126,*ECG262
TIP126	2N6041,2N6042,RCA126,SE9401,SE9402,TIP127,*ECG262
TIP127	2N6042,SE9402
TIP140	TIP141,*ECG270
TIP141	TIP142,*ECG270
TIP142	*ECG270
TIP145	TIP146,*ECG271
TIP146	TIP147,*ECG271
TIP147	*ECG271
TIP501	SDT9010,TIP502,*276-2008
TIP502	SDT4455,SDT9011,*276-2008
TIP503	TIP504,*TR-81,*ECG175,*WEP241
TIP504	
TIP505	TIP506
TIP506	
TIP507	
TIP508	
TIP645	IR645,IR646,TIP646,*ECG246
TIP646	IR646,IR647,TIP647,*ECG246
TIP647	IR647
TIP2955	*PTC168,*ECG219,*WEPS7001
TIP3055	*PTC169,*HEPS5004,*276-2020
TIS03	SEE 2N3702
TIS04	SEE 2N3703
TIS18	2N2368,2N2729,2N3544,2N4137,2SC387AG,2SC601,2SC689,2SC1126,EN918, EN2369A,MM1941,MPS6507,MT1038,MT1038A,MT1039,MT1060,PET3001,TIS49, TIS84,TIS129,*PTC132,*GE-86,*TR-83,*HEPS0016,*SK3039,*EN10,*ECG161, *WEP719,*276-2011,*RT-113
TIS22	*GE-39,*TR-21,*HEPS0011,*SK3122,*EN10,*ECG123A,*WEP736,*BC107B, *276-2013,*RT-102
TIS23	TIS24,*GE-39,*TR-21,*HEPS0011,*SK3122,*EN10,*ECG123A,*WEP736,*BC107B, *276-2013,*RT-102
TIS24	*PTC170,*GE-63,*TR-95,*HEPS0005,*SK3039,*EN10,*ECG108,*WEP56,*276-2008, *RT-113
TIS37	*PTC103,*GE-65,*TR-20,*HEPS0013,*SK3025,*ECG159,*WEP242,*BC177,*276-2034, *RT-115
TIS38	TIS137,*PTC103,*GE-65,*TR-20,*HEPS0013,*SK3114,*ECG159,*WEP242,*BC327, *276-2023,*RT-115
TIS39	SEE TIXS39
TIS44	2N706,2N706A,2N706B,2N783,2N784,2N959,2N2205,2N2318,2N2319,2N2368, 2N2656,2N2710,2N2729,2N3261,2N3510,2N3511,2N3544,2N3605A,2N3606A, 2N3648,2N4137,2N5030,2N5187,2SC54,2SC55,2SC63,2SC68,2SC204,2SC321, 2SC322,2SC356,2SC400-R,2SC576,2SC595,2SC601,2SC689,2SC1126,2SC1293, 40405,GET708,GET914,GET3013,GET3014,GET3646,KT218,MM1941,MPS6507, MPS6511,MPSH19,MT1038,MT1038A,MT1039,MT1060,PET3001,PT2760,TIS45, TIS46,TIS49,TIS51,TIS52,TIS55,TIS84,ZT2368,*PTC132,*GE-61,*TR-21,*HEPS0025, *SK3122,*ECG123A,*WEP736,*BC107B,*276-2015,*RT-102
TIS45	2N708,2N708A,2N784A,2N914,2N2221,2N2242,2N2481,2N2501,2N2539,2N2710, 2N2845,2N2847,2N3013,2N3014,2N3115,2N3210,2N3211,2N3301,2N3510,2N3511, 2N3605A,2N3606A,2N3646,2N3647,2N3648,2N3903,2N3946,2N4123,2N4140, 2N4962,2N5027,2N5144,2N5380,2SC67,2SC169,2SC335,2SC336,2SC577,2SC620, 2SC639,2SC764,2SC1274,BSW82,GET708,GET914,GET2221,GET2369,GET3013,

To Replace	Substitute This Type
(TIS45)	GET3014,GET3646,GI-3641,GI-3642,MM3903,MPS6530,TIS46,TIS52,TIS55,TN-3903, ZT83,ZT708,*GE-20,*TR-21,*HEPS0011,*SK3122,*ECG123A,*WEP736,*BC107B, *276-2016,*RT-102
TIS46	2N708,2N708A,2N784A,2N914,2N2221,2N2242,2N2481,2N2501,2N2539,2N2710, 2N2845,2N2847,2N3013,2N3014,2N3115,2N3210,2N3211,2N3301,2N3510,2N3511, 2N3605A,2N3606A,2N3646,2N3647,2N3648,2N3903,2N3946,2N4123,2N4140, 2N4962,2N5027,2N5144,2N5380,2SC67,2SC169,2SC335,2SC336,2SC577,2SC620, 2SC639,2SC764,2SC1274,BSW82,GET708,GET914,GET2221,GET2369,GET3013, GET3014,GET3646,GI-3641,GI-3642,MM3903,MPS6530,TIS45,TIS52,TIS55,TN-3903, ZT83,ZT708,*GE-20,*TR-21,*HEPS0011,*SK3122,*ECG123A,*WEP736,*BC107B, *276-2016,*RT-102
TIS47	2N834A,2N2368,2N2710,2N3510,2N3648,2SC321,2SC322,2SC576,GET3013, GET3014,GET3646,MPS834,MPS6532,TIS49,TIS52,TIS55,ZT2368,*GE-20,*TR-21, *HEPS0025,*SK3122,*ECG123A,*WEP736,*276-2016,*RT-102
TIS48	2N2369,2N2369A,2N3862,2N5029,2SC169,2SC335,2SC336,2SC577,2SC639,2SC764, MPS2369,ZT2369,*PTC133,*TR-21,*HEPS0011,*SK3122,*ECG123A,*WEP736, *276-2038,*RT-102
TIS49	2N2369,2N2369A,2N2710,2N3511,2N3648,2N3862,2N4137,2N5029,2SC169,2SC335, 2SC336,2SC577,2SC639,2SC764,MPS2369,TIS48,ZT2369,*PTC133,*TR-21, *HEPS0011,*SK3122,*ECG123A,*WEP736,*276-2038,*RT-102
TIS50	2N869A,2N2894,2N3012,2N3209,2N3546,2N3576,2N3640,2N4258A,2N4423,MM2894, *PTC103,*HEPS0013,*SK3122,*ECG159,*WEP242,*276-2021
TIS51	2N2501,2N2710,2N3011,2N3013,2N3014,2N3211,2N3510,2N3511,2N3646,2N3647, 2N3648,2SC67,2SC169,2SC335,2SC577,2SC595,2SC639,2SC764,2SC1274, GET2369,GET3013,GET3014,GET3646,TIS52,TIS55,*GE-20,*TR-21,*HEPS0011, *SK3122,*ECG123A,*WEP736,*276-2015,*RT-102
TIS52	2N708,2N708A,2N784A,2N914,2N2481,2N2501,2N2710,2N3013,2N3014,2N3210, 2N3211,2N3510,2N3511,2N3605A,2N3606A,2N3646,2N3647,2N3648,2N5144,2SC67, 2SC169,2SC335,2SC336,2SC577,2SC639,2SC764,2SC1274,GET708,GET914, GET2369,GET3013,GET3014,GET3646,MPS6530,TIS45,TIS46,TIS55,ZT83,ZT708, *GE-20,*TR-21,*HEPS0025,*SK3122,*ECG123A,*WEP736,*276-2016,*RT-102
TIS53	2N2894A,2N3012,2N3304,2N3546,2N3639,2N3640,2N4257A,2N4258A,2N5056, 2N5057,2N5292,MPS3639,MPS3640,MPSL07,MPSL08,*PTC131,*TR-20,*HEPS0013, *SK3114,*ECG159,*WEP242,*276-2021,*RT-115
TIS54	2N869A,2N2894,2N2894A,2N3012,2N3209,2N3248,2N3545,2N3546,2N3576,2N3640, 2N4258A,2N4423,2N5056,2N5057,2N5292,2N5354,2SA741,MM2894,MPS3640, MPSL08,TIS50,*PTC131,*GE-22,*TR-20,*HEPS0013,*SK3114,*ECG159,*WEP242, *276-2021,*RT-115
TIS55	2N708,2N708A,2N784A,2N914,2N2481,2N2501,2N2710,2N3013,2N3014,2N3210, 2N3211,2N3510,2N3511,2N3605A,2N3606A,2N3646,2N3647,2N3648,2N5144,2SC67, 2SC169,2SC335,2SC336,2SC577,2SC639,2SC764,2SC1274,GET708,GET914, GET2369,GET3013,GET3014,GET3646,MPS6530,TIS45,TIS46,TIS52,ZT83,ZT708, *GE-20,*TR-21,*HEPS0025,*SK3046,*ECG123A,*WEP736,*276-2016,*RT-102
TIS56	2N918,2N5851,2N5852,2SC251,2SC251A,2SC252,2SC253,2SC602,BF232,MT1061, TIS57,*PTC132,*GE-86,*TR-21,*HEPS0017,*SK3122,*ECG123A,*WEP736, *276-2015,*RT-102
TIS57	2N918,2N5851,2N5852,2SC251,2SC251A,2SC252,2SC253,2SC602,BF232,MT1061, TIS56,*PTC132,*GE-86,*TR-21,*HEPS0017,*SK3122,*ECG123A,*WEP736, *276-2015,*RT-102
TIS60	2N909,2N930B,2N2222,2N2350,2N2350A,2N2484A,2N2540,2N3116,2N3227, 2N3242A,2N3302,2N3509,2N3704,2N4386,2N4401,2N4952,2N5107,2N5369,2N5449, 2N5818,2N6004,2N6010,2SC366G-Y,2SC367G-Y,2SC1813,A5T2192,A5T2222, A5T5449,A8T3704,BC337-25,BSW83,D33D26,GET2222,GI-3643,GI-3704,GI-3706, MPS3704,MPS6531,PET3704,PET3706,PET8001,PET8003,TIS60M,TIS90,TIS90-GRY, TIS90-VIO,TIS90M,TIS90M-GRY,TIS90M-VIO,TIS92-GRY,TIS92-VIO,TIS92M-GRY, TIS92M-VIO,TIS109,TIS111,TN60,TN-3904,TZ81,*PTC170,*GE-210,*TR-21, *HEPS0015,*SK3024,*ECG123A,*WEP736,*BC337,*276-2013,*RT-102
TIS60M	2N909,2N930B,2N2222,2N2350,2N2350A,2N2484A,2N2540,2N3116,2N3227, 2N3242A,2N3302,2N3509,2N3704,2N4386,2N4401,2N4952,2N5107,2N5369,2N5449, 2N5818,2N6004,2N6010,2SC366G-Y,2SC367G-Y,2SC1813,A5T2192,A5T2222, A5T5449,A8T3704,BC337-25,BSW83,D33D26,GET2222,GI-3643,GI-3704,GI-3706,

To Replace	Substitute This Type
(TIS60M)	MPS3704,MPS6531,PET3704,PET3706,PET8001,PET8003,TIS60,TIS90,TIS90-GRY, TIS90-VIO,TIS90M,TIS90M-GRY,TIS90M-VIO,TIS92-GRY,TIS92-VIO,TIS92M-GRY, TIS92M-VIO,TIS109,TIS111,TN60,TN-3904,TZ81,*PTC170,*GE-210,*TR-87,*SK3024, *ECG128,*WEP243,*BC337,*276-2013,*RT-114
TIS61	2N2838,2N2907,2N2907A,2N3486,2N3486A,2N3504,2N3505,2N3672,2N3673,2N3912, 2N3915,2N4028,2N4403,2N4415,2N4415A,2N5373,2N5375,2N5817,2N5819,2N6005, 2N6011,2SA467G-Y,2SA720,2SA891,A5T2907,A5T3504,A5T3505,A5T3644, A5T3645,A5T4028,A5T4403,A8T4028,BC327-16,BC327-25,BCW35,BCW37,BSW73, D29E5,D29E6,GI-3644,TIS61M,TIS91,TIS91-BLU,TIS91-GRY,TIS91-VIO,TIS91M, TIS91M-BLU,TIS91M-GRY,TIS91M-VIO,TIS93-BLU,TIS93-GRY,TIS93-VIO, TIS93M-BLU,TIS93M-GRY,TIS93M-VIO,TIS112,TN-3906,TQ60,TZ552,TZ581,TZ582, *PTC103,*GE-82,*TR-20,*HEPS0015,*SK3025,*ECG159,*WEP242,*BC327, *276-2022,*RT-115
TIS61M	2N2838,2N2907,2N2907A,2N3486,2N3486A,2N3504,2N3505,2N3672,2N3673,2N3912, 2N3915,2N4028,2N4403,2N4415,2N4415A,2N5373,2N5375,2N5817,2N5819,2N6005, 2N6011,2SA467G-Y,2SA720,2SA891,A5T2907,A5T3504,A5T3505,A5T3644, A5T3645,A5T4028,A5T4403,A8T4028,BC327-16,BC327-25,BCW35,BCW37,BSW73, D29E5,D29E6,GI-3644,TIS61,TIS91,TIS91-BLU,TIS91-GRY,TIS91-VIO,TIS91M, TIS91M-BLU,TIS91M-GRY,TIS91M-VIO,TIS93-BLU,TIS93-GRY,TIS93-VIO, TIS93M-BLU,TIS93M-GRY,TIS93M-VIO,TIS112,TN-3906,TQ60,TZ552,TZ581,TZ582, *PTC103,*GE-82,*TR-88,*SK3025,*ECG129,*WEP242,*BC327,*276-2022,*RT-115
TIS62	2N2369,2N2369A,2N3862,2N4137,2N4996,2N4997,2N5029,2SC169,2SC335,2SC595, 2SC601,2SC601N,2SC639,2SC764,2SC1123,2SC1128,2SC1188,2SC1189,2SC1240, 2SC1320,2SC1687,2SC1688,BF123,BF125,BF311,MPSH10,MPSH11,MT1060A, TIS48,TIS49,TIS64A,TIS87,TIS129,*PTC132,*GE-86,*TR-95,*HEPS0016,*SK3039, *ECG108,*WEP56,*276-2011,*RT-113
TIS62A	TIS63A,*PTC121,*276-2038
TIS63	2N2368,2N2729,2N4137,2SC302,2SC321,2SC322,2SC595,2SC601,2SC689,2SC1126, MM1941,MPS6507,MT1038,MT1038A,MT1039,MT1060,PET3001,TIS49,TIS84, TIS129,*PTC121,*GE-86,*TR-95,*HEPS0016,*SK3039,*ECG108,*WEP56,*276-2011, *RT-113
TIS63A	TIS62A,*PTC121,*276-2038
TIS64	2N2368,2N2729,2N4137,2SC302,2SC321,2SC322,2SC595,2SC601,2SC689,2SC1126, MM1941,MPS6507,MT1038,MT1038A,MT1039,MT1060,PET3001,TIS49,TIS84, TIS129,*PTC132,*GE-86,*TR-95,*HEPS0016,*SK3039,*ECG108,*WEP56,*276-2011, *RT-113
TIS64A	2N5770,TIS62A,TIS63A,*PTC121,*276-2038
TIS71	*TR-21,*SK3122,*ECG123A,*WEP736,*RT-102
TIS72	*TR-21,*SK3122,*ECG123A,*WEP736,*RT-102
TIS82	2N3735,*GE-215,*ECG186,*WEP245
TIS83	2N2369,2N2369A,2N2710,2N3862,2N4137,2N5029,2SC577,2SC601,2SC601N, 2SC639,2SC764,MPS2369,TIS48,TIS49,TIS87,ZT2369,*PTC132,*GE-61,*TR-21, *HEPS0016,*SK3122,*ECG123A,*WEP736,*276-2038,*RT-102
TIS84	*PTC132,*GE-61,*TR-95,*HEPS0016,*SK3039,*ECG108,*WEP56,*276-2016,*RT-113
TIS85	2N708,2N708A,2N784A,2N914,2N2369,2N2369A,2N2481,2N2710,2N3009,2N3013, 2N3014,2N3210,2N3261,2N3510,2N3511,2N3605A,2N3606A,2N3646,2N3647, 2N3648,2N4137,2N5029,2SC54,2SC67,2SC302,2SC335,2SC336,2SC577,2SC601, 2SC601N,2SC689,2SC1274,GET708,GET914,GET2369,GET3013,GET3014, GET3646,MPS2369,TIS45,TIS46,TIS48,TIS49,TIS52,TIS55,ZT83,ZT708,ZT2369, *PTC132,*GE-61,*TR-95,*SK3039,*ECG108,*WEP56,*276-2016,*RT-113
TIS86	2N5770,*PTC121,*TR-70,*HEPS0016,*SK3039,*ECG107,*WEP56,*276-2038,*RT-108
TIS87	TIS105,*PTC121,*TR-70,*HEPS0016,*SK3039,*ECG107,*WEP56,*276-2015,*RT-108
TIS90	A5T2192,TIS90M,*GE-47,*TR-21,*HEPS0014,*SK3122,*ECG123A,*WEP736,*BC337, *276-2012,*RT-102
TIS90-BLU	2N6426,A5T2192,A5T2222,A5T5449,A8T3704,BC337-25,TIS90,TIS90-GRY,TIS90-VIO, TIS90M,TIS90M-BLU,TIS90M-GRY,TIS90M-VIO,TIS92-BLU,TIS92-GRY,TIS92-VIO, TIS92M-BLU,TIS92M-GRY,TIS92M-VIO,TIS109,TIS111
TIS90-GRN	2N6426,2SC1852,A5T2192,A5T2222,A5T5449,A8T3704,BC337-16,TIS90,TIS90-BLU, TIS90-VIO,TIS90M,TIS90M-BLU,TIS90M-GRN,TIS90M-VIO,TIS92-BLU,TIS92-GRN, TIS92-VIO,TIS92M-BLU,TIS92M-GRN,TIS92M-VIO,TIS111
TIS90-GRY	A5T5451,A8T3706,BC337-40,TIS90M-GRY,TIS92-GRY,TIS92M-GRY

To Replace	Substitute This Type
TIS90-VIO	A5T2192,A5T2222,A5T5449,A5T5451,A8T3704,A8T3706,BC337-25,TIS90,TIS90-GRY, TIS90M,TIS90M-GRY,TIS90M-VIO,TIS92-GRY,TIS92-VIO,TIS92M-GRY,TIS92M-VIO, TIS109,TIS111
TIS90-YEL	2N6426,2SC1852,BC337-16,TIS90-BLU,TIS90-GRN,TIS90M-BLU,TIS90M-GRN, TIS90M-YEL,TIS92-BLU,TIS92-GRN,TIS92-YEL,TIS92M-BLU,TIS92M-GRN, TIS92M-YEL
TIS90M	A5T2192,TIS90
TIS90M-BLU	2N6426,A5T2192,A5T2222,A5T5449,A8T3704,BC337-25,TIS90,TIS90-BLU,TIS90-GRY, TIS90-VIO,TIS90M,TIS90M-GRY,TIS90M-VIO,TIS92-BLU,TIS92-GRY,TIS92-VIO, TIS92M-BLU,TIS92M-GRY,TIS92M-VIO,TIS109,TIS111
TIS90M-GRN	2N6426,2SC1852,A5T2192,A5T2222,A5T5449,A8T3704,BC337-16,TIS90,TIS90-BLU, TIS90-GRN,TIS90-VIO,TIS90M,TIS90M-BLU,TIS90M-VIO,TIS92-BLU,TIS92-GRN, TIS92-VIO,TIS92M-BLU,TIS92M-GRN,TIS92M-VIO,TIS111
TIS90M-GRY	A5T5451,A8T3706,BC337-40,TIS90-GRY,TIS92-GRY,TIS92M-GRY
TIS90M-VIO	A5T2192,A5T2222,A5T5449,A5T5451,A8T3704,A8T3706,BC337-25,TIS90,TIS90-GRY, TIS90-VIO,TIS90M,TIS90M-GRY,TIS92-GRY,TIS92-VIO,TIS92M-GRY,TIS92M-VIO, TIS109,TIS111
TIS90M-YEL	2N6426,2SC1852,BC337-16,TIS90-BLU,TIS90-GRN,TIS90-YEL,TIS90M-BLU, TIS90M-GRN,TIS92-BLU,TIS92-GRN,TIS92-YEL,TIS92M-BLU,TIS92M-GRN, TIS92M-YEL
TIS91	A5T4028,A8T4028,MPS4354,MPS4355,TIS91M,*GE-48,*TR-20,*HEPS0012,*SK3114, *ECG159,*WEP242,*BC327,*276-2025,*RT-115
TIS91-BLU	2SA742,2SA891,A5T2907,A5T3504,A5T3505,A5T3644,A5T3645,A5T4028,A5T4403, A8T4028,BC327-16,BC327-25,MPS4354,MPS4355,TIS91,TIS91-GRN,TIS91-GRY, TIS91-VIO,TIS91M,TIS91M-BLU,TIS91M-GRN,TIS91M-GRY,TIS91M-VIO,TIS93-BLU, TIS93-GRN,TIS93-GRY,TIS93-VIO,TIS93M-BLU,TIS93M-GRN,TIS93M-GRY, TIS93M-VIO,TIS112
TIS91-GRN	2SA742,2SA891,A5T2907,A5T3504,A5T3505,A5T3644,A5T3645,A5T4028,A5T4403, A8T4028,BC327-16,TIS91,TIS91-BLU,TIS91-VIO,TIS91-YEL,TIS91M,TIS91M-BLU, TIS91M-GRN,TIS91M-VIO,TIS91M-YEL,TIS93-BLU,TIS93-GRN,TIS93-VIO,TIS93-YEL, TIS93M-BLU,TIS93M-GRN,TIS93M-VIO,TIS93M-YEL,TIS112
TIS91-GRY	A5T2907,A5T3504,A5T3505,A5T3644,A5T3645,A5T4028,A5T4403,A8T4028,BC327-25, BC327-40,MPS4354,MPS4355,TIS91,TIS91-VIO,TIS91M,TIS91M-GRY,TIS91M-VIO, TIS93-GRY,TIS93-VIO,TIS93M-GRY,TIS93M-VIO,TIS112
TIS91-VIO	2SA891,A5T2907,A5T3504,A5T3505,A5T3644,A5T3645,A5T4028,A5T4403,A8T4028, BC327-16,BC327-25,MPS4354,MPS4355,TIS91,TIS91-BLU,TIS91-GRY,TIS91M, TIS91M-BLU,TIS91M-GRY,TIS91M-VIO,TIS93-BLU,TIS93-GRY,TIS93-VIO, TIS93M-BLU,TIS93M-GRY,TIS93M-VIO,TIS112
TIS91-YEL	2SA742,2SA891,A5T4402,BC327-16,TIS91-BLU,TIS91-GRN,TIS91M-BLU,TIS91M-GRN, TIS91M-YEL,TIS93-BLU,TIS93-GRN,TIS93-YEL,TIS93M-BLU,TIS93M-GRN, TIS93M-YEL
TIS91M	A5T4028,A8T4028,MPS4354,MPS4355,TIS91,*TR-87
TIS91M-BLU	2SA742,2SA891,A5T2907,A5T3504,A5T3505,A5T3644,A5T3645,A5T4028,A5T4403, A8T4028,BC327-16,BC327-25,MPS4354,MPS4355,TIS91,TIS91-BLU,TIS91-GRN, TIS91-GRY,TIS91-VIO,TIS91M,TIS91M-GRN,TIS91M-GRY,TIS91M-VIO,TIS93-BLU, TIS93-GRN,TIS93-GRY,TIS93-VIO,TIS93M-BLU,TIS93M-GRN,TIS93M-GRY, TIS93M-VIO,TIS112
TIS91M-GRN	2SA742,2SA891,A5T2907,A5T3504,A5T3505,A5T3644,A5T3645,A5T4028,A5T4403, A8T4028,BC327-16,TIS91,TIS91-BLU,TIS91-GRN,TIS91-VIO,TIS91-YEL,TIS91M, TIS91M-BLU,TIS91M-VIO,TIS91M-YEL,TIS93-BLU,TIS93-GRN,TIS93-VIO,TIS93-YEL, TIS93M-BLU,TIS93M-GRN,TIS93M-VIO,TIS93M-YEL,TIS112
TIS91M-GRY	A5T2907,A5T3504,A5T3505,A5T3644,A5T3645,A5T4028,A5T4403,A8T4028,BC327-25, BC327-40,MPS4354,MPS4355,TIS91,TIS91-GRY,TIS91-VIO,TIS91M,TIS91M-VIO, TIS93-GRY,TIS93-VIO,TIS93M-GRY,TIS93M-VIO,TIS112
TIS91M-VIO	2SA891,A5T2907,A5T3504,A5T3505,A5T3644,A5T3645,A5T4028,A5T4403,A8T4028, BC327-16,BC327-25,MPS4354,MPS4355,TIS91,TIS91-BLU,TIS91-GRY,TIS91-VIO, TIS91M,TIS91M-BLU,TIS91M-GRY,TIS93-BLU,TIS93-GRY,TIS93-VIO,TIS93M-BLU, TIS93M-GRY,TIS93M-VIO,TIS112
TIS91M-YEL	2SA742,2SA891,A5T4402,BC327-16,TIS91-BLU,TIS91-GRN,TIS91-YEL,TIS91M-BLU, TIS91M-GRN,TIS93-BLU,TIS93-GRN,TIS93-YEL,TIS93M-BLU,TIS93M-GRN, TIS93M-YEL

To Replace	Substitute This Type
TIS92-BLU	2N6426,A5T2192,A5T2222,A5T5449,A8T3704,BC337-25,TIS90,TIS90-BLU,TIS90-GRY, TIS90-VIO,TIS90M,TIS90M-BLU,TIS90M-GRY,TIS90M-VIO,TIS92-GRY,TIS92-VIO, TIS92M-BLU,TIS92M-GRY,TIS92M-VIO,TIS109,TIS111,*PTC170,*GE-81,*ECG123A, *WEP736,*BC337,*276-2012
TIS92-GRN	2N6426,2SC1852,A5T2192,A5T2222,A5T5449,A8T3704,BC337-16,TIS90,TIS90-BLU, TIS90-GRN,TIS90-VIO,TIS90M,TIS90M-BLU,TIS90M-GRN,TIS90M-VIO,TIS92-BLU, TIS92-VIO,TIS92M-BLU,TIS92M-GRN,TIS92M-VIO,TIS111,*PTC170,*GE-81, *ECG123A,*WEP736,*BC337,*276-2012
TIS92-GRY	A5T5451,A8T3706,BC337-40,TIS90-GRY,TIS90M-GRY,TIS92M-GRY,*GE-47, *ECG123A,*WEP736,*BC337,*276-2012
TIS92-VIO	A5T2192,A5T2222,A5T5449,A5T5451,A8T3704,A8T3706,BC337-25,TIS90,TIS90-GRY, TIS90-VIO,TIS90M,TIS90M-GRY,TIS90M-VIO,TIS92-GRY,TIS92M-GRY,TIS92M-VIO, TIS109,TIS111,*PTC170,*GE-81,*ECG123A,*WEP736,*BC337,*276-2012
TIS92-YEL	2N6426,2SC1852,BC337-16,TIS90-BLU,TIS90-GRN,TIS90-YEL,TIS90M-BLU, TIS90M-GRN,TIS90M-YEL,TIS92-BLU,TIS92-GRN,TIS92M-BLU,TIS92M-GRN, TIS92M-YEL,*PTC170,*GE-81,*ECG123A,*WEP736,*BC337,*276-2012
TIS92M-BLU	2N6426,A5T2192,A5T2222,A5T5449,A8T3704,BC337-25,TIS90,TIS90-BLU,TIS90-GRY, TIS90-VIO,TIS90M,TIS90M-BLU,TIS90M-GRY,TIS90M-VIO,TIS92-BLU,TIS92-GRY, TIS92-VIO,TIS92M-GRY,TIS92M-VIO,TIS109,TIS111
TIS92M-GRN	2N6426,2SC1852,A5T2192,A5T2222,A5T5449,A8T3704,BC337-16,TIS90,TIS90-BLU, TIS90-GRN,TIS90-VIO,TIS90M,TIS90M-BLU,TIS90M-GRN,TIS90M-VIO,TIS92-BLU, TIS92-GRN,TIS92-VIO,TIS92M-BLU,TIS92M-VIO,TIS111
TIS92M-GRY	A5T5451,A8T3706,BC337-40,TIS90-GRY,TIS90M-GRY,TIS92-GRY
TIS92M-VIO	A5T2192,A5T2222,A5T5449,A5T5451,A8T3704,A8T3706,BC337-25,TIS90,TIS90-GRY, TIS90-VIO,TIS90M,TIS90M-GRY,TIS90M-VIO,TIS92-GRY,TIS92-VIO,TIS92M-GRY, TIS109,TIS111
TIS92M-YEL	2N6426,2SC1852,BC337-16,TIS90-BLU,TIS90-GRN,TIS90-YEL,TIS90M-BLU, TIS90M-GRN,TIS90M-YEL,TIS92-BLU,TIS92-GRN,TIS92-YEL,TIS92M-BLU, TIS92M-GRN
TIS93-BLU	2SA742,2SA891,A5T2907,A5T3504,A5T3505,A5T3644,A5T3645,A5T4028,A5T4403, A8T4028,BC327-16,BC327-25,MPS4354,MPS4355,TIS91,TIS91-BLU,TIS91-GRN, TIS91-GRY,TIS91-VIO,TIS91M,TIS91M-BLU,TIS91M-GRN,TIS91M-GRY,TIS91M-VIO, TIS93-GRN,TIS93-GRY,TIS93-VIO,TIS93M-BLU,TIS93M-GRN,TIS93M-GRY, TIS93M-VIO,TIS112,*PTC103,*GE-82,*ECG159,*WEP242,*BC327,*276-2025
TIS93-GRN	2SA742,2SA891,A5T2907,A5T3504,A5T3505,A5T3644,A5T3645,A5T4028,A5T4403, A8T4028,BC327-16,TIS91,TIS91-BLU,TIS91-GRN,TIS91-VIO,TIS91-YEL,TIS91M, TIS91M-BLU,TIS91M-GRN,TIS91M-VIO,TIS91M-YEL,TIS93-BLU,TIS93-VIO,TIS93-YEL, TIS93M-BLU,TIS93M-GRN,TIS93M-VIO,TIS93M-YEL,TIS112,*PTC103,*GE-82, *ECG159,*WEP242,*BC327,*276-2025
TIS93-GRY	A5T2907,A5T3504,A5T3505,A5T3644,A5T3645,A5T4028,A5T4403,A8T4028,BC327-25, BC327-40,MPS4354,MPS4355,TIS91,TIS91-GRY,TIS91-VIO,TIS91M,TIS91M-GRY, TIS91M-VIO,TIS93-VIO,TIS93M-GRY,TIS93M-VIO,TIS112,*GE-48,*ECG159,*WEP242, *BC327,*276-2025
TIS93-VIO	2SA891,A5T2907,A5T3504,A5T3505,A5T3644,A5T3645,A5T4028,A5T4403,A8T4028, BC327-16,BC327-25,MPS4354,MPS4355,TIS91,TIS91-BLU,TIS91-GRY,TIS91-VIO, TIS91M,TIS91M-BLU,TIS91M-GRY,TIS91M-VIO,TIS93-BLU,TIS93-GRY,TIS93M-BLU, TIS93M-GRY,TIS93M-VIO,TIS112,*PTC103,*GE-82,*ECG159,*WEP242,*BC327, *276-2025
TIS93-YEL	2SA742,2SA891,A5T4402,BC327-16,TIS91-BLU,TIS91-GRN,TIS91-YEL,TIS91M-BLU, TIS91M-GRN,TIS91M-YEL,TIS93-BLU,TIS93-GRN,TIS93M-BLU,TIS93M-GRN, TIS93M-YEL,*PTC103,*GE-82,*ECG159,*WEP242,*BC327,*276-2025
TIS93M-BLU	2SA742,2SA891,A5T2907,A5T3504,A5T3505,A5T3644,A5T3645,A5T4028,A5T4403, A8T4028,BC327-16,BC327-25,MPS4354,MPS4355,TIS91,TIS91-BLU,TIS91-GRN, TIS91-GRY,TIS91-VIO,TIS91M,TIS91M-BLU,TIS91M-GRN,TIS91M-GRY,TIS91M-VIO, TIS93-BLU,TIS93-GRN,TIS93-GRY,TIS93-VIO,TIS93M-GRN,TIS93M-GRY,TIS93M-VIO, TIS112
TIS93M-GRN	2SA742,2SA891,A5T2907,A5T3504,A5T3505,A5T3644,A5T3645,A5T4028,A5T4403, A8T4028,BC327-16,TIS91,TIS91-BLU,TIS91-GRN,TIS91-VIO,TIS91-YEL,TIS91M, TIS91M-BLU,TIS91M-GRN,TIS91M-VIO,TIS91M-YEL,TIS93-BLU,TIS93-GRN, TIS93-VIO,TIS93-YEL,TIS93M-BLU,TIS93M-VIO,TIS93M-YEL,TIS112
TIS93M-GRY	A5T2907,A5T3504,A5T3505,A5T3644,A5T3645,A5T4028,A5T4403,A8T4028,BC327-25,

To Replace	Substitute This Type
(TIS93M-GRY)	BC327-40,MPS4354,MPS4355,TIS91,TIS91-GRY,TIS91-VIO,TIS91M,TIS91M-GRY, TIS91M-VIO,TIS93-GRY,TIS93-VIO,TIS93M-VIO,TIS112
TIS93M-VIO	2SA891,A5T2907,A5T3504,A5T3505,A5T3644,A5T3645,A5T4028,A5T4403,A8T4028, BC327-16,BC327-25,MPS4354,MPS4355,TIS91,TIS91-BLU,TIS91-GRY,TIS91-VIO, TIS91M,TIS91M-BLU,TIS91M-GRY,TIS91M-VIO,TIS93-BLU,TIS93-GRY,TIS93-VIO, TIS93M-BLU,TIS93M-GRY,TIS112
TIS93M-YEL	2SA742,2SA891,A5T4402,BC327-16,TIS91-BLU,TIS91-GRN,TIS91-YEL,TIS91M-BLU, TIS91M-GRN,TIS91M-YEL,TIS93-BLU,TIS93-GRN,TIS93-YEL,TIS93M-BLU, TIS93M-GRN
TIS94	2N6428,2N6428A,2N6539,TIS97,*PTC123,*GE-210,*TR-21,*HEPS0015,*SK3122, *ECG123A,*WEP736,*BC337,*276-2009,*RT-102
TIS95	2N6540,TIS98,*PTC123,*GE-10,*TR-21,*SK3122,*ECG123A,*WEP736,*276-2030, *RT-102
TIS96	2N6541,TIS99,*PTC123,*GE-20,*TR-21,*HEPS0005,*SK3122,*ECG123A,*WEP736, *RT-102
TIS97	2N6428,2N6428A,2N6539,TIS94,*PTC123,*GE-20,*TR-70,*HEPS0015,*SK3039, *ECG107,*WEP56,*BC337,*276-2009,*RT-108
TIS98	2N6540,TIS95,*PTC123,*GE-63,*TR-70,*HEPS0005,*SK3122,*ECG107,*WEP56, *276-2009,*RT-108
TIS99	2N6541,TIS96,*PTC123,*GE-20,*TR-70,*HEPS0005,*SK3039,*ECG107,*WEP56, *RT-108
TIS100	*PTC170,*GE-27,*TR-78,*HEPS0005,*SK3045,*ECG154,*WEP712,*BF338,*276-2008, *RT-110
TIS101	*PTC170,*GE-27,*TR-78,*HEPS0005,*SK3045,*ECG154,*WEP712,*BF338,*276-2008, *RT-110
TIS105	*PTC121,*HEPS0016,*ECG108,*WEP56,*276-2015
TIS108	TIS84,*PTC132,*GE-61,*TR-70,*HEPS0016,*SK3039,*ECG107,*WEP56,*276-2016, *RT-108
TIS109	*HEPS3001,*ECG195A,*276-2009
TIS110	*HEPS3001,*ECG195A,*276-2009
TIS111	*HEPS3001,*ECG195A,*276-2009
TIS112	A5T2907,A5T3505,A5T3645,*PTC103,*HEPS0019,*ECG159,*276-2023
TIS125	2N834A,2N2368,2SC302,2SC321,2SC322,2SC576,2SC689,BF225J,MPSH20,TIS47, TIS84,ZT2368,*PTC132,*276-2016
TIS126	*PTC121
TIS128	*PTC103
TIS129	2N2369A,2SC764,2SC1188,2SC1189,2SC1320,*PTC132
TIS133	*276-2038
TIS134	TIS133,*276-2038
TIS135	
TIS136	TIS135
TIS137	*PTC103
TIS138	*PTC103,*276-2023
TIX210	SEE 2N3551
TIX211	SEE 2N3552
TIX316	TI-400,TIXM101,*TR-17,*HEPG0003,*ECG160,*WEP637
TIX317	SEE 2N3601
TIX318	SEE 2N3603
TIX888	SEE 2N3554
TIX3016	*TR-17,*ECG160,*WEP637
TIX3016A	*GE-86,*TR-17,*ECG160,*WEP637
TIX3024	
TIX3032	*TR-17,*HEPG0003,*ECG160,*WEP637
TIX3033	SEE 2N3418
TIX3034	SEE 2N3419
TIX3035	SEE 2N3420
TIX3036	SEE 2N3421
TIXM01	2N2587,2N2717,2N2795,2N6365,2N6365A,ASZ21,MM380,*GE-51,*TR-17,*HEPG0003, *ECG160,*WEP637
TIXM02	2N2587,2N2717,2N2795,2N6365,2N6365A,ASZ21,MM380,*TR-17,*HEPG0003, *ECG160,*WEP637

To Replace	Substitute This Type
TIXM03	2N741A,2N2796,*TR-17,*HEPG0003,*ECG160,*WEP637
TIXM04	2N2587,2N2717,2N2795,2N6365,2N6365A,ASZ21,MM380,*TR-17,*HEPG0003, *ECG160,*WEP637
TIXM05	2N2587,2N6365,2N6365A,MM380,*GE-51,*TR-17,*HEPG0003,*HF75,*ECG160, *WEP637
TIXM06	2N2587,2N6365,2N6365A,MM380,*TR-17,*HEPG0003,*ECG160,*WEP637
TIXM07	2N741A,2N2796,*TR-17,*HEPG0003,*ECG160,*WEP637
TIXM08	2N2587,2N6365,2N6365A,MM380,*TR-17,*HEPG0008,*ECG160,*WEP637,*276-2005
TIXM10	2N2587,*TR-17,*HEPG0001,*ECG160,*WEP637
TIXM11	*GE-51,*TR-17,*HEPG0003,*ECG160,*WEP637
TIXM13	*TR-17,*HEPG0003,*ECG160,*WEP637
TIXM14	2N2273,2N3323,2N3324,2N3325,2N3371,*TR-17,*HEPG0002,*ECG160,*WEP637, *276-2005
TIXM15	2N2587,2N2795,2N6365,2N6365A,MM380,*TR-17,*HEPG0003,*ECG160,*WEP637
TIXM16	2N2587,2N2795,2N6365,2N6365A,MM380,*TR-17,*HEPG0002,*ECG160,*WEP637, *276-2005
TIXM17	2N838,2N3371,*TR-17,*HEPG0003,*ECG160,*WEP637
TIXM18	*TR-17,*HEPG0003,*ECG160,*WEP637
TIXM19	*TR-17,*HEPG0003,*ECG160,*WEP637
TIXM101	*TR-17,*HEPG0003,*ECG160,*WEP637
TIXM103	*PTC107,*TR-17,*ECG160,*WEP637
TIXM104	*PTC107,*TR-17,*ECG160,*WEP637
TIXM105	*PTC107,*TR-17,*HEPG0003,*ECG160,*WEP637
TIXM106	*PTC107,*TR-17,*HEPG0003,*ECG160,*WEP637
TIXM107	*PTC107,*TR-17,*HEPG0003,*ECG160,*WEP637
TIXM108	*PTC107,*TR-17,*HEPG0003,*ECG160,*WEP637
TIXM201	2N2587,2N2955,2N6365,2N6365A,*PTC107,*GE-51,*TR-17,*HEPG0003,*HF75, *ECG160,*WEP637
TIXM202	2N2587,2N2955,2N6365,2N6365A,*PTC107,*GE-51,*TR-17,*HEPG0003,*SK3006, *HF35,*ECG160,*WEP637
TIXM203	*PTC107,*GE-51,*TR-17,*HEPG0003,*HF75,*ECG160,*WEP637
TIXM204	*PTC107,*GE-51,*TR-17,*HEPG0003,*SK3006,*HF35,*ECG160,*WEP637
TIXM205	*PTC107,*GE-51,*TR-17,*HEPG0003,*HF35,*ECG160,*WEP637
TIXM206	*PTC107,*GE-51,*TR-17,*HEPG0003,*HF75,*ECG160,*WEP637
TIXM207	*PTC107,*GE-51,*TR-17,*HEPG0009,*HF75,*ECG160,*WEP637
TIXP07	*HEPS5005
TIXS09	A3T3011,A3T929,FX914,FX2369A,FX3013,FX3014,FX4046,MMT2369,MMT3014, *GE-86,*TR-95,*HEPG0005,*SK3039,*EN10,*ECG108,*BC107B,*276-2011,*RT-113
TIXS10	MMT3014,*GE-86,*TR-95,*SK3039,*EN10,*ECG108,*WEP56,*BC107B,*276-2011, *RT-113
TIXS12	*GE-20,*TR-21,*SK3122,*SN80,*ECG123A,*WEP736,*RT-102
TIXS13	*GE-20,*TR-21,*SK3122,*SN80,*ECG123A,*WEP736,*RT-102
TIXS28	2N2369,2N2369A,2N3862,2SC601,2SC764,2SC1188,2SC1189,2SC1320,2SC1687, 2SC1688,TIS129,*PTC132,*GE-61,*TR-95,*HEPS0016,*SK3039,*EN10,*ECG108, *WEP56,*276-2015,*RT-113
TIXS29	2N2369,2N2369A,2N3862,2N4137,2N4295,2N5029,2SC169,2SC335,2SC336,2SC468, 2SC601,2SC601N,2SC639,2SC764,2SC1188,2SC1189,2SC1320,2SC1687,2SC1688, EN2369A,TIS48,TIS49,TIS87,TIS129,*PTC132,*GE-60,*TR-95,*HEPS0016,*SK3039, *EN10,*ECG108,*WEP56,*276-2015,*RT-113
TIXS30	2N2369,2N2369A,2N3862,2N4137,2N4295,2N5029,2SC169,2SC335,2SC336,2SC468, 2SC601,2SC601N,2SC639,2SC764,2SC1188,2SC1189,2SC1320,2SC1687,2SC1688, EN2369A,TIS48,TIS49,TIS87,TIS129,*PTC132,*GE-60,*TR-95,*HEPS0016,*SK3039, *EN10,*ECG108,*WEP56,*276-2015,*RT-113
TIXS31	2N2369,2N2369A,2N3862,2N4137,2N4295,2N5029,2SC169,2SC335,2SC336,2SC468, 2SC601,2SC601N,2SC639,2SC764,2SC1188,2SC1189,2SC1320,2SC1687,2SC1688, EN2369A,TIS48,TIS49,TIS87,TIS129,*PTC132,*GE-60,*TR-95,*HEPS0016,*SK3039, *EN10,*ECG108,*WEP56,*276-2015,*RT-113
TIXS39	2SC651,*276-2038
TN53	2N3725A,2N4961,2SC108A-O,2SC109A-O,*PTC144,*GE-18,*TR-21,*SK3024, *ECG123A,*WEP736,*RT-102
TN54	2N4963,2N5822,D33D30,*PTC144

TRANSISTOR SUBSTITUTES

To Replace	Substitute This Type
TN55	SEE 2N4383
TN56	SEE 2N4384
TN57	SEE 2N4385
TN58	SEE 2N4386
TN59	2N2219,2N2538,2N3300,*GE-47,*TR-21,*SK3024,*ECG123A,*WEP736,*276-2030,*RT-102
TN60	2N2540,2N3242A,2N5818,A5T2222,A5T5449,A5T5451,A8T3704,A8T3706,BC337-25,D33D26,TIS109,TIS111,TN-3904,*PTC136,*GE-20,*TR-21,*HEPS0011,*ECG123A,*WEP736,*BC337,*276-2009,*RT-102
TN61	2N2270,2N2537,2N3724A,2N4960,2SC1384,*PTC144,*GE-47,*TR-21,*SK3024,*ECG123A,*WEP736,*276-2030,*RT-102
TN62	2N2539,2N2897,2N2900,2N3643,2N4962,40458,A5T5450,A8T3705,BC337-16,D33D25,TIS110,TN-3903,*PTC136,*GE-20,*TR-21,*HEPS0011,*SK3122,*ECG123A,*WEP736,*BC337,*276-2009,*RT-102
TN63	2SC147,SE8041,SE8042,*PTC144,*GE-47,*TR-21,*SK3024,*ECG123A,*WEP736,*276-2030,*RT-102
TN64	SE8040,*PTC136,*GE-20,*TR-21,*HEPS0011,*SK3122,*ECG123A,*WEP736,*BC337,*276-2009,*RT-102
TN79	2N4265,BCY58,BCY66,*PTC125,*GE-47,*ECG192,*WEP243,*276-2008
TN80	2N3242A,A5T4124,*PTC123,*GE-20,*ECG123A,*WEP736,*BC337,*276-2009
TN81	2N744A,2N4264,BSY63,*GE-47,*TR-63,*ECG192,*WEP243,*276-2038
TN237	2N3724A,TN61,*PTC144,*GE-47,*TR-21,*HEPS3024,*SK3024,*ECG123A,*WEP736,*276-2030,*RT-102
TN624	2N1973,2N2443,2N2849,2N3107,SDT4456,TN624-1,*276-2008
TN624-1	2N1973,2N2443,2N2849,2N3107,SDT4456,TN624,*276-2008
TN624-2	
TN624-3	
TN-3200	2SC333,2SC334,2SC463,40239,BF168,BF173,BFS62,BFX60,*GE-60,*276-2016
TN-3903	2N4962,2N4963,2N5380,2SC1998,A5T3903
TN-3904	2N5381,A5T3904
TN-3905	2N5382,A5T3905,FT3644,FT3645,TP3644,TP3645
TN-3906	2N5383,A5T3906
TP3566	2N6428,2N6428A,2N6539,BCY59,*PTC170
TP3638	A5T3638,*PTC103,*GE-65,*TR-20,*SK3247,*ECG159,*WEP242,*BC327,*276-2024
TP3638A	2N5378,2N5379,2N5813,2SB598,D29E2,*PTC103,*GE-65,*SK3247,*ECG193,*WEP242,*276-2024
TP3644	2SA891,A5T2907,A5T3504,A5T3505,A5T3644,A5T3645,BCY79,TIS112,TP3645,*PTC103
TP3645	2SA891,A5T2907,A5T3505,A5T3645,TIS112,*PTC103
TP3702	2N3251,2N3251A,2N3702,2N3798,2N3962,2N4034,2N4035,2N5244,2N5373,2N5374,2N5375,2N5379,2N5383,2N5447,2N6005,2SA659,2SA661-GR,2SA661-Y,2SA720,2SA889,2SA891,A5T3906,A5T5447,A8T3702,BC557A,BC557B,BC560A,BC560B,BCY79,BCY79A,BCY79B,BSW73,GET2906,GET2907,MM3906,MM4048,NPS404A,TN-3906,TP3644,TP3645,TZ552,TZ553
TP3703	2N3250,2N3250A,2N3703,2N5372,2N5448,2N6067,2SA661-O,A5T5448,A8T3703,BSW74,GET2904,GET2905,TZ551
TP3704	2N4953,2N5370,2N5376,2N5377,2N5827,2N6006,2N6112,2N6428,2N6428A,2N6539,2SC1166-GR,BC547B,BC550B,TIS94,TIS97,*PTC101
TP3705	2N915,2N3859A,2N3947,2N3974,2N3976,2N4013,2N4951,2N4962,2N5368,2N5380,2N5825,2N6538,2SC1166-O,2SC1318,2SC1781,2SC1852,2SC1999,A5T3903,BCW34,TIS133,TIS134,TN-3903,*PTC101
TP3706	2N4953,2N4954,2N5370,2N5371,2N5376,2N5377,2N5827,2N6006,2N6112,2N6428,2N6428A,2N6539,2SC1166-GR,BC337,BC547B,BC550B,BCY59C,MPSA16,TIS94,TIS97,TP3704,*PTC101
TP3707	2N3707,2N3708,A5T3565,A5T3707,A5T3708,A8T3707,A8T3708,BC547B,BC550B,BCY58B,BCY58C,BCY59B,BCY59C,MPSA16,MPSA17,PET4002,TP3708,*PTC101
TP3708	2N3708,2N3711,A5T3565,A5T3708,A5T3711,A8T3708,A8T3711,BC547B,BC550B,BCY58D,BCY59D,PET4003,TP3711,*PTC101
TP3709	2N3241A,2N3709,2N4074,2SC1999,A5T3709,A8T3709,BCY58A,BCY59A,PET4001,*PTC101
TP3710	2N3707,2N3710,A5T3707,A5T3710,A8T3707,A8T3710,BCY58B,BCY58C,BCY59B,

To Replace	Substitute This Type
(TP3710)	BCY59C,MPSA16,MPSA17,PET4002,TP3707,*PTC101
TP3711	2N3711,A5T3711,A8T3711,BC550C,BCY58D,BCY59D,PET4003,*PTC101
TP4058	2N3582,2N4058,2N4059,2N4061,A5T404A,A5T4058,A5T4059,A5T4061,A8T404A, A8T4058,A8T4059,A8T4061,NPS404A,TP4059,TP4061,*PTC103
TP4059	2N3964,2N4059,2N4062,A5T4059,A5T4062,A8T4059,A8T4062,TP4062,*PTC103
TP4060	2N2946,2N3527,2N3581,2N4060,2N4980,2N4981,A5T4060,A8T4060,TIS37,*PTC103
TP4061	2N3582,2N4058,2N4061,A5T404A,A5T4058,A5T4061,A8T404A,A8T4058,A8T4061, NPS404A,TP4058,*PTC103
TP4062	2N3964,2N4059,2N4062,A5T4059,A5T4062,A8T4059,A8T4062,TP4059,*PTC103
TP4257	2N3012,2N3304,2N3546,2N3639,2N3640,2N4207,2N4208,2N4209,2N4257,2N4257A, 2N4258,2N4258A,2N4872,2N5056,2N5057,2N5292,2N5771,2SA781,A5T4260, A5T4261,MPS3639,MPS3640,MPSH81,TIS53,*PTC131,*SK3118,*ECG159
TP4258	2N3546,2N4208,2N4209,2N4258,2N4258A,2N4872,2N5056,2N5057,2N5292,2N5771, 2SA781,A5T4260,A5T4261,MPSH81,*SK3118,*ECG159,*WEP242
TP4274	2N2475,2N2784,2N3633,2N3959,2N3960,2N5200,2N5201,2SC99,2SC300,2SC301, A5T3571,MPS6568A,MPS6569,MPS6570,*PTC101,*GE-61,*TR-21,*ECG101, *WEP641,*BF173,*276-2016
TP4275	2N2475,2N2615,2N2616,2N2784,2N3011,2N3633,2N3959,2N3960,2N5200,2N5201, 2SC99,2SC300,2SC301,A5T3571,BF123,BF125,BF127,BF377,BF378,MPS3563, MPS6568A,MPS6569,MPS6570,MPSH10,MPSH11,TIS64A,TIS86,*PTC101,*GE-61, *TR-21,*ECG108,*WEP56,*BF173,*276-2016
TP5135	BCY59,TP3566,*PTC170
TP5136	A5T4124,BCY58,BCY66,*PTC170
TP5142	A5T3638,TIS38,TIS137,TIS138
TQ53	2N4031,2N5865,2SA581,2SA606,2SA817,MPS4356,*PTC127,*GE-67,*ECG193, *WEP242
TQ53A	2N4031,2SA581,2SA606,MPS4356,*PTC127
TQ54	2N4027,2N5821,2N5823,2SA817,D29E10,*PTC127,*GE-67;*ECG193,*WEP242
TQ54A	2N4027,*PTC127
TQ55	SEE 2N4412
TQ56	SEE 2N4413
TQ57	SEE 2N4414
TQ58	SEE 2N4415
TQ59	2N2905,2N2905A,2N3502,2N3503,2N4032,2SA503-GR,2SA504-GR,2SA684,A5T2907, A5T3504,A5T3505,A5T3644,A5T3645,A5T4403,BC327-16,BC327-25,MPS4354, MPS4355,TIS112,*PTC127,*GE-48,*ECG193,*WEP242,*BC327
TQ59A	2N2905,2N2905A,2N3503,2N4032,2N4033,2N4405,2SA503-GR,2SA684,2SA817, A5T2907,A5T3505,A5T3645,MPS4354,MPS4355,TIS112,*PTC127,*GE-67,*ECG193, *WEP242
TQ60	2N2907,2N2907A,2N3486,2N3486A,2N3504,2N3505,2N4028,2N5819,A5T2907, A5T3504,A5T3505,A5T3644,A5T3645,A5T4403,BC327-16,BC327-25,D29E6,TIS112, TN-3906,*PTC127,*GE-48,*ECG193,*WEP242,*BC327,*276-2023
TQ60A	2N2907,2N2907A,2N3486,2N3486A,2N3505,2N4028,2N4029,2SA817,A5T2907, A5T3505,A5T3645,TIS112,*PTC127,*GE-67,*ECG193,*WEP242
TQ61	2N2801,2N2904,2N2904A,2N3244,2N3467,2N3671,2N4030,2N4890,2SA503-Y, 2SA504-Y,2SA571,2SA580,2SA684,2SA717,40537,A5T4402,BC327-16,*PTC127, *GE-21,*ECG159,*WEP242,*BC327
TQ61A	2N2904,2N2904A,2N3671,2N4030,2N4031,2N4890,2N5865,2SA503-Y,2SA571, 2SA580,2SA684,2SA717,2SA817,40537,MPS4356,*PTC127,*GE-21,*ECG159, *WEP242
TQ62	2N2838,2N2906,2N2906A,2N3485,2N3485A,2N3672,2N4026,2N5815,2N5817, A5T4402,BC327-16,D29E5,FT3644,FT3645,TN-3905,*PTC127,*GE-21,*ECG159, *WEP242,*BC327,*276-2023
TQ62A	2N2906,2N2906A,2N3485,2N3485A,2N3672,2N4026,2N4027,2N5821,2N5823,2SA817, D29E10,FT3644,*PTC127,*GE-21,*ECG159,*WEP242
TQ63	2N5023,SE8541,SE8542,*PTC127,*GE-21,*TR-88,*ECG129,*WEP242,*BC327, *276-2032,*RT-115
TQ63A	2N2800,2N3245,2N3467,2N3468,2N3762,2N3762A,2N5022,2N5023,2N5042, 2SA504-O,MM3726,SE8541,SE8542,*PTC127,*GE-21,*TR-88,*ECG129,*WEP242, *BC327,*RT-115
TQ64	2N5242,2N5243,SE8540,*PTC127,*GE-21,*TR-88,*ECG129,*WEP242,*BC327,

TRANSISTOR SUBSTITUTES

To Replace	Substitute This Type
(TQ64)	*276-2023,*RT-115
TQ64A	2N2837,2N3764,2N3764A,2N5243,SE8540,*PTC127,*GE-21,*TR-88,*ECG129, *WEP242,*BC327,*276-2023,*RT-115
TR-01	REFER TO SECTION 2
TR-01MP	M.P.TR-01
TR-02	REFER TO SECTION 2
TR-03	REFER TO SECTION 2
TR-05	REFER TO SECTION 2
TR-06	REFER TO SECTION 2
TR-08	REFER TO SECTION 2
TR-09	REFER TO SECTION 2
TR-10	REFER TO SECTION 2
TR-11	REFER TO SECTION 2
TR-12	REFER TO SECTION 2
TR-14	REFER TO SECTION 2
TR-16	REFER TO SECTION 2
TR-17	REFER TO SECTION 2
TR-19	REFER TO SECTION 2
TR-20	REFER TO SECTION 2
TR-21	REFER TO SECTION 2
TR-22	REFER TO SECTION 2
TR-23	REFER TO SECTION 2
TR-24	REFER TO SECTION 2
TR-25	REFER TO SECTION 2
TR-26	REFER TO SECTION 2
TR-27	REFER TO SECTION 2
TR-28	REFER TO SECTION 2
TR-29	REFER TO SECTION 2
TR-30	REFER TO SECTION 2
TR-31	REFER TO SECTION 2
TR-32	REFER TO SECTION 2
TR-33	REFER TO SECTION 2
TR-34	REFER TO SECTION 2
TR-35	REFER TO SECTION 2
TR-36	REFER TO SECTION 2
TR-36MP	M.P.TR-36
TR-37	REFER TO SECTION 2
TR-50	REFER TO SECTION 2
TR-51	REFER TO SECTION 2
TR-52	REFER TO SECTION 2
TR-53	REFER TO SECTION 2
TR-54	REFER TO SECTION 2
TR-55	REFER TO SECTION 2
TR-56	REFER TO SECTION 2
TR-57	REFER TO SECTION 2
TR-58	REFER TO SECTION 2
TR-59	REFER TO SECTION 2
TR-60	REFER TO SECTION 2
TR-61	REFER TO SECTION 2
TR-61MP	M.P.TR-61
TR-62	REFER TO SECTION 2
TR-63	REFER TO SECTION 2
TR-64	REFER TO SECTION 2
TR-65	REFER TO SECTION 2
TR-66	REFER TO SECTION 2
TR-67	REFER TO SECTION 2
TR-68	REFER TO SECTION 2
TR-69	REFER TO SECTION 2
TR-70	REFER TO SECTION 2
TR-72	REFER TO SECTION 2
TR-73	REFER TO SECTION 2

To Replace	Substitute This Type
TR-74	REFER TO SECTION 2
TR-75	REFER TO SECTION 2
TR-76	REFER TO SECTION 2
TR-77	REFER TO SECTION 2
TR-78	REFER TO SECTION 2
TR-79	REFER TO SECTION 2
TR-80	REFER TO SECTION 2
TR-81	REFER TO SECTION 2
TR-81MP	M.P.TR-81
TR-82	REFER TO SECTION 2
TR-83	REFER TO SECTION 2
TR-84	REFER TO SECTION 2
TR-85	REFER TO SECTION 2
TR-86	REFER TO SECTION 2
TR-87	REFER TO SECTION 2
TR-88	REFER TO SECTION 2
TR-91	REFER TO SECTION 2
TR-92	REFER TO SECTION 2
TR-93	REFER TO SECTION 2
TR-94MP	REFER TO SECTION 2
TR-95	REFER TO SECTION 2
TRS301	*PTC117,*GE-27,*TR-78,*SK3045,*ECG154,*WEP712,*276-2008,*RT-110
TRS350	*PTC117,*GE-32
TRS401	*GE-32
TRS450	*GE-32
TRS501	*GE-32
TRS550	*GE-32
TRS601	
TRS701	
TRS750	
TRS801	
TRS3011	*PTC117,*GE-27,*TR-78,*SK3045,*ECG154,*WEP712,*276-2008,*RT-110
TRS3012	*PTC117,*GE-27,*TR-78,*SK3045,*ECG154,*WEP712,*276-2008,*RT-110
TRS3014	*PTC117,*GE-27,*TR-78,*SK3201,*ECG171,*WEPS3021,*276-2008
TRS3015	*GE-32,*SK3220,*ECG198,*WEP244
TRS3501	*PTC117
TRS3502	*PTC117
TRS3504	*PTC117,*GE-32
TRS3505	
TRS4001	
TRS4002	
TRS4004	*GE-32
TRS4005	
TRS4501	
TRS4502	
TRS4504	*GE-32
TRS4505	
TRS5011	
TRS5012	
TRS5014	*GE-32
TRS5501	
TRS5502	
TRS5504	*GE-32
TRS6011	
TRS6012	
TRS6014	
TRS7014	
TRS7015	
TRS7504	
TRS7505	
TRS8014	

To Replace	Substitute This Type
TRS8015	
TS173	2N235A,2N236A,2N285B,2N399,2N401,2N1291,2SB426A,AD130,AD149,AD150, ADY27,*PTC105A,*GE-3,*TR-16,*HEPG6005,*SK3014,*PT40,*ECG121,*WEP232, *OC28,*276-2006
TS601	2N1280,2N1942,*PTC135,*GE-53,*HEPG0002,*AT10N,*AC188/01
TS602	2N660,2N661,2N1705,2N1706,NKT231,NKT232,*PTC135,*GE-53,*HEPG0002, *AT10M,*AC188/01,*276-2005
TS603	2N653,2N1018,2N1942,2SB219A,2SB220A,*PTC135,*GE-53,*HEPG0005,*AT20N, *ECG158,*WEP238,*AC188/01,*276-2005
TS604	2N598,2N660,2N661,2N1706,2N1707,2N1998,2SB222A,2SB223A,2SB327,2SB461, *PTC135,*GE-53,*HEPG0005,*AT20M,*ECG158,*WEP238,*AC188/01,*276-2005
TS612	2N235A,2N285A,2N296,2N399,2N401,AD149,*PTC105A,*GE-3,*TR-01,*HEPG6003, *SK3009,*PT40,*ECG121,*WEP232,*OC28,*276-2006
TS613	AD132,AD163,AUY20,AUY34,*PTC105A,*GE-25,*TR-01,*HEPG6003,*SK3014,*PT40, *ECG121,*WEP232,*OC28,*276-2006
TS615	2N43,2N43A,2N43A+,2N44,2N44A,2N460,2N524,2N524A,2N1614,2N2382,2SB224, *GE-1,*HEPG0002,*HF12M,*276-2005
TS616	2N1495,*GE-53,*HEPG0005,*SK3004,*AT20M,*ECG158,*WEP238,*AC188/01, *276-2005
TS617	2N1495,2N3883,NKT221,NKT228,*GE-53,*HEPG0005,*SK3004,*AT20M,*ECG158, *WEP238,*AC188/01,*276-2005
TS618	2N1008A,*GE-53,*HEPG0005,*SK3004,*AT20M,*ECG158,*WEP238,*AC188/01, *276-2005
TS619	2N59,2N59A,2N60,2N60A,2N241A,2N241A+,2N321,2N360,2N414B,2N414C,2N422A, 2N654,2N654+,2N1175,2N1175A,2N1192,2N1348,2N1349,2N1350,2N1351,2N1374, 2N1382,2N1383,2N1415,2N1478,2N1681,2N1706,2N1707,2N3883,2SA206,2SA207, 2SA407,2SB221,2SB221A,2SB326,2SB496,NKT213,NKT216,SF.T125,SF.T125P, SF.T352,SF.T353,*GE-53,*TR-05,*HEPG0005,*SK3004,*AT20M,*ECG158,*WEP238, *AC126,*276-2005
TS620	2N59,2N59A,2N60,2N60A,2N241A,2N241A+,2N321,2N360,2N414B,2N414C,2N422A, 2N654,2N654+,2N1175,2N1175A,2N1192,2N1348,2N1349,2N1350,2N1351,2N1374, 2N1382,2N1383,2N1415,2N1478,2N1681,2N1706,2N1707,2N3883,2SA206,2SA207, 2SA407,2SB221,2SB221A,2SB326,2SB496,NKT213,NKT216,SF.T125,SF.T125P, SF.T352,SF.T353,*GE-53,*TR-05,*HEPG0002,*SK3008,*AT20M,*AC126,*276-2005
TS621	2N362,2N467,2N508A,2N598,2N599,2N655,2N655+,2N1008A,2N1193,2N1354, 2N1355,2N1356,2N1357,2N1376,2N1381,2N1998,2SB22,2SB186,2SB187,2SB188, 2SB223,2SB223A,2SB327,2SB431,NKT211,NKT218,SF.T223,SF.T323,*GE-53, *TR-05,*HEPG0002,*SK3008,*HF12M,*AC126,*276-2005
TS630	2N60B,2N60C,2N61B,2N61C,2N381,2N525,2N525A,2N526,2N526A,2N586,2N597, 2N650,2N651,2N651A,2N652A,2N1187,2N1447,2N1451,2N1924,2N1925,2SB57A, 2SB225,2SB226,2SB427,MA100,MA882,MA887,SF.T243,*GE-1,*TR-05,*HEPG0002, *SK3123,*HF12M,*ECG160,*WEP637,*276-2005
TV1000	REFER TO SECTION 2
TVCM-551	REFER TO SECTION 2
TW135	*GE-82,*TR-20,*HEPS0019,*SP70,*ECG106,*WEP52,*BC327,*276-2023,*RT-126
TZ81	2N929A,2N930A,2N2484,2N2484A,2N2523,2N5826,2N5827,2N6112,2SC1166-GR, 2SC1850,BC547B,BC550B,BCY59B,BCY59C,MPSA16,MPSA17,TIS94,TIS97,TP3566, TP3704,TP3706,*PTC121,*GE-62,*ECG199,*WEP736,*BC337,*276-2014
TZ82	2N909,2N929A,2N930B,2N956,2N2483,2N2484,2N2484A,2N2645,2N3302,2N4409, 2N5381,2N6540,2N6541,2SC1166-Y,2SC1850,2SC1991,A5T3904,A5T4409,BC546A, GET930,GET2222,GET2222A,MM3904,TIS95,TIS98,TN-3904,*PTC101,*GE-62, *SK3098,*ECG150,*BC337,*276-2014,*RT-255
TZ551	2N3250A,GET2904,GET2905,*PTC103,*GE-65,*TR-20,*HEPS0019,*SK3114,*ECG159, *WEP242,*276-2023
TZ552	2N3251A,2N3962,2N3963,BC556A,BC556B,GET2906,GET2907,TP3645,TZ553, *PTC103,*GE-65,*ECG193,*WEP242,*276-2023
TZ553	2N3962,2N3963,2N3965,BC556B,*GE-65,*ECG193,*WEP242,*276-2023
TZ554	2N3251,2N4035,2N4058,2N4059,2N4061,2N5244,2N5366,2N5383,A5T3906,BC557A, BC557B,BC558A,BC558B,BC559A,BC559B,BC560A,BC560B,BCY78A,BCY78B, BCY78C,BCY79,BCY79A,BCY79B,BCY79C,MM3906,NPS404A,TN-3906,*PTC103, *GE-65,*ECG193,*WEP242,*276-2023

To Replace	Substitute This Type
TZ581	2N2605,2N2605A,2N3251,2N3251A,2N3547,2N3548,2N3582,2N3702,2N3798,2N3962, 2N4035,2N4359,2N5244,2N5383,2N5447,2SA659,2SA661-GR,2SA661-Y,2SA889, A5T2605,A5T3906,A5T4250,A5T5447,A8T3702,BC326,BC557A,BC557B,BC560A, BC560B,BCY79,BCY79A,BCY79B,BCY79C,GET2906,GET2907,MM3906,MM4048, NPS404A,TN-3906,TP3702,TZ552,TZ553,*GE-65,*HEPS3028,*SK3203,*ECG211, *BC327,*276-2023
TZ582	2N2597,2N2605,2N2605A,2N3251A,2N3547,2N3798,2N3962,2N3963,2SA661-Y, A5T2605,BC326,BC556A,GET2906,GET2907,TP3645,TZ552,*PTC103,*GE-65, *HEPS3028,*SK3203,*ECG211,*BC327,*276-2023
U2TA406	U2TA408,U2TB406,U2TB408
U2TA408	U2TA410,U2TB408,U2TB410
U2TA410	U2TB410
U2TA506	U2TA508
U2TA508	U2TA510
U2TA510	
U2TB406	U2TB408,U2TG406,U2TG408
U2TB408	U2TB410,U2TG408,U2TG410
U2TB410	U2TG410
U2TD410	
U2TD420	
U2TD430	
U2TG406	U2TG408
U2TG408	U2TG410
U2TG410	
V415	SEE 2N5762
V417	SEE 2N5761
V575	2SC1039,2SC1673,*276-2017
V643	2SC1038,2SC1039,2SC1673,*276-2017
WEP2	REFER TO SECTION 2
WEP3	REFER TO SECTION 2
WEP10	REFER TO SECTION 2
WEP50	REFER TO SECTION 2
WEP51	REFER TO SECTION 2
WEP52	REFER TO SECTION 2
WEP53	REFER TO SECTION 2
WEP54	REFER TO SECTION 2
WEP55	REFER TO SECTION 2
WEP56	REFER TO SECTION 2
WEP57	REFER TO SECTION 2
WEP67	REFER TO SECTION 2
WEP187	REFER TO SECTION 2
WEP224	REFER TO SECTION 2
WEP230	REFER TO SECTION 2
WEP230MP	M.P.WEP230
WEP231	REFER TO SECTION 2
WEP232	REFER TO SECTION 2
WEP233	REFER TO SECTION 2
WEP235	REFER TO SECTION 2
WEP238	REFER TO SECTION 2
WEP240	REFER TO SECTION 2
WEP241	REFER TO SECTION 2
WEP242	REFER TO SECTION 2
WEP243	REFER TO SECTION 2
WEP244	REFER TO SECTION 2
WEP245	REFER TO SECTION 2
WEP246	REFER TO SECTION 2
WEP247	REFER TO SECTION 2
WEP247MP	M.P.WEP247
WEP250	REFER TO SECTION 2
WEP253	REFER TO SECTION 2
WEP254	REFER TO SECTION 2

To Replace	Substitute This Type
WEP367	REFER TO SECTION 2
WEP371	REFER TO SECTION 2
WEP372	REFER TO SECTION 2
WEP373	REFER TO SECTION 2
WEP380	REFER TO SECTION 2
WEP394	REFER TO SECTION 2
WEP403	REFER TO SECTION 2
WEP405	REFER TO SECTION 2
WEP454	REFER TO SECTION 2
WEP458	REFER TO SECTION 2
WEP460	REFER TO SECTION 2
WEP474	REFER TO SECTION 2
WEP474MP	M.P.WEP474
WEP481	REFER TO SECTION 2
WEP495	REFER TO SECTION 2
WEP517	REFER TO SECTION 2
WEP535	REFER TO SECTION 2
WEP536	REFER TO SECTION 2
WEP537	REFER TO SECTION 2
WEP538	REFER TO SECTION 2
WEP624	REFER TO SECTION 2
WEP628	REFER TO SECTION 2
WEP628MP	M.P.WEP628
WEP630	REFER TO SECTION 2
WEP631	REFER TO SECTION 2
WEP632	REFER TO SECTION 2
WEP635	REFER TO SECTION 2
WEP637	REFER TO SECTION 2
WEP641	REFER TO SECTION 2
WEP641A	REFER TO SECTION 2
WEP641B	REFER TO SECTION 2
WEP642	REFER TO SECTION 2
WEP642MP	M.P.WEP642
WEP643	REFER TO SECTION 2
WEP700	REFER TO SECTION 2
WEP701	REFER TO SECTION 2
WEP703	REFER TO SECTION 2
WEP704	REFER TO SECTION 2
WEP707	REFER TO SECTION 2
WEP709	REFER TO SECTION 2
WEP710	REFER TO SECTION 2
WEP712	REFER TO SECTION 2
WEP715	REFER TO SECTION 2
WEP716	REFER TO SECTION 2
WEP717	REFER TO SECTION 2
WEP719	REFER TO SECTION 2
WEP720	REFER TO SECTION 2
WEP723	REFER TO SECTION 2
WEP724	REFER TO SECTION 2
WEP728	REFER TO SECTION 2
WEP729	REFER TO SECTION 2
WEP735	REFER TO SECTION 2
WEP735A	REFER TO SECTION 2
WEP736	REFER TO SECTION 2
WEP740	REFER TO SECTION 2
WEP740A	REFER TO SECTION 2
WEP740B	REFER TO SECTION 2
WEP772	REFER TO SECTION 2
WEP773	REFER TO SECTION 2
WEP774	REFER TO SECTION 2
WEP777	REFER TO SECTION 2

To Replace	Substitute This Type
WEP778	REFER TO SECTION 2
WEP784	REFER TO SECTION 2
WEP799	REFER TO SECTION 2
WEP828	REFER TO SECTION 2
WEP829	REFER TO SECTION 2
WEP838	REFER TO SECTION 2
WEP900	REFER TO SECTION 2
WEP901	REFER TO SECTION 2
WEP906	REFER TO SECTION 2
WEP907	REFER TO SECTION 2
WEP908	REFER TO SECTION 2
WEP909	REFER TO SECTION 2
WEP910	REFER TO SECTION 2
WEP911	REFER TO SECTION 2
WEP912	REFER TO SECTION 2
WEP913	REFER TO SECTION 2
WEP914	REFER TO SECTION 2
WEP915	REFER TO SECTION 2
WEP917	REFER TO SECTION 2
WEP918	REFER TO SECTION 2
WEP921	REFER TO SECTION 2
WEP922	REFER TO SECTION 2
WEP923	REFER TO SECTION 2
WEP924	REFER TO SECTION 2
WEP956	REFER TO SECTION 2
WEP968	REFER TO SECTION 2
WEP1014	REFER TO SECTION 2
WEP1018	REFER TO SECTION 2
WEP1096	REFER TO SECTION 2
WEP1237	REFER TO SECTION 2
WEP1239	REFER TO SECTION 2
WEP1306	REFER TO SECTION 2
WEP1307	REFER TO SECTION 2
WEP1717	REFER TO SECTION 2
WEP1860	REFER TO SECTION 2
WEP1945	REFER TO SECTION 2
WEPG6001	REFER TO SECTION 2
WEPS3001	REFER TO SECTION 2
WEPS3002	REFER TO SECTION 2
WEPS3003	REFER TO SECTION 2
WEPS3020	REFER TO SECTION 2
WEPS3021	REFER TO SECTION 2
WEPS3023	REFER TO SECTION 2
WEPS3027	REFER TO SECTION 2
WEPS3031	REFER TO SECTION 2
WEPS5003	REFER TO SECTION 2
WEPS5004	REFER TO SECTION 2
WEPS5005	REFER TO SECTION 2
WEPS5007	REFER TO SECTION 2
WEPS7000	REFER TO SECTION 2
WEPS7001	REFER TO SECTION 2
WEPS9100	REFER TO SECTION 2
WX118UA	
WX118UB	
WX118UC	
WX118XA	
WX118XB	
WX118XC	
XA101	2N269,2N404,2N579,2N580,2N580+,2N1017,2N1313,2SA205,2SA206,2SA406, SF.T227,*PTC135,*GE-50,*TR-17,*HEPG0003,*SK3005,*HF35,*ECG160,*WEP637, *AC188/01

To Replace	Substitute This Type
XA102	2N414,2N582,2N584,2N1171,2N1307,2N1309A,2N2953,2SA207,2SA210,2SA407, *DS25,*GE-50,*TR-17,*HEPG0003,*SK3005,*HF35,*ECG160,*WEP637
XA111	2N269,2N404,2N579,2N580,2N580+,2N1017,2N1313,2SA205,2SA206,2SA406, SF.T227,*PTC135,*GE-50,*TR-17,*HEPG0003,*SK3005,*HF35,*ECG160,*WEP637, *AC188/01
XA112	2N414,2N582,2N584,2N1171,2N1307,2N1309A,2N2953,2SA207,2SA210,2SA407, *GE-50,*TR-17,*HEPG0003,*SK3005,*HF35,*ECG160,*WEP637
XA123	2N2496,2N3783,2N3784,2SA234,2SA235,2SA246,A1383,AFZ12,OC170,*PTC107, *GE-50,*TR-17,*HEPG0003,*SK3007,*HF35,*ECG160,*WEP637,*AF125
XA124	2N2496,2N3783,2N3784,2SA234,2SA235,2SA246,A1383,AFZ12,OC170,*PTC107, *GE-50,*TR-17,*HEPG0003,*SK3007,*HF35,*ECG160,*WEP637,*AF125
XA126	2N3783,2N3784,*PTC107,*GE-50,*TR-17,*HEPG0003,*SK3007,*ECG160,*WEP637, *AF125
XA131	2N1023,2N2588,2N3783,2N3784,*PTC107,*GE-51,*TR-17,*HEPG0003,*SK3006, *ECG160,*WEP637
XA141	2N838,*PTC107,*TR-17,*HEPG0001,*ECG160,*WEP637
XA142	2N838,*PTC107,*TR-17,*HEPG0008,*ECG160,*WEP637
XA143	2N838,*PTC107,*TR-17,*HEPG0008,*ECG160,*WEP637
XB102	2N404A,2N465,2N573,2N1056,2N1191,2N1446,2SB397,SF.T226,*PTC135,*GE-53, *TR-85,*HEPG0005,*SK3004,*AT20M,*ECG102A,*WEP250,*276-2005,*RT-121
XB112	2N404A,2N465,2N573,2N1056,2N1191,2N1446,2SB397,SF.T226,*PTC135,*GE-53, *TR-85,*HEPG0005,*SK3004,*AT20M,*ECG102A,*WEP250,*276-2005,*RT-121
XB113	2N217+,2N382,2N422A,2N467,2N1008A,2N1125,2N1192,2N1309A,2N1348,2N1349, 2N1350,2N1351,2N1375,2N1377,2N1448,2SB77A,2SB89A,ACY23,ACY32,*PTC135, *GE-53,*TR-85,*HEPG0005,*SK3004,*AT20M,*ECG102A,*WEP250,*276-2005, *RT-121
XB401	2SC302-M,*GE-28,*TR-76,*ECG186,*WEP245
XB404	*GE-66,*ECG152,*WEP245
XB408	*GE-66,*ECG152,*WEP245,*276-2020
XB433	41027†
XB434	BLY38,XB435,XB436,ZT5590
XB435	XB436,ZT5590
XB436	XB437,ZT5590
XB437	
XB473	2N5947,MT5763
XB474	XB475
XB475	XB476
XB476	*GE-66,*ECG152,*WEP245
XC121	2N1008A,2N1125,2N1375,2N1377,*PTC135,*GE-53,*TR-85,*HEPG0005,*SK3004, *AT20M,*ECG102A,*WEP250,*276-2005,*RT-121
XC131	M.P.XC121
XC141	2N380,2N1168,2N2143,2N2143A,2N2144,2N2144A,*PTC105A,*GE-16,*TR-01, *HEPG6003,*SK3009,*ECG121,*WEP232,*RT-127
XC142	2N380,2N1159,2N2144,2N2144A,2N2145,2N2145A,*PTC105A,*GE-16,*TR-01, *HEPG6005,*SK3009,*ECG121,*WEP232,*RT-127
XC155	*GE-25,*TR-01,*HEPG6003,*SK3009,*PT40,*ECG104,*WEP230,*276-2006,*RT-124
XC156	*DS-503,*GE-25,*TR-01,*HEPG6003,*SK3014,*PT40,*ECG104,*WEP230,*276-2006, *RT-124
XDAR10020	KSP1222,KSP1223,KSP1224,KSP1225,XDAR10025,XGSR15020
XDAR10025	KSP1223,KSP1224,KSP1225,XDAR10030
XDAR10030	KSP1225,XDAR10035
XDAR10035	
XGS7001	
XGS7002	
XGSA3025	XGSA3030,XGSA5025,XGSA5030
XGSA3030	XGSA3035,XGSA3040,XGSA5030,XGSA5035
XGSA3035	XGSA3040,XGSA5035
XGSA3040	
XGSA5025	XGSA5030
XGSA5030	XGSA5035
XGSA5035	

To Replace	Substitute This Type
XGSQ3025	XGSQ3030,XGSQ5025,XGSQ5030
XGSQ3030	XGSQ3035,XGSQ3040,XGSQ5030,XGSQ5035,XGSQ5040
XGSQ3035	XGSQ3040,XGSQ5035,XGSQ5040
XGSQ3040	XGSQ5040
XGSQ5025	XGSQ5030
XGSQ5030	XGSQ5035,XGSQ5040
XGSQ5035	XGSQ5040
XGSQ5040	
XGSQ7530	XGSQ7535,XGSQ7540,XGSR7530,XGSR7535,XGSR7540
XGSQ7535	XGSQ7540,XGSR7535,XGSR7540
XGSQ7540	XGSR7540
XGSR3020	SDT1060,XGSR3025,XGSR5025
XGSR3025	XGSR3030,XGSR5025,XGSR5030
XGSR3030	SDT1061,XGSR3035,XGSR3040,XGSR5030,XGSR5035,XGSR5040
XGSR3035	SDT1061,SDT1062,XGSR3040,XGSR5035,XGSR5040
XGSR3040	SDT1062,XGSR5040
XGSR5025	XGSR5030
XGSR5030	XGSR5035,XGSR5040
XGSR5035	XGSR5040
XGSR5040	
XGSR7530	XGSR7535,XGSR7540
XGSR7535	XGSR7540
XGSR7540	
XGSR10025	KSP1223,KSP1224,KSP1225,XDAR10025,XDAR10030,XGSR10030,XGSR15025, XGSR15030
XGSR10030	2N6579,2N6582,XDAR10030,XDAR10035,XGSR10040,XGSR15030
XGSR10035	2N6579,2N6580,2N6582,2N6583,XDAR10035,XGSR10040,XGSR15035,XGSR15040
XGSR10040	2N6579,2N6580,2N6581,2N6582,2N6583,2N6584
XGSR15020	KSP1222,KSP1223,KSP1224,KSP1225
XGSR15025	KSP1223,KSP1224,KSP1225,XDAR10025,XDAR10030,XGSR15030
XGSR15030	XDAR10030,XDAR10035
XGSR15035	XDAR10035,XGSR15040
XGSR15040	
XGSR50020	
ZDT10	ZDT11,*GE-61,*TR-95,*ECG108,*WEP56,*276-2016,*RT-113
ZDT11	ZDT10,*GE-61,*TR-95,*ECG108,*WEP56,*276-2016,*RT-113
ZDT20	ZDT21,*GE-61,*TR-95,*ECG108,*WEP56,*276-2016,*RT-113
ZDT21	ZDT20,*GE-61,*TR-95,*ECG108,*WEP56,*276-2016,*RT-113
ZDT30	*PTC126,*GE-39,*TR-95,*SK3019,*ECG108,*WEP56,*276-2011,*RT-113
ZDT31	*PTC126,*GE-39,*TR-95,*SK3019,*ECG108,*WEP56,*276-2011,*RT-113
ZDT40	TWO ZT82 TRANSISTORS
ZDT41	TWO ZT84 TRANSISTORS
ZDT42	ZDT44
ZDT44	ZDT42
ZDT45	
ZEN100	REFER TO SECTION 2
ZEN101	REFER TO SECTION 2
ZEN102	REFER TO SECTION 2
ZEN103	REFER TO SECTION 2
ZEN104	REFER TO SECTION 2
ZEN105	REFER TO SECTION 2
ZEN106	REFER TO SECTION 2
ZEN107	REFER TO SECTION 2
ZEN108	REFER TO SECTION 2
ZEN109	REFER TO SECTION 2
ZEN110	REFER TO SECTION 2
ZEN111	REFER TO SECTION 2
ZEN112	REFER TO SECTION 2
ZEN113	REFER TO SECTION 2
ZEN114	REFER TO SECTION 2
ZEN115	REFER TO SECTION 2

To Replace	Substitute This Type
ZEN116	REFER TO SECTION 2
ZEN117	REFER TO SECTION 2
ZEN118	REFER TO SECTION 2
ZEN119	REFER TO SECTION 2
ZEN120	REFER TO SECTION 2
ZEN121	REFER TO SECTION 2
ZEN122	REFER TO SECTION 2
ZEN125	REFER TO SECTION 2
ZEN126	REFER TO SECTION 2
ZEN127	REFER TO SECTION 2
ZEN128	REFER TO SECTION 2
ZEN200	REFER TO SECTION 2
ZEN201	REFER TO SECTION 2
ZEN202	REFER TO SECTION 2
ZEN203	REFER TO SECTION 2
ZEN204	REFER TO SECTION 2
ZEN205	REFER TO SECTION 2
ZEN206	REFER TO SECTION 2
ZEN207	REFER TO SECTION 2
ZEN208	REFER TO SECTION 2
ZEN209	REFER TO SECTION 2
ZEN210	REFER TO SECTION 2
ZEN211	REFER TO SECTION 2
ZEN300	REFER TO SECTION 2
ZEN301	REFER TO SECTION 2
ZEN302	REFER TO SECTION 2
ZEN303	REFER TO SECTION 2
ZEN304	REFER TO SECTION 2
ZEN305	REFER TO SECTION 2
ZEN306	REFER TO SECTION 2
ZEN307	REFER TO SECTION 2
ZEN308	REFER TO SECTION 2
ZEN309	REFER TO SECTION 2
ZEN310	REFER TO SECTION 2
ZEN311	REFER TO SECTION 2
ZEN312	REFER TO SECTION 2
ZEN313	REFER TO SECTION 2
ZEN314	REFER TO SECTION 2
ZEN315	REFER TO SECTION 2
ZEN325	REFER TO SECTION 2
ZEN326	REFER TO SECTION 2
ZEN327	REFER TO SECTION 2
ZEN328	REFER TO SECTION 2
ZEN329	REFER TO SECTION 2
ZEN330	REFER TO SECTION 2
ZEN331	REFER TO SECTION 2
ZT20	*PTC126,*GE-62,*TR-21,*HEPS0014,*SK3122,*ECG123A,*WEP736,*276-2016
ZT21	*PTC126,*GE-62,*TR-21,*HEPS0014,*SK3122,*ECG123A,*WEP736,*276-2016
ZT22	2S102,*PTC123,*GE-62,*TR-21,*HEPS0014,*SK3122,*ECG123A,*WEP736,*276-2009
ZT23	2N3946,2N4962,2N5144,2N5380,2N6538,2S103,A5T3903,MM3903,TN-3903,*PTC123, *GE-62,*TR-21,*HEPS0014,*SK3122,*ECG123A,*WEP736,*276-2009
ZT24	2N3947,2N5381,2S104,2SC1364,2SC1999,A5T3904,BC547A,BCY59A,MM3904, TN-3904,*PTC123,*GE-62,*TR-21,*HEPS0014,*SK3122,*ECG123A,*WEP736, *276-2009
ZT40	2N2571,2N2572,*PTC126,*GE-61,*TR-21,*HEPS0011,*SK3018,*ECG123A,*WEP736, *276-2016
ZT41	2N4418,*PTC126,*GE-61,*TR-21,*HEPS0011,*SK3018,*ECG123A,*WEP736, *276-2016
ZT42	2S102,*PTC123,*GE-212,*TR-21,*HEPS0011,*SK3018,*ECG123A,*WEP736, *276-2009
ZT43	2N3903,2N3946,2N4962,2N5144,2N5380,2N6538,2S103,A5T3903,MM3903,TN-3903,

To Replace	Substitute This Type
(ZT43)	*PTC123,*GE-212,*TR-21,*HEPS0011,*SK3018,*ECG123A,*WEP736,*276-2009
ZT44	2N3904,2N3947,2N5381,2S104,2SC943,2SC1364,2SC1999,A5T3904,BC107A, BC167A,BC171A,BC182A,BC237A,BC547A,BCY59A,MM3904,TN-3904,*PTC123, *GE-212,*TR-21,*HEPS0011,*SK3018,*ECG123A,*WEP735,*276-2009
ZT80	2N2501,ZT87,ZT2369A,ZT2938,*PTC136,*GE-210,*TR-21,*HEPS0011,*SK3122, *ECG123A,*WEP736,*276-2016
ZT81	2N916,2N916A,2N2539,2N3643,2N4013,2N4951,2N4962,2N5368,TN-3903,ZT82,ZT84, *PTC136,*GE-210,*TR-21,*HEPS0011,*SK3122,*ECG123A,*WEP736,*276-2016
ZT82	2N2222,2N2540,2N3116,2N3302,2N4401,2N4952,2N5107,2N5369,2N6010,A5T2222, EN2219,FT3643,GI-3643,MPS6531,SE6020,SE6020A,TIS109,TIS111,TN-3904, *PTC136,*GE-210,*TR-25,*HEPS0025,*SK3122,*ECG149,*276-2016
ZT83	2N2221,2N2221A,2N2539,2N2845,2N2847,2N3115,2N3301,2N4962,2N4963,2N5581, BSW84,FT3641,FT3642,FT3722,GI-3641,GI-3642,MPS6530,TN-3903,*GE-210,*TR-21, *HEPS0011,*SK3122,*ECG123A,*WEP736,*276-2009
ZT84	2N915,2N2222,2N2222A,2N2540,2N3116,2N3643,2N4401,2N4951,2N4952,2N5107, 2N5368,2N5369,2N5582,A5T2222,BSW85,EN2219,GI-3643,MPS6531,SE6020, SE6020A,SE6021,SE6021A,TIS111,TN-3904,ZT89,*GE-210,*TR-21,*HEPS0011, *SK3122,*ECG123A,*WEP736,*276-2009
ZT86	SE8012,*GE-17,*TR-21,*SK3122,*ECG123A,*WEP736
ZT87	2N3227,2N5419,BSW83,*PTC136,*GE-210,*TR-21,*HEPS0011,*SK3122,*ECG123A, *WEP736,*276-2016
ZT88	*GE-17,*TR-21,*SK3122,*ECG123A,*WEP736
ZT89	2N2222A,2N5582,BSW85,SE6021,SE6021A,*GE-17,*TR-21,*SK3122,*ECG123A, *WEP736
ZT90	2N1711A,*PTC144,*TR-87,*SK3024,*ECG128,*WEP243
ZT91	2N2102,2N2102A,2N2405,ZT92,ZT93,ZT2102
ZT92	2N1711B,2N2405
ZT93	2N2102,2N2102A,2N2405,ZT91,ZT92,ZT2102,*PTC144,*TR-87,*SK3024,*ECG128
ZT94	MPS3725,*PTC144,*GE-215,*TR-87,*SK3024,*ECG128,*WEP243
ZT95	2N1711A,2N2270,ZT90,ZT2270
ZT110	ZT117,*PTC136,*GE-210,*TR-25,*SK3099,*ECG151,*276-2016
ZT111	2N4400,MPS3693,MPS-A05,TIS110,ZT112,ZT114,*PTC136,*GE-210,*TR-21, *HEPS0011,*SK3122,*ECG123A,*WEP736,*276-2016
ZT112	2N4401,A5T2222,MPS3694,MPS6531,TIS109,TIS111,*PTC136,*GE-210,*TR-21, *HEPS0011,*SK3122,*ECG123A,*WEP736,*276-2016
ZT113	2N4400,2N5581,MPS6530,TIS110,*GE-210,*TR-21,*HEPS0011,*SK3122,*ECG123A, *WEP736,*276-2009
ZT114	2N4401,2N5582,A5T2222,MPS6531,TIS111,ZT119,*GE-210,*TR-21,*HEPS0011, *SK3122,*ECG123A,*WEP736,*276-2009
ZT116	*GE-17,*TR-21,*ECG123A,*WEP736
ZT117	*PTC136,*GE-210,*TR-21,*HEPS0011,*SK3122,*ECG123A,*WEP736,*276-2016
ZT118	*GE-17,*TR-21,*SK3122,*ECG123A,*WEP736
ZT119	2N5582,*GE-17,*TR-21,*SK3122,*ECG123A,*WEP736
ZT152	*GE-22,*TR-20,*HEPS0013,*SK3114,*ECG159,*WEP242,*276-2022
ZT180	2N869,2N3638A,2N4402,2SA509-O,2SA509G-O,2SA695,2SA719,2SA890,A5T3638A, A5T4402,BSW72,CS9012,GI-3638A,MPS6533,TN-3905,ZT187,ZT280,ZT287, *PTC127,*GE-22,*TR-20,*HEPS0013,*SK3114,*ECG159,*WEP242,*276-2034
ZT181	2N2906,2N2906A,2N3073,2N3135,2N3485,2N3485A,2N3644,2N3645,2N3672,2N3673, 2N4026,2N5372,2SA696,2SA697,2SA720,2SA891,BCW35,BCW37,FT3644,FT3645, TZ551,ZT182,ZT184,ZT281,ZT282,ZT284,*PTC127,*GE-82,*TR-20,*HEPS0019, *SK3114,*ECG159,*WEP242,*276-2023
ZT182	2N2907,2N2907A,2N3136,2N3486,2N3486A,2N3504,2N3505,2N3644,2N3645,2N3672, 2N3673,2N4028,2N5373,2N5819,2N6005,2SA720,2SA891,A5T2907,A5T3504, A5T3505,A5T3644,A5T3645,BCW35,BCW37,D29E6,EN2905,EN3502,FT3644, FT3645,GI-3644,TIS112,TZ552,ZT282,*PTC127,*GE-82,*TR-20,*HEPS0013, *SK3114,*ECG159,*WEP242,*276-2023
ZT183	2N2906,2N2906A,2N3073,2N3121,2N3135,2N3485,2N3485A,2N4026,2N5372,2N5763, 2SA696,2SA697,TZ551,ZT181,ZT281,ZT283,*PTC127,*GE-82,*TR-20,*HEPS0013, *SK3114,*ECG159,*WEP242,*276-2023
ZT184	2N2907,2N2907A,2N3136,2N3486,2N3486A,2N3504,2N3505,2N3644,2N3645,2N3672, 2N3673,2N4028,2N5373,2N6005,2SA696,2SA697,2SA720,2SA891,A5T2907,

TRANSISTOR SUBSTITUTES

To Replace	Substitute This Type
(ZT184)	A5T3504,A5T3505,A5T3644,A5T3645,BCW35,BCW37,EN2905,EN3502,FT3644, FT3645,GI-3644,TIS112,TZ552,ZT181,ZT182,ZT281,ZT282,ZT284,*PTC127,*GE-82, *TR-20,*HEPS0013,*SK3114,*ECG159,*WEP242,*276-2023
ZT187	2N3638A,2N4403,2N5375,2N5999,2N6001,2SA509-Y,2SA509G-Y,2SA719,2SA890, A5T4403,BSW73,CS9012,GI-3638A,MPS3638A,MPS6534,TN-3906,ZT287,*PTC127, *GE-22,*TR-20,*HEPS0013,*SK3114,*ECG159,*WEP242,*276-2034
ZT189	2N4029,2SA817,2SA850,BSW75,*PTC127,*GE-82,*ECG159
ZT202P	2N1839,ZT402P,*PTC123
ZT203P	2N2318,2N2319,2N2320,2N2710,2N3261,2N3510,2N3648,2N5236,2SC343,2SC576, 2SC874,40405,MPS6532,MPS6540,MPSH37,ZT403P,ZT2368,*PTC123
ZT204P	2N916,2N916A,2N1838,2N2237,2N2331,2N2501,2N3211,2N3705,2N4013,2N5144, 2N5450,2N5810,2N6427,2SC509-O,2SC509G-O,2SC577,2SC1210,A5T5450, A8T3705,BSW82,D33D21,GI-3705,MPS2369,MPS3693,MPS3705,PET3705,TN62, ZT81,ZT111,ZT404P,ZT2369,ZT2369A,*PTC123
ZT280	2N4402,A5T3638A,A5T4402,MPS6533,ZT287,*PTC127,*GE-22,*TR-20,*HEPS0013, *SK3114,*ECG159,*WEP242,*276-2034
ZT281	2N3485,2N3485A,2N3673,BCW37,ZT282,ZT284,*PTC127,*GE-82,*TR-20,*HEPS0013, *SK3114,*ECG159,*WEP242,*276-2023
ZT282	2N3486,2N3486A,2N3673,A5T2907,A5T3504,A5T3505,A5T3644,A5T3645,BCW37, TIS112,*PTC127,*GE-82,*TR-20,*HEPS0013,*SK3114,*ECG159,*WEP242, *276-2023
ZT283	2N3485,2N3485A,ZT281,*PTC127,*GE-82,*TR-20,*HEPS0013,*SK3114,*ECG159, *WEP242,*276-2023
ZT284	2N3486,2N3486A,2N3673,A5T2907,A5T3504,A5T3505,A5T3644,A5T3645,BCW37, TIS112,ZT281,ZT282,*PTC127,*GE-82,*TR-20,*HEPS0013,*SK3114,*ECG159, *WEP242,*276-2023
ZT287	2N4403,A5T4403,MPS3638A,MPS6534,*PTC127,*GE-22,*TR-20,*HEPS0013, *SK3114,*ECG159,*WEP242,*276-2034
ZT402P	*PTC123
ZT403P	2N2318,2N2319,2N2710,2N3261,2N3510,2N3648,2SC576,40405,MPS6532,MPS6540, MPSH37,ZT2368,*PTC123
ZT404P	2N916,2N916A,2N2331,2N2501,2N3211,2N3705,2N4013,2N5144,2N5450,2N5810, 2N6427,2SC509-O,2SC509G-O,2SC577,2SC1210,A5T5450,A8T3705,BSW82, D33D21,GI-3705,MPS2369,MPS3693,MPS3705,PET3705,TN62,ZT81,ZT111,ZT2369, ZT2369A,*PTC123
ZT600	*GE-47,*ECG186,*WEP245,*276-2038
ZT696	2N1958,2N1958A,2N2217,2N2477,2N3252,2N3253,2N3444,2N5188,2N6376,2SC97, 2SC109A-R,2SC189,2SC255,2SC344,2SC345,2SC503-O,2SC580,BSX49,BSX60, *PTC123,*GE-81,*TR-21,*HEPS0014,*SK3024,*ECG123A,*WEP736,*276-2030
ZT697	2N1837,2N1837A,2N1959A,2N2218,2N2218A,2N2224,2N2270,2N2410,2N2479, 2N2537,2N2846,2N2848,2N2958,2N3015,2N3299,2N3725,2N3725A,2N4047,2N4960, 2N4961,2SC97A,2SC109A-O,2SC190,2SC307,2SC479,2SC503-Y,2SC560,2SC1218, 40635,BC341-10,BSY51,BSY53,BSY83,TIS110,TIS135,TIS136,TN53,*PTC123, *GE-81,*TR-21,*HEPS0014,*SK3024,*ECG123A,*WEP736,*276-2030
ZT708	2N2501,2N3946,2N4962,2N5380,A5T3903,MM3903,TN-3903,*PTC136,*GE-17,*TR-21, *HEPS0011,*SK3122,*ECG123A,*WEP736,*276-2009
ZT1479	ZT1481,*GE-18,*TR-87,*SK3024,*ECG128,*WEP243,*276-2020
ZT1480	ZT1482,*276-2020
ZT1481	2N1481,*GE-18,*TR-87,*HEPS5014,*SK3024,*ECG128,*WEP243,*276-2020
ZT1482	2N1482,*276-2020
ZT1483	ZT1485,*GE-66,*ECG152,*WEP245,*276-2020
ZT1484	ZT1486,*GE-66,*ECG152,*WEP245,*276-2020
ZT1485	*GE-66,*ECG152,*WEP245,*276-2020
ZT1486	*GE-66,*ECG152,*WEP245,*276-2020
ZT1487	2N1487,2N1489,2SD15,2SD163,BDY23A,HST9201,HST9205,HST9206,HST9801, HST9802,ZT1489,*PTC173,*GE-77,*TR-59,*HEPS7002,*SK3027,*ECG130,*WEP247
ZT1488	2N1488,2N1490,2N3446,2SD16,2SD118-R,2SD119-R,2SD161,2SD164,2SD177, BDY24A,HST9202,HST9203,HST9207,HST9208,HST9803,HST9804,ZT1490, *PTC173,*GE-75,*TR-59,*SK3027,*ECG130,*WEP247
ZT1489	2N1489,2SD15,180T2,BDY23B,*PTC173,*GE-77,*TR-59,*HEPS7002,*SK3027, *ECG130,*WEP247

To Replace	Substitute This Type
ZT1490	2N1490,2N3446,2N3448,2SD16,2SD110-R,2SD111-R,2SD118-R,2SD119-R,181T2, BDY24B,*PTC173,*GE-75,*TR-59,*SK3027,*ECG130,*WEP247
ZT1613	2N1889,2N2193,2N2193A,2N2193B,2N3107,2N3109,BC140-10,BC141-10,BSY83, *PTC144,*GE-216,*TR-87,*SK3024,*ECG128,*WEP243,*276-2008
ZT1700	2N1481,2N2297,2N2594,2N3945,2N4238,2N4239,BSX46,SDT4305,SDT4306, SDT4311,SDT4312,ZT1481,*PTC144,*GE-46,*TR-87,*HEPS5014,*SK3024,*ECG128, *WEP243,*276-2020
ZT1701	2N1485,ZT1485,*GE-66,*ECG152,*WEP245,*276-2020
ZT1711	2N1890,*PTC125,*GE-18,*TR-21,*SK3122,*ECG123A,*WEP736,*276-2008
ZT2102	2N2102,2N2102A,2N2405,ZT91,ZT92,ZT93,*PTC144,*GE-28,*TR-76,*ECG102, *276-2008
ZT2205	2N2242,2N2501,2N3013,2N3014,2N3211,2N4123,2N4418,2S512,2SC67,2SC577, 2SC639,2SC735-O,2SC764,A5T4123,BSW82,PT720,ZT2938,*PTC136,*GE-210, *TR-21,*HEPS0011,*SK3122,*ECG123A,*WEP736,*276-2016
ZT2270	2N1711A,2N2270,*GE-63,*TR-76,*ECG192,*WEP243,*276-2008
ZT2368	2N2710,2SC576,*PTC136,*GE-20,*TR-95,*HEPS0016,*SK3039,*ECG108,*WEP56, *276-2009
ZT2369	2SC577,ZT2369A,*GE-20,*TR-95,*HEPS0016,*SK3039,*ECG108,*WEP56,*276-2009
ZT2369A	*GE-20,*TR-95,*SK3039,*ECG108,*WEP56,*276-2009
ZT2857	2N2808,2N2808A,2N2809A,2N2857,2N3570,2N4252,2N4253,2N5053,2N5054,2N6304, 2SC567,2SC583,2SC1747,A485,MT1061A,*PTC133,*GE-86,*TR-95,*SK3019, *ECG108,*WEP56
ZT2938	*GE-20,*TR-95,*HEPS0016,*SK3039,*ECG108,*WEP56,*276-2016
ZT5589	ZT5590
ZT5590	
ZT5591	
ZTX107	ZTX107B,*PTC133,*GE-212,*SK3245,*ECG199,*WEP736
ZTX107A (RED)	ZTX304,ZTX4401,*PTC121
ZTX107B (GRN)	ZTX107
ZTX108	ZTX107,ZTX107B,ZTX108B,ZTX109B,ZTX223,ZTX223B,ZTX237B,ZTX238B,ZTX238C, ZTX239B,ZTX239C,*GE-212,*SK3245,*ECG199,*WEP736,*276-2016
ZTX108A (RED)	ZTX107A,ZTX223,ZTX223A,ZTX237A,ZTX238A,ZTX303,ZTX3704,ZTX4401
ZTX108B (GRN)	ZTX107,ZTX107B,ZTX108,ZTX109,ZTX109B,ZTX223B,ZTX237B,ZTX238B,ZTX238C, ZTX239B,ZTX239C
ZTX108C (BLU)	ZTX109,ZTX109C
ZTX109	ZTX108C,ZTX109C,ZTX238C,ZTX239C,*GE-212,*SK3245,*ECG199,*WEP736, *276-2016
ZTX109B (GRN)	ZTX107,ZTX107B,ZTX108,ZTX108B,ZTX109,ZTX223B,ZTX237B,ZTX238B,ZTX238C, ZTX239B,ZTX239C
ZTX109C (BLU)	ZTX108C,ZTX109
ZTX114	ZTX239B,ZTX239C,*276-2016
ZTX196	
ZTX197	
ZTX212	ZTX212A,ZTX212B,ZTX3904
ZTX212A (RED)	ZTX212,ZTX212B,ZTX3904,*PTC103
ZTX212B (GRN)	
ZTX213	ZTX212B,ZTX213B,ZTX213C,ZTX214,ZTX214B,ZTX214C,ZTX537B,ZTX537C
ZTX213A (RED)	ZTX212,ZTX212A,ZTX212B,ZTX213B,ZTX214B,ZTX537A,ZTX537B,ZTX3904,*PTC103
ZTX213B (GRN)	ZTX212B,ZTX213,ZTX213C,ZTX214,ZTX214B,ZTX214C,ZTX537B,ZTX537C
ZTX213C (BLU)	ZTX214C,ZTX537C
ZTX214	ZTX212B,ZTX213,ZTX213B,ZTX213C,ZTX214B,ZTX214C,ZTX537C
ZTX214B (GRN)	ZTX212B,ZTX213,ZTX213B,ZTX213C,ZTX214,ZTX214C,ZTX537B,ZTX537C
ZTX214C (BLU)	ZTX213C,ZTX537C
ZTX223	ZTX223B,ZTX337B,ZTX337C
ZTX223A	BFS59,BFS60,ZTX223,ZTX223B,ZTX337A,ZTX337B,ZTX3704,ZTX4401
ZTX223B	ZTX223,ZTX337B,ZTX337C
ZTX237A	ZTX4401
ZTX237B	
ZTX238A	ZTX237A,ZTX4401
ZTX238B	ZTX237B,ZTX239B
ZTX238C	ZTX239C

To Replace	Substitute This Type
ZTX239B	ZTX237B,ZTX238B
ZTX239C	ZTX238C
ZTX300	ZTX301,ZTX302,ZTX338A,ZTX338B,*GE-210,*HEPS0014,*ECG123A,*WEP736, *276-2016
ZTX301	ZTX302,ZTX303,ZTX337A,ZTX337B,*GE-210,*SK3244,*ECG123A,*WEP736, *276-2016
ZTX302	ZTX301,ZTX303,ZTX337A,ZTX337B,*GE-210,*HEPS0015,*SK3244,*ECG128, *WEP243,*276-2016
ZTX303	BFS59,BFS60,ZTX337A,ZTX337B,ZTX4401,*GE-210,*SK3244,*ECG192,*WEP243
ZTX304	*GE-81,*SK3244,*ECG123A
ZTX310	*PTC115,*GE-210,*ECG123A,*WEP736,*276-2016
ZTX311	ZTX338A,*GE-210,*ECG123A,*WEP736,*276-2016
ZTX312	*GE-20,*ECG123A,*WEP736,*276-2016
ZTX313	ZTX314,*276-2038
ZTX314	ZTX313,*276-2038
ZTX320	ZTX321,*PTC132,*GE-61,*ECG108,*WEP56,*276-2015
ZTX321	ZTX320,*PTC132,*GE-61,*ECG108,*WEP56,*276-2015
ZTX325	*PTC132,*GE-86,*ECG316
ZTX326	ZTX325,*PTC132,*GE-86,*ECG316
ZTX327	
ZTX330	ZTX114,ZTX302,ZTX337B,ZTX337C,ZTX338B,ZTX338C,ZTX3704,ZTX3706,*GE-210, *SK3244,*ECG123A,*WEP736,*276-2013
ZTX331	ZTX360,ZTX3705,ZTX4400,*PTC170,*GE-210,*SK3244,*ECG128,*WEP243, *276-2013
ZTX337A	ZTX337B
ZTX337B	ZTX337C
ZTX337C	
ZTX338A	ZTX337A,ZTX337B,ZTX338B
ZTX338B	ZTX337B,ZTX337C,ZTX338C
ZTX338C	ZTX337C
ZTX341	ZTX342,*PTC170,*GE-27,*SK3201,*ECG171,*WEPS3021,*276-2012
ZTX342	*PTC170,*GE-27,*SK3201,*ECG171,*WEPS3021,*276-2012
ZTX360	*GE-47,*HEPS3020,*ECG188,*276-2009
ZTX500	ZTX501,ZTX502,ZTX538A,ZTX538B,ZTX4403,*PTC103,*GE-22,*HEPS0012,*ECG129, *WEP242,*276-2034
ZTX501	ZTX502,ZTX503,ZTX537A,ZTX537B,ZTX4403,*PTC103,*GE-82,*HEPS0019,*ECG159, *WEP242,*276-2034
ZTX502	ZTX501,ZTX503,ZTX537A,ZTX537B,ZTX4403,*PTC103,*GE-82,*HEPS0019,*ECG159, *WEP242,*276-2034
ZTX503	BFS96,BFS97,ZTX537A,ZTX537B,*PTC103,*GE-82,*HEPS0019,*ECG159,*WEP242
ZTX504	*PTC103,*GE-82,*HEPS0019,*ECG159
ZTX510	*PTC103,*HEPS0013,*ECG159,*276-2021
ZTX530	ZTX502,ZTX537B,ZTX537C,ZTX538B,ZTX538C,ZTX4403,*GE-82,*HEPS0019, *SK3114,*ECG159,*WEP242,*276-2022
ZTX531	ZTX3703,*PTC103,*HEPS0019,*SK3114,*ECG159,*WEP242,*276-2022
ZTX537A	ZTX537B
ZTX537B	ZTX537C
ZTX537C	
ZTX538A	ZTX537A,ZTX537B,ZTX538B
ZTX538B	ZTX537B,ZTX537C,ZTX538C
ZTX538C	ZTX537C
ZTX541	ZTX542,*PTC171
ZTX542	*PTC171
ZTX3702	BFS96,BFS97,ZTX537A,ZTX537B,ZTX4403,*PTC103
ZTX3703	*PTC103
ZTX3704	BFS59,BFS60,ZTX337A,ZTX337B,ZTX4401
ZTX3705	ZTX337A,ZTX360,ZTX4400
ZTX3706	ZTX337B,ZTX337C
ZTX3707	ZTX3708,ZTX3710
ZTX3708	ZTX3711
ZTX3709	

To Replace	Substitute This Type
ZTX3710	ZTX3707,ZTX3708
ZTX3711	ZTX3708
ZTX3903	*PTC103
ZTX3904	
ZTX3905	ZTX537A,ZTX3903,*PTC103
ZTX3906	ZTX3904,*PTC103
ZTX4400	
ZTX4401	
ZTX4402	ZTX537A,*PTC103
ZTX4403	ZTX537A,ZTX537B,*PTC103
ZTX4403	*PTC103

SECTION 2

General-Purpose Replacements

This section provides information on the general-purpose replacement transistors shown as substitutes in Section 1. Section 2 is divided into two subsections: (1) Description of Types and (2) Applications. In the first subsection, the left-hand column lists the transistors in numerical and alphabetical order; a second column shows the polarity and material of each transistor. Here the first letter indicates whether it is an npn (N) or a pnp (P) type, and the second letter indicates germanium (G) or silicon (S). The next entry is the abbreviation for the supplier of the transistor. The last entry indicates typical applications for the transistor. In general, the application requiring the highest frequency response is listed, since this can be a limiting factor. A key to the abbreviations follows.

The second listing is alphabetical by application. If there is no suggested substitute in Section 1 of this book, it may be possible to select a replacement from this section on the basis of how the transistor is used in the circuit.

KEY TO MANUFACTURERS

ARC—Radio Shack (Archer), Div. of Tandy Corp.
DEL—Delco Radio Div., General Motors Corp.
GEC—General Electric Co., Semiconductor Products Dept.
INR—International Rectifier
MAL—P. R. Mallory & Co., Inc.
MOT—Motorola Semiconductor Products, Inc.
PHI—Philips Electronics Industries Ltd.
RAY—Raytheon Co., Semiconductor Div.
RCA—Radio Corporation of America, Electronic Components & Devices
SEM—Semitronics Corp.
SPR—Sprague Products Co.
SYL—Sylvania Electric Products, Inc. Semiconductor Div.
WTV—Workman Electronic Products, Inc.
ZEN—Zenith Components & Accessories Div.

KEY TO ABBREVIATIONS

AA—Audio Amplifier
AD—Audio Driver
AM—Used in AM Receivers
AMP—Used in Amplifier Applications
AO—Audio Output
AOD—Darlington Output Transistor for Audio Applications
APA—Audio Power Amplifier
APD—Darlington Audio Preamplifier
APO—Audio Power Output
AR—Used in Automobile Radios, Stereo, and/or Tape Units
AUD—Audio Frequency, Audio Applications
BHO—Black & White TV Horizontal Output Applications
BVO—Black & White TV Vertical Output Applications
CB—Used in CB Applications
CHO—Color TV Horizontal Output Applications
CMC—Used in CATV/MATV/COMM Applications
CMO—Converter, Mixer, Oscillator
CO—Used in Computer Applications
COL—Color Amplifier
CON—Converter, Converter Applications
CTR—Used in Cassette Tape Recorder
CVO—Color TV Vertical Output Applications
DAR—Darlington Transistor
DE—Dual-Emitter Transistor
DF—Drift-Field Transistor
DR—Used in Driver Applications
DUO—Two Transistors in One Package
EUR—General-Purpose Device for European Market
FM—Used in FM Receivers
FX—Frequency Multiplier
G—Germanium Transistor
GEN—General Purpose
HC—High Current
HF—High Frequency, High-Frequency Amplifier
HG—High Gain
HIFI—Used in Hi-Fi Equipment
HOR—Horizontal, Horizontal Amplifier
HORD—Horizontal Deflection
HORR—Horizontal Driver
HP—High Power
HPO—High-Power Output
HS—High Speed
HV—High Voltage
HVA—High-Voltage Amplifier
HVP—High-Voltage Power Transistor
HVR—High-Voltage Regulator
IF—IF Amplifier
IG—Designed for or Used in Ignition Systems/Applications
IND—Used in Industrial Applications
LF—Low Frequency, Low-Frequency Amplifier
LM—Used in Land-Mobile Radio Applications (Low Supply Voltage Use)
LN—Low Noise
LO—Used in 120V Line-Operated Equipment
LP—Low Power
LPO—Low-Power Output
MB—Used in Multiple Band Applications (AM/FM/SW/TV)
MCD—Memory Core Driver Applications
MF—Medium Frequency, Medium-Frequency Amplifier
MIX—Mixer
MLT—Multiple Transistors; Usually Two Connected in Darlington Configuration
MP—Medium Power
MPA—Medium-Power Amplifier
MPD—Medium-Power Driver
MW—Microwave, Microwave Applications
N—NPN Transistor
NIT—Flourescent Numerical Indicator Tube Driver Applications
NTD—Nixie Tube Driver Applications
OSC—Oscillator
OUT—Output
P—PNP Transistor
PRE—Preamplifier
PWR—Power, Power Amplifier
QUAD—Four Transistors in One Package
REG—Regulator Applications
RF—Radio Frequency, RF Amplifier
RFA—RF Amplifier, RF Power Amplifier
RFD—RF Driver
RPO—RF Power Output
S—Silicon Transistor

SCH—Stereo, Communications, and Hi-Fi Applications
SPL PUR—Special Purpose
SS—Small Signal
SSW—Saturated Switch
SW—General-Purpose Switch, Switching Applications
TRO—Transceiver Output Use
TV—Used for Television Circuits
UHF—UHF Amplifier, UHF Applications
UHF OSC—UHF Oscillator
UNI—Universal Transistor for Use in Imported Entertainment-Type Equipment
VA—Voltage Amplifier
VER—Vertical, Vertical Amplifier
VERD—Vertical Deflection
VHF—VHF Amplifier, VHF Applications
VID—Video, Video Amplifier
WBA—Wide-Band Amplifier

DESCRIPTION OF TYPES

Type	Polarity Ge/Si	Mfg	Application
AA1	PG	WTV	CMO,IF
AA2	NS	WTV	CMO,IF
AA3	PG	WTV	HF
AA4	PG	WTV	GEN,MP
AA5	PG	WTV	GEN,HP
AC187/01	NG	PHI	AO
AC188/01	PG	PHI	AO
AT10H	PG	SEM	AD,AO
AT10M	PG	SEM	AD,AO
AT10N	PG	SEM	AD,AO
AT100H	PG	SEM	APA
AT100M	PG	SEM	APA
AT100N	PG	SEM	APA
AT20H	PG	SEM	AD,AO
AT20M	PG	SEM	AD,AO
AT20N	PG	SEM	AD,AO
AT30H	PG	SEM	AD,AO
AT30M	PG	SEM	AD,AO
AT30N	PG	SEM	AD,AO
BDY20	NS	PHI	AO
BD230	NS	PHI	AO
BD231	PS	PHI	AO
BD232	NS	PHI	HV AO
BD237	NS	PHI	AO
BD238	PS	PHI	AO
BU108	NS	PHI	HOR
DS25	PG	DEL	CMO AUD
DS26	PG	DEL	AA
DS41	PG	DEL	FM RF
DS501	PG	DEL	APO
DS503	PG	DEL	APO
DS509	NS	DEL	SPL PUR
DS520	PG	DEL	APO
DS525	PG	DEL	APO
DS56	PG	DEL	FM AMP
DS66	NS	DEL	AUD FM
DS71	NS	DEL	RF AMP
DS72	NS	DEL	CMO,IF
DS74	NS	DEL	FM OSC
DS76	NS	DEL	AA
DS81	NS	DEL	FM RF
DS83	PS	DEL	SPL PUR
ECG100	PG	SYL	CMO,RF
ECG101	NG	SYL	CMO,RF
ECG102	PG	SYL	AD,APO
ECG102A	PG	SYL	AD,APO
ECG103	NG	SYL	AD,APO
ECG103A	NG	SYL	AD,APO
ECG104	PG	SYL	APO
ECG105	PG	SYL	APO
ECG106	PS	SYL	CMO,FM
ECG107	NS	SYL	CMO,IF
ECG108	NS	SYL	CMO,VID
ECG121	PG	SYL	AO,HIFI
ECG123	NS	SYL	AD,VID
ECG123A	NS	SYL	AA,RF
ECG124	NS	SYL	APO LO
ECG126	PG	SYL	CMO,IF
ECG127	PG	SYL	HOR/VER
ECG128	NS	SYL	VID,AO
ECG129	PS	SYL	VID,AO
ECG130	NS	SYL	APA
ECG131	PG	SYL	APO AR
ECG152	NS	SYL	APO
ECG153	PS	SYL	APO
ECG154	NS	SYL	VID OUT
ECG155	NG	SYL	APA
ECG157	NS	SYL	HV APA
ECG158	PG	SYL	APA
ECG159	PS	SYL	APRE,AD
ECG159+	PS	SYL	APRE,AD
ECG160	PG	SYL	CMO,IF
ECG161	NS	SYL	VID IF
ECG162	NS	SYL	VERD
ECG163	NS	SYL	HORD
ECG164	NS	SYL	VERD
ECG165	NS	SYL	HORD
ECG171	NS	SYL	AA,VID
ECG172	NS	SYL	DAR
ECG175	NS	SYL	APA
ECG176	PG	SYL	APA
ECG179	PG	SYL	AA
ECG180	PS	SYL	AA
ECG181	NS	SYL	AA
ECG182	NS	SYL	APA
ECG183	PS	SYL	APA
ECG184	NS	SYL	APA
ECG185	PS	SYL	APA
ECG186	NS	SYL	APA
ECG186A	NS	SYL	APA,CB
ECG187	PS	SYL	APA
ECG187A	PS	SYL	APA
ECG188	NS	SYL	APA,AD
ECG189	PS	SYL	APA,AD
ECG190	NS	SYL	APA
ECG191	NS	SYL	HV VID
ECG192	NS	SYL	APO
ECG192+	NS	SYL	APO
ECG193	PS	SYL	APO
ECG193+	PS	SYL	APO
ECG194	NS	SYL	HV AMP
ECG195	NS	SYL	RPA,CB
ECG195A	NS	SYL	RPA CB
ECG196	NS	SYL	APO
ECG197	PS	SYL	APO
ECG198	NS	SYL	HV A/SW
ECG199	NS	SYL	HG PRE
ECG210	NS	SYL	AO,SW
ECG211	PS	SYL	AO,SW
ECG213	PG	SYL	HC,HP
ECG218	PS	SYL	APO
ECG219	PS	SYL	AO,SW
ECG223	NS	SYL	AO,SW
ECG224	NS	SYL	RPA,CB
ECG225	NS	SYL	AUD,VID
ECG226	PG	SYL	APO
ECG228	NS	SYL	AO,VID
ECG229	NS	SYL	CMO,VID
ECG232	PS	SYL	DAR
ECG233	NS	SYL	VID IF
ECG234	PS	SYL	HG PRE
ECG235	NS	SYL	HP,HF

DESCRIPTION OF TYPES

Type	Polarity Ge/Si	Mfg	Application
ECG236	NS	SYL	HP,HF
ECG237	NS	SYL	RPO CB
ECG238	NS	SYL	HOR OUT
ECG241	NS	SYL	APA,SW
ECG242	PS	SYL	APA,SW
ECG243	NS	SYL	DAR PWR
ECG244	PS	SYL	DAR PWR
ECG245	NS	SYL	DAR PWR
ECG246	PS	SYL	DAR PWR
ECG247	NS	SYL	DAR PWR
ECG248	PS	SYL	DAR PWR
ECG249	NS	SYL	DAR PWR
ECG250	PS	SYL	DAR PWR
ECG251	NS	SYL	DAR PWR
ECG252	PS	SYL	DAR PWR
ECG253	NS	SYL	DAR PWR
ECG254	PS	SYL	DAR PWR
ECG257	NS	SYL	DAR PWR
ECG258	PS	SYL	DAR PWR
ECG259	NS	SYL	DAR PWR
ECG260	PS	SYL	DAR PWR
ECG261	NS	SYL	DAR PWR
ECG262	PS	SYL	DAR PWR
ECG263	NS	SYL	DAR PWR
ECG264	PS	SYL	DAR PWR
ECG265	NS	SYL	DAR PWR
ECG266	NS	SYL	DAR PWR
ECG267	NS	SYL	DAR PWR
ECG268	NS	SYL	DAR PWR
ECG269	PS	SYL	DAR PWR
ECG270	NS	SYL	DAR PWR
ECG271	PS	SYL	DAR PWR
ECG272	NS	SYL	DAR PWR
ECG273	PS	SYL	DAR PWR
ECG274	NS	SYL	DAR PWR
ECG275	PS	SYL	DAR PWR
ECG277	NS	SYL	HOR OUT
ECG278	NS	SYL	RF CMC
ECG280	NS	SYL	APA
ECG281	PS	SYL	APA
ECG282	NS	SYL	RPA,SW
ECG283	NS	SYL	HVHC SW
ECG284	NS	SYL	APA
ECG285	PS	SYL	APA
ECG286	NS	SYL	PWR,SW
ECG287	NS	SYL	HV GEN
ECG288	PS	SYL	HV GEN
ECG289	NS	SYL	APA
ECG290	PS	SYL	APA
ECG291	NS	SYL	PWR,SW
ECG292	PS	SYL	PWR,SW
ECG293	NS	SYL	APA
ECG294	PS	SYL	APA
ECG295	NS	SYL	RPO CB
ECG297	NS	SYL	AD,APA
ECG298	PS	SYL	AD,APA
ECG299	NS	SYL	RFD,PWR
ECG300	NS	SYL	APO
ECG302	NS	SYL	PWR CB
ECG306	NS	SYL	PWR CB
ECG307	PS	SYL	APO

Type	Polarity Ge/Si	Mfg	Application
ECG311	NS	SYL	UHF OSC
ECG313	NS	SYL	RF,VHF
ECG315	NS	SYL	RFDR CB
ECG316	NS	SYL	LN UHF
ECG317	NS	SYL	RPO
ECG318	NS	SYL	RPO
ECG319	NS	SYL	IF TV
ECG320	NS	SYL	RPO
ECG321	NS	SYL	HOR OUT
EN10	NS	SEM	RF,IF
EN30	NS	SEM	AA
EN40	NS	SEM	RF,AUD
EP20	PS	SEM	RF,IF
EP25	PS	SEM	AA
EP35	PS	SEM	RF,AUD
GE-1	PG	GEC	CMO AM
GE-10	NS	GEC	CMO AM
GE-11	NS	GEC	CMO FM
GE-12	NS	GEC	APA LO
GE-14	NS	GEC	APA/HP
GE-16	PG	GEC	APA,SW
GE-17	NS	GEC	FM,TV
GE-18	NS	GEC	AA,AO
GE-19	NS	GEC	AA,OSC
GE-2	PG	GEC	AA
GE-20	NS	GEC	AA,OSC
GE-21	PS	GEC	AA,OSC
GE-210	NS	GEC	AA,RF
GE-211	NS	GEC	RF,IF
GE-212	NS	GEC	VA
GE-213	NS	GEC	IF,TV
GE-214	NS	GEC	UHF/VHF
GE-215	NS	GEC	OUT CB
GE-216	NS	GEC	OUT CB
GE-217	NS	GEC	HVA
GE-218	PS	GEC	HVA
GE-219	NS	GEC	OUT CB
GE-22	PS	GEC	AA,OSC
GE-220	NS	GEC	GEN HVA
GE-221	PS	GEC	GEN HVA
GE-222	NS	GEC	GEN HVA
GE-223	PS	GEC	GEN HVA
GE-224	NS	GEC	TV
GE-225	PS	GEC	TV
GE-226	NS	GEC	HVA
GE-227	PS	GEC	HVA
GE-228	PS	GEC	HVA
GE-229	NS	GEC	VID OUT
GE-23	NS	GEC	APO SCH
GE-230	NS	GEC	GEN HVA
GE-231	NS	GEC	HV VID
GE-232	NS	GEC	AO LO
GE-233	NS	GEC	SW,AMP
GE-234	PS	GEC	SW,AMP
GE-235	NS	GEC	GEN HVA
GE-25	PG	GEC	HVA,TV
GE-26	PS	GEC	APA SCH
GE-27	NS	GEC	VID OUT
GE-28	NS	GEC	APA
GE-29	PS	GEC	APA
GE-3	PG	GEC	APA

DESCRIPTION OF TYPES

Type	Polarity Ge/Si	Mfg	Application
GE-30	PG	GEC	APO SCH
GE-32	NS	GEC	AA LO
GE-35	NS	GEC	VERD,AA
GE-36	NS	GEC	HORD
GE-37	NS	GEC	VERD
GE-38	NS	GEC	HORD
GE-39	NS	GEC	VID IF
GE-4	PG	GEC	APA/HP
GE-40	NS	GEC	VID OUT
GE-43	NG	GEC	APA
GE-44	PG	GEC	APO SCH
GE-45	NS	GEC	PWR SW
GE-46	NS	GEC	RPO CB
GE-47	NS	GEC	APA,AD
GE-48	PS	GEC	APA,AD
GE-49	PG	GEC	APA
GE-5	NG	GEC	CMO AM
GE-50	PG	GEC	FM/TV
GE-51	PG	GEC	IF,RF
GE-52	PG	GEC	LN,AA
GE-53	PG	GEC	AA,AO
GE-54G	PG	GEC	AO,AD
GE-54R	NG	GEC	AO,AD
GE-55	NS	GEC	APA,SW
GE-56	PS	GEC	APA,SW
GE-57	NS	GEC	APA
GE-58	PS	GEC	APA
GE-59	NG	GEC	AA,AO
GE-6	NG	GEC	CMO AM
GE-60	NS	GEC	IF,TV
GE-61	NS	GEC	IF,TV
GE-62	NS	GEC	HG,LN
GE-63	NS	GEC	AA/AO
GE-64	NS	GEC	DAR,LN
GE-65	PS	GEC	HG,LN
GE-66	NS	GEC	APO
GE-67	PS	GEC	AA/AO
GE-69	PS	GEC	APO
GE-7	NG	GEC	IF AM
GE-72	NS	GEC	AA
GE-73	NS	GEC	AA
GE-74	PS	GEC	AA
GE-75	NS	GEC	AA
GE-76	PG	GEC	AA,HVSW
GE-8	NG	GEC	AA
GE-80	PG	GEC	APA
GE-81	NS	GEC	MPD,LPO
GE-82	PS	GEC	MPD,LPO
GE-83	NS	GEC	APO
GE-84	PS	GEC	APO
GE-85	NS	GEC	LN AMP
GE-86	NS	GEC	VHF CMO
GE-88	NS	GEC	AD,AO
GE-89	PS	GEC	AD,AO
GE-9	PG	GEC	CMO,IF
GEMR-6	NS	GEC	PWR TAB
HEPG0001	PG	MOT	RF,LP
HEPG0002	PG	MOT	RF,LP
HEPG0003	PG	MOT	RF,LP
HEPG0005	PG	MOT	GEN,LP
HEPG0006	PG	MOT	GEN AA
HEPG0007	PG	MOT	GEN AA
HEPG0008	PG	MOT	GEN,OSC
HEPG0009	PG	MOT	RF,LP
HEPG0011	NG	MOT	RF,LP
HEPG6000	PG	MOT	HC PWR
HEPG6001	PG	MOT	HC PWR
HEPG6002	PG	MOT	HC PWR
HEPG6003	PG	MOT	HC PWR
HEPG6004	PG	MOT	HC PWR
HEPG6005	PG	MOT	HC PWR
HEPG6006	PG	MOT	HC PWR
HEPG6007	PG	MOT	HC PWR
HEPG6008	PG	MOT	HC PWR
HEPG6009	PG	MOT	HC PWR
HEPG6010	PG	MOT	HC PWR
HEPG6011	PG	MOT	HC PWR
HEPG6012	PG	MOT	HC PWR
HEPG6013	PG	MOT	GEN AA
HEPG6014	PG	MOT	HC PWR
HEPG6015	PG	MOT	HC PWR
HEPG6016	PG	MOT	HC PWR
HEPG6017	PG	MOT	HC PWR
HEPG6018	PG	MOT	HC PWR
HEPS0001	NS	MOT	HV,GEN
HEPS0002	NS	MOT	GEN CMO
HEPS0003	NS	MOT	GEN CMO
HEPS0004	NS	MOT	HF CMO
HEPS0005	NS	MOT	HV,GEN
HEPS0006	PS	MOT	GEN CMO
HEPS0007	NS	MOT	GEN VA
HEPS0008	NS	MOT	VID IF
HEPS0009	NS	MOT	VID IF
HEPS0010	NS	MOT	UHF CMO
HEPS0011	NS	MOT	GEN,LP
HEPS0012	PS	MOT	RF,LP
HEPS0013	PS	MOT	RF,LP
HEPS0014	NS	MOT	RF,LP
HEPS0015	NS	MOT	GEN,LP
HEPS0016	NS	MOT	FM,TV
HEPS0017	NS	MOT	UHF CMO
HEPS0019	PS	MOT	GEN,SW
HEPS0020	NS	MOT	VHF MIX
HEPS0024	NS	MOT	LN,LP
HEPS0025	NS	MOT	GEN,SW
HEPS0026	PS	MOT	AD,AO
HEPS0027	NS	MOT	HV
HEPS0028	PS	MOT	HV
HEPS0029	PS	MOT	HV GEN
HEPS0030	NS	MOT	GEN RF
HEPS0031	PS	MOT	GEN AMP
HEPS0032	PS	MOT	MS SW
HEPS0033	NS	MOT	GEN,LP
HEPS3001	NS	MOT	RF PWR
HEPS3002	NS	MOT	MP
HEPS3003	PS	MOT	MP
HEPS3004	NS	MOT	RF PWR
HEPS3005	NS	MOT	RF PWR
HEPS3006	NS	MOT	RF PWR
HEPS3007	NS	MOT	RF PWR
HEPS3008	NS	MOT	RF PWR
HEPS3009	NS	MOT	RF PWR

DESCRIPTION OF TYPES

Type	Polarity Ge/Si	Mfg	Application
HEPS3010	NS	MOT	MP
HEPS3011	NS	MOT	RF PWR
HEPS3012	PS	MOT	MP
HEPS3013	NS	MOT	RF PWR
HEPS3014	PS	MOT	RF PWR
HEPS3019	NS	MOT	HV AA
HEPS3020	NS	MOT	MP
HEPS3021	NS	MOT	HV
HEPS3022	NS	MOT	HV
HEPS3023	NS	MOT	GEN AMP
HEPS3024	NS	MOT	GEN AMP
HEPS3025	NS	MOT	GEN AMP
HEPS3026	NS	MOT	GEN AMP
HEPS3027	PS	MOT	GEN AMP
HEPS3028	PS	MOT	GEN AMP
HEPS3029	PS	MOT	GEN AMP
HEPS3030	PS	MOT	GEN AMP
HEPS3031	PS	MOT	GEN AMP
HEPS3032	PS	MOT	MP
HEPS3033	NS	MOT	MP
HEPS3034	NS	MOT	MP
HEPS3035	NS	MOT	MP
HEPS3036	NS	MOT	RF PWR
HEPS3037	NS	MOT	RF PWR
HEPS3038	NS	MOT	RF PWR
HEPS3039	NS	MOT	RF PWR
HEPS3040	NS	MOT	RF PWR
HEPS3041	NS	MOT	RF PWR
HEPS5000	NS	MOT	GEN PWR
HEPS5001	NS	MOT	GEN PWR
HEPS5002	PS	MOT	GEN PWR
HEPS5003	NS	MOT	GEN PWR
HEPS5004	NS	MOT	GEN PWR
HEPS5005	PS	MOT	GEN PWR
HEPS5006	PS	MOT	GEN PWR
HEPS5007	PS	MOT	GEN PWR
HEPS5008	PS	MOT	GEN PWR
HEPS5009	PS	MOT	GEN PWR
HEPS5010	PS	MOT	GEN PWR
HEPS5011	NS	MOT	PWR
HEPS5012	NS	MOT	HP
HEPS5013	PS	MOT	MP
HEPS5014	NS	MOT	MP
HEPS5015	NS	MOT	HV,PWR
HEPS5018	PS	MOT	GEN AA
HEPS5019	NS	MOT	GEN AA
HEPS5020	NS	MOT	HV,HP
HEPS5021	NS	MOT	HV,HP
HEPS5022	PS	MOT	LP,RF
HEPS5023	PS	MOT	LFRF,AA
HEPS5024	NS	MOT	GEN AA
HEPS5025	NS	MOT	LF CMO
HEPS5026	NS	MOT	LF CMO
HEPS7000	NS	MOT	HP
HEPS7001	PS	MOT	HP
HEPS7002	NS	MOT	HP
HEPS7003	PS	MOT	HP
HEPS7004	NS	MOT	GEN AA
HEPS7005	NS	MOT	HV,HP
HEPS9100	NS	MOT	DAR
HEPS9101	NS	MOT	DAR

Type	Polarity Ge/Si	Mfg	Application
HEPS9102	NS	MOT	DAR
HEPS9103	NS	MOT	MP DAR
HEPS9120	PS	MOT	DAR
HEPS9121	PS	MOT	DAR
HEPS9122	PS	MOT	DAR
HEPS9123	PS	MOT	MP DAR
HEPS9140	NS	MOT	DAR
HEPS9141	PS	MOT	DAR
HEPS9142	NS	MOT	DAR
HEPS9143	PS	MOT	DAR
HEP1	PG	MOT	LP,RF
HEP2	PG	MOT	LP,RF
HEP200	PG	MOT	MP
HEP230	PG	MOT	MP
HEP231	PG	MOT	HP
HEP232	PG	MOT	MP
HEP233	PG	MOT	HP
HEP234	PG	MOT	HV,PWR
HEP235	PG	MOT	HV,PWR
HEP236	PG	MOT	HC,PWR
HEP237	PG	MOT	HP,HC
HEP238	PG	MOT	MP,GEN
HEP239	PG	MOT	MP,GEN
HEP240	NS	MOT	HV,PWR
HEP241	NS	MOT	HV,PWR
HEP242	PS	MOT	MP
HEP243	NS	MOT	MP
HEP244	NS	MOT	HV,PWR
HEP245	NS	MOT	MP
HEP246	PS	MOT	MP
HEP247	NS	MOT	HP
HEP248	PS	MOT	HP
HEP250	PG	MOT	LP,GEN
HEP251	PG	MOT	LP,GEN
HEP252	PG	MOT	LP,GEN
HEP253	PG	MOT	LP,GEN
HEP254	PG	MOT	LP,GEN
HEP3	PG	MOT	LP,RF
HEP50	NS	MOT	LP,RF
HEP51	PS	MOT	LP,RF
HEP52	PS	MOT	LP,RF
HEP53	NS	MOT	LP,RF
HEP54	NS	MOT	LP,RF
HEP55	NS	MOT	LP,RF
HEP56	NS	MOT	LP,OSC
HEP57	PS	MOT	LP,RF
HEP623	PG	MOT	GEN APA
HEP624	PG	MOT	GEN APA
HEP625	PG	MOT	GEN APA
HEP626	PG	MOT	GEN APA
HEP627	PG	MOT	GEN APA
HEP628	PG	MOT	GEN APA
HEP629	PG	MOT	GEN AA
HEP630	PG	MOT	GEN AA
HEP631	PG	MOT	GEN AA
HEP632	PG	MOT	GEN AA
HEP633	PG	MOT	GEN AA
HEP634	PG	MOT	GEN AA
HEP635	PG	MOT	GEN RF
HEP636	PG	MOT	GEN RF
HEP637	PG	MOT	GEN RF

DESCRIPTION OF TYPES

Type	Polarity Ge/Si	Mfg	Application
HEP638	PG	MOT	GEN DF
HEP639	PG	MOT	GEN DF
HEP640	PG	MOT	GEN DF
HEP641	NG	MOT	GEN AA
HEP642	PG	MOT	GEN APA
HEP643	PG	MOT	GEN APA
HEP644	PG	MOT	HV PWR
HEP700	PS	MOT	GEN APA
HEP701	NS	MOT	GEN APA
HEP702	PS	MOT	GEN APA
HEP703	NS	MOT	GEN APA
HEP704	NS	MOT	GEN APA
HEP705	PS	MOT	GEN APA
HEP706	NS	MOT	GEN APA
HEP707	NS	MOT	HV,HP
HEP708	PS	MOT	MP,RF
HEP709	NS	MOT	UHF CMO
HEP710	PS	MOT	MP,AA
HEP712	NS	MOT	HV,CMO
HEP713	NS	MOT	HV,CMO
HEP714	NS	MOT	HV,AA
HEP715	PS	MOT	GEN,AA
HEP716	PS	MOT	GEN,SW
HEP717	PS	MOT	LP,AA
HEP718	NS	MOT	UHF CMO
HEP719	NS	MOT	UHF CMO
HEP720	NS	MOT	UHF CMO
HEP721	NS	MOT	GEN AA
HEP722	NS	MOT	GEN AA
HEP723	NS	MOT	GEN AA
HEP724	NS	MOT	GEN AA
HEP725	NS	MOT	GEN AA
HEP726	NS	MOT	GEN AA
HEP727	NS	MOT	GEN AA
HEP728	NS	MOT	GEN AA
HEP729	NS	MOT	GEN AA
HEP730	NS	MOT	GEN AA
HEP731	NS	MOT	GEN AA
HEP732	NS	MOT	GEN AA
HEP733	NS	MOT	GEN AA
HEP734	NS	MOT	GEN AA
HEP735	NS	MOT	GEN AA
HEP736	NS	MOT	GEN.AA
HEP737	NS	MOT	GEN AA
HEP738	NS	MOT	GEN AA
HEP739	PS	MOT	LP
HEP740	NS	MOT	HV,HP
HEP75	NS	MOT	RF PWR
HEP76	PS	MOT	RF PWR
HF12H	PG	SEM	CMO AM
HF12M	PG	SEM	CMO AM
HF20H	PG	SEM	CMO AM
HF20M	PG	SEM	CMO AM
HF3H	PG	SEM	IF AM
HF3M	PG	SEM	IF AM
HF35	PG	SEM	CMO MB
HF50	PG	SEM	CMO MB
HF6H	PG	SEM	IF AM
HF6M	PG	SEM	IF AM
HF75	PG	SEM	CMO MB
HN100	NS	SEM	VER

Type	Polarity Ge/Si	Mfg	Application
HN150	NS	SEM	VER
HO300	PG	SEM	HOR
IRTR-50	PG	INR	AA,SW
IRTR-51	NS	INR	GEN TV
IRTR-52	PS	INR	GEN TV
IRTR-53	NS	INR	GEN MB
IRTR-54	PS	INR	GEN MB
IRTR-55	NS	INR	AA,SW
IRTR-56	PS	INR	AA,SW
IRTR-57	NS	INR	AA,SW
IRTR-58	PS	INR	AA,SW
IRTR-59	NS	INR	AA,SW
IRTR-60	NS	INR	MPHV MB
IRTR-61	NS	INR	HPHV MB
IRTR62	NS	INR	RPO CB
IRTR63	NS	INR	RPO CB
IRTR64	NS	INR	RPO CB
IRTR65	NS	INR	RPO CB
IRTR66	NS	INR	RPO CB
IRTR67	NS	INR	HORD,HV
IRTR68	NS	INR	VERD,HV
IRTR69	NS	INR	DAR,PRE
IRTR70	NS	INR	VID IF
IRTR71	NS	INR	VID IF
IRTR72	NS	INR	MP,AD
IRTR73	PS	INR	MP,AD
IRTR74	NS	INR	AO,AD
IRTR75	NS	INR	AA,VID
IRTR76	NS	INR	HP AA
IRTR77	PS	INR	HP AA
IRTR78	NS	INR	VID OUT
IRTR79	NS	INR	AO,VID
IRTR80	NS	INR	UHF CMO
IRTR81	NS	INR	APA
IRTR82	PG	INR	APA
IRTR83	NS	INR	OSC TV
IRTR84	PG	INR	APA
IRTR85	PG	INR	AD,PRE
IRTR86	NS	INR	AUD,TV
IRTR87	NS	INR	AUD,VID
IRTR88	PS	INR	AUD,VID
IRTR91	NG	INR	HP,AA
IRTR92	NS	INR	RPO AR
IRTR93	NS	INR	HOR OUT
IRTR94MP	PG	INR	AA,M.P.
IRTR95	NS	INR	CMO
JR10	PG	SEM	AM/FM
JR100	PG	SEM	CMO MB
JR15	PG	SEM	AO,AD
JR200	PG	SEM	CMO MB
JR30	PG	SEM	CMO MB
JR30X	PG	SEM	CMO MB
JR5	PG	SEM	AO,AD
NA20	NG	SEM	AO,AD
NA30	NG	SEM	AO,AD
NO400	NS	SEM	HOR
NR10	NG	SEM	CMO AM
NR5	NG	SEM	IF AM
NR700	NG	SEM	CMO MB
PAR12	PG	SEM	APA
PN26	NS	SEM	APA

DESCRIPTION OF TYPES

Type	Polarity Ge/Si	Mfg	Application
PN350	NS	SEM	APA
PN66	NS	SEM	APA
PTC101	NS	MAL	SS,PRE
PTC102	PG	MAL	SS,AA
PTC103	PS	MAL	SS,PRE
PTC104	NS	MAL	HV,AO
PTC105	PG	MAL	HP,AO
PTC105A	PG	MAL	HP,APO
PTC106	PG	MAL	HP,AO
PTC107	PG	MAL	RF,IF
PTC108	NG	MAL	RF,IF
PTC109	PG	MAL	LP,AA
PTC110	NS	MAL	MP,AO
PTC111	PS	MAL	MP,AO
PTC112	NS	MAL	MP,AO
PTC113	PS	MAL	MP,AO
PTC114	PG	MAL	HP,AO
PTC115	NS	MAL	SS,HF
PTC116	NS	MAL	HP,AO
PTC117	NS	MAL	VID LPO
PTC118	NS	MAL	VID HPO
PTC119	NS	MAL	HP,AO
PTC120	PG	MAL	HP,AO
PTC120A	PG	MAL	HP,APO
PTC121	NS	MAL	RF,IF
PTC122	PG	MAL	VID,HPO
PTC123	NS	MAL	SS,AA
PTC124	NS	MAL	HV,AO
PTC125	NS	MAL	SS,AA
PTC126	NS	MAL	SS,AA
PTC127	PS	MAL	SS,AA
PTC128	NS	MAL	RF PWR
PTC129	NS	MAL	HOR
PTC129A	NS	MAL	VID HPO
PTC130	NS	MAL	VID,HPO
PTC131	PS	MAL	SS,HF
PTC132	NS	MAL	SS,HF
PTC133	NS	MAL	SS,HF
PTC134	NG	MAL	LP,AA
PTC135	PG	MAL	LP,AA
PTC136	NS	MAL	SS,AA
PTC137	NS	MAL	HP,AO
PTC138	PG	MAL	MP,APO
PTC139	NS	MAL	SS,AA
PTC140	NS	MAL	HP,AO
PTC141	PS	MAL	MP,AO
PTC142	PS	MAL	MP,AO
PTC143	NS	MAL	MP,AO
PTC144	NS	MAL	MP,AO
PTC145	PG	MAL	SS,HF
PTC146	NS	MAL	VID HPO
PTC147	PG	MAL	HP,AO
PTC148	NS	MAL	HP,AO
PTC149	PS	MAL	HP,AO
PTC150	NS	MAL	VID LPO
PTC153	NS	MAL	DAR,PRE
PTC154	NS	MAL	HP,AO
PTC155	PG	MAL	MP,AO
PTC156	PG	MAL	LP,AA
PTC157	PS	MAL	HP,AO
PTC158	NS	MAL	RFA,CB
PTC160	PS	MAL	MP,AO
PTC162	PS	MAL	HP,AO
PTC163	NS	MAL	HP,AO
PTC164	PS	MAL	HP,AO
PTC165	NS	MAL	HP,AO
PTC166	PS	MAL	HP,AO
PTC167	NS	MAL	HP,AO
PTC168	PS	MAL	HP,AO
PTC169	NS	MAL	HP,AO
PTC170	NS	MAL	AA,AO
PTC171	PS	MAL	AA,AO
PTC172	PS	MAL	HP,AO
PTC173	NS	MAL	HP,AO
PTC174	PS	MAL	HP,AO
PTC175	NS	MAL	HP,AO
PTC176	NS	MAL	RF,PWR
PTC177	PS	MAL	LP,AA
PTC178	NS	MAL	LP,AA
PTC179	NS	MAL	RFD CB
PTC180	NS	MAL	RFD CB
PTC186	NS	MAL	RF PWR
PTC191	NS	MAL	RF PWR
PTC192	NS	MAL	RF PWR
PTC193	NS	MAL	APA
PTC194	PG	MAL	MP,APO
PTC195	NS	MAL	GEN,MPA
PTC196	PS	MAL	GEN,MPA
PTC197	NS	MAL	GEN,MPA
PTC198	PS	MAL	GEN,MPA
PTC901	PG	MAL	HP,APO
PTC904	NS	MAL	MP,APA
PTC905	PS	MAL	MP,APA
PTC906	NS	MAL	APA,SW
PT12	PG	SEM	APA
PT150	PG	SEM	APA/HP
PT201	PG	SEM	APA
PT25	PG	SEM	APA
PT250	PG	SEM	APA/HP
PT32	PG	SEM	APA
PT40	PG	SEM	APA
PT50	PG	SEM	APA/HP
PT501	PG	SEM	APA
PT515	PG	SEM	APA/HP
PT6	PG	SEM	APA
RE 1	PG	RAY	CMO AM
RE 10	NS	RAY	CMO
RE 11	PG	RAY	APA
RE 12	NS	RAY	AA,VID
RE 13	NS	RAY	AA,RF
RE 14	NS	RAY	APA
RE 15	PG	RAY	CMO
RE 16	PG	RAY	APA,HOR
RE 17	NS	RAY	AA,VID
RE 18	PS	RAY	AA,VID
RE 19	NS	RAY	APA
RE 191	NS	RAY	HV PWR
RE 192	NS	RAY	LN PRE
RE 193	PS	RAY	LN PRE
RE 194	NS	RAY	HF PWR
RE 2	NG	RAY	CMO AM
RE 20	PG	RAY	APA

DESCRIPTION OF TYPES

Type	Polarity Ge/Si	Mfg	Application
RE 21	NS	RAY	APA
RE 22	PS	RAY	APA
RE 23	NS	RAY	VID OUT
RE 24	NS	RAY	APA
RE 25	PG	RAY	APA
RE 26	PS	RAY	AA
RE 27	PG	RAY	CMO
RE 28	NS	RAY	VID IF
RE 29	NS	RAY	VER AMP
RE 3	PG	RAY	AA
RE 30	NS	RAY	HOR AMP
RE 31	NS	RAY	VER AMP
RE 32	NS	RAY	HOR AMP
RE 33	NS	RAY	APD
RE 34	NS	RAY	APA
RE 35	PG	RAY	APA
RE 36	PG	RAY	APA
RE 37	NS	RAY	APA
RE 38	NS	RAY	APA
RE 39	PS	RAY	APA
RE 4	PG	RAY	AA
RE 40	NS	RAY	APA
RE 41	PS	RAY	APA
RE 42	NS	RAY	APA
RE 43	PS	RAY	APA
RE 44	NS	RAY	APA
RE 5	NG	RAY	AA
RE 53	PS	RAY	CMO
RE 54	NG	RAY	APA
RE 56	PS	RAY	APA
RE 57	PS	RAY	APA
RE 58	PS	RAY	APA
RE 59	NS	RAY	APA
RE 6	NG	RAY	AA
RE 60	NS	RAY	AA,VID
RE 61	PS	RAY	APA
RE 62	PS	RAY	LN AUD
RE 63	PS	RAY	VA,DR
RE 64	NS	RAY	LN AA
RE 66	NS	RAY	RF,IF
RE 67	NS	RAY	CMO
RE 68	PG	RAY	APA
RE 69	PG	RAY	RF,IF
RE 7	PG	RAY	APA
RE 70	NS	RAY	AA
RE 71	NS	RAY	APA
RE 72	PS	RAY	APA
RE 73	NS	RAY	AA,VID
RE 74	PS	RAY	APA
RE 75	NS	RAY	AD,PWR
RE 76	PS	RAY	AD,PWR
RE 77	NS	RAY	AA,VID
RE 78	NS	RAY	GEN AMP
RE 79	NS	RAY	RF PWR
RE 8	PG	RAY	APA
RE 80	NS	RAY	AO,SW
RE 81	PS	RAY	AO,SW
RE 82	PS	RAY	AO,SW
RE 83	PG	RAY	APA
RE 9	NS	RAY	CMO
RS-2001	NG	ARC	HF,SW
RS-2002	PG	ARC	MF AMP
RS-2003	NG	ARC	SS,GEN
RS-2004	PG	ARC	GEN,AA
RS-2005	PG	ARC	GEN,AA
RS-2006	PG	ARC	MP,AA
RS-2007	PG	ARC	HF,SW
RS-2008	NS	ARC	GEN AMP
RS-2009	NS	ARC	GEN AMP
RS-2010	NS	ARC	LN,LP
RS-2011	NS	ARC	LP,SS
RS-2012	NS	ARC	HVP,GEN
RS-2013	NS	ARC	HIFI,LP
RS-2014	NS	ARC	MP,AA
RS-2015	NS	ARC	HF,SS
RS-2016	NS	ARC	LP,SSW
RS-2017	NS	ARC	PWR,SW
RS-2018	NS	ARC	PWR,SW
RS-2019	NS	ARC	PWR,SW
RS-2020	NS	ARC	PWR,SW
RS-2021	PS	ARC	MP,GEN
RS-2022	PS	ARC	LN,HG
RS-2023	PS	ARC	MP,HSSW
RS-2024	PS	ARC	MP,AA
RS-2025	PS	ARC	PWR,SW
RS-2026	PS	ARC	PWR,SW
RS-2027	PS	ARC	PWR,SW
RS-2030	NS	ARC	HC,GEN
RS-2031	NS	ARC	LN,HG
RS-2032	PS	ARC	MP,SW
RS-2033	NS	ARC	AMP,SW
RS-2034	PS	ARC	AMP,SW
RT-100	NS	SPR	AA
RT-100A	NS	SPR	AA,VID
RT-101	PS	SPR	AA
RT-101A	PS	SPR	AA,VID
RT-102	NS	SPR	AA
RT-102A	NS	SPR	AA,VID
RT-103	PS	SPR	AA
RT-103A	PS	SPR	AA,VID
RT-104	NS	SPR	AA,SS
RT-105	NS	SPR	LN,PRE
RT-106	PS	SPR	LN,PRE
RT-107	NS	SPR	FM RF
RT-107A	NS	SPR	RF,OSC
RT-108	NS	SPR	UHF CMO
RT-108A	NS	SPR	PRE,AA
RT-109	NS	SPR	HV,AMP
RT-110	NS	SPR	VID OUT
RT-111	NS	SPR	VID OUT
RT-112	NS	SPR	VID IF
RT-113	NS	SPR	VID IF
RT-114	NS	SPR	AA,AO
RT-115	PS	SPR	AA,AO
RT-116	NS	SPR	BHO
RT-117	PG	SPR	RF,OSC
RT-118	PS	SPR	AM RF
RT-119	NG	SPR	RF CMO
RT-120	PS	SPR	AM AD
RT-121	PS	SPR	AM AO
RT-122	NG	SPR	AD,APO
RT-123	PG	SPR	AD

DESCRIPTION OF TYPES

Type	Polarity Ge/Si	Mfg	Application
RT-124	PG	SPR	APO
RT-125	PG	SPR	APA
RT-126	PS	SPR	RF,IF
RT-126A	PS	SPR	MB AMP
RT-127	PG	SPR	APO
RT-127A	PG	SPR	APA
RT-128	NS	SPR	HV APO
RT-129	NG	SPR	SS
RT-130	PG	SPR	HOR,VER
RT-131	NS	SPR	APA
RT-132	PG	SPR	APO
RT-133	NS	SPR	AUD PWR
RT-134	NS	SPR	RF CMO
RT-135	NS	SPR	HV APA
RT-136	PG	SPR	APA
RT-137	NS	SPR	PWR,SW
RT-139	NS	SPR	HP
RT-140	NS	SPR	HP
RT-141	NS	SPR	AA,AO
RT-143	NS	SPR	RPA,RFD
RT-144	NS	SPR	DAR
RT-145	NS	SPR	APA
RT-146	NS	SPR	PWR OUT
RT-147	PG	SPR	AA
RT-148	PS	SPR	AA
RT-149	NS	SPR	AA
RT-150	NS	SPR	APA
RT-151	PS	SPR	APA
RT-152	NS	SPR	APA
RT-153	PS	SPR	APA
RT-154	NS	SPR	APA
RT-155	PS	SPR	APA
RT-156	NS	SPR	AD,APA
RT-157	PS	SPR	AD,APA
RT-158	NS	SPR	PWR OUT
RT-159	NS	SPR	AA,VID
RT-159A	NS	SPR	AA,HORR
RT-160	NS	SPR	AA,AO
RT-162	NS	SPR	AA,HSSW
RT-163	PS	SPR	AA,HSSW
RT-164	NS	SPR	AUD PWR
RT-164A	NS	SPR	AMP,SW
RT-165	PS	SPR	AUD PWR
RT-165A	PS	SPR	AMP,SW
RT-166	NS	SPR	AMP,SW
RT-167	PS	SPR	AMP,SW
RT-172	NS	SPR	AA
RT-174	NS	SPR	RPO CB
RT-182	NS	SPR	MP AMP
RT-183	PG	SPR	RF
RT-185	PS	SPR	AD,PWR
RT-186	NS	SPR	AD,PWR
RT-187	NS	SPR	VID IF
RT-188	PG	SPR	RF
RT-189	PS	SPR	APD
RT-190	NS	SPR	APA
RT-191	NS	SPR	RPO CB
RT-194	NS	SPR	GEN PWR
RT-195	PS	SPR	GEN PWR
RT-196	PS	SPR	APA
RT-197	NS	SPR	APA

Type	Polarity Ge/Si	Mfg	Application
RT-300	NS	SPR	AMP,SW
RT-301	NS	SPR	AA,RFA
RT-302	NS	SPR	AA,RFA
RT-303	PS	SPR	AA,RFA
RT-304	NS	SPR	AA,RFA
RT-305	PS	SPR	AA,RFA
RT-306	NS	SPR	RPO CB
RT-306F	NS	SPR	RPO CB
RT-307	NS	SPR	RPO CB
RT-308	NS	SPR	CMO CB
RT-309	NS	SPR	RFA CB
RT-310	NS	SPR	RFO CB
RT-311	PS	SPR	AA/RFA
RT-312	NS	SPR	RFD CB
RT-313	NS	SPR	OSC CB
RT-314	PS	SPR	GEN HVA
RT-315	NS	SPR	GEN HVA
SK3003	PG	RCA	AUD,SS
SK3004	PG	RCA	AUD,SS
SK3005	PG	RCA	AM RF
SK3006	PG	RCA	FM,TV
SK3007	PG	RCA	MB RF
SK3008	PG	RCA	AM RF
SK3009	PG	RCA	AUD,HP
SK3010	NG	RCA	AUD,SS
SK3011	NG	RCA	AM RF
SK3012	PG	RCA	AUD,HP
SK3014	PG	RCA	AUD,HP
SK3018	NS	RCA	MB RF
SK3019	NS	RCA	UHF OSC
SK3020	NS	RCA	AUD,MP
SK3021	NS	RCA	AUD,HV
SK3024	NS	RCA	AUD,MP
SK3024+	NS	RCA	AUD,MP
SK3025	PS	RCA	AUD,MP
SK3025+	PS	RCA	AUD,MP
SK3026	NS	RCA	AUD,HP
SK3027	NS	RCA	AUD,HP
SK3034	PG	RCA	TV
SK3035	PG	RCA	HOR OUT
SK3036	NS	RCA	AUD,HP
SK3038	NS	RCA	AUD
SK3039	NS	RCA	RF TV
SK3040	NS	RCA	VID AMP
SK3041	NS	RCA	AUD,HP
SK3044	NS	RCA	TV
SK3045	NS	RCA	AUD/TV
SK3046	NS	RCA	RPO CB
SK3047	NS	RCA	RPO CB
SK3048	NS	RCA	RPO CB
SK3049	NS	RCA	RPO CB
SK3052	PG	RCA	AUD,MP
SK3053	PS	RCA	AUD HV
SK3054	NS	RCA	AUD/VID
SK3079	NS	RCA	VERD
SK3082	PG	RCA	AUD,MP
SK3083	PS	RCA	AUD/VID
SK3084	PS	RCA	AUD,HP
SK3085	PS	RCA	VERD
SK3103	NS	RCA	APO LO
SK3104	NS	RCA	AUD/TV

DESCRIPTION OF TYPES

Type	Polarity Ge/Si	Mfg	Application
SK3111	NS	RCA	HOR OUT
SK3114	PS	RCA	AUD,SS
SK3115	NS	RCA	HOR OUT
SK3117	NS	RCA	VID IF
SK3118	PS	RCA	MB RF
SK3122	NS	RCA	MB RF
SK3123	PG	RCA	AUD,MP
SK3124	NS	RCA	AUD,SS
SK3131	NS	RCA	AUD/TV
SK3132	NS	RCA	TV
SK3132+	NS	RCA	VID IF
SK3133	NS	RCA	VERD
SK3137	NS	RCA	AUD,MP
SK3138	PS	RCA	AUD,MP
SK3156	NS	RCA	AUD,SS
SK3173	PS	RCA	AUD,HP
SK3176	NS	RCA	RPO/VHF
SK3177	NS	RCA	RPO/VHF
SK3178	NS	RCA	AUD,MP
SK3179	PS	RCA	AUD,MP
SK3180	NS	RCA	AUD,HP
SK3181	PS	RCA	HP,HV
SK3182	NS	RCA	AUD,HP
SK3183	PS	RCA	HP,HV
SK3188	NS	RCA	AUD,HP
SK3189	PS	RCA	AUD,HP
SK3190	NS	RCA	AUD,HP
SK3191	PS	RCA	AUD,HP
SK3192	NS	RCA	AUD,HP
SK3193	PS	RCA	AUD,HP
SK3194	NS	RCA	HORD
SK3195	NS	RCA	RPO/VHF
SK3197	NS	RCA	RPO CB
SK3198	PG	RCA	AUD,MP
SK3199	NS	RCA	AUD,MP
SK3200	PS	RCA	AUD,MP
SK3201	NS	RCA	AUD,MP
SK3202	NS	RCA	AUD,MP
SK3203	PS	RCA	AUD,MP
SK3217	NS	RCA	HP,HV
SK3218	NS	RCA	RPO VHF
SK3219	NS	RCA	HORR
SK3220	NS	RCA	HORD
SK3221	PS	RCA	VERD
SK3222	NS	RCA	VERD
SK3232	NS	RCA	TV
SK3239	NS	RCA	RPO CB
SK3241	PS	RCA	AUD,SS
SK3244	NS	RCA	AUD,HV
SK3245	NS	RCA	AUD,SS
SK3247	PS	RCA	AUD,SS
SK3248	NS	RCA	AUD,MP
SK3510	NS	RCA	HP,HV
SK3511	NS	RCA	HP,HV
SK3512	NS	RCA	MP,HV
SK3513	PS	RCA	MP,HV
SK3528	PS	RCA	HV,SW
SK3529	NS	RCA	HV,SW
SK3530	NS	RCA	HP,HV
SK3534	NS	RCA	HP,HV
SK3535	NS	RCA	HP,HV
SK3536	NS	RCA	MP,HV
SK3537	NS	RCA	HV,SW
SK3538	NS	RCA	HP,HV
SK3559	NS	RCA	AUD,HV
SK3560	NS	RCA	AUD,HV
SK3561	NS	RCA	AUD,TV
SK3562	NS	RCA	AUD,TV
SK3563	NS	RCA	HP,HV
SN60	NS	SEM	RF,IF
SN80	NS	SEM	AA
SP70	PS	SEM	AA
SP90	PS	SEM	RF,IF
TCVM-553	NS	SPR	MP
TR-01	PG	INR	HP AA
TR-02	PG	INR	MPLV,AA
TR-03	PG	INR	HPHG,AA
TR-05	PG	INR	MPHG,AA
TR-06	PG	INR	AA
TR-08	NG	INR	MP,CMO
TR-09	NG	INR	AA
TR-10	NG	INR	IF
TR-11	PG	INR	CMO
TR-12	PG	INR	LP,CMO
TR-14	PG	INR	AA,AD
TR-16	PG	INR	HP/AR
TR-17	PG	INR	GEN,CMO
TR-19	PS	INR	HS SW
TR-20	PS	INR	HS SW
TR-21	NS	INR	HS SW
TR-22	NS	INR	HS SW
TR-23	NS	INR	HV,PWR
TR-24	NS	INR	GEN,LN
TR-25	NS	INR	AA/RF
TR-26	NS	INR	HPAA,SW
TR-27	PG	INR	VID,HVP
TR-28	PS	INR	AA,RF
TR-29	PS	INR	HPAA,SW
TR-30	PS	INR	GEN,LN
TR-31	PS	INR	UNI
TR-32	NS	INR	UNI,HV
TR-33	NS	INR	UNI,CMO
TR-34	PG	INR	UNI,HVP
TR-35	PG	INR	UNI,APA
TR-36	NS	INR	HP,AA
TR-37	PS	INR	MP,UNI
TR-50	PG	INR	HPAA,SW
TR-51	NS	INR	GEN,SS
TR-52	PS	INR	GEN,SS
TR-53	NS	INR	GEN
TR-54	PS	INR	GEN
TR-55	NS	INR	MPAA,SW
TR-56	PS	INR	MPAA,SW
TR-57	NS	INR	MPAA,SW
TR-58	PS	INR	MPAA,SW
TR-59	NS	INR	MPAA,SW
TR-60	NS	INR	MP,HV
TR-61	NS	INR	HP,HV
TR-62	NS	INR	OUT/CB
TR-63	NS	INR	OUT/CB
TR-64	NS	INR	OUT/CB
TR-65	NS	INR	OUT/CB

DESCRIPTION OF TYPES

Type	Polarity Ge/Si	Mfg	Application
TR-66	NS	INR	OUT/CB
TR-67	NS	INR	HV,VID
TR-68	NS	INR	HV,VID
TR-69	NS	INR	DAR,LN
TR-70	NS	INR	RF,VID
TR-72	NS	INR	MPAA,AD
TR-73	PS	INR	MPAA,AD
TR-74	NS	INR	MPAA,AD
TR-75	NS	INR	AA,VID
TR-76	NS	INR	HP,AA
TR-77	PS	INR	HP,AA
TR-78	NS	INR	VID OUT
TR-79	NS	INR	VID,AO
TR-80	NS	INR	RF,CMO
TR-81	NS	INR	APA
TR-82	PG	INR	APA
TR-83	NS	INR	OSC,VID
TR-84	PG	INR	APA
TR-85	PG	INR	AD,PRE
TR-86	NS	INR	AUD/VID
TR-87	NS	INR	AUD/VID
TR-88	PS	INR	AUD/VID
TR-91	NG	INR	HP,AA
TR-92	NS	INR	RPO/AR
TR-93	NS	INR	HOR OUT
TR-94MP	PG	INR	M.P./AA
TR-95	NS	INR	CMO,IF
TVCM-551	NS	SPR	AUD PWR
TV1000	PG	SEM	CMO MB
WEPG6001	PG	WTV	AA
WEPS3001	NS	WTV	RPA CB
WEPS3002	NS	WTV	MP,PWR
WEPS3003	PS	WTV	MP,PWR
WEPS3020	NS	WTV	AD,APA
WEPS3021	NS	WTV	APA,HOR
WEPS3023	NS	WTV	APA,MP
WEPS3027	PS	WTV	APA,MP
WEPS3031	PS	WTV	APA,HOR
WEPS5003	NS	WTV	APA,HP
WEPS5004	NS	WTV	APA,HP
WEPS5005	PS	WTV	APA,HP
WEPS5007	PS	WTV	APA,HP
WEPS7000	NS	WTV	AA,HP
WEPS7001	PS	WTV	AA,HV
WEPS9100	NS	WTV	APD,SW
WEP10	PG	MOT	GEN
WEP1014	NS	WTV	APA,RFD
WEP1018	NS	WTV	APA,RFD
WEP1096	NS	WTV	AO,APA
WEP1237	NS	WTV	AO,SW
WEP1239	NS	WTV	RPA CB
WEP1306	NS	WTV	AO,SW
WEP1307	NS	WTV	AO,SW
WEP2	PG	WTV	RF/IF
WEP224	NS	WTV	PWR CB
WEP230	PG	WTV	APO,MP
WEP231	PG	WTV	APO,HP
WEP232	PG	WTV	APO SCH
WEP233	PG	WTV	APO,HP
WEP235	PG	WTV	HOR/VER
WEP238	PG	WTV	APO,MP
WEP240	NS	WTV	APA LO
WEP241	NS	WTV	APA,HP
WEP242	PS	WTV	AO,VID
WEP243	NS	WTV	AO,VID
WEP244	NS	WTV	HV,APA
WEP245	NS	WTV	APA,RFD
WEP246	PS	WTV	APA,APO
WEP247	NS	WTV	APA,HP
WEP250	PG	WTV	AD,PRE
WEP253	PG	WTV	APA,SS
WEP254	PG	WTV	CMO AM
WEP3	PG	WTV	CMO,IF
WEP367	PG	WTV	MP
WEP371	NS	WTV	AA,VID
WEP372	NS	WTV	AA,VID
WEP373	NS	WTV	LP
WEP380	NS	WTV	LP
WEP394	NS	WTV	AMP,CON
WEP403	NS	WTV	LP
WEP405	PG	WTV	LP
WEP454	NS	WTV	LP
WEP458	NS	WTV	LP
WEP460	NS	WTV	AM,FM
WEP474	PG	WTV	HP
WEP50	NS	WTV	AA,RF
WEP51	PS	WTV	AA,VID
WEP517	NS	WTV	RPA CB
WEP52	PS	WTV	CMO MB
WEP53	NS	WTV	AUD,VID
WEP535	NS	WTV	CMO FM
WEP536	NS	WTV	LP
WEP537	NS	WTV	LP
WEP538	NS	WTV	SS
WEP54	NS	WTV	AA,RF
WEP55	NS	WTV	AUD,RF
WEP56	NS	WTV	CMO UHF
WEP57	PS	WTV	AD,PRE
WEP624	PG	WTV	APO,MP
WEP628	PG	WTV	APO SCH
WEP630	PG	WTV	APA,LP
WEP631	PG	WTV	AD,PRE
WEP632	PG	WTV	PRE,SS
WEP635	PG	WTV	CMO MB
WEP637	PG	WTV	CMO MB
WEP641	NG	WTV	CMO AM
WEP641A	NG	WTV	AD,PRE
WEP641B	NG	WTV	AD,PRE
WEP642	PG	WTV	APO AR
WEP643	PG	WTV	APA AR
WEP700	PS	WTV	APO,MP
WEP701	NS	WTV	APO,MP
WEP703	NS	WTV	APA AR
WEP704	NS	WTV	GEN APA
WEP707	NS	WTV	HV,VERD
WEP709	NS	WTV	VID,IF
WEP710	NS	WTV	LN
WEP712	NS	WTV	VID OUT
WEP715	PS	WTV	AD,PRE
WEP716	PS	WTV	AD,PRE
WEP717	PS	WTV	AD,PRE
WEP719	NS	WTV	VID IF

DESCRIPTION OF TYPES

Type	Polarity Ge/Si	Mfg	Application
WEP720	NS	WTV	CMO UHF
WEP723	NS	WTV	CMO TV
WEP724	NS	WTV	AA,RF
WEP728	NS	WTV	AA,RF
WEP729	NS	WTV	AA,RF
WEP735	NS	WTV	AA,RF
WEP735A	NS	WTV	AMP,SW
WEP736	NS	WTV	AA,VID
WEP740	NS	WTV	HV,HORD
WEP740A	NS	WTV	HV,VERD
WEP740B	NS	WTV	HV,HORD
WEP773	NS	WTV	AA,VID
WEP774	NS	WTV	AUD,VID
WEP777	NS	WTV	RPA CB
WEP778	NS	WTV	RPA CB
WEP784	NS	WTV	IF,RF
WEP799	NS	WTV	RPO CB
WEP907	PS	WTV	LN,HG
WEP923	NS	WTV	LN
ZEN100	NS	ZEN	LP,RF
ZEN101	PS	ZEN	LP,RF
ZEN102	NS	ZEN	LP,RF
ZEN103	NS	ZEN	LP,RF
ZEN104	NS	ZEN	UHF OSC
ZEN105	NS	ZEN	UHF CMO
ZEN106	PS	ZEN	GEN,AA
ZEN107	PS	ZEN	GEN AMP
ZEN108	NS	ZEN	UHF CMO
ZEN109	NS	ZEN	UHF CMO
ZEN110	NS	ZEN	GEN AA
ZEN111	NS	ZEN	GEN AA
ZEN112	NS	ZEN	GEN AA
ZEN113	NS	ZEN	GEN AA
ZEN114	NS	ZEN	GEN AA
ZEN115	NS	ZEN	GEN AA
ZEN116	NS	ZEN	GEN AA
ZEN117	NS	ZEN	GEN AA
ZEN118	NS	ZEN	GEN AA
ZEN119	NS	ZEN	GEN AA
ZEN120	NS	ZEN	GEN AA
ZEN121	NS	ZEN	GEN AA
ZEN122	PS	ZEN	LP
ZEN125	NS	ZEN	HV,GEN
ZEN126	NS	ZEN	GEN CMO
ZEN127	NS	ZEN	HF CMO
ZEN128	NS	ZEN	DAR
ZEN200	NS	ZEN	HV,PWR
ZEN201	NS	ZEN	HV,PWR
ZEN202	NS	ZEN	MP
ZEN203	PS	ZEN	MP
ZEN204	NS	ZEN	HV,HP
ZEN205	NS	ZEN	LF CMO
ZEN206	NS	ZEN	HV,HP
ZEN207	NS	ZEN	MP
ZEN208	NS	ZEN	GEN PWR
ZEN209	NS	ZEN	GEN PWR
ZEN210	NS	ZEN	GEN PWR
ZEN211	PS	ZEN	GEN PWR
ZEN300	NG	ZEN	LP,RF
ZEN301	NG	ZEN	LP,RF
ZEN302	PG	ZEN	LP,GEN
ZEN303	PG	ZEN	LP,GEN
ZEN304	PG	ZEN	LP,GEN
ZEN305	PG	ZEN	LP,GEN
ZEN306	PG	ZEN	GEN AA
ZEN307	PG	ZEN	GEN AA
ZEN308	PG	ZEN	GEN AA
ZEN309	PG	ZEN	GEN AA
ZEN310	PG	ZEN	GEN AA
ZEN311	PG	ZEN	GEN RF
ZEN312	PG	ZEN	GEN RF
ZEN313	PG	ZEN	GEN DF
ZEN314	PG	ZEN	GEN DF
ZEN315	NG	ZEN	GEN,AA
ZEN325	PG	ZEN	MP
ZEN326	PG	ZEN	MP
ZEN327	PG	ZEN	HP
ZEN328	PG	ZEN	HV,PWR
ZEN329	PG	ZEN	GEN PWR
ZEN330	PG	ZEN	GEN APA
ZEN331	PG	ZEN	GEN APA
276-2001	NG	ARC	HF SW
276-2002	PG	ARC	LP,AMP
276-2003	NG	ARC	LP,GEN
276-2004	PG	ARC	GEN,AA
276-2005	PG	ARC	GEN,AA
276-2006	PG	ARC	MP/HPAA
276-2007	PG	ARC	HF,SW
276-2008	NS	ARC	GEN AMP
276-2009	NS	ARC	GEN LO
276-2010	NS	ARC	HG,SS
276-2011	NS	ARC	HG,SS
276-2012	NS	ARC	GEN LO
276-2013	NS	ARC	HIFI,LF
276-2014	NS	ARC	GEN AA
276-2015	NS	ARC	SS AMP
276-2016	NS	ARC	GEN SSW
276-2017	NS	ARC	PWR,SW
276-2018	NS	ARC	PWR,SW
276-2019	NS	ARC	PWR,SW
276-2020	NS	ARC	PWR,SW
276-2021	PS	ARC	GEN AMP
276-2022	PS	ARC	HG,SS
276-2023	PS	ARC	GEN AMP
276-2024	PS	ARC	GEN AA
276-2025	PS	ARC	PWR,SW
276-2026	PS	ARC	PWR,SW
276-2027	PS	ARC	PWR,SW
276-2030	NS	ARC	HC,GEN
276-2031	NS	ARC	GEN AMP
276-2032	PS	ARC	GEN,SW
276-2033	NS	ARC	PWR,SW
276-2034	PS	ARC	GEN,SW

APPLICATIONS

Application	Type	Polarity Ge/Si	Mfg
AA	DS26	PG	DEL
AA	DS76	NS	DEL
AA	ECG123A	NS	SYL
AA	ECG171	NS	SYL
AA	ECG179	PG	SYL
AA	ECG180	PS	SYL
AA	ECG181	NS	SYL
AA	EN30	NS	SEM
AA	EP25	PS	SEM
AA	GE-18	NS	GEC
AA	GE-19	NS	GEC
AA	GE-2	PG	GEC
AA	GE-20	NS	GEC
AA	GE-21	PS	GEC
AA	GE-210	NS	GEC
AA	GE-22	PS	GEC
AA	GE-35	NS	GEC
AA	GE-52	PG	GEC
AA	GE-53	PG	GEC
AA	GE-59	NG	GEC
AA	GE-72	NS	GEC
AA	GE-73	NS	GEC
AA	GE-74	PS	GEC
AA	GE-75	NS	GEC
AA	GE-76	PG	GEC
AA	GE-8	NG	GEC
AA	HEPS5023	PS	MOT
AA	HEP710	PS	MOT
AA	HEP714	NS	MOT
AA	HEP715	PS	MOT
AA	HEP717	PS	MOT
AA	IRTR-50	PG	INR
AA	IRTR-55	NS	INR
AA	IRTR-56	PS	INR
AA	IRTR-57	NS	INR
AA	IRTR-58	PS	INR
AA	IRTR-59	NS	INR
AA	IRTR75	NS	INR
AA	IRTR91	NG	INR
AA	IRTR94MP	PG	INR
AA	PTC102	PG	MAL
AA	PTC109	PG	MAL
AA	PTC123	NS	MAL
AA	PTC125	NS	MAL
AA	PTC126	NS	MAL
AA	PTC127	PS	MAL
AA	PTC134	NG	MAL
AA	PTC135	PG	MAL
AA	PTC136	NS	MAL
AA	PTC139	NS	MAL
AA	PTC156	PG	MAL
AA	PTC170	NS	MAL
AA	PTC171	PS	MAL
AA	PTC177	PS	MAL
AA	PTC178	NS	MAL
AA	RE 12	NS	RAY
AA	RE 13	NS	RAY
AA	RE 17	NS	RAY
AA	RE 18	PS	RAY
AA	RE 26	PS	RAY
AA	RE 3	PG	RAY

Application	Type	Polarity Ge/Si	Mfg
AA	RE 4	PG	RAY
AA	RE 5	NG	RAY
AA	RE 6	NG	RAY
AA	RE 60	NS	RAY
AA	RE 70	NS	RAY
AA	RE 73	NS	RAY
AA	RE 77	NS	RAY
AA	RS-2004	PG	ARC
AA	RS-2005	PG	ARC
AA	RS-2006	PG	ARC
AA	RS-2014	NS	ARC
AA	RS-2024	PS	ARC
AA	RT-100	NS	SPR
AA	RT-100A	NS	SPR
AA	RT-101	PS	SPR
AA	RT-101A	PS	SPR
AA	RT-102	NS	SPR
AA	RT-102A	NS	SPR
AA	RT-103	PS	SPR
AA	RT-103A	PS	SPR
AA	RT-104	NS	SPR
AA	RT-108A	NS	SPR
AA	RT-114	NS	SPR
AA	RT-115	PS	SPR
AA	RT-141	NS	SPR
AA	RT-147	PG	SPR
AA	RT-148	PS	SPR
AA	RT-149	NS	SPR
AA	RT-159	NS	SPR
AA	RT-159A	NS	SPR
AA	RT-160	NS	SPR
AA	RT-162	NS	SPR
AA	RT-163	PS	SPR
AA	RT-172	NS	SPR
AA	RT-301	NS	SPR
AA	RT-302	NS	SPR
AA	RT-303	PS	SPR
AA	RT-304	NS	SPR
AA	RT-305	PS	SPR
AA	SN80	NS	SEM
AA	SP70	PS	SEM
AA	TR-02	PG	INR
AA	TR-03	PG	INR
AA	TR-05	PG	INR
AA	TR-06	PG	INR
AA	TR-09	NG	INR
AA	TR-14	PG	INR
AA	TR-28	PS	INR
AA	TR-36	NS	INR
AA	TR-75	NS	INR
AA	TR-76	NS	INR
AA	TR-77	PS	INR
AA	TR-91	NG	INR
AA	WEPG6001	PG	WTV
AA	WEPS7000	NS	WTV
AA	WEPS7001	PS	WTV
AA	WEP371	NS	WTV
AA	WEP372	NS	WTV
AA	WEP50	NS	WTV
AA	WEP51	PS	WTV
AA	WEP54	NS	WTV

APPLICATIONS

Application	Type	Polarity Ge/Si	Mfg
AA	WEP724	NS	WTV
AA	WEP728	NS	WTV
AA	WEP729	NS	WTV
AA	WEP735	NS	WTV
AA	WEP736	NS	WTV
AA	WEP773	NS	WTV
AA	ZEN106	PS	ZEN
AA	ZEN315	NG	ZEN
AA	276-2004	PG	ARC
AA	276-2005	PG	ARC
AA LO	GE-32	NS	GEC
AA/AO	GE-63	NS	GEC
AA/AO	GE-67	PS	GEC
AA/RF	TR-25	NS	INR
AA/RFA	RT-311	PS	SPR
AD	AT10H	PG	SEM
AD	AT10M	PG	SEM
AD	AT10N	PG	SEM
AD	AT20H	PG	SEM
AD	AT20M	PG	SEM
AD	AT20N	PG	SEM
AD	AT30H	PG	SEM
AD	AT30M	PG	SEM
AD	AT30N	PG	SEM
AD	ECG102	PG	SYL
AD	ECG102A	PG	SYL
AD	ECG103	NG	SYL
AD	ECG103A	NG	SYL
AD	ECG123	NS	SYL
AD	ECG159	PS	SYL
AD	ECG159+	PS	SYL
AD	ECG188	NS	SYL
AD	ECG189	PS	SYL
AD	ECG297	NS	SYL
AD	ECG298	PS	SYL
AD	GE-47	NS	GEC
AD	GE-48	PS	GEC
AD	GE-54G	PG	GEC
AD	GE-54R	NG	GEC
AD	GE-88	NS	GEC
AD	GE-89	PS	GEC
AD	HEPS0026	PS	MOT
AD	IRTR72	NS	INR
AD	IRTR73	PS	INR
AD	IRTR74	NS	INR
AD	IRTR85	PG	INR
AD	JR15	PG	SEM
AD	JR5	PG	SEM
AD	NA20	NG	SEM
AD	NA30	NG	SEM
AD	RE 75	NS	RAY
AD	RE 76	PS	RAY
AD	RT-122	NG	SPR
AD	RT-123	PG	SPR
AD	RT-156	NS	SPR
AD	RT-157	PS	SPR
AD	RT-185	PS	SPR
AD	RT-186	NS	SPR
AD	TR-14	PG	INR
AD	TR-72	NS	INR
AD	TR-73	PS	INR
AD	TR-74	NS	INR
AD	TR-85	PG	INR
AD	WEPS3020	NS	WTV
AD	WEP250	PG	WTV
AD	WEP57	PS	WTV
AD	WEP631	PG	WTV
AD	WEP641A	NG	WTV
AD	WEP641B	NG	WTV
AD	WEP715	PS	WTV
AD	WEP716	PS	WTV
AD	WEP717	PS	WTV
AM	WEP460	NS	WTV
AM AD	RT-120	PS	SPR
AM AO	RT-121	PS	SPR
AM RF	RT-118	PS	SPR
AM RF	SK3005	PG	RCA
AM RF	SK3008	PG	RCA
AM RF	SK3011	NG	RCA
AM/FM	JR10	PG	SEM
AMP	GE-233	NS	GEC
AMP	GE-234	PS	GEC
AMP	RS-2033	NS	ARC
AMP	RS-2034	PS	ARC
AMP	RT-109	NS	SPR
AMP	RT-164A	NS	SPR
AMP	RT-165A	PS	SPR
AMP	RT-166	NS	SPR
AMP	RT-167	PS	SPR
AMP	RT-300	NS	SPR
AMP	WEP394	NS	WTV
AMP	WEP735A	NS	WTV
AMP	276-2002	PG	ARC
AO	AC187/01	NG	PHI
AO	AC188/01	PG	PHI
AO	AT10H	PG	SEM
AO	AT10M	PG	SEM
AO	AT10N	PG	SEM
AO	AT20H	PG	SEM
AO	AT20M	PG	SEM
AO	AT20N	PG	SEM
AO	AT30H	PG	SEM
AO	AT30M	PG	SEM
AO	AT30N	PG	SEM
AO	BDY20	NS	PHI
AO	BD230	NS	PHI
AO	BD231	PS	PHI
AO	BD237	NS	PHI
AO	BD238	PS	PHI
AO	ECG121	PG	SYL
AO	ECG128	NS	SYL
AO	ECG129	PS	SYL
AO	ECG210	NS	SYL
AO	ECG211	PS	SYL
AO	ECG219	PS	SYL
AO	ECG223	NS	SYL
AO	ECG228	NS	SYL
AO	GE-18	NS	GEC
AO	GE-53	PG	GEC
AO	GE-54G	PG	GEC
AO	GE-54R	NG	GEC
AO	GE-59	NG	GEC

APPLICATIONS

Application	Type	Polarity Ge/Si	Mfg
AO	GE-88	NS	GEC
AO	GE-89	PS	GEC
AO	HEPS0026	PS	MOT
AO	IRTR74	NS	INR
AO	IRTR79	NS	INR
AO	JR15	PG	SEM
AO	JR5	PG	SEM
AO	NA20	NG	SEM
AO	NA30	NG	SEM
AO	PTC104	NS	MAL
AO	PTC105	PG	MAL
AO	PTC106	PG	MAL
AO	PTC110	NS	MAL
AO	PTC111	PS	MAL
AO	PTC112	NS	MAL
AO	PTC113	PS	MAL
AO	PTC114	PG	MAL
AO	PTC116	NS	MAL
AO	PTC119	NS	MAL
AO	PTC120	PG	MAL
AO	PTC124	NS	MAL
AO	PTC137	NS	MAL
AO	PTC140	NS	MAL
AO	PTC141	PS	MAL
AO	PTC142	PS	MAL
AO	PTC143	NS	MAL
AO	PTC144	NS	MAL
AO	PTC147	PG	MAL
AO	PTC148	NS	MAL
AO	PTC149	PS	MAL
AO	PTC154	NS	MAL
AO	PTC155	PG	MAL
AO	PTC157	PS	MAL
AO	PTC160	PS	MAL
AO	PTC162	PS	MAL
AO	PTC163	NS	MAL
AO	PTC164	PS	MAL
AO	PTC165	NS	MAL
AO	PTC166	PS	MAL
AO	PTC167	NS	MAL
AO	PTC168	PS	MAL
AO	PTC169	NS	MAL
AO	PTC170	NS	MAL
AO	PTC171	PS	MAL
AO	PTC172	PS	MAL
AO	PTC173	NS	MAL
AO	PTC174	PS	MAL
AO	PTC175	NS	MAL
AO	RE 80	NS	RAY
AO	RE 81	PS	RAY
AO	RE 82	PS	RAY
AO	RT-114	NS	SPR
AO	RT-115	PS	SPR
AO	RT-141	NS	SPR
AO	RT-160	NS	SPR
AO	TR-79	NS	INR
AO	WEP1096	NS	WTV
AO	WEP1237	NS	WTV
AO	WEP1306	NS	WTV
AO	WEP1307	NS	WTV
AO	WEP242	PS	WTV

Application	Type	Polarity Ge/Si	Mfg
AO	WEP243	NS	WTV
AO LO	GE-232	NS	GEC
APA	AT100H	PG	SEM
APA	AT100M	PG	SEM
APA	AT100N	PG	SEM
APA	ECG130	NS	SYL
APA	ECG155	NG	SYL
APA	ECG158	PG	SYL
APA	ECG175	NS	SYL
APA	ECG176	PG	SYL
APA	ECG182	NS	SYL
APA	ECG183	PS	SYL
APA	ECG184	NS	SYL
APA	ECG185	PS	SYL
APA	ECG186	NS	SYL
APA	ECG186A	NS	SYL
APA	ECG187	PS	SYL
APA	ECG187A	PS	SYL
APA	ECG188	NS	SYL
APA	ECG189	PS	SYL
APA	ECG190	NS	SYL
APA	ECG241	NS	SYL
APA	ECG242	PS	SYL
APA	ECG280	NS	SYL
APA	ECG281	PS	SYL
APA	ECG284	NS	SYL
APA	ECG285	PS	SYL
APA	ECG289	NS	SYL
APA	ECG290	PS	SYL
APA	ECG293	NS	SYL
APA	ECG294	PS	SYL
APA	ECG297	NS	SYL
APA	ECG298	PS	SYL
APA	GE-16	PG	GEC
APA	GE-28	NS	GEC
APA	GE-29	PS	GEC
APA	GE-3	PG	GEC
APA	GE-43	NG	GEC
APA	GE-47	NS	GEC
APA	GE-48	PS	GEC
APA	GE-49	PG	GEC
APA	GE-55	NS	GEC
APA	GE-56	PS	GEC
APA	GE-57	NS	GEC
APA	GE-58	PS	GEC
APA	GE-80	PG	GEC
APA	IRTR81	NS	INR
APA	IRTR82	PG	INR
APA	IRTR84	PG	INR
APA	PAR12	PG	SEM
APA	PN26	NS	SEM
APA	PN350	NS	SEM
APA	PN66	NS	SEM
APA	PTC193	NS	MAL
APA	PTC904	NS	MAL
APA	PTC905	PS	MAL
APA	PTC906	NS	MAL
APA	PT12	PG	SEM
APA	PT201	PG	SEM
APA	PT25	PG	SEM
APA	PT32	PG	SEM

APPLICATIONS

Application	Type	Polarity Ge/Si	Mfg
APA	PT40	PG	SEM
APA	PT501	PG	SEM
APA	PT6	PG	SEM
APA	RE 11	PG	RAY
APA	RE 14	NS	RAY
APA	RE 16	PG	RAY
APA	RE 19	NS	RAY
APA	RE 20	PG	RAY
APA	RE 21	NS	RAY
APA	RE 22	PS	RAY
APA	RE 24	NS	RAY
APA	RE 25	PG	RAY
APA	RE 34	NS	RAY
APA	RE 35	PG	RAY
APA	RE 36	PG	RAY
APA	RE 37	NS	RAY
APA	RE 38	NS	RAY
APA	RE 39	PS	RAY
APA	RE 40	NS	RAY
APA	RE 41	PS	RAY
APA	RE 42	NS	RAY
APA	RE 43	PS	RAY
APA	RE 44	NS	RAY
APA	RE 54	NG	RAY
APA	RE 56	PS	RAY
APA	RE 57	PS	RAY
APA	RE 58	PS	RAY
APA	RE 59	NS	RAY
APA	RE 61	PS	RAY
APA	RE 68	PG	RAY
APA	RE 7	PG	RAY
APA	RE 71	NS	RAY
APA	RE 72	PS	RAY
APA	RE 74	PS	RAY
APA	RE 8	PG	RAY
APA	RE 83	PG	RAY
APA	RT-125	PG	SPR
APA	RT-127A	PG	SPR
APA	RT-131	NS	SPR
APA	RT-136	PG	SPR
APA	RT-145	NS	SPR
APA	RT-150	NS	SPR
APA	RT-151	PS	SPR
APA	RT-152	NS	SPR
APA	RT-153	PS	SPR
APA	RT-154	NS	SPR
APA	RT-155	PS	SPR
APA	RT-156	NS	SPR
APA	RT-157	PS	SPR
APA	RT-190	NS	SPR
APA	RT-196	PS	SPR
APA	RT-197	NS	SPR
APA	TR-35	PG	INR
APA	TR-81	NS	INR
APA	TR-82	PG	INR
APA	TR-84	PG	INR
APA	WEPS3020	NS	WTV
APA	WEPS3021	NS	WTV
APA	WEPS3023	NS	WTV
APA	WEPS3027	PS	WTV
APA	WEPS3031	PS	WTV

Application	Type	Polarity Ge/Si	Mfg
APA	WEPS5003	NS	WTV
APA	WEPS5004	NS	WTV
APA	WEPS5005	PS	WTV
APA	WEPS5007	PS	WTV
APA	WEP1014	NS	WTV
APA	WEP1018	NS	WTV
APA	WEP1096	NS	WTV
APA	WEP241	NS	WTV
APA	WEP244	NS	WTV
APA	WEP245	NS	WTV
APA	WEP246	PS	WTV
APA	WEP247	NS	WTV
APA	WEP253	PG	WTV
APA	WEP630	PG	WTV
APA AR	WEP643	PG	WTV
APA AR	WEP703	NS	WTV
APA LO	GE-12	NS	GEC
APA LO	WEP240	NS	WTV
APA SCH	GE-26	PS	GEC
APA/HP	GE-14	NS	GEC
APA/HP	GE-4	PG	GEC
APA/HP	PT150	PG	SEM
APA/HP	PT250	PG	SEM
APA/HP	PT50	PG	SEM
APA/HP	PT515	PG	SEM
APD	RE 33	NS	RAY
APD	RT-189	PS	SPR
APD	WEPS9100	NS	WTV
APO	DS501	PG	DEL
APO	DS503	PG	DEL
APO	DS520	PG	DEL
APO	DS525	PG	DEL
APO	ECG102	PG	SYL
APO	ECG102A	PG	SYL
APO	ECG103	NG	SYL
APO	ECG103A	NG	SYL
APO	ECG104	PG	SYL
APO	ECG105	PG	SYL
APO	ECG152	NS	SYL
APO	ECG153	PS	SYL
APO	ECG192	NS	SYL
APO	ECG192+	NS	SYL
APO	ECG193	PS	SYL
APO	ECG193+	PS	SYL
APO	ECG196	NS	SYL
APO	ECG197	PS	SYL
APO	ECG218	PS	SYL
APO	ECG226	PG	SYL
APO	ECG300	NS	SYL
APO	ECG307	PS	SYL
APO	GE-66	NS	GEC
APO	GE-69	PS	GEC
APO	GE-83	NS	GEC
APO	GE-84	PS	GEC
APO	PTC105A	PG	MAL
APO	PTC120A	PG	MAL
APO	PTC138	PG	MAL
APO	PTC194	PG	MAL
APO	PTC901	PG	MAL
APO	RT-122	NG	SPR
APO	RT-124	PG	SPR

APPLICATIONS

Application	Type	Polarity Ge/Si	Mfg
APO	RT-127	PG	SPR
APO	RT-132	PG	SPR
APO	WEP230	PG	WTV
APO	WEP231	PG	WTV
APO	WEP233	PG	WTV
APO	WEP238	PG	WTV
APO	WEP246	PS	WTV
APO	WEP624	PG	WTV
APO	WEP700	PS	WTV
APO	WEP701	NS	WTV
APO AR	ECG131	PG	SYL
APO AR	WEP642	PG	WTV
APO LO	ECG124	NS	SYL
APO LO	SK3103	NS	RCA
APO SCH	GE-23	NS	GEC
APO SCH	GE-30	PG	GEC
APO SCH	GE-44	PG	GEC
APO SCH	WEP232	PG	WTV
APO SCH	WEP628	PG	WTV
APRE	ECG159	PS	SYL
APRE	ECG159+	PS	SYL
AUD	ECG225	NS	SYL
AUD	EN40	NS	SEM
AUD	EP35	PS	SEM
AUD	IRTR86	NS	INR
AUD	IRTR87	NS	INR
AUD	IRTR88	PS	INR
AUD	SK3003	PG	RCA
AUD	SK3004	PG	RCA
AUD	SK3009	PG	RCA
AUD	SK3010	NG	RCA
AUD	SK3012	PG	RCA
AUD	SK3014	PG	RCA
AUD	SK3020	NS	RCA
AUD	SK3021	NS	RCA
AUD	SK3024	NS	RCA
AUD	SK3024+	NS	RCA
AUD	SK3025	PS	RCA
AUD	SK3025+	PS	RCA
AUD	SK3026	NS	RCA
AUD	SK3027	NS	RCA
AUD	SK3036	NS	RCA
AUD	SK3038	NS	RCA
AUD	SK3041	NS	RCA
AUD	SK3052	PG	RCA
AUD	SK3082	PG	RCA
AUD	SK3084	PS	RCA
AUD	SK3114	PS	RCA
AUD	SK3123	PG	RCA
AUD	SK3124	NS	RCA
AUD	SK3137	NS	RCA
AUD	SK3138	PS	RCA
AUD	SK3156	NS	RCA
AUD	SK3173	PS	RCA
AUD	SK3178	NS	RCA
AUD	SK3179	PS	RCA
AUD	SK3180	NS	RCA
AUD	SK3182	NS	RCA
AUD	SK3188	NS	RCA
AUD	SK3189	PS	RCA
AUD	SK3190	NS	RCA
AUD	SK3191	PS	RCA
AUD	SK3192	NS	RCA
AUD	SK3193	PS	RCA
AUD	SK3198	PG	RCA
AUD	SK3199	NS	RCA
AUD	SK3200	PS	RCA
AUD	SK3201	NS	RCA
AUD	SK3202	NS	RCA
AUD	SK3203	PS	RCA
AUD	SK3241	PS	RCA
AUD	SK3244	NS	RCA
AUD	SK3245	NS	RCA
AUD	SK3247	PS	RCA
AUD	SK3248	NS	RCA
AUD	SK3559	NS	RCA
AUD	SK3560	NS	RCA
AUD	SK3561	NS	RCA
AUD	SK3562	NS	RCA
AUD	WEP53	NS	WTV
AUD	WEP55	NS	WTV
AUD	WEP774	NS	WTV
AUD FM	DS66	NS	DEL
AUD HV	SK3053	PS	RCA
AUD PWR	RT-133	NS	SPR
AUD PWR	RT-164	NS	SPR
AUD PWR	RT-165	PS	SPR
AUD PWR	TVCM-551	NS	SPR
AUD/TV	SK3045	NS	RCA
AUD/TV	SK3104	NS	RCA
AUD/TV	SK3131	NS	RCA
AUD/VID	SK3054	NS	RCA
AUD/VID	SK3083	PS	RCA
AUD/VID	TR-86	NS	INR
AUD/VID	TR-87	NS	INR
AUD/VID	TR-88	PS	INR
BHO	RT-116	NS	SPR
CB	ECG186A	NS	SYL
CB	ECG195	NS	SYL
CB	ECG224	NS	SYL
CB	PTC158	NS	MAL
CMO	AA1	PG	WTV
CMO	AA2	NS	WTV
CMO	DS72	NS	DEL
CMO	ECG100	PG	SYL
CMO	ECG101	NG	SYL
CMO	ECG106	PS	SYL
CMO	ECG107	NS	SYL
CMO	ECG108	NS	SYL
CMO	ECG126	PG	SYL
CMO	ECG160	PG	SYL
CMO	ECG229	NS	SYL
CMO	GE-9	PG	GEC
CMO	HEP712	NS	MOT
CMO	HEP713	NS	MOT
CMO	IRTR95	NS	INR
CMO	RE 10	NS	RAY
CMO	RE 15	PG	RAY
CMO	RE 27	PG	RAY
CMO	RE 53	PS	RAY
CMO	RE 67	NS	RAY
CMO	RE 9	NS	RAY

APPLICATIONS

Application	Type	Polarity Ge/Si	Mfg
CMO	TR-08	NG	INR
CMO	TR-11	PG	INR
CMO	TR-12	PG	INR
CMO	TR-17	PG	INR
CMO	TR-33	NS	INR
CMO	TR-80	NS	INR
CMO	TR-95	NS	INR
CMO	WEP3	PG	WTV
CMO AM	GE-1	PG	GEC
CMO AM	GE-10	NS	GEC
CMO AM	GE-5	NG	GEC
CMO AM	GE-6	NG	GEC
CMO AM	HF12H	PG	SEM
CMO AM	HF12M	PG	SEM
CMO AM	HF20H	PG	SEM
CMO AM	HF20M	PG	SEM
CMO AM	NR10	NG	SEM
CMO AM	RE 1	PG	RAY
CMO AM	RE 2	NG	RAY
CMO AM	WEP254	PG	WTV
CMO AM	WEP641	NG	WTV
CMO AUD	DS25	PG	DEL
CMO CB	RT-308	NS	SPR
CMO FM	GE-11	NS	GEC
CMO FM	WEP535	NS	WTV
CMO MB	HF35	PG	SEM
CMO MB	HF50	PG	SEM
CMO MB	HF75	PG	SEM
CMO MB	JR100	PG	SEM
CMO MB	JR200	PG	SEM
CMO MB	JR30	PG	SEM
CMO MB	JR30X	PG	SEM
CMO MB	NR700	NG	SEM
CMO MB	TV1000	PG	SEM
CMO MB	WEP52	PS	WTV
CMO MB	WEP635	PG	WTV
CMO MB	WEP637	PG	WTV
CMO TV	WEP723	NS	WTV
CMO UHF	WEP56	NS	WTV
CMO UHF	WEP720	NS	WTV
CON	WEP394	NS	WTV
DAR	ECG172	NS	SYL
DAR	ECG232	PS	SYL
DAR	GE-64	NS	GEC
DAR	HEPS9100	NS	MOT
DAR	HEPS9101	NS	MOT
DAR	HEPS9102	NS	MOT
DAR	HEPS9120	PS	MOT
DAR	HEPS9121	PS	MOT
DAR	HEPS9122	PS	MOT
DAR	HEPS9140	NS	MOT
DAR	HEPS9141	PS	MOT
DAR	HEPS9142	NS	MOT
DAR	HEPS9143	PS	MOT
DAR	IRTR69	NS	INR
DAR	PTC153	NS	MAL
DAR	RT-144	NS	SPR
DAR	TR-69	NS	INR
DAR	ZEN128	NS	ZEN
DAR PWR	ECG243	NS	SYL
DAR PWR	ECG244	PS	SYL

Application	Type	Polarity Ge/Si	Mfg
DAR PWR	ECG245	NS	SYL
DAR PWR	ECG246	PS	SYL
DAR PWR	ECG247	NS	SYL
DAR PWR	ECG248	PS	SYL
DAR PWR	ECG249	NS	SYL
DAR PWR	ECG250	PS	SYL
DAR PWR	ECG251	NS	SYL
DAR PWR	ECG252	PS	SYL
DAR PWR	ECG253	NS	SYL
DAR PWR	ECG254	PS	SYL
DAR PWR	ECG257	NS	SYL
DAR PWR	ECG258	PS	SYL
DAR PWR	ECG259	NS	SYL
DAR PWR	ECG260	PS	SYL
DAR PWR	ECG261	NS	SYL
DAR PWR	ECG262	PS	SYL
DAR PWR	ECG263	NS	SYL
DAR PWR	ECG264	PS	SYL
DAR PWR	ECG265	NS	SYL
DAR PWR	ECG266	NS	SYL
DAR PWR	ECG267	NS	SYL
DAR PWR	ECG268	NS	SYL
DAR PWR	ECG269	PS	SYL
DAR PWR	ECG270	NS	SYL
DAR PWR	ECG271	PS	SYL
DAR PWR	ECG272	NS	SYL
DAR PWR	ECG273	PS	SYL
DAR PWR	ECG274	NS	SYL
DAR PWR	ECG275	PS	SYL
DR	RE 63	PS	RAY
FM	ECG106	PS	SYL
FM	GE-17	NS	GEC
FM	HEPS0016	NS	MOT
FM	SK3006	PG	RCA
FM	WEP460	NS	WTV
FM AMP	DS56	PG	DEL
FM OSC	DS74	NS	DEL
FM RF	DS41	PG	DEL
FM RF	DS81	NS	DEL
FM RF	RT-107	NS	SPR
FM/TV	GE-50	PG	GEC
GEN	AA4	PG	WTV
GEN	AA5	PG	WTV
GEN	HEPG0005	PG	MOT
GEN	HEPG0008	PG	MOT
GEN	HEPS0001	NS	MOT
GEN	HEPS0005	NS	MOT
GEN	HEPS0011	NS	MOT
GEN	HEPS0015	NS	MOT
GEN	HEPS0019	PS	MOT
GEN	HEPS0025	NS	MOT
GEN	HEPS0033	NS	MOT
GEN	HEP238	PG	MOT
GEN	HEP239	PG	MOT
GEN	HEP250	PG	MOT
GEN	HEP251	PG	MOT
GEN	HEP252	PG	MOT
GEN	HEP253	PG	MOT
GEN	HEP254	PG	MOT
GEN	HEP715	PS	MOT
GEN	HEP716	PS	MOT

APPLICATIONS

Application	Type	Polarity Ge/Si	Mfg
GEN	PTC195	NS	MAL
GEN	PTC196	PS	MAL
GEN	PTC197	NS	MAL
GEN	PTC198	PS	MAL
GEN	RS-2003	NG	ARC
GEN	RS-2004	PG	ARC
GEN	RS-2005	PG	ARC
GEN	RS-2012	NS	ARC
GEN	RS-2021	PS	ARC
GEN	RS-2030	NS	ARC
GEN	TR-17	PG	INR
GEN	TR-24	NS	INR
GEN	TR-30	PS	INR
GEN	TR-51	NS	INR
GEN	TR-52	PS	INR
GEN	TR-53	NS	INR
GEN	TR-54	PS	INR
GEN	WEP10	PG	MOT
GEN	ZEN106	PS	ZEN
GEN	ZEN125	NS	ZEN
GEN	ZEN302	PG	ZEN
GEN	ZEN303	PG	ZEN
GEN	ZEN304	PG	ZEN
GEN	ZEN305	PG	ZEN
GEN	ZEN315	NG	ZEN
GEN	276-2003	NG	ARC
GEN	276-2004	PG	ARC
GEN	276-2005	PG	ARC
GEN	276-2030	NS	ARC
GEN	276-2032	PS	ARC
GEN	276-2034	PS	ARC
GEN AA	HEPG0006	PG	MOT
GEN AA	HEPG0007	PG	MOT
GEN AA	HEPG6013	PG	MOT
GEN AA	HEPS5018	PS	MOT
GEN AA	HEPS5019	NS	MOT
GEN AA	HEPS5024	NS	MOT
GEN AA	HEPS7004	NS	MOT
GEN AA	HEP629	PG	MOT
GEN AA	HEP630	PG	MOT
GEN AA	HEP631	PG	MOT
GEN AA	HEP632	PG	MOT
GEN AA	HEP633	PG	MOT
GEN AA	HEP634	PG	MOT
GEN AA	HEP641	NG	MOT
GEN AA	HEP721	NS	MOT
GEN AA	HEP722	NS	MOT
GEN AA	HEP723	NS	MOT
GEN AA	HEP724	NS	MOT
GEN AA	HEP725	NS	MOT
GEN AA	HEP726	NS	MOT
GEN AA	HEP727	NS	MOT
GEN AA	HEP728	NS	MOT
GEN AA	HEP729	NS	MOT
GEN AA	HEP730	NS	MOT
GEN AA	HEP731	NS	MOT
GEN AA	HEP732	NS	MOT
GEN AA	HEP733	NS	MOT
GEN AA	HEP734	NS	MOT
GEN AA	HEP735	NS	MOT
GEN AA	HEP736	NS	MOT

Application	Type	Polarity Ge/Si	Mfg
GEN AA	HEP737	NS	MOT
GEN AA	HEP738	NS	MOT
GEN AA	ZEN110	NS	ZEN
GEN AA	ZEN111	NS	ZEN
GEN AA	ZEN112	NS	ZEN
GEN AA	ZEN113	NS	ZEN
GEN AA	ZEN114	NS	ZEN
GEN AA	ZEN115	NS	ZEN
GEN AA	ZEN116	NS	ZEN
GEN AA	ZEN117	NS	ZEN
GEN AA	ZEN118	NS	ZEN
GEN AA	ZEN119	NS	ZEN
GEN AA	ZEN120	NS	ZEN
GEN AA	ZEN121	NS	ZEN
GEN AA	ZEN306	PG	ZEN
GEN AA	ZEN307	PG	ZEN
GEN AA	ZEN308	PG	ZEN
GEN AA	ZEN309	PG	ZEN
GEN AA	ZEN310	PG	ZEN
GEN AA	276-2014	NS	ARC
GEN AA	276-2024	PS	ARC
GEN AMP	HEPS0031	PS	MOT
GEN AMP	HEPS3023	NS	MOT
GEN AMP	HEPS3024	NS	MOT
GEN AMP	HEPS3025	NS	MOT
GEN AMP	HEPS3026	NS	MOT
GEN AMP	HEPS3027	PS	MOT
GEN AMP	HEPS3028	PS	MOT
GEN AMP	HEPS3029	PS	MOT
GEN AMP	HEPS3030	PS	MOT
GEN AMP	HEPS3031	PS	MOT
GEN AMP	RE 78	NS	RAY
GEN AMP	RS-2008	NS	ARC
GEN AMP	RS-2009	NS	ARC
GEN AMP	ZEN107	PS	ZEN
GEN AMP	276-2008	NS	ARC
GEN AMP	276-2021	PS	ARC
GEN AMP	276-2023	PS	ARC
GEN AMP	276-2031	NS	ARC
GEN APA	HEP623	PG	MOT
GEN APA	HEP624	PG	MOT
GEN APA	HEP625	PG	MOT
GEN APA	HEP626	PG	MOT
GEN APA	HEP627	PG	MOT
GEN APA	HEP628	PG	MOT
GEN APA	HEP642	PG	MOT
GEN APA	HEP643	PG	MOT
GEN APA	HEP700	PS	MOT
GEN APA	HEP701	NS	MOT
GEN APA	HEP702	PS	MOT
GEN APA	HEP703	NS	MOT
GEN APA	HEP704	NS	MOT
GEN APA	HEP705	PS	MOT
GEN APA	HEP706	NS	MOT
GEN APA	WEP704	NS	WTV
GEN APA	ZEN330	PG	ZEN
GEN APA	ZEN331	PG	ZEN
GEN CMO	HEPS0002	NS	MOT
GEN CMO	HEPS0003	NS	MOT
GEN CMO	HEPS0006	PS	MOT
GEN CMO	ZEN126	NS	ZEN

APPLICATIONS

Application	Type	Polarity Ge/Si	Mfg
GEN DF	HEP638	PG	MOT
GEN DF	HEP639	PG	MOT
GEN DF	HEP640	PG	MOT
GEN DF	ZEN313	PG	ZEN
GEN DF	ZEN314	PG	ZEN
GEN HVA	GE-220	NS	GEC
GEN HVA	GE-221	PS	GEC
GEN HVA	GE-222	NS	GEC
GEN HVA	GE-223	PS	GEC
GEN HVA	GE-230	NS	GEC
GEN HVA	GE-235	NS	GEC
GEN HVA	RT-314	PS	SPR
GEN HVA	RT-315	NS	SPR
GEN LO	276-2009	NS	ARC
GEN LO	276-2012	NS	ARC
GEN MB	IRTR-53	NS	INR
GEN MB	IRTR-54	PS	INR
GEN PWR	HEPS5000	NS	MOT
GEN PWR	HEPS5001	NS	MOT
GEN PWR	HEPS5002	PS	MOT
GEN PWR	HEPS5003	NS	MOT
GEN PWR	HEPS5004	NS	MOT
GEN PWR	HEPS5005	PS	MOT
GEN PWR	HEPS5006	PS	MOT
GEN PWR	HEPS5007	PS	MOT
GEN PWR	HEPS5008	PS	MOT
GEN PWR	HEPS5009	PS	MOT
GEN PWR	HEPS5010	PS	MOT
GEN PWR	RT-194	NS	SPR
GEN PWR	RT-195	PS	SPR
GEN PWR	ZEN208	NS	ZEN
GEN PWR	ZEN209	NS	ZEN
GEN PWR	ZEN210	NS	ZEN
GEN PWR	ZEN211	PS	ZEN
GEN PWR	ZEN329	PG	ZEN
GEN RF	HEPS0030	NS	MOT
GEN RF	HEP635	PG	MOT
GEN RF	HEP636	PG	MOT
GEN RF	HEP637	PG	MOT
GEN RF	ZEN311	PG	ZEN
GEN RF	ZEN312	PG	ZEN
GEN SSW	276-2016	NS	ARC
GEN TV	IRTR-51	NS	INR
GEN TV	IRTR-52	PS	INR
GEN VA	HEPS0007	NS	MOT
HC	ECG213	PG	SYL
HC	HEP236	PG	MOT
HC	HEP237	PG	MOT
HC	RS-2030	NS	ARC
HC	276-2030	NS	ARC
HC PWR	HEPG6000	PG	MOT
HC PWR	HEPG6001	PG	MOT
HC PWR	HEPG6002	PG	MOT
HC PWR	HEPG6003	PG	MOT
HC PWR	HEPG6004	PG	MOT
HC PWR	HEPG6005	PG	MOT
HC PWR	HEPG6006	PG	MOT
HC PWR	HEPG6007	PG	MOT
HC PWR	HEPG6008	PG	MOT
HC PWR	HEPG6009	PG	MOT
HC PWR	HEPG6010	PG	MOT

Application	Type	Polarity Ge/Si	Mfg
HC PWR	HEPG6011	PG	MOT
HC PWR	HEPG6012	PG	MOT
HC PWR	HEPG6014	PG	MOT
HC PWR	HEPG6015	PG	MOT
HC PWR	HEPG6016	PG	MOT
HC PWR	HEPG6017	PG	MOT
HC PWR	HEPG6018	PG	MOT
HF	AA3	PG	WTV
HF	ECG235	NS	SYL
HF	ECG236	NS	SYL
HF	PTC115	NS	MAL
HF	PTC131	PS	MAL
HF	PTC132	NS	MAL
HF	PTC133	NS	MAL
HF	PTC145	PG	MAL
HF	RS-2001	NG	ARC
HF	RS-2007	PG	ARC
HF	RS-2015	NS	ARC
HF	276-2007	PG	ARC
HF CMO	HEPS0004	NS	MOT
HF CMO	ZEN127	NS	ZEN
HF PWR	RE 194	NS	RAY
HF SW	276-2001	NG	ARC
HG	GE-62	NS	GEC
HG	GE-65	PS	GEC
HG	RS-2022	PS	ARC
HG	RS-2031	NS	ARC
HG	WEP907	PS	WTV
HG	276-2010	NS	ARC
HG	276-2011	NS	ARC
HG	276-2022	PS	ARC
HG PRE	ECG199	NS	SYL
HG PRE	ECG234	PS	SYL
HIFI	ECG121	PG	SYL
HIFI	RS-2013	NS	ARC
HIFI	276-2013	NS	ARC
HOR	BU108	NS	PHI
HOR	HO300	PG	SEM
HOR	NO400	NS	SEM
HOR	PTC129	NS	MAL
HOR	RE 16	PG	RAY
HOR	RT-130	PG	SPR
HOR	WEPS3021	NS	WTV
HOR	WEPS3031	PS	WTV
HOR AMP	RE 30	NS	RAY
HOR AMP	RE 32	NS	RAY
HOR OUT	ECG238	NS	SYL
HOR OUT	ECG277	NS	SYL
HOR OUT	ECG321	NS	SYL
HOR OUT	IRTR93	NS	INR
HOR OUT	SK3035	PG	RCA
HOR OUT	SK3111	NS	RCA
HOR OUT	SK3115	NS	RCA
HOR OUT	TR-93	NS	INR
HOR/VER	ECG127	PG	SYL
HOR/VER	WEP235	PG	WTV
HORD	ECG163	NS	SYL
HORD	ECG165	NS	SYL
HORD	GE-36	NS	GEC
HORD	GE-38	NS	GEC
HORD	IRTR67	NS	INR

APPLICATIONS

Application	Type	Polarity Ge/Si	Mfg
HORD	SK3194	NS	RCA
HORD	SK3220	NS	RCA
HORD	WEP740	NS	WTV
HORD	WEP740B	NS	WTV
HORR	RT-159A	NS	SPR
HORR	SK3219	NS	RCA
HP	AA5	PG	WTV
HP	ECG213	PG	SYL
HP	ECG235	NS	SYL
HP	ECG236	NS	SYL
HP	HEPS5012	NS	MOT
HP	HEPS5020	NS	MOT
HP	HEPS5021	NS	MOT
HP	HEPS7000	NS	MOT
HP	HEPS7001	PS	MOT
HP	HEPS7002	NS	MOT
HP	HEPS7003	PS	MOT
HP	HEPS7005	NS	MOT
HP	HEP231	PG	MOT
HP	HEP233	PG	MOT
HP	HEP237	PG	MOT
HP	HEP247	NS	MOT
HP	HEP248	PS	MOT
HP	HEP707	NS	MOT
HP	HEP740	NS	MOT
HP	IRTR91	NG	INR
HP	PTC105	PG	MAL
HP	PTC105A	PG	MAL
HP	PTC106	PG	MAL
HP	PTC114	PG	MAL
HP	PTC116	NS	MAL
HP	PTC119	NS	MAL
HP	PTC120	PG	MAL
HP	PTC120A	PG	MAL
HP	PTC137	NS	MAL
HP	PTC140	NS	MAL
HP	PTC147	PG	MAL
HP	PTC148	NS	MAL
HP	PTC149	PS	MAL
HP	PTC154	NS	MAL
HP	PTC157	PS	MAL
HP	PTC162	PS	MAL
HP	PTC163	NS	MAL
HP	PTC164	PS	MAL
HP	PTC165	NS	MAL
HP	PTC166	PS	MAL
HP	PTC167	NS	MAL
HP	PTC168	PS	MAL
HP	PTC169	NS	MAL
HP	PTC172	PS	MAL
HP	PTC173	NS	MAL
HP	PTC174	PS	MAL
HP	PTC175	NS	MAL
HP	PTC901	PG	MAL
HP	RT-139	NS	SPR
HP	RT-140	NS	SPR
HP	SK3009	PG	RCA
HP	SK3012	PG	RCA
HP	SK3014	PG	RCA
HP	SK3026	NS	RCA
HP	SK3027	NS	RCA

Application	Type	Polarity Ge/Si	Mfg
HP	SK3036	NS	RCA
HP	SK3041	NS	RCA
HP	SK3084	PS	RCA
HP	SK3173	PS	RCA
HP	SK3180	NS	RCA
HP	SK3181	PS	RCA
HP	SK3182	NS	RCA
HP	SK3183	PS	RCA
HP	SK3188	NS	RCA
HP	SK3189	PS	RCA
HP	SK3190	NS	RCA
HP	SK3191	PS	RCA
HP	SK3192	NS	RCA
HP	SK3193	PS	RCA
HP	SK3217	NS	RCA
HP	SK3510	NS	RCA
HP	SK3511	NS	RCA
HP	SK3530	NS	RCA
HP	SK3534	NS	RCA
HP	SK3535	NS	RCA
HP	SK3538	NS	RCA
HP	SK3563	NS	RCA
HP	TR-36	NS	INR
HP	TR-61	NS	INR
HP	TR-76	NS	INR
HP	TR-77	PS	INR
HP	TR-91	NG	INR
HP	WEPS5003	NS	WTV
HP	WEPS5004	NS	WTV
HP	WEPS5005	PS	WTV
HP	WEPS5007	PS	WTV
HP	WEPS7000	NS	WTV
HP	WEP231	PG	WTV
HP	WEP233	PG	WTV
HP	WEP241	NS	WTV
HP	WEP247	NS	WTV
HP	WEP474	PG	WTV
HP	ZEN204	NS	ZEN
HP	ZEN206	NS	ZEN
HP	ZEN327	PG	ZEN
HP AA	IRTR76	NS	INR
HP AA	IRTR77	PS	INR
HP AA	TR-01	PG	INR
HP/AR	TR-16	PG	INR
HPAA	TR-26	NS	INR
HPAA	TR-29	PS	INR
HPAA	TR-50	PG	INR
HPHG	TR-03	PG	INR
HPHV MB	IRTR-61	NS	INR
HPO	PTC122	PG	MAL
HPO	PTC130	NS	MAL
HS SW	TR-19	PS	INR
HS SW	TR-20	PS	INR
HS SW	TR-21	NS	INR
HS SW	TR-22	NS	INR
HSSW	RS-2023	PS	ARC
HSSW	RT-162	NS	SPR
HSSW	RT-163	PS	SPR
HV	HEPS0001	NS	MOT
HV	HEPS0005	NS	MOT
HV	HEPS0027	NS	MOT

APPLICATIONS

Application	Type	Polarity Ge/Si	Mfg
HV	HEPS0028	PS	MOT
HV	HEPS3021	NS	MOT
HV	HEPS3022	NS	MOT
HV	HEPS5015	NS	MOT
HV	HEPS5020	NS	MOT
HV	HEPS5021	NS	MOT
HV	HEPS7005	NS	MOT
HV	HEP234	PG	MOT
HV	HEP235	PG	MOT
HV	HEP240	NS	MOT
HV	HEP241	NS	MOT
HV	HEP244	NS	MOT
HV	HEP707	NS	MOT
HV	HEP712	NS	MOT
HV	HEP713	NS	MOT
HV	HEP714	NS	MOT
HV	HEP740	NS	MOT
HV	IRTR67	NS	INR
HV	IRTR68	NS	INR
HV	PTC104	NS	MAL
HV	PTC124	NS	MAL
HV	RT-109	NS	SPR
HV	SK3021	NS	RCA
HV	SK3181	PS	RCA
HV	SK3183	PS	RCA
HV	SK3217	NS	RCA
HV	SK3244	NS	RCA
HV	SK3510	NS	RCA
HV	SK3511	NS	RCA
HV	SK3512	NS	RCA
HV	SK3513	PS	RCA
HV	SK3528	PS	RCA
HV	SK3529	NS	RCA
HV	SK3530	NS	RCA
HV	SK3534	NS	RCA
HV	SK3535	NS	RCA
HV	SK3536	NS	RCA
HV	SK3537	NS	RCA
HV	SK3538	NS	RCA
HV	SK3559	NS	RCA
HV	SK3560	NS	RCA
HV	SK3563	NS	RCA
HV	TR-23	NS	INR
HV	TR-32	NS	INR
HV	TR-60	NS	INR
HV	TR-61	NS	INR
HV	TR-67	NS	INR
HV	TR-68	NS	INR
HV	WEPS7001	PS	WTV
HV	WEP244	NS	WTV
HV	WEP707	NS	WTV
HV	WEP740	NS	WTV
HV	WEP740A	NS	WTV
HV	WEP740B	NS	WTV
HV	ZEN125	NS	ZEN
HV	ZEN200	NS	ZEN
HV	ZEN201	NS	ZEN
HV	ZEN204	NS	ZEN
HV	ZEN206	NS	ZEN
HV	ZEN328	PG	ZEN
HV A/SW	ECG198	NS	SYL

Application	Type	Polarity Ge/Si	Mfg
HV AA	HEPS3019	NS	MOT
HV AMP	ECG194	NS	SYL
HV AO	BD232	NS	PHI
HV APA	ECG157	NS	SYL
HV APA	RT-135	NS	SPR
HV APO	RT-128	NS	SPR
HV GEN	ECG287	NS	SYL
HV GEN	ECG288	PS	SYL
HV GEN	HEPS0029	PS	MOT
HV PWR	HEP644	PG	MOT
HV PWR	RE 191	NS	RAY
HV VID	ECG191	NS	SYL
HV VID	GE-231	NS	GEC
HVA	GE-217	NS	GEC
HVA	GE-218	PS	GEC
HVA	GE-226	NS	GEC
HVA	GE-227	PS	GEC
HVA	GE-228	PS	GEC
HVA	GE-25	PG	GEC
HVHC SW	ECG283	NS	SYL
HVP	RS-2012	NS	ARC
HVP	TR-27	PG	INR
HVP	TR-34	PG	INR
HVSW	GE-76	PG	GEC
IF	AA1	PG	WTV
IF	AA2	NS	WTV
IF	DS72	NS	DEL
IF	ECG107	NS	SYL
IF	ECG126	PG	SYL
IF	ECG160	PG	SYL
IF	EN10	NS	SEM
IF	EP20	PS	SEM
IF	GE-211	NS	GEC
IF	GE-213	NS	GEC
IF	GE-51	PG	GEC
IF	GE-60	NS	GEC
IF	GE-61	NS	GEC
IF	GE-9	PG	GEC
IF	PTC107	PG	MAL
IF	PTC108	NG	MAL
IF	PTC121	NS	MAL
IF	RE 66	NS	RAY
IF	RE 69	PG	RAY
IF	RT-126	PS	SPR
IF	SN60	NS	SEM
IF	SP90	PS	SEM
IF	TR-10	NG	INR
IF	TR-95	NS	INR
IF	WEP3	PG	WTV
IF	WEP709	NS	WTV
IF	WEP784	NS	WTV
IF AM	GE-7	NG	GEC
IF AM	HF3H	PG	SEM
IF AM	HF3M	PG	SEM
IF AM	HF6H	PG	SEM
IF AM	HF6M	PG	SEM
IF AM	NR5	NG	SEM
IF TV	ECG319	NS	SYL
LF	276-2013	NS	ARC
LF CMO	HEPS5025	NS	MOT
LF CMO	HEPS5026	NS	MOT

APPLICATIONS

Application	Type	Polarity Ge/Si	Mfg
LF CMO	ZEN205	NS	ZEN
LFRF	HEPS5023	PS	MOT
LN	GE-52	PG	GEC
LN	GE-62	NS	GEC
LN	GE-64	NS	GEC
LN	GE-65	PS	GEC
LN	HEPS0024	NS	MOT
LN	RS-2010	NS	ARC
LN	RS-2022	PS	ARC
LN	RS-2031	NS	ARC
LN	RT-105	NS	SPR
LN	RT-106	PS	SPR
LN	TR-24	NS	INR
LN	TR-30	PS	INR
LN	TR-69	NS	INR
LN	WEP710	NS	WTV
LN	WEP907	PS	WTV
LN	WEP923	NS	WTV
LN AA	RE 64	NS	RAY
LN AMP	GE-85	NS	GEC
LN AUD	RE 62	PS	RAY
LN PRE	RE 192	NS	RAY
LN PRE	RE 193	PS	RAY
LN UHF	ECG316	NS	SYL
LP	HEPG0001	PG	MOT
LP	HEPG0002	PG	MOT
LP	HEPG0003	PG	MOT
LP	HEPG0005	PG	MOT
LP	HEPG0009	PG	MOT
LP	HEPG0011	NG	MOT
LP	HEPS0011	NS	MOT
LP	HEPS0012	PS	MOT
LP	HEPS0013	PS	MOT
LP	HEPS0014	NS	MOT
LP	HEPS0015	NS	MOT
LP	HEPS0024	NS	MOT
LP	HEPS0033	NS	MOT
LP	HEPS5022	PS	MOT
LP	HEP1	PG	MOT
LP	HEP2	PG	MOT
LP	HEP250	PG	MOT
LP	HEP251	PG	MOT
LP	HEP252	PG	MOT
LP	HEP253	PG	MOT
LP	HEP254	PG	MOT
LP	HEP3	PG	MOT
LP	HEP50	NS	MOT
LP	HEP51	PS	MOT
LP	HEP52	PS	MOT
LP	HEP53	NS	MOT
LP	HEP54	NS	MOT
LP	HEP55	NS	MOT
LP	HEP56	NS	MOT
LP	HEP57	PS	MOT
LP	HEP717	PS	MOT
LP	HEP739	PS	MOT
LP	PTC109	PG	MAL
LP	PTC134	NG	MAL
LP	PTC135	PG	MAL
LP	PTC156	PG	MAL
LP	PTC177	PS	MAL
LP	PTC178	NS	MAL
LP	RS-2010	NS	ARC
LP	RS-2011	NS	ARC
LP	RS-2013	NS	ARC
LP	RS-2016	NS	ARC
LP	TR-12	PG	INR
LP	WEP373	NS	WTV
LP	WEP380	NS	WTV
LP	WEP403	NS	WTV
LP	WEP405	PG	WTV
LP	WEP454	NS	WTV
LP	WEP458	NS	WTV
LP	WEP536	NS	WTV
LP	WEP537	NS	WTV
LP	WEP630	PG	WTV
LP	ZEN100	NS	ZEN
LP	ZEN101	PS	ZEN
LP	ZEN102	NS	ZEN
LP	ZEN103	NS	ZEN
LP	ZEN122	PS	ZEN
LP	ZEN300	NG	ZEN
LP	ZEN301	NG	ZEN
LP	ZEN302	PG	ZEN
LP	ZEN303	PG	ZEN
LP	ZEN304	PG	ZEN
LP	ZEN305	PG	ZEN
LP	276-2002	PG	ARC
LP	276-2003	NG	ARC
LPO	GE-81	NS	GEC
LPO	GE-82	PS	GEC
M.P.	IRTR94MP	PG	INR
M.P./AA	TR-94MP	PG	INR
MB AMP	RT-126A	PS	SPR
MB RF	SK3007	PG	RCA
MB RF	SK3018	NS	RCA
MB RF	SK3118	PS	RCA
MB RF	SK3122	NS	RCA
MF AMP	RS-2002	PG	ARC
MP	AA4	PG	WTV
MP	HEPS3002	NS	MOT
MP	HEPS3003	PS	MOT
MP	HEPS3010	NS	MOT
MP	HEPS3012	PS	MOT
MP	HEPS3020	NS	MOT
MP	HEPS3032	PS	MOT
MP	HEPS3033	NS	MOT
MP	HEPS3034	NS	MOT
MP	HEPS3035	NS	MOT
MP	HEPS5013	PS	MOT
MP	HEPS5014	NS	MOT
MP	HEP200	PG	MOT
MP	HEP230	PG	MOT
MP	HEP232	PG	MOT
MP	HEP238	PG	MOT
MP	HEP239	PG	MOT
MP	HEP242	PS	MOT
MP	HEP243	NS	MOT
MP	HEP245	NS	MOT
MP	HEP246	PS	MOT
MP	HEP708	PS	MOT
MP	HEP710	PS	MOT

APPLICATIONS

Application	Type	Polarity Ge/Si	Mfg
MP	IRTR72	NS	INR
MP	IRTR73	PS	INR
MP	PTC110	NS	MAL
MP	PTC111	PS	MAL
MP	PTC112	NS	MAL
MP	PTC113	PS	MAL
MP	PTC138	PG	MAL
MP	PTC141	PS	MAL
MP	PTC142	PS	MAL
MP	PTC143	NS	MAL
MP	PTC144	NS	MAL
MP	PTC155	PG	MAL
MP	PTC160	PS	MAL
MP	PTC194	PG	MAL
MP	PTC904	NS	MAL
MP	PTC905	PS	MAL
MP	RS-2006	PG	ARC
MP	RS-2014	NS	ARC
MP	RS-2021	PS	ARC
MP	RS-2023	PS	ARC
MP	RS-2024	PS	ARC
MP	RS-2032	PS	ARC
MP	SK3020	NS	RCA
MP	SK3024	NS	RCA
MP	SK3024+	NS	RCA
MP	SK3025	PS	RCA
MP	SK3025+	PS	RCA
MP	SK3052	PG	RCA
MP	SK3082	PG	RCA
MP	SK3123	PG	RCA
MP	SK3137	NS	RCA
MP	SK3138	PS	RCA
MP	SK3178	NS	RCA
MP	SK3179	PS	RCA
MP	SK3198	PG	RCA
MP	SK3199	NS	RCA
MP	SK3200	PS	RCA
MP	SK3201	NS	RCA
MP	SK3202	NS	RCA
MP	SK3203	PS	RCA
MP	SK3248	NS	RCA
MP	SK3512	NS	RCA
MP	SK3513	PS	RCA
MP	SK3536	NS	RCA
MP	TCVM-553	NS	SPR
MP	TR-08	NG	INR
MP	TR-37	PS	INR
MP	TR-60	NS	INR
MP	WEPS3002	NS	WTV
MP	WEPS3003	PS	WTV
MP	WEPS3023	NS	WTV
MP	WEPS3027	PS	WTV
MP	WEP230	PG	WTV
MP	WEP238	PG	WTV
MP	WEP367	PG	WTV
MP	WEP624	PG	WTV
MP	WEP700	PS	WTV
MP	WEP701	NS	WTV
MP	ZEN202	NS	ZEN
MP	ZEN203	PS	ZEN
MP	ZEN207	NS	ZEN

Application	Type	Polarity Ge/Si	Mfg
MP	ZEN325	PG	ZEN
MP	ZEN326	PG	ZEN
MP AMP	RT-182	NS	SPR
MP DAR	HEPS9103	NS	MOT
MP DAR	HEPS9123	PS	MOT
MP/HPAA	276-2006	PG	ARC
MPA	PTC195	NS	MAL
MPA	PTC196	PS	MAL
MPA	PTC197	NS	MAL
MPA	PTC198	PS	MAL
MPAA	TR-55	NS	INR
MPAA	TR-56	PS	INR
MPAA	TR-57	NS	INR
MPAA	TR-58	PS	INR
MPAA	TR-59	NS	INR
MPAA	TR-72	NS	INR
MPAA	TR-73	PS	INR
MPAA	TR-74	NS	INR
MPD	GE-81	NS	GEC
MPD	GE-82	PS	GEC
MPHG	TR-05	PG	INR
MPHV MB	IRTR-60	NS	INR
MPLV	TR-02	PG	INR
MS SW	HEPS0032	PS	MOT
OSC	GE-19	NS	GEC
OSC	GE-20	NS	GEC
OSC	GE-21	PS	GEC
OSC	GE-22	PS	GEC
OSC	HEPG0008	PG	MOT
OSC	HEP56	NS	MOT
OSC	RT-107A	NS	SPR
OSC	RT-117	PG	SPR
OSC	TR-83	NS	INR
OSC CB	RT-313	NS	SPR
OSC TV	IRTR83	NS	INR
OUT CB	GE-215	NS	GEC
OUT CB	GE-216	NS	GEC
OUT CB	GE-219	NS	GEC
OUT/CB	TR-62	NS	INR
OUT/CB	TR-63	NS	INR
OUT/CB	TR-64	NS	INR
OUT/CB	TR-65	NS	INR
OUT/CB	TR-66	NS	INR
PRE	IRTR69	NS	INR
PRE	IRTR85	PG	INR
PRE	PTC101	NS	MAL
PRE	PTC103	PS	MAL
PRE	PTC153	NS	MAL
PRE	RT-105	NS	SPR
PRE	RT-106	PS	SPR
PRE	RT-108A	NS	SPR
PRE	TR-85	PG	INR
PRE	WEP250	PG	WTV
PRE	WEP57	PS	WTV
PRE	WEP631	PG	WTV
PRE	WEP632	PG	WTV
PRE	WEP641A	NG	WTV
PRE	WEP641B	NG	WTV
PRE	WEP715	PS	WTV
PRE	WEP716	PS	WTV
PRE	WEP717	PS	WTV

APPLICATIONS

Application	Type	Polarity Ge/Si	Mfg
PWR	ECG286	NS	SYL
PWR	ECG291	NS	SYL
PWR	ECG292	PS	SYL
PWR	ECG299	NS	SYL
PWR	HEPS5011	NS	MOT
PWR	HEPS5015	NS	MOT
PWR	HEP234	PG	MOT
PWR	HEP235	PG	MOT
PWR	HEP236	PG	MOT
PWR	HEP240	NS	MOT
PWR	HEP241	NS	MOT
PWR	HEP244	NS	MOT
PWR	PTC176	NS	MAL
PWR	RE 75	NS	RAY
PWR	RE 76	PS	RAY
PWR	RS-2017	NS	ARC
PWR	RS-2018	NS	ARC
PWR	RS-2019	NS	ARC
PWR	RS-2020	NS	ARC
PWR	RS-2025	PS	ARC
PWR	RS-2026	PS	ARC
PWR	RS-2027	PS	ARC
PWR	RT-137	NS	SPR
PWR	RT-185	PS	SPR
PWR	RT-186	NS	SPR
PWR	TR-23	NS	INR
PWR	WEPS3002	NS	WTV
PWR	WEPS3003	PS	WTV
PWR	ZEN200	NS	ZEN
PWR	ZEN201	NS	ZEN
PWR	ZEN328	PG	ZEN
PWR	276-2017	NS	ARC
PWR	276-2018	NS	ARC
PWR	276-2019	NS	ARC
PWR	276-2020	NS	ARC
PWR	276-2025	PS	ARC
PWR	276-2026	PS	ARC
PWR	276-2027	PS	ARC
PWR	276-2033	NS	ARC
PWR CB	ECG302	NS	SYL
PWR CB	ECG306	NS	SYL
PWR CB	WEP224	NS	WTV
PWR OUT	RT-146	NS	SPR
PWR OUT	RT-158	NS	SPR
PWR SW	GE-45	NS	GEC
PWR TAB	GEMR-6	NS	GEC
RF	ECG100	PG	SYL
RF	ECG101	NG	SYL
RF	ECG123A	NS	SYL
RF	ECG313	NS	SYL
RF	EN10	NS	SEM
RF	EN40	NS	SEM
RF	EP20	PS	SEM
RF	EP35	PS	SEM
RF	GE-210	NS	GEC
RF	GE-211	NS	GEC
RF	GE-51	PG	GEC
RF	HEPG0001	PG	MOT
RF	HEPG0002	PG	MOT
RF	HEPG0003	PG	MOT
RF	HEPG0009	PG	MOT
RF	HEPG0011	NG	MOT
RF	HEPS0012	PS	MOT
RF	HEPS0013	PS	MOT
RF	HEPS0014	NS	MOT
RF	HEPS5022	PS	MOT
RF	HEP1	PG	MOT
RF	HEP2	PG	MOT
RF	HEP3	PG	MOT
RF	HEP50	NS	MOT
RF	HEP51	PS	MOT
RF	HEP52	PS	MOT
RF	HEP53	NS	MOT
RF	HEP54	NS	MOT
RF	HEP55	NS	MOT
RF	HEP57	PS	MOT
RF	HEP708	PS	MOT
RF	PTC107	PG	MAL
RF	PTC108	NG	MAL
RF	PTC121	NS	MAL
RF	PTC176	NS	MAL
RF	RE 13	NS	RAY
RF	RE 66	NS	RAY
RF	RE 69	PG	RAY
RF	RT-107A	NS	SPR
RF	RT-117	PG	SPR
RF	RT-126	PS	SPR
RF	RT-183	PG	SPR
RF	RT-188	PG	SPR
RF	SN60	NS	SEM
RF	SP90	PS	SEM
RF	TR-28	PS	INR
RF	TR-70	NS	INR
RF	TR-80	NS	INR
RF	WEP50	NS	WTV
RF	WEP54	NS	WTV
RF	WEP55	NS	WTV
RF	WEP724	NS	WTV
RF	WEP728	NS	WTV
RF	WEP729	NS	WTV
RF	WEP735	NS	WTV
RF	WEP784	NS	WTV
RF	ZEN100	NS	ZEN
RF	ZEN101	PS	ZEN
RF	ZEN102	NS	ZEN
RF	ZEN103	NS	ZEN
RF	ZEN300	NG	ZEN
RF	ZEN301	NG	ZEN
RF AMP	DS71	NS	DEL
RF CMC	ECG278	NS	SYL
RF CMO	RT-119	NG	SPR
RF CMO	RT-134	NS	SPR
RF PWR	HEPS3001	NS	MOT
RF PWR	HEPS3004	NS	MOT
RF PWR	HEPS3005	NS	MOT
RF PWR	HEPS3006	NS	MOT
RF PWR	HEPS3007	NS	MOT
RF PWR	HEPS3008	NS	MOT
RF PWR	HEPS3009	NS	MOT
RF PWR	HEPS3011	NS	MOT
RF PWR	HEPS3013	NS	MOT
RF PWR	HEPS3014	PS	MOT

APPLICATIONS

Application	Type	Polarity Ge/Si	Mfg
RF PWR	HEPS3036	NS	MOT
RF PWR	HEPS3037	NS	MOT
RF PWR	HEPS3038	NS	MOT
RF PWR	HEPS3039	NS	MOT
RF PWR	HEPS3040	NS	MOT
RF PWR	HEPS3041	NS	MOT
RF PWR	HEP75	NS	MOT
RF PWR	HEP76	PS	MOT
RF PWR	PTC128	NS	MAL
RF PWR	PTC186	NS	MAL
RF PWR	PTC191	NS	MAL
RF PWR	PTC192	NS	MAL
RF PWR	RE 79	NS	RAY
RF TV	SK3039	NS	RCA
RF/IF	WEP2	PG	WTV
RFA	PTC158	NS	MAL
RFA	RT-301	NS	SPR
RFA	RT-302	NS	SPR
RFA	RT-303	PS	SPR
RFA	RT-304	NS	SPR
RFA	RT-305	PS	SPR
RFA CB	RT-309	NS	SPR
RFD	ECG299	NS	SYL
RFD	RT-143	NS	SPR
RFD	WEP1014	NS	WTV
RFD	WEP1018	NS	WTV
RFD	WEP245	NS	WTV
RFD CB	PTC179	NS	MAL
RFD CB	PTC180	NS	MAL
RFD CB	RT-312	NS	SPR
RFDR CB	ECG315	NS	SYL
RFO CB	RT-310	NS	SPR
RPA	ECG195	NS	SYL
RPA	ECG224	NS	SYL
RPA	ECG282	NS	SYL
RPA	RT-143	NS	SPR
RPA CB	ECG195A	NS	SYL
RPA CB	WEPS3001	NS	WTV
RPA CB	WEP1239	NS	WTV
RPA CB	WEP517	NS	WTV
RPA CB	WEP777	NS	WTV
RPA CB	WEP778	NS	WTV
RPO	ECG317	NS	SYL
RPO	ECG318	NS	SYL
RPO	ECG320	NS	SYL
RPO AR	IRTR92	NS	INR
RPO CB	ECG237	NS	SYL
RPO CB	ECG295	NS	SYL
RPO CB	GE-46	NS	GEC
RPO CB	IRTR62	NS	INR
RPO CB	IRTR63	NS	INR
RPO CB	IRTR64	NS	INR
RPO CB	IRTR65	NS	INR
RPO CB	IRTR66	NS	INR
RPO CB	RT-174	NS	SPR
RPO CB	RT-191	NS	SPR
RPO CB	RT-306	NS	SPR
RPO CB	RT-306F	NS	SPR
RPO CB	RT-307	NS	SPR
RPO CB	SK3046	NS	RCA
RPO CB	SK3047	NS	RCA

Application	Type	Polarity Ge/Si	Mfg
RPO CB	SK3048	NS	RCA
RPO CB	SK3049	NS	RCA
RPO CB	SK3197	NS	RCA
RPO CB	SK3239	NS	RCA
RPO CB	WEP799	NS	WTV
RPO VHF	SK3218	NS	RCA
RPO/AR	TR-92	NS	INR
RPO/VHF	SK3176	NS	RCA
RPO/VHF	SK3177	NS	RCA
RPO/VHF	SK3195	NS	RCA
SPL PUR	DS509	NS	DEL
SPL PUR	DS83	PS	DEL
SS	PTC101	NS	MAL
SS	PTC102	PG	MAL
SS	PTC103	PS	MAL
SS	PTC115	NS	MAL
SS	PTC123	NS	MAL
SS	PTC125	NS	MAL
SS	PTC126	NS	MAL
SS	PTC127	PS	MAL
SS	PTC131	PS	MAL
SS	PTC132	NS	MAL
SS	PTC133	NS	MAL
SS	PTC136	NS	MAL
SS	PTC139	NS	MAL
SS	PTC145	PG	MAL
SS	RS-2003	NG	ARC
SS	RS-2011	NS	ARC
SS	RS-2015	NS	ARC
SS	RT-104	NS	SPR
SS	RT-129	NG	SPR
SS	SK3003	PG	RCA
SS	SK3004	PG	RCA
SS	SK3010	NG	RCA
SS	SK3114	PS	RCA
SS	SK3124	NS	RCA
SS	SK3156	NS	RCA
SS	SK3241	PS	RCA
SS	SK3245	NS	RCA
SS	SK3247	PS	RCA
SS	TR-51	NS	INR
SS	TR-52	PS	INR
SS	WEP253	PG	WTV
SS	WEP538	NS	WTV
SS	WEP632	PG	WTV
SS	276-2010	NS	ARC
SS	276-2011	NS	ARC
SS	276-2022	PS	ARC
SS AMP	276-2015	NS	ARC
SSW	RS-2016	NS	ARC
SW	ECG210	NS	SYL
SW	ECG211	PS	SYL
SW	ECG219	PS	SYL
SW	ECG223	NS	SYL
SW	ECG241	NS	SYL
SW	ECG242	PS	SYL
SW	ECG282	NS	SYL
SW	ECG286	NS	SYL
SW	ECG291	NS	SYL
SW	ECG292	PS	SYL
SW	GE-16	PG	GEC

APPLICATIONS

Application	Type	Polarity Ge/Si	Mfg
SW	GE-233	NS	GEC
SW	GE-234	PS	GEC
SW	GE-55	NS	GEC
SW	GE-56	PS	GEC
SW	HEPS0019	PS	MOT
SW	HEPS0025	NS	MOT
SW	HEP716	PS	MOT
SW	IRTR-50	PG	INR
SW	IRTR-55	NS	INR
SW	IRTR-56	PS	INR
SW	IRTR-57	NS	INR
SW	IRTR-58	PS	INR
SW	IRTR-59	NS	INR
SW	PTC906	NS	MAL
SW	RE 80	NS	RAY
SW	RE 81	PS	RAY
SW	RE 82	PS	RAY
SW	RS-2001	NG	ARC
SW	RS-2007	PG	ARC
SW	RS-2017	NS	ARC
SW	RS-2018	NS	ARC
SW	RS-2019	NS	ARC
SW	RS-2020	NS	ARC
SW	RS-2025	PS	ARC
SW	RS-2026	PS	ARC
SW	RS-2027	PS	ARC
SW	RS-2032	PS	ARC
SW	RS-2033	NS	ARC
SW	RS-2034	PS	ARC
SW	RT-137	NS	SPR
SW	RT-164A	NS	SPR
SW	RT-165A	PS	SPR
SW	RT-166	NS	SPR
SW	RT-167	PS	SPR
SW	RT-300	NS	SPR
SW	SK3528	PS	RCA
SW	SK3529	NS	RCA
SW	SK3537	NS	RCA
SW	TR-26	NS	INR
SW	TR-29	PS	INR
SW	TR-50	PG	INR
SW	TR-55	NS	INR
SW	TR-56	PS	INR
SW	TR-57	NS	INR
SW	TR-58	PS	INR
SW	TR-59	NS	INR
SW	WEPS9100	NS	WTV
SW	WEP1237	NS	WTV
SW	WEP1306	NS	WTV
SW	WEP1307	NS	WTV
SW	WEP735A	NS	WTV
SW	276-2007	PG	ARC
SW	276-2017	NS	ARC
SW	276-2018	NS	ARC
SW	276-2019	NS	ARC
SW	276-2020	NS	ARC
SW	276-2025	PS	ARC
SW	276-2026	PS	ARC
SW	276-2027	PS	ARC
SW	276-2032	PS	ARC
SW	276-2033	NS	ARC

Application	Type	Polarity Ge/Si	Mfg
SW	276-2034	PS	ARC
TV	GE-17	NS	GEC
TV	GE-213	NS	GEC
TV	GE-224	NS	GEC
TV	GE-225	PS	GEC
TV	GE-25	PG	GEC
TV	GE-60	NS	GEC
TV	GE-61	NS	GEC
TV	HEPS0016	NS	MOT
TV	IRTR86	NS	INR
TV	SK3006	PG	RCA
TV	SK3034	PG	RCA
TV	SK3044	NS	RCA
TV	SK3132	NS	RCA
TV	SK3232	NS	RCA
TV	SK3561	NS	RCA
TV	SK3562	NS	RCA
UHF CMO	HEPS0010	NS	MOT
UHF CMO	HEPS0017	NS	MOT
UHF CMO	HEP709	NS	MOT
UHF CMO	HEP718	NS	MOT
UHF CMO	HEP719	NS	MOT
UHF CMO	HEP720	NS	MOT
UHF CMO	IRTR80	NS	INR
UHF CMO	RT-108	NS	SPR
UHF CMO	ZEN105	NS	ZEN
UHF CMO	ZEN108	NS	ZEN
UHF CMO	ZEN109	NS	ZEN
UHF OSC	ECG311	NS	SYL
UHF OSC	SK3019	NS	RCA
UHF OSC	ZEN104	NS	ZEN
UHF/VHF	GE-214	NS	GEC
UNI	TR-31	PS	INR
UNI	TR-32	NS	INR
UNI	TR-33	NS	INR
UNI	TR-34	PG	INR
UNI	TR-35	PG	INR
UNI	TR-37	PS	INR
VA	GE-212	NS	GEC
VA	RE 63	PS	RAY
VER	HN100	NS	SEM
VER	HN150	NS	SEM
VER	RT-130	PG	SPR
VER AMP	RE 29	NS	RAY
VER AMP	RE 31	NS	RAY
VERD	ECG162	NS	SYL
VERD	ECG164	NS	SYL
VERD	GE-35	NS	GEC
VERD	GE-37	NS	GEC
VERD	IRTR68	NS	INR
VERD	SK3079	NS	RCA
VERD	SK3085	PS	RCA
VERD	SK3133	NS	RCA
VERD	SK3221	PS	RCA
VERD	SK3222	NS	RCA
VERD	WEP707	NS	WTV
VERD	WEP740A	NS	WTV
VHF	ECG313	NS	SYL
VHF CMO	GE-86	NS	GEC
VHF MIX	HEPS0020	NS	MOT
VID	ECG108	NS	SYL

APPLICATIONS

Application	Type	Polarity Ge/Si	Mfg
VID	ECG123	NS	SYL
VID	ECG128	NS	SYL
VID	ECG129	PS	SYL
VID	ECG171	NS	SYL
VID	ECG225	NS	SYL
VID	ECG228	NS	SYL
VID	ECG229	NS	SYL
VID	IRTR75	NS	INR
VID	IRTR79	NS	INR
VID	IRTR87	NS	INR
VID	IRTR88	PS	INR
VID	PTC122	PG	MAL
VID	PTC130	NS	MAL
VID	RE 12	NS	RAY
VID	RE 17	NS	RAY
VID	RE 18	PS	RAY
VID	RE 60	NS	RAY
VID	RE 73	NS	RAY
VID	RE 77	NS	RAY
VID	RT-100A	NS	SPR
VID	RT-101A	PS	SPR
VID	RT-102A	NS	SPR
VID	RT-103A	PS	SPR
VID	RT-159	NS	SPR
VID	TR-27	PG	INR
VID	TR-67	NS	INR
VID	TR-68	NS	INR
VID	TR-70	NS	INR
VID	TR-75	NS	INR
VID	TR-79	NS	INR
VID	TR-83	NS	INR
VID	WEP242	PS	WTV
VID	WEP243	NS	WTV
VID	WEP371	NS	WTV
VID	WEP372	NS	WTV
VID	WEP51	PS	WTV

Application	Type	Polarity Ge/Si	Mfg
VID	WEP53	NS	WTV
VID	WEP709	NS	WTV
VID	WEP736	NS	WTV
VID	WEP773	NS	WTV
VID	WEP774	NS	WTV
VID AMP	SK3040	NS	RCA
VID HPO	PTC118	NS	MAL
VID HPO	PTC129A	NS	MAL
VID HPO	PTC146	NS	MAL
VID IF	ECG161	NS	SYL
VID IF	ECG233	NS	SYL
VID IF	GE-39	NS	GEC
VID IF	HEPS0008	NS	MOT
VID IF	HEPS0009	NS	MOT
VID IF	IRTR70	NS	INR
VID IF	IRTR71	NS	INR
VID IF	RE 28	NS	RAY
VID IF	RT-112	NS	SPR
VID IF	RT-113	NS	SPR
VID IF	RT-187	NS	SPR
VID IF	SK3117	NS	RCA
VID IF	SK3132+	NS	RCA
VID IF	WEP719	NS	WTV
VID LPO	PTC117	NS	MAL
VID LPO	PTC150	NS	MAL
VID OUT	ECG154	NS	SYL
VID OUT	GE-229	NS	GEC
VID OUT	GE-27	NS	GEC
VID OUT	GE-40	NS	GEC
VID OUT	IRTR78	NS	INR
VID OUT	RE 23	NS	RAY
VID OUT	RT-110	NS	SPR
VID OUT	RT-111	NS	SPR
VID OUT	TR-78	NS	INR
VID OUT	WEP712	NS	WTV